UNIVERSITY

PHYSICS

SECOND EDITION

UNIVERSITY

PHYSICS

Second Edition

GEORGE B. ARFKEN

DAVID F. GRIFFING

DONALD C. KELLY

JOSEPH PRIEST

Miami University, Oxford, Ohio

Harcourt Brace Jovanovich, Publishers

and its subsidiary, Academic Press
San Diego   New York   Chicago   Austin   Washington, D.C.
London   Sydney   Tokyo   Toronto

Cover Credit: Visual Horizons/FPG International

Illustration Credits:
Fig. 1-1 Yerkes Observatory Photo; **6-1** Yerkes Observatory Photo; **6-13** Courtesy of University of California Press; **7-16** Photo by Jeff Sabo/Audio Visual Service, Miami University, Oxford, OH; **14-6** Courtesy of NASA; **15-2** Photo by George Holten/Photo Researchers, Inc.; **15-3** Photo by Jeff Sabo/Audio Visual Service, Miami University, Oxford, OH; **16-2** Photo by Bernice Abbott/Photo Researchers, Inc.; **18-17** Photo by Calvin Larsen/Photo Researchers, Inc.; **19-7** Photo by Jeff Sabo/Audio Visual Service, Miami University, Oxford, OH; **23-1** *The Granger Collection*, New York; **25-15** Courtesy of Dr. T. T. Tsong, Pennsylvania State University; **26-13** Courtesy of Central Scientific Company; **29-3** Fundamental Photographs, New York; **29-20** From Marion, J. B. (1981). *Physics in the Modern World* (2nd ed.). Orlando, FL: Academic Press. Used with permission; **30-14** Courtesy of Central Scientific Company; **33-6** Photo by Jeff Sabo/Audio Visual Service, Miami University, Oxford; **35-20** Courtesy of Polaroid Corporation; **36-2** Fundamental Photographs, New York; **36-20(a)** Photo by Jeff Sabo/Audio Visual Service, Miami University, Oxford; **37-1** Fundamental Photographs, New York; **37-3** Courtesy of Joseph Priest; **37-10** Fundamental Photographs, New York; **37-15** Photo by Jeff Sabo/Audio Visual Service, Miami University, Oxford; **38-1** From Rinard, Philip M. (1976). *American Journal of Physics, 44,* 70. Used with permission of the author; **38-9** From Cajent, M., Francon, M., & Thrien, J. (1962). *Atlas of Optical Phenomena.* Heidelberg, West Germany: Springer-Verlag. Used with permission of the publisher; **38-17** Courtesy of Donald C. Kelly; **39-1** AIP Neils Bohr Library; **39-13** Courtesy of Donald C. Kelly; **41-1** AIP Neils Bohr Library; **41-12** AIP Neils Bohr Library; **41-22** AIP Neils Bohr Library; **41-23** AIP Neils Bohr Library.

ISBN: 0-15-592977-1

Library of Congress Catalog Card Number: 88-83505

Printed in the United States of America

PREFACE

In this second edition of *University Physics* we present students with an authoritative and easy-to-use text. Among the features that we deem important for a beginning physics textbook are:

- a sound pedagogical presentation
- the systematic development of problem-solving skills
- a special sensitivity to students and their goals

A SOUND PEDAGOGICAL PRESENTATION

We present the most important physical principles in the least intimidating way. The unity of physics and the universal character of its principles are emphasized. Basic concepts are illustrated with numerous examples, many drawn from such diverse areas as astrophysics, sports, and the environment.

A SYSTEMATIC DEVELOPMENT OF PROBLEM-SOLVING SKILLS

A step-by-step Problem-Solving Guide is introduced in Chapter 3 and extended in Chapters 6 and 28. Students are shown how to approach and solve problems in a systematic fashion. The Guide is illustrated with numerous examples.

The text of each chapter concludes with a challenging Worked Problem, typical of those found in the end-of-chapter problem sets. Each chapter presents a set of Exercises and Problems that allow students to test their grasp of the principles. Single-concept Exercises reinforce ideas developed in the current chapter. The Exercises are followed by a set of substantive Problems that often illustrate the "vertical" structure of physics and require students to draw on concepts learned in earlier chapters. In this edition, we have provided a wide range of problems, with many problems at the challenging end of the spectrum. Instructors can readily match the abilities of their students to the problems.

A SPECIAL SENSITIVITY TO STUDENTS

Our goal is to create a learning environment that inspires student confidence. We are patient with students. For example, we have considered students who are taking calculus concurrently. The first five chapters avoid calculus and allow students to develop problem-solving skills and build confidence before being confronted by calculus-based problems.

Also, the liberal use of examples and illustrations, and our Problem Solving Guide, help to produce the sensitive atmosphere for which we strive.

PATHWAYS THROUGH THIS TEXT

There are many ways to structure physics courses, and this text can be used in a variety of ways to meet that diversity. Here at Miami University we use the text in two slightly different sequences. In class sections open only to entering freshmen, the first semester covers Newtonian Mechanics (Chapters 1–13) and Special Relativity (Chapters 39–40). The special relativity is interwoven with Newtonian mechanics. The second semester is devoted to Electromagnetism (Chapters 23–35). Many of the students continue with a third semester that covers Materials and Fluid Mechanics (Chapters 14–15), Waves (Chapters 16–17), Thermal Physics (Chapters 18–22), Optics (Chapters 36–38), and selections from Contemporary Physics (Chapters 41–42).

In the other sequence here, upperclass students form a more heterogeneous audience. The first semester covers Newtonian mechanics, materials, and waves. Special relativity is omitted. The second semester covers electromagnetism and optics.

SUPPLEMENTS

We provide an Instructor's Answer Book, a Student's Solutions Manual containing solutions or hints to approximately twenty percent of the Exercises and Problems, and a set of transparencies. Instructional software is available for IBM compatibles and Apple II series microcomputers. A Study Guide written by T. William Houk, James E. Poth, and John W. Snider offers additional insights and opportunities for students to sharpen their problem-solving skills.

ACKNOWLEDGMENTS

We are indebted to many reviewers, students, and colleagues for their helpful criticisms during the development of this text. Special thanks go to Bill Adams, Baylor University; Larry Banks, Southwest Missouri State University; James T. Cushing, University of Notre Dame; Patrick Hamill, San Jose State University; Joseph H. Hamilton, Vanderbilt University; James Monroe, Penn State University-Beaver; R. D. Purrington, Tulane University; Eric Sheldon, University of Lowell; K. L. Schick, Union College; and Ken-Hsi Wang, Baylor University. Finally, to Jeff Holtmeier, Debbie Hardin, Chris Nelson, Kim Svetich, Merilyn Britt, Stacy Simpson, and Lynne Bush of Harcourt Brace Jovanovich go our collective thanks for their encouragements, proddings, and zealous attention to detail.

George B. Arfken

David F. Griffing

Donald C. Kelly

Joseph Priest

CONTENTS

This book is available in two versions: a two-volume set (Volume One: Chapters 1–22; Volume Two: Chapters 23–42) and a combined volume (Chapters 1–42). The complete Table of Contents is provided for both versions.

1.1
THE DEVELOPMENT OF PHYSICS

People have always been curious about the world around them, and they have always sought correlations and patterns of behavior in nature. The early Egyptians, for instance, noticed that the Nile River flooded each year when the bright star Sirius rose at dawn, and they wondered whether these events were connected. Such curiosity, and subsequent attempts to find order in the universe, led to the development of physics as a science. Curiosity about the motion of the planets, for example, eventually led to the Copernican revolution (Figure 1.1).

A major goal of physics is to discover the physical laws that govern the patterns of nature in our world as well as in the universe. These laws and principles are concise expressions that enable scientists and engineers to describe many aspects of physical behavior and to predict future physical behavior.

Physics is a diverse and evolving discipline. This text emphasizes the areas of classical physics that lead into developing areas of contemporary physics. These include mechanics, thermodynamics, electromagnetism, optics, sound, and hydrodynamics. In addition you will find introductions to Einstein's theory of special relativity, and quantum and nuclear physics.

Some branches of engineering, including electrical engineering and materials science, developed as branches of physics. The movement of these disciplines illustrates the close alliance of physics and engineering, as well as the changing nature of physics. Some old areas of physics have been revolutionized and some new areas have opened up. For example, the invention of the laser rejuvenated optical physics, and the development of the transistor catalyzed the growth of solid-state physics.

FIGURE 1.1

The Polish astronomer Nicolaus Copernicus (1473–1543) developed the theory that the sun is the center of our solar system. This theory implied a radically new view of the universe and of the place humans held in it.

1.2
SCIENCE AND MEASUREMENT

The laws that unify physics relate particular physical quantities, such as force and position. Before we can understand the laws of physics, we must define and measure these physical quantities. In physics, the definition and measurement of a quantity are interrelated. Physics is built on operational definitions—that is, definitions that are expressed in terms of measurements. For example, average speed is defined as the measured distance traveled divided by the measured time that has elapsed.

We distinguish two types of physical quantities—**fundamental** quantities and **derived** quantities. Fundamental quantities—time and length, for example—are not defined in terms of other quantities. (We will see how the standards of time and length are defined in Sections 1.3 and 1.4.) Derived quantities, on the other hand, are defined in terms of fundamental quantities by means of a relation that is normally an equation. Average speed, for example, is a derived quantity that is defined in terms of the fundamental quantities of length and time

$$\text{average speed} = \text{distance traveled} \div \text{elapsed time}$$

In the nineteenth century European scientists saw the need for a coherent system of fundamental physical quantities and their corresponding standards of measurement. This concern led to the signing of the Treaty of the Meter in

1875 and the establishment of the International Bureau of Weights and Measures. This bureau was created to "establish new metric standards, conserve the international prototypes, and carry out the comparisons necessary to insure the uniformity of measures throughout the world."

The present metric system, called the *Système International d'Unités*, or SI for short, recognizes seven fundamental quantities. In this chapter we consider three of these fundamental quantities—time, length, and mass. Other fundamental quantities will be introduced as they are needed.

1.3
TIME

Time is a fundamental quantity of special importance because it can be measured with great precision. Any repetitive phenomenon can serve as a clock. The Italian scientist Galileo Galilei used his pulse as a clock to measure the time that it took for a swinging chandelier to return to complete one swing. The earth, by rotating on its axis, can serve as a clock. Originally the second was defined as 1/86,400 the duration of one complete revolution of the earth*, averaged over a year. The earth, however, has not proven to be a sufficiently accurate clock for some sciences, for two reasons. First, because of the tidal friction, the rotation of the earth is gradually slowing down. Second, the earth's rate of rotation is slightly irregular, varying with the seasons and from year to year.

The unit of time is now based on the rhythmic behavior of atoms. By international agreement, the second was redefined in 1969 as the duration of 9,192,631,770 periods of the microwave radiation from a cesium–133 atom. A time interval in seconds is measured as the number of cesium–133 periods in that interval divided by 9,192,631,770. This particular number was chosen to match the older standard as closely as possible. With atomic clocks it is possible to measure time intervals to an accuracy of one part in 10^{13}. The cesium clock is reproducible and readily available to scientists and engineers the world over. Table 1.1 shows the wide range of times encountered by physicists.

TABLE 1.1
Range of Times Encountered in Physics (Orders of Magnitude)

Time	Seconds
Time for half of a sample of neodymium–144 nuclei to disintegrate	10^{23}
Age of the universe	10^{18}
Age of the earth	10^{17}
Age of the Great Pyramid	10^{11}
One year	10^{7}
One day	10^{5}
Period of high-frequency sound	10^{-4}
Period of vibration of cesium clock	10^{-10}
Time for a fast-moving proton to cross a nuclear diameter	10^{-23}
Planck time	10^{-43}

---

* The division of a minute into 60 seconds, an hour into 60 minutes, and a day into 24 hours began in ancient Babylon.

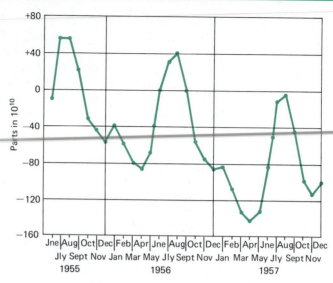

FIGURE 1.2

The variation in the rate of rotation of the earth expressed in parts in 10^{10}. (From L. Essen, *Physics Today 13*, 29 July, 1960)

Figure 1.2 shows the variation in the rate of rotation of the earth as measured by a cesium clock. But how can we be sure that the irregularity is in the earth's rotation and not in the cesium clock? The way that physicists answer this question illustrates how scientists think. First, we must ask ourselves what phenomena might cause irregularities. In the case of the cesium–133 atoms, no such phenomena are known. On the other hand, phenomena such as tidal friction, the seasonal motions of the winds, the melting and refreezing of the polar ice caps, and even major earthquakes could change the earth's rate of rotation. Hence we hypothesize that the earth's rotation is irregular. To confirm this hypothesis we look for other clocks by which to measure the rate of rotation of the earth. Vibrations of other atoms, the movement of the planets around the sun, and the movement of the moon around the earth can all serve as clocks. Although these clocks are not as accurate as the cesium clock, they do confirm that the irregularities shown in Figure 1.2 are rightfully assigned to the earth.

EXAMPLE 1

The Solar Day and the Sidereal Day

We are used to thinking of one day as a span of 24 hours. Each hour is divided into 60 minutes, and each minute is divided into 60 seconds, so the day contains 86,400 seconds. This familiar day is technically a solar day. There is also a second version of "the day," called the **sidereal* day**. Let's see how the two versions are related.

One solar day is defined as the time between successive passages of the sun across the local meridian. Less formally, one solar day is the time between successive high noons. One sidereal day is defined as the time between successive passages of the vernal equinox across the local meridian. Less formally, one sidereal day is the time for the earth to rotate (spin) through 360°.

---

* The word *sidereal* refers to star.

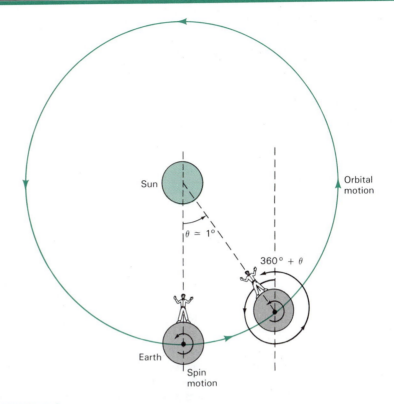

FIGURE 1.3

A view of the earth and sun from above the North Pole. The earth moves counterclockwise in its orbit around the sun, and it spins counterclockwise on its axis. In one solar day the earth spins through an angle of $360° + \theta$, where θ is the angle through which the earth moves in its orbit around the sun. The angle θ is approximately $1°$ and is *not* drawn to scale.

Figure 1.3 lets us see how a solar day differs from a sidereal day. The figure shows the earth at high noon on two successive days, with the sun above an observer. The view is from above the North Pole. The earth moves counterclockwise in its motion about the sun, and it spins in a counterclockwise fashion on its axis. In Figure 1.3 we see that the earth spins through more than $360°$ in one solar day; it spins through $360° + \theta$, where θ is the angle through which the earth moves in its orbit around the sun.

It requires approximately 365.25 solar days for the earth to complete one orbit about the sun. This orbit carries the earth through $360°$, so the angle θ is approximately one degree:

$$\theta = \frac{360°}{365.25} = 0.986°$$

The difference between a solar day and a sidereal day is the time for the earth to spin through the angle θ. The earth spins through $15°$ in approximately one hour, so 4 minutes are required for it to rotate through $1°$. Thus, a sidereal day lasts approximately 23 hours and 56 minutes, 4 minutes less than a solar day.

To obtain a more precise relation for the length of a sidereal day we note that the angle through which the earth spins in one solar day is

$$360° + \theta = 360° \left(1 + \frac{1}{365.25} \right) = 360° \left(\frac{366.25}{365.25} \right)$$

Since a rotation of 360° corresponds to 1 sidereal day we have

$$(1 \text{ sidereal day})\left(\frac{366.25}{365.25}\right) = 1 \text{ solar day} = 86{,}400 \text{ s}$$

The length of a sidereal day is

$$1 \text{ sidereal day} = \left(\frac{365.25}{366.25}\right) 86{,}400 \text{ s}$$

$$= 86{,}164 \text{ s}$$

The difference between a solar day and a sidereal day is 236 s, or about 4 minutes.

1.4
LENGTH

The SI unit of length is the meter. When the metric system was originated in the eighteenth century, the meter (m) was defined as one ten-millionth of the distance from one of the earth's poles to the equator. The attraction of this standard was that it seemed indestructible. However, the direct use of this standard was not feasible. This difficulty was resolved by constructing a bar of an alloy consisting of 90% platinum and 10% iridium, and defining the standard meter as the distance between two lines engraved on the bar when the bar was at the temperature of melting ice. Accurate comparisons between the meter bar and objects of unknown length could then be made by using elaborate and very precise optical techniques. Under the auspices of the International Bureau of Weights and Measures, secondary standards of length were made and distributed to other nations of the world.

The precision of visual comparisons is limited to about one part in ten million (10^7). As science and technology developed, this precision became inadequate. In 1960, by international agreement, the meter was redefined as the length equal to 1,650,763.73 wavelengths (in vacuum) of a particular color of light (a shade of orange) emitted by krypton–86 atoms. This peculiar number was chosen to match the old standard meter as closely as possible. Once again, the standard meter was tied to a property of the natural world. The krypton light source was reproducible and permitted convenient and precise measurements. Remember, though, that all standards are in principle temporary, because they are chosen by consensus.

In 1983 the krypton standard was replaced by an even more precise standard. Because time is the most precisely measurable quantity, length is now defined in terms of time. The definition also involves the speed of light. We recognize that the speed of light in vacuum is a fundamental constant of nature. The **value** of the speed of light is determined by the definition of the meter:

One meter is the distance light travels through vacuum in 1/299,792,458 second.

This definition of the meter *assigns* a value to the speed of light in vacuum.

$$\text{speed of light in vacuum} = 299{,}792{,}458 \text{ m/s}$$

In other words, the speed of light is no longer a *measured* quantity. We recognize it as a fundamental constant and assign it a convenient value. The particular value adopted is the value measured using earlier standards.

TABLE 1.2
Range of Lengths Encountered in Physics (Orders of Magnitude)

Length	Meters
Distance to farthest quasars	10^{26}
Distance to nearby star (Proxima Centauri)	10^{16}
Radius of solar system	10^{13}
Earth–sun distance	10^{11}
Radius of earth	10^{7}
Height of person	10^{0}
Size of air-polluting aerosol particle	10^{-6}
Diameter of influenza virus	10^{-7}
Diameter of hydrogen atom	10^{-10}
Diameter of proton	10^{-15}
Planck length	10^{-35}

In Table 1.2 we list the range of lengths in physics, from the Planck length (about 10^{-35} m) to the distances between the earth and the most remote quasars (perhaps 10^{26} m). When one length is approximately ten times greater than another length, we say that the first length is an **order of magnitude** greater than the second. Orders of magnitude, in other words, refer to powers of 10.

For convenience in handling such a wide range of values, the set of prefixes listed in Table 1.3 is used in conjunction with SI units. Notice that all but one of the exponents are multiples of 3. The only exception, the centimeter (cm), is occasionally used when it is inconvenient to use millimeters or meters.

By law, units of length in the United States (the British system of units) are defined in terms of the standard meter. Thus

$$1 \text{ yard} = 0.9144 \text{ m} \qquad \text{(exactly)}$$

TABLE 1.3
SI Prefixes

Power of 10	Prefix	Symbol
10^{18}	exa	E
10^{15}	peta	P
10^{12}	tera	T
10^{9}	giga	G
10^{6}	mega	M
10^{3}	kilo	k
10^{-2}	centi	c
10^{-3}	milli	m
10^{-6}	micro	μ
10^{-9}	nano	n
10^{-12}	pico	p
10^{-15}	femto	f
10^{-18}	atto	a

Examples: 10^{6} watts = 1 megawatt (1 MW)
10^{3} meters = 1 kilometer (1 km)
10^{-3} gram = 1 milligram (1 mg)
10^{-9} second = 1 nanosecond (1 ns)

and

$$1 \text{ inch} = 2.54 \text{ cm} \qquad \text{(exactly)}$$

From time to time we will use British units in order to give you a better feeling for the physical situation. In most such examples, however, the British units will be converted to SI units before the calculation is carried out.

1.5
MASS

Mass, the third fundamental SI quantity, can be roughly defined as the amount of matter. In Chapter 6 we will discuss mass in terms of its two fundamental properties: (1) gravitational interaction, and (2) resistance to acceleration. For now a standard of mass and a method for comparison of masses is sufficient.

The SI standard of mass is a cylinder of platinum-iridium alloy, which was chosen for its durability and resistance to corrosion. The cylinder is kept at the International Bureau of Weights and Measures in a suburb of Paris, France. The unit of mass, the kilogram, is defined as the mass of this particular metal cylinder. Other masses can be compared to the standard kilogram by using a sensitive balance. If an unknown mass balances the standard kilogram, then the unknown mass also has a mass of 1 kilogram (kg). Objects with masses that are a fraction or a multiple of the standard kilogram can be measured as well, to a precision of one part in 10^8.

Because there is always the remote possibility that the standard platinum-iridium kilogram will be lost or destroyed, there is some preference for adopting an atomic standard of mass comparable to the atomic time standard. Scientists have speculated about taking the mass of a particular atom (such as the mass–12 variety of carbon, ^{12}C) as a standard. At this time, however, there is no pressing need for an atomic standard of mass. The platinum-iridium cylinder standard kilogram is adequate and has no real competitors. Table 1.4 shows the range of masses encountered in physics.

TABLE 1.4

Range of Masses Encountered in Physics (Orders of Magnitude)

Mass	Kilograms
Our galaxy (Milky Way)	10^{41}
The sun	10^{30}
The earth	10^{25}
An asteroid	10^{10}
A weather satellite	10^{3}
A drop of water 1 mm in diameter	10^{-6}
Planck mass	10^{-7}
A uranium atom	10^{-25}
A proton	10^{-27}
An electron	10^{-30}

1.6
DIMENSIONS AND UNITS

You have already been exposed to a variety of units of time: the second, minute, hour, day, year, and century. All of these units are said to have dimensions of time, symbolized by T.

Likewise, all units of length, such as the meter, kilometer, inch, fathom, light-year, and parsec are said to have dimensions of length, symbolized by L. The kilogram and all other units of mass, such as the astrophysicists' solar mass, have dimensions of mass, symbolized by M.

From the three fundamental physical quantities, time, length, and mass, we can develop a variety of useful secondary or derived quantities. Remember that derived quantities have dimensions that can be expressed in terms of the dimensions of fundamental quantities. For instance, area, obtained by multiplying a length times a length, has the dimensions L^2. Volume, such as a cubic meter, has dimensions of L^3. Mass density is defined as mass per unit volume and has dimensions of M/L^3. The SI unit of speed is meters per second (m/s) with the dimensions L/T.

The concept of dimensions is important in both understanding physics and in doing physics problems. For instance, the addition of quantities having different dimensions makes no sense; 1 meter plus 2 seconds is meaningless. Every term in an equation must be dimensionally the same. For example, the equation giving the position of a freely falling body (Chapter 4) is

$$x = v_0 t + \tfrac{1}{2}gt^2 \qquad (1.1)$$

where x is position (length), v_0 is the initial speed (length/time), g is the acceleration of gravity (length/time^2), and t is time. If we analyze the equation dimensionally, we have

$$L = \left(\frac{L}{T}\right)T + \left(\frac{L}{T^2}\right)T^2 = L + L$$

Note that numerical factors such as $\tfrac{1}{2}$ are ignored in dimensional analysis because they have no dimensions. Every term in this equation has dimensions of length, L. If we had written, by mistake,

$$x = v_0 t^2 + \tfrac{1}{2}gt \qquad (1.2)$$

the dimensions of the terms would be

$$L = \left(\frac{L}{T}\right)T^2 + \left(\frac{L}{T^2}\right)T$$

$$= LT + \frac{L}{T}$$

Because this result is dimensionally inconsistent, Equation 1.2 must be incorrect.

EXAMPLE 2

Planetary Acceleration

A planet moving in a circular orbit experiences an acceleration. This acceleration depends only on the speed v of the planet and on the radius r of its orbit. How can we use dimensional analysis to determine how the acceleration, a, is related to v and r?

We assume that the acceleration is proportional to some power of the speed ($a \propto v^p$). Similarly, we assume that the acceleration is proportional to some power of the orbit radius ($a \propto r^q$). Our assumed form for the acceleration can be expressed by

$$a = v^p r^q$$

The exponents p and q are unknown numbers that we want to determine by dimensional analysis. In SI units, the orbit radius is measured in meters and has the dimension of length. Speed is measured in meters per second and has the dimensions of length divided by time. Acceleration is measured in meters per second per second and has the dimensions of length divided by time squared. Thus, the dimensions of a, v, and r separately are

$$[a] = \frac{L}{T^2}$$

$$[v] = \frac{L}{T}$$

$$[r] = L$$

The dimensions of both sides of the equation must be the same. This requirement lets us write

$$LT^{-2} = \left(\frac{L}{T}\right)^p L^q = L^{p+q} T^{-p}$$

We can equate the exponents of T and, separately, the exponents of L

$$-2 = -p$$
$$+1 = p + q$$

This gives $p = 2$ and $q = -1$ and so the acceleration must be proportional to v^2/r.

$$a = \frac{v^2}{r}$$

In this case dimensional analysis gives the exact result. In general, however, dimensional analysis cannot help you infer **dimensionless** numerical factors. For example, dimensional analysis would indicate that an area A has the dimensions of L^2, but it would not distinguish between the area of a circle (πr^2) and the surface area of a sphere ($4\pi r^2$).

A physicist who suspects that an equation may be incorrect generally checks it by using dimensional analysis. You may catch your own algebraic errors by performing dimensional analysis.

Conversion Factors

In physics, and in all the sciences, we frequently have to convert a measurement from one set of units to another. One way to do this is to look up the conversion factor, if it is available. A brief table of conversion factors is given inside the front cover. We can also calculate a conversion factor in terms of conversion factors that we already know. The point to remember is that all conversion factors are intrinsically equal to unity. That is, they are the ratio of two equal quantities. For example, the conversion factor from feet to inches is the ratio

$$\frac{12 \text{ inches}}{1 \text{ foot}}$$

Numerically this ratio is 12, but *intrinsically* it is unity since 12 inches equal 1 foot.

EXAMPLE 3

Conversion of Velocities

Suppose that you are driving at 55 mi/h. How many feet per second are you traveling? How many meters per second? We start with 55 mi/h and multiply by several known conversion factors:

$$55 \text{ mi/h} = (55 \text{ mi/h})\left(\frac{5280 \text{ ft}}{1 \text{ mi}}\right)\left(\frac{1 \text{ h}}{3600 \text{ s}}\right) = 80.7 \text{ ft/s}$$

To convert feet per second to the SI meters per second, we look up the meter in the list of Conversion Factors inside the front cover:

$$1 \text{ m} = 3.281 \text{ ft}$$

The ratio (1 m/3.281 ft) is the conversion factor that converts feet to meters. The speed in m/s is

$$80.7 \text{ ft/s}\left(\frac{1 \text{ m}}{3.281 \text{ ft}}\right) = 24.6 \text{ m/s}$$

One final point about numbers: most of the physical measurements you make will not approach the precision with which the fundamental constants are measured. In this text most of the calculations will carry three significant figures. Using a hand-held calculator, you can easily obtain three-figure accuracy. Throughout this text you can assume that given values for calculations are accurate to three digits unless a higher precision is specified. For example, a velocity of 1 m/s should be taken as 1.00 m/s.

1.7
PHYSICS AND MATHEMATICS

One of the great attractions of physics is its ability to make quantitative predictions. This predictive feature stems from the fact that the laws of physics are expressed by equations. Your understanding of physics will grow as you become more adept at formulating and solving problems expressed in equation form.

We assume that you have a working knowledge of high school mathematics, especially algebra, and have just started a college course in differential and integral calculus. Calculus is very powerful and we are eager to use it. But because you may have just begun to study calculus, we will introduce it very slowly. (Calculus will not really be needed until Chapter 6.)

As you work through this text you will find that physics is more than just a subject to study—physics is also a way of thinking. To excel in physics you need to master this method of thinking, as well as the physics subject matter itself. You should work through the examples in each chapter. The exercises and problems at the end of the chapters are a crucial element of this text, and they are designed to help you master both aspects of learning physics.*

---

* "Just do the exercises diligently. Then you will find out what you have understood and what you have not," Arnold Sommerfeld told his student Werner Heisenberg. Sommerfeld was one of the most famous physics teachers of all time, Heisenberg was one of the founders of modern quantum theory.

We urge you to try all of the problems your teacher assigns. Even if you are unable to solve a problem, you will discover where your understanding fails. One thing is certain—you will learn very little if you do not attempt to solve the problems.

WORKED PROBLEM

Max Planck, who introduced the quantum theory, proposed a fundamental system of units based on three constants: the quantum of action, now called Planck's constant ($h = 6.63 \times 10^{-34}$ kg·m²/s), the speed of light in vacuum ($c = 3.00 \times 10^{8}$ m/s), and Newton's gravitational constant ($G = 6.67 \times 10^{-11}$ m³/kg·s²). Determine the numerical value in seconds of the "Planck time." Express your result as an order of magnitude.

Solution

We seek a combination of h, c, and G having the dimension of time (T). To do this we assign exponents P, Q, and R to h, c, and G and construct the quantity

$$h^P c^Q G^R$$

The dimension of this combination is to be T. Substituting the dimensions of h, c, and G we have

$$\left(\frac{ML^2}{T}\right)^P \left(\frac{L}{T}\right)^Q \left(\frac{L^3}{MT^2}\right)^R = T = M^0 L^0 T^1$$

Collecting terms

$$M^{P-R} L^{2P+Q+3R} T^{-P-Q-2R} = M^0 L^0 T^1$$

Equating exponents on both sides gives three equations relating P, Q, and R

$$P - R = 0$$
$$2P + Q + 3R = 0$$
$$-P - Q - 2R = 1$$

Adding these three equations gives

$$2P = 1$$

so that

$$P = \tfrac{1}{2}$$

This in turn leads to values for the exponents Q and R,

$$Q = -\tfrac{5}{2}$$
$$R = \tfrac{1}{2}$$

So, the combination of h, c, and G having the dimensions of time can be written as

$$\left(\frac{1}{c^2}\right) \sqrt{\frac{hG}{c}} \equiv \text{Planck time}$$

Inserting numerical values gives the Planck time as 1.35×10^{-43} s. As an order of magnitude the Planck time is 10^{-43} s.

EXERCISES

1.3 Time

A. Compare the duration of (a) a microyear and a 1-minute television commercial, (b) a microcentury and a 60-minute television program.

B. The choice of 9,192,631,770 periods of cesium–133 radiation for the definition of the second seems strange. Why didn't scientists pick a round number like 10,000,000,000?

C. Suppose that the earth rotated *clockwise* on its axis (as viewed from above the North Pole), instead of counterclockwise. How would the sidereal day be related to the solar day?

1.4 Length

D. Astronomers, like most scientists, use units that are appropriate for the quantity being measured. For planetary distances they use the astronomical unit (AU), which is equal to the average earth–sun distance, or 1.50×10^{11} m. For stellar distances astronomers use the light-year, which is the distance light travels in 1 year (3.16×10^7 s) at a speed of 3.00×10^8 m/s, and the parsec, which equals 3.26 light-years. Intergalactic distances can be described in megaparsecs. Express the astronomical unit, the light-year, and the megaparsec in meters, each with an appropriate metric prefix.

E. If it is assumed that an atom and its nucleus are each spherical, the ratio of their radii is about 10^5. If the ratio of the radius of the orbit of the moon and the earth's radius were 10^5, how far would the moon be from the earth? How does this distance compare with the actual earth–moon distance?

F. The United States consumes petroleum at a rate of about 6×10^9 barrels per year. Assuming a barrel has a length of 1 meter, compare the length of 6 billion barrels, laid end-to-end, with the coast-to-coast distance of the United States (about 4,000 km).

1.5 Mass

G. A carbon–12 atom ($^{12}_{6}C$) has a mass of 1.99264×10^{-26} kg. How many carbon–12 atoms are there in 12 grams of carbon–12? (This is Avogadro's number.)

H. Density is a derived quantity defined as mass per unit volume. Newton guessed that the average density of the earth was 5.5 grams/cm^3 = 5.5×10^3 kg/m^3. Taking the earth to be a sphere with a radius of 6.38×10^6 m, use Newton's estimate to calculate the mass of the earth.

1.6 Dimensions and Units

I. Work out the dimensions of the following derived quantities: (a) acceleration (speed/time), (b) force (mass × acceleration), (c) work (force × distance), (d) linear momentum (mass × velocity).

J. Newton's law of universal gravitation is given by

$$F = \frac{GMm}{r^2}$$

Here F is the force of gravity, M and m are masses, and r is a length. Force has the units kg·m/s^2. What are the SI units of the proportionality constant G?

K. Consider a system in which the three fundamental quantities are the speed of light c (dimensions LT^{-1}), Planck's constant h (dimensions ML^2T^{-1}), and the mass of the proton, m (dimension M). Develop ratios and/or products of c, h, and m to form a quantity that has the dimension of L.

L. The distance D (in km) traveled by a car is related to the time of travel t (in hours) by

$$D = a + bt + ct^2 + dt^4$$

where a, b, c, and d are constants. What are the units and dimensions of a, b, c, and d?

M. The planet Jupiter has an average radius 10.95 times that of the average radius of the earth and a mass 317.4 times that of the earth. Calculate the ratio of Jupiter's mass density to the mass density of the earth.

N. Assume that it takes 7 minutes to fill a 30 gallon gasoline tank. (a) Calculate the rate at which the tank is filled in gallons per second. (b) Calculate the rate at which the tank is filled in units of cubic meters per second. (c) Determine the time in hours required to fill a one cubic meter volume at the same rate.

P. Imagine that you are driving along a highway in Canada. You see an 80 km/h speed limit sign. What would the sign read in miles per hour? Round your answer to the nearest integer.

Q. Calculate the number of cubic meters in one cubic yard.

R. Calculate the conversion factor for converting meters per second to miles per hour.

S. Under the influence of gravity a freely falling mass accelerates at the rate of 32.2 ft/s^2. Convert this to the SI unit of meters per second per second.

T. To three-figure accuracy, the speed of light is 3.00×10^8 m/s. Convert this to feet per nanosecond.

PROBLEMS

1. Imagine a poison dart experiencing a constant acceleration a over the length x of a blowgun. The speed v of the dart leaving the blowgun depends only on a and x. Find a combination of a and x that has the same dimensions as v.

2. (a) Express your age (approximately) in seconds. (b) A cesium clock can measure time with a precision of one part in 10^{12}. Is such a clock sufficiently precise to measure your age to within one microsecond? To within one millisecond?

3. Because of tidal friction the earth's rate of rotation is decreasing. Early in the history of the earth the length of the day was only one fourth of its present value, and the length of the year was 1461 days (4×365.25). Assume that the earth's initial period of rotation was one fourth of a day and that its period increased at a **constant rate** to its present value. If the age of the earth is 4×10^9 years, determine the age of the earth in days.

4. A copper nucleus has a mass of 1.08×10^{-25} kg and a radius of 4.82×10^{-15} m. Assume the nucleus is a uniform sphere and (a) calculate its mass density. (b) Estimate what the mass of your textbook would be if its density were that of the copper nucleus.

5. Imagine that you are in charge of moving mass from earth to a space colony. The first day you move 2 kg. The second day you move 4 kg. Each day you double the mass moved the day before. (a) How many days are required to move 2.147×10^9 kg? (b) How much mass do you move on the last day?

6. The equatorial radius of the earth is 6.38×10^8 m. You decide to tie several lengths of string together, the cumulative length of which will equal the circumference of the earth. The first string has a length of π meters. The second string has a length of π^2, and so on. Each string has a length π times the length of the preceding shorter string. How many pieces of string will you need?

7. The Reynolds number is a dimensionless quantity used to describe fluid motion. It is formed from products and/or ratios of the density ρ (dimensions ML^{-3}), viscosity η (dimensions $ML^{-1}T^{-1}$), fluid speed u (dimensions LT^{-1}), and a characteristic length D with dimension L. Deduce the form of the Reynolds number by constructing a dimensionless quantity from these four variables that is directly proportional to the fluid speed.

8. A body falling vertically is pulled downward by gravity and held back by air resistance. If the air resistance is negligible, the speed of the body increases by 9.80 m/s each second. (a) Express this change of speed in terms of miles per hour per second. (b) Use dimensional analysis to estimate the time it takes for a diver to reach the water from a 10-meter diving board. (c) Use dimensional analysis to estimate her speed when she hits the water.

9. A stack of N sheets of metal is formed as follows: the bottom sheet has a mass of $m = 3$ kg. Each sheet has a mass that is one-fourth that of the sheet immediately below it. (a) Show that the total mass of the stack is $4m[1 - (1/4)^N]/3$. (b) What is the mass of a stack composed of an infinite number of sheets?

10. Following the technique developed in the *Worked Problem*, determine the Planck length and the Planck mass. Express your results as orders of magnitude and compare them with the values in Table 1.2 and Table 1.4.

11. A **lunar** day is defined as the time between successive passages of the moon across the local meridian; or, it is the time between successive "high moons." It requires 27.32 **sidereal** days for the moon to orbit the earth. The moon moves counterclockwise in its orbit about the earth, as viewed from above the North Pole. Determine the length of a lunar day. Express your result in hours and minutes.

CHAPTER 2

VECTOR ALGEBRA

2.1
SCALARS AND VECTORS

Many quantities in our physical world are characterized by a magnitude only. These physical quantities may be described mathematically by a single number and an appropriate unit. For example, the distance traveled by a projectile might be given as 42.3 meters; and the energy stored in a particular battery can be represented by $E = 2 \times 10^5$ joules. Physical quantities such as these, represented by a magnitude only, are called **scalars**.

Other physical quantities are characterized by a *direction* as well as a magnitude. For instance, the velocity of a car, $\mathbf{v} = 21$ m/s, northeast, or the earth's magnetic field at a particular location, $\mathbf{B} = 4 \times 10^{-5}$ tesla, 4° west of geographic north, are quantities having both magnitude and direction. Such quantities are called **vectors**. Indeed, we may define vectors as quantities having magnitude and direction, and as quantities that combine with each other according to certain rules, which we present in the following sections.

Because a vector differs from a scalar by having a direction associated with it, we represent a vector with a distinctive symbol. In this text, vector quantities are always given in **boldface** type: $\mathbf{v}$, $\mathbf{B}$. Scalar quantities, on the other hand, are given in *italic* type: *m*, *E*. To indicate only the magnitude of vector $\mathbf{v}$, the absolute-value sign may be used, $|\mathbf{v}|$, or the quantity may be printed in italics, *v*. (We prefer the italic designation in this text.)

Vectors may be represented symbolically by an arrow. The direction of the arrow is the direction of the vector, and the length of the arrow is proportional to the vector's magnitude. Vectors may also have dimensions, such as length for a position vector $\mathbf{r}$ and length/time for a velocity vector $\mathbf{v}$.

EXAMPLE 1

Velocity Vector

Velocities relative to the surface of the earth may be described by specifying the magnitude (speed) and the direction. For example, the velocity of an airplane might be described as 50 m/s directed at an angle of 30° north of east. Velocity vectors may also be presented symbolically if we agree that a particular length represents a certain speed. To demonstrate, let's plot a velocity $\mathbf{v}$ of 50 m/s, 30° north of east.

We let 1 cm on the graph in Figure 2.1 correspond to 10 m/s. The magnitude of $\mathbf{v}$ is represented by an arrow 5 cm long. The direction of east is taken to be the positive *x*-axis. The direction of the arrow representing $\mathbf{v}$ is 30° above the positive *x*-axis.

2.2
ADDITION AND SUBTRACTION OF VECTORS

Vectors can be added, subtracted, and multiplied by a scalar. In this section we introduce rules governing these operations.

Addition of Vectors

To understand how vector addition is defined, consider the following situation: A golfer tees off and then walks 120 yards east in search of his ball. He spots his ball several yards away, and walks 16 yards north to reach it. The

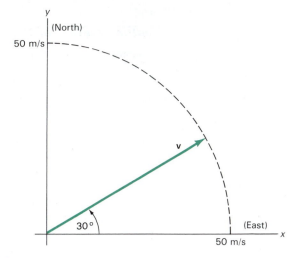

FIGURE 2.1
The velocity vector **v**.

golfer undergoes two changes in position. A change in position is called a **displacement**. We can represent the golfer's two displacements by vectors (Figure 2.2).

$$\mathbf{A} = 120 \text{ yards, east}$$

$$\mathbf{B} = 16 \text{ yards, north}$$

The overall displacement is 121 yards, at an angle of 7.59° north of east. We can also represent the overall displacement by a vector,

$$\mathbf{C} = 121 \text{ yards, } 7.59° \text{ north of east}$$

Because **C** is equivalent to **A** followed by **B**, we *define* **C** to be the **vector sum** of **A** and **B**. Symbolically,

$$\mathbf{C} = \mathbf{A} + \mathbf{B} \tag{2.1}$$

It is important to understand clearly the meaning of this vector equation: The displacement **A** *followed by* the displacement **B** is *equivalent* to the single displacement, **C**. The displacement vector is a model for all vectors. In order to qualify as a vector a quantity must follow the same type of *addition rule* that displacement vectors obey.

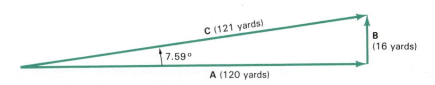

FIGURE 2.2
The two displacements, **A** followed by **B**, are equivalent to the single displacement, **C**.

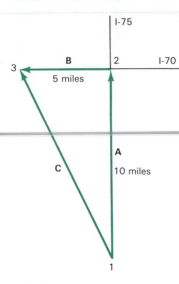

FIGURE 2.3

EXAMPLE 2

Vector Addition of Displacements

A salesman leaves Dayton, Ohio, and drives 10 miles north on I-75 and then 5 miles west on I-70. Where is he relative to his starting point?

We let the vector **A** denote the 10 mile displacement north and we let the vector **B** denote the 5 mile displacement west. We represent these vectors with arrows (Figure 2.3). In Figure 2.3, 1 cm corresponds to 2 miles.

Traveling 10 miles north (from point 1 to point 2) and then 5 miles west (from point 2 to point 3) is equivalent to traveling from point 1 to point 3. The equivalent displacement vector **C** is the vector sum of the displacement vectors **A** and **B**.

$$\mathbf{C} = \mathbf{A} + \mathbf{B}$$

Using a protractor, we find that the direction of **C** is about 26° northwest of Dayton (point 1). We can measure the length of **C** to find it is approximately 5.5 cm long, which corresponds to a distance of 11 miles. Such a graphical solution has the advantage of providing a picture, but it does not yield precise results for the magnitude and direction of **C**.

In our example of a golfer it is clear that the overall displacement of the golfer would have been the same had he first walked 16 yards north and then walked 120 yards east. This means that vector addition is **commutative** (Figure 2.4).

$$\mathbf{A} + \mathbf{B} = \mathbf{B} + \mathbf{A} \tag{2.2}$$

As Figure 2.4 shows, if we form a parallelogram whose sides are the vectors **A** and **B**, then the vector sum **A** + **B** is one diagonal of the parallelogram. Be careful—the parallelogram has two diagonals, only one of which represents the vector sum.

By adding a third vector, **D**, to **A** and **B**, as in Figure 2.5, we see that the way we group the vectors does not affect the final result.

$$(\mathbf{A} + \mathbf{B}) + \mathbf{D} = \mathbf{A} + (\mathbf{B} + \mathbf{D}) \tag{2.3}$$

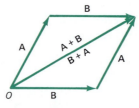

FIGURE 2.4

Vector addition is commutative. The displacement **A** followed by the displacement **B** is equivalent to **B** followed by **A**. The vector sums **A** + **B** and **B** + **A** are represented geometrically by the vector that forms the diagonal of the parallelogram.

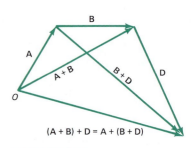

$(A + B) + D = A + (B + D)$

FIGURE 2.5

Vector addition is associative.

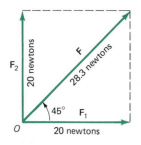

FIGURE 2.6

The force vectors $\mathbf{F}_1$ and $\mathbf{F}_2$ act at point O.

With the tail of **B** at the tip of **A** and the tail of **D** at the tip of **B**, the vector sum $\mathbf{A} + \mathbf{B} + \mathbf{D}$ runs from the tail of **A** to the tip of **D** regardless of the order of addition of the vectors. Vector addition is **associative**, or independent of the grouping of vectors.

The vectors **A** and **B** of Example 2 occur in sequence, one after the other. Vector quantities can also act simultaneously, as we see in Example 3.

EXAMPLE 3

Vector Addition of Forces

Two forces act at a common point on the bumper of a car. One force of 20 newtons (4.50 lb) acts horizontally. A second force, also 20 newtons, acts vertically. What is the *single* force equivalent to the horizontal and vertical forces?

First, we must draw a vector diagram (Figure 2.6) to show the two forces. We let a length of 1 cm correspond to 4 newtons of force. Because F_1 and F_2 in this example are equal, the parallelogram defined by the vectors $\mathbf{F}_1$ and $\mathbf{F}_2$ is a square. The equivalent vector **F**, the vector sum of $\mathbf{F}_1$ and $\mathbf{F}_2$, is given by the diagonal. Because vectors $\mathbf{F}_1$ and $\mathbf{F}_2$ form a right angle in this case, the magnitude of **F** is given by the Pythagorean theorem:

$$|\mathbf{F}| = \sqrt{F_1{}^2 + F_2{}^2} = \sqrt{(20 \text{ newtons})^2 + (20 \text{ newtons})^2}$$
$$= 28.3 \text{ newtons}$$

Thus

$$\mathbf{F} = 28.3 \text{ newtons, } 45° \text{ above the direction of } \mathbf{F}_1$$

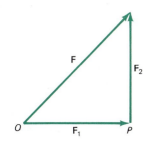

FIGURE 2.7

The vector $\mathbf{F}_2$ is displaced to P in order to add it to $\mathbf{F}_1$.

In Example 3 we have assumed that the result of the two forces $\mathbf{F}_1$ and $\mathbf{F}_2$ acting together is equivalent to that of a single force equal to their vector sum. The justification for treating forces as vectors is based on experiment.

We specified that the two forces $\mathbf{F}_1$ and $\mathbf{F}_2$ act on the same point, P. By adding the forces we can form the triangle shown in Figure 2.7. To form this triangle, $\mathbf{F}_2$ is displaced sideways, parallel to the original $\mathbf{F}_2$. This displacement is a mathematical device for adding the two vectors; *it does not imply that $\mathbf{F}_2$ acts at P.* The resultant force, $\mathbf{F} = \mathbf{F}_1 + \mathbf{F}_2$, whether found as the hypotenuse of a triangle or the diagonal of a parallelogram, acts at the original point, O.

FIGURE 2.8

Subtraction of Vectors

If **B** is a vector shown as an arrow in Figure 2.8, then $-\mathbf{B}$ is defined to be a vector of equal magnitude, but in the opposite direction. The sum of a vector and its negative equals zero. The subtraction of a vector is defined as the addition of a negative vector. For example, we define $\mathbf{A} - \mathbf{B}$ as

$$\mathbf{A} - \mathbf{B} = \mathbf{A} + (-\mathbf{B}) \tag{2.4}$$

Thus, you carry out the subtraction by adding the vector **A** and the vector $-\mathbf{B}$ (Figure 2.9).

Multiplication by a Scalar

A vector **A** may be multiplied by a number or a scalar, α, to give $\alpha\mathbf{A}$. If α is greater than 1, then this multiplication corresponds to a **lengthening** of **A**. If α is less than 1, then the multiplication corresponds to a **shortening** of **A**. If α is positive, the direction of **A** is not affected. If α is negative, the direction

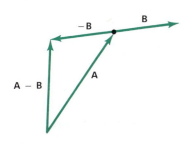

FIGURE 2.9

To form $\mathbf{A} - \mathbf{B}$, first construct the vector $-\mathbf{B}$; then add $-\mathbf{B}$ to **A**.

of **A** is reversed. For instance, $-2\mathbf{A}$ has double the length and is in the opposite direction from **A**.

If the scalar α has dimensions, multiplication will yield a physically different vector. For example, in mechanics the vector velocity **v** is often multiplied by a scalar mass m. The product $m\mathbf{v} = \mathbf{p}$, called **linear momentum**, has the dimensions mass × length/time.

Three important rules of ordinary algebra also apply to vector algebra:

$$\alpha\mathbf{v} = \mathbf{v}\alpha \qquad \text{(commutation)} \qquad (2.5)$$

$$\alpha(\mathbf{v}_1 + \mathbf{v}_2) = \alpha\mathbf{v}_1 + \alpha\mathbf{v}_2 \qquad \text{(distribution)} \qquad (2.6)$$

$$\alpha(\beta\mathbf{v}) = (\beta\alpha)\mathbf{v} \qquad \text{(association)} \qquad (2.7)$$

We can combine the last two of these rules and obtain the result

$$\alpha(\beta\mathbf{v}_1 + \Gamma\mathbf{v}_2) = (\alpha\beta)\mathbf{v}_1 + (\alpha\Gamma)\mathbf{v}_2 \qquad (2.8)$$

Diagrams and graphs help us to picture a physical situation, but offer limited precision. We overcome this limitation in the next section where we develop an algebraic method of vector addition.

2.3
COMPONENTS

When we represent a vector by an arrow, a coordinate system is not needed. Nor do we need a coordinate system to add vectors graphically. But graphical methods are very limited, both in precision and in scope. We now turn to a new vector representation that uses a coordinate system. This new representation will let us add and subtract, and later multiply vectors analytically.

Figure 2.10 depicts a vector **A**, relative to a two-dimensional $x-y$ coordinate system. Consider a line, drawn from the tip of **A** to a point on the x-axis, that intersects this axis perpendicularly. The distance from the origin to this intersection point represents a quantity A_x, which we call the x-component of **A**. Similarly, if we draw a line from the tip of **A** perpendicular to, and intersecting, the y-axis, the distance A_y from the intersection point to the origin is called the y-component of **A**.

As Figure 2.10 shows, the components A_x and A_y form the two sides of a right triangle whose hypotenuse is the vector **A**. The relations between the magnitude of **A** and its components A_x and A_y involve trigonometric functions of the angle θ in Figure 2.10. If we let A denote the magnitude of **A** we can see from Figure 2.10 that A_x and A_y are related to A by

$$A_x = A \cos \theta \qquad (2.9)$$

$$A_y = A \sin \theta \qquad (2.10)$$

EXAMPLE 4

Components of a Velocity Vector

Consider the velocity vector for a car moving at 10 m/s in a direction $35°$ north of east. What are the x- and y-components of the velocity? From Figure 2.11 we see that $\theta = 35°$. With a calculator we find

$$v_x = 10 \text{ m/s } (\cos 35°) = 10 \text{ m/s} \cdot 0.819$$

$$= 8.19 \text{ m/s}$$

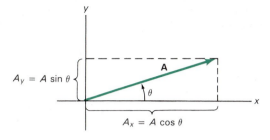

FIGURE2.10

The vector **A** has components A_x and A_y. The trigonometric functions $\sin\theta$ and $\cos\theta$ can be used to express the components in terms of the magnitude (A) and the direction (θ).

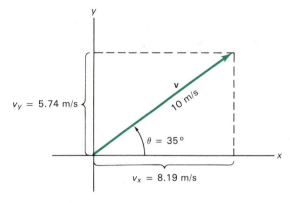

FIGURE 2.11

and

$$v_y = 10 \text{ m/s (sin 35}^\circ) = 10 \text{ m/s} \cdot 0.574$$
$$= 5.74 \text{ m/s}$$

The component of a vector may be negative. For example, in Figure 2.12, the vector **B** is in the second quadrant of the coordinate system. The angle θ running from the positive x-axis to **B** is 140° and $\cos\theta$ is negative.

$$\cos\theta = \cos 140^\circ = -0.766$$

The x-component of **B** is therefore negative

$$B_x = B\cos\theta = -0.766B$$

Often, the measurement of the angles to the nearest axis is more convenient. For instance, if **A** is in the third quadrant, as in Figure 2.13, the angle φ measured from the negative x-axis might be used. The x- and y-components would then be given by

$$A_x = -A\cos\varphi \qquad (2.11)$$

and

$$A_y = -A\sin\varphi \qquad (2.12)$$

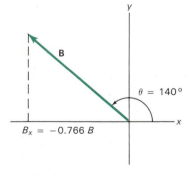

FIGURE 2.12

The vector **B** has a negative x-component. B_x is negative because $\cos\theta = \cos 140^\circ$ is negative.

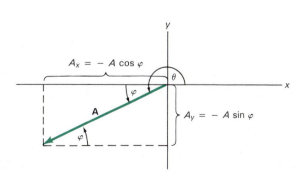

FIGURE 2.13

Angles may be measured from the negative axis. Note the negative signs in Equation 2.11 and Equation 2.12.

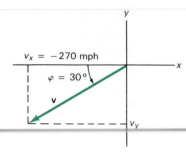

FIGURE 2.14

The velocity components are negative.

The negative signs are inserted because we can see that the components are negative. This "look and see" technique of component calculation can be valuable as verification of formal calculations (Equation 2.9 and Equation 2.10), and will be used frequently.

EXAMPLE 5

Velocity Magnitude and Direction

An airplane flies along a course 30° south of west (Figure 2.14). Its x-component of velocity is -270 mph. We want to determine its speed and its y-component of velocity. Using Equation 2.11 we find the speed of the airplane

$$v = \frac{-v_x}{\cos 30°} = \frac{-(-270 \text{ mph})}{0.866}$$

$$= 312 \text{ mph}$$

From Equation 2.12 we get

$$v_y = -v \sin 30° = -312 \text{ mph} (0.5)$$

$$= -156 \text{ mph}$$

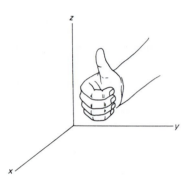

FIGURE 2.15

A right-handed coordinate system. Curl the fingers of your right hand from the positive (+) x-axis toward the positive (+) y-axis; your right thumb points along the positive (+) z-axis.

Three-Dimensional Vectors

So far we have used an x–y coordinate system to examine vectors in two-dimensional space. In order to consider three-dimensional space, we add a z-axis perpendicular to the plane of x- and y-coordinates. Then we choose the positive direction of this z-axis by using what we call the **right-hand rule**. Let the fingers of your right hand curl from the positive x-axis to the positive y-axis. Your extended thumb indicates the positive z-direction for a right-handed system (Figure 2.15).

Figure 2.16 depicts a displacement vector **r** relative to the x–y–z reference frame. We can resolve **r** into components by forming its projections onto each of the three coordinate axes. The components of **r** are denoted by x, y, and z. Imagine that **r** represents the displacement of an ant that crawls from one corner of a room to the farthest corner. Figure 2.16 shows one possible route that the ant could follow: the ant could first move a distance x along the width of the room, then a distance y along its length, and finally a distance z from the floor to the ceiling. A fly could travel in a straight line from one corner to the other along **r**. The displacements of the ant and the fly would be equal.

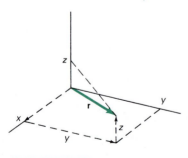

FIGURE 2.16

An ant undergoes the displacements x, y, and z along three mutually perpendicular directions. A fly undergoes an equivalent displacement r along a straight line.

Unit Vectors

An advantage of using vector components is that graphical or geometric manipulation of vector problems can be replaced by algebraic and numerical manipulation. To facilitate these manipulations we introduce three **unit vectors**: **i**, **j**, and **k** (Figure 2.17). The vector **i** is in the positive x-direction; **j** is in the positive y-direction; and **k** is in the positive z-direction. These three vectors are called unit vectors because their magnitudes are unity. They are also dimensionless. The three unit vectors **i**, **j**, and **k** do not change the magnitude or the dimensions of physical quantities; they only indicate directions. We can use the unit vectors to represent a vector as a sum of three mutually perpendicular component vectors. Let's reconsider the displacement vector for the

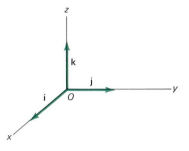

FIGURE 2.17
Unit vectors **i**, **j**, and **k**.

ant. The ant underwent successive displacements of x, y, and z parallel to the corresponding coordinate axes. We can use the vector $\mathbf{i}x$ to describe the displacement along the x-direction. Likewise, the vectors $\mathbf{j}y$ and $\mathbf{k}z$ describe the displacements along the other two directions. By multiplying each component by the corresponding unit vector we get displacement vectors along the three mutually perpendicular axes. From our description of the ant's journey it is clear that the resulting displacement vector **r** equals the vector sum of the three separate displacement vectors

$$\mathbf{r} = \mathbf{i}x + \mathbf{j}y + \mathbf{k}z \qquad (2.13)$$

EXAMPLE 6

Position Vector

A squirrel is on a tree limb, 17.3 m from the origin of a coordinate system. The squirrel's position vector makes equal angles with the three positive coordinate axes. We want to express the squirrel's position vector in the form of Equation 2.13. The square of the length of the position vector is

$$x^2 + y^2 + z^2 = r^2 = (17.3 \text{ m})^2 = 300 \text{ m}^2$$

By symmetry $x = y = z$. This gives us for the components

$$x = y = z = 10 \text{ m}$$

The position vector **r** can be expressed as

$$\mathbf{r} = (10\mathbf{i} + 10\mathbf{j} + 10\mathbf{k}) \text{ m}$$

Any vector **A** can be represented in similar fashion as a sum of three mutually perpendicular component vectors. If **A** has components (A_x, A_y, A_z), it can be expressed as

$$\mathbf{A} = \mathbf{i}A_x + \mathbf{j}A_y + \mathbf{k}A_z \qquad (2.14)$$

Figure 2.18 illustrates the vector addition described by Equation 2.14.

Vector Algebra in Component Form

Let's reexamine some vector operations in terms of unit vectors and components.

Equality. $\mathbf{A} = \mathbf{B}$ when **A** and **B** have the same length, the same direction, and the same physical dimensions. We can express this equality in terms of components as follows:

$$\mathbf{i}A_x + \mathbf{j}A_y + \mathbf{k}A_z = \mathbf{i}B_x + \mathbf{j}B_y + \mathbf{k}B_z$$

This equation implies that $A_x = B_x$, $A_y = B_y$, and $A_z = B_z$. If two vectors are equal, their x-components are equal to each other, their y-components are equal to each other, and their z-components are equal to each other. Here you can begin to see the power of vector algebra. The single vector equation $\mathbf{A} = \mathbf{B}$ is equivalent to three algebraic equations. The vector equation is thus a compact and powerful way of expressing mathematical and physical relations.

Addition.   The addition of two vectors

$$\mathbf{C} = \mathbf{A} + \mathbf{B}$$

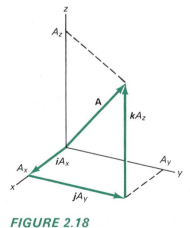

FIGURE 2.18
Components can be used to represent a vector as a sum of three mutually perpendicular vectors.

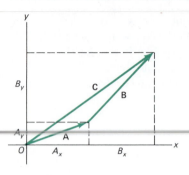

FIGURE 2.19

Vector addition by components. This figure shows only the x- and y-components.

can be expressed as

$$\mathbf{i}C_x + \mathbf{j}C_y + \mathbf{k}C_z = (\mathbf{i}A_x + \mathbf{j}A_y + \mathbf{k}A_z) + (\mathbf{i}B_x + \mathbf{j}B_y + \mathbf{k}B_z)$$
$$= \mathbf{i}(A_x + B_x) + \mathbf{j}(A_y + B_y) + \mathbf{k}(A_z + B_z)$$

Equating corresponding components on both sides of the equation we may write

$$C_x = A_x + B_x \tag{2.15}$$

$$C_y = A_y + B_y \tag{2.16}$$

$$C_z = A_z + B_z \tag{2.17}$$

This method of equating corresponding components can be extended to subtraction. A graphical interpretation of Equation 2.15 and Equation 2.16 is shown in Figure 2.19.

EXAMPLE 7

Vector Addition by Components

A displacement vector $\mathbf{B} = 2\mathbf{i} + 4\mathbf{j} - 3\mathbf{k}$ meters is added to a displacement vector $\mathbf{A} = \mathbf{i} - 2\mathbf{j} + 6\mathbf{k}$ meters. What is the resultant displacement, $\mathbf{C} = \mathbf{A} + \mathbf{B}$?

Adding corresponding components, we write

$$C_x = A_x + B_x = 1\text{ m} + 2\text{ m} = 3\text{ m}$$

$$C_y = A_y + B_y = -2\text{ m} + 4\text{ m} = 2\text{ m}$$

$$C_z = A_z + B_z = 6\text{ m} + (-3\text{ m}) = 3\text{ m}$$

The vector sum of $\mathbf{A}$ and $\mathbf{B}$ can now be written as

$$\mathbf{C} = (3\mathbf{i} + 2\mathbf{j} + 3\mathbf{k})\text{ m}$$

2.4
THE SCALAR PRODUCT

So far the rules of vector algebra have been similar to the rules of ordinary algebra. However, multiplication of two vectors does not correspond to the multiplication of two ordinary numbers. We must therefore define multiplication rules for vectors. The scalar product and the vector product are two forms of vector multiplication that are useful in physics. As you go through this text you will see that these two products have many applications in describing the physical world. It is because they occur so often in applications that we introduce a concise symbolic representation of such products.

The **scalar product** of two vectors $\mathbf{A}$ and $\mathbf{B}$, Figure 2.20, is indicated by a dot between $\mathbf{A}$ and $\mathbf{B}$ and is *defined* as a scalar equal to the product of the two vector magnitudes and the cosine of the included angle.

$$\mathbf{A} \cdot \mathbf{B} = AB \cos \theta \tag{2.18}$$

$\mathbf{A} \cdot \mathbf{B}$ is read "$\mathbf{A}$ dot $\mathbf{B}$." $\mathbf{A} \cdot \mathbf{B}$ is called the scalar product or the **dot product**.

If we form the dot product of $\mathbf{A}$ with $\mathbf{A}$ itself, then $\theta = 0$, and $\cos \theta = 1$. In this case the dot product is the square of the magnitude of that vector.

$$\mathbf{A} \cdot \mathbf{A} = A^2 \cos 0° = A^2$$

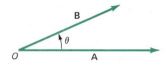

FIGURE 2.20

When **A** and **B** are perpendicular, $\theta = 90°$ and $\cos \theta = 0$. In this case **A** · **B** = 0, even though neither **A** nor **B** are zero.

EXAMPLE 8

Scalar Products of Unit Vectors

Applying the definition of the scalar product to the unit vectors **i**, **j**, and **k**, we see that when one of these unit vectors is dotted into itself the result is 1. Thus,

$$\mathbf{i} \cdot \mathbf{i} = (1)(1) \cos 0° = 1$$

Similarly,

$$\mathbf{j} \cdot \mathbf{j} = 1 \qquad \mathbf{k} \cdot \mathbf{k} = 1$$

When one of these unit vectors is dotted into a different one of these unit vectors we get 0 because the cosine of the included angle (90°) is zero.

$$\mathbf{i} \cdot \mathbf{j} = (1)(1) \cos 90° = 0$$

Similarly,

$$\mathbf{i} \cdot \mathbf{k} = 0 \qquad \mathbf{j} \cdot \mathbf{k} = 0$$

The scalar product of two vectors may be written in terms of the components as

$$\mathbf{A} \cdot \mathbf{B} = (\mathbf{i}A_x + \mathbf{j}A_y + \mathbf{k}A_z) \cdot (\mathbf{i}B_x + \mathbf{j}B_y + \mathbf{k}B_z)$$

The scalar product obeys distributive and associative laws, which we can use to get

$$\begin{aligned}
\mathbf{A} \cdot \mathbf{B} = {} & (\mathbf{i} \cdot \mathbf{i})A_xB_x + (\mathbf{i} \cdot \mathbf{j})A_xB_y + (\mathbf{i} \cdot \mathbf{k})A_xB_z \\
& + (\mathbf{j} \cdot \mathbf{i})A_yB_x + (\mathbf{j} \cdot \mathbf{j})A_yB_y + (\mathbf{j} \cdot \mathbf{k})A_yB_z \\
& + (\mathbf{k} \cdot \mathbf{i})A_zB_x + (\mathbf{k} \cdot \mathbf{j})A_zB_y + (\mathbf{k} \cdot \mathbf{k})A_zB_z
\end{aligned}$$

The six scalar products involving two different unit vectors equal zero. The three scalar products involving the same unit vectors equal unity. This leaves

$$\mathbf{A} \cdot \mathbf{B} = A_xB_x + A_yB_y + A_zB_z \tag{2.19}$$

This form of the scalar product is particularly useful when the vectors are specified in terms of their components. Equation 2.19 also shows that the scalar product is commutative:

$$\mathbf{A} \cdot \mathbf{B} = \mathbf{B} \cdot \mathbf{A}$$

Using the two forms of the scalar product equation, Equation 2.18 and Equation 2.19, we obtain a useful relation for the angle between two vectors.

$$\cos \theta = \frac{\mathbf{A} \cdot \mathbf{B}}{AB} = \frac{A_xB_x + A_yB_y + A_zB_z}{AB}$$

If we express A and B in terms of their components

$$AB = (A_x{}^2 + A_y{}^2 + A_z{}^2)^{1/2}(B_x{}^2 + B_y{}^2 + B_z{}^2)^{1/2}$$

we obtain an equation that lets us use components to determine the angle between two vectors

$$\cos \theta = \frac{A_xB_x + A_yB_y + A_zB_z}{(A_x{}^2 + A_y{}^2 + A_z{}^2)^{1/2}(B_x{}^2 + B_y{}^2 + B_z{}^2)^{1/2}} \tag{2.20}$$

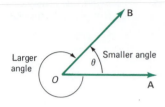

FIGURE 2.21

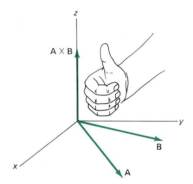

FIGURE 2.22

Right-hand rule for the vector product $\mathbf{A} \times \mathbf{B}$.

Visualizing vectors in three dimensions is not difficult. But, can you visualize a pair of vectors in three dimensions, and the angle between them, and then the problem of determining that angle? Equation 2.20 shows one of the advantages of using vector algebra to solve such problems.

EXAMPLE 9

Angle Between Two Vectors

To illustrate Equation 2.20, let's find the angle between the two vectors $\mathbf{A}$ and $\mathbf{B}$:

$$\mathbf{A} = 3\mathbf{i} - 2\mathbf{j} + \mathbf{k}$$

$$\mathbf{B} = 2\mathbf{i} + 4\mathbf{j} - 3\mathbf{k}$$

You might want to sketch $\mathbf{A}$ and $\mathbf{B}$ to get a rough idea of the angle between them. Equation 2.20 gives

$$\cos \theta = \frac{(3)(2) + (-2)(4) + (1)(-3)}{(3^2 + 2^2 + 1^2)^{1/2}(2^2 + 4^2 + 3^2)^{1/2}}$$

$$= \frac{-5}{\sqrt{(14)(29)}} = -0.248$$

The angle θ is

$$\theta = \cos^{-1}(-0.248) = 104°$$

2.5
THE VECTOR PRODUCT

It is useful to define an alternate way of multiplying two vectors so that their product yields a *vector*. This is accomplished with the **vector product**, or **cross product**, of two vectors. The symbolism $\mathbf{A} \times \mathbf{B}$ is used to denote the vector product which is read as "$\mathbf{A}$ cross $\mathbf{B}$." The magnitude of $\mathbf{A} \times \mathbf{B}$ is **defined** as the product of the magnitudes of $\mathbf{A}$ and $\mathbf{B}$ and the magnitude of the sine of the smaller angle between $\mathbf{A}$ and $\mathbf{B}$ (Figure 2.21):

$$|\mathbf{A} \times \mathbf{B}| = AB \sin \theta \qquad (2.21)$$

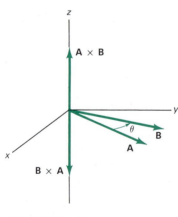

FIGURE 2.23

Interchanging the order of the vectors changes the sign of their vector product; $\mathbf{A} \times \mathbf{B} = -\mathbf{B} \times \mathbf{A}$. The vectors $\mathbf{A}$ and $\mathbf{B}$ lie in the x–y plane. The vector product $\mathbf{A} \times \mathbf{B}$ is directed in the positive z-direction. The vector product $\mathbf{B} \times \mathbf{A}$ is directed in the negative z-direction.

The direction of $\mathbf{A} \times \mathbf{B}$ is **defined** to be perpendicular to both $\mathbf{A}$ and $\mathbf{B}$ and is therefore perpendicular to the plane defined by $\mathbf{A}$ and $\mathbf{B}$. There are, however, two possible directions perpendicular to this plane. We remove this ambiguity by using the right-hand rule. As shown in Figure 2.22, let the fingers of your right hand curl from the first vector $\mathbf{A}$ to the second vector $\mathbf{B}$ through the smaller angle joining them. Your extended thumb then points in the direction of $\mathbf{A} \times \mathbf{B}$. If $\mathbf{A}$ and $\mathbf{B}$ both lie in the x–y plane, as shown in Figure 2.22, the vector product $\mathbf{A} \times \mathbf{B}$ will be in the positive z-direction.

Interchanging $\mathbf{A}$ and $\mathbf{B}$ reverses the sign of the cross product. To see why, again use the right-hand rule, letting the fingers of your right hand curl from the first vector $\mathbf{B}$ to the second vector $\mathbf{A}$ through the smaller angle. Your extended thumb now points along the negative z-axis and gives the direction of $\mathbf{B} \times \mathbf{A}$ (Figure 2.23). The magnitude of $\mathbf{B} \times \mathbf{A}$ is $BA \sin \theta$, which also equals the magnitude of $\mathbf{A} \times \mathbf{B}$. Thus, $\mathbf{A} \times \mathbf{B}$ and $\mathbf{B} \times \mathbf{A}$ are equal in magnitude but opposite in direction.

$$\mathbf{A} \times \mathbf{B} = -(\mathbf{B} \times \mathbf{A}) \qquad (2.22)$$

This definition of $\mathbf{A} \times \mathbf{B}$ may seem complicated and you may rightfully ask, "Why invent $\mathbf{A} \times \mathbf{B}$?" The pragmatic answer is that the vector product is very useful. It provides a concise representation of many physical quantities—including torque, angular momentum, magnetic force, and the flow of energy in an electromagnetic wave.

The cross products of the unit vectors $\mathbf{i}$, $\mathbf{j}$, and $\mathbf{k}$ with themselves and with one another are basic to evaluating cross products of vectors in general. Let's start by considering the cross products $\mathbf{i} \times \mathbf{i}$, $\mathbf{j} \times \mathbf{j}$, and $\mathbf{k} \times \mathbf{k}$. The vector product of any vector with itself is zero because the included angle is zero and $\sin 0° = 0$. Thus the vector product of any unit vector, $\mathbf{i}$, $\mathbf{j}$, or $\mathbf{k}$, with itself is zero.

$$\mathbf{i} \times \mathbf{i} = 0 \qquad \mathbf{j} \times \mathbf{j} = 0 \qquad \mathbf{k} \times \mathbf{k} = 0 \qquad (2.23)$$

The vector product of any two different of these unit vectors is the other unit vector. For example, $\mathbf{i} \times \mathbf{j} = \mathbf{k}$. To prove this, use the right-hand rule to see that $\mathbf{i} \times \mathbf{j}$ is directed along the positive z-axis. The magnitude of $\mathbf{i} \times \mathbf{j}$ is

$$|\mathbf{i} \times \mathbf{j}| = (1)(1) \sin 90° = 1$$

Thus, $\mathbf{i} \times \mathbf{j}$ is a vector of unit magnitude along the positive z-axis, which is the unit vector $\mathbf{k}$.

There are six nonzero vector products involving $\mathbf{i}$, $\mathbf{j}$, and $\mathbf{k}$:

$$\begin{aligned} \mathbf{i} \times \mathbf{j} = \mathbf{k} \qquad \mathbf{j} \times \mathbf{k} = \mathbf{i} \qquad \mathbf{k} \times \mathbf{i} = \mathbf{j} \\ \mathbf{j} \times \mathbf{i} = -\mathbf{k} \qquad \mathbf{k} \times \mathbf{j} = -\mathbf{i} \qquad \mathbf{i} \times \mathbf{k} = -\mathbf{j} \end{aligned} \qquad (2.24)$$

We can use these vector products of the unit vectors to develop the component form of the vector product

$$\mathbf{A} \times \mathbf{B} = (\mathbf{i}A_x + \mathbf{j}A_y + \mathbf{k}A_z) \times (\mathbf{i}B_x + \mathbf{j}B_y + \mathbf{k}B_z)$$

The vector product obeys associative and distributive laws that we can use to write $\mathbf{A} \times \mathbf{B}$ as

$$\begin{aligned} \mathbf{A} \times \mathbf{B} = \mathbf{i} \times \mathbf{i}A_xB_x + \mathbf{i} \times \mathbf{j}A_xB_y + \mathbf{i} \times \mathbf{k}A_xB_z \\ + \mathbf{j} \times \mathbf{i}A_yB_x + \mathbf{j} \times \mathbf{j}A_yB_y + \mathbf{j} \times \mathbf{k}A_yB_z \\ + \mathbf{k} \times \mathbf{i}A_zB_x + \mathbf{k} \times \mathbf{j}A_zB_y + \mathbf{k} \times \mathbf{k}A_zB_z \end{aligned}$$

Three of the vector products are zero and the other six give unit vectors. Equation 2.23 and Equation 2.24 yield

$$\begin{aligned} \mathbf{A} \times \mathbf{B} = \mathbf{i}(A_yB_z - A_zB_y) + \mathbf{j}(A_zB_x - A_xB_z) \\ + \mathbf{k}(A_xB_y - A_yB_x) \end{aligned} \qquad (2.25)$$

This shows that $\mathbf{A} \times \mathbf{B}$ can be expressed in terms of the components of $\mathbf{A}$ and $\mathbf{B}$ as

$$(\mathbf{A} \times \mathbf{B})_x = A_yB_z - A_zB_y \qquad (2.26)$$

$$(\mathbf{A} \times \mathbf{B})_y = A_zB_x - A_xB_z \qquad (2.27)$$

$$(\mathbf{A} \times \mathbf{B})_z = A_xB_y - A_yB_x \qquad (2.28)$$

Equations 2.26, 2.27, and 2.28 may seem formidable, but in many cases they can be used to greatly simplify a problem. For example, if $\mathbf{A}$ and $\mathbf{B}$ lie in the x–y plane (or if we orient the coordinates so that the x–y plane coincides with the plane determined by $\mathbf{A}$ and $\mathbf{B}$), then $A_z = 0$ and $B_z = 0$. Substituting these values into Equation 2.26 and Equation 2.27, we get $C_x = 0$, and $C_y = 0$. Only the z-component, C_z, is nonzero.

$$(iA_x + jA_y) \times (iB_x + jB_y) = k(A_xB_y - A_yB_x) \tag{2.29}$$

The cross-product vector is perpendicular to the plane of the two vectors (the x–y plane).

The vector product of two vectors has a useful geometric interpretation: *area*. As shown in Figure 2.24, **A** and **B** are two vectors in the x-y plane. The magnitude of the vector product **A** × **B** may be written

$$\left| \mathbf{A} \times \mathbf{B} \right| = A(B \sin \theta)$$

But $B \sin \theta$ is the perpendicular distance from the tip of **B** to **A** (Figure 2.24). If A is the base of the shaded parallelogram, then $B \sin \theta$ is its height and $AB \sin \theta$ is its area. Thus, we have the useful relation,

$$\left| \mathbf{A} \times \mathbf{B} \right| = \text{area of parallelogram with sides } \mathbf{A} \text{ and } \mathbf{B} \tag{2.30}$$

EXAMPLE 10

Triangle Area

Consider a triangle (Figure 2.25) with vertices at (0, 0, 0), (2, 4, 3), and (4, 0, 4). We want to use a vector product to determine the area of this triangle.

From Figure 2.25 we see that the area of the triangle is half the area of the parallelogram having sides **A** and **B**, where

$$\mathbf{A} = \text{vector from (0, 0, 0) to (2, 4, 3)} = 2\mathbf{i} + 4\mathbf{j} + 3\mathbf{k}$$

and

$$\mathbf{B} = \text{vector from (0, 0, 0) to (4, 0, 4)} = 4\mathbf{i} + 4\mathbf{k}$$

If we denote the vector product **A** × **B** as the vector **C** then

$$\mathbf{C} = \mathbf{A} \times \mathbf{B}$$

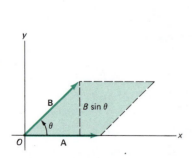

FIGURE 2.24

The magnitude of the cross product **A** × **B** equals the area of the parallelogram with sides formed by **A** and **B**.

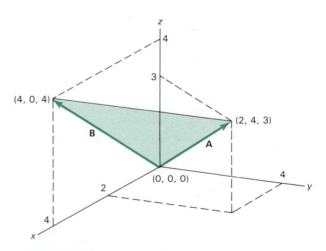

FIGURE 2.25

A triangle with two sides formed by the vectors **A** and **B**. Its area is half the magnitude of **A** × **B**.

and

$$\text{Area} = \tfrac{1}{2}\left|\mathbf{A} \times \mathbf{B}\right| = \tfrac{1}{2}C$$

For $\mathbf{C} = \mathbf{A} \times \mathbf{B}$ we find

$$C_x = A_y B_z - A_z B_y = 4 \cdot 4 - 3 \cdot 0 = 16$$
$$C_y = A_z B_x - A_x B_z = 3 \cdot 4 - 2 \cdot 4 = 4$$
$$C_z = A_x B_y - A_y B_x = 2 \cdot 0 - 4 \cdot 4 = -16$$

The magnitude of $\mathbf{C}$ is

$$C = (C_x{}^2 + C_y{}^2 + C_z{}^2)^{1/2} = (16^2 + 4^2 + 16^2)^{1/2} = 23.0$$

The area of the triangle is

$$\text{Area} = \tfrac{1}{2}\left|\mathbf{A} \times \mathbf{B}\right| = \tfrac{1}{2}C = 11.5$$

WORKED PROBLEM

Find a unit vector in the x–y plane perpendicular to a line drawn from the point $(2, 5, 0)$ to the point $(-3, 8, 0)$ (Figure 2.26).

Solution

A vector from the origin to $(2, 5, 0)$ may be written as $\mathbf{A} = 2\mathbf{i} + 5\mathbf{j}$; a vector from the origin to $(-3, 8, 0)$ may be written as $\mathbf{B} = -3\mathbf{i} + 8\mathbf{j}$. The vector

$$\mathbf{C} = \mathbf{B} - \mathbf{A} = (-3 - 2)\mathbf{i} + (8 - 5)\mathbf{j} = -5\mathbf{i} + 3\mathbf{j}$$

is directed from $(2, 5, 0)$ to $(-3, 8, 0)$. Let

$$\mathbf{D} = x\mathbf{i} + y\mathbf{j}$$

denote the unit vector we want to determine. There are two unknowns, x and y, so we need two equations. The two equations relating x and y describe the fact that $\mathbf{D}$ is a unit vector and that $\mathbf{D}$ is perpendicular to $\mathbf{C}$.

For $\mathbf{D}$ to be perpendicular to $\mathbf{C}$ we must have $\mathbf{C} \cdot \mathbf{D} = 0$. In terms of the components of $\mathbf{C}$ and $\mathbf{D}$,

$$\mathbf{C} \cdot \mathbf{D} = -5x + 3y = 0$$

Therefore, $y = (5/3)x$.

For $\mathbf{D}$ to be a unit vector we must have $\mathbf{D} \cdot \mathbf{D} = 1$. In terms of the components of $\mathbf{D}$,

$$\mathbf{D} \cdot \mathbf{D} = x^2 + y^2 = 1$$

Setting $y = (5/3)x$ and solving for x gives

$$x = \frac{3}{\sqrt{34}} = 0.5145$$

The value of y then follows as

$$y = \frac{5}{\sqrt{34}} = 0.8575$$

Hence the vector of interest is

$$\mathbf{D} = 0.5145\mathbf{i} + 0.8575\mathbf{j}$$

There is a second solution because $x^2 + y^2 = 1$ is a quadratic equation. The second solution is the negative of the first solution.

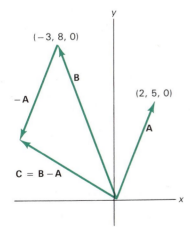

FIGURE 2.26

EXERCISES

2.2   Addition and Subtraction of Vectors

A. Using a graphical technique, find the vector sum $\mathbf{A} + \mathbf{B}$ and the vector difference $\mathbf{A} - \mathbf{B}$ of the two force vectors shown in Figure 1. (Note: A graphical solution means that you draw the vectors with accurately measured magnitude and direction and then measure from your drawing the magnitude and direction of the resultant vector.)

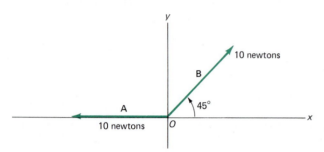

FIGURE 1

B. The magnitudes of vectors $\mathbf{C}$ and $\mathbf{D}$ are different. Is it possible to have $\mathbf{C} + \mathbf{D} = 0$? If you think so, illustrate with a diagram.

C. Is it possible to add three vectors of equal magnitude but different directions to get zero? If you think so, illustrate with a diagram.

D. A force vector $\mathbf{F}_1$ of magnitude 6 N acts at the origin in a direction $30°$ above the positive x-axis. A second force vector $\mathbf{F}_2$ of magnitude 5 N acts at the origin in the direction of the positive y-axis. Find graphically the magnitude and direction of the resultant force vector $\mathbf{F}_1 + \mathbf{F}_2$.

E. The velocity of sailboat A relative to sailboat B is defined by the equation $\mathbf{v}_{AB} = \mathbf{A} - \mathbf{B}$, where $\mathbf{A}$ is the velocity of sailboat A, and $\mathbf{B}$ is the velocity of sailboat B. Determine $\mathbf{v}_{AB}$ if

$$\mathbf{A} = 30 \text{ km/h east} \qquad \mathbf{B} = 40 \text{ km/h north}$$

F. Two students exert forces having magnitudes of 100 N and 50 N on the arms of their physics teacher. Using the principles of vector addition, draw vector diagrams showing the situations that give the largest and smallest net force they can exert. Determine the magnitudes of the maximum and minimum forces.

2.3   Components

G. In a certain experiment two electric fields (vectors) are present. The first field $\mathbf{P}$ has components $P_x = 21$, $P_y = 68$, $P_z = -18$, and the second field $\mathbf{R}$ has components $R_x = 10$, $R_y = -16$, $R_z = 0$, in units of volts per meter. Find the components and the magnitude of the combined electric field $\mathbf{E} = \mathbf{P} + \mathbf{R}$.

H. A barrel of beer rolls down a plane inclined at $35°$ to the horizontal. The force of gravity on the barrel is 450 N vertically downward. Find the component of this gravitational force parallel to the inclined plane.

I. A football leaves the foot of a punter at an angle of $36°$ with respect to the vertical (positive y-direction) at a speed of 21 m/s. Determine the horizontal and vertical components of the velocity.

2.4   The Scalar Product

J. Given the vectors $\mathbf{A} = 2\mathbf{i} + 4\mathbf{j} - \mathbf{k}$ and $\mathbf{B} = 3\mathbf{i} + 2\mathbf{j} - 3\mathbf{k}$, calculate the scalar product $\mathbf{A} \cdot \mathbf{B}$.

K. Vector $\mathbf{A}$ has a magnitude 2 and is directed along the positive x-axis. Vector $\mathbf{B}$ has a magnitude 4 and makes an angle of $45°$ with respect to the positive x-axis (Figure 2). Find the scalar product $\mathbf{A} \cdot \mathbf{B}$.

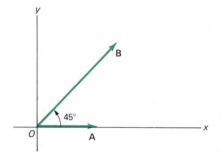

FIGURE 2

L. The force of gravity $\mathbf{F}$ on the beer barrel (Exercise H) is vertically downward. Calculate the scalar product of $\mathbf{F}$ and the displacement vector $\mathbf{s}$ of the barrel when the barrel moves 2 m down the inclined plane.

2.5   The Vector Product

M. Given $\mathbf{M} = 6\mathbf{i} + 2\mathbf{j} - \mathbf{k}$ and $\mathbf{N} = 2\mathbf{i} - \mathbf{j} - 3\mathbf{k}$, calculate the vector product $\mathbf{M} \times \mathbf{N}$.

N. At a certain instant the position of a stone in a sling is given by $\mathbf{r} = 1.7\mathbf{i}$ m. The linear momentum $\mathbf{p}$ of the stone is $12\mathbf{j}$ kg·m/s. Calculate the angular momentum $\mathbf{L} = \mathbf{r} \times \mathbf{p}$.

PROBLEMS

1. The driver of a car, obviously lost, drives 3 km north, 2 km northeast ($45°$ east of north), 4 km west, and then 3 km southeast ($45°$ east of south). Where does he end up relative to his starting point? Work out your answer graphically. Check by using components. (The car is not near the North Pole or the South Pole.)

2. Vector **B** has x-, y-, and z-components of 4, 6, and 3, respectively. Calculate B and the angles that **B** makes with the coordinate axes.

3. The vertices of a triangle, A, B, and C, are given by the points $(1, 2, 1)$, $(3, 4, 0)$, and $(-1, 5, -1)$, respectively. All lengths are in meters. Find the components of the vector that starts at A and ends at B.

4. A police radar detector pointing $25°$ away from the direction of the highway records the speed of a car at point P as 50 mph (Figure 3). This is actually the component of the car's velocity along the direction of the radar beam. What is the actual speed of the car along the highway?

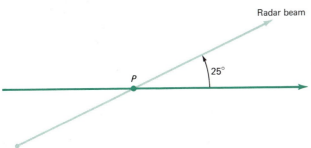

FIGURE 3

5. Position vectors $\mathbf{r}_1$ and $\mathbf{r}_2$ join the origin and the points $(4, -1, 2)$ and $(2, 3, 1)$. Calculate the scalar product $\mathbf{r}_1 \cdot \mathbf{r}_2$.

6. Find the angle between $\mathbf{F} = 2\mathbf{i} + 3\mathbf{j} - \mathbf{k}$ and $\mathbf{G} = 3\mathbf{j} - \mathbf{k}$.

7. Certain theorems of trigonometry can be proved easily using vector algebra. Expand the right side of the equation

$$\mathbf{C} \cdot \mathbf{C} = (\mathbf{A} + \mathbf{B}) \cdot (\mathbf{A} + \mathbf{B})$$

and derive the law of cosines, $C^2 = A^2 + B^2 + 2AB \cos \theta$ (Figure 4).

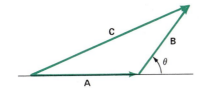

FIGURE 4

8. Compute the area of the triangle in Problem 3 using the geometric interpretation of the vector product.

9. The equation $\mathbf{F} = q(\mathbf{v} \times \mathbf{B})$ gives the force $\mathbf{F}$ on an electric charge q moving with velocity $\mathbf{v}$ in a magnetic field $\mathbf{B}$. Calculate the force vector for a proton ($q = 1.60 \times 10^{-19}$ coulomb) moving with a velocity $(2.1\mathbf{i} + 0.2\mathbf{j} + 1.3\mathbf{k}) \times 10^5$ m/s in a magnetic field of $0.62\mathbf{k}$ tesla. These SI units yield the force in newtons.

10. A rectangular parallelepiped has edges of 4 m, 2 m, and 1 m. Use vector methods to determine the angles (α, β, γ) between the edges and the main body diagonal (Figure 5).

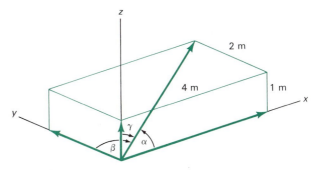

FIGURE 5

11. Given a vector $\mathbf{A} = 2\mathbf{i} + 6\mathbf{j}$, find a vector $\mathbf{B}$ such that $\mathbf{A} \cdot \mathbf{B} = 2$ and $\mathbf{A} \times \mathbf{B} = 2\mathbf{k}$.

12. $N - 1$ unit vectors are produced by rotating the unit vector $\mathbf{i}$ counterclockwise in angular steps of $2\pi/N$. When these $N - 1$ unit vectors are added to $\mathbf{i}$ the vector sum is zero. Prove this geometrically. We suggest that you start with a specific case, such as $N = 3$. This specific case may clue you on how to approach the general case.

13. The tetrahedron $ABCD$ (Figure 6) is characterized by an edge length L. Its base lies in the x–y plane with one edge on the y-axis. (a) Treat the edges as vectors and express the edges A to B, A to C, and A to D in terms of the unit vectors $\mathbf{i}$, $\mathbf{j}$, and $\mathbf{k}$. (b) Use the geometric interpretation of the cross product to prove that the faces ABD and ACD have equal areas.

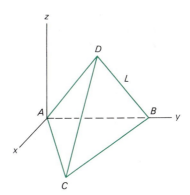

FIGURE 6

14. The volume of a parallelepiped is given by

$$V = \mathbf{A} \cdot (\mathbf{B} \times \mathbf{C})$$

where **A**, **B**, and **C** are vectors along the edges of the parallelepiped, starting at one of its corners. The primitive cell for a body-centered cubic lattice forms a parallelepiped with edge vectors

$$\mathbf{A} = \frac{L(\mathbf{i}+\mathbf{j})}{2} \qquad \mathbf{B} = \frac{L(\mathbf{j}+\mathbf{k})}{2} \qquad \mathbf{C} = \frac{L(\mathbf{i}+\mathbf{k})}{2}$$

where L is a length. In terms of L, determine the volume of the primitive cell.

15. Given that the three vectors **a**, **b**, and **c** describe a crystal lattice, the reciprocal lattice vectors are defined by

$$\mathbf{A} = \frac{2\pi(\mathbf{b} \times \mathbf{c})}{V} \qquad \mathbf{B} = \frac{2\pi(\mathbf{c} \times \mathbf{a})}{V}$$

$$\mathbf{C} = \frac{2\pi(\mathbf{a} \times \mathbf{b})}{V}$$

where $V = \mathbf{a} \cdot (\mathbf{b} \times \mathbf{c})$ is a volume. The hexagonal lattice is described by the vectors

$$\mathbf{a} = \frac{(\sqrt{3}\,\mathbf{i}+\mathbf{j})L}{2} \qquad \mathbf{b} = \frac{(-\sqrt{3}\,\mathbf{i}+\mathbf{j})L}{2}$$

$$\mathbf{c} = W\mathbf{k}$$

(a) Prove that $V = \sqrt{3}\,L^2W/2$. (b) Calculate the reciprocal lattice vectors **A**, **B**, and **C**.

16. Given $\mathbf{A} = \mathbf{i}+\mathbf{j}$, $\mathbf{B} = \mathbf{i}-\mathbf{j}$, and $\mathbf{C} = \mathbf{i}+\mathbf{k}$, determine (a) $(\mathbf{A} \times \mathbf{B}) \cdot \mathbf{C}$, (b) $\mathbf{A} \times (\mathbf{B} \times \mathbf{C})$

17. Use the scalar product to prove that $\mathbf{A} = \mathbf{i}+\mathbf{j}+\sqrt{2}\,\mathbf{k}$ is perpendicular to $\mathbf{B} = \mathbf{i}+\mathbf{j}-\sqrt{2}\,\mathbf{k}$.

18. Given $\mathbf{A} = \mathbf{i}+\mathbf{j}+\mathbf{k}$ and $\mathbf{B} = \mathbf{i}+\mathbf{j}-\mathbf{k}$, determine (a) A, (b) B, (c) $\mathbf{A} \cdot \mathbf{B}$, (d) the angle between **A** and **B**.

19. Construct a unit vector that makes equal angles with the positive x, y, and z axes. Express this vector in terms of **i**, **j**, and **k**.

CHAPTER 3

EQUILIBRIUM OF RIGID BODIES

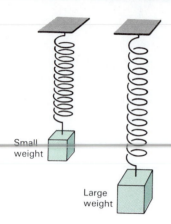

FIGURE 3.1

A coiled spring stretches more when a large weight is attached than when a small weight is attached.

3.1
FORCE

The word **force** has different meanings for different people. Debaters use forceful arguments; the local police force maintains law and order. Intuitively, you probably think of a force as a push or a pull. This is a good starting point. Now we need to sharpen it.

In physics *force* represents the interaction of a body with its environment. Physicists recognize four basic or **fundamental** forces:

1. gravitational force
2. weak nuclear force
3. electromagnetic force
4. strong nuclear force

Forces in nature, such as those associated with friction, water, and wind, can be understood in terms of these four fundamental forces. Much research going on now is aimed at increasing our understanding of these forces. This research seeks to find a common basis, or a unification, for all forces.*

Force is defined in terms of its effects. When a force acts on a body, the effect is a deformation of the body and/or a change in the state of motion of the body. In Chapter 6 we develop an operational definition of force in terms of Newton's second law. In this chapter we introduce a method for *measuring* force.

Force can be measured in terms of the stretch of a spring. When a coiled spring hangs vertically, its length depends on the weight attached, as suggested by Figure 3.1.† The force that stretches the spring is a result of gravity acting on the suspended weight. Each time a particular weight is attached the spring stretches to a definite and reproducible length. Different stretches correspond to different forces or weights.

Suppose the object we attach to the spring is the standard kilogram. Let F denote the force experienced by the spring when the standard kilogram is attached. Any object that produces the same stretch as the standard kilogram exerts the same force.

We are free to define the magnitude of the force exerted by the standard kilogram. We choose the value

$$F = 9.80 \text{ newtons}$$

This choice calibrates the spring in newtons, the SI unit of force. The reason for choosing the value 9.80 will become apparent when we introduce Newton's second law in Chapter 6.

If we assume that mass is an additive property, we can produce multiples and submultiples of the standard kilogram. Using these multiples and submultiples we can calibrate a spring scale of force in newtons (Figure 3.2).

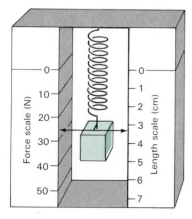

FIGURE 3.2

With no weight attached the spring is relaxed and both pointers indicate zero readings. When the spring stretches 3.3 cm the scale pointer indicates a 25-N force.

---

* Experimental and theoretical efforts by physicists have demonstrated that electromagnetism and the weak nuclear force are different aspects of a single more fundamental interaction, the **electroweak force**. The theory behind this effort is called the **grand unified theory** (GUT). Physicists are also trying to explain the strong nuclear force and the electroweak force as aspects of a single basic force.

† The force of gravity exerted on a body is called the **weight** of the body.

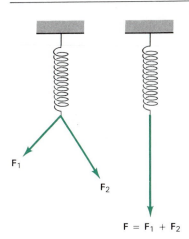

FIGURE 3.3

A pair of forces (F_1 and F_2) acting on the spring produce the same effect as a single force (F) equal to their vector sum. This experimental result shows that force is a vector quantity.

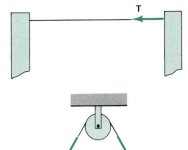

FIGURE 3.4

(a) The force **T** is exerted on the post by the string. We call this force the tension in the string. A force of equal magnitude and opposite direction is exerted on the string by the post.
(b) A frictionless pulley changes the direction of the tension, but not its magnitude.

FIGURE 3.5

An ice skater coasting without friction experiences only an upward thrust by the ice.

EXAMPLE 1

From Pounds to Newtons

The newton (abbreviated N) is the SI unit of force. You are likely to be more familiar with the pound, the British unit of force. The relationship between these units is

$$1 \text{ pound} = 4.448 \text{ N}$$

A student weighs 113 pounds on her bathroom scale. Let's determine her weight in newtons:

$$113 \text{ lb} \left(\frac{4.448 \text{ N}}{1 \text{ lb}} \right) = 503 \text{ N}$$

The weight of an object in newtons is about four and a half times its weight in pounds.

Force has both magnitude and direction. Is force a vector quantity? Experiments prove that forces are vector quantities because they add as vectors. Figure 3.3 shows how experiments demonstrate the vector character of force. A pair of forces, F_1 and F_2, act to stretch the spring. If these two forces are replaced by a single force equal to the vector sum of F_1 and F_2, then the stretch of the spring is found to be the same as when the two forces acted. Thus a single force **F** given by

$$\mathbf{F} = \mathbf{F}_1 + \mathbf{F}_2$$

is equivalent to F_1 and F_2 acting together, showing that force is a vector quantity.

Contact Forces

Contact forces arise when two bodies touch. A child pushes, but is unable to move, the family refrigerator. The motion of the refrigerator is opposed by contact forces exerted by the floor.

Contact forces are electromagnetic in nature. Two electrically neutral molecules that are far apart exert little force on each other. When they come close together, however, strong mutually repulsive forces arise, preventing the molecules from coming any closer together. It is impractical to deal with contact forces on a molecule-by-molecule basis. The vector character of force lets us deal with a single force—the vector sum of the many microscopic molecular forces.

When two objects are in contact, the direction of the forces they exert on each other is not always obvious. In some instances the direction of a contact force can be inferred from the nature of the objects. For example, a flexible string can exert only a pull in a direction parallel to its length, Figures 3.4a, 3.4b. We say that the string is under **tension** or that it exerts a tensile force. A smooth or frictionless surface like a sheet of ice can exert a force perpendicular to and away from the surface (Figure 3.5). We call such a force a **push** or a **thrust**. Often, however, you do not know the direction of the contact force in advance. In such cases it is often advisable to resolve the contact force into its components parallel to and perpendicular to the surface. For example, a runner's shoes experience contact forces (Figure 3.6). These forces may be resolved into a thrust-like component perpendicular to the surface and a frictional component parallel to the surface. We often refer to the perpendicular

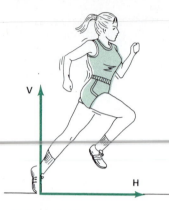

The runner experiences a contact force exerted on her foot by the track. The normal component of this force is the thrust, **V**. The tangential component is the frictional force, **H**.

component as the **normal** force. The component parallel to the surface is often referred to as a **tangential** force. Both are merely components of a single contact force.

Rigid Bodies

If forces act on an object it is deformed. For example, a steel beam may bend slightly because of the weight it supports. If the deformation of the object is negligible, it is called a **rigid body**. The notion of a rigid body will prove especially useful in our studies.

3.2
THE FIRST CONDITION OF EQUILIBRIUM

States of Equilibrium

If forces act on a body they may change its state of motion. **Translational motion** causes a change in position. **Rotational motion** causes a change in orientation. If certain conditions are satisfied, the forces maintain states of **translational equilibrium** and **rotational equilibrium**.

Translational equilibrium is defined as a state of rest or a state of motion with constant velocity. A car at rest or moving down a road in a straight line with a constant speed is in translational equilibrium. Rotational equilibrium is defined as a state of no rotation or a state of rotation at a constant rate. A bicycle wheel that is not rotating, or that is rotating about a fixed axis at a constant number of revolutions per second, is in rotational equilibrium.

The principles and techniques we develop in this chapter are applicable to bodies in translational and rotational equilibrium. However, in this chapter we deal only with rigid bodies that are at rest in a state of **static equilibrium**.

Experiment shows that the forces acting on a body in translational equilibrium satisfy the *first condition of equilibrium*:

A body is in translational equilibrium if and only if the *net* force exerted *on* the body equals zero.

The net force is the vector sum of the forces acting on the body. When you stand at rest you are in translational equilibrium under the action of three forces: the downward pull of gravity and the upward thrust of the contact forces exerted on your feet (Figure 3.7).

We can express the first condition in symbolic form by the vector equation

$$\sum \mathbf{F} = 0 \qquad (3.1)$$

The symbol $\sum$ means "the sum of." In Equation 3.1, $\sum \mathbf{F}$ denotes the vector sum of all forces acting on the body. For the person standing at rest in Figure 3.7, the summation implied by $\sum \mathbf{F}$ reads

$$\mathbf{C}_1 + \mathbf{C}_2 + \mathbf{W} = 0$$

Equation 3.1 is a vector equation. It is equivalent to three independent algebraic equations involving the components of the forces.

$$\sum F_x = 0$$
$$\sum F_y = 0 \qquad (3.2)$$
$$\sum F_z = 0$$

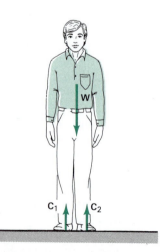

A person standing at rest experiences three forces. When he is in equilibrium, the vector sum of the forces is zero.

Although the first condition of equilibrium also applies to objects in motion with a constant velocity, we are concerned in this chapter with statics, that is, the equilibrium of objects at rest.

In order to apply the first condition of equilibrium, we imagine that the universe is divided into two parts. The part we choose to study is called the **body** and everything else is called the **environment**, which is the part of the universe that interacts with the body. The first condition of equilibrium deals only with **external forces**—forces that are exerted *on* the body by the environment. Forces exerted on the environment *by* the body are not included in the formulation of the first condition of equilibrium.

EXAMPLE 2

Solution by Inspection

A bowl of fruit weighing 4 pounds rests on the top of a table. What is the magnitude of the contact force exerted on the bowl by the table? The answer, by inspection, is 4 pounds. Of course we can prove this by using the first condition of equilibrium. When a body in equilibrium experiences only two forces, the first condition of equilibrium requires

$$\mathbf{F}_1 + \mathbf{F}_2 = 0$$

This shows that the two forces must be equal in magnitude and opposite in direction.

$$\mathbf{F}_1 = -\mathbf{F}_2$$

The bowl of fruit experiences a downward gravitational force of 4 pounds and an upward contact force of 4 pounds.

Problem Solving Guide

Most problems cannot be solved simply by inspection. In physics it pays to be systematic. For many students the most difficult part of physics is the formulation of the equations needed to solve a problem. You may have no trouble solving equations formulated by others, but you may experience difficulty converting word problems into equations. The following guide will help you attack word problems:

Guide to Problem Solving I

1. What's given, what's not? (What information is given explicitly or by implication? What are you asked to determine?)
2. Make an educated guess at the answer. (Right or wrong, this practice will help you develop physical insight.)
3. Make a simple line drawing.
4. Pick a body. (Pick one acted on by one or more of the unknown forces you seek to determine.)
5. Draw a force diagram showing the forces that act on the body you picked. (This may suggest a solution.)
6. Choose coordinate axes and resolve the forces into their components. (Exploit symmetry where possible.)
7. Apply the first condition of equilibrium. (You need one equation for each unknown.)
8. Decide whether or not the solution makes sense.

The following examples elaborate and illustrate the Problem Solving Guide.

EXAMPLE 3

Tension in a Rope

A rope of negligible weight hanging vertically supports a crate weighing 500 N. What is the tension in the rope?

Let's follow the Problem Solving Guide.

(1) The weight of the crate is given and we are asked to determine the tension in the rope. There is one unknown so we will need to apply the first condition of equilibrium only once.

(2) Try to guess the answer. Go ahead, guess!

(3) Make a simple line drawing.

(4) Pick a body. An obvious choice here is the crate.

(5) The line drawing and force diagram that we will use for this problem are shown in Figure 3.8. Forces acting on the crate—the body we picked—include the weight of the crate (**W**), and the tension exerted by the rope (**T**).

(6) We take up as the positive y-direction. The y-component of the tension **T** is $+T$, and it is positive because **T** is directed upward. The y-component of the weight **W** is $-W = -500$ N, and it is negative because **W** is directed downward.

(7) The first condition of equilibrium, Equation 3.2, requires that

$$\sum F_y = T - W = 0$$

Solving for the tension gives

$$T = W$$
$$= 500 \text{ N}$$

The tension in the rope is 500 N, the weight of the crate.

(8) Does the answer make sense? If you didn't guess correctly, try to understand where your thinking went wrong. As a beginner you will not always see that the correct solution is also reasonable. Experience is a great teacher.

Let's consider a variation of this example by specifying that the rope is uniform and has a weight of 20 N. We want to determine the tension in the rope at its midpoint.

Following the Problem Solving Guide, we now have added information; the rope weighs 20 N and it is uniform. We still need to solve for just one quantity so we will

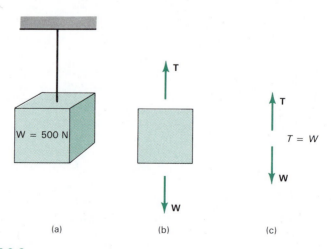

(a)   (b)   (c)

FIGURE 3.8

(a) We take the crate to be the body. (b) The cable exerts an upward force on the crate. Gravity exerts a downward force on the crate. (c) Force diagram.

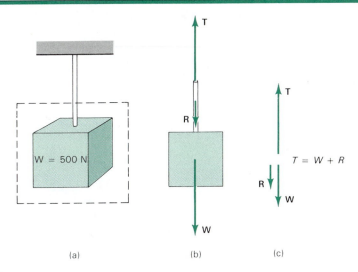

FIGURE 3.9
The tension in the cable is increased if we take into account the weight of the cable. (a) We take the body to be the crate and the lower half of the rope. (b) The upper half of the rope exerts an upward force **T**. Gravity exerts downward forces **R** and **W**. (c) Force diagram.

need to apply the first condition only once. We want to determine the tension in the rope at its midpoint so we pick as a body the lower half of the rope and the crate. Figure 3.9 shows the rope and crate and the forces acting on the body we chose. The upward force **T** is the tension in the rope. The downward forces are **W**, the weight of the crate, and **R**, the weight of the lower half of the rope. Since the rope is uniform the magnitude of **R** is 10 N. If we again take up as the positive y-direction, the first condition of equilibrium takes the form

$$\sum F_y = T - R - W = 0$$

The tension is

$$T = R + W = 10\,\text{N} + 500\,\text{N}$$
$$= 510\,\text{N}$$

If this makes sense to you then you can understand the general result: The tension at any point in a vertical strand of rope in equilibrium equals the total weight suspended beneath that point, including the weight of the portion of the rope beneath it.

In the preceding example, the first condition of equilibrium led to a single equation because the forces all acted vertically. The problem was one-dimensional. In a two-dimensional problem, the forces all lie in a plane. The forces may have horizontal and/or vertical components. In this case two equations follow from the first condition of equilibrium, as we see in the following example.

EXAMPLE 4

Holding a Sled on a Hill

A young girl wants to demonstrate her strength by pulling her brother and his sled up a 30° hill (Figure 3.10). The combined weight of the sled and her brother is 785 N. Although she fails to move him up the hill, she pulls with sufficient force parallel to the slope to prevent her brother and the sled from going downhill. The girl

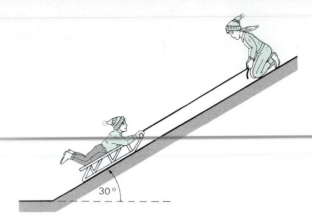

FIGURE 3.10

A sled in equilibrium on a 30° slope.

is at rest on a bare spot, but the sled is on an icy (frictionless) surface. We want to determine the tension in the cord, assuming that the weight of the cord is negligible.

To determine the cord tension we must pick a body on which this force acts. We have two choices: one, the girl, or two, the sled plus the brother. The girl's weight is not known, nor is the force exerted on the girl by the hill. The combined weight of the sled plus the brother *is* known. In addition, we know the direction of the force exerted on the sled by the hill. Because we have more information regarding forces acting on the sled plus the brother, we choose this as the body. We consider the sled *and* the brother as a single body because only their combined weight is given, and because we have no information about the forces that the sled and the brother exert on each other.

There are two unknown forces, the cord tension and the thrust exerted by the hill. Since we want to determine only the tension in the cord it is conceivable that we could do so with just one equation. In Figure 3.11 the three vectors represent the weight (**W**), cord tension (**C**), and thrust of the hill (**H**). The thrust **H** is perpendicular to the surface because the ice is assumed to be frictionless.

Coordinate axes may be selected in any direction, but it is best to choose axes that simplify rather than complicate the solution. For this problem, it is advantageous to choose axes parallel to the slope and perpendicular to the slope. In this coordinate system, **H** and **C** have only one nonzero component each.

To obtain the equations of the first condition of equilibrium we must find the components of each vector along the chosen axes. Because **C** is parallel to the x-axis and **H** is parallel to the y-axis, we obtain the following:

$$C_x = C \qquad H_x = 0$$

and

$$C_y = 0 \qquad H_y = H$$

Figure 3.11 shows that the components of **W** are negative:

$$W_x = -W \sin 30°$$

$$W_y = -W \cos 30°$$

Substituting these components into the first condition of equilibrium gives

$$\sum F_x = C - W \sin 30° = 0$$

$$\sum F_y = H - W \cos 30° = 0$$

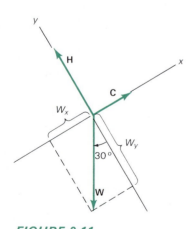

FIGURE 3.11

Forces acting on the combined system of sled plus brother.

We can determine the tension in the cord from the first equation

$$C = W \sin 30° = (785\ N)(0.5)$$
$$= 393\ N$$

Remembering the Problem Solving Guide, ask yourself if this result for C makes sense? In retrospect we see that the girl must pull with a force that balances $W \sin 30°$, the component of the weight *parallel* to the slope. The equation $C = W \sin 30°$ describes this balance.

Other choices of axes lead to the same result, but the algebra involved is more difficult. Because the solution is independent of the coordinate axes you chose, try to select axes that simplify the algebra. As you gain experience in problem solving, you will find it increasingly easy to select the best axes.

Newton's Third Law

Observation shows us that forces come in pairs. When a car crashes into a tree the force exerted by the car damages the tree. But the tree, in turn, exerts a force that damages the car. Observation and experiment reveal a very simple and fundamental relationship between such pairs of forces. This relationship is known as **Newton's third law**:

To every action there is always an equal but opposite reaction.

The words *action* and *reaction* mean *forces*. Newton's third law recognizes that forces always occur in pairs. For every force there is another force equal in magnitude and oppositely directed. We refer to these forces as an **action-reaction pair**.

When a car collides with a tree, the force that the car exerts on the tree and the force that the tree exerts on the car form an action-reaction pair. If you push down on a bed with a force of 500 N, it pushes up on you with a force of 500 N. If earth's gravity exerts a downward force of 600 N on you, then your gravity exerts an upward force of 600 N on the earth.

Either force of a pair may be regarded as the action, making the other force the reaction. Newton's third law involves forces constituting the mutual interaction of two bodies. The two forces of an action-reaction pair never act on the same body. This is an important point to remember when you apply the first condition of equilibrium. The first condition relates forces acting *on* a body. The formulation of the first condition for one body can never include both members of an action-reaction pair.

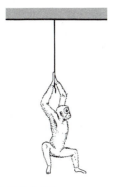

FIGURE 3.12

EXAMPLE 5

The Monkey and the Rope

To illustrate Newton's third law and the treatment of internal forces, we consider a monkey weighing 98.0 N hanging on the end of a rope weighing 10.0 N (Figure 3.12). What are the tensions, T and B, at the ends of the rope?

We are given two weights and we are asked to determine two forces, so we anticipate that we will need two equations. Because the forces act in the vertical direction we will need to apply the first condition of equilibrium to two different bodies.

We pick the monkey as a body. Figure 3.13 shows the two forces acting on the monkey. His weight **W** acts downward and the tension **B** exerted by the rope acts upward. The first condition of equilibrium relates the vertical components $+B$ and $-W$

$$\sum F_y = B - W = 0$$

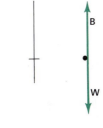

FIGURE 3.13
Forces acting on the monkey.

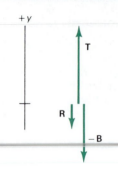

FIGURE 3.14

Forces acting on the rope.

that gives

$$B = W$$
$$= 98.0 \text{ N}$$

Thus, the tension at the bottom of the rope is 98.0 N, which is the total weight suspended beneath it.

To determine the tension at the top of the rope we pick the rope as a body. Figure 3.14 shows the three forces acting on the rope. The tension **T** at the top acts upward; the weight of the rope **R** acts downward; the force −**B** exerted by the monkey acts downward also. We have made use of Newton's third law by recognizing that the rope and the monkey exert equal but oppositely directed forces on each other. The action-reaction forces are labeled **B** and −**B** in Figures 3.13 and 3.14.

For the rope, the first condition of equilibrium gives

$$\sum F_y = T - R - B = 0$$

or

$$T = B + R = 98.0 \text{ N} + 10.0 \text{ N}$$
$$= 108 \text{ N}$$

The problem can be solved in more than one way. Suppose we regard the rope and the monkey as a single body. Then three external forces act on this choice of body: the upward force T and the two weights W and R. Equilibrium requires

$$T = W + R = 98.0 \text{ N} + 10.0 \text{ N}$$
$$= 108 \text{ N}$$

By choosing the monkey-plus-rope as the body we simplify the solution for T. But this new choice of body does not let us determine B, the force exerted on the rope by the monkey, because B becomes an *internal* force, balanced by the equal but opposite force exerted on the monkey by the rope. No *internal* force can be determined via the first condition of equilibrium—only *external* forces enter the formulation of the first condition of equilibrium.

In general, the forces that maintain equilibrium do not lie along the same line. Equilibrium problems become more difficult conceptually and mathematically when the forces involved have two or three nonzero components. The next two examples illustrate the first condition of equilibrium in two-dimensions.

EXAMPLE 6

The Tension in a High Wire

Consider a tightrope walker who slips while attempting to walk a high wire across the chasm downstream from Niagara Falls. Prevented from falling only by a safety rope attached to his waist and to the high wire, he hangs motionless (Figure 3.15). The wire makes the given angles θ and φ with the horizontal. The tightrope walker weighs 800 N.

Let's find the tensions in the wire on each side of the point where the safety rope is attached. These tensions are labeled **L** and **R** in Figure 3.15. We consider only the weight of the tightrope walker and neglect the weight of the rope. We take the angles θ and φ to be given, although we will not specify numerical values.

The angles θ and φ give the directions of **L** and **R**. The magnitudes of **L** and **R** are the unknowns. The first condition of equilibrium gives the two necessary equations to solve for the two unknowns.

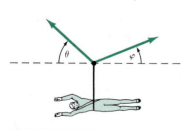

FIGURE 3.15

A tightrope artist hanging by a safety rope.

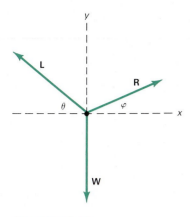

FIGURE 3.16

The forces **L**, **R**, and **W** act on the tie point.

We must choose a body to which we can apply the first condition of equilibrium. We pick the tie point, the place where the rope contacts the wire. Figure 3.16 shows the three forces acting on the tie point. These forces are the tensions **L** and **R** and the tension in the safety rope. From earlier examples we know that the tension in the safety rope equals the weight of the tightrope walker (W). Resolving the forces into x- and y-components and applying the first condition of equilibrium yields two equations relating the unknowns L and R,

$$\sum F_x = -L \cos \theta + R \cos \varphi = 0$$

$$\sum F_y = L \sin \theta + R \sin \varphi - W = 0$$

One way of solving for the tensions L and R is to multiply the first equation by $\sin \theta$, and the second equation by $\cos \theta$. Then we add equations (equals) and use the trigonometric identity

$$\sin (\theta + \varphi) = \sin \theta \cos \varphi + \sin \varphi \cos \theta$$

to find

$$R = \frac{W \cos \theta}{\sin (\theta + \varphi)}$$

L follows by symmetry as

$$L = \frac{W \cos \varphi}{\sin (\theta + \varphi)}$$

You can verify that L and R are solutions by substituting these results back into the original equilibrium equations.

Complicated solutions, like the expressions for L and R, generally cannot be determined to be correct simply by inspection. Often we can use other methods. One powerful technique is the use of a **symmetry argument**. For example, when $\theta = \varphi$, the force vector diagram in Figure 3.16 becomes a symmetric "Y." It seems reasonable that $L = R$ in this symmetric case. You should check to see that the algebraic solution reduces to $L = R$ when $\theta = \varphi$.

3.3
TORQUE

When the first condition of equilibrium is satisfied, a body is in **translational equilibrium**. A second condition must be satisfied to insure **rotational equilibrium**. To see why we need the second condition, imagine that you and a friend approach opposite sides of a revolving door (Figure 3.17). You both exert equal but oppositely directed forces on the door and it is set into rotation. The net force exerted on the door is zero, but it is not in equilibrium. The second condition of equilibrium is formulated in terms of a quantity called **torque** that *measures the tendency of a force to cause rotation.* Before we introduce the second condition of equilibrium, we develop the concept of torque.

Consider the force **F** in Figure 3.18 acting on a rod that is free to rotate about an axis through the point O. The dashed line lying along **F** is called the **line of action** of the force. The force exerts a torque τ *defined* with respect to the point O by

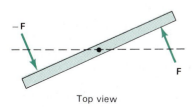

Top view

FIGURE 3.17

Equal but oppositely directed forces on the revolving door cause it to rotate. The first condition of equilibrium is satisfied, but the door is not in rotational equilibrium.

$$\tau = (\text{force}) \cdot (\text{lever arm}) = F \cdot \ell \tag{3.3}$$

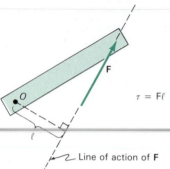

FIGURE 3.18

The magnitude of the torque is the product of the magnitude of the force F and the lever arm ℓ.

The lever arm, ℓ in Figure 3.18, is *defined* as the perpendicular distance from the point O to the line of action of the force. If you increase the lever arm you increase the torque and thereby increase the tendency to cause rotation. The handle on a door is placed well away from the hinges in order to give a large lever arm.

If the force is applied at the point O then the torque is zero because ℓ is zero, therefore the rod will not rotate. If the line of action of the force does not pass through O then the torque is not zero and the rod will rotate. We will study the dynamic effects of torque in Chapter 12. In this section we are concerned with the relationship between the torques that act on a body in rotational equilibrium.

Torque has a vector character related to the two possible directions of rotation. If a torque* tends to cause a clockwise rotation, we treat it as a negative quantity. If a torque tends to cause a counterclockwise rotation, we treat it as a positive quantity. In Figure 3.18 the force F sets up a positive torque relative to O, because if it were the only force acting it would cause a counterclockwise rotation.

EXAMPLE 7

Teeter-Totter Torques

Figure 3.19 shows two children seated on a balanced teeter-totter. The weight of the child on the left is $L = 421$ N (95 lb). The weight of the child on the right is $R = 340$ N (76 lb). The children exert downward forces on the teeter-totter. The fulcrum exerts an upward force, U. Figure 3.20 shows the three forces and the lever arms ℓ_L and ℓ_R measured relative to the fulcrum. The lever arms are

$$\ell_L = 1.30 \text{ m} \qquad \ell_R = 1.61 \text{ m}$$

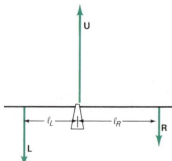

FIGURE 3.19

Two persons of unequal weight balance on a teeter-totter.

Let's evaluate the torques exerted by each force. We choose the fulcrum as a torque reference point. The torque exerted by U is zero because its lever arm is zero. The torques exerted by L and R are:

$$\tau_L = +L \cdot \ell_L = +(421 \text{ N})(1.30 \text{ m}) = +547 \text{ N·m}$$

$$\tau_R = -R \cdot \ell_R = -(340 \text{ N})(1.61 \text{ m}) = -547 \text{ N·m}$$

Note the sign convention. The torque τ_L is positive because it tends to cause counterclockwise rotation. The torque τ_R is negative because it tends to cause clockwise rotation. There is no rotation of the teetor-totter; it is in static equilibrium.

The numerical value and the sign of the torque depend on the location of the torque reference point. In the following example we consider the torques exerted by a set of forces relative to three different torque reference points.

EXAMPLE 8

Torques on a Beam

Consider a horizontal beam 4 m long and of negligible weight, as shown in Figure 3.21. The left end of the beam is in contact with a smooth vertical wall. The wall exerts a horizontal force **H** on the beam. The right end of the beam experiences forces exerted by two cables of negligible weight. One cable is 5 m long and exerts a force **T**, as shown. The vertical cable exerts a force **W**. The weight of the sign is 300 N.

FIGURE 3.20

Forces acting on a balanced teeter-totter.

---

* In static equilibrium there is no rotation. To decide whether a particular torque is positive or negative you must imagine that it is the *only* torque acting on the body.

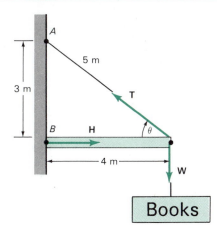

FIGURE 3.21

The "Books" sign weighs 300 N. By inspection, the tension W in the vertical cable equals 300 N.

We want to determine the torques associated with these three forces. We will evaluate the torques for three different reference points—the points A, B, and C in Figure 3.21.

The given information is the weight of the sign and the lengths of the beam and one of the cables. Knowing the lengths lets us figure out the direction of the tension in the cable and in the lever arms.

First we determine the magnitudes of **H**, **T**, and **W**. Because we neglect the weights of the cables, W equals 300 N, or the weight of the sign. Figure 3.22 shows the force diagram that lets us determine T and H. Applying the first condition of equilibrium to the beam we have

$$\sum F_x = H - T \cos \theta = 0$$
$$\sum F_y = T \sin \theta - W = 0$$

From Figure 3.21 we can see that $\cos \theta = 4/5$ and $\sin \theta = 3/5$. Solving for the second equation for T we get

$$T = \frac{W}{\sin \theta} = 300 \text{ N} \left(\frac{5}{3}\right)$$
$$= 500 \text{ N}$$

With T determined we can solve the first equation for H,

$$H = T \cos \theta = 500 \text{ N} \left(\frac{4}{5}\right)$$
$$= 400 \text{ N}$$

To evaluate torques we must determine the lever arms for each force. Figure 3.23 helps with the needed geometry. First, consider point C as a torque reference point. The lines of action of all three forces pass through C. Hence, the lever arms for **T**, **W**, and **H** are zero relative to C, and the torque associated with each of these forces is zero. (The results are listed in the table that follows.)

Next, consider B as a torque reference point. The line of action of the thrust force **H** passes through B, so the lever arm and torque for **H** are zero relative to B. The lever arm for **W** is the length of the beam, 4 m. The weight tends to cause clockwise rotation of the beam about the point B, so the torque due to W is negative

$$\tau_W = -W\ell_W = -(300 \text{ N})(4 \text{ m})$$
$$= -1200 \text{ N} \cdot \text{m}$$

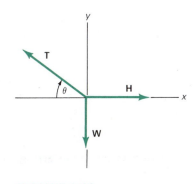

FIGURE 3.22

Forces and geometry for the system shown in Figure 3.21.

The lever arm for the tension $\mathbf{T}$ is the perpendicular distance from the point B to the line of action of $\mathbf{T}$. In Figure 3.23, this lever arm is the distance from B to D and has the value

$$\ell_T = 4 \text{ m} \sin \theta = 4 \text{ m} \left(\frac{3}{5} \right) = 2.4 \text{ m}$$

The torque set up by $\mathbf{T}$ is positive because it tends to cause counterclockwise rotation about B,

$$\tau_T = + T\ell_T = + (500 \text{ N})(2.4 \text{ m}) = + 1200 \text{ N} \cdot \text{m}$$

Finally, consider A as a torque reference point. The line of action of $\mathbf{T}$ passes through A so its lever arm and torque are zero. The lever arm for the thrust force $\mathbf{H}$ is 3 m and its torque is

$$\tau_H = H\ell_H = (400 \text{ N})(3 \text{ m})$$
$$= 1200 \text{ N} \cdot \text{m}$$

For the force $\mathbf{W}$, the lever arm is 4 m, and the torque is negative,

$$\tau_W = - W\ell_W = - (300 \text{ N})(4 \text{ m})$$
$$= - 1200 \text{ N} \cdot \text{m}$$

These torques are listed in the table below.

Torque Reference Point	Force (N)	Lever Arm (m)	Torque (N·m)
A	$T = 500$	0	0
	$W = 300$	4	-1200
	$H = 400$	3	$+1200$
B	$T = 500$	2.40	$+1200$
	$W = 300$	4	-1200
	$H = 400$	0	0
C	$T = 500$	0	0
	$W = 300$	0	0
	$H = 400$	0	0

Notice that the sum of the torques about each reference point is zero. This is not coincidental. It illustrates the second condition of equilibrium that we will develop in Section 3.4.

Now that we have had some practice evaluating torques, let's see how we can represent torque in terms of a **vector product**. In Figure 3.24 we again consider a force $\mathbf{F}$ acting on a rod. We let $\mathbf{r}$ denote the vector from the torque reference point O to the point P where $\mathbf{F}$ acts. The angle θ is the angle between the vectors $\mathbf{r}$ and $\mathbf{F}$. As you can see in Figure 3.24, the lever arm is related to r and θ by

$$\ell = r \sin \theta \tag{3.4}$$

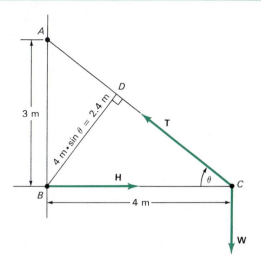

FIGURE 3.23

Points *A*, *B*, and *C* are torque reference points. The lever arms for **T**, **H**, and **W** depend on which reference point is selected.

It follows that the torque $\tau = F\ell$ can be expressed as

$$\tau = rF \sin \theta \tag{3.5}$$

From our study of the vector product in Chapter 2 we recognize $rF \sin \theta$ as the magnitude of a torque vector $\boldsymbol{\tau}$ defined as

$$\boldsymbol{\tau} = \mathbf{r} \times \mathbf{F} \tag{3.6}$$

For example, when the vectors **r** and **F** are in the plane of this page the torque vector is perpendicular to the page. A torque vector directed up from the page tends to produce a counterclockwise rotation. A torque vector directed down into the page tends to produce clockwise rotation. You can use a right-hand rule to relate the direction of the torque vector to the sense of rotation the torque tends to produce. Curl the fingers of your right hand in the direction of rotation that the torque tends to cause. Your extended right thumb points in the direction of the torque. In Figure 3.24, your fingers would curl counterclockwise and $\boldsymbol{\tau}$ would be directed upward out of the page.

3.4
THE SECOND CONDITION OF EQUILIBRIUM

A body is in **translational** equilibrium if the vector sum of external **forces** acting on it is zero. For **rotational** equilibrium, a condition involving **torques** must be satisfied. The teeter-totter in Example 7 and the beam in Example 8 are illustrations of bodies in rotational equilibrium. For both the teeter-totter and the beam, the sum of external torques is zero. As we said before, this is not coincidental. Experiment shows that it is a general requirement for rotational equilibrium. The **second condition of equilibrium** is a formal statement of this zero-torque condition:

A body that is in translational equilibrium is also in rotational equilibrium if and only if the *net* torque exerted *on* the body is zero.

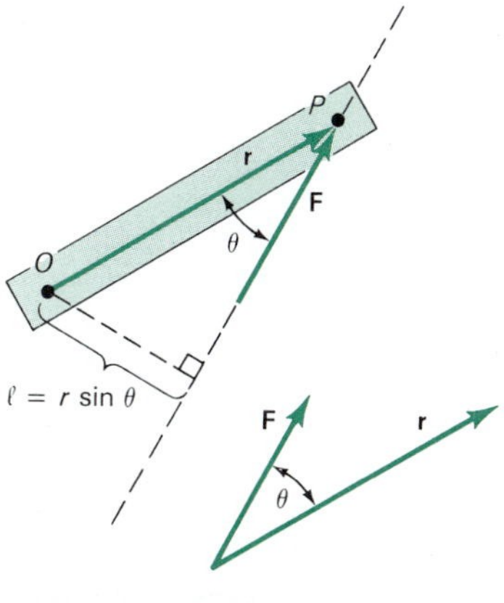

FIGURE 3.24

The torque vector equals **r** × **F** and is directed upward, out of the page.

The second condition of equilibrium can be expressed mathematically by

$$\sum \tau = 0 \qquad\qquad (3.7)$$

The summation includes the torques exerted by all external forces.

Before illustrating the second condition of equilibrium we add three items to the Problem Solving Guide of Section 3.2.

Guide to Problem Solving II

9. Choose a torque reference point. (A particular choice may simplify the problem by eliminating an unknown force.)
10. Determine the lever arm and torque for each external force.
11. Apply the second condition of equilibrium.

This advice is put to use in the following examples.

EXAMPLE 9

Which Hand Exerts the Greater Force on the Pole?

A vaulter holds a 29.4-N pole in equilibrium by exerting an upward force, **U**, with his leading hand, and a downward force, **D**, with his trailing hand, as shown in Figure 3.25. If we assume that the weight of the pole acts at its midpoint, what are the values of U and D?

The given information includes the weight of the pole and the distances between the points where the three forces act. These distances will serve as lever arms.

Before reading further, you should make qualitative guesses about the nature of the solution. Should U be larger than D? Smaller? Would equilibrium be maintained if the directions of **U** and **D** were reversed?

We want to choose a torque reference point that will simplify the solution. If the reference point is located at either A or B, then only one of the two forces, either **U** or **D**, will appear in the torque equation. If the reference point is located elsewhere along the pole, then both **U** and **D** will appear in the torque equation. We first choose A as the torque reference point. This choice eliminates the unknown force **D** that acts at A.

The forces **U**, **D**, and **W** are perpendicular to the pole, therefore their lever arms equal the distances measured along the pole from A to the points where the forces act.

$$\ell_D = 0 \qquad \ell_U = 0.75 \text{ m} \qquad \ell_W = 2.25 \text{ m}$$

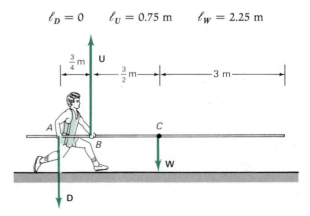

FIGURE 3.25
The vaulter's pole is in equilibrium.

Inserting these values into the second condition of equilibrium yields an equation in which U is the only unknown

$$\sum \tau_A = D\ell_D + U\ell_U - W\ell_W$$
$$= D(0 \text{ m}) + U(0.75 \text{ m}) - (29.4 \text{ N})(2.25 \text{ m}) = 0$$

This gives

$$U = 88.2 \text{ N}$$

In order to determine D we choose B as a torque reference point. This choice eliminates U and gives

$$\sum \tau_B = U(0 \text{ m}) + D(0.75 \text{ m}) - (29.4 \text{ N})(1.5 \text{ m}) = 0$$

Solving for D we get

$$D = 58.8 \text{ N}$$

How do these results compare with your conjectures?

We can verify that these values of U and D are compatible with the first condition of equilibrium. Taking the positive y-axis to be up we get

$$\sum F_y = U - D - W = 29.4 \text{ N} - 58.8 \text{ N} - 29.4 \text{ N}$$
$$= 0$$

Thus, the solution deduced using the second condition of equilibrium is consistent with the first condition of equilibrium.

We obtained values for U and D without invoking the first condition of equilibrium. Often it is unclear in advance whether both conditions of equilibrium are needed, or if they are not, which of the two should be used. In some instances a solution can be obtained using either the first or second condition. Sometimes, both conditions of equilibrium are needed. Insight comes with experience. Keep in mind that you need one equation for each unknown.

Suppose that we want to determine a particular force **F**. Unless **F** acts on the chosen body, **F** will not appear in the equilibrium equations. To determine a particular force, a body must be selected on which that force acts. Choosing a body was relatively simple in Examples 8–9. As the structure becomes complex, however, making a good choice becomes more difficult. To illustrate how choosing a body determines which forces appear in the equilibrium equations, we consider an A-frame structure. In this example it is necessary to choose more than one body.

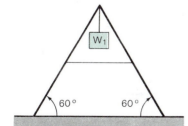

FIGURE 3.26

The A-frame has two beams fastened together at the apex. The steel cable connecting the beams is under tension.

EXAMPLE 10

The A-Frame

An A-frame Swiss chalet is constructed with beams 10 m long, weighing 5000 N each. The beams are joined at the apex of an equilateral triangle, as indicated in Figure 3.26. Halfway to the top a horizontal cable of negligible weight is attached to the beams. A chandelier weighing 1000 N is hung from the apex. The tension in the cable is adjusted so that the forces exerted on the beams by the ground have no horizontal components.

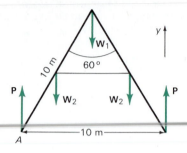

FIGURE 3.27

External forces acting on the
A-frame.

We want to determine the tension in the cable, the forces exerted on the beams by the ground, and the contact force between the beams at the apex. We assume that the weight of each beam acts at its center.

There are four unknown forces so we need to apply the conditions of equilibrium more than once. In problems where it is necessary to determine several forces you may have to apply the conditions of equilibrium to more than one body as well.

First we pick the entire assembly as a body. This choice gives us the force vector diagram shown in Figure 3.27. Since the A-frame is symmetric, we assume that the two forces exerted by the ground are equal and denote their magnitudes by P. Applying the first condition of equilibrium we write

$$\sum F_y = +2P - 2W_2 - W_1 = 0$$

where $W_2 = 5000$ N is the weight of each beam and $W_1 = 1000$ N is the weight of the chandelier. Solving for P gives

$$P = W_2 + \tfrac{1}{2}W_1 = 5000 \text{ N} + \tfrac{1}{2}(1000 \text{ N})$$
$$= 5500 \text{ N}$$

We could also determine P by applying the second condition of equilibrium. We use point A as a reference point for torques. The lever arms are shown in Figure 3.27. The second condition of equilibrium is expressed by

$$\sum \tau_A = P(0 \text{ m}) - W_2(2.5 \text{ m}) - W_1(5 \text{ m}) - W_2(7.5 \text{ m}) + P(10 \text{ m}) = 0$$

This also gives $P = 5500$ N.

No information about the tension in the cable or the contact force between the beams at the apex may be obtained by using the full A-frame as a body because the tension and contact force are internal forces. To determine these forces we must pick a body on which these forces act.

We pick the left beam as a body and obtain the force vector diagram shown in Figure 3.28. The vector **T** represents the force exerted on the left beam by the cable. The force exerted on the left beam by the right beam has been resolved into horizontal and vertical components, **H** and **V**. But how do we know that the vertical component **V** is downward as shown, rather than upward? In fact, we do not *need* to know in advance whether **V** is up or down. If we assume it is down and our solution gives a negative value for **V** this means that our choice was wrong and that **V** is directed upward.

Since we have already determined the value of P we need three equations to solve for the tension T, and H and V. Applying the first condition of equilibrium to the vertical components gives

$$\sum F_y = P - W_2 - V = 0$$

Solving for V we get

$$V = P - W_2 = 5500 \text{ N} - 5000 \text{ N}$$
$$= 500 \text{ N}$$

Next, we apply the first condition to the horizontal components.

$$\sum F_x = T - H = 0$$

This gives $T = H$, but does not determine the value of T or H. In order to obtain T and H we must apply the second condition of equilibrium. We choose the apex point C as a torque reference point. This choice makes the lever arms for H and V zero. The lever arms for P, W_2, and T are shown in Figure 3.28. The second condition gives

$$\sum \tau_C = -P\ell_P + T\ell_T + W_2\ell_2 = 0$$

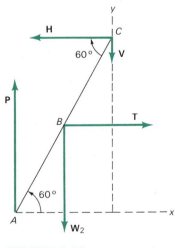

FIGURE 3.28

Forces acting on the left beam of
the A-frame in Figure 3.26.

Solving for the tension T gives

$$T = \frac{P\ell_P - W_2\ell_2}{\ell_T}$$

$$= \frac{5500 \text{ N} \cdot 5 \text{ m} - 5000 \text{ N} \cdot 2.5 \text{ m}}{4.33 \text{ m}}$$

$$= 3460 \text{ N}$$

Since we have shown that $T = H$ we have

$$H = 3460 \text{ N}$$

This completes the solution. We have determined five forces by applying the first and second conditions of equilibrium and a symmetry argument.

3.5
CENTER OF GRAVITY

In Examples 9–10 we assumed that the total weight of an extended body acted at a point. This point is called the **center of gravity** of a body. Its location may be determined experimentally or can be deduced from the conditions of equilibrium. We will use the latter method before we describe an experimental procedure.

Each particle in an extended body has a weight, $\mathbf{W}_i$. The vector sum of the weights of all of the particles is the total weight of the body.

$$\mathbf{W} = \sum \mathbf{W}_i$$

It is possible to place an object in equilibrium by applying a single force, called the **equilibrant**, in just the right direction at just the right point. The magnitude and direction of the equilibrant, $\mathbf{E}$, is determined by the first condition of equilibrium.

$$\mathbf{W} + \mathbf{E} = 0$$

Thus,

$$\mathbf{E} = -\mathbf{W} \tag{3.8}$$

showing that the equilibrant must be equal and opposite to the weight of the object in order to satisfy the first condition of equilibrium. The point of application of the equilibrant must be chosen to satisfy the second condition of equilibrium, thereby guaranteeing rotational equilibrium. To locate the center of gravity we consider a laminar body that lies in the x–z plane. Figure 3.29 shows an edge-on view of the body. We locate the coordinate origin at some arbitrary point in the body. Let's see what is required to have rotational equilibrium about the z-axis. In Figure 3.29 we see that the equilibrant sets up a torque about the z-axis given by $E\bar{x}$. Since $E = W$ the equilibrant torque is

$$E\bar{x} = W\bar{x}$$

The particle of weight W_i exerts a torque $-W_i x_i$ about the z-axis. The combined z-torque due to all particles is $-\sum W_i x_i$, where the summation extends over all particles. The second condition of equilibrium requires that

$$\sum \tau_z = W\bar{x} - \sum W_i x_i = 0$$

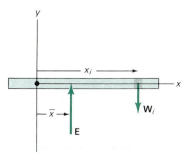

FIGURE 3.29

The torque exerted by the equilibrant is $E\bar{x}$. The torque exerted by the particle of weight W_i is $-W_i x_i$.

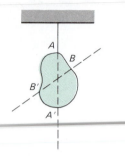

FIGURE 3.30

When an object is suspended by a string from point A, the center of gravity lies below A on the vertical line AA'. A second line, BB', similarly obtained, locates the center of gravity as the point where AA' and BB' intersect.

The x-component of the center of gravity follows as

$$\bar{x} = \frac{\sum W_i x_i}{W} \tag{3.9}$$

This equation is valid no matter what the shape of the body. We can generalize this result to three dimensions and write down the equations for the y- and z-coordinates of the center of gravity.

$$\bar{y} = \frac{\sum W_i y_i}{W} \tag{3.10}$$

$$\bar{z} = \frac{\sum W_i z_i}{W} \tag{3.11}$$

The center of gravity of a thin, flat object can be located experimentally by hanging the object by a string. The string exerts the equilibrant force at the point A, as shown in Figure 3.30, so the center of gravity must lie somewhere along the line AA'. Repeating the procedure by using the suspension point B establishes that the center of gravity lies along the line BB'. The center of gravity must be at the intersection of these two lines.

The center-of-gravity concept is valuable because it simplifies many problems. Extended objects may be replaced by a particle whose weight acts at the center of gravity. We did this in Examples 9–10 and it simplified our calculations.

There are other ways to locate the center of gravity of an object. For example, when the object is highly symmetric we may locate the center of gravity by inspection. The center of gravity of a homogeneous symmetric object coincides with the center of symmetry. In the following example, we use Equation 3.10 and symmetry to locate the center of gravity of a symmetric array of billiard balls.

EXAMPLE 11

Center of Gravity of a Billiard Ball Array

A common initial arrangement in pocket billiards has 15 object balls, each weighing 1.64 N, distributed symmetrically in a triangle, as suggested by the dot arrangement in Figure 3.31. The center-to-center distance of the billiard balls is 5.72 cm. What are the coordinates of the center of gravity of the array?

By symmetry, each ball has its center of gravity at its geometric center, so the array of centers adequately represents the balls themselves. The array of centers is symmetric about the x-axis, so $\bar{x} = 0$. For $\bar{y}$, Equation 3.10 yields

$$\bar{y} = [5 \cdot (1.64 \text{ N})(0 \text{ cm}) + 4(1.64 \text{ N})(4.95 \text{ cm}) + 3(1.64 \text{ N})(9.91 \text{ cm})$$

$$+ 2(1.64 \text{ N})(14.9 \text{ cm}) + (1.64 \text{ N})(19.8 \text{ cm})]/[15 \cdot (1.64 \text{ N})]$$

$$= 6.61 \text{ cm}$$

The factor 4.95 cm = (5.72 cm) cos 30° is y_i for the row of four balls. Symmetry indicates that $\bar{y}$ should be at the intersection of the perpendicular bisectors of the edges of the formation. Our calculated value of 6.61 cm confirms this numerically. The center of gravity is marked by an asterisk in Figure 3.31.

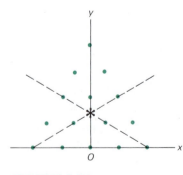

FIGURE 3.31

The dots represent the centers of gravity of 15 billiard balls arranged in a triangular array. The asterisk marks the center of gravity of the 15-ball array.

The center of gravity location does not depend on the weight of the body. To convince yourself of this, mentally double the weight of each particle in

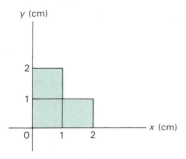

FIGURE 3.32

A 3-cube Soma piece.

the body. This doubles both the numerators and the denominators in Equations 3.9–3.11, leaving $\bar{x}$, $\bar{y}$, and $\bar{z}$ unchanged.

When a body lacks sufficient symmetry to allow its center of gravity to be determined by inspection, we may be able to use Equations 3.9–3.11. In Example 12 we use Equation 3.9 to find the center of gravity of a Soma puzzle piece.

EXAMPLE 12

The Soma Puzzle

The Soma puzzle consists of 27 small cubes organized into six pieces, which are composed of 4 cubes each and one piece composed of 3 cubes. The 3-cube piece is illustrated in Figure 3.32. The three separate cubes are 1 cm on an edge. Where is the center of gravity of the piece?

By symmetry we know that the center of each cube is the location of its center of gravity. We can also see by symmetry that $\bar{x} = \bar{y}$ for the 3-cube piece. Let the weight of one cube be C. From Figure 3.32 we see that $x_1 = x_2 = 0.5$ cm and $x_3 = 1.5$ cm. Using Equation 3.9 we get

$$\bar{x} = \frac{\sum W_i x_i}{W} = \frac{[C(0.5 \text{ cm}) + C(0.5 \text{ cm}) + C(1.5 \text{ cm})]}{3C}$$

$$= 0.83 \text{ cm}$$

From the symmetry relation $\bar{x} = \bar{y}$ we get

$$\bar{y} = 0.83 \text{ cm}$$

The location of the center of gravity in Figure 3.32 is marked with an asterisk.

Negative Weight Method

We now reconsider Example 12 to develop a negative weight procedure, which is useful in some center-of-gravity calculations. One more cube, placed in the position suggested by the dotted lines in Figure 3.33, would change the 3-cube piece into a symmetric 4-cube piece. With this in mind, we rewrite $\bar{x}$ for Example 12:

$$\bar{x} = \frac{[2C \cdot (0.5 \text{ cm}) + 2C \cdot (1.5 \text{ cm}) - C \cdot (1.5 \text{ cm})]}{[4C - C]}$$

Terms have been added and subtracted in both the numerator and the denominator, leaving $\bar{x}$ unchanged. A slight rearrangement gives

$$\bar{x} = \frac{[4C \cdot (1.0 \text{ cm}) - C \cdot (1.5 \text{ cm})]}{[4C - C]}$$

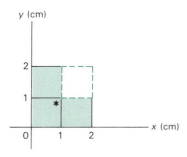

FIGURE 3.33

The dotted lines indicate the position where a fourth cube could be placed to create a symmetric object.

This formula represents a two-particle system. One particle ($W_1 = +4C$) is at $x_1 = 1.0$ cm, and the other particle ($W_2 = -C$) is at $x_2 = 1.5$ cm.

This result may be generalized as follows: If an unsymmetric object can be converted into a symmetric object by adding or subtracting one or more symmetric pieces, then the negative weight procedure will yield the correct coordinates of the center of gravity. In the case of Example 12, for instance, a symmetric one-cube object can be added to the unsymmetric object to make a symmetric four-cube object. The positive weight four-cube object coupled with the negative weight one-cube object results in the same center of gravity as the original three-cube object. The positive four-cube piece has a center of gravity at (1.0 cm, 1.0 cm), and the negative one-cube piece has a center of gravity at (1.5 cm, 1.5 cm). We thus calculate from these data

$$\bar{x} = \frac{[4C \cdot (1.0 \text{ cm}) - C \cdot (1.5 \text{ cm})]}{[4C - C]}$$

$$= 0.83 \text{ cm}$$

Again we have by symmetry

$$\bar{y} = 0.83 \text{ cm}$$

in agreement with our previous result. The following example further illustrates the negative weight procedure.

EXAMPLE 13

Center of Gravity of a Lunate Area

Consider the lunate (crescent-shaped) area bounded by circles having radii R and $R/2$ shown in Figure 3.34. We wish to locate the center of gravity, assuming that weight is proportional to area. By symmetry $\bar{y} = 0$, so only $\bar{x}$ needs to be calculated.

If the hole is filled in, a symmetric circle is constructed from the lunate shape. It may be viewed, therefore, as a combination of one circular area ($W_1 = +4C$) with $x_1 = 0$ and another circular area ($W_2 = -C$) with $x_2 = +R/2$. This interpretation leads to the calculation

$$\bar{x} = \frac{[4C \cdot (0) - C \cdot (R/2)]}{[4C - C]}$$

$$= -R/6$$

which is the x-coordinate of the center of gravity of the original lunate figure. An asterisk marks the center of gravity in Figure 3.34.

Stability of Equilibrium

A cone balanced on its point is in equilibrium, just as it is when firmly seated on its base. A basketball delicately balanced on top of a player's finger is in equilibrium, as it is when it is motionless on the floor. In each situation two kinds of equilibrium are described, and they are not difficult to distinguish. We classify types of equilibrium according to their stability.

When a body is in equilibrium the net external force acting on it is zero. The net external torque acting on the body is also zero. Let's consider what happens when the object undergoes a *small displacement* from equilibrium. There are three possible consequences of this small displacement:

1. The displaced object remains in equilibrium.
2. A net force and/or torque is created, urging the body still further from equilibrium.
3. A net force and/or torque is created, urging the body back toward its original equilibrium.

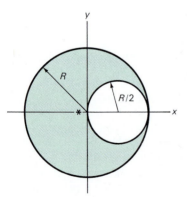

FIGURE 3.34

A lunate-shaped area bounded by two circles. The asterisk marks the center of gravity.

We consider each of these three possibilities in turn.

When a body in equilibrium is given a slight displacement and the new position is an equilibrium position, then the original position is said to be one of **neutral equilibrium**. For example, consider a book at rest on a level surface. If the book is given a small displacement, both the new and the old position are equilibrium positions. Hence the original position is one of neutral equilibrium.

When the net force or torque acts to urge the body further away from the original equilibrium position, then the original equilibrium is an **unstable equilibrium**. The gymnast shown in Figure 3.35, poised above the bar balanced on his hands, exemplifies an unstable equilibrium. If he rotates through a small angle away from equilibrium, then the net torque created as a result of the small rotation will act to move him still further away.

When the displaced position gives rise to a net force and/or torque that acts to urge the body back toward its original equilibrium position, then the original equilibrium is a **stable equilibrium**. For example, picture a child in a swing. If the swing is rotated through a small angle away from the bottom equilibrium position and then released, it will be urged back toward the original position by the torque generated by the weight of the child and swing.

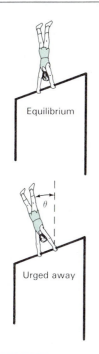

Equilibrium

Urged away

FIGURE 3.35

WORKED PROBLEM

Figure 3.36 shows a window washer standing on a horizontal plank suspended at its ends by cables. The window washer weighs 800 N. The plank is uniform, 5 m long, and weighs 500 N. The weights of the cables are negligible. Find the minimum and maximum values of the tensions in the cables.

Solution

Pick the plank as the body in equilibrium. The two cables pull upward on the plank and gravity pulls down on the plank. The downward force exerted on the plank by the window washer equals his weight.

Choose the left end of the plank as a torque reference point (A in Figure 3.37). The downward forces of the window washer and the weight of the plank produce negative (clockwise) torques. The upward force of T_2 produces a positive torque. The force T_1 produces zero torque because its line of action passes through the torque reference point.

Applying the second condition of equilibrium we have

$$\sum \tau_A = -(800 \text{ N})x - (500 \text{ N})(2.5 \text{ m}) + T_2(5 \text{ m}) = 0$$

Solving for T_2 gives

$$T_2 = 250 \text{ N} + (160 \text{ N/m})x$$

The minimum value of T_2 occurs when the window washer stands at the left end of the plank, making $x = 0$ m: $T_2(\text{min}) = 250$ N. The maximum value of T_2 occurs when the window washer stands at the right end of the plank, making $x = 5$ m: $T_2(\text{max}) = 1050$ N.

The symmetry of the structure suggests that the maximum values of T_1 and T_2 are equal, and that their minimum values are also equal. This can be verified by using the first condition of equilibrium or by applying the second condition of equilibrium about a different reference point.

Choose up as the positive y-direction and apply the first condition of equilibrium. This gives

$$\sum F_y = T_1 + T_2 - 800 \text{ N} - 500 \text{ N} = 0$$

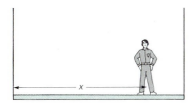

FIGURE 3.36

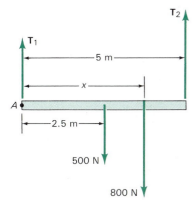

FIGURE 3.37

Solving for T_1 gives

$$T_1 = 1300 \text{ N} - T_2$$

Inserting the maximum and minimum values of T_2 verifies the expectation that T_1 and T_2 have the same maximum and minimum values.

EXERCISES

3.1 Force

A. The *Guinness Book of Records* (1980 edition) lists the weight of the heaviest human of all time as 1069 pounds. Express this weight in newtons.

3.2 The First Condition of Equilibrium

B. An ideal pulley changes the direction of the tension in a rope without changing its magnitude. Such a pulley weighing 100 N is used with a rope of negligible weight to hoist a large fish weighing 4000 N, as shown in Figure 1. Determine the force F and the tension T in the rope supporting the pulley.

FIGURE 1

C. Two forces act on the pendulum bob shown in Figure 2, the tension exerted by the cord and the weight of the bob. Give an argument based on the first condition of equilibrium showing that the bob is not in equilibrium.

FIGURE 2

D. Two forces act on a person standing on a level surface; the person's weight **W** acts downward and the contact

force **P** acts upward. In order for the person to be in translational equilibrium, the first condition of equilibrium requires that $P = W$. Explain why **P** and **W** do not constitute an action-reaction pair of forces.

E. A spider descends along its self-spun thread at a constant speed. The spider weighs 0.01 N. What is the tension in the thread if the weight of the thread is negligible? (Be sure to guess at an answer and then check it by applying the remaining steps in the problem solving guide.)

F. A 12-N weight is suspended as shown in Figure 3. The system is held motionless by the frictional force exerted on the weight W by the horizontal table top. What is the magnitude of the frictional force?

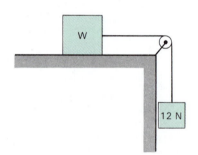

FIGURE 3

G. A 9000-N crate is being loaded on a ship with the symmetric cable shown in Figure 4. Assume that the weight acts at the center of the crate and neglect the cable weight. Determine the tensions T_1, T_2, and T_3.

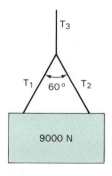

FIGURE 4

H. Figure 5 shows a frictionless plane inclined at an angle of 15° relative to the horizontal. The cords attached to the blocks are of negligible weight and the pulley is frictionless. One block weighs 20 N, the other weighs 40 N. Determine the tension in the cord attached to the left side of the 20-N block.

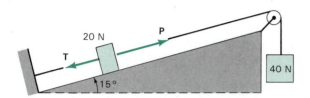

FIGURE 5

3.4 The Second Condition of Equilibrium

I. A baseball player holds a 36-oz bat (weight = 10.0 N) with one hand at the point O (Figure 6). The bat is in equilibrium. The weight of the bat acts along a line 60 cm to the right of O. Determine the force and the torque exerted on the bat by the player.

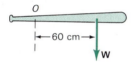

FIGURE 6

J. To demonstrate his strength a student holds a 50-N chair by the bottom of one leg. The weight **W** acts along a line 20 cm to the right of the left side of the chair (Figure 7). Determine the torque of **W** relative to the point O.

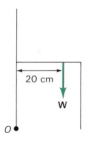

FIGURE 7

K. A rod 4 m long is hinged at the point P at its left end (Figure 8). It is horizontal and in equilibrium. A 10 N force is applied at its right end at an angle of 30° above the horizontal. (a) Calculate the torque of the 10 N force relative to P. (b) Determine the weight W of the rod, given that it acts at its midpoint.

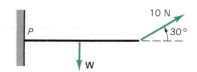

FIGURE 8

L. Determine the lever arms and torques for the three forces acting on the weightless beam in Figure 9 relative to the point A.

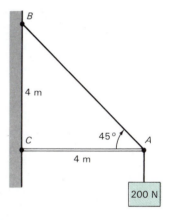

FIGURE 9

M. A sign weighing 300 N is hung from the end of a uniform beam 4 m long that weighs 100 N (Figure 10). A wire 5 m long connects the building to the end of the beam. Determine the torque for each of the five forces relative to the point A.

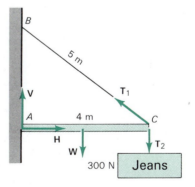

FIGURE 10

N. Two children, one at each end of a teeter-totter, are in static equilibrium. One child, weighing 340 N, is 3 m to the right of the pivot. The other child is 2.3 m to the left of the pivot. (a) Neglecting friction, calculate the weight of the second child. (b) Determine the upward force exerted by the pivot on the teeter-totter.

P. Figure 11 shows a truss that supports a downward force of 1000 N applied at the point B. Neglect the weight of the truss. Apply the conditions of equilibrium to prove that $N_A = 366$ N and $N_C = 634$ N.

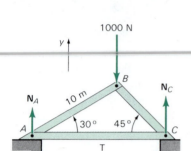

FIGURE 11

Q. A yo-yo weighing 5 N has an outer radius 10 times larger than its axial radius (Figure 12). The yo-yo is placed in equilibrium by hanging it from one string wound around its axis, and by hanging a weight W from a second string wound around the outer part as shown. Determine the tensions T_1 and T_2 in the two strings.

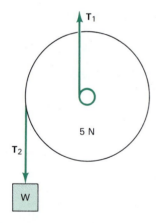

FIGURE 12

3.5 Center of Gravity

R. A pendulum-collision toy consists of five steel balls in contact (Figure 13). Each ball is of weight W and radius R. Where is the center of gravity when (a) the balls are in contact? (b) the leftmost ball is pulled a distance $4R$ to the left? (c) the first two balls on the left are pulled a distance $4R$ to the left while in contact?

S. The minimum center-to-center distance between adjacent bowling pins is 12 inches. When initially set up, the ten pin centers form a symmetric array that is rep-

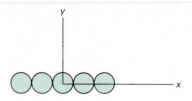

FIGURE 13

resented in Figure 14 by dots and asterisks. A common arrangement for three pins left standing after the first ball has been bowled is that of the three dots. Find the x- and y-coordinates of the center of gravity of these three pins.

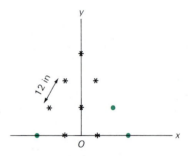

FIGURE 14

T. Locate the center of gravity of the plane Soma puzzle piece shown in Figure 15.

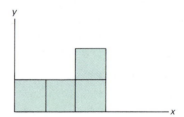

FIGURE 15

U. A triangular piece of wire is shown in Figure 16. Assuming the wire is uniform, locate the center of gravity.

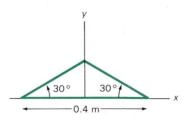

FIGURE 16

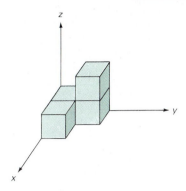

FIGURE 17

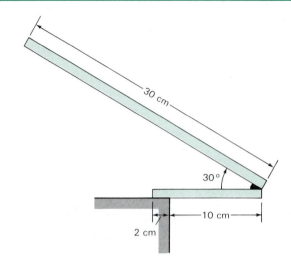

FIGURE 18

V. Locate the coordinates of the center of gravity of the four-cube Soma puzzle piece shown in Figure 17.

W. A rigid wooden structure consists of two rods glued together at an angle of 30° (Figure 18). The longer segment has a weight of 0.500 N and a length of 30 cm. The shorter segment has a mass of 0.250 N and a length of 12 cm. The structure is carefully placed on the edge of a desk as shown. Does it remain stationary or topple off? Explain.

PROBLEMS

1. A child weighing 400 N sits on a playground swing of negligible weight. Normally, the two supporting ropes are attached at the top (Figure 19a) and each rope has a tension of 200 N. Suppose, however, that the two ropes pass over pulleys at the top (Figure 19b) and the ends are held by the child. Apply the first condition of equilibrium and a symmetry argument to determine the tension in the ropes. Assume the ropes are vertical, and that the child sits in the middle of the seat. Disregard friction.

weight parallel to and perpendicular to the plane. (b) Apply the first condition of equilibrium to the 100-N weight and calculate the string tension, T, and the force P exerted by the plane on the 100-N weight. (c) Determine the weight W.

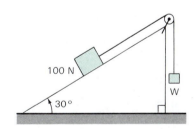

FIGURE 20

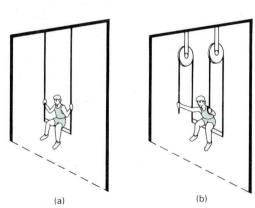

(a)   (b)

FIGURE 19

2. A weight W just balances the 100-N weight shown in Figure 20. The pulley and inclined plane are frictionless. (a) Obtain the components of the 100-N

3. A beam 4 meters long and of negligible weight is perpendicular to a frictionless wall at point C (refer to Figure 9). The beam is in equilibrium. It experiences a thrust, P, perpendicular to the wall at C, a tension T, directed along the cable from A to B, and a downward tension at A due to the 200-N weight. Determine the magnitudes of P and T.

4. Two 200-N traffic lights are suspended from a single cable as shown in Figure 21. Neglect the cable weight and (a) prove that if $\theta_1 = \theta_2$, then $T_1 = T_2$. (b) Determine the three tensions if $\theta_1 = \theta_2 = 8°$.

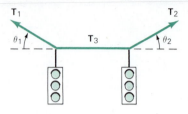

FIGURE 21

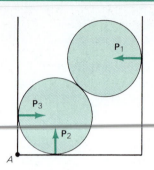

FIGURE 24

5. A bottle of wine weighing 6.00 N is stored on its side in a diamond-shaped rack as shown in Figure 22. Assume the walls of the rack are smooth. (a) Determine the thrust exerted on the bottle by each wall when the receptacle is symmetrically oriented ($\theta_1 = \theta_2 = 30°$). (b) Determine the thrusts when $\theta_1 = 25°$ and $\theta_2 = 35°$.

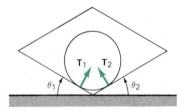

FIGURE 22

6. A beam 4 m long and of negligible weight is perpendicular to a wall (Figure 23). The beam is supported by the wall and a cable that makes an angle of 30° with the horizontal. A 500-N weight is suspended as shown. Determine the force exerted on the beam by the wall and the tension in the cable. Ignore the weight of the cable.

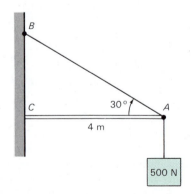

FIGURE 23

7. Two racket balls are placed in a glass as shown in Figure 24. Their centers and the point A lie on a straight line. (a) Assume that the walls are smooth, and determine P_1, P_2, and P_3. (b) Determine the magnitude of the force exerted on the right ball by the left ball. Assume each ball weighs 1.64 N.

8. To get a heavy barrel off the street a worker wraps a rope around it and pulls horizontally, as shown in Figure 25. The curb height is half the radius of the barrel. What is the minimum tension in the rope that will pull the barrel onto the sidewalk? Express your answer in terms of the barrel weight, W.

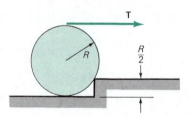

FIGURE 25

9. One type of automobile jack, shown in Figure 26, consists of a four-sided figure ($ABCD$) with the corners B and D connected by a screw. To raise one corner of a car, the screw is turned, shortening DB and raising point A. Suppose the car exerts a downward force of 1000 N at A, and the jack forms two equilateral triangles. Determine the tension in the screw and the compression in each of the four sides.

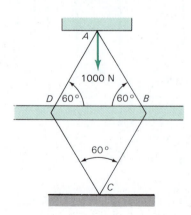

FIGURE 26

10. Three strings are tied at the point A as shown in Figure 27. A 39.3-N weight hangs from one string. There is a tension of 50 N in the string that makes an angle of $37°$ with the horizontal. The system is in equilibrium. Determine the magnitude (T) and direction (θ) of the tension in the third string.

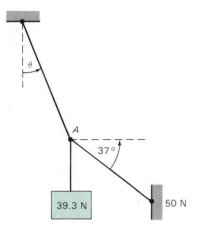

FIGURE 27

11. The packing crate shown in Figure 28 is of weight W, height b, and width a. Its center of gravity is at its center of symmetry. Consider its stability with respect to rotation about an axis through the point O. Express $\tan \theta$ in terms of a and b, where θ is the angle through which the crate must rotate in order to move it into a position of unstable equilibrium.

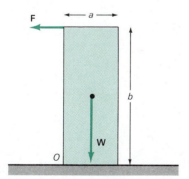

FIGURE 28

12. Four billiard balls of weight W each are in contact forming a tetrahedron as shown in Figure 29. (a) In terms of W, determine the horizontal (H) and vertical (V) components of the contact force exerted on each of the lower balls by the table. (b) Determine the magnitude of the contact force (C) exerted on the upper ball by each of the lower balls.

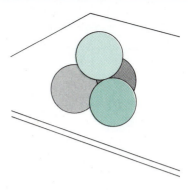

FIGURE 29

13. A slab of steel weighing 200 N has an equilateral triangle cross section (Figure 30). It rests in a smooth open-bottom groove as shown in the figure. Determine the magnitudes of the two normal forces L and R.

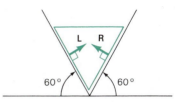

FIGURE 30

14. A 1000-N weight hangs from a rope 5 meters long. A student attaches a second rope at a point 3 meters from the top of the rope suspending the weight (Figure 31). She pulls horizontally and holds the weight stationary. The weight is raised 20 cm. (a) What is the force she applies and, (b) how far horizontally did the weight move?

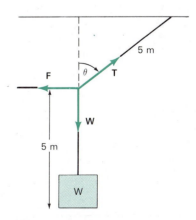

FIGURE 31

15. A uniform rod of weight W and length L is supported at its ends by a frictionless trough (Figure 32).

(a) Prove that the center of gravity of the rod C is directly above the point O when the rod is in equilibrium. (b) Determine the equilibrium value of the angle θ.

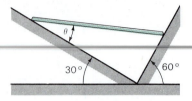

FIGURE 32

16. A sign is formed from two uniform pieces of wood, each of weight W, held together by a horizontal wire. Viewed from the side (Figure 33) the sign resembles the letter A. The sign rests on a smooth surface that can exert only upward forces on the sign. The wire is connected to the midpoints of the sides, which form a 60° angle. Show that the tension in the wire equals $W/\sqrt{3}$.

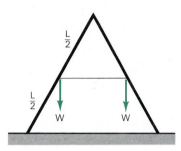

FIGURE 33

17. An acrobat weighing 580 N suspends himself horizontally by a rope held in his teeth and by the forces exerted on his shoe soles by a nearby wall (Figure 34). The distance from his shoe soles to his center of gravity and his mouth are 1.0 m and 1.8 m, respectively. Determine the tension in the rope and the frictional force exerted on his shoe soles.

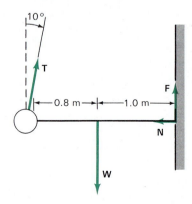

FIGURE 34

18. A uniform ladder leaning against a frictionless wall is held in place by a box on the floor (Figure 35). The ladder weighs 196 N. The floor exerts only a normal force on the ladder. The maximum horizontal force that the box can exert is 200 N. What is the minimum value the angle θ can have such that the ladder does not fall?

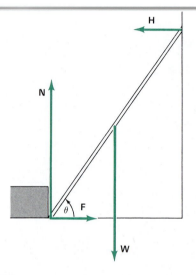

FIGURE 35

19. A uniform plank 8 m long and weighing 490 N is suspended horizontally by a rope attached at a point 3 m from one end (Figure 36). A man weighing 500 N walks along the plank. If the rope is likely to break if its tension exceeds 2200 N, how far out can the man safely walk?

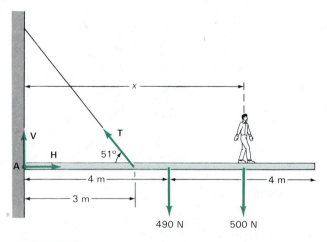

FIGURE 36

20. A boom and cable (Figure 37) are used to suspend a crate. The weight of the boom is 800 N and the weight of the crate is 5000 N. Determine the tension in each cable, and the horizontal and vertical compo-

nents of the force exerted on the bottom of the boom. Assume that the weight of the boom acts at its midpoint and neglect the weight of the cables.

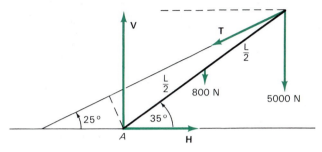

FIGURE 37

21. A meter stick weighing 1.27 N is supported horizontally by small wedges at points 5 cm from each end (Figure 38). A 1.96-N weight is suspended beneath the 30 cm mark and a 4.90-N weight is suspended beneath the 75 cm mark. Determine the support forces exerted by the wedges.

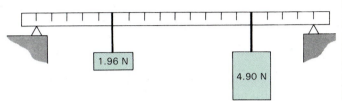

FIGURE 38

CHAPTER

4

MOTION IN
ONE DIMENSION

4.1
FRAMES OF REFERENCE

Kinematics is the study of motion without reference to its causes. The logical development of kinematics begins with the concept of a **frame of reference**:

> A frame of reference is a coordinate system attached to a body or moving relative to a body in a specified way.

In most of our studies we will use coordinate systems attached to the earth. In certain instances it is more convenient to use a coordinate system moving relative to the earth in a specified way. For example, we might use a coordinate system attached to a car moving west at 50 mph relative to the earth.

An object moving relative to a frame of reference undergoes a **displacement**—that is, a **change of position**. Displacements and **time intervals** are the fundamental quantities in kinematics. All other kinematic quantities are defined directly or indirectly in terms of displacements and time intervals.

EXAMPLE 1

Halfback Displacements

A football halfback catches a pass and races down the sideline. He crosses the 50-yard line with the clock showing just 9 seconds left in the game. He reaches the 20-yard line with 5 seconds left, stops for 1 second to avoid an opponent, and then crosses the goal line as the game ends. During the first 4-second time interval his displacement was 30 yards along a direction parallel to the sideline. His displacement was zero over the next 1-second interval. His displacement during the final 4-second interval was 20 yards, also along a direction parallel to the sideline.

In this chapter we consider motion along a straight line. This is called *linear motion*. We take the x-axis to define the straight line. All vector quantities are oriented along the positive or the negative x-direction. Because all motion is along a single axis, we will *not* use boldface symbols to denote vectors in this chapter.

4.2
AVERAGE VELOCITY

Imagine a group of sprinters crouching in their starting positions. The gun starts the automatic timer and the sprinters race toward the finish line 100 m away. It is a close race. First one runner is slightly ahead, and then another surges into the lead, only to be overtaken near the finish by a third runner who wins in a time of 11.34 s.

The winner is the sprinter who travels the 100 m distance in the least time. She is also the sprinter with the largest **average velocity**. The average velocity is denoted by $\bar{v}$ and is defined as

$$\bar{v} = \frac{\text{displacement}}{\text{time interval}}$$

(4.1)

The displacement of the sprinter was 100 m and the time interval for that displacement was 11.34 s. The average velocity of the winner was

$$\bar{v} = \frac{100 \text{ m}}{11.34 \text{ s}} = 8.82 \text{ m/s}$$

As a vector quantity, $\bar{v}$ is in the direction of the displacement. If the position changes from x_i at time t_i to position x_f at time t_f, then the displacement over the time interval $t_f - t_i$ is $x_f - x_i$. We can express the average velocity as

$$\bar{v} = \frac{x_f - x_i}{t_f - t_i} \qquad (4.2)$$

A positive value of $\bar{v}$ indicates that x increases $(x_f > x_i)$. A negative value of $\bar{v}$ means that x decreases $(x_f < x_i)$.

We denote the displacement $x_f - x_i$ by Δx and the time interval $t_f - t_i$ by Δt. In terms of Δx and Δt, the average velocity is

$$\bar{v} = \frac{\Delta x}{\Delta t} \qquad (4.3)$$

The average velocity can be calculated for any time interval, large or small, for which values of Δx and Δt are known.

EXAMPLE 2

Average Velocity of a Sprinter

Table 4.1 lists the positions and the corresponding times for one of the sprinters in the 100 m race. Let's determine her average velocity over the time interval from $t_i = 0.25$ s to $t_f = 1.03$ s.

Her position changes from 0.91 m to 5.49 m, a displacement of 4.58 m. The displacement occurs over a time interval of 0.78 s. Her average velocity is .

$$\bar{v} = \frac{\Delta x}{\Delta t} = \frac{4.58 \text{ m}}{0.78 \text{ s}}$$

$$= 5.87 \text{ m/s}$$

Interpolation and Extrapolation

We cannot calculate the average velocity for arbitrary time intervals, such as the one from $t_i = 0.50$ s to $t_f = 1.00$ s, because there are no position data for these times in Table 4.1. However, if we plot the data in Table 4.1 to produce the set of points shown in Figure 4.1, we can then draw a smooth curve through the data points and select values of x and t that fall on this curve as data points. This procedure, called **interpolation**, is useful and often necessary.

In many experiments computers often record data "on line" (as they occur). The computer output is often in graph form, like the x versus t relation in Figure 4.2. Even when the computer-produced graph is a solid line, interpolation is involved. We can determine the average velocity by first measuring Δx and Δt from the graph and then using Equation 4.3 to determine $\bar{v}$.

In some instances we may want to estimate the average velocity over a time interval *outside* the range of measured values. This requires an *extension* of the x versus t data beyond the range of measured values. The process of

TABLE 4.1

Position–Time Data for Sprinter

x (m)	t (s)
0.00	0.00
0.31	0.11
0.91	0.25
2.13	0.48
5.49	1.03
9.14	1.53

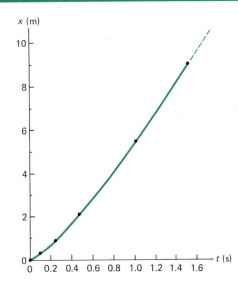

FIGURE 4.1
The points represent data given in Table 4.1 for a runner. The solid lines connect the points to provide a graphical *interpolation* of the data. The dashed line is an *extrapolation* of the data.

extending beyond the measured data range is called **extrapolation**. In Figure 4.1, the dashed line indicates an extrapolation, and is simply a "smooth curve" extension representing an educated guess.

In addition to tables and graphs, we often use equations to relate positions and times. A theory of motion may provide us with an equation relating position and time. By using the equation specifying position as a function of time, $x = x(t)$, we can evaluate the average velocity. For example, an object released from rest near the surface of the earth falls under the influence of gravity. The equation describing the distance it falls in a time t is

$$x = \tfrac{1}{2}gt^2 \tag{4.4}$$

The constant g is called the acceleration of gravity. On the surface of the earth its numerical value varies slightly from one location to the next. We will take its value to be 9.80 m/s^2. Let's see how we can use Equation 4.4 to determine the average velocity.

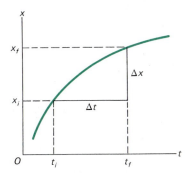

FIGURE 4.2
The solid curve represents the relation $x(t)$. The displacement $\Delta x = x_f - x_i$ occurs during the time interval $\Delta t = t_f - t_i$.

EXAMPLE 3

Average Velocity of a Sinker

An accurate method of measuring the time at which an object passes a point uses photodiodes. A photodiode produces an electrical signal when an object interrupts a light beam.

Using a photodiode sensitive to infrared radiation, students measure the time for a lead sinker to travel past two photodiodes. The sinker passes the photodiodes at the times $t_i = 0.127$ s and $t_f = 0.501$ s. The time interval is

$$\Delta t = t_f - t_i = 0.501 \text{ s} - 0.127 \text{ s} = 0.374 \text{ s}$$

We use Equation 4.4 to determine the corresponding positions

$$x_i = \tfrac{1}{2}(9.80 \text{ m/s}^2)(0.127 \text{ s})^2 = 0.08 \text{ m}$$

$$x_f = \tfrac{1}{2}(9.80 \text{ m/s}^2)(0.501 \text{ s})^2 = 1.23 \text{ m}$$

The displacement is

$$\Delta x = x_f - x_i = 1.23 \text{ m} - 0.08 \text{ m} = 1.15 \text{ m}$$

The average velocity is

$$\bar{v} = \frac{\Delta x}{\Delta t} = \frac{1.15 \text{ m}}{0.374 \text{ s}}$$

$$= 3.07 \text{ m/s}$$

The average velocity is positive because the displacement of the object is positive.

Average Velocity and Average Speed

There are situations in which the average velocity is not a meaningful quantity. Suppose a swimmer in a 50 meter pool covers 100 m in 52 s by swimming from one end to the other and back again. The displacement Δx is zero over the round trip. Hence, the average velocity over the 52 s interval is zero. The vector character of average velocity is responsible for this result. Average speed, a scalar, is more appropriate for such a problem. Average speed is determined by dividing the total distance traveled by the elapsed time.

$$\text{average speed} = \frac{\text{distance covered}}{\text{time interval}}$$

The average speed of the swimmer is therefore 1.92 m/s, *not* zero.

4.3
INSTANTANEOUS VELOCITY

When you glance at the speedometer of your car and see that it reads 52 mph, you have learned to regard this value as the speed of the car at that instant. The concepts of instantaneous velocity and instantaneous speed are easy to grasp because we have wide experience with motions that are continuous.

Instantaneous velocity is defined at a moment in time—at some *instant*. Its calculation requires evaluating a limit. We can conceptualize the limit process as follows: To evaluate the velocity at a time t we start by considering a time interval from t to $t + \Delta t$. If Δx is the displacement over the interval then $\bar{v} = \Delta x / \Delta t$ is the average velocity over the interval. Then we repeat the procedure using the same t but with a smaller Δt. As we shorten the time interval the displacement Δx also decreases. The calculation is repeated, again and again, each time with smaller Δt and Δx. Thus we get a sequence of average velocities, all defined over intervals that start at the time t. In this sequence, both Δt and Δx approach zero. The ratio $\Delta x / \Delta t$ approaches a definite limiting value as Δt approaches zero. This limiting value of the average value is defined as the instantaneous velocity at time t. For linear motion the instantaneous velocity is denoted by v and is defined by the limit

$$v = \lim_{\Delta t \to 0} \frac{\Delta x}{\Delta t} \tag{4.5}$$

The notation "$\lim_{\Delta t \to 0}$" is read as "the limit as Δt goes to zero." To illustrate this limiting procedure we will use three methods. First we will develop a geometric interpretation using a graph. Then we will show how to perform a

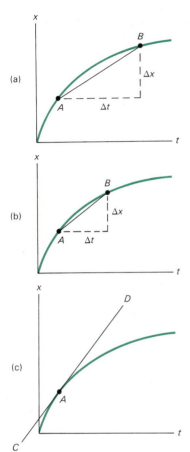

FIGURE 4.3

(a) The slope of the secant line *AB* equals the average velocity. (b) As the time interval approaches zero the secant line approaches the tangent line. (c) The slope of the tangent line *CAD* equals the instantaneous velocity.

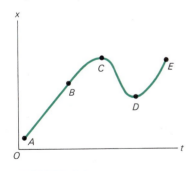

FIGURE 4.4

The slope of the curve at time *t* equals the instantaneous velocity at time *t*. Inspection of the curve determines whether the velocity is positive, zero, or negative. The velocity is zero at points *C* and *D*.

numerical evaluation using a hand calculator or a personal computer. Finally, we will determine the instantaneous velocity via an analytic method using an equation that relates x and t.

Geometric Interpretation of Instantaneous Velocity

To develop a geometric interpretation of instantaneous velocity we first consider the geometric interpretation of average velocity. In Figure 4.3 we show graphically three members of the sequence of average velocities that define the instantaneous velocity. For each graph the time t is the same, but the time interval Δt becomes progressively shorter from one graph to the next. In Figures 4.3a and 4.3b we see that the average velocity $\Delta x/\Delta t$ equals the *slope of the secant line* connecting the points x and $x + \Delta x$. As Δt becomes smaller and smaller the secant line approaches the line tangent to the curve at time t. Figure 4.3c shows that the tangent line is the limit of the secant line. Because the instantaneous velocity is the limit of the average velocity it equals the slope of the tangent line.

This geometric interpretation allows us to make some general observations about the x versus t graph:

1. If the graph is a straight line, such as *AB* in Figure 4.4, the instantaneous velocity is a constant.
2. Points where the slope is zero correspond to positions and times where the instantaneous velocity is zero. In Figure 4.4 points *C* and *D* correspond to zero-velocity points, where the particle is momentarily at rest.
3. A positive slope corresponds to a positive velocity and shows that x is increasing. In Figure 4.4 the instantaneous velocity is positive between *A* and *C* and between *D* and *E*. Between *C* and *D*, x is decreasing, and the instantaneous velocity is negative.

EXAMPLE 4
Instantaneous Velocity: Graphical Solution

Figure 4.5 is a graph of the relation $x = \frac{1}{2}gt^2$ where $g = 9.80$ m/s². Let's determine the instantaneous velocity at the time $t = 3$ s. The tangent line has been drawn with a straight edge to "eyeball" accuracy. The slope of the tangent line appears to pass through the point $x = 0$, $t = 1.6$ s. At $t = 3$ s the value of x is 44 m. The slope of the tangent line through these two points gives us a graphical value of the instantaneous velocity at $t = 3$ s.

$$v = \frac{\Delta x}{\Delta t} = \frac{(44 - 0)\ \text{m}}{(3.0 - 1.6)\ \text{s}}$$

$$= 31\ \text{m/s}$$

As we will show later, the exact value of the instantaneous velocity at $t = 3$ s is 29.4 m/s.

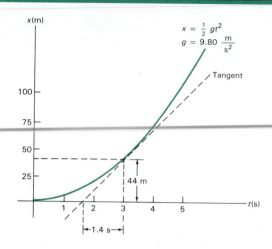

FIGURE 4.5
The instantaneous velocity at $t = 3$ s is the slope of the tangent line that passes through the $t = 3$ s point.

The accuracy of a graphical determination is limited by the precision with which the tangent line can be drawn, and by the precision with which values for Δx and Δt can be read off the graph. However, when an equation relating x and t is known, we can determine the instantaneous velocity numerically or analytically, rather than graphically.

EXAMPLE 5

Numerical Determination of Instantaneous Velocity

Let's determine the instantaneous velocity at $t = 3$ s for the motion described by $x = \frac{1}{2}gt^2$. The values of x in Table 4.2 were obtained from this equation using a hand calculator ($g = 9.80$ m/s^2). The sequence of average velocities is apparently approaching 29.4 m/s. We cannot set $\Delta t = 0$, because that would give us the indeterminate expression 0/0. Nevertheless we have some confidence that the trend established for small values of Δx and Δt represents the limiting behavior.

TABLE 4.2
Numerical Evaluation of Average Velocity

t_f (s)	t_i (s)	x_f (m)	x_i (m)	Δx (m)	Δt (s)	$\Delta x/\Delta t$ (m/s)
3.0	2.9	44.1	41.209	2.891	.1	28.91
3.0	2.99	44.1	43.80649	0.29351	.01	29.351
3.0	2.999	44.1	44.070605	0.029395	.001	29.395
3.0	2.9999	44.1	44.09706	0.00294	.0001	29.4

Analytic Evaluation of Instantaneous Velocity

To remove any doubt as to the limit of the sequence of average velocities we now determine the instantaneous velocity by evaluating the limit in Equation 4.5. For $x = \frac{1}{2}gt^2$ we have

$$\Delta x = \tfrac{1}{2}g(t + \Delta t)^2 - \tfrac{1}{2}gt^2 = gt \cdot \Delta t + \tfrac{1}{2}g(\Delta t)^2$$

and

$$\Delta x/\Delta t = gt + \tfrac{1}{2}g \cdot \Delta t$$

As Δt is made smaller the term $\frac{1}{2}g \cdot \Delta t$ becomes smaller. In the limit as Δt approaches zero the term $\frac{1}{2}g \cdot \Delta t$ approaches zero. It follows that the limit of $\Delta x / \Delta t$ as Δt approaches zero is gt. Thus, the instantaneous velocity at time t is

$$v = gt$$

With $g = 9.80$ m/s^2 and $t = 3$ s we get for the instantaneous velocity

$$v = (9.8 \text{ m/s}^2)3 \text{ s} = 29.4 \text{ m/s}$$

By now you may have recognized the limit defined by Equation 4.5 as a **time derivative**. The instantaneous velocity v is the time derivative of the position $x(t)$.

$$v = \frac{dx}{dt} \tag{4.6}$$

In later chapters we will make liberal use of derivatives. For now we do not use derivatives explicitly.*

We saw earlier that the magnitude of the average velocity can differ from the average speed. However, the magnitude of the instantaneous velocity equals the instantaneous speed.

instantaneous speed = magnitude of instantaneous velocity

Thus, if two objects have instantaneous velocities of $+3$ m/s and -3 m/s, they are traveling in opposite directions with equal speeds of 3 m/s.

4.4
ACCELERATION

The rate of change of velocity is called **acceleration**. A car changes its velocity when it speeds up, slows down, or changes direction. The acceleration of the passengers in the car has a sensory impact—they feel **forces** when they are accelerated. In Chapter 6 we will see how force and acceleration are related by Newton's second law. Here, we are concerned with the kinematic aspects of acceleration. Just as we introduced average and instantaneous velocity, we can define two kinds of acceleration—average and instantaneous. The **average acceleration** is a vector quantity *defined* by

$$\bar{a} = \frac{\text{change in velocity}}{\text{time interval}} \tag{4.7}$$

If we denote the change in velocity by Δv and the time interval by Δt, the average acceleration is

$$\bar{a} = \frac{\Delta v}{\Delta t} \tag{4.8}$$

---

* Fashions in mathematical notation change from time to time. You may have been introduced to the derivative through the following definition:

$$\frac{dx}{dt} = \lim_{h \to 0} \frac{[x(t + h) - x(t)]}{h}$$

This notation differs from that displayed in Equation 4.5, but the definitions are identical. The notation used in Equation 4.5 conveys both the geometrical and physical significance of the derivative.

The direction of $\bar{a}$ is the direction of Δv. For linear motion a positive value of $\bar{a}$ indicates that v is increasing. Two types of motion can give a positive value of $\bar{a}$: one, the object could be traveling in the positive x-direction and increasing its speed, or two, it could be traveling in the negative x-direction and decreasing its speed. You should try to figure out the two types of motion that give a negative average acceleration.

The SI unit of acceleration is m/s². Other units are in common use. An automobile that reaches a speed of 60 miles per hour in 10 s would be credited with an average acceleration of 6 miles per hour per second.

EXAMPLE 6

A Drag Race

Drag racing is an acceleration competition for automobiles over a track 1320 feet long. Cars start from rest and attempt to reach the end of the quarter-mile track in the shortest possible time. In a typical race a Camaro was timed in 11.91 s and achieved a final velocity of 117 mi/h. What was the average acceleration for the Camaro?

Using British units we have

$$\Delta v = 117 \text{ mi/h} - 0 \text{ mi/h} = 117 \text{ mi/h} \qquad \Delta t = 11.91 \text{ s}$$

The average acceleration is

$$\bar{a} = \frac{\Delta v}{\Delta t} = \frac{(117 \text{ mi/h})}{11.91 \text{ s}}$$

$$= 9.83 \text{ mi/h·s}$$

To convert this to SI units we convert each unit separately.

$$9.83 \text{ mi/h·s} = 9.83 \text{ mi/h·s} (1610 \text{ m/mi}) \left(\frac{1 \text{ h}}{3600 \text{ s}} \right)$$

$$= 4.40 \text{ m/s}^2$$

Instantaneous Acceleration

Just as with instantaneous velocity, the definition of instantaneous acceleration requires the use of a limit. For linear motion the **instantaneous acceleration** is defined as the time derivative of the velocity

$$a = \lim_{\Delta t \to 0} \frac{\Delta v}{\Delta t} = \frac{dv}{dt} \qquad (4.9)$$

We will stress applications for which the instantaneous acceleration is constant. For these cases $a = \bar{a}$.

The geometric interpretation of acceleration closely parallels that of velocity. A graph giving v versus t is drawn in Figure 4.6. The average acceleration during the time interval Δt is the slope of the secant line AB, in appropriate units. The instantaneous acceleration at the point B is the slope of the tangent line at point B.

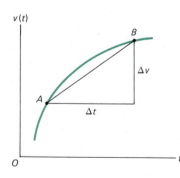

FIGURE 4.6
Average acceleration is $\Delta v/\Delta t$. Geometrically this is the slope of the secant line AB.

EXAMPLE 7

Acceleration of a Runner

Figure 4.7 shows the velocity-versus-time graph of a runner during the first 5 s of a race. We want to determine the instantaneous acceleration of the runner at $t = 2$ s.

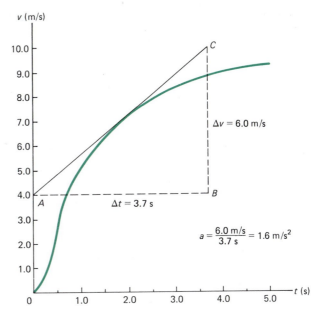

FIGURE 4.7

The colored curve represents the velocity-time relation for a sprinter. The instantaneous acceleration at $t = 2$ s equals the slope of the tangent to the line AC.

First we draw the tangent line AC at $t = 2$ s. This line becomes the hypotenuse of a right triangle (ABC) that has Δv and Δt as its other two sides. From the graph we obtain

$$\Delta v = 6.0 \text{ m/s} \qquad \Delta t = 3.7 \text{ s}$$

The instantaneous acceleration at $t = 2$ s is

$$a = \frac{6.0 \text{ m/s}}{3.7 \text{ s}}$$

$$= 1.6 \text{ m/s}^2.$$

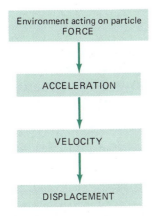

FIGURE 4.8

The laws of dynamics determine the acceleration of a particle. Given the acceleration, kinematics can be used to determine the velocity and displacement.

4.5
THE PROGRAM OF PARTICLE KINEMATICS

Now that we have defined velocity and acceleration it is useful to pause and ask, "What is the purpose of kinematics?" The primary application of kinematics is to start with the acceleration of a particle and then *predict* its future velocity and position. Figure 4.8 illustrates the chain of steps in the program of particle kinematics.

The ability to predict the future has always been a fascinating aspect of physics. But how can we predict the acceleration? The dynamic laws of nature contain the answer to this question. In Chapter 6 we will see how the acceleration of a particle is related to the external force by Newton's second law.

To illustrate this program of particle kinematics we will consider linear motion with a **constant acceleration**. This type of acceleration is both common and mathematically simple. It arises frequently because the earth's gravity can produce a constant acceleration.

In the remainder of this chapter we deal with a particle undergoing linear motion with a constant acceleration. You will learn how to predict future velocities and positions and to solve other types of problems involving motion with a constant acceleration. In the next chapter we will extend the analysis to two-dimensional motion with constant acceleration, and study projectile motion in a plane.

4.6
LINEAR MOTION WITH CONSTANT ACCELERATION

We begin our development of the kinematics of motion with a constant acceleration by establishing a geometrical relation between velocity and displacement. Consider a particle whose instantaneous velocity is a constant 8 m/s. We want to determine its displacement over a time interval of 12 s. Evidently the particle's average velocity is 8 m/s and we can use Equation 4.3 to determine its displacement.

$$\Delta x = \bar{v} \cdot \Delta t = (8 \text{ m/s})(12 \text{ s})$$
$$= 96 \text{ m}$$

This result can be interpreted graphically. In the velocity-versus-time graph of Figure 4.9 the constant velocity plots as a horizontal line. The shaded area in the figure corresponds to the 96-m displacement. These conclusions are valid for all motions with constant velocity: the average velocity equals the instantaneous velocity, and the displacement equals the area under the velocity-time curve. Keep in mind that the units of the "area" under the velocity-time curve are length units.

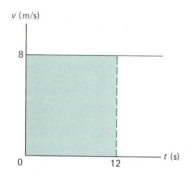

FIGURE 4.9

The horizontal line is the velocity-time relation for an object moving at a constant velocity of 8 m/s. In 12 s the object moves 96 m, which is represented by the shaded area.

A similar geometrical relation relates the change in velocity to the acceleration. Consider a particle whose instantaneous acceleration is a constant 9.8 m/s². Let's find its change in velocity over a 12 s interval. The average acceleration equals 9.8 m/s² and we can use Equation 4.8 to determine the change in velocity

$$\Delta v = \bar{a} \cdot \Delta t = (9.8 \text{ m/s}^2)(12 \text{ s})$$
$$= 118 \text{ m/s}$$

The graph in Figure 4.10 plots instantaneous acceleration versus time. The constant acceleration plots as a horizontal line. The shaded area in the figure corresponds to the 118 m/s change in velocity. These conclusions are valid for all motions with a constant acceleration: The average acceleration equals the instantaneous acceleration, and the change in velocity equals the "area" under the acceleration-time curve in units of length/time.

In fact, both of our area interpretations are generally valid. They apply even when the velocity and acceleration are not constants:

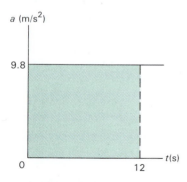

FIGURE 4.10

The horizontal line is the acceleration-time relation for an object moving with a constant acceleration of 9.8 m/s². In 12 s the velocity of the object changes by 118 m/s, which is represented by the shaded area.

The area under a velocity-time curve equals the displacement, and the area under an acceleration-time curve equals the change in velocity.

Proving these statements requires the use of integral calculus. We will not attempt such a proof here. Instead, we simply assert the validity of the statements and apply them to the important special case in which the acceleration is constant.

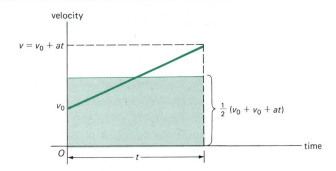

FIGURE 4.11

A velocity-time graph for motion with a constant acceleration. The trapezoid area corresponds to the displacement of the object over the time interval from 0 to t. This area is $v_0 t + \frac{1}{2}at^2$.

We consider a particle that experiences a constant acceleration a. We take the origin of the x-axis to be at the position of the particle at the initial time $t = 0$. The initial velocity is denoted as v_0. We want to deduce equations for the velocity v and for the position x at some arbitrary time t. From Equation 4.8 we have

$$a = \frac{\Delta v}{\Delta t} = \frac{(v - v_0)}{t}$$

This gives the velocity-time relation

$$v = v_0 + at \qquad (4.10)$$

We can use the area interpretation of displacement to derive the position-time relation (Figure 4.11 shows a graph of Equation 4.10). The trapezoidal area equals the displacement over the time interval from 0 to t. The trapezoid area equals the product of its base and its average height. Thus, the displacement x in time t is

$$x = \frac{t(v_0 + v_0 + at)}{2}$$

or

$$x = v_0 t + \frac{1}{2}at^2 \qquad (4.11)$$

Equations 4.10 and 4.11 let us predict the position and velocity at any time t, provided we know the initial velocity v_0 and the constant acceleration a.

FIGURE 4.12

The baseball travels straight up, rising for a time of 3 s.

EXAMPLE 8

Foul Tip Physics

A baseball is hit so that it travels straight up (Figure 4.12). A fan observes that it requires 3 s to reach its maximum height. We want to find both its initial velocity and the height it reached. We ignore the effects of air resistance.

We orient the positive x-axis vertically upward with the origin at the point where the ball left the bat. The acceleration is downward, in the negative x-direction. Because air resistance is neglected, we take the acceleration to be constant

$$a = -g = -9.80 \text{ m/s}^2$$

Equations 4.10 and 4.11 become

$$v = v_0 - gt$$

$$x = v_0 t - \tfrac{1}{2}gt^2$$

The given information is that the ball rises for 3 s. From this we *infer* that the velocity of the ball is zero at time $t = 3$ s. Setting $v = v_0 - gt = 0$ with $t = 3$ s lets us determine the value of v_0.

$$v_0 = gt = (9.8 \text{ m/s}^2)(3 \text{ s})$$

$$= 29.4 \text{ m/s}$$

To determine the height the ball reached we use this value of v_0 in $x = v_0 t - \tfrac{1}{2}gt^2$ and take $t = 3$ s. This gives

$$x = (29.4 \text{ m/s})(3 \text{ s}) - \tfrac{1}{2}(9.80 \text{ m/s}^2)(3 \text{ s})^2$$

$$= 44.1 \text{ m}$$

Symmetries

The equations describing motion with constant acceleration exhibit a number of *symmetries*. For example, it took 3 s for the ball to rise 44.1 m. It takes the same time for it to fall 44.1 m. We can demonstrate this most simply by choosing a coordinate origin at the point where the ball is momentarily at rest. The ball starts from rest at this new origin. With $v_0 = 0$ and $a = -g$, Equation 4.10 gives

$$x = -\tfrac{1}{2}gt^2$$

Setting $x = -44.1$ m and $g = 9.80$ m/s^2 gives $t = 3$ s, verifying the time symmetry.

The instantaneous speed also exhibits symmetry. At any point along the path the speed of the ball is the same on the way up as it is on the way down. To prove this statement we use Equation 4.10 and Equation 4.11 to derive a relation between v and x that does not involve the time t. We can do this by using Equation 4.10 to express t as

$$t = \frac{v - v_0}{a}$$

Inserting this expression into Equation 4.11 leads to

$$v^2 = v_0{}^2 + 2ax \tag{4.12}$$

Equation 4.12 is the desired equation: it relates v and x without reference to the time. For given values of v_0, a, and x the two solutions of Equation 4.12 are

$$v = \pm\sqrt{v_0{}^2 + 2ax}$$

This shows that for *a* given x the velocity can have two values of equal magnitude, or, in other words, equal *speed*. The object passes a given position on the way up with a particular velocity. It passes the same position on the way down with an equal but oppositely directed velocity.

Equations 4.10, 4.11, and 4.12 relate the variables x, v, and t for motion with a constant acceleration. Deciding which equation is best suited to solve a particular problem requires experience and a systematic approach. In general you should size up a problem by noting what quantities are given—either *explicitly* or *implicitly*—and what quantities you are asked to figure out. You must then sort through Equations 4.10–4.12 and select those that relate only the given and unknown quantities. Let's put this advice to work by solving some typical problems.

EXAMPLE 9

Stopping Distance of an Automobile

By applying the brakes without causing a skid, the driver of a car is able to achieve a constant acceleration of -4.0 m/s^2. How far does the car travel after the brakes are applied at 35 mi/h?

The given information is

$$v_0 = 35 \text{ mi/h} = 51.3 \text{ ft/s}$$

$$a = -4.0 \text{ m/s}^2 = -13.1 \text{ ft/s}^2$$

We must *infer* that the stopping distance corresponds to the value of x for which the velocity reaches zero. We see that Equation 4.12 relates the unknown x to the known quantities v, v_0, and a. With $v = 0$, Equation 4.12 gives

$$x = \frac{-v_0^2}{2a} = \frac{-(51.3 \text{ ft/s})^2}{2(-13.1 \text{ ft/s}^2)}$$

$$= 100 \text{ ft}$$

Note that if the initial velocity is *doubled* to 70 mi/h, the stopping distance *quadruples*, becoming 400 ft.

Police routinely measure the length of skid marks after automobile accidents to estimate the speed of the car when the brakes were applied. The acceleration limits for different highway conditions are well known to the police. The acceleration and stopping distance together establish the initial velocity of the car. This example reverses the analysis used by the police to determine the initial velocity.

Example 9 also illustrates how careful we must be with units. The velocities v and v_0 may have units of (m/s) or (mi/h), and the same units must be used for both. The acceleration a and the distance x must combine to give compatible units. As originally defined, the units are mixed. We must convert units before we can obtain meaningful numerical values. A useful conversion factor is

$$15 \text{ mi/h} = 22 \text{ ft/s}$$

Equations 4.10–4.12 are interrelated, so it may happen that there is more than one way to solve a problem.

EXAMPLE 10

Time of Flight

A baseball is thrown straight downward from a rooftop with an initial velocity of 26 m/s. It reaches the ground 24.5 m below traveling at 34 m/s. We want to determine its time of flight. We neglect the effects of air resistance, and assume a constant acceleration of 9.8 m/s^2. We are given

$$v = 34 \text{ m/s} \qquad v_0 = 26 \text{ m/s} \qquad a = 9.8 \text{ m/s}^2$$

and asked to find t. Equation 4.10 is an obvious candidate to help us solve the problem. Taking the positive direction as down it gives

$$t = \frac{v - v_0}{a} = \frac{(34 - 26) \text{ m/s}}{9.8 \text{ m/s}^2}$$

$$= 0.82 \text{ s}$$

We could also use Equation 4.11 because x, v_0, and a are all specified. Equation 4.11 is a *quadratic* equation in t and so has two solutions, one of which is $t = 0.82$ s. You should determine the other solution of the quadratic equation, and explain its physical significance.

The acceleration due to gravity is the same for all objects at a particular location. The equations that we have developed apply equally well to the constant acceleration at the surface of the earth and at the surface of the moon. On the moon, however, this acceleration is roughly one-sixth of its value on earth. Does this mean that an object requires six times as long to fall through 1 m on the moon as to fall through 1 m on the earth? The following example addresses this question.

EXAMPLE 11

Free-Fall on the Moon

An astronaut standing on the moon drops a hammer, letting it fall 1 m to the surface. The lunar gravity produces a constant acceleration of 1.62 m/s². Upon returning to earth the astronaut again drops the hammer, letting it fall to the ground with an acceleration of 9.8 m/s². How do the times of fall compare? In both cases the initial velocity is zero and $x = 1$ m. We want to determine the time of fall, so we use Equation 4.11 to find

$$t = \sqrt{\frac{2x}{a}}$$

On the moon,

$$t = \sqrt{\frac{2(1 \text{ m})}{1.62 \text{ m/s}^2}}$$

$$= 1.11 \text{ s}$$

and on the earth

$$t = \sqrt{\frac{2(1 \text{ m})}{9.8 \text{ m/s}^2}}$$

$$= 0.452 \text{ s}$$

The time of free-fall is longer on the moon, but not by the factor of 6 as suggested earlier. For equal distances the times are inversely proportional to the *square root* of the acceleration.

WORKED PROBLEM

A ball is thrown vertically upward with an initial speed of 12 m/s. A second ball is thrown vertically upward 0.500 s later with an initial speed of 16 m/s. At what height above the release point do the balls pass, and is the first ball moving upward or downward when they pass? Disregard air friction.

Solution

We orient the x-axis along the paths of the balls, call up the positive direction, and take the origin at the position where the balls leave the thrower's hand. With this convention, the acceleration of each ball is $a = -g = -9.80 \text{ m/s}^2$.

To solve this problem, we first write down the position versus time relation for each ball. We can then equate the positions of the balls and find the time at which the balls pass. Having determined the moment at which they pass, we can calculate the position at which they pass.

A time t after the first ball is thrown, its position is given by

$$x_1 = (12 \text{ m/s})t - (4.90 \text{ m/s}^2)t^2$$

The second ball is released 0.500 s later with an initial speed of 16 m/s. At time t it has been moving for a time $t - 0.500$ s, so its position at time t is given by

$$x_2 = (16 \text{ m/s})(t - 0.500 \text{ s}) - (4.90 \text{ m/s}^2)(t - 0.500 \text{ s})^2$$

Equating x_1 and x_2 and solving for t gives

$$t = 1.04 \text{ s}$$

The height above the release point at which they pass is the value of x_1 (and x_2) at $t = 1.04$ s. We find

$$x_1 = (12 \text{ m/s})(1.04 \text{ s}) - (4.90 \text{ m/s}^2)(1.04 \text{ s})^2$$
$$= 7.18 \text{ m}$$

The velocity of the first ball at time t is given by

$$v_1 = 12 \text{ m/s} - (9.80 \text{ m/s}^2)t$$

At $t = 1.04$ s we find v_1 equals $+1.8$ m/s. The positive value of v_1 means that the first ball is moving *upward* at the moment the balls pass.

Thus, the balls pass at a height 7.18 m above the release point and the first ball is still traveling upward at that moment.

EXERCISES

(Disregard air resistance in all Exercises.)

4.1   Frames of Reference

A. Two observers, A and B, are positioned at $x_A = 2$ m, and $x_B = 12$ m. At time t_i they observe a bluebird at $x_i = 10$ m. At time t_f they observe the bird at $x_f = 8$ m. Determine the displacement of the bluebird for each observer.

4.2   Average Velocity

B. Using the data in Table 4.1, obtain the average velocity for (a) $x_i = 0.31$ m, $x_f = 0.91$ m; (b) $x_i = 0.31$ m, $x_f = 2.13$ m; (c) $x_i = 0.91$ m, $x_f = 2.13$ m.

C. A graph of position versus time is given for a swimmer in Figure 1. Obtain her average velocity for (a) $t_i = 0$ s, $t_f = 4$ s; (b) $t_i = 3$ s, $t_f = 4$ s.

D. A freely falling ball has a position described by $x = -\frac{1}{2}gt^2$, where $g = 9.80 \text{ m/s}^2$. Determine the average velocity over the time interval from $t_i = 1$ s to $t_f = 3$ s.

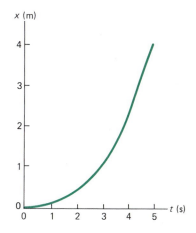

FIGURE 1

E. The position of a feather dropped on the moon by Astronaut David Scott is described by

$$x = -\tfrac{1}{2}gt^2 \qquad g = 1.62 \text{ m/s}^2$$

Find the average velocity for $t_f = 1.00$ s and (a) $t_i = 0.00$ s, (b) $t_i = 0.50$ s, (c) $t_i = 0.90$ s.

F. *Plate tectonic* motion refers to the motion of the earth's crustal plates. Measurements indicate northward plate tectonic motions of 2.5 cm per year for coastal portions of southern California. Estimate how long it would take for this motion to carry southern California to Alaska. What is assumed in your calculation?

4.3   Instantaneous Velocity

G. From the graph in Figure 1, estimate the instantaneous velocity at $t = 3$ s.

H. The position of a plane accelerating for takeoff down a runway is given by

$$x = bt^2 \qquad b = 1.200 \text{ m/s}^2$$

Estimate the plane's instantaneous velocity at $t = 4$ s by evaluating a sequence of average velocities. Take $t_i = 4.000$ s and (a) $t_f = 4.100$ s, (b) $t_f = 4.010$ s, (c) $t_f = 4.001$ s.

I. The velocity of an object is given by $v = at + b$ where $a = 2.4$ m/s^2 and $b = 30$ m/s. Plot this velocity on graph paper from $t = 0$ to $t = 10$ s, and calculate the distance covered during that time interval by determining the area under the curve.

J. A subway train moves in a straight line from one station to another. The position of the front of the train as a function of time is shown in Figure 2. Estimate the instantaneous velocity at 0, 2, 4, and 6 minutes.

4.4   Acceleration

K. The velocity of a rocket during the launch stage is given by $v = bt - kt^2$, where $b = 30$ m/s^2 and $k = 0.50$ m/s^3. Determine the average acceleration of the rocket from (a) $t_i = 0$ s to $t_f = 1$ s, (b) $t_i = 5$ s to $t_f = 6$ s.

L. Using the formula for velocity given in Exercise K, estimate the instantaneous acceleration at $t = 5$ s by taking $t_i = 5$ s and successively letting $t_f = 5.1$ s, 5.01 s, and 5.001 s.

M. The advertised characteristics of the 1980 Porsche Turbo sportscar are shown in Table 1. For which gear is the acceleration largest and what is the largest acceleration?

TABLE 1

Gear	Time starting from rest (s)	Final speed (mi/h)
1	2.3	30
2	9.3	52
3	21.3	78
4	44.5	100
5	93.4	120

N. (a) A car moving in a straight line at 10 mi/h accelerates for 3 s. The average acceleration over the 3 s interval is 20 mi/h·s. What is its speed at the end of the 3 s acceleration interval? (b) The car then decelerates, coming to a stop in 7 s. What is the average acceleration over the 7 s interval

4.6   Linear Motion with Constant Acceleration

P. In this exercise, a, v, and x denote an acceleration, a velocity, and a displacement. A quiz problem asks that you identify the time of flight of a projectile. The following multiple-choice answers are presented: (a) a/v, (b) ax/v, (c) $x/(v - x)$, (d) x/v, (e) x/a. Use dimensional analysis to identify the correct answer.

Q. In a drag race one car reaches a maximum velocity of 90 m/s. The car's acceleration was 10 m/s^2 throughout the race. How far did the car travel?

R. A research rocket starting from rest at $t = 0$ climbs to an altitude of 9600 m in 80 s. Its acceleration is constant. Calculate the rocket's (a) acceleration, (b) velocity at $t = 80$ s, (c) average velocity over the 80 s interval.

S. A drag racer covers the quarter-mile distance in 10 s. (a) What was his acceleration, assuming it was constant? (b) What was his final velocity? (Express answers in SI units.)

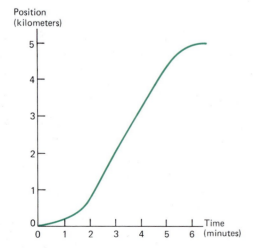

FIGURE 2

PROBLEMS

(Disregard air resistance in all Problems.)

1. A student jumps straight upward. He rises for 0.30 s and descends in equal time. (a) What was his initial velocity? (b) How high did he rise?

2. A child drops stones from a bridge at regular intervals of 1 second. At the moment the fourth stone is released, the first strikes the water below. (a) How high is the bridge? (b) How far above the water are each of the other falling stones at the moment when the first stone strikes?

3. Imagine that you are in charge of a camera crew that is filming a miniaturized scene in which a bridge explodes and pieces fly in all directions. The scale is 1/9; that is, all linear dimensions are one-ninth their true size. However, you do not have the liberty to scale g, the acceleration of gravity. It will be 9.80 m/s² when the miniature pieces begin flying about. If a full-scale scene were photographed at 60 frames/s, and lasted 6 s, how many frames per second would be required to make the scene look natural—that is, to make it last 6 s when projected at the normal speed of 60 frames/s?

4. A rubber ball is released from a height of 4.90 m above the floor. It falls with an acceleration of 9.80 m/s² and strikes the floor 1 s later. It bounces repeatedly, always rising to 81/100 of the height through which it falls. (a) Treat the ball as a point mass and show that it travels a distance of 46.7 m during its infinite number of bounces. (Suggestion: Work out the first few distances and then set up and sum the infinite series. If you are unable to do this algebraically, try to sum the first few terms of the series arithmetically, perhaps using a calculator.) (b) Determine the time required for the infinite number of bounces.

5. The horizontal motion of a football punt proceeds at a constant rate of 12 yd/s. A player begins running in the same direction as the football with a velocity of 8 yd/s. If both start at the same time, and the player is 5 yd behind the ball when it lands, how far does the football travel horizontally?

6. A group of physics students devised a modified game of darts. They dropped the dart board to the ground from a window. One second after dropping the dart board they threw darts at it. By trial and error they found that the darts arrived at the board at the instant the board touched the ground if the initial dart speed was 20 m/s. Determine (a) the time of flight for the dart, (b) the height of the window.

7. An object is projected upward in a vacuum (Figure 3) past elevations B and A (labeled B' and A' on the trip downward). The object rises to an unknown height and then returns to its starting point, passing A' and B' on the way. The distance between the two elevations is measured to be exactly 1 meter. A laser accurately

triggers a timer, which records the times when the object is located at x_B, x_A, $x_{A'}$, and $x_{B'}$. These times are $t_B = 0.1982$ s, $t_A = 0.2207$ s, $t_{A'} = 9.2651$ s, and $t_{B'} = 9.2876$ s. Treat the initial velocity and the acceleration of gravity as unknowns. (a) Obtain a value for g (to three significant figures) using the given data. (b) Obtain the initial velocity. (c) Determine the height to which the object rises relative to its starting position.

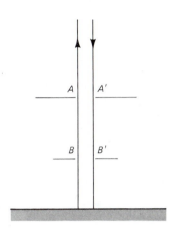

FIGURE 3

8. A daring cowboy sitting on a tree limb wishes to drop vertically onto a horse galloping under the tree. The speed of the horse is 10 m/s and the distance from the limb to the saddle is 3 m. (a) What must be the horizontal distance between the saddle and limb when the cowboy makes his move? (b) How long is the cowboy in the air?

9. The fuel consumption of a car (G), expressed as gallons consumed per kilometer of travel, depends on the speed (v) of the car. (a) Drawing on your own experience or from talking with a friend, does the relation $G = A + Bv$ represent a reasonable description of the fuel consumption of the car? $(A = 0.05$ gal/km, $B = 0.01$ gal·h/km²$)$ (b) If a car starts from rest, accelerates at a constant rate, and achieves a speed of 100 km/h in 1 minute, how much fuel was consumed by the car? Assume that $G = A + Bv$ holds during the acceleration.

10. Two cars are at rest 1 km apart. The first car starts and accelerates at 2 m/s² toward the second car. Five seconds later the second car starts and accelerates toward the first at 3 m/s². Where do the cars meet relative to the starting position of the first car?

11. A car starts from rest, accelerates to 20 m/s at the constant rate of 3.5 m/s², and then moves at a constant speed of 20 m/s. Two seconds after the first car

starts, a second car starts from rest at the same point, accelerates to a speed of 24 m/s at the constant rate of 4.0 m/s^2, and then moves at a constant speed of 24 m/s. (a) What is the time interval between the start of the first car and the time when the second car catches the first? (b) How far do the cars travel during this time?

12. To execute a triple-somersault dive a diver springs upward 1.5 m and then falls 4.5 m into the water. If she changes into the tuck position 0.1 s after leaving the board and emerges from it 0.1 s before entering the water, how much time will she have for the rotation in the tuck position?

13. At the conclusion of an experiment on the motion of a particle a student finds the velocity and position of the particle are related by the graph in Figure 4. The position of the particle at time $t = 0$ is $x = 0$. Determine equations describing (a) velocity as a function of position, (b) position as a function of time, (c) velocity as a function of time.

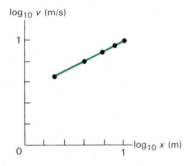

FIGURE 4

14. The cables supporting an elevator snap and it falls freely a distance of 19.6 m. (a) Determine its speed just before impact. (b) A cat inside the elevator senses its precarious situation. One-half second before impact the cat leaps upward with an initial speed of 12 m/s relative to the elevator. At what speed does the cat impact the floor? (c) At what speed would the cat have to leap in order to minimize its impact speed? (d) What is the minimum impact speed?

15. An acrobat leaves a trampoline moving upward at a speed of 8 m/s. One-half second later she throws a small ball downward. The ball leaves her hand at a speed of 2 m/s relative to her hand. The release point is 1 m above her feet. (a) Determine the maximum height reached by the ball. (b) Determine the time that elapses between the release of the ball and its impact on the trampoline.

16. A car and train move together along parallel paths at 25 m/s. The car stops for a red light, decelerating at 2.5 m/s^2. It remains at rest for 45 s, then accelerates back to a speed of 25 m/s at a rate of 2.5 m/s^2. How far behind the train is the car when it reaches the speed of 25 m/s, assuming that the train speed has remained at 25 m/s?

17. Two mirrors in outer space move toward each other. Each travels at half the speed of light relative to a fixed point midway between the mirrors. A light pulse starts from one mirror when the mirrors are 10 light-years apart. The pulse reaches the second mirror, is reflected back to the first, and so on. What distance has the light pulse traveled when the mirrors collide?

18. A baseball is thrown vertically upward. It passes an overhead wire 0.4 s after being released. On the way down it passes the same wire 3.2 s after being released. Determine the initial speed of the ball.

19. One car moving at a constant velocity of 50 km/h passes a second car at rest. At the instant it is passed, the second car begins accelerating at a constant rate of 5 km/h·s. The second car maintains this constant acceleration until it catches the first car. How much time is required for the second car to catch the first car, and what distance is covered in the process?

CHAPTER

5

MOTION IN

A PLANE

5.1
PROJECTILE MOTION IN A PLANE

Position, velocity, and acceleration are vector quantities. In our study of one-dimensional kinematics the direction of these vectors was indicated by a plus sign or a minus sign. The vector nature of position, velocity, and acceleration is more apparent when we deal with motion in a plane since these vectors are not directed along a single line. We begin by considering the kinematics of **projectile** motion because we all have some experience with projectiles like tennis balls and baseballs.

No new mathematics is needed to describe two-dimensional kinematics. We can carry over the ideas and techniques developed for linear motion. The mathematical simplicity of projectile motion stems from two facts. First, in the absence of air resistance, acceleration is constant. Second, motions perpendicular to and parallel to the direction of the acceleration are *independent*. Your everyday experience serves as experimental support for this observation. For example, suppose you watch a friend seated in a classroom as he tosses a ball straight upward. The ball experiences a constant acceleration in the downward direction. It rises and falls in accord with the one-dimensional kinematics we developed in Chapter 4. When the class ends, your friend rises from his seat and walks toward the door, still tossing the ball with the same motion. The vertical part of the ball's motion is *unaffected* by its horizontal motion. Its vertical motion is still described by the equations developed in Chapter 4.

For as little as two cents you can demonstrate for yourself the independence of the horizontal and vertical motions. Simply place two coins near the corner of a table (Figure 5.1). Use your finger to simultaneously push both coins off the table, giving one a relatively large horizontal velocity and giving the other virtually zero horizontal velocity. One coin will fly off the table; the other will barely topple off. Your ears will tell you the result of this demonstration. The coins strike the floor at the same moment. Their vertical motions are identical. Therefore, their different horizontal motions have no effect on their vertical motions.

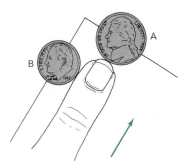

FIGURE 5.1

Two coins can be used to demonstrate the independence of the horizontal and vertical components of motion. The tip of the finger imparts a horizontal velocity to coin *A*. Simultaneously, the side of the finger nudges coin *B* off the edge of the table.

Component Equations for Projectile Motion

We want to describe the motion of a projectile that experiences a constant acceleration in the downward direction. We use the x–y coordinate system shown in Figure 5.2. The positive y-axis is upward. At any point along the projectile's path we can resolve the position vector $\mathbf{r}$, the velocity vector $\mathbf{v}$, and the acceleration vector $\mathbf{a}$, into their x- and y-components. The coordinate origin is placed at the position of the projectile at time $t = 0$. In Figure 5.2a, the initial velocity has a magnitude v_0 and is directed at the launch angle θ_0 above the positive x-axis. A zero subscript is used to identify values at the initial time $t = 0$. An x or y subscript identifies a component along the corresponding axis.

The constant acceleration of the projectile is in the negative y-direction.

$$a_y = -g = \text{constant} \tag{5.1}$$

Near the surface of the earth $g = 9.80 \text{ m/s}^2$.

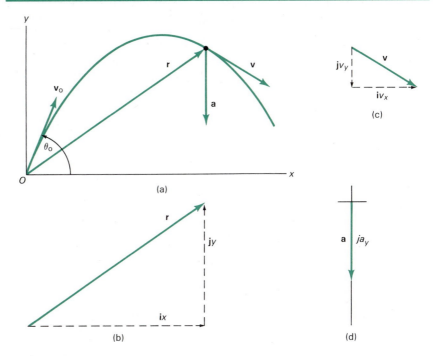

FIGURE 5.2

The heavy curve in (a) represents the trajectory of a projectile moving in the x–y plane. The vectors **r**, **v**, and **a** denote the position, velocity, and acceleration at the same moment. In (b), (c), and (d) the components of **r**, **v**, and **a** are shown. The vector $\mathbf{v}_0$ denotes the initial (launch) velocity and θ_0 is the launch angle.

We use the results developed in Chapter 4 to write the y-components of the velocity and position.

$$v_y = v_{y0} - gt \tag{5.2}$$

$$y = v_{y0}t - \tfrac{1}{2}gt^2 \tag{5.3}$$

v_{y0} is the y-component of the initial velocity. Eliminating t from Equation 5.2 and Equation 5.3 we obtain an equation relating v_y and y.

$$v_y{}^2 = v_{y0}{}^2 - 2gy \tag{5.4}$$

Motion in the horizontal direction is unaccelerated. The x-component of the velocity is constant and equal to its initial value

$$v_x = v_{x0} = \text{constant} \tag{5.5}$$

The x-displacement in time t is

$$x = v_{x0}t \tag{5.6}$$

If you are given the initial velocity (v_{x0}, v_{y0}) and acceleration $(-g)$, Equations 5.2–5.6 enable you to predict the velocity (v_x, v_y) and position (x, y) at any time t. Thus, Equations 5.2–5.6 comprise a kinematic solution of projectile motion in a plane. The following example illustrates these results.

EXAMPLE 1

Vertical Drop of a Pitched Baseball

A baseball is thrown horizontally by a pitcher toward a batter. Assuming that the ball has the velocity of a good major league fastball (90 mph or 40 m/s), and travels 17 m (somewhat less than the distance between the pitcher's mound and home plate), what vertical distance does the ball drop? Disregard air resistance effects.

We use Equation 5.6 to obtain the time of flight

$$t = \frac{x}{v_{x0}} = \frac{17 \text{ m}}{40 \text{ m/s}} = 0.425 \text{ s}$$

The initial velocity is horizontal, so $v_{y0} = 0$. With $g = -9.80 \text{ m/s}^2$, Equation 5.3 gives the vertical drop

$$y = -\tfrac{1}{2}gt^2 = -4.90 \text{ m/s}^2 \,(0.425 \text{ s})^2 = -0.885 \text{ m}$$

The minus sign indicates that the baseball arrives 0.885 m below the release level. Remember that our calculation did not take air resistance into account. If air resistance is taken into account, the drop is about 10 percent larger than what we calculated.

Parabolic Trajectory

The path followed by a projectile is called its **trajectory**. For a constant acceleration the trajectory is a **parabola**. We can derive the equation describing the trajectory by eliminating the time t from the equations for x and y. Let's show that this procedure leads to a parabolic trajectory.

We solve Equation 5.6 for t:

$$t = \frac{x}{v_{x0}}$$

Substituting this expression into Equation 5.3 gives

$$y = \left(\frac{v_{y0}}{v_{x0}}\right)x - \left(\frac{g}{2v_{x0}{}^2}\right)x^2 \tag{5.7}$$

You should recall that a parabola is described by the quadratic form

$$y = bx^2 + cx \qquad b \neq 0$$

with b and c as constants. Equation 5.7 has this form, illustrating that the trajectory is parabolic.

Often, the launch angle θ_0 and the magnitude of the initial velocity v_0 are given instead of v_{x0} and v_{y0}. In these situations, it is helpful to use the *polar forms* of v_{x0} and v_{y0}. Figure 5.2a shows that the polar forms of v_{x0} and v_{y0} are

$$v_{x0} = v_0 \cos \theta_0$$
$$v_{y0} = v_0 \sin \theta_0 \tag{5.8}$$

Substituting these expressions into Equation 5.7 yields a new form for the trajectory of the projectile

$$y = x \tan \theta_0 - \frac{1}{2}\left(\frac{g}{v_0{}^2 \cos^2 \theta_0}\right)x^2 \tag{5.9}$$

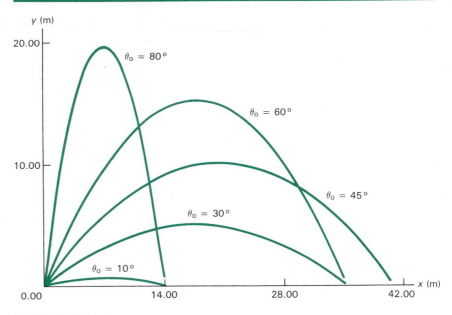

FIGURE 5.3

The parabolic trajectories correspond to the same initial speed ($v_0 = 20$ m/s), and different launch angles (θ_0). The launch angle is shown in degrees for each trajectory.

Trajectories that correspond to several launch angles are given in Figure 5.3. These parabolic paths share the same values of v_0 and g. The maximum height, the distance from launch to landing point, and the launch angle vary from one trajectory to the next.

Time Symmetry

The parabolic trajectory exhibits the same symmetries we found in Chapter 4 for one-dimensional motion. In particular, the time for the projectile to reach its maximum height equals the time for it to return to its initial level. In the case of a parabolic trajectory, time symmetry persists because the time for the projectile to reach its peak is independent of its horizontal motion. Let's verify this claim and then use the result to derive an equation for the range of the projectile.

To find the time it takes for the projectile to reach its peak we set $v_y = 0$ in Equation 5.2. This gives

$$t_{\text{rise}} = \frac{v_{y0}}{g} \tag{5.10}$$

To find the total time of flight we set $y = 0$ in Equation 5.3. This gives

$$t(v_{y0} - \tfrac{1}{2}gt) = 0$$

The solution $t = 0$ corresponds to the initial position of the projectile. The other solution gives the time of flight, t_f

$$t_f = \frac{2v_{y0}}{g} = 2t_{\text{rise}} \tag{5.11}$$

This shows that the projectile rises and falls for equal times. To determine the maximum height reached by the projectile we set $t = t_{rise}$ in Equation 5.3:

$$y_{max} = v_{y0}\left(\frac{v_{y0}}{g}\right) - \frac{1}{2}g\left(\frac{v_{y0}}{g}\right)^2 = \frac{v_{y0}^2}{2g} \qquad (5.12)$$

This shows that the maximum height is independent of the horizontal motion. If we use Equation 5.8 to express v_{y0} in terms of the launch angle we get

$$y_{max} = \left(\frac{v_0^2}{2g}\right)\sin^2\theta_0 \qquad (5.13)$$

Projectile Range

The horizontal distance the projectile travels is called its **range**. Assuming that the projectile lands at the same level ($y = 0$) from which it was launched, the range is simply v_{x0} times the time of flight. We denote the range by R. Equation 5.11 yields

$$R = v_{x0} \cdot t_f = v_{x0} \cdot \left(\frac{2v_{y0}}{g}\right) \qquad (5.14)$$

If we use the polar forms of v_{x0} and v_{y0} given in Equation 5.8, and the trigonometric identity $\sin 2\theta_0 = 2\sin\theta_0\cos\theta_0$, we get

$$R = \left(\frac{v_0^2}{g}\right)\sin 2\theta_0 \qquad (5.15)$$

The range depends on the initial speed v_0, the launch angle θ_0, and the acceleration, g. This form of the range equation lets us prove an important result: If v_0 and g are constant, the range is a maximum for $\theta_0 = 45°$. From Equation 5.15 we see that the range is a maximum when $\sin 2\theta_0 = 1$, which is satisfied by $\theta_0 = 45°$.

We see from Equation 5.14 that the value of g determines the time of flight. A projectile launched with a given initial speed and launch angle would travel approximately 6 times farther on the moon than on the earth. For golfers who yearn to be "long drivers," the moon is the place to tee up.

EXAMPLE 2
Two Golf Drives

Let's compare the ranges of two golf drives, one on the earth and the other on the moon. We take the same initial speed $v_0 = 80$ m/s, and the same launch angle $\theta_0 = 8°$, for both. The range follows from Equation 5.15. On the earth

$$R_{earth} = \left(\frac{v_0^2}{g}\right)\sin 2\theta_0 = \left[\frac{(80\ \text{m/s})^2}{9.80\ \text{m/s}^2}\right]\sin 16°$$

$$= 180\ \text{m}$$

So, the range is approximately 200 yards. The value of g on the moon is 1.62 m/s^2, about six times smaller than the value of g on the earth. As a consequence the range on the moon is about six times greater than the range on the earth

$$R_{moon} = \left[\frac{(80\ \text{m/s})^2}{1.62\ \text{m/s}^2}\right]\sin 16°$$

$$= 1090\ \text{m}$$

This is approximately 1200 yards—over two-thirds of a mile.

Elevated Launch

A launch angle of $45°$ gives the maximum range only if the impact point is at the same level as the launch point. In general, the range is a maximum when the angle between the initial velocity $\mathbf{v}_0$ and the final velocity $\mathbf{v}$ is $90°$. To prove this statement we consider the elevated launch shown in Figure 5.4. The projectile is launched at an angle θ_0 from a height h above the level at which it impacts. The relation between the initial and final speeds follows from Equation 5.4. At impact $y = -h$, and Equation 5.4 gives

$$v^2 = v_0{}^2 + 2gh \tag{5.16}$$

Calling β the angle between $\mathbf{v}_0$ and $\mathbf{v}$, we observe that the **magnitude** of the vector product $\mathbf{v} \times \mathbf{v}_0$ is

$$|\mathbf{v} \times \mathbf{v}_0| = vv_0 \sin \beta \tag{5.17}$$

We can obtain another expression for $|\mathbf{v} \times \mathbf{v}_0|$ by using the component forms of $\mathbf{v}$ and $\mathbf{v}_0$. In terms of the unit vectors $\mathbf{i}$ and $\mathbf{j}$ we can write

$$\mathbf{v}_0 = \mathbf{i}v_{x0} + \mathbf{j}v_{y0}$$

$$\mathbf{v} = \mathbf{i}v_{x0} + \mathbf{j}(v_{y0} - gt)$$

Forming the vector product $\mathbf{v} \times \mathbf{v}_0$ and noting $\mathbf{i} \times \mathbf{j} = \mathbf{k}$ gives

$$\mathbf{v} \times \mathbf{v}_0 = \mathbf{k}gv_{x0}t$$

So that

$$|\mathbf{v} \times \mathbf{v}_0| = gv_{x0}t \tag{5.18}$$

But $v_{x0}t$ equals R, the range of the projectile. Equating these two expressions for $|\mathbf{v} \times \mathbf{v}_0|$ gives

$$vv_0 \sin \beta = gR$$

Using Equation 5.16 to express v in terms of v_0 and h gives

$$R = \left(\frac{v_0}{g}\right) \sqrt{v_0{}^2 + 2gh} \sin \beta \tag{5.19}$$

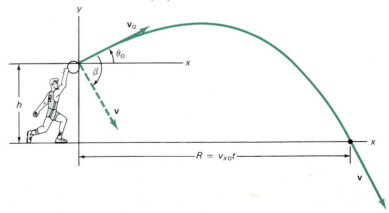

FIGURE 5.4

A projectile is launched from a height h above its impact level. For a given launch speed v_0 the range R is a maximum when the initial and final velocities are perpendicular ($\beta = 90°$).

By inspection, R is a maximum when $\sin \beta = 1$, which requires $\beta = 90°$. Thus the range is a maximum when the angle β between the initial and final velocity vectors is $90°$. The maximum range is given by

$$R_{\max} = \left(\frac{v_0}{g}\right)\sqrt{v_0^2 + 2gh} \tag{5.20}$$

The launch angle which produces this value of R is given by

$$\tan \theta_0 = \frac{v_0}{\sqrt{v_0^2 + 2gh}} \tag{5.21}$$

(See Problem 5.17.) For $h = 0$ this gives the result $\theta_0 = 45°$ as determined earlier.

EXAMPLE 3

Optimum Launch Angle in the Shot Put

In the shot put event (refer to Figure 5.4) an athlete releases the shot from a point 2 m above ground level with a speed of 14.1 m/s. At what angle should the shot be launched to achieve maximum distance and what is the maximum distance? From Equation 5.21,

$$\tan \theta_0 = \frac{v_0}{\sqrt{v_0^2 + 2gh}} = \frac{14.1 \text{ m/s}}{\sqrt{(14.1 \text{ m/s})^2 + 2(9.80 \text{ m/s}^2)(2 \text{ m})}} = 0.914$$

This gives $\theta_0 = 42.4°$.

The maximum range is given by Equation 5.20:

$$R_{\max} = \left(\frac{v_0}{g}\right)\sqrt{v_0^2 + 2gh} = \frac{(14.1 \text{ m/s})(15.4 \text{ m/s})}{9.80 \text{ m/s}^2}$$

$$= 22.2 \text{ m}$$

5.2
ACCELERATION IN CIRCULAR MOTION

There are many examples of circular motion in nature. When a particle moves along a circular path its velocity vector continually changes direction. There is an acceleration associated with this change in direction. To focus on this particular aspect, we consider a particle undergoing **uniform circular motion**—that is, circular motion at a *constant speed*. The velocity vector changes direction continuously while its magnitude remains constant. Examples include moons in circular orbits about a planet, satellites in circular orbits around the earth, and charged particles in circular orbits in accelerator storage rings.

Acceleration in Uniform Circular Motion

We want to demonstrate that a particle undergoing uniform circular motion experiences an acceleration of constant magnitude directed toward the center of the circular path.

To establish the relationships between **a**, **v**, and **r**, consider the two triangles in Figure 5.5. The triangle in Figure 5.5a relates the velocity and position vectors. During a short time interval Δt the position changes from

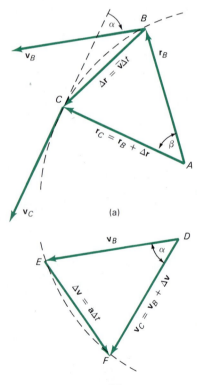

FIGURE 5.5

Two vector diagrams illustrate the geometrical relationships
(a) between **r** and **v** and
(b) between **v** and **a**.

$\mathbf{r}_B$ to $\mathbf{r}_C$. The change in position is $\Delta\mathbf{r} = \bar{\mathbf{v}}\,\Delta t$, where $\bar{\mathbf{v}}$ is the average velocity over the time interval Δt. As shown in Figure 5.5a, $\bar{\mathbf{v}}$ is directed along the secant line BC. The triangle ABC is isosceles because the radius of the circular orbit is constant. In the limit as $\Delta t \to 0$, $\bar{\mathbf{v}}$ approaches the instantaneous velocity $\mathbf{v}$, which is perpendicular to $\mathbf{r}$ and directed along the tangent to the circular path.

The triangle in Figure 5.5b relates the velocity and acceleration vectors. We can see the two velocity vectors $\mathbf{v}_B$ and $\mathbf{v}_C$ from Figure 5.5a. These vectors have the same magnitude because the particle moves at a constant speed, v. Geometrically, the velocity vector sweeps around a circle of "radius" v as the particle moves around the circular path.

During the time Δt the velocity changes from $\mathbf{v}_B$ to $\mathbf{v}_C$. The change in velocity is $\Delta\mathbf{v} = \bar{\mathbf{a}}\,\Delta t$, where $\bar{\mathbf{a}}$ is the average acceleration over the interval Δt. As shown in Figure 5.5b, $\bar{\mathbf{a}}$ is directed along the secant line EF. The triangle DEF is isosceles because the speed of the particle is constant. In the limit as $\Delta t \to 0$, $\bar{\mathbf{a}}$ approaches the instantaneous acceleration, $\mathbf{a}$. You can see from Figure 5.5b that $\Delta\mathbf{v} = \bar{\mathbf{a}}\,\Delta t$ becomes perpendicular to $\mathbf{v}_B$ in the limit $\Delta t \to 0$. This shows that the instantaneous acceleration is perpendicular to the velocity $\mathbf{v}$. Since $\mathbf{v}$ is directed along the tangent to the path, $\mathbf{a}$ must be **radial**. A moment's reflection shows that $\mathbf{a}$ is directed radially *inward*, toward the center of the circular path. We can express the magnitude of $\mathbf{a}$ in terms of v and r by comparing the two triangles in Figure 5.5. Because Δt is the same for both triangles, the angles α and β are equal and so the triangles are similar. It follows that the ratios of corresponding sides are equal. Thus

$$\frac{\bar{v}\,\Delta t}{r} = \frac{\bar{a}\,\Delta t}{v}$$

In the limit $\Delta t \to 0$ we can replace $\bar{v}$ by v and $\bar{a}$ by a. Canceling the factor Δt gives

$$a = \frac{v^2}{r} \tag{5.22}$$

The acceleration can also be expressed in terms of the **period** (T) of the motion, the time for one round trip of the circle. In one round trip the particle moves at a speed v for a time T and covers a distance $2\pi r$, the circumference of the path. Thus,

$$vT = 2\pi r$$

Using this relation to eliminate v from Equation 5.22 gives

$$a = \frac{4\pi^2 r}{T^2} \tag{5.23}$$

Because v, r, and T are constant, a is also constant. Thus, in uniform circular motion the particle experiences an acceleration of constant magnitude directed toward the center of the circular path. This "center-seeking" acceleration is called **centripetal**, from the Latin *petere* (to seek).

What causes the centripetal acceleration of a particle in uniform circular motion? Kinematics cannot provide an answer. The answer comes from Newton's second law that relates *force* and *acceleration*. Forces "cause" accelerations. However, the physical nature of the force cannot be inferred from the centripetal acceleration. Many different types of forces can produce uniform circular motion. For a moon in uniform circular motion around a planet, the force is gravitational. In accelerator storage rings, the force is magnetic. In a spinning flywheel, the circular motion is maintained by electromagnetic forces acting between atoms.

EXAMPLE 4

Centripetal Acceleration for the Crab Pulsar

The Crab Nebula is the remnant of a stellar explosion, a supernova. At the center of the Crab is a stellar corpse, a **pulsar** in the form of a **neutron star** that rotates 30 times per second. A direct measurement of the radius of the neutron star is not possible, but theoretical calculations predict a radius of about 10 km. Let's estimate the magnitude of the centripetal acceleration of a point on the equator of the neutron star.

Using Equation 5.23, with $r = 10^4$ m and $T = 1/30$ s, we get

$$a = \frac{4\pi^2 \cdot 10^4 \text{ m}}{(0.0333 \text{ s})^2} = 3.56 \times 10^8 \text{ m/s}^2$$

This is over thirty million times the acceleration of gravity at the surface of the earth.

In solving a problem, the choice of reference frame often influences the complexity of the solution. For example, in Chapter 3 the orientation and origin of a reference frame were chosen to simplify the solution of statics problems.

In the following example we consider a bicycle tire. As observed from the ground, the tire undergoes a combination of linear and circular motion. From a reference frame at rest at the center of the tire, the motion is observed to be circular. The description of the motion is simpler in the second reference frame.

EXAMPLE 5

Radial Acceleration of a Tire

A bicycle moves in a straight line at a speed of 30 mi/h (13.4 m/s). The diameter of the tires is 26 inches (0.660 m). We want to determine the radial acceleration of a point on the circumference of the tire.

Imagine that you are viewing points on the circumference from a position on the axle of the wheel. You are moving forward at 13.4 m/s relative to the road. You see points on the circumference moving about you in uniform circular motion at a speed of 13.4 m/s. The magnitude of the radial acceleration of points on the circumference is given by Equation 5.22,

$$a = \frac{v^2}{r} = \frac{(13.4 \text{ m/s})^2}{0.330 \text{ m}}$$
$$= 544 \text{ m/s}^2$$

This is more than 55 times as large as g!

Tangential Acceleration in Circular Motion

If the speed varies in circular motion there are changes in both the direction and magnitude of the velocity. The particle travels on a circular path, but speeds up or slows down as it moves. A pendulum exhibits this type of motion.

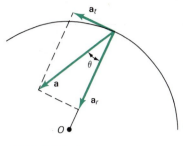

FIGURE 5.6

The acceleration, **a**, of a particle in non-uniform circular motion has a tangential component, **a**$_t$, and a radial component, **a**$_r$.

If the speed changes, the acceleration has both radial and tangential components (Figure 5.6). As before, the radial acceleration is given by Equation 5.22. The **tangential acceleration** is determined by the rate at which the speed v changes.

$$a_t = \lim_{\Delta t \to 0} \frac{\Delta v}{\Delta t} = \frac{dv}{dt} \tag{5.24}$$

The sign of a_t indicates whether v is increasing or decreasing. If $a_t > 0$, then v is *increasing*; if $a_t < 0$, then v is *decreasing*. In uniform circular motion, v is constant and Equation 5.24 yields $a_t = 0$. Thus, in non-uniform circular motion the acceleration can be resolved into two mutually perpendicular components, a_t and a_r. The magnitude of the total acceleration is

$$a = \sqrt{a_t^2 + a_r^2} \tag{5.25}$$

Figure 5.6 shows that the direction of the total acceleration relative to the direction of the radial component a_r is given by

$$\tan \theta = \frac{a_t}{a_r} \tag{5.26}$$

EXAMPLE 6

Tangential Acceleration

A child uses her finger to spin the wheel of a toy car. During a one-second interval the speed of the periphery of the wheel increases from 10 cm/s to 120 cm/s. The child then releases her finger and friction causes the speed to decrease to 90 cm/s during the next one-second interval. Let's determine the average tangential acceleration over these one-second intervals. Over the first one-second interval the tangential speed increases and the average tangential acceleration is positive,

$$\bar{a}_t = \frac{\Delta v}{\Delta t} = \frac{(120 - 10) \text{ cm/s}}{1 \text{ s}}$$
$$= 1.10 \text{ m/s}^2$$

Over the next one-second interval the tangential speed decreases and the average tangential acceleration is negative

$$\bar{a}_t = \frac{\Delta v}{\Delta t} = \frac{(90 - 120) \text{ cm/s}}{1 \text{ s}}$$
$$= -0.30 \text{ m/s}^2$$

5.3
RELATIVE MOTION: TWO FRAMES OF REFERENCE

So far, our descriptions of motion have been made relative to observers in a single frame of reference. Now let's investigate the relationship between descriptions of motion made by observers in different reference frames.

Consider the two reference frames shown in Figure 5.7. One frame (S) is attached to the ground. The second frame (S') is attached to a railway flatcar. The flatcar moves at a constant speed v relative to the ground. Observers on the ground and flatcar are in relative motion. For convenience we refer to S as the "rest" frame of reference and S' as the "moving" frame of reference. As

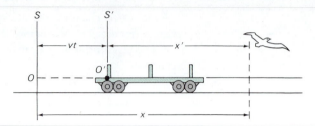

FIGURE 5.7
The reference frame S' attached to the flatcar moves with velocity v relative to the reference frame S attached to the ground.

shown in Figure 5.7, we take the x and x' axes to lie along the direction of relative motion. We want to establish the relationships between the positions, the velocities, and the accelerations measured in these two reference frames. We use primes to denote quantities referred to S'. Unprimed symbols refer to measurements made in S.

Observers in S and S' each carry identical clocks that read zero when the two origins coincide. This means that their initially synchronized clocks run at the same rate and so always display equal times. Mathematically this means

$$t = t' \tag{5.27}$$

The observers in S and S' measure the position of a bird, assigning the values x and x' at time t. As Figure 5.7 shows, the position values for the bird are related by

$$x = x' + vt \tag{5.28}$$

We note two special cases of Equation 5.28. First, if $v = 0$ there is no relative motion. Equation 5.28 reduces to $x = x'$, showing that the two reference frames coincide. Second, suppose that the bird is perched on the origin of S'. In this case $x' = 0$ and Equation 5.28 gives $x = vt$. This simply describes the fact that the S' origin moves away from the S origin at a speed v.

Equations 5.27 and 5.28 are referred to as the **Galilean transformation** equations. They constitute a recipe for changing, or, transforming, the values of t and x, measured in S, into the corresponding quantities, t' and x', measured in S'.

Galilean Velocity Addition

Now we establish the relationship between velocities measured in S and S'. Consider a short time interval, Δt, during which the bird undergoes a displacement Δx, as observed in S. The corresponding displacement, as observed in S', is denoted $\Delta x'$. We write

$$u = \frac{\Delta x}{\Delta t} = \text{velocity of bird relative to } S \tag{5.29}$$

$$u' = \frac{\Delta x'}{\Delta t} = \text{velocity of bird relative to } S' \tag{5.30}$$

The relationship between Δx, $\Delta x'$, and Δt follows from Equation 5.28,

$$\Delta x = \Delta x' + v\,\Delta t$$

Dividing both sides by Δt and using Equations 5.29 and 5.30 converts this into the **Galilean velocity addition relation,**

$$u = u' + v \qquad (5.31)$$

We can test Equation 5.31 by considering two special cases. First, if $v = 0$, there is no relative motion of the two frames of reference, and Equation 5.31 reduces to $u = u'$, showing that both S' and S measure the same velocity.

As a second special case, suppose that the bird moves at 30 mph relative to S (relative to the ground) and that the flatcar also moves at 30 mph relative to the ground. Since both S' and the bird move at the same velocity relative to S, the bird is at rest relative to S'. Equation 5.31 confirms this: with $u = v = 30$ mph, we find

$$u' = u - v = 30 \text{ mph} - 30 \text{ mph} = 0$$

Example 7 illustrates the Galilean velocity addition relation in a more general context.

EXAMPLE 7

Relative Velocity of a Golf Ball and a Club Head

A golf club head moving at 100 mph approaches a golf ball at rest (Figure 5.8). After the collision the club head moves at 69 mph and the ball travels at 139 mph. All motion is in the $+x$-direction. We want to determine the velocity of the ball relative to the club head before and after the collision.

Let

$$u = \text{velocity of ball relative to ground}$$

$$u' = \text{velocity of ball relative to club head}$$

$$v = \text{velocity of club head relative to ground}$$

Before the collision

$$u = 0 \qquad v = 100 \text{ mph}$$

and Equation 5.31 gives

$$u' = u - v$$
$$= -100 \text{ mph}$$

Velocities relative to ground   Velocities relative to club head

(a)
Before

100 mi/h   At rest   100 mi/h

(b)
After

69 mi/h   139 mi/h   70 mi/h

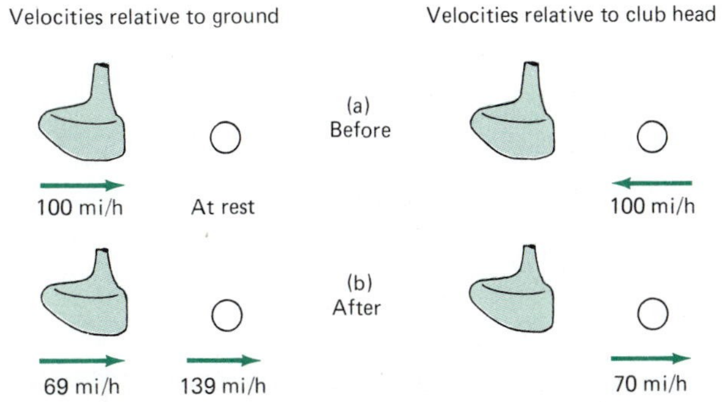

FIGURE 5.8

(a) The relative speed of the club head and the golf ball is 100 mi/h before they collide.
(b) After the collision the ball and club head move to the right at a relative speed of 70 mi/h.

The minus sign indicates that the ball is moving opposite to the direction the club head is moving.

After the collision

$$u = 139 \text{ mph} \qquad v = 69 \text{ mph}$$

and

$$u' = u - v = 139 \text{ mph} - 69 \text{ mph}$$
$$= +70 \text{ mph}$$

With the ball moving at 139 mph and the club head moving at 69 mph, the ball moves away from the club head at 70 mph. The plus sign indicates that the relative motion is in the positive x-direction.

Throughout this section we have assumed that v is fixed, or in other words, that the relative velocity of the two frames of reference is constant. However, no such restriction applies to the velocity of objects moving relative to the two frames of reference. Both u and u' may vary with time. Let Δu and $\Delta u'$ denote the change in u and u', respectively, during the interval Δt. Using Equation 5.31 we have, for the relation connecting changes in velocity,

$$\Delta u = \Delta u' + \Delta v$$

However, $\Delta v = 0$, because v is constant, therefore

$$\Delta u = \Delta u' \tag{5.32}$$

The changes in velocity measured by observers in two different frames of reference are equal even though the velocities themselves are unequal. It follows that the two observers will measure the same acceleration because the equal changes of velocity take place in equal times. Thus,

$$a = \frac{\Delta u}{\Delta t} = \frac{\Delta u'}{\Delta t} = a'$$

confirming that

$$a = a' \tag{5.33}$$

The two observers, moving with constant velocity relative to each other, will therefore record equal accelerations.

Vector Form of Galilean Transformation Equations

The relations just developed for one dimension are readily extended to motion in two or three dimensions. Position, velocity, and acceleration are vectors, and Equations 5.28, 5.31, and 5.33 are the x-components of vector equations.

The vector form of Equation 5.28 is

$$\mathbf{r} = \mathbf{r}' + \mathbf{v}t \tag{5.34}$$

Figure 5.9 shows the geometric relation between $\mathbf{r}$, $\mathbf{r}'$, and $\mathbf{v}t$. The origins of O and O' coincide at time $t = 0$. Figure 5.9 shows that O' has undergone a displacement $\mathbf{v}t$ in time t. The vectors $\mathbf{r}$ and $\mathbf{r}'$ locate the position of an object relative to O and O'.

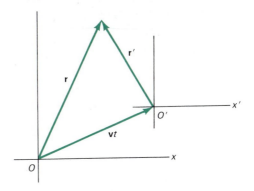

FIGURE 5.9

Two reference frames in relative motion. O' moves at a constant velocity **v** relative to O. The origins coincide at time $t = 0$. At time t they are separated by a vector displacement **v**t. The vectors **r** and **r'** locate an object relative to O and O'.

The generalization of Equation 5.31 is

$$\mathbf{u} = \mathbf{u'} + \mathbf{v} \qquad (5.35)$$

where **u** and **u'** denote the velocity of an object relative to O and O'. Figure 5.10 shows the triangular relation between the velocities **u**, **u'**, and **v**.

The generalization of Equation 5.33 is

$$\mathbf{a} = \mathbf{a'} \qquad (5.36)$$

Relative motion at a constant velocity does not affect accelerations. Observers in O and O' will measure the same acceleration for an object.

Equations 5.34–5.36 describe a theory of relative motion called **Galilean relativity**.

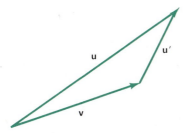

FIGURE 5.10

The relative velocity triangle embodies Equation 5.35, **u** = **u'** + **v**.

Relative Velocity

If $\mathbf{v}_1$ and $\mathbf{v}_2$ are the velocities of objects 1 and 2 relative to the *same* frame of reference, the *velocity of 1 relative to 2* is given by

$$\text{velocity of 1 relative to 2} = \mathbf{v}_1 - \mathbf{v}_2 \qquad (5.37)$$

This useful result is essentially a special case of Equation 5.35.

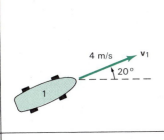

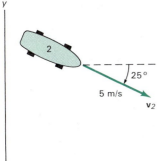

EXAMPLE 8

Relative Speed of Skateboards

Two skateboards travel with the velocities indicated in Figure 5.11. We want to determine their relative speed.

The relative speed is the *magnitude* of the relative velocity. Let

$$\mathbf{v} = \mathbf{v}_1 - \mathbf{v}_2$$

denote the relative velocity of the skateboards. From Figure 5.11 we see that the components of **v** are

$$v_x = v_{1x} - v_{2x} = (4 \text{ m/s}) \cos 20° - (5 \text{ m/s}) \cos 25°$$
$$= -0.77 \text{ m/s}$$

$$v_y = v_{1y} - v_{2y} = (4 \text{ m/s}) \sin 20° - (5 \text{ m/s})(-\sin 25°)$$
$$= 3.48 \text{ m/s}$$

FIGURE 5.11

The skateboards move relative to the ground at velocities $\mathbf{v}_1$ and $\mathbf{v}_2$. Their relative speed is the magnitude of their relative velocity, $\mathbf{v}_1 - \mathbf{v}_2$.

The relative speed is

$$v = \sqrt{v_x{}^2 + v_y{}^2} = \sqrt{(-0.77 \text{ m/s})^2 + (3.48 \text{ m/s})^2}$$
$$= 3.56 \text{ m/s}$$

WORKED PROBLEM

A ball is launched straight up from a position 10 m above the ground. Its initial speed is 10 m/s. A second ball is launched simultaneously from the ground at a point 30 m away horizontally (Figure 5.12). The balls collide just as they reach the ground. Determine the initial speed and launch angle of the second ball. Disregard air resistance.

Solution

We first recognize that the time of flight is the same for both balls. The information given for the ball launched vertically lets us determine its time of flight. Once we have the time of flight we can use the given final position of the second ball to find its initial speed and launch angle.

Take the origin at the point where the second ball is launched (Figure 5.12). The displacement of the ball launched vertically is given by

$$y_1 = 10 \text{ m} + (10 \text{ m/s})t - (4.90 \text{ m/s}^2)t^2$$

Setting $y_1 = 0$ gives the time of flight,

$$t = \frac{10 \text{ m/s}}{9.80 \text{ m/s}^2} + \sqrt{\left(\frac{10 \text{ m/s}}{9.80 \text{ m/s}^2}\right)^2 + \frac{10 \text{ m}}{4.90 \text{ m/s}^2}}$$
$$= 2.78 \text{ s}$$

For the second ball, we know that $x_2 = 30$ m at $t = 2.78$ s. Equation 5.6 gives the x-component of its initial velocity,

$$v_{x0} = \frac{x_2}{t} = \frac{30 \text{ m}}{2.78 \text{ s}}$$
$$= 10.8 \text{ m/s}$$

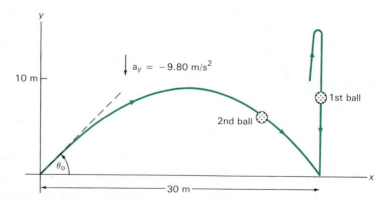

FIGURE 5.12

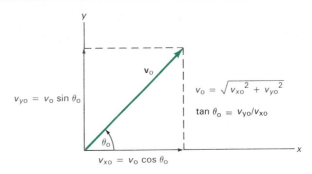

FIGURE 5.13

We set $y_2 = 0$ at $t = 2.78$ s in Equation 5.3 to find the y-component of the initial velocity

$$v_{y0} = \tfrac{1}{2}gt = \tfrac{1}{2}(9.80 \text{ m/s}^2)(2.78 \text{ s}) = 13.6 \text{ m/s}$$

The launch speed is the magnitude of the initial velocity (Figure 5.13),

$$v_0 = \sqrt{v_{x0}^2 + v_{y0}^2} = \sqrt{(10.8 \text{ m/s})^2 + (13.6 \text{ m/s})^2}$$
$$= 17.4 \text{ m/s}$$

The launch angle is determined from $\tan \theta_0 = v_{y0}/v_{x0}$ (Figure 5.13). Hence,

$$\theta_0 = \tan^{-1}\left(\frac{v_{y0}}{v_{x0}}\right) = \tan^{-1}\left(\frac{13.6 \text{ m/s}}{10.8 \text{ m/s}}\right)$$
$$= 51.5°$$

EXERCISES

5.1 Projectile Motion in a Plane

A. A particle moves in a plane with the velocity components

$$v_x = 1 \text{ m/s} \qquad v_y = -1 \text{ m/s}$$

It passes the origin ($x = 0$, $y = 0$) at $t = 0$. Sketch the particle's position versus time over the time interval $t = 0$ to $t = 3$ s.

B. A projectile lands at the same level at which it was launched. Sketch a few trajectories having the same time of flight but different ranges. For different trajectories, compare (a) the vertical component of initial velocity, (b) the height to which the projectile rises, (c) the horizontal component of the initial velocity.

C. A particle has velocity components

$$v_x = +4 \text{ m/s} \qquad v_y = -(6 \text{ m/s}^2)t + 4 \text{ m/s}$$

Calculate the speed of the particle and the direction $\theta = \tan^{-1}(v_y/v_x)$ of the velocity vector at $t = 2$ s.

D. The position of a particle is given by

$$x = 2 \text{ m} + (3 \text{ m/s})t \qquad y = x - (5 \text{ m/s}^2)t^2$$

How far from the origin is the particle at (a) $t = 0$; (b) $t = 2$ s?

E. What initial speed would a football need in order to travel 60 m horizontally if it is kicked at an angle of 45°?

F. Determine the range and time of flight of the golf ball in Example 2 if the ball is launched with the same speed (80 m/s) at 23° on the moon.

G. In Example 1 a baseball thrown horizontally at 40 m/s was shown to fall 0.885 m vertically while traveling 17 m horizontally. For the same speed and horizontal distance, determine the launch angles that will cause the ball to reach home plate at its launch elevation.

H. A sling is used to hurl a rock a horizontal distance of 80 m. The initial speed is 38 m/s. Find the two possible launch angles.

5.2 Acceleration in Circular Motion

I. For what duration of the sidereal day would the radial acceleration at the equator have the magnitude 9.80 m/s²?

J. A drop of water rides on the edge of a moving $33\tfrac{1}{3}$-rpm record. The diameter of the record is 30 cm. Determine the magnitude of the radial acceleration of the drop.

K. A passenger in a car travels around a 90° arc of a circle of radius 20 m when the car turns a corner. At a speed of 60 mph what is the magnitude of her radial acceleration?

L. The Syncom satellite, at a distance of 4.21×10^7 m from the center of the earth, completes one orbit in one sidereal day. Determine the magnitude of its radial acceleration.

M. A discus thrower moves the 1-kg implement along a circular path of radius 1.06 m. The maximum speed of the discus is 20 m/s. Determine the magnitude of the maximum radial acceleration of the discus.

N. Astronauts training for the weightlessness of space flights are in an airplane that travels along a parabolic trajectory. Near the peak this parabola closely approximates the arc of a circle. If the plane is moving at 156 m/s (350 mi/h), what is the radius of its circular arc?

P. The speed of the bob of a pendulum of length 10 m is given by $v = v_0 \cos(2\pi t/T)$ where $v_0 = 1.00$ m/s and $T = 2\pi$ s. Determine the magnitude of the radial acceleration of the bob at (a) $t = \frac{1}{4}\pi$ s and (b) $t = \frac{1}{2}\pi$ s.

Q. Determine the magnitudes of the radial and tangential components of the acceleration of the pendulum bob in Exercise P at time $t = \pi/4$ s.

5.3   Relative Motion: Two Frames of Reference

R. A man standing on the shore observes that a raft floats with the river at 2 m/s (Figure 1). Relative to an ob-

server on the raft, a motorboat moves directly across the stream with a velocity $v = at$, where $a = 2$ m/s^2. Determine the magnitudes and directions of the motorboat velocity and acceleration relative to the shore observer at $t = 3$ s.

S. The bases are loaded in a baseball game when the batter hits a line drive single. The three base runners (A, B, C in Figure 2) move at equal speeds of 8 m/s. (a) Sketch their velocity vectors. (b) Consider the relative velocity vectors, for A relative to B, for B relative to C, and for C relative to A. Which two runners have the largest relative speed? Determine their relative speed.

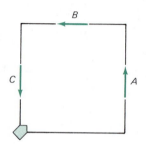

FIGURE 2

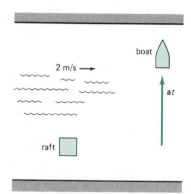

FIGURE 1

T. Car A travels east at 30 mph while car B travels south at 40 mph (Figure 3). Determine the velocity of A relative to B.

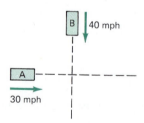

FIGURE 3

PROBLEMS

1. A coin slides across a table and falls to the floor 1.1 m below (Figure 4). (a) For what length of time does it fall? (b) If the coin strikes the floor 1.1 m beyond the edge of the table, what was its horizontal velocity when it left the table? (c) What is the coin's horizontal velocity just before impact?

2. A punter wants a football to travel 60 m horizontally and stay airborne for 5 s. Determine the required initial speed and launch angle.

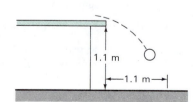

FIGURE 4

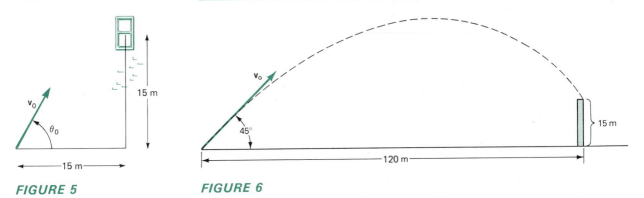

FIGURE 5

FIGURE 6

3. A rock is thrown through a window that is 15 m higher than the launch point and 15 m away horizontally (Figure 5). The rock is moving horizontally when it hits the window. Determine the initial speed and launch angle of the rock.

4. A batted ball is launched at $45°$ and just clears a fence 15 m higher than the launch point and 120 m away (Figure 6). Determine (a) the launch speed (v_0), (b) the time for the ball to reach the fence.

5. Show that the range of a projectile is the same for $\theta_0 = 45° + \varphi$ as for $\theta_0 = 45° - \varphi$, provided $\varphi \leq 45°$.

6. A ball bearing is dropped from the point $x = 4$ m, $y = 2$ m. At the same moment, a second bearing is launched from $x = 0$, $y = 0$ at an angle of $20°$ above the $+x$-axis at a speed of 6 m/s. Determine (a) the minimum distance between the bearings, (b) the time at which this minimum occurs. *Suggestion*: If you are unable to solve this problem analytically, you may wish to write and run a short computer program that locates the minimum of the *square* of the distance between the bearings.

7. A ball is thrown horizontally from a rooftop, 20 m above the ground at a speed of 8 m/s. A second ball is thrown horizontally out of a first floor window, 3 m above the ground. The balls collide as they strike the ground. (a) What is the launch speed of the second ball? (b) What is the time between the release of the balls?

8. A plane moving at 360 m/s pursues a plane directly ahead moving at 180 m/s. Each pilot can withstand a maximum centripetal acceleration of 49 m/s². Just as the faster plane catches up, the pilot of the slower plane executes a $180°$ turn in a horizontal plane. The pursuing pilot reacts quickly and also executes a $180°$ turn. (a) Calculate the minimum turn radius for each plane, (b) the time each plane requires to complete the $180°$ turn, (c) the distance between the planes when the slower plane finishes the $180°$ turn. Assume the planes start turning together.

9. During the 1968 Olympics in Mexico City, Bob Beamon executed a record long jump. The horizontal distance he achieved was 8.90 m. His center of gravity started at an elevation of 1.0 m, reached a maximum height of 1.90 m, and finished at 0.15 m. From these data determine (a) his time of flight, (b) his horizontal and vertical velocity components at the takeoff time, (c) his takeoff angle.

10. A race car starts from rest on a circular track of radius 400 m. Its speed increases at the constant rate of 0.5 m/s². At the point where the magnitudes of the radial and tangential accelerations are equal, determine (a) the speed of the race car, (b) the distance traveled, (c) the elapsed time.

11. An ultracentrifuge spins a test tube in a circle (effective radius 0.08 m) at 40,000 revolutions per minute. This produces a large centripetal acceleration that functions as an artificial gravity. With normal gravity, a serum-filled test tube clears in roughly 30 days. (a) Calculate the centripetal acceleration of the ultracentrifuge. (b) *Estimate* the settling time for the serum. (Assume the sedimentation velocity is proportional to the centripetal acceleration.)

12. The position of a particle is represented by the vector

$$\mathbf{r} = 10 \text{ m } (\mathbf{i} \sin \omega t - \mathbf{j} \cos \omega t)$$

$$\omega = 5 \text{ rad/s}$$

(a) Show that the particle moves counterclockwise in a circle of 10 m radius. (b) How long does it take for the particle to complete 10 revolutions? (c) Show that the magnitudes of the position, velocity, and acceleration vectors are related by $a = v^2/r$.

13. The position of a particle as a function of time t is described by

$$\mathbf{r} = (bt)\mathbf{i} + (c - dt^2)\mathbf{j} \qquad b = 2 \text{ m/s}$$

$$c = 5 \text{ m} \qquad d = 1 \text{ m/s}^2$$

(a) Express y in terms of x and sketch the trajectory. What is the shape of the trajectory? (b) Derive a vector relation for the velocity. (c) At what time $(t > 0)$ is the velocity vector perpendicular to the position vector?

14. A sailboat sails for 1 hour at 4 km/h on a steady compass heading of 40° east of north. The sailboat is simultaneously carried along by a current. At the end of the hour the boat is 6.12 km from its starting point. The line from the starting point to its location lies 60° east of north. Find the components of the velocity of the water.

15. A sailor aims his rowboat toward an island located 2 km east and 3 km north of his starting position. After an hour of rowing he sees the island due west. He then aims the boat in the opposite direction from which he was rowing, rows for another hour, and ends up 4 km east of his starting position. He correctly deduces that the current is from west to east. (a) What is the speed of the current? (b) Show that the boat's velocity relative to the shore for the first hour can be expressed as $\mathbf{u} = (4\ \text{km/h})\mathbf{i} + (3\ \text{km/h})\mathbf{j}$, where $\mathbf{i}$ is directed east and $\mathbf{j}$ is directed north.

16. A sailboat goes 2 km east and then turns and goes along a straight line in a different direction. The sailboat arrives at a point 4 km northeast of its starting point. (a) Draw a clear diagram indicating north, east, the initial 2 km run, and the final 4 km northeast position. (b) Find the components of the displacement vector for the second segment of the trip. Find (c) the magnitude, (d) and the direction of the second segment.

17. Derive Equation 5.21. *Hint*: The horizontal component of the velocity remains constant throughout the motion. Use a sketch of the perpendicular initial and final velocity vectors to show that $v_0 \cos \theta_0 = v \sin \theta_0$.

CHAPTER 6

NEWTON'S LAWS

6.1
INTRODUCTION TO NEWTON'S LAWS

The laws of physics are man-made attempts to describe nature in a coherent and quantitative fashion. In order to rank as a law of physics, a relationship must be wide in scope, predict new phenomena, and stand the test of time. Newton's laws of motion and his law of universal gravitation are monuments to an incredibly inventive mind.

Sir Isaac Newton (1642–1727) discovered the law of universal gravitation. He incorporated this law, and his three laws of motion, into a comprehensive dynamic theory that explained celestial and terrestrial motion. He also developed calculus and made important contributions to other areas of physics and mathematics. In 1687 Newton published the *Principia*. This treatise includes (1) Newton's three laws of motion, (2) his law of universal gravitation, and (3) his mathematical development of calculus. The *Principia* is a supreme intellectual achievement.

In this chapter we formulate Newton's three laws of motion. We will examine their logical content and implications and show how they are applied. Most of our attention will be devoted to Newton's second law. This law is the heart of *dynamics,* the theory that describes how external forces act to cause accelerations.

We also introduce Newton's law of universal gravitation. Prior to Newton, gravity had been considered a local force affecting only objects near the surface of the earth. Newton proposed that the force of gravity is not limited to short distances, but rather that every piece of matter attracts every other piece of matter, to some extent, no matter how great their separation.

Newton's laws of motion and his law of gravitation are **universal laws**. They describe phenomena on the earth, on the moon, and in the farthest reaches of the universe. Since Newton's pioneering effort, physics has developed as a science concerned with universal relationships.

FIGURE 6.1
Sir Isaac Newton

Inertial Frames of Reference

Newton's laws of motion have an especially simple form when they are formulated in an **inertial frame of reference**. An inertial frame is one that is not accelerated; that is, it is at rest or moving with a constant velocity. But, what frame of reference can we use to decide if some other frame of reference is at rest or moving with a constant velocity? Experiment shows that for many purposes the earth is a satisfactory approximation to an inertial frame. The surface of the earth is accelerated by virtue of the earth's daily rotation and its orbital motion about the sun. However, for most practical purposes these accelerations are negligibly small. If we require a better approximation to an inertial frame, we use the reference frame provided by the stars. Their measured positions change almost imperceptibly over long time intervals. Conceptually, and in practice, they can serve as an inertial frame of reference. Keep in mind, as we develop and apply Newton's laws of motion, that the results refer to *inertial frames of reference*.

6.2
NEWTON'S FIRST LAW

In Chapter 3 we adopted the pragmatic viewpoint that a force is a push or a pull. To set up a quantitative measure of force we used a spring balance to measure weight, the force of gravity. In order to develop a more complete and logical viewpoint of force we now turn to the concept of **inertia**. Inertia is related to the fact that objects in motion *tend* to remain in motion, and objects at rest *tend* to remain at rest. We describe this tendency by saying that objects possess inertia.

The concept of inertia is not obvious. In fact, our everyday experiences often obscure it. The superficial observation that objects left to themselves come to rest led the early Greek philosophers to identify the state of rest as the "natural state" of matter. It seemed clear to them that a push or pull was needed to keep an object moving but that a force was not needed to keep an object at rest. Our childhood experiences seem to bear out the idea that a force is needed to maintain motion. We learn that a tricycle comes to rest unless we keep pedaling, a rolling ball will eventually come to rest.

It was Galileo who first recognized the inertia of matter. He swept away faulty Greek notions with a brilliant argument. He noted that when a ball rolled *down* a nearly frictionless inclined plane the speed of the ball *increased* slightly. On the other hand, when the same ball rolled *up* the same inclined plane its speed *decreased* slightly. By considering the in-between case with no friction and *no* inclination of the plane, he argued that, once in motion, the speed of the ball would not change. Therefore it should be just as natural for the ball's velocity to remain constant as for the ball to remain at rest (Figure 6.2).

Galileo could imagine a frictionless environment only as a limiting case. Today, air tracks, magnetic suspensions, and orbiting satellites are examples of nearly zero-friction situations. Galileo's observations are formalized by **Newton's first law**, the law of inertia.

A body, at rest or in motion, remains at rest or in motion with a constant velocity, unless acted upon by a net force exerted by its environment.

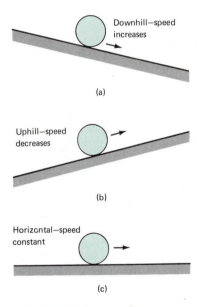

Downhill—speed increases

(a)

Uphill—speed decreases

(b)

Horizontal—speed constant

(c)

FIGURE 6.2

The ball moves on a flat surface with negligible friction. The speed of the ball (a) increases as the ball moves downhill, (b) decreases as the ball moves uphill, and (c) remains constant as the ball moves horizontally.

$\theta_1 = \theta_2 = \theta_3$ for a collision course

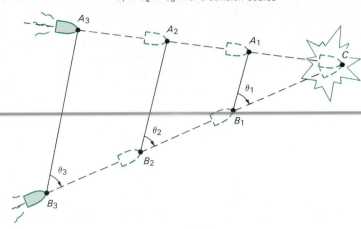

FIGURE 6.3

Two ships moving at constant speeds collide at C. The direction of the line of sight from one ship to the other remains constant when the ships are on a collision course.

EXAMPLE 1

Collision Course?

Two ships move along paths that intersect (see Figure 6.3). The ships experience forces, but the **net** force on each is zero. As a consequence the ships move along straight lines at constant speeds. How can sailors aboard the two ships decide if they are on a collision course? The sailors do not know the velocity of either ship; they know only that the velocities are constant.

Let's suppose that the ships *do* collide at point C where their paths intersect. In Figure 6.3, the pairs of points (A_1, B_1), (A_2, B_2), and (A_3, B_3) mark the ships' positions at corresponding times of 1 minute, 2 minutes, and 3 minutes before the collision. Because the speeds are constant, the lengths A_3C, A_2C, and A_1C are in the ratio $3:2:1$. Likewise, the lengths B_3C, B_2C, and B_1C are in the ratio $3:2:1$. It follows that the triangles CA_3B_3, CA_2B_2, and CA_1B_1 are congruent. This means that the angles θ_1, θ_2, and θ_3 are equal. Observe that these angles define the line of sight from ship B to ship A. Thus, for a collision course, the angle at which sailors aboard ship B observe ship A remains constant. This criterion also applies for the sailors aboard ship A.

Conceptually, Newton's first law says, "matter has inertia—it tends to resist changes in velocity." From a logical viewpoint, Newton's first law is an operational definition of zero net force. If the velocity of a body remains constant, then, by definition, no net force acts on it.

The concept of inertia is qualitative. The quantity we call *mass* is used to measure inertia. An object with a large mass has more inertia than an object with a small mass. A party balloon has a small mass compared to the mass of a Mercedes-Benz. As children we learn we can stop a moving party balloon without much effort. We are warned not to try to stop moving autos.

When the velocity is constant, the acceleration is zero. Newton's first law relates zero force to zero acceleration. The implication is that there is a direct relationship between force and acceleration. This relationship is made precise by Newton's second law.

6.3
NEWTON'S SECOND LAW

Newton's first law implies that a body that is accelerated experiences a net force. This idea fits in with our everyday experience that a force is a push or a pull. When a pool ball is struck by the tip of a cue it accelerates. The *force causes the acceleration.* Furthermore, the acceleration is in the direction of the push exerted by the cue. So, force and acceleration are vectors having the same direction.

On the other hand, a body resists acceleration because of its inertia. Suppose you push a tricycle and cause it to accelerate at 1 m/s^2. If you wanted to cause the same acceleration of an auto you would have to push a lot harder. For a given acceleration, the force must increase as the mass increases.

These observations suggest that any quantitative relationship between force, mass, and acceleration should satisfy two criteria:

1. For a given mass force should increase as acceleration increases and be in the same direction as the acceleration.
2. For a given acceleration a larger mass should require a larger force.

Newton's second law of motion establishes just such a relation:

> The net external force acting on a body equals the product of its mass and its acceleration.

Newton's second law can be written as an equation

$$\mathbf{F} = m\mathbf{a} \tag{6.1}$$

where m is the mass of the body, $\mathbf{a}$ is its acceleration, and $\mathbf{F}$ is the net external force acting on the body. Because we already have operational definitions of mass and acceleration, Newton's second law gives us an operational definition of force.

The SI unit of force, the newton, is defined by setting $m = 1$ kg and $a = 1 \text{ m/s}^2$ in Equation 6.1. Thus,

$$1 \text{ newton} = 1 \text{ kg} \cdot \text{m/s}^2$$

A 1-kg mass accelerating at 1 m/s^2 experiences a net force of 1 newton (abbreviated 1 N). The equivalence between newtons and pounds is

$$1 \text{ N} = 0.2248 \text{ lb} \qquad 1 \text{ lb} = 4.448 \text{ N}$$

The dimensions of force are

$$[F] = MLT^{-2}$$

We will have many opportunities to apply Newton's second law.

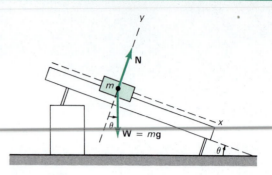

FIGURE 6.4

An air car travels without friction along the inclined air track. Its acceleration along the track equals $g \sin \theta$.

EXAMPLE 2

Air Track Acceleration

An air track is inclined at an angle θ (Figure 6.4). We want to determine the acceleration of an air car gliding along the track. Two external forces act on the car, its weight mg and a normal force N. The air friction force is negligible. We choose axes parallel and perpendicular to the track. The acceleration of the air car is along the track, that is, the x-axis, in Figure 6.4. Expressing Newton's second law in component form, we have

$$\sum F_x = mg \sin \theta = ma_x$$
$$\sum F_y = N - mg \cos \theta = 0$$

The first equation yields the solution we seek

$$a_x = g \sin \theta$$

We can check the plausibility of this solution. When $\theta = 0$ the track is horizontal and we expect $a_x = 0$. Our solution confirms this expectation because $\sin \theta = 0$. When $\theta = 90°$ we expect $a_x = g$ because the track would be vertical and the car would be in free-fall. Again, our solution confirms this expection because $\sin 90° = 1$.

The second equation determines the normal force

$$N = mg \cos \theta$$

Does this expression for N give the correct values for the special cases, $\theta = 0°$ and $\theta = 90°$?

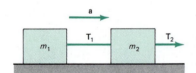

FIGURE 6.5

Two air cars move without friction on a horizontal air track. The string connecting the air cars is under tension T_1. The string accelerating the system to the right is under tension T_2.

EXAMPLE 3

Acceleration of Coupled Objects

Two objects having masses m_1 and m_2 move horizontally on an air track (Figure 6.5). One string (under tension T_1) connects the objects, and a second string (under tension T_2) is used to accelerate the system to the right. The acceleration, a, and the masses are given. We want to determine the two tensions in terms of m_1, m_2, and a. You might first want to guess the answers to T_1 and T_2. In particular, does your intuition tell you that $T_1 = T_2$?

First, we choose m_2 as a body. Two horizontal forces act on this body. The force vector diagram is given in Figure 6.6, with the x-axis horizontal. Applying Newton's second law to the horizontal components, we obtain

$$\sum F_x = +T_2 - T_1 = m_2 a$$

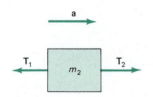

FIGURE 6.6

Horizontal forces acting on the right air car in Figure 6.5 are the tensions T_1 and T_2.

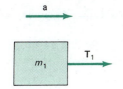

FIGURE 6.7

The only horizontal force acting on the left air car in Figure 6.5 is the tension T_1.

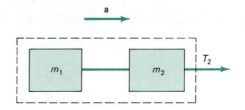

FIGURE 6.8

The only external horizontal force acting on the two-car system is the tension T_2.

This tells us that T_2 exceeds T_1 by the product m_2a. The two tensions are not equal. Neither T_1 nor T_2 can be determined using this equation alone. We must pick another body in order to obtain a second relation between the acceleration and the tensions.

We select m_1 as a body. One horizontal force acts on m_1. The corresponding force vector diagram is shown in Figure 6.7. Newton's second law yields

$$\sum F_x = T_1 = m_1a$$

This equation gives T_1 in terms of m_1 and a, as desired. We complete the solution by using this result to solve the first equation for T_2. The result is

$$T_2 = (m_1 + m_2)a$$

This result for T_2 can also be obtained by picking the combined system of the two masses and the connecting string as a body. Such a choice is permissible provided all parts of an object move as a unit—that is, they have the same acceleration. One horizontal force acts on this composite body. The force vector diagram is shown in Figure 6.8. Applying Newton's second law, we have

$$\sum F_x = T_2 = (m_1 + m_2)a$$

This confirms our first solution for T_2. No information about T_1 is obtained because T_1 is an internal force.

Insight is needed to decide how to choose a body. In the problem just described, three choices were available. Any two of these choices leads to a complete solution. As you work through examples and end-of-chapter problems, this sort of insight will come more easily.

6.4
NEWTON'S THIRD LAW

Newton's second law relates the acceleration of a body to forces exerted on it by the environment—**external forces**. Internal forces exerted between different portions of a body do not affect its acceleration. For example, a pencil at rest on a desk remains motionless despite the enormous number of internal forces between its atoms. **Newton's third law** relates this fact to the force concept by observing the give and take nature of force:

> To every action there is always an equal but opposite reaction.

The *action* and the *reaction* are *forces*. Newton's third law recognizes that forces always occur in pairs. An action-reaction pair is composed of forces equal in magnitude and opposite in direction. The vector sum of the action-reaction forces between each pair of atoms in a pencil is zero. It follows that the vector

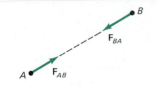

FIGURE 6.9

The force $\mathbf{F}_{AB}$ exerted on the particle A by the particle B is equal in magnitude and oppositely directed to the force $\mathbf{F}_{BA}$ exerted on the particle B by the particle A.

sum of all internal forces in a pencil is zero. Internal forces alone cannot set the pencil into motion.

Internal forces are important. They hold objects together. But Newton's third law tells us that we can ignore them when we want to determine the acceleration of a body; only external forces need to be considered.

We can formalize Newton's third law as follows. Consider the interaction of the particles A and B in Figure 6.9. The force exerted on A by B, $\mathbf{F}_{AB}$, and the force exerted on B by A, $\mathbf{F}_{BA}$, satisfy the relation

$$\mathbf{F}_{AB} = -\mathbf{F}_{BA}$$

This relation holds whether A (or B) experiences other forces or not.

You must be careful not to confuse Newton's second and third laws. For example, how would you explain the following dilemma? The force you exert on the ground and the force the ground exerts on you are equal in magnitude but opposite in direction. The vector sum of these two forces is zero. If this net force is always zero, then how could you ever jump off the ground? The question accurately labels two forces as belonging to an action-reaction pair. But they do not act on the same object! Whether you jump off the ground or not is determined by the net force acting on *you*. The force you exert on the ground is *not* a part of the net force exerted on you.

EXAMPLE 4

Ice Skater Action-Reaction

Two ice skaters ($m_1 = 60$ kg and $m_2 = 80$ kg) face each other on the ice (Figure 6.10). They stand motionless and exchange greetings. Suddenly, the 60-kg skater pushes the 80-kg skater, causing him to move away at a speed of 0.625 m/s. What happens to the 60-kg skater?

According to Newton's third law, the skaters exert equal but opposite forces on each other. While these forces act, the skaters accelerate away from each other. Using the third law, along with Newton's second law for each skater, we have

$$+F = m_1 a_1$$

$$-F = m_2 a_2$$

Solving for the acceleration of the 60-kg skater we get

$$a_1 = -\left(\frac{m_2}{m_1}\right) a_2$$

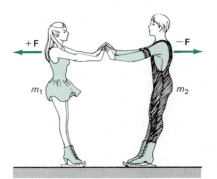

FIGURE 6.10

Ice skaters exerting forces on one another.

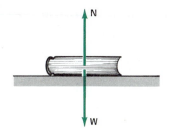

FIGURE 6.11

A book at rest experiences two forces. Its weight, **W**, acts downward and the contact force, **N**, acts upward. Although they are equal in magnitude and oppositely directed, the forces do not constitute an action-reaction pair because they act on the same body.

The accelerations of the skaters are inversely proportional to their masses; the smaller mass experiences the larger acceleration.

Because both forces act for the same time, the final velocities of the skaters are also inversely proportional to their masses. The final velocity of the 60-kg skater is

$$v_1 = -\left(\frac{m_2}{m_1}\right)v_2 = -\left(\frac{80 \text{ kg}}{60 \text{ kg}}\right)0.625 \text{ m/s}$$
$$= -0.833 \text{ m/s}$$

The minus sign shows that the skaters move in opposite directions.

It is also easy to confuse Newton's first and third laws. Consider a book at rest on a desk (Figure 6.11). Two forces act on the book. Its weight **W** acts downward, and the normal force **N** exerted on it by the desk acts upward. These forces are equal in magnitude and oppositely directed. Do they constitute an action-reaction pair? No, they do not. Both forces act on the book, whereas the forces in an action-reaction pair must act on different bodies. Yet we know that every force belongs to an action-reaction pair. Where, then, are the other two forces? The first is the downward push on the desk by the book. This contact force, together with **N** in Figure 6.11, provides one action-reaction pair of forces. The second is the gravitational attraction exerted on the earth by the book. This force, together with **W**, provides the other action-reaction pair of forces.

6.5
NEWTON'S LAW OF UNIVERSAL GRAVITATION

We noted before that there are four fundamental forces in nature: gravitational, electromagnetic, and the weak and strong nuclear forces. We have first hand experience with gravity and with electromagnetic forces. Your weight is just the gravitational force exerted on you by the earth. We begin our study of forces with gravity for two reasons: The gravitational force is fundamental and our everyday experience gives us a feel for gravity.

In Newton's day the word *gravity* referred to the force exerted on objects by the earth. As we have said, gravity was thought to be a local force, and only those objects near the earth were believed to be affected by it. Galileo had already formulated the laws of kinematics, which described the motion of objects influenced by gravity. However, the explanation of these gravitational effects remained unknown.

Prior to Newton, scientists believed that the laws describing motions on the earth differed from the laws describing motions in the heavens. The motion of falling apples, for example, was never connected to the motion of an orbiting planet. In Newton's hands, however, **universal gravitation** and a **universal set of dynamic laws** evolved together.

Newton's realization that gravity is universal had important implications. He recognized that the laws of motion must also be universal in character. The same laws of motion—Newton's laws of motion—govern both celestial and terrestrial motion.

Newton benefited from the research of Johannes Kepler and Tycho Brahe. It was Brahe who accumulated precise data on planetary motions. Kepler analyzed the data and through trial and error formulated three empirical relations that correctly describe planetary motion. These are known as **Kepler's laws of planetary motion**:

1. Planets move in elliptic orbits with the sun at one focus.
2. During equal time intervals, equal areas are swept out by the line connecting the planet and the sun.
3. The square of the orbital period is proportional to the cube of the average orbital radius.

Figure 6.12 illustrates certain aspects of Kepler's laws. We will show later how Kepler's second and third laws follow from Newton's laws.

Newton recognized that any theory of planetary motion should account for Kepler's three laws. It was Kepler's first law that gave Newton a hint as to the nature of the force acting between the sun and a planet. Newton had proved that an elliptical orbit required that the force between the sun and the planet be an **inverse square law** force. Specifically, the force must be proportional to the inverse square of the sun-to-planet distance, r.

$$F_{grav} \propto \frac{1}{r^2}$$

Near the surface of the earth gravity causes a constant acceleration that is the same for all bodies. We can use Newton's second law to see that the force of gravity must be proportional to the mass of the body. Thus, Newton's second law ($F = ma$) shows that the gravitational acceleration of a body of mass m is

$$a = \frac{F_{grav}}{m}$$

If a is the same for all bodies, regardless of their mass, then the force of gravity must be directly proportional to the mass of the body that *experiences* the force. In order to satisfy Newton's third law the force of gravity must also be proportional to the mass of the body that *exerts* the force. Collecting these results, one sees that the gravitational force between two masses m_1 and m_2 separated by a distance r is proportional to $m_1 m_2/r^2$

$$F_{grav} \propto \frac{m_1 m_2}{r^2}$$

The equation form of this relation is obtained by introducing a proportionality factor, G.

$$F_{grav} = \frac{G m_1 m_2}{r^2} \tag{6.2}$$

The gravitational constant G must be determined experimentally. To three-figure accuracy

$$G = 6.67 \times 10^{-11} \ \text{N·m}^2/\text{kg}^2$$

Equation 6.2 is **Newton's law of universal gravitation**. Even though we have presented arguments showing how the gravitational force must depend on mass and distance, such reasoning does not constitute a derivation of the law of gravity. Even today we have no clear explanation of why gravity exists. Gravity is simply a fundamental force that we can describe quantitatively.

Newton proved that the gravitational force between two spherically symmetric objects is given by Equation 6.2, with r equal to the distance between their centers. The gravitational force exerted on an object by the earth is also explained by Equation 6.2, with r equal to the distance from the object to the center of the earth.

When an apple falls from a tree it experiences a constant gravitational acceleration. The value of this acceleration, which we denote as g, depends on the location. Over much of the earth it has the value

$$g = 9.80 \text{ m/s}^2$$

You may have measured g in a laboratory experiment. Let's use Newton's second law and the law of gravity, together with this experimental value of g, to determine the mass of the earth.

EXAMPLE 5

Determining the Mass of the Earth

An object of mass m falling freely near the surface of the earth experiences a force of magnitude

$$F = \frac{GM_Em}{R_E^2}$$

where R_E is the radius of the earth. If we substitute this force into Newton's second law we get

$$\frac{GM_Em}{R_E^2} = ma = mg$$

(with $a = g$). We can solve the resulting equation for M_E, the mass of the earth,

$$M_E = \frac{gR_E^2}{G}$$

Inserting numerical values we get

$$M_E = \frac{(9.80 \text{ m/s}^2)(6.38 \times 10^6 \text{ m})^2}{6.67 \times 10^{-11} \text{ N·m}^2/\text{kg}^2}$$
$$= 5.98 \times 10^{24} \text{ kg}$$

We are all aware that gravity is an attractive force. In the case of planetary motion the force acts along the line from the planet to the sun. Kepler arrived at his laws of planetary motion by trial and error. He had no valid theory to guide him. We can use Newton's second law of motion and the law of universal gravitation to derive Kepler's third law of planetary motion.

Kepler's Third Law

Although the planetary orbits are ellipses they differ only slightly from circles. Accordingly, we consider a planet of mass m in a circular orbit of radius r about the sun. We use Newton's second law to relate the acceleration of the planet to the gravitational force exerted on it by the sun.

$$ma = \frac{GM_sm}{r^2} \tag{6.3}$$

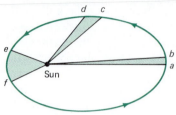

FIGURE 6.12

The planets travel in elliptical orbits about the sun. A line drawn from the sun to the planet sweeps out equal areas in equal times. Thus, if equal times are required for the planet to travel from *a* to *b* and from *c* to *d* and from *e* to *f*, then the three shaded areas are equal.

Here, M_s is the mass of the sun. The gravitational force is always directed toward the sun. The resulting acceleration is also directed radially inward. In Chapter 5 we derived a relation between the centripetal acceleration and the period T, the time for one orbit.

$$a = \frac{4\pi^2 r}{T^2}$$

Inserting this result into Equation 6.3 and rearranging gives

$$T^2 = \left(\frac{4\pi^2}{GM_s}\right) r^3 \tag{6.4}$$

The factor $4\pi^2/GM_s$ is a constant. It has the same value for all satellites of the sun. Thus, Equation 6.4 shows that the square of the orbital period is proportional to the cube of the orbit radius. This is Kepler's third law.

For elliptical orbits the radius r in Equation 6.4 is replaced by the semi-major axis (Figure 6.12). Kepler based his work on observations of the "naked-eye" planets, Mercury, Venus, Mars, Jupiter, and Saturn. Astronomical observations have since confirmed Kepler's third law for the other planets, the asteroids, and the comets, all of which are satellites of the sun.

Newton also used the law of gravity and his second law of motion to derive Kepler's first and second laws of planetary motion. What were simply empirical facts for Kepler became theoretical deductions for Newton.

The Newtonian Synthesis

Newton also developed a detailed theory explaining celestial motions. The sun's gravity maintains planetary orbits; the earth's gravity holds the moon in its orbit. We have come to accept these facts. However, our everyday experience certainly does not suggest that a falling apple has anything in common with the moon sailing majestically through the sky. Newton used a diagram to show the plausibility of the connection (Figure 6.13).

Looking at the figure we imagine a cannon placed on top of a high mountain. If a cannonball is simply dropped, it falls to the ground with a constant acceleration—just like an apple. Next, we imagine a series of shots of the cannon, all aimed to fire parallel to the earth's surface. Trajectories are drawn for several cannonball speeds; all assume no air friction. For small speeds the ball lands near the mountain, as expected. At large speeds, however, because of the curvature of the earth, the cannonball will land partway around the earth. Evidently a critical speed exists at which the ball could travel around the globe in a circle; it would then be in orbit. With this clever drawing Newton bridged the gap between Kepler's celestial orbits and Galileo's terrestrial trajectories. With just a slight stretch of our imagination we can picture the moon traveling around the earth like a gigantic cannonball in an orbit maintained by gravity.

Newton's unification of terrestrial and celestial motions is called the **Newtonian synthesis**. Once we accept the picture of the moon tied to the earth by gravity, it is easy to believe that gravity can bind planets to the sun, and that we can use Newton's theory to describe how the sun circles our galaxy under the gravitational action of other stellar matter.

Newton's laws of motion and his law of gravity have proved themselves on a universal scale. He set a precedent and thereby established a goal for scientists involved in all areas of fundamental research—namely, to search for *universally valid* relationships.

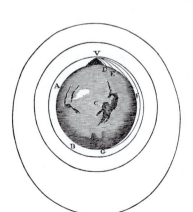

FIGURE 6.13

Newton's drawing suggests the universal character of gravity.

Weight and Mass

The force of gravity exerted on an object is called its *weight*. We signify the weight by W. For a body of mass m a distance r from the center of the earth, the weight is given by

$$W = \frac{GM_E m}{r^2} \tag{6.5}$$

This expression for W assumes that the gravitational forces exerted on the object by the moon, the sun, and other bodies, are negligible in comparison to the force exerted by the earth. For objects on the surface of the earth, we write,

$$W = mg \tag{6.6}$$

where g denotes the acceleration of gravity at the surface of the earth. Comparing Equation 6.5 ($r = R_E = 6.38 \times 10^6$ m) with Equation 6.6, we find

$$g = \frac{GM_E}{R_E{}^2} = 9.80 \text{ m/s}^2$$

The weight of a 1-kg mass at the surface of the earth is

$$W = 1 \text{ kg} \cdot 9.80 \text{ m/s}^2 = 9.80 \text{ N}$$

Recall in Chapter 3 that we arbitrarily assigned a force of 9.80 N to the weight of the standard kilogram. Here we see the reason for the choice of the number 9.80.

The mass of a body is the same no matter where the body is located, but its weight can change because W is proportional to $1/r^2$. If you walk up a flight of stairs, climb a mountain, or fly in a plane you "lose weight" because W decreases as r increases.

EXAMPLE 6

Weight Loss in the Sky

A student weighs 116 lb when she boards an airplane. How much weight does she "lose" when the plane rises to its cruising altitude of 30,000 ft?

The change in r is

$$30{,}000 \text{ ft} \times \left(\frac{0.3048 \text{ m}}{1 \text{ ft}} \right) = 9{,}140 \text{ m}$$

Because the change in r is small compared to the radius of the earth we can treat it mathematically as a differential, dr. The change in weight is correspondingly small and is related to dr by

$$dW = \left(\frac{dW}{dr} \right) dr$$

where (dW/dr) is the derivative of W, evaluated at the surface. From Equation 6.5 we find $(dW/dr) = -2W/R_E$. With $dr = 9{,}140$ m, the change in weight is

$$dW = -2(116 \text{ lb}) \left(\frac{9{,}140 \text{ m}}{6.38 \times 10^6 \text{ m}} \right)$$

$$= -0.333 \text{ lb}$$

This weight loss is slightly over 5 ounces.

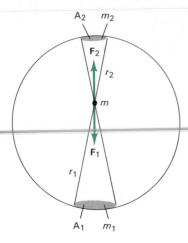

FIGURE 6.14

The masses m_1 and m_2 exert gravitational forces on the mass m.

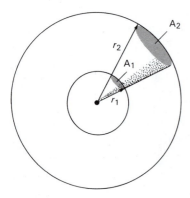

FIGURE 6.15

The surface areas A_1 and A_2, defined by the cone, are proportional to the squares of their radii.

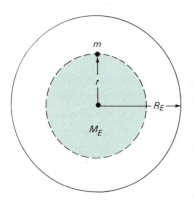

FIGURE 6.16

The mass farther from the center of the earth than the particle of mass m does not exert a net gravitational force on the particle.

Your weight also decreases if you descend below the surface of the earth. This would seem to contradict the $1/r^2$ dependence of the gravitational force. To understand this point requires that we investigate the gravitational force experienced by a particle *inside* a spherical shell of mass.

Gravitational Force Inside a Spherical Shell of Mass

A particle of mass m is located inside a uniform shell of mass (Figure 6.14). We want to show that the *net gravitational force exerted on the particle by the shell is zero*, no matter where it is positioned inside the shell.

The proof depends on a geometric property, related to the fact that the surface area of a sphere is proportional to the square of its radius. Figure 6.15 shows a pair of concentric spheres of radii r_1 and r_2. Surfaces of area A_1 and A_2 on the surfaces of the spheres are defined by a cone. Because A_1 and A_2 are defined by the *same* cone, we have the equality

$$\frac{A_1}{r_1{}^2} = \frac{A_2}{r_2{}^2}$$

This is the geometric result we need.

In Figure 6.14 the mass m is positioned at an arbitrary point inside the spherical mass shell. A *narrow* double-ended cone defines the *small* areas A_1 and A_2 on the surface of the shell. The masses of the areas A_1 and A_2 can be expressed as

$$m_1 = \sigma A_1 \qquad m_2 = \sigma A_2$$

where σ is the mass per unit area of the shell. The *uniformity* of the shell means that σ is constant, that is, the same at all positions on the shell.

In Figure 6.14 $\mathbf{F}_1$ and $\mathbf{F}_2$ denote the gravitational forces exerted on the mass m by the masses m_1 and m_2. The directions of $\mathbf{F}_1$ and $\mathbf{F}_2$ are opposite because the masses m_1 and m_2 are located on opposite sides of the mass m. The magnitudes of $\mathbf{F}_1$ and $\mathbf{F}_2$ are

$$F_1 = \frac{Gmm_1}{r_1{}^2} = \frac{Gm\sigma A_1}{r_1{}^2}$$

$$F_2 = \frac{Gmm_2}{r_2{}^2} = \frac{Gm\sigma A_2}{r_2{}^2}$$

Using the geometric result $A_1/r_1{}^2 = A_2/r_2{}^2$ shows that $F_1 = F_2$. Thus, $\mathbf{F}_1$ and $\mathbf{F}_2$ are equal but opposite forces, and their vector sum is zero. The two masses exert zero net gravitational force on the mass m. The complete shell can be divided into such pairs of masses and so the entire shell exerts zero gravitational force on the mass m.

Weight Beneath the Surface of the Earth

Let's apply the result for a spherical shell of mass to derive an equation for the weight of an object beneath the surface of the earth. We treat the earth as a uniform sphere of mass M_E and radius R_E. We want to determine the weight of a particle of mass m positioned beneath the surface, a distance r from the center (Figure 6.16).

The portion of the earth's mass *farther* from the center than the particle comprises a series of concentric shells. Each shell exerts zero gravitational force on the particle. Thus, all mass located farther from the center of the shell than the particle can be ignored—this mass exerts zero gravitational force on the particle.

The net gravitational force on the particle—its *weight*—is

$$W = \frac{GmM}{r^2} \tag{6.7}$$

where M is the *portion* of the earth's mass inside a sphere of radius r. For a uniform sphere, the *mass density* is constant, so

$$\frac{M}{4\pi r^3/3} = \frac{M_E}{4\pi R_E{}^3/3}$$

This gives

$$M = M_E \left(\frac{r}{R_E}\right)^3$$

Using this in Equation 6.7, we find that the weight of the particle is

$$W = m\left(\frac{GM_E}{R_E{}^3}\right)r \tag{6.8}$$

This equation shows that the weight of the particle is zero at the center of the earth, a result to be expected on the basis of symmetry.

We have established that if an object moves upward from the surface of the earth, its weight decreases; if it moves inward toward the center, its weight also decreases. Thus, the weight of an object is a *maximum* at the surface of the earth.

Apparent Weight in an Accelerated Frame of Reference

The reading on a scale does not disclose the true weight (mg) of a person who is accelerating. To see why it does not, consider an astronaut standing on a scale in a rocket that is leaving the launching pad (Figure 6.17). We pick the astronaut as a body and write Newton's second law.

$$\sum F = ma$$

Two forces contribute to the net force. One is the weight, mg, of the astronaut, and the other is the normal force, N, exerted on the astronaut by the scale. Thus we have

$$\sum F = N - mg = ma$$

Solving for the force N gives the result

$$N = m(g + a) \tag{6.9}$$

The contact force N exerted on the astronaut equals the scale reading. We call the scale reading the **apparent weight**, to distinguish it from the true weight, mg. When the rocket is at rest on the launching pad, $a = 0$, $N = mg$, and the apparent weight equals the true weight. When the rocket accelerates upward, $a > 0$ and $N > mg$; the apparent weight exceeds the weight of the astronaut.

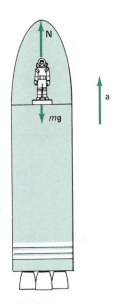

FIGURE 6.17

In an accelerating rocket the scale reading, N, differs from the true weight, mg.

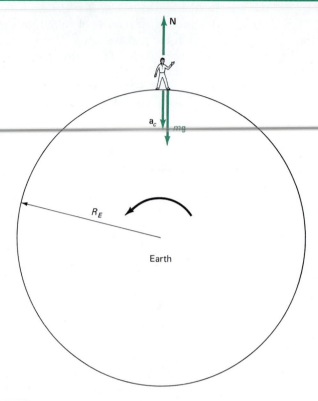

FIGURE 6.18
A view from above the North Pole. A person standing on the equator experiences a centripetal acceleration. This acceleration causes the apparent weight of the person to differ very slightly from the true weight.

In Figure 6.17 the acceleration of the astronaut is along a straight line. However, the effect just described is also observed with centripetal acceleration. In Figure 6.18, a person is shown standing at the equator. The diagram is drawn from the viewpoint of someone above the North Pole. The earth spins counterclockwise, and the person on the equator travels in a circular orbit whose radius equals the earth's radius. Two forces act on him. His weight is directed toward the center of the earth, and the normal force is directed outward. Applying Newton's second law gives

$$\sum F = mg - N = ma_c$$

where a_c is the person's centripetal acceleration. We solve for N to get

$$N = m(g - a_c) \tag{6.10}$$

Clearly, N is less than the weight. The apparent weight N differs little from the true weight mg because a_c is less than one percent of g. If $mg = 200$ lb, the scale would read about $199\frac{1}{3}$ lb because of the centripetal acceleration.

Astronauts in earth orbit experience "weightlessness," meaning that $N = 0$. This weightless condition is described by

$$mg = ma_c$$

This means that gravity supplies the full force needed to maintain the centripetal acceleration.

Gravitation and Inertia

Mass measures two remarkable properties of matter. In Newton's second law ($\mathbf{F} = m\mathbf{a}$), mass is a measure of inertia. In the law of gravity, mass is a measure of the strength with which an object exerts and experiences gravitational forces. There is no obvious connection between these two aspects of mass. However, experiments of great precision have been unable to distinguish the inertial and gravitational aspects of mass. Is this a coincidence, or is there some underlying connection between gravitation and inertia? We don't know! Einstein and others attempted, unsuccessfully, to establish a quantitative link between gravitation and inertia. Although most scientists believe that there *is* a fundamental relation between gravitation and inertia, there is no satisfactory theory that establishes the details of such a connection.

6.6
MECHANICAL FORCE LAWS

Logically, Newton's second law defines force. In practice, we generally use Newton's second law to determine the acceleration of a body subjected to known forces. Once the acceleration is specified we can use kinematics to determine the changes in velocity and position.

The contact forces we often deal with are electromagnetic in character. They arise because of the electrical structure of atoms and molecules. It is not necessary to deal with these forces at the atomic level in order to describe the net force that acts at our sensory level. Instead we describe the net force by some empirical (experimental) relation. For example, when you stretch a spring, it exerts a restoring force on you. The observable quantity is the distance you stretch the spring, and it is possible to establish an empirical relation between the stretch and the restoring force. Frictional forces are also electromagnetic in character. The frictional force between two surfaces can be described in terms of directly measurable quantities. It is not necessary or useful to deal with the forces acting at the atomic level.

In this section we describe several types of mechanical forces and use them to illustrate Newton's second law.

Linear Spring

An *elastic* material is one that resumes its original shape after a distorting force is removed. Springs are formed from elastic materials. When a spring is stretched it exerts a *restoring* force (Figure 6.19b). This force opposes the stretch. If a spring is compressed it also exerts a *restoring* force (Figure 6.19c). This force opposes the compression.

For some springs the force is directly proportional to the change in length of the spring. In this case we write

$$F = -kx \tag{6.11}$$

F is the restoring force and x is the change in length of the spring. The constant k is called the **spring constant**; it is a positive quantity characteristic of the spring material and geometry. The minus sign shows that F and x are in opposite directions; the force acts to oppose the change in length.

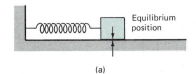

(a)

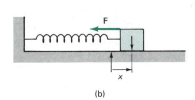

(b)

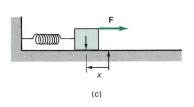

(c)

FIGURE 6.19

(a) A horizontal spring in its equilibrium configuration.
(b) The stretched spring exerts a force to the left. (c) The compressed spring exerts a force to the right.

Because F and x are linearly related, any spring described by Equation 6.11 is called a **linear spring**. In order to use the spring to measure force, we must measure the spring constant.

EXAMPLE 7

Measuring the Spring Constant

A linear spring is stretched by hanging it vertically (Figure 6.20). With no weight attached the length of the spring is 8.20 cm. A 200-gram mass is attached and its weight is allowed to slowly stretch the spring. The 200-gram mass reaches an equilibrium position when its weight is balanced by the spring force,

$$kx = mg$$

The equilibrium length of the stretched spring is 9.65 cm, so

$$x = 9.65 \text{ cm} - 8.20 \text{ cm} = 1.45 \text{ cm}$$

For k we find

$$k = \frac{mg}{x} = \frac{(0.200 \text{ kg})(9.80 \text{ m/s}^2)}{0.0145 \text{ m}}$$

$$= 135 \text{ N/m}$$

Sliding Friction

When two solid surfaces touch they experience contact forces. In general, the contact force has one component perpendicular to the surface and another component parallel to the surface. The component perpendicular to the surface is called the **normal force**; it prevents the two surfaces from penetrating into one another. The parallel component is called the **frictional force**; it acts to oppose relative motion of the two surfaces. When the surfaces are in relative motion we speak of *kinetic friction*; when there is no relative motion, we speak of *static friction*.

Experiment shows that the kinetic friction force (f) depends on the nature of the two surfaces and is directly proportional to the normal force (N). We relate f and N by the equation

$$f = \mu_k N \tag{6.12}$$

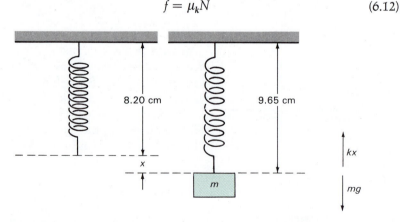

FIGURE 6.20

The weight of the mass stretches the linear spring. At equilibrium, the spring force *kx* equals the weight *mg*.

Material	μ_k
waxed ski on dry snow (0°C)	0.04
waxed ski on dry snow (-10°C)	0.18
waxed ski on dry snow (-40°C)	0.40
ice on ice (-10°C)	0.035
ebonite on ice (-10°C)	0.05
brass on ice (-10°C)	0.075
rubber on dry cement	1.02
rubber on wet cement	0.97

The quantity μ_k is called the **coefficient of kinetic friction**. Experiment shows that μ_k depends on the nature of the two surfaces in contact but not on their relative speed or the contact area. Typical values of μ_k are given in Table 6.1.

EXAMPLE 8

Skid Distance

Consider a van moving down an icy hill (Figure 6.21). At a moment when the speed of the van is 42 mi/h the driver spots a deer 360 feet ahead. He slams on the brakes and slides to a stop. The deer is paralyzed by fright and unable to move. Will the deer be hit or will the van stop before reaching it?

We take the coefficient of kinetic friction to be $\mu_k = 0.26$. The angle θ is 5°. The contact force acting on the sliding van has a normal component N and a frictional component $f = \mu_k N$. We choose axes parallel to and perpendicular to the hill. The acceleration of the van is along the x-axis. The components of Newton's second law are

$$\sum F_x = mg \sin \theta - \mu_k N = ma_x$$
$$\sum F_y = N - mg \cos \theta = 0$$

Eliminating N we find for the acceleration

$$a_x = g(\sin \theta - \mu_k \cos \theta)$$

Inserting numerical values we get

$$a = (9.8 \text{ m/s}^2)(\sin 5° - 0.26 \cdot \cos 5°)$$
$$= -1.68 \text{ m/s}^2$$

The negative value indicates that the van slows down. To determine how far the van slides we use a result derived in Chapter 3.

$$v^2 = v_0{}^2 + 2ax$$

With $v_0 = 42$ mi/h $= 18.8$ m/s and $a = -1.68$ m/s², we set $v = 0$ and solve for x. This gives the distance the van slides before coming to rest.

$$x = \frac{-v_0{}^2}{2a} = \frac{-(18.8 \text{ m/s})^2}{2(-1.68 \text{ m/s}^2)} = 105 \text{ m}$$
$$= 345 \text{ ft}$$

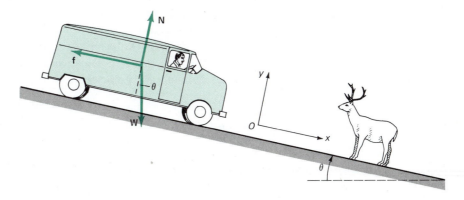

FIGURE 6.21
The van skids down the hill toward the deer.

In equilibrium

$$f_s = F$$

FIGURE 6.22

An external horizontal force **F** exerted on the book is opposed by a static friction force, **f**$_s$. So long as the book remains at rest, the net force on the book is zero, and **f**$_s$ must equal −**F**.

Because the deer was 360 ft away when the slide began, the van stops 15 ft short of the deer.

Static Friction

A frictional force can exist even when there is no relative motion between surfaces in contact. In this case we speak of **static** friction. The static friction force can take on values ranging from zero up to a maximum whose value depends on the normal force and the nature of the surfaces. For example, suppose that a book is at rest on a table (Figure 6.22). If no horizontal force is exerted on the book the frictional force also must be zero. Otherwise, equilibrium would not be preserved. Suppose you push weakly on the book. The book will remain at rest because the static friction force takes on a value just large enough to balance your applied force. If you push harder and harder the book eventually moves. There is a limit to the static friction force.

Experiment shows that the maximum static friction force $(f_s)_{max}$ can be expressed by

$$(f_s)_{max} = \mu_s N \tag{6.13}$$

where N is the normal force and μ_s is the **coefficient of static friction**. In most situations $\mu_s > \mu_k$; static friction is "stronger" than kinetic friction. This relation has implications in braking an automobile.

Drivers are routinely taught to stop their cars without allowing it to skid. If the car skids there is relative motion between the pavement and the tires; the friction is kinetic and the frictional force is $f_k = \mu_k N$. If the car rolls to a stop without skidding there is no relative motion between the pavement and the tires*; the friction is static and the frictional force can take on a maximum value equal to $\mu_s N$. Because $\mu_s > \mu_k$ the braking force can be greater—and the stopping distance shorter—when no skidding occurs.

Fluid Friction

When a body moves through a fluid (a gas or a liquid) it experiences a frictional force opposing its motion. This fluid friction force is often called a **viscous force** because it depends on a property of the fluid called *viscosity*. In aerodynamics the fluid friction force is referred to as the *drag*. The fluid friction force depends strongly on the size and shape of the body, on its speed, and on the type of fluid.

Experiment shows that the force increases as the speed increases. At very low speeds the drag is directly proportional to the speed of the body. For spherical objects the fluid friction force is well represented by *Stokes' law*, which expresses the drag, F_D, on a sphere of radius r and speed v as

$$F_D = 6\pi\eta r v \tag{6.14}$$

The constant η (Greek eta) is an empirical quantity called the dynamic viscosity of the fluid. Stokes' law is adequate for describing the drag on tiny water droplets in fog and mists. Ordinary raindrops travel at relatively high speeds

---

* This may seem perplexing! The car is moving and the roadway is at rest. However, so long as there is no skidding, the part of the tire in contact with the road has no velocity component parallel to the road.

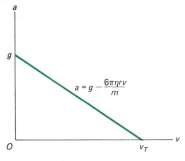

FIGURE 6.23

The acceleration of the fog droplet, *a*, is plotted versus the speed of the droplet. The acceleration decreases as the speed increases. When the acceleration reaches zero, the droplet has reached its terminal speed, v_T.

and Stokes' law does not give a reliable estimate of the drag force they experience.

Terminal Velocity

As we said before, fog and mist are collections of tiny water droplets. Experiment reveals that these droplets fall very slowly. They experience a viscous drag described by Stokes' law, Equation 6.14.

Consider the vertical motion of a fog droplet subject to the forces of gravity and drag. Two forces act on the droplet. Its weight, *mg*, is downward, and the drag, $6\pi\eta rv$, is upward. The droplet's acceleration downward is determined from Newton's second law

$$\sum F = mg - 6\pi\eta rv = ma$$

Dividing by *m* we obtain

$$a = g - \frac{6\pi\eta rv}{m}$$

Perhaps the most striking feature of this result is the velocity dependence. Freely falling objects share a common acceleration *g*. The droplet acceleration (Figure 6.23) is *g* initially ($v = 0$), but then approaches zero. The velocity increases until the two forces become equal in magnitude. At this point the velocity is at a maximum called the **terminal velocity**, v_T. Setting $a = 0$, we obtain for the terminal velocity

$$v_T = \frac{mg}{6\pi\eta r} \tag{6.15}$$

We see from this relation that the terminal velocity of an object depends on the object's mass. The more massive the sphere, the faster it falls through a fluid. To determine the size dependence of the terminal velocity we introduce the mass density, ρ,

$$m = \rho(\tfrac{4}{3}\pi r^3)$$

where $\tfrac{4}{3}\pi r^3$ is the volume of the droplet. Using this expression for the mass in Equation 6.15 we find for the terminal velocity

$$v_T = \frac{2\rho g r^2}{9\eta} \tag{6.16}$$

The terminal velocity of a sphere is directly proportional to the square of the radius; doubling the radius produces a fourfold increase in terminal velocity.

EXAMPLE 9

Terminal Velocity of a Fog Droplet

Using a microscope we find that the radius of a small fog droplet is 5.1×10^{-6} m, or about five-thousandths of a millimeter. (This radius, typical for droplets found in fog and clouds, is roughly one-tenth of the radius of the smallest droplet visible to the human eye.) We can use this measurement to obtain the terminal velocity of the droplet.

For air at a temperature of 20° C,

$$\eta = 1.81 \times 10^{-5} \text{ N} \cdot \text{s/m}^2$$

Using Equation 6.16, with $\rho = 10^3$ kg/m^3 for water, we obtain

$$v_T = \frac{2\rho g r^2}{9\eta}$$

$$= \frac{2(10^3 \text{ kg/m}^3)(9.8 \text{ m/s}^2)(5.1 \times 10^{-6} \text{ m})^2}{9(1.81 \times 10^{-5} \text{ kg/m} \cdot \text{s})}$$

$$= 3.1 \times 10^{-2} \text{ m/s}$$

A droplet falling with this speed would require about one minute to fall from your head to your feet.

WORKED PROBLEM

A ball ($m = 200$ grams) is thrown vertically upward with an initial speed of 30 m/s. During its flight the ball experiences an air resistance force given by $\mathbf{f} = -b\mathbf{v}$, where $\mathbf{v}$ is the velocity and $b = 0.03$ N$\cdot$s/m. Determine the time for the ball to reach its maximum height, and compare this with the time required if there were no air resistance.

Solution

Whether or not air resistance is accounted for, the ball leaves the thrower's hand with a speed of 30 m/s and decreases to zero at the maximum height. With no air resistance, only gravity slows the ball. But with air resistance, both gravity and air resistance act to reduce the speed of the ball. Because we know the forces acting on the ball, we can use Newton's second law to formulate a solution.

We take the positive direction for vectors to be upward. The net force is $-mg - bv$ because both forces on the rising ball act downward. Equating mass times acceleration to net force yields

$$m\frac{dv}{dt} = -mg - bv$$

This can be rewritten as

$$\frac{dv}{g + bv/m} = -dt$$

We integrate this equation to find how the velocity changes with time. Let $t = 0$ and $v = v_0$ be the time and velocity when the ball is released. Let $t = T$ and $v = 0$ denote the time and velocity at maximum height. These values of t and v serve as limits on the integrals. Thus,

$$\int_{v_0}^{0} \frac{dv}{g + bv/m} = -\int_{0}^{T} dt$$

The integral on the left gives a natural logarithm. Doing the integrations and performing some algebra gives

$$T = \left(\frac{m}{b}\right) \ln\left(1 + \frac{bv_0}{mg}\right)$$

Substituting the given numerical values we find,

$$bv_0 = (0.03 \text{ N} \cdot \text{s/m})(30 \text{ m/s}) = 0.9 \text{ N}$$

$$mg = (0.200 \text{ kg})(9.80 \text{ m/s}^2) = 1.96 \text{ N}$$

and

$$T = \left(\frac{0.2 \text{ kg}}{0.03 \text{ N·s/m}}\right) \ln\left(1 + \frac{0.9 \text{ N}}{1.96 \text{ N}}\right)$$
$$= 2.52 \text{ s}$$

The acceleration equals $-g$ when there is no air friction. The velocity after the release at $t = 0$ is related to the initial velocity by

$$v = v_0 - gt$$

The time for the ball to reach $v = 0$ at its peak is

$$T = \frac{v_0}{g} = \frac{30 \text{ m/s}}{9.80 \text{ m/s}^2}$$
$$= 3.06 \text{ s}$$

Air resistance causes the maximum height to be reached 0.54 s sooner than if air resistance were absent. The maximum height is, of course, less than when air resistance is present.

EXERCISES

6.2 Newton's First Law

A. A particle moves in a straight line. Its positions (x) at various moments of time are displayed in Table 1.

x (m)	t (s)
1.1	0
2.2	1
3.4	2
4.8	3
6.0	4
7.0	5

TABLE 1

Determine whether or not the particle remains free of a force over the entire time interval.

B. A particle moves in a circle at a constant speed of 30 m/s. Does it experience a net force? Explain.

6.3 Newton's Second Law

C. A 1-kg mass is observed to accelerate at 10 m/s² in a direction 30° north of east. One of the two forces acting on the kilogram has a magnitude of 5 N and is directed north. Determine the magnitude and direction of the second force acting on the mass.

D. Forces $\mathbf{F}_1$ (east) and $\mathbf{F}_2$ (north) are applied to a 1-kg mass (Figure 1). Its acceleration is 10 m/s² at 30° north of east. (a) Give the magnitude and direction of the net force acting on the mass. (b) Determine the magnitudes of $\mathbf{F}_1$ and $\mathbf{F}_2$.

E. A 1000-kg car accelerates at 4 m/s². What is the magnitude of the net force on the car?

F. A 0.2-kg mass attached to the end of a string is whirled in a vertical circle by a student. At some position the mass experiences a downward force of $-62\mathbf{j}$ N and a horizontal force of $38\mathbf{i}$ N. Determine the acceleration at this position.

G. Forces of 10 N north, 20 N east, and 15 N south are simultaneously applied to a 4-kg mass. Obtain its acceleration.

H. A net force applied to a 2-kg mass gives it an acceleration of 3 m/s² east. The same net force results in an acceleration of 1 m/s² east when applied to a different mass m. Determine m and the net force.

I. A constant force changes the speed of an 80-kg sprinter from 3 m/s to 4 m/s in 0.5 s. (a) Calculate the magnitude of the acceleration of the sprinter. (b) Obtain the magnitude of the force. (c) Determine the magnitude of the acceleration of a 50-kg sprinter experiencing the same force. (Assume linear motion.)

J. Two blocks connected by a string move as a unit on a frictionless horizontal surface (Figure 2). The acceleration is 3 m/s² with $m = 2$ kg and $M = 4$ kg. (a) Determine the tensions in the strings. (b) For the same acceleration, repeat the calculation if the two blocks are interchanged.

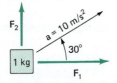

FIGURE 1

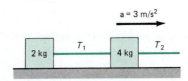

FIGURE 2

K. A worker piles one box on top of another and pushes on the bottom box (Figure 3). The two boxes move together with an acceleration of 1 m/s². What horizontal force does the bottom box exert on the upper box?

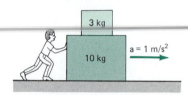

FIGURE 3

6.4   Newton's Third Law

L. The mass of the earth is 5.98×10^{24} kg. If the earth's gravitation force causes a 60-kg student to accelerate downward at 9.8 m/s², calculate the upward acceleration of the earth resulting from the student's gravitational reaction force acting on the earth.

M. Newspapers and television news frequently report accidents between large trucks and small cars. Based on your knowledge of such reports and Newton's third law, present an argument to support the idea that damage in a collision is proportional to *acceleration, not force*.

6.5   Newton's Law of Universal Gravitation

N. The force of gravity is proportional to the mass of a body, yet all freely falling bodies experience the same acceleration at the same location. Explain.

P. On the earth a weight lifter can lift a mass of 200 kg. If we assume no change in his strength, what mass could he lift on the moon?

Q. Calculate the gravitational force between two bowling balls, each with a mass of 7.27 kg, whose centers are 1 m apart. Compare this force with the weight of one bowling ball.

R. At a point along the line joining the centers of the earth and the moon, the gravitational forces they exert on a third body are equal in magnitude but oppositely directed. Determine the distance of this point from the center of the moon. Take the mass of the earth to be 81 times the mass of the moon and express your answer as a fraction of the earth–moon distance.

S. If you weigh 160 pounds at the surface of the earth, how far would you have to descend beneath the surface of the earth to reduce your weight to 159 pounds?

T. If you weigh 120 pounds at the surface of the earth, how much do you weigh at an altitude of 6380 km? (The radius of the earth is 6380 km.)

U. Imagine an 80-kg man standing on a spring scale (calibrated in newtons) in an elevator. (a) The elevator accelerates downward at 3 m/s². What is the scale reading? (b) When the scale reading is 1000 N, what is the acceleration of the elevator?

6.6   Mechanical Force Laws

V. A horizontal spring with an unstretched length of 0.20 m stretches to a length of 0.25 m when a force of 5 N is applied. (a) Determine the spring constant, k. (b) A 5-kg mass is attached to the spring and the spring is stretched and released. Obtain the acceleration of the mass when the spring length is 0.15 m.

W. A sprinter accelerates uniformly from 6.5 m/s to 9.0 m/s in a distance of 7.75 m. Assuming the normal force equals the weight of the sprinter, determine the coefficient of fricton.

X. Two identical linear springs have a spring constant k. Determine the spring constant of a new spring constructed by putting the two springs (a) end to end, (b) side by side.

Y. A fireman whose mass is 75 kg slides down a 6-m pole in a time of 2 s. The force of friction exerted on the fireman by the pole is constant. Determine (a) the acceleration of the fireman, (b) the force of friction.

Z. A block slides down a plane at a constant speed (Figure 4). Prove that the angle θ and the coefficient of friction μ_k are related by $\mu_k = \tan \theta$.

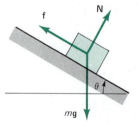

FIGURE 4

PROBLEMS

1. A 1000-kg car moving at 20 m/s is brought to rest in a distance of 80 m by a constant net force, **F**, directed opposite to the velocity of the car. Determine the (a) magnitude of the acceleration of the car, (b) the magnitude of **F**.

2. An air-hockey puck ($m = 0.1$ kg) moves without friction in a circle of radius 0.5 m on a horizontal table. The puck is attached to the center of rotation by a string of negligible mass. The speed of the puck is 6 m/s. Determine (a) the centripetal acceleration of the puck, and (b) the string tension. (c) The 0.1-kg puck is replaced by a 0.4-kg puck. The speed of the new

puck is adjusted until the string tension equals the tension in part (b). What is the speed of the new puck?

3. At the moment of a total solar eclipse the moon lies along a line from the earth to the sun (Figure 5). Your weight, as indicated by a scale, is affected by the gravitational pulls of the sun and moon. Compare the gravitational forces exerted on you by (a) the sun and (b) the moon, with the gravitational force exerted on you by the earth. Express your answers as ratios of the forces rather than as the forces themselves. (c) If the earth exerts a gravitational force of 800 N on you, what is the combined pull of the sun and moon?

FIGURE 5

4. A group of people decides to construct the world's tallest brick tower. As a rigid body, the tower shares in the rotation of the earth. At what altitude will the ratio of the weight of a brick to its mass be equal to its centripetal acceleration? (At this point, the topmost brick is in orbit, or, in other words, it experiences no contact force with the bricks beneath it.)

5. A hypothetical planet has five moons that move in circular orbits. Table 2, listing orbit radii and periods, contains one *erroneous radius* entry. Use Kepler's third law to discover and correct the error.

TABLE 2

Moon	r (10^6 m)	T (days)
I	$\frac{1}{4}$	$\frac{1}{8}$
II	1	1
III	2	2.83
IV	3	6
V	4	8

6. Some stars evolve into small dense objects called neutron stars. Consider a neutron star of mass $m = 2.8 \times 10^{30}$ kg and radius $r = 1.2 \times 10^4$ m. (a) Calculate the acceleration of gravity at the surface of the star. (b) Assume the star spins with period T. Determine the value of T for which the acceleration of gravity equals the centripetal acceleration for a particle on the equator. (c) Obtain the speed of this particle. (For comparison, the speed of light is 3×10^8 m/s.)

7. Two identical spherical masses ($m = 6 \times 10^7$ kg) are tied together by a weightless cord. The centers of the two spheres are 200 m apart. The spheres move in uniform circular motion about a point midway between

them with a period of 36 hours. Determine the (a) radial acceleration of one of the masses, (b) the gravitational force exerted on your chosen mass, (c) the tension in the cord.

8. A car moving at 20 m/s brakes to a stop without skidding. The driver in the car behind the first, moving at 30 m/s, sees the brake lights, applies his brakes after a 0.1-s delay to react, and brakes to a stop without skidding. Assume that $\mu_s = 0.75$ for both cars. Calculate the minimum distance between the cars at the instant the driver of the lead car applies the brakes if a rear-end collision is to be avoided. (Assume constant accelerations.)

9. A crate of mass m slides across a warehouse floor. The acceleration, a, of the crate is opposite in direction to its motion. Assume the force of friction is constant and prove that $\mu_k = a/g$.

10. About 200 years ago Coulomb invented the tribometer, a device employed to investigate static friction. The instrument is represented schematically in Figure 6. To determine the coefficient of static friction, the hanging mass M is increased or decreased as necessary until m is on the verge of sliding. Prove that $\mu_s = M/m$.

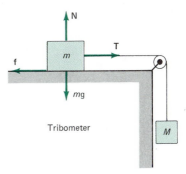

FIGURE 6

11. On a dry road a 1000-kg car moving at 30 m/s requires a minimum distance of 50 m to come to rest without skidding. Determine (a) the magnitude of the car's acceleration; (b) the magnitude of the frictional force exerted by the road.

12. A spherical oil droplet (density = 900 kg/m³) whose radius is 5×10^{-8} m is emitted from the top of a smokestack 250 m above the ground. Estimate the time for the oil droplet to descend to the ground if its descent is governed by Stokes' law. (Take $\eta = 1.81 \times 10^{-5}$ N·s/m² for air.)

13. Particulates emerging from a smokestack settle vertically in accord with Stokes' law while at the same time they are carried horizontally by a 10-mph wind (Figure 7). Consider two spherical particulates (density = 2000 kg/m³) after they emerge from a smokestack

150 m high. Assume a perfectly flat earth and estimate the horizontal distance from the smokestack to the points where the spheres strike the ground. Take the radii of the spheres to be (a) 1.5×10^{-6} m and (b) 2×10^{-4} m. Use $\eta = 1.81 \times 10^{-5}$ N·s/m² for air.

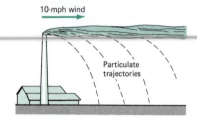

FIGURE 7

14. To determine the viscosity of a liquid, a steel sphere 1 mm in radius is timed as it falls in the liquid at its terminal velocity through a distance of 0.10 m. The weight of the sphere (corrected for buoyancy) is shown in the following table. Assume Stokes' law is obeyed and determine the viscosities of the three liquids listed in the table.

Liquid	Weight in Liquid	Time to Fall 0.1 m
Water	2.9×10^{-4} N	6.5×10^{-3} s
Glycerin	2.8×10^{-4} N	5.0 s
Sucrose	2.7×10^{-4} N	1.9×10^{4} s

15. In an acceleration test a van (Figure 8) is brought from rest to a speed of 25 m/s in 10 s. A small mass is suspended from the ceiling of the van. At constant acceleration (a) what angle will the cord make with the vertical? (b) What acceleration time for the same final speed would give rise to an angle of 30°?

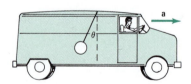

FIGURE 8

16. Two ice skaters are at rest on a frozen canal 50 m wide, a distance x from one side (Figure 9). They push each other apart and then coast without friction, arriving at the same time on opposite sides of the canal. Their masses are 80 kg and 50 kg, and during the acceleration period of 0.5 s they push on each other with constant forces of 400 N. Determine (a) the magnitudes

of their accelerations during the push, (b) their final speeds, (c) the distance over which each accelerates. (d) Locate the starting point. (e) Calculate the total time (accelerating time plus coasting time) required for the skaters to reach the sides of the canal.

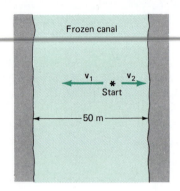

FIGURE 9

17. A 60-kg skier moving initially at 10 m/s glides up a 15° slope ($\mu_k = 0.18$). She coasts up the slope to a stop, and then glides back down to the starting point. Determine (a) the time to coast up the slope and (b) the time to coast back to the starting point.

18. A 1000-kg car is moving up a 10° hill at 25 m/s. It strikes an icy stretch and coasts to a stop without friction. Determine (a) the time required to come to rest, (b) the stopping distance.

19. The mean orbital distance of Pluto from the sun is 39.44 AU. Calculate its period in years.

20. Use Kepler's third law to calculate the altitude of a synchronous satellite (a satellite with a period of 1 sidereal day.)

21. Show that the weight of a particle of mass m a distance r from the center of the earth can be expressed as

$$W = mg\left(\frac{r}{R_E}\right) \qquad r \le R_E$$

$$W = mg\left(\frac{R_E}{r}\right)^2 \qquad r \ge R_E$$

where g is the acceleration of gravity at the surface of the earth.

22. In Larry Niven's science fiction novel *Ringworld*, a solid ring of material rotates about a star (Figure 10). The rotational speed of the ring is 1.25×10^6 m/s. Its radius is 1.53×10^{11} m. The inhabitants of this ring world experience a normal contact force, N. Acting alone, this normal force would produce an inward acceleration of 9.90 m/s². Additionally, the star at the center of the ring exerts a gravitational force on the ring and its inhabitants. (a) Show that the total centripetal acceleration of the inhabitants is 10.2 m/s². (b) The difference between the total acceleration and the acceleration provided by the normal force is due

to the gravitational attraction of the central star. Show that the mass of the star is approximately 10^{32} kg.

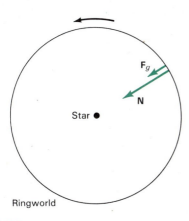

FIGURE 10

23. A person stands on a scale in an elevator (Figure 11). The maximum and minimum scale readings are 591 N and 391 N. Assume the magnitude of the acceleration is the same during starting and stopping, and determine (a) the weight of the person, (b) the person's mass, (c) the elevator acceleration.

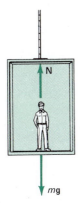

FIGURE 11

24. Figure 12 shows Atwood's machine. Assume that the pulley is frictionless and has no effect other than to change the direction of the tension in the cord. (a) Write down Newton's second law for m as a body, and for M as a body. Eliminate the tension from the resulting pair of equations and show that the magnitude of the acceleration of the masses is (assume $M > m$)

$$a = \frac{(M - m)g}{M + m}$$

(b) Derive an equation for the tension in the cord.

FIGURE 12

25. A conical pendulum is a bob moving in a horizontal circle at the end of a long wire (Figure 13). The angle between the wire and the vertical does not change. Consider a conical pendulum with an 80-kg bob on a 10-m wire making an angle of $5°$ with the vertical. (a) Determine the tension in the wire and its horizontal and vertical components. (b) Determine the radial acceleration of the bob.

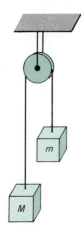

FIGURE 13

26. A ball ($m = 200$ grams) is dropped into a fluid. The velocity of the ball as a function of time is represented by the graph in Figure 14. Deduce the magnitude of the net upward force acting on the ball.

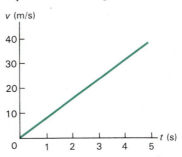

FIGURE 14

27. A block is released from rest at the top of a plane inclined at an angle of 45°. The coefficient of kinetic friction varies along the plane according to the relation $\mu_k = \sigma x$, where x is the distance along the plane measured in meters from the top and where $\sigma = 0.50 \text{ m}^{-1}$. Determine (a) how far the block slides before coming to rest, (b) the maximum speed it attains.

28. A van accelerates down a hill (Figure 15), going from rest to 30 m/s in 6 s. During the acceleration, a toy ($m = 100$ grams) hangs by a string from the ceiling. The acceleration is such that the string remains perpendicular to the ceiling. Determine (a) the angle θ, (b) the tension in the string.

FIGURE 15

29. A hockey stick accelerates a puck (not touching the ice) horizontally (Figure 16). Determine the minimum acceleration needed to provide an upward frictional force equal to the weight of the puck. Take the coefficient of static friction to be $\mu_s = 0.40$.

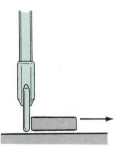

FIGURE 16

30. The bob of a conical pendulum 5 m long moves in a horizontal circle, completing one revolution in 4 s (refer to Figure 13). Determine the angle θ between the wire and the vertical.

31. A flagpole of mass m and length L is vertical at the North Pole. The flagpole has a uniform mass per unit length (m/L). Show that the weight of the flagpole is $mgR/(R + L)$, where R is the radius of the earth and $g = GM_E/R^2$ is the acceleration of gravity at the base of the flagpole.

32. A long thin rod of mass m and length $2L$ has a uniform mass per unit length ($m/2L$). The rod is balanced with its center at the North Pole (Figure 17). (a) Show that the weight of the rod is $mgR/(R^2 + L^2)^{1/2}$, where R is the radius of the earth and $g = GM_E/R^2$ is the acceleration of gravity at the North Pole. (b) Show that the weight of the rod is finite even when its length and mass approach infinite values (the mass per unit length remaining constant).

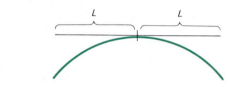

FIGURE 17

CHAPTER

7

WORK ENERGY AND POWER

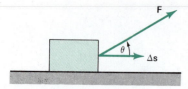

FIGURE 7.1

A constant force acts on an object. The work done by this force equals $\mathbf{F} \cdot \Delta \mathbf{s} = F\,\Delta s \cos \theta$.

7.1
WORK DONE BY A CONSTANT FORCE

The word "work" has many connotations in our everyday language: people *work* for a living; the TV set doesn't *work*; my physics professor *works* me too hard. In physics, the meaning of "work" is very specific: work relates the concepts of force and energy.

Consider the special case of a body that experiences a constant force **F** while undergoing a displacement $\Delta \mathbf{s}$ (Figure 7.1). The force has the same magnitude and direction at every point. We define the work W done by **F** over the displacement $\Delta \mathbf{s}$ to be the scalar product of **F** and $\Delta \mathbf{s}$.

$$W = \mathbf{F} \cdot \Delta \mathbf{s} = F\,\Delta s \cos \theta \qquad (7.1)$$

where θ is the angle between **F** and $\Delta \mathbf{s}$ (Figure 7.2). The work can be positive or negative, depending on the sign of $\cos \theta$. Note that $F_s = F \cos \theta$ is the component of **F** along the direction of $\Delta \mathbf{s}$; hence an alternative form of W is,

$$W = F_s\,\Delta s \qquad (7.2)$$

According to Equation 7.2, one unit of work is the product of a force of 1 newton and a displacement of 1 meter. The SI unit of work, 1 newton·meter, is called 1 joule (1 J);

$$1 \text{ joule} = 1 \text{ newton·meter} \qquad \text{or} \qquad 1\,\text{J} = 1\,\text{N·m}$$

Because $1 \text{ newton} = 1 \text{ kg·m/s}^2$ we may also write

$$1\,\text{J} = 1\,\text{kg·m}^2/\text{s}^2$$

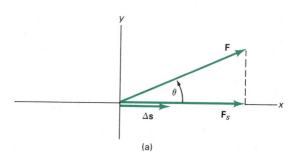

(a)

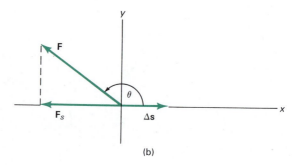

(b)

FIGURE 7.2

F_s is the component of **F** along the direction of the displacement. In (a) F_s is parallel to $\Delta \mathbf{s}$ and W is positive. In (b) F_s and $\Delta \mathbf{s}$ are in opposite directions and W is negative.

EXAMPLE 1

Work Done by Gravity

A sky diver (mass = 66 kg) falls vertically through a distance of 1000 m. What is the work done by the force of gravity (presumed constant) on the sky diver during the fall?

The magnitude of the force vector is

$$F = mg = (66 \text{ kg})(9.80 \text{ m/s}^2)$$
$$= 647 \text{ N}$$

and the magnitude of the displacement vector is

$$\Delta s = 1000 \text{ m}$$

The force and displacement vectors are parallel, so $\theta = 0$, and $\cos \theta = +1$. Using Equation 7.1, the work done by gravity becomes

$$W = F \, \Delta s = (647 \text{ N})(1000 \text{ m})$$
$$= 6.47 \times 10^5 \text{ J}$$

The work done by the force of gravity does not depend on the speed of the sky diver or on the presence or absence of other forces.

According to Equation 7.1, the work W is zero if any of the factors (F, Δs, $\cos \theta$) equals zero. If $F = 0$, then no force acts, so it is not surprising that the work is zero.

If $\Delta s = 0$, then $W = 0$ regardless of the size of the force that is acting. In other words, if a force does not cause a displacement of a body, then it does not do work on that body. This idea may contradict your everyday understanding of work. For example, a weight lifter exerting an upward force on a barbell does work as he lifts the barbell from the floor ($\Delta s \neq 0$). However, he does no work while he holds the barbell motionless above his head after the lift ($\Delta s = 0$). This does not mean that holding a barbell overhead is an easy task! It means only that work, as we define it in physics, is not performed.

The work W is also zero if $\cos \theta = 0$. This occurs when $\theta = 90°$, and $\mathbf{F}$ is perpendicular to $\Delta\mathbf{s}$. A planet moving in a circular orbit under the influence of the gravitational force of the sun is an example of motion with $\mathbf{F}$ perpendicular to $\Delta\mathbf{s}$. The displacement is along the circumference of the orbit while the force is radial (Figure 7.3). Clearly the force influences the motion in these cases, and yet the work done is zero. This result is a direct consequence of our definition of work, and illustrates the fact that work does not *completely* represent the effect of the environment on particle motion.

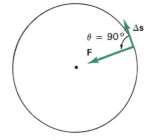

FIGURE 7.3

For a particle moving at constant speed in a circle the work done equals zero because the force and displacement are perpendicular and $\cos 90° = 0$.

EXAMPLE 2

Work Done on a Truck

Consider a pickup truck of mass m moving up a ramp as shown in Figure 7.4. The forces acting on the truck include the normal force $\mathbf{N}$ and the weight $m\mathbf{g}$. The work done by the normal force equals zero because $\mathbf{N}$ and $\Delta\mathbf{s}$ are perpendicular. What is the work done by gravity? The angle between the force, $m\mathbf{g}$, and $\Delta\mathbf{s}$ is $90° + \theta$, and we have from Equation 7.1

$$W = mg \, \Delta s \, (90° + \theta)$$

or

$$W = -mg \, \Delta s \sin \theta$$

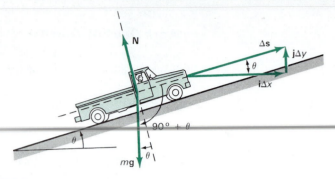

FIGURE 7.4
The displacement of the truck, $\Delta\mathbf{s}$, makes an angle of $90° + \theta$ with the weight of the truck, $m\mathbf{g}$.

From the geometry of Figure 7.4 the angle θ is both the angle of the ramp and the angle between $m\mathbf{g}$ and a line perpendicular to that ramp. Figure 7.4 also shows that the quantity $\Delta s \sin \theta$ is the vertical component of $\Delta\mathbf{s}$.

$$\Delta s \sin \theta = \Delta y$$

Hence

$$W = -mg \, \Delta y$$

This is consistent with the result of Example 1 except for the minus sign. Here, the vertical part of the displacement is opposite in direction to the force, whereas in Example 1 they were in the same direction. To bring out this aspect of the calculation we resolve $\Delta\mathbf{s}$ into components parallel and perpendicular to $m\mathbf{g}$, and recall that the scalar product $\mathbf{F} \cdot \Delta\mathbf{s}$ can be expressed in terms of the components of the vectors $\mathbf{F}$ and $\Delta\mathbf{s}$. Thus,

$$\begin{aligned} W &= \mathbf{F} \cdot \Delta\mathbf{s} \\ &= (iF_x + jF_y) \cdot (\mathbf{i} \, \Delta x + \mathbf{j} \, \Delta y) \\ &= F_x \, \Delta x + F_y \, \Delta y \end{aligned}$$

In this case we have

$$\mathbf{F} = -mg\mathbf{j} \qquad F_x = 0 \qquad F_y = -mg$$

This gives the previous result for the work

$$\begin{aligned} W &= (0) \, \Delta x - (mg) \, \Delta y \\ &= -mg \, \Delta y \end{aligned}$$

7.2
WORK DONE BY A VARIABLE FORCE

The work W done by a constant force F acting parallel to a displacement Δx is

$$W = F \, \Delta x \tag{7.3}$$

As long as the force acting on a particle is constant, Equation 7.3 can be used to calculate the work done by that force, whether the displacement of the particle is large or small.

Force

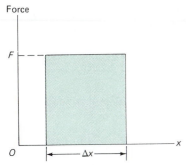

FIGURE 7.5
Work done by a constant force corresponds to the shaded rectangular area, expressed in appropriate units.

Force

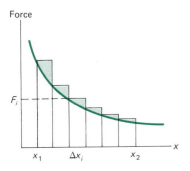

FIGURE 7.6
Work done by a variable force corresponds to the area under the force curve. This area may be estimated as the area of a set of rectangles.

Figure 7.5 shows that W corresponds to the shaded rectangular area, measured in appropriate units. As we will now show, work can be represented by an area under the force-versus-position graph, even if the force is not constant. Figure 7.6 shows a graph of a force that is not constant. The work done by this force over the displacement from $x = x_1$ to $x = x_2$ can be approximated by the work done by constant forces that act over short displacements. Let F_i denote the force acting in the x-direction at $x = x_i$. The rectangular area $F_i \Delta x_i$ approximates the work done between x_i and $x_i + \Delta x_i$. The total work W_{12} done between x_1 and x_2 is approximated by the sum of the rectangular areas.

$$W_{12} \approx \sum F_i \Delta x_i$$

The summation spans the displacement from x_1 to x_2. The approximation becomes exact as we shrink the displacements Δx_i toward zero while increasing their number.

$$W_{12} = \lim_{\Delta x_i \to 0} \sum F_i \Delta x_i$$

The right side defines the area under the F versus x curve between x_1 and x_2. This limit is called the **definite integral** and is represented symbolically by $\int_{x_1}^{x_2} F \, dx$.

The general definition of work for a force acting parallel to the x-axis is

$$W_{12} = \int_{x_1}^{x_2} F \, dx \qquad (7.4)$$

Let's summarize the development of the work-integral concept:

1. The work done by a force F over a displacement from x_1 to x_2 equals the area under a graph of the force-versus-position relation, expressed in appropriate units.
2. To determine this work (area) we evaluate the definite integral $\int_{x_1}^{x_2} F \, dx$.

Examples 3 and 4 illustrate the evaluation of the work integral for two important variable forces.

EXAMPLE 3

Work by a Linear Spring Force

The restoring force exerted by a linear spring is

$$F = -kx$$

where x denotes the stretch or compression of the spring and k is the spring constant. Figure 7.7 shows a graph of F versus x. Let's take $k = 10^4$ N/m and evaluate the work for three different displacements

A. $x_1 = -0.1$ m to $x_2 = 0$
B. $x_1 = -0.1$ m to $x_2 = +0.1$ m
C. $x_1 = -0.1$ m to $x_2 = +0.2$ m

Because the graph of F versus x is a straight line the work "areas" are triangular. For example, in case A the work is the area of a triangle of height kx_1 and base length x_1. Thus,

$$W_A = \tfrac{1}{2}kx_1{}^2$$

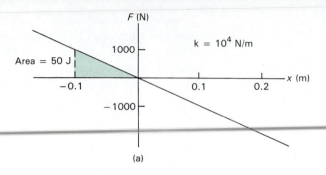

(a)

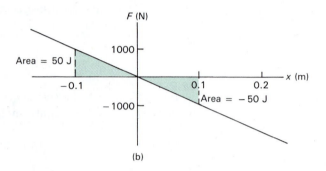

(b)

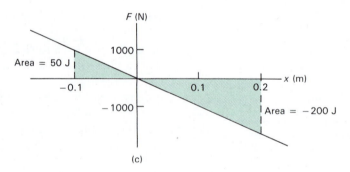

(c)

FIGURE 7.7
Work done by a linear spring force is (a) positive, (b) zero, and (c) negative.

The same result is obtained by evaluating the work integral. In each case we have

$$W = \int_{x_1}^{x_2} (-kx)\, dx = -\tfrac{1}{2}kx^2 \Big|_{x_1}^{x_2}$$

$$W = \tfrac{1}{2}kx_1{}^2 - \tfrac{1}{2}kx_2{}^2$$

For case A, $x_1 = -0.1$ m, $x_2 = 0$ and

$$W_A = \tfrac{1}{2}(10^4\ \text{N/m})(-0.1\ \text{m})^2 = 50\ \text{N·m}$$
$$= 50\ \text{J}$$

The work done is positive because the compressed spring exerts a force in the same direction as the displacement.

For case B, $x_1 = -0.1$ m, $x_2 = +0.1$ m and

$$W_B = \tfrac{1}{2}(10^4\ \text{N/m})(-0.1\ \text{m})^2 - \tfrac{1}{2}(10^4\ \text{N/m})(0.1\ \text{m})^2$$
$$= 0$$

This case illustrates an important aspect of the work-area interpretation. The total area consists of two triangular areas, one positive and one negative. Negative work is done over the range $x = 0$ to $x = x_2 = +0.1$ m. For this range of x the spring is being stretched and the restoring force acts in the direction opposite to the displacement. In general, the work will be negative whenever the force and displacement are in opposite directions.

In case B the positive work done between $x = x_1$ and $x = 0$ is exactly canceled by the negative work done between $x = 0$ and $x = x_2$. In general we can determine the total work by adding the work done over each portion of the interval. Let's adopt this viewpoint to evaluate the work for case C. We can interpret W_C as the sum of the work done from x_1 to 0 and the work done from $x = 0$ to $x = x_2$.

$$W_C = \int_{x_1}^{x_2} F \, dx = \int_{x_1}^{0} (-kx) \, dx + \int_{0}^{x_2} (-kx) \, dx$$

The integral from x_1 to 0 is positive

$$\int_{x_1}^{0} (-kx) \, dx = \tfrac{1}{2}kx_1{}^2 = \tfrac{1}{2}(10^4 \text{ N/m})(-0.1 \text{ m})^2 = +50 \text{ J}$$

The integral from 0 to x_2 is negative

$$\int_{0}^{x_2} (-kx) \, dx = -\tfrac{1}{2}kx_2{}^2 = -\tfrac{1}{2}(10^4 \text{ N/m})(0.2 \text{ m})^2 = -200 \text{ J}$$

The total work is negative

$$W_C = +50 \text{ J} - 200 \text{ J}$$
$$= -150 \text{ J}$$

We saw in Example 2 that negative work was done by the force of gravity when the truck moved upward. We treated the force of gravity as constant, a legitimate step because the displacement was small compared to the radius of the earth. In Example 4 we evaluate the work done by gravity over a displacement that is so large that we must take into account the inverse square nature of the force.

EXAMPLE 4

Work by Universal Force of Gravitation

Let's evaluate the work done by gravity on an astronaut (Figure 7.8) who moves from a point A on the surface of the earth to a point B whose altitude is 2 earth radii ($r = 3R_E$). The gravitational force acting on the astronaut when she is a distance r from the center of the earth is given by Newton's law of universal gravitation

$$F = \frac{GM_E m}{r^2}$$

M_E is the mass of the earth and m is the mass of the astronaut. This force is directed radially inward and the displacement of the astronaut is radially outward, so the work done is negative. The work done as she moves from r to $r + dr$ (Figure 7.8) is

$$dW = -F \, dr = \frac{-GM_E m \, dr}{r^2}$$

The total work over the displacement from A to B is

$$W_{AB} = -GM_E m \int_{R_E}^{3R_E} \frac{dr}{r^2}$$

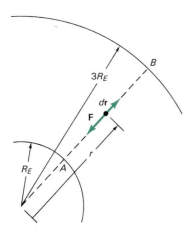

FIGURE 7.8

The work done by the force of gravity is negative when the astronaut moves upward because the force **F** and the displacement d**r** are in opposite directions.

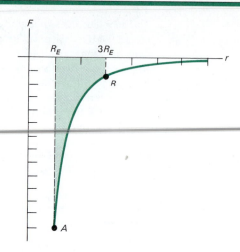

FIGURE 7.9

The shaded area represents the negative work done by the force of gravity as the astronaut moves from the surface of the earth to an altitude of $2R_E$.

In Figure 7.9 the shaded area represents W_{AB}. The integral is

$$\int_{R_E}^{3R_E} \frac{dr}{r^2} = -\left(\frac{1}{r}\right)\Bigg|_{R_E}^{3R_E} = \frac{-1}{3R_E} + \frac{1}{R_E} = \frac{2}{3R_E}$$

The work is

$$W_{AB} = \frac{-\frac{2}{3}GM_E m}{R_E}$$

The acceleration of gravity at the surface of the earth is given by

$$g = \frac{GM_E}{R_E{}^2}$$

In terms of g

$$W_{AB} = -\frac{2}{3}mgR_E$$

This is one-third of the result we would have obtained had we treated the gravitational force as a constant equal to the weight (mg) at the surface of the earth.

Work and Path Integrals

When external forces act on a particle, the particle may not move in a straight line. Therefore, we extend our definition of work to include motion along a curved path. Consider a particle moving along the curved path shown in Figure 7.10. We can take a path segment, Δs_i, small enough so that Equation 7.1 approximates the work done on the particle for that segment.

$$W_i = \mathbf{F}_i \cdot \Delta \mathbf{s}_i$$

The smaller $\Delta \mathbf{s}_i$, the more accurate the approximation of W_i will be. We divide the path from the initial position of the particle (1) to its final position

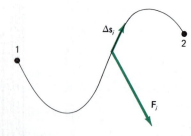

FIGURE 7.10

For an arbitrary path connecting points 1 and 2, the work done by the force $\mathbf{F}_i$ over the small displacement $\Delta \mathbf{s}_i$ is approximately $\mathbf{F}_i \cdot \Delta \mathbf{s}_i$.

(2) into a large number of segments. Each segment contributes a term like W_i to the work. Using a limiting procedure analogous to the one used in this section, we obtain the general definition of work.

$$W = \int_1^2 \mathbf{F} \cdot d\mathbf{s} \qquad (7.5)$$

Both the integral in Equation 7.5 and the integral in Equation 7.4 are called **path integrals** or **line integrals**, because they are evaluated along a path or line. Equation 7.5 is more general than Equation 7.4 because it includes both straight and curved paths.

EXAMPLE 5

Roadway Path Integral

Let's return to the pickup truck in Example 2. The work performed by gravity can be expressed as a path integral

$$W = \int_1^2 \mathbf{F} \cdot d\mathbf{s}$$

The displacement $d\mathbf{s}$ (Figure 7.11) can be expressed as

$$d\mathbf{s} = \mathbf{i}\,dx + \mathbf{j}\,dy$$

The gravitational force is

$$\mathbf{F} = -mg\mathbf{j}$$

The scalar product is

$$\mathbf{F} \cdot d\mathbf{s} = -mg(\mathbf{j} \cdot \mathbf{i}\,dx + \mathbf{j} \cdot \mathbf{j}\,dy) = -mg\,dy$$

Only the vertical displacement results in the performance of work.

$$W = \int_{y_1}^{y_2} -mg\,dy = -mg(y_2 - y_1)$$

If we write $\Delta y = y_2 - y_1$ we recover the result obtained in Example 2

$$W = -mg\,\Delta y$$

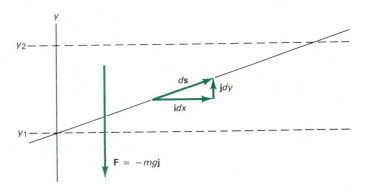

FIGURE 7.11

The displacement along the roadway can be resolved into horizontal and vertical components. Only the vertical component of the displacement results in gravitational work because the force of gravity is vertical.

7.3
KINETIC ENERGY AND THE WORK-ENERGY PRINCIPLE

When a particle experiences a net force it accelerates. If the net force is in the direction of motion it performs positive work and the particle speed increases (Figure 7.12). If the net force is opposite to the direction of motion it performs negative work and the particle speed decreases. We introduce a quantity K, called the **kinetic energy** of the particle, which measures the effect of the work done by the net force acting on the particle.

The kinetic energy of a particle of mass m moving at speed v is defined by

$$K = \tfrac{1}{2}mv^2 \tag{7.6}$$

Kinetic energy is the energy of *motion*; it increases as the speed increases. The SI unit of kinetic energy is the joule, the same unit as the SI unit of work.

EXAMPLE 6

Kinetic Energy of a Skier

A 52-kg skier moves down a slope at a speed of 14.0 m/s (31.3 mph). Determine the kinetic energy of the skier. Using Equation 7.6 we have

$$K = \tfrac{1}{2}mv^2 = \tfrac{1}{2}(52 \text{ kg})(14.0 \text{ m/s})^2$$
$$= 5100 \text{ J}$$

Similarly, a 52-kg platform diver falling 10 m down to the water is accelerated by gravity from rest to a speed of 14 m/s. Thus the kinetic energy of the 52-kg platform diver at impact is also 5100 joules. So, although the diver and skier may travel along quite different paths, both start from rest and acquire the same speed.

When the net force acting on a particle does work, the kinetic energy of the particle changes. The **Work-Energy Principle** relates the work done to the change in kinetic energy:

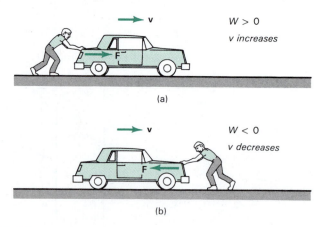

(a)

(b)

FIGURE 7.12

(a) The force acts in the same direction as the displacement and does positive work, causing the speed to increase. (b) The force acts opposite to the displacement and does negative work, causing the speed to decrease.

The work done on a particle by the net force equals the change in the kinetic energy of the particle.

In symbolic form the work-energy principle states

$$W = K_f - K_i \qquad (7.7)$$

W is the work done by the net force and K_f and K_i are the final and initial kinetic energies. We can derive Equation 7.7 most simply by considering the one-dimensional case. The work done by the net force is

$$W = \int_i^f F_{net} \, dx$$

The integral limits i and f are symbolic. They indicate arbitrary initial and final states of motion. To relate the work to the kinetic energy, we start with Newton's second law

$$F_{net} = m \frac{dv}{dt}$$

The work done over the displacement dx is

$$dW = F_{net} \, dx = m \left(\frac{dv}{dt} \right) dx$$

The displacement dx is related to the velocity by

$$dx = v \, dt$$

Replacing dx by $v \, dt$ we can express the work dW as

$$dW = m \left(\frac{dv}{dt} \right) v \, dt = mv \, dv$$

The last step lets us perform the integration

$$W = \int_i^f F_{net} \, dx = \int_{v_i}^{v_f} mv \, dv = \frac{1}{2} mv^2 \Big|_{v_i}^{v_f}$$
$$= \frac{1}{2} mv_f^2 - \frac{1}{2} mv_i^2$$

This confirms our statement of the work-energy principle

$$W = K_f - K_i$$
$$= \frac{1}{2} mv_f^2 - \frac{1}{2} mv_i^2 \qquad (7.7)$$

Examples 7–9 illustrate different applications of the work-energy principle.

EXAMPLE 7

Galileo's Cannonball

Galileo first deduced a relation for the speed of a cannonball that falls vertically from rest under the influence of gravity. Let's use the work-energy principle to deduce Galileo's result.

Consider a cannonball of mass m that starts from rest and falls vertically through a distance h. We assume that only the constant force of gravity, mg, acts on the cannonball. The work done by this constant force is mgh

$$W = mgh$$

The cannonball starts from rest and accelerates to a speed v. Its change in kinetic energy is

$$K_f - K_i = \tfrac{1}{2}mv^2 - 0$$

The work-energy principle gives

$$mgh = \tfrac{1}{2}mv^2$$

or

$$v = \sqrt{2gh}$$

which is Galileo's result.

For very large displacements the gravitational force cannot be treated as a constant. We must evaluate the gravitational work using Newton's inverse square law form of the gravitational force. The following example illustrates an application of the work-energy principle to such a situation.

EXAMPLE 8

Lunar Impact Speed

Let's reconsider Galileo's cannonball experiment. Imagine an astronaut, hovering high above the lunar surface. She drops a cannonball from an altitude equal to twice the moon's radius. We want to determine its impact speed. We can use the calculation performed in Example 4 to see that the work performed by gravity is

$$W = \frac{\tfrac{2}{3}GM_M m}{R_M}$$

M_M is the mass of the moon, m is the mass of the cannonball, and R_M is the lunar radius. The impact speed follows from the work-energy principle

$$\tfrac{1}{2}mv^2 = W = \frac{\tfrac{2}{3}GM_M m}{R_M}$$

or

$$v = \sqrt{\frac{\tfrac{4}{3}GM_M}{R_M}}$$

Inserting numerical values

$$M_M = 7.35 \times 10^{22}\ \text{kg} \qquad R_M = 1.74 \times 10^6\ \text{m}$$

we get

$$v = \sqrt{\frac{\tfrac{4}{3}(6.67 \times 10^{-11}\ \text{N·m}^2/\text{kg}^2)(7.35 \times 10^{22}\ \text{kg})}{1.74 \times 10^6\ \text{m}}}$$
$$= 1.94 \times 10^3\ \text{m/s}$$

This converts to 4,340 mi/h.

As a third example of the work-energy principle we consider a situation where work is performed by a linear spring force.

EXAMPLE 9

Driving into a Brick Wall

An automobile ($m = 10^3$ kg) is driven into a brick wall in a safety test. The bumper behaves like a spring ($k = 5 \times 10^6$ N/m), and is observed to compress a distance of 3.16 cm as the car is brought to rest. What was the initial speed of the automobile?

As the spring is being compressed, the spring force opposes the compression and so the work performed is negative.

$$W = \int_0^x (-kx)\, dx = -\tfrac{1}{2}kx^2$$
$$= -\tfrac{1}{2}(5 \times 10^6 \text{ N/m})(0.0316 \text{ m})^2 = -2.50 \times 10^3 \text{ J}$$

The final kinetic energy equals zero, and the initial kinetic energy is given by

$$K_i = \tfrac{1}{2}mv^2 = \tfrac{1}{2}(10^3 \text{ kg})v^2$$

Substituting these expressions into $W = K_f - K_i$, we get

$$-2.50 \times 10^3 \text{ J} = 0 - \tfrac{1}{2}(10^3 \text{ kg})v^2$$

This gives

$$v = \sqrt{\frac{2(2500 \text{ J})}{1000 \text{ kg}}}$$
$$= 2.24 \text{ m/s}$$

This converts to 5.00 mi/h.

We made use of Newton's second law in the derivation of the work-energy principle. It follows that results found using the work-energy principle can also be obtained by using Newton's second law. If so, why invent work and kinetic energy? Here are two reasons:

1. Calculations involving the work-energy principle are generally simpler than those using Newton's second law. The work-energy principle is a scalar relation, whereas Newton's second law is a vector relation.
2. The concepts of work and kinetic energy offer a new viewpoint, a new approach. As you continue your study of physics you will encounter many different forms of energy. The energy concept unifies many areas of physical science.

7.4
POWER

Often we are interested in *power*, which is defined as the *rate* at which work is done. If work ΔW is done in a time interval Δt, then the average power $\bar{P}$ during Δt is defined as

$$\bar{P} = \frac{\Delta W}{\Delta t} \tag{7.8}$$

The instantaneous power P is defined as

$$P = \frac{dW}{dt} \tag{7.9}$$

The SI unit of power is the watt. From Equation 7.9

$$1 \text{ watt} = 1 \text{ joule/second} \quad \text{or} \quad 1 \text{ W} = 1 \text{ J/s}$$

Another frequently used unit of power is the horsepower (hp). The conversion factor is:

$$1 \text{ hp} = 746 \text{ W}$$

Any power unit multiplied by a time unit is a valid unit of energy or work. For example, electric energy is usually measured in kilowatt hours (kWh).

EXAMPLE 10

Weight-lifter Power

A weight lifter executes a lift by raising a 200-kg barbell a distance of 2.2 m in a time of 0.82 s. Let's calculate the average power developed. The work performed by the weight lifter is

$$\Delta W = mgh$$

The average power is

$$\bar{P} = \frac{\Delta W}{\Delta t} = \frac{200 \text{ kg } (9.80 \text{ m/s}^2)(2.2 \text{ m})}{0.82 \text{ s}} = 5300 \text{ J/s}$$
$$= 5.3 \text{ kW}$$

This is approximately 7 horsepower.

Another useful expression relates power to force and velocity. We have

$$\Delta W = \mathbf{F} \cdot \Delta \mathbf{s}$$

Dividing by Δt and taking the limit as $\Delta t \to 0$ gives

$$\frac{dW}{dt} = \mathbf{F} \cdot \frac{d\mathbf{s}}{dt}$$

Recognizing dW/dt and $d\mathbf{s}/dt$ as the power and velocity we can write

$$P = \mathbf{F} \cdot \mathbf{v} \tag{7.10}$$

EXAMPLE 11

Auto Friction

When an automobile moves with constant velocity the power developed by the engine is used to overcome the frictional forces exerted by the air and the road. If the power developed in an engine is 50 hp, what total frictional force acts on that car at 55 mph (24.6 m/s)? One horsepower equals 746 W, so

$$P = 50 \text{ hp} \cdot 746 \text{ W/hp} = 3.73 \times 10^4 \text{ W}$$

The force follows from Equation 7.10

$$F = \frac{P}{v} = \frac{3.73 \times 10^4 \text{ W}}{24.6 \text{ m/s}} = 1520 \text{ N}$$

This equals 341 pounds and illustrates why automotive engineers strive for designs that reduce air resistance.

7.5
SIMPLE MACHINES

Machines normally serve one of two functions. They either increase the applied force, or they increase the applied velocity. For example, by using a claw hammer a carpenter easily pulls nails that he couldn't move with his bare hands. The claw hammer increases the applied force. The velocity of a tennis racket in the hand of a tennis player is much greater than the velocity of the player's hand. The tennis racket increases the applied velocity. In a simple machine, the increase or multiplication of force or velocity is governed by a work principle, which we will now derive from the work-energy principle.

We assume that the kinetic energy of the machine does not change, and disregard internal friction. It follows from the work-energy principle that the net work done on the machine is zero. It is useful to distinguish between the positive and negative parts of the work done on the machine. Let W_i (i for input) be the positive work, and $-W_o$ (o for output) be the negative work. Because zero net work is done on the machine, work input equals work output, or

$$W_i = W_o \tag{7.11}$$

In most simple machines we deal with a single input force and a single output force. Schematically we can think of the machine as a **work transmitter**. Any work done on the machine by the environment, $+W_i$, is transmitted, or returned, to the environment as work done by the machine, $+W_o$. We can express Equation 7.11 as

$$F_i s_i = F_o s_o \tag{7.12}$$

where F_i is the input force, s_i is the input displacement, F_o is the output force, and s_o is the output displacement. Equations 7.11 and 7.12 can be read as *work in equals work out.*

EXAMPLE 12

A Pry Bar

A 3-ft rod is used as a pry bar (Figure 7.13). An output force of 1400 lb is required to lift the edge of a crate. The rod pivots about a point 1 inch from the end of the rod. What input force must the worker apply?

From Equation 7.12 the input force is

$$F_i = F_o \left(\frac{s_o}{s_i} \right)$$

From Figure 7.13 we see that the ratio s_o/s_i equals the ratio of the lever arms L_o/L_i. Thus,

$$F_i = F_o \left(\frac{L_o}{L_i} \right) = 1400 \text{ lb} \left(\frac{1 \text{ in}}{35 \text{ in}} \right) = 40 \text{ lb}$$

The worker is able to lift a 1400-pound crate by applying only a 40-pound force.

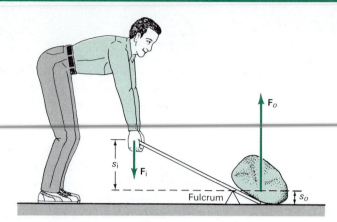

FIGURE 7.13
A person using a rod as a pry bar.

The input work and the output work in Example 12 are equal and they are done in equal times. This makes the rate at which input work is done equal to the rate at which output work is performed. This in turn means that the input power equals the output power. We can express this equality as

$$F_i v_i = F_o v_o \tag{7.13}$$

where v_i and v_o are the input and output velocities.

The **mechanical advantage** (*MA*) of a machine is defined as the force multiplication factor F_o/F_i,

$$MA = \frac{F_o}{F_i} \tag{7.14}$$

Using Equations 7.12 and 7.13 we can express the factor F_o/F_i as

$$\frac{F_o}{F_i} = \frac{s_i}{s_o} = \frac{v_i}{v_o} \tag{7.15}$$

provided friction is disregarded. In the presence of friction the *MA* will be less than the ratios s_i/s_o and v_i/v_o. To obtain the *MA* of a machine we must measure the ratio of output force to input force. Equation 7.15 shows that force multiplication is attained at the expense of velocity (or distance), and vice versa.

EXAMPLE 13
Wheel and Axle

For centuries the wheel and axle combination has been used to raise water from a well. This system (Figure 7.14) consists of a drum of radius r_1 and a handle that moves in a circle of radius r_2. As the handle is turned, the rope supporting the bucket wraps around the drum. What is the mechanical advantage of this system?

The input force is the force applied on the handle and the output force is the force applied by the drum. For one complete revolution of the drum and handle

$$s_i = 2\pi r_2$$

$$s_o = 2\pi r_1$$

Equation 7.15 gives the *MA*

$$MA = \frac{s_i}{s_o} = \frac{2\pi r_2}{2\pi r_1}$$

$$= \frac{r_2}{r_1}$$

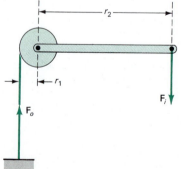

FIGURE 7.14
The wheel and axle combination is a force multiplier. In this version of the wheel and axle combination, the input force is applied to the handle (the wheel). The output force is the tension in the rope wound around the drum (the axle).

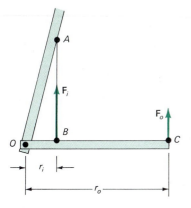

FIGURE 7.15

The elbow joint is a lever that multiplies velocity. The elbow lever can be modeled as two sticks hinged at *O*, with an input force supplied by the biceps muscle at *B*. The output force is exerted by the hand at *C*.

Suppose $MA = 10$ for the wheel and axle. This means a worker is able to raise a 40-lb bucket of water by applying a 4-lb force. But, gaining a mechanical advantage always has drawbacks. Because the well is 40 ft deep, the worker's hand must move through a distance of 400 ft. Furthermore, the bucket moves at a speed of only one-tenth that of the worker's hand.

A **lever** is a simple machine that consists of a rigid body pivoted on a fixed point called a **fulcrum**. The pry bar considered in Example 12 is a simple lever. The human body can also be seen as a complex system of levers, some acting to multiply force and others acting to multiply velocity.

EXAMPLE 14

The Elbow as a Lever

A schematic representation of the human arm is shown in Figure 7.15. The shoulder, elbow, and hand are points *A*, *O*, and *C*, respectively. The forearm pivots about *O* as a result of the force, $\mathbf{F}_i$, exerted by the muscle at point *B*. Let's calculate the *MA* for this system.

When the forearm, *OC*, rotates about *O*, the distances moved by the points *B* and *C* are proportional to their radii. Thus we have

$$\frac{s_i}{s_o} = \frac{r_i}{r_o}$$

so that

$$MA = \frac{r_i}{r_o}$$

You can estimate this ratio by looking at your own elbow. The *MA* of the elbow is less than 1. Suppose the *MA* of the elbow equals 1/8. The muscle in your forearm must exert a 400-lb force if your hands are to support 50 lb. Clearly the arrangement of your elbow is designed as a velocity multiplier, not as a force multiplier. (Note that forces applied at the pivot do not require consideration, since no work is done by them.)

You can demonstrate the velocity multiplier action of the elbow as follows: Raise your elbow to eye level and place a small object on it (Figure 7.16). Then sweep your hand toward the object and catch it. The velocity lever action of the elbow and the inertia of the object make this an easy trick to perform.

FIGURE 7.16

An object placed on the elbow can be caught by sweeping the hand toward it.

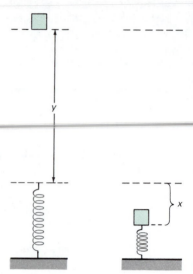

FIGURE 7.17

WORKED PROBLEM

A mass m falls freely through a distance y, and then strikes a vertical spring having a spring constant k (Figure 7.17). The spring is compressed a distance x as the mass is momentarily brought to rest. Use the work-energy principle to show that (a) the speed is reduced to zero for

$$x = \frac{mg}{k} + \sqrt{\left(\frac{mg}{k}\right)^2 + \frac{2mgy}{k}}$$

and (b) the maximum speed attained is given by

$$v_{\max} = \sqrt{2gy + \frac{mg^2}{k}}$$

Solution

(a) As the mass descends, gravity does positive work on it. The force exerted on the mass by the spring opposes its motion and does negative work. The total distance the mass falls in coming to rest is $y + x$, so the work done by gravity is

$$W_g = mg(y + x)$$

The work done by the spring is

$$W_s = -\tfrac{1}{2}kx^2$$

The total work done on the mass is the sum of W_g and W_s.

The mass started from rest ($K_i = 0$) and came to rest momentarily ($K_f = 0$), so the change in kinetic energy is also zero.

The work-energy principle states that the work done equals the change in kinetic energy. Equating net work to change in kinetic energy we have

$$mgy + mgx - \tfrac{1}{2}kx^2 = 0$$

We are seeking x, the maximum compression of the spring. The solutions of the quadratic equation are,

$$x = \frac{mg}{k} \pm \sqrt{\left(\frac{mg}{k}\right)^2 + \frac{2mgy}{k}}$$

Algebraically x can be positive or negative. The positive root corresponds to compression of the spring. Rejecting the negative root leaves the desired result for part (a).

(b) You might think that the maximum speed occurs at the moment the mass strikes the spring. If so, your first impression is wrong. Once contact is made, the net force on the mass is

$$F = mg - kz$$

where z is the compression of the spring. As long as the gravitational force is greater than the spring force, the mass continues to accelerate and its speed increases. The speed reaches a maximum value when the acceleration drops to zero. Zero acceleration means zero net force. Hence, the speed is a maximum when

$$z = \frac{mg}{k}$$

The work-energy principle gives the relation between speed (v) and compression (z):

$$K = \tfrac{1}{2}mv^2 = W_{\text{net}} = mg(y + z) - \tfrac{1}{2}kz^2$$

Setting $z = mg/k$ we find the desired result for the maximum speed

$$v_{max} = \sqrt{2gy + \frac{mg^2}{k}}$$

In physics there is seldom only one way to solve a problem. If you can figure out more than one solution, your understanding will increase. In the present instance, you can also find the maximum speed by observing that the kinetic energy, K, is a maximum when the speed, v, is a maximum. The maximum of K occurs when the derivative dK/dz equals zero. Here's an opportunity for you to apply your calculus. Using the expression for K above, set dK/dz equal to zero and double-check to see that it produces $z = mg/k$.

EXERCISES

7.1   Work Done by a Constant Force

A.   A raindrop ($m = 3.35 \times 10^{-5}$ kg) falls vertically at constant speed under the influence of the forces of gravity and drag (Figure 1). In falling through 100 m, what work is done by (a) gravity, (b) drag?

FIGURE 1

B.   A 4-kg bowling ball skids horizontally for the first 4 m down a lane. Forces acting on the ball include the normal force **N**, the frictional force **f**, and the gravitational force m**g**, as shown in Figure 2. If $\mu_k = 0.1$, what work is done by (a) gravity, (b) the normal force, (c) friction?

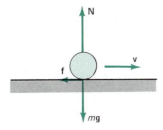

FIGURE 2

C.   Two 4-kg objects are connected by a string passing over a frictionless pulley of negligible mass (Figure 3). The vertically moving object falls at constant speed through a distance of 2 m. Determine the work done

by (a) gravity on m_1, (b) gravity on m_2, (c) tension on m_1, (d) tension on m_2, (e) friction on m_1.

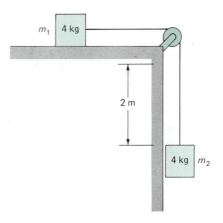

FIGURE 3

7.2   Work Done by a Variable Force

D.   In Figure 4 the force exerted by gravity on a 945-kg object is plotted versus its position. To achieve an upward displacement of 8 m, determine the work done by gravity (a) graphically (by measuring the area), (b) analytically.

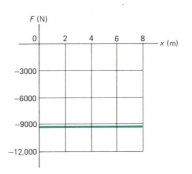

E. A child on a pogo stick bounces up and down, repeatedly reaching the same height. Consider the work done on the child by gravity from the peak of one jump to the next peak. Is it positive, negative, or zero? How would your answer change if the height of the jumps steadily decreased?

F. The free end of a linear spring ($k = 10$ N/m) is pulled 10 cm from its equilibrium position. Calculate the work done by the spring in the displacement from (a) $x = 0$ to $x = 5$ cm, (b) $x = 5$ cm to $x = 10$ cm.

G. The force exerted by a spring is plotted versus the change in its length in Figure 5. Use the area under the curve to determine the work done from (a) $x = -0.3$ m to $x = -0.1$ m, (b) $x = -0.2$ m to $x = +0.1$ m, (c) $x = -0.3$ m to $x = 0$, (d) $x = -0.2$ m to $x = +0.2$ m.

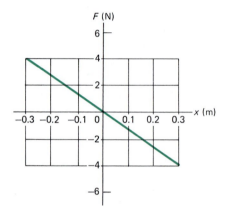

FIGURE 5

H. An athlete exercises his legs on a machine by repeatedly bending and then extending them against a large resistive force. The force he exerts is represented in Figure 6. From the graph determine the work done by his legs during one extension.

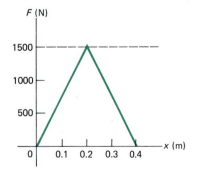

FIGURE 6

I. Imagine a very large frictionless plane tangent to the surface of the earth at point C, as shown in Figure 7.

The points A and B are equidistant from C. Consider the block as it slides along the plane. Will the work done by gravity be positive, zero, or negative if the block moves from (a) A to C, (b) C to B, (c) A to B?

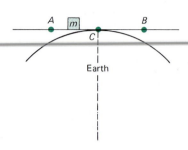

FIGURE 7

J. A charged particle moving in the vicinity of a magnet experiences a magnetic force that is perpendicular to its velocity. Explain why the work done on the charge by the magnetic force is zero.

K. A 50-N weight is taken on a closed rectangular path $1 \rightarrow 2 \rightarrow 3 \rightarrow 4 \rightarrow 1$, as shown in Figure 8. The height of the rectangle is 2 m and its width is 3 m. Obtain numerical values for the work done by gravity for a displacement of the weight from (a) 1 to 2, (b) 2 to 3, (c) 3 to 4, (d) 4 to 1. (e) What is the net work done by gravity around the complete path?

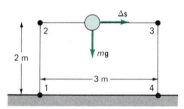

FIGURE 8

7.3 Kinetic Energy and the Work-Energy Principle

L. According to kinetic theory (Section 22.3), a typical gas molecule in thermal equilibrium at room temperature has a kinetic energy $K = 6 \times 10^{-21}$ J, regardless of mass. Estimate the speed at room temperature of (a) a hydrogen molecule ($m = 3.3 \times 10^{-27}$ kg), (b) a xenon atom ($m = 2.0 \times 10^{-25}$ kg).

M. An astronaut fires a distress signal packet ($m = 0.5$ kg) upward from the lunar surface. The position of the packet is described by

$$x = v_0 t + \tfrac{1}{2}at^2$$

with $v_0 = +32.4$ m/s and $a = -1.62$ m/s^2. Calculate (a) the velocity of the packet at $t = 10$ s, 20 s, and 30 s, (b) the kinetic energy of the packet at the same times.

N. A 50-gram mass is at rest on a horizontal frictionless air track. This mass is held against a linear spring compressed a distance of 10 cm. The mass is released, allowing the spring to expand toward its relaxed position. Determine the velocity of the mass at the moment the spring is relaxed ($k = 100$ N/m).

P. Determine the stopping distance for a skier with a speed of 20 m/s (Figure 9). Assume $\mu_k = 0.18$ and $\theta = 5°$.

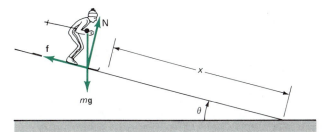

FIGURE 9

7.4 Power

Q. How many joules of energy are used by a homeowner whose electric bill is $65.00 and who pays $0.10 per kilowatt hour?

R. A gymnast ($m = 80$ kg) pulls himself up a 6-m rope in 6 s. What power does he develop if he moves at a constant velocity? Express your result in kW and in hp.

7.5 Simple Machines

S. A man ($m = 80$ kg) raises himself at constant velocity by pulling downward on a rope that passes around a fixed pulley overhead and is tied around his waist. (a) How much work is done by the man if a 10-m length of the rope passes through his hands? (b) What power is expended if 4 s are required?

T. What force F is required to lift the weight using the block and tackle pulley arrangement shown in Figure 10? (Disregard friction.)

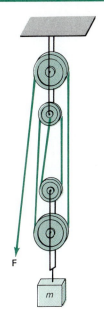

FIGURE 10

U. An external force $F = 50$ N raises a mass $m = 12$ kg (Figure 11) at constant velocity. Neglecting friction, calculate r_2 if $r_1 = 0.1$ m.

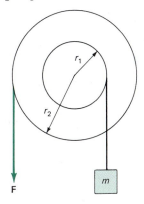

FIGURE 11

PROBLEMS

1. A group of students pushes a stalled car ($m = 1000$ kg) up a 3° incline for a distance of 100 m measured parallel to the incline. How much work is done by (a) the normal force during the trip, (b) gravity during the trip, (c) gravity as the car retraces its path?

2. The force acting on a particle is shown in Figure 12 for the interval -4 m $\leq x \leq +4$ m. Determine the work done on the particle for displacements from (a) $x = -4$ m to $x = -2$ m, (b) $x = -4$ m to $x = 0$ m, (c) $x = 0$ to $x = +4$ m, (d) $x = -4$ m to $x = +4$ m, (e) $x = -2$ m to $x = +2$ m.

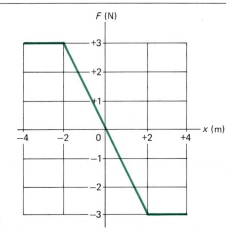

FIGURE 12

3. The repulsive force exerted on an alpha particle by a gold nucleus is given by

$$F = \frac{A}{r^2}$$

where F is in newtons, r is in meters, and $A = +3.65 \times 10^{-26}$ N·m². This function is plotted in Figure 13. Determine the work done over the interval from $r = 4 \times 10^{-14}$ m to $r = 12 \times 10^{-14}$ m by evaluating the work integral.

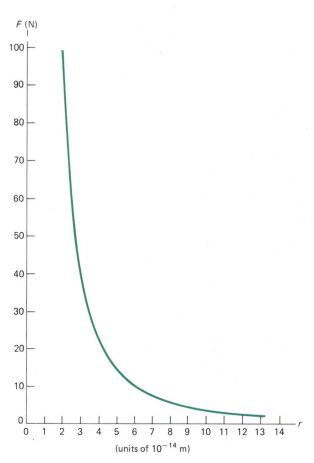

F (N)

(units of 10^{-14} m)

FIGURE 13

4. For small displacements, the horizontal force on a pendulum may be approximated by the expression $F = -m(g/L)x$, where x is the horizontal displacement, m is the mass of the bob, and L is the length of the pendulum. Calculate the work done by this force as the pendulum bob changes from position $x = x_0$ to position $x = 0$.

5. In diatomic molecules, the constituent atoms exert attractive forces at large distances and repulsive forces

at short distances. The Lennard-Jones force is a good approximation for many molecules and is given by

$$F = F_0\left[2\left(\frac{\sigma}{r}\right)^{13} - \left(\frac{\sigma}{r}\right)^{7}\right]$$

where r is the center-to-center distance of the atoms in the molecule and σ is a length parameter. For an oxygen molecule $F_0 = 9.6 \times 10^{-11}$ N and $\sigma = 3.5 \times 10^{-10}$ m. Determine the work done by this force from $r = 4 \times 10^{-10}$ m to $r = 9 \times 10^{-10}$ m by evaluating the work integral.

6. When a certain spring is stretched a distance x it exerts a restoring force represented by

$$F = -kx + k'x^3$$

with $k = 10$ N/m and $k' = 100$ N/m³. Calculate the work done by this force when the spring is stretched by 0.1 m.

7. In Olympic competition, an athlete throws a 16-lb hammer as far as possible horizontally. Assume a 45° launch angle and an 80-m throw (near the World Record). Determine the initial (a) speed, (b) kinetic energy of the hammer. For comparison, the kinetic energy of a runner ranges from approximately 1000 J in the marathon to 4000 J in a sprint.

8. In two dimensions a spring force may be represented by the equation

$$\mathbf{F} = -k(\mathbf{i}x + \mathbf{j}y)$$

with x and y in meters and $k = 5$ N/m. Evaluate the path integral for the work done by this force by using $d\mathbf{s} = \mathbf{i}\,dx + \mathbf{j}\,dy$. Consider the three paths shown in Figure 14: (a) $(0, 0) \rightarrow (0, 1) \rightarrow (1, 1)$, (b) $(0, 0) \rightarrow (1, 0) \rightarrow (1, 1)$, and (c) $(0, 0) \rightarrow (1, 1)$ along the straight line $x = y$.

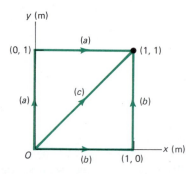

y (m)

FIGURE 14

9. A linear spring ($k = 19.6$ N/m) hanging vertically has an equilibrium length of 0.25 m (Figure 15). A 100-gram particle is carefully attached to the spring and then allowed to drop. As the particle falls the spring increases in length from 0.25 m to 0.35 m. The particle is brought to rest momentarily when the spring attains

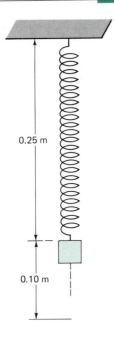

0.25 m

0.10 m

FIGURE 15

the 0.35 m length, and subsequently moves upward as the spring contracts. Calculate the work done on the particle during the fall by (a) gravity, (b) the spring, (c) the net force.

10. With a 100-gram particle attached, the equilibrium length of the vertical spring in Problem 9 differs from its original equilibrium length of 0.25 m (Figure 16).

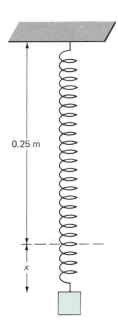

0.25 m

x

FIGURE 16

(a) Calculate this new equilibrium length. Note that the force exerted on the spring is up if the spring is longer than 0.25 m. Now let the particle be pushed 0.05 m above the new equilibrium position determined in (a) and then released. It returns to and moves through the new equilibrium position. Calculate the work done on the particle for the first 0.05 m of this downward move by (b) gravity, (c) the spring, (d) the net force. (e) Calculate the kinetic energy of the particle as it moves through its new equilibrium position.

11. A 50-gram yo-yo moves in a vertical circle 1 m in radius (Figure 17). When the yo-yo is at point A, the string tension equals the weight of the yo-yo. (a) Calculate the yo-yo's speed at point A. (b) Use the work-energy principle to determine its speed at point B. (c) If the string breaks when the yo-yo is at point A, determine the yo-yo's speed when it arrives at the elevation of point B.

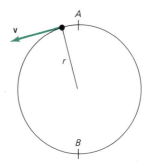

FIGURE 17

12. The force F exerted by one oxygen atom on another as a function of their separation is represented by the Lennard-Jones force (see equation in Problem 5). (a) Determine the change in kinetic energy of one oxygen atom as two oxygen atoms move from an infinite distance to a separation of 4.35×10^{-10} m. (Assume the two atoms share the kinetic energy equally.) (b) Using $m = 2.70 \times 10^{-26}$ kg for the mass of an oxygen atom, calculate the speed of one of the atoms at a separation of 4.35×10^{-10} m, assuming the kinetic energy at infinity is negligible.

13. The force $F = A/r^2$ is exerted on an alpha particle with $A = 3.65 \times 10^{-26}$ N·m² and r in meters. Suppose the speed of the alpha particle at a great distance from the force center (nucleus) is 1.70×10^7 m/s. As the alpha particle approaches the nucleus, the repulsive force F does work on the alpha particle, reducing its speed. Either graphically (refer to Figure 13, page 152) or by integration, calculate the work done by F for motion of the alpha particle from $r \to \infty$ to $r = r_0$. Use the work-energy principle to obtain the alpha-particle speed when r_0 equals (a) 10×10^{-14} m, (b) 8×10^{-14} m, (c) 6×10^{-14} m. (d) Determine the value of r_0 for which the alpha-particle speed equals zero.

14. A simple pendulum of mass m moves along a circular arc of radius R in a vertical plane (Figure 18). (a) Prove that in a downward swing from the angle θ to $0°$ the work done by gravity is $mgR(1 - \cos \theta)$. (b) For the same swing, prove that the work done by the tension **T** equals zero. (c) Use the work-energy principle to determine the speed of the pendulum bob at the bottom when the bob is released from rest at $\theta = 90°$. (d) Compare the answer to part (c) with the speed acquired by the same bob in falling from rest through a vertical distance R.

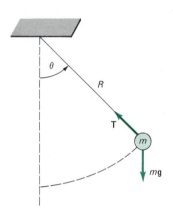

FIGURE 18

15. A 1000-kg automobile undergoes constant acceleration from rest to 25 m/s in 10 s. (a) Obtain the speed as a function of time. (b) Use $P = Fv$ to calculate the power developed by the net force at $t = 4$ s. (c) Calculate the kinetic energy as a function of time. (d) Use the work-energy principle to determine the work done by the net force during the first 8 s.

16. A lever 2 m long is used to lift one edge of a 9000-N commercial refrigerator (Figure 19). The maximum available external force F is 500 N. (a) Assuming the lever supports half the weight, how far from the point A should the fulcrum be placed? (b) For what fulcrum distance from A will the $MA = 49$?

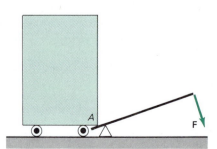

FIGURE 19

17. To operate one of the generators in the Grand Coulee Dam in Washington 7.24×10^8 W of mechanical power are required. This power is provided by gravity as it performs work on the water while it falls 87 m to the generator. The kinetic energy acquired during the fall is given up turning the generator. (a) Prove that the available water power is mgh/t where m is the mass of water falling through the height h during the time t. (b) Calculate the flow rate in kilograms per second. (c) Determine the volume of water needed to power this generator for one day. (d) If the water for one day were stored in a circular lake 10 m deep, what would the radius of the lake be?

18. Power windmills turn in response to the force of high-velocity drag. For a sphere, $F_D \sim r^2 v^2$, where r is the radius of the sphere and v is the fluid speed. The power developed, $P = F_D v$, is proportional to $r^2 v^3$. The power developed by a windmill can be expressed as $P = ar^2 v^3$, where r is the windmill radius, v is the wind speed, and $a = 2$ W·s^3/m^5. For a home windmill with $r = 1.5$ m, calculate the power delivered to the generator (this representation ignores windmill efficiency—which is about 25%) if (a) $v = 8$ m/s, (b) $v = 24$ m/s. For comparison, a typical home needs about 3 kW of electric power.

19. A jackscrew is part of the automobile jack, a once-common device used to lift a car while changing a tire. Each revolution of the screw raises the screw a distance equal to its pitch (0.5 cm in Figure 20). If the handle of the automobile jack moves in a circle of 25 cm radius, then (a) determine the MA for this machine. (b) What force is required (neglecting friction) to raise a 1000-kg car?

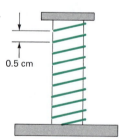

FIGURE 20

20. Three identical wooden dowels are at rest on a horizontal surface. Each has a weight of 0.82 N and a diameter of 2.0 cm. A child lifts one dowel and places it in the trough formed by the other two, giving the arrangement shown in Figure 21 (end view). Determine the work done on the dowel by gravity.

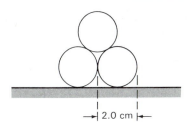

FIGURE 21

21. A variable force acting in the $+x$-direction is given by

$$F = 200(1 - bx^2)^{1/2} \text{ N} \qquad b = \tfrac{1}{4} \text{ m}^{-2}$$

Calculate the work performed by this force over the displacement from $x = 0$ to $x = 2$ meters.

22. A hydrodynamic force is described by the equation

$$\mathbf{F} = -k(\mathbf{i}y + \mathbf{j}b)$$

with y in meters, $b = 1$ m, and $k = 3$ N/m. Using $d\mathbf{s} = \mathbf{i}\, dx + \mathbf{j}\, dy$, evaluate the path integral for the work done by this force along the two paths shown in Figure 22: (a) $(0, 0) \rightarrow (0, 2) \rightarrow (3, 2)$, (b) $(0, 0) \rightarrow (3, 0) \rightarrow (3, 2)$. Note that the x-component of the force is proportional to y.

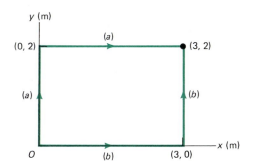

FIGURE 22

23. Two men lift a uniform 12-kg board. The board remains horizontal and rises 0.8 m. They hold the board (4 m long) at the points A and B (Figure 23), and exert constant forces F_1 and F_2. Determine the work done by each man, and the work done by gravity.

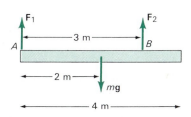

FIGURE 23

24. A shot (7.27 kg) is launched 2.13 m above the ground with a speed of 13.4 m/s (Figure 24). It rises to a maximum height of 4.11 m before falling to the ground. Determine the work done by gravity during (a) the rise, (b) the subsequent fall from its maximum height to impact. Use the work-energy principle to obtain the speed at (c) the highest point, (d) at impact.

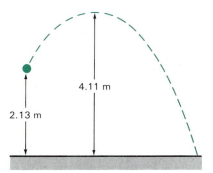

FIGURE 24

25. The 3-kg mass of an Atwood's machine is held against the floor in the position shown in Figure 25, and then released. (a) Determine the work done by gravity on the 3-kg mass during the subsequent 5-m displacement. (b) Determine the impact speed of the 15-kg mass. (Disregard friction.)

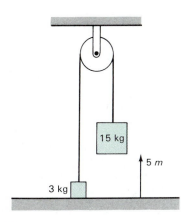

FIGURE 25

26. Water falls over a dam of height H meters at a rate of R kilograms/second. (a) Show that the power available from the water is

$$P = RgH$$

where g is the acceleration of gravity. (b) Each hydroelectric unit at the Grand Coulee Dam discharges water at a rate of 8.5×10^5 kg/s from a height of 87 m. The power developed by the falling water is converted to electric power with an efficiency of 85%. How much electric power is produced by each hydroelectric unit?

27. If a steady wind of speed v impinges on a wind turbine having a blade length d, the maximum power delivered to the turbine is

$$P = \left(\frac{\pi}{8}\right)\rho d^2 v^3$$

where ρ is the density of air. (a) Derive this result by assuming that the blade completely fills a circle of diameter d. Consider that in a time t all of the air within a cylinder of length vt and diameter d will strike the blade. Use this idea to derive an expression for the *rate* at which kinetic energy reaches the turbine. (b) The speed of a fairly strong wind is about 8 m/s and the density of air is about 1.1 kg/m^3. If a wind turbine having a blade diameter of 5 m converts 25% of the wind power reaching it into electric power, how much electric power is produced?

28. A tidal electric power plant is a form of a hydroelectric power plant. Water is captured in a basin when the tide changes from low to high tide. When the tidal conditions reverse, some of the kinetic energy of the water is converted to electric energy by a turbine. The power available from the outrushing water is given by

$$P = \tfrac{1}{2}RgH$$

where H is the maximum depth of the water in the basin. (a) The power expression for the tidal system is identical to that for a hydroelectric system except for the factor of $\frac{1}{2}$ (See Problem 26). What is the origin of the factor $\frac{1}{2}$ in the tidal power expression? (b) The change from high tide to low tide takes 6 hours. This 6 hour time determines the rate at which the tidal basin can empty. Suppose that the tidal power is to be 10^9 watts (roughly that of a large coal-burning electric power plant) and the maximum depth in the tidal basin is 8 meters. What area is required for the tidal basin? If the surface of the tidal basin were square, what would be the length of a side?

29. The force required to stretch a nonlinear spring by x meters is given by

$$F = bx^3 \qquad b = 6000 \text{ N/m}^3$$

Determine the work performed in stretching the spring from (a) $x = 0$ to $x = 0.1$ m, (b) $x = 0.1$ m to $x = 0.2$ m.

30. A 65-kg sprinter accelerates from 0 to 5 m/s in a time of 1.2 s. (a) Assuming her acceleration is a constant, calculate the maximum power she develops over the 1.2 s interval. (b) Verify that

$$\int P\, dt = \text{kinetic energy acquired}$$

CHAPTER 8

CONSERVATION OF ENERGY

8.1
ENERGY AND ENERGY CONSERVATION

In physics, energy is defined as the capacity to perform work. In Chapter 7 we saw that a particle possessing kinetic energy is capable of performing work—that is, converting kinetic energy into work. In this chapter we enlarge the concept of energy to include a new form of energy, **potential energy**. A system possessing potential energy is also capable of performing work.

The law of **conservation of energy** is one of the most fruitful results to emerge from Newtonian mechanics. Since its beginning, the law has been extended to include thermal energy, chemical energy, electromagnetic energy, and nuclear energy. The word *conservation* often means the wise use of resources. In physics, however, the conservation of a physical quantity means that it *does not change with time*.

Conservative and Nonconservative Forces

Path integrals representing work are used to classify forces as *conservative* or *nonconservative*. If the work done by a force between any two points is the same for *all paths* between the points, then the force is conservative.

The gravitational force is a good example of a conservative force. When you move from the first floor to the second floor of a building, the force of gravity acting on you does work. This work depends on your weight and your vertical displacement, not on your path between floors. Therefore, the path integral representing the work done by gravity is the same if you take the stairs or if you take the elevator.

Frictional forces are nonconservative. The work done by a frictional force does depend on the path. For example, suppose you get in a car and drive 20 miles west from point A to point B (Figure 8.1). The frictional forces acting on the car are always directed opposite to its motion. Hence, the work done by friction over the route $A \rightarrow B$ is negative. Imagine that from point B you drive north 20 miles to a point C, turn around, and return to point B. Over the round trip from B to C and back to B the frictional forces do additional negative work. The total work over the route $A \rightarrow B \rightarrow C \rightarrow B$ is clearly different from that of the work done over the first route $A \rightarrow B$. Both paths take you from A to B but the work done by friction is not the same for the two paths. This fact identifies friction as a nonconservative force.

Because the focus of this chapter is potential energy and the law of conservation of energy, and because potential energy can be defined only for conservative forces, we now restrict ourselves to conservative forces.

8.2
POTENTIAL ENERGY

You can think of potential energy as energy that is stored in a system by virtue of its position or configuration. For example, a raised weight has gravitational potential energy. If the weight falls it can convert its potential energy into work.

Let's look at another example of potential energy. Suppose you compress the spring of a toy dart gun by inserting the dart. The compressed spring now

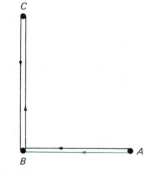

FIGURE 8.1

Two paths that begin at A and end at B. The work done by friction along $A \rightarrow B \rightarrow C \rightarrow B$ is different from the work done along $A \rightarrow B$.

has **elastic** potential energy. If you pull the trigger, the spring uncoils. The elastic potential energy is converted into work and then into the kinetic energy of the dart.

For any conservative force we can define a corresponding potential energy. For a conservative force, the potential energy *difference* between two configurations is defined by the statement

Difference in potential energy = −Work done by the force

For a conservative force **F** this definition of potential energy difference is expressed symbolically by

$$U_f - U_i = -\int_i^f \mathbf{F} \cdot d\mathbf{s} \tag{8.1}$$

where U_f and U_i denote the potential energy at the positions f and i. The letters i and f are symbols that represent arbitrary initial and final configurations. Note that Equation 8.1 defines only the *difference* in potential energy. This means we can freely choose the $U = 0$ configuration or position.

Let's illustrate these ideas with specific forces. We start by evaluating the potential energy difference associated with a linear spring force.

Linear Spring Potential Energy

In Figure 8.2 the force exerted by a linear spring stretched a distance x is

$$F = -kx$$

The difference in potential energy follows from the definition in Equation 8.1:

$$U_f - U_i = -\int_{x_i}^{x_f} (-kx)\, dx = \tfrac{1}{2}kx_f{}^2 - \tfrac{1}{2}kx_i{}^2$$

We observe that this relation is satisfied by choosing

$$U = \tfrac{1}{2}kx^2 \tag{8.2}$$

With this form of U, the unstretched spring $(x = 0)$ becomes the $U = 0$ configuration. Figure 8.3 shows a graph of $U = \tfrac{1}{2}kx^2$. Note that the spring stores potential energy in both the stretched $(x > 0)$ and the compressed $(x < 0)$ configurations.

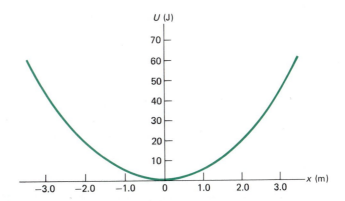

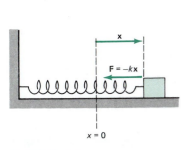

FIGURE 8.2

The linear spring exerts a force opposing the displacement.

FIGURE 8.3

The potential energy of a linear spring, $U = \tfrac{1}{2}kx^2$, is graphed versus the spring extension. The value of the spring constant is $k = 10$ N/m.

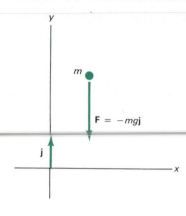

FIGURE 8.4

A constant gravitational force acts in the negative y-direction. The vector $\mathbf{j}$ is a unit vector in the positive y-direction.

The gravitational force is conservative, so we can define a gravitational potential energy. We do this first for the case where the gravitational force may be treated as a constant.

Gravitational Potential Energy: Constant Gravitational Force

In Figure 8.4 we take the positive y-axis upward. The constant gravitational force acting on a mass m can be expressed as

$$F = -mg\,\mathbf{j}$$

For an arbitrary displacement

$$d\mathbf{s} = \mathbf{i}\,dx + \mathbf{j}\,dy + \mathbf{k}\,dz$$

we have

$$\mathbf{F} \cdot d\mathbf{s} = -mg\,dy$$

Using the definition in Equation 8.1 we can calculate the difference in potential energy between the levels y_i and y_f.

$$U_f - U_i = -\int_i^f \mathbf{F} \cdot d\mathbf{s} = mg \int_{y_i}^{y_f} dy$$

This gives

$$U_f - U_i = mgy_f - mgy_i$$

This equation is satisfied by the relation

$$U = mgy \tag{8.3}$$

The level $y = 0$ corresponds to $U = 0$. Figure 8.5 shows a graph of $U = mgy$ for an 80-kg mass near the surface of the earth ($g = 9.80$ m/s^2). Note that U is negative for $y < 0$. This is a consequence of our choice that $U = 0$ at $y = 0$. If the mass is below the $y = 0$ level then we would have to do a positive amount of work on it to raise it to the $U = 0$ level. When you use $U = mgy$

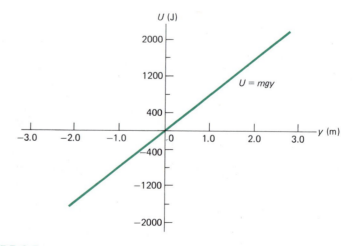

FIGURE 8.5

The potential energy $U = mgy$ of an 80-kg mass is shown versus height y near the surface of the earth.

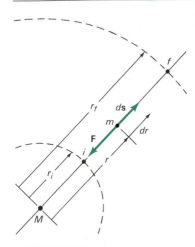

FIGURE 8.6

The path of integration used to evaluate the change in gravitational potential energy is radially away from M, from point i to point f.

in problems you can choose the $y = 0$ level. Often, a particular choice will simplify the algebraic solution of a problem.

When the gravitational force acts over large displacements we cannot treat the force as a constant. If we take account of the inverse square form of the gravitational force, the gravitational potential energy is no longer described by Equation 8.3.

Gravitational Potential Energy: Universal Gravitational Force

The magnitude of the gravitational force acting between point masses M and m separated by a distance r is

$$F = \frac{GMm}{r^2}$$

Figure 8.6 shows the path along which we carry out the integral that defines the potential energy difference. The force exerted on m by M acts opposite to the displacement over the path from i to f.

$$\mathbf{F} \cdot d\mathbf{s} = -F\,dr = \frac{-GMm\,dr}{r^2}$$

The difference in gravitational potential energy between the positions r_i and r_f is

$$U_f - U_i = -\int_i^f \mathbf{F} \cdot d\mathbf{s} = GMm \int_{r_i}^{r_f} \frac{dr}{r^2}$$

Carrying out the integration gives

$$U_f - U_i = \frac{-GMm}{r_f} + \frac{GMm}{r_i}$$

This equation is satisfied by the relation

$$U = \frac{-GMm}{r} \qquad (8.4)$$

Figure 8.7 shows a graph of $U = -GMm/r$ versus r for $m = 80$ kg and $M = 5.98 \times 10^{24}$ kg (mass of the earth). Note that U is negative for all finite values of r. This is a consequence of our choice that $U = 0$ when r is infinite. Note also that U increases toward zero as r increases.

8.3
THE LAW OF CONSERVATION OF ENERGY

Our understanding of energy is widened when we consider potential energy and kinetic energy together. Our starting point is the work-energy principle:

$$\int_i^f \mathbf{F} \cdot d\mathbf{s} = W = K_f - K_i \qquad (8.5)$$

Suppose that a conservative force, $\mathbf{F}$, acts on a particle. Then, Equation 8.1 shows that the work W is related to the potential energy change by

$$W = U_i - U_f \qquad (8.6)$$

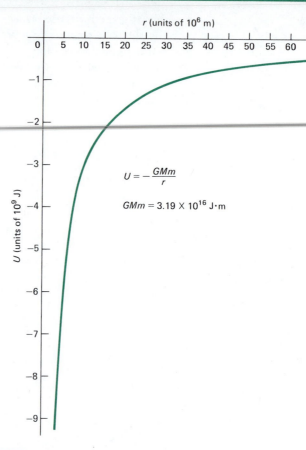

FIGURE 8.7

The gravitational potential energy $U = -GMm/r$ between an 80-kg object and the earth is shown versus their separation r.

Using this expression for W in Equation 8.5 gives

$$K_i + U_i = K_f + U_f \qquad (8.7)$$

This result shows that the sum of the kinetic energy and the potential energy $(K + U)$ does not change—it is therefore a *conserved* quantity. The sum $K + U$ is called the **mechanical energy** of the system. Equation 8.7 represents the law of conservation of mechanical energy, which can be stated as follows:

> When only conservative forces act, the sum of the kinetic and potential energies is conserved.

If the mechanical energy of the system is conserved, kinetic energy and potential energy change in such a way that their sum is constant. Kinetic energy is converted to potential energy and vice versa.

Let's illustrate the law of conservation of energy for each of the three potential energies developed in Section 8.2.

EXAMPLE 1

Dart Gun Muzzle Speed

A toy dart ($m = 30$ g) is propelled by a linear spring ($k = 800$ N/m). If the spring is compressed 5 cm and then released, what is the muzzle speed of the dart?

The only force that performs work on the dart is the conservative spring force. The total mechanical energy of the spring and dart is

$$K + U = \tfrac{1}{2}mv^2 + \tfrac{1}{2}kx^2$$

Initially, the dart is at rest ($K_i = 0$) and the spring is compressed. The dart is launched at the point where the spring potential energy is zero ($U_f = 0$). With $K_i = 0$ and $U_f = 0$, the conservation of mechanical energy reduces to

$$U_i = K_f$$

This expresses the transformation of the elastic potential energy of the spring into the kinetic energy of the dart. Inserting numerical values we have

$$\tfrac{1}{2}(800 \text{ N/m})(0.05 \text{ m})^2 = \tfrac{1}{2}(0.030 \text{ kg})v_f{}^2$$

This gives the muzzle speed of the dart

$$v_f = \sqrt{\frac{(800 \text{ N/m})(0.05 \text{ m})^2}{0.030 \text{ kg}}}$$

$$= 8.16 \text{ m/s}$$

In Chapter 7 we used the work-energy principle to derive Galileo's result for the speed of an object that falls from rest through a vertical distance h. The conservation of mechanical energy also leads to the same result. Equation 8.3 gives the potential energy, $U = mgy$. Here we take $y_i = h$ and $y_f = 0$, so that $U_i = mgh$ and $U_f = 0$. With $K_i = 0$ and $U_f = 0$, the energy conservation relation reduces to

$$U_i = K_f$$

or

$$mgh = \tfrac{1}{2}mv_f{}^2$$

This gives Galileo's result

$$v_f = \sqrt{2gh} \tag{8.8}$$

This equation gives the speed achieved when the object falls from rest vertically through a distance h. However, if the object had slid down a frictionless incline through a *vertical* distance h (Figure 8.8), the speed still would be given by Equation 8.8. Horizontal displacements do not result in any change in potential energy.

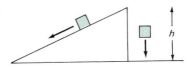

FIGURE 8.8

The vertical displacement of the object determines its speed. If the object falls vertically through the distance h it reaches a speed $v_f = \sqrt{2gh}$. It reaches the same speed when it slides along the frictionless incline.

EXAMPLE 2

Loop-the-Loop

Figure 8.9 shows a small block that slides without friction along an incline and then around a circular loop of radius R. The block will maintain contact with the loop provided it starts from a sufficiently great height h above the base. If h is too small the block will break contact with the loop. We want to determine the *minimum* value of h needed to keep the block in contact with the surface throughout the loop.

We determine the energy at the start of the slide ($y_i = h$; $v_i = 0$)

$$U_i = mgh \qquad K_i = 0$$

and at the top of the loop ($y_f = 2R$; $v_f = v$)

$$U_f = mg \cdot 2R \qquad K_f = \tfrac{1}{2}mv^2$$

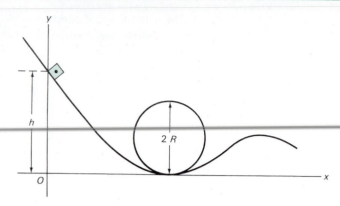

FIGURE 8.9

The block slides along a frictionless track that carries it around the circular loop.

The energy conservation relation

$$K_i + U_i = K_f + U_f$$

becomes

$$0 + mgh = \tfrac{1}{2}mv^2 + mg \cdot 2R$$

This gives

$$v^2 = 2g(h - 2R) \tag{8.9}$$

At this point you might be tempted to conclude that $h = 2R$ is the minimum height required. This is wrong because the block cannot reach the top of the loop with zero speed. (If it did, it would then fall vertically downward along the diameter of the loop.) In order to reach the top of the loop the block must travel with a certain minimum speed.

We can deduce this minimum speed by using Newton's second law. At the top of the loop two forces act on the block: its weight, mg, and the contact force, N (Figure 8.10). Both forces are directed toward the center of the loop. The acceleration of the block is also directed toward the center of the loop. With the block moving at a speed v in a circle of radius R this radial acceleration of the block is v^2/R. Newton's second law takes the form

$$N + mg = \frac{mv^2}{R} \tag{8.10}$$

The minimum speed required for the block to reach the top of the loop corresponds to the limit where N drops to zero. When N drops to zero the block just breaks contact with the loop. With $N = 0$, Equation 8.10 shows that the minimum speed is given by

$$v_{\min}^2 = gR \tag{8.11}$$

Using Equation 8.9 then gives the minimum value of h needed to achieve this minimum speed. Thus,

$$2g(h_{\min} - 2R) = gR$$

or

$$h_{\min} = \frac{5R}{2} \tag{8.12}$$

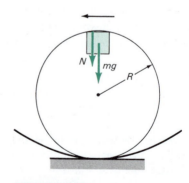

FIGURE 8.10

At the top of the loop the block experiences two forces, its weight mg and the contact force N. Both forces are directed toward the center of the loop and give rise to the radial acceleration v^2/R.

Many decades before our astronauts landed on the moon, Jules Verne wrote a tale called *De la Terre à la Lune* (*From the Earth to the Moon*) about a voyage to the moon and back. Verne's fictitious spaceship was launched vertically upward out of a huge cannon buried in the ground. Disregarding air resistance, let's find the minimum launch velocity that a projectile requires in order to escape the earth's gravity. By escape we mean that the projectile velocity and kinetic energy approach zero as $r \to \infty$.

EXAMPLE 3

Escape from Earth

Take the initial point of the spaceship to be at the earth's surface ($r_i = R_E$) and the final point at infinity ($r_f \to \infty$), and let v_i and v_f be the corresponding projectile speeds. We have

$$K_i = \tfrac{1}{2}mv_i^2 \qquad \text{and} \qquad K_f = \tfrac{1}{2}mv_f^2$$

From Equation 8.4 we have for the gravitational potential energies

$$U_i = \frac{-GM_Em}{R_E}$$

$$U_f = \lim_{r \to \infty} \frac{-GM_Em}{r} = 0$$

The law of conservation of energy requires

$$K_i + U_i = K_f + U_f$$

or

$$\tfrac{1}{2}mv_i^2 - \frac{GM_Em}{R_E} = \tfrac{1}{2}mv_f^2$$

Solving this equation for the launch speed, we get

$$v_i = \sqrt{\frac{2GM_E}{R_E} + v_f^2}$$

The minimum launch speed is called the **escape speed**, and it corresponds to $v_f = 0$. Using

$$g = \frac{GM_E}{R_E^2}$$

we can express the escape speed as

$$v_{\text{escape}} = \sqrt{2gR_E}$$

With

$$R_E = 6.38 \times 10^6 \text{ m}$$

$$g = 9.80 \text{ m/s}^2$$

we find

$$v_{\text{escape}} = \sqrt{2(9.80 \text{ m/s}^2)(6.38 \times 10^6 \text{ m})} = 1.12 \times 10^4 \text{ m/s}$$
$$= 11.2 \text{ km/s}$$

This is slightly more than 25,000 mph.

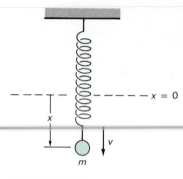

FIGURE 8.11

The mass is released from $x = 0$ and acquires kinetic energy. The total potential energy of the system is the sum of a positive elastic potential energy and a negative gravitational potential energy.

If more than one conservative force acts on a body, then U is the sum of the corresponding potential energies. The following example illustrates energy conservation when gravity and a linear spring force act simultaneously.

EXAMPLE 4

Combined Elastic and Gravitational Potential Energy

A 120-gram mass is attached to the free end of a linear spring ($k = 40$ N/m) that hangs vertically. The mass is held with the spring in the relaxed position ($x = 0$), and then dropped (Figure 8.11). We want to determine the maximum speed of the falling mass.

As the mass falls it acquires kinetic energy ($\frac{1}{2}mv^2$). The spring is stretched and stores elastic potential energy ($\frac{1}{2}kx^2$). The gravitational potential energy (mgx) is negative when the mass is below the initial position.

The total mechanical energy can be written as

$$E = \tfrac{1}{2}mv^2 + \tfrac{1}{2}kx^2 + mgx$$

At the moment the mass is dropped, $v = 0$ and $x = 0$, so the total energy is zero. Because energy is conserved, $E = 0$ for all configurations. It follows that

$$v^2 = -2gx - \left(\frac{k}{m}\right)x^2$$

Setting $d(v^2)/dx$ equal to zero and solving for x gives the value of x for which v^2 (and thus v) is a maximum.

$$\frac{d(v^2)}{dx} = -2g - 2\left(\frac{k}{m}\right)x = 0$$

This gives

$$x = \frac{-mg}{k} = \frac{-(0.120 \text{ kg})(9.80 \text{ m/s}^2)}{(40 \text{ N/m})} = -0.0294 \text{ m}$$

$$= -2.94 \text{ cm}$$

Note that this is the value of x that gives zero net force on the mass ($mg + kx = 0$). This is expected because zero force implies zero acceleration, which is achieved when v reaches a maximum. The maximum value of v follows by inserting $x = -mg/k$ into the equation for v^2. The result is

$$v_{max} = \sqrt{\frac{mg^2}{k}} = \sqrt{\frac{(0.120 \text{ kg})(9.80 \text{ m/s}^2)^2}{(40 \text{ N/m})}}$$

$$= 0.537 \text{ m/s}$$

8.4

FORCE FROM POTENTIAL ENERGY

In Section 8.2 we saw how to start with a given force and determine the corresponding potential energy. Now we want to learn how to reverse this procedure, that is, how to determine the force corresponding to a given potential energy.

The change in potential energy is defined by Equation 8.1

$$U_f - U_i = -\int_i^f \mathbf{F} \cdot d\mathbf{s}$$

The differential form of this equation is

$$dU = -\mathbf{F} \cdot d\mathbf{s} \qquad (8.13)$$

For the case where $\mathbf{F}$ lies along the x-axis we can write

$$\mathbf{F} = \mathbf{i}F_x$$

$$d\mathbf{s} = \mathbf{i}\, dx$$

in which case Equation 8.13 becomes

$$dU = -F_x\, dx$$

This result shows that we can determine F_x by *differentiating* U.

$$F_x = -\frac{dU}{dx} \qquad (8.14)$$

To check this relation, consider the linear spring. We found

$$U = \tfrac{1}{2}kx^2$$

We readily verify that $-dU/dx$ gives the linear spring force $(-kx)$

$$-\frac{dU}{dx} = -kx$$

In general, the force is not directed along the x-axis, and U may depend on all three coordinates. In this case the derivative in Equation 8.14 becomes a *partial derivative*

$$F_x = -\frac{\partial U}{\partial x} \qquad (8.15)$$

The y- and z-components of $\mathbf{F}$ are given by the corresponding partial derivatives.

$$F_y = -\frac{\partial U}{\partial y} \qquad (8.16)$$

$$F_z = -\frac{\partial U}{\partial z} \qquad (8.17)$$

For example, for a constant gravitational force we found the potential energy

$$U = mgy$$

The force, as given by Equation 8.16, is

$$F_y = -\frac{\partial U}{\partial y} = -mg$$

In terms of the unit vector $\mathbf{j}$, the gravitational force is

$$\mathbf{F} = \mathbf{j}F_y = -mg\,\mathbf{j}$$

This is the form of $\mathbf{F}$ we used in Section 8.2 to derive $U = mgy$. Bear in mind that potential energy is a scalar quantity, whereas force is a vector. In many instances the simplest way to determine the force is to differentiate the potential energy function.

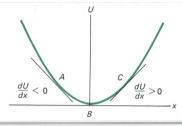

FIGURE 8.12

The potential energy $U = \frac{1}{2}kx^2$ is shown for a linear spring. At the equilibrium position B, the slope is zero and the force is zero. When the spring is compressed, $x < 0$, the slope is negative, and the force is positive. When the spring is stretched, $x > 0$, the slope is positive, and the force is negative.

8.5
ENERGY GRAPHS

When U depends on only one variable (x, for instance) we can draw a graph of U versus x. Such a graph is useful in analyzing the motion of the system. For example, the relation $F_x = -dU/dx$ shows that the force is the negative slope of the U versus x graph. At positions where the slope is zero the force vanishes. These are equilibrium configurations. Figure 8.12 shows a graph of U versus x for the linear spring

$$U = \tfrac{1}{2}kx^2$$

The slope is zero at $x = 0$, which corresponds to the equilibrium configuration (zero force) in which the spring is relaxed.

Recall that in one dimension the direction of the force vector is determined by its sign. Thus for $F > 0$ the force is in the positive x-direction, and for $F < 0$ the force is in the negative x-direction. In Figure 8.12, the slope dU/dx is positive for $x > 0$ and so the force $F = -dU/dx$ is negative. Thus, for $x > 0$ the spring force acts in the negative x-direction. For $x < 0$ the slope is negative and F is positive, showing that the force acts in the positive x-direction. In general, the linear spring exerts a **restoring** force that urges any attached object toward the $x = 0$ equilibrium position. This means that $x = 0$ is a position of **stable** equilibrium.

For the linear spring, the slope is zero at a *minimum* of the U versus x graph. In general, stable equilibrium configurations correspond to *minima* of the U versus x graph. Thus, in Figure 8.13 the positions x_1, x_3, and x_5 locate positions of stable equilibrium. Similar reasoning shows that *maxima* of the U versus x graph correspond to *unstable* equilibrium positions (positions x_2 and x_4). In Figure 8.13, the potential energy U is a maximum at x_2 and x_4. Any slight displacement away from x_2 or x_4 gives rise to forces that urge the system still further away.

The behavior of a ball rolling on a hilly surface provides an easy way to visualize the force-potential energy relationship. Suppose that the ball in Figure 8.14 is free to roll under the influence of gravity on a contour having the U versus x shape. At $x = x_A$, for example, the ball is urged to the right because the force is to the right. At $x = x_C$, the ball is urged to the left. The point $x = x_B$ is a position of stable equilibrium for the ball, whereas $x = x_D$ is a position of unstable equilibrium.

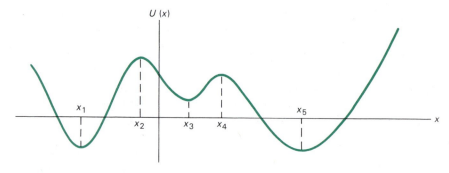

FIGURE 8.13

The potential energy minima at $x = x_1$, x_3, and x_5 are positions of stable equilibrium. The potential energy maxima at $x = x_2$ and x_4 are positions of unstable equilibrium.

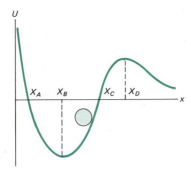

FIGURE 8.14

To remember the force-potential energy relation, imagine a ball free to move under the influence of gravity on a surface having the shape of the U versus x graph.

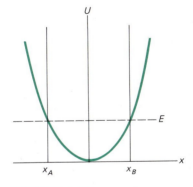

FIGURE 8.15

The heavy solid curve is the potential energy, U, of a particle whose total energy $E = K + U$ is constant. Motion is possible only between x_A and x_B, the range of positions for which the kinetic energy, K, is positive.

In addition to the location of the equilibrium points, much can be learned about the motion of a particle by considering an energy graph. Let's consider the case of a particle that moves along the x-axis under the influence of a conservative force. Its mechanical energy E is a constant

$$E = K + U = \text{constant} \tag{8.18}$$

Figure 8.15 shows a graph of U versus x for a particle that experiences a linear spring force. The dashed horizontal line identifies the constant total energy, E. The kinetic energy, $K = \frac{1}{2}mv^2$, must be positive or zero. From Equation 8.18 we see that we must have $E \geq U$ in order to have $K \geq 0$. Regions where $E < U$ correspond to negative kinetic energies, and therefore represent inaccessible regions on the graph. Regions where $E \geq U$ correspond to non-negative kinetic energies, and therefore represent accessible regions.

In Figure 8.15, for example, the kinetic energy is positive between the points $x = x_A$ and $x = x_B$. At $x = x_A$ and $x = x_B$ the kinetic energy is zero. When the particle reaches $x = x_A$ or $x = x_B$ it is momentarily at rest. Because F is a restoring force at both points, the particle acceleration is directed toward $x = 0$, the equilibrium point. If x is slightly less than x_B, for example, the particle is slowing down if moving to the right or speeding up if moving to the left. Its direction of motion changes at $x = x_B$. A similar argument holds for x_A. These points are called **turning points** of the trajectory.

Qualitatively we can see from the energy graph that the particle oscillates back and forth between the turning points at x_A and x_B. Because $K = E - U$, we can infer that the kinetic energy is a maximum when the potential energy is a minimum. In Figure 8.15 the kinetic energy is a maximum at $x = 0$, where the spring is relaxed. The range of positions where K is positive or zero defines an accessible region for particle motions. The turning points where $K = 0$ mark boundaries between accessible and inaccessible regions. Whether or not a region is inaccessible on a particular U versus x graph depends on the total energy E.

In Figure 8.16, for example, the region between x_B and x_C is inaccessible when the total energy is E_1. If a particle with energy E_1 starts in the region between x_A and x_B.it oscillates back and forth between the turning points at x_A and x_B. Likewise, if a particle with energy E_1 starts in the region between x_C and x_D, it oscillates back and forth between the turning points at x_C and

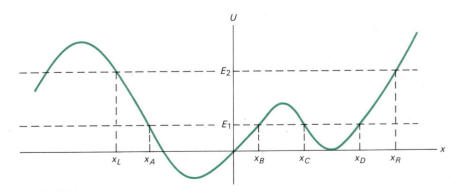

FIGURE 8.16

For a particle with a total energy E_1 the region between x_B and x_C is inaccessible. For a particle with a total energy E_2 the region between x_B and x_C is accessible; the particle oscillates between turning points at x_L and x_R.

x_D. The two accessible regions are separated by the inaccessible region between x_B and x_C. If the total energy is E_2 then the region between x_B and x_C is accessible, and so the particle oscillates between the turning points at x_L and x_R.

EXAMPLE 5
Turning Points

The potential energy of a particle free to move along the x-axis is given by

$$U = \left(\frac{x}{b}\right)^4 - 5\left(\frac{x}{b}\right)^2_j \text{ J} \qquad b = 1 \text{ m}$$

Figure 8.17 shows a graph of U versus x. Let's find the turning points for a particle whose total energy is $E = 36$ joules. The kinetic energy is

$$K = E - U = 36 - \left(\frac{x}{b}\right)^4 + 5\left(\frac{x}{b}\right)^2_j \text{ J}$$

At the turning points $K = 0$. This gives a quadratic equation for $(x/b)^2$

$$\left[\left(\frac{x}{b}\right)^2\right]^2 - 5\left(\frac{x}{b}\right)^2 - 36 = 0$$

The solution is

$$\left(\frac{x}{b}\right)^2 = \frac{5}{2} + \sqrt{\left(\frac{5}{2}\right)^2 + 36} = 9$$

With $b = 1$ m, the turning points are located at

$$x = \pm 3 \text{ m}$$

When the total energy is 36 joules the particle oscillates back and forth in the accessible region between $x = \pm 3$ m.

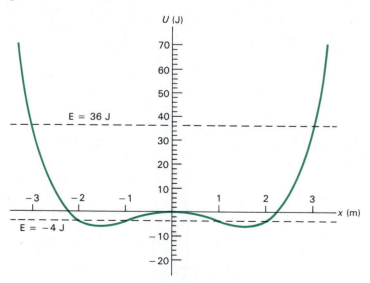

FIGURE 8.17
The locations of the turning points depend on the total energy.

Next, let's consider a particle with a total energy of -4 joules. As before, the turning points follow from the equation $K = 0$. This gives

$$\left(\frac{x}{b}\right)^4 - 5\left(\frac{x}{b}\right)^2 + 4 = 0$$

and

$$\left(\frac{x}{b}\right)^2 = \frac{5}{2} \pm \sqrt{\left(\frac{5}{2}\right)^2 - 4} = \frac{5}{2} \pm \frac{3}{2}$$

There are four turning points. With $b = 1$ m, they are located at

$$x = \pm 2 \text{ m}$$
$$x = \pm 1 \text{ m}$$

As Figure 8.17 shows, the two accessible regions are separated by an inaccessible region between $x = \pm 1$ m.

It is remarkable how much information can be obtained from a study of the potential energy of a system. Its scalar nature makes potential energy easier to use than force, which is a vector concept. In advanced classical and quantum mechanics, potential energy is used almost exclusively, and force is deemphasized.

WORKED PROBLEM

A child slides without friction from a height h along a curved water slide (Figure 8.18). She is launched from a height $h/5$ into the pool. In terms of h and the launch angle θ, determine the maximum height reached by the child.

Solution

This problem can be solved by using the energy conservation principle and concepts developed for projectile motion. Our interest in position is a key to recognizing the appropriateness of energy conservation. Had position versus time or velocity versus time been called for, we would need a different approach.

Since friction is not involved, mechanical energy is conserved. For any positions along the slide, $K + U$ is constant. At the top of the slide

$$K_1 = 0 \qquad U_1 = mgh$$

When the child is launched from the slide her gravitational potential energy has decreased to $U_2 = mgh/5$. It follows from energy conservation that her kinetic energy ($\frac{1}{2}mv^2$) is

$$K_2 = \frac{4}{5} mgh$$

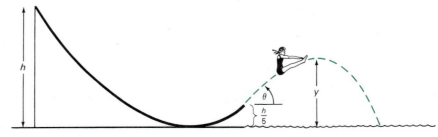

FIGURE 8.18

As the child leaves the slide, the horizontal component of her velocity is

$$v_x = v \cos \theta$$

where v is her speed. Because gravity is the only force acting on the airborne child, v_x remains constant. Furthermore, her total mechanical energy also remains constant. As she rises, her kinetic energy decreases and her gravitational potential energy increases. At the peak of her trajectory, the vertical component of her velocity is zero and her kinetic energy is

$$K_3 = \tfrac{1}{2}mv_x{}^2 = \tfrac{1}{2}m(v \cos \theta)^2 = K_2 \cos^2 \theta = \frac{4}{5} mgh \cos^2 \theta$$

Her potential energy at the peak is

$$U_3 = mgy$$

Mechanical energy conservation ($K_3 + U_3 = K_1 + U_1$) gives

$$mgy + \frac{4}{5} mgh \cos^2 \theta = mgh$$

Solving for y yields,

$$y = h\left(1 - \frac{4}{5} \cos^2 \theta\right)$$

To check the result, note that it has the correct dimensions of length and that it reduces to expected results in two special cases: with $\theta = 0$, $y = h/5$, and with $\theta = 90°$, $y = h$.

EXERCISES

8.1   Energy and Energy Conservation

A.   A conservative force performs -2.2 J of work over the path $A \rightarrow B \rightarrow C$ in Figure 1. How much work is done by the force (a) along the path $A \rightarrow D \rightarrow E \rightarrow C$, (b) along the path $C \rightarrow E \rightarrow B \rightarrow A$?

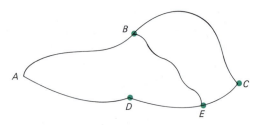

FIGURE 1

B.   A crane lifts a 450-kg load straight upward through 22 m, moves the load sideways for a distance of 10 m, lowers the load 22 m, and then moves the load sideways 10 m to its original position. Determine the work done by the force of gravity during the entire trip.

C.   A constant force of $+275$ N acts in the $+x$-direction on an object moving along the x-axis. The position of the object varies with time according to the equation

$$x = -bt^2 + ct \qquad b = 3 \text{ m/s}^2 \qquad c = 6 \text{ m/s}$$

How much work is done by the constant force during the time interval $t = 1$ s to $t = 2$ s?

8.2   Potential Energy

D.   A linear spring is stretched 0.1 m in excess of its equilibrium length. The potential energy stored is 5 J. What is the spring constant?

E.   A pair of identical linear springs ($k = 500$ N/m) are connected. As seen in Figure 2, the springs are relaxed. Calculate the potential energy of the system if the point where they are connected is displaced 0.1 m from its equilibrium position (a) in the x-direction, (b) in the y-direction.

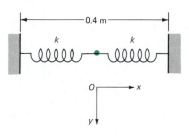

FIGURE 2

F. A 5-kg mass and a 10-kg mass are connected by an inextensible string of negligible mass (Figure 3). The connecting string passes over a pulley so that a rise by one mass is accompanied by an equal fall of the other. Calculate the change that occurs in the total potential energy of the system when the 10-kg mass falls 2 m.

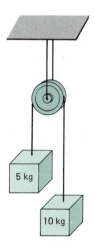

FIGURE 3

G. Cable cars ($m = 2000$ kg) that carry passengers between two mountain peaks are suspended from steel cables strung between the peaks. Pairs of cars operate together, so that while one travels up from A to B (Figure 4) the other travels down from B to A. If B is 1 km higher than A, calculate the change in potential energy (a) of the rising empty car in a trip from A to B, (b) of the falling empty car in a trip from B to A, (c) of the pair of empty cars in the trips defined in (a) and (b). (d) Twenty passengers ride from A to B while five ride from B to A. If $m = 70$ kg for each passenger, calculate the change in the potential energy of the system for the trip.

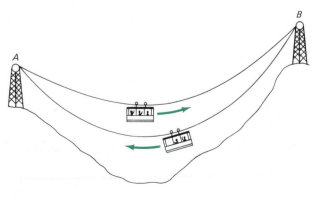

FIGURE 4

H. A swimmer experiences a drag force. Explain why no potential energy can be associated with this drag force.

8.3 The Law of Conservation of Energy

I. A block ($m = 2$ kg) collides with a horizontal linear spring ($k = 10$ N/m). The spring (Figure 5) is compressed a distance $x = 0.2$ m before bringing the block to rest. Determine the original speed of the block.

J. If the two masses in Exercise F start from rest, what is their speed at the moment the 10-kg mass has dropped 2 m?

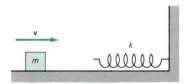

FIGURE 5

K. A pendulum bob (Figure 6) is dropped when the supporting wire ($L = 5$ m) is horizontal. Determine the speed of the bob when $\theta = 45°$.

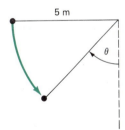

FIGURE 6

L. An object is whirled in a vertical circle 1 m in radius. The tension in the string is 6 times the weight of the object at the top and 12 times the weight of the object at the bottom of the swing. Determine the speed of the object at (a) the top and (b) the bottom of the swing.

M. A meteorite with negligible initial velocity starts to fall toward the earth from a great distance. What is the impact speed of the meteorite at the earth's surface? (Disregard atmospheric effects.)

N. A satellite of mass m is in a circular orbit of radius r about the earth. Use Newton's second law together with Equation 8.4 to prove that the kinetic energy and total energy of the satellite can be expressed in the form

$$K = \frac{\frac{1}{2}GM_Em}{r} \qquad E = \frac{-\frac{1}{2}GM_Em}{r}$$

P. A scientific satellite is sent radially toward the sun from a point on the earth's orbit. Assume that the satellite's initial speed is 30 km/s, and calculate its speed at the sun's surface. ($M_s = 1.99 \times 10^{30}$ kg; $R_s = 6.96 \times 10^8$ m.)

Q. A frictionless water slide is shaped as shown in Figure 7. A person starts from position A with a speed of 4 m/s. Determine the speed of the slider at the positions B, C, D, and E.

What force acts on the sky diver at (a) $x = 5000$ m, (b) $x = 10$ m?

T. A 10-kg block is free to slide on a fractionless plane elevated 40° above the horizontal (Figure 9). (a) Express the gravitational potential energy of the block in terms of its position s along the plane. (b) Use the derivative relation between force and potential energy to calculate the component of the gravitational force directed along the plane.

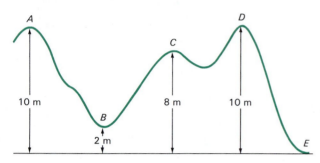

FIGURE 7

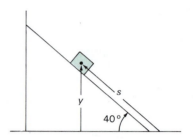

FIGURE 9

8.4 Force from Potential Energy

R. The potential energy of a particle is shown in Figure 8 as a function of r. (a) Estimate the value of r for which the attractive force acting on the particle is a maximum. (b) Estimate the maximum attractive force graphically. (c) Estimate the force at $r = 1$ m. Is it attractive or repulsive?

U. A particular nonlinear spring has an elastic potential energy $U = Cx^3$, which C is a constant. Find the restoring force exerted by this spring.

8.5 Energy Graphs

V. The graph in Figure 10 represents the potential energy of a particle versus x. (a) Give the intervals for which the force is positive. (b) List the points where the force is zero. (c) Give the intervals for which the force is negative. (d) List the points of stable equilibrium, (e) list the points of unstable equilibrium.

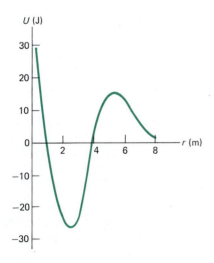

FIGURE 8

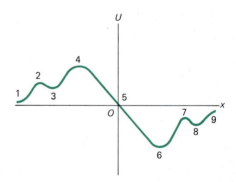

FIGURE 10

S. The potential energy of a sky diver is described by the equation

$$U = Ax + B \qquad A = 588 \text{ N} \qquad B = 3200 \text{ J}$$

W. Construct a graph of U versus θ for a pendulum (Figure 11) covering the interval $-250° < \theta < +250°$. Locate all points of equilibrium and determine whether they are stable or unstable. (Take $m = 10$ kg and $R = 2$ m.)

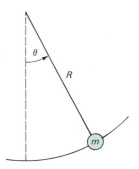

FIGURE 11

X. The potential energy of a particle satisfies the equation

$$U = C\left(\sqrt{1 + \left(\frac{x}{b}\right)^2} - 1\right)^2 \qquad \begin{array}{l} C = 0.500 \text{ J} \\ b = 1 \text{ m} \end{array}$$

(a) Prove that $x = 0$ is an equilibrium position. (b) Characterize the stability of this equilibrium position by plotting U versus x and examining the graph.

Y. The potential energy of a system is described by

$$U = Bx^4 - Cx^2 \qquad B = 1 \text{ J/m}^4 \qquad C = 4 \text{ J/m}^2$$

(a) Find the points of equilibrium. (b) Determine which of the points of equilibrium are stable and which are unstable.

Z. A particle subject to the potential energy in Figure 8 has a total energy $E = +10$ J. (a) Determine the turning points. (b) Give the intervals defining accessible and inaccessible regions of motion.

PROBLEMS

1. The planet Mongo has a mass of 7×10^{26} kg and a radius of 3×10^7 m. A projectile launched from the surface of Mongo travels radially outward, rising to an *altitude* of 6×10^7 m, where it comes to rest momentarily before it falls back. Determine the launch speed. (Do *not* assume a constant acceleration of gravity.)

2. A particle whose potential energy U is described by the equation given in Exercise X has a total energy $E = 1$ J. (a) What is the minimum value of U? (b) What is the maximum value of the kinetic energy K? (c) Calculate the turning points of the motion. (d) Determine K for $x = \sqrt{3}$ m.

3. The system of identical springs ($k = 1000$ N/m) and masses ($m_1 = m_2 = m_3 = 0.1$ kg) in Figure 12 is held at rest in the configuration shown, and abruptly released. (a) Assume that m_1 remains at rest after the release, and determine the speed of m_2 at the instant when all three angles are $120°$. When all angles are $120°$, all 3 springs are relaxed. (b) Determine a second

set of values for θ_1, θ_2, and θ_3 for which the three masses are at rest *momentarily*. Do this by inspection.

4. A child's pogo stick (Figure 13) stores energy in a spring ($k = 2.5 \times 10^4$ N/m). At the position A ($x_1 = -0.1$ m) the spring compression is a maximum, and the child is momentarily at rest. At position B ($x = 0$) the spring is relaxed, and the child is moving upward. At position C the child is again momentarily at rest at the top of the jump. Assume that the combined mass of the child and the pogo stick is 25 kg, and (a) calculate the total energy of the system if both potential energies are zero at $x = 0$. (b) Determine x_2. (c) Calculate the speed of the child at $x = 0$. (d) Determine the value of x for which the kinetic energy of the system is a maximum. (e) Obtain the child's maximum speed.

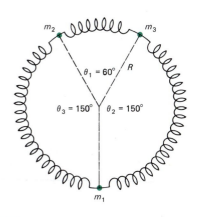

FIGURE 12

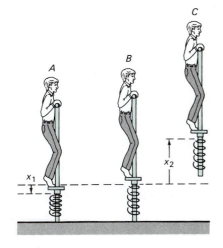

FIGURE 13

5. A particle of mass 3 kg moving in the $+x$-direction passes $x = 0$ at 6 m/s. It moves under the influence of a conservative force with a potential energy given by

$$U = A\left(\frac{x}{b}\right)^2 + C\left(\frac{x}{b}\right)^4$$

$$A = -\sqrt{40}\,\text{J} \qquad C = 1\,\text{J} \qquad b = 1\,\text{m}$$

Determine (a) the turning points, (b) the maximum speed attained by the particle.

6. A 15-kg mass (Figure 14) is attached to a vertical linear spring ($k = 1000$ N/m). It is held motionless with the spring in its relaxed position, and then dropped. How far will it drop before coming to rest momentarily?

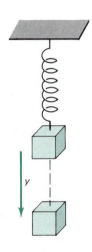

FIGURE 14

7. A rocket ($m = 1000$ kg) is launched from the earth's surface and moves completely out of the solar system. Consider only the earth-rocket system, and (a) prove that when the rocket passes a point a distance r from the center of the earth its gravitational potential energy has changed by

$$\Delta U = mgR_E\left(1 - \frac{R_E}{r}\right)$$

where $R_E = 6.38 \times 10^6$ m is the radius of the earth. (b) Calculate ΔU for the complete flight ($r = R_E$ to $r \to \infty$). (c) Determine the value of r for which the potential energy change is half of its final value.

8. An astronaut of mass m moves from a point P (at a distance r from the center of the earth) to a point Q (at a distance $r + y$ from the center of the earth) (see Figure 15). (a) Prove that the change in the gravitational potential energy of the astronaut equals

$$\Delta U = \frac{mgy(R_E/r)^2}{1 + y/r}$$

where g is the acceleration of gravity at the surface ($g = GM_E/R_E^2$). (b) Show that $\Delta U \approx mgy$ if $r = R_E$ and $y \ll r$.

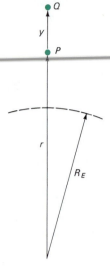

FIGURE 15

9. A 1000-kg satellite orbits the earth at an altitude of 100 km. It is desired to increase the altitude of the orbit to 200 km. How much energy must be added to the system to effect this change in altitude?

10. A weightless cord passing over a frictionless pulley has 3-kg and 7-kg masses attached to its ends. The masses start from rest with the 7-kg mass 1.2 m above the 3-kg mass. Calculate the speed of the masses when they pass.

11. A small mass starts from rest and slides along a frictionless cylindrical surface (Figure 16). If the mass starts from a position $\theta = 30°$ from the vertical, show that it leaves the surface when $\theta = 54.7°$.

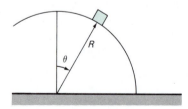

FIGURE 16

12. The effective potential energy of an electron near a proton may be written in the form

$$U = \frac{-A}{r} + \frac{B}{r^2}$$

(A and B are positive constants). Show that (a) the equilibrium separation is $r_{eq} = 2B/A$, (b) the potential energy at this separation is $U_{eq} = -A^2/4B$, (c) sketch U and explain why the equilibrium is stable.

13. Three identical springs are attached to three identical masses as shown in Figure 12. The system of springs and masses is free to move on a circle of radius R without friction. The springs are relaxed when the arrangement is symmetric. Determine the potential energy of the system as drawn. (Take $k = 1000$ N/m and $R = 0.1$ m.)

14. The mechanical energy of a particle moving along the x-axis is conserved. (a) Show that the time required for the particle to change positions from x_0 to x is

$$t = \int_{x_0}^{x} \frac{dx}{\sqrt{\left(\dfrac{2}{m}\right)[E - U]}}$$

where E is the mechanical energy and U is the potential energy. (b) Determine the time for a particle with a mass of 1 kg and a total energy of 2 J to move from $x_0 = 2$ m to $x = 3$ m, if its potential energy is given by

$$U = E\left[2\left(\frac{x}{b}\right) - \left(\frac{x}{b}\right)^2\right] \qquad b = 4m$$

15. (a) Using energy conservation principles, derive an expression for the escape velocity of a mass that starts from a position a distance H above the surface of the earth. (b) Imagine that a mass is in a "tree top" orbit (that is, its orbit radius is equal to the radius of the earth). Derive an expression for the speed of the mass in its orbit. (c) Determine the value of H for which the escape speed calculated in (a) equals the orbital speed found in (b). Express your answer in terms of the radius of the earth.

16 Two springs having equal spring constants, k, and negligible mass, are attached to a mass m as shown in Figure 17. Each spring has the relaxed length L. Show that at equilibrium the mass sags a distance y_{eq} given approximately by

$$y_{eq} = \left(\frac{mgL^2}{k}\right)^{1/3}$$

The approximation is valid provided the sag is small compared to L. Hint: Start with the expression for potential energy and assume $y \ll L$.

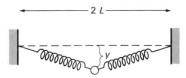

FIGURE 17

CHAPTER 9

CONSERVATION OF LINEAR MOMENTUM

9.1
LINEAR MOMENTUM

In this chapter we introduce and develop the concept of **linear momentum**. An important reason for adding linear momentum to our repertoire of physical variables is that it provides the basis for another conservation law, the **law of conservation of linear momentum**.

The linear momentum of a particle of mass m and velocity $\mathbf{v}$ is defined as the product $m\mathbf{v}$ and is denoted by the symbol $\mathbf{p}$.

$$\mathbf{p} = m\mathbf{v} \tag{9.1}$$

Let's see how linear momentum is related to Newton's second law.

Newton orginally formulated the second law in the form

$$\mathbf{F} = \frac{d(m\mathbf{v})}{dt} \tag{9.2}$$

If the mass m is constant

$$\frac{d(m\mathbf{v})}{dt} = m\frac{d\mathbf{v}}{dt} = m\mathbf{a}$$

and Equation 9.2 reduces to

$$\mathbf{F} = m\mathbf{a} \tag{9.3}$$

Because linear momentum is defined by $\mathbf{p} = m\mathbf{v}$, we see that Newton's original formulation can be written

$$\mathbf{F} = \frac{d\mathbf{p}}{dt} \tag{9.4}$$

This form of Newton's second law is generally valid, but Equation 9.3 is limited to situations in which the mass remains constant. If the mass varies, as it does in a rocket, for example, because fuel is burned and forcefully ejected, then $\mathbf{F} = m\mathbf{a}$ does not apply and we use $\mathbf{F} = d\mathbf{p}/dt$ instead.

Remember that linear momentum is a vector quantity. It is particularly important to keep this in mind when evaluating changes in linear momentum.

EXAMPLE 1

Superball Momentum Change

A superball with a mass of 60 grams is dropped from a height of 2 m. It rebounds to a height of 1.80 m (Figure 9.1). We want to evaluate the change in its linear momentum.

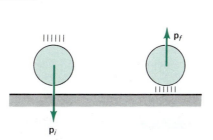

Just before impact   Just after impact   Linear momentum change

FIGURE 9.1

The linear momentum of the superball is changed by the impact. The change in linear momentum, $\mathbf{p}_f - \mathbf{p}_i$, is directed upward.

We disregard air resistance and calculate the linear momentum of the superball just before impact. Applying the law of conservation of energy to the fall we obtain

$$\tfrac{1}{2}mv_i{}^2 = mgh$$

Solving for mv_i we get

$$p_i = mv_i = m\sqrt{2gh} = +(0.060\ \text{kg})\sqrt{2(9.80\ \text{m/s}^2)(2\ \text{m})}$$
$$= +0.376\ \text{kg·m/s}$$

By choosing the positive root, we have taken *down* to be the *positive* direction for vectors.

The linear momentum just after the ball rebounds follows from the same energy conservation relation, using 1.80m for the height to which the ball rises.

$$p_f = mv_f = -(0.060\ \text{kg})\sqrt{2(9.80\ \text{m/s}^2)(1.80\ \text{m})}$$
$$= -0.356\ \text{kg·m/s}$$

The minus sign indicates the upward direction of p_f, opposite to our choice of the positive direction. The change in linear momentum resulting from the force exerted on the superball is

$$p_f - p_i = -0.356\ \text{kg·m/s} - 0.376\ \text{kg·m/s}$$
$$= -0.732\ \text{kg·m/s}$$

The minus sign indicates that the *change* in linear momentum is directed upward.

Average Force and Linear Impulse

When a baseball is struck by a bat the force lasts only a fraction of a second, increasing and then decreasing abruptly as suggested in Figure 9.2. The value of the force at any particular moment is not a good indicator of its effect on the ball. The net effect of the force is to change the linear momentum of the ball. The momentum change follows from the time integral of Newton's second law.

Let Δt denote the time interval over which the force acts and let p_f and p_i denote the linear momentum at the beginning and end of this time interval. Then

$$\int_0^{\Delta t} F\,dt = \int_{p_i}^{p_f} dp = p_f - p_i \tag{9.5}$$

The integral $\int_0^{\Delta t} F\,dt$ is called the **linear impulse** delivered by the force. *Geometrically*, the integral $\int_0^{\Delta t} F\,dt$ corresponds to the *area* under a graph of force versus time.

The average force, $\bar{F}$, is defined in terms of the momentum change

$$\bar{F} = \frac{p_f - p_i}{\Delta t} = \frac{1}{\Delta t}\int_0^{\Delta t} F\,dt \tag{9.6}$$

From its definition we can see that $\bar{F}$ is the value of a constant force that would deliver the same linear impulse as the variable force in the time Δt.

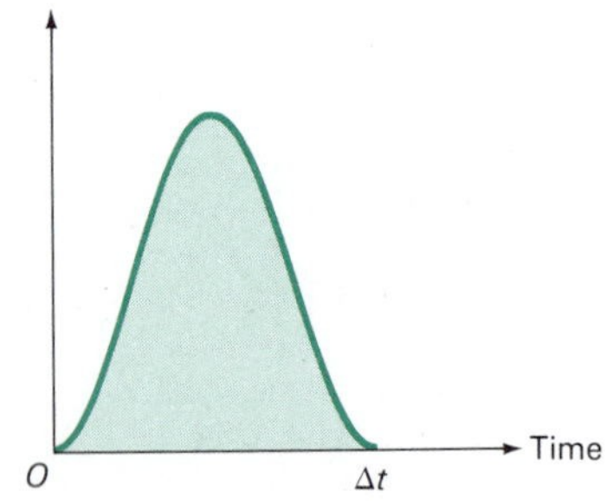

Force

→ Time

O Δt

FIGURE 9.2

Force versus time for an impulsive force. The linear impulse corresponds to the shaded area.

EXAMPLE 2

Linear Impulse and Average Force

A car is stopped for a traffic signal. When the light turns green, the car accelerates, increasing its speed from zero to 5.20 m/s over a time interval of 0.832 s. Let's evaluate the linear impulse and the average force a 70-kg passenger in the car experiences.

From Equation 9.5, the linear impulse equals the change in linear momentum,

$$\text{Linear impulse} = p_f - p_i = mv_f - mv_i$$
$$= 70 \text{ kg } (5.20 \text{ m/s}) - 70 \text{ kg } (0)$$
$$= 364 \text{ kg} \cdot \text{m/s}$$

The average force is given by Equation 9.6

$$\bar{F} = \frac{p_f - p_i}{\Delta t} = \frac{364 \text{ kg} \cdot \text{m/s}}{0.832 \text{ s}}$$
$$= 438 \text{ N}$$

The magnitude of this average force is slightly more than half the weight of the passenger.

Kinetic Energy–Linear Momentum Relationship

It is often useful to express the kinetic energy of a particle

$$K = \tfrac{1}{2}mv^2 \tag{9.7}$$

in terms of the magnitude of the linear momentum rather than speed. Replacing v by p/m gives

$$K = \frac{p^2}{2m} \tag{9.8}$$

This expression for kinetic energy is especially convenient in situations where linear momentum is conserved.

EXAMPLE 3

Superball Kinetic Energy

Let's illustrate Equation 9.8 by evaluating the kinetic energy of the superball in Example 1. The magnitude of the linear momentum of the superball as it rebounds from the floor is

$$p = 0.356 \text{ kg} \cdot \text{m/s}$$

So, its kinetic energy is

$$K = \frac{p^2}{2m} = \frac{(0.356 \text{ kg} \cdot \text{m/s})^2}{2 \cdot 0.060 \text{ kg}}$$
$$= 1.06 \text{ J}$$

9.2
THE LAW OF CONSERVATION OF LINEAR MOMENTUM

Remember that in physics we say that a quantity is conserved if it remains constant. For a single particle, linear momentum is conserved when zero net force acts. This follows from Newton's second law. Thus,

$$\mathbf{F} = \frac{d\mathbf{p}}{dt} = 0$$

implies **p** is a constant. For a single particle, linear momentum conservation amounts to a restatement of Newton's first law, the law of inertia. We can discover the power of the law of conservation of linear momentum by studying a system of two particles.

For example, consider two air hockey pucks that are free to move on a horizontal surface. The pucks collide and exert forces on each other (Figure 9.3). No other net force acts on either puck. If $\mathbf{f}_1$ and $\mathbf{f}_2$ denote the forces that the pucks exert on one another, then Newton's third law specifies that

$$\mathbf{f}_2 = -\mathbf{f}_1$$

We define the combined linear momentum, **p**, of the pucks to be the *vector sum* of the linear momenta of the two pucks,

$$\mathbf{p} = \mathbf{p}_1 + \mathbf{p}_2 \tag{9.9}$$

The rate of change of the total linear momentum is

$$\frac{d\mathbf{p}}{dt} = \frac{d\mathbf{p}_1}{dt} + \frac{d\mathbf{p}_2}{dt} = \mathbf{f}_1 + \mathbf{f}_2$$

We have used Newton's second law to relate $d\mathbf{p}_1/dt$ and $d\mathbf{p}_2/dt$ to the forces $\mathbf{f}_1$ and $\mathbf{f}_2$. But, as we have already observed, Newton's third law requires $\mathbf{f}_2 = -\mathbf{f}_1$. So,

$$\frac{d\mathbf{p}}{dt} = 0 \tag{9.10}$$

The time derivative of **p** is zero, showing that **p** is constant. We conclude that the total linear momentum of the two-particle system is conserved.

$$\mathbf{p} = \mathbf{p}_1 + \mathbf{p}_2 = \text{constant}$$

The linear momentum of each individual puck is changed by the collision. The conservation of linear momentum requires only that the vector sum of the linear momenta remains constant. If we know the linear momentum of both pucks before the collision and the linear momentum of one puck after the collision

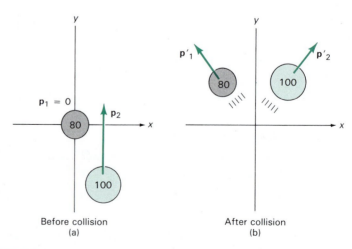

Before collision
(a)

After collision
(b)

FIGURE 9.3
Air puck linear momenta before and after they collide.

we can use the conservation principle to determine the linear momentum of the second puck after the collision.

EXAMPLE 4

Air Puck Linear Momentum and Velocity

Figure 9.3 shows a pair of air hockey pucks before and after they collide on a frictionless horizontal surface. Their masses are

$$m_1 = 80 \text{ grams} \qquad m_2 = 100 \text{ grams}$$

The 80-gram puck is at rest before the collision. The 100-gram puck is moving in the positive y-direction at a speed of 1.10 m/s before the collision

$$\mathbf{v}_1 = 0 \qquad \mathbf{v}_2 = 1.10\mathbf{j} \text{ m/s}$$

The linear momenta before the collision are

$$\mathbf{p}_1 = m_1\mathbf{v}_1 = 0 \qquad \mathbf{p}_2 = m_2\mathbf{v}_2 = 0.110\mathbf{j} \text{ kg·m/s}$$

After the collision the 100-gram puck is observed to move with a velocity

$$\mathbf{v}_2' = (0.300\mathbf{i} + 0.700\mathbf{j}) \text{ m/s}$$

Its linear momentum is

$$\mathbf{p}_2' = m_2\mathbf{v}_2' = (0.030\mathbf{i} + 0.070\mathbf{j}) \text{ kg·m/s}$$

We want to apply the conservation of linear momentum to determine the linear momentum and velocity of the 80-gram puck after the collision. The total linear momentum of the two-puck system before the collision is $\mathbf{p}_1 + \mathbf{p}_2$. Because the 80-gram puck is at rest the total linear momentum is carried entirely by the 100-gram puck,

$$\mathbf{p}_1 + \mathbf{p}_2 = 0.110\mathbf{j} \text{ kg·m/s}$$

Let $\mathbf{p}_1'$ denote the linear momentum of the 80-gram puck after the collision. Linear momentum conservation is expressed by

$$\mathbf{p}_1' + \mathbf{p}_2' = \mathbf{p}_1 + \mathbf{p}_2$$

Figure 9.4 shows the graphical expression of linear momentum conservation. Solving for $\mathbf{p}_1'$ gives

$$\mathbf{p}_1' = \mathbf{p}_1 + \mathbf{p}_2 - \mathbf{p}_2' = (0.110\mathbf{j} - 0.030\mathbf{i} - 0.070\mathbf{j}) \text{ kg·m/s}$$
$$= (-0.030\mathbf{i} + 0.040\mathbf{j}) \text{ kg·m/s}$$

The velocity of the 80-gram puck after the collision is

$$\mathbf{v}_1' = \frac{\mathbf{p}_1'}{m_1} = \frac{(-0.030\mathbf{i} + 0.040\mathbf{j}) \text{ kg·m/s}}{0.080 \text{ kg}}$$
$$= (-0.375\mathbf{i} + 0.500\mathbf{j}) \text{ m/s}$$

The velocity assigned to the 100-gram puck after the collision is not the only possible velocity. There is a range of possible velocities, determined by the details of how the pucks interact during the collision. But, for each possible velocity of the 100-gram puck, the velocity and linear momentum of the 80-gram puck must be such as to satisfy linear momentum conservation.

The conservation of linear momentum principle is a consequence of Newton's third law and does not depend on the nature of the forces. For example, if we mounted small magnets on the air pucks they could exert forces on each other without touching. The magnetic forces obey Newton's third law and the linear momentum of the two-puck system would be conserved.

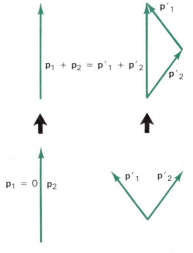

FIGURE 9.4

Linear momentum diagrams for air pucks before and after collision. The vector sum of the momenta is the same before and after the collision.

Is linear momentum always conserved? For a single particle the answer is "no." When a net force acts on a single particle its linear momentum changes, as described by Newton's second law, Equation 9.4. For a system composed of more than one particle the answer is also "no." Linear momentum is not always conserved. What equation would govern the change of linear momentum of a 2-particle system? The total linear momentum is

$$\mathbf{p} = \mathbf{p}_1 + \mathbf{p}_2$$

Let

$$\mathbf{F}_1 = \text{net external force on particle 1}$$

$$\mathbf{f}_1 = \text{force exerted on particle 1 by particle 2}$$

$$\mathbf{F}_2 = \text{net external force on particle 2}$$

$$\mathbf{f}_2 = \text{force exerted on particle 2 by particle 1}$$

The time rate of change of the total linear momentum is

$$\frac{d\mathbf{p}}{dt} = \frac{d\mathbf{p}_1}{dt} + \frac{d\mathbf{p}_2}{dt} = \mathbf{F}_1 + \mathbf{f}_1 + \mathbf{F}_2 + \mathbf{f}_2$$

Newton's third law, $\mathbf{f}_2 = -\mathbf{f}_1$, causes the internal forces to cancel and gives the result

$$\frac{d\mathbf{p}}{dt} = \mathbf{F}_1 + \mathbf{F}_2 \qquad (9.11)$$

The vector sum

$$\mathbf{F} = \mathbf{F}_1 + \mathbf{F}_2$$

is the net external force acting on the 2-particle system. It follows that the form of Newton's second law for the 2-particle system is

$$\mathbf{F} = \frac{d\mathbf{p}}{dt} \qquad (9.12)$$

This equation is identical in form to the single-particle version, Equation 9.4. From Equation 9.12 we conclude that if the net external force is zero, then the momentum* of the 2-particle system is conserved.

9.3
TWO-PARTICLE COLLISIONS

When a bowling ball collides with a bowling pin each object exerts a large force on the other for a short time. The changes in the velocities of the colliding bodies are associated with their mutual interaction, and not with the influence of the rest of their environment. A collision between two particles is an interaction in which the forces between the particles are large compared with all other forces acting on either particle. Under these conditions, the colliding particles can be considered free of external forces and their combined linear momentum is conserved.

---

* When no confusion is likely we use the terms *momentum* and *linear momentum* interchangeably.

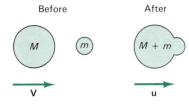

Before   After

M m $M + m$

V   u

FIGURE 9.5

In a perfectly inelastic collision, the colliding objects stick together.

Perfectly Inelastic Collisions

When two colliding objects stick together, the collision is termed **perfectly inelastic**. Consider a perfectly inelastic collision in which a particle of mass M moving at velocity $\mathbf{V}$ strikes a particle of mass m at rest. The two particles stick together and move off as a single particle of mass $M + m$ and velocity $\mathbf{u}$. Figure 9.5 illustrates the situation before and after the collision. The conservation of linear momentum is expressed by

$$(M + m)\mathbf{u} = M\mathbf{V} \tag{9.13}$$

which gives

$$\mathbf{u} = \frac{M\mathbf{V}}{M + m} \tag{9.14}$$

Because $M/(M + m)$ is less than unity we can see that the composite particle moves more slowly than the incident particle.

EXAMPLE 5

What a Catch!

An outfielder sees a high drive about to sail over his head. He throws his glove straight upward. Just as the glove reaches its peak it is struck by the ball. The ball and glove move off together at a speed of 5.8 m/s.* We want to determine the speed of the incoming ball. The mass of the baseball is 0.160 kg and the mass of the glove is 0.625 kg. Linear momentum is conserved in the collision and Equation 9.13 gives

$$V = \frac{(M + m)u}{M} = \frac{(0.160\ \text{kg} + 0.625\ \text{kg})(5.8\ \text{m/s})}{0.160\ \text{kg}}$$

$$= 28\ \text{m/s}$$

This converts to a speed of about 63 mi/h.

Although linear momentum is conserved in a perfectly inelastic collision, *kinetic energy is not conserved.* There is a simple relationship between the kinetic energies before and after the collision. We use Equation 9.8 to express the kinetic energies before and after the collision in terms of the linear momentum

$$K_i = \frac{p^2}{2M} \qquad \text{(before collision)} \tag{9.15}$$

$$K_f = \frac{p^2}{2(M + m)} \qquad \text{(after collision)} \tag{9.16}$$

Note that we have made use of momentum conservation by taking the momentum to be the same before and after the collision. Eliminating p^2 from Equations 9.15 and 9.16 gives

$$K_f = \frac{M}{M + m} K_i \tag{9.17}$$

The ratio $M/(M + m)$ is less than unity, showing that $K_f < K_i$. The kinetic energy after the collision is less than the kinetic energy before the collision.

* Baseball fans will recognize that such a stroke of luck would not count as a catch of the ball.

Kinetic energy is not conserved. The difference between the initial kinetic energy and the final kinetic energy is not lost. It is transformed into other forms of energy, including work performed as the two colliding objects deform each other.

EXAMPLE 6

Inelastic Car-Truck Collision

Suppose an 18-wheel truck (mass $M = 8000$ kg) strikes a compact car (mass $m = 900$ kg) at rest. Equation 9.17 relates the kinetic energy of the combined car and truck to the initial kinetic energy of the truck.

$$K_f = \frac{M}{M + m} K_i = \left(\frac{8000 \text{ kg}}{8900 \text{ kg}} \right) K_i = 0.899 K_i$$

Nearly 90% of the initial kinetic energy is retained.

Now let's interchange the vehicles. Let the moving compact car strike the truck at rest. Do you expect the same fraction of the initial kinetic energy to be retained? With $M = 900$ kg and $m = 8000$ kg, Equation 9.17 gives

$$K_f = \left(\frac{900 \text{ kg}}{8900 \text{ kg}} \right) K_i = 0.101 K_i$$

With the vehicles interchanged, only about 10% of the initial kinetic energy is retained.

Elastic Collisions

Linear momentum is conserved in collisions whether or not the particles stick together, as they do in a perfectly inelastic collision. A perfectly inelastic collision is a limiting case in which the fraction of the kinetic energy transformed into other forms is a maximum. So, in general, kinetic energy is *not* conserved in a collision. At the other extreme are collisions in which there is no change in the combined kinetic energy of the colliding particles.

If the combined kinetic energy of the colliding particles is conserved, the collision is termed **elastic**. You can demonstrate an elastic collision using two coins. If you slide one coin across a smooth surface so that it strikes a second coin the collision will be elastic. During the collision the shapes of the coins are altered slightly. Kinetic energy is transformed into elastic potential energy—very briefly—and then it goes entirely back into kinetic energy.

Let's examine an elastic collision between two particles and see how the conservation of linear momentum and kinetic energy restrict the possible outcomes. We begin with a one-dimensional (head-on) elastic collision in which a particle of mass M and speed V_0 strikes a particle of mass m at rest (Figure 9.6). Linear momentum conservation is expressed by

$$P_0 = P + p \tag{9.18}$$

where P_0 and P denote the linear momentum of the particle of mass M before and after the collision, and p denotes the linear momentum of the recoiling particle of mass m. With the aid of Equation 9.8 ($K = p^2/2m$), the conservation of kinetic energy can be expressed by

$$\frac{P_0{}^2}{2M} = \frac{P^2}{2M} + \frac{p^2}{2m} \tag{9.19}$$

Before

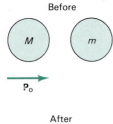

P_0

After

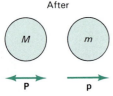

P   p

FIGURE 9.6

Head-on elastic collision. The two-headed arrow indicates that the mass M can move to the right or to the left after the collision, depending on the mass ratio M/m.

We envision a situation in which P_0 is given and we seek a solution for the momentum P in terms of P_0.

Using Equation 9.18 to set $p = P_0 - P$ we can eliminate p from Equation 9.19. This gives a quadratic equation for P

$$P^2\left(\frac{1}{2M} + \frac{1}{2m}\right) - 2P\frac{P_0}{2m} - P_0{}^2\left(\frac{1}{2M} - \frac{1}{2m}\right) = 0$$

The solutions are

$$P = P_0$$

and

$$P = \frac{M - m}{M + m}P_0 \tag{9.20}$$

The solution $P = P_0$ is trivial—it corresponds to a miss in which the particles pass without colliding. The solution corresponding to a collision is described by Equation 9.20.

Let's think of P_0 as a quantity under our control. Then Equation 9.20 shows us that there is just one possible value of the after-collision momentum of the incident particle. The conservation of linear momentum and kinetic energy impose severe constraints. In a head-on elastic collision these two conservation laws allow just one possible outcome.

The momentum of the recoiling target particle can be expressed in terms of P_0 by using Equation 9.20 in Equation 9.18 to eliminate P. This gives

$$p = \frac{2m}{M + m}P_0 \tag{9.21}$$

Again, you should note that there is only one possible value of p for a given P_0, a consequence of the constraints imposed by the conservation of linear momentum and kinetic energy.

EXAMPLE 7

Two-Penny Collision

A penny slides across a desk top and suffers a head-on elastic collision with another penny at rest. The incident coin moves at 1.2 m/s before the collision. We want to determine the speeds of the two coins after the collision.

Linear momentum is conserved, so we can use Equations 9.20 and 9.21. The masses are equal so we set $M = m$. Equation 9.20 gives

$$P = \frac{m - m}{m + m}P_0 = 0$$

which means that the incident coin comes to rest. You can verify this for yourself through experimentation.

The struck coin carries off all of the linear momentum after the collision. Equation 9.21 verifies this:

$$p = \frac{2m}{m + m}P_0 = P_0$$

Having the same mass as the incident coin, the struck coin must move at 1.2 m/s after the collision.

Two-Dimensional Elastic Collisions

When two colliding particles undergo a glancing collision, they end up moving along different directions. The conservation of linear momentum and kinetic energy still impose important restrictions but do not uniquely determine the final momenta. Figure 9.7 shows the geometry of an elastic collision in which an incident particle of mass M strikes a target particle of mass m at rest. The incident particle scatters through an angle θ measured relative to its incident direction. The target particle recoils at an angle β. We use $\mathbf{P}_0$ and $\mathbf{P}$ to denote the linear momenta of the incident particle before and after the collision, respectively. The linear momentum of the recoiling target is denoted by $\mathbf{p}$. Figure 9.8 shows the triangular relationship that embodies the conservation of linear momentum

$$\mathbf{P}_0 = \mathbf{P} + \mathbf{p} \tag{9.22}$$

We want to express P, the magnitude of the momentum of the scattered particle, in terms of P_0 and θ, quantities that can be controlled or measured. Such an equation lets us predict the value of P for specified values of P_0 and θ.

We use the conservation of linear momentum and kinetic energy to eliminate p. From linear momentum conservation, Equation 9.22, we have

$$\mathbf{p} = \mathbf{P}_0 - \mathbf{P}$$

Next, we form the scalar product $\mathbf{p} \cdot \mathbf{p}$

$$\mathbf{p} \cdot \mathbf{p} = p^2 = (\mathbf{P}_0 - \mathbf{P}) \cdot (\mathbf{P}_0 - \mathbf{P}) = P_0{}^2 + P^2 - 2\mathbf{P}_0 \cdot \mathbf{P}$$

Using

$$\mathbf{P}_0 \cdot \mathbf{P} = P_0 P \cos\theta$$

gives

$$p^2 = P_0{}^2 + P^2 - 2P_0 P \cos\theta \tag{9.23}$$

The conservation of kinetic energy relation also contains p^2

$$\frac{P_0{}^2}{2M} = \frac{P^2}{2M} + \frac{p^2}{2m} \tag{9.24}$$

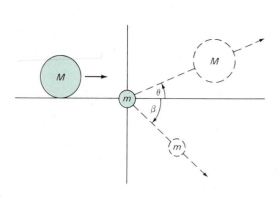

FIGURE 9.7

Elastic collision in two dimensions.

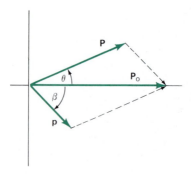

FIGURE 9.8

Vector triangle showing linear momentum conservation in an elastic collision.

Using Equation 9.23 to eliminate p^2 from Equation 9.24 gives a quadratic equation for P:

$$(M + m)P^2 - (2P_0 M \cos \theta)P + (M - m)P_0^2 = 0$$

Since P is the *magnitude* of the momentum it cannot be negative. If the mass of the incident particle is less than the mass of the target particle, only one of the two roots of the quadratic equation is positive.

$$P = \frac{\{M \cos \theta + \sqrt{m^2 - M^2 \sin^2 \theta}\}}{M + m} P_0 \qquad M \leq m \qquad (9.25)$$

If the mass of the incident particle is greater than the mass of the target particle, both solutions of the quadratic equation describe physically possible outcomes of the collision

$$P = \frac{\{M \cos \theta \pm \sqrt{m^2 - M^2 \sin^2 \theta}\}}{M + m} P_0 \qquad M > m \qquad (9.26)$$

The two roots mean that two values of P are permitted for a given angle of scatter. This can be understood by thinking of situations that result in the incident particle being scattered through a *small* angle. There are two ways of achieving a small-angle scattering: a *near-miss collision* and a *nearly head-on collision*. The plus sign root in Equation 9.26 corresponds to the near-miss collision. Such a collision results in a relatively small transfer of kinetic energy and momentum to the target particle. Likewise, the root with the minus sign in Equation 9.26 corresponds to the nearly head-on collision.

If the mass of the incident particle is greater than the mass of the target particle ($M > m$) there is a maximum angle of scatter, θ_{max}. This maximum angle of scatter follows from Equation 9.26 by noting that the momentum P must be a real quantity and thus the radicand

$$m^2 - M^2 \sin^2 \theta$$

must not be negative. The maximum angle of scatter is therefore defined by

$$\sin \theta_{max} = \frac{m}{M} \qquad (M > m) \qquad (9.27)$$

In 1911–1912 one of the cornerstones of nuclear physics was laid when Geiger and Marsden performed their famous alpha-particle scattering experiment. Under the supervision of Ernest Rutherford they directed a beam of alpha particles toward a thin gold foil. Geiger and Marsden observed the angles through which the alpha particles were scattered. Most of the observed scattering angles were quite small and could be understood in terms of collisions between an alpha particle and an electron in the gold atom. The masses of the electron and alpha particle are

$$m = 9.11 \times 10^{-31} \text{ kg} \qquad \text{(electron)}$$

$$M = 6.69 \times 10^{-27} \text{ kg} \qquad \text{(alpha)}$$

From Equation 9.27, the maximum angle of scatter is

$$\theta_{max} = \sin^{-1}\left(\frac{m}{M}\right) = \sin^{-1}\left(\frac{9.11 \times 10^{-31} \text{ kg}}{6.69 \times 10^{-27} \text{ kg}}\right)$$

$$= 0.0078°$$

Allowing for multiple scatterings by electrons enabled Geiger and Marsden to explain most of the observations. But a tiny fraction of the scatterings resulted in large deflections—through angles as great as 150°. Rutherford realized that such a large scattering angle was possible only if the alpha particle collided with a *more massive* particle.

Furthermore, the small fraction of collisions that produced large-angle deflections implied that the target was small compared to the dimensions of the entire atom. Rutherford concluded that most of the mass of an atom is concentrated in a small region, the **nucleus** as it is now called.

EXAMPLE 8

Neutron Scatter by a Carbon Nucleus

A neutron with a kinetic energy of 6.2 MeV* collides elastically with a carbon–12 nucleus and scatters through an angle of 60°. We want to determine the kinetic energies of the neutron and carbon nucleus after the collision.

With $\theta = 60°$,

$$\cos \theta = \cos 60° = 0.500 \qquad \sin \theta = \sin 60° = 0.866$$

The carbon nucleus is 11.9 times as massive as the neutron. Because only mass *ratios* enter Equation 9.26, the actual masses in kilograms need not be used. Setting $m = 1$ and $M = 11.9$ in Equation 9.26 gives

$$P = \frac{\{0.500 + \sqrt{(11.9)^2 - (0.866)^2}\}}{1 + 11.9} P_0$$

$$= 0.959 \, P_0$$

The kinetic energy of the scattered neutron is

$$K = \frac{P^2}{2M} = \frac{(0.959)^2 P_0{}^2}{2M}$$

The factor $P_0{}^2/2M$ is the kinetic energy of the incident neutron, so the kinetic energy of the scattered neutron is

$$K = (0.959)^2 K_0 = (0.920)6.20 \text{ MeV}$$

$$= 5.70 \text{ MeV}$$

Because kinetic energy is conserved in an elastic collision, the recoiling carbon nucleus has a kinetic energy of

$$K_{\text{carbon}} = (6.20 - 5.70) \text{ MeV}$$

$$= 0.50 \text{ MeV}$$

9.4
CENTER OF MASS

In Chapter 3 we introduced the concept of the *center of gravity*. We saw that we could regard all of the weight of a body as being located at its center of gravity. We now introduce and develop the closely related concept of the **center of mass**.

---

* One electron volt (1 eV) equals 1.60×10^{-19} J. One MeV equals one **million electron** volts and is a convenient unit of energy at the nuclear level.

Figure 9.9 shows a diver executing a forward one-and-a-half somersault dive. An observer viewing the dive from poolside would be aware of the intricate rotational motion. But an observer five blocks away would see only a particle-like speck moving along a parabolic trajectory. For the observer nearby, just one point—the center of mass—moves along a parabola.

For a discrete set of particles, the center of mass position vector $\mathbf{r}_{cm}$ is defined by

$$\mathbf{r}_{cm} = \frac{m_1\mathbf{r}_1 + m_2\mathbf{r}_2 + \cdots}{M}$$

$$= \frac{\sum m_i\mathbf{r}_i}{M} \tag{9.28}$$

Here, M is the total mass of the set of particles

$$M = \sum m_i$$

and

$$\mathbf{r}_i = \mathbf{i}x_i + \mathbf{j}y_i + \mathbf{k}z_i \tag{9.29}$$

is the position vector of the ith particle, whose mass is m_i (Figure 9.10). The summations extend over all particles in the set. If we use Equation 9.29 in Equation 9.28 we can express the center of mass position vector in component form

$$\mathbf{r}_{cm} = \mathbf{i}x_{cm} + \mathbf{j}y_{cm} + \mathbf{k}z_{cm} \tag{9.30}$$

The components are given by

$$x_{cm} = \frac{\sum m_i x_i}{M}$$

$$y_{cm} = \frac{\sum m_i y_i}{M} \tag{9.31}$$

$$z_{cm} = \frac{\sum m_i z_i}{M}$$

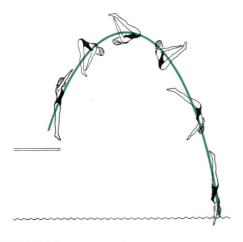

FIGURE 9.9

The center of mass of the diver follows a parabolic path, even though the diver rotates while moving through the air.

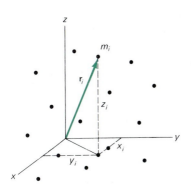

FIGURE 9.10

The position vector, $\mathbf{r}_i$, for the ith particle in a collection, has components x_i, y_i, and z_i.

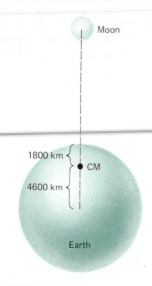

FIGURE 9.11

The center of mass (CM) of the earth-moon system is located beneath the earth's surface.

For symmetric systems the center of mass is at the center of symmetry. For example, in a binary star system composed of two equal-mass stars, the center of mass is halfway between the two star centers.

Numerical values of the center of mass coordinates depend on the origin chosen. For example, relative to an origin at the center of the earth, the center of mass of the earth-moon system is located a distance of 4.6×10^6 m away (Figure 9.11). For an origin at the surface of the earth the center of mass is located a distance of 1.8×10^6 m away, beneath the surface. The physical location of the center of mass is the same in both cases, but the numbers used to describe its position differ. Often it is convenient to place the coordinate origin at the center of mass. In this case all components of $\mathbf{r}_{cm}$ are zero.

EXAMPLE 9

Center of Mass Located By Symmetry

Figure 9.12 shows a symmetric arrangement of six identical pool balls. Three balls are located at the vertices of an equilateral triangle. The other three balls are located at the midpoints of the sides of the triangle. The length of each side of the triangle is 2 m.

The center of mass of each ball is at its geometric center, *by symmetry*. We can replace the array of balls by an array of point masses located at the centers of the balls.

For the coordinate axes shown in Figure 9.12 we can see that $x_{cm} = 0$ *by symmetry*. The y-coordinate of the center of mass is given by Equation 9.31. If we let m denote the mass of each ball,

$$y_{cm} = \frac{\{3m(0) + 2m(0.866 \text{ m}) + m(1.732 \text{ m})\}}{6m}$$

$$= 0.577 \text{ m}$$

The center of mass is marked by an asterisk in Figure 9.12. The value of y_{cm} can also be determined by symmetry. To do so, note that, by symmetry, the center of mass must lie on the line drawn from a vertex to the center of the side opposite. There are three such lines and they have one point in common. This point marks the center of mass.

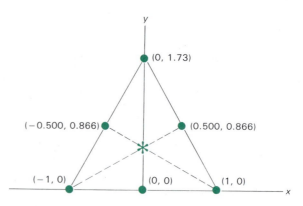

FIGURE 9.12

A symmetric arrangement of six identical pool balls. The asterisk marks the center of mass.

In many situations it is legitimate and desirable to ignore the rotation and internal motions of a system. In these situations the center of mass concept greatly simplifies the analysis of the motion because the many-particle system can be treated as a single particle located at the center of mass. For example, astronomers study the motion of a galaxy as a whole while disregarding separate individual motions of the 100 billion stars in that galaxy. We begin with the defining equation of the center of mass, Equation 9.28, and derive a series of equations that show how a collection of particles can be treated as a single particle located at the center of mass.

Velocity of the Center of Mass

The velocity of the center of mass is defined as the time derivative of the center of mass position vector,

$$\mathbf{v}_{cm} = \frac{d\mathbf{r}_{cm}}{dt} \tag{9.32}$$

Forming the time derivative of Equation 9.28 we obtain

$$\mathbf{v}_{cm} = \frac{\sum m_i \mathbf{v}_i}{M} \tag{9.33}$$

which expresses the center of mass velocity in terms of the individual particle velocities, $\mathbf{v}_i = d\mathbf{r}_i/dt$.

Linear Momentum

The vector sum $\sum m_i \mathbf{v}_i$ in Equation 9.33 is the total linear momentum of the system. We designate it by $\mathbf{p}$.

$$\mathbf{p} = \sum m_i \mathbf{v}_i \tag{9.34}$$

Substituting $\mathbf{p}$ for $\sum m_i \mathbf{v}_i$ in Equation 9.33 we can write

$$\mathbf{p} = M\mathbf{v}_{cm} \tag{9.35}$$

This equation has the following interpretation. If the mass (M) of the system were concentrated at the center of mass and moved with the center of mass velocity ($\mathbf{v}_{cm}$), then the linear momentum of the mass would equal the sum of the linear momenta of all the particles. This is the justification for treating a system, a galaxy for instance, as a particle located at the center of mass.

Newton's Second Law

We now want to show how the center of mass concept is related to Newton's second law. Newton's second law for a particle can be written

$$\mathbf{F}_i = \frac{d\mathbf{p}_i}{dt} \tag{9.4}$$

where $\mathbf{F}_i$ is the net force acting on the ith particle. We sum the forces for all particles in the system and make use of the fact that the sum of linear momenta gives the total linear momentum

$$\sum \mathbf{F}_i = \frac{\sum d\mathbf{p}_i}{dt} = \frac{d(\sum \mathbf{p}_i)}{dt} = \frac{d\mathbf{p}}{dt}$$

Using Equation 9.35 we can replace $\mathbf{p}$ by $M\mathbf{v}_{cm}$

$$\sum \mathbf{F}_i = \frac{d(M\mathbf{v}_{cm})}{dt} \tag{9.36}$$

The sum $\sum \mathbf{F}_i$ includes external forces and internal forces acting between particles. The internal forces cancel in pairs because of Newton's third law, and so the sum $\sum \mathbf{F}_i$ equals the net external force acting on the system. We denote this net force by $\mathbf{F}$:

$$\sum \mathbf{F}_i = \mathbf{F} \tag{9.37}$$

Comparing Equations 9.36 and 9.37 gives Newton's second law for the system

$$\mathbf{F} = \frac{d(M\mathbf{v}_{cm})}{dt} \tag{9.38}$$

This form of Newton's second law states that the net external force determines the time rate of change of the total linear momentum of the system. Because Equation 9.38 has the same form as Equation 9.4 for a single particle it shows that the system as a whole moves like a particle of mass M located at the system's center of mass.

Conservation of Linear Momentum

One of the most significant and useful consequences of the center of mass concept follows from Equation 9.38 when the net external force is zero. With $\mathbf{F} = 0$, Equation 9.38 shows that the total linear momentum of the system is conserved.

$$M\mathbf{v}_{cm} = \text{constant} \qquad \mathbf{F} = 0 \tag{9.39}$$

Equation 9.39 applies to all force-free systems, no matter how complex their makeup.

If the center of mass is at rest and linear momentum is conserved, the center of mass remains at rest even though the particles making up the system are in motion. In such cases the conservation of linear momentum helps us deduce features of the relative motion of parts of the system.

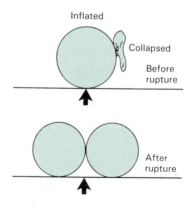

Inflated

Collapsed

Before rupture

After rupture

FIGURE 9.13

Identical balloons are connected by a thin membrane. Initially the left balloon is inflated and the right balloon is collapsed. The membrane breaks, producing a final situation where the gas fills both balloons equally. If the balloon mass is neglected, the center of mass of the gas will not move.

EXAMPLE 10

Siamese-Twin Balloons

Two identical balloons are joined by a thin membrane (Figure 9.13). Initially one balloon is filled with gas while the other is evacuated and collapsed. The mass of the balloons is negligible in comparison with the mass of the gas. At a certain moment the membrane ruptures, allowing the gas to fill the balloons equally. Disregarding friction, and if only horizontal motion is permitted, where is the final location of the balloons?

In the case of the balloons, the net external force is zero and so linear momentum is conserved. The center of mass is at rest before the membrane ruptures and so the center of mass *remains at rest*. The arrows in Figure 9.13 indicate the center of mass position before and after the rupture. Because the only mass in the system is that of the gas, the center of mass before rupture is at the center of the gas-filled balloon. By symmetry, the center of mass after rupture is at the point between the balloons. Hence the balloons must move to the left to maintain the fixed position of the center of mass.

9.5
VARIABLE MASS SYSTEMS

In formulating Newton's second law for a system of particles we have been careful to emphasize the "force-equals-time rate of change of linear momentum" version rather than the more restrictive "force-equals-mass-times-acceleration" form. For systems whose mass remains constant we have seen that the two formulations are equivalent. For systems whose mass changes we must use

$$\mathbf{F} = \frac{d\mathbf{p}}{dt}$$

where $\mathbf{p}$ is the total linear momentum of the system and $\mathbf{F}$ is the net external force acting on the system. We start by considering a simple example of a system of variable mass. Then we will consider the rocket, an important example of a variable mass system.

EXAMPLE 11

How to Accelerate a Handcar

A railroad brakeman (Figure 9.14) of mass m stands on a handcar of mass M moving at speed v. Taken together, the brakeman and handcar constitute a system of mass $M + m$ and linear momentum

$$p_i = (M + m)v$$

The brakeman runs along the car at a speed u_0 relative to the car and then jumps off. While he runs and jumps, the brakeman exerts a force on the handcar. The handcar exerts an equal but oppositely directed reaction force on the brakeman. This action-reaction pair of forces is *internal* to the brakeman-handcar *system* and does not change the total linear momentum of that system.

We want to determine the change in the velocity of the handcar, assuming that friction between the rails and the handcar is negligible. Disregarding friction, the net external force is zero and therefore linear momentum is conserved. By ejecting mass (the brakeman) the handcar increases its linear momentum. By the time the brakeman leaps from the car its velocity has increased to $v + \Delta v$. The brakeman's velocity relative to the earth is $v + \Delta v - u_0$. The final linear momentum of the handcar-plus-brakeman system is

$$p_f = M(v + \Delta v) + m(v + \Delta v - u_0)$$

Setting $p_i = p_f$ gives

$$(M + m)v = Mv + M\,\Delta v + mv + m\,\Delta v - mu_0$$

Solving for the change in velocity gives

$$\Delta v = \frac{mu_0}{M + m} \tag{9.40}$$

From a force viewpoint, the handcar acceleration is a consequence of the force exerted by the brakeman. From a linear momentum viewpoint the handcar is accelerated because the brakeman carries off linear momentum. The handcar velocity increases to maintain overall momentum conservation.

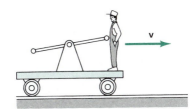

$\mathbf{v}$

$\mathbf{v} + \Delta\mathbf{v} - \mathbf{u}_0$

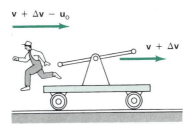

$\mathbf{v} + \Delta\mathbf{v}$

FIGURE 9.14

A brakeman and a handcar are initially moving to the right at speed v. When the brakeman runs to the left and jumps from the car, the handcar recoils ahead, increasing its speed by Δv.

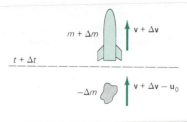

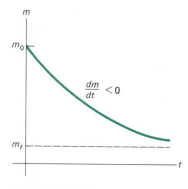

FIGURE 9.15

A force-free rocket system. At time t the rocket system comprises a mass m moving with velocity v. At time $t + \Delta t$ the same system consists of a rocket of mass $m + \Delta m$ moving with velocity $v + \Delta v$, and ejected gases of mass $-\Delta m$ moving with velocity $v + \Delta v - u_0$.

A Force-Free Rocket

Like the handcar and brakeman system just considered, a rocket ejects mass. In the process the rocket exerts a force on the mass being ejected. It follows from Newton's third law that this mass exerts a reaction force on the rocket. It is this reaction force that accelerates the rocket.

Consider a rocket (Figure 9.15) that ejects mass from its engines. The velocity of the ejected gases is constant *relative to the rocket*, and is equal to u_0. We want to determine the acceleration of the rocket in the absence of any net external force.

In Figure 9.15, the mass of the rocket at time t is m and it is moving at a velocity v. The linear momentum of the rocket system at time t is

$$p_i = mv$$

A moment later, at time $t + \Delta t$, the rocket mass has changed to $m + \Delta m$ and the rocket velocity has increased to $v + \Delta v$. Note that the mass of the rocket decreases, so Δm is *negative*. Additionally, a positive mass $(-\Delta m)$ of gas has been ejected by the engine and moves with a velocity $v + \Delta v - u_0$. At time $t + \Delta t$ the linear momentum of the system is

$$p_f = (m + \Delta m)(v + \Delta v) + (-\Delta m)(v + \Delta v - u_0)$$
$$= mv + m\Delta v + u_0 \Delta m$$

The change in linear momentum of the system in the time interval Δt is

$$\Delta p = p_f - p_i = m\Delta v + u_0 \, \Delta m$$

In the force-free case we know that linear momentum is conserved, so $\Delta p = 0$, and we can write

$$\Delta v = -u_0 \frac{\Delta m}{m} \tag{9.41}$$

This equation shows that the change in the velocity of the rocket is proportional to the mass ejected and to the velocity of the escaping gases, and is inversely proportional to the rocket mass. This is exactly the same situation as in the handcar and brakeman example. Note that Δv is positive because Δm is negative. The rocket accelerates itself by literally throwing part of itself (the ejected gases) in a direction opposite to that in which it accelerates.

The mass of the rocket during the fuel burn decreases with time, as suggested in Figure 9.16. If we divide Equation 9.41 by Δt and take the limit as Δt approaches zero we obtain

$$\frac{dv}{dt} = -\left(\frac{u_0}{m}\right)\frac{dm}{dt} \tag{9.42}$$

We introduce the acceleration, $a = dv/dt$, and the *rate* at which mass is ejected

$$\mu = -\frac{dm}{dt} \tag{9.43}$$

Note that μ is positive because m decreases as mass is ejected. Equation 9.42 takes the form

$$a = \frac{u_0 \mu}{m} \tag{9.44}$$

FIGURE 9.16

A rocket, whose initial total mass is m_0, loses mass as the engine burns fuel. The slope of the mass versus time curve is negative.

Note that the acceleration is directly proportional to μ, the rate of mass ejection, and to u_0, the exhaust gas velocity. These two quantities are determined by the design of the rocket engine. For maximum acceleration, both quantities must be large.

In Equation 9.44 the quantity

$$T = u_0 \mu \qquad (9.45)$$

is called the *thrust* of the rocket engine. In terms of T we can write Equation 9.44 in the form

$$T = ma$$

which resembles the constant-mass form of Newton's second law. Rockets are often rated in terms of the thrust their engines deliver. The *Saturn V* rocket that carried American astronauts to the moon had a thrust of over 34 million newtons, about 4000 tons.

We obtain the velocity of the rocket as a function of its mass by integrating Equation 9.42

$$\int_{v_0}^{v} dv = -u_0 \int_{m_0}^{m} \frac{dm}{m}$$

This gives

$$v = v_0 + u_0 \ln \left(\frac{m_0}{m} \right) \qquad (9.46)$$

Here, v_0 and m_0 are the initial values of v and m.

EXAMPLE 12

A Rocket to the Stars

We can use Equation 9.46 to answer an intriguing question. Can we accelerate a conventional chemical rocket to a speed comparable to the speed of light? Let's explore this possibility.

What mass ratio would be required to accelerate a rocket from rest to a speed of $v = 3 \times 10^6$ m/s (one percent of the speed of light)? We take $u_0 = 5000$ m/s, a very large velocity for a chemical rocket engine.

From Equation 9.46, with $v_0 = 0$, we have

$$\ln \left(\frac{m_0}{m} \right) = \frac{v}{u_0} = \frac{3 \times 10^6 \text{ m/s}}{5 \times 10^3 \text{ m/s}} = 600$$

This makes

$$\frac{m_0}{m} = e^{600} = 10^{260}$$

This ratio is prohibitively large. If the empty rocket consisted of just *one molecule*, the mass of the fuel would exceed the mass of the visible universe! Clearly, chemical rockets are not viable vehicles for interstellar travel.

To accelerate a rocket to a speed comparable to the speed of light requires reactions that liberate more energy than chemical reactions. A study called Project Orion, initiated in 1958, showed that hydrogen bombs could be used to safely accelerate space vehicles to speeds approaching one-tenth the speed of light. Project Orion is discussed in Chapter 40.

WORKED PROBLEM

A particle of mass M and kinetic energy K_0 collides head-on with a particle of mass μ, initially at rest. The particle of mass μ then collides head-on with a particle of mass m, initially at rest. Both collisions are elastic. (a) Determine the value of μ that maximizes the linear momentum of the mass m. Express μ in terms of M and m. (b) If $M = 4m$ and $K_0 = 100$ J, what is the maximum kinetic energy that can be transferred to m?

Solution

This problem is a variation of the two-particle elastic collision discussed in Section 9.3. We can apply Equation 9.21 to both two-particle collisions.

(a) For a mass M, making a head-on elastic collision with a mass μ initially at rest, Equation 9.21 reduces to

$$P = \frac{2\mu}{M + \mu} P_0$$

where P is the momentum of μ and P_0 is the momentum of M. When μ collides with m, Equation 9.21 reduces to

$$p = \frac{2m}{m + \mu} P$$

where p is the momentum of m. Eliminating P yields

$$p = 4mP_0\mu(M + \mu)^{-1}(m + \mu)^{-1}$$

We take M, m, and P_0 as constant, and set the derivative $dp/d\mu$ equal to zero in order to determine the value of μ that maximizes p. You can verify that this procedure gives

$$\frac{dp}{d\mu} = 4mP_0(m + \mu)^{-2}(M + \mu)^{-2}(mM - \mu^2) = 0$$

Solving for μ gives

$$\mu = \sqrt{mM}$$

This result shows that the value of μ that maximizes the momentum transfer is the geometric mean of m and M.

(b) The kinetic energy of m is related to its linear momentum by

$$k = \frac{p^2}{2m}$$

This shows that k is a maximum when p is a maximum. Using $K_0 = P_0^2/2M$, and the relation $p = 4mP_0\mu(M + m)^{-1}(m + \mu)^{-1}$ established above, gives an expression for the maximum kinetic energy of m

$$k_{max} = 16mM\mu^2(M + \mu)^{-2}(m + \mu)^{-2}K_0$$

Setting

$$M = 4m$$

$$K_0 = 100 \text{ J}$$

$$\mu = \sqrt{mM} = 2m$$

gives

$$k_{max} = \left(\frac{256}{324}\right)K_0 = (0.790)100 \text{ J}$$

$$= 79.0 \text{ J}$$

EXERCISES

9.1 Linear Momentum

A. The linear momentum of a particle moving along the x-axis is given by $p = A \cos \omega t$, where A and ω are constants. Express the force acting on the particle in terms of A and ω at (a) $\omega t = 90°$, (b) $\omega t = 180°$.

B. A child bounces a superball on the sidewalk. The linear impulse delivered by the sidewalk to the superball is 2 N·s during the 1/800 s of contact. What is the magnitude of the average force exerted on the superball by the sidewalk?

C. Figure 1 is a graph showing the force exerted by a boxer on a punching bag. Determine the linear impulse delivered over the time interval from $t = 0$ to $t = 0.10$ s.

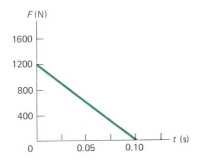

FIGURE 1

D. A 75-kg ice skater moving at 10 m/s crashes into a stationary skater of equal mass. After the collision the two skaters move as a unit at 5 m/s. The average force that a skater can experience without breaking a bone is 1000 lb. If the impact time is 0.1 s, does a bone break?

E. A 5-kg particle experiences the force shown in Figure 2 for 6 s. (a) Calculate the linear impulse delivered during the 6 s. (b) Determine the particle speed at $t = 6$ s if it started from rest.

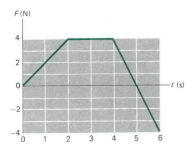

FIGURE 2

9.2 The Law of Conservation of Linear Momentum

F. A 70-kg astronaut is floating freely in space several meters from her spacecraft. She has a 1.60 kg wrench that she can throw with a speed of 22.0 m/s. (a) In what direction should she throw the wrench so that she will move toward the spacecraft? (b) What will be her speed toward the spacecraft if she is initially at rest relative to the spacecraft?

G. A $^{238}_{92}$U nucleus (mass = 238 units) decays, transforming into an alpha particle (mass = 4 units) and a residual $^{234}_{90}$Th nucleus (mass = 234 units). If the uranium nucleus was at rest, and the alpha particle has a speed of 1.5×10^7 m/s, determine the recoil speed of the thorium nucleus.

H. Astronaut A ($m = 60$ kg) is motionless a distance x from one end of a space capsule 20 m long (Figure 3). She holds an 8-kg mass that she then throws to astronaut B. It arrives at B simultaneously with the arrival of A at the other end of the capsule. Determine the value of x.

FIGURE 3

I. Two hockey players ($m_1 = 70$ kg, $m_2 = 90$ kg) move together with velocity $\mathbf{v}$ as they fight. At a certain moment they push each other apart. As they move apart, the 70-kg player's velocity is $15\mathbf{i}$ m/s and the 90-kg player's velocity is $10\mathbf{j}$ m/s. Determine the magnitude and direction of $\mathbf{v}$.

9.3 Two-Particle Collisions

J. A runaway string of 90 freight cars moving at a speed of 5 km/h collides with a string of 10 freight cars at rest. What is the speed of the 100-car string just after impact if the cars move as a unit? Take each car to have the same mass.

K. The danger to a car driver involved in a collision with a stationary car is related to the change in the speed of his car. To assess the role of the masses of the cars, consider a perfectly inelastic collision of a moving car with a car initially at rest. Determine the change in speed for the car initially in motion at 50 km/h for the following situations:

Moving Car	Stationary Car
(a) mass M	Mass $2M$
(b) mass $2M$	Mass M

L. A neutron (mass = 1 unit) collides elastically with a nucleus at rest. Calculate the fractional energy loss of the neutron $(K_i - K_f)/K_i$ in a head-on collision if the struck nucleus is (a) a proton (mass = 1 unit), (b) carbon (mass = 12 units).

M. A car makes a perfectly inelastic collision with an identical car at rest. (a) Prove that the kinetic energy of the system after the collision is one-half of the kinetic energy before the collision. (b) What happened to the "missing" kinetic energy?

N. An alpha particle (mass = 4 units) scatters from a proton at rest (mass = 1 unit). Determine the maximum scattering angle of the proton.

P. A 10-gram bullet is stopped in a block of wood ($m = 5$ kg). The speed of the bullet-plus-wood combination immediately after the collision is 0.60 m/s. What was the original speed of the bullet?

9.4   Center of Mass

Q. The three large dots in Figure 4 represent equal masses. The points labeled 1, 2, and 3 are prospects for the center of mass. Identify the center of mass and give a qualitative reason why each of the other two points cannot be the center of mass.

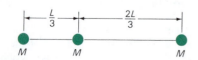

FIGURE 4

R. A uniform rod of length L and mass M has three point masses, each of mass M, attached as shown in Figure 5. Determine the location of the center of mass of the system.

FIGURE 5

S. Locate the center of mass of the model of the human body shown in Figure 6, assuming that all parts have the same density and the same thickness.

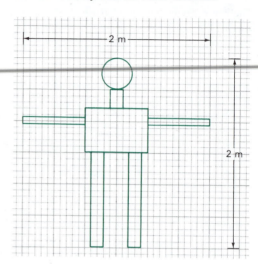

FIGURE 6

T. Three objects are located in the x–y plane as shown in Figure 7. Determine the location of a fourth object having mass $4M$ such that the center of mass of the four objects is at the origin.

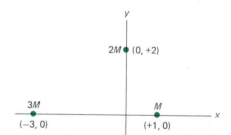

FIGURE 7

U. A boomerang (Figure 8) is constructed in the shape of a 90° angle with each side uniform and 40 cm in length. Locate the center of mass by using (a) a symmetry argument, (b) integration.

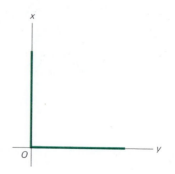

FIGURE 8

V.   Using the center of the earth as an origin, calculate the center of mass of the earth-moon system, treating both objects as points. Use the average center-to-center distance

$$r = 3.84 \times 10^8 \text{ m}$$

and the mass ratio

$$\frac{M_{moon}}{M_{earth}} = 0.0123.$$

9.5   Variable Mass Systems

W.   A rocket in outer space (force free) starts from rest and ejects mass at a constant rate with speed u_0 relative to the rocket. When the rocket achieves a speed $v = u_0$, what fraction of its initial mass has been ejected?

X.   A large rocket with an exhaust velocity $u_0 = 3000$ m/s develops a thrust of 24 million newtons. How much mass is being blasted out of the rocket exhaust per second?

Y.   What is the maximum velocity a rocket can acquire if it starts from rest in a force-free environment with $u_0 = 3$ km/s and 90% of its initial mass is fuel?

Z.   A force-free two-stage rocket is constructed so that 90% of the total original mass is fuel for the first stage. Of the remaining 10% of the original mass, half becomes the total mass of the second stage and half is jettisoned when the first-stage burn is complete. The second-stage rocket is also 90% (by mass) fuel. Each rocket engine has an exhaust speed of 2500 m/s. Determine the maximum speed of the second stage, assuming that its initial speed equals the final speed of the first stage and that the first stage starts from rest.

PROBLEMS

1.   Is the collision between air pucks in Example 4 elastic? Explain.

2.   Three cars are lined up a few inches apart, waiting for a traffic signal to change. A fourth car moving at 20 mph plows into the rear car. A series of perfectly inelastic collisions follows, ending with four stuck-together cars. The cars have equal masses. Determine the speed of the cars just after the final collision.

3.   The average velocity $\bar{v}$ is defined by

$$\bar{v} = \left(\frac{1}{t_f - t_i}\right) \int_{t_i}^{t_f} v \, dt$$

Prove that this reduces to $\bar{v} = \frac{1}{2}(v_i + v_f)$ for motion in which the acceleration is constant.

4.   A particle of mass 2.6 kg moving with a velocity of $(7.2\mathbf{i} - 6.1\mathbf{j})$ m/s is brought to rest in 3.2 s by a constant force. Determine this force.

5.   A particle of mass 2.0 kg moving with a velocity of $(3\mathbf{i} + 2\mathbf{j})$ m/s experiences a constant force $(8\mathbf{i} - 3\mathbf{j})$ N for 2.2 s. Determine its linear momentum at the end of this 2.2-s interval.

6.   A bullet $(m = 0.050 \text{ kg})$ moving horizontally at 400 m/s embeds itself in a 2-kg block of wood initially at rest. The block-and-bullet combination subsequently slides along a horizontal surface. How far will they slide if the coefficient of friction is 0.8?

7.   A bullet $(m = 0.01 \text{ kg})$ moving at 300 m/s strikes a wooden fence 2 cm thick and comes out the other side at a speed of 200 m/s. Assuming constant acceleration while passing through the fence, determine the force experienced by the fence during contact with the bullet.

8.   A highly excited carbon nucleus at rest spontaneously disintegrates into three alpha particles. One alpha particle moves in the $+x$-direction with speed of 2×10^7 m/s and another moves in the $+y$-direction with speed 4×10^7 m/s. Determine the velocity of the third alpha particle.

9.   Three ice skaters of equal mass move north as a unit at 6 m/s. The middle skater pushes the rear skater southward. Subsequently the rear skater moves north at 2 m/s slower than the other two skaters. Calculate the velocity of (a) the rear skater, (b) the other two skaters.

10. The speed of a bullet (mass m) may be measured by firing it into a block of wood (mass M). The bullet embeds itself in the wood. The system of wood plus bullet then swings like a pendulum a horizontal distance d (Figure 9). In terms of d, L, g, and the two masses, obtain an expression for the speed of the bullet before impact.

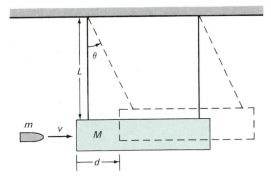

FIGURE 9

11. A mass M slides across a frictionless horizontal surface at speed v. It strikes a linear spring (spring

constant k) attached to an identical mass as shown in Figure 10. Assuming a perfectly inelastic collision, show that the maximum compression of the spring is given by $\sqrt{Mv^2/2k}$.

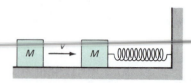

FIGURE 10

12. A small mass ($m = 0.2$ kg) starts from rest at the top of a frictionless incline ($M = 0.6$ kg). The incline is initially stationary on a frictionless horizontal surface, as shown in Figure 11. As the mass accelerates down the incline, the incline recoils. In moving from the top of the incline to the bottom, the smaller mass moves through a vertical distance of 16 cm. (a) Determine the magnitude of the normal force acting between the two masses. (b) Determine the time for the mass to reach the bottom of the incline. (c) Determine the recoil speed of the incline when the mass reaches the bottom.

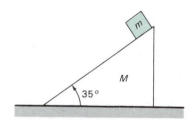

FIGURE 11

13. A particle of mass m_1 collides elastically with a particle of mass m_2 at rest. After the collision the particles recoil from the impact point with equal speeds. Determine the mass ratio m_2/m_1.

14. In a nuclear reaction, a nucleus at rest splits into a neutron (mass = 1 unit) and an alpha particle (mass = 4 units). The total kinetic energy of the two particles is 14 MeV. What is the kinetic energy of the neutron?

15. Two particles of mass M and $2M$ traveling at 3 m/s collide head-on. Determine their speeds after the collision if the collision (a) is perfectly inelastic, (b) is elastic.

16. To execute a vertical jump, a 50-kg woman first crouches and then accelerates upward. Her center of mass moves upward 20 cm with a constant acceleration before her feet break contact with the ground. What downward force must she exert on the ground to raise her center of mass 60 cm above its take-off position?

17. An air hockey puck moving in the x-direction strikes an identical puck at rest. After the elastic collision the

pucks move as shown in Figure 12. The pucks make angles θ and β with the x-axis. Neglect friction and use the conservation of linear momentum and kinetic energy to prove that $\theta + \beta = 90°$.

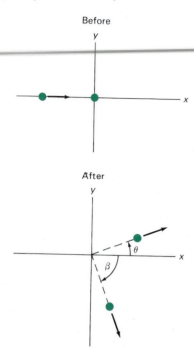

FIGURE 12

18. Water in a glass forms a truncated cone 12 cm high (Figure 13). The radius is 2 cm at the base and 3 cm at the top. Locate the center of mass of the water.

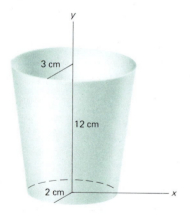

FIGURE 13

19. The spring-mass system shown in Figure 14 is at rest on a frictionless surface. The linear spring ($k = 300$ N/m) is compressed 8.2 cm when the retaining string snaps, allowing the masses to recoil. Determine the maximum speeds attained by the masses.

20. A Volkswagen (mass = 900 kg) moves at 20 m/s toward a Cadillac (mass = 1600 kg) traveling in the

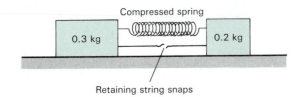

Compressed spring

0.3 kg   0.2 kg

Retaining string snaps

FIGURE 14

opposite direction at 30 m/s. Determine the velocity of their center of mass.

21. A boy and a girl, each of mass m, are standing on a sled of mass $2m$. They run and jump off the rear with a speed u_0 relative to the sled. Neglecting friction, calculate the final speed of the recoiling sled for two distinct cases: (a) The boy and girl jump off simultaneously, (b) the boy remains stationary while the girl runs and jumps, and then the boy follows her.

22. An asteroid of mass 3×10^{15} kg has a moon of mass 2×10^{14} kg. They are separated by 3,000 km when their orbital motion is halted. They subsequently fall toward their center of mass. How far does the moon fall before colliding with the asteroid? Treat the objects as point masses.

23. A particle of mass m moving in the $+x$-direction at a speed v collides with a particle of mass $4m$ at rest. Three particles emerge from the collision, moving as shown in Figure 15. Two particles (mass $= m$) move in opposite directions along the y-axis. The third particle (mass $= 3m$) moves along the $+x$-axis. Assume conservation of both kinetic energy and linear momentum and prove that $v_1 = v_2 = v/\sqrt{3}$ and $v_3 = v/3$

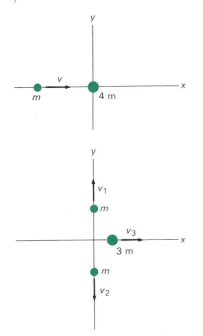

FIGURE 15

24. A 100-kg man stands motionless at point A on a 900-kg raft at rest in the water (Figure 16). He walks across the raft, coming to rest at B. Neglecting water resistance, determine how far (a) the man moved from his initial location, (b) the raft moved. Take the distance $AB = 10$ m.

FIGURE 16

25. Locate the center of mass of a homogeneous right circular cone (Figure 17) having an altitude a and a base of radius b.

FIGURE 17

26. A uniform chain of mass M and length L lies at rest on a table. At $t = 0$ an upward force is applied to one end so that thereafter the end of the chain rises at a constant speed v (Figure 18). Show that the lifting force may be expressed as

$$F = \frac{Mv(v + gt)}{L}$$

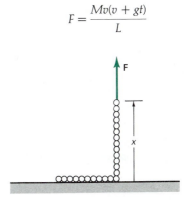

FIGURE 18

27. The ammonia molecule consists of three hydrogen atoms and one nitrogen atom situated at the corners of a pyramid. The hydrogen atoms are situated at (8.14, 0, 0), (−8.14, 0, 0), and (0, 14.1, 0). The nitrogen atom is located at (0, 4.70, 3.83). All lengths are expressed in units of 10^{-11} m. The mass of the nitrogen atom is 14 times the mass of a hydrogen atom. Locate the center of mass of the ammonia molecule.

28. A steel bar (Figure 19) is bent into the shape of one-fourth of a circle of radius 40 cm. Locate its center of mass.

29. Identical air cars ($m = 200$ grams) are equipped with identical linear springs ($k = 3,000$ N/m). They move toward each other with speeds of 3.0 m/s on a horizontal air track and collide, compressing the springs (Figure 20). Find the maximum compression of a spring.

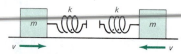

FIGURE 20

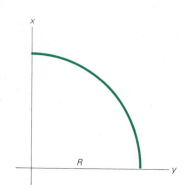

FIGURE 19

CHAPTER 10

ROTATIONAL MOTION

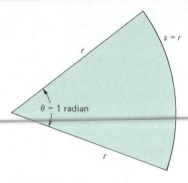

FIGURE 10.1

The angle subtended by an arc that is equal to the radius equals one radian.

10.1
ROTATIONAL KINEMATICS

In Chapter 9 we developed general principles for treating the motion of systems that consist of many particles. When there is no relative motion of the particles, we call the system a **rigid body**. Many objects, ranging from atomic nuclei to planets, can be approximated as rigid bodies. The motion of a rigid body reduces to a translation of its center of mass and a rotation relative to its center of mass. We developed the variables needed to describe the center-of-mass motion in Chapter 9. Now we introduce the variables required to describe rotational motions.

Radian Measure

Two common methods are used to establish a unit of angle. One method divides a circle into 360 degrees. The second method defines one unit of angle to be the angle subtended by an arc equal in length to the radius, as shown in Figure 10.1. Because the arc length for a complete circle equals 2π times the radius, it follows that the angle subtended by the circumference is 2π units. The value of this unit of angle, called the **radian**, is given in degrees (to three-significant-figure accuracy) by

$$1 \text{ radian} = \frac{360°}{2\pi} = 57.3°$$

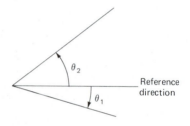

FIGURE 10.2

An angle is positive, like θ_2, when measured counterclockwise from a fixed reference direction. An angle is negative, like θ_1, when measured clockwise from a fixed direction.

Many pocket calculators have built-in conversion factors that let you use radians or degrees.

For a circle of radius r, the angle θ subtended by an arc length s is given in radians by

$$\theta = \frac{s}{r} \tag{10.1}$$

Thus, $s = r$ means $\theta = 1$ radian. For a full circle, $s = 2\pi r$, and $\theta = 2\pi$ radians. Thus, 360° correspond to 2π radians, and 90° to $\pi/2$ radians, and so on.

By convention, an angle measured counterclockwise is positive. In Figure 10.2, θ_2 is a positive angle and θ_1 is a negative angle. Angular displacement, $\Delta\theta$, is the difference between two angles measured relative to the same direction:

$$\Delta\theta = \theta_2 - \theta_1 \tag{10.2}$$

A positive angular displacement corresponds to $\theta_2 > \theta_1$ (Figure 10.3).

Angular Velocity

The **instantaneous linear velocity** is defined as the time derivative of the linear displacement, $v = dx/dt$. Analogously, we define the **instantaneous angular velocity**, ω, as the time derivative of the angular displacement

$$\omega = \frac{d\theta}{dt} \tag{10.3}$$

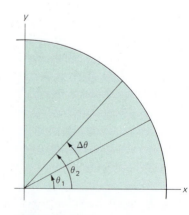

FIGURE 10.3

Angular displacement, $\Delta\theta = \theta_2 - \theta_1$, equals the difference between two angles measured from the same fixed direction, the x-axis in this diagram.

If time is expressed in seconds and angular displacement in radians, then the units for ω are radians per second (rad/s). Equation 10.3 can be integrated to obtain,

$$\int_{\theta_0}^{\theta} d\theta = \int_0^{\Delta t} \omega \, dt \tag{10.4}$$

For constant ω this equation can be integrated to give

$$\Delta\theta = \omega \, \Delta t \tag{10.5}$$

where $\Delta\theta = \theta - \theta_0$ is the angular displacement during the time interval Δt. When ω is not constant, Equation 10.5 defines the average angular velocity over the time interval Δt.

EXAMPLE 1

The Spinning Earth

The earth spins about its axis, completing a $360°$ revolution in one sidereal day. We want to determine the spin angular velocity of the earth. For one revolution we have

$$\Delta\theta = 2\pi \text{ rad}$$

$$\Delta t = 1 \text{ sidereal day} = 86{,}164 \text{ s}$$

To three significant figures the spin angular velocity of the earth is

$$\omega = \frac{\Delta\theta}{\Delta t} = \frac{2\pi \text{ rad}}{86{,}164 \text{ s}} = 7.29 \times 10^{-5} \text{ rad/s}$$

Angular Velocity and Tangential Velocity Relation

In Figure 10.4, the displacement $\mathbf{v} \, \Delta t$ is composed of two perpendicular components; a radial component $v_r \, \Delta t$, and a tangential component $v_t \, \Delta t$. The radial component of the velocity is given by

$$v_r = \frac{dr}{dt} \tag{10.6}$$

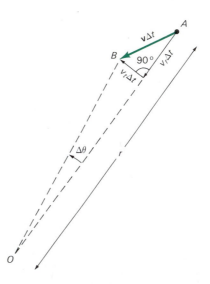

FIGURE 10.4

A particle moving from A to B has a displacement $\mathbf{v} \, \Delta t$ composed of two perpendicular components. The component $v_r \, \Delta t$ is parallel to r and the component $v_t \, \Delta t$ is perpendicular to r.

and measures the rate at which the particle approaches or recedes from the observer. We can use Figure 10.4 to relate the angular velocity ω to the tangential velocity, v_t, and the position, r. The tangential component of the displacement, $v_t \, \Delta t$, is related to the angular displacement, $\Delta \theta$, by

$$r \, \Delta\theta = v_t \, \Delta t \tag{10.7}$$

Dividing both sides of this equation by Δt and noting that the angular velocity is $\Delta\theta/\Delta t$ we arrive at the general relation between ω, r, and the tangential velocity, v_t,

$$r\omega = v_t \tag{10.8}$$

EXAMPLE 2

Angular Velocity with Linear Motion

A car travels in a straight line down a highway at 80 km/h. A farmer located 100 meters away observes the car's motion (Figure 10.5). Let's determine the angular velocity of the car relative to the farmer.

The tangential velocity of the car is

$$v_t = v \sin 60° = (80 \text{ km/h})0.866$$
$$= 69.3 \text{ km/h}$$

Converting to m/s gives

$$v_t = 69.3 \times 10^3 \, \frac{\text{m}}{\text{h}} \, \frac{1 \text{ h}}{3600 \text{ s}} = 19.2 \text{ m/s}$$

With $r = 100$ m, Equation 10.8 gives

$$\omega = \frac{v_t}{r} = \frac{19.2 \text{ m/s}}{100 \text{ m}} = 0.192 \text{ rad/s}$$

Note that the unit of angular velocity is the rad/s. Because the radian is dimensionless it may not appear explicitly at each stage of a calculation. Thus, in evaluating ω we divided v_t (units m/s) by r (units m), which seems to give ω in units of s^{-1}. But you must remember that the radian unit is implicit and that ω is expressed in units of rad/s.

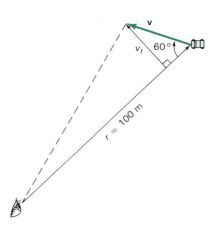

FIGURE 10.5
The tangential component of the car's velocity is $v_t = v \sin 60°$.

Angular Acceleration

Analogous to the linear acceleration a, defined by $a = dv/dt$, the angular acceleration, α, is defined as the rate of change of angular velocity

$$\alpha = \frac{d\omega}{dt} \qquad (10.9)$$

The SI unit for α is the radian per second per second (rad/s^2).

Equation 10.9 can be integrated to yield

$$\int_{\omega_0}^{\omega} d\omega = \int_{0}^{\Delta t} \alpha \, dt$$

For constant α the integrations yield

$$\Delta\omega = \omega - \omega_0 = \alpha \, \Delta t \qquad (10.10)$$

When α is not constant, Equation 10.10 defines the average angular acceleration over the time interval Δt. This relation is analogous to the one-dimensional motion equation for constant acceleration, $\Delta v = v - v_0 = a \, \Delta t$.

For a particle moving in a circular path, Equation 10.8 relates the tangential and angular velocities. To obtain the corresponding relation between the tangential and angular accelerations we differentiate the equation $v_t = r\omega$. Keeping r constant we get

$$\frac{dv_t}{dt} = r \frac{d\omega}{dt}$$

Using the definitions $a_t = dv_t/dt$ and $\alpha = d\omega/dt$, we obtain the desired result:

$$a_t = r\alpha \qquad (10.11)$$

The tangential acceleration, a_t, of a particle moving in a circular path is associated with a change in the magnitude of the tangential velocity of the particle. If the tangential speed is constant, then $a_t = 0$.

You may recall from Chapter 5 that a particle moving at speed v in a circle of radius r experiences a *radial* acceleration, $a_r = v^2/r$, associated with the changing direction of its velocity. This radial acceleration should not be confused with the *tangential* acceleration, a_t. The tangential acceleration arises only if the magnitude of the velocity changes.

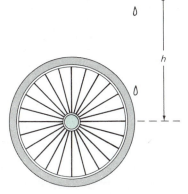

FIGURE 10.6

Drops of water rise vertically from the spinning tire.

EXAMPLE 3

Flying Water Drops and Angular Acceleration

A bicycle is turned upside down while its owner repairs a flat tire. A friend spins the other wheel and observes drops of water fly off tangentially. She measures the height reached by drops moving vertically (Figure 10.6). A drop that breaks loose from the tire on one turn rises 54 cm above the tangent point. A drop that breaks loose on the next turn rises 51 cm above the tangent point. The radius of the wheel is 0.381 m. She correctly infers that the height to which the drops rise decreases because the angular velocity of the wheel decreases. She decides to estimate the average angular acceleration of the wheel, using only the observed heights and the radius of the wheel. Let's follow her analysis.

The average angular acceleration is

$$\bar{\alpha} = \frac{\omega - \omega_0}{\Delta t}$$

Here, Δt is the time for one revolution, during which the angular velocity changes from ω_0 to ω. She estimates Δt by using the average angular velocity,

$$\bar{\omega} = \tfrac{1}{2}(\omega + \omega_0)$$

in the relation

$$\bar{\omega}\,\Delta t = 2\pi$$

This gives

$$\bar{\alpha} = \frac{\omega^2 - \omega_0{}^2}{4\pi}$$

Using the conservation of energy ($\tfrac{1}{2}mv^2 = mgh$) she relates the height, h, to which a drop rises to the tangential speed of the wheel

$$v^2 = 2gh$$

Finally, using $v = r\omega$, she obtains an equation for α in terms of the measured heights (h, h_0) and the radius of the wheel (r).

$$\bar{\alpha} = \frac{2g(h - h_0)}{4\pi r^2}$$

Her measured values of h, h_0, and r give

$$\bar{\alpha} = \frac{2(9.80 \text{ m/s}^2)(0.51 \text{ m} - 0.54 \text{ m})}{4\pi(0.381 \text{ m})^2}$$

$$= -0.32 \text{ rad/s}^2$$

The negative value of $\bar{\alpha}$ signifies that the angular velocity is decreasing.

We have introduced angular displacement, angular velocity, and angular acceleration. In each case the defining equation was analogous to the corresponding equation for linear motion. The kinematic relations between the angular variables are also analogous to their linear counterparts. Table 10.1 displays these analogous linear and rotational relations.

TABLE 10.1

Linear		Angular
s	displacement	θ
v	velocity	ω
a	acceleration	α
$v = at + v_0$		$\omega = \alpha t + \omega_0$
$s = s_0 + v_0 t + \tfrac{1}{2}at^2$		$\theta = \theta_0 + \omega_0 t + \tfrac{1}{2}\alpha t^2$
$v^2 = v_0{}^2 + 2as$		$\omega^2 = \omega_0{}^2 + 2\alpha\theta$

EXAMPLE 4

The Bicycle Wheel Revisited

Assuming that the angular acceleration of the bicycle wheel in Example 3 is constant, through how many revolutions does it turn before coming to rest?

From Table 10.1, we use $\omega^2 = \omega_0^2 + 2\alpha\theta$. From Example 3

$$\alpha = -0.32 \text{ rad/s}^2$$

$$\omega_0^2 = \left(\frac{v_0}{r}\right)^2 = \frac{2gh_0}{r^2} = \frac{2(9.80 \text{ m/s}^2)(0.54 \text{ m})}{(0.381 \text{ m})^2}$$

$$= 73 \text{ (rad/s)}^2$$

Setting $\omega = 0$, we get the value of θ for which the wheel has stopped rotating.

$$\theta = \frac{-\omega_0^2}{2\alpha} = \frac{73 \text{ (rad/s)}^2}{2(-0.32 \text{ rad/s}^2)}$$

$$= 110 \text{ radians}$$

Dividing by 2π shows that this angle corresponds to approximately 18 revolutions.

We make one final comment regarding angular motion variables. It is useful to define the angular velocity as a vector. The magnitude has already been defined by $\omega = d\theta/dt$*. The direction of the vector $\boldsymbol{\omega}$ is defined by a right-hand rule: Curl the fingers of the right hand in the direction of particle motion. The direction of the extended right thumb is defined as the direction of $\boldsymbol{\omega}$. Thus, in Figure 10.7 the particle moves in the x–y plane in a counterclockwise sense. The vector $\boldsymbol{\omega}$ is along the positive z-axis.

If the radius and tangential velocity of the particle are labeled $\mathbf{r}$ and $\mathbf{v}_t$, then the three vectors are related by

$$\mathbf{v}_t = \boldsymbol{\omega} \times \mathbf{r} \tag{10.12}$$

This is the vector form of Equation 10.8.

10.2
ROTATIONAL KINETIC ENERGY

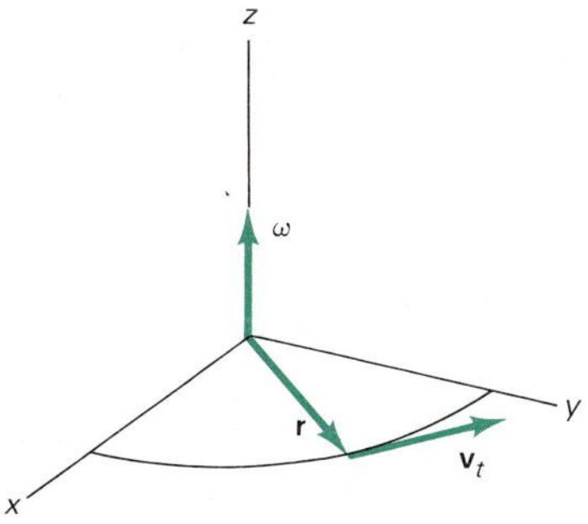

The kinetic energy of a particle is given by $K = \frac{1}{2}mv^2$. For a rotating rigid body—for example, a diskette in a microcomputer—the kinetic energy is not given by $K = \frac{1}{2}mv^2$ because all particles in the body do not have the same speed. The speed of a particle rotating about a fixed axis is directly proportional to its distance from the axis of rotation. The quantity that *is* the same for all particles is ω, the angular velocity. Let's see how the kinetic energy of a rigid body rotating about a fixed axis is related to the angular velocity of the body.

For a collection of particles, the kinetic energy is defined by

$$K = \sum \frac{1}{2}m_i v_i^2 \tag{10.13}$$

FIGURE 10.7

A particle moves along a counterclockwise circular path in the x-y plane. The angular velocity vector for the particle, $\boldsymbol{\omega}$, is in the positive z-direction.

---

* The angular velocity vector cannot be defined as the time derivative of an angular displacement vector because angular displacements do not qualify as vectors. (See Exercise H at the end of this chapter.)

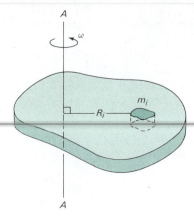

FIGURE 10.8

The mass m_i is located a distance R_i from the axis of rotation, A–A.

where $\frac{1}{2}m_i v_i^2$ is the kinetic energy of the ith particle. In a rigid body rotating about a fixed axis, each particle moves with the same angular velocity, ω (Figure 10.8). The speed of a particle moving in a circle of radius R_i is

$$v_i = R_i \omega \tag{10.14}$$

Substituting Equation 10.14 into Equation 10.13 gives

$$K = \tfrac{1}{2}\left(\sum m_i R_i^2\right)\omega^2$$

The quantity in parentheses is called the **moment of inertia** of the body for the given axis and it is designated by the symbol I.

$$I = \sum m_i R_i^2 \tag{10.15}$$

From this definition we see that the units of I are $\text{kg}\cdot\text{m}^2$. In terms of I the kinetic energy becomes

$$K = \tfrac{1}{2}I\omega^2 \tag{10.16}$$

A comparison of Equation 10.16 with the equation for the kinetic energy of a particle, $K = \frac{1}{2}mv^2$, shows that the moment of inertia, I, is the rotational analog of mass, m.

EXAMPLE 5

Tidal-Friction Energy

The rotational energy of the earth is decreasing steadily because of tidal friction. We want to estimate the change in the rotational energy of the earth in *one sidereal day*, given that the rotational period of the earth decreases by about 10 microseconds each *year*.

The moment of inertia of the earth about its spin axis is

$$I = 8.04 \times 10^{37} \text{ kg}\cdot\text{m}^2$$

and the rotational period is

$$T = 86{,}164 \text{ s}$$

The rotational energy, $K = \frac{1}{2}I\omega^2$, can be expressed in terms of T by using the relation $\omega T = 2\pi$. This gives

$$K = \frac{2\pi^2 I}{T^2}$$

The relative changes in K and T are small in comparison to K and T themselves. This lets us treat the changes as *differentials*. Thus,

$$dK = 2\pi^2 I(-2T^{-3}\,dT)$$
$$= -\frac{4\pi^2 I\,dT}{T^3}$$

The change in T in one day is

$$dT = \frac{10\ \mu s}{365} = \frac{10^{-5}\text{ s}}{365}$$
$$= 2.7 \times 10^{-8} \text{ s}$$

The change in the rotational kinetic energy in one day is

$$dK = -\frac{4\pi^2(8.04 \times 10^{37}\text{ kg}\cdot\text{m}^2)(2.7 \times 10^{-8}\text{ s})}{(8.62 \times 10^4\text{ s})^3}$$
$$= -1.3 \times 10^{17} \text{ J}$$

The minus sign indicates that the rotational energy is decreasing. This dissipation is roughly one-half of the daily energy consumption in the United States.

10.3
CALCULATION OF THE MOMENT OF INERTIA

The moment of inertia for a collection of point masses is defined by

$$I = \sum m_i R_i{}^2 \tag{10.15}$$

To convert this equation into an expression for the moment of inertia of a body whose mass is distributed continuously, we replace the sum by an integral. Thus, in place of Equation 10.15 we get

$$I = \int R^2 \, dm \tag{10.17}$$

This integration is over the total mass of the body.

When the object is sufficiently symmetric, it is possible to evaluate the moment of inertia analytically. The goal of such a calculation is to express I in terms of the total mass and the dimensions of the body. (Table 10.2 displays moment-of-inertia formulas for several objects.)

TABLE 10.2
Moments of Inertia about Axes through the Center of Mass

Hoop or cylindrical shell (Axis through the center perpendicular to the plane of the hoop)		MR^2
Hoop (Axis along a diameter)		$\dfrac{1}{2} MR^2$
Disk or solid cylinder (Axis through the center, perpendicular to the plane of disk)		$\dfrac{1}{2} MR^2$
Thin disk (Axis along a diameter)		$\dfrac{1}{4} MR^2$
Spherical shell (Axis along a diameter)		$\dfrac{2}{3} MR^2$
Solid sphere (Axis along a diameter)		$\dfrac{2}{5} MR^2$
Thin rod (Axis through the midpoint, perpendicular to the rod)		$\dfrac{1}{12} ML^2$
Rectangular lamina (Axis perpendicular to figure, and through the center)		$\dfrac{1}{12} M(a^2 + b^2)$

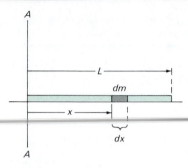

FIGURE 10.9

A long, thin rod of length L and mass M. A small section of the rod of width dx and mass dm is located at a distance x from the axis relative to which I is evaluated.

We illustrate Equation 10.17 by evaluating the moment of inertia of a long, thin rod.

EXAMPLE 6

The Long, Thin Rod

We want to calculate the moment of inertia for a rod of length L and mass M with respect to an axis perpendicular to the length of the rod and through one end of the rod (Figure 10.9). The rod has a constant density and a uniform cross-sectional area. Let dm denote the mass between x and $x + dx$. Because the rod is uniform,

$$\frac{dm}{M} = \frac{dx}{L}$$

Using this in Equation 10.17 gives

$$I = \int_0^L x^2 \, dm = \frac{M}{L} \int_0^L x^2 \, dx = \frac{M}{L} \frac{x^3}{3} \Big|_0^L$$

$$= \tfrac{1}{3}ML^2$$

Note that I is directly proportional to M and depends on the geometry through the length L.

Note also that the moment of inertia depends on the location and orientation of the reference axis. If we move the axis from the end of the rod to the center of mass, we get the same form for the moment-of-inertia integral, but with different limits. With the origin at the center of the rod, we integrate from $x = -\tfrac{1}{2}L$ to $x = +\tfrac{1}{2}L$. This gives

$$I_c = \tfrac{1}{12}ML^2$$

Parallel Axis Theorem

The moment of inertia for an axis through the center of mass of a body is related to the moment of inertia for any parallel axis by the **parallel axis theorem**. We prove this theorem for an axis perpendicular to a lamina (Figure 10.10).

Let I_c be the moment of inertia about an axis passing through the center of mass of the lamina, and let I be the moment of inertia about a parallel axis. The parallel axis theorem states

$$I = I_c + Mh^2 \tag{10.18}$$

where M is the mass of the lamina and h is the perpendicular distance between the two axes.

To prove this theorem, consider the arbitrarily shaped lamina in Figure 10.10. The points O and O' lie on parallel axes through the origin and center of mass. The vectors $\mathbf{R}_i$ and $\mathbf{r}_i$ are from O and O' to an element of mass m_i, and the vector $\mathbf{h}$ is perpendicular to the two axes. We see from the figure that $\mathbf{R}_i = \mathbf{r}_i + \mathbf{h}$. Since

$$R_i^2 = \mathbf{R}_i \cdot \mathbf{R}_i$$

we have

$$R_i^2 = (\mathbf{r}_i + \mathbf{h}) \cdot (\mathbf{r}_i + \mathbf{h}) = r_i^2 + h^2 + 2\mathbf{r}_i \cdot \mathbf{h}$$

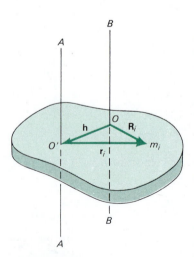

FIGURE 10.10

Two parallel axes, A–A and B–B. One passes through the center of mass at O'. The vector $\mathbf{h}$ denotes the perpendicular separation of the axes.

Using this in the defining equation for I,

$$I = \sum m_i R_i^2 \qquad (10.15)$$

gives

$$I = \sum m_i r_i^2 + \left(\sum m_i\right)h^2 + 2\left(\sum m_i \mathbf{r}_i\right) \cdot \mathbf{h}$$

The last term is zero because

$$\mathbf{r}_{cm} = \frac{1}{M}\sum m_i \mathbf{r}_i = 0$$

defines the position of the center of mass relative to the center of mass, and therefore equals zero. The second term equals Mh^2 and the first term defines I_c, thus proving the parallel axis theorem for a lamina,

$$I = I_c + Mh^2$$

To generalize this theorem, observe that the arbitrarily shaped object shown in Figure 10.11 can be viewed as a stack of parallel laminae. The parallel axis theorem holds for each lamina and hence for the entire stack.

Note also that the term Mh^2 is never negative, so the moment of inertia is a minimum for the axis passing through the center of mass.

FIGURE 10.11

A three-dimensional object can be considered to be a stack of parallel laminae.

EXAMPLE 7

The Long, Thin Rod—Again

In Example 6 we calculated the moment of inertia of a long thin rod about an axis perpendicular to the length of the rod through one end of the rod, and about a parallel axis through its center of mass. Let's use the parallel axis theorem to verify our result for the moment of inertia about the axis through the center of mass.

The two parallel axes are separated by a distance $h = \frac{1}{2}L$. With

$$I = \tfrac{1}{3}ML^2$$

the parallel axis theorem confirms our earlier calculation,

$$I_c = I - Mh^2 = \frac{1}{3}ML^2 - M\left(\frac{L}{2}\right)^2 = \left(\frac{1}{3} - \frac{1}{4}\right)ML^2$$

$$= \tfrac{1}{12}ML^2$$

WORKED PROBLEM

A 15-kg mass and a 10-kg mass are suspended by a pulley with a radius of 10 cm and a mass of 3 kg (Figure 10.12). The cord has a negligible weight and causes the pulley to rotate without slipping. The pulley rotates without friction. The masses start from rest 3 meters apart. Treat the pulley as a uniform disk and determine the speeds of the two masses as they pass each other.

Solution

Apply conservation of mechanical energy, taking account of the rotational energy of the pulley. Take the zero level of gravitational potential energy at the initial position of the 10-kg mass. Then

$$K_i = 0 \qquad U_1 = m_1 g h_1 \qquad (m_1 = 15 \text{ kg} \qquad h_1 = 3 \text{ m})$$

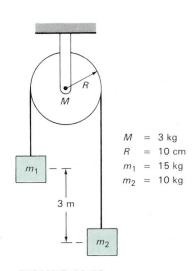

$M = 3$ kg
$R = 10$ cm
$m_1 = 15$ kg
$m_2 = 10$ kg

FIGURE 10.12

The masses pass at a point 1.5 m below the initial position of the 15-kg mass. The two masses have the same speed throughout the motion because the cord does not stretch. At the moment the masses pass

$$K_f = \tfrac{1}{2}(m_1 + m_2)v^2 + \tfrac{1}{2}I\omega^2 \qquad (I = \tfrac{1}{2}MR^2 \qquad M = 3 \text{ kg})$$

$$U_f = (m_1 + m_2)gh_f \qquad (m_2 = 10 \text{ kg} \qquad h_f = 1.5 \text{ m})$$

Energy conservation, $K_i + U_i = K_f + U_f$, takes the form

$$m_1gh_1 = \tfrac{1}{2}(m_1 + m_2)v^2 + \tfrac{1}{2}I\omega^2 + (m_1 + m_2)gh_f$$

The pulley rotates without allowing the cord to slip, so the angular velocity of the pulley is related to the speed of the masses by

$$\omega = \frac{v}{R}$$

Using this relation converts the rotational energy of the pulley to

$$\tfrac{1}{2}I\omega^2 = \tfrac{1}{4}M(R\omega)^2 = \tfrac{1}{4}Mv^2$$

Solving for v gives

$$v = \sqrt{\frac{2g(m_1h_1 - m_1h_f - m_2h_f)}{m_1 + m_2 + \tfrac{1}{2}M}} = \sqrt{\frac{2(9.80 \text{ m/s}^2)(5 \text{ kg})(1.50 \text{ m})}{26.5 \text{ kg}}}$$

$$= 2.36 \text{ m/s}$$

EXERCISES

10.1   Rotational Kinematics

A. Determine the difference in angle between 3 radians and $180°$ in (a) degrees, (b) radians.

B. What arc length subtends an angle of $1°$ if the radius is 1 meter?

C. You walk at constant speed once around a circle of radius 10 m in a time of 10 s. What is your angular velocity about the center of the circle?

D. The moon always presents the same hemisphere to an observer on the earth. Use this fact to determine the spin angular velocity of the moon on its axis. (The period of the moon's orbital motion about the earth is 27.3 sidereal days.)

E. A model airplane is flying in a horizontal circle ($R = 40$ m) at 20 mi/h when the engine stops. The plane travels through half the circumference of the circle while coming to rest. Assume that the angular acceleration during this coasting period is constant, and calculate (a) the angular velocity of the plane when the engine stops, (b) the angular acceleration of the plane while it coasts, (c) the time required to coast to a stop.

F. A student accelerates her sports car from rest at a constant rate. After 5 s of motion the tachometer indicates an engine rotation of 1500 rpm. Determine the angular acceleration of the engine, assuming that it is constant.

G. A discus thrower accelerates the discus from rest to a speed of 25 m/s by whirling it through 1.25 revolu-

tions (Figure 1). Assume the discus moves on the arc of a circle 1 m in radius and (a) calculate the final angular velocity of the discus. (b) Determine the angular acceleration of the discus, assuming it to be constant. (c) Calculate the acceleration time.

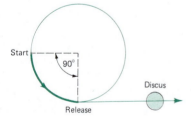

FIGURE 1

H. Remember that vectors satisfy the commutative law of addition. For example, the displacement **A** followed by the displacement **B** leads to the same total displacement as **B** followed by **A**. We describe this fact by the equation **A** + **B** = **B** + **A**.

Figure 2 shows three views of a die. The number of dots on the three hidden faces in each view can be inferred from the fact that the sum of the dots on opposite faces is seven. In Figure 2a the die is shown in its initial orientation relative to the axes. In Figure 2b the same die is shown after a rotation of $\theta_1 = 90°$ about the x-axis. In Figure 2c the same die is shown following a second rotation of $\theta_2 = 90°$ about the y-axis. Is the final orientation of the die the same if the two 90° rotations are performed in reverse order—that is, first about the y-axis and then about the x-axis? What does this experiment suggest about the vector character of rotations?

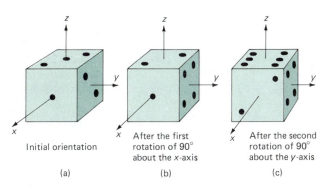

Initial orientation   After the first   After the second
                             rotation of 90°   rotation of 90°
                             about the x-axis   about the y-axis

(a)   (b)   (c)

FIGURE 2

10.2 Rotational Kinetic Energy

I. The moment of inertia of the earth about its spin axis is 8.04×10^{37} kg·m². Use the fact that the earth rotates through 2π radians in 1 sidereal day to determine its rotational kinetic energy.

J. The center of mass of a pitched baseball (radius = 3.8 cm) moves at 38 m/s. It spins about an axis through its center of mass with an angular velocity of 125 rad/s. Calculate the ratio of the spin kinetic energy to the translational kinetic energy. Treat the ball as a uniform sphere.

K. An amusement park ride consists of a rotating cylinder (radius = 3m). Passengers stand inside with their backs to the wall and when the cylinder rotates fast enough, the floor drops out and the passengers stick to the sides of the cylinder. When the rotational speed is 25 rpm, what is the rotational kinetic energy of an 80-kg person? (Treat the person as a point mass.)

L. Treat the earth and moon as homogeneous spheres with $I = (2/5)MR^2$, and calculate their spin kinetic energies. For comparison, their orbital kinetic energies are $K_E = 4.60 \times 10^{26}$ J and $K_M = 3.75 \times 10^{28}$ J.

10.3 Calculation of the Moment of Inertia

M. The moment of inertia of a long, thin rod of mass M and length L is $(1/3)ML^2$ for an axis perpendicular to the length of the rod and through one end of it. Derive an equation for the moment of inertia about a parallel axis a distance x from the center of the rod.

N. A cylinder, a sphere, and a cube are homogeneous and have equal masses. Their dimensions are as shown in Figure 3. For the axes $A–A$ through the center of mass, which has the largest moment of inertia? Which has the smallest?

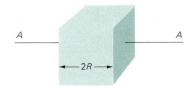

FIGURE 3

P. Derive an equation for the moment of inertia of a hoop of mass M and radius R about an axis in the plane of the hoop tangent to the circumference.

Q. Determine the moment of inertia of a hoop of mass 3 kg and radius 0.38 m about an axis perpendicular to the plane of the hoop and passing through the circumference.

R. Use the parallel axis theorem to deduce the moment of inertia of a homogeneous sphere of mass M and radius R about an axis tangent to its surface.

PROBLEMS

1. One bicycle wheel rotates with a constant angular velocity of 20.6 rad/s. A second wheel starts from rest and undergoes a constant angular acceleration of 1.12 rad/s². (a) How long does it take for the second wheel to rotate through the same angle as the first? (b) Through what angle, in radians, have the wheels rotated in part (a)?

2. One fan blade rotating at 1600 rpm begins to slow down at the same moment a second blade begins to rotate. Four seconds later the two blades have equal angular velocities. The constant angular accelerations of the two blades have equal magnitudes. Determine (a) the magnitude of their angular acceleration, (b) their angular velocity at $t = 4$ s.

3. A 1500-kg car that has been traveling at 16 m/s comes to a stop. Imagine that all of its kinetic energy has been transformed into the rotational kinetic energy of a flywheel. The flywheel is a uniform disk with a radius of 40 cm and a mass of 50 kg. Calculate its angular velocity.

4. Determine the moment of inertia of the L-shaped rod (Figure 4) and the cross (Figure 5) about an axis through O perpendicular to the plane of each object. Treat each of the arms as long thin rods of mass M and length L.

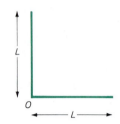

FIGURE 4

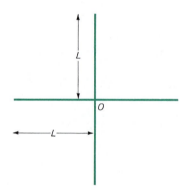

FIGURE 5

5. A wheel is formed from a hoop and 12 spokes (Figure 6). The mass of the hoop is six times the mass of a spoke. Calling M and R the mass and length of a spoke, determine the moment of inertia of the wheel about an axis through its center and perpendicular to a plane containing the wheel.

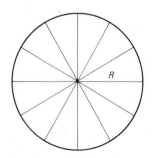

FIGURE 6

6. The density of the earth, a distance r from its center, is given approximately by

$$\rho = \left[14.2 - 11.6\left(\frac{r}{R}\right) \right] \times 10^3 \text{ kg/m}^3$$

where R is the radius of the earth. Show that this density leads to a moment of inertia about an axis through the center given by $I = 0.330\ MR^2$, where M is the mass of the earth. (The value deduced from astronomical observations is $0.331\ MR^2$.)

7. A thin rod of mass M is bent into the shape of a semicircle of radius R (Figure 7). Determine the three moments of inertia about axes through the center of mass and parallel to the x-, y-, and z-axes.

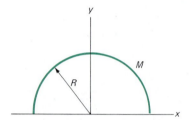

FIGURE 7

8. Three identical point masses (mass M) are attached to a thin rod of length L and mass M, as shown in Figure 8. (a) Determine the moment of inertia about an axis A–A a distance x from the center of the rod and perpendicular to its length. Express your result in terms of M, L, and x. (b) Determine the value of x that minimizes the moment of inertia and express the minimum value in terms of M and L.

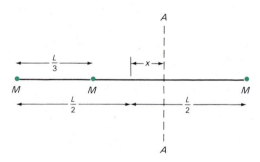

FIGURE 8

9. Calculate the moment of inertia of a spherical shell of mass M about an axis through the center of mass. The inner radius of the shell is R_1; its outer radius is R_2. Express your result in terms of M, R_1, and R_2.

10. As a result of friction the angular velocity of a wheel changes with time according to

$$\frac{d\theta}{dt} = \omega_0 e^{-\sigma t}$$

where ω_0 and σ are constants. The angular velocity changes from 3.5 rad/s at time $t = 0$ to 2.0 rad/s at time $t = 9.3$ s. Use this information to determine σ and ω_0. Then determine (a) the angular acceleration at $t = 3$ s, (b) the number of revolutions the wheel makes in the first 2.5 s, (c) the number of revolutions the wheel makes before coming to rest.

11. Tidal friction causes the rotational energy of a planet to decrease at a rate given by

$$\frac{dK}{dt} = -bK^2 \qquad (b = \text{constant})$$

Let K_0 denote the energy at time $t = 0$ and take

$$bK_0 = \frac{1}{10^9 \text{ years}}$$

(a) Integrate to show that the energy K varies with time according to

$$K = \frac{K_0}{1 + \dfrac{t}{\tau}} \qquad (\tau = 10^9 \text{ years})$$

(b) The planet spins with a period of 10 hours at $t = 0$. Determine its period 3×10^9 years later.

CHAPTER 11

CONSERVATION OF ANGULAR MOMENTUM

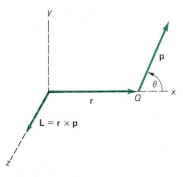

FIGURE 11.1

The linear momentum of the particle is $\mathbf{p}$ when it passes the point Q. The angular momentum of the particle is defined as $\mathbf{L} = \mathbf{r} \times \mathbf{p}$.

11.1
ANGULAR MOMENTUM OF A PARTICLE

As we saw in Chapter 9, the concept of linear momentum is of special importance because it is associated with a conservation law. In this chapter we will see that angular momentum is the rotational analog of linear momentum and it is also associated with a conservation law. The law of conservation of angular momentum is used in studies of a wide range of phenomena, including the emission of light by atoms, spinning ice skaters, and the formation of galaxies. We begin by defining the angular momentum of a particle.

Consider a particle whose linear momentum is $\mathbf{p}$ at the moment it passes the point Q in Figure 11.1. The vector $\mathbf{r}$ locates the position of the particle as it passes the point Q. The angular momentum of the particle is denoted by the vector $\mathbf{L}$ and it is defined by the vector product of $\mathbf{r}$ and $\mathbf{p}$:

$$\mathbf{L} = \mathbf{r} \times \mathbf{p} \tag{11.1}$$

The magnitude of the angular momentum is given by

$$L = rp \sin \theta \tag{11.2}$$

where θ is the smaller angle between the vectors $\mathbf{r}$ and $\mathbf{p}$. The angular momentum vector is perpendicular to the plane containing $\mathbf{r}$ and $\mathbf{p}$, and in a direction determined by a right-hand rule for the vector product. In Figure 11.1 the particle moves in the x–y plane and the vector $\mathbf{L}$ is directed in the positive z-direction.

For a particle of mass m and velocity $\mathbf{v}$, the linear momentum is $\mathbf{p} = m\mathbf{v}$. Therefore, we can write

$$\mathbf{L} = \mathbf{r} \times m\mathbf{v}$$

From this equation we see that the units of angular momentum are kg·m²/s. Because 1 joule·second equals 1 kg·m²/s, the J·s is an equivalent unit of angular momentum.

The angular momentum of a particle depends on the location of the origin through the dependence of $\mathbf{L}$ on $\mathbf{r}$. Thus, in Figure 11.2 the angular momentum vector $\mathbf{L'}$ of the particle relative to the origin at O' differs from the angular momentum $\mathbf{L}$ of the same particle relative to the origin at O. Note that if $\mathbf{p}$ is directed toward or away from the origin, then the angular momentum is zero relative to this origin.

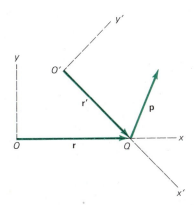

FIGURE 11.2

A particle passing the point Q has a linear momentum $\mathbf{p}$. The angular momentum relative to the origin O' is $\mathbf{L'} = \mathbf{r'} \times \mathbf{p}$. The angular momentum relative to the origin O is $\mathbf{L} = \mathbf{r} \times \mathbf{p}$. In general $\mathbf{L'}$ and $\mathbf{L}$ will not be equal; the magnitude and direction of the angular momentum depend on the coordinate reference frame.

Angular Momentum for Uniform Linear Motion

A particle moving with constant velocity has a constant linear momentum. Equation 11.1 seems to suggest that the angular momentum $\mathbf{r} \times \mathbf{p}$ of a particle with constant $\mathbf{p}$ might vary. However, as we now demonstrate, such a particle has a constant angular momentum.

Consider the particle at the two positions in Figure 11.3 labelled $\mathbf{r}_1$ and $\mathbf{r}_2$. We take $\mathbf{r}_1$ to denote the position along the particle's path nearest to O, and $\mathbf{r}_2$ to denote an arbitrary position along the path. We assign the symbol b to the magnitude of the vector $\mathbf{r}_1$. Because the particle has constant linear momentum, $p_1 = p_2 = mv$. From the definition of angular momentum, $\mathbf{L}_1 = \mathbf{r} \times \mathbf{p}_1$ and $\mathbf{L}_2 = \mathbf{r}_2 \times \mathbf{p}_2$. Using the right-hand rule we find that both $\mathbf{L}_1$ and $\mathbf{L}_2$ are directed into the page. Indeed, $\mathbf{L}$ is directed into the page for all particle positions along the path, and thus $\mathbf{L}$ is constant in direction.

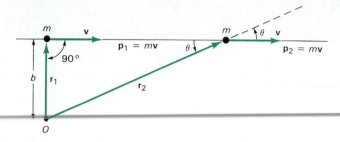

FIGURE 11.3

A particle of mass m and speed v moves in a straight line. Its angular momentum vector **L** is directed into the page and has the magnitude $L = mvb$.

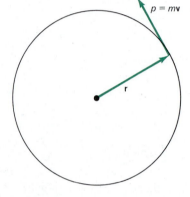

FIGURE 11.4

A particle moves in a circular path with constant speed v. Its angular momentum vector **L** is directed out of the page and has the magnitude $L = mvr$.

Next, we consider the magnitude of **L**. To obtain the magnitude of $\mathbf{L}_1$ and $\mathbf{L}_2$ we use Equation 11.2 to get

$$L_1 = r_1 p_1 \sin 90° = mvr_1 = mvb$$

$$L_2 = r_2 p_2 \sin \theta = mvr_2 \sin \theta$$

Figure 11.3 shows that $r_2 \sin \theta = b$. Therefore

$$L_2 = mvb = L_1$$

Hence the magnitude of **L** is also constant.

Because both the magnitude and direction of the angular momentum are constant, we have shown that the vector **L** is constant.

Angular Momentum for Uniform Circular Motion

Next, we consider a particle moving in a circular path at a constant speed (Figure 11.4). Such a particle will have a constant angular momentum relative to an origin at the center of the circle. Both **r** and **p** have constant magnitudes, and vary in direction with a constant angle of 90° between them. The vector $\mathbf{L} = \mathbf{r} \times \mathbf{p}$ has a constant direction, perpendicular to the plane of the path. The vector **L** is directed out of the page for counterclockwise motion. The constant magnitude of **L** is given by

$$L = rp = mvr \tag{11.3}$$

EXAMPLE 1

Orbital Angular Momentum of the Moon

We can use Equation 11.3 to evaluate the angular momentum of the moon relative to the earth. We take $r = 3.84 \times 10^8$ m and $m = 7.35 \times 10^{22}$ kg. The orbital period of the moon is $T = 2.36 \times 10^6$ s.

The orbital speed of the moon is

$$v = \frac{2\pi r}{T} = \frac{2\pi \cdot 3.84 \times 10^8 \text{ m}}{2.36 \times 10^6 \text{ s}} = 1.02 \times 10^3 \text{ m/s}$$

The magnitude of the moon's angular momentum is

$$L = mvr = (7.35 \times 10^{22} \text{ kg})(1.02 \times 10^3 \text{ m/s})(3.84 \times 10^8 \text{ m})$$
$$= 2.88 \times 10^{34} \text{ kg·m/s}^2$$

The vectors **r** and **p** lie in the moon's orbital plane. The vector $\mathbf{L} = \mathbf{r} \times \mathbf{p}$ is perpendicular to the orbital plane of the moon and is directed (approximately) toward Polaris, the North Pole star.

As we will show later the orbital angular momentum of the moon is not strictly constant. It slowly changes both its magnitude and direction because of the gravitational action of the earth.

There are many examples of systems for which angular momentum is conserved. What criterion determines whether the angular momentum will be constant or will change with time? To answer this question we must determine what causes changes in angular momentum. We start with the definition $\mathbf{L} = \mathbf{r} \times \mathbf{p}$ and differentiate this equation to get $d\mathbf{L}/dt$, the time rate of change of angular momentum of a particle.

$$\frac{d\mathbf{L}}{dt} = \mathbf{r} \times \frac{d\mathbf{p}}{dt} + \frac{d\mathbf{r}}{dt} \times \mathbf{p}$$

Because $\mathbf{v} = d\mathbf{r}/dt$ and $\mathbf{p} = m\mathbf{v}$, we can write the second term as

$$\frac{d\mathbf{r}}{dt} \times \mathbf{p} = \mathbf{v} \times m\mathbf{v}$$

But $\mathbf{v} \times m\mathbf{v} = m(\mathbf{v} \times \mathbf{v}) = 0$, so the second term equals zero, leaving

$$\frac{d\mathbf{L}}{dt} = \mathbf{r} \times \frac{d\mathbf{p}}{dt}$$

Newton's second law relates $d\mathbf{p}/dt$ to the net force $\mathbf{F}$ acting on the particle

$$\mathbf{F} = \frac{d\mathbf{p}}{dt}$$

Replacing $d\mathbf{p}/dt$ by $\mathbf{F}$ gives

$$\frac{d\mathbf{L}}{dt} = \mathbf{r} \times \mathbf{F}$$

We observe that $\mathbf{r} \times \mathbf{F}$ is the torque acting on the particle

$$\boldsymbol{\tau} = \mathbf{r} \times \mathbf{F}$$

Thus, the time rate of change of angular momentum equals the torque acting on the particle

$$\boldsymbol{\tau} = \frac{d\mathbf{L}}{dt} \tag{11.4}$$

Equation 11.4 is the rotational analog of Newton's second law. Force determines the rate of change of *linear momentum*; torque determines the rate of change of *angular momentum*. The analogies between the linear and rotational forms of Newton's second law are summarized in Table 11.1.

TABLE 11.1
Linear and Rotational Analogs

Linear		Rotational
$\mathbf{p}$	← momentum →	$\mathbf{L}$
$\mathbf{F}$	← force torque →	$\boldsymbol{\tau}$
$\mathbf{F} = \dfrac{d\mathbf{p}}{dt}$	← law of motion →	$\boldsymbol{\tau} = \dfrac{d\mathbf{L}}{dt}$

Conservation of Angular Momentum for a Particle

When the net torque acting on a particle is zero, Equation 11.4 shows that $d\mathbf{L}/dt$ equals zero; the angular momentum of the particle does not change and we say that angular momentum is conserved. Thus,

$$\mathbf{L} = \text{constant} \qquad \text{when} \qquad \tau = 0 \qquad (11.5)$$

Equation 11.5 is the law of conservation of angular momentum for a particle. The criterion that must be satisfied if angular momentum is to be conserved is that the net torque equals zero.

One situation where the net torque is zero, even though a net force acts, is the case of a **central force**. A central force is one whose direction is always toward or away from a fixed point (Figure 11.5). If $\mathbf{r}$ is the vector from the force center to the particle, then $\mathbf{F}$ is along $\mathbf{r}$. Electrical and gravitational forces provide important examples of central forces. When $\mathbf{F}$ and $\mathbf{r}$ are parallel their vector product is zero. Therefore, for a central force $\mathbf{r} \times \mathbf{F} = 0$ and $\tau = 0$. The angular momentum of a particle moving under the influence of a central force is conserved.

Kepler's Second Law

Just prior to the invention of the telescope the astronomer Tycho Brahe recorded precise naked-eye observations of planetary positions. Working with Brahe's data for Mars, Johannes Kepler found that the orbit of Mars was an ellipse with the sun at one focus. This result, subsequently generalized to all planets, is now known as Kepler's first law of planetary motion:

Planets move in elliptic orbits with the sun at one focus.

Kepler's second law of planetary motion states:

During equal time intervals, equal areas are swept out by the line connecting the planet and the sun.

Kepler deduced these laws empirically through trial and error. We can show that Kepler's second law is a consequence of angular momentum conservation.

Consider the part of the planetary orbit exhibited in Figure 11.6. The planet moves from P to Q with velocity $\mathbf{v}$ in a short time interval Δt. The radius vector $\mathbf{r}$ sweeps out the shaded area in Figure 11.6 in the time Δt. This area is one-half the area of the parallelogram whose sides are formed by the vectors $\mathbf{r}$ and $\mathbf{v}\,\Delta t$. The area of the parallelogram formed by $\mathbf{r}$ and $\mathbf{v}\,\Delta t$ equals the

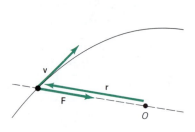

FIGURE 11.5

As the particle moves along the curved path it experiences a central force **F**, directed toward the fixed point O. The torque $\tau = \mathbf{r} \times \mathbf{F}$ is zero so the angular momentum of the particle does not change.

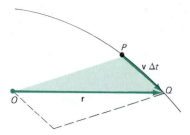

FIGURE 11.6

As the planet moves from P to Q, the line from the sun at O to the planet sweeps out the shaded area. This area is one half the area of the parallelogram $OPQRO$.

magnitude of their *vector product* (see Section 2.5). Thus, the area ΔA swept out in time Δt is given by

$$A = \tfrac{1}{2}|\mathbf{r} \times \mathbf{v}\,\Delta t|$$

Factoring Δt and multiplying both numerator and denominator by the mass m gives

$$\Delta A = \frac{\Delta t}{2m}|\mathbf{r} \times m\mathbf{v}|$$

We recognize that $|\mathbf{r} \times m\mathbf{v}| = L$ is the magnitude of the angular momentum. Thus the area swept out in the time Δt is

$$\Delta A = \frac{L\,\Delta t}{2m}$$

The rate at which area is swept out is $\Delta A/\Delta t$

$$\frac{\Delta A}{\Delta t} = \frac{L}{2m} \tag{11.6}$$

The gravitational force exerted on the planet by the sun is a central force, so the torque is zero. It follows that $\mathbf{L}$ is constant and therefore it also follows that $L/2m$ is constant. Kepler's second law is a statement that $\Delta A/\Delta t$ is constant. Equation 11.6 shows that Kepler's second law is a consequence of angular momentum conservation.

EXAMPLE 2

Halley's Comet

Halley's comet moves about the sun in an elliptic orbit (Figure 11.7) with perihelion (closest to the sun) distance $r_p = 0.59$ AU and aphelion (farthest from the sun) distance $r_a = 35$ AU (1 AU $= 1.5 \times 10^{11}$ m). We can use the conservation of angular momentum to find the aphelion speed if we know the perihelion speed is $v_p = 5.4 \times 10^4$ m/s.

The angular momentum of the comet is conserved because the sun exerts a central force on it. The angular momentum at perihelion equals the angular momentum at aphelion. Applying angular momentum conservation at perihelion and aphelion we write

$$mr_pv_p = mr_av_a$$

The aphelion speed is

$$v_a = v_p\left(\frac{r_p}{r_a}\right) = (5.4 \times 10^4 \text{ m/s})\left(\frac{0.59 \text{ AU}}{35 \text{ AU}}\right) = 910 \text{ m/s}$$

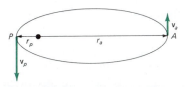

FIGURE 11.7

The angular momentum of Halley's comet may be written $L = mr_av_a$ in terms of aphelion quantities or $L = mr_pv_p$ in terms of perihelion quantities. The two expressions for L are equal because angular momentum is conserved.

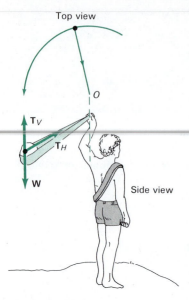

FIGURE 11.8

If the thrower's hand is motionless, the stone in the sling moves with constant speed in a horizontal circle.

To illustrate a physical situation where the angular momentum of a particle may or may not be conserved we next consider the motion of an ancient sling, such as that used by David to slay Goliath. For centuries this weapon competed favorably with the bow and arrow.

EXAMPLE 3

The Sling

A sling consists of a stone and strap. The geometry of this device is shown in Figure 11.8. The stone moves in a horizontal circle. The angular momentum of the stone is vertical, perpendicular to the plane of motion.

Initially we consider motion in which the thrower's hand remains essentially fixed, as Figure 11.8 suggests. The strap exerts a force whose vertical component, T_V, balances the weight of the stone. The horizontal component of the force exerted by the strap, T_H, is central, being directed toward the motionless hand of the thrower (point O in Figure 11.8). Relative to a torque reference point at O, the net torque exerted on the stone is zero, so its angular momentum remains constant.

Next, we suppose that the thrower moves his hand in a horizontal circle of radius b (Figure 11.9). The horizontal component, T_H, acts along the strap. In general, the line of action of this force does not pass through the point O, and the stone experiences a net torque. Angular momentum is not conserved. The thrower coordinates the motion so that the net torque increases the angular momentum of the stone. The increase in the angular momentum appears as an increase in the speed of the stone.

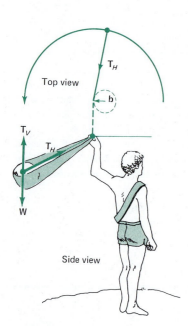

FIGURE 11.9

If the thrower's hand moves in a small circle (radius b), a torque is developed that increases the angular momentum and the speed of the stone.

11.2
ANGULAR MOMENTUM OF A SYSTEM OF PARTICLES

In the preceding section we defined the angular momentum for a single particle. Now we extend the angular momentum principles to a system of particles. We define the angular momentum of a system of particles to be the vector sum of the individual angular momentum vectors. Symbolically,

$$\mathbf{L} = \sum \mathbf{L}_i$$

To determine the dynamical equation governing $\mathbf{L}$ we form the time derivative of $\mathbf{L}$:

$$\frac{d\mathbf{L}}{dt} = \sum \frac{d\mathbf{L}_i}{dt}$$

We use Equation 11.4 to set $\boldsymbol{\tau}_i = d\mathbf{L}_i/dt$. This gives

$$\frac{d\mathbf{L}}{dt} = \sum \boldsymbol{\tau}_i$$

In general the torque acting on each particle is composed of external torques and internal torques exerted by other particles in the system. The sum of the internal torques is zero, a consequence of Newton's third law. The sum $\sum \boldsymbol{\tau}_i$ is therefore the vector sum of external torques. We denote this sum simply as $\boldsymbol{\tau}$.

$$\sum \boldsymbol{\tau}_i = \boldsymbol{\tau}$$

The resulting equation for $d\mathbf{L}/dt$ is

$$\tau = \frac{d\mathbf{L}}{dt} \qquad (11.7)$$

You will notice that Equation 11.7, which describes the rotational dynamics of a system of particles, has the same form as Equation 11.4, which describes a single particle. In Equation 11.7, $\mathbf{L}$ is the vector sum of the angular momenta of the particles comprising the system, and τ is the vector sum of the external torques acting on the system.

The Conservation of Angular Momentum

If the external torque is zero, $d\mathbf{L}/dt$ is zero and $\mathbf{L}$ is constant. Thus, just as we found for a single particle, there is a law of angular momentum conservation for a system of particles:

> If the net external torque acting on a system is zero, then the total angular momentum of the system is conserved.

In equation form

$$\mathbf{L} = \text{constant} \qquad \text{when} \qquad \tau = 0 \qquad (11.8)$$

A system experiencing zero external torque can exhibit striking changes in its rotational motion through the action of its internal forces, even though the angular momentum of the system is conserved. Insight into this behavior is provided by the expressions for the angular momentum of a particle moving in a circle

$$L = mvr \qquad (11.3)$$

This equation shows that if r decreases, then v must increase in order to keep L constant.

For example, a figure skater begins her final spin about a vertical axis with her arms and legs as far from the axis of rotation as possible (Figure 11.10). Her angular momentum about a vertical axis is nearly constant because the ice exerts only a weak frictional torque. As she draws her arms and legs closer to the spin axis, reducing r, her average tangential speed must increase. Her speed of rotation increases just enough to keep her angular momentum constant.

The conservation of linear momentum and the conservation of angular momentum are mutually independent conditions. It is possible for a system to have its linear momentum conserved while its angular momentum changes, and vice versa. As an example, consider the view of the top of a revolving door, shown in Figure 11.11. The forces $\mathbf{F}_1$ and $\mathbf{F}_2$ exerted on the door are equal in magnitude and opposite in direction, so the net force acting on the door is zero. The linear momentum of the door remains zero because the net force is zero.

However, the torques τ_1 and τ_2 set up by $\mathbf{F}_1$ and $\mathbf{F}_2$ are in the same direction—that is up, out of the page. Their vector sum is not zero, so there is a net torque acting on the door. This net torque sets the door into rotation and changes its angular momentum.

Large r, small v

Small r, large v

FIGURE 11.10

To initiate her final spin, a figure skater extends both arms and one leg as far from the spin axis as possible. To complete the spin she brings all parts of her body as close to the axis as possible.

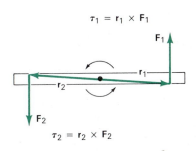

FIGURE 11.11

Top view of a revolving door. Equal but opposite forces are exerted on the door. The vector sum of the forces is zero, so the net force on the door is zero. The torques produced by the two forces are in the same direction—up, out of the page—so their vector sum is not zero. The door is set into rotation because the net torque is not zero.

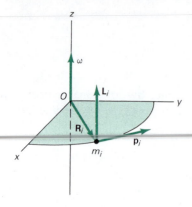

FIGURE 11.12

A particle moving in the x–y plane has linear momentum $\mathbf{p}_i$ and moves in a circle of radius $\mathbf{R}_i$. With the directions shown for $\mathbf{p}_i$ and $\mathbf{R}_i$, the angular momentum vector $\mathbf{L}_i = \mathbf{R}_i \times \mathbf{p}_i$ is in the positive z-direction.

Spin Angular Momentum

When a rigid body rotates about a fixed axis all parts of the body have the same angular velocity. Under these circumstances the angular momentum of the body can be expressed in the vector form

$$\mathbf{L} = I\boldsymbol{\omega} \tag{11.9}$$

where I is the moment of inertia of the body about the axis of rotation and $\boldsymbol{\omega}$ is its angular velocity vector.

To derive Equation 11.9 we consider a particle in a rigid body rotating about the z-axis with an angular velocity ω. As Figure 11.12 shows, the ith particle moves in a circle of radius R_i. Its mass is m_i and its linear momentum is

$$p_i = m_i v_i = m_i R_i \omega$$

where $v_i = R_i \omega$ expresses the speed of the particle in terms of R_i and the angular velocity ω. The angular momentum of the particle is

$$L_i = m_i R_i^2 \omega$$

As Figure 11.12 shows, the vector $\mathbf{L}_i$ is along the positive z-direction. The vector form of $L_i = m_i R_i^2 \omega$ is

$$\mathbf{L}_i = m_i R_i^2 \boldsymbol{\omega}$$

The vector $\boldsymbol{\omega}$ is directed along the positive z-axis. The total angular momentum is obtained by summing over all particles comprising the body

$$\mathbf{L} = \sum \mathbf{L}_i = \left(\sum m_i R_i^2 \right) \boldsymbol{\omega} \tag{11.10}$$

The sum in parentheses is the moment of inertia about the axis of rotation

$$I = \sum m_i R_i^2 \tag{11.11}$$

Inserting this result into Equation 11.10 shows that $\mathbf{L} = I\boldsymbol{\omega}$, as given by Equation 11.9. The relation $\mathbf{L} = I\boldsymbol{\omega}$ is the rotational analog of the linear momentum relation $\mathbf{p} = m\mathbf{v}$. Comparing these two equations we see that the moment of inertia is the rotational analog of mass, a conclusion we also drew in our discussion of rotational kinetic energy.

Because $\mathbf{L}$ is directly proportional to the angular velocity of the spinning body, the quantity $\mathbf{L} = I\boldsymbol{\omega}$ is often referred to as the **spin angular momentum.**

EXAMPLE 4

The Conservation of Spin Angular Momentum

A circular disk spins on a thin air film. The moment of inertia of the disk about its axis of rotation is 0.062 kg·m². The disk rotates with a period of 0.36 s. The magnitude of the spin angular momentum is given by

$$L = I\omega$$

The angular velocity ω is related to the period T by

$$\omega = \frac{2\pi}{T}$$

Thus

$$L = \frac{2\pi l}{T} = \frac{2\pi \cdot 0.062 \ \text{kg} \cdot \text{m}^2}{0.36 \ \text{s}}$$
$$= 1.08 \ \text{kg} \cdot \text{m}^2/\text{s}$$

A small mass is dropped onto the rotating disk and subsequently rotates with the disk. The moment of inertia of the mass about the axis of rotation is 0.046 kg·m². We want to determine the final period of the rotating disk and mass.

When the disk and mass collide they exert equal and opposite torques on each other. No external torques act, so the angular momentum of the combined system is conserved. The initial angular momentum of the mass is zero, so the total angular momentum of the system is $L = 1.08$ kg·m²/s. The final angular momentum is given by

$$L_f = (I_{\text{disk}} + I_{\text{mass}})\omega_f = (I_{\text{disk}} + I_{\text{mass}}) \frac{2\pi}{T_f}$$

where T_f is the final period. The conservation of angular momentum lets us set $L_f = L = 1.08$ kg·m²/s. Solving for T_f gives

$$T_f = \frac{2\pi(I_{\text{disk}} + I_{\text{mass}})}{L} = \frac{2\pi(0.062 + 0.046) \ \text{kg} \cdot \text{m}^2}{1.08 \ \text{kg} \cdot \text{m}^2/\text{s}}$$
$$= 0.63 \ \text{s}$$

Qualitatively we can see how angular momentum conservation requires that the final period be greater than the initial period. From

$$L = I\omega = \frac{2\pi I}{T}$$

it follows that if I increases then T must also increase in order for L to remain constant.

11.3
COMBINED SPIN AND ORBITAL ANGULAR MOMENTUM

Many systems possess two distinctive types of angular momentum. For example, the angular momentum associated with the earth's orbital motion about the sun is called its **orbital angular momentum**. The angular momentum associated with the spinning of the earth about its own axis is its spin angular momentum.

It can be shown that the total angular momentum of any system can be expressed as the vector sum of its orbital and spin angular momenta

$$\mathbf{L} = \mathbf{L}_o + \mathbf{L}_s$$

Here, $\mathbf{L}$ is the total angular momentum of the system, $\mathbf{L}_o$ is the orbital angular momentum, and $\mathbf{L}_s$ is the spin angular momentum.

When the external torque acting on the system equals zero, the angular momentum of the system is conserved. In this situation, any change in the spin angular momentum must be compensated for by an equal but opposite change in the orbital angular momentum. The earth-moon system provides an example of such a situation.

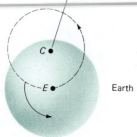

FIGURE 11.13

The earth and moon orbit their center of mass at *C*. Additionally, each spins about its own axis.

Angular Momentum Conservation in the Earth–Moon System

Consider the earth and its moon as an isolated system. The earth and moon each orbit the center of mass *C*, as suggested in Figure 11.13. Additionally, the earth and the moon each spin about their own axes. The total angular momentum of the system has two spin contributions and two orbital contributions. The spin angular momentum of the moon and the orbital angular momentum of the earth are relatively small, and can be neglected. The total angular momentum of the earth–moon system can be represented by

$$\mathbf{L} = \mathbf{L}_o + \mathbf{L}_s \tag{11.12}$$

where $\mathbf{L}_o$ is the orbital angular momentum of the moon and $\mathbf{L}_s$ is the spin angular momentum of the earth.

Because **L** is constant, any change in the spin angular momentum of the earth is accompanied by an equal but opposite change in the orbital angular momentum of the moon. The spin angular momentum of the earth and the orbital angular momentum of the moon are not precisely parallel, but to simplify the analysis we take them to be parallel. The conservation of angular momentum, then, requires that the sum of the magnitudes of $\mathbf{L}_o$ and $\mathbf{L}_s$ remains constant

$$L_o + L_s = \text{constant}$$

The spin angular momentum of the earth can be written as

$$L_s = \frac{2\pi I_E}{T_E} \tag{11.13}$$

where I_E is the moment of inertia of the earth about its spin axis and T_E is its rotational period (1 sidereal day).

The orbital angular momentum of the moon can be written as

$$L_o = M_M v r$$

Using Newton's second law and the law of gravitation we get

$$\frac{M_M v^2}{r} = \frac{GM_E M_M}{r^2}$$

which leads to the result

$$L_o = \sqrt{GM_M^2 M_E r} \tag{11.14}$$

The angular momentum of the earth–moon system, a conserved quantity, is

$$L_o + L_s = \sqrt{GM_M^2 M_E r} + \frac{2\pi I_E}{T_E} = \text{constant} \tag{11.15}$$

The two quantities in Equation 11.15 that can undergo systematic changes are T_E, the earth's rotational period, and *r*, the earth–moon distance. You will remember that the earth's rotational motion is systematically slowing down because of *tidal friction*. We can understand this slowing down if we think of the earth as a giant flywheel rotating inside a brake drum. The tidal friction in the earth's crust, atmosphere, and oceans acts like a brake and transforms rotational kinetic energy into heat. As the earth's rate of rotation decreases, the time it takes for it to rotate increases. The time it takes for the earth to rotate is the duration of the day, T_E. Thus, tidal friction causes a systematic increase in T_E. As Equation 11.15 shows, an increase in T_E requires an increase in r, the earth–moon distance. The increase in T_E causes a decrease in the earth's spin angular momentum. The increase in r causes an increase in the orbital angular momentum of the moon. The changes in T_E and r are related by the angular momentum conservation condition, Equation 11.15.

The earth–moon distance is measured regularly with great precision by bouncing laser beams off reflectors placed on the moon by United States astronauts. These measurements confirm that the earth–moon distance is increasing a few centimeters each year. The earth's rotational period increases by about one millisecond per century. Over a time of hundreds of millions of years this minuscule variation accumulates into a significant change in the length of the day. Geologic studies indicate that the day was about 20 hours long 700 million years ago (Figure 11.14). At that time the earth–moon distance was approximately 3.53×10^8 m, or 92% of its present value. Ultimately the earth and moon will reach a *tidal lock* condition in which the rotational period of the earth equals the orbital period of the moon. This condition is referred to as a lock because once it is reached, the tidal friction that produced it drops to zero, preventing further change. When the tidal lock is achieved, the length of the day will be about 47 times its present value, and the earth–moon distance will be approximately 1.44 times its present value.

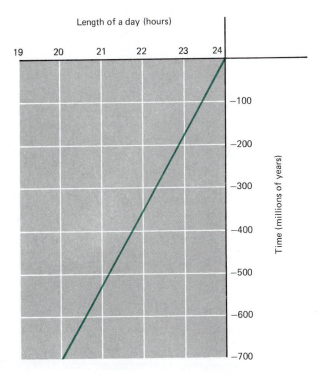

FIGURE 11.14

Geologic studies indicate that the duration of the day has increased systematically. The day was only 20 hours long 700 million years ago.

EXAMPLE 5

Tidal Lock

We can use the conservation of angular momentum to determine the tidal-lock configuration of the earth–moon system. The total angular momentum of the earth–moon system can be evaluated using Equation 11.15. With

$$M_M = 7.35 \times 10^{22} \text{ kg}$$

$$r = 3.84 \times 10^8 \text{ m}$$

$$T_E = 8.62 \times 10^4 \text{ s}$$

$$M_E = 5.98 \times 10^{24} \text{ kg}$$

$$I_E = 8.04 \times 10^{37} \text{ kg·m}^2$$

we find

$$L_o = \sqrt{GM_M{}^2 M_E r} = 2.88 \times 10^{34} \text{ kg·m}^2/\text{s}$$

$$L_s = \frac{2\pi I_E}{T_E} = 0.586 \times 10^{34} \text{ kg·m}^2/\text{s}$$

The total angular momentum of the earth–moon system is

$$L_o + L_s = 3.47 \times 10^{34} \text{ kg·m}^2/\text{s}$$

Angular momentum conservation requires that the combined spin and orbital angular momenta have this same value when tidal lock is achieved

$$L_o(\text{lock}) + L_s(\text{lock}) = 3.47 \times 10^{34} \text{ kg} \cdot \text{m}^2/\text{s} \qquad (11.16)$$

As the earth–moon distance increases, the orbital angular momentum of the moon increases. The tidal-lock value of the orbital angular momentum can be related to its present value. Thus, with r_L denoting the tidal-lock value of the earth-moon distance, and r denoting its present value,

$$L_o(\text{lock}) = \sqrt{GM_M{}^2 M_E r_L} = \sqrt{GM_M{}^2 M_E r}\ \sqrt{\frac{r_L}{r}}$$

Numerically,

$$L_o(\text{lock}) = 2.88 \times 10^{34} \text{ kg·m}^2/\text{s}\ \sqrt{\frac{r_L}{r}}$$

We can use angular momentum conservation to determine r_L/r. Although the earth will still be rotating when tidal lock is achieved, its spin angular momentum will be negligible in comparison to the moon's orbital angular momentum. With

$$L_s(\text{lock}) \approx 0$$

the conservation of angular momentum, Equation 11.16, becomes

$$2.88 \times 10^{34} \text{ kg·m}^2/\text{s}\ \sqrt{\frac{r_L}{r}} = 3.47 \times 10^{34} \text{ kg·m}^2/\text{s}$$

This gives

$$r_L = \left(\frac{3.47}{2.88}\right)^2 r = 1.44\ r \qquad (11.17)$$

This means that at tidal lock the earth–moon distance will be 1.44 times its present value.

Having determined r_L/r we can evaluate the tidal-lock period of the earth's rotation. First, we use Kepler's third law to write

$$T_L = \left(\frac{r_L}{r}\right)^{3/2} T_M \qquad (11.18)$$

where T_L is the orbital period of the moon at tidal lock and where

$$T_M = 27.3 \text{ sidereal days}$$

is the moon's present orbital period. With $r_L/r = 1.44$

$$T_L = (1.44)^{3/2}\ 27.3 = 47.2 \text{ sidereal days}$$

The condition for tidal lock is that the *earth's spin period equal the moon's orbital period.* The earth will then always present the same face to the moon. This means that the length of the day at tidal lock is related to its present value by

$$T_E(\text{lock}) = T_L = 47.2 \text{ sidereal days}$$

At tidal lock the day will be 47.2 times its present duration.

These calculations do not take into account the effect of the sun, which exerts significant tidal torques on the earth and moon.

Tidal friction has also affected the moon's rotation. The earth and moon exert equal but oppositely directed tidal torques on each other. Furthermore, the moment of inertia of the moon is approximately 1000 times smaller than that of the earth. The dynamic relation $\tau = I\alpha$ then explains why the angular deceleration caused by the tidal torques has slowed the moon's rotation more rapidly than it has slowed the earth's rotation.

The tidal torque exerted on the moon by the earth has brought the moon into a tidal lock with the earth. As a consequence the moon always presents the same face toward the earth. The moon's spin period equals its orbital period. The tidal torque exerted on the earth by the moon is still driving the earth toward a tidal lock. When the earth reaches a tidal lock, it will always present the same face toward the moon.

WORKED PROBLEM

Two astronauts (Figure 11.15), each having a mass of 75 kg, are connected by a 10-m rope of negligible mass. They are isolated in space, orbiting their center of mass at speeds of 5 m/s. (a) Calculate the magnitude of the angular momentum of the system by treating the astronauts as particles. (b) Calculate the kinetic energy of the system.

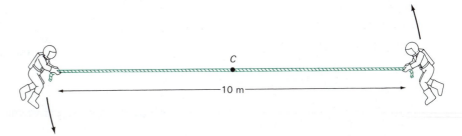

C

—10 m—

FIGURE 11.15

By pulling in on the rope, the astronauts shorten the distance between them to 5 m. (c) What is the new angular momentum of the system? (d) What are their new speeds? (e) What is the new kinetic energy of the system? (f) How much work is done by the astronauts in shortening the rope?

Solution

Solution of this problem involves an understanding of the law of angular momentum conservation, and the work-energy principle.

(a) The angular momentum vectors of the astronauts are parallel and equal in magnitude. Treating the astronauts as particles, the angular momentum of the system is

$$L = 2(mvr) = 2(75 \text{ kg})(5 \text{ m/s})(5 \text{ m})$$
$$= 3.75 \times 10^3 \text{ kg·m}^2/\text{s}$$

(b) The kinetic energy of the system is

$$K = 2(\tfrac{1}{2} mv^2) = (75 \text{ kg})(5 \text{ m/s})^2$$
$$= 1.88 \text{ kJ}$$

(c) The forces exerted on the astronauts by the rope produce no net torque. They are equal and opposite *internal* forces that act to hold the two astronauts together. Because the net torque acting on the system is zero, its angular momentum remains conserved and remains 3.75×10^3 kg· m^2/s.

(d) Calling V and R the new orbital speed and radius, and invoking conservation of angular momentum, we have

$$2(mVR) = 3.75 \times 10^3 \text{ kg·m}^2/\text{s}$$

Substituting $m = 75$ kg and $R = 2.5$ m we find $V = 10$ m/s. The speed is doubled when the radius is halved.

(e) The new total kinetic energy is,

$$K = 2(\tfrac{1}{2}mV^2) = (75 \text{ kg})(10 \text{ m/s})^2$$
$$= 7.5 \text{ kJ}$$

When the speed doubles, the kinetic energy increases by a factor of four. Kinetic energy is not conserved; the astronauts do work as they shorten their separation.

(f) We invoke the work-energy principle to compute the work done by the astronauts:

$$W = \Delta K = 7.50 \text{ kJ} - 1.88 \text{ kJ}$$
$$= 5.62 \text{ kJ}$$

EXERCISES

11.1   Angular Momentum of a Particle

A.   A 1000-kg car travels on a circular track 4 km in circumference. The driver completes one lap at 300 km/h. What is the magnitude of the angular momentum of the car relative to the center of the track?

B.   A 1500-kg car travels down a highway at constant velocity ($v = 80$ km/h). A farmer located at a point 100 m off the highway observes the car. Determine the magnitude of the car's angular momentum (a) relative to a car ahead, moving along the same line at 90 km/h, (b) relative to the farmer.

C.   An alpha particle has a kinetic energy of 6.60 MeV. It starts along a path that would carry it past a nucleus at a distance of 3.12 fm if it were not deflected by the electrical force exerted by the nucleus (Figure 1). Determine the magnitude of its angular momentum. Express your result as a multiple of $\hbar$ ($\hbar = 1.05 \times 10^{-34}$ J·s).

D.   Assume the orbit of the earth is circular. Determine the area per second swept out by the earth's orbital radius.

E.   A comet orbits the sun in an elliptic orbit. The ratio of aphelion radius to perihelion radius is $r_a/r_p = 10$.

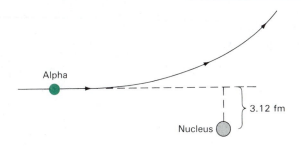

FIGURE 1

If the perihelion speed is $v_p = 3 \times 10^5$ m/s, what is its aphelion speed?

F. A 50-gram particle passes the position $x = +2$m, $y = -3$ m with the velocity $v_x = +6$ m/s, $v_y = +2$ m/s. Determine its angular momentum (magnitude and direction).

G. A 1.5-kg particle moves with velocity $\mathbf{v} = (4.2\mathbf{i} - 3.6\mathbf{j})$ m/s. Determine its angular momentum when its position is $\mathbf{r} = (1.5\mathbf{i} + 2.2\mathbf{j})$ m.

H. A particle ($m = 1.2$ kg) located at $x = 2.5$ m, $y = 4.3$ m, $z = 0$ moves in the $+x$-direction at 12 m/s. Determine its angular momentum relative to the point $x = 5.0$ m, $y = -5.0$ m, $z = 0$.

I. A simple pendulum swings in a fixed plane. (a) Is the magnitude of its angular momentum relative to the support point constant? Explain. (b) Is the direction of its angular momentum constant? Explain. (c) Is its angular momentum conserved?

11.2 Angular Momentum of a System of Particles

J. Calculate the magnitude of the combined angular momentum of the earth and moon as they orbit their common center of mass. Take each body to be a point mass located at its own center of mass. The center of mass of the earth–moon system is located 4.66×10^6 m from the center of the earth.

K. Two 80-kg ice skaters (Figure 2) move in opposite directions along parallel lines separated by 2 m at speeds of 15 m/s. Determine the magnitude of the angular momentum of the system relative to origins at points A, B, and C.

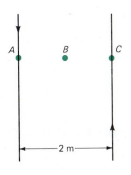

FIGURE 2

L. Two astronauts connected by a rope orbit their center of mass. Assume that only the rope exerts forces on them. By pulling on the rope, the astronauts move closer to each other. (a) Is the total angular momentum conserved? Explain. (b) Is the kinetic energy of the astronauts conserved? Explain.

M. Three identical particles each have an angular momentum of magnitude $\hbar = 1.05 \times 10^{-34}$ J · s. Draw figures to show how the three angular momenta can be oriented to give a total angular momentum (vector sum) with magnitudes of 0, $\hbar$, $2\hbar$, and $3\hbar$.

N. Two 2-kg particles move with velocities of $+2\mathbf{j}$ m/s and $-2\mathbf{j}$ m/s at the moment when they have the positions $x = +2$ m and $x = -2$ m. Calculate the angular momentum of the system relative to (a) the origin, (b) the point $x = +1$ m, $y = 0$, (c) the point $x = +2$ m, $y = 0$.

P. Two particles confined to the same plane move in circular paths about a common center. The first particle ($m = 5.2$ grams) has an orbital speed of 31.0 m/s and an orbit radius of 2.1 m. The second particle ($m = 8.1$ grams) has an orbital speed of 26.1 m/s. (a) Determine the orbit radius of the second particle if the total angular momentum of the system is zero. (b) Determine the magnitude of the total angular momentum of the system if the second particle reverses its direction.

Q. The velocity of a particle is $\mathbf{v} = (3.12\mathbf{i} + 4.82\mathbf{j})$ m/s. When the particle crosses the x-axis at $x = 2.13$ m, another identical particle traveling parallel to the x-axis crosses the y-axis at $y = 1.32$ m. What is the velocity of the second particle if the angular momentum of the system is zero?

R. A pair of astronauts are in space at opposite ends of a 20-m cord. They rotate about their center of mass with a period of 32 s. They begin shortening the cord. What is the period of their motion when they are 12 m apart? Treat the astronauts as equal point masses.

11.3 Combined Spin and Orbital Angular Momentum

S. The electron has a spin angular momentum $L_s = h/4\pi = 0.527 \times 10^{-34}$ J·s. (a) Determine the magnitude of its orbital angular momentum as it moves in a circular orbit about the proton in a hydrogen atom. Take $r = 0.529 \times 10^{-10}$ m and $v = 2.20 \times 10^6$ m/s. (b) If the spin and orbital angular momentum vectors must be either parallel or antiparallel, what are the possible values for the magnitude of the total angular momentum?

T. Treat the moon as a uniform sphere with a mass of 7.35×10^{22} kg and a radius of 1.74×10^6 m that spins on its axis with a period of 2.36×10^6 seconds. Determine the magnitude of its spin angular momentum. Compare its spin angular momentum to its orbital angular momentum (see Example 5). Is it reasonable to neglect the moon's spin angular momentum in comparison to its orbital angular momentum?

PROBLEMS

1. Prove that the orbit of a particle lies in a fixed plane if its angular momentum is constant.

2. The orbital angular momentum of the earth, relative to the center of mass of the earth–moon system, is given by

$$L_E = M_E v r_E = \frac{2\pi M_E r_E^2}{T}$$

where M_E is the mass of the earth, r_E is the distance of the center of the earth from the center of mass of the earth–moon system, and T is the orbital period of the earth–moon system about its center of mass. A similar expression can be written for L_M, the angular momentum of the moon. Show that

$$\frac{L_E}{L_M} = \frac{M_M}{M_E}$$

and use the tabulated masses (inside rear cover) of the moon and earth to evaluate the ratio L_E/L_M.

3. A pair of 2-kg particles are attached by a light string so that they rotate about an axis perpendicular to the plane of the page and through their *moving* center of mass. At a certain moment their positions and velocities are as shown in Figure 3. Calculate the total angular momentum of the system (a) relative to the origin, (b) relative to the center of mass.

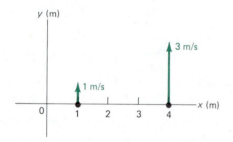

FIGURE 3

4. Two identical stars (mass m) move in a circular orbit about their center of mass. Show that the magnitude of the total angular momentum is given by

$$L^2 = Gm^3r$$

where r is the radius of the orbit and G is the universal gravitation constant.

5. The earth possesses orbital angular momentum by virtue of its motion around the sun, and spin angular momentum due to its daily rotation. If its orbital angular momentum could be completely transformed into spin angular momentum, what would be its spin period? Could the earth withstand the stresses set up by such a spin period? The moment of inertia of the

earth about its spin axis is 8.04×10^{37} kg·m². Ignore the effects of the moon.

6. A conical pendulum consists of a bob of mass m in motion in a circular path in a horizontal plane as shown in Figure 4. During the motion, the supporting wire of length ℓ maintains the constant angle θ with the vertical. Prove that

$$L^2 = m^2g\ell^3\left(\frac{\sin^4\theta}{\cos\theta}\right)$$

where L is the magnitude of the angular momentum relative to the support point.

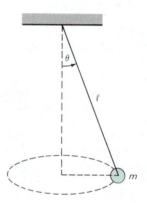

FIGURE 4

7. A particle of mass m moves along a circular path described by

$$\mathbf{r} = \mathbf{i}R\cos\omega t + \mathbf{j}R\sin\omega t$$

Prove that the angular momentum relative to a point located by the position vector $\mathbf{d}$ is given by

$$\mathbf{L} = mR^2\omega\mathbf{k} + mR\omega[-(\mathbf{d}\times\mathbf{i})\sin\omega t + (\mathbf{d}\times\mathbf{j})\cos\omega t]$$

This shows that the angular momentum is constant only if $\mathbf{d}$ lies along the z-axis ($d_x = 0$, $d_y = 0$).

8. If the orbital radius of the moon increases by 5%, by how much does the length of the day increase?

9. The period of a 150-kg satellite orbiting the earth in a circular orbit equals the rotational period of the earth. Determine (a) the magnitude of the linear momentum of the satellite, (b) the magnitude of the angular momentum of the satellite, (c) the time rate at which area is swept out by a line from the center of the earth to the satellite.

10. Two identical disks separately rotate without friction on a common axis (Figure 5). One disk rotates with an angular velocity ω. The other rotates in the same sense with an angular velocity of 4ω. The disks

then collide and stick together. (a) Calculate their final angular velocity in terms of ω. (b) Calculate the ratio

$$\frac{\text{final rotational energy}}{\text{initial rotational energy}}$$

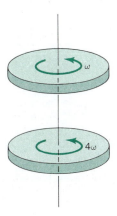

FIGURE 5

11. A planetary moon moves in an elliptical orbit. The semi-major and semi-minor axes have lengths of 32,000 km and 29,500 km. The orbital period is 12.2 sidereal days. The mass of the moon is 6.32×10^{20} kg. Determine the magnitude of the angular momentum of the moon.

12. A particle of mass m slides without friction on a wire bent in the shape of the parabola $y = ax^2$. The particle starts from rest at $y = H$ as shown in Figure 6. (a) Show that the angular momentum of the particle relative to the point O may be written

$$\mathbf{L} = \left(\frac{2g[H - y]}{1 + 4ay}\right)^{1/2} my\mathbf{k} \qquad \text{for } y < H$$

(b) Use a geometric argument to prove that this result holds for $v_x > 0$, but that if $v_x < 0$ a minus sign is required.

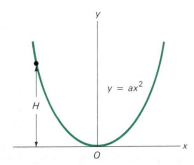

FIGURE 6

13. A 20-kg pendulum bob is attached to an 18-m cord. The bob is pulled aside through an angle of $10°$ and released from rest. Calculate the magnitude of the angular momentum of the bob relative to the point of support as the bob passes the vertical position.

14. The bob of a pendulum is released from rest at A when the supporting string is horizontal (Figure 7). The bob moves along the circular path AB (radius R_1) until the string strikes a peg of negligible thickness at O'. Thereafter the bob moves along the circular path BC (radius R_2). (a) Show that the points O and O' are separated by the distance $3R_1/5$ if the tension in the string drops to zero just as the bob reaches point C. (b) Show that the magnitude of the angular momentum of the bob at B, relative to points O and O', is

$$L_O = m\sqrt{2g}\, R_1{}^{3/2} \qquad L_{O'} = m\sqrt{0.32g}\, R_1{}^{3/2}$$

(c) If the separation of O and O' is $3R_1/5$, show that the magnitude of the angular momentum of the bob at C is

$$L_O = m\sqrt{0.016g}\, R_1{}^{3/2} \qquad L_{O'} = m\sqrt{0.064g}\, R_1{}^{3/2}$$

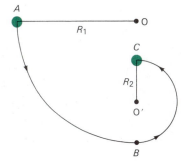

FIGURE 7

15. A solid disk ($M = 250$ grams) with a radius of 45 cm is mounted on a shaft through its center (Figure 8). A wad of gum (mass $= 10$ grams) moving horizontally at a speed of 25 m/s sticks to the perimeter of the disk, setting it into rotation about the shaft. Disregard frictional torques exerted by the shaft, and determine (a) the angular velocity of the combination, (b) the number of revolutions the combination makes in 10 seconds. Take the moment of inertia of the disk to be $\frac{1}{2}MR^2$.

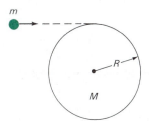

FIGURE 8

16. A uniform disk (mass = 1.2 kg, radius = 0.33 m) rotates about an axis perpendicular to the disk and through its center of mass. When its angular velocity equals 12 rad/s a 1.2-kg mass initially at rest becomes attached to the rim of the disk. No external torques act on the disk–particle system. Treat the particle as a point mass and determine (a) the new angular velocity of the system, (b) the new kinetic energy of the system. (c) Is kinetic energy conserved?

17. An 80-kg diver has a moment of inertia of 30 kg·m² in the layout position and 4 kg·m² in the tuck position. During a dive his angular momentum is such that he has 0.6 s in which to complete three somersaults in the tuck position. Determine (a) the maximum angular velocity he attains, (b) his angular momentum, (c) the time needed to complete one revolution in the layout position.

18. Early in the history of the earth–moon system, the moon was much closer to the earth than it is now, and the earth was spinning much faster than it is today. Although there was no tidal lock, there was a possible spin-orbit synchronism for which the spin period of the earth equaled the orbital period of the moon. The equation describing tidal lock in Example 5 also describes spin-orbit synchronism. The conservation of angular momentum for the earth–moon system can be expressed as

$$2.88 \sqrt{\frac{r_s}{r}} + \frac{0.586}{27.3 \left[\dfrac{r_s}{r}\right]^{3/2}} = 3.47$$

where each term represents angular momentum measured in units of 10^{34} kg·m²/s; r denotes the present earth–moon distance and r_s is the earth–moon distance at the time of synchronism. (a) Verify that this equation is satisfied by $r_s/r = 0.038$. (b) Determine the duration of the day corresponding to $r_s/r = 0.038$. (c) Determine the angular diameter of the moon, as viewed from the earth, for $r_s/r = 0.038$. The present value is 0.519°.

CHAPTER 12

ROTATIONAL DYNAMICS

12.1
DYNAMICS OF RIGID BODY MOTION

Dynamics is the study of accelerated motion and its causes. The dynamics of particle motion is governed by Newton's second law, $\mathbf{F} = d\mathbf{p}/dt$. For a system of constant mass, we have seen that Newton's second law reduces to the form $\mathbf{F} = m\mathbf{a}$. In Chapter 11 we found that the rotational analog of $\mathbf{F} = d\mathbf{p}/dt$ relates the time rate of change of the angular momentum ($d\mathbf{L}/dt$) to the net torque acting on the system ($\boldsymbol{\tau}$),

$$\boldsymbol{\tau} = \frac{d\mathbf{L}}{dt} \tag{12.1}$$

This equation is the starting point for developing the dynamics of rigid body motion. Under certain rather restrictive conditions, Equation 12.1 reduces to $\tau = I\alpha$, where α is the angular acceleration. The equation $\tau = I\alpha$ is the rotational analog of $F = ma$.

Consider a rigid body that rotates about an axis *fixed in direction*. The axis of rotation can move so long as it does not change direction. For example, the wheels of a car moving in a straight line rotate about axes that are fixed in direction.

The angular momentum of the wheels, or any rigid body, rotating about an axis can be expressed by

$$L = I\omega \tag{12.2}$$

where ω is the angular velocity of the body. For a rigid body rotating about an axis fixed in direction, the moment of inertia is constant. The time derivative of Equation 12.2 gives

$$\frac{dL}{dt} = I\frac{d\omega}{dt} = I\alpha$$

where $\alpha = d\omega/dt$ is the angular acceleration of the body. Using Equation 12.1 to equate dL/dt with the torque gives

$$\tau = I\alpha \tag{12.3}$$

Note the analogy with $F = ma$. Just as a net force causes *linear* acceleration, a net torque causes *angular* acceleration.

Equation 12.3 is the basis of dynamic methods for measuring a moment of inertia of a rigid body. Such methods require a measurement of the angular acceleration of the body and of the net torque acting on it.

EXAMPLE 1

Measurement of the Moment of Inertia of a Disk

The moment of inertia of a disk can be determined in a simple experiment. A sensitized paper tape is wrapped around the disk (Figure 12.1), and a small mass m is attached to it. As the tape unwinds, a constant torque acts on the disk, causing it to undergo a constant angular acceleration. Position measurements on the tape at known time intervals allow us to determine its tangential acceleration. Measuring the external torque is accomplished indirectly, permitting I to be expressed in terms of measured quantities.

Let T denote the tension in the tape. Newton's second law for the attached mass m is

$$\sum F = mg - T = ma$$

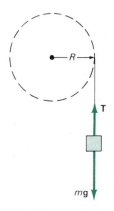

FIGURE 12.1

A particle of mass m is attached to the paper tape that unwinds from the cylinder.

where a is the linear acceleration of the mass. All points on the tape experience an acceleration whose magnitude also equals a because the tape does not stretch. The moment of inertia is introduced by applying Equation 12.3 to the disk. We take the center of the disk as a torque reference point. Ignoring frictional torques, the net torque is due to the tension in the tape. The lever arm for T is R, the radius of the disk, so the net torque is TR. Equation 12.3 gives

$$\sum \tau = TR = I\alpha$$

Furthermore, the tape unwinds without slipping. This means that the acceleration of a point on the circumference of the disk equals the acceleration of the tape in contact with the disk at that point. The magnitude of the tangential acceleration of points on the circumference of the disk equals $R\alpha$. Thus, we have

$$a = R\alpha$$

Combining these three equations we can express I in terms of the measured quantities m, R, and a:

$$I = \frac{mR^2(g - a)}{a}$$

For a uniform disk of mass M the theoretical value is $I = \frac{1}{2}MR^2$. A measurement of M and R gives a value for I that can be compared with the result just obtained.

The comprehensive nature of physics is illustrated in the next example. The linear and rotational forms of Newton's second law are used, together with several results from rotational kinematics developed in Chapter 10.

EXAMPLE 2

The Yo-Yo

A yo-yo is released (Figure 12.2) and at the same time the string is pulled upward with a constant tension T so that the center of mass remains fixed in position as the string unwinds. What is the angular velocity when the 1-m string has unwound, assuming that the initial angular velocity is zero?

The axial radius of the yo-yo is 0.90 cm, its moment of inertia is 1.60×10^{-5} kg·m^2, and its mass is 40 grams.

For zero vertical acceleration the net force must equal zero. Hence the string tension equals the yo-yo weight.

$$T = mg = (4 \times 10^{-2} \text{ kg})(9.80 \text{ m/s}^2)$$
$$= 0.392 \text{ N}$$

The net torque relative to the center of mass is due to the tension alone. Hence

$$\tau = Tr = (0.392 \text{ N})(0.009 \text{ m})$$
$$= 3.53 \times 10^{-3} \text{ N·m}$$

Using Equation 12.3 we obtain the angular acceleration

$$\alpha = \frac{\tau}{I} = \frac{3.53 \times 10^{-3} \text{ N·m}}{1.60 \times 10^{-5} \text{ kg·m}^2}$$
$$= 221 \text{ rad/s}^2$$

The angle through which the yo-yo has turned equals the string length divided by the radius ($\theta = s/r$)

$$\theta = \frac{1 \text{ m}}{0.009 \text{ m}} = 111 \text{ rad}$$

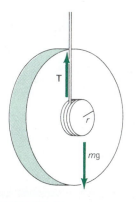

FIGURE 12.2

A yo-yo unwinds under the influence of two forces. The string exerts an upward tension **T**, and gravity exerts a downward force mg.

Because the angular acceleration is constant we can use the relation $\omega^2 = \omega_0{}^2 + 2\alpha\theta$ given in Table 10.1. With $\omega_0 = 0$,

$$\omega^2 = 2\alpha\theta = 2(221 \text{ rad/s}^2)(111 \text{ rad})$$

For the angular velocity this gives

$$\omega = 221 \text{ rad/s}$$

This is about 35 revolutions per second.

~~The angular velocity achieved may be increased by giving the yo-yo an initial~~ downward velocity. In this case the tension must exceed mg to decelerate the yo-yo, causing an increase in τ, α, and ω.

12.2
PRECESSION

We now consider motion where the axis of rotation changes direction. Spinning bodies such as gyroscopes, toy tops, and the earth exhibit precession, which is *a regular rotation of the spin axis.*

A spinning bicycle wheel is often used in classroom demonstrations of precession. One end of the axle is supported with a rope while the other end is left free to move. In Figure 12.3 such a wheel is shown with its spin angular velocity vector **ω** along the positive *y*-axis. The *z*-axis is vertical and passes through the supported end of the axle. If the wheel is properly started, the spin axis will rotate in the horizontal (*x–y*) plane.

To the extent that friction in the bearings can be ignored, the spin angular velocity of the wheel remains constant in magnitude. But, as the spin axis changes direction, the vector **ω** changes direction. The spin axis and **ω** move together around a circle in the *x–y* plane. The angular velocity at which the spin axis moves around the circle is called the **angular velocity of precession**, and is denoted by **Ω**. The vector relationship between **ω** and **Ω** is shown in Figure 12.4. The figure shows that **ω** lies in the *x–y* plane and rotates from the positive *x*-axis toward the positive *y*-axis. The vector **Ω** is along the positive *z*-axis.

We can use the equation of rotational dynamics, $\tau = d\mathbf{L}/dt$, to derive a relation between ω and Ω. Figure 12.5 shows a top view of the precessing wheel. The spin angular momentum, $\mathbf{L} = I\boldsymbol{\omega}$, is parallel to $\boldsymbol{\omega}$. As $\boldsymbol{\omega}$ precesses, $\mathbf{L}$ precesses with it. In a time interval dt the spin axis precesses through an angle $d\varphi$. The angular velocity of precession is Ω, so

$$d\varphi = \Omega \, dt \tag{12.4}$$

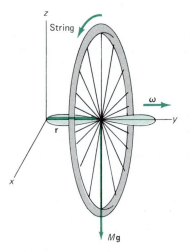

FIGURE 12.3

The bicycle wheel is supported by a cord. The spin axis and angular velocity vector *ω* precess in the horizontal plane.

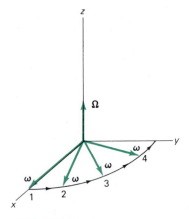

FIGURE 12.4

The angular velocity of precession **Ω** is directed upward along the positive *z*-axis. The spin angular velocity *ω* lies in the *x–y* plane. As the wheel precesses, the spin angular velocity vector precesses in the sense shown, from 1 to 2 to 3 to 4, and so on.

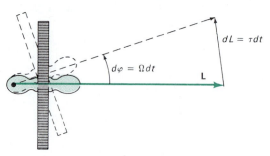

FIGURE 12.5

Top view of a precessing wheel. In a time interval *dt* the angular momentum **L** precesses through the angle $d\varphi = \Omega \, dt$. The magnitude of **L** remains constant, but it changes direction. The magnitude of the change in **L** is $dL = \tau \, dt$.

The angular momentum vector also precesses through the angle $d\varphi$. Because I and ω are constant in magnitude, the angular momentum vector **L** changes direction, but not magnitude. From Figure 12.5 we see that the change dL is related to L and Ω by

$$dL = L\, d\varphi = L\Omega\, dt \qquad (12.5)$$

The weight of the wheel sets up a torque that causes **L** to change direction. From the dynamic relation $\boldsymbol{\tau} = d\mathbf{L}/dt$ we obtain

$$dL = \tau\, dt \qquad (12.6)$$

Comparing Equations 12.5 and 12.6 shows that

$$dL = L\Omega\, dt = \tau\, dt$$

It follows that the angular velocity of precession is given by

$$\Omega = \frac{\tau}{L} \qquad (12.7)$$

The torque set up by the weight Mg is given by $\tau = rMg$, where r is the distance from the center of gravity to the point of support of the precessing wheel (Figure 12.3). The spin angular momentum is given by $L = I\omega$. Substituting $L = I\omega$ and $\tau = rMg$ into Equation 12.7 gives an alternative expression for Ω:

$$\Omega = \frac{rMg}{I\omega} \qquad (12.8)$$

The reciprocal relation between ω and Ω indicates that the rate of precession increases as ω decreases. As rotational energy is lost through friction, ω will decrease. Equation 12.8 suggests why a top or gyroscope precesses faster and faster as its spin velocity decreases.

EXAMPLE 3

A Toy Gyroscope

A toy gyroscope is made in the shape of a uniform disk with an axle of negligible mass. The radius of the disk is 5.0 cm. The point of support is 3.0 cm from the center of gravity. We want to determine the precessional angular velocity when the gyroscope spins at 12 revolutions per second. The moment of inertia for the uniform disk is $I_c = \frac{1}{2}MR^2$. Equation 12.8 gives

$$\Omega = \frac{rMg}{\frac{1}{2}MR^2\omega} = \frac{2rg}{R^2\omega}$$

With

$$\omega = (12\ \text{rev/s})(2\pi\ \text{rad/rev})$$
$$= 75.4\ \text{rad/s}$$

we get

$$\Omega = \frac{2(0.030\ \text{m})(9.80\ \text{m/s}^2)}{(0.050\ \text{m})^2(75.4\ \text{rad/s})}$$
$$= 3.1\ \text{rad/s}$$

This corresponds to a precessional period of about 2 seconds.

We have assumed that the axis of the precessing wheel remains horizontal. This requires that the vertical component of the support force be equal to the weight of the wheel. If this condition is not satisfied, the precessing axis may exhibit a wobbling or nodding motion called **nutation**.*

Precession of the Equinoxes

Precession is not limited to tops and gyroscopes. Planets and stars spin and thus can precess if subjected to a net torque. More than 2000 years ago Hipparchus discovered the precession of the equinoxes. The equinoxes mark positions in the earth's orbit about the sun at which day and night are of equal length. The equinox line (Figure 12.6) is defined by the intersection of the earth's **equatorial plane** and the earth's **orbital plane** (the *ecliptic plane*). The precession of the equinoxes is caused by gravitational torques exerted on the equatorial bulges of the earth by the moon and the sun.

The earth's spin axis and thus its angular momentum **L** are perpendicular to the equatorial plane. The equinox line lies *in* the equatorial plane and so the vector **L** and the spin axis are perpendicular to the equinox line. As the spin axis precesses, the equinox line precesses at the same rate. This maintains a 90° angle between the spin axis and the equinox line and gives rise to the precession of the equinoxes. The earth's spin axis is known to precess through 360° once every 26,000 years, tracing a cone with an angular radius of 23.5°. Thus, experimentally, the angular velocity for the precession of the equinoxes is

$$\Omega = \frac{2\pi \text{ rad}}{26,000 \text{ years}}$$
$$= 7.7 \times 10^{-12} \text{ rad/s}$$

---

* From the Latin *nutare*: to nod

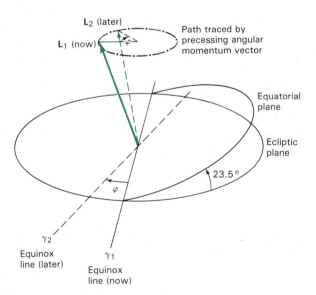

FIGURE 12.6

The equinox line is defined by the intersection of the earth's equatorial plane and the orbital plane (ecliptic). The equinox line is perpendicular to the earth's spin angular momentum, **L**. As **L** precesses the equinox line also precesses.

By way of comparison, the spin angular velocity of the earth is about ten million times greater:

$$\omega = 7.27 \times 10^{-5} \text{ rad/s}$$

12.3
WORK, ENERGY, AND POWER IN ROTATION

The work-energy principle, $\Delta K = W$, relates the change in kinetic energy ΔK of a system to the work W done by the net force acting on the system. In particular, the work-energy principle applies to rigid bodies. To restate this principle using variables suitable for rotation, we must first derive an expression for the work done on a rigid body by a single force F. We assume that the axis of rotation is fixed in direction. A consequence of this restriction is that only forces having components perpendicular to the axis of rotation need to be considered because only those components can produce torques parallel to the axis of rotation.

Referring to Figure 12.7, we see that the work dW done by the force $\mathbf{F}$ acting through the displacement $d\mathbf{s}$ is

$$dW = \mathbf{F} \cdot d\mathbf{s} = F \, ds \cos \theta$$

Substituting $ds = r \, d\theta$ into this equation we get

$$dW = Fr \cos \theta \, d\theta = (Fr \sin \varphi) \, d\theta$$

The quantity in parentheses is the magnitude of the torque, τ, relative to the reference point O. Therefore, we have the result

$$dW = \tau \, d\theta$$

This is the rotational analog of the one-dimensional relation $dW = F \, dx$.

For a finite rotation about an axis fixed in direction, the integral of Equation 12.9 gives

$$W = \int_{\theta_1}^{\theta_2} \tau \, d\theta \tag{12.10}$$

If more than one external torque acts, then the resultant torque is the sum of the individual torques.

$$\tau = \sum \tau_i$$

The work is given by Equation 12.10 with the torque τ being the sum of the individual torques.

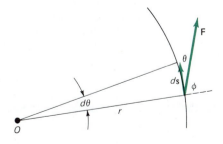

FIGURE 12.7
The tangential displacement $d\mathbf{s}$ subtends the angle $d\theta$. The angles θ and φ are complementary.

The work done on a system of particles by the net force equals the increase in kinetic energy of the system, provided the internal forces do no work. In a rigid body the interparticle distances are fixed, guaranteeing that internal forces do no work. Thus we can write

$$W = \text{change in rotational kinetic energy} = K_f - K_i$$

The rotational kinetic energy of a rigid body is given by $\frac{1}{2}I\omega^2$. Thus, the rotational form of the work-energy principle becomes

$$W = \tfrac{1}{2}I\omega_f{}^2 - \tfrac{1}{2}I\omega_i{}^2 \tag{12.11}$$

If the work done is *positive*, the rotational kinetic energy *increases*. If the work done is *negative*, then the kinetic energy *decreases*.

Negative work is illustrated by the work done by frictional torques. For example, the tidal torques acting on the earth as it rotates perform negative work, causing a decrease in the earth's spin velocity and in its rotational kinetic energy about its spin axis.

Rotational Power

The rate at which work is performed, dW/dt, is the power P. A rotational form of the equation for power delivered to a body by a torque τ follows directly from Equation 12.9, which gives the work done by the torque τ acting through the angle $d\theta$.

$$dW = \tau \, d\theta \tag{12.9}$$

The power is

$$P = \frac{dW}{dt} = \tau \frac{d\theta}{dt}$$

Noting that $d\theta/dt = \omega$ is the angular velocity, we conclude that the power can be expressed as

$$P = \tau\omega \tag{12.12}$$

This is the rotational analog of the linear relation $P = Fv$.

EXAMPLE 4

An Electric Power Generator

Electric power is generated by applying an external torque to a generator rotor. This torque causes the rotor to turn at a predetermined angular velocity. Such generators are so efficient ($\geq 99\%$) that virtually all of the mechanical power input is converted to electric power output. A typical generator running at 1800 rpm yields an electric power of 0.840×10^9 watts. What torque is required?

The angular velocity corresponding to 1800 rpm = 30 rev/s is

$$\omega = (30 \text{ rev/s})(2\pi \text{ rad/rev})$$
$$= 188 \text{ rad/s}$$

Using Equation 12.12 we obtain for the torque

$$\tau = \frac{P}{\omega} = \frac{0.840 \times 10^9 \text{ W}}{188 \text{ rad/s}}$$
$$= 4.46 \times 10^6 \text{ N·m}$$

In power stations this torque is often supplied by a large steam turbine that is equivalent in its action to a wheel and axle. What driving force would be needed to develop this torque by using a wheel and axle comparable in size to the turbine? For a radius of 2 m, the torque of 4.46×10^6 N·m corresponds to a force of 2.23×10^6 N, or roughly 250 tons.

12.4
COMBINED TRANSLATION AND ROTATION

It is useful to separate the kinetic energy of a rigid body into two terms. We want to show that the kinetic energy is given by

$$K = \tfrac{1}{2}Mv_{cm}^2 + \tfrac{1}{2}I_c\omega_c^2 \qquad (12.13)$$

The center of mass speed is denoted by v_{cm}. The quantity I_c is the moment of inertia about the center of mass and ω_c is the angular velocity relative to an axis through the center of mass. The term $\tfrac{1}{2}Mv_{cm}^2$ is the kinetic energy the body would have if its mass were all to move at the center of mass velocity. The term $\tfrac{1}{2}I_c\omega_c^2$ represents the rotational kinetic energy of the body relative to an axis through the center of mass. The fact that the kinetic energy can be expressed as the sum of translational and rotational contributions reflects the fact that any displacement of a rigid body can be expressed as a *displacement of the center of mass* and *a rotation about the center of mass*. Equation 12.13 is not restricted to situations in which the body rotates about its center of mass.

To derive Equation 12.13 we begin with the definition of the kinetic energy of a collection of particles

$$K = \sum \tfrac{1}{2}m_i v_i^2$$

We introduce the center of mass velocity $\mathbf{v}_{cm}$ and write

$$\mathbf{v}_i = \mathbf{v}_{cm} + \mathbf{V}_i$$

where $\mathbf{V}_i$ is the velocity of the particle relative to the center of mass. Noting that $v_i^2 = \mathbf{v}_i \cdot \mathbf{v}_i$ we can write the kinetic energy as

$$\begin{aligned}
K &= \sum \tfrac{1}{2}m_i(\mathbf{v}_{cm} + \mathbf{V}_i) \cdot (\mathbf{v}_{cm} + \mathbf{V}_i) \\
&= \tfrac{1}{2}\sum m_i(v_{cm}^2 + V_i^2 + 2\mathbf{v}_{cm} \cdot \mathbf{V}_i) \\
&= \tfrac{1}{2}(\sum m_i)v_{cm}^2 + \tfrac{1}{2}\sum m_i V_i^2 + \mathbf{v}_{cm} \cdot \sum m_i \mathbf{V}_i \qquad (12.14)
\end{aligned}$$

The third term is zero because

$$\frac{\sum m_i \mathbf{V}_i}{M}$$

represents the velocity of the center of mass *relative to the center of mass*. Since this is zero, $\sum m_i \mathbf{V}_i = 0$.

To simplify the form of the second term in Equation 12.14, we introduce the angular velocity of the body relative to an axis through the center of mass, ω_c. Then

$$V_i = R_i\omega_c$$

where R_i is the radius of the circle on which the particle moves. The second term in Equation 12.14 becomes

$$\tfrac{1}{2}\sum m_i V_i^2 = \tfrac{1}{2}\left(\sum m_i R_i^2\right)\omega_c^2$$
$$= \tfrac{1}{2}I_c\omega_c^2$$

where I_c is the moment of inertia of the body about an axis through the center of mass.

The first term in Equation 12.14 is simplified by noting that the total mass of the body is $M = \sum m_i$. Hence,

$$\tfrac{1}{2}\left(\sum m_i\right)v_{cm}^2 = \tfrac{1}{2}Mv_{cm}^2$$

This completes the derivation and confirms Equation 12.13 for the kinetic energy of a rigid body:

$$K = \tfrac{1}{2}Mv_{cm}^2 + \tfrac{1}{2}I_c\omega_c^2 \tag{12.13}$$

EXAMPLE 5

The Great Race

A solid disk, a hoop, and a solid sphere roll without slipping down an inclined plane (Figure 12.8). They have equal radii and masses and start from rest. Which one reaches the bottom of the incline first?

We take the finish line to be a vertical distance y below the starting line. The winner is the object whose center of mass finishes with the greatest speed. We use the law of energy conservation to write

$$Mgy = \tfrac{1}{2}Mv_{cm}^2 + \tfrac{1}{2}I_c\omega_c^2$$

For rolling without slipping

$$\omega_c = \frac{v_{cm}}{R}$$

so

$$\tfrac{1}{2}Mv_{cm}^2 + \tfrac{1}{2}I_c\omega_c^2 = \tfrac{1}{2}Mv_{cm}^2\left(1 + \frac{I_c}{MR^2}\right)$$

The conservation of energy relation gives

$$v_{cm}^2 = \frac{2gy}{1 + \dfrac{I_c}{MR^2}}$$

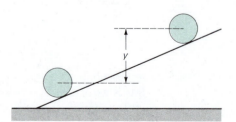

FIGURE 12.8
An object with a circular cross section rolls without slipping down the plane. The center of gravity of the object falls through a vertical distance y.

This shows that v_{cm} is largest for the object with the smallest value of I_c. From Table 10.2 we have

$$I_c = \tfrac{2}{5}MR^2 \qquad \text{(sphere)}$$

$$I_c = \tfrac{1}{2}MR^2 \qquad \text{(disk)}$$

$$I_c = MR^2 \qquad \text{(hoop)}$$

So, the sphere has the smallest value of I_c and is therefore the winner of the race.

Dynamically, the smallest I_c presents the smallest resistance to angular acceleration. The angular acceleration will be largest for the object with the smallest I_c. This in turn implies that the linear acceleration, and therefore v_{cm}, will be largest for the object with the smallest I_c. We can also understand the result of the race in terms of energy. The smallest I_c corresponds to the smallest rotational energy. Coupled with a constant total energy, this implies a larger translational kinetic energy, and hence a larger v_{cm}.

WORKED PROBLEM

A marble of radius r starts from rest at the top of a circular pipe of radius R and rolls without slipping along a path perpendicular to the axis of the pipe (Figure 12.9). Determine the angular position (θ) of the point where the marble breaks contact with the surface and becomes airborne.

Solution

We can use the law of energy conservation and Newton's second law to obtain two equations relating the speed of the center of mass (v_{cm}) to the angle θ. Eliminating v_{cm} from this pair of equations will give a solution for the angle θ.

Take $U = mgy$, with $y = 0$ at the center of the pipe. With $y = R + r$,

$$U_i = mg(R + r) \qquad K_i = 0$$

When the marble rolls to a point where its angular position is θ,

$$U_f = mg(R + r) \cos \theta$$

and

$$K_f = \tfrac{1}{2}mv_{cm}{}^2 + \tfrac{1}{2}I_c\omega_c{}^2$$

For rolling without slipping, $\omega_c = v_{cm}/r$. The marble is a uniform sphere, so $I_c = (2/5)mr^2$, and

$$K_f = \tfrac{7}{10}mv_{cm}{}^2$$

Energy conservation is expressed by

$$0 + mg(R + r) = \tfrac{7}{10}mv_{cm}{}^2 + mg(R + r) \cos \theta$$

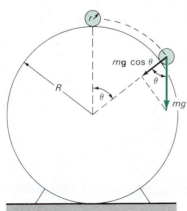

FIGURE 12.9

This gives one of the equations relating v_{cm} and θ:

$$v_{cm}{}^2 = \tfrac{10}{7}g(R + r)(1 - \cos \theta)$$

The second equation relating v_{cm} and θ follows from Newton's second law. The *radial* component of the net force acting on the marble is $mg \cos \theta - N$, where N is the normal component of the contact force. The radial component of the acceleration of the marble is $v_{cm}{}^2/(R + r)$. Newton's second law gives

$$\frac{mv_{cm}{}^2}{R + r} = mg \cos \theta - N$$

The normal force becomes zero when the marble breaks contact with the pipe. With $N = 0$, Newton's second law gives

$$v_{cm}^2 = g(R + r) \cos \theta$$

Equating the two expressions for v_{cm}^2 gives

$$\tfrac{10}{7} g(R + r)(1 - \cos \theta) = g(R + r) \cos \theta$$

The factor $g(R + r)$ cancels, leaving the result

$$\cos \theta = \tfrac{10}{17} = 0.588$$

The angle θ is

$$\theta = \cos^{-1}(0.588)$$
$$= 54°$$

EXERCISES

12.1   Dynamics of Rigid Body Motion

A. A bicycle wheel ($M = 18$ kg, $R = 40$ cm) is set into rotation by applying a force of 150 N tangent to the circumference. Treat the wheel as a hoop and calculate its angular acceleration.

B. A child rolls a uniform disk at a constant linear speed. What fraction of the total kinetic energy of the disk is due to the translational motion of the center of mass?

C. A superball dropped with the correct spin and direction of velocity will bounce back and forth as suggested in Figure 1. Explain.

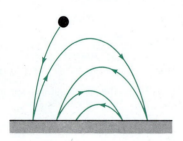

FIGURE 1

D. The experiment described in Example 1 uses a mass $m = 200$ grams. The measured acceleration is $a = 1.12$ m/s^2. To three significant figure accuracy, it is found that the measured moment of inertia equals the theoretical value, $\frac{1}{2}MR^2$. Determine the mass (M) of the disk.

E. A meter rod is held horizontally, with its tip resting on the edge of a table top. The other end is then released. (a) Determine the angular acceleration of the rod at the moment of release. (b) Determine the linear acceleration of the free end at the moment of release.

12.2   Precession

F. Suppose that the toy gyroscope in Example 2 has a mass of $M = 0.50$ kg, a radius $R = 0.04$ m, and spins at 20 rev/s. Calculate (a) the spin angular momentum, (b) the torque due to gravity, (c) the angular momentum associated with the precessional motion about the z-axis. Take $I = (5/4)MR^2$.

G. A wheel ($I = 1.50$ kg·m^2 and $M = 8$ kg) spins at 25 rad/s with its axle horizontal. If the axle is pivoted about a point O (Figure 2) located 10 cm from the plane of the wheel, calculate the angular velocity of precession of the axle.

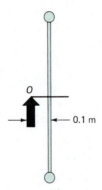

FIGURE 2

H. The moment of inertia of a precessing wheel about its spin axis is 1.20 kg·m^2. The gravitational torque causing its precession is 20.7 N·m. Its spin angular velocity is 14 times its angular velocity of precession. Determine its precessional period.

12.3   Work, Energy, and Power in Rotation

I. A constant force of 150 N is applied tangentially to the circumference of a wheel ($M = 8$ kg, $R = 40$ cm).

Starting from rest the wheel rotates through 90°. Treat the wheel as a hoop and calculate (a) the applied torque, (b) the work done by the applied torque, (c) the angular velocity at the end of the 90° rotation.

J. A flywheel energy storage unit being proposed for a small bus consists of 6 disks, each of radius 0.48 m and mass 15 kg. When moving at maximum speed (that is, when the flywheel is charged) each flywheel rotates at 16,000 rev/min. If the bus requires 4.1 MJ for each kilometer of travel, how many kilometers can the bus travel on a single charge of the flywheel?

K. Gasoline has an energy content of 133 MJ per gallon. What is the angular velocity of a 200-kg flywheel with a radius of 1 meter if its kinetic energy is equivalent to 1 gallon of gasoline? (Regard the flywheel as a uniform disk.)

12.4 Combined Translation and Rotation

L. A pool ball (radius = 2.88 cm, mass = 165 grams) rolls without slipping. Its center of mass moves at a speed of 2.1 m/s. Determine (a) the translational kinetic energy $(\frac{1}{2}Mv_{cm}^2)$, (b) the rotational kinetic energy $(\frac{1}{2}I_c\omega_c^2)$, and (c) the total kinetic energy.

M. All uniform disks rolling without slipping on an inclined plane reach the bottom at the same time,

regardless of their mass or radius, when started together from rest. Suppose that the density of a disk is proportional to the distance measured from its center. If this disk is raced against a uniform disk, which will win? Give a qualitative reason to justify your answer.

N. As a car rounds a corner it loses a wheel. The wheel (mass = 14.6 kg) rolls without slipping at a speed of 30 miles per hour. Its moment of inertia about an axis through its center is given by $I_c = (2/3)MR^2$, where M is its mass and R is its radius. Determine the total kinetic energy of the wheel.

P. Explain why the yo-yo in Figure 3 can be made to roll toward you, slide toward you, or roll away from you, despite the fact that the tension in the cord is always directed away from the yo-yo.

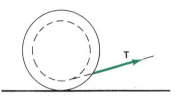

FIGURE 3

PROBLEMS

1. A long, thin rod of mass M and length L is hinged at one end (Figure 4). It is released and rotates about the hinged end. (a) Show that its angular acceleration is

$$\alpha = \left(\frac{3g}{2L}\right)\cos\theta$$

(b) Show that the tangential acceleration of the free end exceeds g when the angle θ falls below 48°.

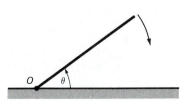

FIGURE 4

2. A uniform disk is set in rotation as described in Example 1. The disk has a radius of 10 cm and a mass of 2.10 kg. The mass $m = 0.16$ kg experiences a downward acceleration of 1.30 m/s². Calculate the moment of inertia of the disk using (a) the theoretical formula, (b) the acceleration of the small mass.

3. Refer to Example 1. Show that the acceleration of the mass m is independent of the radius of the uniform

disk and depends on the ratio of the masses. Specifically, show that

$$a = \frac{2mg}{M + 2m}$$

Check the limiting values of a for $m \to 0$ and $m \to \infty$ and argue that these limits are to be expected.

4. A student working on an upended bicycle sets a wheel in motion and then slows it down by applying a tangential force on the tire. The force is applied in such a way that it increases according to $F = bt$, where $b = 1.21$ N/s. If the moment of inertia of the wheel is 0.2 kg·m² and the radius is 0.3 m, what is the angular acceleration 2 s after application of the force?

5. A thin rod (mass = M, length = L) is held in a horizontal position. One end of the rod is attached to the ceiling by a weightless string (Figure 5). At a certain

FIGURE 5

moment the rod is released. At the instant of release (a) show that the angular acceleration of the rod is $3g/2L$ about an axis through the point A, (b) show that the magnitude of the linear acceleration of the rod's center of mass is $3g/4$, (c) show that the tension in the string is $mg/4$.

6. A thin hoop (mass $= M$, radius $= R$) is caused to rotate in place on a frictionless incline by a thin tape of negligible weight wrapped around and attached to the hoop (Figure 6). The tape, parallel to the plane as it unwinds, passes over a frictionless peg and is attached to a mass m. The mass m falls in such a way that the hoop undergoes an angular acceleration while its center of mass remains fixed in place. Show that (a) the tension in the tape is $T = mg/(1 + m/M)$, (b) the acceleration of m is $a = mg/(m + M)$, (c) the angle θ is determined by $\sin \theta = m/(m + M)$, (d) assuming the system starts from rest, the speed of m after one complete revolution of the hoop is given by $v^2 = 4\pi mgR/(m + M)$.

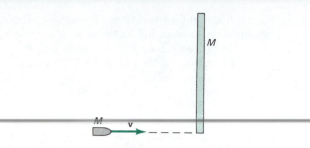

FIGURE 7

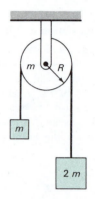

FIGURE 8

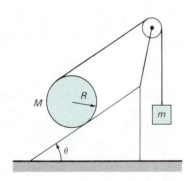

FIGURE 6

7. A bullet (mass $= M$) moving with speed v, as shown in Figure 7, embeds itself in the end of a long, thin rod (mass $= M$, length $= L$) initially at rest. In terms of L and v calculate (a) the subsequent velocity of the center of mass, (b) the angular velocity of the rod, (c) the fraction of the kinetic energy that remains after impact.

8. A string passing over a uniform disk (mass $= m$, radius $= R$) connects the masses m and $2m$ as shown in Figure 8. Friction prevents slippage between the string and disk, which rotates on a frictionless bearing. The system is released from rest. Show that the speed

of the masses is given by $v = \sqrt{8gR/7}$ when the mass $2m$ has dropped a distance $2R$.

9. A hammer thrower develops a torque that varies with time according to the equation

$$\tau = At - Bt^2$$

$$A = 120 \text{ N·m/s} \qquad B = 32 \text{ N·m/s}^2$$

The angular velocity of the hammer varies with time according to the relation $\omega = Ct$, where $C = 20$ rad/s. These equations apply over the interval $t = 0$ to $t = 2.8$ s. Determine (a) the time at which the power is a maximum, (b) the maximum power developed.

10. A uniform disk and a uniform sphere, each of mass M and radius R, start from rest and race down an incline in the fashion described in Example 5. The incline makes an angle of 20° with the horizontal and the race covers a distance of 1.60 m along the plane. Determine the elapsed times for both racers.

11. A billiard ball is given a linear velocity of 10 m/s without spin. It begins to roll without slipping after traveling a distance of 2 m. Determine the coefficient of friction between the table and the billiard ball.

12. Prove that all uniform spheres, independent of mass or radius, arrive at the bottom of the incline simultaneously under the conditions of the race described in Example 5.

13. A uniform ladder (mass = 10.3 kg, length = 5.1 m) rests against a wall and makes an angle of 60° with the horizontal. The wall and floor are frictionless, and the ladder is held in place with a string. At a certain moment the string breaks and the ladder slides down to the floor. At the moment of impact with the floor, determine (a) the total kinetic energy of the ladder, (b) the magnitude of its center-of-mass velocity, (c) its angular velocity relative to its center of mass.

14. A uniform sphere moving horizontally with a center of mass speed v encounters an inclined plane. It moves up the plane a distance D (measured parallel to the plane) before coming to rest. Let D_1 denote the value of D for motion along a frictionless incline and let D_2 denote the value of D for rolling without slipping along the incline. (a) Show that $D_2/D_1 = 7/5$. (b) Explain whether or not the spin angular momentum of the sphere changes in the two cases.

15. A uniform disk of radius R and mass M rolls without slipping on a plane inclined at the angle θ (Figure 9). A flat tape is wound around the cylinder and attached to a mass m after passing over a frictionless pulley. The tape is parallel to the plane. (a) Show that the angular acceleration of the disk equals zero about an axis through the point O if $\sin\theta = 2m/M$. (b) For $\theta = 30°$, show that the linear acceleration of the center of mass is $g/7$ up the incline if $M = 2m$.

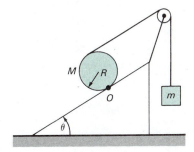

FIGURE 9

16. A sling accelerates a 100-gram projectile from rest to a speed of 70 m/s. (a) Determine the work performed on the projectile. (b) The projectile experiences a constant torque and the acceleration takes place over an angular displacement of $\pi/3$ radians. Determine the magnitude of the torque that acts. (c) The angular velocity of the projectile just before the moment of release is 100 rad/s. Determine the power delivered at that moment.

17. Flywheels weighing several hundred tons have been proposed as mechanical energy storage devices. Consider a 100-ton flywheel that stores 10^4 kWh of energy at an angular speed of 3500 rev/min. Determine (a) the moment of inertia of the flywheel, (b) the time required to bring it to rest while it delivers 3000 kW of power, (c) the initial torque delivered externally by the flywheel while it is losing energy at the rate of 3000 kW.

18. Tidal friction exerts a retarding torque that slows the earth's rate of rotation and thereby increases the length of the day. We showed in Example 5 of Chapter 10 that tidal friction reduces the earth's rotational kinetic energy by 1.3×10^{17} J per sidereal day. Determine the average frictional torque exerted on the tidal bulges.

19. A father hears a cry for help and discovers that his daughter has fallen into a well. He tells her to hold on to the bucket so that he can pull her up with the hand winch. If the father is capable of a maximum power of 60 W and the mass of the daughter is 40 kg, (a) what is the maximum upward velocity of the daughter? (b) If the father rotates the handle at 1 rev/s, what torque does he exert?

20. A bowling ball is started down a bowling lane without spin. What fraction of the ball's original kinetic energy is lost by the time it starts to roll without slipping?

21. In Example 2 of Chapter 8 we considered a loop-the-loop in which an object slides around a vertical circle. Suppose instead, that a ball bearing rolls without slipping around the loop (Figure 10). The radius of the loop is R and the radius of the bearing is r. The bearing is to maintain contact with the loop throughout its motion. Show that $h = 2.7R - 1.7r$ is the minimum height above the base of the circle from which the center of gravity of the bearing can be released.

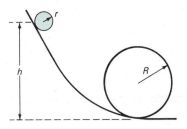

FIGURE 10

22. A long, thin rod (mass M, length L) hangs vertically, free to rotate about an axis through one end (Figure 11). A wad of gum of mass m moving horizontally at speed v strikes the lower end of the rod and sticks to it. (a) Determine the speed of the lower end of the rod immediately after the collision with the gum. (b) Determine the maximum angular displacement of the rod. Express your results in terms of m, M, L, and v.

23. A uniform sphere of radius r and mass m rolls without slipping down the track shown in Figure 12. At the end of its run at point Q its center-of-mass velocity is directed vertically upward. To what height above the base of the track will the sphere rise? Express your results in terms of R.

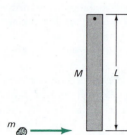

FIGURE 11

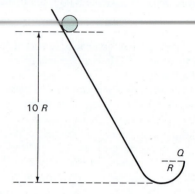

FIGURE 12

CHAPTER 13

OSCILLATORY MOTION

Horizontal Spring

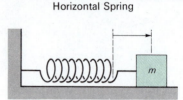

FIGURE 13.1

The mass undergoes simple harmonic motion if the spring is linear and if friction is absent.

13.1
SIMPLE HARMONIC MOTION

A periodic motion is one that repeats with a characteristic time. For example, the motions of a jump rope, a pendulum, and an orbiting planet are all periodic. An oscillation is a special type of periodic motion in which the system moves back and forth through an equilibrium position. The motion of a pendulum is both periodic and oscillatory.

The time for one cycle of an oscillatory motion is called the **period**, denoted by T. The **frequency** of the motion is the number of cycles per unit time, designated by f. By definition, T and f are mutually reciprocal:

$$f = \frac{1}{T} \tag{13.1}$$

If time is expressed in seconds, frequency is expressed in reciprocal seconds (s^{-1}). This frequency unit is called the **hertz** and is abbreviated Hz.

$$1 \text{ Hz} = 1 \text{ cycle/s} = 1 \text{ oscillation/s}$$

In order for a system to oscillate, any net force (or torque) that acts on it must be **restoring** in character. A restoring force is one that urges a body toward a position of stable equilibrium. A spring-mass system (Figure 13.1) is a good example of a system that oscillates in response to a restoring force. When certain conditions are satisfied, the oscillatory motion is called **simple harmonic motion**:

When the net force (or torque) results in an *acceleration* that is *directly proportional* to the *displacement* from equilibrium and in the opposite direction the resulting motion is called **simple harmonic motion (SHM)**.

Spring-Mass Oscillator

Figure 13.1 shows a spring-mass oscillator. The mass oscillates along the x-axis on a frictionless horizontal surface under the influence of the spring. When the spring is *linear*, the mass experiences a restoring force directly proportional to x, the displacement of the mass from equilibrium. For a displacement $+x$ the force is $-kx$, where k is the spring constant. Using Newton's second law we have

$$F = -kx = ma$$

The acceleration is

$$a = -\left(\frac{k}{m}\right)x \tag{13.2}$$

The ratio k/m is constant. Hence the acceleration is directly proportional to the displacement and it is in the opposite direction, so the mass executes SHM.

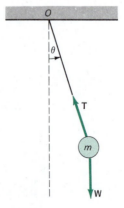

FIGURE 13.2

Two forces act on the pendulum bob. For small angular displacements the bob executes simple harmonic motion.

The Simple Pendulum

A simple pendulum (Figure 13.2) consists of a bob supported by a cord of negligible mass. The pendulum oscillates in a plane. Consider the pendulum bob and the cord as a body undergoing rotational oscillations. We choose the point of suspension O as a torque reference point and apply the rotational form of Newton's second law, $\tau = I\alpha$. Forces acting on the bob are its weight, $W = mg$, and the tension in the supporting cord, T. The moment of inertia of the bob about O is $I = m\ell^2$. The torque due to T is zero, leaving only the torque due to the weight. When the angular displacement of the bob is θ, the lever arm for the weight is $\ell \cdot \sin\theta$ and the torque is $\tau = -mg\ell \cdot \sin\theta$. The minus sign accounts for the restoring character of the torque.

The relation $\tau = I\alpha$ becomes

$$-mg\ell \sin\theta = m\ell^2\alpha$$

Solving for the angular acceleration, we find

$$\alpha = -\left(\frac{g}{\ell}\right)\sin\theta$$

As it stands this equation does not represent SHM because the angular acceleration α is not directly proportional to the angular displacement, θ. However, if θ is small we can use the approximation $\sin\theta \approx \theta$ to write

$$\alpha = -\left(\frac{g}{\ell}\right)\theta \qquad\qquad (13.3)$$

For small angular displacements, the angular acceleration α is directly proportional to θ and opposite in sign, and the simple pendulum executes SHM.

Any system that undergoes SHM is called a **harmonic oscillator**. To determine the general form of the equation describing a harmonic oscillator we reconsider Equations 13.2 and 13.3.

Recall that acceleration is the second derivative of displacement,

$$a = \frac{d^2x}{dt^2} \qquad \alpha = \frac{d^2\theta}{dt^2}$$

In terms of the second derivative, Equations 13.2 and 13.3 can be expressed as

$$\frac{d^2x}{dt^2} = -\left(\frac{k}{m}\right)x \qquad \frac{d^2\theta}{dt^2} = -\left(\frac{g}{\ell}\right)\theta$$

Mathematically, these two equations are particular forms of the same general equation

$$\frac{d^2u}{dt^2} = -\omega^2 u \qquad\qquad (13.4)$$

in which u denotes a displacement from equilibrium and ω^2 is a constant determined by the physical constants in Equations 13.2 and 13.3:

$$\omega^2 = \frac{k}{m} \qquad \text{(spring-mass)} \qquad\qquad (13.5)$$

$$\omega^2 = \frac{g}{\ell} \qquad \text{(simple pendulum)} \qquad\qquad (13.6)$$

As we will see, the units of ω are radians per second.

Equation 13.4 is called the *harmonic oscillator equation*. Let's examine the solution of this equation.

Solution of the Harmonic Oscillator Equation

Equation 13.4 involves the second derivative, so its solution, u, will contain two constants of integration. Calling A and φ the constants of integration, we propose the following solution of the harmonic oscillator equation:

$$u = A \cos(\omega t - \varphi) \tag{13.7}$$

To verify that this form of u does in fact satisfy Equation 13.4, we differentiate $u = A \cos(\omega t - \varphi)$ twice. The first derivative is

$$\frac{du}{dt} = -\omega A \sin(\omega t - \varphi)$$

The second derivative is

$$\frac{d^2 u}{dt^2} = -\omega^2 A \cos(\omega t - \varphi) \tag{13.8}$$

Substituting Equations 13.7 and 13.8 into Equation 13.4 gives an identity

$$-\omega^2 A \cos(\omega t - \varphi) = -\omega^2 A \cos(\omega t - \varphi)$$

This identity confirms that $u = A \cos(\omega t - \varphi)$ is a solution of Equation 13.4, the harmonic oscillator equation.

Period of Simple Harmonic Motion

We have already noted that oscillatory motion repeats with a characteristic period, T. The period of simple harmonic motion is related to the quantity ω appearing in Equation 13.4. To deduce the relation between T and ω we first consider the periodicity of the cosine function. Figure 13.3a shows that the

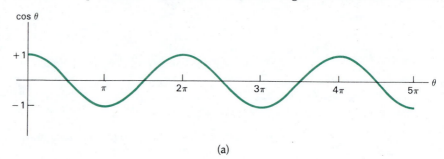

(a)

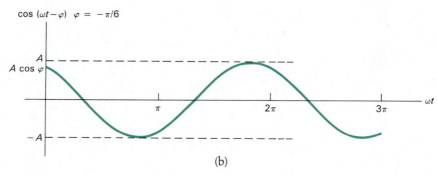

(b)

FIGURE 13.3

(a) The function $\cos \theta$ is periodic, repeating when θ changes by 2π. (b) The function $A \cos(\omega t - \varphi)$ is periodic, repeating when t changes by 2π. The time for ωt to change by 2π is the period, T.

cosine function is periodic, with a period of 2π radians. Mathematically, this periodicity is expressed by

$$\cos(\theta + 2\pi) = \cos\theta$$

In the solution $u = A\cos(\omega t - \varphi)$, the argument of the cosine function $(\omega t - \varphi)$ is called the **phase**. As t changes, the phase changes. During one complete cycle, the phase changes by 2π. If the time increases from t to $t + T$, the phase increases by ωT, from $\omega t - \varphi$ to $\omega t - \varphi + \omega T$. In order for the displacement to be the same at times t and $t + T$, the phase increase must equal 2π. Thus, ω is related to the period T by

$$\omega T = 2\pi \qquad (13.9)$$

For a spring-mass oscillator, Equation 13.5 shows that $\omega = \sqrt{k/m}$, and the period is given by

$$T = 2\pi\sqrt{\frac{k}{m}} \qquad (13.10)$$

For a simple pendulum, Equation 13.6 gives $\omega = \sqrt{g/\ell}$, so

$$T = 2\pi\sqrt{\frac{\ell}{g}} \qquad (13.11)$$

We can also relate ω to the frequency $f = 1/T$. Thus,

$$\omega = \frac{2\pi}{T} = 2\pi f \qquad (13.12)$$

This equation shows that the three quantities ω, T, and f contain the same information. With the system executing f oscillations per second, the phase changes by $2\pi f$ radians per second. For this reason, $\omega = 2\pi f$ is called the **angular frequency**.

EXAMPLE 1

A Seconds Pendulum

Many antique clocks use a pendulum to keep time. The bob of a seconds pendulum requires 1 s for each swing, so its period is 2 s. What are the values of f and ω for a simple pendulum that has a period of 2 s, and what is its length?

Given that $T = 2$ s, Equation 13.1 gives the frequency

$$f = \frac{1}{T} = \frac{1}{2\text{ s}}$$
$$= 0.500\text{ Hz}$$

Equation 13.12 gives the angular frequency

$$\omega = \frac{2\pi}{T} = \frac{2\pi\text{ rad}}{2\text{ s}}$$
$$= 3.14\text{ rad/s}$$

To obtain the pendulum length we use Equation 13.11 to get

$$\ell = g\left(\frac{T}{2\pi}\right)^2 = 9.80\text{ m/s}^2\left(\frac{2\text{ s}}{2\pi}\right)^2$$
$$= 0.994\text{ m}$$

Construct such a pendulum for yourself and see how close you can come to a period of 2 seconds.

Let's summarize the key features of simple harmonic motion:

1. Simple harmonic motion results when the net force or torque results in an acceleration directly proportional to the displacement from equilibrium and in the opposite direction.
2. Mathematically, the equation describing the relation between the acceleration a and the displacement u for SHM is

$$a = -\omega^2 u \qquad \omega^2 = \text{constant}$$

3. The period of SHM is related to the constant ratio of the displacement and acceleration by

$$T = 2\pi \cdot \sqrt{\frac{-\text{displacement}}{\text{acceleration}}} = \frac{2\pi}{\omega}$$

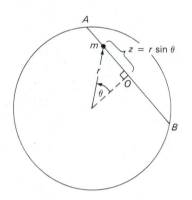

FIGURE 13.4

The mass m is free to oscillate along the frictionless tunnel between points A and B. The z-axis lies along the tunnel with its origin at the midpoint O.

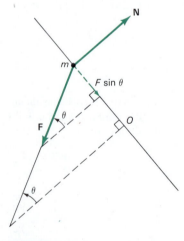

FIGURE 13.5

Two forces act on the mass m; the normal force **N** and the gravitational force **F**. The normal force is balanced by $F \cos \theta$, the component of the gravitational force perpendicular to the tunnel. The component $F \sin \theta$ parallel to the tunnel gives rise to the acceleration of the mass.

EXAMPLE 2

The Gravity Transit System

Imagine that the earth is a uniform, nonrotating spherical mass. A narrow, frictionless tunnel is drilled along a straight line between the surface points A and B (Figure 13.4). We want to show that objects released into the tunnel from A or B will undergo SHM as they oscillate inside. The unusual aspect of the motion is that the period of the SHM is the same no matter where we choose to drill A and B.

Figure 13.5 shows the two forces acting on the object (mass m) in the tunnel: the gravitational force **F** and the normal force **N**. The component $F \cos \theta$ of the gravitational force is balanced by the normal force. The net force on the object is the component of **F** parallel to the tunnel.

We take the z-axis to lie along the tunnel. The origin O is at the midpoint of the tunnel. The net force acting on the object is zero when it is at the origin. When the object is on either side of the origin the net force acting on it urges it toward the origin. Thus, the object experiences a restoring force when it is displaced from the origin. To prove that the object undergoes SHM we must show that the restoring force is *directly* proportional to the displacement from O.

In Chapter 6 (Equation 6.8) we showed that the gravitational force exerted on an object beneath the surface of the earth is directly proportional to its distance from the center of the earth (r). The force on an object of mass m is given by

$$F = \left(\frac{GM_E m}{R_E^3} \right) r$$

where M_E and R_E are the mass and radius of the earth. Figure 13.5 shows that the net force on the object is $F \sin \theta$, the component of **F** parallel to the tunnel. Inserting a minus sign to account for the restoring character we have

$$F_z = -F \sin \theta = -\left(\frac{GM_E m}{R_E^3} \right) r \sin \theta$$

$$= -\left(\frac{GM_E m}{R_E^3} \right) z$$

The last step follows from Figure 13.4, which shows that $z = r \sin \theta$ is the displacement of the object from equilibrium. Newton's second law gives

$$ma_z = F_z = -\left(\frac{GM_E m}{R_E{}^3}\right) z$$

The acceleration along the tunnel is

$$a_z = -\left(\frac{GM_E}{R_E{}^3}\right) z$$

This acceleration is directly proportional to the displacement from equilibrium and it is in the opposite direction. Thus, the mass executes SHM. The period of the oscillations is

$$T = 2\pi \sqrt{\frac{-z}{a_z}}$$
$$= 2\pi \sqrt{\frac{R_E{}^3}{GM_E}}$$

Inserting numerical values gives

$$T = 2\pi \sqrt{\frac{(6.38 \times 10^6 \text{ m})^3}{(6.67 \times 10^{-11} \text{ N·m}^2/\text{kg}^2)(5.98 \times 10^{24} \text{ kg})}}$$
$$= 5.07 \times 10^3 \text{ s}$$
$$= 84.5 \text{ minutes}$$

Note that the period is independent of the length of the tunnel. The transit time between any two points on the surface is $\frac{1}{2}T$, or slightly less than 43 minutes.

Initial Conditions

In Figure 13.3b, the displacement $x = A \cos(\omega t - \varphi)$ is plotted versus ωt. From the figure we can see the physical significance of A. Because the values of $\cos(\omega t - \varphi)$ oscillate between maxima and minima of $+1$ and -1, the constant A equals the maximum displacement. It is called the **amplitude** of the motion. The angle φ is the initial $(t = 0)$ value of the phase.

The harmonic oscillator equation of motion is satisfied by $x = A \cos(\omega t - \varphi)$ for *all* values of A and φ. This means that, *mathematically*, there are an infinite number of solutions. *Physically*, however, just one solution will satisfy the *initial conditions*—or, in other words, conditions describing some particular motion at $t = 0$. We take the initial conditions to be the position and velocity at $t = 0$. We write

$$x_0 = x \big|_{t=0} \qquad \text{(initial position)}$$

$$v_0 = \frac{dx}{dt} \bigg|_{t=0} \qquad \text{(initial velocity)}$$

From $x = A \cos(\omega t - \varphi)$ we find

$$x_0 = A \cos(-\varphi) = A \cos \varphi \qquad (13.13)$$

and

$$v = \frac{dx}{dt} = -\omega A \sin(\omega t - \varphi)$$

Setting $t = 0$ gives the initial velocity

$$v_0 = -\omega A \sin(-\varphi) = \omega A \sin \varphi \qquad (13.14)$$

We can solve Equations 13.13 and 13.14 to express A and φ in terms of the initial conditions. Thus,

$$x_0{}^2 + \left(\frac{v_0}{\omega}\right)^2 = A^2 \cos^2 \varphi + A^2 \sin^2 \varphi = A^2$$

which gives

$$A = \pm \sqrt{x_0{}^2 + \left(\frac{v_0}{\omega}\right)^2} \qquad (13.15)$$

Similarly,

$$\frac{v_0/\omega}{x_0} = \frac{A \sin \varphi}{A \cos \varphi} = \tan \varphi$$

which gives

$$\varphi = \tan^{-1}\left(\frac{v_0}{\omega x_0}\right) \qquad (13.16)$$

Equations 13.15 and 13.16 allow us to construct a solution of the harmonic oscillator equation that matches a given set of initial conditions.

EXAMPLE 3

Air Track Harmonic Oscillator

An air car (mass $m = 200$ grams) is attached to one end of a linear spring ($k = 50$ N/m). The other end of the spring is attached to one end of a horizontal air track. This arrangement allows the car to oscillate on a frictionless surface. The air car is started with an initial displacement of 3.0 cm and an initial velocity of 25 cm/s. We want to determine the form of x.

The initial conditions are

$$x_0 = 3.0 \text{ cm} \qquad v_0 = 25 \text{ cm/s}$$

The angular frequency of the oscillations is

$$\omega = \sqrt{\frac{k}{m}} = \sqrt{\frac{50 \text{ N/m}}{0.20 \text{ kg}}}$$
$$= 16 \text{ rad/s}$$

Using Equation 13.15 gives

$$A = \sqrt{(3.0 \text{ cm})^2 + \left[\frac{25 \text{ cm/s}}{16 \text{ rad/s}}\right]^2}$$
$$= 3.4 \text{ cm}$$

(We choose the positive root to make x_0 positive.)

The initial phase follows from Equation 13.16

$$\tan \varphi = \frac{25 \text{ cm/s}}{16 \text{ s}^{-1} (3.0 \text{ cm})} = 0.53$$

This gives

$$\varphi = 0.49 \text{ radian}$$

Thus, the solution that satisfies the initial conditions is

$$x = 3.4 \cos(\omega t - 0.49) \text{ cm} \qquad \omega = 16 \text{ rad/s}$$

Note that the maximum displacement is 3.4 cm, which is greater than the initial displacement of 3 cm. This is because the mass was started with an initial velocity.

Energy Conservation in SHM

Systems that undergo SHM invariably experience conservative forces, so their total mechanical energy is conserved. Their oscillations can be viewed as the cyclic transformation of kinetic energy into potential energy and then back into kinetic energy.

For example, imagine a pendulum pulled aside and released from rest (Figure 13.6). Initially its kinetic energy is zero. Its total energy is in the form of gravitational potential energy. As it swings downward its gravitational potential energy (mgy) decreases and its kinetic energy increases. As it passes through the equilibrium position (when the cord is vertical) its gravitational potential energy is a minimum and its kinetic energy is a maximum. As it swings upward its kinetic energy decreases and its gravitational potential energy increases.

If we take the gravitational potential energy to be mgy, with y measured vertically from the equilibrium position, then the total energy of the pendulum is

$$E = \tfrac{1}{2}mv^2 + mgy$$

This energy can be expressed in terms of the angular displacement θ and the angular velocity $d\theta/dt$. The speed of the pendulum bob is

$$v = \ell\omega = \ell \frac{d\theta}{dt} \tag{13.17}$$

For an angular displacement θ, Figure 13.6 shows that the elevation of the bob is

$$y = \ell(1 - \cos\theta)$$

For small θ (sufficiently small for the motion to be SHM),

$$\cos\theta \approx 1 - \tfrac{1}{2}\theta^2$$

so

$$mgy = \tfrac{1}{2}mg\ell\theta^2 \tag{13.18}$$

The total energy is

$$E = \tfrac{1}{2}m\ell^2 \left(\frac{d\theta}{dt}\right)^2 + \tfrac{1}{2}mg\ell\theta^2 \tag{13.19}$$

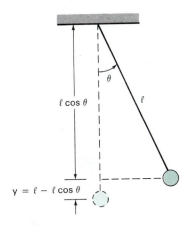

FIGURE 13.6

When the pendulum's angular displacement is θ the bob is a distance $y = \ell(1 - \cos\theta)$ above the equilibrium position.

$\ell \cos\theta$

θ

ℓ

$y = \ell - \ell\cos\theta$

EXAMPLE 4

Pendulum Energetics

A simple pendulum ($m = 11.3$ kg, $\ell = 5.76$ m) is released from rest. Its initial angular displacement is 0.300 radian. We want to find its maximum kinetic energy, and its maximum angular velocity.

The conservation of energy, $K + U = E = $ constant, shows that the maximum kinetic energy is attained when the potential energy is a minimum. Because

$$U_{\min} = K_{\min} = 0$$

we have

$$K_{\max} = U_{\max} = \tfrac{1}{2}mg\ell\theta_0{}^2$$

where θ_0 is the initial angular displacement. Thus,

$$K_{\max} = \tfrac{1}{2}(11.3 \text{ kg})(9.80 \text{ m/s}^2)(5.76 \text{ m})(0.300)^2$$
$$= 28.7 \text{ J}$$

The angular velocity is related to K by

$$K = \tfrac{1}{2}m\ell^2\left(\frac{d\theta}{dt}\right)^2$$

The maximum angular velocity is attained when K is a maximum.

$$\left(\frac{d\theta}{dt}\right)_{\max} = \sqrt{\frac{2K_{\max}}{m\ell^2}} = \sqrt{\frac{2(28.7 \text{ J})}{11.3 \text{ kg } (5.76 \text{ m})^2}}$$
$$= 0.391 \text{ rad/s}$$

This converts to about 22 degrees per second.

Energy is also conserved for a spring-mass system undergoing SHM. The moving mass carries the kinetic energy ($\tfrac{1}{2}mv^2$) and the linear spring stores the elastic potential energy ($\tfrac{1}{2}kx^2$). The total energy

$$E = \tfrac{1}{2}mv^2 + \tfrac{1}{2}kx^2 \tag{13.20}$$

remains constant throughout the oscillations.

13.2
APPLICATIONS OF SIMPLE HARMONIC MOTION

The period of simple harmonic motion depends on the physical properties of the oscillating system and on the system's environment. A precise measurement of the period is possible by counting many vibrations. Measurement of the period often provides a dynamic method of determining characteristics of the system or of its environment. For instance, for the spring-mass system Equation 13.5 shows that

$$m = \frac{k}{\omega^2}$$

Using

$$\omega = \frac{2\pi}{T}$$

we can relate the mass to the period of the oscillations

$$m = \frac{kT^2}{4\pi^2} \qquad (13.21)$$

If we use a calibrated spring then the value of k is known and a measurement of T can be used to determine m. This technique was used in Skylab where ordinary balances could not be used.

In Example 1 the measured period of a seconds pendulum, together with the known value of g, enabled us to determine the pendulum length. We can reverse the logic and use a pendulum of measured length to infer a value for g. Using $\omega = 2\pi/T$ in Equation 13.6 gives

$$g = \frac{4\pi^2\ell}{T^2} \qquad (13.22)$$

EXAMPLE 5

Measurement of the Acceleration of Gravity with a Simple Pendulum

A pendulum that is 1.00 m in length is observed to require 100.6 s to execute 50 complete oscillations. These data give a period

$$T = \frac{100.6 \text{ s}}{50} = 2.01 \text{ s}$$

Using Equation 13.22 we calculate for the acceleration of gravity

$$g = \frac{4\pi^2 \cdot 1 \text{ m}}{(2.01 \text{ s})^2} = 9.77 \text{ m/s}^2$$

In Newton's day the pendulum was used to obtain values for g at different latitudes over the surface of the earth. The systematic variations observed for g confirmed Newton's conjecture that the earth was an oblate spheroid.

For a simple pendulum T can be determined with accuracy, but it is not possible to obtain an equally accurate value of ℓ. This in turn limits the accuracy of the values of g obtained by using a simple pendulum.

The Physical Pendulum

A physical pendulum is a body free to oscillate under the influence of gravity (Figure 13.7). As we will show, the moment of inertia of a physical pendulum is related to its period. This makes it possible to determine the moment of inertia of a body by measuring its period while it oscillates as a physical pendulum.

In Figure 13.7 the body oscillates about the point P. The restoring torque is caused by the weight of the body. Using P as a reference point for torque we get

$$\tau = -mgh \sin \theta$$

For small angles we set $\sin \theta \approx \theta$ and

$$\tau = -mgh\theta$$

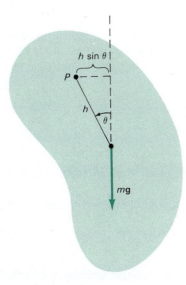

FIGURE 13.7
A physical pendulum is a body free to oscillate under the influence of gravity.

From the equation of rotational dynamics

$$\tau = I_P \alpha$$

we find that the angular acceleration is

$$\alpha = -\left(\frac{mgh}{I_P}\right)\theta \qquad (13.23)$$

where I_P is the moment of inertia about the axis through P. This equation describes SHM with the period given by

$$T^2 = 4\pi^2 \left(\frac{I_P}{mgh}\right) \qquad (13.24)$$

Solving for I_P gives

$$I_P = \frac{mghT^2}{4\pi^2} \qquad (13.25)$$

Measurements of m, h, and T enable I_P to be determined. The parallel axis theorem (Section 10.3) relates I_P to I_c, the moment of inertia about a parallel axis through the center of mass.

$$I_P = I_c + mh^2 \qquad (13.26)$$

Using Equation 13.26 in Equation 13.25 we get

$$T^2 = \frac{4\pi^2}{g}\left(\frac{I_c}{mh} + h\right) \qquad (13.27)$$

A graph of T^2 versus h reveals that T^2 has a minimum. The value of h for which the period is a minimum can be determined by setting $d(T^2)/dh = 0$. The minimum value of T occurs for

$$h = \sqrt{\frac{I_c}{m}} \qquad (13.28)$$

Figure 13.8 shows data for an experiment in which a 2-meter rod is used as a physical pendulum. The theoretical points are obtained using Equation 13.27. If the 2-meter rod is long and thin, $I_c = (1/12)m\ell^2$, and we find that the minimum period should occur at

$$h = \frac{\ell}{\sqrt{12}} = \frac{200 \text{ cm}}{3.46}$$

$$= 57.7 \text{ cm}$$

By substituting this value of h into Equation 13.27 you can verify that the minimum period equals 2.16 s.

The data are consistent with this prediction, although the minimum is not sharply defined.

The Torsion Pendulum

A torsion pendulum consists of an object suspended on the end of a wire or rod that can undergo twisting oscillations (Figure 13.9). The period of such oscillations depends on the moment of inertia of the system. A torsion pendulum can be used to measure the moment of inertia of an object.

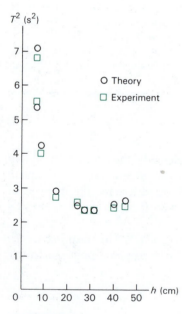

FIGURE 13.8

T^2 versus h data for a physical pendulum in the form of a 2-meter rod.

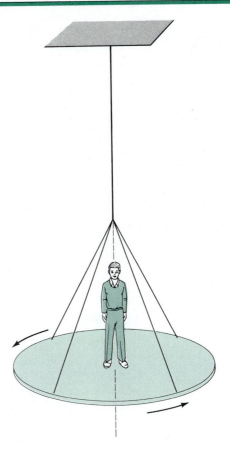

FIGURE 13.9

A torsion pendulum is free to undergo twisting oscillations. Here the torsion pendulum is used to measure the moment of inertia of a person.

A vertical torsion pendulum, arranged to measure the moment of inertia of a standing person about an axis through the center of mass, is shown in Figure 13.9. We now show how a measurement of T for torsional oscillations can be used to determine the moment of inertia of the person.

First, we must calibrate the torsion pendulum. A known torque, τ, is applied to twist the rod until it comes to equilibrium at the angle θ. We assume that torque is directly proportional to the twist angle. This assumption permits us to write

$$\tau = -\beta\theta \tag{13.29}$$

The minus sign is needed because τ is a restoring torque. A static measurement of τ and θ establishes the value of the twist constant, β.

The pendulum is set into oscillation without the person on board. If I_0 is the moment of inertia of the empty pendulum, the rotational form of Newton's second law enables us to write

$$\tau = -\beta\theta = I_0\alpha$$

or

$$\alpha = -\left(\frac{\beta}{I_0}\right)\theta$$

showing that the pendulum executes SHM. The period is related to the moment of inertia by

$$I_0 = \frac{\beta T_0^2}{4\pi^2} \tag{13.30}$$

Then, the person climbs onto the pendulum platform, and the pendulum is again set into oscillation. If I_P is the moment of inertia of the person alone, the total moment of inertia $I_0 + I_P$ is related to the new period T by

$$I_0 + I_P = \frac{\beta T^2}{4\pi^2} \tag{13.31}$$

Substituting the value of I_0 obtained in Equation 13.30 into Equation 13.31, we can solve for the moment of inertia of the person in terms of the three measured quantities β, T, and T_0:

$$I_P = \frac{\beta(T^2 - T_0^2)}{4\pi^2} \tag{13.32}$$

EXAMPLE 6

Moment of Inertia of a Person

Let's use the torsion pendulum to determine the moment of inertia of a person who stands with the symmetry line of the apparatus passing through their center of mass. We measure $\beta = 125$ N·m/rad, $T_0 = 2.011$ s, and $T = 2.152$ s. The person's moment of inertia follows from Equation 13.32:

$$I_P = 125 \text{ N·m} \frac{[(2.152 \text{ s})^2 - (2.011 \text{ s})^2]}{4\pi^2}$$

$$= 1.86 \text{ kg·m}^2$$

13.3
DAMPED MOTION

When a pendulum is set into oscillation, friction gradually causes this oscillation to stop. So far we have neglected frictional effects in oscillatory motion. We now consider this kind of decaying oscillation, which we call **damped motion**.

To account for friction analytically we need an expression for the frictional force. We limit our consideration of frictional forces to linear motion in one dimension along the x-axis. Experiment shows that for small velocities the frictional force exerted on many objects is directly proportional to the object's velocity. Such a force can be represented by the equation

$$f = -\sigma v \tag{13.33}$$

where σ is an empirically determined constant. The minus sign indicates that the force and the velocity are in opposite directions. The units of the constant σ are kg/s.

Let's add the frictional force to our treatment of the spring-mass oscillator. The oscillator now consists of a mass m subjected to the forces $-kx$ and $-\sigma v$. Applying Newton's second law, we write

$$F = -kx - \sigma v = ma$$

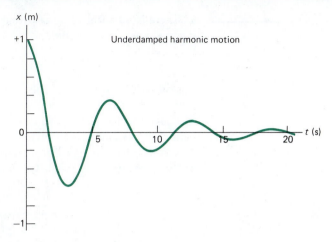

FIGURE 13.10

In underdamped harmonic motion the system oscillates with a steadily decreasing amplitude.

With $v = dx/dt$ and $a = d^2x/dt^2$, this equation can be written in the form

$$m \frac{d^2x}{dt^2} + \sigma \frac{dx}{dt} + kx = 0 \qquad (13.34)$$

Solutions of this equation describe damped harmonic motion. It turns out that there are two distinct types of damped motion:

1. For sufficiently small values of σ, the frictional force on a harmonic oscillator will be weak and the motion approximates SHM. The system oscillates with a steadily decreasing amplitude. This motion is illustrated in Figure 13.10. It is called **underdamped harmonic motion**.
2. For sufficiently large values of σ the frictional force will nearly balance the restoring force of the spring. As Equation 13.34 indicates, the acceleration will be small throughout the motion. If the mass is displaced from equilibrium and released, we expect it to creep back to equilibrium. In other words, no oscillation will occur. This motion, as illustrated in Figure 13.11, is called **overdamped motion**.

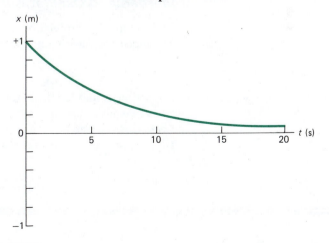

FIGURE 13.11

With overdamped motion, no oscillation occurs. The displacement steadily decreases toward zero.

There is a critical value of σ that defines the borderline between the underdamped harmonic motion and the overdamped motion. This critical value is

$$\sigma_c = 2\sqrt{mk} \tag{13.35}$$

For this critical value of σ, the mass returns to equilibrium without oscillating. Any smaller value of σ leads to oscillatory motion.

Equations like 13.34 arise in many physical situations. For example, electric currents in certain types of circuits are described by a similar equation. Therefore, it is convenient to develop a standardized form of Equation 13.34 that will apply to many situations. If we divide the equation by m we get

$$\frac{d^2x}{dt^2} + \frac{\sigma}{m}\frac{dx}{dt} + \frac{k}{m}x = 0$$

We set

$$\omega^2 = \frac{k}{m} \tag{13.5}$$

and introduce

$$\tau = \frac{2m}{\sigma} \tag{13.36}$$

Dimensionally, τ is a time. It should not be confused with torque, a quantity represented by the same symbol. Equation 13.34 becomes

$$\frac{d^2x}{dt^2} + \frac{2}{\tau}\frac{dx}{dt} + \omega^2 x = 0 \tag{13.37}$$

Here, ω is the angular frequency of the oscillator in the absence of damping. We will find that τ is the parameter that characterizes the time needed for the oscillations to decay.

Underdamped Oscillations

To develop an interpretation of τ we start by comparing the motion of the underdamped oscillator with SHM. An equation describing SHM is

$$x = A \cos \omega t$$

This function is plotted in Figure 13.12 with $A = 1$ m and $\omega = 1$ rad/s. The curve begins at $x = +1$ m and oscillates between limits set by $x = \pm 1$ m. The period is calculated using $T = 2\pi/\omega$. In this case, $T = 2\pi$ s $= 6.28$ s.

The corresponding equation that describes underdamped harmonic motion is

$$x = Ae^{-t/\tau} \cos \omega_1 t \tag{13.38}$$

where

$$\omega_1 = \omega \left(1 - \frac{1}{\omega^2 \tau^2}\right)^{1/2} \tag{13.39}$$

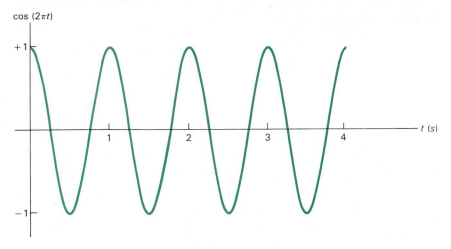

FIGURE 13.12
When there is no damping the maximum displacement of a harmonic oscillator started at $x = +1$ m with zero speed is bounded by $x = +1$ m and $x = -1$ m.

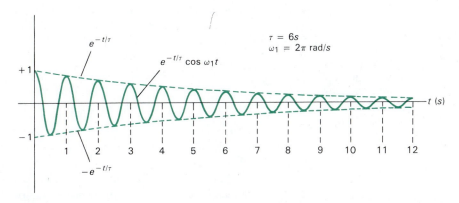

FIGURE 13.13
With underdamped harmonic motion the displacement of an oscillator started at $x = +1$ m with zero speed is bounded by the two exponentials $x = +e^{-t/\tau}$ and $x = -e^{-t/\tau}$ (here $\tau = 6$ s).

is the angular frequency of the oscillation. Figure 13.13 is a plot of the function $Ae^{-t/\tau} \cos \omega_1 t$, with $A = 1$ m and $\tau = 6$ s. The curve begins at $x = +1$ and oscillates between decreasing limits set by the enveloping curves $x = \pm e^{-t/\tau}$. In this case, the time between alternate zeros is

$$T = \frac{2\pi}{\left(\omega^2 - \dfrac{1}{\tau^2}\right)^{1/2}} \tag{13.40}$$

Let's see graphically how τ controls $Ae^{-t/\tau}$. In Figure 13.14 we have plotted the exponential $e^{-t/\tau}$ versus t for three values of τ. All of the curves approach zero as $t \to \infty$. For large τ the approach to zero is slow. For small τ the approach to zero is rapid. The time τ determines how rapidly the exponential approaches zero, and thereby how rapidly damped oscillations die out. Accordingly, we call τ the **damping time**.

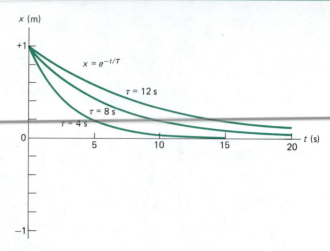

FIGURE 13.14

The time τ determines the rate at which the exponential $e^{-t/\tau}$ decreases to zero. The greater the value of τ, the more slowly the exponential drops toward zero.

We can also see numerically how rapidly $e^{-t/\tau}$ decreases with time.

$$t = 0 \qquad e^{-t/\tau} = e^0 = 1$$

$$t = 2\tau \qquad e^{-t/\tau} = e^{-2} = 0.135$$

$$t = 4\tau \qquad e^{-t/\tau} = e^{-4} = 0.00674$$

$$t = 10\tau \qquad e^{-t/\tau} = e^{-10} = 0.0000454$$

The rapid decline of $e^{-t/\tau}$ shows that τ is the characteristic time for damping to essentially kill the motion.

For small σ, the time τ is large and the damping is slow. Underdamped motion occurs only for large values of τ. But how large must τ be for oscillations to occur? Inserting the critical value of σ given by Equation 13.35 into Equation 13.36 defines a corresponding critical value of τ given by

$$\tau_c = \frac{2m}{\sigma_c} = \sqrt{\frac{m}{k}} \qquad (13.41)$$

But $\omega = \sqrt{k/m}$ is the angular frequency of undamped motion. Thus

$$\omega\tau_c = 1 \qquad (13.42)$$

If $\omega\tau$ is greater than unity the motion is underdamped and oscillatory.

EXAMPLE 7

A Damped Simple Pendulum

A long pendulum with a heavy bob will oscillate for several hours before coming to rest. This system is a good illustration of underdamped harmonic motion. Consider a pendulum 20 m long that requires 1 hour for its amplitude to be reduced to half its original value. What are the values of ω_1 and τ?

We assume that $\omega\tau \gg 1$ so that we can approximate the angular frequency ω_1 by the frequency for undamped motion, ω.

$$\omega_1 = \omega\left(1 - \frac{1}{\omega^2\tau^2}\right)^{1/2} \approx \omega$$

From the undamped simple pendulum, $\omega = \sqrt{g/\ell}$, so

$$\omega_1 \approx \omega = \sqrt{\frac{9.80 \text{ m/s}^2}{20 \text{ m}}}$$
$$= 0.700 \text{ rad/s}$$

To evaluate τ we use Equation 13.38, noting that the amplitude at time t is $Ae^{-t/\tau}$. In order for the amplitude at time t to be half its initial ($t = 0$) value we must have

$$Ae^{-t/\tau} = \tfrac{1}{2}A$$

Setting $t = 1$ hour and solving for the damping time τ gives

$$\tau = \frac{1 \text{ hour}}{0.693} = 1.44 \text{ hours}$$

We assumed initially that $\omega\tau \gg 1$. This assumption was justified because

$$\omega\tau = (0.700 \text{ rad/s})(1.44 \text{ hour})(3600 \text{ s/hour}) = 3630 \gg 1$$

Energy Considerations in Damped Motion

One consequence of the friction that causes damped motion is that mechanical energy is not conserved: it is converted to thermal energy. To show this, we write the total mechanical energy, $E = K + U$, in the form

$$E = \tfrac{1}{2}mv^2 + \tfrac{1}{2}kx^2 \qquad (13.43)$$

Differentiating and using $v = dx/dt$ and $a = dv/dt$ we can write the rate of change of E as

$$\frac{dE}{dt} = (ma + kx)v \qquad (13.44)$$

If there is no damping, the simple harmonic oscillator equation, $ma + kx = 0$, shows that $dE/dt = 0$, and mechanical energy is conserved. With the damped harmonic oscillator, however, we have $(ma + kx) = -\sigma v$, so Equation 13.44 can be written

$$\frac{dE}{dt} = -\sigma v^2 \qquad (13.45)$$

The damping constant σ is positive, as is v^2; thus, dE/dt is negative, showing that the mechanical energy decreases with time. Mechanical energy is converted into thermal energy by friction.

13.4
DRIVEN OSCILLATOR: RESONANCE

In damped motion, no external forces act to replace the mechanical energy that is continuously being converted to thermal energy through friction. Consequently, the motion eventually stops. In Section 13.3 we assumed that the moving body experienced an elastic force ($-kx$) and a frictional force ($-\sigma v$). We now study the behavior of a damped harmonic oscillator that is also subjected to a third force that varies sinusoidally with time. We call such a system a **driven oscillator**.

Our prototype of a driven oscillator consists of a mass m that is subjected to a spring force, $-kx$, a frictional force, $-\sigma v$, and a periodic driving force, $F_0 \cos \Omega t$. The motion is governed by the equation

$$ma = -\sigma v - kx + F_0 \cos \Omega t$$

Making the same substitutions as in the previous section, we obtain a standardized form of the driven oscillator equation

$$\frac{d^2x}{dt^2} + \frac{2}{\tau}\frac{dx}{dt} + \omega^2 x = a_0 \cos \Omega t \qquad (13.46)$$

In this equation $\omega^2 = k/m$ and $\tau = 2m/\sigma$, as in Section 13.3, Ω is the angular frequency of the driving force, $F_0 \cos \Omega t$, and $a_0 = F_0/m$.

This equation has a general solution consisting of a sum of two parts. There is a transient part describing underdamped, critically damped, or overdamped motion, depending on the value of τ, as discussed in Section 13.3. The transient part dies out exponentially with time. The other part of the general solution to Equation 13.46 is called the **steady-state solution**.

At the onset of forced vibrations the motion of an oscillator can be described by a superposition of transient and steady-state motions. Any transient motion initially present will die out exponentially with time, but the steady-state motion will persist. We assume here that the system executes steady-state motion without the presence of transients.

In the steady state, the system is forced to execute SHM with an angular frequency Ω determined by the driving force. Therefore we write the solution in the form

$$x = A \cos (\Omega t - \varphi) \qquad (13.47)$$

to resemble the harmonic oscillator solution given by Equation 13.7. The most significant difference between the two solutions concerns the angular frequency of the oscillations. In Equation 13.7 the angular frequency of oscillation (ω) is determined by k and m, whereas in Equation 13.47 the frequency Ω is a characteristic of the driving force.

Much of our interest in the forced oscillator relates to the response of the system. (By response we mean the dependence of A and φ on Ω.)

The variation of A with Ω is illustrated by the following example. A car with nylon cord tires is left parked for a prolonged period of time. Consequently the tires develop flat spots. Then, when the car is first driven again the passengers feel a periodic bumping each time that these flat spots hit the road. The amplitude of vibration of the car is small provided the speed of the car is either slow or fast. But, at 38 mph the amplitude becomes large and the car shakes violently. This large amplitude at a particular frequency is an example of **resonance**.

To express A and φ in terms of Ω, ω, τ, and a_0, we substitute $x = A \cos (\Omega t - \varphi)$ into Equation 13.46 and use the identities.

$$\cos (\Omega t - \varphi) = \cos \Omega t \cos \varphi + \sin \Omega t \sin \varphi$$

$$\sin (\Omega t - \varphi) = \sin \Omega t \cos \varphi - \cos \Omega t \sin \varphi$$

to obtain an equation of the form

$$\left[A(\omega^2 - \Omega^2) \cos \varphi + \left(\frac{2A\Omega}{\tau} \right) \sin \varphi - a_0 \right] \cos \Omega t$$

$$+ \left[A(\omega^2 - \Omega^2) \sin \varphi - \left(\frac{2A\Omega}{\tau} \right) \cos \varphi \right] \sin \Omega t = 0$$

This equation is satisfied only if the coefficients of cos Ωt and sin Ωt separately vanish. This requirement gives two equations relating A and φ. The solutions are

$$\tan \varphi = \frac{2\Omega/\tau}{\omega^2 - \Omega^2} \tag{13.48}$$

and

$$A = \frac{a_0}{\sqrt{(\omega^2 - \Omega^2)^2 + \dfrac{4\Omega^2}{\tau^2}}} \tag{13.49}$$

We regard ω and τ as fixed parameters describing the oscillator. We treat the driving frequency Ω as a parameter that we can vary to study the response of the oscillator.

Graphs of φ versus Ω and A versus Ω are shown in Figures 13.15 and 13.16. Figure 13.16 shows that the amplitude A is relatively small when the driving frequency Ω is very small or very large. With a large Ω, the inertia of the mass impairs its response to the rapidly changing driving force. The phase angle φ approaches π (180°) for a large Ω, showing that the displacement reaches a maximum in one direction when the driving force reaches a maximum in the opposite direction. With a small Ω, the phase angle φ approaches zero, showing that the displacement is virtually in step with the driving force.

For small damping, the amplitude shows a sharp maximum at $\Omega = \omega$. This is *resonance*, the strong response that results when a system is driven at its *natural* (resonant) frequency.

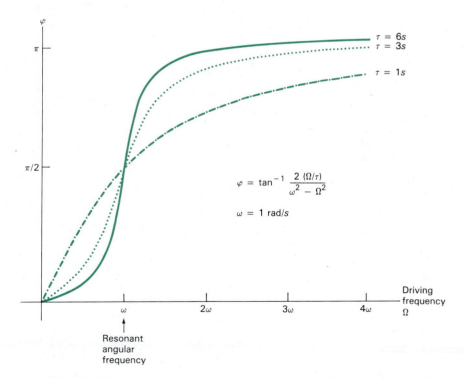

FIGURE 13.15

The relative phase angle φ between the displacement and the driving force. At resonance, the phase angle equals $\pi/2 = 90°$.

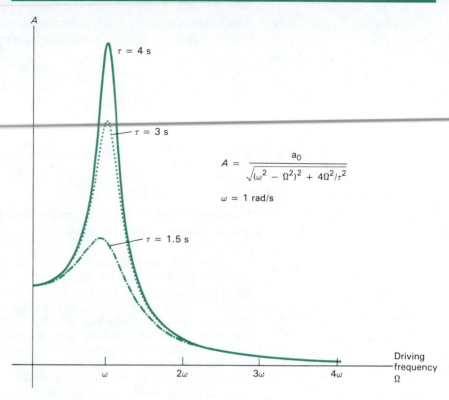

$$A = \frac{a_0}{\sqrt{(\omega^2 - \Omega^2)^2 + 4\Omega^2/\tau^2}}$$

$$\omega = 1 \text{ rad/s}$$

FIGURE 13.16

The amplitude A of the motion is plotted versus the angular driving frequency, Ω. Near resonance the amplitude becomes large. The resonant amplitude becomes larger as the damping is reduced.

EXAMPLE 8

Spring-Mass Resonance

A 1-kg mass connected to a linear spring ($k = 3600$ N/m) experiences a weak damping force with $\sigma = 6 \times 10^{-3}$ kg/s. The mass is driven by a very weak force, with $F_0 = 0.02$ N. The parameters a_0 and τ have the values

$$a_0 = \frac{F_0}{m} = 0.02 \text{ m/s}^2$$

$$\tau = \frac{2m}{\sigma} = 333 \text{ s}$$

We want to determine the resonant frequency and show that a sizable resonant amplitude can result even though the driving force is quite weak. The resonant angular frequency of the oscillator is

$$\omega = \sqrt{\frac{k}{m}} = \sqrt{\frac{3600 \text{ N/m}}{1 \text{ kg}}} = 60 \text{ rad/s}$$

Resonance occurs when $\Omega = \omega = 60$ rad/s. The resonance frequency in Hz is

$$f = \frac{1}{2\pi}\sqrt{\frac{k}{m}} = \frac{60 \text{ Hz}}{2\pi}$$

$$= 9.55 \text{ Hz}$$

The resonant amplitude follows from Equation 13.49

$$A = \frac{a_0 \tau}{2\Omega} = \frac{(0.02 \text{ m/s}^2)(333 \text{ s})}{2(60 \text{ s}^{-1})} = 0.0555 \text{ m}$$

$$= 5.55 \text{ cm}$$

For comparison, the amplitude at a very low driving frequency approaches the $\Omega = 0$ limit

$$A(\Omega = 0) = \frac{a_0}{\omega^2} = \frac{0.02 \text{ m/s}^2}{(60 \text{ s}^{-1})^2}$$

$$= 5.55 \times 10^{-4} \text{ cm}$$

This is exactly ten thousand times smaller than the resonant amplitude. Resonance can excite relatively gigantic responses.

WORKED PROBLEM

A circular wire hoop, hanging on a small nail, is set into oscillation (Figure 13.17). The mass of the hoop is M and its radius is R. (a) For small oscillations, show that the motion is SHM and derive an expression for the period. (b) Calculate the period for $R = 10$ cm.

Solution

To demonstrate SHM we must show that the acceleration is directly proportional to the displacement and in the opposite direction. We can use the angular version of Newton's second law, $\tau = I\alpha$, to determine the angular acceleration.

(a) We measure the angular displacement of the hoop relative to a vertical line through the nail. Two forces act on the hoop; the contact force exerted by the nail, and the weight of the hoop, Mg. Relative to a torque reference point at the nail, the contact force exerts zero torque. The weight exerts a restoring torque. From Figure 13.17 we identify the lever arm as $R \sin \theta$. The torque exerted by the weight is,

$$\tau = -MgR \sin \theta$$

The minus sign indicates the restoring nature of the torque. Using Newton's second law in angular form we have

$$I_p \alpha = -MgR \sin \theta$$

For small angular displacements, $\sin \theta \approx \theta$, and α is directly proportional to θ. The hoop undergoes SHM with a period given by

$$T = 2\pi \sqrt{\frac{-\alpha}{\theta}} = 2\pi \sqrt{\frac{I_p}{MgR}}$$

This same result follows directly from Equation 13.24, the equation giving the period of a physical pendulum. In the case of the hoop, the distance from the center of mass to the point about which the pendulum oscillates equals the radius of the hoop.

We can use the parallel axis theorem to determine I_p. This gives

$$I_p = I_c + mh^2 = MR^2 + MR^2 = 2MR^2$$

The period is,

$$T = 2\pi \sqrt{\frac{2R}{g}}$$

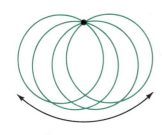

FIGURE 13.17

Interestingly, the period is equal to that of a simple pendulum of length $2R$.
(b) For $R = 10$ cm $= 0.10$ m, the period is

$$T = 2\pi \sqrt{\frac{0.20 \text{ m}}{9.80 \text{ m/s}^2}} = 0.90 \text{ s}$$

EXERCISES

13.1   Simple Harmonic Motion

A. A particle starts from rest and executes simple harmonic motion along the x-axis. It returns to its starting position every 0.25 s. Determine the period, frequency, and angular frequency.

B. A simple harmonic oscillator in the form of a linear spring and a 0.150-kg mass has a period of 2.10 s. Determine the spring constant k.

C. A mass m is oscillating freely on a linear spring (Figure 1). When $m = 0.81$ kg the period is 0.91 s. An unknown mass on the same spring is observed to have a period of 1.16 s. Determine (a) the spring constant k, (b) the unknown mass.

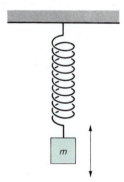

FIGURE 1

D. A horizontal spring-mass system is started by pulling the mass aside a distance A, thereby stretching the spring a distance A, and then releasing the mass from rest. Determine the stretch of the spring for which the potential and kinetic energies are the same. Express your answer in terms of the length A.

E. The displacement of a simple harmonic oscillator is described by the equation

$$x = 0.073 \cos \omega t \text{ m} \qquad \omega = 2 \text{ rad/s}$$

(a) Where are the turning points? (b) What is the period?

F. The angular displacement of a pendulum is represented by the equation $\theta = 0.32 \cos \omega t$ ($\omega = 4.43$ rad/s). Determine the period and the length of the pendulum.

G. The velocity of a 300-gram mass attached to the end of a linear spring is represented by $v = 1.60 \sin \omega t$ m/s, with $\omega = 2.83$ rad/s. Determine the total energy of the system.

13.2   Applications of Simple Harmonic Motion

H. A meter stick undergoes SHM about an axis that is perpendicular to its length and that passes through a point 20 cm from one end. Determine the period of the oscillations.

I. A 120-gram particle attached to the end of a linear spring oscillates horizontally on a frictionless surface. Its position as a function of time is given by $x = 0.35 \cos \omega t$ m. Determine the spring constant if the particle's maximum acceleration equals 9.80 m/s².

J. (a) Explain why you would expect the period of a simple pendulum to increase when it is moved to an elevated position above the earth's surface. (b) If a simple pendulum has a period of 1 s when on the earth's surface, at what distance above the earth's surface is its period 2 s?

K. A simple pendulum of length ℓ has a period T. If the length is increased by a small amount $\Delta\ell$, show that the change in the period is

$$\Delta T = T\left(\frac{\Delta\ell}{2\ell}\right)$$

L. A simple pendulum of length 1.45 m is observed to take 242 s to complete 100 oscillations. Calculate g. (Note: The experiment is carried out at high altitude.)

M. An object undergoing uniform circular motion in a plane has a position given by

$$\mathbf{r} = 0.31(\mathbf{i} \cos \omega t + \mathbf{j} \sin \omega t) \text{ m} \qquad \omega = 4 \text{ rad/s}$$

(a) Determine the radius of the circle. (b) Determine the period of the motion.

13.3   Damped Motion

N. The oscillation of a damped pendulum is described by Equation 13.34, with $m = 10$ kg, $\sigma = 6.25 \times 10^{-3}$ kg/s, and $k = 23.5$ N/m. Calculate (a) the angular frequency, ω; (b) the period, T; (c) the time for the amplitude to fall to half of its original value.

13.4   Driven Oscillator: Resonance

P. A spring-mass oscillator is driven at its resonant frequency of 60 Hz. The parameters are $m = 1.21$ kg, $k = 1.72 \times 10^5$ N/m, $\sigma = 3.10 \times 10^{-2}$ kg/s,   and

$F_0 = 3.68 \times 10^{-3}$ N. Calculate the damping time τ and the resonant amplitude A.

Q. A spring-mass oscillator has $k = 4.72$ N/m, $m =$ 0.93 kg. If it is driven by a periodic driving force, what must be the period of the force if it is to excite resonance?

PROBLEMS

1. When a 5-gram mass is attached to a vertical linear spring, the spring's length increases by 2 cm. The same spring is then arranged to oscillate horizontally on a frictionless surface. A mass of 2 grams is attached and the spring is pulled a distance x_0 from its equilibrium position and released from rest. The initial acceleration is 50 m/s^2. Determine the spring constant, k, and the initial displacement, x_0.

2. An athlete of mass 80 kg assumes a prone position with his center of mass on the symmetry line of the torsion pendulum shown in Figure 13.9. The measured period of oscillation of the pendulum with the athlete is $T = 3$ s; without the athlete it is $T_0 = 2$ s. The value of the twist constant is $\beta = 125$ N·m/rad. Determine the moment of inertia of the athlete about the axis of rotation.

3. Two particles of equal mass are connected to separate springs having spring constants of k and $4k$. With one end of each spring fixed, the particles are given identical displacements and then released. They undergo SHM on a horizontal frictionless surface. If the smaller period of the two springs is 1 s, (a) what is the minimum time required for the two particles to return simultaneously to their starting positions? (b) How many oscillations has each particle made during this time?

4. A damped simple harmonic oscillator has

$$\sigma = 0.11 \text{ kg/s} \qquad k = 1.20 \text{ N/m} \qquad m = 0.130 \text{ kg}$$

(a) Is its motion overdamped or underdamped? (b) Determine the value of σ for critically damped motion.

5. Students in an undergraduate physics laboratory recorded the data shown in Figure 2 for a damped harmonic oscillator. Show that these data are described by Equation 13.38 by finding values for A, τ, and ω_1, and by plotting the displacement x versus time t.

6. A wire frame in the form of an equilateral triangle undergoes SHM about an axis through one vertex. (The axis is perpendicular to the plane of the triangle.) The sides of the triangle have lengths of 1 m. Determine the period of the small-amplitude oscillations.

7. A floating bottle bobs up and down with a period of 0.52 s and an amplitude of 4.1 cm. Its motion is simple harmonic and its mass is 0.28 kg. Determine the magnitude of the maximum force acting on it.

8. A horizontal spring-mass oscillator undergoes SHM with an amplitude of 3.2 cm and an angular frequency of 12.6 rad/s. The spring constant is 5.12 N/m. Just as the mass reaches its equilibrium position a student stops it with her finger. The impulsive force delivered to the student's finger lasts for 0.01 s. Determine the average force exerted on her finger.

9. A long, thin rod of mass M and length L oscillates about its center on a cylinder of radius R (Figure 3). Show that small displacements give rise to SHM with a period given by $\pi L/\sqrt{3gR}$.

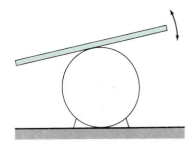

FIGURE 3

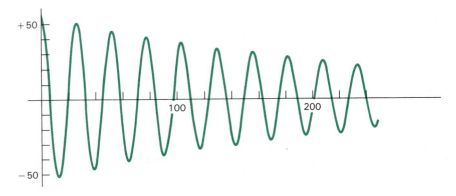

FIGURE 2

10. A solid sphere (radius = R) rolls without slipping in a cylindrical trough (radius = $5R$) as shown in Figure 4. Show that for small displacements from equilibrium, perpendicular to the length of the trough, the sphere executes SHM with a period

$$T = 2\pi \sqrt{\frac{13R}{2g}}$$

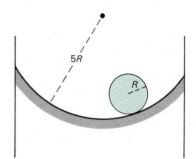

FIGURE 4

11. When a mass is attached to a linear spring the period of oscillation is found to be 1 s. Then the spring is cut in half and the mass is set into oscillation with both halves attached (Figure 5). The springs are relaxed when the mass is in its equilibrium position. (a) What is the new period? (b) A new mass replaces the first mass and oscillates with a period of 1 s. Compare the two masses.

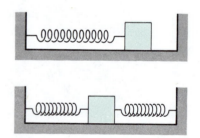

FIGURE 5

12. A glass tube in the shape of the letter U has a cross-sectional area A and contains a volume V of mercury (Figure 6). The mercury is displaced slightly from its equilibrium position and released. Find the oscillation period (disregard friction) in terms of V, A, and g.

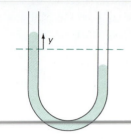

FIGURE 6

13. Identical linear springs (relaxed length = L) are attached to a mass m so that both springs are relaxed in the equilibrium position (Figure 7). The system will execute vibrational motion in the y-direction. The motion is not SHM because the restoring force is not directly proportional to the displacement. Prove that for small displacements from equilibrium the restoring force in the y-direction is given approximately by ky^3/L^2.

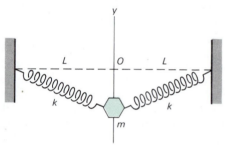

FIGURE 7

14. A simple pendulum with a length of 2.23 m and a mass of 6.74 kg is given an initial speed of 2.06 m/s at its equilibrium position. Assume it undergoes SHM and determine its (a) period, (b) total energy, (c) maximum angular displacement.

15. A simple pendulum (length = 1.73 m, mass = 1.21 kg) is started from $\theta = 0.324$ radian with a speed of 1.42 m/s. Its initial motion is toward the equilibrium position. Assume it undergoes SHM and that its angular displacement has the form

$$\theta = \theta_{max} \cos(\omega t - \varphi)$$

Determine (a) ω, (b) θ_{max}, (c) φ. (d) Find the initial phase φ for the case where the initial motion is away from the equilibrium position.

CHAPTER 14

MECHANICAL PROPERTIES OF MATTER

14.1
STATES OF MATTER

We are going to analyze the properties of matter in this chapter. Our analysis is at the macroscopic level—that is, the level at which measuring instruments and our senses can respond. A *macroscopic* quantity reflects the combined actions of many atoms or molecules. A *microscopic* quantity, on the other hand, describes conditions at the atomic level. The mass of your body, for example, is a macroscopic quantity; the mass of a single carbon atom in your body is a microscopic quantity.

Physical states of matter can be divided into two broad classes: solids and fluids. A solid tends to maintain its shape: it resists the actions of external forces that try to change it. In contrast, external forces can easily change the shape of a fluid.

Gases and liquids are fluids. The distinction between a gas and a liquid is made on the basis of compressibility. Under ordinary conditions, gases are relatively easy to compress, whereas liquids are virtually incompressible.

One characteristic of solids can be demonstrated by splitting a log; it will easily split when an axe strikes it parallel to the grain. However, the same blow applied across the grain fails to split the wood. We describe this type of directional dependence by saying that wood is **anisotropic**. Many crystalline solids display anisotropies in their mechanical, thermal, and electromagnetic properties. The directional dependence of these properties of bulk matter can be traced to the anisotropic crystalline structure at the atomic level. Materials that do not exhibit any such directional preferences are said to be **isotropic**. Gases and most liquids are isotropic because the motions of their atoms are not strongly correlated. Noncrystalline solids, called **amorphous** solids, are isotropic. We limit our discussion in this text to isotropic materials.

14.2
STRESS AND STRAIN

External forces can change the shape of a solid slightly. The solid is said to be **elastic** if it reverts to its original shape when the forces are removed. **Plastic** materials, on the other hand, remain permanently distorted when the deforming forces are removed.

When an elastic solid is deformed, its response is measured by an **elastic modulus**. The deforming force is described in terms of the **stress** that it exerts. The deformation of the solid is described in terms of the **strain** that results. In this section we define and illustrate the primary types of stresses and strains. In the sections that follow we will see how stress and strain are related by the elastic moduli.

Tensile Stress and Tensile Strain

Consider a rod of length L and cross-sectional area A. Suppose that a force T is applied at both ends, parallel to the length, stretching the rod by ΔL (Figure 14.1). The **tensile stress** is defined as the ratio of the force of tension to the cross-sectional area

$$\text{tensile stress} = \frac{T}{A} \tag{14.1}$$

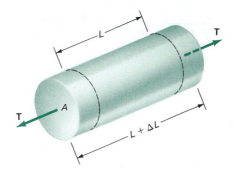

FIGURE 14.1

Equal but opposite forces of magnitude T applied to the ends of the rod cause its length to increase from L to $L + \Delta L$. The rod is said to be in tension.

The resulting **tensile strain** is defined as a ratio that compares the change in length to the original length:

$$\text{tensile strain} = \frac{\Delta L}{L} \tag{14.2}$$

Note that tensile stress has the dimensions of force per unit area and is measured in newtons per square meter (N/m^2). Tensile strain is dimensionless.

EXAMPLE 1

Stress and Strain

A television commercial shows a 47,000-lb (210,000 N) railroad car being lifted by a cord made of a material used to strengthen automobile tires. The diameter of the cord is 9.1 cm. The tensile stress is

$$\frac{T}{A} = \frac{210,000 \text{ N}}{\frac{1}{4}\pi(0.091 \text{ m})^2} = 3.2 \times 10^7 \text{ N/m}^2$$

The unstretched length of the cord is 8 m. The length increases by 2.60 cm when the railroad car is raised. The tensile strain is

$$\frac{\Delta L}{L} = \frac{2.60 \text{ cm}}{800 \text{ cm}} = 0.00325$$

The large applied stress produces a relatively small change in length.

Pressure and Volume Strain

An object that is submerged in a fluid experiences compressive forces over its entire surface. If these compressive forces change the object's volume from V to $V + \Delta V$, the **volume strain** is defined as $\Delta V/V$. Because the compressive forces cause the volume to decrease, ΔV is negative. Consequently, the volume strain is negative. The **compressive stress** associated with volume strain is called **pressure.*** If F_p denotes the compressive force normal to a surface element of area A (Figure 14.2) the pressure is defined as

$$\text{pressure} \equiv P = \frac{F_P}{A} \tag{14.3}$$

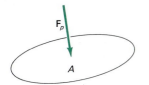

FIGURE 14.2

The pressure P is defined to be the ratio of the magnitude of the normal force $\mathbf{F}_p$ to the area A on which the force acts: $P \equiv F_P/A$.

---

* For liquids, the terminology *hydrostatic pressure* is often used, even in situations where the pressure is not exerted by water.

Several units of pressure are commonly encountered in the sciences. The N/m^2 is called the **pascal** (Pa). The pound per square inch (psi), the atmosphere (atm), the bar, and the torr are other pressure units. The **atmosphere** is defined as

$$1 \text{ atm} = 1.01325 \times 10^5 \text{ N/m}^2$$

The bar is defined as

$$1 \text{ bar} = 10^5 \text{ N/m}^2$$

so it is very nearly equal to one atmosphere. Newspaper weather maps often show **isobars**, lines of constant pressure, with the pressure indicated in millibars ($1 \text{ mb} = 10^{-3}$ bar).

The torr is defined by the relation

$$760 \text{ torr} = 1 \text{ atm}$$

The pressure of a confined gas results from impacts of the gas molecules on the container walls. In dealing with gas pressures, we often measure the **gauge pressure**, which is the difference between the total or **absolute** pressure and the surrounding atmospheric pressure. For example, if the gauge pressure for an auto tire reads 32 psi at a time when the atmospheric pressure is 15 psi, the absolute pressure in the tire is 47 psi.

EXAMPLE 2

How to Weigh Your Car with a Ruler

The four tires of an automobile are inflated to a gauge pressure of $2.0 \times 10^5 \text{ N/m}^2$ (29 psi). Each of the four tires has an area of 0.024 m^2 that is in contact with the ground. We want to determine the weight of the automobile.

The weight, W, is balanced by the contact forces exerted on the tires by the ground. The total contact force is $4PA$, where P is the gauge pressure and A is the contact area for one tire. We have for the weight

$$W = 4PA = 4(2.0 \times 10^5 \text{ N/m}^2)(0.024 \text{ m}^2)$$
$$= 19,000 \text{ N}$$

This converts to approximately 4300 pounds.

Shear Stress and Shear Strain

The compressive stress that we call pressure changes the volume but not the shape of an object. A cube remains cubical when it is compressed by a uniform pressure. A **shear stress**, on the other hand, tends to change the shape of a body. Referring to Figure 14.3, we see that a force F_S parallel to the surface of area A deforms the rectangular area xy into a parallelogram—the shape of the body is altered. The shear stress is defined as

$$\text{shear stress} \equiv S = \frac{F_S}{A} \tag{14.4}$$

and the **shear strain** is defined as

$$\text{shear strain} \equiv \frac{\Delta x}{y} = \tan \theta \approx \theta \tag{14.5}$$

The approximation $\tan \theta \approx \theta$ is valid for small strains.

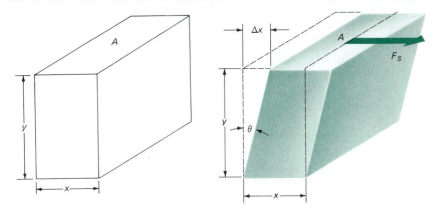

FIGURE 14.3

The tangential force F_s distorts the solid. The cross section is changed from a rectangle to a parallelogram. The shear stress S is defined to be the ratio of the magnitude of the tangential force F_s to the area A on which the force acts.

14.3
HOOKE'S LAW AND ELASTIC MODULI

Many situations reveal that stress and strain are related. For example, as a balloon is inflated, two quantities increase: the balloon's volume and the pressure of the air inside. The change in volume is a measure of the strain, and the air pressure is a measure of the stress. As you put more air into the balloon there is an increase in the stress, which results in an increase in the strain as the balloon expands. Similarly, the extension of a spring balance is a measure of its strain. The suspended weight establishes the stress. The greater the weight, the longer the extension, and, therefore the greater the stress and the strain.

For many elastic solids, stress and strain are linked by **Hooke's law**, which states that *stress is directly proportional to strain*. The proportionality factor between stress and strain is called an **elastic modulus**. Hooke's law may be expressed as an equation:

$$\frac{\text{stress}}{\text{strain}} = \text{elastic modulus} \tag{14.6}$$

We will first introduce the elastic modulus that relates tensile stress and tensile strain.

Young's Modulus

In the particular case of tensile stress and strain, the elastic modulus is known as **Young's modulus**, denoted by Y. For tensile stress and strain, Hooke's law is expressed by

$$Y = \frac{\text{tensile stress}}{\text{tensile strain}} = \frac{T/A}{\Delta L/L} \tag{14.7}$$

To emphasize the fact that the larger Y is, the more rigid or unyielding the material, Hooke's law may be expressed as

$$\frac{\Delta L}{L} = \frac{T/A}{Y} \tag{14.8}$$

TABLE 14.1
Physical Properties of Solids

Material	Density (units of 10^3 kg/m^3)	Young's Modulus Y	Shear Modulus μ (units of 10^{10} N/m^2)	Bulk Modulus B	Poisson's Ratio (σ)
Carbon (diamond)	3.3	83	34	50	0.22
Aluminum	2.7	6.9	2.6	6.6	0.33
Iron (Cast)	7.9	15	6.0	10	0.25
Copper	9.0	12	4.6	10	0.30
Zinc	7.1	9.2	3.7	6.0	0.24
Tin	7.3	5.4	2.0	6.0	0.35
Lead	11.4	1.6	0.56	3.7	0.43
Steel	7.8	21	8.0	19	0.31
Glass	2.5	6.1	2.5	3.6	0.22
Brass	8.6	9.0	3.5	7.0	0.29

Table 14.1 presents values of Young's modulus for several solids. Other elastic constants that will be introduced later are also included in Table 14.1

EXAMPLE 3

Stretching Steel

A Foucault pendulum has an 80-kg mass suspended from a steel wire. The wire is 18 m long and 3 mm in diameter. We want to determine the increase in the length of the wire when the pendulum is not in motion. In this situation, the tensile force equals the weight of the suspended mass

$$T = mg = (80 \text{ kg})(9.80 \text{ m/s}^2)$$
$$= 784 \text{ N}$$

The cross-sectional area is

$$A = \frac{\pi}{4}(0.003)^2 \text{ m}^2$$
$$= 7.07 \times 10^{-6} \text{ m}^2$$

The tensile stress is

$$\frac{T}{A} = \frac{784 \text{ N}}{7.07 \times 10^{-6} \text{ m}^2} = 1.11 \times 10^8 \text{ N/m}^2$$

From Table 14.1, Young's modulus for steel is 21×10^{10} N/m^2. Using Equation 14.8, we get the tensile strain

$$\frac{\Delta L}{L} = \frac{T/A}{Y} = \frac{1.11 \times 10^8 \text{ N/m}^2}{21 \times 10^{10} \text{ N/m}^2} = 0.00053$$

With $L = 18$ m we find $\Delta L = 9.5$ mm. The wire is stretched nearly one centimeter.

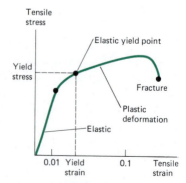

FIGURE 14.4

Tensile stress-strain relation for a metal rod.

Figure 14.4 shows the nature of the stress-strain relation for a metal. Hooke's law is valid only for very small strains. The relation between stress and strain generally ceases to be linear for a strain somewhere between 0.001 and 0.01. For example, if a metal rod is subjected to increasing tension, it eventually undergoes a plastic deformation—that is, it "gives"—at a characteristic **yield stress**. Further increases in the strain eventually lead to **fracture**—the metal snaps. The fracture stress lies in the range of 10^8 N/m^2 to 10^9 N/m^2 for many solids.

Maximum Size of Man-Made Objects

Man-made structures cover a wide range of sizes, from tiny solid-state electronic circuits to the Great Wall of China. The shapes of these structures, as well as the shapes of most solids, are maintained by electrical forces between atoms. These forces act to oppose any change in shape—compression, extension, shearing, and the like.

Gravitational forces acting between different segments of typical man-made objects are negligible in comparison to these electrical forces. Electrical forces, not gravity, hold together desks and chairs and other relatively small objects. Gravity holds together the matter forming the sun and the planets. If an object is large enough, gravity not only holds it together, but also determines its shape. The crush of gravity overwhelms the electrical forces in a large object and compacts the matter into a nearly spherical form.*

We want to show that the limiting size for a body whose shape is determined by electrical forces is a few hundred kilometers. The limiting size may be inferred by using a dimensional argument and by noting that 10^9 N/m^2 is a typical fracture stress for solids. Gravity will fracture a solid whenever the gravitational stress exceeds the fracture stress.

The attractive gravitational force between two particles having masses m_1 and m_2 and separated by a distance r is

$$F = \frac{Gm_1m_2}{r^2} \tag{14.9}$$

Consider a sphere of mass M and radius R (Figure 14.5). On purely dimensional grounds the characteristic gravitational force that the two hemispheres exert on each other is roughly GM^2/R^2. This force acts over an area of approximately πR^2. Our order-of-magnitude approach does not justify inclusion of factors of π, 2, and the like, so we take R^2 as the area over which the gravitational force acts. The gravitational stress, or pressure, is therefore

$$P_{\text{gravity}} = \frac{F_{\text{gravity}}}{\text{area}} \approx \frac{GM^2/R^2}{R^2} = \frac{GM^2}{R^4}$$

It is convenient to eliminate M in favor of the mass density, ρ. The volume is roughly R^3, so

$$M \approx \rho R^3$$

In terms of ρ and R,

$$P_{\text{gravity}} \approx GR^2\rho^2 \tag{14.10}$$

When $P_{\text{gravity}} > P_{\text{fracture}}$, gravity determines the shape of the object. Taking $\rho = 10^4$ kg/m^3 and $P_{\text{fracture}} = 10^9$ N/m^2 as typical of solids, we find that gravity wins when

$$GR^2\rho^2 > 10^9 \text{ N/m}^2$$

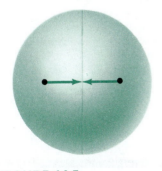

FIGURE 14.5

The two hemispheres exert equal but opposite gravitational forces on each other, and thereby set up a pressure within the solid.

---

* Rotation causes deviations from a spherical shape. The earth, for example, is an oblate spheroid because of its rotation.

or

$$R > \frac{1}{\rho} \sqrt{\frac{10^9 \ \text{N/m}^2}{G}}$$

$$\approx (10^{-4} \ \text{m}^3/\text{kg}) \cdot \sqrt{\frac{10^9 \ \text{N/m}^2}{6.67 \times 10^{-11} \ \text{N} \cdot \text{m}^2/\text{kg}^2}} \approx 400 \ \text{km}$$

The radius of Mercury, the smallest planet, is 2432 km. So, its shape is determined by gravity, as are the shapes of all the planets in our solar system.

Most asteroids have sizes of a few kilometers. We therefore expect their shapes to be determined by electrical forces. If this is so, it is unlikely that they are spherical. This expectation is borne out by observations of the sunlight that is reflected by asteroids. The light intensity varies as the asteroids rotate, and the shapes of the asteroids can be determined by studying light intensity patterns. Most asteroids are found to be oblong bodies. Phobos, one of the two moons of Mars, has an irregular shape—a shape determined by electrical forces rather than gravity (Figure 14.6). This is not surprising because its largest dimension is less than 30 km.

Elastic Potential Energy

An elastic solid behaves like a linear spring because the force required to change its length is directly proportional to this change. For an elastic solid, $F = (YA/L) \ \Delta L$. For a linear spring, $F = kx$. We see that ΔL corresponds to x and that the spring constant for an elastic solid of length L and cross-sectional area A is

$$k = \frac{YA}{L} \tag{14.11}$$

For a linear spring the elastic potential energy is given by $U = \frac{1}{2}kx^2$. By analogy, the elastic potential energy stored in an elastic solid is

$$U = \frac{1}{2}\left(\frac{YA}{L}\right)(\Delta L)^2 \tag{14.12}$$

EXAMPLE 4

Gravitational Wave Detector

Gravitational waves are extremely weak ripples in the strength of gravity that travel at the speed of light. When a gravitational wave passes through an elastic solid it sets up tiny strains that store elastic potential energy. To appreciate the difficulty of detecting gravitational waves, we will estimate this elastic potential energy.

An aluminum cylinder used in the earliest attempts to detect gravitational waves was 4 m long and 0.80 m in diameter. The gravitational waves were expected to change the length by no more than $\Delta L = 10^{-16}$ m. This minuscule distance is comparable to the dimension of an atomic nucleus. For aluminum, $Y = 6.9 \times 10^{10}$ N/m², and

$$U = \frac{1}{2}\left(\frac{YA}{L}\right)(\Delta L)^2$$

$$= \frac{1}{2}\left[\frac{6.9 \times 10^{10} \ \text{N/m}^2)\pi(0.40 \ \text{m})^2}{4 \ \text{m}}\right](10^{-16} \ \text{m})^2$$

$$= 4.3 \times 10^{-23} \ \text{J}$$

This energy is comparable to the kinetic energy of *one* oxygen molecule in the air.

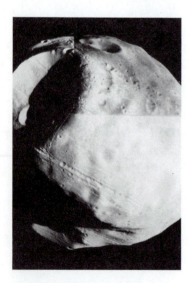

FIGURE 14.6

A view of Phobos from the Viking Orbiter 1 (range 612 km). The largest crater at the top lies near the North Pole (NASA photo).

Bulk Modulus and Compressibility

For volume stress and strain the elastic modulus is called the **bulk modulus**, and is denoted by B. Hooke's law is expressed by

$$B = -\left(\frac{\text{volume stress}}{\text{volume strain}}\right) = \frac{-P}{\Delta V/V} \tag{14.13}$$

The change in volume, ΔV, is negative. The minus sign has been inserted in Equation 14.13 to make B positive. The units of bulk modulus, like those of Young's modulus, are N/m^2. Values of B for several materials are given in Table 14.1. Typically, values of B for solids are in the neighborhood of 10^{11} N/m^2, or 10^6 atm.

The reciprocal of B is sometimes a more convenient parameter. It is called the **compressibility** and is denoted by the symbol K:

$$K \equiv \frac{1}{B} = -\frac{(\Delta V/V)}{P} \tag{14.14}$$

The smaller the value of K, the larger the pressure needed to produce a specified volume strain must be. The compressibility is a meaningful quantity for fluids as well as solids. In fact, liquids can transmit hydraulic pressure effectively because of their very small compressibilities.

EXAMPLE 5

Compression of Water and Steel

From experiment, the compressibilities of water and steel are found to be

$$K_{\text{water}} = 5 \times 10^{-5}/\text{atm}$$

and

$$K_{\text{steel}} = 6 \times 10^{-7}/\text{atm}$$

We see that the compressibility of water is about 80 times that of steel—it is 80 times easier to compress water than it is to compress steel. However, both water and steel are difficult to compress. To demonstrate this, we rewrite Equation 14.14 as

$$-\frac{\Delta V}{V} = KP \tag{14.15}$$

A 1% decrease in volume ($\Delta V/V = -0.01$) requires a pressure of

$$P = \frac{0.01}{K}$$

Solving this equation for water we obtain

$$P_{\text{water}} = \frac{0.01}{5 \times 10^{-5}} \text{ atm}$$

$$= 200 \text{ atm}$$

Thus, to compress water by 1% requires a pressure of 200 atm, which is about 100 times the pressure in an automobile tire. To compress steel by the same amount one needs a pressure

$$P_{\text{steel}} = \frac{0.01}{6 \times 10^{-7}} \text{ atm}$$

$$\approx 16,000 \text{ atm}$$

One of the consequences of the small compressibility of water is a slight increase in the density of water with depth. At great depths in the oceans, water is compressed by the weight of the water above it, and its density is thereby increased.

The bulk modulus, B, is defined by Equation 14.13 in terms of the volume strain, $\Delta V/V$. The volume strain $\Delta V/V$ is related to the fractional change in mass density through the conservation of mass. Thus, it follows from

$$\rho V = \text{constant}$$

that

$$-\frac{\Delta V}{V} = \frac{\Delta \rho}{\rho} \tag{14.16}$$

Inserting this relation into Equation 14.13, $-P(\Delta V/V) = B$, gives us an alternative expression for B in terms of the pressure and fractional density change,

$$B = \frac{P}{\Delta \rho / \rho} \tag{14.17}$$

Shear Modulus

For shear stress and strain the elastic modulus is called the **shear modulus**, and is denoted by μ. Hooke's law takes the form

$$\mu = \frac{\text{shear stress}}{\text{shear strain}} = \frac{F_s/A}{\theta} \tag{14.18}$$

As Table 14.1 reveals, for many materials the values of μ are comparable to those of Y and B.

Shear stresses and strains arise in rotating or twisted structures. For example, the twisted rod in Figure 14.7 exhibits a shear strain.

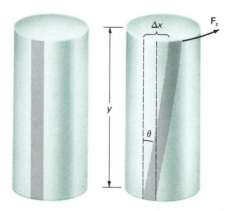

FIGURE 14.7

Shear stress and strain for a cylindrical rod. The base of the rod is anchored. The vertical strip (gray area) is distorted when the tangential twisting force F_s is applied. The shear strain is $\Delta x/y \approx \theta$, with the angle θ expressed in radians.

FIGURE 14.8
As the shot-putter drives herself forward the sole of her shoe experiences a shear stress and undergoes a shear strain.

EXAMPLE 6

The Shear Strain of a Shot-Putter's Shoe Sole

A shot-putter exerts a shear stress on the sole of her shoe as she drives her body forward (Figure 14.8). What is the resulting shear strain on the sole? The athlete exerts a shearing force of 700 N, which is distributed over an area of 0.015 m². The shear modulus of the rubber sole is $\mu = 1.5 \times 10^6$ N/m². We substitute these values into Equation 14.18 and solve for the shear strain, θ.

$$\theta = \frac{F_s/A}{\mu} = \frac{700 \text{ N}/0.015 \text{ m}^2}{1.5 \times 10^6 \text{ N/m}^2}$$

$$= 0.03 \text{ rad}$$

This is an angle of approximately 2 degrees. This strain is rather large because the shear modulus μ for rubber is much smaller than it is for typical elastic solids.

Poisson's Ratio

When an elastic solid is stretched, its transverse dimensions decrease. For example, if a cylinder of length L and diameter w is stretched, L increases and the transverse dimension w decreases. The transverse strain is defined as the fractional change in a transverse dimension. For example, if the cylinder diameter changes by Δw, the transverse strain is $\Delta w/w$.

The relationship between the tensile strain and the transverse strain is embodied in **Poisson's ratio**. Let $\Delta L/L$ denote the tensile strain and let $\Delta w/w$ denote the transverse strain. For small strains, experiment shows that $\Delta w/w$ and $\Delta L/L$ are directly proportional

$$\frac{\Delta w}{w} = -\sigma \left(\frac{\Delta L}{L} \right) \tag{14.19}$$

The dimensionless quantity σ is Poisson's ratio. The value of σ is a characteristic property of an elastic solid. As Table 14.1 shows, typical values of σ lie in the range from 0.2 to 0.4.

EXAMPLE 7

Volume Change for a Stretched Wire

When an object is stretched, its transverse dimensions decrease. Does its volume increase or decrease? Make a guess!

Suppose we take the solid to be a right circular cylinder of diameter w and length L. The volume is

$$V = \frac{\pi w^2 L}{4}$$

Treating small changes (ΔV, Δw, ΔL) as mathematical differentials we find

$$\Delta V = \frac{\pi (2w \, \Delta w \, L + w^2 \, \Delta L)}{4}$$

which gives

$$\frac{\Delta V}{V} = \frac{2 \, \Delta w}{w} + \frac{\Delta L}{L}$$

Poisson's ratio enters through Equation 14.19, allowing us to eliminate $\Delta w/w$ and obtain

$$\frac{\Delta V}{V} = (1 - 2\sigma) \frac{\Delta L}{L} \qquad (14.20)$$

We find experimentally that $\sigma < 1/2$, and so $(1 - 2\sigma)$ is positive. It follows that $\Delta V/V$ has the same sign as $\Delta L/L$. When the solid is stretched, ΔL is positive. Therefore the volume increases when the solid is stretched.

To illustrate the size of the volume changes involved, consider the steel wire of Example 3. We found $\Delta L/L = 0.00053$. From Table 14.1, we have $\sigma = 0.28$ for steel, giving

$$\frac{\Delta V}{V} = (1 - 2\sigma) \frac{\Delta L}{L} = (1 - 0.56)(0.00053)$$

$$= 0.00023$$

The volume of the wire increases by approximately two parts in ten thousand.

Only two of the four quantities, Y, B, μ, and σ, are independent. For example, B and μ can be expressed in terms of Y and σ. The reason for this interdependence is that any strain can be expressed in terms of a tensile strain and σ, and any stress can be expressed in terms of a tensile stress and σ. Thus, any stress-strain relationship can be expressed in terms of σ and the tensile stress-strain relationship, which brings in Young's modulus, Y. The bulk modulus and Poisson's ratio are related to Y and μ by

$$B = \frac{Y}{9 - \dfrac{3Y}{\mu}} \qquad (14.21)$$

$$\sigma = \frac{Y}{2\mu} - 1 \qquad (14.22)$$

Keep in mind that these relations apply only to isotropic materials.

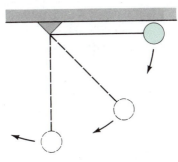

FIGURE 14.9

WORKED PROBLEM

A pendulum bob of mass m is suspended from a wire of length L (unstretched). The bob is released with the wire in a horizontal position (Figure 14.9). Show that the increase in length of the wire at the bottom of the swing is given by $\Delta L = (3mg/YA)L$, where Y is Young's modulus for the wire, A is its cross-sectional area, and g is the acceleration of gravity.

Solution

We can determine L by applying Hooke's law, $\Delta L/L = T/YA$. To do so, we need to determine T, the tension in the wire. If the bob were at rest with the wire vertical, the tension would equal the weight of the bob. When the bob is in motion, the tension is larger because the wire exerts the force that keeps the bob moving along a circular arc. Because the bob accelerates, we suspect that Newton's second law plays a role in the solution.

Two forces act on the bob; gravity and the tension in the wire. The net force on the bob at the bottom of the swing is $T - mg$, where we have taken the direction down as negative. Newton's second law gives

$$ma = T - mg$$

The acceleration of the bob at the bottom of the swing is centripetal, being directed upward toward the point of support:

$$a = \frac{v^2}{L + \Delta L}$$

Solving for the tension gives

$$T = \frac{mv^2}{L + \Delta L} + mg$$

The presence of the expression mv^2 suggests kinetic energy and the conservation of energy principle. The total mechanical energy of the bob is conserved. We take the gravitational potential energy to be zero at the bottom of the swing. Remembering that the bob starts from rest with the wire horizontal, the conservation of energy principle gives

$$\tfrac{1}{2}mv^2 + 0 = 0 + mg(L + \Delta L)$$

This in turn yields

$$\frac{mv^2}{L + \Delta L} = 2mg$$

With this result the tension becomes $T = 2mg + mg = 3mg$, and Hooke's law gives the desired result:

$$\Delta L = \left(\frac{3mg}{YA}\right)L$$

EXERCISES

14.1 States of Matter

A. Table 1 lists the mass and radius values for the planets in the solar system. (Pluto is not included because its radius is uncertain.) Determine the mass density of each planet. The four inner planets (Mercury, Venus, Earth, and Mars) are called the terrestrial, or Earth-like, planets. The outer four are referred to as the Jovian, or Jupiter-like, planets. What property of the planets explains these divisions?

TABLE 1

Planet	Mass (kg)	Radius (m)
Mercury	3.3×10^{23}	2.43×10^6
Venus	4.87×10^{24}	6.06×10^6
Earth	5.98×10^{24}	6.37×10^6
Mars	6.4×10^{23}	3.38×10^6
Jupiter	1.90×10^{27}	6.97×10^7
Saturn	5.69×10^{26}	5.79×10^7
Uranus	8.69×10^{25}	2.37×10^7
Neptune	1.03×10^{26}	2.25×10^7

14.2 Stress and Strain

B. (a) Calculate the total force exerted by the atmospheric pressure (1 atm) on 1 cm^2 of your skin. (b) If an object that weighed 10.1 N were placed on top of your head, would you notice it? (c) Why don't you ordinarily notice the forces exerted on your body by the atmosphere?

C. The gauge pressure of the air in a car tire is found to be 29 psi. The atmospheric pressure is 14.8 psi. Calculate the absolute pressure of the air in the tire in (a) pounds per square inch, (b) atmospheres, (c) kilopascals (kPa).

D. A single ice skate contacts the ice over an area 10 cm long and 3 mm wide. Assume that the weight of the skater (mass = 60 kg) is uniformly distributed over this area, and calculate the resulting pressure exerted on the ice (a) in pascals, (b) in atmospheres.

E. A mass of 4 kg is suspended from the end of a copper wire that has a diameter of 1 mm. What tensile stress does the wire experience?

F. A steel cube is subjected to a hydrostatic pressure that shortens each edge of the cube by 1 percent. What is the approximate volume strain?

G. A physics textbook weighing 15 N slides across a desk. The coefficient of sliding friction is 0.40 and the area of the book that is in contact with the desk is 0.05 m^2. Determine the shear stress exerted on the book.

H. The strain at which a solid yields and undergoes plastic deformation is typically 0.01. This may seem like a small strain. In fact, it is quite large. To sense just how large, imagine that you experience a tensile strain of 0.01. How much taller would you be? Do you think your body could withstand such a stress without undergoing a plastic deformation?

I. The weight of a 50-kg student is distributed over an area of 400 cm^2 of the soles of her shoes. Determine the compressive stress exerted on the soles. Express the pressure in pascals, pounds per square inch, and atmospheres.

14.3 Hooke's Law and Elastic Moduli

J. A student records the following data from an experiment designed to determine Young's modulus for a steel wire:

Unstressed length = 2.06 m

Stressed length = 2.10 m

Diameter = 0.22 mm

Tensile force = 133 N

Using these data, determine the value of Young's modulus.

K. A copper wire 2 m long and 1 mm in diameter is subjected to a tensile force of 125 N. How much will it stretch?

L. A steel wire 100 m long and 3 mm in diameter hangs down alongside a skyscraper under construction. How much would the wire stretch if one of the workers (mass = 100 kg) grabbed the bottom of the wire and hung freely?

M. A copper rod 1 m long and 4 mm in diameter is stretched 0.15 mm by pulling on each end. (a) What tensile force acts on each end? (b) How much elastic energy is stored?

N. A 40-kg mass is hung from a 16-m length of wire, stretching it by 1.25 cm. If the diameter of the wire is 2 mm determine the tensile stress, the tensile strain, and Young's modulus.

P. A steel wire 15 m long and 4 mm in diameter experiences a tensile strain of 4.8×10^{-4}. Determine the elastic potential energy stored in the wire.

Q. A uniform wire can act as a linear spring. Prove that cutting the wire in half doubles its spring constant.

R. The pressure of seawater increases by about 1 atm for each 10-m increase in depth. By what percentage is the density of water increased in the deepest oceanic trenches, which have depths of approximately 12 km? The compressibility of water is 50×10^{-6}/atm.

S. A woman who weighs 500 N (112 pounds) wears spike heels. The contact area of one heel is 2×10^{-6} m^2. Her weight causes a normal force of 250 N to act on this area. (a) What pressure does one heel exert on the floor? (b) If a floor covering has a bulk modulus of 10^{10} N/m^2 and is damaged by volume strains larger than -0.003, will the woman's spike heel damage the floor?

T. A pedestrian leaps sideways to avoid being run down by an automobile. Her rubber shoe sole experiences a shear force of 800 N on an area of 0.021 m^2. The shear modulus of the sole is 1.8×10^7 N/m^2. Determine the shear strain of the sole. Express your result in degrees.

U. To polish a rectangular block of brass the polishing tool moves back and forth repeatedly, pushing to the right or left on the top surface, AB (Figure 1). During the polishing process the bottom surface CD is held rigidly fixed in position. If polishing gives rise to a shear strain of 0.001 radian, calculate (a) the accompanying shear stress, (b) the shear force exerted on the polished surface by the polisher. Assume a top surface area equal to 10 cm².

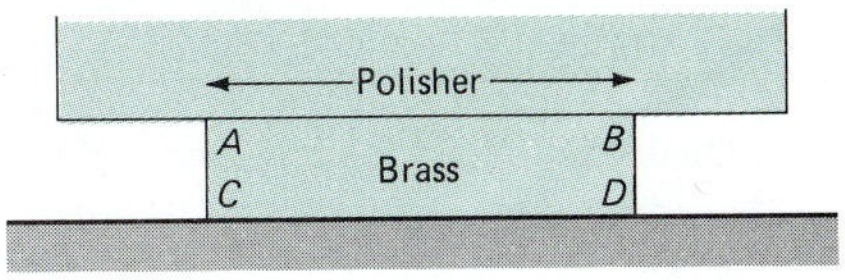

FIGURE 1

V. What value would Poisson's ratio need to have in order for a solid to undergo no change in density when it is compressed?

PROBLEMS

1. The radius of the earth is approximately 6300 km. The pressure of the atmosphere at the surface of the earth is approximately 10^5 N/m². The thickness of the atmosphere is small compared to the radius of the earth. Consequently the acceleration of gravity may be treated as a constant ($g = 9.80$ m/s²) throughout the atmosphere. (a) What is the surface area of the earth? (b) What is the weight of the earth's atmosphere? (c) What is the mass of the earth's atmosphere? (d) If the average mass of one atmospheric molecule is 4.8×10^{-26} kg, how many molecules are there in the earth's atmosphere?

2. To pluck a steel guitar string (length = 1 m), a musician pulls it sideways a distance of 2 mm, forming the triangle shown in Figure 2. If the point B is originally 10 cm from one end of the string, calculate the tensile strain produced by plucking. (Watch significant figures!)

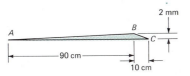

FIGURE 2

3. A cylindrical aluminum rod has a length of 120 cm and a diameter of 3 cm. The rod is held vertically with its lower end 2 m above a rigid steel plate and then released. Assume that the steel plate does not deform, and that gravitational potential energy is converted into elastic energy. (a) Show that $(\Delta L/L)^2 = 2\rho gh/Y$, where h is the distance the center of gravity of the rod falls. (b) Estimate the maximum strain experienced by the rod. (c) Show that the strain-producing force is more than 1000 times the weight of the rod.

4. The pressure on an object submerged in seawater increases in direct proportion to its depth. For every 10-meter increase in depth, the pressure on the object increases by about 1 atm. At what depth would a typical solid be compressed to 99.9% of its volume

at the surface? (Volume strain = -0.001.) Consult Table 14.1 and choose a reasonable value for the bulk modulus of a typical solid.

5. Two climbers are connected by a nylon safety rope (Figure 3). One climber loses his footing and plummets over the edge of an overhang. His quick-thinking companion winds their connecting line around a tree, and thus avoids being pulled over the cliff. The first climber ends up oscillating at the end of the line, which is elastic. The mass of the oscillating climber is 75 kg. The *unstretched* length of the line from tree to climber is 30 m. The diameter of the line is 1 cm and its Young's modulus is 10^8 N/m². (a) Calculate the spring constant (k) for the line. (b) Determine the maximum length of the line. (*Hint:* The maximum length of the line is the distance the climber falls. The overall result of this fall is to convert gravitational potential energy into elastic potential energy. A mathematical statement of this fact leads to a quadratic equation for the stretch of the line.)

FIGURE 3

6. A rod stretched from its equilibrium length possesses elastic potential energy that is distributed throughout the volume of the rod. (a) Calling e the strain in the rod show that the energy per unit volume (u) is

$$u = \tfrac{1}{2}Ye^2$$

where Y is Young's modulus. (b) Determine the energy density in a steel rod experiencing a strain of 0.001. (c) Compare the energy density determined in part (b) with the electric energy density in the battery of an automobile. Assume that the car battery has a total energy of 4×10^6 joules and dimensions 0.2 m × 0.15 m × 0.10 m.

7. A shearing force on the upper surface of a parallelepiped deforms the solid as shown in Figure 4. Show that if forces are applied only to the upper and lower faces they cannot both be parallel to the surface and have the solid in rotational equilibrium.

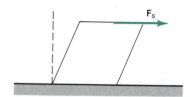

FIGURE 4

8. In an undergraduate laboratory, a wire is suspended from a ceiling and a small mass is attached to the free end to keep the wire taut. With a sensitive measuring instrument, students measure the extension of the wire as selected masses are added to the free end. A summary of the data for the experiment is presented in Table 2. (a) Graphically verify that Hooke's law is valid. Deduce Young's modulus for the wire and speculate on the composition of the wire by comparing your value of Y with the data presented in Table 14.1.

9. (a) A plot of stress versus strain is a straight line for a material obeying Hooke's law. Show that the elastic potential energy per unit volume is the area under the straight line of the stress-strain relation. (b) If an elastic material does not obey Hooke's law the elastic potential energy per unit volume is still the area under the curve of the stress-strain relation. Figure 5 displays the stress-strain relation for a bone. Determine the elastic potential energy per unit volume for a tensile strain of 0.015 and for a compressive strain of −0.015.

10. A 70-kg mass is suspended by a steel wire 3 mm in diameter. When oscillating as a simple pendulum its period is 7 s. Calculate the period for vertical oscillations.

11. A steel wire 1 mm in diameter is connected to fixed points 2 m apart (Figure 6). Initially straight and under no tension, the wire sags when a 100-g mass is attached to the middle. Calculate the vertical sagging distance, assuming the strain to be small.

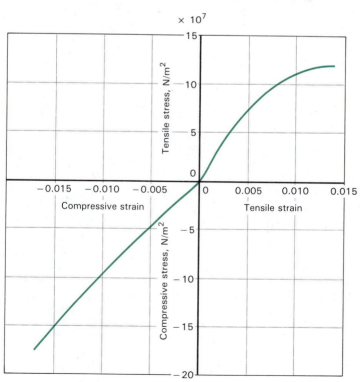

FIGURE 5

TABLE 2

unstretched length of wire = 2.21 m
diameter of wire = 1.05 mm

mass (kg)	extension (mm)
1	0.13
2	0.26
3	0.34
4	0.48
5	0.58
6	0.73
7	0.80
8	0.92
9	1.10
10	1.19

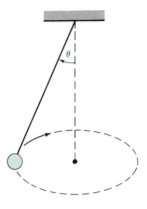

FIGURE 7

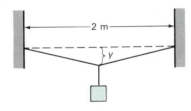

FIGURE 6

12. A 1-kg mass hangs on a steel wire 2 m long (relaxed length) with a diameter of 0.1 mm. The system is set into motion as a conical pendulum with an apex angle θ (Figure 7). Calculate (a) the strain in the wire, (b) the apex angle θ, (c) the period of rotational motion when the wire tension is twice the weight of the mass.

13. A solid cylinder having radius r and length L is twisted so that the bottom rotates through an angle φ relative to the top (Figure 8). Prove that the twisting torque is given by

$$\tau = \left(\frac{\pi r^4}{2L}\right)\mu\varphi$$

where μ is the shear modulus for the cylinder.

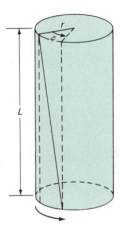

FIGURE 8

14. A textbook has a thickness of 10 cm and a weight of 16 N. When 10 identical textbooks are stacked on top of each other the height of the pile is less than 1 m because of compression. Suppose that the thickness of a book compressed by a weight W is $10e^{-x}$ cm, where $x = W/(160 \text{ N})$. Determine the height of a stack of 10 textbooks.

CHAPTER 15

FLUID MECHANICS

15.1
FLUID STATICS

Fluid mechanics is concerned with the behavior of fluids at the *macroscopic* level—that is, the level at which measurements are made with pressure gauges, thermometers, and flow meters. Accordingly, when we speak of a **fluid particle** or **fluid element** we are not referring to a single molecule. Rather, we mean some volume of fluid that is small by macroscopic standards, but nonetheless contains many molecules. The properties of such a fluid particle do not depend on the behavior of any one molecule.

Fluid statics is concerned with fluids in which the center of mass of each fluid particle is at rest relative to its container. Although the fluid particles are at rest, individual molecules move about incessantly. A fluid at rest is said to be in **hydrostatic equilibrium**.

Archimedes' Principle

An important property of fluids in hydrostatic equilibrium is described by **Archimedes' principle**:

> An object experiences a buoyant force equal to the weight of the fluid it displaces.

To understand Archimedes' principle, consider a fluid particle in hydrostatic equilibrium. The particle experiences a gravitational force equal to its weight and contact forces exerted on its surface by the surrounding fluid. In equilibrium, the net force on the particle is zero, so these surface forces balance the particle's weight. The surface force is a buoyant force—it opposes the tendency to sink. If a solid object displaces the fluid particle, the same forces act on the solid's surface that once acted on the now displaced fluid. Hence, the object experiences a buoyant force equal to the weight of the fluid it displaced (Figure 15.1).

You may have noticed that it is relatively easy to carry someone in a swimming pool who is partially submerged. This is because you support only part of the person's weight. The buoyant force supports the remainder. Water-skiers notice a change in buoyancy as they go from a nearly submerged starting position to an upright stance. While submerged a skier experiences a large buoyant force, and the vertical force on the skis is small. In an upright skiing position the skier is no longer submerged. The buoyant force is negligible and the skis must support nearly all the skier's weight.

We can use Archimedes' principle to prove that an object will float in any liquid whose density is greater than the average density of the object. Ice floats in water, for example, because the density of ice is less than that of water. If an object floats, the net force on it is zero. This means that its weight equals the weight of the fluid it displaces. If M_O is the mass of the object and M_f is the mass of the fluid it displaces we have

$$M_O g = M_f g$$

We can express the masses in terms of the mass densities (ρ_O, ρ_f) and the volumes (V_O, V_f)

$$M_O = \rho_O V_O \qquad M_f = \rho_f V_f$$

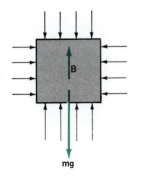

FIGURE 15.1

The net effect of the surface forces is a buoyant force, *B*, equal to the weight of the displaced fluid.

It follows that the volume of the fluid displaced and the volume of the object are related by

$$\frac{\rho_O}{\rho_f} = \frac{V_f}{V_O}$$ (15.1)

For an object that floats V_f/V_O is the fraction of its volume that is submerged. Hence, Equation 15.1 shows that an object floats if its average density is less than that of the fluid in which it is placed.

EXAMPLE 1

Just the Tip of the Iceberg

The density of ice is 920 kg/m³ and the density of seawater is 1020 kg/m³. From Equation 15.1, the fraction of an iceberg that is submerged is

$$\frac{V_f}{V_O} = \frac{920 \text{ kg/m}^3}{1020 \text{ kg/m}^3}$$

$$= 0.90$$

Thus, only about 10% of an iceberg—its tip—is above the surface of the water (Figure 15.2).*

The Cartesian diver is an interesting demonstration that involves Archimedes' principle. It consists of a hollow glass or plastic "diver" placed in a water-filled bottle (Figure 15.3). Small holes in the diver allow it to be partially

FIGURE 15.2
Only about 10% of the volume of an iceberg is above water.

---

* The expression, "That's just the tip of the iceberg," has come to mean that the enormity of a situation is suspected, even though only a tiny part is evident.

FIGURE 15.3

The Cartesian diver in action. An eye dropper can be used for the diver.

filled with water to the point where it has an average density slightly less than the density of the water. This enables the diver to float with a small amount of air trapped inside. The bottle is filled, leaving little or no air space. If the bottle is squeezed, the diver plummets to the bottom. Squeezing the bottle increases the pressure in the water, and forces water into the diver. This increases the diver's average density. The diver sinks when its average density rises above that of the water. When the pressure on the bottle is reduced, the pressure inside the diver is momentarily greater than the pressure in the surrounding water. This causes water to be forced out of the diver and it rises.

Pressure-Depth Relation

Each square meter of the earth's surface experiences a force of approximately 10^5 N (over 11 tons). This force is the total weight of the atmosphere above each square meter of the surface. The atmospheric weight per square meter of surface area is called the **atmospheric pressure**. If you were to ascend through the atmosphere, let's say by climbing a mountain, you would find that the atmospheric pressure *decreases* as you move upward. The atmospheric pressure at any level reflects the weight of the air above that level. At an altitude of 5 km the pressure is approximately one-half of the pressure at the surface. In other words, approximately one-half of the earth's atmosphere lies within 5 km of the surface. At an altitude of 31 km the pressure is less than 1% of the pressure at the surface. We literally live at the bottom of an ocean of air, the weight of which produces the atmospheric pressure. Similarly, when divers descend in the ocean they experience a pressure increase. The pressure at any depth reflects the weight of the fluids—water and air—above each square meter at that level.

We can derive the pressure-depth relation by considering the forces acting on a thin slab of fluid at a depth z (Figure 15.4). Let A denote the cross-sectional area of the slab and dz its thickness. The volume of the slab is $A\,dz$ and its mass is $dm = \rho A\,dz$, where ρ is the mass density of the fluid. The weight of the fluid slab is

$$dW = g\,dm = \rho g A\,dz$$

For a fluid in hydrostatic equilibrium, the net force on the slab must be zero. The upward pressure on the bottom face exceeds the downward pressure by

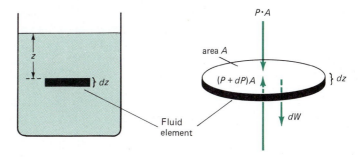

FIGURE 15.4

Three vertical forces act on the fluid element. The weight of the element is dW. The fluid force on the top is PA. The fluid force on the bottom is $(P + dP)A$. The difference between these two fluid forces, dPA, constitutes the buoyant force. For hydrostatic equilibrium the buoyant force balances the weight.

an amount dP. In hydrostatic equilibrium the buoyant force, $A\,dP$, just balances the weight dW. Thus,

$$A\,dP = dW = \rho g A\,dz$$

Canceling the common factor A gives

$$dP = \rho g\,dz \tag{15.2}$$

This equation relates the changes in pressure and depth. If P_0 denotes the pressure at the surface ($z = 0$), then the pressure P at a depth z is determined by integrating Equation 15.2:

$$\int_{P_0}^{P} dP = \int_{0}^{z} \rho g\,dz$$

For an incompressible fluid ρ is constant. If g is also taken to be constant, the integration gives

$$P = P_0 + \rho g z \tag{15.3}$$

This is the **pressure-depth relation**. Note that if the pressure at the surface (P_0) is changed, an equal change in pressure is felt at all depths, that is, throughout the fluid. This fact is called **Pascal's principle**:

> Pressure applied to an enclosed fluid is transmitted undiminished throughout the fluid.

EXAMPLE 2

The Barometer

A barometer is used to measure the pressure of the atmosphere. A mercury-filled barometer consists of a narrow tube closed at one end. The tube is first filled with mercury, then inverted in a mercury reservoir (Figure 15.5). The pressure at the surface of the reservoir is atmospheric (P_{atm}). Inside the tube the pressure at the level of the reservoir is $P_0 + \rho g h$, where h is the height of the mercury column. The pressure P_0 at the top of the mercury column is not quite zero because a small amount of mercury vapor fills the otherwise evacuated space. This vapor produces the pressure P_0, which is negligible at ordinary temperatures, except in precise scientific work. Hydrostatic equilibrium is achieved when the pressure at the reservoir surface ($P_0 + \rho g h$) equals the atmospheric pressure (P_{atm}).

$$P_{atm} = P_0 + \rho g h$$

In practice, this relation lets us determine the atmospheric pressure by measuring the height of the mercury column. Let's find the height of the mercury column when the pressure is 1 atm. For mercury, $\rho = 13.6 \times 10^3 \text{ kg/m}^3$. With $P_{atm} = 1$ atm we get

$$h = \frac{P_{atm}}{\rho g} = \frac{1.01 \times 10^5 \text{ N/m}^2}{13.6 \times 10^3 \text{ kg/m}^3 \cdot 9.80 \text{ m/s}^2}$$

$$= 0.758 \text{ m}$$

$$= 75.8 \text{ cm}$$

Thus, the atmosphere exerts a pressure that can support a column of mercury nearly 76 cm in height.

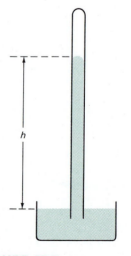

FIGURE 15.5

In a mercury barometer, the atmospheric pressure exceeds $\rho g h$ slightly, because of the vapor pressure P_0 in the region above the mercury.

Equation 15.3 is restricted to fluids in which g and ρ are constant. In the case of planetary atmospheres, for example, both g and ρ decrease as the altitude increases, so Equation 15.3 cannot be used. However, in liquids and solids the density variations are usually minor enough so that Equation 15.3 is applicable. In particular, the pressure at modest depths in the earth's crust can be estimated using Equation 15.3. This estimation enables us to answer a fascinating question: How deep can you dig a hole?

EXAMPLE 3

How Deep Can You Dig a Hole?

There is a limit to how far we can dig—or drill—a hole in the ground because the pressure of the earth increases with depth. At a sufficient depth the pressure becomes so great that it will crush any material, whether man-made steel or nature's most rigid rock.

We can estimate the maximum depth for a hole in the ground by combining the pressure-depth relation, Equation 15.3,

$$P = P_O + \rho g z$$

with Hooke's law for volume stress and strain (Section 14.3),

$$P = -B\frac{\Delta V}{V}$$

If we equate these expressions for P and ignore P_O (1 atm) we obtain a relation between the depth (z) and the volume strain ($-\Delta V/V$):

$$z = \frac{-B\dfrac{\Delta V}{V}}{\rho g}$$

Because we want to estimate the maximum depth, we set $\Delta V/V$ equal to the volume strain at which solids fracture. Values appropriate to the crust of the earth are as follows:

$$\rho = 3 \times 10^3 \text{ kg/m}^3 \qquad \text{density of crustal rocks}$$

$$g = 9.8 \text{ m/s}^2 \qquad \text{acceleration of gravity}$$

$$B = 6 \times 10^{10} \text{ N/m}^2 \qquad \text{bulk modulus of crustal rocks}$$

$$\frac{\Delta V}{V} = -0.02 \qquad \begin{array}{l}\text{volume strain at which}\\ \text{crustal rock crumbles (estimate)}\end{array}$$

Substituting these values into our equation, we get

$$z = \frac{(0.02)(6 \times 10^{10} \text{ N/m}^2)}{(3 \times 10^3 \text{ kg/m}^3)(9.8 \text{ m/s}^2)}$$

$$= 4 \times 10^4 \text{ m}$$

$$= 40 \text{ km}$$

This is a depth of approximately 25 miles. If we compare this value of z with the radius of the earth ($R_E = 6380$ km), we see that the deepest hole possible would extend less than 1% of the way to the center of the earth.

Our rough calculation shows that rigid material cannot exist at depths greater than about 40 km. It follows that this same depth, 40 km, should represent the thickness of the earth's rigid crustal layer. Seismic studies bear out this expectation. The earth's crustal thickness ranges from about 5 km in the ocean floors to 35 km in the continental blocks. At such depths the earth's makeup changes from a rigid solid to a plastic-like material.

15.2
FLUID DYNAMICS

We can describe the macroscopic motions of a fluid in terms of the motions of a fluid particle. In particular, the velocity **u** of a *fluid particle* may vary both with position and with time. Thus, in general

$$\mathbf{u} = \mathbf{u}(x, y, z, t)$$

We limit our study of fluid dynamics to a special, but very important, class of fluid flows. This allows us to deal with the more important concepts of fluid dynamics without using sophisticated mathematical techniques.

First, fluid flow may be **rotational** or **irrotational**. If we place a small paddle wheel in a fluid where the flow is irrotational, the paddle wheel does not rotate (Figure 15.6a). In a rotational flow it rotates (Figure 15.6b). We will restrict ourselves to irrotational flows.

Second, we distinguish gases and liquids on the basis of their abilities to withstand compressive stresses. There are many situations in which the flow of a gas can be regarded as incompressible. These are flows where the gas is not subjected to large pressure changes. In an incompressible fluid, the *mass density* (ρ) *is a constant*. We will deal only with incompressible fluid flow.

Third, fluids exhibit **viscosity**, a form of friction. Viscosity can be ignored in many situations, just as rolling or sliding friction often can be neglected in treating the motions of rigid bodies. In this section we restrict ourselves to **nonviscous** flow.

Fourth, and finally, fluid flow is termed **steady** if the fluid velocity **u** does not change with time. This restriction does not require that the fluid velocity be the same at all positions. For example, in the steady flow of water in a stream, the flow velocity is greatest at the surface. The velocity decreases with depth, falling to zero at the bottom, where the water adheres to the streambed. The velocity is different at different positions, but **u** does not change with time. We restrict our analysis to *steady flows*.

To begin our study of steady, irrotational, incompressible flow we introduce the concept of a **streamline**. If we draw the velocity vector **u** of a fluid particle

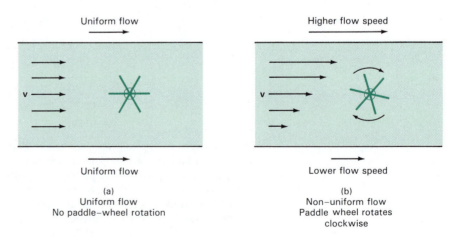

FIGURE 15.6

(a) A small paddle wheel placed in an irrotational flow undergoes *translation* but not *rotation*.
(b) In a rotational flow the paddle wheel rotates.

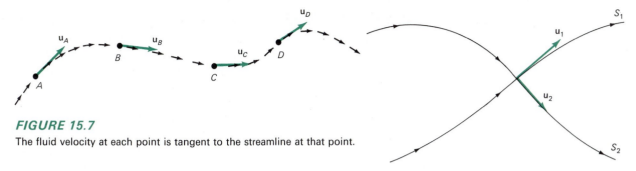

FIGURE 15.7

The fluid velocity at each point is tangent to the streamline at that point.

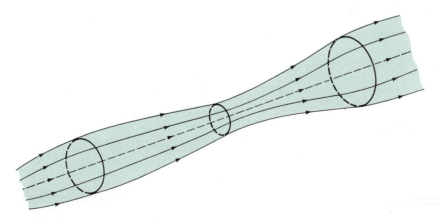

FIGURE 15.8

Streamlines never cross. If the streamlines S_1 and S_2 crossed, the velocities $\mathbf{u}_1$ and $\mathbf{u}_2$ would be in different directions at the point of intersection. This in turn would imply that the fluid velocity has two different values at one point. This is impossible. Hence, streamlines never cross.

at successive points (A, B, C, etc.) and connect these points, we construct a line called a streamline (Figure 15.7). Fluid particles move along streamlines. Streamlines have interesting and useful properties. By definition, they are parallel to the fluid velocity at all points. As a consequence, no two streamlines can cross. We can prove this by assuming that two streamlines *do* cross (Figure 15.8). If two streamlines crossed, the fluid velocities associated with them would be in different directions at the point of intersection. This in turn would imply that the fluid velocity has two different values at one point. This is clearly impossible. Hence, streamlines never cross.

The fact that fluid particles move along streamlines that never cross enables us to keep track of any given volume of fluid. We need only sketch a few representative streamlines on the periphery of the fluid volume of interest. These streamlines outline a surface. The region bounded by this surface is called a **tube of flow** (see Figure 15.9). Fluid particles inside a tube of flow remain inside. Fluid particles outside never enter. Either sort of crossing would entail the crossing of streamlines.

FIGURE 15.9

Segment of a tube of flow generated by streamlines.

The Equation of Continuity

We can use the tube-of-flow concept to derive the equation of continuity, a statement of mass conservation. We focus our attention on one segment of a tube of flow (see Figure 15.10). For steady flow the streamlines are fixed and the volume of the tube segment remains constant. For incompressible flow the density remains constant. The total mass in the segment, the product of density and volume, is therefore constant. It follows that equal masses must flow into and out of the ends of the segment.

In particular, consider points ① and ② at the ends of the segment, where the flow speeds are u_1 and u_2 and the cross-sectional areas are A_1 and A_2 (Figure 15.10). In a time t, a fluid volume A_1u_1t flows past point ① into the segment. A volume A_2u_2t flows past point ② and out of the segment. The mass of each volume is equal to ρ times the volume. Thus,

$$m_1 = \rho A_1 u_1 t = \text{mass into segment}$$

$$m_2 = \rho A_2 u_2 t = \text{mass out of segment}$$

In order for the mass inside the segment to remain constant, the mass flowing into the segment must equal the mass flowing out. Setting $m_1 = m_2$ and canceling common factors, we get

$$A_1 u_1 = A_2 u_2 \tag{15.4}$$

This is one form of the **equation of continuity** for steady incompressible flow. According to this equation, if the cross-sectional area of the flow is decreased, the fluid velocity is increased. Many common devices illustrate the equation of continuity. For example, when you slowly squeeze the trigger of a water pistol, water is forced through a very small opening, and it emerges at a high speed. Likewise, in a syringe, the plunger has a relatively large cross-sectional area and moves slowly. At the tiny opening in the tip of the needle the fluid emerges at a much higher speed.

The quantity Au is called the **discharge rate**, Q, and gives the volume of fluid that crosses the area A per unit time:

$$\text{Discharge rate} \equiv Q = Au \tag{15.5}$$

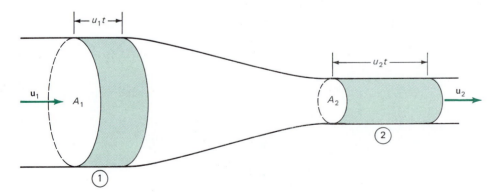

FIGURE 15.10

For steady incompressible flow, mass in equals mass out.

Multiplying the discharge rate by the mass density gives the rate of mass flow. Thus

$$\text{rate of mass flow} \equiv \alpha = \frac{\text{mass}}{\text{volume}} \cdot \frac{\text{volume}}{\text{second}}$$

$$= (\text{mass density}) \cdot (\text{discharge rate})$$

or

$$\alpha = \rho A u \qquad (15.6)$$

Let's apply Equation 15.6 to the flow of natural gas through a pipeline.

EXAMPLE 4

Flow Speed in a Pipeline

A natural gas pipeline has a diameter of $D = 0.25$ m. The rate of mass flow is $\alpha = 1.4$ kg/s. The density of the gas is $\rho = 0.90$ kg/m^3. We wish to find the flow speed, u. Using Equation 15.6 for the rate of mass flow, we write

$$\alpha = \frac{\rho \pi D^2 u}{4}$$

where $\pi D^2/4 = A$. Solving for the flow speed and substituting the values given, we get

$$u = \frac{4\alpha}{\rho \pi D^2} = \frac{4(1.4 \text{ kg/s})}{0.90 \text{ kg/m}^3 \cdot 3.14(0.25 \text{ m})^2}$$

$$= 32 \text{ m/s}$$

This is over 70 mi/h!

Bernoulli's Equation

Because fluid particles obey Newton's second law, the work-energy principle can be applied to their motions (see Section 7.4). For nonviscous flow, the work-energy principle leads to Bernoulli's equation.

Consider a section of a narrow tube of flow (Figure 15.11). Let's apply the work-energy principle to the fluid that, at some instant, occupies the region between the cross sections labeled ① and ②. We follow the motion of the fluid over a short time interval Δt during which the fluid moves in response to the combined actions of the fluid forces $P_1 A_1$ and $P_2 A_2$, and gravity. Each fluid particle in the system advances along the tube of flow, but net work and net energy transfers involve only the mass elements at the ends. The energy of the intervening mass is unchanged because the density is constant and the flow is steady. For a steady flow of constant density, equal masses (m) flow past ① and ② during the interval Δt. The net result of the flow during the interval Δt is to transfer a mass m from the left end of the tube to the right end. The change in the kinetic energy of the system is the change in the kinetic energy of this mass m. Thus,

$$\text{change in kinetic energy} = \tfrac{1}{2}mu_2{}^2 - \tfrac{1}{2}mu_1{}^2 \qquad (15.7)$$

Gravitational work is performed because the mass m moves through a vertical distance $z_2 - z_1$. The net work done on the system by gravity is thus

$$W_g = -mg(z_2 - z_1) \qquad (15.8)$$

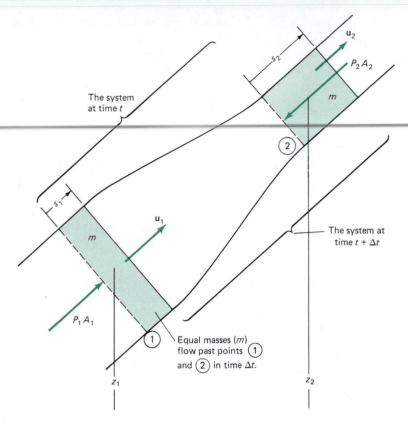

FIGURE 15.11

The system consists of the fluid between points 1 and 2 at time t. In a time interval t the system moves forward under the influence of the forces P_1A_1, P_2A_2, and gravity. Applying the work-energy principle to the system relates the work done by these forces to the change in kinetic energy of the system.

At the left end, the fluid force P_1A_1 acts through the displacement s_1 and does positive work $P_1A_1s_1$. At the right end, the fluid force P_2A_2 and the displacement s_2 are oppositely directed and the negative work done is $-P_2A_2s_2$. The net work done on the system by the fluid forces is

$$W_f = P_1A_1s_1 - P_2A_2s_2 \qquad (15.9)$$

Note that A_1s_1 and A_2s_2 are the volumes of the equal masses m and that the density is constant. Because volume is equal to mass divided by density, we can substitute m/ρ for both A_1s_1 and A_2s_2:

$$A_1s_1 = A_2s_2 = \frac{m}{\rho}$$

whereupon Equation 15.9 becomes

$$W_f = (P_1 - P_2)\frac{m}{\rho} \qquad (15.10)$$

A verbal statement of the work-energy principle for the system is

$$\begin{matrix} \text{net work done} & & \text{change in} \\ \text{on system by} & = & \text{kinetic energy} \\ \text{external forces} & & \text{of system} \end{matrix}$$

Collecting the results of Equations 15.7, 15.8, and 15.10, we can translate this verbal statement into

$$\underbrace{(P_1 - P_2)m/\rho}_{\text{work by fluid}} + \underbrace{[-mg(z_2 - z_1)]}_{\text{work by gravity}} = \underbrace{\tfrac{1}{2}mu_2{}^2 - \tfrac{1}{2}mu_1{}^2}_{\substack{\text{change in} \\ \text{kinetic energy}}}$$

which can be rearranged to read

$$P_2 + \rho g z_2 + \tfrac{1}{2}\rho u_2{}^2 = P_1 + \rho g z_1 + \tfrac{1}{2}\rho u_1{}^2 \qquad (15.11)$$

This is a standard form of Bernoulli's equation. An equivalent statement is

$$P + \rho g z + \tfrac{1}{2}\rho u^2 = \text{constant}$$

that is, the quantity $P + \rho g z + \tfrac{1}{2}\rho u^2$ has the same value at all points along a streamline.

Each term in Bernoulli's equation has the dimensions of energy per unit volume. The term $\tfrac{1}{2}\rho u^2$ is the kinetic energy per unit volume associated with the macroscopic fluid motions. The term $\rho g z$ is the gravitational potential energy per unit volume. We call the pressure, P, the **flow energy** per unit volume. Flow energy is work done by the fluid against its surroundings as a result of its motions. The sum of the three terms

$$P + \rho g z + \tfrac{1}{2}\rho u^2$$

is the *total energy per unit volume*. Bernoulli's equation is therefore a statement that the total energy per unit volume remains constant along any given tube of flow. The various forms of energy (kinetic energy, gravitational potential energy, and flow energy) can be transformed into one another, but the total energy per unit volume remains constant. We now consider two examples of Bernoulli's equation.

EXAMPLE 5

Hydrostatic Limit

When a fluid is at rest, $u_1 = u_2 = 0$. In this special case, which we call the **hydrostatic limit**, Bernoulli's equation reduces to the pressure-depth relation

$$P_1 = P_2 + \rho g(z_2 - z_1)$$

This equation was derived earlier (Section 15.1). Here, however, we see it in a new light. We see that change in flow energy is balanced by an opposite change in the gravitational energy. This equation is an example of energy conservation. Energy can be changed from one form to another, but there can be no change in the total energy.

In certain circumstances, the total energy per unit volume has the same value along *all* streamlines, and thus is a constant throughout the fluid. The following situation illustrates this point.

Torricelli's Theorem

We can use Bernoulli's equation to determine the speed at which water escapes through a small orifice at the base of a reservoir. Both the top of the

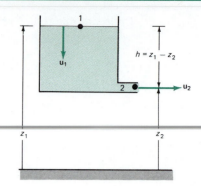

FIGURE 15.12
Torricelli's theorem follows from Bernoulli's equation when the speed of the surface, u_1, is ignorable compared to the exit speed, u_2.

reservoir and the orifice are open to the atmosphere (Figure 15.12). Bernoulli's equation is applied at the upper surface (point 1) and in the stream emerging from the orifice (point 2). The total energy is the same at all points on the upper surface. Because all tubes of flow originate at this surface, the total energy is the same along all streamlines. We can therefore regard the entire volume of water between 1 and 2 as a single tube of flow.

The Bernoulli equation, Equation 15.11, is simplified by noting that the height of the water column is

$$h = z_1 - z_2$$

and that $P_1 = P_2 = P_{atm}$, since both points 1 and 2 are open to the atmosphere. Bernoulli's equation reduces to

$$\tfrac{1}{2}u_2{}^2 - \tfrac{1}{2}u_1{}^2 = gh \tag{15.12}$$

This form of Bernoulli's equation illustrates the transformation of gravitational potential energy into kinetic energy.

We could eliminate u_1 in our equation by using the equation of continuity, Equation 15.4. Instead, however, we make use of the fact that the surface area A_1 is much greater than the cross-sectional area of the emerging stream, A_2. This makes u_1 small in comparison with u_2. Therefore, we drop $u_1{}^2$ from the right side of Equation 15.12 to get

$$u_2 = \sqrt{2gh} \tag{15.13}$$

This result is known as Torricelli's theorem. Notice that the speed of flow is the same as would be acquired in free-fall from the surface of the reservoir to the level of the orifice.

There are many other situations in which the gravitational potential energy term is ignorable. In these situations Bernoulli's equation relates pressure changes and kinetic energy changes, as we see in Example 6.

EXAMPLE 6

Fluid Dynamics of Vaccination

Vaccinations are now routinely administered using high-pressure guns rather than hypodermic needles. The guns force the vaccine through a tiny opening in a sapphire tip. Typically guns exert a pressure of 500 psi to force vaccine through an orifice 0.005 in. in diameter. We can use Bernoulli's equation to determine the speed at which the vaccine emerges. We ignore the gravitational energy term in Bernoulli's equation and approximate the flow speed inside the gun as zero. In this case, Bernoulli's equation describes the transformation of flow energy into kinetic energy. Equation 15.11 assumes the form

$$P_i - P_o = \tfrac{1}{2}\rho u_o{}^2$$

where the subscripts i and o mean inside and outside. We take

$$\rho = 1.1 \times 10^3 \text{ kg/m}^3$$

$$P_i = 500 \text{ psi} = 3.45 \times 10^6 \text{ N/m}^2$$

$$P_o = 15 \text{ psi} = 0.10 \times 10^6 \text{ N/m}^2$$

The speed of the emerging vaccine is

$$u_o = \sqrt{\frac{2(P_i - P_o)}{\rho}} = \sqrt{\frac{2(3.35 \times 10^6) \text{ N/m}^2}{1.1 \times 10^3 \text{ kg/m}^3}}$$

$$= 78 \text{ m/s}$$

which is approximately 175 mph, or about one quarter of the speed of sound in air.

15.3
VISCOSITY

Although fluids cannot support shearing stresses they do *resist* shearing motions. The property of a fluid that measures its resistance to shearing motions is called **viscosity**.

Viscosity is the fluid analog of the shear modulus for an elastic solid. A shearing force F_v acting on a fluid surface of area A sets the fluid in motion and produces relative motion of the fluid layers (Figure 15.13). Because of this relative motion, the flow velocity u changes with position. The velocity profile is a graph or equation that relates the fluid velocity to position. The derivative du/dy is called the **velocity gradient**. Geometrically, du/dy measures how u changes with position. Dynamically du/dy measures the fluid's response to the

FIGURE 15.13
In a liquid, a shearing stress causes adjacent layers to be in relative motion. The velocity gradient, du/dy, is a measure of the distortion produced by the stress.

TABLE 15.1
Viscosities of Selected Liquids and Gases*

Fluid	Dynamic Viscosity (η) ($N \cdot s/m^2$)	Kinematic Viscosity (v) (m^2/s)
Liquids		
Benzene	6.6×10^{-4}	7.5×10^{-7}
Gasoline	2.9×10^{-4}	4.3×10^{-7}
Jet Fuel (JP–4)	8.7×10^{-4}	1.1×10^{-6}
Mercury	1.55×10^{-3}	1.2×10^{-7}
Ethyl alcohol	1.7×10^{-3}	2.2×10^{-6}
Crude Oil	7.2×10^{-3}	8.4×10^{-6}
Water	1.0×10^{-3}	1.0×10^{-6}
Glycerin	1.5	1.2×10^{-3}
Gases		
Hydrogen	0.90×10^{-5}	1.1×10^{-4}
Helium	1.97×10^{-5}	1.2×10^{-4}
Argon	2.2×10^{-5}	1.3×10^{-5}
Methane	1.34×10^{-5}	2.0×10^{-5}
Ethane	0.90×10^{-5}	0.70×10^{-5}
Air	1.81×10^{-5}	1.5×10^{-5}
Nitrogen	1.75×10^{-5}	1.4×10^{-5}
Oxygen	2.01×10^{-5}	1.5×10^{-5}
Carbon Dioxide	1.47×10^{-5}	0.80×10^{-5}

* Values refer to a temperature of 20°C and a pressure of 1 atm. Adapted in part from J. K. Vannard and R. L. Street, *Elementary Fluid Mechanics,* 6th ed., with permission of the publisher, John Wiley & Sons, New York: 1982.

shear stress F_v/A. Experiment shows that the stress and velocity gradient are directly proportional. Their ratio is termed the **dynamic viscosity** and is denoted by the Greek letter eta (η)

$$\eta = \frac{\text{shear stress}}{\text{velocity gradient}} = \frac{F_v/A}{du/dy} \tag{15.14}$$

The SI unit of viscosity is the $N \cdot s/m^2$; (This unit suffers the indignity of having no name.) The cgs unit of viscosity is the $dyne \cdot s/cm^2$ and is called the **poise**. You can verify,

$$1 \text{ poise} \equiv 1 \text{ dyne} \cdot s/cm^2 = 0.1 \text{ N} \cdot s/m^2$$

The viscosities of liquids and gases typically differ by a factor of roughly 100. For example, the viscosities of water and air at 20°C are $1.0 \times 10^{-3} \, N \cdot s/m^2$ and $1.8 \times 10^{-5} \, N \cdot s/m^2$, respectively. Table 15.1 gives viscosities for several fluids at 20°C.

The viscous force F_v in Equation 15.14 is the fluid analog of the sliding friction force between two solid surfaces. For this reason, viscosity is often referred to as **fluid friction**. Like other frictional forces, viscous forces oppose the relative motion of adjacent fluid layers.

Velocity Profile for Viscous Flow in a Pipe

Consider a steady fluid flow in a horizontal pipe. If the fluid were nonviscous, a constant flow speed could be maintained without any variation in pressure along the pipe. This conclusion is consistent with Bernoulli's equation,

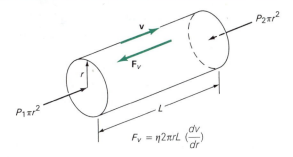

FIGURE 15.14
Forces acting on a cylindrical volume of fluid.

Equation 15.11. However, when viscous forces are taken into account, a pressure variation along the pipe is needed to maintain a steady flow.

We consider the steady flow of a fluid with a viscosity η along a horizontal pipe. The pressure decreases uniformly along the direction of flow. We want to determine the **velocity profile** of the flow, the relation that relates the flow speed (v) to the distance (r) from the center of the pipe.

Figure 15.14 shows the forces acting on a cylindrical volume of the moving fluid. The cylinder radius is r and its length is L. The forces acting on the ends are $P_1 \pi r^2$ and $P_2 \pi r^2$, where P_1 and P_2 are the pressures on the ends. The viscous force exerted on the surface of the cylinder by the surrounding fluid follows from Equation 15.14. The viscous shear stress is F_v/A, where $A = 2\pi rL$ is the surface area of the cylindrical volume of the fluid. The viscous force is

$$F_v = \eta A \frac{dv}{dr} = \eta 2\pi rL \frac{dv}{dr}$$

For steady flow, the net force on the cylinder is zero:

$$\sum F = P_1 \pi r^2 - P_2 \pi r^2 + \eta 2\pi rL \frac{dv}{dr} = 0$$

This equation can be integrated to express the flow speed v as a function of r. Taking the flow speed to be zero at the surface where the water and the pipe are in contact gives

$$\int_0^v dv = -\frac{(P_1 - P_2)}{2\eta L} \int_R^r r \, dr$$

Hence the velocity profile is,

$$v = \frac{(P_1 - P_2)(R^2 - r^2)}{4\eta L} \tag{15.15}$$

Figure 15.15 shows the parabolic shape of the velocity profile. The flow speed is a maximum at the center of the pipe ($r = 0$), and falls to zero at the inner surface of the pipe ($r = R$). We note that v is directly proportional to the **pressure gradient**,

$$\text{pressure gradient} \equiv \frac{(P_1 - P_2)}{L}$$

The pressure gradient is a measure of how sharply the pressure varies with position along the flow.

Parabolic Velocity Profile

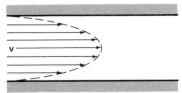

FIGURE 15.15
Velocity profile for viscous fluid flow in a pipe of radius R. The fluid velocity is zero at the wall and rises to a maximum on the central axis.

EXAMPLE 7

Pressure Gradient for a Water Line

A horizontal pipe with an inner diameter of 4.0 cm carries water. We want to determine the pressure gradient necessary to maintain a maximum flow speed (on the axis of the pipe) of 15 m/s.

From Equation 15.15, with $r = 0$, we get the pressure gradient

$$\frac{P_1 - P_2}{L} = \frac{4\eta v}{R^2}$$

From Table 15.1, $\eta = 1.0 \times 10^{-3}$ N·s/m², and

$$\frac{P_1 - P_2}{L} = \frac{4(1.0 \times 10^{-3} \text{ N·s/m}^2)(15 \text{ m/s})}{(0.02 \text{ m})^2}$$

$$= 6.0 \text{ N/m}^3$$

A pressure of 1 pound per square inch equals 6.89×10^3 N/m². A pressure gradient of 6.0 N/m³ converts to 0.26 pound per square inch *per thousand feet*. Thus, a very modest pressure gradient is adequate to maintain the flow.

Poiseuille's Law

We developed an expression for the discharge rate of a nonviscous fluid in Section 15.2, Equation 15.5. When the flow speed is limited by viscosity, a relation known as **Poiseuille's law** describes the discharge rate. We can use Equation 15.15 for the flow speed to derive Poiseuille's law.

In Figure 15.16 we consider viscous flow in a pipe. The fluid that crosses the annular ring area $2\pi r \, dr$ in a time dt occupies a cylindrical sleeve whose volume is

$$dV = 2\pi r \, dr \, v \, dt$$

The speed (v) of the fluid is given by Equation 15.15. The *rate* of flow (in m³/s) is given by dV/dt.

$$\frac{dV}{dt} = \frac{2\pi r \, dr \, (P_1 - P_2)(R^2 - r^2)}{4\eta L}$$

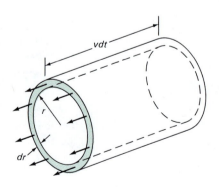

FIGURE 15.16

The fluid that crosses the annular ring in a time dt fills the cylindrical sleeve. The volume of the fluid is $2\pi r \, dr \, v \, dt$.

Integrating over the cross-sectional area of the pipe accounts for the flow across all annular rings and gives the total volume rate of flow, the discharge rate,

$$Q = \left[\frac{\pi(P_1 - P_2)}{2\eta L} \right] \int_0^R (R^2 - r^2)r \, dr$$

The result gives Poiseuille's law:

$$Q = \frac{\pi R^4 (P_1 - P_2)}{8\eta L} \tag{15.16}$$

Note that the discharge rate is directly proportional to the pressure gradient and inversely proportional to the viscosity.

Kinematic Viscosity

The viscosity η is generally referred to as the *dynamic viscosity* to distinguish it from a related quantity, the **kinematic viscosity**. The kinematic viscosity, v, is defined as the ratio of the dynamic viscosity and the mass density,

$$v \equiv \frac{\eta}{\rho} \tag{15.17}$$

Both v and η measure the resistance to shearing motions. The kinematic viscosity is introduced because in many situations it is the ratio η/ρ rather than η alone that determines the importance of viscous forces. The cgs unit of kinematic viscosity is the *stoke*:

$$1 \text{ stoke} = 1 \text{ cm}^2/\text{s}$$

We will make use of the kinematic viscosity in the next section in which we study fluid turbulence.

Molecular Origin of Viscosity

The origin of the shearing force in a fluid is not immediately evident. Why should there be a shearing force between two layers of fluid in relative motion? There are, in fact, two sources of viscous forces. This is suggested by the observation that the viscosity of a liquid *decreases* as temperature increases, whereas the viscosity of a gas *increases* as its temperature rises.

In liquids, viscosity originates with the cohesive forces between molecules in adjacent layers. The viscous force results because the attractive forces between molecules tend to prevent the relative motion of molecules. In low-density gases cohesive forces are weak and make only a minor contribution to the viscosity. A different mechanism is responsible for viscosity. In a gas, molecules constantly are being exchanged between adjacent layers. If molecules of mass m are traded between layers where the flow speeds are u and $u + du$, there will be a net transfer of momentum from the faster layer to the slower layer. The slower layer is accelerated—it receives an impulse of $+m \, du$ when it catches a molecule from the faster layer. The faster layer receives a negative impulse, $-m \, du$, in the trade. Overall, these momentum transfers tend to reduce the relative velocity of the layers. At the macroscopic level this momentum transfer shows itself as the viscous force that opposes relative motion.

15.4
TURBULENCE

The stable streamline flow in a viscous fluid is called **laminar flow**. Adjacent layers, or laminae, of fluid slide smoothly over each other. The stability of laminar flow is maintained by viscous forces. However, experiment shows that laminar flow becomes *unstable* and is disrupted if the speed of flow is sufficiently great. A random and irregular motion of the fluid replaces the smooth laminar flow. This unstable fluid motion is called turbulent flow, or *turbulence*. The flow speed at which turbulence develops depends on the viscosity of the liquid, and on the geometry of the flow.

There are countless examples of turbulence around us. There may be turbulence in the smoke rising from a smokestack, or in the buildup of a cumulus cloud on a summer afternoon. The twinkle of starlight is the result of turbulence in our atmosphere. The Gulf Stream is a turbulent flow. The wakes in the air and in the water left by autos, airplanes, and ships are turbulent.

We can think of fluid turbulence as groups of swirling eddies or currents. Viscous forces gradually dissipate the energy of individual eddies, causing them to die out. However, new eddies will form as long as the basic fluid instability persists.

Turbulence can be good or bad, depending on the circumstances. Turbulent flows are often undesirable because they invariably waste energy. As eddies form and decay, energy is dissipated uselessly as heat. As a consequence, when the flow through a pipe changes from laminar to turbulent, the pressure must be increased in order to maintain a constant flow speed. On the other hand, turbulence can be very desirable because it causes rapid mixing and increases rates of transfer of mass, momentum, and energy. For example, the blades of an eggbeater promote turbulent flow and rapid mixing.

Reynolds Number

The Reynolds number is a dimensionless parameter that allows us to estimate whether a flow is steady or turbulent. The **Reynolds number** is defined as

$$R = \frac{\rho u L}{\eta} \tag{15.18}$$

where ρ is the fluid density, u is the flow speed, η the dynamic viscosity, and L some *characteristic length* associated with the flow. For flow past a sphere, L would be the diameter of the sphere. For flow through a pipe, L would be the diameter of the pipe.

The Reynolds number can also be written as

$$R = \frac{uL}{v} \tag{15.19}$$

where v is the kinematic viscosity.

Experiment shows that geometrically similar flows become turbulent at the same value of R. For example, oil and water have different kinematic viscosities, but if we pump oil and water through pipes with the same type of cross section, the flows become unstable and turbulent at approximately the same value of R; this value is called the *critical* Reynolds number. For many fluids the critical Reynolds number lies between 1000 and 10,000.

EXAMPLE 8

A Classroom Determination of the Critical Reynolds Number

The critical Reynolds number for water can be measured in the following simple experiment. The flow velocity of tap water is increased until the stream changes its appearance. Laminar flow is characterized by a smooth, transparent stream. When turbulence sets in, the stream becomes twisted and translucent. The volume of water that flows out of the pipe in a time t is $V = ut \cdot \pi L^2/4$, where L is the diameter of the pipe. If the water is collected in a cylindrical vessel of diameter D, the volume V can be expressed as

$$V = \frac{h\pi D^2}{4}$$

where h is the height of the water. Equating these two expressions for V gives

$$uL = \frac{D^2 h}{Lt}$$

The Reynolds number is $R = uL/v$. Thus,

The following data were obtained in a classroom experiment:

$$D = 0.167 \text{ m}$$

$$L = 0.014 \text{ m}$$

$$h = 0.089 \text{ m}$$

$$t = 40 \text{ s}$$

Taking $v = 1.0 \times 10^{-6} \text{ m}^2/\text{s}$ gives

$$R = \frac{(0.167 \text{ m})^2(0.089 \text{ m})}{(1.0 \times 10^{-6} \text{ m}^2/\text{s})(0.014 \text{ m})(40 \text{ s})}$$

$$= 4400$$

You can try the experiment yourself and then compare your result with this one.

We have stressed the idea that turbulence results from a fluid instability. The stability of fluid particles moving along a tube of flow depends on the viscous friction exerted by the surrounding fluid. If the viscous friction is disturbed, instability may result.

The following analogy animates the idea of unstable flow and illustrates the significance of the Reynolds number. Consider a group of students running to their physics class on a cold winter morning. They run at a speed u, taking strides of length L. The coefficient of friction between their shoes and the sidewalk is μ. We define the analog of the Reynolds number for this flow of students as

$$R_s = \frac{uL}{\mu}$$

Consider what happens when the students move from a bare sidewalk to one covered with ice. The value of μ decreases abruptly, causing the student Reynolds number to increase. While friction prevents any slipping on the bare sidewalk, the students must be careful on the ice—any slight imbalance may result in *unstable* motion. To recover stability, the students slow down and take shorter strides. By decreasing u and L, the students lower their Reynolds number and thereby tend to stabilize their motions.

WORKED PROBLEM

A thin wooden rod of length L and density $\rho_r = 640 \text{ kg/m}^3$ has a cross-sectional area A. The rod hangs freely from a pivot (Figure 15.17) so that it is partially submerged in water (density $\rho_w = 1000 \text{ kg/m}^3$). The water surface is a distance D below the pivot, and the rod makes an angle θ with the vertical.

(a) Show that $\theta = 0$ is a stable equilibrium orientation if $D > L\sqrt{1 - \dfrac{\rho_r}{\rho_w}}$.

(b) Show that $\theta = \cos^{-1}\left(\dfrac{D}{L\sqrt{1 - (\rho_r/\rho_w)}}\right)$ is an equilibrium orientation if $D < L\sqrt{1 - \dfrac{\rho_r}{\rho_w}}$.

Solution

In problems involving rotational equilibrium we want to examine the torques acting on the system. We choose the pivot point O as a reference point for torques. The two forces that exert torques on the rod are its weight W and the buoyant force B. We take the buoyant force to act at the center of the submerged portion of the rod. From Figure 15.17 we can see that the torque due to the weight tends to cause clockwise rotation. The torque due to the buoyant force tends to produce counterclockwise rotation. Thus, we have the possibility of rotational equilibrium in which these two torques balance.

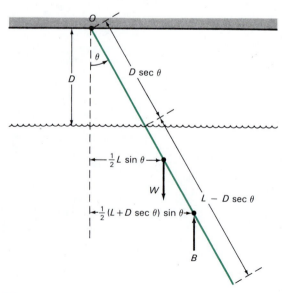

FIGURE 15.17

Let's look qualitatively at why there are two possible equilibrium orientations, and then we can work out the details. Imagine a limiting case in which D is almost equal to L, so that only the tip of the rod is submerged. In this case the buoyant force is nearly zero and we expect that equilibrium will occur for $\theta = 0$, the orientation for a stable equilibrium when the rod does not reach the water.

At the other extreme, imagine that D is much less than L. In this case the rod is nearly horizontal and we expect that there is a stable equilibrium solution in which D approaches zero; the rod is horizontal and floating. Intuitively, we expect this will be a stable equilibrium.

Let's verify these qualitative observations by applying the second condition of equilibrium to the rod. The lever arms for the forces W and B are shown in Figure 15.17. The net torque τ is the sum of the positive (counterclockwise) torque exerted by the buoyant force and the negative torque exerted by the weight of the rod.

$$\tau = B \cdot \tfrac{1}{2}(L + D \sec \theta) \sin \theta - W \cdot \tfrac{1}{2}L \sin \theta$$

The weight of the rod is given by $W = \rho_r ALg$. The buoyant force equals the weight of the water displaced, and is given by

$$B = \rho_w A(L - D \sec \theta)g$$

Substituting for W and B we find

$$\tau = \tfrac{1}{2}Ag \cdot \sin \theta \, [\rho_w(L^2 - D^2 \sec^2 \theta) - \rho_r L^2]$$

Equilibrium corresponds to $\tau = 0$. There are three equilibrium orientations. Two orientations make the bracketed expression zero. They give orientations with equal angles on either side of the vertical. The third equilibrium angle is $\theta = 0$ (the rod is vertical), which makes $\sin \theta = 0$.

(a) For $\theta = 0$ to be a stable equilibrium, the net torque must be a restoring torque for values of θ near zero. This is the case provided the gravitational torque exceeds the buoyant torque, a condition that is satisfied if the term inside the brackets is negative for $\theta = 0$. For $\theta = 0$, $\sec^2 \theta = 1$, and the bracketed term is negative, provided

$$\rho_r L^2 > \rho_w(L^2 - D^2)$$

This gives the desired relation

$$D > L \sqrt{1 - \frac{\rho_r}{\rho_w}}$$

(b) The other equilibrium orientations correspond to

$$[\rho_w(L^2 - D^2 \sec^2 \theta) - \rho_r L^2] = 0$$

Solving for θ with the specified densities gives

$$\theta = \cos^{-1} \frac{D}{L\sqrt{1 - (\rho_r/\rho_w)}}$$

As D approaches zero, the equilibrium angle approaches $90°$, the limit in which the rod is horizontal and floating.

A curious property of this equilibrium solution is that the length of the rod above water is the same regardless of the distance D between the pivot and the surface of the water. Thus, the distance along the rod from the pivot to the surface of the water is $D/\cos \theta$. Using the relation for the equilibrium value of $\cos \theta$ shows that $D/\cos \theta = L\sqrt{1 - (\rho_r/\rho_w)}$, a value that is independent of D.

You may find it fun to test this final conclusion. Use a thread to suspend a small wooden rod from one end. Holding the thread, dip the vertical rod into the water. You should observe that the rod remains stable until a critical depth is reached, at which point the rod will assume an equilibrium orientation at an angle with the vertical.

Continue lowering the rod and you should see the orientation angle increase, but the same length of rod stays above water.

When you pull the rod from the water you can see the high water mark from which you can measure the length of the rod that remains above water. Since many woods have densities near 640 kg/m³ you should find that the distance from the pivot to the surface of the water is about 0.60L, or, in other words, about six-tenths the length of the rod. Try it!

EXERCISES

15.1   Fluid Statics

A. One mole of water contains 6.02×10^{23} molecules and occupies a volume of 18×10^{-6} m³. Determine the average volume per molecule.

B. A 425-N pig marooned on a wooden board floats down a flooded river. If the board is 15 cm thick and 3 m long, and has a density of 600 kg/m³, what is the width of the board if the top surface is level with the water?

C. An oil well 1600 m deep (about 1 mi) is filled with a fluid whose density is 950 kg/m³. What is the pressure at the bottom of the well? Express your answer in atmospheres.

D. Experimentalists measuring pressures in the range of a few torr sometimes use a barometer (Figure 15.5) filled with oil rather than mercury. If the density of the oil is 920 kg/m³, what is the change in the level of oil when the pressure changes by 1 torr?

E. The gravitational acceleration at the surface of the earth's moon is 1.62 m/s². The average mass density of the moon is 3.33×10^3 kg/m³ and the bulk modulus of its crustal rocks is comparable to the bulk modulus of the earth's crustal rocks. Estimate the depth to which one could dig a hole in the moon.

F. A diver descends to a depth of 20 m in search of abalone. What pressure acts on her at that depth? Express your result in both pascals and atmospheres. Is your answer an absolute pressure or a gauge pressure?

G. A block of wood is held beneath the surface of a bucket of water by a string. The bucket rests on a scale. The string breaks and the wood floats to the surface. After things settle down, has the scale reading increased, decreased, or remained unchanged? Explain your answer.

H. When you weigh yourself, the scale reading understates your true weight because of the buoyant force exerted by the air you displace. *Estimate* the buoyant force on a 140-lb person. The density of air is approximately 1 kg/m³.

15.2   Fluid Dynamics

I. A particular garden hose has an inside diameter of 5/8 inch. The nozzle opening has a diameter of 1/4 inch. When the flow speed of the water in the hose is 0.43 m/s, what is the flow speed through the nozzle?

J. An electromagnet used in a research laboratory is cooled with water flowing through copper tubes (inside diameter = 0.50 cm). A pump that maintains the circulation is connected to the copper tubes by a flexible tube having an inside diameter of 1 cm. The flow speed of water in the copper tube is 5.0 m/s. Determine (a) the flow speed in the flexible tubing, (b) the discharge rate of the pump, (c) the rate of mass flow from the pump.

K. Oil flows out of a pipe at a speed of 1.15 m/s. The diameter of the pipe is 5.12 cm. Determine the discharge rate.

L. Water flows through a horizontal pipe that gradually narrows so that the final inside diameter is one-half the original diameter. In the wide section the flow speed is 14.1 m/s and the pressure is 10.2 atm. Determine (a) the flow speed in the narrow section, (b) the pressure in the narrow section. Express the pressure in atmospheres.

M. A J-shaped glass tube is connected to a pipe as shown in Figure 1. The tube is filled with mercury and the open end is exposed to the atmosphere in which the pressure is $P_0 = 1.05 \times 10^5$ N/m². If the difference in levels of the mercury is 28 cm, what is the pressure of the air in the pipe?

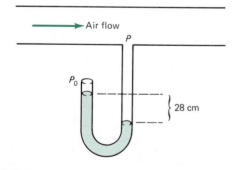

FIGURE 1

N. When a water tower is filled, the surface is 14 m above the exit pipe. (a) Assuming that Torricelli's theorem applies, determine the exit flow speed. (b) How far above the exit pipe must the surface be in order to give an exit flow speed of 8.3 m/s?

P. Water squirts from a syringe at a speed of 30 m/s. The diameter of the opening is 0.01 cm. Determine the pressure difference between the inside of the syringe magazine and the outside. Assume that the flow speed inside is zero. Express your result in pascals and atmospheres.

Q. A toy squirt gun ejects water with a nozzle speed of 10 m/s. The diameter of the nozzle is 1 mm. The diameter of the water magazine is 1 cm. Determine the pressure inside the gun. Perform two determinations. In the first, ignore the flow speed inside the magazine. Express the pressure in pascals and in atmospheres. In the second, include the internal flow speed in your calculation. Do the results differ if you work to three-significant-figure accuracy?

15.3 Viscosity

R. A sphere of radius r moving through a fluid of dynamic viscosity η at a speed v experiences a viscous force given by Stokes' law

$$F = 6\pi\eta rv$$

Show that this equation is dimensionally consistent.

S. Determine the discharge rate for the water flow in Example 7. Express your result in m^3/s and in gallons per minute.

15.4 Turbulence

T. The flow of helium through a particular pipe becomes turbulent at a flow speed of 30 m/s. At what speed will the flow become turbulent for oxygen (through the same pipe at the same temperature and pressure)?

U. Oil flows through a pipe at a speed of 6 m/s. The kinematic viscosity of the oil is 8.0×10^{-6} m^2/s. The diameter of the pipe is 12 cm. (a) Determine the Reynolds number for the flow. (b) Would you expect the flow to be streamline or turbulent?

V. Air flows through a tube at a speed of 10 m/s. If the flow must be laminar, what is the maximum diameter that the pipe can have? (Take the critical Reynolds number to be 2300.)

W. A golf ball with a radius of 2 cm moves through air at 55 m/s. Calculate the Reynolds number. Would you expect the flow to be laminar or turbulent?

X. Stokes' law correctly describes the viscous drag force only for laminar flows with a Reynolds number of no more than 10. Experiment shows that for higher Reynolds numbers the drag force can be represented by

$$F_D = \tfrac{1}{2}C_D A \rho v^2$$

where A is the cross-sectional area of the object, ρ is the density of the fluid, v is the speed of the object relative to the fluid, and C_D is the drag coefficient. (a) Determine the dimensions of the drag coefficient. The drag force is often expressed as a function of the Reynolds number (R) rather than as a function of the speed of flow. (b) Argue that F_D can be represented in the form

$$F_D = \frac{B\eta^2 R^2}{\rho}$$

where η is the viscosity and B is a dimensionless constant.

PROBLEMS

1. A body of mass m and density ρ_0 is completely submerged in a fluid of density ρ_f, where $\rho_f < \rho_0$. (a) Show that the apparent weight of the submerged body (mg − buoyant force) is

$$W = mg\left(1 - \frac{\rho_f}{\rho_0}\right)$$

(b) A 10-kg rock is submerged in water. If the density of the rock is 3.1 g/cm³, what is its apparent weight under water?

2. A pousse-café is a drink made of six different liqueurs that lie in layers, one on top of the other. The colors and densities of these liqueurs are indicated in Figure 2. Each layer is 2 cm thick. A small plastic cube with a volume of 1 cm³ and a density of 1.056 g/cm³ is carefully lowered into the drink. (a) Between which layers does it float? (b) Show that five-sevenths of its volume is in one layer and two-sevenths is in another.

3. A plastic sphere floats in water with 0.50 of its volume submerged. This same sphere floats in oil with 0.40

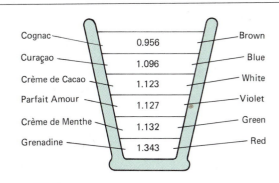

Cognac	0.956	Brown
Curaçao	1.096	Blue
Crème de Cacao	1.123	White
Parfait Amour	1.127	Violet
Crème de Menthe	1.132	Green
Grenadine	1.343	Red

FIGURE 2

of its volume submerged. Determine the densities of the oil and the sphere.

4. The molecular weight of helium is 4.00 g/mole. The molecular weight of air is 28.9 g/mole. A helium-filled balloon rises because it displaces a weight of air in excess of its own weight. The molar volume is approximately the same for all gases, 0.0224 m³ at a temperature of 0°C and a pressure of 1 atm. (a) Estimate the

mass and volume of helium required to lift a mass of 100 kg. (b) In view of your answer for part (a), would it be feasible to market miniblimps? In particular, would such a blimp fit into a one-car garage?

5. Torricelli was the first to realize that we live at the bottom of an ocean of air. He correctly surmised that the pressure of our atmosphere is due to the weight of the air. The density of air at the earth's surface is 1.3 kg/m^3. The density decreases with increasing altitude as the atmosphere thins out. Suppose that the density were a constant (1.3 kg/m^3) up to some altitude h, and zero above. Then h would represent the depth of the ocean of air, or the thickness of our atmosphere. Determine the value of h that gives a pressure of 1 atm at the surface of the earth. Would the peak of Mt. Everest rise above the surface of such an ocean?

6. The velocity profile for laminar flow in a pipe of radius R is given by

$$u = u_0 \left(1 - \frac{r^2}{R^2}\right)$$

where r is the distance from the axis of the pipe and u_0 is the flow speed along the axis. (a) Where is the fluid velocity greatest? (b) At what position(s) does the velocity gradient have the largest magnitude? (c) Where is the shearing stress greatest?

7. Imagine an open water tank filled to a depth H (Figure 3). If a hole is punched in the tank just slightly below

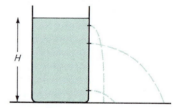

FIGURE 3

the surface of the water, the stream of water through this hole will have a low speed and will strike the ground relatively close to the base of the tank. If a hole is punched in the tank near its base, the water will emerge from the hole with a higher speed, but it will still strike the ground relatively close to the base of the tank. Presumably there is a point at which a hole can be punched so as to maximize the horizontal distance that the water travels. (a) Show that the horizontal distance the water travels is a maximum when the hole is punched at a point $H/2$ above the base of the tank. (b) Show that the maximum horizontal distance is equal to H. (Hint: Observe that $R = vt$ expresses the horizontal distance of travel in terms of the speed of the emerging water, v, and the time of flight, t, of the water. It is algebraically advisable to start with the equation $R^2 = v^2t^2$, and then use Bernoulli's equation and the free-fall equation (relating the time of flight, t, and

the distance of vertical fall, y) to obtain the equation $R^2 = 4(H - y)y$. The condition

$$\frac{d(R^2)}{dy} = 0 \qquad \text{for maximum } R^2$$

leads to the value of y that maximizes R.)

8. A particle falling through air accelerates until it reaches what is called the **terminal speed**. The terminal speed is reached when the downward gravitational force on the particle (its weight) equals the upward viscous force. Take the particle to be a sphere of radius r and assume that the viscous force is given by Stokes' law. Show that the terminal speed is given by

$$v = \frac{2(\rho - \rho_a)gr^2}{9\eta}$$

where ρ is the mass density of the particle, ρ_a is the density of air, and η is the viscosity of air.

9. The Reynolds number for flow through a pipe can be related to the discharge rate Q. Show that, for a pipe having a circular cross section of diameter d,

$$R = \frac{4Q}{\pi v d}$$

where v is the kinematic viscosity.

10. A brass sphere having a diameter of 2.0 cm is connected to the end of a linear spring having a spring constant of 20.0 N/m. The spring is suspended so that the sphere can move up and down in a container of water. If the sphere is set into motion with an initial vertical displacement of 3.0 cm, (a) estimate the maximum speed of the sphere. You may neglect the buoyant force and assume that the drag force is proportional to the velocity. (b) For what range of speeds is the fluid flow laminar? (c) Explain why the fluid flow changes from laminar to turbulent during the oscillations.

11. (a) Figure 4 depicts a dam holding back water of depth H. Determine the force exerted on the dam by the water. (b) Find the torque exerted by the water about an axis parallel to the bottom of the dam. (c) If the total were concentrated along a horizontal line across the dam, how high would the line be if the torque was the same as calculated in part (b)?

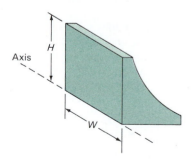

FIGURE 4

12. Pressure can be measured by using a transducer that produces an electrical signal related to the pressure. Imagine that two pressure transducers record the pressures labeled P_1 and P_2 in the pipe in Figure 5. (a) Show that the speed of the fluid in the larger section is

$$v = \sqrt{\frac{2(P_1 - P_2)}{\rho\left[\left(\dfrac{D}{d}\right)^4 - 1\right]}}$$

where D is the diameter of the larger section and d is the diameter of the smaller section. (b) Determine the pressure difference if the pipe contains water and $v = 25$ m/s, $D = 2$ cm, and $d = 0.3$ cm.

FIGURE 5

13. (a) The tension recorded by a spring balance (Figure 6) is less than the weight of the object because of the buoyant force. Given the density of the fluid surrounding the object and the volume of the object, express the spring tension in terms of the weight of the object. (b) Show how to determine the density of a liquid from spring balance tensions, taken when the object is immersed first in water and then in the liquid.

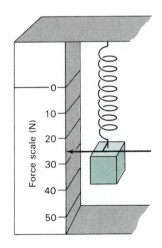

FIGURE 6

14. A rectangular block, 4 cm by 4 cm by 10 cm, floats in water (Figure 7). When slightly depressed and released it executes simple harmonic motion. What is the period of the motion if the density of the block is 800 kg/m³?

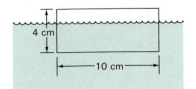

FIGURE 7

15. A thin, rigid rod held in the form of a semicircle of radius R is free to rotate about a horizontal axis through the point O (Figure 8). As the rod rotates, its ends alternately break free of the liquid surface. Prove that the equilibrium orientation, $\theta = 0$, is stable, neutral, or unstable, depending on whether the ratio of the rod density to the fluid density is greater than, equal to, or less than 0.5. (Neglect the supporting wire.)

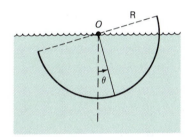

FIGURE 8

16. A rectangular block has dimensions $b \times a \times a$. The block is floating with the length b vertical (Figure 9). It is given a small angular displacement in such a way that the point P on the block remains fixed at the surface of the liquid. You can think of P as an axis of rotation. This displacement does not change the volume of fluid displaced, so the buoyant force remains equal to the weight of the block. The net force remains zero but the net torque is *not* zero. Prove that the equilibrium is stable or unstable according to whether the ratio a/b is greater than or less than $\sqrt{6\Gamma(1 - \Gamma)}$ where Γ is the ratio of the density of the block to the density of the fluid. Note that the net buoyant force acts at the center of buoyancy, which is the center of gravity of the displaced fluid.

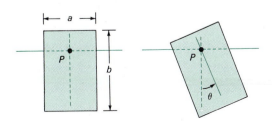

FIGURE 9

17. Assume that the viscous force on a sky diver is given by

$$F = kv^3$$

where v is the speed in m/s. Evaluate the constant k by making use of the fact that the terminal velocity for an 80-kg sky diver is 58 m/s.

18. A wooden dowel has a diameter of 1.2 cm. It floats in water with 0.4 cm of its diameter above water (Figure 10). Determine the density of the dowel.

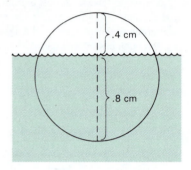

.4 cm

.8 cm

FIGURE 10

19. The masses of the United States penny, nickel, and dime coins are 2.7 grams, 5.2 grams, and 2.4 grams, respectively. A wooden cube (volume 18.6 cm³) floats on water with 0.72 of its volume submerged. A coin placed on top of the cube causes the cube to just float, or, in other words, the cube's top surface is level with the water surface, with no part of the coin submerged. Determine whether the coin is a penny, a nickel, or a dime.

CHAPTER 16

WAVE KINEMATICS

16.1
WAVE CHARACTERISTICS

We learn about waves at an early age. As children we watch water waves sweep across ponds. We are taught that the sun sends out light waves that illuminate our planet and provide the energy to nourish life. Shock waves from supersonic aircraft may rattle our windows. We may feel the effects of seismic waves produced by earthquakes that travel thousands of miles along the surface of the earth and through its interior. Physicists actively search for gravitational waves—faint wrinkles in the strength of gravity.

This chapter is about the **kinematics** of waves. You will learn how to describe wave motions without reference to the particular physical nature of the wave. Sound waves, light waves, and water waves are quite different physically. Yet, each displays certain kinematic features that are identical, or very similar.

A wave is a disturbance that travels at a definite speed. For example, a stone thrown into a still pond disturbs the surface. This disturbance moves across the surface of the pond as a circular wave, transferring energy and momentum as it moves. It is the ability of waves to transfer energy and momentum that interests scientists and engineers. In the example just mentioned, the waves transfer energy and momentum from the point where the stone strikes the surface to the edge of the pond. The surface of the pond returns to its undisturbed condition after the waves pass. There is a transfer of energy and momentum but no transfer of mass.

A water wave is one type of **mechanical** wave. In a mechanical wave, matter experiences an oscillatory displacement from equilibrium. Forces acting on and between adjacent elements of the matter cause the disturbance to advance at a definite speed. The matter is the **medium** through which the wave travels.

Some waves do not require a material medium to support their travel. Electromagnetic waves travel through empty space between the earth and the stars. In electromagnetic waves the disturbances are oscillatory **electric** and **magnetic fields**. Gravitational waves can also travel through empty space. Gravitational waves are oscillatory variations of the **gravitational field**, and they travel at the speed of light. As yet there is no clear-cut experimental evidence that gravitational waves exist. A number of scientific research teams are currently trying to detect gravitational waves, which presumably are generated in various astrophysical events, such as the supernova explosion of an aging star.

Many waves may be classified as either **transverse** or **longitudinal**. In a transverse mechanical wave the displacement of the medium is perpendicular to the direction of wave travel (Figure 16.1a). A wave moving across a pond appears to be transverse. We see the surface of the water rise and fall as the wave form moves horizontally. However, water waves are not purely transverse. The surface of the water also moves to and fro in the horizontal direction, although this component of the motion is not usually apparent to the casual observer.

In a longitudinal wave the displacement is parallel to the direction of wave travel (Figure 16.1b). Sound waves in air are longitudinal. The displacements are variations in the air pressure. These pressure variations are parallel to the direction in which the sound travels. A helical spring (a Slinky® toy) can be used to demonstrate longitudinal wave motion. The wave is seen as moving variations in the spacing of the turns of the helix (Figure 16.2).

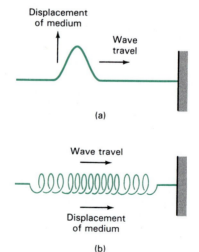

FIGURE 16.1

(a) In a transverse mechanical wave, such as a wave on a stretched string, the displacement of the medium is perpendicular to the direction of wave travel. (b) In a longitudinal mechanical wave, such as a wave along the length of a helical coil, the displacement of the medium is parallel to the direction of the wave.

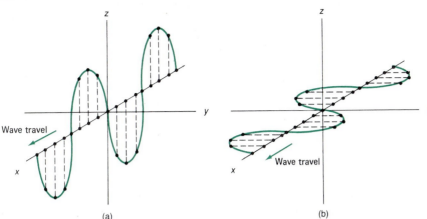

Wave travel

(a)

Wave travel

(b)

FIGURE 16.3

A wave on a string is transverse. The motion of points along the string is indicated by dashed lines and is perpendicular to the direction of wave travel. In both (a) and (b) the wave travels along the x-axis. In (a) the wave is polarized along the z-axis. In (b) the wave is polarized along the y-axis.

Another aspect of a wave is its **polarization**. A wave is said to be polarized along the direction of the disturbance. The polarization of a longitudinal wave is parallel to the direction of wave motion, and is of little practical importance because it cannot be changed or controlled. In a transverse wave the disturbance (and thus the polarization) lies in a plane perpendicular to the direction of wave travel. The direction of polarization can be controlled in some transverse waves. For example, a transverse wave can be generated by shaking one end of a flexible cord or string (Figure 16.3). For a wave on a string advancing along the x-axis the polarization lies in the y–z plane. Figure 16.3 shows two possible polarizations, one along the z-axis and one along the y-axis.

To avoid making our study of wave motion unnecessarily abstract, we frequently refer to the type of wave called a **wave on a string**. This is the kind of wave that is generated by shaking one end of a flexible cord or string (Figure 16.3). Waves on a string propagate in one dimension, that is, along the length of the string. The description of one-dimensional waves requires just one space variable. In contrast, the water waves generated by throwing a stone into a pond are two-dimensional waves—they spread out over a surface. The flash of light generated by a firefly travels outward in all directions as a three-dimensional pulse. In general, the mathematical description of two- and three-dimensional waves requires two and three space variables.

In our study of waves we ignore the effects of absorption and dispersion. Absorption reduces the wave height, and dispersion changes the wave shape. Thus, we consider waves that maintain their shape and their height.

Suppose a wave on a string travels in the positive x-direction at a constant speed (see Figure 16.4). Let $\psi(x, t)$ denote the height of the wave at a point x at time t. The shape of the wave form describes how ψ depends on x at each moment of time. There is a special relationship between x and t for all waves: for a wave traveling in the positive x-direction at speed v the variables x and t *must* occur in $\psi(x, t)$ *only* as the combination $x - vt$. Symbolically,

$$\psi(x, t) = \psi(x - vt) \tag{16.1}$$

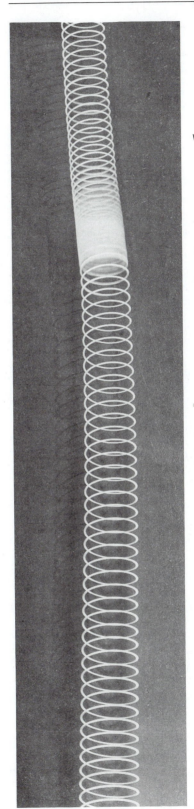

FIGURE 16.2

A longitudinal wave travels along the helical spring of a Slinky.

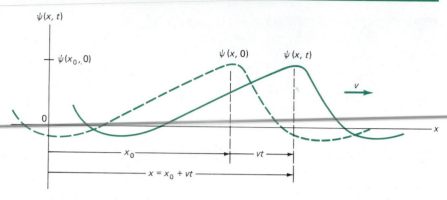

FIGURE 16.4

The wave form advances at speed v. If the crest is at x_0 at time $t = 0$, it will advance to $x_0 + vt$ in time t.

This relation between x and t describes a *kinematic* aspect of waves: It has nothing to do with the physical nature of the waves, and would be present in descriptions of water waves, light waves, and sound waves.

EXAMPLE 1

Traveling Wave Pulse

Figure 16.5 illustrates how the $x - vt$ combination results in wavelike behavior for a pulse described by

$$\psi(x, t) = \frac{a^3}{a^2 + (x - vt)^2}$$

where $a = 1$ cm and $v = 3$ cm/s. Note that x and t occur only in the combination $x - vt$. Figure 16.5 shows that the crest of the wave is at $x = 0$ at $t = 0$. The height of the crest is

$$\psi(x = 0, t = 0) = a = 1 \text{ cm}$$

With the wave form moving at a speed of $v = 3$ cm/s, the entire wave form moves 3 cm to the right in one second. We can verify this by showing that $\psi = a = 1$ cm at $x = 3$ cm and $t = 1$ s. Setting $x = 3$ cm, $v = 3$ cm/s, and $t = 1$ s gives

$$\psi(x = 3 \text{ cm}, t = 1 \text{ s}) = \frac{a^3}{a^2 + (3 \text{ cm} - 3 \text{ cm})^2} = a = 1 \text{ cm}$$

At $t = 2$ s the crest is at $x = 6$ cm. In general, the crest is at $x = vt$ at time t.

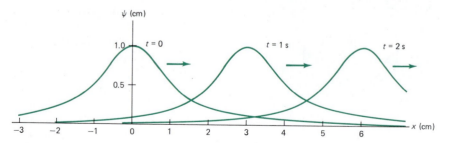

FIGURE 16.5

The wave form $\psi(x, t) = a^3/[a^2 + (x - vt)^2]$ with $a = 1$ cm and $v = 3$ cm/s is shown at the times $t = 0$, $t = 1$ s, and $t = 2$ s. The wave moves to the right without change of shape.

16.2
SINUSOIDAL WAVES

A **wave form** is described by the function $\psi(x, t)$. **Sinusoidal** wave forms are particularly important. They have the forms

$$\psi(x, t) = A \sin\left[\frac{2\pi}{\lambda}(x - vt)\right] \tag{16.2}$$

$$\psi(x, t) = B \cos\left[\frac{2\pi}{\lambda}(x - vt)\right] \tag{16.3}$$

The quantity λ is the wavelength, and it is defined below. The importance of sinusoidal wave forms stems from two facts:

1. Sinusoidal wave forms adequately describe many types of waves observed in nature.
2. More complicated wave forms can be synthesized through the superposition (addition) of sinusoidal waves.

Our immediate concern is with the kinematic aspects of sinusoidal waves. Figure 16.6 shows the wave form $\psi = A \sin[(2\pi/\lambda)(x - vt)]$ at $t = 0$. From the figure, or the equation

$$\psi(x, t = 0) = A \sin\left(\frac{2\pi x}{\lambda}\right) \tag{16.4}$$

we see that A is the maximum value of ψ. We define A to be the **amplitude** of the wave. Geometrically, A is the height of the wave crests above the x-axis. The **wavelength**, λ, is the distance over which the wave form repeats itself. The **period**, T, of the wave can be defined as the time between the passage of successive crests, or equivalently, as the time required for the wave form to advance through a distance λ. Since the wave form moves at speed

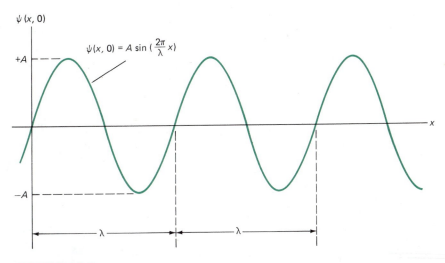

FIGURE 16.6
The wave form $\psi(x, t) = A \sin[(2\pi/\lambda)(x - vt)]$ at time $t = 0$ has the form $\psi(x, 0) = A \sin(2\pi x/\lambda)$.

v, it advances a distance vT in time T. Thus, the wavelength λ is related to the speed and period by

$$\lambda = vT \qquad (16.5)$$

Equation 16.5 is the basic kinematic relation for sinusoidal waves.

EXAMPLE 2

Determining $\psi(x, t)$ for a Sinusoidal Wave

A transverse sinusoidal wave travels along a rope at a speed of $v = 5$ m/s. The wavelength is $\lambda = 4$ m and the amplitude is $A = 0.05$ m. The height of the wave is zero at $x = 0$, $t = 0$. The height of the wave at any point and time is found by using Equation 16.2.

Let's determine the height of the wave at $x = 6$ m, $t = 1$ s. We have

$$\frac{2\pi}{\lambda}(x - vt) = \frac{2\pi}{4 \text{ m}}(6 \text{ m} - 5 \text{ m/s} \cdot 1 \text{ s}) = \frac{\pi}{2}$$

Thus,

$$\psi(6 \text{ m}, 1 \text{ s}) = A \sin\left[\frac{2\pi}{\lambda}(x - vt)\right] = 0.05 \sin\left(\frac{\pi}{2}\right) \text{ m}$$

$$= 0.05 \text{ m}$$

Because $\psi = A = 0.05$ m locates a wave crest, this result shows that there is a crest at the position $x = 6$ m at the time $t = 1$ s.

The **frequency** of the wave, f, is the number of crests that pass an observer per second. If the time between crests (T) is 0.1 s, ten crests pass per second. Thus, the frequency is the reciprocal of the period:

$$f = \frac{1}{T} \qquad (16.6)$$

The SI unit of frequency, 1 cycle per second (cps), is called the *hertz* (abbreviated Hz). We can substitute the reciprocal of the frequency, $1/f$, for T in the basic kinematic relation $\lambda = vt$. This gives us a second form of the kinematic relation,

$$f\lambda = v \qquad (16.7)$$

The two forms of the kinematic relation can be used to represent sinusoidal waves in alternative forms. A particularly useful representation of the sinusoidal wave form is obtained by introducing the **wave number** (k),

$$k = \frac{2\pi}{\lambda} \qquad (16.8)$$

measured in radians per meter, and the **angular frequency** (ω),

$$\omega = 2\pi f \qquad (16.9)$$

measured in radians per second. Expressed in terms of k and ω, Equation 16.2 becomes

$$\psi(x, t) = A \sin(kx - \omega t) \qquad (16.10)$$

EXAMPLE 3

Determining T, f, k, and ω for a Sinusoidal Wave

For the wave of Example 2 we can use Equation 16.5 to establish the period

$$T = \frac{\lambda}{v} = \frac{4 \text{ m}}{5 \text{ m/s}} = 0.800 \text{ s}$$

The frequency of the wave is

$$f = \frac{1}{T} = \frac{1}{0.800 \text{ s}} = 1.25 \text{ Hz}$$

The wave number, k, is

$$k = \frac{2\pi}{\lambda} = \frac{2\pi \text{ rad}}{4 \text{ m}} = 1.57 \text{ rad/m}$$

The angular frequency, ω, is

$$\omega = 2\pi f = 2\pi \text{ rad} (1.25/\text{s}) = 7.85 \text{ rad/s}$$

According to Equation 16.5 the wave speed is given by $v = \lambda/T$. By combining Equations 16.8 and 16.9 we obtain yet another form of the kinematic relation, namely

$$v = \frac{\omega}{k} \tag{16.11}$$

Table 16.1 summarizes the relationships among the many wave properties.

Phase and Phase Difference

For the sinusoidal wave form $\psi(x, t) = A \sin(kx - \omega t)$, the argument of the sine function

$$kx - \omega t \equiv \varphi \tag{16.12}$$

TABLE 16.1
Relationships among Wave Properties

Wavelength	Wave Number	Period	Frequency	Angular Frequency	Velocity
λ	k	T	f	ω	v
$\dfrac{2\pi}{\lambda} = k$			$\dfrac{1}{T} = f$ $2\pi f = \omega$		
Basic Kinematic Relation	$\lambda = vT$		$f\lambda = v$		$\dfrac{\omega}{k} = v$
Equivalent Sinusoidal Wave forms		$\psi(x, t) = A \sin\left[\dfrac{2\pi}{\lambda}(x - vt)\right]$ $\psi(x, t) = A \sin(kx - \omega t)$			

is called the **phase**. At any given time the phase is different for each point along the wave form. For example, at a fixed time t, the points x_1 and x_2 differ in phase by

$$\text{phase difference} = \varphi_2 - \varphi_1 = (kx_2 - \omega t) - (kx_1 - \omega t)$$
$$= k(x_2 - x_1)$$

In terms of the wavelength,

$$\text{phase difference} = \frac{2\pi(x_2 - x_1)}{\lambda} \tag{16.13}$$

If we let $\Delta\varphi$ stand for the phase difference $\varphi_2 - \varphi_1$, and let Δx stand for the spatial separation along the wave form $(x_2 - x_1)$, we can write

$$\Delta\varphi = 2\pi\left(\frac{\Delta x}{\lambda}\right)$$

This relationship between phase difference and spatial separation along the wave form is used extensively in the study of optics.

EXAMPLE 4

Phase Difference and Path Difference

A tuning fork generates sound waves with a frequency of 246 Hz. The waves travel in opposite directions along a hallway, are reflected by walls, and return. What is the phase difference between the reflected waves when they meet? The corridor is 47 m long and the tuning fork is located 14 m from one end. The speed of sound in air is 343 m/s.

The kinematic relation $f\lambda = v$ lets us determine the wavelength of the sound waves

$$\lambda = \frac{v}{f} = \frac{343 \text{ m/s}}{246 \text{ s}^{-1}} = 1.29 \text{ m}$$

One wave travels 14 m, is reflected, and returns 14 m. The other travels 33 m, is reflected, and returns 33 m. The difference in their path lengths is

$$\Delta x = 66 \text{ m} - 28 \text{ m} = 38 \text{ m}$$

The phase difference between the two paths is

$$\Delta\varphi = 2\pi\frac{\Delta x}{\lambda} = \frac{2\pi \text{ rad } (38 \text{ m})}{1.29 \text{ m}} = 185 \text{ rad}$$

It follows from the relation $\Delta\varphi = 2\pi(\Delta x/\lambda)$ that two points separated by one wavelength differ in phase by 2π radians (Figure 16.7). Because the wave form repeats over a distance of one wavelength, the corresponding phase difference of 2π leaves the wave form unchanged. This reasoning can be extended to establish the following:

1. Phase differences that are *integral multiples* of 2π are equivalent to zero phase difference.
2. Any phase difference can be increased or reduced by an integral multiple of 2π without affecting subsequent calculations and results.

For example, the phase difference in Example 4 was $2\pi(29.5)$. We can ignore the $2\pi(29)$ and concern ourselves only with $2\pi(0.5) = \pi$, without affecting subsequent calculations involving $\Delta\varphi$.

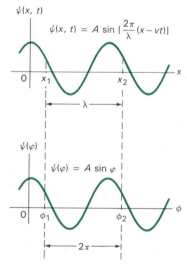

FIGURE 16.7

The wave represented by $\psi(x, t) = A \sin[(2\pi/\lambda)(x - vt)]$ is shown in the upper graph. Using the definition of phase, $\varphi = (2\pi/\lambda)(x - vt)$, the wave can be represented as $\psi(\varphi) = A \sin \varphi$. A plot of $A \sin \varphi$ is shown in the lower graph. As shown, a spatial separation of one wavelength is equivalent to a phase difference of 2π radians.

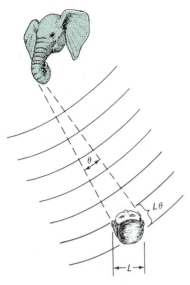

FIGURE 16.8

Turning the head through a small angle θ results in a path difference of approximately $L\theta$ between waves reaching the two ears simultaneously. For sound with a frequency of 330 Hz the minimum detectable phase difference corresponds to an angle $\theta \approx 3°$.

EXAMPLE 5

How to Find a Wounded Elephant

Imagine that you are an African game warden searching for a wounded elephant. You wander aimlessly until you hear the elephant's distress call. Then, confident that you will find the animal, you set off in the direction from which the sound came. If you have normal hearing, you will be able to determine the direction of a source of sound rather accurately. When you face the source, sound waves entering your ears have zero phase difference. If you turn your head as little as 3° to either side of the direction of the source there is a noticeable phase difference (Figure 16.8). If your ears are separated by a distance L, then rotating your head through a small angle θ will result in a path difference of $L\theta$. The corresponding phase difference is

$$\text{phase difference} = 2\pi\left(\frac{L\theta}{\lambda}\right)$$

If we take $\theta = 3°$ we get the minimum detectable phase difference. If the elephant trumpets at a wavelength of 1 m (frequency = 330 Hz) and we take $L = 20$ cm, $\theta = 3° = 0.052$ rad, we find that the minimum detectable phase difference is 0.068 rad. An alternative way of emphasizing this high degree of directional sensitivity is to note that $L\theta/\lambda$ is the path difference expressed in wavelengths. For these data, $L\theta/\lambda = 0.0104$. In other words, at a wavelength of 1 m, your ears can detect a path difference of about 1 cm.

16.3
THE PRINCIPLE OF SUPERPOSITION FOR WAVES

There are many instances in which two or more sources generate waves that subsequently meet. For example, raindrops strike the surface of a pond and generate circular waves that meet each other. Experimentally, we find that the effect of these meetings is described by a **principle of superposition**:

A meeting of two or more waves produces a wave form that is the sum of the wave forms produced by each wave acting separately.

The principle of superposition can be stated mathematically as follows:

If ψ_1 and ψ_2 are the wave forms produced by two sources of waves acting separately, then $\psi_1 + \psi_2$ is the wave form produced when the two sources act together.

Not all waves obey the principle of superposition. Large-amplitude waves in fluids and solids interact in a complicated fashion. Shock waves generated by jet aircraft and ocean surf do not obey the principle of superposition. In general, only experiment can determine whether or not the principle of superposition is applicable to a given situation.

Interference of Waves

The superposition of waves can result in **interference**. At points where the waves tend to cancel one another we say that there is ***destructive*** **interference**. At points where the waves reinforce one another we say that there is ***constructive*** **interference**.

Various types of mirrors and lenses, both optical and acoustical, focus waves at a point (ideally) where they exhibit constructive interference. The parabolic

FIGURE 16.9
A receiver for television signals transmitted from a communications satellite. Waves reflected by the dish interfere constructively at the horn.

dishes in evidence along the sidelines of televised football games reflect sound waves to a microphone. The microphone is mounted at a point where the waves interfere constructively. Similar reflecting dishes are used to collect television signals beamed from communication satellites (Figure 16.9). Waves reflected from the dish interfere constructively at the point where the receiving horn is positioned.

Beats

The superposed sound waves produced by two identical tuning forks are heard by the human ear as a single pure tone. However, if the frequency of one fork is altered slightly by loading it with a piece of wax, the resulting sound exhibits periodic variations in loudness called **beats**. The beats are a result of the periodic swings between constructive interference and destructive interference. Beat phenomena are not restricted to sound waves; they are produced whenever two waves with slightly different frequencies are superposed.

We can analyze the beat phenomenon by considering the superposition of the waves

$$\psi_1 = A \cos(k_1 x - \omega_1 t)$$

and

$$\psi_2 = A \cos(k_2 x - \omega_2 t)$$

At the fixed point $x = 0$ (chosen for convenience), the wave form is

$$\psi = \psi_1 + \psi_2 = A(\cos \omega_1 t + \cos \omega_2 t) \qquad (16.14)$$

From this equation we can see that at $t = 0$ the waves interfere constructively and give a wave of amplitude $2A$. If we suppose that $\omega_2 > \omega_1$, then the phase $\omega_2 t$ increases more rapidly than the phase $\omega_1 t$.

The quantity

$$\Delta\varphi = (\omega_2 - \omega_1)t \qquad (16.15)$$

is the phase difference between the waves. This phase difference increases linearly with time. Constructive interference occurs at times when

$$(\omega_2 - \omega_1)t = 0, 2\pi, 4\pi, \ldots$$

The time between successive moments of constructive interference is called the **beat period**, T_{beat}. In a time interval T_{beat} the phase difference increases by 2π. The number of beats per second, called the **beat frequency**, is the reciprocal of the beat period

$$\text{beat frequency} = f_{\text{beat}} = \frac{\omega_2 - \omega_1}{2\pi}$$

Recall that $\omega = 2\pi f$ relates the ordinary and angular frequencies. This relation allows us to replace $(\omega_2 - \omega_1)/2\pi$ by $f_2 - f_1$. Thus,

$$f_{\text{beat}} = f_2 - f_1 \qquad (16.16)$$

The beat frequency equals the difference between the two superposed frequencies.

Equation 16.16 shows that the beat frequency equals the difference between the frequencies of the superposed waves. This fact makes it possible to measure an unknown frequency by superposing a wave of known frequency and then measuring the beat frequency. The advantage of this method is that the beat frequency can be made much lower than either of the two superposed frequencies. This relatively low beat frequency can then be measured by audio or electronic methods that are too slow to measure the individual frequencies. For example, guitar players who lack a precise sense of pitch rely on beats to tune their instrument.

16.4
THE DOPPLER EFFECT

Relative motion between a source of waves and an observer of waves results in the **Doppler effect**, a change in the *observed* frequency or wavelength. We measure this change by comparing the frequency or wavelength when there *is* relative motion to the frequency or wavelength when there is *no* relative motion.

A motorist waiting at a railroad crossing can observe the Doppler effect for sound waves. As the train passes, the motorist notices a sudden decrease in the frequency of the train whistle. If the motorist then speeds away from the crossing in an effort to make up for the delay, he may discover an application of the Doppler effect—police radar. A police radar unit sends out a beam of microwaves. These are electromagnetic waves with a wavelength of a few centimeters. An approaching vehicle reflects portions of the microwave beam back to the radar unit, which also acts as a receiver. Portions of the transmitted and received waves are superposed, giving a signal at the beat frequency—the difference between the two frequencies. As we will show, the difference between the two frequencies is directly proportional to the speed of the vehicle. Hence, a device capable of measuring the beat frequency is readily converted into a "speed meter."

The shifts in frequency and wavelength caused by relative motion depend on the speed of the observer and on the speed of the waves. For waves traveling in a material medium such as sound waves we can distinguish two different situations:

1. wave source in motion, observer at rest
2. observer in motion, wave source at rest

Let's see how the frequency and wavelength are affected when the observer is at rest and the wave source is in motion.

Source in Motion

Consider a tuning fork that emits waves of frequency f_0. If the waves travel at speed v, an observer at rest with respect to the fork would find that their wavelength is

$$\lambda_0 = \frac{v}{f_0} = vT_0 \tag{16.17}$$

It is helpful to think of T_0, the period of the wave, as the time between the emission of successive wave crests. If the fork *recedes* from the observer at speed u, the distance between successive crests is increased by uT_0, the distance the fork moves between the emission of successive crests (see Figure 16.10). The wavelength is thereby increased from λ_0 to $\lambda_0 + \Delta\lambda$, where

$$\Delta\lambda = uT_0 \tag{16.18}$$

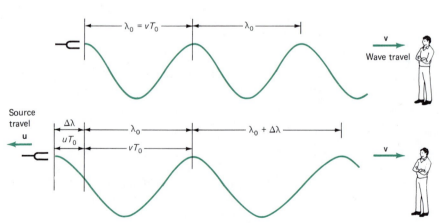

FIGURE 16.10
When the source of waves recedes from the observer the wavelength is increased, resulting in a decrease in frequency for the observer.

Combining Equations 16.17 and 16.18 gives us one version of the Doppler formula

$$\frac{\Delta\lambda}{\lambda_0} = \frac{u}{v} = \frac{\text{speed of recession}}{\text{speed of wave}} \qquad (16.19)$$

Note that the wavelength *increases* when the relative motion *increases* the distance between the source of waves and the observer.

Waves pass the observer at a constant speed v whether or not the fork is in motion, because v is determined by the medium, not by the motion of the source. Consequently, the increased wavelength also registers as a decreased frequency. To show how the frequency is reduced, we note that the speed of the wave can be expressed by

$$v = f(\lambda_0 + \Delta\lambda)$$

where f is the frequency corresponding to the wavelength $\lambda_0 + \Delta\lambda$. The wave speed is also given by

$$v = f_0\lambda_0$$

Equating these two expressions for v and using Equation 16.19 gives an expression for the Doppler-shifted frequency

$$f = \frac{f_0}{1 + \dfrac{u}{v}} \qquad \text{(source in motion)} \qquad (16.20)$$

When the source moves toward the observer the wavelength is decreased and the frequency is increased. The relation between the frequencies f and f_0 when the source moves toward the observer is obtained by replacing u by $-u$ in Equation 16.20. The following example illustrates the Doppler effect for a source that first moves toward an observer and then moves away from the observer.

EXAMPLE 6

Doppler Shift of the Frequency of an Air Horn

A student standing at the side of a roadway hears the air horn of an approaching truck. She notices that the frequency of the horn suddenly decreases as the truck passes her. The truck moves at a speed of 90 km/h (25 m/s), and the speed of sound is 330 m/s. The frequency of the horn as observed by the truck driver is 1200 Hz. What frequencies does she hear as the truck approaches and recedes?

The truck driver is at rest relative to the horn so the frequency he hears is f_0. The frequency heard by the student as the truck approaches is

$$f = \frac{f_0}{1 - \dfrac{u}{v}} = \frac{1200 \text{ Hz}}{1 - \dfrac{25}{330}} = 1300 \text{ Hz}$$

The frequency she hears as the truck moves away is

$$f = \frac{f_0}{1 + \dfrac{u}{v}} = \frac{1200 \text{ Hz}}{1 + \dfrac{25}{330}} = 1120 \text{ Hz}$$

As the truck moves toward her and then away from her the student hears the frequency drop from 1300 Hz to 1120 Hz. You may have observed a similar decrease in the frequency of the whine of automobile tires as cars move past you.

Observer in Motion

When the wave source is at rest and the observer is in motion Equations 16.19 and 16.20 do not apply. If the observer moves away from a stationary source at speed u the time between the passage of successive crests (T) is increased over its value (T_0) when there is no relative motion (Figure 16.11). The distance between successive crests, as perceived by the moving observer, can be expressed by two equations

$$\lambda = vT$$

and

$$\lambda = vT_0 + uT$$

Using $f = 1/T$ and $f_0 = 1/T_0$ leads to the relation

$$f = f_0\left(1 - \frac{u}{v}\right) \qquad \text{(observer in motion)} \qquad (16.21)$$

Although Equation 16.21 differs quantitatively from Equation 16.20, both describe the same qualitative result: the frequency *decreases* when their relative motion increases the distance between the source and observer.

The Doppler Effect for Electromagnetic Waves

One of the basic postulates of Einstein's Special Relativity Theory (Chapter 39) is that the speed of light is a constant, independent of any relative motion between the source of light and the observer. As a consequence, the Doppler effect for light waves depends only on the *relative* speed of the source and observer.

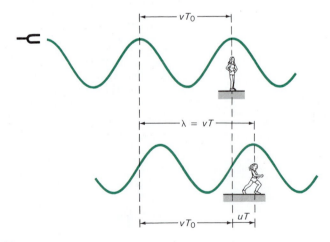

FIGURE 16.11

For an observer moving away from a source, the time between the passage of successive crests is increased, lowering the frequency.

If an observer and a source of light or any other form of electromagnetic waves *recede* from each other at a speed v, then the observed frequency is given by

$$f = f_0 \sqrt{\frac{1 - \dfrac{v}{c}}{1 + \dfrac{v}{c}}} \qquad (16.22)$$

where c is the speed of light, and f_0 is the frequency of the waves for an observer at rest relative to the source. The corresponding relation between wavelengths follows from the kinematic relation

$$f\lambda = f_0\lambda_0 = c$$

Thus,

$$\lambda = \lambda_0 \sqrt{\frac{1 + \dfrac{v}{c}}{1 - \dfrac{v}{c}}} \qquad (16.23)$$

Here, λ_0 is the wavelength for an observer at rest relative to the source. Replacing v by $-v$ in Equations 16.22 and 16.23 describes the case where the source and observer *approach* one another.

Doppler Radar

Police radar waves are one type of electromagnetic wave. The transmitted radar wave strikes an approaching vehicle (Figure 16.12) and is reflected. The radar unit functions as a transmitter and a receiver. It compares the frequencies of the transmitted wave and the wave that is reflected by the vehicle. We can show that the difference between the transmitted and received frequencies is proportional to the speed of the vehicle. This makes it possible to calibrate the radar unit to measure speeds.

In Figure 16.12, the transmitted radar wave has a frequency f_0. The frequency received by the oncoming vehicle is

$$f_1 = f_0 \sqrt{\frac{1 + \dfrac{v}{c}}{1 - \dfrac{v}{c}}}$$

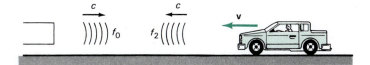

FIGURE 16.12
The approaching car reflects the radar waves. The Doppler effect operates twice; the car acts like a moving listener. When it reflects the waves the car acts like a moving source.

The vehicle acts as a source in motion. The waves that it reflects are Doppler shifted a second time so that the frequency of the waves returned to the radar unit is

$$f_2 = f_1 \sqrt{\frac{1 + \dfrac{v}{c}}{1 - \dfrac{v}{c}}} = f_0 \left[\frac{1 + \dfrac{v}{c}}{1 - \dfrac{v}{c}} \right]$$

The difference between the frequencies transmitted and received by the radar unit is

$$f_2 - f_0 = \frac{2f_0 \dfrac{v}{c}}{1 - \dfrac{v}{c}} \tag{16.24}$$

Because $v \ll c$ we can replace the factor $1 - v/c$ by unity. Solving for the vehicle speed gives

$$v = \frac{1}{2} \frac{c(f_2 - f_0)}{f_0} \tag{16.25}$$

This shows that the vehicle speed is directly proportional to the difference between the transmitted and received wave frequencies.

EXAMPLE 7

Police Radar

A police radar transmits waves with a frequency of 20 GHz. An approaching sports car reflects the radar wave. The radar unit detects a frequency that is 3990 Hz greater than the transmitted frequency. What is the speed of the car?

In Equation 16.25 we set

$$f_2 - f_0 = 3990 \text{ Hz} \qquad f_0 = 2 \times 10^{10} \text{ Hz}$$

This gives for the speed of the car

$$v = \frac{(3 \times 10^8 \text{ m/s})(3990 \text{ Hz})}{2(2 \times 10^{10} \text{ Hz})} = 29.9 \text{ m/s}$$

This converts to 68 miles per hour.

WORKED PROBLEM

A tuning fork vibrating at 512 Hz falls from rest and accelerates at 9.80 m/s². How far below the point of release is the tuning fork when waves of frequency 485 Hz reach the release point? Take the speed of sound to be 330 m/s.

Solution

Solving this problem involves an understanding of the Doppler effect and the motion of a freely falling object. Figure 16.13 shows the tuning fork at three moments: at release, at time t_1 when the waves that reach the release point with a frequency of

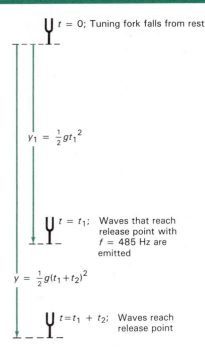

$t = 0$; Tuning fork falls from rest

$y_1 = \frac{1}{2}gt_1{}^2$

$t = t_1$; Waves that reach release point with $f = 485$ Hz are emitted

$y = \frac{1}{2}g(t_1 + t_2)^2$

$t = t_1 + t_2$; Waves reach release point

FIGURE 16.13

485 Hz are *emitted*, and at time $t_1 + t_2$ when these waves *reach* the release point. We can figure out the distance y from $y = \frac{1}{2}g(t_1 + t_2)^2$. Let's see how we can find t_1 and t_2.

The source is moving away from the stationary listener, so Equation 16.20 is appropriate. Solving Equation 16.20 for the speed of the object gives

$$u = v\left(\frac{f_0}{f} - 1\right) = 330 \text{ m/s} \left(\frac{512 \text{ Hz}}{485 \text{ Hz}} - 1\right)$$
$$= 18.4 \text{ m/s}$$

The tuning fork is moving at 18.4 m/s when it emits the waves that are heard at the release point with a frequency of 485 Hz. The time for the tuning fork to reach this speed follows from $u = gt_1$. Thus,

$$t_1 = \frac{u}{g} = \frac{18.4 \text{ m/s}}{9.80 \text{ m/s}^2}$$
$$= 1.87 \text{ s}$$

The tuning fork continues to fall as the Doppler-shifted waves travel back to the listener. Their transit time is given by

$$t_2 = \frac{y_1}{v} = \frac{\frac{1}{2}gt_1{}^2}{330 \text{ m/s}} = \frac{4.90 \text{ m/s}^2 \, (1.87 \text{ s})^2}{330 \text{ m/s}}$$
$$= 0.052 \text{ s}$$

When the sound waves are heard, the tuning fork has been falling for a time $t_1 + t_2$ and has fallen a distance

$$y = \frac{1}{2}g(t_1 + t_2)^2 = 4.90 \text{ m/s}^2 \, (1.87 \text{ s} + 0.05 \text{ s})^2$$
$$= 18.1 \text{ m}$$

EXERCISES

16.1   Wave Characteristics

A. Which of the following are functions of $x - vt$?
(a) $e^{a(x-vt)}$, (b) $(x - vt)^3$, (c) $x^2 - 2xvt + v^2t^2$,
(d) $x^2 - v^2t^2$, (e) $x^4 - v^4t^4$, (f) $\sin(kx) \cdot \cos(kvt)$?

B. Ocean waves with a crest-to-crest distance of 10 m can be described by

$$\psi(x, t) = 0.8 \sin[0.63(x - vt)] \text{ m} \qquad v = 1.2 \text{ m/s}$$

(a) Sketch $\psi(x, t)$ at $t = 0$. This corresponds to a snapshot of the waves at $t = 0$. (b) Sketch $\psi(x, t)$ at $t = 2$ s. Note how the entire wave form has shifted 2.4 m in the positive x-direction.

16.2   Sinusoidal Waves

C. A sinusoidal wave form has a maximum displacement (ψ_{max}) of 1.80 cm at $x = 2.11$ m and a (nearest) minimum displacement of -1.80 cm at $x = 2.62$ m. Determine its amplitude and wavelength.

D. A wave form is described by

$$\psi = 2.0 \sin(kx - \omega t) \text{ cm} \qquad k = 2.11 \text{ rad/m}$$

$$\omega = 3.62 \text{ rad/s}$$

where x is the position along the wave form (in meters) and t is the time (in seconds). Determine the amplitude, wave number, wavelength, angular frequency, and speed of the wave.

E. A sinusoidal wave on a string is described by

$$\psi = 0.51 \sin(kx - \omega t) \text{ cm} \qquad k = 3.1 \text{ rad/cm}$$

$$\omega = 9.3 \text{ rad/s}$$

How far does a wave crest move in 10 s? Does it move in the positive x-direction or in the negative x-direction?

F. Light travels through a vacuum at a speed of 3.00×10^8 m/s. Light with a wavelength of 650 nm ($1 \text{ nm} = 10^{-9}$ m) is red in color. Determine the frequency of red light waves.

G. A sinusoidal wave has a wavelength of 6 m and a speed of 10 m/s. Determine its period, frequency, wave number, and angular frequency.

H. A transverse sinusoidal wave travels to the right along a rope at a speed of 10 m/s. The wavelength is 2 m and the amplitude is 0.1 m. The height of the wave is zero at $x = 0$, $t = 0$. (a) Write an expression for the displacement $\psi(x, t)$. (b) Determine the period of the wave motion.

I. A longitudinal sound wave traveling along a steel rod is described by

$$\psi(x, t) = \psi_0 e^{-(ax-bt)^2} \qquad a = 0.500 \text{ m}^{-1}$$

$$b = 2900 \text{ s}^{-1}$$

Determine the speed of the wave.

J. A calculation shows a phase difference of 63.7 rad. What is the smallest (positive) *equivalent* phase difference?

K. Use the addition theorem,

$$\sin(x + y) = \sin x \cos y + \sin y \cos x$$

to prove that

$$\sin(\varphi + 2\pi N) = \sin \varphi$$

where N is an integer. This shows that any phase angle can be altered by an integral multiple of 2π without changing the value of a sinusoidal wave property.

L. A beam of red light ($\lambda = 650$ nm) moves in the $+x$-direction. Calculate Δx corresponding to a phase difference of π radians.

M. A light wave undergoes a phase change of $\frac{1}{4}\pi$ rad in passing through a thin film of oil. If the wavelength of the light as it moves through the film is 520 nm, how thick is the film?

N. A wave has the form

$$\psi(x, t) = 0.26 \sin(kx - \omega t + 1.6) \text{ cm}$$

$$k = 2.3 \text{ rad/cm} \qquad \omega = 8.1 \text{ rad/s}$$

Determine how the origin on the x-axis can be shifted so that the wave has the form

$$\psi(x', t) = 0.26 \sin(kx' - \omega t) \text{ cm}$$

where x' denotes position measured from the new origin.

P. Two cellos in an orchestra are lined up, one behind the other, with a listener (Figure 1). When they sound a note ($f = 220$ Hz) simultaneously, the sounds arrive at the listener with a phase difference of 180°. Determine the smallest separation of the cellos that will give this phase difference (sound speed = 344 m/s).

Q. A pair of tuning forks are slightly out of tune: One vibrates at 513 Hz, the other at 511 Hz. What is the beat frequency of their superposed waves?

R. Two tuning forks exhibit beats at a beat frequency of 3 Hz. The frequency of one fork is known to be 256 Hz. The frequency of this fork is then lowered slightly by adding a bit of tacky wax to one tine. The two forks then exhibit a beat frequency of 1 Hz. Determine the final frequencies of the two forks.

S. Determine the beat frequency resulting from the superposition of two waves represented by

$$\psi_1 = \sin(kx - \omega_1 t)$$
$$\psi_2 = \sin(kx - \omega_2 t)$$

$$k = 3.1 \text{ rad/m}$$
$$\omega_1 = 4320 \text{ rad/s}$$
$$\omega_2 = 4330 \text{ rad/s}$$

T. Beat frequencies in the range of $2 - 8$ Hz are considered musically pleasant. To take advantage of this,

FIGURE 1

100 km/h

FIGURE 2

an organ designer will sometimes include two pipes constructed to play the same note, and then deliberately detune one of them to produce beats in the pleasant range. If one pipe is to produce an A ($f = 440$ Hz), what frequencies are available for the second pipe if pleasant beats are to result?

16.4 The Doppler Effect

U. A police radar unit emits waves with a frequency of 4.80×10^9 Hz. A motorist moves toward the radar unit at a speed of 100 km/h (Figure 2). What is the difference between the broadcast frequency of 4.80×10^9 Hz and the frequency his radar warning unit detects?

V. A train whistle generates sound waves with a frequency of 486 Hz. (a) What frequency will you hear if the train moves toward you at a speed of 30.1 m/s? Take the speed of sound to be 345 m/s. (b) What frequency will you hear if the train moves away from you at 30.1 m/s?

W. Starting with Equation 16.20, show that $\Delta f \equiv f - f_0$ is given by

$$\Delta f = \frac{-f_0 \left(\dfrac{u}{v} \right)}{1 + \dfrac{u}{v}}$$

If the speed of the source (u) is small compared to the speed of the waves (v), this reduces to $\Delta f = -(u/v)f_0$, showing that the Doppler shift is directly proportional to the speed of the source.

X. A toy train moves counterclockwise around a circular track at a constant speed. The train generates sound waves that reach a child at O (Figure 3). The four frequencies received by the child when the train is at positions 1, 2, 3, and 4 are labeled f_1, f_2, f_3, and f_4. Arrange the frequencies in order of increasing magnitude, with the lowest frequency first.

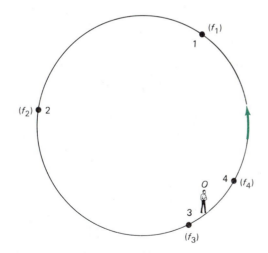

FIGURE 3

Y. A police car approaches an accident scene with the siren blaring. The siren has a frequency of 2.80 kHz. An observer at the accident scene hears a frequency of 3.12 kHz. Determine the speed of the car. Take the speed of sound to be 343 m/s.

PROBLEMS

1. Two traveling waves are represented by

$$\psi_1 = A \sin(kx + \omega t)$$

$$\psi_2 = A \sin(kx + \omega t + \pi/2)$$

Show that the amplitude of the superposed waves is $\sqrt{2}A$.

2. A physics professor, willing to go to any limit to provide memorable demonstrations, goes one step too far in his demonstration of the Doppler effect. With vibrating tuning fork in hand, he invites his students to note its frequency. A music major who possesses perfect pitch correctly identifies the frequency as 300 Hz. The professor then runs away from his class at top speed, urging them "to listen to the Doppler ef . . ." His last word is only partly audible because his last step carries him out a window (Figure 4). Unfortunately, the classroom is located on the 25th floor. The students rush to the window, eager to observe the Doppler effect. (a) Do the students standing at the window hear a frequency greater than or less than 300 Hz? (The professor remains silent so as not to spoil the demonstration.) (b) Does the frequency they hear remain constant throughout the demonstration? Explain. (c) The acceleration of gravity is 9.80 m/s², and the professor demonstrates for 5.5 s. Determine the approximate frequency that will be heard during the final moments of the demonstration. The speed of sound in air is 343 m/s. (d) One student with exceptional hearing is able to hear the waves arriving from the tuning fork and those that travel to the ground and are reflected upward. What is the maximum beat frequency that he hears?

3. Two students carry identical 256-Hz tuning forks. They approach each other at equal speeds (Figure 5). What relative speed must they have to experience a beat frequency of 1 Hz? Take the speed of sound to be 345 m/s.

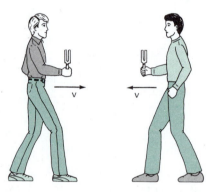

FIGURE 5

4. A train whistle ($f_0 = 400$ Hz) sounds higher or lower in pitch depending on whether it approaches or recedes. (a) Prove that the difference in frequency between the approaching and receding train whistle is

$$\Delta f = \frac{2f_0 \dfrac{u}{v}}{1 - \dfrac{u^2}{v^2}} \qquad \begin{array}{l} u = \text{speed of train} \\ v = \text{speed of sound} \end{array}$$

and, (b) calculate this difference for a 130-km/h train speed. Take the speed of sound to be 330 m/s.

5. The position of a particle moving in the x–y plane is described by the equations

$$x = a \cos(\omega t)$$

$$y = b \sin(\omega t + \varphi)$$

(a) Prove that the path of the particle is an ellipse if $\varphi = 0$. (b) Does the particle rotate clockwise or counterclockwise in the x–y plane? (Take $\varphi = 0$ and $\omega > 0$.)

6. A student demonstrates the Doppler effect by attaching a sound source to the end of a string and then twirling it in a horizontal plane (Figure 6). The frequency of the source is 3 kHz, the string is 1 m long, and the source makes 5 revolutions each second. Determine the maximum and minimum frequencies heard by a listener 20 meters away. Take the speed of sound to be 342 m/s.

FIGURE 4

FIGURE 6

7. Two transverse waves are initiated in a stretched string aligned along the z-axis of a coordinate system. One wave is described by $U = A \sin (kz - \omega t)$, and produces transverse displacements in the x-direction. The other wave is described by $V = A \sin (kz - \omega t)$, and produces transverse displacements in the y-direction.

(a) Make three-dimensional sketches of the individual waves and the superposition of the two waves at $t = 0$. (b) Describe the state of polarization of the superposed waves.

8. Two traveling waves in a stretched string are described by

$$\psi_1 = A_1 \sin (k_1 x - \omega_1 t)$$

$$\psi_2 = A_2 \sin (k_2 x - \omega_2 t)$$

Sketch the two waves and their superposition at $t = 0$ if $A_2 = A_1/2$ and $k_2 = k_1$.

9. A string supports a transverse wave traveling in the positive x-direction. At $t = 0$ the wave function describing the transverse displacement of the string is given by

$$\psi = 10 \, e^{-ax^2} \text{ cm} \qquad a = 1.00 \text{ cm}^{-2}$$

where x is the distance in centimeters along the string. (a) Make an accurate sketch of the wave at $t = 0$. (b) One second later the crest of the wave has moved 10 cm. What is the wave velocity? (c) Deduce the form of the function $\psi(x, t)$ describing this transverse wave.

CHAPTER 17

MECHANICAL WAVES

17.1
WAVES ON A STRING

In this chapter we are concerned primarily with **mechanical waves** such as sound waves and water waves. Mechanical waves involve the displacement of ordinary matter. We can send a mechanical wave along a flexible string by anchoring one end of the string and shaking the other end (Figure 17.1). The movement of the string is perpendicular to the direction of the wave; therefore, the wave is a transverse wave.

The variable that we use to describe this wave is the displacement of the string from its equilibrium shape. We denote this displacement as $\psi(x, t)$. We can derive an equation governing $\psi(x, t)$ by applying Newton's second law to a small segment of the string.

We assume that the tension in the string, T, remains constant even when waves are present. Figure 17.2 shows three external forces acting on the segment located between x and $x + \Delta x$. These forces are the weight of the segment and the forces exerted on each end. We ignore the weight of the segment because the dynamics of the string are controlled by the tension T, not by gravity. As Figure 17.3 indicates, the forces pulling on the ends of the segment have equal magnitudes. The directions of these two forces are different (in general), and so a net vertical force acts on the segment. This net vertical force $T_V(x + \Delta x) - T_V(x)$ results in a vertical acceleration of the segment, as prescribed by Newton's second law, $ma = F_{\text{net}}$.

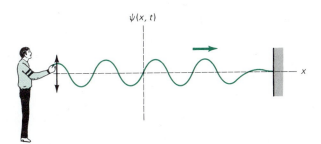

FIGURE 17.1

A series of waves is generated by raising and lowering the left end of the string. $\psi(x, t)$ denotes the displacement of the string from its equilibrium shape (horizontal). The figure depicts $\psi(x, t)$ for different values of x at a fixed moment of time t, just before the wave reaches the wall.

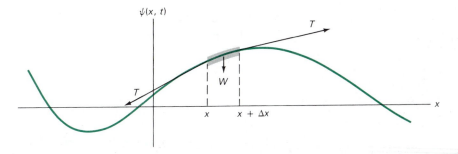

FIGURE 17.2

Three external forces act on the segment located between x and $x + \Delta x$: the weight W and the forces acting on the ends.

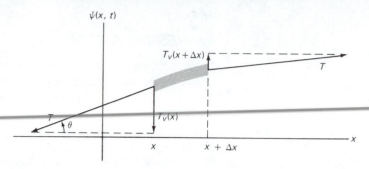

FIGURE 17.3

The forces acting on the ends of the segment have equal magnitudes but different directions. As a result there is a net vertical force on the segment.

We consider a uniform string of mass M and length L. The mass per unit length is M/L, which is denoted by σ:

$$\sigma \equiv \frac{M}{L} \tag{17.1}$$

The mass of the segment of length Δx is $\sigma \cdot \Delta x$. The vertical acceleration of the segment is

$$a = \frac{\partial^2 \psi}{\partial t^2}$$

Newton's second law for the segment is

$$\sigma \, \Delta x \, \frac{\partial^2 \psi}{\partial t^2} = T_V(x + \Delta x) - T_V(x) \tag{17.2}$$

Next, we divide through by Δx and consider the limit as $\Delta x \to 0$. The right side is recognized as the partial derivative of T_V:

$$\lim_{\Delta x \to 0} \frac{T_V(x + \Delta x) - T_V(x)}{\Delta x} = \frac{\partial T_V}{\partial x}$$

Newton's second law is thereby transformed into

$$\sigma \, \frac{\partial^2 \psi}{\partial t^2} = \frac{\partial T_V}{\partial x}$$

The vertical component T_V is related to the tension T by

$$T_V = T \sin \theta$$

For small-amplitude waves θ is small and we can replace $\sin \theta$ by $\tan \theta$, which in turn equals the slope, $\partial \psi / \partial x$.

$$\sin \theta \approx \tan \theta = \frac{\partial \psi}{\partial x}$$

Newton's second law for the rope takes the form

$$\frac{\partial^2 \psi}{\partial t^2} = \frac{T}{\sigma} \frac{\partial^2 \psi}{\partial x^2} \tag{17.3}$$

The quantity $\sqrt{T/\sigma}$ has the dimensions of *speed*. We therefore introduce

$$v = \sqrt{\frac{T}{\sigma}} \qquad (17.4)$$

whereupon Equation 17.3 becomes

$$\frac{\partial^2 \psi}{\partial t^2} = v^2 \frac{\partial^2 \psi}{\partial x^2} \qquad (17.5)$$

Equation 17.5 is a standard form of the wave equation for small-amplitude waves on a string. These waves travel at the speed v given by Equation 17.4.

EXAMPLE 1

Clothesline Waves

A clothesline has a mass of 0.80 kg. The line stretches between two supports that are 12 m apart. The tension in the line is 100 N. What is the time required for a wave to travel from one end of the line to the other and then return? The mass per unit length is

$$\sigma = \frac{M}{L} = \frac{0.80 \text{ kg}}{12 \text{ m}}$$

$$= 0.067 \text{ kg/m}$$

The speed of the waves is given by Equation 17.4,

$$v = \sqrt{\frac{100 \text{ N}}{0.067 \text{ kg/m}}}$$

$$= 38.7 \text{ m/s}$$

The round trip travel time is $2L/v$,

$$t = \frac{24 \text{ m}}{38.7 \text{ m/s}}$$

$$= 0.620 \text{ s}$$

The tension in a string is a measure of the string's elasticity, and the mass per unit length of the string is an inertial factor. Therefore, according to Equation 17.4, the speed of a wave on a string depends on the ratio of an elastic factor (T) and an inertial factor ($\sigma = M/L$). Equation 17.4 is just one example of the general relation

$$v = \sqrt{\frac{\text{elastic factor}}{\text{inertial factor}}} \qquad (17.6)$$

We will see that Equation 17.6 holds for other types of mechanical waves.

17.2
MECHANICAL WAVES: A SAMPLING

In this section we present descriptions and discussions of several types of mechanical waves, both to expose you to unfamiliar types of waves and to reveal some interesting features of familiar waves. The presentations are brief and **many results are stated without proof.**

All of the waves studied in this section can be described by the wave equation, Equation 17.5:

$$\frac{\partial^2 \psi}{\partial t^2} = v^2 \frac{\partial^2 \psi}{\partial x^2}$$

(17.5)

However, the physical significance of $\psi(x, t)$ and the dependence of the wave speed on the properties of the medium differ from one type of wave to the next.

1. Waves on a Liquid Surface (Water Waves)

Water waves are among the more familiar wave forms encountered in our everyday experience. These waves appear to be transverse because the liquid rises and falls as the wave form travels along the surface. But in fact water waves have both transverse and longitudinal components. You will see this if you watch a small floating object such as a cork or bottle. The cork not only bobs up and down but also moves back and forth along the surface. A detailed theoretical analysis reveals that the fluid particles execute *circular* motions. The radius of the circular paths (the wave amplitude) is greatest at the surface. At a depth of a few wavelengths, the wave amplitude has diminished to a negligible value (Figure 17.4).

Gravity and surface tension tend to restore displacements of the water. Accordingly, two elastic factors will enter the equation for wave speed. There are also two characteristic lengths to be considered. These are the wavelength and the depth of the water. When the depth (h) is much smaller than the wavelength (λ), we are dealing with **shallow-water waves**. The speed of shallow-water waves is

$$v = \sqrt{gh} \qquad h \ll \lambda$$

(17.7)

Equation 17.7 can be viewed as an expression of Equation 17.6, in which gravity supplies the elastic factor ($\rho g h$) and in which the mass density (ρ) serves as the inertial factor.

When the depth of the water is much greater than the wavelength, the waves are called **deep-water waves**. The speed of deep-water waves is given by

$$v = \sqrt{\frac{g\lambda}{2\pi} + \frac{2\pi S}{\rho\lambda}} \qquad h \gg \lambda$$

(17.8)

In Equation 17.8, g is the acceleration of gravity, ρ is the density of the liquid, and S is the surface tension, which is a measure of the force required to stretch

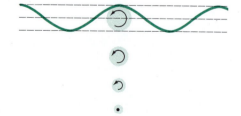

FIGURE 17.4

In water waves, the fluid particles execute circular motions. The radii of the circular motions decrease as the depth increases, becoming negligible at a depth of a few times the wavelength. Because the wave motion is confined to layers near the surface we classify water waves as surface waves.

the surface, thereby increasing its area. This expression shows that waves of very short wavelength (ripples) are controlled by the surface tension, whereas those of long wavelength are controlled by gravity. There is a minimum speed for deep-water waves. We can determine this minimum by setting $dv/d\lambda = 0$, where v is given by Equation 17.8. We find that the minimum speed occurs for the wavelength

$$\lambda = 2\pi \sqrt{\frac{S}{\rho g}} \tag{17.9}$$

We then substitute this expression for λ into Equation 17.8 to obtain the equation for minimum speed:

$$v_{min} = \left(\frac{4gS}{\rho}\right)^{1/4} \tag{17.10}$$

The surface tension of water varies slightly with temperature, ranging from 0.0756 N/m at 0°C to 0.0589 N/m at 100°C. At 15°C (59°F)

$$S = 0.07359 \text{ N/m} \quad \text{and} \quad \rho = 1002 \text{ kg/m}^3$$

Taking $g = 9.80$ m/s^2, we get 0.232 m/s for the minimum speed and 1.72 cm as the wavelength at which this minimum speed occurs.

The existence of this minimum wave speed for deep-water waves has an interesting consequence. When air moves over the surface of water at a speed less than the minimum wave speed, it is unable to transfer energy to the water. Any aspiring wave finds itself fighting a headwind and dies out. Ripples are generated only for wind speeds greater than $(4gS/\rho)^{1/4}$. You can demonstrate this by blowing on the surface of a glass of water. The fact that ripples do not arise for low air speeds can be used to measure the surface tension of a liquid.

2. Acoustic Gravity Waves

The density of our atmosphere decreases with altitude. This variation in density allows the propagation of transverse density variations called **acoustic gravity waves**. The acoustic gravity waves travel parallel to the earth's surface.

Two forces combine to produce acoustic gravity waves. These are the force of gravity and a buoyant force. Consider a parcel of air—that is, some arbitrary chunk of the atmosphere. In equilibrium, a parcel of air experiences a buoyant force equal to its weight. If the parcel is displaced upward, the buoyant force acting on it is reduced because the parcel displaces air that is less dense. Because the parcel's weight is unchanged, the net force is downward, and the air parcel is urged downward toward its equilibrium altitude. If the parcel is displaced downward, the buoyant force increases because the parcel displaces air that is more dense. The buoyant force then exceeds the weight and the net force is an upward restoring force. Thus, a vertical displacement either upward or downward gives rise to a restoring force. Parcels of air can undergo oscillations about equilibrium, just as segments of a string can oscillate vertically about equilibrium. These oscillatory displacements of air parcels propagate horizontally as acoustic gravity waves.

Acoustic gravity waves are of considerable meteorological interest because they appear to be related to clear-air turbulence. Strong winds can cause the waves to become unstable. Their amplitudes grow until they break like surf on a beach. The breaking of an acoustic gravity wave leaves a wake of turbulent eddies. These swirling eddies produce the annoying, sometimes dangerous, bumps that are felt by airplane passengers.

3. Longitudinal and Transverse Elastic Waves in Solids and Fluids

A homogeneous, isotropic solid can support both longitudinal and transverse elastic waves. The longitudinal waves in a solid correspond to the acoustic waves in a fluid. Both are traveling pressure and density variations. Transverse shear waves can exist because a solid can successfully resist shearing stresses.

The speeds of the longitudinal and transverse waves are given by the following equations:

$$v_L = \sqrt{\frac{B}{\rho} + \frac{4}{3}\frac{\mu}{\rho}} \tag{17.11}$$

$$v_T = \sqrt{\frac{\mu}{\rho}} \tag{17.12}$$

In these equations, ρ is the mass density, B is the bulk modulus, and μ is the shear modulus. The longitudinal waves travel at a greater speed than the transverse waves. This can be seen by comparing Equation 17.11 with Equation 17.12, keeping in mind that B and μ are always positive. Typical values of v_L and v_T fall in the range from 1 km/s to 10 km/s. Values of B, μ, and ρ for several solids and liquids are displayed in Table 17.1, along with values of v_T and v_L computed from Equations 17.11 and 17.12.

TABLE 17.1
Mechanical Wave Speeds

Densities are expressed in units of 10^3 kg/m^3. The elastic moduli (Chapter 14), Y, B, and μ are expressed in units of 10^{10} N/m^2. Wave speeds are expressed in units of km/s. All values are approximate.

Material	ρ	Y	B	μ	v_L	v_T	v_{rod}
Aluminum	2.70	6.9	6.6	2.6	6.1	3.1	5.1
Iron (cast)	7.9	15	10	6.0	4.8	2.8	4.4
Copper	9.0	12	10	4.6	4.2	2.3	3.7
Lead	11.4	1.6	3.7	0.56	2.0	0.70	1.2
Steel	7.8	21	19	8.0	6.2	3.2	5.2
Glass	2.5	6.1	3.6	2.5	5.3	3.2	4.9
Brass	8.6	9.0	7.0	3.5	3.7	2.0	3.2
Benzene	0.88	—	0.22	—	1.6	—	—
Ethyl alcohol	0.79	—	0.09	—	1.1	—	—
Water	1.0	—	0.22	—	1.5	—	—
Mercury	13.5	—	0.24	—	0.42	—	—

Solids: $\quad v_L = \sqrt{\frac{B}{\rho} + \frac{4\mu}{3\rho}} \qquad v_T = \sqrt{\frac{\mu}{\rho}} \qquad v_{rod} = \sqrt{\frac{Y}{\rho}}$

Liquids: $\quad v_L = \sqrt{\frac{B}{\rho}}$

EXAMPLE 2

Pipeline Waves

A workman strikes a steel pipeline with a hammer, generating both longitudinal and transverse waves. Reflected waves return 2.4 s apart. How far away is the reflection point?

Let L denote the distance to the reflection point. The round-trip travel times for the two waves are

$$t_L = \frac{2L}{v_L} \qquad t_T = \frac{2L}{v_T}$$

Solving for L in terms of $t_T - t_L$ gives

$$L = \frac{1}{2} \frac{(t_T - t_L)}{\left(\dfrac{1}{v_T} - \dfrac{1}{v_L}\right)}$$

From Table 17.1, $v_T = 3.2$ km/s and $v_L = 6.2$ km/s. The distance to the reflection point is

$$L = \frac{2.4 \text{ s}}{2\left(\dfrac{1}{3.2 \text{ km/s}} - \dfrac{1}{6.2 \text{ km/s}}\right)}$$

$$= 7.9 \text{ km}$$

In an ideal fluid there is no resistance to shearing motions and the shear modulus (μ) is zero. The speed of longitudinal waves in an ideal fluid is given by Equation 17.11, with $\mu = 0$.

$$v_L = \sqrt{\frac{B}{\rho}} \tag{17.13}$$

The speed of transverse waves drops to zero when $\mu = 0$, showing that transverse waves do not propagate in an ideal fluid.

EXAMPLE 3

The Speed of Underwater Sound

We can use Equation 17.13 to calculate the speed of sound in water. The bulk modulus of water at 20°C is 2.18×10^9 N/m². The density of water at 20°C is 998 kg/m³. The speed of sound in water at 20°C is therefore

$$v = \sqrt{\frac{2.18 \times 10^9 \text{ N/m}^2}{998 \text{ kg/m}^3}} = 1.48 \times 10^3 \text{ m/s}$$

$$= 1.48 \text{ km/s}$$

This is nearly five times the speed of sound in air at 20°C.

In more sophisticated treatments of acoustic waves, the ratio B/ρ appearing in Equation 17.13 is replaced by the partial derivative $\partial P/\partial \rho$, where P denotes the fluid pressure. The wave speed as given by Equation 17.13 then becomes

$$v_{\text{long}} = \sqrt{\frac{\partial P}{\partial \rho}} \tag{17.14}$$

EXAMPLE 4

The Speed of Sound in Air

For ordinary sound waves in a gas, the variations in pressure and density are related by

$$P = C\rho^{\gamma} \qquad C, \gamma = \text{constants}$$

We can use this relation and Equation 17.14 to determine the speed of sound in air. The constant γ has the value 1.40 for air. We find

$$\frac{\partial P}{\partial \rho} = \gamma C\rho^{\gamma - 1} = \frac{\gamma P}{\rho}$$

Thus, from Equation 17.14 the speed of sound in air is

$$v_{\text{sound}} = \sqrt{\frac{\gamma P}{\rho}}$$

Using values corresponding to standard conditions (0° C, 1 atm)

$$P = 1 \text{ atm} = 1.01 \times 10^5 \text{ N/m}^2 \qquad \rho = 1.29 \text{ kg/m}^3 \qquad \gamma = 1.40$$

we find

$$v_{\text{sound}} = 331 \text{ m/s}$$

This is approximately 740 miles per hour.

Ultrasonic waves are used widely in the study of solids and liquids. Measurements of the wave speeds (v_L, v_T) yield values of the elastic constants (B, μ) of solids and liquids. Ultrasonic waves have many practical applications. Reflections of ultrasound by cracks and impurities make such defects "visible," much like reflected light reveals a crack in a window pane. Medical applications of ultrasonic waves include the inspection of a fetus within the womb and the detection of tumors. The mechanical properties of metal and concrete structures can be improved by passing ultrasonic waves through the material before it solidifies. The waves pulverize tiny impurities, remove gas bubbles, and homogenize the material.

The solid portions of the earth support both longitudinal and transverse elastic waves. Much of our knowledge about the interior of the earth has been inferred from measurements on seismic waves generated by earthquakes or man-made explosions. An earthquake generates both longitudinal and transverse waves. Seismologists refer to the faster-moving longitudinal disturbances as primary, or "P" waves. The slower, transverse waves are called secondary, or "S" waves. Because of the difference in their speeds, the P waves reach a seismic station before the S waves. The time delay between their arrival can be used to determine the distance from the station to the epicenter.*

The speed of elastic waves may be affected by the geometry of the solid. For example, Equation 17.11 gives the speed of longitudinal waves in a medium whose dimensions are large compared to the wavelength. For elastic waves traveling along a long thin rod the wavelength may be much larger than the

---

* The epicenter is the point on the surface of the earth directly above the origin of the earthquake.

diameter of the rod. When the wavelength is large compared to the diameter of the rod, longitudinal waves can travel along the rod at a speed given by

$$v_{\text{rod}} = \sqrt{\frac{Y}{\rho}} \qquad (17.15)$$

where Y is Young's modulus.

4. Elastic Surface Waves (Rayleigh Waves)

In addition to the longitudinal and transverse waves that can propagate through the body of an elastic solid, various types of surface waves can also be transmitted across the surface of a solid. Among these are what we call **Rayleigh waves**.

Like surface waves on a liquid, the amplitude of a Rayleigh wave diminishes with depth below the surface, and is ignorably small at depths of more than a few wavelengths. The disturbance is therefore confined to the surface layers of a solid. The surface layers experience both transverse and longitudinal displacements. Rayleigh waves are well-suited for use in a variety of transducers* that process radar, television, and radio signals. Of primary importance in considering Rayleigh waves is the fact that, for a given frequency, the wavelength of an electromagnetic wave is about 10^5 times greater than the wavelength of a Rayleigh wave. This follows from the basic kinematic relation

$$f\lambda = v$$

Electromagnetic waves travel at the speed of light, $c = 3 \times 10^8$ m/s. Rayleigh waves travel approximately 100,000 times more slowly. This means that the wavelength of a Rayleigh wave is approximately 100,000 times shorter than an electromagnetic wave of the same frequency.

If we want to transmit a wave without causing severe attenuation, the length of the transmitting device must be comparable to or greater than the wavelength. Because Rayleigh waves of a given frequency are 10^5 times shorter than electromagnetic waves of the same frequency, devices that transmit Rayleigh waves can be made much smaller than their electromagnetic counterparts. This reduction in size reduces the weight, which is an important consideration for applications in aircraft and space vehicles. Another desirable property of Rayleigh waves is that they are **nondispersive**—that is, their speed is independent of frequency. Therefore, a Rayleigh-wave device can process a multifrequency signal without any change in the wave form (zero distortion).

17.3
ENERGY FLOW AND WAVE INTENSITY

An important feature of wave motion is the transfer of energy. The **intensity** of a wave measures the rate at which it transfers energy.

Suppose that a steady progression of waves emanates from a source. The wave energy E falling on a detector of collecting area A perpendicular to the direction of energy flow in a time t is proportional to both A and t. If we

---

* A transducer is a device that converts electrical energy into mechanical energy and/or vice versa.

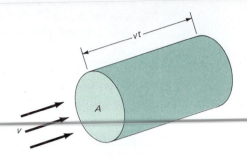

FIGURE 17.5

The waves that carry energy E across area A travel a distance vt in time t and thus fill a volume Avt.

divide E by t and A, we get a quantity that is independent of A and t, an intrinsic measure of energy flow in the wave. The rate of energy flow per unit area, $(E/t)/A$, is called the intensity of the wave and is represented by the symbol I:

$$I \equiv \frac{E/t}{A} \qquad (17.16)$$

The SI unit of intensity is the watt per square meter. The intensity of sunlight reaching the upper levels of the earth's atmosphere is 1.38×10^3 W/m². This intensity is called the **solar constant** (see also Section 20.3). The minimum intensity of sound waves (at a frequency of 1000 Hz) to which the human ear is sensitive is approximately 10^{-12} W/m². This is called the **threshold of hearing**.

We obtain a very useful expression for I by multiplying the numerator and denominator of the right side of Equation 17.16 by the wave velocity, v. The result is

$$I = \left(\frac{E}{Avt}\right)v$$

As Figure 17.5 indicates, the quantity Avt is the *volume* that contains the wave energy E. The quantity E/Avt, then, is the wave energy per unit volume. We call this quantity the **energy density** of the wave, u:

$$u = \text{wave energy per unit volume} = \frac{E}{Avt} \qquad (17.17)$$

The intensity can now be written as

$$I = uv \qquad (17.18)$$

We can use this relation to estimate the energy density in a wave.

EXAMPLE 5

Laser Beam Energy Density

A high-power laser develops a power of 1.2 TW over a cross-sectional area of 10^{-12} m². We want to determine the energy density of the beam. The intensity is

$$I = \frac{E/t}{A} = \frac{\text{power}}{\text{area}} = \frac{1.2 \times 10^{12} \text{ W}}{10^{-12} \text{ m}^2}$$

$$= 1.2 \times 10^{24} \text{ W/m}^2$$

The laser energy travels at the speed of light. The energy density follows from Equation 17.18,

$$u = \frac{I}{c} = \frac{1.2 \times 10^{24} \text{ W/m}^2}{3 \times 10^8 \text{ m/s}} = 4 \times 10^{15} \text{ J/m}^3$$

We can use Equation 17.18 to relate the intensity of a sinusoidal wave to the wave amplitude and frequency. Consider a sinusoidal wave traveling through an elastic medium, such as a longitudinal sound wave traveling along a metal rod. The atoms that make up the rod execute simple harmonic motion about their equilibrium positions. Their total energies are a sum of kinetic and potential energies. The total energy of an individual particle undergoing simple harmonic motion is a constant, equal to the maximum kinetic energy. For sinusoidal waves of the form,

$$\psi = D \cos (kx - \omega t)$$

the particle velocity is

$$\frac{\partial \psi}{\partial t} = \omega D \sin (kx - \omega t)$$

The maximum speed of the particle is ωD. If m is the mass of some segment of the rod undergoing this motion, the maximum kinetic energy is $\frac{1}{2}m(\omega D)^2$. Dividing by the volume of the segment gives the energy density, u,

$$u = \frac{1}{2}\rho\omega^2 D^2 \qquad (17.19)$$

where ρ is the mass density of the rod. Using this expression in Equation 17.18 gives the intensity for a sinusoidal wave in an elastic medium:

$$I = \frac{1}{2}\rho\omega^2 D^2 v \qquad (17.20)$$

Note that the intensity is proportional to D^2, the square of the wave amplitude, and proportional to the square of the wave frequency ($\omega^2 = (2\pi f)^2$).

Wave Fronts

When you watch water waves travel across a pond your eye tends to follow a wave crest. The visual impression of a wave crest leads to the idea of a **wave front**. A wave front is defined as a surface over which the phase of the wave is constant. In a particular wave front, at a given moment of time, all particles of the medium are undergoing the same motion.

Plane wave fronts and spherical wave fronts are particularly important. If a wave travels in one fixed direction, we call it a **plane wave**. For example, a sound wave in which the pressure differences $P(x, t)$ vary sinusoidally,

$$P(x, t) = P_0 \sin (kx - \omega t)$$

is a plane wave traveling in the x-direction. The wave fronts of plane waves are planes that are perpendicular to the direction of propagation (Figure 17.6a).

Spherical wave fronts can be produced by a point source—a source whose dimensions are small compared with the distance to an observer. If waves travel outward in all directions from a point source, the phase of one of these waves will be the same at points equidistant from the source, that is, on the surface of a sphere centered on the source (see Figure 17.6b). Therefore, the wave fronts are spherical surfaces and the waves are called **spherical waves**.

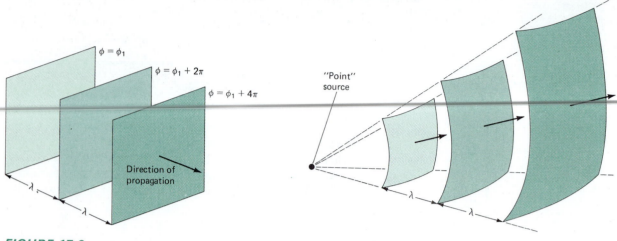

$\phi = \phi_1$

$\phi = \phi_1 + 2\pi$

$\phi = \phi_1 + 4\pi$

"Point" source

Direction of propagation

λ

FIGURE 17.6

(a) Plane wave fronts. (b) Spherical wave fronts.

If a wave front diverges, the energy carried by the wave spreads over a larger and larger area, and consequently the intensity decreases. The intensity of a plane wave is constant because plane wave fronts neither converge nor diverge.* The intensity of a spherical wave, on the other hand, is inversely proportional to the square of the distance from the source. This is so because the outgoing waves must spread over larger and larger surface areas, while obeying the energy conservation law. To understand this inverse square law variation of intensity, consider spherical waves leaving the point source of Figure 17.7. Next, imagine a series of spherical surfaces concentric with the source. These are conceptual *geometric* surfaces, not real *physical* surfaces. The rate of energy flow across each such spherical surface must be the same because no energy is absorbed or released between surfaces. Thus, the rate of energy flow, E/t, is the same for each spherical surface. The surface area of a sphere of radius r is $4\pi r^2$. The rate of energy flow is proportional to the product of the intensity and the surface area.

$$I(4\pi r^2) = \frac{E}{t} = \text{constant}$$

Because E/t is constant, it follows that the intensity of a spherical wave varies inversely as the square of the distance from the source

$$I \propto \frac{1}{r^2} \tag{17.21}$$

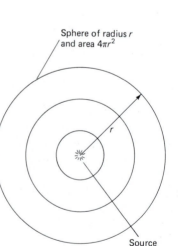

Sphere of radius r and area $4\pi r^2$

r

Source

FIGURE 17.7

The same energy passes through all spherical surfaces surrounding the source. The areas of the spherical surfaces increase as the square of their radii. This causes the energy per unit area to be inversely proportional to the square of the radius.

In *surface* waves, particularly seismic waves, the intensity is proportional to $1/r$, making such waves destructive over much greater distances than if their intensity followed Equation 17.21.

---

* This assertion assumes that the medium through which the wave travels does not absorb energy from the wave and does not feed energy into the wave.

Sound Wave Intensity and Intensity Level

The intensity of ordinary sound waves in air and the auditory sensation of loudness are not linearly related. Doubling the intensity does not double the loudness of the sound. Instead, there is a roughly *logarithmic* relation between intensity and the sensation of loudness. For frequencies near 1000 Hz, in the frequency range where human hearing is best, the loudness seems twice as great if the intensity is increased by a factor of about 10. This qualitative notion of loudness is measured by a quantity called the **intensity level**, which *by definition*, exhibits a logarithmic relation to the intensity. Specifically, the intensity levels of two waves of intensity I and I_0 are defined to differ by β decibels (dB), where

$$\beta = 10 \log_{10} \frac{I}{I_0} \text{ dB} \qquad (17.22)$$

Ten decibels equal 1 bel, a unit that is not widely used. Thus, if I is twice I_0, then

$$\log_{10} \frac{I}{I_0} = \log_{10} 2 = 0.301$$

and their intensity levels differ by 3.01 dB. In the case of ordinary sound waves, I_0 is taken to be 10^{-12} W/m². This serves as a standard of intensity that is near the threshold of hearing for most people (at 1000 Hz). The intensity of ordinary speech is on the order of 10^{-6} W/m², and gives intensity levels of about 60 dB. The *threshold of pain* for sound is set at an intensity level of 120 dB, which corresponds to an intensity of 1 W/m².

The intensity of sinusoidal waves is proportional to the square of the wave amplitude. For sound waves in air, the wave amplitude is the maximum pressure, measured relative to the undisturbed atmospheric pressure. Thus,

$$\frac{I}{I_0} = \frac{P^2}{P_0{}^2}$$

where P and P_0 are the pressure amplitudes. An alternative definition of the intensity level for sound waves is therefore

$$\beta = 20 \log_{10} \left(\frac{P}{P_0} \right) \text{ dB} \qquad (17.23)$$

The standard for sound wave pressure is $P_0 = 2 \times 10^{-5}$ Pa, the threshold for hearing at 1000 Hz. Compared to the typical atmospheric pressure (1 atm = 1.013×10^5 Pa), P_0 is less than one billionth of an atmosphere. Ordinary sound waves are indeed small-amplitude waves.

EXAMPLE 6

Sound Intensities at Rock Concerts

The intensity level of the sound generated by some rock musicians frequently reaches 120 dB. The intensity level of ordinary speech is typically 60 dB. The ratio of the actual intensities of rock music and ordinary speech follows from Equation 17.22:

$$\beta_{\text{rock}} - \beta_{\text{speech}} = 10 \log_{10} \frac{I_R}{I_0} - 10 \log_{10} \frac{I_S}{I_0}$$

Recalling that $\log a - \log b = \log (a/b)$ gives

$$120 - 60 = 10 \log_{10} \frac{I_R}{I_S}$$

and thus

$$\log_{10} \frac{I_R}{I_S} = 6$$

Therefore

$$\frac{I_R}{I_S} = 10^6$$

The intensity of rock music can be a million times greater than that of ordinary speech. Since medical studies show that prolonged exposure to high-intensity sound causes hearing loss, the conclusion should be obvious. Don't sit near the speakers at a rock concert!

17.4
STANDING WAVES AND BOUNDARY CONDITIONS

In Chapter 16 we introduced the basic kinematic relation for sinusoidal waves,

$$f\lambda = v \qquad (17.24)$$

As we have seen, the speed v is determined by the nature of the wave and the medium through which the wave moves. Depending on the physical constraints imposed on the medium, either f or λ may be regarded as the independent variable. Up to now, we have envisioned situations where a source generates a steady succession of waves. The frequency of the source can be set at any value, within limits. We have therefore made frequency the independent variable. The wavelength becomes the dependent variable, and its value is dictated by Equation 17.24. Let's turn to a different situation, where physical constraints on the medium fix the value of λ. The frequency, as given by Equation 17.24, then becomes the dependent variable.

The physical constraints on a medium are called **boundary conditions**. To illustrate the idea of boundary conditions, imagine a uniform string stretched between two rigid supports a distance L apart (Figure 17.8). No matter how the string moves, its displacement must be zero at both ends, where the string is anchored. We can describe these boundary conditions mathematically as follows: Label the positions of the ends of the string $x = 0$ and $x = L$, and let $\psi(x, t)$ denote the transverse displacement of the string. The boundary conditions are then

$$\psi(0, t) = 0$$
$$\psi(L, t) = 0 \qquad (17.25)$$

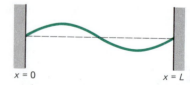

$x = 0$ $x = L$

FIGURE 17.8

A standing wave on a string. The only physically acceptable solutions of the wave equation for the string are those for which the displacement is zero at both ends.

These boundary conditions can be satisfied by a superposition of two sinusoidal waves with equal amplitudes, traveling in opposite directions. Such a superposition results in **standing waves**, like those shown in Figure 17.9. At any given position x the string undergoes simple harmonic motion in the vertical direction, just as it does in a traveling wave. However, in a standing wave the

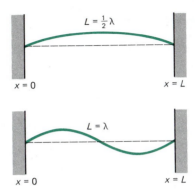

$L = \frac{1}{2}\lambda$

$x = 0$ $x = L$

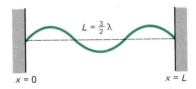

$L = \lambda$

$x = 0$ $x = L$

$L = \frac{3}{2}\lambda$

$x = 0$ $x = L$

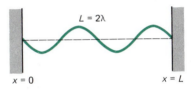

$L = 2\lambda$

$x = 0$ $x = L$

FIGURE 17.9

Standing wave forms on a string.

amplitude varies with position along the string. The wave form gives no impression of advancing crests—hence the designation *standing* wave.

Positions of zero displacement in a standing wave are called **nodes**. Only sinusoidal wave forms that have nodes at $x = 0$ and $x = L$ *at all times* can satisfy the boundary conditions. If we examine a sinusoidal wave form, we discover that the distance between adjacent nodes is one-half wavelength. Therefore, acceptable sinusoidal wave forms are those for which an *integral* number of half wavelengths fit into the distance L,

$$L = \tfrac{1}{2}\lambda, \, \lambda, \, \tfrac{3}{2}\lambda, \, 2\lambda, \ldots \tag{17.26}$$

or, in a more succinct notation,

$$L = n \cdot \tfrac{1}{2}\lambda \qquad n = \text{positive integer} \tag{17.27}$$

Figure 17.9 shows the wave forms at one instant of time for $n = 1, 2, 3, 4$.

The condition imposed by Equation 17.27 can be derived more formally. It is worthwhile to learn how this formal derivation is carried out for standing waves on a string because the techniques carry over to many types of boundary value problems. First, we note that no *single* traveling wave can satisfy the boundary conditions. For example, the traveling wave $\psi = A \sin(kx - \omega t)$ would satisfy the conditions $\psi(0, t) = 0$ at time $t = 0$ but not an instant earlier or an instant later. However, a standing wave form that does satisfy the boundary conditions can be built up by a superposition of two waves traveling in opposite directions. The superposition of the two waves

$$\psi_R(x, t) = A \sin(kx - \omega t)$$

$$\psi_L(x, t) = A \sin(kx + \omega t)$$

gives a standing wave of the form

$$\psi(x, t) = \psi_R + \psi_L = 2A \sin kx \cdot \cos \omega t \tag{17.28}$$

Equation 17.28 is the mathematical description of a standing wave. It describes a simple harmonic motion ($\cos \omega t$) with an amplitude ($2A \cdot \sin kx$) that varies with position along the string. This wave form clearly satisfies the boundary condition at $x = 0$ because $\sin 0 = 0$. The boundary condition at $x = L$,

$$\psi(L, t) = 0 = 2A \sin kL \cdot \cos \omega t$$

is also satisfied for all times t, provided that

$$\sin kL = 0$$

The condition $\sin kL = 0$ is satisfied for special values of kL, namely

$$kL = n\pi \qquad n = 1, 2, 3, \ldots \tag{17.29}$$

The substitution $k = 2\pi/\lambda$ converts Equation 17.29 into Equation 17.27, the standing wave condition deduced earlier.

Equations 17.27 and 17.29 illustrate our earlier statement that "physical constraints on the medium fix the value of λ." Only certain values of λ satisfy the boundary conditions and produce a standing wave form.

We can find the frequencies of the standing waves by using Equation 17.27 and the kinematic relation, $f\lambda = v$. Thus,

$$f = \frac{v}{\lambda} = n\left(\frac{v}{2L}\right) \qquad n = 1, 2, 3, \ldots \tag{17.30}$$

For waves on a uniform string the speed v is given by Equation 17.4,

$$v = \sqrt{\frac{T}{\sigma}}$$

Substituting this expression into Equation 17.30 we get

$$f = \frac{n\sqrt{\frac{T}{\sigma}}}{2L} \qquad n = 1, 2, 3, \ldots \qquad (17.31)$$

for the standing-wave frequencies. Equation 17.31 shows that the *spectrum* of standing-wave frequencies consists of integral multiples of the **fundamental frequency** ($n = 1$),

$$f_{\text{fund}} = \frac{\sqrt{\frac{T}{\sigma}}}{2L} \qquad (17.32)$$

The guitar provides a familiar example of this relation. Increasing the tension, T, in a guitar string raises the string's fundamental frequency. We hear this increase in frequency as a higher note when we pluck the string. Decreasing the length L through fingering accomplishes the same result.

EXAMPLE 7

Standing Waves on a String

A uniform string of mass 0.1 kg is stretched between two supports separated by a distance of 3 m. The tension in the string is 2000 N. What are the speed, the fundamental frequency, and the corresponding wavelength of the waves along the string? The fundamental frequency corresponds to the longest wavelength, which is twice the length of the string, so $\lambda = 6$ m.

The wave speed is given by Equation 17.4:

$$v = \sqrt{\frac{T}{\sigma}} = \sqrt{\frac{2000 \text{ N}}{0.1 \text{ kg/3 m}}}$$
$$= 245 \text{ m/s}$$

The fundamental frequency follows from Equation 17.32,

$$f_{\text{fund}} = \frac{v}{2L} = \frac{245 \text{ m/s}}{6 \text{ m}}$$
$$= 40.8 \text{ Hz}$$

This frequency is near the lower end of the audible range of frequencies.

Our treatment of standing waves on a string illustrates a general consequence of boundary conditions.

Boundary conditions identify the physically acceptable solutions contained within the larger group of mathematically acceptable solutions of the wave equation.

To see this more clearly, consider the wave equation

$$\frac{\partial^2 \psi}{\partial t^2} = v^2 \frac{\partial^2 \psi}{\partial x^2}$$

One set of mathematically acceptable solutions to this equation has the form

$$\psi(x, t) = A \sin(kx - \omega t) + B \sin(kx + \omega t)$$

The only constraint imposed by the wave equation is that $\omega/k = v$. In particular, A and B can have any values. However, for standing waves on a string, only solutions with $A = B$ are admissible, and the values of k are restricted by the condition $kL = n\pi$.

Fourier Synthesis of Wave Forms

These restrictions do not limit the physically acceptable solutions to sinusoidal wave forms. For example, a triangular wave form can be set up by plucking the string at its midpoint. Such a triangular wave form can be synthesized by a *superposition* of sinusoidal wave forms. In Figure 17.10 the dashed line shows the single sinusoidal wave form

$$\psi_1 = \left(\frac{8}{\pi^2}\right) \sin\frac{\pi x}{L}$$

The dotted line shows the sum (superposition) of three sinusoidal wave forms

$$\psi_3 = \frac{8}{\pi^2}\left[\sin\frac{\pi x}{L} - \frac{1}{9}\sin\frac{3\pi x}{L} + \frac{1}{25}\sin\frac{5\pi x}{L}\right]$$

Even better fits to the triangular wave form can be achieved by adding properly chosen sinusoidal wave forms to ψ_3. Figure 17.10 shows how the superposition of two sinusoidal wave forms (one dashed and one dotted) approximates the triangular wave form. The technique of building up wave forms by superposition of sinusoidal wave forms is called **Fourier synthesis**. The sum of sinusoidal wave forms that describes the synthesis is called a **Fourier series**.

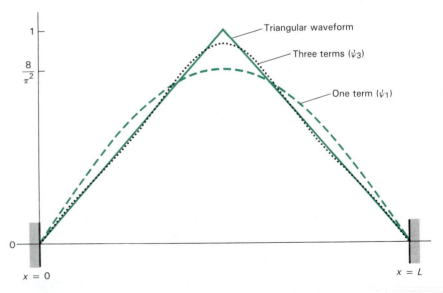

FIGURE 17.10

The triangular wave form can be synthesized by superposing standing waves. The dashed line indicates the approximation achieved with a standing wave of the fundamental frequency. The dotted line shows the approximation achieved by superposing three standing waves having appropriate amplitudes.

Fourier synthesis also can be used to synthesize various types of wave forms in electrical circuits.

The effect of boundary conditions—which is to sort out physically acceptable solutions—is not limited to waves on a string, nor to waves in material media. Electromagnetic waves and the probability waves of quantum mechanics are frequently subjected to boundary conditions. Thus, the ideas and techniques developed in connection with standing waves on a string can be extended to many other types of wave phenomena.

You don't need to be a mathematical wizard to gain insight into standing waves. In fact, it is possible to *estimate* the fundamental frequency for any oscillatory system without solving the wave equation.* The fundamental frequency is the *lowest* frequency for a standing wave form. From $f = v/\lambda$ it follows that the fundamental frequency corresponds to the *maximum* wavelength. Thus,

$$f_{\text{fund}} = \frac{v}{\lambda_{\text{max}}} \tag{17.33}$$

where λ_{max} is the maximum standing wavelength and v is the speed of the wave. The precise value of λ_{max} depends on geometric factors and the boundary conditions. However, we can estimate λ_{max} quite simply, and thereby obtain an approximate value for f_{fund}. The estimate for λ_{max} is

$$\lambda_{\text{max}} \approx \text{largest dimension of system} \equiv \mathcal{L} \tag{17.34}$$

This estimate is readily justified because it is the only possible choice. No other length could enter the problem. With $\lambda_{\text{max}} = \mathcal{L}$ we have

$$f_{\text{fund}} \approx \frac{v}{\mathcal{L}} \tag{17.35}$$

This estimate for the fundamental frequency is generally within a factor of 2 or 3 of the precise value obtained by solving the wave equation. Thus, Equation 17.35 gives an *order of magnitude estimate* for the fundamental frequency of standing waves. You can use this equation to double-check your formal solution of a given wave equation. If your solution indicates a fundamental frequency that differs by a large factor from your order of magnitude calculation, go back and check your mathematics.

EXAMPLE 8

Seismic Wave Frequencies

Earthquakes can excite standing waves that involve the entire earth. The approximate fundamental frequency of such waves is given by Equation 17.35,

$$f_{\text{fund}} = \frac{v_{\text{seismic}}}{D}$$

where v_{seismic} is the speed of the seismic wave and D is the diameter of the earth. The seismic wave speed varies with depth and direction. However, an estimate of the seismic wave speed, which is typical of the speed of sound in solids, is

$$v_{\text{seismic}} \approx 11 \text{ km/s}$$

---

* P. A. M. Dirac, a Nobel Laureate in Physics, said, "I understand what an equation means if I can figure out the characteristics of its solution without actually solving it."

Taking $D \approx 12,000$ km gives

$$f_{\text{fund}} \approx 10^{-3} \text{ Hz}$$

The reciprocal of f_{fund}, the *period* of the fundamental mode of vibration, is about 1000 s, or 16 min. The motion to which this period refers is called the *breathing mode* because the radius of the earth expands and contracts slightly, as if the planet were breathing. How good is our estimate? A rigorous calculation sets the period at 1228.8 s. Experimental data fix the period at 1227.6 s. Because our value, 1000 s, is within a factor of 2 of these figures, it is a satisfactory order of magnitude estimate.

WORKED PROBLEM

A string of length L is held horizontal between two walls under a tension T. The mass per unit length varies along the string according to

$$\sigma = \frac{\sigma_0 x}{L}$$

where x denotes the distance measured from one end of the string, and σ_0 is a constant. Determine the time for a transverse wave to travel the length of the string. Take $L = 2.00$ m, $T = 8.62$ N, and $\sigma_0 = 2.57 \times 10^{-2}$ kg/m.

Solution

Recall that the speed of a transverse wave of a string depends on the mass per unit length according to $v = \sqrt{T/\sigma}$. Because σ varies along the string, the speed v also changes. To determine the transit time, we cannot simply divide its length by the speed, because the speed changes. Instead, we must integrate the defining relation for speed, taking account of the variation in speed.

We start with the definition,

$$\frac{dx}{dt} = v$$

and set

$$v = \sqrt{\frac{T}{\sigma}} = \sqrt{\frac{TL}{\sigma_0 x}}$$

Separating the variables gives

$$x^{1/2} \, dx = \sqrt{\frac{TL}{\sigma_0}} \, dt$$

Calling τ the time for the wave to travel the length of the string we integrate:

$$\int_0^L x^{1/2} \, dx = \sqrt{\frac{TL}{\sigma_0}} \int_0^\tau dt$$

This gives

$$\frac{2L^{3/2}}{3} = \sqrt{\frac{TL}{\sigma_0}} \, \tau$$

Solving for the transit time gives

$$\tau = \frac{2L}{3} \sqrt{\frac{\sigma_0}{T}}$$

Evaluating τ for the given data we obtain,

$$\tau = \frac{2(2.00 \text{ m})}{3} \sqrt{\frac{2.57 \times 10^{-2} \text{ kg/m}}{8.62 \text{ N}}}$$

$$= 0.0728 \text{ s}$$

Had the mass per unit length been constant (σ_0) along the string, the transit time would have been longer by a factor of 3/2.

EXERCISES

17.1   Waves on a String

A. (a) The tension in a cord is 60 N. The mass per unit length is 0.01 kg/m. What is the wave speed along the rope? (b) If the rope is 8 m long, what time is required for a wave to travel from end to end?

B. A violin string has a length of 0.76 m and a mass of 6 grams. The tension in the string is 104 N. Compare the speed of waves along the string with the speed of sound in air (330 m/s).

C. A bucket of water weighing 80 N hangs at the end of a rope. The rope is 6 m long and has a mass of 0.60 kg. (a) Determine the tension in the rope at both ends. (b) Determine the wave speed at both ends.

D. A rule of thumb for estimating the distance between yourself and a lightning stroke is this: count the seconds that elapse between the flash and the arrival of the thunder, and divide this number by three. The result is the distance in kilometers. For example, a 6-s delay indicates a distance of 2 km. Explain the basis for this rule.

17.2   Mechanical Waves: A Sampling

E. Transverse elastic waves travel through a solid at a speed of 2 km/s. Longitudinal waves travel at 4 km/s. The shear modulus of the solid is 9×10^{10} N/m². Determine the bulk modulus.

F. The speed of sound in air equals $\sqrt{1.4P/\rho}$, an expression that is independent of frequency. As a consequence, sound waves should not exhibit dispersion. Waves of all frequencies should travel at the same speed. What elementary observations on sound are consistent with the absence of dispersion?

G. Do you think that you can hear sound whose wavelength is larger than the size of your ear? Calculate the wavelengths corresponding to the auditory limits of 20 Hz and 20 kHz.

H. The bulk modulus for mercury at 40°C is 2.40×10^9 N/m². The density of mercury at 40°C is 13,450 kg/m³. Calculate the speed of sound in mercury at 40°C.

I. Determine the speed of deep-water waves having a wavelength of 20 m. Take $g = 9.80$ m/s². Does your answer differ significantly if you ignore the surface tension term in Equation 17.8? (Surface tension $S = 0.0736$ N/m; seawater density $\rho = 1002$ kg/m³.)

J. Imagine that you are standing in water with a depth of 1.6 m near the shore. Ocean waves are coming in toward the beach. A surfer riding a wave collides with you. What is the speed of impact in meters per second and in miles per hour?

K. P waves and S waves in the outer core of the earth move at speeds of 8.15 km/s and 1.40 km/s. The density is 9400 kg/m³. Determine the bulk modulus and shear modulus for the outer core.

L. One end of a long steel beam is struck with a hammer. The blow generates longitudinal and transverse waves that travel the length of the beam, are reflected, and return. The round trip times for the waves are 3.2 ms and 5.8 ms. Determine the ratio B/μ for steel. Compare your answer with the value of B/μ computed from the data in Table 17.1.

M. A Rayleigh wave with a frequency of 100 MHz travels at a speed of 5.2 km/s. Determine its wavelength.

17.3   Energy Flow and Wave Intensity

N. A laser beam transfers an energy of 6 J across an area of 2×10^{-6} m² in a time of 1 ns. (a) What is the beam intensity? (b) Determine the energy density of the beam, assuming that the wave travels at the speed of light, 3×10^8 m/s.

P. The intensity of a sound wave is raised, first from 10^{-12} W/m² to 6×10^{-12} W/m², and then to 10^{-11} W/m². Determine the sequential changes in intensity level (in decibels) for the two changes.

Q. Near a jet engine, the intensity level can reach 140 dB. (a) What is the corresponding sound wave intensity? Determine the corresponding sound pressure in (b) N/m², (c) atmospheres.

R. An SST produces sonic booms in which the pressure changes by about 100 N/m². Use the relation $\beta = 20 \log_{10}(P/P_0)$ to estimate the boom intensity. (The sonic boom is a shock wave, and quite different from the sinusoidal wave forms for which $\beta = 20 \log_{10}(P/P_0)$ strictly applies.)

S. A laser used in thermonuclear fusion research produces a beam power of 27 PW. The beam diameter is 300

μm. (a) Determine the beam intensity. (b) If the laser pulse lasts for 120 ps determine the energy delivered in one pulse.

T. A radio antenna radiates energy with a power of 50 kW. Determine the intensity of the wave at a distance of 100 km, assuming that the energy spreads out in all directions, with (a) the half that strikes the ground being completely absorbed; (b) the half that strikes the ground being completely reflected, thereby enhancing the energy radiated directly outward.

U. A light meter records an intensity of 16 W/m^2 at an unknown distance from a point source. Increasing the distance by 1 meter reduces the intensity to 9 W/m^2. What is the distance from the source to the point where the intensity is 9 W/m^2?

17.4 Standing Waves and Boundary Conditions

V. The head of an Indian tom-tom drum has a diameter of 30 cm. It is made of leather stretched so that the surface tension is $S = 1.8 \times 10^4$ N/m. The inertial property of the leather is measured by its mass per unit area, $\beta = 0.32$ kg/m^2. (a) Show that the quantity $\sqrt{S/\beta}$ has the dimensions of speed. (b) Assume that $\sqrt{S/\beta}$ is the characteristic speed of waves on the drum head. Estimate the fundamental frequency of the standing waves that the tom-tom can support. Is your calculated frequency in the audible range?

W. Estimate the fundamental frequency for longitudinal sound waves (standing waves) in your skull. Assume a wave speed of 4 km/s.

X. Make an order-of-magnitude estimate of the fundamental frequency for standing waves on a lake 6 km long and 12 m deep. Assume shallow-water waves.

Y. Estimate the fundamental frequency of sound waves in an aluminum rod 80 cm long. Is this frequency in the audible range?

PROBLEMS

1. A uniform rope of weight Mg and length L hangs vertically. The tension at the free end is zero. The tension increases linearly with distance above the free end, reaching the value Mg at the point of support. (a) Show that the wave speed a distance x from the free end is given by

$$v = \sqrt{gx}$$

(b) Show that the time for a wave to travel the length of the rope is

$$t = 2\sqrt{\frac{L}{g}}$$

(Hint: The speed varies along the length of the rope because the tension varies. Start with

$$\frac{dx}{dt} = \sqrt{gx}$$

and separate the variables before integrating to relate t and L.)

2. A jet plane takes off in Hawaii and flies to California at an average speed of 800 km/h. At the moment of takeoff a volcanic eruption occurs, generating a tidal wave (tsunami) in the ocean and sound waves in the air. Determine the order of arrival in California for the plane, tsunami, and sound waves. Take $h = 5$ km for the depth of the Pacific Ocean. Take the pressure and density of air to be 1.06×10^5 N/m^2 and 1.31 kg/m^3.

3. The pressure amplitude of a wave is raised from 2×10^{-5} N/m^2 to 4×10^{-5} N/m^2. (a) By how many decibels is the intensity level raised? (b) By what factor is the intensity raised? (c) If the pressure amplitude is raised to 6×10^{-5} N/m^2, will the intensity level rise by the same number of decibels as in part (a)? Explain.

4. A student wants to establish a standing wave on a wire 1.8 m long clamped at both ends. The wave speed is 509 m/s. (a) What is the minimum frequency she should apply to set up standing waves? (b) What next higher frequency should she apply to create a different standing wave?

5. The intensity level of an orchestra is 85 dB. A single violin achieves an intensity level of 70 dB. How does the intensity of the full orchestra compare with that of the violin?

6. The intensity of sunlight incident at the top of the earth's atmosphere is 1.38 kW/m^2. (a) The absolute luminosity of a star is defined as the total energy it radiates per second. Show that the absolute intensity of the sun is 3.9×10^{26} W. (The earth–sun distance is 1.50×10^{11} m.) (b) If 39% of the incident energy is reflected back into space, determine the total solar energy absorbed by the earth in 1 year. (c) What fraction of this solar energy would have to be utilized in order to match the annual United States electrical energy consumption of 1.5×10^{12} kWh? (The radius of the earth is 6.38×10^6 m.)

7. A surface wave that spreads out from a point source generates cylindrical wave fronts. Present an argument that shows that the intensity of a cylindrical wave varies inversely as the first power of the distance from the source ($I \propto 1/r$). Earthquakes generate surface waves as well as the P and S waves that travel through the body of the earth. The surface waves are particularly destructive, even at great distances from the

origin of the disturbance, because the intensity diminishes only as $1/r$ rather than as $1/r^2$.

8. A string is anchored at $x = 0$ and $x = 2.12$ m. Transverse standing waves on the string are described by

$$\psi(x, t) = 0.0348 \sin kx \cdot \cos \omega t \qquad \text{m}$$

where $k = 5.93$ rad/m and $\omega = 60$ rad/s. Determine (a) the wavelength and (b) the locations of the nodes.

9. Treat the outer ear as a pipe 3 cm in length, open at one end and closed at the other end. Assume that the ear hears sound most efficiently for frequencies close to the fundamental frequency of the pipe. Estimate the most sensitive frequency of the human ear.

10. A water wave traveling in a lake 9.2 m deep has a wavelength of 1.22 m. (a) Is it a shallow-water wave or a deep-water wave? (b) Determine the speed of the wave. (Ignore surface tension.)

11. Standing waves are generated in a string with one end tied to a wall. The other end is oscillated up and down with a mechanical vibrator to form an antinode at the driven end. Present an analysis parallel to that in Section 17.4, for a string anchored at both ends. (a) Write down the boundary conditions. (b) Sketch the standing wave forms for the three lowest frequencies. (c) Obtain the equation that determines the fundamental frequency.

12. The equation for the speed of deep-water waves is given by

$$v = \sqrt{\frac{g\lambda}{2\pi} + \frac{2\pi S}{\rho\lambda}}$$

(a) Rewrite this expression as a quadratic equation having the form

$$a\lambda^2 + b\lambda + c = 0$$

(b) Use the equation for the roots of a quadratic equation to show that the physically meaningful solutions

require a minimum speed given by

$$v_{min} = \left(\frac{4gS}{\rho}\right)^{1/4}$$

13. (a) Show that

$$\psi(x, t) = A \sin(600x - 50t) + B \cos(600x - 50t)$$

with A and B constant, is a solution of the wave equation (Equation 17.5). (b) If the wave function in part (a) describes a wave on a string, evaluate A and B, given that the transverse velocity of the string is 5.0 m/s at $t = 0$, $x = 0$, and the displacement of the string is 1.2 cm at $t = 0$, $x = 0$.

14. A wave pulse has the form

$$\psi = Ae^{-(x - ut)^2/a^2}$$

(a) Prove that the point where the slope vanishes ($\partial\psi/\partial x = 0$) travels according to $x = ut$. (b) Determine the maximum value of ψ. (c) Calculate the velocities, $\partial\psi/\partial t$, of points located at distances $x = \pm a$ from the center of the pulse at time t.

15. The displacement of a vibrating string is given by the expression

$$\psi = A[e^{-(x + a - vt)^2/a^2} - e^{-(x - a + vt)^2/a^2}]$$

(a) Prove that $\psi = 0$ for all x for $t = a/v$. (b) Sketch ψ versus x at $t = a/2v$. (c) Interpret your sketch to explain why $\psi = 0$ when $t = a/v$.

16. A wave traveling on a string is described by

$$\psi = 2 \cos\left(kx - \omega t + \frac{\pi}{6}\right)$$

$$k = 3 \text{ rad/m} \qquad \omega = 5 \text{ rad/s}$$

Determine the form of the traveling wave that produces a standing wave when superposed with ψ.

CHAPTER

18

THE FIRST LAW OF THERMODYNAMICS

18.1
INTRODUCTION TO THERMODYNAMICS

Thermodynamics deals with the properties of bulk matter under conditions in which the effects of heat and temperature are significant. Thermodynamics often seems rather abstract, partly because of the universal character of its laws. The generality of the laws of thermodynamics requires that they be independent of the detailed workings of any particular physical system. Thermodynamics did not evolve from atomic models of matter that let you see how it works. Such developments came later with **kinetic theory** and **statistical mechanics**. A taste of the statistical approach to thermodynamics is presented in Chapter 22, where we will study the kinetic theory of gases.

18.2
TEMPERATURE

In mechanics we rely on three fundamental quantities: mass, length, and time. In thermodynamics, we must introduce a fourth fundamental quantity: *temperature*. In this chapter we give temperature an operational definition that allows us to measure it. In Chapter 22 we will uncover the kinetic interpretation of temperature, and find that temperature measures the average kinetic energy of molecular motions.

Qualitatively, temperature is a property based on our subjective sense of hot and cold. An object that feels hot is said to be at a higher temperature than one that feels cold. We know from experience that if a hot object is placed in contact with a cold object, their temperatures change. If the objects are isolated from other objects, changes in their measurable properties eventually subside and the objects are said to be in **thermal equilibrium** or at **equal temperature**. This definition of temperature equality is not tied to any particular scale of temperature. At this point in our discussion we do not have an operational way of assigning temperatures—we have simply recognized temperature as a property that indicates whether or not two objects are in thermal equilibrium.

The Zeroth Law of Thermodynamics

Suppose a warm watermelon is placed in a mountain stream. It cools off and comes to thermal equilibrium with the stream. A bottled soft drink placed in the stream also comes into thermal equilibrium with the stream. When the watermelon and the soft drink are placed in contact, neither undergoes any change of temperature. They are already in thermal equilibrium. This result may not strike you as surprising, but it is important. It is called the **zeroth law of thermodynamics** and may be stated formally as

> Two objects, each in thermal equilibrium with a third object, are in thermal equilibrium with each other.

The zeroth law is significant because it recognizes that equality of a *single* physical property—temperature—is both necessary and sufficient to ensure thermal equilibrium.

FIGURE 18.1

An elementary form of optical pyrometer. [The color of the hot coals is a temperature indicator.]

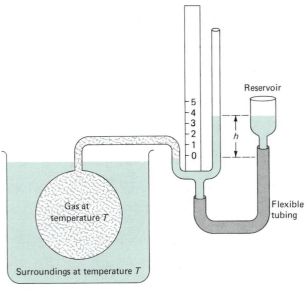

FIGURE 18.2

Schematic of a constant volume gas thermometer. The gas is maintained at a constant volume by means of the mercury reservoir. If the temperature T of the bath rises, the level of the reservoir must be increased to maintain the same volume.

Temperature Measurement

In principle, any substance whose properties change with temperature can be used as a thermometer. In the optical pyrometer, the property is color (Figure 18.1). Other properties used to indicate temperature include the pressure of a confined gas, the electrical resistance of a wire, and the length of a column of alcohol or mercury. In certain medical thermometers the color of a liquid crystal is used to register body temperature.

Constant Volume Gas Thermometer

An empirical temperature scale is obtained by simply *defining* a temperature in terms of some measurable property. In the constant volume gas thermometer (Figure 18.2), temperature (T) is defined in terms of gas pressure (P) by

$$T = bP \qquad (18.1)$$

where b is a proportionality constant. The choice of a linear relation between temperature and pressure is arbitrary but convenient. The scale defined by Equation 18.1 is one example of an **absolute scale**. On an absolute scale, the lowest temperature is numerically zero.*

---

* The fact that temperature has a lower limit is embodied in the third law of thermodynamics. The fact that such a lower limit exists is not important here. For details, see J. Wilks, *The Third Law of Thermodynamics*. London and New York: Oxford Univ. Press, 1961.

The value of the constant b in Equation 18.1 is fixed by assigning a temperature to some standard *fixed point*. By international agreement, the temperature of water at its triple point* has been assigned the value 273.16 K, where K is the symbol for the temperature unit called the **kelvin**.[†] Accessibility and ease of reproducibility are the primary reasons for using the triple point of water as the fundamental fixed point.

If P_3 denotes the gas pressure at the triple point of water, we have

$$273.16 \text{ K} = bP_3$$

The constant volume gas temperature scale becomes

$$T = 273.16 \left(\frac{P}{P_3} \right) \text{ kelvins} \tag{18.2}$$

The freezing and boiling of water are familiar phenomena. The **ice point** of pure water is the temperature at which the solid and liquid are in thermal equilibrium at a pressure of 1 atm. The **boiling point** is the temperature at which a liquid and its vapor are in thermal equilibrium. The **normal boiling point** is the equilibrium temperature when the pressure of the vapor is exactly 1 atm. The normal boiling point of pure water is also called the **steam point**.

EXAMPLE 1

Steam Point Measurement

Using a constant volume gas thermometer of modest precision, a student finds the values $P_3 = 2.68 \times 10^4 \text{ N/m}^2$ and $P_s = 3.67 \times 10^4 \text{ N/m}^2$ for the triple point and steam point. The corresponding steam point is

$$T_s = 273.16 \text{ K} \left(\frac{3.67 \times 10^4 \text{ N/m}^2}{2.68 \times 10^4 \text{ N/m}^2} \right)$$

$$= 374 \text{ K}$$

The temperature recorded by a constant volume gas thermometer depends on the amount and the type of gas. Thermometers that use different gases generally do not agree, except at the triple point of water. Furthermore, if we remove some of the gas in a thermometer and then repeat the measurements, we record a lower triple-point pressure and a different temperature. A sequence of such measurements in which progressively lower triple-point pressures are used enables us to extrapolate the results toward the limit $P_3 \to 0$. The salient feature of constant volume gas thermometers is that this extrapolated temperature is the *same for all gases*. Figure 18.3 shows how sequences of measurements using different gases and different triple-point pressures extrapolate to a common value for the steam point.

---

* The solid, liquid, and gaseous phases of water can coexist in thermodynamic equilibrium at a unique pressure and temperature. This is referred to as the **triple point of water**.
[†] At one time temperatures on the Kelvin scale were designated °K, read "degrees kelvin." In 1968 the designation was changed to K, read "kelvins." Thus, 100 K is the temperature "one hundred kelvins," *not* "one hundred degrees kelvin."

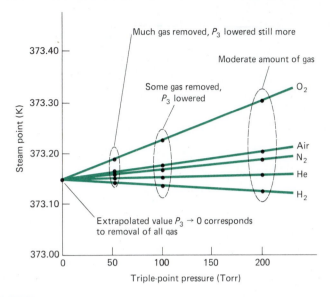

FIGURE 18.3

The steam point recorded by a constant volume gas thermometer. For any particular gas, the steam point depends on the gas used. However, when the data are extrapolated to zero pressure, the unique steam point 373.15 K is obtained for all gases. (Adapted from *Heat and Thermodynamics*, 5th ed., by Mark Zemansky; McGraw-Hill, New York, 1968. Used with permission of McGraw-Hill Book Company.)

The fact that the extrapolated temperature is the same for all gases leads to the important concept of an *ideal gas*. The extrapolated temperature is called the **ideal gas temperature**,

$$T = 273.16 \lim_{P_3 \to 0} \left(\frac{P}{P_3}\right) \text{kelvins} \tag{18.3}$$

Other features of the ideal gas are developed in subsequent chapters.

In addition to the Kelvin scale, three other prominent scales are used. These are the Celsius, Fahrenheit, and Rankine scales. On the Celsius scale, temperature t_C is related to the Kelvin temperature T by

$$t_C = (T - 273.15)°C \tag{18.4}$$

where °C is the abbreviation for degrees Celsius. For example, the Celsius temperature of the triple point of water is 0.01°C because $T_3 = 273.16$ K. Notice that the Celsius degree and the kelvin have the same size. The scales differ only in the location of their zeros.*

The symbol °F stands for degrees Fahrenheit. Temperatures on the Fahrenheit scale are related to those of the Celsius scale by the equation

$$t_F = (32 + \tfrac{9}{5}t_C)°F \tag{18.5}$$

This places the ice and steam points at 32°F and 212°F, respectively. So, the Fahrenheit and Celsius scales differ in the size of their units and in the location of their zeros.

---

* It is meaningful to distinguish the units of *temperature* and the units of *temperature change* on the Celsius scale. We denote temperatures on the Celsius scale by the symbol °C (read "degrees Celsius"). For example, 20°C is a temperature and 20 C° is a temperature difference.

EXAMPLE 2

Celsius and Fahrenheit Temperatures of Liquid Nitrogen

The normal boiling point of nitrogen is 77.4 K. The corresponding Celsius temperature follows from Equation 18.4,

$$t_C = (77.4 - 273.15)°C$$
$$= -195.8°C$$

We can insert this result into Equation 18.5 to find the equivalent Fahrenheit temperature,

$$t_F = [32 + \tfrac{9}{5}(-195.8)]°F$$
$$= -320.4°F$$

Thus, 77.4 K, $-195.8°C$, and $-320.4°F$ are equivalent temperatures expressed on different scales.

The Rankine temperature scale is used in some branches of engineering. The Rankine and Kelvin scales have a common zero, but their units differ in size. The Rankine degree is equal to the Fahrenheit degree, and thus is only 5/9 as large as the Celsius degree and the kelvin,

$$1\,R° = \tfrac{5}{9}\,K \qquad (18.6)$$

Consequently, a temperature on the Rankine scale is 9/5 as great as the corresponding Kelvin temperature

$$t_R = \tfrac{9}{5}T°R \qquad (18.7)$$

For example, the Rankine temperature of the ice point of water is

$$\frac{9}{5}(273.15)°R = 491.67°R$$

Figure 18.4 compares temperatures on the four scales. Figure 18.5 shows the wide range of temperatures encountered in physics.

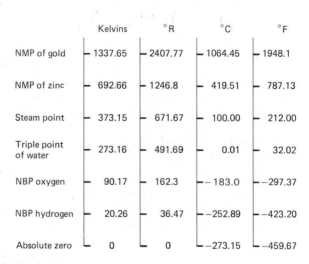

	Kelvins	°R	°C	°F
NMP of gold	1337.65	2407.77	1064.45	1948.1
NMP of zinc	692.66	1246.8	419.51	787.13
Steam point	373.15	671.67	100.00	212.00
Triple point of water	273.16	491.69	0.01	32.02
NBP oxygen	90.17	162.3	-183.0	-297.37
NBP hydrogen	20.26	36.47	-252.89	-423.20
Absolute zero	0	0	-273.15	-459.67

FIGURE 18.4

A comparison of four temperature scales. Numerical values indicate the temperatures of several standard fixed points. NBP stands for normal boiling point, the melting point at a pressure of one atmosphere. NMP stands for normal melting point.

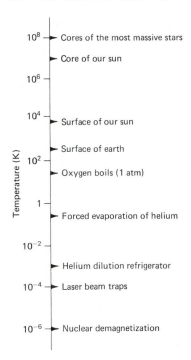

10^8 — Cores of the most massive stars

— Core of our sun

10^6 —

10^4 — Surface of our sun

— Surface of earth

10^2 —

— Oxygen boils (1 atm)

1 —

— Forced evaporation of helium

10^-2 —

— Helium dilution refrigerator

10^-4 — Laser beam traps

10^-6 — Nuclear demagnetization

Temperature (K)

FIGURE 18.5

A sampling of the temperatures encountered in nature (Kelvin scale). Note that the scale is logarithmic.

18.3
HEAT

Today we recognize heat as energy transferred by virtue of a temperature difference. We can set up an operational definition of heat that relates it to other forms of energy. The **kilocalorie** is a unit of heat defined as follows:

One kilocalorie (1 kcal) is the amount of heat required to raise the temperature of 1 kg of water from 14.5°C to 15.5°C.

Because heat is a form of energy, the kilocalorie can be expressed in kilojoules (kJ).

$$1.0000 \text{ kcal} = 4.1858 \text{ kJ} \tag{18.8}$$

Many engineering applications of thermodynamics employ the British thermal unit (Btu) of heat. The equivalences are

$$1 \text{ Btu} = 0.252 \text{ kcal} = 1.05 \text{ kJ}$$

In dietetics, the Calorie (abbreviation Cal) is used as a unit of energy. One Calorie equals 1000 calories, or 1 kilocalorie.

$$1 \text{ Cal} = 1 \text{ kcal}$$

Experiment reveals two significant attributes of heat:

1. The amount of heat required to produce a specified change of temperature of a material is directly proportional to the mass of the material.
2. For a given mass of material the amount of heat absorbed is directly proportional to the temperature increase. This direct proportionality holds true only for small changes of temperature and only as long as the heated material does not undergo a change of phase, such as melting.

These attributes lead to the concept of specific heat capacity.

Specific Heat Capacity

Let Q stand for the heat absorbed or ejected when a body of mass m undergoes a temperature change ΔT. The observation that Q is proportional to m and ΔT lets us introduce a proportionality factor (C) and write the equation

$$Q = mC \, \Delta T \tag{18.9}$$

The factor C is called the **average specific heat capacity** over the temperature range ΔT. The value of the average specific heat capacity depends on the physical and chemical composition of the material. We shorten "average specific heat capacity" to "**specific heat**." The specific heats for a number of substances are presented in Table 18.1. The tabulated values refer to measurements made with the pressure held constant at 1 atm. The subscript on C_P signifies constant pressure.

EXAMPLE 3

The Specific Heat of Water

Table 18.1 lists the specific heat of water as 4.19 kJ/kg·C°. To see that this value conforms to the definition of 1 kcal, we use Equation 18.9 with $m = 1$ kg,

$C = 4.19$ kJ/kg·C°, and $\Delta T = 1$ C°. This gives the amount of heat needed to raise the temperature of 1 kg of water by 1 C°:

$$Q = (1 \text{ kg})(4.19 \text{ kJ/kg·C°})(1 \text{ C°})$$
$$= 4.19 \text{ kJ}$$

This confirms the definition of the kilocalorie and the equivalence stated in Equation 18.8.

Equation 18.9 can be expressed in differential form as

$$dQ = mC \, dT \qquad (18.10)$$

In this equation dT is the temperature change resulting from the exchange of the amount of heat dQ. The total heat absorbed or ejected in a finite change of temperature is the sum (integral)

$$Q = \int_{T_i}^{T_f} dQ = m \int_{T_i}^{T_f} C \, dT \qquad (18.11)$$

In general, the specific heat varies with T. If C does not change significantly over the temperature range T_i to T_f, it can be treated as a constant in the integration. In this event the integration results in

$$Q = mC(T_f - T_i) = mC \, \Delta T \qquad (18.12)$$

which is identical with Equation 18.9. Happily, many substances have nearly constant specific heat capacities over sizable temperature ranges, permitting us to use the relation $Q = mC \, \Delta T$ instead of the more involved integral relation of Equation 18.11.

TABLE 18.1
Specific Heat Capacity at Constant Pressure (C_P)*

Substance	C_P(kJ/kg·C°)	Substance	C_P(kJ/kg · C°)
Solids		**Liquids**	
Silver (Ag)	0.234	Benzene (C_6H_6)	1.62
Aluminum (Al)	0.900	Bromine (Br)	0.448
Gold (Au)	0.130	Ethyl alcohol (C_2H_5OH)	2.43
Bismuth (Bi)	0.123	Methyl alcohol (CH_3OH)	2.52
Copper (Cu)	0.385	Water (H_2O)	4.19
Iron (Fe)	0.448		
Ice (H_2O, 0° C)	2.10	**Gases**	
Brick	0.837	Steam (100° C)	2.05
Concrete	0.879	Argon (Ar)	0.523
		Carbon dioxide (CO_2)	0.95
Magnesium (Mg)	1.04	Chlorine (Cl_2)	0.481
Sodium (Na)	1.23	Helium (He)	5.24
Nickel (Ni)	0.431	Hydrogen (H_2)	14.2
Lead (Pb)	0.130	Nitrogen (N_2)	1.04
Zinc (Zn)	0.352	Neon (Ne)	1.03
Wood (Pine)	2.81	Oxygen (O_2)	0.917

* Except where indicated otherwise, values refer to a pressure of 1 atmosphere and a temperature of 25°C for solids and liquids; 15°C for gases. Most values are adapted with permission from International Critical Tables of Numerical Data: Physics, Chemistry, and Technology, National Research Council, Washington, D.C., 1923–1933, and E. H. Kennard, *Kinetic Theory of Gases.* New York, McGraw-Hill, 1938.

EXAMPLE 4

Constant Specific Heat

The specific heat of zinc is 0.352 kJ/kg·C° for temperatures near 25°C. Let's determine the amount of heat required to raise the temperature of 0.500 kg of zinc from 20°C to 30°C. Taking the specific heat to be a constant, we can use Equation 18.12 to find

$$Q = mC\,\Delta T = (0.50 \text{ kg})(0.352 \text{ kJ/kg·C°})(10 \text{ C°})$$
$$= 1.76 \text{ kJ}$$

Equation 18.12 is not applicable when the specific heat changes significantly over the temperature range of interest. For a variable C, we must use Equation 18.11 to determine Q.

EXAMPLE 5

Variable Heat Capacity

At cryogenic temperatures the heat capacities of many crystalline solids are proportional to T^3. For zinc at temperatures below 100 K, the specific heat is given by

$$C = (1.02 \times 10^{-6} \text{ kJ/kg·K}^4)T^3$$

Let's determine the heat needed to raise the temperature of 0.50 kg of zinc from 10 K to 20 K. The temperature change is 10 kelvins, equal to the 10 C° temperature change in Example 4. Here we must use Equation 18.11 because the heat capacity varies with temperature:

$$Q = m \int_{T_i}^{T_f} C\,dT$$

$$= 0.50 \text{ kg } (1.02 \times 10^{-6} \text{ kJ/kg·K}^4) \int_{T_i}^{T_f} T^3\,dT$$

giving

$$Q = (1.28 \times 10^{-7} \text{ kJ/K}^4)(T^4{}_f - T_i{}^4)$$

Setting $T_f = 20$ K and $T_i = 10$ K gives $Q = 0.0192$ kJ. The heat required is more than a hundred times smaller than the heat required for the same change of temperature in Example 4. This is because the average specific heat is far lower over the range from 10 K to 20 K than it is over the range from 20°C to 30°C.

The Method of Mixtures

The *method of mixtures* is a technique for measuring specific heats. For example, suppose we want to measure the specific heat of lead. We place 0.600 kg of lead, initially at a temperature of 25°C, in an aluminum can that holds 0.10 kg of water. The can is called a **calorimeter**. The initial temperature of the water and calorimeter is 80°C. The mass of the calorimeter is 0.020 kg and its specific heat is 0.900 kJ/kg·C°. We find that the final temperature of the mixture is 72°C. We can use these data to determine the specific heat of lead.

The basic assumption in the method of mixtures is that all of the heat ejected by the initially warmer material is absorbed by the initially cooler material. This assumes there is negligible heat transfer to or from the system via radiation, conduction, and convection. The heat absorbed by the lead is

$$Q_A = m_{Pb} C_{Pb} \Delta T_{Pb} = (0.60 \text{ kg})(47 \text{ C}°)C_{Pb}$$

The heat given up by the water and calorimeter is

$$Q_E = (m_{A1} C_{A1} + m_w C_w) \Delta T$$
$$= [(0.020 \text{ kg})(0.900 \text{ kJ/kg·C}°) + (0.10 \text{ kg})(4.18 \text{ kJ/kg·C}°)] \text{ 8 C}°$$
$$= 3.50 \text{ kJ}$$

Equating Q_A and Q_E gives for the specific heat of lead

$$C_{Pb} = 3.50 \text{ kJ}/(0.600 \text{ kg})(47 \text{ C}°)$$
$$= 0.124 \text{ kJ/kg·C}°$$

At ordinary temperatures the specific heats of solids and liquids depend only slightly on the pressure and volume. In contrast, the specific heats of gases depend strongly on pressure and volume. In particular, the specific heat of a gas maintained at a constant pressure is greater than the specific heat measured when the gas is held at a constant volume. This curious fact led Robert Mayer to hypothesize that heat is a form of energy transfer and that thermal energy can be converted into other forms of energy. In 1842 he observed that the amount of heat required to change the temperature of a gas was greater when the *pressure* of the gas was held constant than when the *volume* of the gas was held constant. When heat is added at constant pressure the gas expands and performs work. When heat is added at constant volume no work is performed. Mayer realized that the extra heat required under constant-pressure conditions was converted into the work of expansion.

The convertibility of heat into work means that heat and work are both forms of energy transfer. In the next section we develop the concept of work as it is used in thermodynamics.

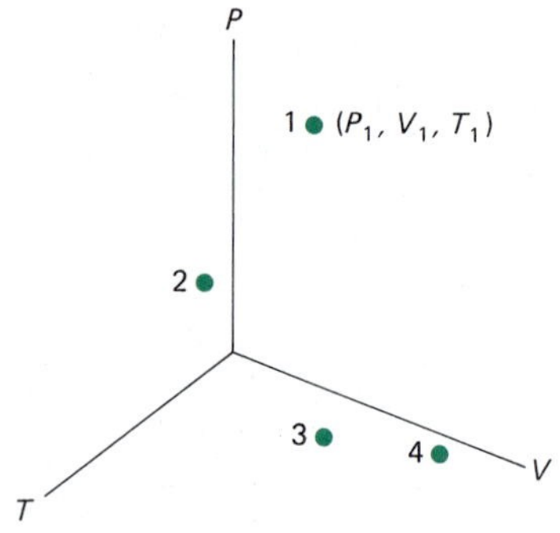

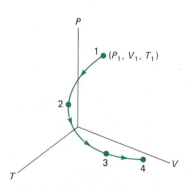

FIGURE 18.6

Equilibrium states of a gas can be represented by points defined by values of pressure (P), volume (V), and temperature (T).

18.4
THERMODYNAMIC WORK

An **equilibrium state** of a system is defined by a set of values of the thermodynamic variables. For a gas enclosed in a cylinder fitted with a piston, a set of values of pressure, volume, and temperature (P, V, T) specifies an equilibrium state. If we use a Cartesian coordinate system to plot P, V, and T, different equilibrium states correspond to different points (Figure 18.6).

A **quasi-static** process is one in which the surroundings and thermodynamic variables of the system change so slowly that the process can be viewed as one in which the system passes through a succession of equilibrium states. Geometrically, a quasi-static process can be represented by a path that joins the initial and final equilibrium states by a succession of intermediate equilibrium states (Figure 18.7). Figure 18.8 suggests how a gas might be compressed quasi-statically. A cylinder filled with gas has a piston at one end. Sand piled on the piston helps to compress the gas. By adding sand slowly, one grain at a time, the gas can be compressed slowly.

We can use the example of a gas being compressed quasi-statically to introduce the concept of a **reversible process**. Consider first the behavior of the gas-piston system when there is no friction between the piston and the

FIGURE 18.7

A quasi-static process may be represented by a thermodynamic path that joins the initial and final states by a succession of intermediate equilibrium states.

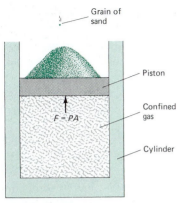

FIGURE 18.8

The gas confined in the cylinder is compressed quasi-statically by slowly piling sand on the piston.

cylinder walls. Adding or removing a grain of sand constitutes an *infinitesimal change* in the surroundings. At any stage we can halt the compression and initiate expansion by removing a grain of sand, provided there is no friction. If the sand is removed grain by grain, so as to reverse the order in which it was added, the gas retraces the thermodynamic path (the same sequence of equilibrium states) it followed during the compression.

The quasi-static, frictionless compression just described is an example of a reversible process:

A **reversible process** is a quasi-static process that can be reversed by an infinitesimal change in the surroundings.

Please note that a reversible process does not necessarily go one way and then back again. The word "reversible" implies that the process *can be* reversed, not that it *is* reversed.

In contrast, an **irreversible process** is one that cannot be reversed by an infinitesimal change in the surroundings. Examples of irreversible processes include the following:

(a) a chemical reaction in which a precipitate forms
(b) the explosive combination of substances, for example, the formation of water from hydrogen and oxygen
(c) the diffusion of perfume vapor in the air

Irreversible processes occur spontaneously within a system that is not in thermodynamic equilibrium. Heat flow is another type of irreversible process. Heat flows spontaneously—that is, without any outside help—from regions of higher temperature to regions of lower temperature. Irreversible processes tend to produce a state of mutual equilibrium between the system and its surroundings.

Friction destroys reversibility. If there is friction between the piston and the cylinder shown in Figure 18.8, we can still compress the gas quasi-statically by adding sand slowly. However, the compression cannot be reversed by an infinitesimal change in the surroundings. Removing one grain of sand—an infinitesimal change—does not reverse the process.

Although friction cannot be eliminated completely, the frictional force can often be rendered negligible by comparison with the force exerted by the gas. In such circumstances compressions and expansions may be regarded as reversible processes.

The work done by the system during a reversible process can be expressed in terms of its thermodynamic variables and the changes in these variables. For the gas-piston system (Figure 18.9), the force exerted on the piston by the gas is

$$F = PA \qquad (18.13)$$

where P is the gas pressure and A is the cross-sectional area of the piston. Suppose that the gas expands, pushing the piston outward a distance ds. The work done by the gas is

$$dW = F\,ds = PA\,ds \qquad (18.14)$$

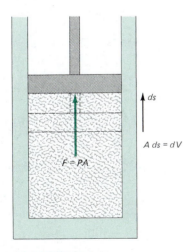

FIGURE 18.9

When the piston moves upward a distance ds the volume of the gas increases by $dV = A\,ds$ and the gas does work $P\,dV$.

However, $A\,ds = dV$ is the change in the volume of the gas. Thus, the work done by the gas dW during a quasi-static change of volume dV is

$$dW = P\,dV \qquad (18.15)$$

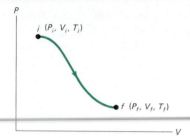

FIGURE 18.10

P-V diagram showing the expansion of a gas.

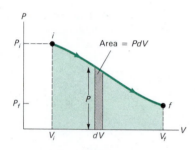

FIGURE 18.11

Geometrically, the work $dW = P\,dV$ corresponds to the grey area. The total work W done by the gas during the expansion from V_i to V_f corresponds to the aqua shaded area.

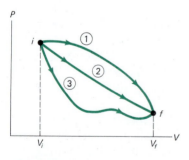

FIGURE 18.12

Three different paths connecting the same initial and final states. The work done by the gas is greatest for the path marked ①.

Equation 18.15 expresses the work solely in terms of the thermodynamic variables of the system.

The work $dW = P\,dV$ is often referred to as **thermodynamic work**, because it is expressed entirely in terms of the thermodynamic variables. We will deal only with thermodynamic work, and therefore will refer to it simply as work. For the gas-piston system shown in Figure 18.9, we find the total work done during a finite change of volume by integrating Equation 18.15 between some initial state (i) and some final state (f):

$$W = \int_i^f dW = \int_{V_i}^{V_f} P\,dV \qquad (18.16)$$

V_i and V_f denote the initial and the final volumes of the gas. In order to evaluate the integral it is necessary to specify a relationship between P and V. Using a P-V diagram (Figure 18.10) to label the sequence of states through which the gas passes, we get a graphical description of the process. We can think of the sequence of points on the P-V diagram as a path from the initial state to the final state.

The P-V diagram in Figure 18.11 shows the path of an expansion process. The quantity $P\,dV$ is the work done during the volume change dV and is represented by the area under the P-V path. (Keep in mind that although $P\,dV$ can be represented by this area, its units are those of work, not of area.) The total work, W, done by the gas corresponds to the shaded area beneath the P-V path.

Different paths between V_i and V_f can result in different amounts of work, even though they connect the same initial and final states. We say that the work, W, is a **path-dependent** quantity. These different paths can be achieved by controlling the temperature of the gas during the expansion.

Figure 18.12 shows three different paths connecting the same initial and final states. The work done by the gas is greatest for the path marked "1." The work done by the gas is smallest for the path marked "3."

EXAMPLE 6

Path-Dependent Work

We can demonstrate the path dependence of thermodynamic work numerically as follows: Consider a gas that starts at a pressure of 10^6 N/m² and a volume of 1 m³ and expands to a final pressure of 5×10^5 N/m² and a final volume of 2 m³. Figure 18.13 shows two possible paths connecting the initial and final states. The work (W_1) performed along path 1 corresponds to a rectangular area. The work (W_2) performed along path 2 corresponds to a trapezoidal area. It is clear from the figure that $W_1 > W_2$. Direct calculation verifies this. The work W_1 corresponds to the area of a rectangle of height P_i and width $V_f - V_i$:

$$W_1 = P_i(V_f - V_i) = (10^6 \text{ N/m}^2)(2 - 1) \text{ m}^3$$
$$= 10^6 \text{ J}$$

The work W_2 corresponds to the area of a trapezoid of average height $\frac{1}{2}(P_i + P_f)$ and width $V_f - V_i$:

$$W_2 = \tfrac{1}{2}(P_i + P_f)(V_f - V_i) = \tfrac{1}{2}(10^6 + 5 \times 10^5) \text{ N/m}^2 \cdot (2 - 1) \text{ m}^3$$
$$= 7.5 \times 10^5 \text{ J}$$

The fact that $W_1 \neq W_2$ shows the thermodynamic work done by a system depends on the path of the process connecting the initial and final states.

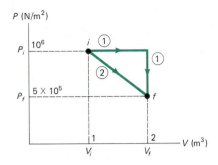

FIGURE 18.13
The work performed along path ① is greater than the work performed along path ②.

The work integral can be negative. For example, if the expansion described by Figure 18.11 is reversed, the magnitude of the work done is still represented by the area under the P-V path. However, the work is negative because dV is negative—the volume is decreasing.

The mechanical work performed by a force is the most familiar form of work. However, the work performed by a thermodynamic system can take other forms. In thermodynamics, *work includes all forms of energy transfer except heat*. For example, when a battery is being charged, electrical work is performed and stored in chemical form. Another important nonmechanical form of work is magnetic work. When a piece of steel is magnetized, magnetic work is performed.

Just as there are nonmechanical forms of thermodynamic work, the energy a system stores internally need not be purely mechanical. Chemical and magnetic energy can be stored internally. The concept of **internal energy** is introduced in thermodynamics to cover the many possible forms of energy that can be stored within a system. In the next section we show how internal energy is operationally defined.

18.5
INTERNAL ENERGY

Consider a system that undergoes a transformation during which no heat is absorbed or ejected. Such a process is called an **adiabatic process**. In an adiabatic process a system can perform work only at the expense of its stored energy. If energy is conserved and the system performs a positive amount of work on the environment, then the energy stored in the system decreases. This observation can be used to set up an operational definition of **internal energy**, the energy stored within a system. Let W denote the work done by a system during an adiabatic process that carries the system from an initial equilibrium state of internal energy U_i to a final equilibrium state of internal energy U_f. Energy conservation is described by the relation

$$\left\{\begin{array}{c}\text{The change in the}\\\text{internal energy}\\\text{of the system}\end{array}\right\} + \left\{\begin{array}{c}\text{the work done}\\\text{by the system on}\\\text{its environment}\end{array}\right\} = 0$$

In symbolic form this equation becomes

$$(U_f - U_i) + W = 0 \tag{18.17}$$

We already have an operational definition of work, so Equation 18.17 defines the change in the internal energy of the system, $U_f - U_i$.

EXAMPLE 7

Adiabatic Compression of a Gas

A person exerts a force of 1000 N in compressing the gas in a pump cylinder. The force acts over a distance of 0.500 m. Because gases are poor heat conductors the rapid expansion or compression of a gas can be regarded as an adiabatic process; it is over before the gas has time to absorb or eject a significant amount of heat. Let's calculate the work done by the gas and the change in its internal energy.

The work done *on* the gas by the piston is positive. The work done *by* the gas on the piston (force × displacement) is negative,

$$W = -(1000 \text{ N})(0.500 \text{ m})$$
$$= -500 \text{ J}$$

The increase in the internal energy of the gas follows from the adiabatic form of the first law, Equation 18.17:

$$U_f - U_i = -W$$
$$= +500 \text{ J}$$

The temperature of the gas rises as a result of this increase in the internal energy. If you have ever inflated a tire using a hand pump you may have noticed the temperature rise that results from an adiabatic compression.

Our operational definition does not give us much of a feel for internal energy. If we want to see inside a system and understand what makes it tick, we must construct a **microscopic model** and a theory to go with it. We must believe in atoms and puzzle over the forces that bind them together to form molecules and solid objects. The challenge of understanding thermodynamics from a microscopic viewpoint inspired men like Rudolph Clausius, James Clerk Maxwell, Ludwig Boltzmann, and Josiah Willard Gibbs to develop the **kinetic theory of gases** and **statistical mechanics**. These two disciplines permit events to be followed at the microscopic level in a statistical fashion. We will get a taste of the ideas involved when we study kinetic theory in Chapter 22.

18.6
THE FIRST LAW OF THERMODYNAMICS

Energy is a fruitful concept with two primary attributes: energy can be converted from one form to another, and energy is conserved. The first law of thermodynamics quantifies these two attributes of energy. All forms of energy (mechanical, thermal, electrical, chemical, nuclear, mass, and so on) are recognized, together with the possibility of conversions from one type to another. In particular, the recognition of heat as a form of energy transfer is necessary in order to preserve the conservation of energy. A statement of the first law of thermodynamics that is adequate for our purposes follows:

Energy is conserved in all processes. It cannot be created or destroyed.

The experimental foundation of the first law is the fact that no one has been able to build a *perpetual motion machine of the first kind*—a device that, operating in a cycle, can produce more energy than it takes in.

Credit for formulating the first law of thermodynamics is shared by Robert Mayer and James Prescott Joule. They independently realized that heat is a form of energy transfer and that energy is conserved.

The first law of thermodynamics is formulated in terms of heat, internal energy, and work:

$$\left\{ \begin{array}{c} \text{heat added} \\ \text{to a system} \end{array} \right\} \begin{array}{c} \text{appears} \\ \text{as} \end{array} \left\{ \begin{array}{c} \text{a change in the} \\ \text{internal energy} \end{array} \right\} \text{and/or} \left\{ \begin{array}{c} \text{is used to} \\ \text{do work} \end{array} \right\}$$

The equation expressing this statement of the first law of thermodynamics is

$$Q = U_f - U_i + W \qquad (18.18)$$

where Q is the name for the net amount of heat absorbed by the system. If Q is negative, it simply means that the system *ejected* a net amount of heat. In an adiabatic process $Q = 0$ and Equation 18.18 reverts to the adiabatic form, Equation 18.17.

EXAMPLE 8

Internal Energy Change

A piston-fitted cylinder contains superheated steam. The steam absorbs 12 MJ of heat as it expands, driving the piston. The expanding steam does 17 MJ of work against its environment. We want to determine the change in the internal energy of the steam. Heat is absorbed by the steam, so $Q = +12$ MJ. The expanding steam does positive work so $W = +17$ MJ. Inserting these values of Q and W gives

$$U_f - U_i = 12 \text{ MJ} - 17 \text{ MJ}$$
$$= -5 \text{ MJ}$$

The minus sign means that the final internal energy is 5 MJ less than the initial internal energy. Internal energy decreased because the work performed exceeded the heat absorbed.

When a thermodynamic process proceeds smoothly we envision it as a continuous sequence of small changes. Mathematically we represent the quantities appearing in the first law as differentials. The differential form of the first law is

$$dQ = dU + dW \qquad (18.19)$$

The physical content of the first law is unaltered by rewriting it in this form. However, Equation 18.19 presents a pitfall of which you must be aware. Representing heat and work as dQ and dW *does not imply* the existence of properties that measure the heat and work content of a system. *There are no such properties.* In other words, dQ and dW are not true differentials: they simply denote small amounts of heat and work.

The first law serves two purposes: In its adiabatic form (Equation 18.17) it gives an operational definition of internal energy change. In the general form (Equation 18.18) it recognizes heat as a form of energy transfer and states that energy is conserved.

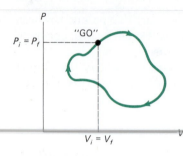

FIGURE 18.14

P–V diagram for a cyclic process.

18.7
APPLICATIONS OF THE FIRST LAW

We illustrate the first law by showing how it is applied to various types of thermodynamic processes.

Cyclic Process

In a **cyclic process** (Figure 18.14), the initial and final states of the system are the same. This means that $U_f = U_i$, and the first law reduces to

$$Q = W \tag{18.20}$$

In this form the first law states that the net work done by the system over the cycle equals the net heat absorbed over the cycle.

In the case of a fluid system, where $dW = P\,dV$, we can give the net work a geometric interpretation. The net work done over one complete cycle can be written

$$W_{\text{cycle}} = \oint P\,dV \tag{18.21}$$

The symbol $\oint$ means that the integration is over a full cycle. Geometrically, $\oint P\,dV$ is represented by the area enclosed by the path describing the cycle on a *P-V* diagram. Figure 18.15 illustrates this point. During expansion the volume increases so that dV and thus $dW = P\,dV$ are positive. The total work

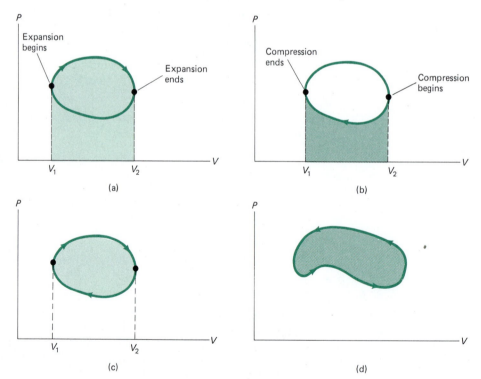

FIGURE 18.15

(a) The work of expansion is positive; it corresponds to the light aqua area. (b) The work of compression is negative; its magnitude corresponds to the darker aqua area. (c) The net work is positive. It corresponds to the light aqua area enclosed by the clockwise path. (d) A cycle for which the work done is negative; its magnitude corresponds to the enclosed area. Note that the cycle generates a counterclockwise path.

of expansion is thus positive and its magnitude corresponds to the shaded area of Figure 18.15a. The compression from V_2 to V_1 reduces the volume so that dV and dW are negative. The total work of compression is therefore negative, and its magnitude is given by the shaded area of Figure 18.15b. The net work done over a full cycle is evidently the sum of the positive work of expansion and the negative work of compression. Therefore, the net work corresponds geometrically to the *difference* between the corresponding areas. Figure 18.15c shows that this difference is the *area enclosed* by the path of the cycle. In Figure 18.15c the net work done by the system is positive because the pressure during expansion is greater than the pressure during compression. Note that the cyclic process in Figure 18.15c generates a clockwise path on the *P-V* diagram. A *clockwise* path on a *P-V* diagram corresponds to *positive* work.

Figure 18.15d shows a cyclic process for which the net work is negative. The work of compression is negative and its magnitude exceeds the positive work of expansion. Note that the cycle traces out a counterclockwise path. On a *P-V* diagram, a *counterclockwise* path corresponds to *negative* work. The magnitude of the net work corresponds to the enclosed area.

EXAMPLE 9

Cyclic Work

Example 6 illustrates the path dependence of work. Now consider a cycle that uses the two paths of that example (Figure 18.16). Starting at i the cycle proceeds in clockwise fashion along the two-legged path 1 to f, and returns to it along the path labeled "2." The net work performed over the cycle is positive (the cycle traces a clockwise path) and corresponds geometrically to the triangular area enclosed. Using the numerical values given in Example 6, we find that the net work performed is

$$W = \tfrac{1}{2}(P_i - P_f)(V_f - V_i) = \tfrac{1}{2}(10^6 - 5 \times 10^5)\,\mathrm{N/m^2}(2 - 1)\,\mathrm{m^3}$$
$$= 2.5 \times 10^5\,\mathrm{J}$$

In this cycle the net heat absorbed also equals 2.5×10^5 J. This is not a coincidence: the first law requires that $Q = W$ for a cyclic process.

Adiabatic Process

An adiabatic process is defined as one in which *no heat is absorbed or ejected by the system*. For an adiabatic process the first law of thermodynamics states that

$$U_f - U_i + W = 0 \qquad (18.17)$$

If positive work is performed by a system in an adiabatic process, the internal energy decreases ($U_f < U_i$). In most instances a decrease in internal energy is reflected by a decrease in temperature.

The earth's atmosphere provides a majestic illustration of the temperature drop that accompanies an adiabatic expansion. In the summer, parcels of moist air near the surface of the earth are heated. This causes them to expand. The expansion decreases the density of these parcels, with the result that they become buoyant and then rise. Such rising parcels of air are called **thermals**. The pressure of the surrounding atmosphere decreases as one moves upward, and so the thermals float upward into surroundings of decreasing pressure. The thermals therefore expand and perform positive work. Very little heat is transferred because gases are poor heat conductors and because the thermals rise swiftly. Therefore, the expansion of a thermal may be treated as an

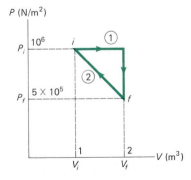

FIGURE 18.16

A cyclic process. The net work done by the system corresponds to the area of the triangle.

FIGURE 18.17
A thermal develops when
moisture-laden air expands
adiabatically and cools.

adiabatic process. The internal energy of an expanding thermal decreases—it is converted into the work of expansion. This decrease of internal energy is accompanied by a drop in the temperature of the thermal. In fact, the temperature may decrease to the point where the water vapor present in the thermal condenses to form a billowing cloud (Figure 18.17).

Isobaric Process

An **isobaric process** is one that takes place at *constant pressure*. In general, the first law does not assume any special form for an isobaric process. That is, W, Q, and $U_f - U_i$ are all nonzero. The work done by a system that expands or contracts isobarically has a simple form. In order to evaluate the work integral

$$W = \int_{V_i}^{V_f} P \, dV$$

the relationship between pressure and volume must be specified. For an isobaric process the pressure is a constant factor in the integrand, and the work integral gives

$$W_{\text{isobaric}} = P(V_f - V_i) \tag{18.22}$$

Isothermal Process

An **isothermal process** is one that takes place at *constant temperature*. As is the case for an isobaric process, the first law does not assume any special form for an isothermal process. In general, W, Q, and $U_f - U_i$ are all nonzero. However, there is an important type of isothermal process in which W is much less than Q. This is the process of **melting**, in which a solid changes to the liquid phase. If heat is added to a crystalline solid at its melting point, the melting proceeds at constant temperature. The heat required to produce melting after a solid has reached its melting point is called the **heat of fusion**. The reverse process, in which a liquid changes to the solid phase, is called **freezing**. The amount of heat released during freezing equals the heat of fusion. **Phase changes** like melting and freezing are discussed in detail in the next chapter (Section 19.4).

Under ordinary atmospheric pressure, melting changes the volume of a substance by only a few percent. The corresponding work is much smaller than the heat of fusion. Thus, to a good approximation, the melting process is described by the first law with W set equal to zero:

$$Q \approx U_f - U_i \tag{18.23}$$

We interpret this form of the first law by saying that the added heat of fusion is used to increase the internal energy of the system.

EXAMPLE 10

A Working Ice Cube

We wish to show that $W \ll Q$ for the case of melting ice. The heat of fusion of ice is 334 kJ/kg. The density of ice at its melting point is 920 kg/m³, and the density of water is 1000 kg/m³. Thus melting changes the volume of 1 kg of ice by

$$\Delta V = \left(\frac{1}{1000} - \frac{1}{920} \right) \text{m}^3$$
$$= -8.70 \times 10^{-5} \text{ m}^3$$

The work done at a constant pressure of 1 atm is

$$W = P \Delta V = (1.01 \times 10^5 \text{ N/m}^2)(-8.70 \times 10^{-5}) \text{ m}^3$$
$$= -8.8 \text{ J}$$

The minus sign appears because ice contracts when it melts. Negative work is done by the ice. If we compare $W = -8.8$ J with $Q = 334 \times 10^3$ J, we see that W is less than 0.01% of Q. This result is typical—the mechanical work is negligible in comparison to the heat of fusion. Virtually all of the added heat of fusion shows up as an increase in the internal energy.

The first law of thermodynamics applies to all conceivable processes that connect equilibrium states. However, not every conceivable process that is consistent with the first law is actually possible. Many are not. For example, we can imagine a power plant that extracts heat from the atmosphere and converts it into an equivalent amount of electrical energy. Such a device would not violate the first law and it would solve our energy problems. Unfortunately, technology will never be able to fabricate such a device because there are other restrictions on the kinds of thermodynamic processes that can occur. The *second law* of thermodynamics, introduced in Chapter 21, describes the *limits* on processes that convert heat into work.

WORKED PROBLEM

One cubic meter of a gas at an initial pressure of 1 atmosphere expands, doubling its volume. During the expansion the pressure P and the volume V are related by

$$P = e^{-(bV - 1)} \text{ atm} \qquad b = 1 \text{ m}^{-3}$$

At the end of the expansion, the pressure is decreased at constant volume until the pressure drops to half the pressure achieved at the end of the expansion. The gas is then compressed and returned to its initial pressure and volume. During the compression, the pressure and volume are related by

$$P = P_0 e^{a(bV - 2)} \qquad b = 1 \text{ m}^{-3}$$

where P_0 and a are constants. Figure 18.18 shows the cycle. (a) Determine the pressure P_0. (b) Calculate the work done during the cycle. (c) Determine the net heat absorbed or ejected by the gas during the cycle.

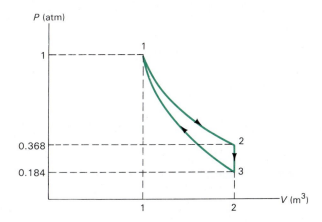

FIGURE 18.18

Solution

(a) The pressure at the end of the expansion is

$$P_2 = e^{-(2-1)} \text{ atm} = e^{-1} \text{ atm}$$
$$= 0.368 \text{ atm}$$

This is reduced by a factor of two to $P_3 = \frac{1}{2}e^{-1}$ atm, which is the pressure at the beginning of the compression

$$P_3 = \frac{1}{2}e^{-1} \text{ atm} = P_0 e^{a(2-2)} = P_0$$

Thus,

$$P_0 = \frac{1}{2}e^{-1} \text{ atm}$$
$$= 0.184 \text{ atm}$$

(b) The net work done over the cycle equals the area enclosed by the path describing the cycle on the P-V diagram. This can be expressed by

$$W = \int_{1 \text{ m}^3}^{2 \text{ m}^3} (P_{\text{expansion}} - P_{\text{compression}}) \, dV$$
$$= 1 \text{ atm} \int_{1 \text{ m}^3}^{2 \text{ m}^3} (e^{-(bV-1)} - \frac{1}{2}e^{-1}e^{a(bV-2)}) \, dV$$

With $b = 1 \text{ m}^{-3}$ and 1 atm = 1.01×10^5 N/m^2, this gives

$$W = 1.01 \times 10^5 \left[1 - e^{-1} - \frac{\frac{1}{2}e^{-1}(1 - e^{-a})}{a} \right] \text{ J}$$

The constant a is evaluated by using the fact that the compression returns the pressure to 1 atm when the volume reaches $V = 1 \text{ m}^3$,

$$1 \text{ atm} = P_0 e^{a(b \cdot 1\text{m}^3 - 2)} = (\frac{1}{2}e^{-1})e^{-a} \text{ atm}$$

With $b = 1 \text{ m}^{-3}$, this gives

$$e^{-a} = 2e \quad \text{and} \quad a = -\ln(2e) = -1.693$$

The work done over the cycle is

$$W = 1.01 \times 10^5 \left[1 - e^{-1} + \frac{\left(\frac{e^{-1}}{2} - 1\right)}{1.693} \right] \text{ J}$$
$$= 1.52 \times 10^4 \text{ J}$$

(c) In a cyclic process the net heat absorbed equals the work performed, so the net heat absorbed by the gas over the cycle is 1.52×10^4 J.

EXERCISES

18.2   Temperature

A. Using a constant volume gas thermometer, values of 1.374, 1.370, and 1.368 are observed for the ratio of the steam-point pressure to the triple-point pressure. (a) Determine the corresponding steam points. (b) The three values correspond to triple-point pressures of 26.67×10^3 N/m^2, 13.33×10^3 N/m^2, and 6.67×10^3 N/m^2. Extrapolate the data (graphically or otherwise) to determine the $P_3 \to 0$ limit of the steam point.

B. Express a body temperature of 98.6°F on the Celsius and Kelvin scales.

18.3   Heat

C. The specific heat of gold is 0.130 kJ/kg·C°. Determine the amount of heat required to raise the temperature of 100 grams of gold from 270 K to 300 K.

D. The temperature of a silver bar rises by 10.0 C° when it absorbs 1.23 kJ of heat. The mass of the bar is 525 grams. Determine the specific heat of silver.

E. Lake Erie contains roughly 4×10^{11} m^3 of water. (a) How much heat is required to raise the temperature of that volume of water from 62°C to 63°C? (b) Approximately how many years would it take to supply this amount of heat by using the full output of a 1000-MW electric power plant?

F. A student inhales air at a temperature of 22°C and exhales air at 37°C. The average volume of air in one breath is 200 cm³. Make a rough estimate of the amount of heat absorbed in one day by the air breathed by the student. Consult Table 18.1 to estimate the specific heat of air. The density of air is approximately 1 kg/m³.

G. Equal masses of silver and chlorine at room temperature absorb equal amounts of heat at constant pressure. (a) Which undergoes the larger temperature increase? (b) If the temperature of the silver rises 4 C°, determine the temperature increase of the chlorine.

H. The Joule constant, J = 4.1858 kJ/kcal, is often referred to as the *mechanical equivalent of heat*. At the end of his 1842 essay, Mayer expressed his results for the mechanical equivalent of heat by stating that the warming of a given weight of water from 0°C to 1°C is equivalent to the mechanical energy acquired by the same weight of water in falling from a height of 365 m. Show that the value of J for these data is approximately 3.58 kJ/kcal. Assume that the specific heat of water is 1 kcal/kg·C°.

I. Soil has a specific heat of 0.80 kJ/kg·C°. Equal masses of soil and water absorb equal quantities of heat. The temperature of water increases 4 C°. Determine the temperature increase of the soil.

18.4 Thermodynamic Work

J. A block of aluminum in the form of a cube with a volume of 1 m³ expands. The surrounding pressure is 1 atm. The length of each edge increases by 1 mm. Determine the work done by the block.

K. A fluid expands at a constant pressure of 6.2 × 10⁵ N/m², increasing its volume by 3.1 m³. Determine the work done by the fluid.

L. A gas expands from an initial volume of 1.0 m³ to a final volume of 2.0 m³. The pressure decreases as shown in Figure 1. Determine the work done by the gas.

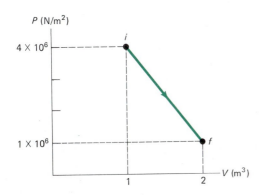

FIGURE 1

M. (a) Determine the work done by a fluid that expands from *i* to *f* as indicated in Figure 2. (b) How much work is performed by the fluid if it is compressed from *f* to *i* along the same path?

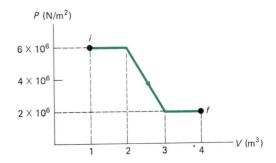

FIGURE 2

N. A certain gas obeys the equation of state

$$\frac{PV}{T} = \text{constant}$$

(a) Use a P-V diagram to trace the path of processes in which the gas expands, doubling its volume, (i) isobarically and (ii) isothermally. (b) Use the geometric interpretation of work to determine in which process the gas performs more work.

18.5 Internal Energy

P. The internal energy of a system is doubled as the result of a thermodynamic process. Is there any way to decide whether the increase came about as a result of the absorption of heat or through the performance of work?

18.6 The First Law of Thermodynamics

Q. Electric utilities charge for energy usage on the basis of the kilowatt hour (kWh). Show that 1 kWh is approximately equal to 3600 kJ.

R. The quad is an energy unit

$$1 \text{ quadrillion Btu} = 10^{15} \text{ Btu} \equiv 1 \text{ quad}$$

The megaton is another energy unit. One megaton is the energy released in the explosion of 1 million tons of TNT.

$$1 \text{ megaton} \equiv 4 \times 10^{12} \text{ kJ}$$

The yearly energy consumption in the United States totals nearly 100 quads. Show that the yearly energy consumption is approximately 25,000 megatons.

S. The environment performs 17 MJ of work on the system of Example 8. If the system absorbs 5 MJ of heat, determine its change in internal energy.

T. For many years the U.S. Patent Office was deluged with applications for patents on various types of perpetual motion machines. This flood subsided abruptly when the rules were amended to require that a working model accompany an application. Suppose that you were employed in the Patent Office prior to this ruling and that you received an application for a patent on a device that claimed the following performance figures:

Net heat consumption per cycle = 40 kJ

Mechanical work output per cycle = 100 kJ

Electrical work output per cycle = 10^5 W·s

Would it be necessary to give detailed consideration to the application? Explain.

U.   A pond of water absorbs 40 MJ of heat. The temperature of the pond rises and the water expands slightly performing 600 kJ of work against the atmosphere. Determine the change in the internal energy of the pond.

18.7 Applications of the First Law

V.   A fluid is carried through the cycle indicated in Figure 3. How much work (in kJ) is done by the fluid during (a) the expansion from 1 to 2, (b) the constant-volume decompression from 2 to 3, (c) the compression from 3 to 4? (d) What is the net amount of heat (in kJ) transformed into work during the cycle?

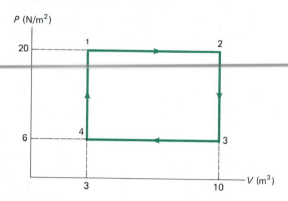

FIGURE 3

W.   The wreckage of a submarine shows charring and other evidence of high temperatures. If the hull cracked at great depth, what would happen to the air inside the submarine? Does the air absorb any heat?

PROBLEMS

1.   The electrical resistance, R, of a thermistor is related to its kelvin temperature, T, by

$$R = R_0 e^{B/T}$$

where R_0 and B are constants. (a) Show that the relation between R and T can also be expressed as

$$\ln R = \ln R_0 + \frac{B}{T}$$

(b) Use the data below for a thermistor to determine the constants R_0 and B.

Temperature (°C)	Resistance (ohms)
0	5700
25	2000
50	810
100	185
150	59
200	25

(c) Using your values of R_0 and B determine the temperature corresponding to a resistance of 900 ohms.

2.   The paths of three cyclic processes are indicated in Figure 4. Indicate for each part whether the net work done is positive, negative, or zero. Explain the basis for your choice.

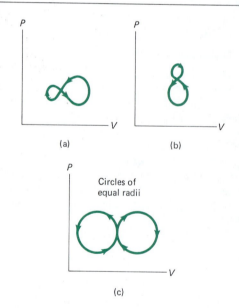

FIGURE 4

3.   A fluid undergoes a cyclic process. The P-V path is a "circle," as shown in Figure 5. Determine the net work done by the fluid.

4.   A gas undergoes a controlled expansion during which its pressure, P, is related to its volume, V, by

$$P = a \cdot V^{3/2} \qquad a = 10^5 \text{ N·m}^{-13/2}$$

Calculate the work done by the gas when its volume increases from 1 m³ to 2 m³.

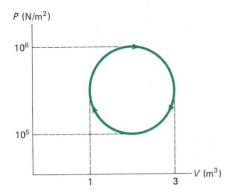

P (N/m^2)

10^6

10^5

1   3

V (m^3)

FIGURE 5

5. Helium with an initial volume of 1 liter (10^{-3} m^3) and an initial pressure of 10 atm expands to a final volume of 1 m^3. The relationship between pressure and volume during the expansion is $PV =$ constant. Determine (a) the value of the constant, (b) the final pressure, (c) the work done by the helium during the expansion.

6. The condenser of a steam turbine in a large electric power plant is cooled by water in contact with the condenser. Water flows continuously through the condenser and warms in the process. Typically, heat flows into the water at a rate of 1500 megawatts. (a) If the temperature rise of the water is held to 10 C$^\circ$, what is the rate of flow of water (in cubic meters per second) through the condenser? (b) The use of water in a city equals 150 gallons per person per day. What is the population of a city having a water-usage rate equal to the rate of water flow through the condenser in part (a)? (100 gallons = 0.379 cubic meter.)

7. A rubber ball dropped from a height of 1 m rebounds to a height of 0.5 m. The decrease in gravitational potential energy is accompanied by a rise in the ball temperature of 6×10^{-3} C$^\circ$. Determine the specific heat of the ball.

8. At very low temperatures solids exhibit temperature-dependent specific heats. For example, the specific heat of solid argon at constant volume is given by

$$C_v = (0.063 \text{ J/kg} \cdot \text{K}^4)T^3$$

with T expressed in kelvins. Determine the heat required to raise the temperature of one mole of solid

argon from 0.001 K to 2 K at constant volume. The atomic weight of argon is 39.95.

9. Three trays of ice cubes at 0°C are placed in a container of water at 25°C. All of the ice (1.08 kg) melts, resulting in a final temperature of 5°C. How many quarts of water were originally in the container? (Ignore the heat ejected by the container.)

10. The specific heat of a substance is given by

$$C = a + bT$$

where $a = 1.12$ kJ/kg$\cdot$K and $b = 0.016$ kJ/kg$\cdot$K^2. Determine the amount of heat required to raise the temperature of 1.20 kg of the material from 280 K to 312 K.

11. A student obtains the following data in a method-of-mixtures experiment designed to measure the specific heat of aluminum:

Initial temperature of water and calorimeter: 70°C
Mass of water: 0.400 kg
Mass of calorimeter: 0.040 kg
Specific heat of calorimeter: 0.63 kJ/kg$\cdot$C$^\circ$
Initial temperature of aluminum: 27°C
Mass of aluminum: 0.200 kg
Final temperature of mixture: 66.3°C

Use these data to determine the specific heat of aluminum. Your result should be within 15% of the value listed in Table 18.1.

12. The air temperature above coastal areas is profoundly influenced by the large specific heat of water (4.19 kJ/kg$\cdot$C$^\circ$). One reason is that the heat released when 1 cubic meter of water cools by 1 C$^\circ$ will raise the temperature of an enormously larger volume of air by 1 C$^\circ$. Estimate this volume of air. The specific heat of air is approximately 1.0 kJ/kg$\cdot$C$^\circ$. Take the density of air to be 1.3 kg/m^3.

13. During a controlled expansion, the pressure of a gas is given by

$$P = 12e^{-bV} \text{ atm} \qquad b = \frac{1}{12 \text{ m}^3}$$

where the volume V is expressed in m^3. Determine the work performed when the gas expands from $V = 12$ m^3 to $V = 36$ m^3.

CHAPTER 19

THERMAL PROPERTIES OF MATTER

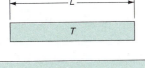
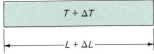

FIGURE 19.1

A small temperature change, ΔT, results in a change in length, ΔL. Experiment shows that ΔL is directly proportional to ΔT and directly proportional to L.

19.1
THERMAL EXPANSION

Have you ever noticed that electric power lines expand and sag most noticeably in hot weather? This is an example of thermal expansion, which is where we will begin our study of thermal properties.

If a rod of length L and temperature T experiences a small change in temperature ΔT, its length changes by an amount proportional to both ΔT and L (Figure 19.1). The change in length, ΔL, is proportional to the product $L\,\Delta T$. We introduce a proportionality factor α characteristic of the solid and we write

$$\Delta L = \alpha L\,\Delta T \tag{19.1}$$

The quantity α is called the **coefficient of linear expansion**. If we set $\Delta T = 1\ \mathrm{C}^\circ$, we can see from Equation 19.1 that α represents the fractional change in length for a temperature change of 1 Celsius degree.

Table 19.1 presents values of α for a variety of materials. All of these values are close to $10^{-5}/\mathrm{C}^\circ$. We can get a feeling for thermal expansion by noting that a solid 1 km long with $\alpha = 10^{-5}/\mathrm{C}^\circ$ increases its length by 1 cm for each 1-degree rise in temperature.

Materials with near-zero values of α maintain their dimensions and their shapes very well in spite of temperature changes. Such materials can be very useful. For example, the mirrors used in astronomical telescopes must have a paraboloidal reflecting surface in order to focus light sharply. Thermal expansion or contraction alters the shape of the surface. The glass used for the 200-inch Hale telescope on Mount Palomar has $\alpha = 0.325 \times 10^{-5}/\mathrm{C}^\circ$. The glass to be used in the mirror of the Space Telescope, which will be put in orbit in 1990, has a value of α at least 100 times smaller than that of the Hale mirror.

TABLE 19.1
Coefficients of Expansion for Selected Liquids and Solids*

Solids	α (units of $10^{-5}/\mathrm{C}^\circ$)	Liquids	β (units of $10^{-3}/\mathrm{C}^\circ$)
Silver (Ag)	1.89	Acetic acid	3.21
Aluminum (Al)	2.80	Acetone	4.47
Gold (Au)	1.42	Benzene	3.72
Calcium (Ca)	2.50	Carbon tetrachloride	3.72
Iron (Fe)	1.17	Ether	4.98
Sodium (Na)	7.1	Mercury	0.54
Lead (Pb)	2.91	Pentane	4.83
Rubidium (Rb)	9.0	Water	0.63
Zinc (Zn)	3.3		
Carbon steel	1.15		
Glass	0.90		
Concrete	1.2		
Brass	2.0		
Copper (Cu)	1.7		
Magnesium (Mg)	2.5		

* The column entry for α expresses the coefficient of linear expansion in units of $10^{-5}/\mathrm{C}^\circ$. Thus, the value of α for silver is $1.89 \times 10^{-5}/\mathrm{C}^\circ$. For the coefficient of volume expansion, β, the units are $10^{-3}/\mathrm{C}^\circ$. Thus, the value of β for acetic acid is $3.21 \times 10^{-3}/\mathrm{C}^\circ$. Values refer to measurements at a pressure of 1 atm and for a temperature near 20° C. Values for liquids are adapted from J. H. Perry, *Chemical Engineer's Handbook*, McGraw-Hill, New York, 1950. Used with permission of McGraw-Hill Book Company. Values for most solids are adapted from *International Critical Tables of Numerical Data: Physics, Chemistry, and Technology*, with permission of the National Academy of Sciences, Washington, D.C.

EXAMPLE 1

Why Telephone Calls Travel Farther in Warm Weather

The value of α for copper is $1.7 \times 10^{-5}/\text{C}°$. A copper telephone wire 10 km long at a temperature of 16°C (60°F) lengthens when the temperature rises. At 36°C (96°F), the increase in length is

$$\Delta L = \alpha L \, \Delta T = (1.7 \times 10^{-5}/\text{C}°)(10^4 \text{ m})(20 \text{ C}°)$$

$$= 3.4 \text{ m}$$

Differences in the thermal expansion of two solids can be demonstrated using a bimetallic strip. This consists of two pieces of metal bound together, each having a different value of α (Figure 19.2). An increase in temperature produces unequal changes in their lengths, and results in a curvature of the bimetallic strip. A decrease in temperature results in a curvature in the opposite direction. Bimetallic strips are used to activate electrical switches in home thermostats.

Temperature changes that alter the dimensions of an object change its volume. If a substance of volume V experiences a small change in temperature, ΔT, its volume changes by an amount proportional to V and ΔT. We express the change in volume as

$$\Delta V = \beta V \, \Delta T \tag{19.2}$$

where β is called the **coefficient of volume expansion**.

Typical values of β for liquids are listed in Table 19.1. The tabulated values of β for liquids cluster near $10^{-3}/\text{C}°$. This means that a liquid expands by about one part in 10^3 for each degree of change in temperature. A few substances exhibit negative values of β over certain ranges of temperature. Most notable of these is water, for which β is negative over the narrow range of temperature from 0°C to 3.98°C. Thus, if water at 1°C is warmed to 2°C, its volume decreases and its density increases. The maximum density of water occurs at a temperature of 3.98°C (at a pressure of 1 atm). Figure 19.3 shows a graph of the density of water over a narrow range of temperature.

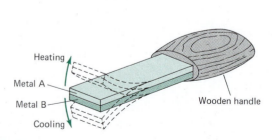

FIGURE 19.2

A bimetallic strip is used to demonstrate differences in thermal expansion. Two metals with different values of α are bound together. An increase in temperature produces unequal changes in their lengths and results in a curvature. A decrease in temperature produces a curvature in the opposite direction.

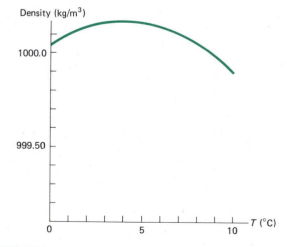

FIGURE 19.3

The density of water has a maximum at 3.98°C.

Table 19.1 lists values of β only for liquids. For isotropic solids, β and α are related by

$$\beta = 3\alpha \tag{19.3}$$

(See Problem 19.)

19.2
THE IDEAL GAS

Solids and liquids expand only slightly when heated. By comparison, gases expand greatly when heated. Experiments show that the thermal properties of many different gases are strikingly similar. These similarities are embodied in the concept of an **ideal gas**. From an experimental viewpoint, the ideal gas model is an extrapolation that describes basic features common to all real gases.

The ideal gas model offers both practical and conceptual advantages. The engineer who wants a rough idea of how a temperature change will affect the pressure of a particular gas can rely on the ideal gas model for an approximate answer. The ideal gas model also makes it relatively easy to grasp certain thermodynamic concepts. The ideal gas plays the same role in thermodynamics as does the point mass in mechanics. It can be used to illustrate the principles of thermodynamics at an intuitive level.

The ideal gas is characterized by two relations: (1) an equation of state and (2) an equation for the heat capacity.

1   Equation of State   In general, an equation of state relates two or more measurable properties of a system. The *ideal gas equation of state* is

$$PV = vRT \tag{19.4}$$

where P, V, and T are the pressure, volume, and temperature of the gas, v denotes the number of moles* of gas, and R is the universal gas constant,

$$R = 8.31434 \text{ joules/kelvin} \tag{19.5}$$

For a fixed amount of gas ($v = $ constant) the ideal gas equation of state shows that only two of the three variables P, V, and T are independent. For example, if P and T are used as independent variables, then V is determined by Equation 19.4.

EXAMPLE 2
Molar Volume

Consider 1 mole of gas at the standard temperature and pressure:

$$P = 1 \text{ atm} = 1.01 \times 10^5 \text{ N/m}^2$$

$$T = 0°C = 273 \text{ K}$$

---

* By definition, 1 mole is the amount of a substance that contains the same number of molecules (or atoms in the case of monatomic substances) as there are carbon atoms in 12 grams of carbon 12. This number is Avogadro's number, $N_A = 6.022 \times 10^{23}$. The mass (in grams) of 1 mole of a substance is called its *molecular weight* (see Section 22.1).

The volume occupied by 1 mole of a gas at this pressure and temperature is given approximately by the ideal gas equation of state:

$$V = \frac{vRT}{P} = \frac{1(8.31 \text{ J/K})(273 \text{ K})}{1.01 \times 10^5 \text{ N/m}^2}$$

$$= 2.24 \times 10^{-2} \text{ m}^3$$

$$= 22{,}400 \text{ cm}^3$$

This value is approximate because real gases do not conform precisely to the ideal gas equation of state under standard conditions. Real gases are better described by more complicated equations of state. However, even in these equations only two of the three variables P, V, and T are independent.

2   Heat Capacity Equation

The second relation that characterizes the ideal gas is the equation for the specific heat. Experiment shows that the specific heat of a gas is nearly constant over a moderate range of temperature. For an ideal gas the *specific heat* is taken to be *constant*.

We now assume a constant specific heat and use the first law of thermodynamics to show that the internal energy of the ideal gas increases linearly with its Kelvin temperature. The differential form of the first law of thermodynamics is

$$dQ = dU + dW \tag{19.6}$$

where dQ is the heat absorbed by the gas, dU is the change in the internal energy of the gas, and dW is the work performed by the gas. We can express dQ as

$$dQ = mC \, dT$$

where m is the mass of the gas, C is its specific heat, and dT is the change in its temperature. The work done by the gas is given by

$$dW = P \, dV$$

Substituting these expressions for dQ and dW into Equation 19.6 gives

$$mC \, dT = dU + P \, dV \tag{19.7}$$

If heat is added with the volume of the gas held constant, then $dV = 0$, in which case Equation 19.7 reduces to

$$mC_v \, dT = dU \tag{19.8}$$

The subscript v is a reminder that Equation 19.8 applies to the condition of heat exchange at constant volume. For the ideal gas the heat capacity C_v is independent of temperature. This permits us to integrate Equation 19.8 to obtain

$$U = mC_v T + \text{constant}$$

This is the desired result: The internal energy of the ideal gas increases linearly with its temperature. We take the integration constant to be zero and we write

$$U = mC_v T \tag{19.9}$$

Heat added to a gas at constant volume is used entirely to increase the internal energy because no work is done. The same amount of heat added at constant

TABLE 19.2
Molar Heat Capacities of Gases

Gas	MC_p(J/K)	MC_v(J/K)	$MC_p - MC_v$ (J/K)	$\gamma = C_p/C_v$
Helium (He)	20.93	12.61	8.32	1.66
Argon (Ar)	20.93	12.53	8.40	1.67
Hydrogen (H_2)	28.76	20.39	8.37	1.41
Nitrogen (N_2)	29.05	20.75	8.30	1.40
Oxygen (O_2)	29.43	21.02	8.41	1.40

pressure produces a smaller rise in temperature because the gas expands and converts some of the heat into work. As a consequence, the specific heat is larger at constant pressure than at constant volume.

Let C_p denote the specific heat at constant pressure and let dQ_p be the heat absorbed by the gas at constant pressure. Then

$$dQ_p = mC_p\, dT \tag{19.10}$$

For heat added at constant pressure the first law has the form

$$mC_p\, dT = dU + P\, dV \tag{19.11}$$

With pressure constant we can form the differential $P\, dV$ using the ideal gas equation of state, $PV = vRT$. This gives

$$P\, dV = vR\, dT \tag{19.12}$$

Using Equations 19.8 and 19.12 in Equation 19.11 gives

$$mC_p = mC_v + vR \tag{19.13}$$

We can conclude from this equation that $C_p > C_v$, as we asserted earlier. Equation 19.13 also shows that C_p is a constant for an ideal gas, because C_v and R are constants. For 1 mole of gas ($v = 1$, $m \equiv M$, the molecular weight), Equation 19.13 shows that the *molar heat capacities*, MC_p and MC_v, differ by $R = 8.31$ J/K.

$$MC_p - MC_v = R \tag{19.14}$$

Table 19.2 shows that the molar heat capacities of various real gases are in good agreement with Equation 19.14.

Reversible Adiabatic Process

An adiabatic process is one in which there is no exchange of heat. Many processes occur so rapidly that there is little heat transfer. The rapid expansions and compressions of gases in internal-combustion engines constitute one example. Gases are poor conductors of heat and the expansions and compressions occur more rapidly than heat can be absorbed or ejected, allowing the process to be treated as adiabatic. Furthermore, such expansions and compressions are approximately *reversible* because pressure variations within the gas adjust themselves at the speed of sound. So long as the piston moves at subsonic speeds the gas pressure remains essentially uniform and equal to the pressure exerted by the piston.

Let's consider a reversible adiabatic transformation of an ideal gas. The fact that there is no heat exchange will let us establish another relation among P, V, and T, in addition to the equation of state. An adiabatic process is defined by the condition $dQ = 0$, and the first law becomes

$$dU + P \, dV = 0 \tag{19.15}$$

For the ideal gas

$$dU = mC_v \, dT$$

and

$$P = \frac{vRT}{V}$$

Inserting these expressions into Equation 19.15 and dividing through by mC_v gives

$$dT + \frac{vRT}{mC_v}\left(\frac{dV}{V}\right) = 0 \tag{19.16}$$

From Equation 19.13 we find

$$\frac{vR}{mC_v} = \frac{m(C_p - C_v)}{mC_v} = \frac{C_p}{C_v} - 1$$

The ratio of specific heats C_p/C_v is denoted by γ:

$$\frac{C_p}{C_v} = \gamma \tag{19.17}$$

The quantity γ is sometimes referred to as the **adiabatic exponent**, for reasons that will become apparent shortly.

For an ideal gas C_p and C_v are constants, so γ is a constant. In terms of γ,

$$\frac{vR}{mC_v} = \gamma - 1$$

We can see from this relation that $\gamma - 1$ is a positive number. Replacing vR/mC_v by $\gamma - 1$ in Equation 19.16 and multiplying through by $V^{\gamma-1}$ leads to

$$V^{\gamma-1} \, dT + T(\gamma - 1)V^{\gamma-2} \, dV = d(TV^{\gamma-1}) = 0$$

This shows that the differential of $TV^{\gamma-1}$ is zero during an adiabatic process. It follows that $TV^{\gamma-1}$ is a constant for a reversible adiabatic transformation of an ideal gas:

$$TV^{\gamma-1} = \text{constant} \tag{19.18}$$

Equation 19.18 can be used to show in a quantitative way how an adiabatic compression raises the temperature of a gas. To illustrate, let (T_0, V_0) denote the initial values of temperature and volume. The final temperature, T_f, follows from Equation 19.18 as

$$T_f = T_0 \left(\frac{V_0}{V_f}\right)^{\gamma-1} \tag{19.19}$$

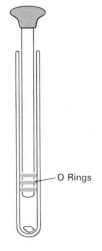

FIGURE 19.4

A device for demonstrating the temperature increase that occurs during an adiabatic compression. A glass tube serves as a transparent cylinder in which air is compressed by a plunger fitted with rubber O-rings. A small piece of tissue paper placed in the cylinder will flash to incandescence when the air is suddenly compressed.

where V_f is the final volume. For a compression, V_0/V_f is greater than unity, and so T_f exceeds T_0.

EXAMPLE 3

Just Hot Air

Adiabatic compression can lead to a substantial increase in temperature. The piston-fitted cylinder device shown in Figure 19.4 can be used to demonstrate this fact. A small piece of tissue paper is placed inside the cylinder. If the piston is forced downward rapidly, the air is compressed adiabatically. The piston moves at a subsonic speed and so the compression is nearly reversible. To the extent that the air behaves like an ideal gas, the temperature increases in accord with Equation 19.19. The tissue paper flashes to incandescence, revealing the rapid increase in temperature.

In a typical compression the volume decreases by a factor of 20. Both nitrogen and oxygen, the primary constituents of air, have $\gamma = 1.4$. If we take $T_0 = 300$ K and $V_0/V_f = 20$, Equation 19.19 gives

$$T_f = 300 \text{ K} \cdot (20)^{0.4}$$
$$= 994 \text{ K}$$

a temperature well above the flash point of tissue. Note that absolute temperatures must be used in Equation 19.19.

In an ordinary automobile engine, an air-fuel mixture is ignited by a spark. In a diesel engine there are no spark plugs. The air-fuel mixture is raised to ignition temperature by a nearly adiabatic compression. The high temperatures at which ordinary automobile engines operate often cause dieseling—that is, the engine continues to run for a few moments after the ignition has been turned off.

We can obtain a relationship between the pressure and temperature during a reversible adiabatic process by noting that the ideal gas equation of state for a fixed amount of gas can be expressed as

$$\frac{P_f V_f}{T_f} = \frac{P_0 V_0}{T_0}$$

Using this relation to eliminate V_0/V_f from Equation 19.19 gives

$$\frac{T_f}{T_0} = \left(\frac{P_f}{P_0}\right)^{(\gamma - 1)/\gamma} \tag{19.20}$$

EXAMPLE 4

Adiabatic Expansion and Gas Liquefaction

In one method used to liquefy a gas the temperature is lowered by causing the gas to expand and thereby to perform PdV work against its surroundings. If the expansion is adiabatic or nearly so, the work is done at the expense of the internal energy of the gas, and its temperature decreases. That is, $U = mC_v T$ so if U decreases so must T. We can get a rough idea of the cooling produced by treating the gas as ideal. We take $\gamma = 1.40$, a value appropriate for diatomic gases (Table 19.2). For initial-state values we set

$$P_0 = 50 \text{ atm} \qquad T_0 = 300 \text{ K}$$

For the final-state pressure we take

$$P_f = 1 \text{ atm}$$

With $P_f/P_0 = 1/50$, the final temperature follows from Equation 19.20 as

$$T_f = T_0 \left(\frac{P_f}{P_0}\right)^{(\gamma-1)/\gamma} = 300 \text{ K} \left(\frac{1}{50}\right)^{0.283}$$

$$\approx 99 \text{ K}$$

Real gases depart significantly from the ideal behavior at a pressure of 50 atm. Nevertheless the result is not unrealistic. Adiabatic expansion engines are prominent features of gas liquefiers.

19.3
P-V-T SURFACES

A *pure* substance is one that is composed of a single chemical species rather than a mixture of chemically distinct components. Water, for example, consists only of the compound H_2O, and is a pure substance. The equilibrium states of a fixed amount of a pure substance can be described by three variables: pressure, volume, and temperature (P, V, T). It is convenient to take the fixed amount to be one mole so that V becomes the volume of one mole, or the **molar volume**.

The equation of state specifies one relation among $P, V,$ and T so that only two are independent variables. For example, the equation of state for one mole of an ideal gas is $PV = RT$. From a geometric viewpoint the equation of state defines a *surface* in a space in which $P, V,$ and T are the coordinates. Figure 19.5 shows the P-V-T surface for an ideal gas. Figures 19.6 and 19.7 show the P-V-T surface for carbon dioxide. Using a P-V-T surface implicitly restricts us to reversible processes. Only for reversible processes can we view the states of the system as a sequence of equilibrium states, related by the equation of state embodied by the P-V-T surface.

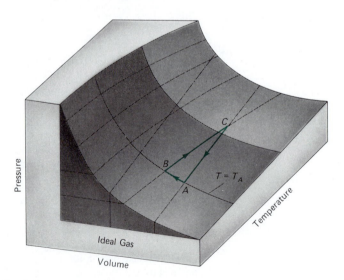

FIGURE 19.5

The P-V-T surface for an ideal gas. The path from A to B describes an isothermal (constant temperature) compression of the gas. The path from B to C describes an isobaric (constant pressure) expansion. The path from C to A describes an isovolumic (constant volume) decompression.

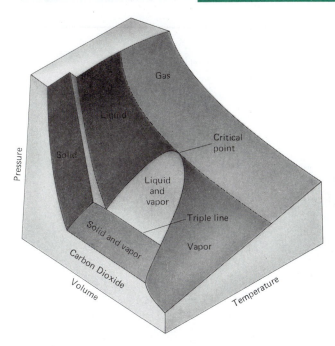

FIGURE 19.6

The *P-V-T* surface for carbon dioxide. The dashed line marks the critical isotherm. The appearance of the surface, when viewed along the volume axis, is shown in Figure 19.8.

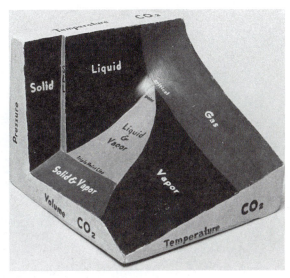

FIGURE 19.7

A model of the *P-V-T* surface for carbon dioxide.

P-V-T surfaces are helpful in following the progress of various reversible processes. Figure 19.5 shows a closed path $A \rightarrow B \rightarrow C \rightarrow A$ on the ideal gas surface. The path $A \rightarrow B$ describes an isothermal compression (T = constant). The path $B \rightarrow C$ describes an isobaric expansion (P = constant). The segment $C \rightarrow A$ describes a process in which the volume remains constant while pressure and temperature decrease.

EXAMPLE 5

Ideal Gas Cycle

Consider 1 mole of air (treated as an ideal gas) that is carried through the reversible cycle of Figure 19.5. Let point A correspond to the conditions

$$P_A = 1 \text{ atm} \qquad T_A = 273 \text{ K} \qquad V_A = 10 \text{ m}^3$$

Suppose that the isothermal compression $A \rightarrow B$ reduces the volume to 5 m³. It follows from the equation of state $PV = RT$ that the pressure doubles if V is reduced to half its initial value while T remains constant. Thus, the conditions at point B are

$$P_B = 2 \text{ atm} \qquad V_B = 5 \text{ m}^3 \qquad T_B = T_A = 273 \text{ K}$$

The path from B to C is an expansion that proceeds at constant pressure. The volume doubles, becoming 10 m³ at C. The equation of state shows that the temperature must also double if pressure remains constant. Thus,

$$P_C = 2 \text{ atm} \qquad V_C = 10 \text{ m}^3 \qquad T_C = 546 \text{ K}$$

Finally, $C \rightarrow A$ proceeds at a constant volume of 10 m³. Heat is ejected and the pressure and temperature return to their original values of 1 atm and 273 K.

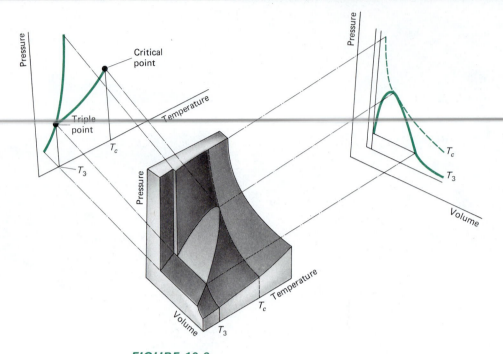

FIGURE 19.8
Projections of the P-V-T surface onto a P-V diagram and a P-T diagram.

Figures 19.6 and 19.7 reveal that certain portions of the P-V-T surface correspond to states where two phases coexist in thermal equilibrium. An ice cube in equilibrium with water is a familiar example of such a condition. The **triple line** marks states where liquid, solid, and vapor* simultaneously exist in equilibrium.

The **critical point** marks the upper terminus of the liquid-vapor coexistence region. It is located at a unique pressure, temperature, and molar volume characteristic of the substance (P_C, T_C, V_C). For example, for water $P_C = 217.7$ atm, $T_C = 647.3$ K, and $V_C = 5.7 \times 10^{-5}$ m^3. If you look at a liquid confined in a test tube, the meniscus marks the liquid-vapor boundary. This meniscus disappears at the critical point. If a gas is compressed at a temperature in excess of the critical temperature, it never undergoes condensation—it remains a gas. If a gas is compressed at a temperature less than the critical temperature, it separates into liquid and vapor phases.

In many situations it is convenient to view a projection of the P-V-T surface. Figure 19.8 shows projections of a P-V-T surface onto the P-T plane and onto the P-V plane. Figure 19.9 shows the P-T diagram for carbon dioxide. The triple line projects onto the P-T diagram as the triple point. The portion of the P-V-T surface labeled "liquid and vapor" in Figure 19.6 appears as a line on the P-T diagram—the vaporization curve. The other coexistence regions likewise appear as lines on the P-T plane. The solid-liquid region projects onto the P-T plane as the line designated "melting curve," or the fusion line. The solid-gas region projects onto the P-T plane as the sublimation curve. The physical significance of these various lines on the P-T diagram is discussed in the next section.

---

* The word *vapor* is often used to describe a gas that is in equilibrium with the liquid or solid phase. We use the words *vapor* and *gas* interchangeably.

FIGURE 19.9

P-T diagram for carbon dioxide (not to scale). The aqua lines define conditions for stable equilibrium between two phases. They intersect at the triple point (5.1 atm, 216.6 K). Carbon dioxide is typical of most pure substances in that it expands upon melting. One consequence is that the slope of the melting curve is positive ($dP/dT > 0$). If the pressure is raised above 5.1 atm, heating the solid may result in melting. The values at the critical point are 72.9 atm, and 309.1 K.

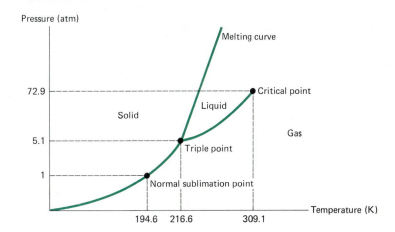

19.4
CHANGE OF PHASE

A solid can be heated until it melts and becomes a liquid. The liquid in turn can be heated until it reaches its boiling point and vaporizes. For pure substances, such phase changes occur at a fixed temperature. There is an exchange of heat between the substance and its environment, but the temperature remains constant during the change of phase. The **heat of fusion**, L_f, is the amount of heat per kilogram absorbed in converting a solid to a liquid. The **heat of vaporization**, L_v, is the amount of heat per kilogram absorbed in converting a liquid to a vapor. The process called **sublimation** is one in which a solid is converted to a vapor without passing through the liquid phase. The sublimation of "dry ice" (solid CO_2) is a familiar example. The sublimation process involves a **heat of sublimation**, L_s, the amount of heat per kilogram absorbed in the solid-to-vapor conversion. Table 19.3 lists latent heats for various substances.

Sublimation is a process that has found its niche in our high-tech world. Each year floods and fires result in water damage to thousands of books and documents. Permanent damage can be minimized by *freezing* the water-soaked materials. The ice is then removed by forcing it to *sublime* in a reduced-pressure enclosure. A similiar technique is used to produce freeze-dried coffee crystals.

A heating curve illustrates the relationship between temperature and heat absorbed. Figure 19.10 shows the heating curve for ice and water near the melting point. It is a plot of temperature versus Q, the amount of heat absorbed. If heat is added at a constant rate, the amount of heat absorbed is directly

FIGURE 19.10

The heating curve for a sample of *m* kilograms of ice. The initial temperature (point *A*) is $-5°C$. Heat is added at a constant rate, making the absorbed heat (Q) directly proportional to the heating time, *t*. At B the ice has reached the melting point. During the change of phase (melting) the temperature remains constant. At *C* the ice has melted completely. The specific heat of ice is one half that of water. This is reflected in the slope of the heating curve. The slope from *A* to *B* is twice the slope from *C* to *D*.

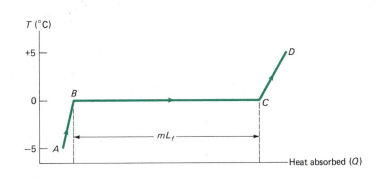

TABLE 19.3
Latent Heats

Substance	Normal Melting Point (K)	Latent Heat of Fusion (kJ/kg)	Normal Boiling Point (K)	Latent Heat of Vaporization (kJ/kg)
Water	273.16	334	373.16	2,260
Silver	1234	111	2485	2,360
Aluminum	931.7	399	2600	10,500
Gold	1336.16	64.4	2933	62,300
Cesium	301.9	15.7	963	514
Copper	1356.2	205	2868	4,800
Germanium	1232	479	2980	3,920
Lithium	459	416	1640	19,600
Sodium	371	115	1187	4,260
Lead	600.6	23.1	2023	859
Silicon	1683	1650	2750	10,600
Zinc	692.7	102	1180	1,760
Helium-4	—	—	4.2	20.5
Hydrogen (H_2)	14.0	57.8	20.3	452
Neon	24.5	13.8	27.1	85.8
Nitrogen (N_2)	63.2	25.8	77.4	199
Argon	83.8	29.6	87.3	163
Oxygen (O_2)	54.4	13.9	90.2	213
Xenon	161	17.6	165	396

proportional to the heating time, t. The heating curve shows that the sample starts as ice at $-5°C$. The addition of heat at a constant rate results in a uniform temperature rise until the ice reaches $0°C$. No further increase of temperature results until all of the ice has melted. For a sample of m kg, the total heat absorbed during the phase change is mL_f, as indicated in Figure 19.10. Once all of the ice has melted, the continued addition of heat causes the temperature rise to resume. The slope of the heating curve is inversely proportional to the specific heat. The temperature rises most slowly when the specific heat is greatest.

EXAMPLE 6

The Heat of Fusion for Ice

A student performs an experiment in which a heating curve is used to determine the latent heat of ice. Heat is added at a constant rate of 100 W to 20 grams of ice. The time during which the temperature remains constant (melting time) is 66 s. The total heat absorbed during melting may be expressed as

$$mL_f = \text{(rate at which heat is added)} \cdot \text{(heating time)}$$

For the latent heat of fusion this gives

$$L_f = \frac{100 \text{ J/s (66 s)}}{0.02 \text{ kg}}$$

$$= 330 \text{ kJ/kg}$$

This compares favorably with the accepted value of 333.6 kJ/kg.

The temperature at which a solid melts is called the **melting point**. The melting point varies with pressure. For example, Figure 19.9 shows that the melting point of carbon dioxide increases with pressure. The melting point at a pressure of 1 atm is called the **normal melting point**. Similar terms are used for boiling and sublimation. The temperature at which a liquid boils is called the **boiling point** and the temperature at which a solid sublimes is called the **sublimation point**. The boiling point at a pressure of 1 atm is called the **normal boiling point**. The heats of fusion, vaporization, and sublimation also vary with pressure. For example, as a substance approaches the critical point, two phases become one and the heat of vaporization drops to zero (Figure 19.6).

EXAMPLE 7

A Heat Calculation Involving Melting

A mixture composed of 0.160 kg of ice and 0.320 kg of water is initially at 0°C. What is the amount of heat required to melt the ice and then raise the water to a temperature of 16°C? The latent heat of fusion of ice is 334 kJ/kg (Table 19.3). The heat required to melt the ice is

$$Q_{\text{melt}} = m_{\text{ice}}L_f = 0.160 \text{ kg } (334 \text{ kJ/kg})$$
$$= 53.4 \text{ kJ}$$

The heat required to raise the 0.480 kg of water from 0°C to 16°C is

$$Q_{\text{water}} = m_{\text{water}}C_{\text{water}} \, \Delta T$$
$$= 0.480 \text{ kg } (4.19 \text{ kJ/kg} \cdot \text{C}°)16 \text{ C}°$$
$$= 32.2 \text{ kJ}$$

The total heat required is 85.6 kJ.

Most of the heat absorbed during a phase change is used to alter the microscopic structure of the substance. For example, when a solid melts most of the heat absorbed (the heat of fusion) is used to break the bonds between atoms. A tiny fraction of the heat of fusion is converted into the $P \, dV$ work associated with a change of volume. In sublimation and vaporization, however, the volume of the substance increases substantially and a significant portion of the heat absorbed is used to perform the work of expansion.

EXAMPLE 8

P dV *Work during Vaporization*

The heat of vaporization for water at a pressure of 1 atm is 2256 kJ/kg. The vaporization of water at the boiling point (373.15 K) increases its volume by a factor of 1761. Thus, 1 kg of liquid, which occupies 0.001 m³, expands as a vapor to a volume of approximately 1.76 m³. A portion of the heat of vaporization goes into the $P \, dV$ work of expansion. The remainder is used to break molecular bonds and shows up as a change of internal energy. We apply the first law of thermodynamics,

$$Q = U_f - U_i + W$$

to the process in which 1 kg of water at its boiling point absorbs 2256 kJ and is vaporized.

$$Q = mL_v = 2256 \text{ kJ} = \text{heat of vaporization of 1 kg}$$

$$U_f - U_i = U_{\text{Vapor}} - U_{\text{Liquid}}$$
$$= \text{internal energy change between liquid and vapor phases}$$

$$W = \int_{V_L}^{V_V} P \, dV = \text{work of expansion against atmosphere}$$

Since the vapor expands at constant pressure the work expansion is given by

$$W = \int_{V_L}^{V_V} P \, dV = P(V_V - V_L)$$

Taking $P = 1$ atm $\approx 10^5$ N/m^2 and $V_V - V_L \approx 1.76$ m^3 gives

$$W \approx 176 \text{ kJ}$$

Comparing $W = 176$ kJ with $mL_v = 2256$ kJ, we conclude that about 8% of the heat of vaporization goes into $P \, dV$ work. The remaining 92% is accounted for by the increase in internal energy.

Vaporization

The vaporization of a liquid is readily understood from the microscopic viewpoint. By virtue of their kinetic energy some molecules are able to overcome the attractive intermolecular forces, escape the liquid, and become part of the vapor atmosphere above the liquid surface. When the vapor and liquid are in equilibrium we say the vapor is **saturated**.

In general, the pressure just above the surface of a liquid that is open to the atmosphere is partially due to the surrounding air and partially due to the vapor. For example, suppose you set a small pan of water on a heating element in a room where the atmospheric pressure is 1 atm, and slowly heat the water. The pressure throughout the room, including the region just above the surface of the water, will remain 1 atm. At 60°C the pressure of the water vapor is about $\frac{1}{4}$ atm. At this temperature, water vapor exerts a pressure of $\frac{1}{4}$ atm and air molecules exert a pressure of 3/4 atm. At 82°C the vapor pressure of water is about $\frac{1}{2}$ atm. Thus, when the temperature of the water reaches 82°C the 1-atm pressure just above the liquid surface is due in equal parts to the water vapor and the air molecules. As the temperature approaches 100°C, the water vapor contributes an increasingly larger fraction of the total pressure of 1 atm. When the water reaches 100°C, the vapor pressure equals 1 atm. The atmosphere immediately above the liquid contains only water vapor, and we say that the water has reached its normal boiling point. If heating continues, the water will boil away. Or, in other words, the liquid will continue to escape into the surrounding atmosphere while maintaining a pressure of 1 atm at the surface of the liquid.

The saturated vapor pressure depends on two things: (1) the physical nature of the molecules, and (2) the temperature. The lower the temperature, the lower the saturated vapor pressure is. In general, the boiling point of a liquid is the temperature at which the vapor pressure equals the surrounding atmospheric pressure. Consequently, if the atmospheric pressure is reduced, the boiling point is reduced. For example, if a pan of water is placed in a sealed container and the air is pumped out, the water begins to boil even though no heat is added.

Pumping away the air reduces the surrounding pressure until it equals the saturated vapor pressure, at which time boiling occurs. One consequence of this pressure dependence of the boiling point is that mountain climbers may find it impossible to cook certain foods properly. The atmospheric pressure decreases as they climb upward, and at high elevations the boiling point may be reduced to a temperature too low to cook foods properly.

Cooling via Evaporation

The relationship between vapor pressure and temperature is important in cryogenics, the science of low-temperature phenomena. If the vapor above a cryogenic liquid is pumped away, liquid evaporates in an attempt to maintain equilibrium. But vaporization requires heat and the only source of the heat of vaporization is the liquid itself. The vapor carries away thermal energy and the liquid is cooled. Using a pump to maintain a low vapor pressure, we can lower the temperature of a liquid—that is, we can cool it by forced evaporation. For example, the normal boiling point of liquid helium is 4.2 K. The temperature of liquid helium may be lowered to less than 1 K by forced evaporation.

The motion of the "drinking duck" (Figure 19.11) relies on cooling by evaporation and on the fact that the saturated vapor pressure depends strongly on temperature. The beak of the duck is covered with a layer of an absorptive fabric, much like a paper towel. If the beak is moistened and atmospheric conditions permit the liquid to evaporate, the duck begins to sway on its perch, and finally tips forward to "drink" from a cup of water. It then rights itself. So long as the beak is moist the drinking action continues. The fluid filling the duck is highly volatile (usually ethyl ether). The duck's behavior is a consequence of the cooling action of evaporation. Initially, the vapor trapped in the body of the duck is at the same temperature and pressure as that in the head. Evaporation cools the head and the ensuing condensation reduces the vapor pressure there. The vapor pressure in the body is then greater than

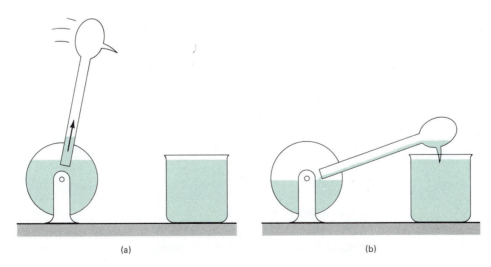

(a)   (b)

FIGURE 19.11

The drinking duck illustrates cooling by evaporation.

the vapor pressure in the head. The greater pressure in the body forces liquid upward in the stem and causes the center of gravity of the duck to rise. As the duck loses its balance and tips forward, the lower end of the stem breaks contact with the fluid in the body. The vapor itself then rises through the tube, allowing the liquid to drain and permitting the duck to right itself.

Plastic Flow, Melting, and Mountains

In Section 15.1 we showed that there is a limit to the depth of a hole in the ground. This limit exists because the pressure increases with depth, eventually reaching a value sufficient to crush any material. A similar effect limits the height of mountains. The limiting mountain height can be estimated by the following energy argument. Imagine that you set out to build a mountain by stacking blocks of granite one on top of the other. As the height of your mountain increases, the pressure and temperature at its base increase. Eventually, the pressure and temperature reach a value at which plastic flow occurs— that is, the bottom block turns into something resembling silly putty. Once plastic flow occurs at the base, the mountain has achieved its maximum height.

The amount of energy required to induce plastic flow is roughly the same as the heat of fusion, and we shall assume that they are equal. The energy required to melt a block of mass m is mL_f. The work done to lift a block of mass m to a height h is mgh. As long as

$$mgh < mL_f \tag{19.21}$$

it is energetically favorable to stack up blocks. Nature favors the configuration of minimum energy, and so the height of the mountain can be increased as long as the inequality of Equation 19.21 is satisfied. The limiting height is therefore set by the condition that the work done to lift a block to the top equals the energy required to melt a block at the base. Thus, we set

$$mgh_{max} = mL_f \tag{19.22}$$

The heat of fusion for rocks is roughly 250 kJ/kg. The resulting limit is

$$h_{max} = \frac{L_f}{g} = \frac{250 \text{ kJ/kg}}{9.80 \text{ m/s}^2}$$

$$\approx 25 \text{ km}$$

The highest mountains reach about one-half this height. Our calculated limit is too high, primarily because the energy required to produce plastic flow is less than L_f.

WORKED PROBLEM

The period of a pendulum is proportional to the square root of its length. A seconds pendulum has a period of precisely 2 s and executes 43,200 complete oscillations in one day. The pendulum coefficient of linear expansion is $1.2 \times 10^{-5}/\text{C}°$. The pendulum is calibrated at a temperature of 20°C. How many oscillations does it execute in one day when the temperature is held fixed at 40°C?

Solution

The number of oscillations in one day is given by

$$N = \frac{86,400 \text{ s}}{P}$$

where P is the period of the pendulum, expressed in seconds. The period is proportional to the square root of the length,

$$P = CL^{1/2} \qquad (L = \text{length of pendulum})$$
$$(C = \text{constant})$$

The length varies with temperature changes according to

$$L = L_0(1 + \alpha\,\Delta T)$$

The quantity $\alpha\,\Delta T$ is small compared to unity, so we can use the binomial expansion and write

$$L^{1/2} = L_0^{1/2}(1 + \alpha\,\Delta T)^{1/2} \approx L_0^{1/2}(1 + \tfrac{1}{2}\alpha\,\Delta T)$$

and

$$P = CL^{1/2} = CL_0^{1/2}(1 + \tfrac{1}{2}\alpha\,\Delta T)$$

Take L_0 to be the length of the pendulum at 20°C, for which the period is 2 s. Setting $CL_0^{1/2} = 2$ s gives

$$N = \frac{86{,}400 \text{ s}}{2 \text{ s }(1 + \tfrac{1}{2}\alpha\,\Delta T)} = \frac{43{,}200}{1 + \tfrac{1}{2}\alpha\,\Delta T}$$

With $\alpha = 1.2 \times 10^{-5}/\text{C}°$ and $T = 40°C$ ($\Delta T = 20 \text{ C}°$), the number of oscillations in one day is

$$N = \frac{43{,}200}{1 + \tfrac{1}{2}(1.2 \times 10^{-5}/\text{C}°)(20 \text{ C}°)}$$
$$= 43{,}195$$

EXERCISES

19.1 Thermal Expansion

A. A gold ring has an inner diameter of 2.168 cm at a temperature of 37°C. Determine its diameter at 122°C.

B. Explain why ice forms first at the surface of a lake rather than at the bottom.

C. A block of aluminum (volume 0.48 m³) absorbs heat and expands. Its temperature rises by 22 C°. Determine the work done against the surrounding atmosphere (pressure 1 atm).

D. A brass sphere ($\alpha = 1.9 \times 10^{-5}/\text{C}°$) has a diameter of 5.000 cm at 25°C. A steel ring ($\alpha = 1.1 \times 10^{-5}/\text{C}°$) has a diameter of 4.995 cm at 25°C. At what temperature will the sphere just be able to pass through the ring (a) if the ring is heated, but not the sphere, (b) if both the ring and sphere have the same temperature?

E. Figure 1 shows a circular iron casting with a gap. If the casting is heated, the iron expands ($\alpha = 1.17 \times 10^{-5}/\text{C}°$). (a) Does the width of the gap increase or decrease? (b) The gap width is 1.600 cm when the temperature is 30°C. Determine the gap width when the temperature is 190°C.

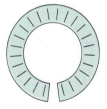

FIGURE 1

F. A bimetallic strip consisting of strips of aluminum and zinc is riveted together as shown in Figure 2. When the strip is heated does it curl upward or downward? Explain.

FIGURE 2

G. (a) The 200-inch diameter Hale telescope mirror is made of a borosilicate glass ($\alpha = 3.25 \times 10^{-6}/\text{C}°$).

Calculate the change in diameter (in millimeters) for a temperature change of $10C°$. (b) A titanium silicate glass being used for a space telescope mirror has an expansion coefficient of less than $3 \times 10^{-8}/C°$. Calculate the maximum change in diameter (in millimeters) of a 200-inch mirror of this glass for a $10C°$ change in temperature.

19.2   The Ideal Gas

H. A gas at a pressure of 2.11×10^5 N/m² occupies a volume of 0.116 m³. The gas temperature is 284 K. Assuming the gas is ideal, how many moles are present?

I. One mole of an ideal gas expands reversibly and isothermally at a temperature of 294 K. The initial pressure is 2.20×10^5 N/m². The final pressure is 1.24×10^5 N/m². Determine the (a) initial volume, (b) the final volume.

J. Two moles of an ideal gas occupy a volume of 0.017 m³ at a temperature of 288 K. (a) Determine the gas pressure. (b) The gas is compressed reversibly and isobarically to half its initial volume. Determine its final temperature. (c) Does the internal energy increase or decrease?

K. One mole of helium expands reversibly from 2 atm pressure to 1 atm pressure. The initial temperature is 20°C. Treat the helium as an ideal gas. (a) Calculate the initial volume. (b) Assuming the expansion to be isothermal, calculate the final volume. (c) Assuming the expansion to be adiabatic, calculate the final volume, (d) the final temperature.

L. Assume that helium behaves like an ideal gas. Consult Table 19.2 for values of MC_v and MC_p. Determine the amount of heat that one mole of helium must absorb in order to raise its temperature from 280 K to 290 K when (a) the pressure is held constant, (b) the volume is held constant. (c) Determine the internal energy increases for (a) and (b).

M. Air expands reversibly and adiabatically from $T_0 = 300$ K, $P_0 = 100$ atm to a final pressure of 1 atm. Treat the gas as ideal and determine the final temperature (Take $\gamma = 1.40$). Compare your result with the normal boiling points of oxygen (90.2 K) and nitrogen (77.4 K). Does your result necessarily mean that some air would liquefy?

N. Equal amounts of heat are added to 1 mole of helium and 1 mole of hydrogen, both at constant pressure. (a) Which gas undergoes the larger temperature change? (b) Would your answer be different if equal amounts of heat were added to the helium at constant pressure and to the hydrogen at constant volume?

P. Air initially at 27°C expands reversibly and adiabatically, tripling its volume. Its initial pressure is 4 atm. Treat the gas as ideal with $\gamma = 1.40$ and determine its final pressure and temperature.

Q. Why does a playground basketball bounce noticeably higher on a hot day than it does on a cold day?

19.3   P-V-T Surfaces

R. Figure 3 shows the P-T diagram for water. Use the diagram to explain how ice at $-20°C$ can melt under pressure even though its temperature remains constant. (This is called **regelation** and is the mechanism by which glaciers flow around large obstacles.)

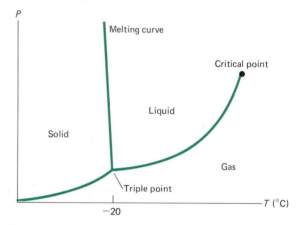

FIGURE 3

S. The weight of an ice skater spread over a narrow skate blade produces a *moderate* pressure. The ice beneath the blade experiences a *sudden* increase in pressure. The ice beneath the blade melts, then refreezes as the pressure is removed. The compression of the ice takes place *adiabatically*—that is, too quickly for heat to flow. Use the P-T diagram for water to suggest how ice initially at $-20°C$ can melt under adiabatic compression.

19.4   Change of Phase

T. A block of ice in a picnic cooler maintains a temperature of 0°C. The ice melts at the rate of 6 grams/minute. At what rate (watts) is heat leaking into the cooler?

U. A mixture composed of 0.10 kg of ice and 0.25 kg of water is initially at a temperature of 0°C. Enough heat is added to melt the ice completely and raise the temperature of the 0.35 kg of water to 15°C. Determine the amount of heat added.

V. A drop of liquid sodium with a mass of 0.130 gram is heated until it reaches its boiling point. How much additional heat must it absorb in order to vaporize?

W. A sample of dry ice at the normal sublimation point of 194.6 K absorbs heat at a rate of 100 W. Its mass decreases via sublimation from 200 grams to 100 grams in 9 min 21 s. Determine the heat of sublimation.

X. An undergraduate laboratory experiment is designed to measure the latent heat of fusion of ice. One hundred grams of ice at 0°C are to be melted by the heat developed by an electric current. Determine the rate at which heat must be absorbed to melt the ice in 15 minutes.

PROBLEMS

1. A surveyor's tape made of Invar ($\alpha = 7 \times 10^{-7}/C°$) is calibrated at a temperature of $25°C$. When the temperature is $85°F$, the tape indicates a distance of 61.16 m. Is the actual distance greater or smaller? By what amount? Compare the thermal expansion distance with the width of a scale marking, which is 0.04 mm.

2. An ideal gas expands reversibly, doubling its volume. Among the many ways this can be accomplished are (i) isobarically, (ii) isothermally, and (iii) adiabatically. Determine which of these types of reversible expansion (a) results in the largest and smallest changes in the gas temperature (regard a temperature decrease as smaller than no change), (b) results in the largest and smallest amounts of work being performed against the surroundings, (c) requires the largest and smallest heat absorption by the gas. Where possible, use qualitative arguments based on the P-V diagram, knowledge of the temperature dependence of U, and so on, rather than quantitative calculations.

3. Use the pressure and temperature data given in Example 4 to calculate the final temperature for an ideal gas with $\gamma = 5/3$, a value appropriate for monatomic gases. (a) Is the cooling larger or smaller for a monatomic gas than it is for a diatomic gas ($\gamma = 1.4$ for a diatomic gas)? (b) What would the final temperature be in the limit $\gamma \rightarrow 1$?

4. An immersion heater is placed in a mixture of 100 grams of ice and 150 grams of water at $0°C$. The heater converts electrical energy into heat at a rate of 1 kW. Assume that all of the heat is transferred to the ice-water mixture. Determine the time intervals required to (a) melt the ice, (b) bring the water to its normal boiling point, (c) completely evaporate the water.

5. An insulated glass holds 300 grams of water and 100 grams of ice at $0°C$. The ice melts completely in 31 minutes. Calculate the thermal power input to the ice.

6. A company ships meat in insulated chests packed with dry ice. Heat leaks into the chests at a rate of 0.7 W. Estimate the amount (in kg) of dry ice needed to protect the meat for 1 week. The heat of sublimation of dry ice is 573 kJ/kg.

7. Comets are believed to be composed largely of ice and frozen organic molecules. Assume that a comet composed entirely of ice strikes the earth moving at a speed of 30 km/s (the orbital speed of the earth). Assume further that one-half of its kinetic energy is converted into heat that is absorbed by the comet. (a) Determine the heat absorbed by 1 kg of the comet. The temperature of the comet before impact is $-20°C$. Is the heat absorbed sufficient to (b) melt the comet, (c) to vaporize it? (The average specific heat of ice is 2.1 kJ/kg·C°.) (d) Describe what you would expect to happen if a comet 10 km in radius struck a populated area of the earth.

8. The data in the table below record the length of an aluminum rod as a function of temperature. (a) Beginning with the lowest temperature, determine the coefficient of linear expansion for each of the 10 C° temperature changes. (b) Determine the average coefficient of linear expansion and compare your results with the value given for aluminum in Table 19.1. If you have access to a personal computer and the programs that accompany the text you can work through this problem and discover the optimum method for determining the coefficient of linear expansion.

Length (m)	Temperature (°C)
1.00023	10
1.00049	20
1.00077	30
1.00107	40
1.00139	50
1.00174	60
1.00211	70
1.00250	80
1.00291	90

9. Glauber's salt (sodium sulfate, $Na_2SO_4 \cdot 10H_2O$) melts at $32°C$ with a latent heat of 251 kJ/kg. (a) How does the transition temperature compare with a typical room temperature in a house? (b) Glauber's salt is a possible material for storing thermal energy in a solar home. Energy is stored when the salt converts from solid to liquid and is released when the salt reverts to a solid. Determine the amount of Glauber's salt (in kg) required to store 125,000 Btu, the thermal energy needed to maintain the overnight temperature of a solar home. (c) Compare the volume of the salt required in (b) with the volume of a household refrigerator having dimensions of 1 m × 1 m × 2 m. The density of Glauber's salt is 1460 kg/m³.

10. The variation of temperature with altitude in the earth's atmosphere is called the lapse rate. Normally, temperature decreases as altitude increases. The temperature at an altitude y can be approximated by $T = T_0 - ry$, where T_0 is the temperature at the surface and r is the lapse rate. (a) Assuming that the air is an ideal gas with molecular weight M, show that the atmospheric pressure at an altitude y can be computed from

$$\ln\left(\frac{P_0}{P}\right) = \left(\frac{Mg}{Rr}\right)\ln\left(\frac{T_0}{T_0 - ry}\right)$$

(b) Take the lapse rate to be 6 K/km, $T_0 = 300$ K, and $M = 28.8$ grams. Compare the pressure at the earth's surface with the pressure at an altitude of 2 km.

11. A diesel cycle (Figure 4) consists of an isobaric process $(a \rightarrow b)$, an isovolumic process $(c \rightarrow d)$, and two adiabatic processes $(b \rightarrow c$ and $d \rightarrow a)$. Calculate the ratio of the work done over $a \rightarrow b \rightarrow c \rightarrow d$ to the magnitude of the work done over $d \rightarrow a$. Assume an ideal gas with $\gamma = 1.4$. Take $V_1/V_3 = 5$ and $V_1/V_2 = 15$.

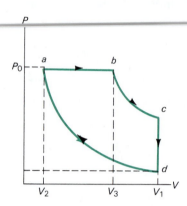

FIGURE 4

12. An ideal diatomic gas undergoes a reversible isobaric expansion. Prove that the work done is 0.4 times as large as the change in internal energy of the gas.

13. Wagon wheels of pioneering days had iron tires. To make a wheel the iron was heated so that when it cooled it clamped tightly over the wooden framework of the wheel. For a tire 1.5 m in diameter, what increase in temperature is necessary for its diameter to be 5 mm larger than that of the wheel?

14. A total mass of 720 grams of ice and steam is placed in an insulated container. Initially, the ice is at 0°C and the steam is at 100°C. When equilibrium is attained the final temperature is 50°C. Determine the original masses of ice and steam.

15. A sample of dry ice is at a temperature of 194.6 K and a pressure of 5.1 atm. It is subjected to a sudden increase of 60 atm, causing its temperature to rise to 300 K. Use Figure 19.9 to determine what happens to the sample.

16. A pond containing 4800 m³ of water absorbs 40 MJ of heat. Its temperature rises 4 C° and it expands, performing work against the atmosphere (pressure = 1 atm). The coefficient of volume expansion for water is $\beta = 6.3 \times 10^{-4}/C°$. Determine the change in internal energy of the water.

17. An ideal gas expands reversibly and adiabatically, tripling its volume and cooling from 450 K to 300 K. Determine the value of γ for the gas.

18. The van der Waals equation of state for one mole of gas is

$$(P + a/V^2)(V - b) = RT$$

where a and b are constants. For oxygen (O_2)

$$a = 0.084 \text{ J} \cdot \text{m}^3 \qquad b = 2.5 \times 10^{-5} \text{ m}^3$$

The van der Waals equation is a refinement of the ideal gas equation. Calculate the work done by one mole of oxygen during an isothermal ($T = 150$ K) compression from a volume of 225 cm³ to a volume of 125 cm³ using (a) the ideal gas equation of state, (b) the van der Waals equation of state.

19. Prove that the coefficients of volume expansion (β) and linear expansion (α) for an isotropic solid are related by Equation 19.3 ($\beta = 3\alpha$). (Hint: Start by considering a solid in the form of a cube with sides of length L that undergoes a temperature change from T to $T + dT$. Treat the accompanying changes in its volume and the length of its edges as differentials, dV and dL, and prove that $dV/V = 3\ dL/L$.)

CHAPTER 20

HEAT TRANSFER

20.1
INTRODUCTION TO HEAT TRANSFER

A transfer of heat occurs whenever an object is not in thermal equilibrium with its surroundings. Applications of heat transfer are found in virtually every phase of engineering and in many areas of the life sciences. There are three forms of heat transfer: conduction, convection, and thermal radiation.

Conduction is a transfer of heat without any flow of matter. The handle of a spoon is heated by conduction when the bowl of the spoon is submerged in hot coffee. We understand heat conduction well, and we can give it a concise mathematical formulation. **Convection** is a form of heat flow that can occur only in fluids. Heat is transported by currents of fluid that circulate between hotter and cooler regions. Heat is transferred between the currents and their surroundings by conduction. The mathematical description of convection can be very complicated because convection involves both fluid motions and heat conduction. We will avoid most of the mathematical complications by restricting ourselves to those aspects of convection described by **Newton's law of cooling**. **Thermal radiation** involves the transfer of energy in the form of **electromagnetic waves**. The rate at which thermal radiation is emitted depends strongly on the temperature of the emitting object.

20.2
CONDUCTION

The flow of heat via conduction through a pane of glass is an example of heat flow in one dimension. We let the x-axis denote the direction of heat flow and consider the flow of heat between the faces of a flat plate of thickness Δx and face area A (Figure 20.1). Let ΔT denote the temperature difference that is maintained across the thickness Δx. This temperature difference gives rise to the heat flow. Let q denote the rate of heat flow across the slab. Experiment shows that q is directly proportional to the temperature difference and inversely proportional to the slab thickness:

$$q \propto \frac{\Delta T}{\Delta x}$$

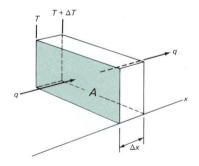

FIGURE 20.1

Heat flow by conduction. The rate of heat flow is proportional to $A \, \Delta T/\Delta x$. For heat flow in the direction shown, ΔT is negative.

TABLE 20.1
*Thermal Conductivity**

Material	k (W/m·C°)
Gases	
Air	0.0237
Carbon dioxide	0.0145
Oxygen	0.0246
Hydrogen	0.167
Helium	0.142
Methane	0.0305
Liquid	
Water	0.569
Solids	
Styrofoam®	0.040
Fiberglass	0.045
Corrugated cardboard	0.064
Rubber	0.12
Wood	0.19
Concrete	1.0
Glass	0.75−1.2
Ice	2.21
Steel	15−50
Iron	73
Brass	120
Magnesium	160
Aluminum	228
Copper	386
Silver	417

* Values refer to a temperature of 0°C and a pressure of 1 atmosphere.

Further, experiment confirms the expectation that the rate of heat flow should be directly proportional to the face area:

$$q \propto A$$

It follows that the rate of heat flow is proportional to $A\,\Delta T/\Delta x$:

$$q \propto A\frac{\Delta T}{\Delta x}$$

This relation can be converted into an equation by introducing a proportionality factor, k, called the **thermal conductivity**:

$$q = -kA\frac{\Delta T}{\Delta x} \tag{20.1}$$

Equation 20.1 is known as **Fourier's law**. The minus sign accounts for the fact that heat flows from a higher temperature to a lower temperature. The units of k are W/m·C°. The value of the thermal conductivity primarily depends on the physical composition of the material. A material with a large value of k would be described as a good thermal conductor. Metals are among the best conductors. A material with a small value of k would be classified as a poor thermal conductor, or a good thermal insulator. Gases and various porous plastics such as Styrofoam® are among the best insulators. Table 20.1 lists the thermal conductivities of some common materials.

Many types of insulation make use of the low thermal conductivity of air. The fluffy insulation used in the walls of homes traps air and provides a thick layer of material with a low thermal conductivity. The fluffy composition also restricts convective heat flow by limiting circulation of the trapped air. Thermal underwear likewise makes use of the low thermal conductivity of air. The numerous small holes in thermal underwear trap air between the body and outer garments.

Throughout this section we assume steady-state conditions, meaning that the rate of heat flow and the temperature do not change with time. Under steady-state conditions the slab does not absorb a net amount of heat. It follows that the rate of heat flow into one face of the slab is equal to the rate of heat flow out of the opposite face.

Thermal Resistance

The temperature change ΔT can be expressed by

$$\Delta T = -R\frac{q}{A} \tag{20.2}$$

where the quantity

$$R = \frac{\Delta x}{k} \tag{20.3}$$

is called the **thermal resistance** of the slab. The concept of thermal resistance is especially useful in dealing with heat flow through a series of layers of materials (Figure 20.2). The rate of heat flow q and the cross-sectional area A are the same for each layer. The overall temperature change across the series of layers ΔT is the sum of individual temperature changes,

$$\Delta T = \Delta T_1 + \Delta T_2 + \cdots + \Delta T_N$$

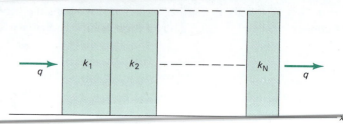

FIGURE 20.2
The temperature change across the series of slabs is the sum of the temperature changes across each slab. The effective value of R for the series of slabs is the sum of R values for the individual slabs.

Using Equation 20.2 we can set $\Delta T_i = R_i(q/A)$ and obtain

$$\Delta T = -(R_1 + R_2 + \cdots + R_N)\frac{q}{A}$$

$$= -R\frac{q}{A}$$

where

$$R = R_1 + R_2 + \cdots + R_N \qquad (20.4)$$

is the effective thermal resistance of the series of layers. The series of layers is equivalent to a single layer with a thermal resistance equal to the sum of the individual thermal resistances.

EXAMPLE 1

R *Value for Fiberglass Insulation*

The thermal conductivity of fiberglass is 0.045 W/m·C°. Let's determine the thermal resistance of a 6-inch-thick slab of fiberglass insulation. With

$$\Delta x = 6 \text{ in } (0.0254 \text{ m/in})$$
$$= 0.152 \text{ m}$$

we get

$$R = \frac{0.152 \text{ m}}{0.045 \text{ W/m·C}^\circ} = 3.4 \text{ m}^2 \cdot \text{C}^\circ/\text{W}$$

Values of R are often expressed in units of ft²·F°·h/Btu. Let's convert the fiberglass value of R. We get

$$R = 3.4 \text{ m}^2 \cdot \text{C}^\circ \cdot \text{s/J} \left(\frac{1 \text{ ft}}{0.348 \text{ m}}\right)^2 (1.8 \text{ F}^\circ/\text{C}^\circ)\left(\frac{1 \text{ h}}{3600\text{s}}\right) (1055 \text{ J/Btu})$$

$$= 19 \text{ ft}^2 \cdot \text{F}^\circ \cdot \text{h/Btu}$$

(Perhaps you have seen television commercials that call attention to an "R value" of 19 for fiberglass insulation.)

Differential Form of Fourier's Law

We obtain the differential form of Fourier's law by replacing $\Delta T/\Delta x$ in Equation 20.1 by the derivative dT/dx. This replacement produces

$$q = -kA\frac{dT}{dx} \qquad (20.5)$$

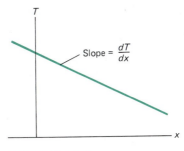

FIGURE 20.3

The temperature gradient dT/dx is the slope of a graph of temperature versus position. For one-dimensional heat flow, dT/dx is constant and the T versus x graph is a straight line.

The derivative dT/dx is called the **temperature gradient** and measures how temperature changes with position. Geometrically, dT/dx is the slope of a graph of T versus x. For one-dimensional heat flow the temperature gradient is a constant, provided that the material is uniform ($k = $ constant):

$$\frac{dT}{dx} = -\frac{q}{kA} = \text{constant}$$

A constant value of dT/dx means that the temperature decreases *linearly* with distance along the direction of heat flow, giving a straight-line graph of T versus x (Figure 20.3).

EXAMPLE 2

Temperature Variation for One-Dimensional Heat Flow

A plate glass window 0.50 cm thick has an inner temperature of 20°C and an outer temperature of -15°C. The temperature gradient across the window is

$$\frac{dT}{dx} = \frac{\Delta T}{\Delta x} = \frac{-35 \ \text{C}°}{0.005 \ \text{m}} = -7000 \ \text{C}°/\text{m}$$

This equation can be integrated to give

$$T = -(7000 \ \text{C}°/\text{m})x + \text{constant}$$

If the inner face of the window is taken to be at $x = 0$, the constant of integration must equal the temperature at that position. Thus, $T_{\text{inner}} = 20$°C gives

$$T = 20°\text{C} - (7000 \ \text{C}°/\text{m})x$$

We can test this equation by checking to see that it gives the correct temperature for the outer face. With $x = 0.50 \ \text{cm} = 5 \times 10^{-3}$ m we find

$$T_{\text{outer}} = 20 \ \text{C}° - (7000 \ \text{C}°/\text{m})(5 \times 10^{-3} \ \text{m})$$
$$= -15°\text{C}$$

Heat Flow in Two Dimensions

We can apply Fourier's law to situations in which the heat flow is two-dimensional by allowing for variation in the area A. Consider a steam pipe in which heat flows radially outward. This type of heat flow is called **cylindrical heat flow**. The geometry of cylindrical heat flow is shown in Figure 20.4. We conceptually divide the cylinder into a series of concentric cylindrical sleeves. The rate of heat flow through a sleeve of radius r and thickness dr is given by Fourier's law:

$$q = -kA\frac{dT}{dr} \tag{20.6}$$

where

$$A = 2\pi r L$$

is the surface area of the sleeve. For steady-state conditions we can rewrite Equation 20.6 as

$$dT = -\frac{q}{2\pi kL}\frac{dr}{r}$$

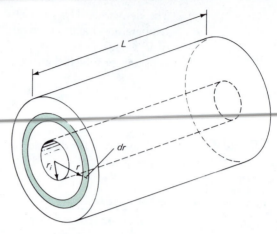

FIGURE 20.4

Geometry for cylindrical heat flow. The inner and outer radii of the pipe are r_i and r_o. The imaginary cylindrical sleeve has a radius r and thickness dr. Its inner surface area is $2\pi rL$.

and integrate from a point r_i where the temperature is T_i to some arbitrary point r where the temperature is $T(r)$,

$$\int_{T_i}^{T(r)} dT = -\frac{q}{2\pi kL} \int_{r_i}^{r} \frac{dr}{r}$$

The result is

$$T(r) - T_i = -\frac{q}{2\pi kL} \ln \frac{r}{r_i} \tag{20.7}$$

This equation shows that the temperature decreases **logarithmically** with r for cylindrical heat flow.

EXAMPLE 3

Reducing Heat Loss with Insulation

A stainless steel pipe has an inner radius of 2.0 cm and an outer radius of 2.5 cm (Figure 20.5a). The pipe carries hot water at a temperature of 130°F (54.4°C). The temperature at the pipe's outer surface is 126°F (52.2°C). What is the rate of heat flow per unit length of the pipe? Solving Equation 20.7 for q/L gives

$$\frac{q}{L} = \frac{-2\pi k[T(r) - T_i]}{\ln \dfrac{r}{r_i}}$$

Setting

$$
\begin{array}{ll}
r = 2.5 \text{ cm} & r_i = 2.0 \text{ cm} \\
T(r) = 52.2°C & T_i = 54.4°C \\
k = 19 \text{ W/m·C°} & \text{(for stainless steel)}
\end{array}
$$

gives

$$\frac{q}{L} = \frac{-2\pi(19 \text{ W/m·C°})(52.2 - 54.4) \text{ C°}}{\ln \dfrac{2.5}{2.0}} = 1200 \text{ W/m}$$

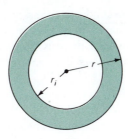
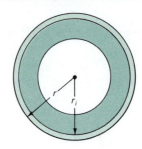

FIGURE 20.5

(a) Cross section of a steel pipe ($r_i = 2.0$ cm; $r = 2.5$ cm). (b) Cross section of a pipe wrapped with insulation. Here r_i denotes the inner radius of the insulating sleeve ($r_i = 2.5$ cm) and r denotes the outer radius of the sleeve.

Now let's determine the thickness of insulation required to reduce this heat loss by a factor of 10 and acheive a temperature of 100°F (37.8°C) at the outer surface of the insulation. For the *pipe*, a reduction in q/L by a factor of 10 requires that the temperature difference between the inner and outer surfaces be reduced by the same factor. Thus, the original 4 F° difference is reduced to 0.4 F° so that the outer surface of the pipe is at 129.6°F (54.2°C). We can use Equation 20.7 again, this time for a cylindrical shell of insulation, with an inner radius of 2.5 cm and an outer radius (r) that remains to be determined (Figure 20.5b). Solving Equation 20.7 for $\ln(r/r_i)$ gives

$$\ln \frac{r}{r_i} = -2\pi k[T(r) - T_i]\frac{q}{L}$$

Setting

$$T(r) = 37.8°C \qquad T_i = 54.2°C$$

$$\frac{q}{L} = 120 \text{ W/m} \qquad k = 0.15 \text{ W/m·C°}$$

gives

$$\ln \frac{r}{r_i} = \frac{-2\pi(0.15 \text{ W/m·C°})(37.8 - 54.2) \text{ C°}}{120 \text{ W/m}} = 0.129$$

and

$$r = 1.14r_i = 2.85 \text{ cm}$$

The required thickness of the insulation is

$$r - r_i = 3.5 \text{ mm}$$

20.3
CONVECTION

Conduction can take place in both solids and fluids. Convection, on the other hand, occurs only in fluids. In convective heat flow, circulating currents transfer parcels of fluid between regions of different temperature. The parcels from a higher temperature region are gradually cooled by conduction as they move through regions of lower temperature.

If the forces that drive the convection currents arise from a lack of thermal equilibrium, the heat flow is called **free convection**. A **forced convection**, on the other hand, is a convective flow driven by a pump or some other mechanical means.

Radiation, conduction, and convection compete as heat transfer mechanisms. If radiation and conduction transport heat swiftly enough, thermal instabilities do not occur and there is no natural convection. If radiation and conduction cannot remove heat fast enough, a thermal instability arises and convection results.

Convection is commonplace in our atmosphere because gases are poor heat conductors. We can compare atmospheric convection to convection in a pan of water. Water in a pan is heated from below by a flame or by electric heating coils. Our atmosphere is heated from below by the earth. The heating of both the water and the atmosphere is uneven. This causes some parcels of fluid to become hotter than the surrounding fluid. Conduction cannot transfer this heat through the fluid quickly enough, and so these parcels expand, lowering their density. This lowered density makes the parcels buoyant and they begin to rise, passing through the cooler, more dense fluid above them. As the warmer parcels rise, cooler parcels descend and take their places, and thus we have convection.

Newton's Law of Cooling

Convection often involves a fluid in contact with a solid surface. Your body may cool by convection through the air currents in contact with your skin. Figure 20.6 suggests how the temperature changes near the surface. Most of the temperature change takes place by conduction across a thin **boundary layer** of the fluid. This boundary layer is like a fluid skin that clings to the solid surface. The thickness of the boundary layer is not a fixed quantity. It depends on such things as viscosity, flow speed, and whether the fluid flow is laminar or turbulent. For air or water a typical boundary layer thickness is about 0.1 mm. Beyond the boundary layer the temperature changes more gradually.

The fact that most of the temperature change occurs by conduction through the boundary layer allows us to make a very useful, though approximate,

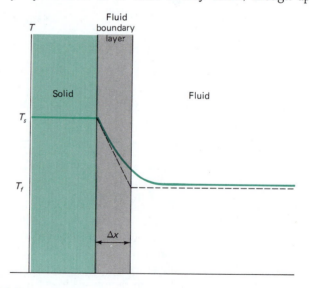

FIGURE 20.6

The solid aqua line suggests the actual way temperature varies with position near a solid surface. The dashed line that approximates the solid line is embodied in Newton's law of cooling.

analysis of convective heat transfer. Let's assume that the full surface-to-fluid temperature change occurs across the boundary layer. Such heat transfer could be described by Fourier's law:

$$q = A \frac{k}{\Delta x} (T - T_f) \qquad (20.8)$$

where Δx is the boundary layer thickness, T is the temperature of the cooling body, and T_f is the temperature of the surrounding fluid. However, as we already noted, the thickness of the boundary layer is not a fixed quantity. For this reason the combination $k/\Delta x$ in Equation 20.8 is replaced by a single empirical factor, h, called the **heat transfer coefficient** or **film coefficient**. The resulting equation for the rate of convective heat flow is

$$q = Ah(T - T_f) \qquad (20.9)$$

Equation 20.9 is known as **Newton's law of cooling**. When it was originally proposed by Isaac Newton, this law was meant to describe all cooling because at that time there was no distinction between convection and conduction. Even though Equation 20.9 is only an approximate description, it is widely used in many applications where great precision is not required.

The value of h depends on the type of fluid, the nature of the surface, the speed of fluid flow, and whether the flow is laminar or turbulent. Some representative values of h are given in Table 20.2. The wide variations in these values reflect the wide range of flow conditions encountered in convective heat transfer.

EXAMPLE 4

Highway Heat Flux

A highway surface is heated to a temperature of 140°F (60°C) by the summer sun. The air temperature several feet above the surface is 88°F (31°C). The heat transfer coefficient for air flowing over the pavement is 25 W/m²·C°. What is the rate of heat flow per square meter of highway surface? From Equation 20.9,

$$\frac{q}{A} = h(T - T_f) = (25 \text{ W/m}^2\cdot\text{C}^\circ)(60 - 31)\text{ C}^\circ$$

$$= 700 \text{ W/m}^2$$

TABLE 20.2

Fluid	Type of Convection	$h(W/m^2 \cdot C^\circ)$
air	free	5–50
air	forced	25–250
water	forced	100–10,000
boiling water	forced	2500–25,000
condensing steam	forced	5000–50,000

* Values are approximate. The wide range of values attest to the complex nature of convection. The value of h depends on various geometrical factors and the nature of the fluid flow.

Cooling Time

We can use Newton's law of cooling to study how an object cools by convection. Consider an object that no longer receives heat and let dQ denote the amount of heat convected from the object in a time dt. The rate of convective heat flow, q, is related to dQ and dt by

$$dQ = q\,dt$$

Using Newton's law of cooling (Equation 20.9) we can replace q with $Ah(T - T_f)$ to get

$$dQ = Ah(T - T_f)\,dt \tag{20.10}$$

The heat loss is related to the temperature decrease of the object dT by

$$dQ = -mC\,dT \tag{20.11}$$

where m is the mass of the object and C is its specific heat. Combining Equations 20.10 and 20.11 gives

$$\frac{dT}{T - T_f} = \left(\frac{Ah}{mC}\right)dt$$

This can be integrated to give

$$\frac{T - T_f}{T_0 - T_f} = e^{-Aht/mC} \tag{20.12}$$

This equation relates the temperature of the body (T) to the time (t) that the body has been cooling. In Equation 20.12, T_0 is the initial temperature of the body, and T_f is the temperature of the fluid that carries away the heat. As Figure 20.7 shows, the temperature of the body decreases steadily toward T_f.

The exponent on the right side of Equation 20.12 is dimensionless, so the quantity mC/Ah must have the dimensions of time. We refer to mC/Ah as the **cooling time**:

$$t_c = \frac{mC}{Ah} \tag{20.13}$$

In a time t_c the temperature difference $T - T_f$ is reduced by a factor of $e^{-1} = 0.368$.

The cooling time dependence on m, C, and A is reasonable: The larger the mass m and specific heat C of an object, the more slowly the object cools. Equation 20.13 verifies this by showing that $t_c \propto mC$. The larger the surface area A the more rapidly an object cools. Equation 20.13 also verifies this because it shows that $t_c \propto 1/A$. Finally, t_c is inversely proportional to the heat transfer coefficient. The larger the h, the faster the heat is convected away and the more rapidly cooling proceeds.

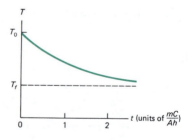

FIGURE 20.7

Temperature (T) versus time (in units of the cooling time mC/Ah) according to Newton's law of cooling. Cooling begins at $t = 0$ with the body at a temperature T_o. As cooling proceeds the body temperature approaches T_f, the temperature of the surrounding fluid.

EXAMPLE 5

Cooling Fins

A power transistor is mounted on a base equipped with cooling fins. Heat is convected to the surrounding air. How much time is required for the fins to cool to a temperature of 32°C?

The temperature of the fin when the transistor is switched off is 80°C. The temperature of the surrounding air is 27°C. The other data are:

$$A = 6.0 \text{ cm}^2 = 6.0 \times 10^{-4} \text{ m}^2$$

$$h = 38 \text{ W/m}^2 \cdot \text{C}°$$

$$m = 1.2 \text{ grams} = 1.2 \times 10^{-3} \text{ kg}$$

$$C = 1.2 \text{ kJ/kg} \cdot \text{C}°$$

Inserting these values into Equation 20.13 we get $t_c = 63$ s for the characteristic cooling time. We can anticipate that the time for the fins to cool from 80°C to 32°C will be comparable to 63 s. We base this expectation on the *exponential* decrease of $T - T_f$ with time. Setting $t_c = 63$ s, $T = 32$°C, $T_f = 27$°C, and $T_0 = 80$°C in Equation 20.12 leads to

$$t = -(63 \text{ s}) \cdot \ln\left[\frac{(32 - 27) \text{ C}°}{(80 - 27) \text{ C}°}\right] \text{ s}$$

$$= 149 \text{ s}$$

As expected, this is comparable to the cooling time.

20.4
THERMAL RADIATION

Every object loses energy by the emission of electromagnetic waves at a rate that depends strongly on the object's temperature. This type of energy transfer is called **thermal radiation**. An object also absorbs a fraction of the thermal radiation incident from its environment. The temperature of an object tends to rise when the rate of absorption exceeds the rate of emission. The temperature tends to drop when the rate of emission exceeds the rate of absorption. Radiative equilibrium exists when the rates of absorption and emission are equal.

An ordinary object that absorbs most of the visible light reaching it appears black. A **blackbody** is an idealized system that absorbs *all* incident radiant energy. The emission of thermal radiation by a blackbody is described by the **Stefan-Boltzmann law**. This law states that the rate at which energy is radiated per unit surface area is proportional to the fourth power of the Kelvin temperature of the blackbody. In equation form the Stefan-Boltzmann law reads

$$F = \sigma T^4 \tag{20.14}$$

where F is the energy radiated per second per square meter and is called the **surface flux**, T is the Kelvin temperature, and σ is a universal constant called **Stefan's constant**,

$$\sigma = 5.6703 \times 10^{-8} \text{ W/m}^2 \cdot \text{K}^4 \tag{20.15}$$

The surface flux of thermal radiation from an object that is not a blackbody is usually expressed as

$$F = \epsilon \sigma T^4 \tag{20.16}$$

where ϵ is called the *emissivity* of the surface. The emissivity depends on the physical composition of the surface and on its temperature. For a blackbody $\epsilon = 1$, but for all other objects $0 < \epsilon < 1$.

The total rate at which an object radiates energy is called its **luminosity**, L. The luminosity of a blackbody that radiates energy according to Equation 20.14 is

$$L = FA = \sigma T^4 A \qquad (20.17)$$

where A is the radiating area.

EXAMPLE 6

Stellar Luminosity

The star Antares is one of a class of stars called **red giants**, appropriately named after their red color and enormous size. The radius of Antares is 500 times that of our sun. The surface temperature of Antares is 3800 K. Antares and other stars are among nature's closest approximations to blackbodies. Therefore we can take $\epsilon = 1$ for Antares. What is the luminosity of Antares? The luminosity is related to its surface flux and radius (R) by Equation 20.17:

$$L = FA = F \cdot 4\pi R^2$$

The radius of our sun is 6.96×10^8 m. Because the radius of Antares is 500 times that of our sun, we set $R = 500 \times (6.96 \times 10^8$ m). Substituting this value of R into Equation 20.17 we get

$$L = (5.67 \times 10^{-8} \text{ W/m}^2 \cdot \text{K}^4)(3800 \text{ K})^4 \cdot 4\pi(500 \times 6.96 \times 10^8 \text{ m})^2$$
$$= 1.8 \times 10^{31} \text{ W}$$

This figure is truly astronomical. The luminosity of our sun is small in comparison—*only* 3.0×10^{26} W. Antares emits thermal radiation at a rate 44,000 times that.

If an object absorbs and radiates energy at equal rates it is in **radiative equilibrium** with its environment. However, radiative equilibrium generally does not require that the object have the same temperature as the source from which it receives radiant energy.* For example, the earth is in radiative equilibrium, even though its temperature differs greatly from that of the sun, the source of most of the radiation that the earth absorbs. We can use the fact that the earth is in radiative equilibrium to estimate the sun's surface temperature.

EXAMPLE 7

Surface Temperature of the Sun

The rate at which the earth receives energy from the sun depends on the temperature of the sun. The rate at which the earth emits thermal radiation depends on the temperature of the earth. The condition of radiative equilibrium relates these two temperatures. By treating both the earth and the sun as blackbodies, we can estimate the sun's surface temperature.

The luminosity of the sun is given by Equation 20.17, with $A = 4\pi R_S^2$ for the surface area of the sun,

$$L_S = \sigma T_S^4 \cdot 4\pi R_S^2$$

T_S is the solar surface temperature and R_S is the sun's radius. The energy radiated by the sun spreads out uniformly in all directions. A tiny fraction is intercepted by our

* If the source is maintained at a fixed temperature and completely surrounds the object, then radiative equilibrium does require that the object have the same temperature as the source.

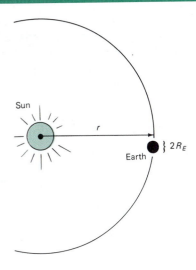

FIGURE 20.8
The earth intercepts a small fraction of the energy radiated by the sun.

planet (Figure 20.8). As viewed from the sun, the earth appears as a circular disk. Its absorbing area is πR_E^2, where R_E is the earth's radius. At a distance r from the sun the radiation is spread uniformly over a sphere of area $4\pi r^2$. If we take r equal to the earth-to-sun distance, the fraction of the solar luminosity absorbed by the earth is $\pi R_E^2/4\pi r^2$. The energy absorbed per second is therefore

$$\sigma T_S^4 \cdot 4\pi R_S^2 \cdot \pi R_E^2/4\pi r^2 = \text{energy absorbed per second}$$

The temperature variations over the earth are only a small fraction of the average earth temperature. We therefore regard the earth as a spherical blackbody of temperature T_E. It emits thermal radiation at a rate

$$\sigma T_E^4 \cdot 4\pi R_E^2 = \text{energy emitted per second}$$

For radiative equilibrium the rates of absorption and emission are equal. Therefore we can equate the rates of energy absorption and emission,

$$\sigma T_E^4 \cdot 4\pi R_E^2 = \sigma T_S^4 \cdot 4\pi R_S^2 \cdot \pi R_E^2/4\pi r^2$$

Solving for the temperature of the sun we get

$$T_S = \frac{2T_E}{\sqrt{\dfrac{2R_S}{r}}} \tag{20.18}$$

We can determine the numerator, $2T_E$, simply by using a thermometer. Let's take $T_E = 285$ K (65°F) as a representative value of the earth's temperature. The quantity $2R_S/r$ is the *angular diameter* of the sun as viewed from the earth (Figure 20.9). The angular diameter is $2R_S/r \approx 0.00934$ rad. Using Equation 20.18, our estimate of the surface temperature of the sun is

$$T_S = \frac{2(285 \text{ K})}{\sqrt{0.00934}} \approx 5900 \text{ K}$$

More precise measurements give a value of 5800 K.

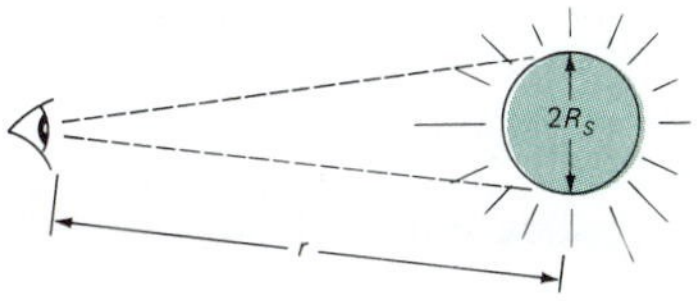

FIGURE 20.9
The angular diameter of the sun is $2R_S/r$.

In some instances, thermal radiation represents an undesirable energy loss. Engineers have devised ingenious ways to reduce such losses. The Thermos® bottle is a prime example (Figure 20.10). The space between the double glass

FIGURE 20.10
Thermos® bottle. The silvering blocks the flow of radiation. The vacuum prevents heat flow by convection and conduction.

walls is evacuated, and the two surfaces facing the evacuated space are silvered. These silvered surfaces reflect thermal radiation, thereby preventing its inward flow. In addition, because the space between the walls is a vacuum, there can be no heat flow across it by convection and conduction. The primary heat flow takes place by conduction through the stopper and the glass.

WORKED PROBLEM

Consider the heat conducted from the surface of the earth's core (radius = 3500 km) to the earth's surface. Estimate the temperature at the surface of the core, assuming that heat is conducted at a rate of 10^{12} W. Take $k = 2.0$ W/m·C°.

Solution

The one-dimensional heat flow equation can be used to estimate the temperature difference between the surface of the core and the surface of the earth,

$$\Delta T = -\frac{q\,\Delta x}{kA}$$

Set Δx equal to the core-to-surface distance:

$$\Delta x = 6380 \text{ km} - 3500 \text{ km} = 2880 \text{ km}$$
$$\approx 2.9 \times 10^6 \text{ m}$$

Set A equal to the surface area of a sphere 5,000 km in radius. This corresponds to a sphere about halfway between the core and the surface,

$$A = 4\pi R^2 = 4\pi (5 \times 10^6 \text{ m})^2$$
$$\approx 3.1 \times 10^{14} \text{ m}^2$$

This gives

$$\Delta T = T_{\text{surface}} - T_{\text{core}} = -\frac{q\,\Delta x}{kA}$$
$$= -\frac{(10^{12} \text{ W})(2.9 \times 10^6 \text{ m})}{(2.0 \text{ W/m·C°})(3.1 \times 10^{14} \text{ m}^2)}$$
$$= -4,700 \text{ C°} = -4,700 \text{ K}$$

Taking the surface temperature to be 300 K gives a value of 5,000 K for the core temperature,

$$T_{\text{core}} = T_{\text{surface}} - \Delta T = 300 \text{ K} - (-4,700 \text{ K})$$
$$= 5,000 \text{ K}$$

EXERCISES

20.2   Conduction

A. A temperature difference of 110 C° is maintained across a copper plate 2 cm thick. The plate is 12 cm long and 8 cm wide. Calculate the rate at which heat is conducted across the plate.

B. A large plate glass window measures 3.0 m × 2.5 m × 0.80 cm. Determine the rate of heat flow through the window when the temperature drop across it is 10 C°.

C. A large picture window consists of a sandwich of air between two panes of glass. The thermal conductivity of air is much lower than that of glass, so a thin layer of air provides better insulation than an equal thickness of glass. What thickness of glass would give the same heat flux (q/A) and temperature drop (ΔT) as a 0.50-cm thickness of air?

D. A window 1.2 cm thick has an inside temperature of 27°C and an outside temperature of 15°C. Determine the temperature gradient across the window.

E. Homeowners are urged to lower their thermostats in winter in order to conserve energy. If the outside temperature is 5°C, how does the conductive heat flow through a window compare for thermostat settings of 21°C and 20°C?

F. The heat flow through the steel pipe in Example 3 is 120 W/m. The pipe is wrapped with a cylindrical sleeve of insulation ($k = 0.15$ W/m·C°) 1 cm thick. If the temperature of the inner surface of the insulation is 129.6°F (54.2°C), determine the temperature of the outer surface.

20.3 Convection

G. The heat transfer coefficient for boiling water in a tea-kettle is $h = 1500$ W/m²·C°. The temperature difference across the boundary layer is 0.50 C°. Find the rate of heat flow through an area of 0.12 m².

H. How long will it take the power transistor of Example 5 to cool to (a) 30°C, (b) 28°C?

I. Certain types of solar energy collectors have a transparent glass plate above a black absorbing surface. A portion of the incident sunlight is reflected by the plate and never reaches the absorbing surface. Nevertheless, the presence of the glass plate raises the temperature of the absorber and increases the overall efficiency of the collector. What is the purpose of the plate?

J. A swimmer emerges from the water and experiences a chill because a breeze increases the convective heat flow. If $h = 250$ W/m²·C° and the temperature difference between the air and wet skin is 4.4 C°, determine the rate of heat flow from a swimmer with an exposed area of 0.62 m².

K. A person's body ejects heat to it surroundings via convection ($h = 10$ W/m²·C°). If the surface area is 0.7 m² and the temperature difference is 15 C°, determine the heat convected in one day. Compare this with the caloric intake of 2500 kilocalories.

L. Estimate the characteristic cooling time of a 30-gram aluminum cooling fin with a surface area of 200 cm². Take $h = 10$ W/m²·C°.

20.4 Thermal Radiation

M. Assume that a nudist's body radiates like a blackbody at a temperature of 32°C and calculate the *net* radiation loss per day to an environment at 27°C. Compare this with the typical caloric intake of 3000 kcal/day. Take the radiating area to be 0.8 m².

N. A solar collector used for heating water has an area of 3.0 m² and an emissivity of 0.86. Determine the rate at which it radiates energy when its temperature is 120°C.

P. A 100-W light bulb emits approximately 10 W of radiation. The wire filament of the bulb is 2.5 cm long and 0.1 mm in diameter. Estimate the temperature of the filament, assuming it behaves as a blackbody.

Q. Pluto is roughly 40 times as far from the sun than the earth. Estimate the surface temperature of Pluto.

PROBLEMS

1. The intensity of solar radiation impinging on the earth's atmosphere is 1.38 kW/m². Averaged over time the energy received from the sun is balanced by energy radiated by the earth. (a) Assume that all solar energy is absorbed and the earth radiates as a blackbody. By equating energy absorbed to energy radiated, show that the equilibrium temperature of the earth is

$$T = \left(\frac{S}{4\sigma}\right)^{1/4}$$

where $S = 1.38$ kW/m² and σ is Stefan's constant. (b) Evaluate the equilibrium temperature with this model.

2. Three flat materials having thermal conductivities k_1, k_2, and k_3, and equal cross-sectional areas and thicknesses, form a sandwich. The temperatures of the exposed faces are T_1 and T_2. Determine the temperature at each of the two interfaces in terms of k_1, k_2, k_3, T_1, and T_2.

3. A house has a wooden roof three inches thick and 1500 ft² in area. During a cold spell lasting two weeks the average temperature outside is 7°F and inside is 70°F. (a) What does it cost the owner at $0.08/kWh to provide for the heat conduction through the roof during the cold snap? (b) How much would it cost the owner if the roof had 3 inches of insulation whose thermal conductivity is one-fourth that of wood?

4. Take the surface of the sun to be at 5800 K. (a) How much power must be provided by nuclear reactions to provide for the energy radiated away? Treat the sun as a blackbody. (b) How many 1000-MW power plants would be needed to supply this power?

5. Suppose that in the far reaches of our Milky Way galaxy a planet orbits a star that is ten thousand times as luminous as our sun and whose radius is ten times the radius of our sun. (a) If the surface temperature of our sun is 5800 K, determine the surface temperature of the star. (b) If the planetary surface temperature range that will allow life to develop is 260 K to 320 K, determine the corresponding range of planet-to-star distances.

6. A sphere of radius R is maintained at a surface temperature T by an internal heat source (Figure 1). The

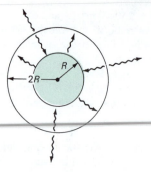

FIGURE 1

sphere is surrounded by a thin concentric shell of radius $2R$. Both objects absorb and emit as blackbodies. Show that the temperature of the shell is $T/(8^{1/4}) = 0.595T$.

7. (a) Show that the flow of heat in a system with spherical symmetry (radially outward from a spherical source) is governed by the equation

$$q = -4\pi k r^2 \frac{dT}{dr}$$

where r is the distance from the center of the source to a point where the temperature is T. (b) Rearrange this equation and integrate to show that the temperature difference between the points r_1 and r_2 (Figure 2) is given by

$$T_1 - T_2 = \frac{q}{4\pi k}\left(\frac{1}{r_1} - \frac{1}{r_2}\right)$$

(c) A spherical pressure vessel made of steel surrounds a small nuclear reactor. The inner and outer diameters of the vessel are 0.88 m and 1.22 m. The outward rate of heat flow is 3700 W. The temperature of the inner surface is 650°F (343°C). Determine the temperature of the outer surface..

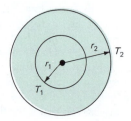

FIGURE 2

8. Long-distance swimmers stop at regular intervals to consume high-caloric foods (typically, several thousand calories each hour). A significant fraction of the energy obtained from these foods is used to compensate the body heat loss to the water. Assume that the heat transfer coefficient for the swimmer is 100 W/m²·C°. Take the surface area to be 1 m². If the temperature of the water is 21°C and the body temperature is 37°C, determine the heat loss in kilocalories in 1 hour. Round your answer to the nearest 100 kcal.

9. Heat is conducted radially outward through the walls of a cylindrical pipe. If the inner and outer walls of the pipe are r_i and r_0, respectively, prove that $r = (r_i r_0)^{1/2}$ is the radius for which the temperature is halfway between the inside and outside wall temperatures.

10. A cylindrical layer of insulation is 4 cm thick. The inner face has a radius of 7 cm and is at a temperature of 140°C. Halfway through the layer the temperature is 120°C. Determine the temperature at the outer surface.

11. In Section 20.3 we carried through an analysis that showed how the temperature decreases steadily for a body cooled by convection. Perform a similar analysis for a body that cools by thermal radiation alone. (a) Show that the temperature T a time t after cooling begins is given by

$$T = \frac{T_0}{\left(1 + \dfrac{t}{\tau}\right)^{1/3}}$$

where T_0 is the initial ($t = 0$) temperature and τ is a characteristic cooling time given by

$$\tau = \frac{mC}{3A\epsilon\sigma T_0^{\,3}}$$

in which m is the mass of the body, C its specific heat, ϵ its emissivity, and A its surface area. Assume that the body radiates but does not absorb. (b) Initially an iron ingot cools primarily by radiation. Its initial temperature is 2000 K. How much time is required for the ingot to cool to a temperature of 1400 K? Take $m = 3 \times 10^4$ kg; $C = 0.45$ kJ/kg·C°; $\epsilon = 0.09$; $A = 12$ m².

12. The star Betelgeuse is a red giant with a surface temperature of 3700 K. Its luminosity is $10^5 L_S$, where $L_S = 3.9 \times 10^{26}$ W is the luminosity of our sun. Assume that $\epsilon = 1$ (blackbody). (a) Calculate the surface flux, F. (b) Calculate the radius of Betelgeuse. Express your result as a multiple of the solar radius, $R_S = 7.0 \times 10^8$ m.

13. The star Sirius B is a white dwarf with a surface temperature of 6600 K. Its luminosity is $1.1 \times 10^{-3} L_S$, where $L_S = 3.9 \times 10^{26}$ W is the luminosity of our sun. Assuming that $\epsilon = 1$ (blackbody), calculate the radius of Sirius B. Express your result as a multiple of the earth's radius ($R_E = 6.38 \times 10^6$ m).

14. A lamp filament reaches 2500 K. The surrounding glass bulb is at 400 K. The emissivity of the filament is 0.32 and its surface area is 0.04 cm². At what rate is electrical energy supplied to the filament?

15. A pipe with an outer radius of 1 cm carries a hot fluid. With a 1-cm thickness of insulation the temperature difference across the insulation is 100 C°. The rate of heat loss per meter length of the pipe (q/L) is found to be 100 W/m. Calculate the value of k, the thermal conductivity of the insulation.

16. A pair of metal plates having equal areas and equal thicknesses are in thermal contact (Figure 3). One is made of iron, and the other is aluminum. Assume that the thermal conductivity of aluminum is exactly three times that of iron. The outer face of the iron plate is maintained at 0°C. The outer face of the aluminum plate is maintained at 60°C. Determine the temperature of the surface where the two plates are in contact.

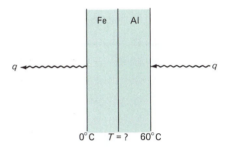

0°C   T = ?   60°C

FIGURE 3

17. Consider a metal bar of length L and uniform cross section A. The sides of the bar are heavily insulated and the ends of the bar have been maintained at temperatures T_1 and T_2 $(T_2 < T_1)$ for a long time. The end of the bar at temperature T_1 is at $x = 0$ and the other end is at $x = L$. (a) Make a sketch showing how the temperature decreases from T_1 at $x = 0$ to T_2 at $x = L$. (b) Suppose that the temperature of the bar is initially T_2 throughout. Then the temperature of the end at $x = 0$ is suddenly increased to temperature T_1. Make qualitative sketches of temperature versus x for different times showing how the bar approaches the steady state of part (a). (c) The time required for the bar to achieve a steady state is influenced by the thermal conductivity k, the specific heat C, the density ρ, and the length L, of the bar. Combine k, C, ρ, and L to produce a quantity having the units of seconds. (d) A detailed theory of heat flow shows that the temperature a distance x from the warm end (T_1) of the bar is given by

$$T(x, t) = T_1 + (T_2 - T_1)\frac{x}{L}$$
$$+ \sum_{n=1}^{\infty} \left[\frac{2(T_1 - T_2)}{n\pi}\right] e^{\frac{-n^2\pi^2 at}{L^2}} \cdot \sin\left(\frac{n\pi x}{L}\right)$$

where $a^2 = k/\rho C$ and t is the time. Show that this equation is consistent with your ideas presented in part (a).

18. A copper pipe has an inner radius R and an outer radius $2R$. It is surrounded by an iron pipe with an inner radius $2R$ and an outer radius $3R$ (Figure 4). The temperature of the inner surface of the copper pipe is 120°C. The temperature of the outer surface of the iron pipe is 30°C. Determine the temperature of the surface where the two pipes are in contact.

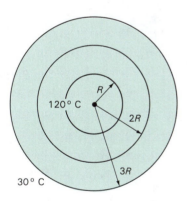

FIGURE 4

CHAPTER 21

THE SECOND LAW OF THERMODYNAMICS

21.1
INTRODUCTION TO THE
SECOND LAW OF THERMODYNAMICS

Most of the laws of physics take the form of equations that describe precisely what must happen in a given situation. Newton's second law states that an object subjected to a net force must experience an acceleration, and this law determines precisely the magnitude and direction of the acceleration. The first law of thermodynamics states that energy must be conserved; energy remains precisely constant. But the second law of thermodynamics is a different kind of physical law. It is an inequality that describes what *cannot* happen. With certain qualifications, the second law recognizes that heat cannot be converted completely into work. The second law places restrictions on heat engines—devices that transform heat into electrical, mechanical, and chemical energy. The upshot of the second law is that the complete conversion of heat into work is impossible. This law is of great practical importance because the conversion of heat into work is basic to many areas of technology.

21.2
HEAT ENGINES AND THERMODYNAMIC EFFICIENCY

We must introduce the concepts of a heat reservoir and a heat engine in order to formulate the second law. A heat reservoir is a body of uniform temperature throughout, the mass of which is sufficiently large so that its temperature is unchanged by the absorption or ejection of heat. For example, if an ice cube is tossed into the Atlantic Ocean, it produces no observable change in the temperature of the ocean. Therefore, the Atlantic Ocean qualifies as a heat reservoir.

Our industrialized society is powered by many types of **heat engines**. A heat engine is any device, operating in a cycle, that absorbs heat from a high-temperature reservoir, converts part of the heat into work, and then ejects the remainder into a low-temperature reservoir. Some sort of **working substance** is used to exchange heat and perform work. In the steam turbine, water is the working substance. Water is heated in a boiler and converted into high-pressure steam. The steam expands and does work on the turbine blades, is condensed by ejecting heat into the atmosphere or a river, and is then pumped back to the boiler. These steps are shown in Figure 21.1.

FIGURE 21.1

Thermodynamic aspects of steam turbine operation. (a) Water is converted to steam by absorbing heat from a coal-fired heat reservoir. (b) Steam expands and does work on the turbine blades. Heat is converted into rotational kinetic energy. (c) Steam condenses in the condenser and the heat liberated is ejected to the environment. (d) Cycle is completed by pumping water back to boiler.

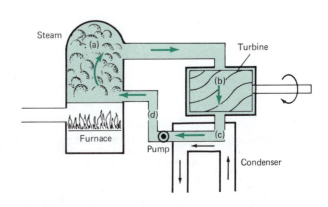

The fact that the earth's supply of fossil fuels is dwindling has made people energy conscious. The ejection of heat to the environment seems wasteful, and perhaps harmful. In recent years, engineers have found a variety of beneficial uses for the thermal discharges from heat engines. For example, the warm water ejected by electric power plants can be used to heat homes, and the hot air ejected by refrigeration units can heat the interiors of supermarkets.

However, no clever or innovative engineering design can overcome the fundamental limitation imposed on heat engines by the second law of thermodynamics. The practical consequence of the second law is that no heat engine can absorb a given amount of heat and convert it completely into work.

Thermodynamic Efficiency

Figure 21.2 shows the energy transfers common to all types of heat engines. Heat Q_H is absorbed from a high temperature reservoir. A portion of Q_H is converted into work, and the remainder, Q_L, is ejected to a low-temperature reservoir.* The working substance is carried through a cycle and is left unchanged. For a cyclic process there is no change in the internal energy, and so the first law of thermodynamics requires that the net work done, W, equal the net heat absorbed:

$$W = Q_H - Q_L \qquad (21.1)$$

A heat engine is efficient if it converts a large fraction of its heat intake into work. The thermodynamic efficiency of a heat engine, denoted by the Greek letter eta (η), is defined as the fraction of Q_H converted into work:

$$\eta \equiv \frac{\text{work out}}{\text{heat absorbed}} = \frac{W}{Q_H} \qquad (21.2)$$

An alternative expression for η follows if we use Equation 21.1 to replace W by $Q_H - Q_L$,

$$\eta = \frac{Q_H - Q_L}{Q_H} \qquad (21.3)$$

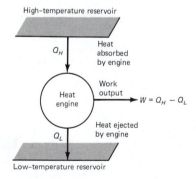

FIGURE 21.2

Energy transfers in a heat engine. Heat Q_H is absorbed from a high-temperature reservoir. A portion of Q_H is converted into work W and the remainder Q_L is ejected to a low-temperature reservoir. The first law of thermodynamics requires that $W = Q_H - Q_L$.

* Both Q_H, heat absorbed, and Q_L, heat ejected, are *positive* quantities.

TABLE 21.1
Thermodynamic Efficiencies

Device	η
Liquid-fuel rocket	0.48
Steam turbine	0.46
Diesel engine	0.37
Automobile engine	0.25
Steam locomotive	0.08
Thermocouple	0.07

Insofar as the first law of thermodynamics is concerned, η can take on any value between 0 and 1. In practice the values of η are well below unity in all areas of energy conversion technology. At electric power plants, steam turbines convert thermal energy into work, which in turn is converted into rotational kinetic energy of the turbine. The thermodynamic efficiency for these steam turbines is approximately 0.46. This means that only 46% of the heat absorbed by the water is converted into work. Over half of the heat absorbed is ejected. Other heat engines in wide-spread use today, like the automobile engine, have even smaller efficiencies. Table 21.1 lists thermodynamic efficiencies for several contemporary heat engines.

EXAMPLE 1
Thermodynamic Efficiency

A series of reversible processes carry a fluid through the cycle shown in Figure 21.3. Arrows directed into the closed figure indicate heat absorbed. Arrows directed out of the closed figure indicate heat ejected. Let's determine the thermodynamic efficiency of the cycle.

The total heat absorbed is $Q_H = 20$ J. The total heat ejected is 9 J. The work done over the cycle is

$$W = Q_H - Q_L = 11 \text{ J}$$

The efficiency is

$$\eta = \frac{W}{Q_H} = \frac{11 \text{ J}}{20 \text{ J}} = 0.55$$

The thermodynamic efficiency η refers only to processes in which heat is transformed into work. Many other energy conversion processes are important. In an electric motor, electric energy is converted into rotational kinetic energy. The conversion efficiency (mechanical energy output/electric energy input) of such processes is generally quite high (over 0.90) in comparison with typical thermodynamic efficiencies.

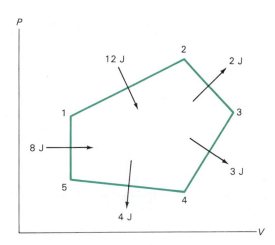

FIGURE 21.3

Refrigerators, Heat Pumps, and Coefficient of Performance

A heat engine extracts heat from a high-temperature reservoir, converts part of it into work, and ejects the remaining heat to a low-temperature reservoir, normally our environment. A refrigerator extracts heat from a low-temperature reservoir, has positive work performed on it by the environment, and ejects heat at a higher temperature. It pumps heat uphill, from a lower to a higher temperature.

Because a refrigerator ejects heat, one method of heating a building is to refrigerate the surrounding ground or atmosphere. Heat pumps are devices based on this idea. They are reversible heat engines that cool a home in the summer (by ejecting heat to the atmosphere) and warm it in the winter (by refrigerating the atmosphere).

A useful measure of the capabilities of a refrigerator or heat pump is the coefficient of performance, or *COP*. It is defined as:

$$COP \equiv \frac{\text{heat extracted from low-temperature reservoir}}{\text{net work performed on refrigerant to extract heat}}$$

or

$$COP = \frac{Q_L}{W} = \frac{Q_L}{Q_H - Q_L} \tag{21.4}$$

Q_H is the heat ejected to the high-temperature reservoir and Q_L is the heat extracted from the low-temperature reservoir. Optimally, a refrigerator extracts a sizable amount of heat from a low-temperature reservoir with a minimal expenditure of work; that is, it has a large coefficient of performance. Typical values of the *COP* for refrigerators and heat pumps range from 2 to 5.

The economic appeal of heat pumps is apparent when we consider the following: Suppose that $COP = 2$, a modest value. We can replace Q_L by $Q_H - W$ in Equation 21.4 to obtain

$$COP = \frac{Q_H - W}{W} = 2$$

from which it follows that $Q_H = 3W$. This means that the amount of heat ejected to the high-temperature reservoir is three times the work done. Suppose you were given the choice of heating your home with (a) a heat pump that supplies 3 kJ of heat to your home for every 1 kJ of electric energy used, or (b) electric heating elements that supply 1 kJ of heat for every 1 kJ of electric energy used. On the basis of this comparison alone, it seems clear that a heat pump is to be preferred. However, there are other factors such as initial cost and maintenance that might favor electric heat.

Several generations of intelligent, dedicated people have worked to improve the efficiency of heat engines. Why, then, are the efficiencies listed in Table 21.1 still under 50%? The answer is that nature has established a fundamental limit on all processes that convert heat into work. The second law of thermodynamics is a formal statement that such a limit exists, and the relation $\eta = (Q_H - Q_L)/Q_H$ is a quantitative measure of this limit. This limit is *not related to friction* within a heat engine. It is true that friction may reduce the actual work output of heat engines and thereby reduce their thermodynamic efficiencies, but only in a minor way. In a steam turbine, friction reduces the thermodynamic efficiency by less than 1%. The fundamental limit to efficiency is thermodynamic in origin.

What then is the optimal design for a heat engine? This question was answered by the French engineer Sadi Carnot. Carnot's research is especially noteworthy because it was completed in 1824, nearly 20 years before Mayer and Joule formulated the first law of thermodynamics. Because of his brilliant analysis of the problem, Carnot is given credit for discovering the second law of thermodynamics.

There are two remarkable aspects of Carnot's research. First, he found that the maximum efficiency with which heat can be converted into other forms of energy depends on the nature of the cyclic process rather than on the working substance. Second, he discovered what is now known as the Carnot cycle, the most efficient cyclic process for converting heat into other forms of energy.

21.3
THE CARNOT CYCLE

Consider a system composed of an ideal gas confined in a cylinder fitted with a piston (Figure 21.4). The ideal gas is the working substance, the piston and cylinder walls are its environment. The gas can exchange heat with its surroundings via conduction through the walls of the cylinder. Heat exchange can be eliminated if desired by suitably insulating the cylinder.

Figure 21.5 shows a P-V diagram for a Carnot cycle in which an ideal gas serves as the working substance. The Carnot cycle consists of two adiabatic paths (no heat exchange) connected by a pair of isothermal paths (constant temperature). We can analyze the Carnot cycle shown in Figure 21.5 by using the following four relations:

1. the equation of state for an ideal gas

$$PV = vRT \qquad (21.5)$$

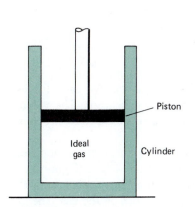

FIGURE 21.4

The ideal gas is confined in a cylinder fitted with a piston. The ideal gas is the working substance. The piston and cylinder walls are its environment.

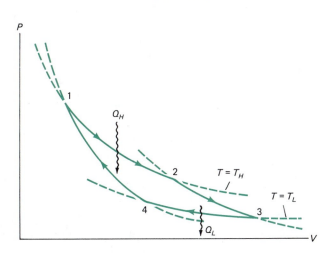

FIGURE 21.5

P-V diagram for the Carnot cycle for a gas. Heat Q_H is absorbed from the high-temperature reservoir and heat Q_L is ejected to the low-temperature reservoir. The four segments of the cycle consist of an isothermal expansion (1 to 2), an adiabatic expansion (2 to 3), an isothermal compression (3 to 4), and an adiabatic compression (4 to 1).

State 1   State 2

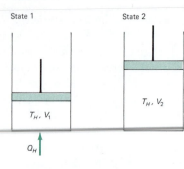

T_H, V_1 T_H, V_2

Q_H

FIGURE 21.6

The gas absorbs heat Q_H and expands from a volume V_1 to a volume V_2. The process proceeds at a rate that maintains the gas at a constant temperature, T_H.

State 2   State 3

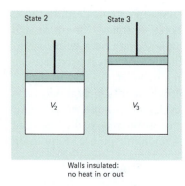

V_2 V_3

Walls insulated:
no heat in or out

FIGURE 21.7

An adiabatic expansion from volume V_2 to volume V_3 results in a drop in temperature for the gas.

State 3   State 4

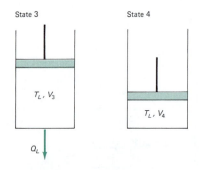

T_L, V_3 T_L, V_4

Q_L

FIGURE 21.8

Heat Q_L is ejected during the isothermal compression from volume V_3 to volume V_4. The gas temperature remains constant at T_L during the compression.

2. the equation for the internal energy (U) of an ideal gas

$$U = mC_v T \qquad (21.6)$$

3. the relation describing the adiabatic transformation of an ideal gas

$$TV^{\gamma - 1} = \text{constant} \qquad \gamma = \frac{C_p}{C_v} \qquad (21.7)$$

4. the first law of thermodynamics

$$Q = \Delta U + W \qquad (21.8)$$

The four legs of the Carnot cycle described in Figure 21.5 consist of an isothermal expansion ($1 \rightarrow 2$), an adiabatic expansion ($2 \rightarrow 3$), an isothermal compression ($3 \rightarrow 4$), and an adiabatic compression ($4 \rightarrow 1$). All four of these processes must proceed *reversibly* (Section 18.5). During the isothermal expansion from 1 to 2, heat Q_H is absorbed from a high-temperature reservoir (Figure 21.6). The expanding gas performs a positive amount of work W_{12} against its environment. Because $U = mC_v T$ for the ideal gas, $\Delta U = 0$ along an isotherm. Thus, there is no change in the internal energy over the path from 1 to 2, and the first law reduces to

$$Q_H = W_{12}$$

All of the heat absorbed from 1 to 2 is used to perform work against the environment.

The expansion of the gas continues from 2 to 3 under adiabatic conditions; that is, the gas is insulated from its surroundings during the expansion (Figure 21.7). During the adiabatic expansion the gas performs a positive amount of work, W_{23}. This work is done at the expense of the internal energy of the gas because no heat is absorbed. The internal energy decreases during the adiabatic expansion (2 to 3). The logic is

$$Q = 0 \qquad \text{(adiabatic process)}$$

$$W_{23} > 0 \qquad \text{(expansion)}$$

$$\Delta U = U_3 - U_2 = -W_{23} < 0 \qquad \text{(first law; work performed at expense of internal energy)}$$

$$U_3 < U_2 \qquad \text{(internal energy decreases from 2 to 3)}$$

Because $U = mC_v T$, a decrease in U means a decrease in temperature, $T_L < T_H$. Therefore, the gas cools during the adiabatic expansion. We can also use Equation 21.7, $TV^{\gamma - 1} = \text{constant}$, to reach the same conclusion. The volume increases as the gas expands, so the temperature must decrease to satisfy Equation 21.7.

The foregoing arguments are reversed for the isothermal compression from 3 to 4 (Figure 21.8) and the adiabatic compression from 4 to 1 (Figure 21.9), completing the cycle. During the isothermal compression from 3 to 4 the environment (piston) performs a positive amount of work on the gas. The gas must eject heat (Q_L) to the low-temperature reservoir in order to avoid a temperature increase. The adiabatic compression from 4 to 1 completes the cycle. There is no heat transfer during this compression, and the positive work performed by the environment therefore increases the internal energy of the gas. This energy increase shows up as a temperature rise from T_L to T_H.

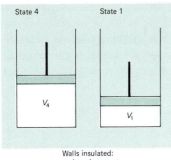

Walls insulated:
no heat in or out

FIGURE 21.9

An adiabatic compression from volume V_4 to volume V_1 results in a temperature rise for the gas.

We chose an ideal gas as the working substance in the Carnot cycle simply for convenience. We could have used any number of other working substances. The *essential* features of the Carnot cycle are that (1) it consists of two isothermal paths connected by two adiabatic paths, and (2) the adiabatic and isothermal processes proceed *reversibly*.

The overall result of the Carnot cycle is the conversion into work of some of the heat extracted from the high-temperature reservoir. The particular Carnot cycle just described acts as a heat engine, converting the net heat $Q_H - Q_L$ into work. We often refer to such a cycle as a *Carnot engine*.

The Carnot engine is the *most efficient* heat engine that can be conceived. This fact is known as **Carnot's theorem**, and it is proved in Section 21.4. It makes the Carnot engine the ultimate standard of comparison for all heat engines.

One special feature of the Carnot engine is that heat is absorbed and ejected only during isothermal processes. Thus, a single temperature is associated with Q_H and likewise with Q_L. This makes it possible to define temperature in terms of heat. In Section 21.5 we show how the efficiency of the Carnot engine is used to define the Kelvin temperature scale.

21.4
THE SECOND LAW OF THERMODYNAMICS

Precise statements of the second law of thermodynamics were formulated by William Thomson—who later became Lord Kelvin—and by Rudolph Clausius, in the early 1850s. Although worded differently, their statements are equivalent.

The Kelvin statement of the second law is this:

> It is impossible to devise a process whose only result is to convert heat, extracted from a single reservoir, entirely into work.

The word "only" in Kelvin's statement is an important qualifier. Many processes can convert heat completely into work. However, all such processes result in a final state that differs from the initial state. Thus, there is some other result besides the conversion of heat into work. In the isothermal expansion of the ideal gas considered in the Carnot cycle, the heat absorbed was converted entirely into work. At the end of the expansion, however, the thermodynamic state of the gas had changed. Its pressure had decreased and its volume had increased.

The word "single" is also a key word in Kelvin's statement. Converting the *net* heat absorbed into work does not violate the second law, and is in fact required in order to satisfy the first law of thermodynamics. A violation of Kelvin's statement of the second law would result if a heat engine could extract heat from a single reservoir and convert it entirely into work without the ejection of a net amount of heat to other reservoirs.

The Clausius statement of the second law is this:

> It is impossible to devise a process whose only result is to extract heat from a reservoir and eject it to a reservoir at a higher temperature.

According to the Clausius statement, the spontaneous flow of heat from a low-temperature reservoir to a high-temperature reservoir is impossible. The Clausius statement recognizes that heat flow is an irreversible process. Heat flows from higher to lower temperatures. It *never* spontaneously flows in the other direction. Of course, a refrigerator transfers heat from a low-temperature reservoir to a high-temperature reservoir, but work must be performed to accomplish this transfer: it is not spontaneous.

Heat flow is one very obvious example of the intimate relation between irreversible processes and the second law. The second law is a consequence of irreversibility. It is possible, although not always enlightening, to show that *if* a particular irreversible process were to become reversible, *then* the reversed process would violate the second law.

The evidence supporting the second law of thermodynamics is the failure of all attempts to construct a perpetual motion machine of the second kind*, that is, a machine that would contradict either the Kelvin or the Clausius statements of the second law.

Equivalence of Kelvin and Clausius Formulations

The Kelvin and Clausius formulations of the second law are equivalent. This equivalence is demonstrated by showing that if Kelvin's statement is false, then so is Clausius' statement, and vice versa.

Let us suppose that Kelvin's statement is false. This would mean that a process is possible whose only result is to convert heat extracted from a reservoir completely into work (Figure 21.10). The work obtained could then be completely converted back into heat at any desired temperature, via friction. In particular, it could be deposited in a heat reservoir whose temperature *exceeds* that of the reservoir from which the heat was extracted. The only result of this process would be to transfer a quantity of heat from a lower temperature to a higher temperature. But this is a violation of Clausius' statement. Thus, *if* Kelvin's statement is false, *then* Clausius' statement is false.

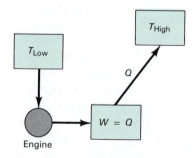

If Kelvin's statement were false we could implement this procedure:

which is equivalent to

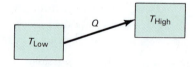

which means Clausius' statement is false.

FIGURE 21.10

Sequence of processes that show Clausius' statement is false *if* Kelvin's statement is false.

---

* Recall that a perpetual motion machine of the first kind is one that would manufacture energy—it would violate the first law of thermodynamics.

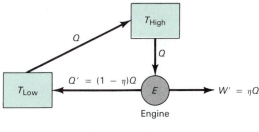

which is equivalent to

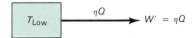

which means Kelvin's statement is false.

FIGURE 21.11

Sequence of processes that show Kelvin's statement is false *if* Clausius' statement is false.

We can also show that *if* Clausius' statement is false, *then* Kelvin's statement must also be false. Suppose that Clausius' statement is false, so that there is a process whose only result is to extract heat from one reservoir and eject it to another reservoir at a higher temperature (Figure 21.11). We can transfer heat from a low-temperature reservoir to a high-temperature reservoir and then extract an equal amount of heat to run a Carnot engine through an integral number of cycles. The Carnot engine converts some of the heat into work and ejects the remainder. The heat ejected by the Carnot engine is returned to the low-temperature reservoir.

The high-temperature reservoir is left unchanged: it absorbs and ejects equal amounts of heat. The Carnot engine is left unchanged because it goes through an integral number of cycles. The net result of the series of processes is to convert heat, extracted from the low-temperature reservoir, entirely into work, which violates Kelvin's statement of the second law.

Proving the equivalence of the Kelvin and Clausius statements *does not prove the second law*. Only experiment can confirm the validity of the laws of nature—the second law of thermodynamics included.

Carnot's Theorem

Carnot's theorem provides the basis for a **thermodynamic temperature scale**. This is a scale that is independent of the materials actually used to fabricate a thermometer. Carnot's theorem is this:

No heat engine operating between two heat reservoirs can be more efficient than a Carnot engine operating between the same two reservoirs.

The proof of the theorem follows these lines: we assume that the second law is valid. Next, two reversible heat engines are introduced. One is a Carnot engine and the other is an engine we assume to be more efficient than the Carnot engine. *If* Carnot's theorem were false it would be possible to use the two engines in a cycle that violates the second law. By assumption it is not possible to circumvent the second law. Thus, Carnot's theorem is not false.

Carnot's theorem is important. It imposes a fundamental limit on the conversion of heat into work. No amount of ingenuity can devise a heat engine more efficient than the Carnot engine.

21.5
THE KELVIN TEMPERATURE SCALE

We saw in Section 21.4 that the efficiency of a Carnot engine is independent of the working substance used in the engine. In Chapter 18 we observed that the Kelvin scale of temperature is based on the laws of thermodynamics—it is independent of the thermometric substance used to construct a thermometer. The fact that the efficiency of a Carnot engine is independent of the working substance makes it possible to use the Carnot cycle to define the Kelvin temperature scale. Let's see how this possibility can be achieved.

Consider a Carnot engine that absorbs heat Q along one isothermal path and ejects Q_L along the other isothermal path. A thermodynamic temperature T is *defined* by taking $Q_L/Q = T_L/T$ or, equivalently,

$$\frac{Q}{T} = \frac{Q_L}{T_L} = \text{constant} \tag{21.9}$$

We emphasize that this is an arbitrary, but operational, definition of temperature. This definition tells us that the thermodynamic temperature T is proportional to Q, but leaves open the size of the temperature unit. As noted in Chapter 18, the temperature of the triple point of water has been assigned the value 273.16 K. Suppose that we operate a Carnot engine with the low-temperature reservoir at 273.16 K. Let Q_3 denote the heat ejected to the low-temperature reservoir. If the heat absorbed from the high-temperature reservoir is Q, its Kelvin temperature T is given by Equation 21.9,

$$\frac{Q}{T} = \frac{Q_3}{273.16 \text{ K}}$$

or

$$T = \left(\frac{Q}{Q_3}\right) 273.16 \text{ K} \tag{21.10}$$

This relation *defines* the Kelvin temperature scale.

The efficiency of a Carnot engine operating between reservoirs at temperatures T_L and T_H can be expressed in terms of T_L and T_H. We start by rearranging

$$\eta = \frac{Q_H - Q_L}{Q_H} = 1 - \frac{Q_L}{Q_H} \tag{21.11}$$

and use Equation 21.9 to write

$$\frac{Q_L}{Q_H} = \frac{T_L}{T_H} \tag{21.12}$$

Replacing Q_L/Q_H by T_L/T_H in Equation 21.11 gives an equation that expresses the efficiency of a Carnot engine in terms of T_L and T_H,

$$\eta_C = 1 - \frac{T_L}{T_H} \tag{21.13}$$

We call $1 - T_L/T_H$ the *Carnot efficiency* and denote it by η_C. The Carnot efficiency is an *upper limit* for the efficiency of heat engines operating between the temperatures T_L and T_H.

EXAMPLE 2

Carnot Efficiency of a Steam Turbine

A steam turbine operates between a boiler temperature of 550° F (561 K) and a condenser temperature of 100° F (311 K). We want to determine the Carnot efficiency, an upper limit to the actual efficiency

$$\eta_C = 1 - \frac{T_L}{T_H}$$

Taking $T_H = 561$ K and $T_L = 311$ K in Equation 21.13 gives

$$\eta_C = 1 - \left(\frac{311 \text{ K}}{561 \text{ K}}\right)$$

$$= 0.446$$

This efficiency of less than 100% illustrates the second law of thermodynamics. Some of the absorbed heat must be ejected and consequently is not available for conversion into work.

A glance at the relation $\eta_C = 1 - T_L/T_H$ suggests that to raise the efficiencies of heat engines we must lower T_L and/or raise T_H. But the only convenient low-temperature reservoirs that nature provides (large bodies of water and the atmosphere) have $T_L \approx 290$ K. Despite the best efforts of scientists there is still a disappointingly low limit on T_H. The thermal properties of working substances and materials used to fabricate high-temperature reservoirs have so far prevented significant increases in T_H. In particular, prolonged operation at high temperatures reduces the strength of metals and increases their rate of corrosion.

As reserves of abundant and cheap fuels dwindle, scientists and engineers strive to increase the efficiency of existing heat engines, and seek to develop more efficient new heat engines. In the quest for such new heat engines, the second law of thermodynamics provides both a goal and a limit.

21.6
ENTROPY

In this section we introduce a new physical property called **entropy**. Let's see how entropy is defined and what it measures. Consider the differential form of the first law of thermodynamics.

$$dQ = dU + dW$$

We found that the work could be expressed as

$$dW = P \, dV$$

provided the process took place **reversibly**. If heat exchange occurs during a **reversible process**, then dQ can be expressed in terms of the temperature (T) and the **entropy change** dS. The relation between dQ and dS is

$$dQ = T \, dS$$

We can rewrite this as

$$dS = \frac{dQ}{T} \tag{21.14}$$

to stress that it is a *definition* of the entropy change of a system that exchanges heat dQ. Remember that Equation 21.14 is valid only for reversible processes. The total entropy change in a reversible process is obtained by integrating Equation 21.14 to obtain the entropy change, ΔS:

$$\Delta S \equiv S_f - S_i = \int_i^f \frac{dQ}{T} \tag{21.15}$$

where S_f and S_i are the entropies of the final and initial states.

Isothermal Entropy Change

A special case of Equation 21.15 is noteworthy. If the process is isothermal, the right-hand side of Equation 21.15 becomes

$$\int_i^f \frac{dQ}{T} = \frac{1}{T} \int_i^f dQ = \frac{Q}{T}$$

where Q is the total heat exchanged by the system. Thus, for an isothermal process the entropy change is

$$\Delta S = \frac{Q}{T} \tag{21.16}$$

The units of entropy and entropy change are joules per kelvin (J/K). As the following demonstrates, entropy is a property of a system.

EXAMPLE 3

Entropy Change for Carnot Cycle

If entropy is a property of a system, it must be unchanged in any *cyclic* process. Let's check to see if entropy is unchanged in the Carnot cycle. Figure 21.6 shows the Carnot cycle for a gas-piston system. There is no entropy change along either of the reversible **adiabats** ($S = $ constant) because $dQ = 0$ and thus

$$dS = \frac{dQ}{T} = 0$$

Along the high-temperature isotherm ($1 \rightarrow 2$), heat Q_H is absorbed. The entropy change is given by Equation 21.16, with $Q = Q_H$ and $T = T_H$.

$$\text{Entropy change along } T_H \text{ isotherm} = \frac{Q_H}{T_H}$$

Along the low-temperature isotherm ($3 \rightarrow 4$), heat Q_L is ejected at $T = T_L$. Equation 21.16 applies, with $T = T_L$ and $Q = -Q_L$. The minus sign enters because an amount of heat Q_L is ejected along the low-temperature isotherm.

$$\text{Entropy change along } T_L \text{ isotherm} = -\frac{Q_L}{T_L}$$

The minus sign shows that the entropy decreases when heat is ejected. The total entropy change over the Carnot cycle is

$$\Delta S_{\text{cycle}} = \oint \frac{dQ}{T} = \frac{Q_H}{T_H} - \frac{Q_L}{T_L}$$

However, Equation 21.9 shows that $Q_H/T_H = Q_L/T_L$ for the Carnot cycle, and thus

$$\Delta S_{\text{cycle}} = 0 \qquad\qquad (21.17)$$

The entropy change for the Carnot cycle is zero because entropy is a property of a physical system. The entropy depends on the thermodynamic state and hence is unchanged in any cyclic process.

Entropy and Disorder

Equation 21.15 provides a way to measure entropy changes, but does not explain what entropy itself measures. Entropy is important because it is useful, but it will help our understanding if we can relate entropy to some aspect of our day-to-day experience.

Entropy is a measure of **disorder**. If the entropy of a system increases, then its microscopic structure becomes more disordered. The examples that follow should help you to understand the idea that entropy measures disorder.

EXAMPLE 4

Entropy Change for Melting

Consider a 1-kg chunk of ice at its normal melting point. If we add the heat of fusion, 334 kJ, the ice melts. What is the change in entropy? Because this change of phase takes place at constant temperature and is reversible, we can use Equation 21.16 to obtain

$$\Delta S = S_{\text{liquid}} - S_{\text{solid}} = \frac{mL_f}{T_m}$$

Taking $mL_f = 334$ kJ and $T_m = 273$ K gives

$$\Delta S = 1.22 \text{ kJ/K}$$

The value of ΔS is positive, showing that melting results in an **entropy increase**. This entropy increase is also in accord with the idea that entropy measures disorder. The molecular structure of the final state (liquid) is more disordered than the molecular structure of the initial state (solid). In a crystalline solid the molecules form an ordered, periodic structure. In a liquid, only a short-range order is evident. By short-range order we mean that any given molecule is able to influence only the behavior of nearby molecules. Melting results in an entropy increase and molecular disordering. Entropy measures disorder.

The computed value $\Delta S = 1.22$ kJ/K obtained in Example 4 will mean more if we compare it with the entropy changes for other processes. To provide a basis for comparison we determine the entropy change when 1 kg of water is heated isobarically to the boiling point.

Isobaric Entropy Increase

If heat is added to a substance at constant pressure, its temperature and entropy increase. The heat added in changing the temperature by an amount dT is

$$dQ = mC_p \, dT$$

where C_p is the specific heat at constant pressure. The corresponding entropy change is

$$dS = \frac{dQ}{T} = mC_p \frac{dT}{T}$$

If C_p is treated as constant, the total entropy change between the initial and final temperatures is

$$\Delta S = mC_p \int_{T_i}^{T_f} \frac{dT}{T} = mC_p \ln\left(\frac{T_f}{T_i}\right) \tag{21.18}$$

To illustrate, suppose we heat 1 kg of water from 0°C to 100°C. For 1 kg of water, $mC_p = 4.19$ kJ/K. With $T_i = 273$ K and $T_f = 373$ K, we find $\ln(T_f/T_i) = 0.312$, and

$$\Delta S = 1.31 \text{ kJ/K}$$

This change in entropy is only slightly larger than the entropy of melting. Raising the temperature of liquid water increases the molecular disorder by increasing the average molecular speed. The molecules move about more chaotically and disorder rises.

Ideal Gas Entropy

It is useful to have an expression for the entropy of an ideal gas. Starting with the first law we can write $dQ = dU + P \, dV$, obtaining for dS,

$$dS = \frac{dQ}{T} = \frac{dU + P \, dV}{T}$$

For an ideal gas

$$dU = mC_v \, dT$$

and

$$PV = vRT$$

The specific heat C_v is a constant. The expression for dS converts to

$$dS = mC_v \frac{dT}{T} + vR \frac{dV}{V}$$

This relation can be integrated to give the entropy change in a process that carries the ideal gas from (T_1, V_1) to (T_2, V_2). The result is

$$S_2 - S_1 = mC_v \ln\left(\frac{T_2}{T_1}\right) + vR \ln\left(\frac{V_2}{V_1}\right) \tag{21.19}$$

Notice that the entropy change has been expressed in terms of the temperatures and volumes of the two states, showing that entropy is a property of a physical system.

We have stressed that the relation $dS = dQ/T$ is valid only for reversible processes. The following discussion of the free expansion of a gas illustrates this fact, and the idea that entropy measures disorder.

Free Expansion of a Gas

Consider the experiment represented in Figure 21.12. A vessel with rigid insulated walls is divided by a partition. Initially, one side contains an ideal gas and the other side is a vacuum. Consider what happens when the partition ruptures, allowing the gas to expand and fill the entire container. The insulation prevents heat exchange. Thus, $dQ = 0$ for each stage of the expansion. The rigid walls guarantee that there can be no change in the volume of the system, and thus no $P\,dV$ work. The process is called a **free expansion** because of the absence of any external work as gas moves into the evacuated region. We therefore have $dQ = 0$ and $dW = 0$ at each stage of the expansion. It follows from the first law, $dU = 0$, that the internal energy of the gas does not change. For an ideal gas, $U = mC_vT$; the internal energy depends on T alone. If U remains constant, so must the temperature, and thus $T_2 = T_1$. The final volume of the gas V_2 is clearly greater than the initial volume V_1, so that $\ln(V_2/V_1)$ is a positive number. From Equation 21.19 we then have

$$S_2 - S_1 = vR \ln\left(\frac{V_2}{V_1}\right) > 0$$

The final entropy exceeds the initial entropy. The gas has expanded but its temperature has not changed. The gas spreads out over a larger volume, resulting in further disorder. The predicted entropy increase is in agreement with the idea that entropy measures disorder.

It all seems very straightforward, but is it? Let's return momentarily to the point where we said that the container was insulated. We noted that the insulation guaranteed that $dQ = 0$ *at each stage* of the expansion. From $dS = dQ/T$ it might *seem* that we should conclude that $dS = 0$ at each stage, and therefore that the entropy of the gas remains constant. But this is not so! The entropy *does* change. The expression $dS = dQ/T$ does *not apply* to the free expansion because the free expansion is an irreversible process. In other words, it cannot be reversed by any infinitesimal change in the surroundings. The relation $dS = dQ/T$ holds only for reversible transformations.

The result

$$S_2 - S_1 = vR \ln\left(\frac{V_2}{V_1}\right) > 0$$

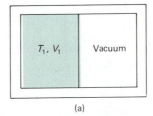 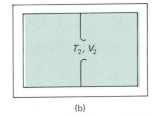

(a)   (b)

FIGURE 21.12

A free expansion. Initially, (a) one portion of the rigid, insulated container is a vacuum and an ideal gas occupies the other section. After the partition ruptures, (b) the gas expands to fill the container. The insulation and rigid walls insure that the system is isolated.

obtained via Equation 21.19 is valid here even though the process leading from state 1 to state 2 is irreversible. This is because entropy is a property of the thermodynamic state. The entropy change $S_2 - S_1$ is independent of the process that carries the system from one state to another.

Entropy Formulation of the Second Law

Although the energy of an isolated system remains constant, its entropy need not remain constant. The free expansion of an ideal gas considered in the preceding section shows that the entropy of an isolated system increases as the result of an irreversible process. Entropy, unlike energy, is not a conserved quantity. In general, we can say that the irreversible processes that drive a system toward equilibrium generate entropy. The diffusion of perfume vapor is an irreversible process that leads to increased disorder, and so the entropy increases. This connection between entropy and irreversibility led Rudolph Clausius to an entropy formulation of the second law:

The entropy of an isolated system never decreases.

Stated mathematically, this version of the second law is

$$\Delta S \geq 0 \qquad (21.20)$$

where ΔS denotes the change in entropy brought about by any thermodynamic processes that connect equilibrium states of an isolated system. If an isolated system is not in equilibrium, irreversible processes operate to drive it toward equilibrium. Irreversible processes generate entropy. They mix, disorder, or randomize. The inequality in Equation 21.20 refers to nonequilibrium situations in which irreversible processes generate entropy. From these remarks it follows that equilibrium corresponds to a state of maximum entropy. The entropy increases until the system attains equilibrium, after which the entropy remains constant.

Keep in mind that the entropy principle applies *only* to isolated systems. For example, a 10-year-old child converts mashed potatoes into muscle—disorder into order. But the child is not an isolated system, and the entropy principle does not apply.

WORKED PROBLEM

A heat engine uses an ideal gas ($\gamma = 1.40$) that undergoes the reversible cycle shown in Figure 21.13. Show that the thermodynamic efficiency for the engine is 0.24.

Solution

The efficiency can be determined by using

$$\eta = \frac{Q_A - Q_E}{Q_A}$$

where Q_A is the heat absorbed over the cycle and Q_E is the heat ejected over the cycle.

Heat is absorbed or ejected over each of the four legs of the cycle. The heat involved can be evaluated by using

$$Q = mC\,\Delta T$$

Over the legs $1 \to 2$ and $3 \to 4$, the specific heat, C, equals C_v, the specific heat at constant volume. Over the legs $2 \to 3$ and $4 \to 1$, C equals C_p, the specific heat at constant pressure.

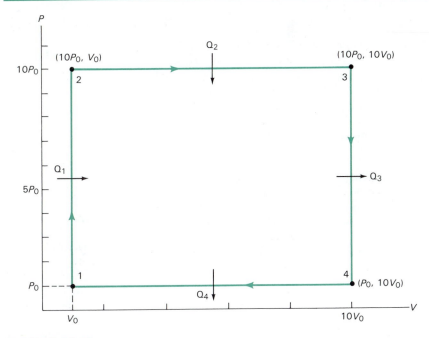

FIGURE 21.13

Each leg involves a change by a factor of ten in either the pressure or the volume. It follows from the ideal gas equation of state, $PV = \nu RT$, that the temperature also changes by a factor of ten over each leg. This conclusion leads to the assignment of temperatures shown in Figure 21.13.

$$T_2 = 10T_1 \qquad T_3 = 100T_1 \qquad T_4 = 10T_1$$

The temperature increases along $1 \rightarrow 2$ and $2 \rightarrow 3$ and heat is absorbed.

$$Q_1 = mC_v\,\Delta T = mC_v(10T_1 - T_1) = mC_v(9T_1)$$

$$Q_2 = mC_p(100T_1 - 10T_1) = mC_p(90T_1)$$

The temperature decreases along $3 \rightarrow 4$ and $4 \rightarrow 1$ and heat is ejected

$$Q_3 = mC_v(100T_1 - 10T_1) = mC_v(90T_1)$$

$$Q_4 = mC_p(10T_1 - T_1) = mC_p(9T_1)$$

The heat absorbed is

$$Q_A = Q_1 + Q_2 = m(9C_v + 90C_p)T_1$$

The heat ejected is

$$Q_E = Q_3 + Q_4 = m(90C_v + 9C_p)T_1$$

The efficiency is

$$\eta = \frac{Q_A - Q_E}{Q_A} = \frac{81(C_p - C_v)}{90C_p + 9C_v}$$

Dividing numerator and denominator by C_v and using $\gamma = C_p/C_v = 1.40$ gives the desired result,

$$\eta = \frac{81(\gamma - 1)}{90\gamma + 9} = \frac{81(0.40)}{90 \cdot 1.40 + 9}$$

$$= 0.24$$

EXERCISES

21.2   Heat Engines and Thermodynamic Efficiency

A.   An inventor claims that his heat engine absorbs 38 kJ of heat, performs 42 kJ of work, and ejects 4 kJ of heat per cycle. Would you be inclined to invest money in this engine? Explain.

B.   A heat engine absorbs 18 kJ of heat per cycle. Its thermodynamic efficiency is 0.36. (a) How much heat does this engine eject per cycle? (b) How much work does it do each cycle?

C.   A heat engine performs 24 kJ of work and ejects 36 kJ of heat during one cycle. (a) Determine the heat absorbed by the heat engine over one cycle. (b) Determine the thermodynamic efficiency.

D.   Determine the thermodynamic efficiency for the cycle shown in Figure 1.

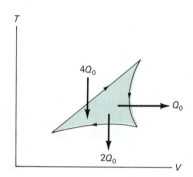

FIGURE 1

E.   Show that the efficiency of the cycle shown in Figure 2 is 1/3.

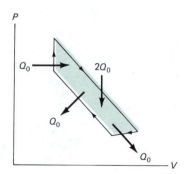

FIGURE 2

F.   A heat engine operates on the cycle shown in Figure 3. It absorbs heat along two legs and ejects heat along two legs, as shown. It performs 10 kJ of work in each cycle. (a) Determine the heat ejected along the lower leg of the cycle. (b) Determine the efficiency of the engine.

G.   A refrigerator has a $COP = 2.7$. The net work performed on the refrigerant in 2 hours in 7 kWh.

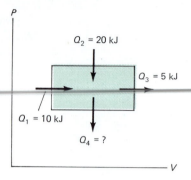

FIGURE 3

Determine the heat extracted from the low-temperature reservoir in 2 hours.

H.   The thermodynamic efficiency of an air conditioner is 0.23. (a) Determine its coefficient of performance. (b) If the air conditioner performs 3×10^7 J of work in 1 h, how much heat does it extract from the low-temperature reservoir in that time?

I.   During each cycle of its operation, a refrigerator extracts 360 kJ of energy from the low-temperature reservoir and ejects 480 kJ of energy to the high-temperature reservoir. Determine (a) the work done on the refrigerant, (b) the coefficient of performance.

21.3   The Carnot Cycle

J.   Can you air-condition your kitchen by opening the refrigerator door? How would you explain your answer to (a) your physics classmates, (b) someone who has never studied physics?

K.   A ship driven by steam turbines consumes fuel at a lower rate when it is in the cold North Atlantic Ocean than when it is in the warm South Pacific Ocean, even though its speed is the same in both. Suggest an explanation based on thermodynamic efficiency.

L.   Why is it necessary to condense the steam on the low-temperature side of a steam turbine? It seems somewhat wasteful to cool the water, return it to the boiler, and then convert it back into superheated steam. Why not just let the steam move through the turbine without condensing?

21.4   The Second Law of Thermodynamics

M.   One mole of an ideal gas expands isothermally. (a) If the gas doubles its volume, show that the work of expansion is

$$W = RT \ln 2$$

(b) Since the internal energy of an ideal gas depends solely on its temperature, there is no change in U during the expansion. It follows from the first law that the heat absorbed by the gas during the expansion is converted completely into work. Why does this *not* violate the second law?

21.5 The Kelvin Temperature Scale

N. A Carnot cycle operates between reservoirs at temperatures of 180°C and 30°C. Determine its Carnot efficiency.

P. A heat engine absorbs heat from a reservoir at 60°C and ejects heat to the surroundings at 20°C. Its thermodynamic efficiency is 0.08. Determine its Carnot efficiency.

Q. A heat pump ejects 6 kJ of heat to a high-temperature reservoir for every 5 kJ of heat it extracts from the low-temperature reservoir. Determine its coefficient of performance.

R. Over one cycle a Carnot engine absorbs 24 MJ of heat along an isotherm at 140°C and ejects heat along a 14°C isotherm. Determine the amount of heat ejected and the Carnot efficiency.

S. There are proposals to tap the thermal energy of the ocean by operating a heat engine utilizing $T_H = 32°C$ of warm surface water in the tropics and $T_L = 14°C$ of cooler deep water. Calculate the Carnot efficiency of a heat engine operating between these temperatures.

T. An inventor comes to you with the claim that her heat engine, which employs water as a working substance, has a thermodynamic efficiency of 0.61. She explains that it operates between heat reservoirs at 4°C and 0°C. It is a very complicated device, with many pistons, gears, and pulleys, and the cycle involves freezing and melting. Does her claim that $\eta = 0.61$ warrant serious consideration? Explain.

U. The Carnot efficiency is given by Equation 21.13. For the same value of T_L, which produces the larger increase in the Carnot efficiency, a rise in T_H from 500 K to 510 K or a rise in T_H from 1000 K to 1010 K?

V. If T_L for a Carnot cycle is limited to 290 K, what value of T_H is needed for a Carnot efficiency of 0.70?

21.6 Entropy

W. At a pressure of 1 atm, liquid helium boils at 4.2 K. The latent heat of vaporization is 20.5 kJ/kg. Determine the entropy change (per kilogram) resulting from vaporization.

X. The heat of vaporization of ethyl alcohol is 1120 kJ/kg at 27°C. Determine the entropy change of 1 gram of ethyl alcohol when it evaporates at room temperature (27°C).

Y. How much water must be vaporized at the normal boiling point to produce the same entropy change as occurs in the melting of 1 kg of ice?

PROBLEMS

1. A substance undergoes a Carnot cycle during which its temperature changes by a factor of 2 and its entropy changes by a factor of 4. (a) Sketch a graph of the cycle using temperature-entropy (T-S) axes. (b) What is the physical significance of the area enclosed by the graph?

2. The isothermal expansion of an ideal gas carries it from $P = 10 \times 10^5$ N/m², $V = 1$ m³ to $P = 5 \times 10^5$ N/m², $V = 2$ m³ at a temperature of 400 K. Calculate the entropy change.

3. The rigid insulated container used in the free-expansion experiment (Section 21.6) is prepared with 1 mole of chlorine gas on one side of the partition and 1 mole of bromine gas on the other. The partition ruptures, allowing the two gases to mix. The volumes V_1 and V_2 are equal. Treat the two gases as ideal and show that the entropy increase of the system is

$$\Delta S = 2R \ln 2$$

This quantity is known as the **entropy of mixing**. How has disorder increased?

4. The molar heat of fusion of lead is 4.69 kJ/mole. The normal melting point is 327.5°C. Determine the entropy change of 100 grams of lead when it melts at a pressure of 1 atm. The atomic weight of lead is 207.2.

5. One mole of an ideal gas ($C_p = 5R/2$; $C_v = 3R/2$) in a Carnot engine executes the following cycle: starting at 600 K, a pressure of 1 atm, and volume of 1 m³, the gas undergoes

  i. an isothermal expansion until $V = 2$ m³ followed by an adiabatic expansion until $T = 300$ K
  ii. an isothermal compression followed by an adiabatic compression to the initial state

(a) What is the efficiency of this engine? (b) Suppose that the adiabatic expansion in Step i is replaced by an isobaric change followed by an isovolumic change until the temperature drops to 300 K, and the adiabatic compression in Step ii is replaced by an isobaric change and an isovolumic change until the temperature rises to 600 K (Figure 4). What is the efficiency of this cycle?

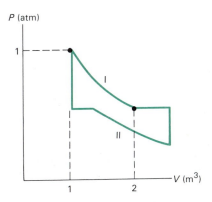

FIGURE 4

6. One mole of an ideal gas ($\gamma = 5/3$) at an initial pressure of 1 atm and volume 0.01 m³ is the working substance in a heat engine that executes the cycle depicted in Figure 5. The first part of the cycle is an isothermal expansion. For each part of the cycle, determine (a) the work, (b) the heat exchanged, (c) the change in internal energy. (d) Determine the efficiency of the engine.

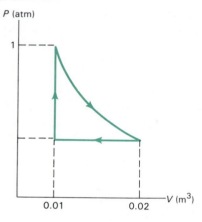

P (atm)

1

0.01   0.02   V (m³)

FIGURE 5

7. (a) Calculate the net work done over the cycle shown in Figure 6. The expansion from 2 to 3 and the compression from 4 to 1 are adiabatic. The arrows indicate the absorption of heat from 1 to 2 and the ejection of heat from 3 to 4. (b) Calculate the thermodynamic efficiency.

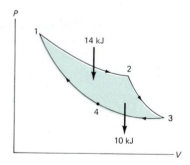

P

1

14 kJ

2

4

3

10 kJ

V

FIGURE 6

8. Which has the greater thermodynamic efficiency, a heat engine for which the heat absorbed from the high-temperature reservoir is three times the work output, or a heat engine for which the heat absorbed from the high-temperature reservoir is twice the heat ejected to the low-temperature reservoir?

9. Prove that the coefficient of performance (COP) and thermodynamic efficiency (η) are related by

$$\eta = \frac{1}{1 + COP}$$

10. Which change would produce a greater increase in the Carnot efficiency? (a) A 25-K increase in the temperature of the high-temperature reservoir; (b) A 25-K decrease in the temperature of the low-temperature reservoir. Explain.

11. Suppose that an ideal gas is used as the working substance in the Carnot cycle depicted in Figure 21.5. Remember that the internal energy of an ideal gas depends on the temperature alone. (a) If the gas absorbs $Q_H = 3$ kJ during the isothermal leg from 1 to 2, how much work is performed by the gas? (b) If the gas performs 1 kJ of work during the adiabatic expansion from 2 to 3, what is the change in its internal energy? (c) How much work must be performed on the gas from 3 to 4 for it to eject $Q_L = 2$ kJ of heat? (d) How much work must be performed on the gas to take it from 4 to 1? (e) What is the net amount of work performed by the gas during the cycle? (f) What is the thermodynamic efficiency of the cycle?

12. (a) If the temperature of the cold reservoir of a Carnot engine is changed by some small amount ΔT, show that the fractional change in efficiency is

$$\frac{\Delta \eta}{\eta} = \frac{-\Delta T}{T_H - T_L}$$

(b) If the temperature of the hot reservoir of a Carnot engine is changed by some small amount $\Delta T \ll T_H$, show that

$$\frac{\Delta \eta}{\eta} \approx \frac{\left(\dfrac{T_L}{T_H}\right) \Delta T}{T_H - T_L}$$

(c) Given the choice of decreasing the cold temperature by an amount ΔT or increasing the hot temperature by the same amount, which would produce the larger change in efficiency?

13. A "figure of merit" (FOM) for a heat pump may be defined as

$$FOM = \frac{\text{heat transferred to}}{\text{high-temperature reservoir in each cycle}}{\text{work done each cycle}}$$

(a) Using the first law of thermodynamics, show that for a heat pump

$$FOM = \frac{Q_H}{Q_H - Q_L}$$

(b) Show that the maximum FOM for a heat pump is

$$FOM \text{ (maximum)} = \frac{T_H}{T_H - T_L}$$

where the temperatures are expressed in kelvins. (c) Determine the maximum FOM if the interior of a house is 70°F and the exterior is 40°F.

14. An ideal gas undergoes the Carnot cycle shown in Figure 21.5. The P, V, and T values at points 1, 2, 3, and 4 are as follows:

1: $P = 10 \times 10^5 \, \text{N/m}^2$ $V = 1 \, \text{m}^3$ $T = 400 \, \text{K}$
2: $P = 5 \times 10^5 \, \text{N/m}^2$ $V = 2 \, \text{m}^3$ $T = 400 \, \text{K}$
3: $P = 2.44 \times 10^5 \, \text{N/m}^2$ $V = 3.08 \, \text{m}^3$ $T = 300 \, \text{K}$
4: $P = 4.87 \times 10^5 \, \text{N/m}^2$ $V = 1.54 \, \text{m}^3$ $T = 300 \, \text{K}$

(a) Calculate the heat absorbed or ejected over each of the four legs of the cycle. (b) Calculate the thermodynamic efficiency of the cycle.

15. One kg of water at 10°C is mixed with 1 kg of water at 30°C. The process proceeds at constant pressure. When the mixture has reached equilibrium, (a) what is the final temperature? (b) Take $C_p = 4.19 \, \text{kJ/kg·K}$ for water and show that the entropy of the system increases by

$$\Delta S = 4.19 \ln\left[\left(\frac{293}{283}\right) \cdot \left(\frac{293}{303}\right)\right] \text{kJ/K}$$

(c) Verify numerically that $\Delta S > 0$. (d) Is the mixing process irreversible?

16. It is an empirical fact known as **Trouton's rule** that the entropy change per mole associated with boiling is roughly the same for many liquids. Test this rule for the liquids listed below by calculating the entropy change per *kilomole*.

Liquid	Latent Heat of Vaporization (kJ/kg)	Molecular Weight (kg/kmol)	Normal Boiling Point (K)
H_2O	2256	18	373
O_2	213	32	90
^4He	20.5	4	4.2
C_2H_5 (ethyl alcohol)	854	46	351
C_6H_6 (benzene)	402	78	353

17. One mole of helium absorbs 126 J of heat at constant volume. Its initial temperature is 80 K. Determine its final temperature and its change in entropy.

18. One mole of helium at 283 K absorbs 252 J of heat at constant volume. One mole of oxygen at 283 K also absorbs 252 J at constant volume. (a) Which gas undergoes the smaller temperature change? (b) Which gas undergoes the smaller entropy change?

19. At a party a guest claims that his glass of water (0.25 kg at 23°C) spontaneously separated into an ice cube (0.02 kg at 0°C) and 0.23 kg of water. (a) Calculate the final water temperature required to satisfy the first law of thermodynamics. (b) Calculate the entropy change for the entire process. (c) Interpret your answer for (b) using the second law of thermodynamics.

20. An ideal gas is carried from the state a to the state b reversibly. Starting with the entropy change

$$\Delta S_{ab} = mC_v \ln\left(\frac{T_b}{T_a}\right) + vR \ln\left(\frac{V_b}{V_a}\right)$$

derive the adiabatic relation, $P_a V_a^{\gamma} = P_b V_b^{\gamma}$, by requiring that $\Delta S_{ab} = 0$ for a reversible adiabatic process.

22.1
THE ATOMIC MODEL OF MATTER

A model is a conceptual picture coupled with a set of assumptions. Kinetic theory is based on an atomic model that pictures matter as composed of atoms and molecules. Kinetic theory assumes that the measurable properties of gases, liquids, and solids reflect the combined actions of countless numbers of atoms and molecules. For example, the pressure exerted on the walls of a bicycle tire is produced by the impacts of an enormous number of air molecules.

The **mole**, one of the seven fundamental quantities in the SI system, is related to the atomic model of matter. One mole is the amount of a substance that contains the same number of **fundamental entities** (atoms or molecules) as are contained in 12 grams of carbon–12 (^{12}C). The number of fundamental entities in 1 mole is called **Avogadro's number** and has the value

$$N_A = 6.02205 \times 10^{23}$$

In water, the fundamental entities are H_2O molecules. In helium, the fundamental entities are helium atoms. One mole of water contains N_A water molecules and 1 mole of helium contains N_A helium atoms.

The mass of 1 mole of a substance is called its **gram molecular weight** (often shortened to **molecular weight**). By definition, then, the gram molecular weight of ^{12}C is 12 grams. Table 22.1 lists the gram molecular weights of several elements.

TABLE 22.1
Characteristic Atomic Dimension

Element	Gram Molecular Weight (grams)	Mass Density (grams/cm³)	Atomic Dimension (Å)
Solids			
Lithium (Li)	6.9	0.53	2.8
Neon (Ne)	20.2	1.00	3.2
Sodium (Na)	23.0	0.97	3.4
Aluminum (Al)	27.0	2.70	2.6
Phosphorus (K)	29.1	0.86	4.2
Argon (Ar)	39.9	1.65	3.4
Copper (Cu)	63.5	8.92	2.3
Zinc (Zn)	65.4	7.1	2.5
Rubidium (Rb)	85.5	1.53	4.5
Silver (Ag)	107.9	10.5	2.6
Cesium (Cs)	132.9	1.87	4.9
Platinum (Pt)	195.2	21.4	2.5
Gold (Au)	197.0	19.3	2.6
Lead (Pb)	207.2	11.3	3.1
Liquids			
Helium (He)	4.0	0.12	3.8
Neon (Ne)	20.2	1.21	2.9
Argon (Ar)	39.9	1.39	3.6
Krypton (Kr)	83.8	2.61	3.7
Xenon (Xe)	131.3	3.06	4.1
Mercury (Hg)	200.6	13.6	2.9

The molecules of a gas are very small, but just how small are they? What is the size of a typical atom? A reliable estimate of atomic size can be made by using the fact that solids and liquids are virtually *incompressible*.

Let's estimate atomic sizes by assuming that the volume of a solid or liquid equals the total volume of the constituent atoms. We ignore the space between atoms. This is a reasonable assumption because if there were much space between atoms in a solid or liquid it would be possible to squeeze them closer together by applying pressure. However, as we already noted, squeezing a solid or a liquid has virtually no effect on its volume.

Consider 1 mole of a solid. Let V denote its volume and let M signify its gram molecular weight. The number of molecules in this molar volume is $N_A \approx 6 \times 10^{23}$. Now imagine that the volume is composed of N_A cubical subvolumes of equal size—one for each molecule. If the length of the edge of such a cube is d, the cube's volume is d^3. Thus d is a measure of molecular size. Taken collectively, the N_A cubes comprise the total volume V and

$$V = N_A d^3$$

Next we observe that the mass density ρ may be expressed by

$$\rho = \frac{\text{mass of 1 mole}}{\text{volume of 1 mole}} = \frac{M}{V} = \frac{M}{N_A d^3}$$

Solving for d gives

$$d = \left(\frac{M}{N_A \rho}\right)^{1/3} \tag{22.1}$$

We can use Equation 22.1 to estimate the size of a water molecule. For water, $M = 18$ grams and $\rho = 1$ g/cm^3. Equation 22.1 gives

$$d = \left(\frac{18 \text{ g}}{(6 \times 10^{23})(1 \text{ g/cm}^3)}\right)^{1/3}$$

$$= 3 \times 10^{-8} \text{ cm} = 3 \times 10^{-10} \text{ m}$$

The angstrom (abbreviated Å) is a convenient unit of length for atomic and molecular sizes.

$$1 \text{ angstrom} \equiv 10^{-10} \text{ m} = 1 \text{ Å}$$

Evidently, the characteristic dimension of a water molecule is 3 Å.

An interesting pattern emerges when Equation 22.1 is used to evaluate d for many solids and liquids. As Table 22.1 shows, $d = (M/N_A \rho)^{1/3}$ is very near 3 Å for many elemental solids and liquids, even though the values for M and ρ differ widely. Other materials furnish similar results. Measurements of atomic spacing made by scattering X rays from solids give values within 10% of those in Table 22.1. We conclude that the angstrom characterizes molecular and atomic sizes.

22.2
MEAN FREE PATH

The atoms of a gas are constantly in motion. When atoms collide with each other and with the walls of their container they are deflected. As a result, they travel in irregular zigzag paths. The distance that an atom travels between

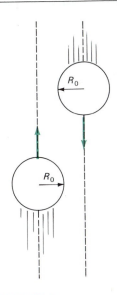

FIGURE 22.1

The effective radius of an atom (R_0) is a measure of the range of interatomic forces. A collision occurs if parallel lines through the centers of two approaching atoms are separated by less than $2R_0$.

collisions is called its **free path length**. Since collisions occur randomly, the length of individual free paths cannot be predicted. A useful measure of free path length is the mean distance between collisions. This distance is called the **mean free path** and is denoted as λ. We can estimate λ as follows: let the effective radius of an atom be denoted by R_0. By "effective" we mean that a collision occurs if parallel lines through the centers of two approaching atoms are separated by less than $2R_0$ (Figure 22.1). The effective radius R_0 is a length that measures the range of interatomic forces—that is, the distance over which the atoms exert forces on one another.

Corresponding to R_0 there is an area, πR_0^2. Geometrically, πR_0^2 represents the **cross-sectional area** of a sphere of radius R_0. Physically, πR_0^2 is a measure of the target area that each atom presents to other atoms. This area is called the **collision cross section**, or simply the cross section, and is denoted by σ. If we take $R_0 \approx 3$ Å the corresponding cross section is

$$\sigma = \pi R_0^2 \approx 3 \times 10^{-19} \text{ m}^2 \tag{22.2}$$

As atoms move about between collisions, their circular cross sections sweep out cylindrical volumes. On the average the cross-sectional area σ sweeps out a volume $\sigma\lambda$, where λ is the mean free path (Figure 22.2). For N atoms in a container of volume V, the volume per atom is V/N, so we can write

$$\sigma\lambda = \frac{V}{N} \tag{22.3}$$

The number of atoms per unit volume is called the **number density**, and is denoted as n,

$$n = \frac{N}{V} \tag{22.4}$$

Setting $V/N = 1/n$, we can rewrite Equation 22.3 as

$$n\sigma\lambda = 1 \tag{22.5}$$

This is one of the most useful of all kinetic theory relations, and it can be applied to many processes besides atomic collisions.

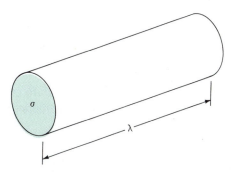

FIGURE 22.2

The cross-sectional area σ sweeps out a cylindrical volume. Over one mean free path length λ the volume swept out is $\sigma\lambda$.

EXAMPLE 1

Mean Free Path for Air Molecules

Let's use Equation 22.5 to estimate the mean free path for a molecule in our atmosphere. The number density (n) can be estimated by using the fact that 1 mole of gas under standard conditions occupies a volume of approximately 0.0224 m^3. The corresponding number density is

$$n = \frac{6.02 \times 10^{23}}{0.0224 \text{ m}^3} \approx 2.7 \times 10^{25} \text{ m}^{-3}$$

For the cross section we use the value $\sigma = 3 \times 10^{-19} \text{ m}^2$, given by Equation 22.2. The mean free path is

$$\lambda = \frac{1}{n\sigma} = \frac{1}{2.7 \times 10^{25} \text{ m}^{-3} \cdot 3 \times 10^{-19} \text{ m}^2} = 1.2 \times 10^{-7} \text{ m}$$
$$= 1200 \text{ Å}$$

Our estimate for σ used $R_0 = 3 \text{ Å}$. The mean free path of 1200 Å is 400-times as large, confirming the view that atmospheric molecules are far apart and spend most of their time in free flight.

22.3
THE IDEAL GAS: KINETIC INTERPRETATION OF TEMPERATURE

We want to use the atomic model of matter to derive the ideal gas equation of state

$$PV = vRT \tag{22.6}$$

We first rewrite this equation to obtain a form that emphasizes the microscopic viewpoint. We introduce the **Boltzmann constant** (k) in terms of the gas constant R and Avogadro's number, N_A:

$$k = \frac{R}{N_A} = 1.38062 \times 10^{-23} \text{ J/K}$$

For a gas containing N molecules the number of moles is given by

$$v = \frac{N}{N_A}$$

Using these results in Equation 22.6 we obtain an alternative form of the ideal gas equation of state,

$$PV = NkT \tag{22.7}$$

This equation of state can be derived by introducing a microscopic model of an ideal gas. The model makes four basic assumptions:

1. The gas consists of a large number of atoms* that obey Newton's laws of motion.
2. The motion of the atoms is random.

---

* For the present we consider a monatomic gas such as helium.

3. The volume of the atoms themselves is negligible in comparison with the volume of the container. The atoms of a monatomic gas are treated as point masses.

4. The atoms collide elastically with the container walls, but do not collide with one another.

Let's consider these four assumptions one by one and see what evidence supports each:

1   Experimental evidence for assuming that the gas contains many atoms is related to determinations of Avogadro's number, $N_A \approx 6 \times 10^{23}$. At 0°C and 1 atm, 1 cm^3 of gas contains approximately 3×10^{19} atoms, an enormous number by any standard.

The assumption that the atoms obey Newton's laws of motion is adequate for most aspects of kinetic theory. Quantum theory is needed to fully explain the internal structures of atoms and molecules. However, the most basic features of kinetic theory are independent of the internal structure of the atoms and molecules. Historically, kinetic theory evolved within a classical framework, and we follow this historical path.

2   Experimental evidence for the random motion of atoms is based on observations of *Brownian motion*. In 1827, the Scottish botanist Robert Brown used a microscope to observe tiny spores of pollen suspended in water. The pollen appeared to dance about in erratic fashion. At first, Brown thought that the pollen was alive and that its motion was some sort of fertility dance! Subsequent studies of liquid suspensions of various inanimate particles convinced observers that the liquid itself was responsible for the erratic motions.

Eventually, Einstein explained the irregular motion of the suspended particles in detail by assuming that the liquid was composed of molecules in random motion. The random motion of atoms in a gas was demonstrated by Fletcher and Millikan in 1911, by observing the Brownian motion of electrically charged oil droplets.

3   An atom behaves like a point mass in only a few major respects. However, it is these same few characteristics that are crucial to determining the equation of state. The translational kinetic energy of the atom can be written as $\frac{1}{2}mv^2$, where m is the mass of the atom and v is the speed of the center of mass. The linear momentum is given by mv. With respect to translational kinetic energy and linear momentum, the structure of the atom is irrelevant—it behaves like a point mass.

Liquid-to-vapor transition data justify the neglect of the atomic volume in a rarified gas. For example, 1 g of water at the normal boiling point occupies a volume of approximately 1 cm^3. After vaporization it occupies a volume of over 1700 cm^3, suggesting that the volume of the molecules themselves is much less than the volume occupied by the vapor.

4   The assumption of elastic collisions is a simplification. Some atoms collide inelastically with the walls of the container. An inelastic collision may result in the transfer of energy from the atom to the wall, or vice versa. However, when thermal equilibrium prevails, inelastic collisions collectively do not result in any net transfer of energy to or from the gas. Hence, ignoring inelastic collisions does not affect our conclusions about the equilibrium states of the gas.

The assumption that the atoms do not collide with one another is supported by our estimate of the mean free path λ in Example 1 in which we found that $\lambda = 1200$ Å. Compared with an atomic size of 3 Å it is evident that atoms spend most of their time in free flight. Thus, we ignore collisions between atoms.

Pressure Calculation

The ideal gas equation of state, $PV = NkT$, relates the gas pressure to the thermodynamic variables N, T, and V. Collisions of the atoms with the container walls give rise to a force and thus to a pressure. To determine the pressure we calculate the average momentum transferred to a wall per second. According to Newton's second law, the momentum change per second equals the force. This force divided by the area of the wall is the gas pressure.

We first calculate the momentum transfer to a wall by an atom. Let v_x denote the x-component of the velocity of an atom of mass m that subsequently collides with the wall (see Figure 22.3). The atom rebounds with an x-component of velocity equal to $-v_x$. The change in its x-component of linear momentum is

$$(p_x)_{\text{after}} - (p_x)_{\text{before}} = -2mv_x$$

A momentum change of equal magnitude and opposite direction is transferred to the container wall.

$$\genfrac{}{}{0pt}{}{\text{momentum transfer to}}{\text{wall in one collision}} \equiv \Delta p_x = +2mv_x$$

The gas is enclosed in a rectangular container having sides of length L_x, L_y, and L_z (Figure 22.4). An atom with velocity component v_x travels the length L_x of the container in a time L_x/v_x. The time interval between collisions with a particular wall Δt is twice as great:

$$\Delta t = \frac{2L_x}{v_x}$$

The average rate at which an atom transfers momentum to the wall is

$$\frac{\Delta p_x}{\Delta t} = \frac{2mv_x}{2L_x/v_x} = \frac{mv_x^2}{L_x}$$

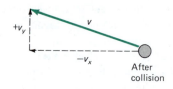

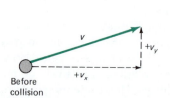

FIGURE 22.3

An atom undergoes an elastic collision with the wall. The normal component of its velocity (v_x) changes sign. Its speed (v) is unchanged.

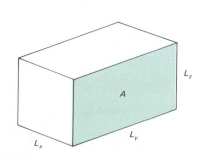

FIGURE 22.4

The rectangular container has a volume $L_xL_yL_z$. The area A of the face perpendicular to the x-direction is L_yL_z.

By Newton's second law, the rate of change of momentum equals force, f_x. Thus, the average force exerted by the impacts of a single atom having a velocity component v_x is

$$f_x = \frac{m v_x{}^2}{L_x}$$

The force exerted by a single atom is not observable: a pressure gauge responds to the combined effects of many such forces. The total force on the wall is obtained by summing over all atoms:

$$F_x = \sum f_x = \frac{\sum m v_x{}^2}{L_x}$$

The area of the container wall perpendicular to the x-axis is $A = L_y L_z$ (Figure 22.4). The pressure P on this area is the force F_x divided by the wall area $A = L_y L_z$,

$$P = \frac{F_x}{A} = \frac{\sum m v_x{}^2}{L_x L_y L_z}$$

Noting that $L_x L_y L_z$ is the volume (V) of the container, we can write

$$PV = \sum m v_x{}^2$$

The pressure of the gas is the same on every wall; *there is no preferred direction of motion.* Thus we can also write

$$PV = \sum m v_y{}^2 = \sum m v_z{}^2$$

Further, since

$$v_x{}^2 + v_y{}^2 + v_z{}^2 = v^2$$

we can write

$$PV = \tfrac{1}{3} \sum m(v_x{}^2 + v_y{}^2 + v_z{}^2) = \tfrac{1}{3} \sum m v^2 \qquad (22.8)$$

The total kinetic energy of the gas is

$$K = \sum \tfrac{1}{2} m v^2 \qquad (22.9)$$

Comparing Equations 22.8 and 22.9 we see that

$$PV = \tfrac{2}{3} K \qquad (22.10)$$

By introducing the average kinetic energy per atom we can maintain the particle point of view and express Equation 22.10 in a form directly comparable with the equation of state, $PV = NkT$. We write K as N times the average kinetic energy per atom, $\langle K \rangle$:

$$K = N \langle K \rangle$$

whereupon Equation 22.10 becomes

$$PV = \tfrac{2}{3} N \langle K \rangle \qquad (22.11)$$

Equation 22.11 is the equation of state that the ideal gas model has produced. Experimentally, the equation of state that is approximately obeyed by all gases is

$$PV = NkT \qquad (22.7)$$

Comparing Equations 22.7 and 22.11, we can conclude that the temperature of an ideal gas is proportional to the average kinetic energy of the atoms. Specifically, if the model is to reproduce the experimentally determined equation of state we must have

$$\langle K \rangle = \langle \tfrac{1}{2}mv^2 \rangle = \tfrac{3}{2}kT \tag{22.12}$$

We emphasize that Equation 22.12 is *not* an operational definition of temperature because we cannot measure the average kinetic energy of an atom. The Kelvin temperature T is defined operationally in terms of heat and the Carnot cycle. However, our model of the ideal gas does give us a **microscopic interpretation** of temperature. It shows that the temperature of a gas is a direct measure of the kinetic energy of the thermal motions of its atoms.

Our derivation of the equation of state applies equally well to diatomic and polyatomic gases because our considerations involved only the translational kinetic energy and linear momentum of the center of mass. Because we follow only the center-of-mass motion, any structure of the atoms or molecules is irrelevant insofar as the equation of state is concerned. Thus, ideal gases obey the same equation of state regardless of any microscopic structure attributed to their atoms and molecules.

EXAMPLE 2

Average Thermal Energy of Atoms

What is the average kinetic energy for atoms in a gas at room temperature ($T \approx 300$ K)? Using Equation 22.12 we have

$$\langle K \rangle = \tfrac{3}{2}kT = \tfrac{3}{2}(1.38 \times 10^{-23} \text{ J/K})(300 \text{ K})$$
$$= 6.21 \times 10^{-21} \text{ J}$$

This value of $\langle K \rangle$ is more conveniently expressed in **electron volt** units,

$$1 \text{ eV} = 1.60 \times 10^{-19} \text{ J}$$

Thus

$$\langle K \rangle = 6.21 \times 10^{-21} \text{ J} \left(\frac{1 \text{ eV}}{1.60 \times 10^{-19} \text{ J}} \right)$$
$$= 0.0388 \text{ eV}$$

or approximately 1/25 eV. Note that the mean kinetic energy is independent of the mass of the atom.

rms Speed

The speed characteristic of molecular motions can be inferred from Equation 22.12, $\tfrac{1}{2}m\langle v^2 \rangle = (3/2)kT$. The **mean square** speed is

$$\langle v^2 \rangle = \frac{3kT}{m} \tag{22.13}$$

The square root of $\langle v^2 \rangle$ is called the **root-mean-square** (rms) speed. Thus,

$$v_{\text{rms}} \equiv \sqrt{\langle v^2 \rangle} = \sqrt{\frac{3kT}{m}} \tag{22.14}$$

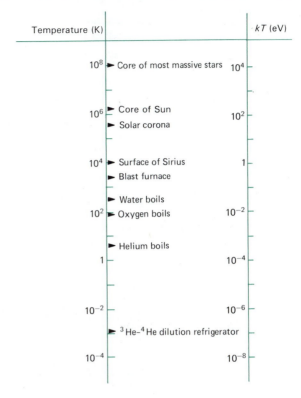

FIGURE 22.5

Characteristic temperatures and thermal energies. Note that the thermal energy (kT) is expressed in electron volts, a convenient unit of energy.

An equivalent expression for the rms speed, in terms of the molecular weight M and the gas constant R is

$$v_{\text{rms}} = \sqrt{\frac{3RT}{M}} \tag{22.15}$$

At ordinary temperatures the rms speeds of atoms and molecules are typically several hundreds of meters per second. In fact, the rms speed is comparable to the **speed of sound**. The near equality of the rms speed and the speed of sound is not accidental because sound waves travel by virtue of collisions between atoms.

When it is necessary to *estimate* the energy or speed characteristic of thermal motions we can take

$$\textbf{thermal energy} \approx \textbf{kT}$$

$$\textbf{thermal speed} \approx \sqrt{\frac{\textbf{kT}}{\textbf{m}}} \tag{22.16}$$

Figure 22.5 indicates the wide range of temperatures and thermal energies encountered in nature.

Quantum Zero-Point Energy

There is a potential pitfall in the relation

$$\langle K \rangle = \tfrac{3}{2}kT \tag{22.12}$$

Taken alone, it implies that the kinetic energy drops to zero as the temperature approaches absolute zero. This in turn suggests that absolute zero is a state in which there is no atomic or molecular motion. However, such an inference *is not valid*. The kinetic energy does not drop to zero as the temperature approaches absolute zero because there is another contribution to the kinetic energy of the atoms and molecules besides thermal kinetic energy. This contribution to the total kinetic energy, ignorably small at ordinary temperatures and densities, is quantum mechanical in nature. The quantum contribution to the kinetic energy is called the **zero-point energy** because it is the energy that would remain at absolute zero, $T = 0$.

There are several physical systems for which the zero-point energy outweighs the ordinary thermal kinetic energy. Liquid helium is perhaps the best-known example of such a system. At atmospheric pressure all but one liquid solidifies as the temperature is lowered. The only exception is helium, which remains liquid even at the lowest temperature. The attractive force between He atoms is too weak to overcome the quantum zero-point motion.

22.4
THE DISTRIBUTION OF MOLECULAR SPEEDS

In deriving the ideal gas equation of state, we dealt with quantities such as the average kinetic energy and the average force per unit area. These average values enter the kinetic theory because molecular speeds are distributed over a wide range of values.

The quantity used to describe the distribution of molecular speeds is called a **distribution function**. We define the distribution function $f(v)$ by the statement

$$\left[\begin{array}{l}\text{number of molecules with speeds} \\ \text{in the range } v \text{ to } v + dv\end{array}\right] = dN = f(v)\,dv \qquad (22.17)$$

Figure 22.6 is a graph of $f(v)$ versus v that describes the distribution of molecular speeds for 1 mole of oxygen molecules at a temperature of 293 K (68°F). The distribution function for the oxygen molecules in your classroom is likely

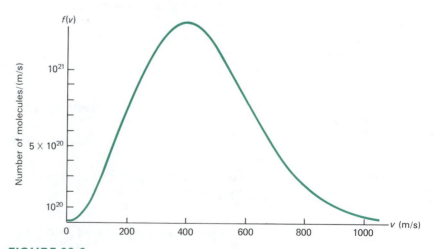

FIGURE 22.6

The distribution function $f(v)$ for 1 mole of oxygen molecules in thermal equilibrium at a temperature of 293 K (68°F).

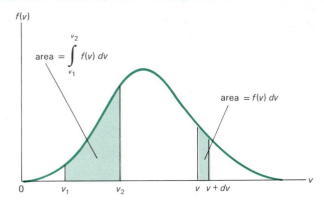

FIGURE 22.7

The number of molecules with speeds between v and $v + dv$ corresponds to the "area" under the $f(v)$ graph between the speeds v and $v + dv$. The number of molecules with speeds between v_1 and v_2 corresponds to the area under the $f(v)$ graph between v_1 and v_2. The total number of molecules in the gas corresponds to the total area under the $f(v)$ graph.

to be very similar to the one shown in Figure 22.6. The quantity $dN = f(v)\,dv$ has a geometric interpretation—it corresponds to the "area" under the graph of $f(v)$ between v and $v + dv$ (Figure 22.7). The number of molecules with speeds in the range from v_1 to v_2 is given by the integral

$$\int_{v_1}^{v_2} f(v)\,dv = \left[\begin{array}{l}\text{number of molecules with speeds}\\ \text{in the range from } v_1 \text{ to } v_2\end{array}\right] \qquad (22.18)$$

In Figure 22.7 we see that this integral corresponds to the area under the graph of $f(v)$ between v_1 and v_2. The total area under the $f(v)$ graph corresponds to the total number of molecules in the gas. Thus the integral of $f(v)\,dv$ over the full range of molecular speeds* equals N, the total number of molecules:

$$N = \int_0^\infty f(v)\,dv \qquad (22.19)$$

Over 100 years ago James Clerk Maxwell determined the form of $f(v)$ that describes a gas in thermal equilibrium. The Maxwellian distribution function for N molecules, each of mass m, in thermal equilibrium at a temperature T is

$$f(v) = 4\pi N\left(\frac{m}{2\pi kT}\right)^{3/2} v^2 e^{-(mv^2/2kT)} \qquad (22.20)$$

The only intrinsic property of the molecules that appears in the Maxwellian $f(v)$ is their mass, m. In particular, the Maxwellian distribution is independent of the forces that colliding molecules exert on one another.

EXAMPLE 3

Maxwellian Distribution of Oxygen Molecule Speeds

Consider a gas composed of 10^6 oxygen molecules with a Maxwellian distribution of speeds at a temperature of 300 K. We want to determine the number of molecules with speeds between (a) 100 m/s and 101 m/s, (b) 300 m/s and 301 m/s, (c) 1000 m/s and 1001 m/s, (d) 3000 m/s and 3001 m/s.

---

* The actual upper limit of molecular speeds is the speed of light, c. A relativistic theory takes this fact into account. Taking the upper limit of molecular speeds to be infinite rather than equal to c has a negligible effect under ordinary conditions.

For the small range of speeds considered (1 m/s) we can use Equation 22.17, which relates the number of molecules (dN) for a specified range of speeds (dv) to the distribution function $f(v)$:

$$dN = f(v)\, dv = \text{number of molecules with speeds in range } v \text{ to } v + dv$$

We set $dv = 1$ m/s as the range of speeds. For 10^6 oxygen molecules at a temperature of 300 K,

$$N = 10^6$$

$$T = 300 \text{ K}$$

$$m = 5.31 \times 10^{-26} \text{ kg}$$

Inserting these values of N, T, and m into the Maxwellian form of $f(v)$, Equation 22.20, with $dv = 1$ m/s, gives

$$dN = f(v)\, dv = (3.67 \times 10^{-2} v^2) e^{-(6.42 \times 10^{-6} v^2)}(1 \text{ m/s})$$

By consistently using SI units throughout, we make certain that the exponent will be dimensionless provided the speed v is expressed in the SI unit of m/s. We set $v = 100$ m/s to find dN for the range $v = 100$ m/s to 101 m/s. This gives

$$dN = 3.67 \times 10^{-2} (100)^2 e^{-[6.42 \times 10^{-6}(100)^2]} = 340$$

We have rounded the value of dN to two significant figures. In the same way we can evaluate dN for $v = 300$ m/s, 1000 m/s, and 3000 m/s to find (rounded to two significant figures):

$$dN = 1900 \quad (v = 300 \text{ m/s to 301 m/s})$$

$$dN = 60 \quad\quad (v = 1000 \text{ m/s to 1001 m/s})$$

$$dN = 0 \quad\quad\; (v = 3000 \text{ m/s to 3001 m/s})$$

Even though we considered equal ranges of speeds ($dv = 1$ m/s) for each of the four intervals, the numbers of molecules are quite different because of the $v^2 e^{-(mv^2/2kT)}$ dependence of the Maxwellian distribution function. In particular, it is the exponential factor $e^{-(mv^2/2kT)}$ that causes the Maxwellian distribution function to drop sharply toward zero at high speeds.

The distribution function is a central concept in more advanced treatments of kinetic theory. Such advanced theories relate deviations of $f(v)$ from the Maxwellian form to phenomena such as conductive heat flow and molecular viscosity.

WORKED PROBLEM

Determine the most probable speed for oxygen molecules with a Maxwellian distribution of speeds at a temperature of 293 K.

Solution

The most probable speed is the speed for which $f(v)$ is a maximum. Mathematically, the peak of $f(v)$ is determined by the condition $df(v)/dv = 0$. For the Maxwellian distribution the condition $df/dv = 0$ leads to

$$2v\left(1 - \frac{mv^2}{2kT}\right) e^{-(mv^2/2kT)} = 0$$

There are three solutions to this equation. Two of them ($v = 0$ and $v \to \infty$) correspond to minima of $f(v)$. The third solution comes from

$$1 - \frac{mv^2}{2kT} = 0$$

and determines the most probable speed, v_{mp}:

$$v_{mp} = \sqrt{\frac{2kT}{m}}$$

For oxygen molecules ($m = 5.31 \times 10^{-26}$ kg) at a temperature of 293 K, the most probable speed is

$$v_{mp} = \sqrt{\frac{2(1.38 \times 10^{-23} \text{ J/K})(293 \text{ K})}{5.31 \times 10^{-26} \text{ kg}}}$$

$$= 390 \text{ m/s}$$

EXERCISES

22.1   The Atomic Model of Matter

A. Calculate the mass in kilograms of (a) one atom of ^{12}C, (b) one atom of hydrogen.

B. The size characteristic of the neutron is 10^{-15} m. The mass of the neutron is 1.67×10^{-27} kg. A star composed of closely packed neutrons has a mass of 2×10^{30} kg (the mass of our sun). Estimate the radius of such a neutron star.

C. A 1000-page textbook has a total thickness of approximately 4 cm. Use this information to estimate the thickness of one page.

D. A laboratory meter rod has a width of 2.5 cm and a thickness of 0.70 cm. Assuming that the characteristic dimension of a molecule is 3 Å, estimate the number of molecules in the rod.

22.2   Mean Free Path

E. A small vacuum chamber has a width of 20 cm. If the mean free path in the chamber is 1000 Å when the pressure is 10^5 N/m^2, at what pressure will the mean free path equal 20 cm? (Assume that the temperature remains constant.)

F. In some regions of intergalactic space the number density may be roughly 1 atom per cubic meter. Estimate the mean free path in kilometers and in light-years.

G. The pressure of a gas is doubled by compressing it at constant temperature. Assuming that the gas obeys the ideal gas equation of state, what change, if any, is there in the mean free path of the gas molecules?

H. The standard billiard table measures 4 ft × 8 ft. The diameter of a billiard ball is 2.25 inches. Estimate the mean free path for collisions between balls when there are four balls moving about on the table.

I. Photons are particle-like quanta of light. Photons produced by nuclear reactions in the core of the sun are scattered by electrons. The scattering cross section is 2.4×10^{-29} m^2. If there are 10^{31} electrons per cubic meter in the sun, determine the photon mean free path.

J. In using Equation 22.5 to estimate λ, we ignored the atomic volume ($4\pi R_0^3/3$), claiming that it is small compared to the volume per particle, V/N. Use the data of Example 2 to justify ignoring the atomic volume.

22.3   The Ideal Gas: Kinetic Interpretation of Temperature

K. Three atoms move at speeds of 1, 2, and 3 km/s. Determine the mean speed and the rms speed.

L. The escape speed from earth's gravity is 11.2 km/s. Compare this with the rms speed of a hydrogen atom for a temperature of 200 K. Over many millions of years, would you expect any particular hydrogen atom in the upper atmosphere to have a good chance to escape earth's gravity?

M. Approximately how long does it take a nitrogen molecule moving at the rms speed to travel across your classroom, assuming it does not suffer any collisions?

N. A neutron emitted in a nuclear fission reaction has a kinetic energy of 2 MeV. (a) If the neutron mass is 1.67×10^{-27} kg, what is the neutron speed? Fission neutrons transfer their kinetic energy to moderating materials via collisions. A *thermal* neutron is one whose kinetic energy equals the mean thermal energy of the atoms in the material through which it moves. (b) Determine the speed of a thermal neutron moving through a gas with a temperature of 290 K.

P. In our treatment of the ideal gas we ignored the effects of gravity. Compare the mean thermal energy of an oxygen molecule with the change in its gravitational potential energy between the ceiling and the floor of your classroom. Use your own values for room temperature and height. Explain how your calculation justifies ignoring gravity in the derivation of the ideal gas equation of state.

Q. Particles in a plasma (an ionized gas) have an average kinetic energy of E electron volts. Assume that the average kinetic energy of a plasma particle is given by $(3/2)kT$ and derive the conversion factor that will let you express this energy in terms of the Kelvin temperature:

$$E \text{ (eV)} = \text{(conversion factor)} \cdot T$$

What temperature corresponds to an energy of 1 eV?

22.4 The Distribution of Molecular Speeds

R. Gaseous helium is in thermal equilibrium with liquid helium at a temperature of 4.20 K. Determine the most probable speed of a helium atom (mass $= 6.70 \times 10^{-27}$ kg).

S. Smoke particles with a mass of 10^{-17} kg are suspended in an oxygen atmosphere. The oxygen molecules and the smoke particles are in thermal equilibrium. Calculate the ratio of the most probable speeds of an oxygen molecule and a smoke particle.

PROBLEMS

1. A gas composed of 1000 oxygen molecules has a Maxwellian distribution of speeds. Use graph paper to plot the Maxwellian distribution function for two temperatures:

$$T_1 = 100 \text{ K} \qquad T_2 = 300 \text{ K}$$

Make your plots on the same speed axis and use them to answer the following questions: (a) Which temperature has the larger most probable speed? (b) Which temperature has the larger maximum value of $f(v)$?

2. For a gas in thermal equilibrium we considered three characteristic thermal speeds: the rms speed, the most probable speed, and the speed of sound in the gas. For a diatomic gas, arrange these three speeds in order, with the lowest speed first. Take $\gamma = 1.40$.

3. Figure 22.6 shows the Maxwellian distribution function for 1 mole of molecular oxygen at a temperature of 293 K. How would it compare to the Maxwellian distribution function for (a) 1 mole of atomic sulfur vapor at 293 K, (b) 1 mole of acetylene (C_2H_2) at 238 K, (c) 1 mole of methane (CH_4) at 146.5 K?

4. Figure 22.6 shows the distribution of molecular speeds for oxygen molecules. Using the fact that area under the $f(v)$ graph corresponds to the number of molecules, estimate the fraction of oxygen molecules with speeds that are two or more times the speed of a molecule traveling at the most probable speed.

5. A gas composed of one million carbon–12 atoms has a Maxwellian distribution of speeds at a temperature of 300 K. Determine the number of atoms with speeds between (a) 100 m/s and 101 m/s, (b) 300 m/s and 301 m/s, (c) 1500 m/s and 1501 m/s. Round your calculated numbers to two significant figures.

6. A gas of argon atoms is at a temperature of 300 K and a pressure of 10^{-6} atm. Estimate the average time between collisions of atoms.

7. As number density increases, the mean free path decreases. When the free path falls to a value comparable to the size of an atom one expects the material to be

a liquid or a solid. Set the mean free path equal to 3 Å and estimate the corresponding number density. Compare your estimate with the number density in water.

8. Assuming a Maxwellian distribution of speeds, the mean square speed is given by

$$\langle v^2 \rangle = \frac{\int_0^\infty v^2 f(v) \, dv}{\int_0^\infty f(v) \, dv}$$

Use the integral

$$\int_0^\infty x^4 \exp(-ax^2) \, dx = \left(\frac{3\sqrt{\pi}}{8} \right) a^{-5/2}$$

and the Maxwellian form of $f(v)$ to show that the rms speed is given by

$$v_{\text{rms}} = \sqrt{\frac{3kT}{m}}$$

9. Ordinary table salt has a structure in which sodium and chlorine atoms are positioned on adjacent corners of cubes. Given that the molecular weight is 58.5 and the density is 2165 kg/m^3, verify that the separation of atoms in salt is about 3 Å.

10. (a) A molecule with sufficient kinetic energy at the surface of the earth can escape the earth's gravitation. Using energy conservation, show that the minimum kinetic energy at the surface needed to escape is mgR, where m is the mass of the molecule, g is the acceleration of gravity at the surface, and R is the radius of the earth. (b) Calculate the temperature for which the minimum escape kinetic energy equals 10 times the average kinetic energy of an oxygen molecule.

11. Make a careful plot of the Maxwellian distribution of molecular speeds for molecular oxygen at a temperature of 300 K and determine the fraction of the molecules that have a speed between 100 m/s and 300 m/s.

12. Two hydrogen atoms traveling in opposite directions

at equal speeds undergo a perfectly inelastic collision, causing both atoms to ionize. (a) If 13.6 eV are required to ionize each atom, what minimum speed must the atoms have? (b) Would you expect hydrogen on the surface of the sun to be ionized or not? Explain.

13. Estimate the mean time between collisions for the atmospheric molecules in Example 2.

14. Consider the mean value of the quantity $(v - \langle v \rangle)^2$ to prove that the rms speed is never less than the mean speed.

15. A gas has a Maxwellian distribution of speeds. Determine what fraction of the molecules have speeds less than the rms speed. You will need to consult integral tables or evaluate the integral numerically.

16. The relative velocity of atoms moving with velocities $\mathbf{v}_1$ and $\mathbf{v}_2$ is $\mathbf{v}_1 - \mathbf{v}_2$. For identical atoms their relative kinetic energy is $\frac{1}{2}m(\mathbf{v}_1 - \mathbf{v}_2) \cdot (\mathbf{v}_1 - \mathbf{v}_2)$. Show that the average relative kinetic energy of the atoms equals $3kT$ for a gas in which the atoms have a Maxwellian distribution of speeds.

17. Using multiple laser beams, physicists have been able to cool and trap sodium atoms. In one experiment the temperature of the atoms was reduced to 2.4×10^{-4} K. (a) Determine the rms speed of the sodium atoms at this temperature. The atoms can be trapped for about 1 second. The trap has a linear dimension of roughly 1 centimeter. (b) Approximately how long would it take an atom to wander out of the trap region if there were no trapping action?

18. Estimate the number of air-molecule impacts each second on 1 cm^2 of your forehead.

CHAPTER 23

ELECTRIC CHARGE

23.1
ELECTRIC CHARGE

Mankind was introduced to electricity by a variety of natural phenomena—lightning, for example. Another phenomenon was the strange blue glow that early sailors saw atop their ship's masts. In Christian times this became known as St. Elmo's Fire, named after the patron saint of sailors (Figure 23.1). We now know this blue glow is the result of an electric discharge.

You have probably experienced the effects of **static electricity**. On a dry day, if you shuffle across a rug and then bring your finger close to a friend's ear, there will be a small spark and a large reaction. Less dramatically, your hair may tend to stand on end after you brush it.

It was found that many different materials could be electrified—or charged—by friction. Furthermore, it became apparent that charged objects could exert forces on other charged objects. Unlike the always-attractive gravitational force, electrical forces can be attractive or repulsive.

Progress toward understanding electrical forces was slow. Early in the nineteenth century the atomic model of matter gained recognition. Late in the nineteenth century it was discovered that atoms have an electrical structure. Today we know that atoms are composed of three basic types of particles: protons, electrons, and neutrons. Two of these are electrically charged: a proton carries a positive charge* and an electron carries a negative charge. A neutron is uncharged; it is neutral. The magnitude of the electron charge is exactly equal to the magnitude of the proton charge to within the experimental error of one part in 10^{22}.

The mass of a proton is nearly equal to the mass of a neutron. Both the neutron and proton are over 1800 times as massive as an electron. Table 23.1 summarizes the charge and mass properties of these three particles.

Quantization

Eighteenth-century scientists thought of electricity as a continuous fluid. From a wide variety of experiments, including electro-chemistry and high-energy particle physics, we now know that electric charge is not a continuous fluid, but rather, it is a quantity that occurs only in certain discrete amounts. We describe this by saying that electric charge is **quantized**. In particular, the net electric charge for any group of charges is an integral multiple of the electron or proton charge. In atoms, for example, the net charge is zero because there are equal numbers of electrons and protons.

FIGURE 23.1

St Elmo's fire is an electrical discharge at the tip of a mast.

TABLE 23.1

Particle	Symbol	Charge	Mass
Proton	p	$+1.60 \times 10^{-19}$ C	1.673×10^{-27} kg
Neutron	n	0	1.675×10^{-27} kg
Electron	e	-1.60×10^{-19} C	9.110×10^{-31} kg

---

* The symbols + and − are just labels. They are convenient in that they suggest and facilitate algebraic addition of charge.

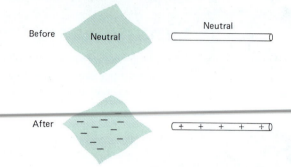

FIGURE 23.2

Separation of electric charge when a glass rod is rubbed with a silk cloth.

Conservation of Electric Charge

The concept that a net electric charge can never be created or destroyed goes back at least to Benjamin Franklin. When a glass rod is rubbed with silk, electrons are transferred from the glass to the silk, giving the silk a negative charge and leaving the glass rod with an equal positive charge. This transfer of electrons is illustrated in Figure 23.2. No change in the total charge of glass-plus-silk occurs. Conservation of charge has been tested repeatedly in the realm of high-energy physics and has been found to hold without exception. The principle of conservation of electric charge can be stated as follows:

The net electric charge remains constant in all processes.

The next two examples illustrate the conservation of electric charge in nuclear processes.

EXAMPLE 1

Uranium 238 Alpha Decay

The radioactive uranium 238 nucleus $^{238}_{92}\text{U}$ disintegrates by emitting an alpha particle (helium nucleus). This nuclear reaction may be written as

$$^{238}_{92}\text{U} \rightarrow {}^{4}_{2}\text{He} + {}^{234}_{90}\text{Th}$$

The superscripts give the combined number of neutrons and protons in each nucleus. The subscripts give the number of protons in each nucleus and therefore measure the positive nuclear charge. There are 92 protons in the uranium nucleus. The decay products contain a total of 92 protons, 2 in helium and 90 in thorium. The balancing of the subscripts, 92 for both sides, describes the exact conservation of electric charge in this nuclear reaction.

In Example 1, electric charge in the form of protons is simply rearranged. Sometimes electric charges are created. When this occurs, positive and negative charges are created in equal amounts, keeping the net charge unchanged.

EXAMPLE 2

Carbon 14 Beta Decay

Carbon 14 ($^{14}_{6}\text{C}$) has six protons in its nucleus and is formed in our atmosphere by cosmic ray bombardment of nitrogen. Carbon 14 is unstable and transforms into nitrogen 14 by emitting an electron and an antineutrino (zero mass, zero charge).

$$^{14}_{6}\text{C} \rightarrow {}^{14}_{7}\text{N} + {}_{-1}e + \bar{\nu}$$

In this process one of the 8 neutrons in the carbon 14 nucleus is transformed into three particles: a positively charged proton, a negatively charged electron, and a neutral antineutrino. The proton, the electron, and the antineutrino are *created* in the reaction. Although both positive and negative charges are created, the net charge remains the same $(+6 \text{ before} = -1 + 7 = +6 \text{ after})$.

23.2
COULOMB'S LAW

A quantitative breakthrough in electrostatics occurred in 1785 when the French scientist Charles Augustin de Coulomb measured the force between two small electrically charged spheres. Coulomb found that the force between the charged spheres was inversely proportional to the square of the distance between them and directly proportional to the product of their charges:

$$F \propto \frac{q_1 q_2}{r^2}$$

This proportionality is converted into an equation by introducing a proportionality constant. The result is known as **Coulomb's law of electrostatic force** and may be written as

$$F = k_e \frac{q_1 q_2}{r^2} \tag{23.1}$$

The SI unit of charge is the coulomb (symbol C). The operational definition of the coulomb, based on magnetic effects of electric currents, is presented in Chapter 30. The proportionality constant k_e is

$$k_e = 8.98755 \times 10^9 \text{ N·m}^2/\text{C}^2 \tag{23.2}$$

Like all forces, the electrostatic force obeys Newton's third law. That is, Equation 23.1 describes the magnitude of the equal but oppositely directed forces that the charges q_1 and q_2 exert on each other (Figure 23.3). The Coulomb force is repulsive for like charges and attractive for unlike charges.

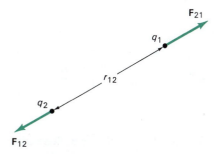

FIGURE 23.3

$\mathbf{F}_{12}$ is the force exerted *by* q_1 *on* q_2 and $\mathbf{F}_{21}$ is the equal but oppositely directed force exerted *by* q_2 *on* q_1. If q_1 and q_2 have the same sign, the forces are repulsive, as shown here. If q_1 and q_2 have opposite signs, the forces are attractive.

EXAMPLE 3

Electrostatic Repulsion

Let's calculate the force of repulsion between two 1 C charges 1 m apart. From Coulomb's law (Equation 23.1) we have

$$F = (8.99 \times 10^9 \text{ N·m}^2/\text{C}^2) \cdot \frac{(1 \text{ C})^2}{(1 \text{ m})^2}$$

$$= 8.99 \times 10^9 \text{ N}$$

This is a force of about 1 million tons. Clearly, one coulomb is an enormous charge. In fact, this example is unrealistic in the sense that we could not get charges of 1 C to stay on small surfaces separated by 1 meter.

Coulomb's law of electrostatic interaction and Newton's law of universal gravitation have the same mathematical form; both are inverse square laws. But what about the relative strength of these two fundamental forces? Let's calculate the ratio of the electrostatic force and the gravitational force between an electron and a proton. These forces are described by Coulomb's law (Equation 23.1) and Newton's law of universal gravitation. The ratio is

$$\frac{F_{\text{elec}}}{F_{\text{grav}}} = \frac{k_e \left(\dfrac{q_e q_p}{r^2} \right)}{G \left(\dfrac{m_e m_p}{r^2} \right)} = \frac{k_e q_e q_p}{G m_e m_p}$$

Note that the distance factor cancels out. The value of the ratio is

$$\frac{F_{\text{elec}}}{F_{\text{grav}}} \approx 2 \times 10^{39}$$

This is an enormous number. Imagine grains of sand so fine that you can pack 10^6 grains in 1 cm^3: 10^{39} of these grains would occupy the volume of a million earths! Clearly, the electrostatic force is far stronger than gravity.

The fact that Newton's law of gravitation and Coulomb's electrostatic law have the same $1/r^2$ distance dependence has impressed many scientists, including Einstein, as more than mere coincidence. So far, no profound relationship or common origin has been discovered.

23.3

SUPERPOSITION

Coulomb's law specifies the force between a pair of point charges. When more than two charges interact, experiment shows that the net force on any particular charge is the vector sum of the Coulomb forces exerted on it by the other charges.

EXAMPLE 4

Superposition

Charges of 3 μC, 4 μC, and 6 μC are placed along a line (Figure 23.4). Let's use Coulomb's law to calculate the two separate forces exerted on the 6 μC charge. First,

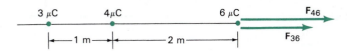

FIGURE 23.4

The total force on the 6-μC charge is the sum of the forces exerted by the 3-μC and 4-μC charges.

consider the force exerted by the 3-μC charge. From Coulomb's law (Equation 23.1) the force exerted on the 6-μC charge is

$$F_{36} = \frac{8.99 \times 10^9 \text{ N·m}^2/\text{C}^2 \, (3 \times 10^{-6} \text{ C})(6 \times 10^{-6} \text{ C})}{(3 \text{ m})^2}$$

$$= 1.80 \times 10^{-2} \text{ N} \qquad \text{(directed to the right)}$$

Next, we consider the force exerted by the 4 μC charge.

$$F_{46} = \frac{8.99 \times 10^9 \text{ N·m}^2/\text{C}^2 \, (4 \times 10^{-6} \text{ C})(6 \times 10^{-6} \text{ C})}{(2 \text{ m})^2}$$

$$= 5.39 \times 10^{-2} \text{ N} \qquad \text{(directed to the right)}$$

Superposing F_{36} and F_{46} yields the total force on the 6 μC charge:

$$F_6 = F_{36} + F_{46} = 7.19 \times 10^{-2} \text{ N}$$

To within the limits of experimental accuracy the total force on the 6 μC charge has been confirmed to be the sum of F_{36} and F_{46}, or 7.19×10^{-2} N. In other words, experiment shows that the presence of a third charge does not influence the Coulomb force between the other two charges.

We can generalize the experimental result stated in Example 4 by saying that electrical forces obey a **principle of superposition**:

The net force exerted by two or more charges on a single charge Q is the vector sum of the individual forces exerted on Q.

Keep in mind that this principle is the result of experiment.

Vector Form of Coulomb's Law

The electrostatic force is a vector quantity—it has direction as well as magnitude. We can write Coulomb's law in vector form by introducing a unit **vector** to indicate direction. In Figure 23.5, $\hat{\mathbf{r}}_{12}$ is a unit vector directed from q_1 toward q_2. The force $\mathbf{F}_{12}$ exerted by q_1 on q_2 is

$$\mathbf{F}_{12} = k_e \frac{q_1 q_2}{r_{12}^2} \hat{\mathbf{r}}_{12} \qquad (23.3)$$

If the charges q_1 and q_2 are both positive or both negative, the force is repulsive, and $\mathbf{F}_{12}$ is parallel to $\hat{\mathbf{r}}_{12}$ (Figure 23.5a). If q_1 and q_2 have opposite signs, then the force is attractive, indicating that q_1 is urged toward q_2 (Figure 23.5b).

Example 4 involves only parallel forces. Now let's apply Equation 23.3 and the principle of superposition to a system where the forces are not parallel. We build in enough symmetry so that we can check our results.

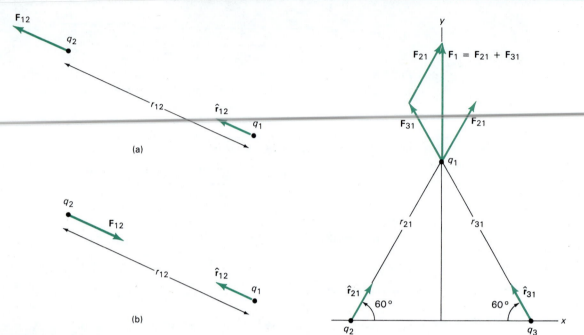

FIGURE 23.5

$\mathbf{F}_{12}$ denotes the force exerted *by* q_1 *on* q_2. The unit vector $\hat{\mathbf{r}}_{12}$ is directed from q_1 toward q_2. (a) The direction of $\mathbf{F}_{12}$ indicates a repulsive force between like charges. (b) When q_1 and q_2 have opposite signs the force is attractive.

FIGURE 23.6

A 1-μC charge is located at each vertex of the equilateral triangle. The net force $\mathbf{F}_1$ on q_1 is the vector sum of the forces $\mathbf{F}_{31}$ and $\mathbf{F}_{21}$ exerted by q_3 and q_2.

EXAMPLE 5

Vector Addition for Coulomb Forces

Consider three 1 μC charges at the vertices of an equilateral triangle, 1 m on a side (Figure 23.6). What is the net force that the two bottom charges exert on the top charge (q_1)?

Figure 23.6 shows that the array of three equal charges has left-right symmetry relative to a vertical line through q_1. We know from this symmetry that the net force on q_1 will be vertical and in the upward direction. (All charges have the same sign; all forces are repulsive.) The net force on q_1 is the vertical component of the force exerted by q_2 plus the vertical component of the force exerted by q_3. Since the two vertical components are equal by symmetry, the magnitude of the force on q_1 is

$$F_1 = 2k_e \left(\frac{q_1 q_2}{r^2} \right) \cdot \cos 30°$$

$$= 2(8.99 \times 10^9 \text{ N·m}^2/\text{C}^2) \left[\frac{10^{-6} \text{ C} \cdot 10^{-6} \text{ C}}{1 \text{ m}^2} \right] \cdot (0.866)$$

$$= 1.56 \times 10^{-2} \text{ N}$$

Now, let's go through the calculation in detail using the vector form of Coulomb's law. The unit vectors are illustrated in Figure 23.7. For $\mathbf{F}_{21}$—the force that q_2 exerts on q_1—we have

$$\mathbf{F}_{21} = k_e \left(\frac{q_1 q_2}{r_{21}^2} \right) \hat{\mathbf{r}}_{21}$$

$$= (8.99 \times 10^9 \text{ N·m}^2/\text{C}^2) \left[\frac{10^{-6} \text{ C} \cdot 10^{-6} \text{ C}}{(1 \text{ m})^2} \right] \hat{\mathbf{r}}_{21}$$

$$= 8.99 \times 10^{-3} \hat{\mathbf{r}}_{21} \text{ N}$$

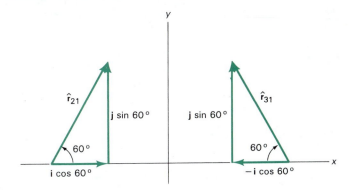

FIGURE 23.7
The unit vectors $\hat{\mathbf{r}}_{21}$ and $\hat{\mathbf{r}}_{31}$ can be resolved into x- and y-components.

We can resolve the unit vector $\hat{\mathbf{r}}_{21}$ into its Cartesian components (Figure 23.7)

$$\hat{\mathbf{r}}_{21} = \mathbf{i}\cos 60° + \mathbf{j}\sin 60°$$

This gives

$$\mathbf{F}_{21} = (8.99 \times 10^{-3}\text{ N})\cos 60°\,\mathbf{i} + (8.99 \times 10^{-3}\text{ N})\sin 60°\,\mathbf{j}$$
$$= (4.49 \times 10^{-3}\text{ N})\mathbf{i} + (7.78 \times 10^{-3}\text{ N})\mathbf{j}$$

The force $\mathbf{F}_{31}$ is given by

$$\mathbf{F}_{31} = k_e\left(\frac{q_1 q_3}{r_{31}^2}\right)\hat{\mathbf{r}}_{31}$$

The unit vector $\hat{\mathbf{r}}_{31}$ is given by

$$\hat{\mathbf{r}}_{31} = -\mathbf{i}\cos 60° + \mathbf{j}\sin 60°$$

This gives

$$\mathbf{F}_{31} = (-4.49 \times 10^{-3}\text{ N})\mathbf{i} + (7.78 \times 10^{-3}\text{ N})\mathbf{j}$$

The net force on q_1 is the vector sum of $\mathbf{F}_{21}$ and $\mathbf{F}_{31}$. The horizontal components cancel each other because of the symmetry of the system. The result is

$$\mathbf{F}_1 = \mathbf{F}_{21} + \mathbf{F}_{31}$$
$$= (1.56 \times 10^{-2}\text{ N})\mathbf{j}$$

This shows that the net force is vertically upward with a magnitude of 1.56×10^{-2} N. This solution agrees with the first calculation.

Using symmetry made it easier to find a solution. If symmetry is not present, however, we can still find the net force by using the vector form of Coulomb's law and the principle of superposition.

Continuously Distributed Charge

All electric charge distributions are collections of discrete charges such as electrons and protons. However, when we consider a large number of closely packed charges, we can treat the distributed discrete charges as continuous. To determine forces exerted by continuous distributions of charge the principle of superposition may be applied, but integrations replace discrete sums.

Figure 23.8 shows a **line charge**, a collection of charges spread continuously along a line. A point charge Q located at the point P experiences forces exerted

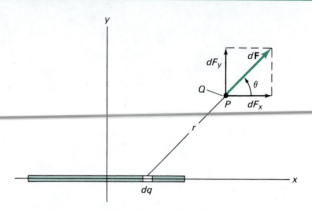

FIGURE 23.8

The charge element dq exerts a force $d\mathbf{F}$ on the charge Q. The total force on the charge Q is obtained by integrating the $d\mathbf{F}$'s exerted by each charge element along the line charge.

by charges distributed along the line. Figure 23.8 shows the force dF exerted on Q by the charge element dq. Each element of charge along the line exerts a similar force on Q.

The magnitude of the force exerted on Q by the charge element dq is given by Coulomb's law, Equation 23.1,

$$dF = \frac{k_eQ \, dq}{r^2}$$

Since $d\mathbf{F}$ is a vector we must integrate each component separately. In Figure 23.8, the x- and y-components of $d\mathbf{F}$ are given by

$$dF_x = dF \cos \theta = \frac{k_eQ \, dq \cos \theta}{r^2}$$

$$dF_y = dF \sin \theta = \frac{k_eQ \, dq \sin \theta}{r^2}$$

The total force $\mathbf{F}$ has components F_x and F_y expressed as integrals over the charge distribution.

$$F_x = \int dF_x = k_eQ \int \frac{dq}{r^2} \cos \theta \tag{23.4}$$

$$F_y = \int dF_y = k_eQ \int \frac{dq}{r^2} \sin \theta \tag{23.5}$$

In order to carry out the integrations it is necessary to specify how the charge is distributed.

EXAMPLE 6

Force Exerted on a Point Charge by a Line Charge

A thin rod on the x-axis extends from $x = -a$ to $x = +a$ in Figure 23.9. The rod has an electric charge q, distributed *uniformly* over its length. The charged rod constitutes a line charge with a charge per unit length given by

$$\beta \equiv \frac{\text{charge}}{\text{length}} = \frac{q}{2a}$$

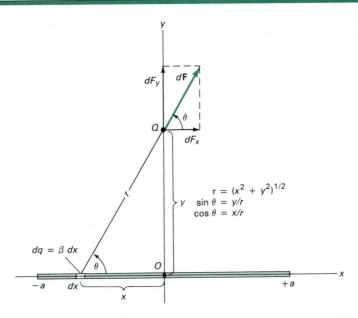

FIGURE 23.9

Electric charge is distributed uniformly along the rod from $x = -a$ to $x = +a$. Each element of charge dq exerts a force on the point charge Q.

We want to use Equations 23.4 and 23.5 to calculate the force on a point charge Q at a point a distance y above the midpoint of the rod. Here we have the same left-right symmetry as we did in Example 5, and just as in Example 5, the net force exerted on Q will be vertical. Thus, we expect to find that F_x is zero.

In Figure 23.9, dq denotes the charge on a segment of length dx. The relation between dq and dx is

$$dq = (\text{charge per unit length})(\text{length}) = \beta \, dx$$

From Figure 23.9 we see that the distance r is given by

$$r = (x^2 + y^2)^{1/2}$$

The functions $\sin \theta$ and $\cos \theta$ are also variables that must be expressed in terms of x and y before carrying out the integrations. From Figure 23.9,

$$\sin \theta = \frac{y}{r} = \frac{y}{(x^2 + y^2)^{1/2}}$$

$$\cos \theta = \frac{x}{r} = \frac{x}{(x^2 + y^2)^{1/2}}$$

Inserting these results into Equations 23.4 and 23.5 we get

$$F_x = k_e Q \beta \int_{-a}^{a} \frac{x \, dx}{(x^2 + y^2)^{3/2}}$$

$$F_y = k_e Q \beta y \int_{-a}^{a} \frac{dx}{(x^2 + y^2)^{3/2}}$$

The integrations range from $x = -a$ to $x = +a$ because this is the range of x over which the charge is distributed.

The integral for F_x gives zero, confirming our expectation based on symmetry. The integral involved in F_y is

$$\int_{-a}^{a} \frac{dx}{(x^2 + y^2)^{3/2}} = \frac{2a}{y^2 \sqrt{a^2 + y^2}}$$

Noting that $q = 2\beta a$, the net force exerted by the charged rod on Q becomes

$$F_y = \frac{k_e Qq}{y\sqrt{a^2 + y^2}} \qquad (23.6)$$

We can make two consistency checks. First, F_y has the proper units, ($k_e \cdot$ charge^2/length^2). Second, when the charge Q is far enough away, the line charge looks like a point charge q, and we expect the force between them to be given by the Coulomb force between point charges separated by a distance y. For $y \gg a$, Equation 23.6 has the limiting form

$$F_y \approx \frac{k_e Qq}{y^2} \qquad y \gg a$$

which is the Coulomb force between point charges Q and q separated by a distance y.

WORKED PROBLEM

Three charges of equal magnitude q are fixed in position at the vertices of an equilateral triangle (Figure 23.10). The charge located at the origin in Figure 23.10 is negative; the other two are positive. A fourth charge Q is free to move along the positive x-axis under the influence of the forces exerted by the three fixed charges. Locate an equilibrium position for Q.

Solution

At an equilibrium position, the net force on the charge Q is zero. The equilibrium position can be located by first determining the angle θ corresponding to equilibrium. In terms of the lengths s, $\frac{1}{2}a\sqrt{3}$, and r, shown in Figure 23.10, the charge at the origin exerts an attractive force $k_e Qq/(s + \frac{1}{2}a\sqrt{3})^2$. The other two charges exert equal repulsive forces of magnitude $k_e Qq/r^2$. The horizontal components of the two repulsive forces add, balancing the attractive force,

$$F_{\text{net}} = k_e Qq \left\{ \frac{2\cos\theta}{r^2} - \frac{1}{(s + \frac{1}{2}a\sqrt{3})^2} \right\} = 0$$

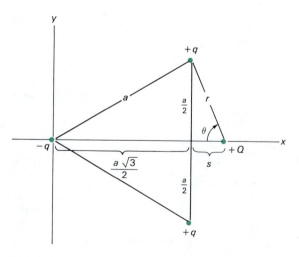

FIGURE 23.10

From Figure 23.10,

$$r = \frac{\frac{1}{2}a}{\sin \theta} \qquad s = \tfrac{1}{2}a \cot \theta$$

The equilibrium condition is expressed in terms of θ by

$$F_{net} = \left(\frac{4}{a^2}\right) k_e Qq \left\{ 2 \cos \theta \sin^2 \theta - \frac{1}{(\sqrt{3} + \cot \theta)^2} \right\} = 0$$

This gives an equation for the equilibrium value of θ,

$$2 \cos \theta \sin^2 \theta \, (\sqrt{3} + \cot \theta)^2 = 1$$

One method for solving for θ is to tabulate the left side. To three significant figures the value of θ corresponding to equilibrium is $81.7°$. The distance from the origin to the equilibrium position is

$$x = \tfrac{1}{2}a(\sqrt{3} + \cot 81.7°)$$
$$= 0.939a$$

θ	$2 \cos \theta \sin^2 \theta (\sqrt{3} + \cot \theta)^2$
$60°$	4
$70°$	2.654
$80°$	1.226
$90°$	0
$81°$	1.091
$81.5°$	1.024
$81.7°$	0.997

EXERCISES

23.2   Coulomb's Law

A. (a) From Coulomb's law show that the proportionality constant k_e has units of $N \cdot m^2/C^2$. (b) Given that one coulomb equals one ampere·second $(A \cdot s)$, show that the units of k_e may also be written as $kg \cdot m^3/A^2 \cdot s^4$.

B. (a) Calculate the number of electrons in a small silver pin, electrically neutral, with a mass of 10 grams. Silver has 47 electrons per atom. The atomic weight of silver is 107.87. (b) Electrons are added to the pin until the net charge is 1 mC. How many electrons are added for every 10^9 electrons already present?

C. Two small neutral spheres are 12 cm apart. Electrons are removed from one sphere and deposited on the other. The result is an attractive force of 1.03×10^{-3} N exerted by each sphere on the other. How many electrons were transferred?

D. Atomic nuclei contain protons and neutrons. A typical proton-proton separation is 1 fm. Calculate the repulsive force between two protons separated by this distance. (In stable nuclei this disruptive Coulomb force is counterbalanced by the strong attractive nuclear force.)

E. A hydrogen atom may be pictured as an electron revolving around a fixed proton at a distance of 0.529 Å.

(a) Calculate the attractive electrical force the proton exerts on the orbiting electron. (b) This electrical force is centripetal. Determine the speed of the electron and compare it to the speed of light.

23.3   Superposition

F. Four equal charges $(Q = 1 \mu C)$ are located as shown in Figure 1. Determine the net electrical force on the charge at position 3, midway between the charges at positions 2 and 4.

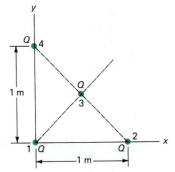

FIGURE 1

G. Two equal charges (magnitude q) are fixed in position 1 m apart on the x-axis. (a) Where could you place a third charge Q having the same sign so that the net force on it is zero? (b) Is Q in a position of stable equilibrium? Assume that Q is free to move only along the x-axis. (c) How would your answers change if Q and q have opposite signs?

PROBLEMS

1. ~~Two identical small metal spheres attract each other~~ with a force of 0.0853 N. The distance between the spheres is 1.19 m. The spheres are brought into electrical contact with each other so that the net charge is shared equally. When returned to a separation of 1.19 m, the spheres repel each other with a force of 0.0196 N. Find the charge originally on each sphere.

2. In the novel *Ringworld*, science-fiction writer Larry Niven describes a device that separates the positive charges from the negative charges in atoms. Suppose that the electrons and protons of 1 microgram of hydrogen were separated into two small spheres 0.1 m apart. What would be the attractive force exerted by each sphere on the other? Express your result in newtons and in tons.

3. Two small insulating spheres are suspended by insulating cords 50 cm long. Each sphere carries a net charge $+q$ uniformly distributed. Each sphere has a mass of 20 grams. When the spheres are hung from the same point, it is found that each cord makes an angle of 5° with the vertical. (a) Determine q. (b) What angle will each cord make with the vertical if one charge is halved and the other charge doubled?

4. Two small spheres contain charges Q and q. Charge is removed from one sphere and transferred to the other. Show that the electrostatic force between the two spheres is a maximum when each sphere carries a charge $\frac{1}{2}(Q + q)$.

5. Three identical positive charges, all q, are fixed in position at the vertices of an equilateral triangle, as shown in Figure 2. A fourth charge, Q, is free to move along the x-axis under the influence of the Coulomb forces exerted by the other three charges. Prove that an equilibrium position exists for Q at $x = +a/\sqrt{3}$, $y = 0$.

6. Each of the four corners of a square 2 m on a side is occupied by a $+1\,\mu C$ charge (Figure 3). (a) Write down an algebraic equation for the force on a $+1\,\mu C$ charge at an arbitrary point on the x-axis within the square. (b) Show that the points $x = \pm 0.773$ m are equilibrium positions.

7. A uniform line of charge (β C/m) is formed in the shape of a semicircle of radius R and placed in the x–y plane with the ends at $x = \pm R$ (Figure 4). (a) Calculate the force on a charge Q placed at the origin. (b) If the charge on the semicircle and Q each equal 2 μC and the force is 0.173 N, determine the radius of the semicircle.

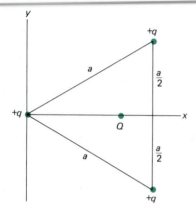

FIGURE 2

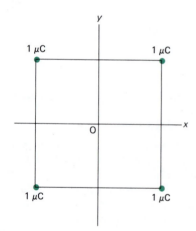

FIGURE 3

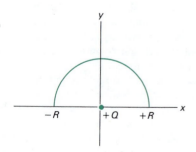

FIGURE 4

8. A thin rod carrying a charge q distributed uniformly along its length is placed on the x-axis with one end at the origin and the other end at $x = +L$. Determine the magnitude of the force on a point charge Q located at $x = a$ $(a > L)$.

9. A charge of $+9 \, \mu C$ is fixed at the origin in Figure 5. A second charge, q, is fixed on the x-axis at $x = a$. A positive charge Q on the x-axis at $x = 3a$ is in stable equilibrium. Determine the magnitude and sign of q.

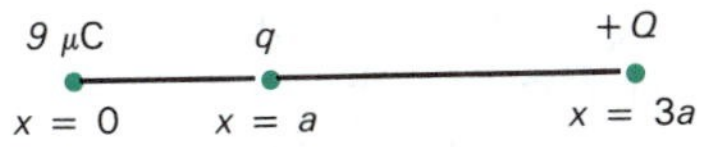

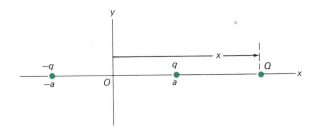

FIGURE 5

10. The electric dipole shown in Figure 6 exerts a force on a test charge Q placed on the x-axis. Show that for a position $x \gg a$ the force is given by

$$F = i \left(\frac{4k_e q Q a}{x^3} \right)$$

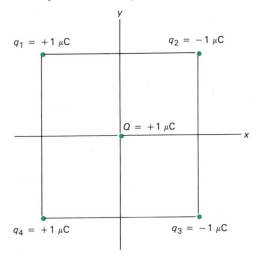

FIGURE 6

11. Four charges, $q_1 = q_4 = +1 \, \mu C$, $q_2 = q_3 = -1 \, \mu C$, are placed on the corners of a square with a side length of 1 m (Figure 7). A fifth charge $Q = +1 \, \mu C$ is placed at the center of the square. (a) Sketch the force vectors exerted on Q by each of the four charges and sketch their vector sum. (b) Determine the net force on Q by completing the table below:

Charge	F	F_x	F_y
q_1			
q_2			
q_3			
q_4			
		Net $F_x =$	Net $F_y =$

Net F = _____ i + _____ j

(c) Compare the calculated net force vector with the vector you sketched in part (a).

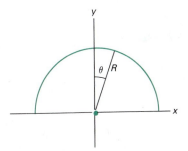

FIGURE 7

12. A line of positive charge is formed into a semicircle of radius $R = 60$ cm as shown in Figure 8. The charge per unit length along the semicircle is described by the expression

$$\beta = \beta_0 \cos \theta$$

The total charge on the semicircle is 12 μC. Calculate the total force on a charge of 3 μC placed at the center of curvature of the semicircle.

FIGURE 8

13. Identical parallel line charges ($\beta = +1 \, \mu C/m$) of infinite length lie in the x–z plane (Figure 9). The line charges are parallel to the z-axis and pass through the points $(+a, 0, 0)$ and $(-a, 0, 0)$. A point charge $(q = 1 \, \mu C, m = 2$ grams$)$ is in equilibrium on the y-axis under the combined electric force exerted by the line charges and its weight. (a) Show that there are two equilibrium positions given by the solutions of the equation

$$y^2 - 2by + a^2 = 0$$

provided $a < b$, where $b = 2k_e q\beta/mg$. (b) Show that for displacements along the y-axis, one of the equilibrium positions is stable and one is unstable.

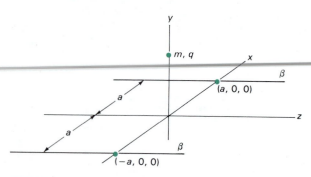

FIGURE 9

14. Identical thin rods of length $2a$ carry equal charges $+Q$ uniformly distributed along their lengths. The rods lie along the x-axis with their centers separated by a distance $b > 2a$ (Figure 10). Show that the force exerted on the right rod is given by

$$F = \left(\frac{k_e Q^2}{4a^2}\right) \ln\left(\frac{b^2}{b^2 - 4a^2}\right)$$

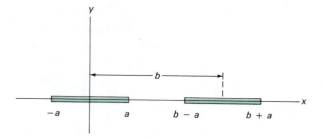

FIGURE 10

15. A positive charge $+q$ of mass M is free to move along the x-axis. It is in equilibrium at the origin, midway between a pair of identical point charges, $+q$, located on the x-axis at $x = +a$ and $x = -a$ (Figure 11). The charge at the origin is displaced a small

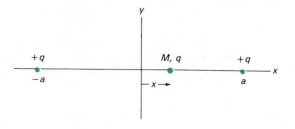

FIGURE 11

distance $x \ll a$ and released. Show that it can undergo simple harmonic motion with an angular frequency

$$\omega = \left(\frac{4k_e q^2}{Ma^3}\right)^{1/2}$$

16. A point charge ($q = -1\ \mu C$, weight $= 0.1$ N) free to move along the y-axis is in equilibrium under the combined influence of its weight and the electrical attraction of a pair of identical charges ($Q = +1\ \mu C$) located on the x-axis at $x = +13.4$ cm and $x = -13.4$ cm (Figure 12). (a) Show that the two equilibrium angles are $\theta = 5.8°$ and $\theta = 71°$. (b) Why is the equilibrium stable for the smaller angle and unstable for the larger angle?

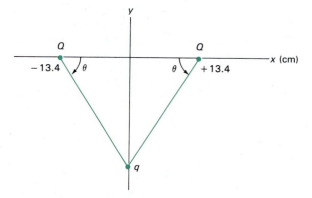

FIGURE 12

17. A negative point charge $-q$ of mass m moves in a circular path of radius r in the y–z plane under the influence of identical point charges $+q$ located on the x-axis at $x = +a$ and $x = -a$ (Figure 13). Prove that for $r \ll a$ the angular frequency of the motion is given by

$$\omega = \left(\frac{2k_e q^2}{ma^3}\right)^{1/2}$$

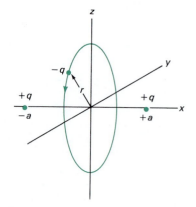

FIGURE 13

18. A charge of 50 μC is distributed uniformly along a U-shaped form that comprises three sides of a 20-cm square (Figure 14). Determine the force on a 1 μC charge placed at the center of the form.

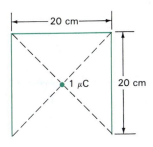

FIGURE 14

19. A thin circular metal disk carries electric charge on both faces. The charge density on both faces increases with distance from the center according to

$$\sigma(r) = \sigma_0 \left[1 + \left(\frac{r}{b} \right)^2 \right]^{1/2} \qquad r \le b$$

$$\sigma(r) = 0 \qquad\qquad\qquad r > b$$

where σ_0 is a constant and $b = 3$ cm is the disk radius. The total charge on the disk is 20 nC. Show that

$$\sigma_0 = 2.90 \ \mu C/m^2$$

CHAPTER 24

ELECTRIC FIELD

24.1
ELECTRIC FIELD

Coulomb's law describes the forces between electric charges, but offers no explanation of or mechanism for these forces. This made many scientists uncomfortable. Other forces with which they were familiar required contact between objects, such as the contact between two colliding billiard balls. They reasoned that if a charge experienced a force then there should be something directly affecting that charge.

Michael Faraday conceived the idea of an electric field to explain how charges could exert and experience forces without touching. According to Faraday's view, *an electric charge modifies the space around it*, or, in other words, the charge establishes an **electric field**. Other charges placed in this field experience forces.

Consider a collection of electric charges fixed in position. If we place a positive test charge q_0 in the vicinity of this collection, the test charge will experience a force. This force is the vector sum of the Coulomb forces exerted by the charges in the collection. Each of the Coulomb forces is directly proportional to the test charge q_0. It follows that the total force $\mathbf{F}$ experienced by q_0 is directly proportional to q_0. This means that the ratio $\mathbf{F}/q_0$ is independent of q_0. This ratio, the force per unit charge, is defined to be the electric field set up by the collection of charges. We denote the electric field by the symbol $\mathbf{E}$:

$$\mathbf{E} = \frac{\text{electric force}}{q_0} = \frac{\mathbf{F}}{q_0} \qquad (24.1)$$

In general, if we move the test charge from one position to another, the force on it will change. That is, the electric field $\mathbf{E}$ generally has a magnitude and direction that changes from point to point. Conceptually, it is often helpful to think of the test charge as sampling the electric field. In particular, the direction of $\mathbf{E}$ is the direction of the force acting on the test charge. From Equation 24.1, the SI unit of electric field is the newton per coulomb (N/C).

Electric Field of a Point Charge

We start our study of electric fields with the field set up by a single point charge q. From Coulomb's law, the force exerted by q on a test charge q_0 is given by

$$\mathbf{F} = k_e \frac{q q_0}{r^2} \hat{\mathbf{r}}$$

where r is the distance from q to q_0 and $\hat{\mathbf{r}}$ is a unit vector directed from q to q_0 (Figure 24.1). The electric field set up by q is $\mathbf{F}/q_0$

$$\mathbf{E} = \frac{k_e q}{r^2} \hat{\mathbf{r}} \qquad (24.2)$$

The unit vector $\hat{\mathbf{r}}$ is directed away from q, toward the point where the field is evaluated. If q is positive, $\mathbf{E}$ is directed away from it. If q is negative, $\mathbf{E}$ is directed toward it. The magnitude of $\mathbf{E}$ depends on the magnitude of q and the distance from the charge r, but not on direction. In other words, $\mathbf{E}$ is **spherically symmetric**. The electric field set up by a point charge is often

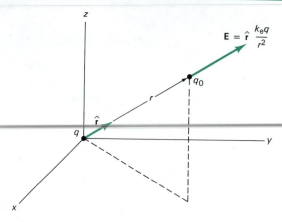

FIGURE 24.1

The positive point charge q sets up an electric field $\mathbf{E} = (k_e q / r^2)\hat{\mathbf{r}}$ at the position of the test charge q_0. The test charge experiences a force $q_0 \mathbf{E}$. If q were negative $\mathbf{E}$ would be opposite to the direction shown.

called a **Coulomb field** because it is associated with the force described by Coulomb's law.

EXAMPLE 1

Electric Field of a Hydrogen Nucleus

The nucleus of a hydrogen atom, a proton, sets up an electric field. Ordinarily the distance between the proton and the electron is 5.30×10^{-11} m. Let's calculate the magnitude of the electric field at this distance from the proton. From Equation 24.2, with $q = 1.60 \times 10^{-19}$ C and $r = 5.30 \times 10^{-11}$ m,

$$E = \frac{k_e q}{r^2}$$

$$= \frac{(8.99 \times 10^9 \ \text{N} \cdot \text{m}^2/\text{C}^2)(1.60 \times 10^{-19} \ \text{C})}{(5.3 \times 10^{-11} \ \text{m})^2}$$

$$= 5.12 \times 10^{11} \ \text{N/C}$$

This is an enormous electric field. Even though q is the smallest unit of electric charge we can isolate, the distance r is also very small. Because E is inversely proportional to r^2 the electric field is very large.

Faraday's concept of electric field lets us shift our attention from forces between separated charges to local interactions between charges and fields. This subtle shift of emphasis has profound consequences. The concept of a field finds wide application in physics—it is not limited to electric fields. In fact, the most esoteric theories dealing with the structure of matter and the nature of the universe are **field theories**.

Electric Field Lines

The electric field is a vector, specified at each point in space by the magnitude and direction of **E**. To help visualize this abstract concept Faraday introduced **electric field lines**. The *direction* of an electric field line at any point is the

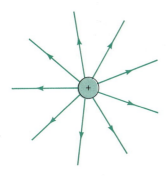

FIGURE 24.2

Electric field lines for an isolated positive charge. Note the closer spacing where the field is relatively strong and wider spacing where the field is relatively weak.

direction of **E** at that point. The *spacing* of the field lines gives a qualitative indication of the *magnitude* of **E**: the closer together the field lines, the stronger the field.

Figure 24.2 shows representative electric field lines for an isolated positive charge. They are directed radially away from the charge in agreement with Equation 24.2. Remember, the direction of **E** and the direction of the field lines is the direction in which a positive test charge would be urged. In this case the test charge would be urged away from the positive charge. In Equation 24.2, r denotes the distance from the charge to the point where the field has the magnitude $k_e q/r^2$. The magnitude of the field increases as r decreases and as Figure 24.2 shows, the spacing of the field lines decreases as r decreases. For an isolated negative charge the field lines are directed radially inward, as shown in Figure 24.3. We say that electric field lines originate on positive charges and terminate on negative charges.

Principle of Superposition

We learned in Chapter 23 that the Coulomb force obeys a principle of superposition. This same principle holds for electric fields. To see why, we need only recall that the electric field is defined as the force per unit charge. If the force obeys a principle of superposition, so also will the force per unit charge. Thus, if charges q_1, q_2, q_3, ... set up electric fields $\mathbf{E}_1$, $\mathbf{E}_2$, $\mathbf{E}_3$, ..., the total electric field is the vector sum

$$\mathbf{E} = \mathbf{E}_1 + \mathbf{E}_2 + \mathbf{E}_3 + \cdots \qquad (24.3)$$

FIGURE 24.3

Electric field lines for an isolated negative charge.

We can illustrate the principle of superposition of electric fields by evaluating the field produced by a pair of 1-μC charges 2 m apart on the x-axis (Figure 24.4). Let's determine the electric field at a point P on the x-axis a distance x from the left charge. The field created by the left charge is

$$\mathbf{E}_1 = k_e \left(\frac{10^{-6}\ \text{C}}{x^2} \right) \mathbf{i} = \frac{8.99 \times 10^3\ \mathbf{i}}{x^2}\ \text{N/C}$$

The field created by the right charge is

$$\mathbf{E}_2 = k_e \left[\frac{10^{-6}\ \text{C}}{(x - 2)^2} \right] (-\mathbf{i}) = \frac{-8.99 \times 10^3\ \mathbf{i}}{(x - 2)^2}\ \text{N/C}$$

For our chosen field point, between the two charges, $\mathbf{E}_1$ and $\mathbf{E}_2$ are oppositely directed. The unit vectors $\mathbf{i}$ and $-\mathbf{i}$ account for the opposite directions. Adding $\mathbf{E}_1$ and $\mathbf{E}_2$ gives the net electric field at the point P,

$$\mathbf{E} = \mathbf{E}_1 + \mathbf{E}_2 = 8.99 \times 10^3 \left[\frac{1}{x^2} - \frac{1}{(x - 2)^2} \right] \mathbf{i}\ \text{N/C}$$

FIGURE 24.4

The net electric field at the point P is obtained by superposing the Coulomb fields produced by each of the 1-μC charges.

If the point P is midway between the charges, the superposed electric fields cancel and $\mathbf{E} = 0$. The net force on a test charge placed midway between the two equal charges is zero. Such a test charge would experience equal and opposite forces.

Electric Field of an Electric Dipole

We have just seen how the principle of superposition can be used to determine the electric field set up by two equal, positive charges. Now we use the same principle to examine the electric field produced by two charges of equal magnitude but opposite sign. This combination of charges is called an **electric dipole**. Figure 24.5 shows an electric dipole consisting of charges $+q$ and $-q$. We call the field produced by the positive charge $\mathbf{E}_+$ and the field produced by the negative charge $\mathbf{E}_-$. By the principle of superposition the net electric field is

$$\mathbf{E} = \mathbf{E}_+ + \mathbf{E}_- = \frac{k_e q}{r_+{}^2}\hat{\mathbf{r}}_+ - \frac{k_e q}{r_-{}^2}\hat{\mathbf{r}}_- \tag{24.4}$$

where $\hat{\mathbf{r}}_+$ and $\hat{\mathbf{r}}_-$ are unit vectors directed from $+q$ and $-q$ to the point where the electric field is evaluated. At points on the y-axis where x equals zero,

$$r_+{}^2 = r_-{}^2 = a^2 + y^2$$

and the form of Equation 24.4 simplifies to

$$\mathbf{E} = \frac{k_e q(\hat{\mathbf{r}}_+ - \hat{\mathbf{r}}_-)}{a^2 + y^2}$$

Figure 24.6 shows that

$$\hat{\mathbf{r}}_+ - \hat{\mathbf{r}}_- = \frac{2a(-\mathbf{i})}{\sqrt{a^2 + y^2}}$$

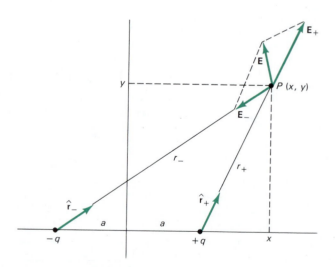

FIGURE 24.5

An electric dipole consists of a pair of charges of equal magnitude and opposite sign. The electric field set up by the dipole is calculated by superposing the Coulomb fields $\mathbf{E}_+$ and $\mathbf{E}_-$ set up by the charges $+q$ and $-q$.

The electric field of the dipole can be expressed as

$$\mathbf{E} = \frac{(-\mathbf{i})k_e p}{(a^2 + y^2)^{3/2}} \tag{24.5}$$

where

$$p = 2qa \tag{24.6}$$

is called the **electric dipole moment**. Equation 24.5 indicates that $\mathbf{E}$ is in the negative x-direction at all points on the y-axis. Figure 24.7 depicts the electric field lines for the electric dipole and confirms that $\mathbf{E}$ is in the negative x-direction at points on the y-axis.

Next, let's examine the behavior of the electric dipole field at large distances from the center of the dipole ($y \gg a$). In Equation 24.5 we can set

$$(a^2 + y^2)^{3/2} \approx y^3 \qquad y \gg a$$

whereupon $\mathbf{E}$ becomes

$$\mathbf{E} = \frac{(-\mathbf{i})k_e p}{y^3}$$

This limiting form of $\mathbf{E}$ illustrates a general feature of the electric field of the dipole: At large distances from the dipole the electric field is *inversely* proportional to the *cube* of the distance from the center of the dipole.

$$E_{\text{dipole}} \propto \frac{1}{r^3} \qquad r \gg a \tag{24.7}$$

The inverse cube (r^{-3}) dependence of E is the **signature** of the dipole field.

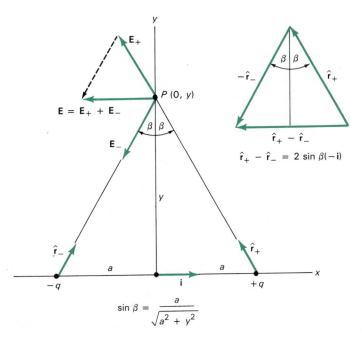

FIGURE 24.6

At points on the y-axis the dipole electric field is directed along the negative x-axis.

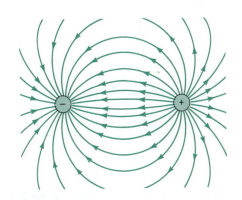

FIGURE 24.7

Electric field lines for an electric dipole.

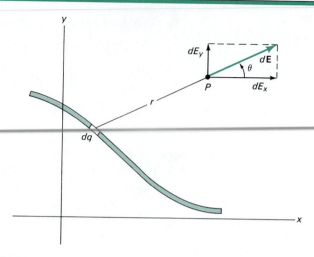

FIGURE 24.8

Charge is spread continuously along a line. The electric field set up at the point P is the vector sum of the electric fields set up by all of the charge elements.

Electric Field of a Continuous Charge Distribution

Figure 24.8 shows a collection of charges spread continuously along a line. We picture the distribution as a collection of point-like charge elements. In Figure 24.8 one such charge element is labeled dq. The net electric field set up at the field point P is the vector sum of the electric fields set up by all of the charge elements spread along the line. In practice the vector sum is evaluated by an integration.

The magnitude of the field set up at P by the charge element dq is

$$dE = \frac{k_e \, dq}{r^2}$$

Because $d\mathbf{E}$ is a vector, we must integrate each component separately. In Figure 24.8, the x- and y-components of $d\mathbf{E}$ are

$$dE_x = dE \cos \theta = k_e \left(\frac{dq}{r^2} \right) \cos \theta$$

$$dE_y = dE \sin \theta = k_e \left(\frac{dq}{r^2} \right) \sin \theta$$

The components E_x and E_y are expressed as integrals over the charge distribution

$$E_x = k_e \int \left(\frac{dq}{r^2} \right) \cos \theta \qquad (24.8)$$

$$E_y = k_e \int \left(\frac{dq}{r^2} \right) \sin \theta \qquad (24.9)$$

In order to carry out the integrations it is necessary to specify how the charge is distributed.

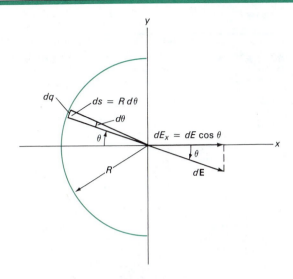

FIGURE 24.9

Charge is spread uniformly along the semicircle. The electric field at the center of the semi-circle is evaluated by integrating the electric fields of all charge elements.

EXAMPLE 2

Electric Field Set Up by a Semicircular Line Charge

Figure 24.9 shows a charge q spread *uniformly* along a semicircle of radius R. We want to determine the electric field set up at the center of the semicircle.

In Equations 24.8 and 24.9 we express the charge on an arc length ds as $dq = \beta \, ds$ where β is the charge per unit length

$$\beta = \frac{q}{\pi R}$$

All charge elements are equidistant from the center of the semicircle, so $r = R$ and $ds = R \, d\theta$. The integral for E_x (Equation 24.8) becomes

$$E_x = k_e \int \left(\frac{dq}{r^2}\right) \cos \theta = k_e \int_{-\pi/2}^{+\pi/2} \left(\frac{\beta R \, d\theta}{R^2}\right) \cos \theta$$

$$= \frac{k_e \beta}{R} \int_{-\pi/2}^{+\pi/2} \cos \theta \, d\theta = \frac{2k_e \beta}{R}$$

Setting $\beta = q/\pi R$ gives

$$E_x = \left(\frac{2}{\pi}\right) \frac{k_e q}{R^2}$$

The y-component of the electric field at the center of the semicircle is zero because of the symmetry of the charge distribution.

24.2
ELECTRIC FLUX

The concept of electric field lines is qualitative at this point; the lines are just a pictorial way of describing the electric field. We can use the concept of **electric flux** to quantify the idea of electric field lines.

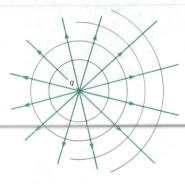

FIGURE 24.10

A series of spherical surfaces surrounding an isolated positive charge. The number of electric field lines through each surface is the same.

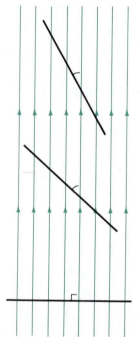

Side view of 3 surfaces of equal area

FIGURE 24.11

The number of electric field lines passing through a given area depends on three factors: field strength, area, and orientation of the area.

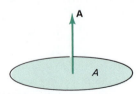

FIGURE 24.12

Vector representation of area.

Consider an isolated positive charge and its radial pattern of electric field lines. Furthermore, imagine a series of spherical surfaces enclosing the charge (Figure 24.10). These are geometric surfaces, not physical surfaces. As the distance from the charge to the surface increases, the surface area increases, in direct proportion to the *square* of the distance from the charge. The electric field strength is *inversely* proportional to the *square* of the distance from the charge. Because of the inverse square nature of the Coulomb electric field, the number of electric field lines through each surface is the same. Each electric field line passes through all of the surfaces because there are no negative charges for it to end on, nor are there any other positive charges to give rise to new field lines.

The spherical surfaces in Figure 24.10 are perpendicular to the field lines. In general the number of field lines passing through a surface depends on the orientation of the surface relative to the direction of the field (Figure 24.11). For a given surface area, the number of field lines is a maximum when the surface is perpendicular to the field. When the surface is parallel to the field then no field lines pass through it. In general the number of electric field lines through an area A is directly proportional to $A \cos \theta$, where θ is the angle between the electric field and a line perpendicular to the surface.

The **electric flux Φ_E** through a surface area A in a region where the electric field magnitude is E is defined by

$$\Phi_E = EA \cos \theta \qquad (24.10)$$

This definition makes the electric flux directly proportional to the number of electric field lines through the surface. The units of electric flux are $N \cdot m^2/C$.

It is useful to represent the area A by a vector, **A**, whose magnitude equals the area A and whose direction is perpendicular to the surface (Figure 24.12). With this definition we can express Φ_E as the scalar product of **E** and **A**.

$$\Phi_E = \mathbf{E} \cdot \mathbf{A} \qquad (24.11)$$

When **A** is part of a closed surface the direction of **A** is defined to be **outward**. As defined by Equation 24.11, the electric flux Φ_E can be positive or negative. Figure 24.13 shows an electric field line that enters a closed surface through the area $\mathbf{A_2}$ and emerges through the area $\mathbf{A_1}$. The flux is negative for the

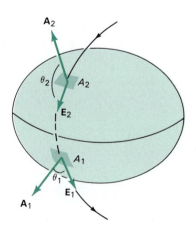

FIGURE 24.13

Field lines entering a closed surface make negative contributions to the flux. Emerging field lines make positive contributions.

area where the field line enters the surface because the angle θ_2 is between $90°$ and $180°$ and its cosine is negative,

$$\Phi_2 = E_2 A_2 \cos \theta_2 < 0$$

The flux is positive over the area where the field line emerges because the angle θ_1 is between $0°$ and $90°$ and its cosine is positive:

$$\Phi_1 = E_1 A_1 \cos \theta_1 > 0$$

In general, field lines entering a surface make negative contributions to the electric flux and emerging field lines make positive contributions.

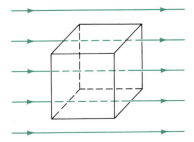

FIGURE 24.14

The field lines for a uniform electric field are parallel straight lines. The cubic surface is oriented so that four of its faces are parallel to the field. The other two faces are normal to **E**.

EXAMPLE 3

Electric Flux Through a Cubic Surface

To illustrate the fact that Φ_E can be positive or negative, consider a uniform electric field. The field lines (Figure 24.14) are parallel straight lines, indicating that the field is not changing in magnitude or direction. Let's evaluate the flux through the cubic surface shown in Figure 24.14. Four of the six sides are parallel to the electric field. The flux through these sides is zero because **E** is perpendicular to **A**:

$$\mathbf{E} \cdot \mathbf{A} = EA \cos 90° = 0$$

Geometrically, the field lines do not enter or leave the cube through these four surfaces. The flux through the right face of the cube is positive because **E** and **A** are parallel:

$$\mathbf{E} \cdot \mathbf{A} = EA \cos 0° = +EA \qquad \text{(right end)}$$

The flux through the left end face is negative because the outwardly directed surface vector **A** is antiparallel to **E**:

$$\mathbf{E} \cdot \mathbf{A} = EA \cos 180° = -EA \qquad \text{(left end)}$$

The total electric flux, the sum of the fluxes through the six faces, is zero. This simply reflects the fact that equal numbers of electric field lines enter and emerge from the cube. The entering field lines make a negative contribution to the flux. The emerging field lines make an equal but positive contribution to the total flux.

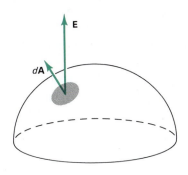

FIGURE 24.15

The flux through a finite surface area can be obtained by integrating $d\Phi_E = \mathbf{E} \cdot d\mathbf{A}$ over the surface.

We can also formulate the flux definition in differential form to handle curved surfaces and situations where **E** varies with position:

$$d\Phi_E = \mathbf{E} \cdot d\mathbf{A} \tag{24.12}$$

The flux through a finite area can be obtained by integration (Figure 24.15).

EXAMPLE 4

Flux Through a Surface Enclosing a Point Charge

Let's use Equation 24.12 to evaluate the total electric flux through a closed spherical surface surrounding an isolated positive charge, q. As we argued earlier (Figure 24.10), the same electric field lines pass through all surfaces enclosing the charge. This means the flux is the same for all surfaces enclosing the charge. We simplify the flux calculation by choosing the surface to be a sphere centered on the charge. Figure 24.16 shows that **E** and $d\mathbf{A}$ are parallel at every point on the surface of the sphere. This is because **E** is radial and $d\mathbf{A}$ is normal to the surface. The scalar product $\mathbf{E} \cdot d\mathbf{A}$ equals $E\,dA$, and the flux as given by Equation 24.12 is

$$d\Phi_E = \mathbf{E} \cdot d\mathbf{A} = E\,dA$$

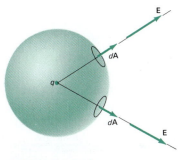

FIGURE 24.16

For a spherical surface centered on the point charge q, the field **E** and the surface vector $d\mathbf{A}$ are parallel over the entire surface.

The total electric flux through the sphere is obtained by integrating over the spherical surface,

$$\Phi_E = \int d\Phi_E = \int E \, dA$$

The electric field of the point charge is spherically symmetric: it has the same magnitude at all points on the spherical surface,

$$E = \frac{k_e q}{r^2}$$

This fact lets us remove E from the integral. The integral that remains gives the surface area of the sphere

$$\Phi_E = E \int dA = \left(\frac{k_e q}{r^2}\right) A_{\text{sphere}} = \left(\frac{k_e q}{r^2}\right) 4\pi r^2$$

The r^2 factors cancel, giving for the total electric flux

$$\Phi_E = 4\pi k_e q \qquad (24.13)$$

As we noted earlier, Φ_E is independent of the shape of the surface. Hence, Example 4 establishes a very general result: the electric flux through *any* surface enclosing a point charge q is given by $\Phi_E = 4\pi k_e q$. In the next section we will generalize this result and obtain Gauss' law.

One change in notation is in order. Up to this point we have used the Coulomb proportionality constant, k_e. It is convenient to introduce another constant ϵ_0, defined by

$$\epsilon_0 = \frac{1}{4\pi k_e} = 8.85419 \times 10^{-12} \text{ C}^2/\text{N}\cdot\text{m}^2 \qquad (24.14)$$

ϵ_0 is called the **permittivity of free space**.

In terms of ϵ_0 the flux through a surface enclosing a point charge q is

$$\Phi_E = \frac{q}{\epsilon_0} \qquad (24.15)$$

In the next section we generalize this result to an arbitrary collection of electric charges.

24.3
GAUSS' LAW

Most of the physical laws that we have studied so far are expressed in differential form. In this section we develop an integral relation—Gauss' law. The differential laws, such as Newton's second law ($F = ma$), are sometimes called **local** laws because they describe what happens at a point. Certain integral laws, on the other hand, can be termed **global** because they describe events over some broad region. Gauss' law is a very powerful and useful global law.

Gauss' law is a generalization of the result derived in the preceding section: the electric flux through a surface enclosing a point charge q equals q/ϵ_0. To derive Gauss' law we use the *principle of superposition*. Consider a collection of point charges q_1, q_2, q_3, ... Each of these charges sets up an electric field, $\mathbf{E}_1$, $\mathbf{E}_2$, $\mathbf{E}_3$, ... The principle of superposition specifies that the net electric field set up by the collection of charges is the vector sum of the individual electric fields

$$\mathbf{E} = \mathbf{E}_1 + \mathbf{E}_2 + \mathbf{E}_3 + \cdots$$

Consider an arbitrary surface enclosing the collection of charges. Remember, the surface is geometrical, not physical. We refer to it as a **Gaussian surface** (GS). Because **E** is the vector sum of the individual point charge fields, the net electric flux is the sum of the fluxes due to the individual fields. Thus,

$$\Phi_E = \int_{GS} \mathbf{E} \cdot d\mathbf{A} = \int_{GS} \mathbf{E}_1 \cdot d\mathbf{A} + \int_{GS} \mathbf{E}_2 \cdot d\mathbf{A} + \cdots$$

We have already shown that the flux of the Coulomb electric field equals $1/\epsilon_0$ times the enclosed charge (Equation 24.15). The net flux may therefore be written

$$\Phi_E = \frac{q_1}{\epsilon_0} + \frac{q_2}{\epsilon_0} + \frac{q_3}{\epsilon_0} + \cdots$$

The right side is

$$\frac{(q_1 + q_2 + q_3 + \cdots)}{\epsilon_0} = \frac{q}{\epsilon_0}$$

where q is the **net** electric charge enclosed by the Gaussian surface. The result is Gauss' law:

$$\int_{GS} \mathbf{E} \cdot d\mathbf{A} = \frac{q}{\epsilon_0} \qquad (24.16)$$

In words, Gauss' law states that the net electric flux through any closed surface equals $1/\epsilon_0$ times the *net* charge enclosed by that surface. We emphasize that it is the net electric charge that is important. For example, the net flux would be zero if there were no charges inside the Gaussian surface. The net flux would also be zero if there was just one electron and one proton inside the Gaussian surface. Any collection of charges which is neutral gives zero net flux.

EXAMPLE 5

Absence of Net Charge Inside a Conductor

A conductor consists of an array of positively charged atomic nuclei bound in place relative to one another. Electrons provide an equal amount of negative charge, and so the conductor is electrically neutral. Most of the electrons are bound to the nuclei, but in a metallic conductor some electrons (perhaps one per atom) are free to move through the metal. The electrons can move in response to an electric field, and in motion they constitute an electric current. When a conductor is in electrostatic equilibrium there is no electric current, and by implication, no electric field.

Although the conduction electrons are free to move about inside the conductor they cannot move to or away from the surface. The reason is that if a net number of electrons move out of any volume they leave behind a net positive charge. This positive charge exerts strong attractive forces that urge the electrons to return. Thus, the interior of a conductor is a region where the ease of charge motion operates to preserve electrical neutrality.

We can add electric charge to a conductor so that it then carries a net or excess charge. We want to show that there can be no net charge in the interior of a conductor that is in electrostatic equilibrium. In other words, any *excess* charge must reside on the surface of the conductor. Physical reasoning will convince you that this is true.

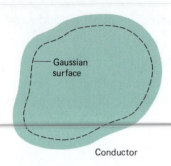

FIGURE 24.17

A Gaussian surface just inside the actual surface of the conductor.

Suppose we started with a neutral conductor and then injected a number of electrons into its interior. These electrons would exert repulsive forces on one another. They would get away from one another as far as possible—by moving to the surface of the conductor. Thus, any excess charge resides on the surface. Now let's see how we can use Gauss' law to prove this result.

We must remember that in electrostatic equilibrium the electric field in the interior of the conductor must be zero. If **E** were not zero the mobile charges would move. Moving charges constitute an electric current, and consequently we would not have electrostatic equilibrium.

Now we choose a Gaussian surface just beneath the actual physical surface of the conductor (Figure 24.17). The electric field is zero at every point on this surface. Integrating over this Gaussian surface, Gauss' law gives

$$\frac{q}{\epsilon_0} = \int_{GS} \mathbf{E} \cdot d\mathbf{A} = 0$$

Thus Gauss' law proves that $q = 0$: There can be no net charge inside the conductor when it is in electrostatic equilibrium. Any excess charge must lie on the surface of the conductor.

Note that charge on the surface of an isolated conductor is actually trapped on the surface. The charge is not free to wander off into the surrounding space. In fact, an extremely large electric force must be exerted to remove electrons from the surface of an isolated conductor in a vacuum. We will explore this point in the next chapter.

Electric Field at the Surface of a Charged Conductor

If there is charge on the surface of a conductor it sets up an electric field outside the conductor. We can use Gauss' law to relate the electric field at the surface of a conductor to the surface density of charge (coulombs per square meter).

Consider a conductor such as a piece of metal, perhaps, that is in electrostatic equilibrium with a net electric charge on its surface. This charge sets up an electric field outside the conductor. The electric field must be perpendicular to the conductor at its surface. If the electric field had a component parallel to the surface the charges would experience an electric force and be set in motion. This motion of charge would constitute an electric current—a nonequilibrium condition. Thus, if we sprayed some charges onto a conducting surface they would move around until they achieved an equilibrium in which the electric field has no component parallel to the surface.

We can use Gauss' law to evaluate the electric field at the surface. We pick the Gaussian surface shown in Figure 24.18. It is a short cylinder of cross-sectional area A. One end of the surface is buried inside the conductor where $E = 0$. The other end is just above the surface of the conductor. The sides of the cylinder are perpendicular to the surface and so are parallel to the electric field. If we let σ denote the charge per unit area on the surface, then the net charge enclosed by the Gaussian cylinder is σA. Gauss' law specifies

$$\int_{GS} \mathbf{E} \cdot d\mathbf{A} = \frac{\sigma A}{\epsilon_0}$$

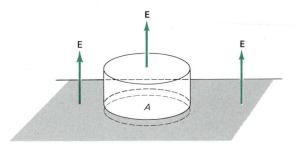

FIGURE 24.18

A cylindrical Gaussian surface. The base of the cylinder is buried in the conductor.

The flux integral is easy to evaluate. The bottom of the Gaussian cylinder contributes zero to the flux integral because $E = 0$ inside the conductor

$$\int_{\text{bottom}} \mathbf{E} \cdot d\mathbf{A} = 0 \qquad E = 0$$

The sides of the Gaussian cylinder also contribute zero because $\mathbf{E}$ is parallel to the Gaussian surface and thus perpendicular to the surface area vector $d\mathbf{A}$;

$$\int_{\text{sides}} \mathbf{E} \cdot d\mathbf{A} = 0 \qquad \mathbf{E} \perp d\mathbf{A}$$

The net flux is due entirely to the electric field lines that emerge through the top of the Gaussian cylinder. The field is parallel to the area vector $d\mathbf{A}$ on the top of the cylinder and so $\mathbf{E} \cdot d\mathbf{A} = E \, dA$. We can always choose the area A small enough so that E may be considered constant. The flux integral over the top of the Gaussian cylinder, the total flux in this case, is

$$\int_{\text{top}} \mathbf{E} \cdot d\mathbf{A} = \int_{\text{top}} E \, dA = EA$$

Setting $\int \mathbf{E} \cdot d\mathbf{A} = EA$ and cancelling a factor A gives

$$E = \frac{\sigma}{\epsilon_0} \tag{24.17}$$

This result shows that the electric field at the surface of a charged conductor is determined by the surface charge density. Equation 24.17 is a very general result. It applies to all charged conducting surfaces regardless of their shape. In practice, Equation 24.17 can be used to evaluate E numerically only when the surface has enough symmetry to make it possible to determine σ. For example, if the conductor is a sphere of radius r and the net charge on its surface is q, then we can argue, by symmetry, that the charge density will be constant,

$$\sigma = \frac{q}{4\pi r^2}$$

Equation 24.17 then gives the spherically symmetric result

$$E = \frac{\sigma}{\epsilon_0} = \frac{1}{4\pi\epsilon_0} \frac{q}{r^2} \tag{24.18}$$

Notice that Equation 24.17 determines E only *at* the surface. If we move away from the surface Gauss' law can be used to determine E only when the field

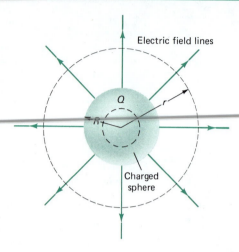

Electric field lines

Charged
sphere

FIGURE 24.19
The charge Q is spread uniformly throughout the volume of the sphere of radius R. The
dashed lines indicate the surfaces of two spherical Gaussian surfaces.

has a high degree of symmetry. Any symmetry of E reflects a symmetry of
the charge distribution. We can illustrate this remark by using Gauss' law to
determine the electric field set up by a positive charge Q that is distributed uniformly throughout the *volume* of a sphere of radius R (Figure 24.19).

EXAMPLE 6

Electric Field for a Uniform Sphere of Charge

Imagine a sphere of radius R with a charge Q spread *uniformly throughout its volume*.
The symmetry of the charge distribution results in a spherically symmetric electric field.
The electric field is *radial*; **E** is directed away from the center of the sphere if Q is posi-
tive. Its magnitude E depends only on the distance r from the center of the sphere.

To determine E in terms of r we use a Gaussian surface that exploits the symmetry.
We choose a spherical Gaussian surface that is concentric with the spherical charge distribution. In Figure 24.19 we use dashed lines to indicate two such Gaussian surfaces.
One encloses the entire charge Q. Its radius r is greater than R. The other Gaussian
sphere has a radius r that is less than R. It encloses only a fraction of the total charge
Q. Regardless of whether we have $r < R$ or $r > R$, the algebraic expression for E
is the same. The electric field is radial, which makes it parallel to $d\mathbf{A}$, the surface area
vector. Thus

$$\mathbf{E} \cdot d\mathbf{A} = E\, dA$$

and the flux integral is

$$\Phi_E = \int \mathbf{E} \cdot d\mathbf{A} = \int E\, dA$$

Since E depends only on r it is constant over the Gaussian sphere surface,

$$\Phi_E = \int E\, dA = E \int dA$$

The integral gives $4\pi r^2$, the surface area of the Gaussian sphere. Thus, the electric flux
is given by

$$\Phi_E = 4\pi r^2 E \qquad (24.19)$$

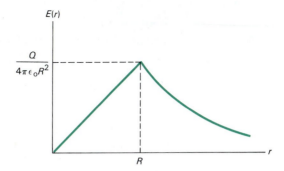

FIGURE 24.20

The magnitude of the electric field set up by a uniformly-charged sphere. The field strength is a maximum at the surface of the sphere.

To determine E via Gauss' law we must evaluate the charge enclosed by the Gaussian sphere. For $r > R$ the enclosed charge is simply Q, the total charge. Setting $Q = q$ in Gauss' law gives

$$4\pi r^2 E = \frac{Q}{\epsilon_0}$$

showing that the electric field is

$$E = \frac{1}{4\pi\epsilon_0} \frac{Q}{r^2} \qquad r \geq R \qquad (24.20)$$

Note that this is the same electric field that would be produced at r by a point charge Q located at the center of the sphere.

To complete our solution for E we need to determine the enclosed charge when the radius of the Gaussian surface is less than R. We make use of the fact that the charge is spread uniformly throughout the *volume* of the sphere. The fraction enclosed by the Gaussian sphere is

$$\frac{q}{Q} = \frac{\frac{4}{3}\pi r^3}{\frac{4}{3}\pi R^3} = \frac{r^3}{R^3}$$

Setting $q = Qr^3/R^3$ in Gauss' law gives

$$4\pi r^2 E = \frac{Qr^3}{\epsilon_0 R^3}$$

The electric field for $r \leq R$ is therefore

$$E = \frac{1}{4\pi\epsilon_0} \frac{Qr}{R^3} \qquad r \leq R \qquad (24.21)$$

Figure 24.20 is a graph of E versus r. Note that E is zero at the center of the sphere. This also follows from symmetry—a test charge placed at the center would experience equal forces from all directions, but no net force.

To this point our focus has been on stationary charges as a source of electric field. But charges also experience forces when placed in an existing electric field. In the next section we consider the motion of electric charges in a specified electric field.

24.4
MOTION OF A CHARGED PARTICLE IN AN ELECTRIC FIELD

A charged particle experiences a force when it is subjected to an electric field. The force on a charge q in a field $\mathbf{E}$ is

$$\mathbf{F} = q\mathbf{E}$$

If this is the only force acting, the charged particle undergoes an acceleration. In accord with Newton's second law, $\mathbf{F} = m\mathbf{a}$, the acceleration is

$$\mathbf{a} = \frac{q\mathbf{E}}{m} \tag{24.22}$$

In situations where the electric field is uniform (E = constant) the acceleration is constant and we can apply the results derived in Chapters 4 and 5 to draw conclusions about its motion.

EXAMPLE 7

Proton Penetration into Electric Field

A proton moving at a speed of 10^5 m/s enters an electric field $E = 1000$ N/C. The proton's velocity and the field are initially antiparallel (Figure 24.21). How far into this field will the proton penetrate before momentarily coming to rest?

We choose the x-axis in the direction of the proton velocity. The acceleration is given by Equation 24.22,

$$a = \frac{-qE}{m}$$

The minus sign enters because the electric field is in the negative x-direction. Physically, the field is opposing the motion of the proton. Because the field is uniform the acceleration is constant. We can use the results of Chapter 4:

$$v^2 - v_0{}^2 = 2ax = \frac{-2qEx}{m}$$

The speed of the proton when it enters the field at $x = 0$ is denoted as v_0. We want to find x when the proton's speed is zero. Setting $v = 0$ we get

$$x = \frac{mv_0{}^2}{2qE}$$

Using

$$m = 1.67 \times 10^{-27} \text{ kg} \qquad v_0 = 10^5 \text{ m/s}$$

$$E = 1000 \text{ N/C} \qquad q = +1.60 \times 10^{-19} \text{ C}$$

gives $x = 5.22$ cm for the penetration distance.

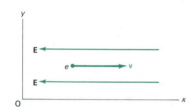

FIGURE 24.21

The proton velocity and the electric field are oppositely directed.

The Millikan oil-drop experiment is another illustration of the motion of charged particles in an electric field. Millikan observed the motion of electrically charged oil droplets as they moved up and down under the combined influence of gravity, a buoyant force, a viscous drag force, and the force exerted by a uniform electric field (Figure 24.22).

The oil drops experienced a buoyant force equal to the weight of the air they displaced. Their effective weight was reduced by this buoyant force.

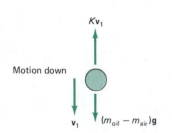

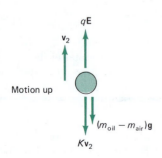

FIGURE 24.22

Forces on a moving oil droplet.

Because of air drag, Millikan's droplets quickly reached a terminal velocity. Falling at a terminal velocity v_1, the effective gravitational force $(m_{oil} - m_{air})g$ was balanced by the viscous drag force Kv_1.

$$(m_{oil} - m_{air})g = Kv_1$$

Applying an electric field E, Millikan introduced an upward force qE. This force caused the oil drops to stop and then move upward. They reached an upward terminal velocity v_2. The balance of forces for the upward motion is described by

$$(m_{oil} - m_{air})g + Kv_2 = qE$$

Eliminating the effective gravitational force gives for the charge q

$$q = \left(\frac{K}{E}\right)(v_1 + v_2)$$

Repeating the experiment many times, Millikan found that the values of q clustered closely about the values 1.6×10^{-19} C, 3.2×10^{-19} C, 4.8×10^{-19}C, ... These data led Millikan to conclude that the charge of the electron is

$$-(1.603 \pm 0.002) \times 10^{-19} \text{ C}$$

and that the charges on all oil drops are integral multiples of this value. In other words, Millikan concluded that electric charge occurs only in discrete amounts—it is **quantized**.

Motion Perpendicular to an Electric Field

The motion of a charged particle initially moving perpendicular to a uniform electric field is like that of a ball thrown horizontally in a uniform gravitational field—both move along a parabolic path.

The motion of electrons in a field perpendicular to their initial velocity is illustrated by the oscilloscope. Electrons that have been accelerated and focused into a narrow beam are moving horizontally at a speed v_x when they arrive at a pair of horizontal metal plates (Figure 24.23). The metal plates are charged,

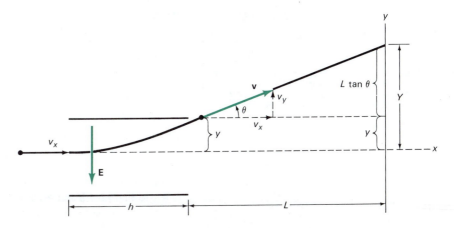

FIGURE 24.23

Electrons are deflected upward by a uniform electric field during the time they are between the horizontal plates. Beyond the field region they travel in a straight line.

and set up a uniform electric field E. This field is vertically downward in Figure 24.23. The negatively charged electrons experience an upward acceleration given by Equation 24.22

$$a = \frac{eE}{m}$$

where e is the magnitude of the electron charge. The horizontal component of velocity, v_x, is unchanged as the electron moves through the electric field region. The time for the electron to cross the field region is

$$t = \frac{h}{v_x}$$

where h is the length of a plate. The electrons emerge from the field region with a vertical velocity component

$$v_y = at = \left(\frac{eE}{m}\right)\left(\frac{h}{v_x}\right)$$

and a vertical deflection

$$y = \frac{1}{2}at^2 = \frac{1}{2}\left(\frac{eE}{m}\right)\left(\frac{h}{v_x}\right)^2$$

Once beyond the field region the electrons travel in a straight line. Overall, they undergo a deflection through an angle θ given by

$$\tan\theta = \frac{v_y}{v_x} = \left(\frac{eE}{mv_x^2}\right)h \tag{24.23}$$

They strike a screen a distance L beyond the plates where they reveal their position by converting their kinetic energy into light. The total deflection of the beam is given by

$$Y = y + L\tan\theta = \left(\frac{eE}{mv_x^2}\right)(L + \tfrac{1}{2}h)h \tag{24.24}$$

WORKED PROBLEM

A point charge Q is located on the axis of a ring of radius R at a distance b from the plane of the ring (Figure 24.24). Show that if one-fourth of the electric flux from the charge threads the ring, then $R = \sqrt{3}\,b$.

Solution

The total flux through a surface enclosing the charge Q is Q/ϵ_0. The flux through the ring is

$$\Phi_{\text{ring}} = \int \mathbf{E} \cdot d\mathbf{A}$$

where the integration covers the area of the ring. We must evaluate this integral and set it equal to $\tfrac{1}{4}Q/\epsilon_0$ to find how b and R are related.

In Figure 24.24, take dA to be the area of an annular ring of radius s and width ds. The flux through dA is

$$\mathbf{E} \cdot d\mathbf{A} = E\,dA\cos\theta = E(2\pi s\,ds)\cos\theta$$

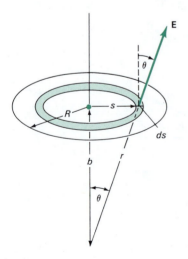

FIGURE 24.24

The magnitude of the electric field has the same value at all points within the annular ring,

$$E = \frac{1}{4\pi\epsilon_0}\frac{Q}{r^2} = \frac{1}{4\pi\epsilon_0}\frac{Q}{s^2 + b^2}$$

and

$$\cos\theta = \frac{b}{r} = \frac{b}{(s^2 + b^2)^{1/2}}$$

Integrate from $s = 0$ to $s = R$ to get the flux through the entire ring.

$$\Phi_{\text{ring}} = \frac{Qb}{2\epsilon_0}\int_0^R \frac{s\,ds}{(s^2 + b^2)^{3/2}}$$

$$= \frac{Qb}{2\epsilon_0}[-(s^2 + b^2)^{-1/2}]\Big|_0^R$$

$$= \frac{Q}{2\epsilon_0}\left[1 - \frac{b}{(R^2 + b^2)^{1/2}}\right]$$

The flux through the ring equals $\frac{1}{4}Q/\epsilon_0$ provided

$$\frac{b}{(R^2 + b^2)^{1/2}} = \frac{1}{2}$$

This is satisfied if $R = \sqrt{3}\,b$.

EXERCISES

.24.1 Electric Field

A. An electric field is adjusted to just counteract the force of gravity on an electron. What field is required?

B. An electric field of magnitude 100 N/C acts on an alpha particle, a doubly ionized helium atom with a charge twice that of a proton and a mass of 6.64×10^{-27} kg. (a) What is the magnitude of the electric force exerted on the alpha particle? (b) What is the magnitude of the acceleration of the alpha particle?

C. An electron inside a TV picture tube experiences a force of 8.0×10^{-14} N directed toward the front of the tube. What is the magnitude and direction of the electric field that produces this force?

D. A charge of 1.5 μC and a charge of 2.8 μC are 1.62 m apart. (a) Find the magnitude of the electric field each charge produces at the position of the other. (b) What is the magnitude of the force each charge exerts on the other? Is Newton's third law satisfied?

E. Four charges, $q_1 = q_4 = +1\ \mu$C, $q_2 = q_3 = -1\ \mu$C, are placed at the corners of a square 1 m on a side (Figure 1). (a) Draw arrows representing the electric field set up at the center of the square by each charge and an arrow representing the net electric field. (b) Determine the net electric field by completing the table provided. In the column labeled "E" enter the *magnitudes* of the electric fields set up by each charge.

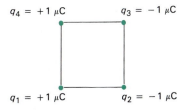

FIGURE 1

Charge	E	E_x	E_y
q_1			
q_2			
q_3			
q_4			
		Net $E_x =$	Net $E_y =$

Net $E =$ _____$\mathbf{i} +$ _____$\mathbf{j}$

(c) Sketch the vector calculated for the net electric field. Is it consistent with the net electric field vector drawn in (a)?

F. Three identical positive point charges lie at the vertices of an equilateral triangle with sides of length a (Figure 2). The origin is at the geometric center of the triangle. Determine the direction of the electric field on the x-axis for points (a) to the right of the origin and (b) to the left of the origin.

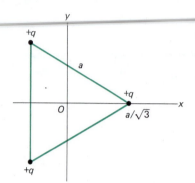

FIGURE 2

G. Two oppositely directed dipoles are placed side by side (Figure 3). This is one form of electric quadrupole. Sketch the electric field lines for this configuration of charges. Include the interior of the square and a portion of the exterior space.

FIGURE 3

H. Two dipoles that are back to back form a linear electric quadrupole (Figure 4). Sketch the electric field lines for this configuration of charges.

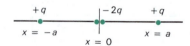

FIGURE 4

24.2   Electric Flux

I. A 10 μC charge located at the origin of a Cartesian coordinate system is surrounded by a non-conducting hollow sphere of radius 10 cm. A drill with a radius of 1 mm is aligned along the z-axis, and a hole is drilled in the sphere. Calculate the electric flux through the hole.

24.3   Gauss' Law

J. A sphere of radius R contains a total charge Q. The charge density decreases with distance from the center of the sphere according to

$$\rho = \frac{K}{r^2} \qquad r \le R$$

$$\rho = 0 \qquad r > R$$

Show that the constant K is related to Q and R by $K = Q/4\pi R$.

K. A point charge of 0.0462 μC is inside a pyramid. Determine the total electric flux through the surface of the pyramid.

L. The electric flux threading a spherical Gaussian surface is 1.53×10^4 N·m²/C. Determine the net charge inside the surface.

M. A charge of 12 μC is at the geometric center of a cube. What is the electric flux through one of the cube's faces?

N. A charge of 6.0×10^{-10} C is spread uniformly over the surface of a hollow metal sphere. The sphere has a radius of 10 cm. Determine the magnitude of the electric field (a) 5 cm from the center of the sphere; (b) 20 cm from the center of the sphere.

P. A battery connected across two parallel metal plates 1 cm apart produces an electric field of 150 N/C between the plates. (a) Find the surface charge density (in C/m²) on the negative plate. (b) How many electrons per square centimeter are represented by your value of the charge density?

Q. Two horizontal metal plates are separated by a small distance (Figure 5). Positive electric charge with a surface density of 10.0 μC/m² is distributed over the upper surface of the lower plate. On the lower surface of the upper plate, negative charge is distributed with equal density. (a) Calculate the magnitude of the electric field between the plates produced by this distribution of charge. (b) Is the field directed up or down?

FIGURE 5

R. Determine the electric flux through the closed surfaces A and B in Figure 6.

S. The electric flux through a Gaussian surface is 5.42×10^{-8} N·m²/C. It is known that there are three charged particles inside the surface, and that the particles are not all the same. The particles can be electrons, protons, or alpha particles. Determine the particles inside the surface.

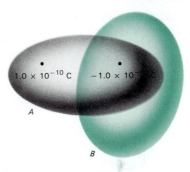

1.0 × 10⁻¹⁰ C   −1.0 × 10 C

A

B

FIGURE 6

24.4 Motion of a Charged Particle in an Electric Field

T. An electron placed in an electric field experiences an upward acceleration of 3.6×10^{13} m/s². Determine the acceleration of a proton placed in the same electric field.

U. An electron with a speed of 3×10^6 m/s moves into a uniform electric field of 1000 N/C. The field is parallel to the electron's velocity and acts to decelerate the electron. How far does the electron travel before it is brought to rest?

PROBLEMS

1. Consider the electric dipole shown in Figure 7. Calculate the electric flux through the y–z plane as follows: (a) show that the electric field on the y–z plane is in the +x-direction and is given by

$$E_x = \frac{q}{2\pi\epsilon_0(R^2 + a^2)^{3/2}}$$

(b) evaluate the electric flux by taking the area element dA to be a ring of radius R and width dR, centered on the origin. Integrate $d\Phi_E = E_x \, dA$ to show that the flux through the y–z plane is

$$\Phi_E = \frac{q}{\epsilon_0}$$

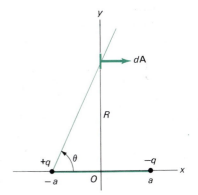

FIGURE 7

2. An infinite plane sheet of charge has a uniform surface charge density σ. Apply Gauss' law to show that the magnitude of the electric field above or below the sheet is given by

$$E = \frac{\sigma}{2\epsilon_0}$$

3. The nucleus of lead-208, $^{208}_{82}$Pb, has 82 protons within a sphere of radius 6.34 fm. Assume that the electric charge is spread uniformly throughout the spherical volume. Calculate the electric field produced by this nuclear charge (a) at the surface of the nucleus, (b) at a distance of 10^{-12} m from the center of the nucleus.

4. Suppose that you have two concentric conducting spherical shells. A charge $+Q$ is placed on the inner shell (radius a), and a charge $-(Q + q)$ on the outer shell (radius b). Let r denote the distance from the center of the shells and use Gauss' law to express the electric field E as a function of r in the three regions $r < a$, $a < r < b$, and $r > b$.

5. An early model of the hydrogen atom consisted of a positive charge uniformly distributed over a spherical volume, and a point electron free to move under the influence of the electric field set up by the positive charge. Consider such a model having a total positive charge e, radius $R = 0.500$ Å, and an electron of charge $-e$. (a) From Gauss' law, show that the Coulomb force on the electron a distance r from the center of the spherical distribution of positive charge is

$$F = -\left(\frac{e^2}{4\pi\epsilon_0}\right)\left(\frac{r}{R^3}\right) \qquad r \leq R$$

(b) Assuming this is the net force, show that the electron can undergo simple harmonic motion radially with an angular frequency of 4.50×10^{16} rad/s.

6. The charge density inside a sphere of radius R is given by

$$\rho = kr \qquad r \leq R$$
$$\rho = 0 \qquad r > R$$

where k is a constant and r is the distance from the center of the sphere. Use Gauss' law to deduce the electric field inside and outside the sphere. Express your results in terms of k, R, and r.

7. A straight line of length L is uniformly charged. Imagine a surrounding Gaussian surface in the shape of a

concentric cylinder of radius R and length L (Figure 8). Show that if half the flux from the line charge passes through the ends of the cylinder, then

$$L = \frac{4R}{3}$$

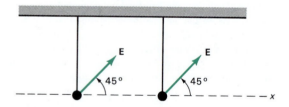

FIGURE 8

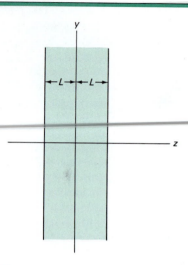

FIGURE 10

8. An electron starts from rest at $x = 0$ and is accelerated in the $+x$-direction by an electric field

$$\mathbf{E} = -bx^7\mathbf{i} \qquad b = 40 \text{ N} \cdot \text{C}^{-1} \cdot \text{m}^{-7}$$

Calculate the kinetic energy of the electron when it reaches $x = 3$ m. Express your result in keV units.

9. Two small spheres hang vertically at the ends of cords 1 m long (Figure 9). The mass of each sphere is 100 grams. The left sphere carries a charge of $-1\ \mu C$ and the right sphere carries a charge of $+1\ \mu C$. A uniform electric field is directed at $45°$ above the $+x$-axis. The tension in the right cord is zero. Determine (a) the distance between the spheres, (b) the tension in the left cord, (c) the magnitude of the electric field.

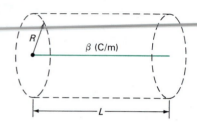

FIGURE 9

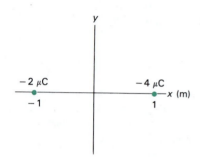

FIGURE 11

10. An infinitely long slab of thickness $2L$ is charged uniformly throughout its volume (charge density $= 10^{-3}$ C/m^3). The x–y plane is parallel to the slab faces and located midway between them (Figure 10). (a) Determine the electric field in the x–y plane. (b) Determine E at a point outside the slab a distance z from its center. (c) Determine E at a point inside the slab, a distance z from its center.

11. Negative charges ($-2\ \mu C$ and $-4\ \mu C$) are located on the x-axis at $x = -1$ m and $x = +1$ m, as shown in Figure 11. (a) Determine $\mathbf{E}$ on the x-axis at $x = -2$ m, $x = 0$, and $x = +2$ m. (b) Prove that the electric field cannot be zero for points not on the x-axis. (Exclude the point at infinity.) (c) Determine the value(s) of x for which $E = 0$.

12. A uniformly charged vertical plane with a surface charge density σ repels a point charge q of mass m that is attached to the plane by a thread of length L (Figure 12). (a) Show that the charge can be in equilibrium at an angle

$$\theta = \tan^{-1}\left(\frac{mg\epsilon_0}{\sigma q}\right)$$

with the horizontal, under the combined electrical and gravitational forces. (b) Prove that the charge can exe-

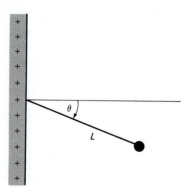

FIGURE 12

cute simple harmonic motion in a vertical plane about its equilibrium position with an angular frequency (ω) given by

$$\omega^2 = \left(\frac{1}{mL}\right)\left[\left(\frac{\sigma q}{\epsilon_0}\right)^2 + (mg)^2\right]^{1/2}$$

13. A proton in a Van de Graaff accelerator starts from rest and is accelerated by an electric field of 1.2×10^6 N/C over a distance of 4 m. (a) What is the acceleration of the proton? (b) What is the velocity of the proton at the end of the 4-m acceleration path? (c) How long does it take the proton to travel this 4-m distance?

14. An electron is ejected with a kinetic energy of 1.6×10^{-19} J from a metal surface. The metal is left with a net positive charge, e, which can be considered to be a point charge located beneath the surface directly below the electron. The positive charge is at a depth r, where r is the distance from the surface to the electron. Assume that the force between the electron and the charge e begins to act when $r = 0.5$ Å. Determine the maximum value of r.

15. A long straight wire carries a charge density of β C/m. Use Gauss' law to show that the magnitude of the electric field a distance r from the wire is

$$E = \frac{\beta}{2\pi\epsilon_0 r}$$

16. An infinitely long solid cylinder of radius R is uniformly charged throughout its volume. The total charge per unit length is β C/m. Use Gauss' law to show that the magnitude of the electric field inside the cylinder a distance r from the axis is

$$E = \left(\frac{\beta}{2\pi\epsilon_0}\right)\left(\frac{r}{R^2}\right) \qquad r \leq R$$

17. A long straight wire has a charge density of 6.21×10^{-9} C/m. Use Gauss' law to calculate the magnitude of the electric field a distance 0.182 m away from the wire.

18. A coaxial cable consists of a cylindrical inner conductor surrounded by a cylindrical conducting shell, as suggested by Figure 13. The net charge on the two conductors is zero. Assume that the charge per unit length on the central conductor is a positive constant, β. (a) Use Gauss' law to prove that the charge per unit length on the surrounding conductor is $-\beta$, and that the negative charge resides on the inner surface. (b) Calculate the electric field in the region between the conductors. Express the field as a function of the radial distance from the center of the cable.

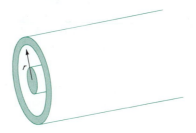

FIGURE 13

19. A pair of identical charges ($+Q$) are located on the x-axis at ($+a$, 0) and ($-a$, 0). A single charge ($-Q$) is located on the y-axis at (0, $+a$). (a) Explain why E cannot be zero for a point on the y-axis with $0 < y < a$. (b) Explain why E cannot be zero at any point on the x-axis. (c) Determine the coordinates of the two points on the y-axis where $E = 0$.

20. A "V" is formed by two uniform line charges, as shown in Figure 14. The angle between the lines is $30°$ and the length of each side is 20 cm. The charge on each side is 12 μC. Determine the electric field at the origin.

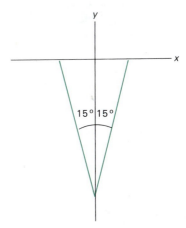

FIGURE 14

21. A photon transfers its energy of 5.78×10^{-19} J to an electron on the surface of a piece of cesium. The electron loses 3.04×10^{-19} J in escaping from the surface. The electron moves directly against a retarding electric field of 100 N/C. How far does the electron travel before its speed is reduced to zero?

22. Two equal masses (m) are suspended from a common point by threads of length L (Figure 15). The right mass has a positive charge ($+q$) and the left mass has an equal but negative charge. A uniform electric field (E) is directed horizontally to the right. In addition to the tension in the string, each mass experiences three

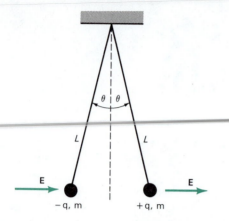

FIGURE 15

forces: the weight (mg), the constant electric field force (qE), and the attractive Coulomb force (k_eq^2/r^2). Take

$$mg = qE = 1 \text{ N} \qquad \frac{k_eq^2}{L^2} = 0.25 \text{ N} \qquad L = 1 \text{ m}$$

(a) Show that if the masses are in equilibrium, then the angle θ is determined by

$$\tan \theta = 1 - \frac{1}{16 \sin^2 \theta}$$

(b) Graph or tabulate both $\tan \theta$ and $1 - 1/(16 \sin^2 \theta)$ versus θ to locate the two solutions. (c) Repeat parts (a) and (b), but with $mg = qE = k_eq^2/L^2 = 1$ N. Show that equilibrium requires

$$\tan \theta = 1 - \frac{1}{4 \sin^2 \theta}$$

Use a graph or tabulations to show that there are no solutions, or, in other words, that equilibrium is not possible.

23. A charge q is spread uniformly around a ring of radius R. (a) Show that the electric field a distance x above the center of the ring has a magnitude

$$E = \frac{k_eqx}{(x^2 + R^2)^{3/2}}$$

(b) Find the value(s) of x for which E has its maximum value.

CHAPTER 25

ELECTRIC POTENTIAL

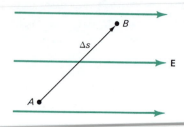

FIGURE 25.1

The test charge undergoes a displacement Δs from A to B in a uniform field **E**.

25.1
POTENTIAL DIFFERENCE

The electric field idea pictures space as being modified by the presence of electric charges. In this chapter we introduce the electric potential, another field concept. We begin by defining the potential difference for a uniform electric field.

Uniform Electric Field

Consider a uniform electric field **E**, constant in magnitude and in direction. Suppose we place a positive test charge q_0 in this field. The charge experiences a force

$$\mathbf{F} = q_0 \mathbf{E}$$

Suppose the charge undergoes a displacement Δs from point A to point B (Figure 25.1). The work performed on the charge by this force is

$$W_{AB} = \mathbf{F} \cdot \Delta \mathbf{s} = q_0 \mathbf{E} \cdot \Delta \mathbf{s} \tag{25.1}$$

In keeping with the field viewpoint we say that the electric field performs the work. Equation 25.1 shows that the work done by **E** is directly proportional to the test charge q_0. To obtain a quantity independent of q_0 and characteristic of the field we divide W_{AB} by q_0. This gives the work per unit charge, W_{AB}/q_0. The **potential difference**, $V_B - V_A$, between the points A and B is *defined* as

$$V_B - V_A = \frac{-W_{AB}}{q_0} \tag{25.2}$$

In terms of the electric field **E** and the displacement $\Delta \mathbf{s}$ the potential difference is

$$V_B - V_A = -\mathbf{E} \cdot \Delta \mathbf{s} \tag{25.3}$$

The unit of potential difference is the joule/coulomb (J/C). One joule per coulomb is called one **volt** (symbol, V).

$$1 \text{ volt} = 1 \text{ joule/coulomb}$$

The definition, Equation 25.2, shows that only *differences* in potential are measurable. This means that we are free to assign zero potential to any convenient point. For many electric circuits the zero potential point is called **ground**. In electronic circuits the zero potential point is called the **common**.

EXAMPLE 1
Potential Difference in the Atmospheric Electric Field

Electrical discharges during rainstorms leave the earth with a negative charge. This charge gives rise to an electric field near the surface of the earth. The magnitude of this atmospheric electric field is approximately 100 N/C and it is directed toward the earth's surface. Let's calculate the potential difference between two points in this field that are separated by a vertical distance of 1.7 m (Figure 25.2).

From Equation 25.3, with **E** and $\Delta \mathbf{s}$ antiparallel,

$$V_B - V_A = -\mathbf{E} \cdot \Delta \mathbf{s} = -E \, \Delta s \cos 180° = -100 \text{ N/C} \cdot 1.7 \text{ m} \, (-1)$$
$$= 170 \text{ N·m/C} = 170 \text{ J/C}$$
$$= 170 \text{ V}$$

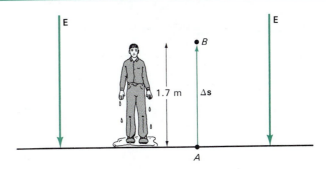

FIGURE 25.2
The electric field near the surface of the earth gives rise to a potential difference between the points A and B.

If we take the surface of the earth to be ground, then the potential 1.7 m above the surface is 170 volts.

Nonuniform Electric Field

If the field $\mathbf{E}$ varies in either magnitude or direction the work done by the field is described in differential form. The work dW done by the field when the test charge q_0 moves a distance $d\mathbf{s}$ is

$$dW = \mathbf{F} \cdot d\mathbf{s} = q_0 \mathbf{E} \cdot d\mathbf{s}$$

The total work done by the field as q_0 moves from A to B (Figure 25.3) is given by the integral of dW

$$W_{AB} = q_0 \int_A^B \mathbf{E} \cdot d\mathbf{s}$$

The potential difference between the points A and B, defined as $-W_{AB}/q_0$ in Equation 25.2, is

$$V_B - V_A = -\int_A^B \mathbf{E} \cdot d\mathbf{s} \qquad (25.4)$$

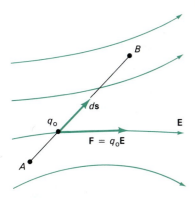

FIGURE 25.3
In a nonuniform electric field the work must be calculated by integrating $\mathbf{F} \cdot d\mathbf{s} = q_0 \mathbf{E} \cdot d\mathbf{s}$ along the path from A to B.

Potential Difference in the Field of a Point Charge

An isolated point charge q sets up an electric field

$$\mathbf{E} = \frac{1}{4\pi\epsilon_0} \frac{q}{r^2} \hat{\mathbf{r}}$$

Let's evaluate the potential difference between the points B and A located at distances b and a from the charge,

$$V(b) - V(a) = -\int_a^b \mathbf{E} \cdot d\mathbf{s}$$

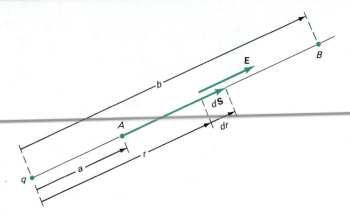

FIGURE 25.4

The integration of $\mathbf{E} \cdot d\mathbf{s}$ is carried out along a radial path from A to B.

We integrate along the radial path shown in Figure 25.4. The displacement along the path is $ds = dr$ and is parallel to $\mathbf{E}$. Thus

$$\mathbf{E} \cdot d\mathbf{s} = E\,dr = \frac{1}{4\pi\epsilon_0} \frac{q\,dr}{r^2}$$

and

$$V(b) - V(a) = -\frac{1}{4\pi\epsilon_0} q \int_a^b \frac{dr}{r^2}$$

Carrying out the integration gives an expression for the potential difference between two points in the Coulomb field of a point charge,

$$V(b) - V(a) = \frac{1}{4\pi\epsilon_0} \left(\frac{q}{b} - \frac{q}{a} \right) \tag{25.5}$$

EXAMPLE 2

Coulomb Potential Difference

To get an idea of the magnitude of potential differences at the atomic level, we calculate the potential difference between two points in the electric field of a proton. Let

$$a = 1 \text{ Å} = 10^{-10} \text{ m} \qquad b = 2 \text{ Å} = 2 \times 10^{-10} \text{ m}$$

With $q = 1.60 \times 10^{-19}$ C we get

$$V(b) - V(a) = \frac{1}{4\pi\epsilon_0} (q) \left(\frac{1}{b} - \frac{1}{a} \right)$$

$$= (8.99 \times 10^9 \text{ N·m}^2/\text{C}^2)(1.6 \times 10^{-19} \text{ C})(5 \times 10^9 - 10^{10}) \text{ m}^{-1}$$

$$= -7.2 \text{ volts}$$

The fact that the potential difference is negative indicates that point A, closer to the proton, is at a higher potential than point B.

25.2
ELECTRIC POTENTIAL

Equation 25.5 gives the potential difference between two points in the electric field of a point charge. We are free to pick a convenient position of zero potential. Because it yields the simplest expression for the potential, we assign zero potential to a position *infinitely* distant from the point charge. In Equation 25.5 we set

$$\lim_{b \to \infty} V(b) = 0$$

Equation 25.5 then gives for the potential a distance $a = r$ from a point charge q

$$V(r) = \left(\frac{1}{4\pi\epsilon_0}\right)\frac{q}{r} \tag{25.6}$$

Note the $1/r$ dependence. This potential is often referred to as the **Coulomb potential**.

Effects of Potential Difference

A potential difference of several thousand volts between you and the earth can result from combing your hair, taking off your sweater, or shuffling your feet across a rug. In the case of the latter, touching a doorknob afterward can result in an electrical discharge. The amount of charge involved is very low (fractions of microcoulombs), and you feel only a minor shock. The lethal effects of electricity come not from potential difference but from **current**, or charge in motion, passing through the body. The effect of even a small current (millicoulombs per second) passing through the heart can be lethal.

To illustrate the importance of potential difference, consider a squirrel walking along a 4400-volt power line. As long as the squirrel stays on one wire, everything is fine. The potential of the one wire is irrelevant. There is no potential difference across any part of the squirrel. But if the squirrel straddles two wires between which there is a large potential difference, the ensuing flow of charge may result in charcoal squirrel.

Principle of Superposition

Any distribution of electric charge can be viewed as a collection of point charges. In dealing with the electric field we saw that a principle of superposition applies. The electric field set up by a collection of point charges is the sum of the electric fields set up by each charge. The potential is related to the electric field by Equation 25.4. The relationship is **linear**. As a consequence, we also have a principle of superposition for potential. The potential set up by a collection of charges $q_1, q_2, q_3, \ldots$ is given by the algebraic sum of Coulomb potentials,

$$V = \frac{1}{4\pi\epsilon_0}\left(\frac{q_1}{r_1} + \frac{q_2}{r_2} + \frac{q_3}{r_3} + \cdots\right). \tag{25.7}$$

where r_i denotes the distance from the charge q_i to the point P at which the potential has the value V (Figure 25.5).

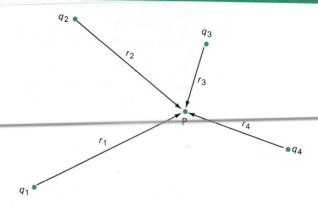

FIGURE 25.5
The potential at point *P* is the algebraic sum of the Coulomb potentials set up by each of the point charges.

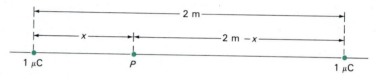

FIGURE 25.6
The principle of superposition lets us calculate the potential set up at *P* by the two 1-μC charges.

EXAMPLE 3

Potential for Two Equal Charges

In the preceding chapter we determined the electric field set up between a pair of 1 μC charges 2 meters apart (Figure 25.6). Let's determine the potential at a point between the charges. From Equation 25.7 we have

$$V = \frac{1}{4\pi\epsilon_0}\left(\frac{q_1}{r_1} + \frac{q_2}{r_2}\right)$$

$$= (8.99 \times 10^9 \text{ N·m}^2/\text{C}^2) \cdot 10^{-6} \text{ C}\left(\frac{1}{x} + \frac{1}{2-x}\right)$$

This gives *V* in volts when *x* is expressed in meters. This algebraic expression is valid at points between the charges, where *x* and 2 − *x* are positive. When *x* is close to zero or close to 2 meters the potential at that point is dominated by one of the charges. At the point midway between the charges (*x* = 1 m) the potential is

$$V = 18{,}000 \text{ volts}$$

At the midpoint each charge contributes 9,000 volts to the potential. The total potential is the sum of these two equal potentials. Note that the electric field is zero at the midpoint whereas the potential is not zero.

Electric Dipole Potential

Another system we studied in the preceding chapter was the electric dipole, a pair of charges of equal magnitude but opposite sign. Let's use the principle of superposition to determine the potential set up by an electric dipole (Figure 25.7). The potential at a point *P* in the *x*–*y* plane is

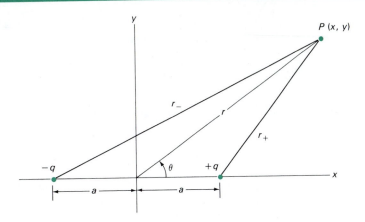

FIGURE 25.7
An electric dipole consists of a pair of point charges of equal magnitude but opposite sign.

$$V(x, y) = \frac{1}{4\pi\epsilon_0}\left(\frac{q}{r_+} - \frac{q}{r_-}\right) \tag{25.8}$$

where

$$r^2_+ = y^2 + (x - a)^2$$

and

$$r^2_- = y^2 + (x + a)^2$$

The potential is zero at all points on the y-axis, where $x = 0$ and $r_+ = r_-$. Let's examine the form of V when r_+ and r_- are both much greater than a, the dimension characteristic of the dipole. We may write without approximation

$$V = \frac{q}{4\pi\epsilon_0}\frac{(r_- - r_+)}{r_- r_+}$$

When r_+ and r_- are both large compared to a, Figure 25.8 shows that we can set

$$r_+ r_- \approx r^2$$

$$r_- - r_+ \approx 2a\cos\theta$$

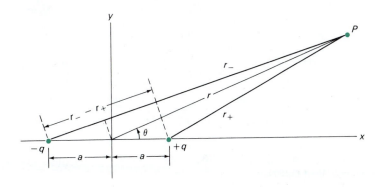

FIGURE 25.8
For r_+ and r_-, both large compared to $2a$, $r_- - r_+$ can be replaced by $2a\cos\theta$ and $r_+ r_-$ can be replaced by r^2.

where r and θ are the polar coordinates of the point P. With these replacements the potential becomes

$$V(r, \theta) = \frac{2qa \cos \theta}{4\pi\epsilon_0 r^2} \tag{25.9}$$

We note that the product $2qa$ in the numerator of Equation 25.9 is the electric dipole moment introduced in the preceding chapter

$$p = 2qa$$

An alternative form of the dipole potential is therefore

$$V(r, \theta) = \frac{p \cos \theta}{4\pi\epsilon_0 r^2} \tag{25.10}$$

EXAMPLE 4

Water Molecule Dipole Potential

A water molecule has an electric dipole moment of 6.17×10^{-30} C·m. Let's use Equation 25.10 to evaluate the potential at a distance of $10 \text{ Å} = 10^{-9}$ m from the center of the dipole, along the dipole axis ($\theta = 0$):

$$V = \frac{p}{4\pi\epsilon_0 r^2} = \frac{6.17 \times 10^{-30} \text{ C·m} \, (8.99 \times 10^9 \text{ N·m}^2/\text{C}^2)}{(10^{-9} \text{ m})^2}$$

$$= 0.0555 \text{ volt}$$

This relatively feeble potential is typical of molecules. Their range of interaction extends only a few angstroms.

In the preceding chapter we remarked that the signature of an electric dipole was an electric field that varied inversely as the cube of the distance (r^{-3}) for $r \gg a$. In terms of the potential, Equations 25.9 and 25.10 show that the dipole character is identified by an inverse square (r^{-2}) behavior of the potential for $r \gg a$.

Continuous Charge Distribution

Equation 25.7 gives the potential for a discrete collection of charges. When charges are distributed continuously, the sum in Equation 25.7 is replaced by an integral. The Coulomb potential at a point P (Figure 25.9) due to the charge dq a distance r away is given by

$$dV = \frac{1}{4\pi\epsilon_0} \frac{dq}{r^2}$$

Integrating over the charge distribution gives the potential at P

$$V = \frac{1}{4\pi\epsilon_0} \int \frac{dq}{r^2} \tag{25.11}$$

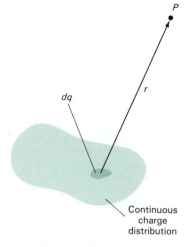

FIGURE 25.9

The potential at the point P is the superposition of the Coulomb potentials set up by the charge elements that comprise the entire charge distribution.

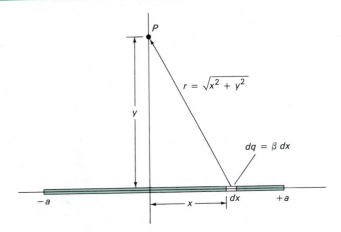

FIGURE 25.10
The potential at the point P on the y-axis is obtained by integrating the Coulomb potentials set up by each charge element.

EXAMPLE 5

Potential of a Line of Charge

Consider a charge q distributed uniformly along the x-axis from $x = -a$ to $x = +a$. We want to use the principle of superposition (Equation 25.11) to determine the potential at a point on the y-axis, a distance y from the origin (Figure 25.10). To carry out the integration in Equation 25.11 we note that the charge per unit length is

$$\beta = \frac{q}{2a}$$

The charge on a segment of length dx is $dq = \beta \, dx$. Furthermore,

$$r = \sqrt{x^2 + y^2}$$

The integration in Equation 25.11 ranges from $x = -a$ to $x = +a$. The explicit form of the integral for $V(y)$ is

$$V(y) = \frac{1}{4\pi\epsilon_0}\beta \int_{-a}^{+a} dx \, (x^2 + y^2)^{-1/2}$$

Carrying out the integration and using $\beta = q/2a$ gives

$$V(y) = \frac{1}{4\pi\epsilon_0}\left(\frac{q}{a}\right)\ln\left(\frac{a}{y} + \sqrt{1 + \left(\frac{a}{y}\right)^2}\right)$$

One way to check this result is to consider a limiting case. At a point far from the origin ($y \gg a$) the line charge appears pointlike and we expect that $V \approx q/4\pi\epsilon_0 y$. When $(a/y)^2 \ll 1$ the logarithmic expression may be approximated by

$$\ln\left(\frac{a}{y} + \sqrt{1 + \left(\frac{a}{y}\right)^2}\right) \approx \ln\left(1 + \frac{a}{y}\right) \approx \frac{a}{y}$$

and the potential approaches the point-charge limit,

$$V(y) \approx \frac{1}{4\pi\epsilon_0}\frac{q}{y} \qquad y \gg a$$

This confirms our expectation: at large distances from the line of charge, it appears pointlike and V reduces to the potential of a point charge.

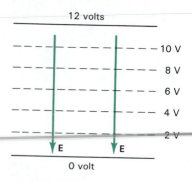

FIGURE 25.11

Side view of two parallel metal plates. Dashed lines indicate equipotential surfaces between the plates.

25.3
EQUIPOTENTIAL SURFACES

Any surface, planar or curved, over which the potential is constant is called an **equipotential surface**. The equipotential surface may coincide with a physical surface or it may simply be a geometrical surface. Figure 25.11 shows a pair of parallel metal plates separated by one centimeter. The top plate is an equipotential surface with $V = 12$ volts. The bottom plate is an equipotential surface with $V = 0$ volts. The uniform electric field between the plates has a magnitude of 12 V/cm. The equipotential surfaces between the two plates are planes, parallel to the two plates. The potential changes linearly between the plates. For example, the equipotential surface with $V = 6$ volts is the plane midway between the two plates.

Because the potential is constant on an equipotential surface, the change in potential between any two points on the surface is zero. The general relation between potential difference and field is

$$\Delta V = -\mathbf{E} \cdot \Delta \mathbf{s}$$

If $\Delta \mathbf{s}$ is a displacement between two adjacent points on the equipotential surface, $\Delta V = 0$ and

$$\mathbf{E} \cdot \Delta \mathbf{s} = 0$$

The scalar product of two vectors is zero when the vectors are perpendicular. Since $\Delta \mathbf{s}$ lies on the equipotential surface, the electric field $\mathbf{E}$ is perpendicular to the equipotential surface. This is a general result:

The electric field is perpendicular to the equipotential surfaces.

Furthermore, the electric field is directed from higher potential to lower potential. Figure 25.11 illustrates this relationship for planar equipotential surfaces.

For a point charge q the Coulomb potential is

$$V(r) = \frac{1}{4\pi\epsilon_0} \frac{q}{r}$$

The equipotential surfaces are spheres ($r = $ constant) centered on the charge. The electric field is radial, perpendicular to the spherical equipotentials, and directed from higher potential to lower potential (Figure 25.12).

More complicated charge systems will have more complicated equipotential surfaces. Knowing that the electric field lines are perpendicular to the equipotentials helps you envision them. Figure 25.13 shows the field lines and equipotential surfaces for an electric dipole.

The surface of any isolated conductor is an equipotential surface. For example, the surface of a conducting sphere of radius R carrying a charge q is at a potential

$$V = \frac{1}{4\pi\epsilon_0} \frac{q}{R}$$

The magnitude of the electric field at the surface of the conducting sphere is

$$E = \frac{1}{4\pi\epsilon_0} \frac{q}{R^2}$$

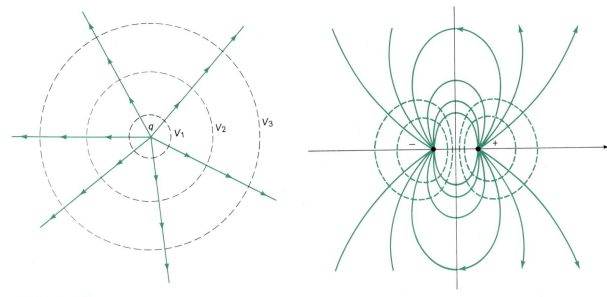

FIGURE 25.12

Equipotential surfaces and electric field lines for a point charge. Dashed lines are cross-sections of the spherical equipotential surfaces. Solid lines indicate the electric field pattern.

FIGURE 25.13

Equipotential surfaces and electric field lines for an electric dipole. Dashed lines are cross-sections of equipotential surfaces. Solid lines show the electric field pattern.

Thus, for the conducting sphere the surface values of E and V are related by

$$E = \frac{V}{R} \qquad \text{(conducting sphere)} \qquad (25.12)$$

This relationship between E and V will help us understand the phenomenon known as corona discharge.

Corona Discharge

When a conductor in air has such a large surface charge that the electric field at the surface exceeds 3×10^6 V/m, the oxygen and nitrogen molecules in the air will be ionized and an electric discharge will occur. This is a **corona discharge**. St. Elmo's fire is a harmless and beautiful form of corona discharge. However, corona discharges in electrical equipment may do physical damage and are to be avoided. Figure 25.14 shows a conducting surface that is blunt at one end and has a sharp tip at the other. Qualitatively, we can regard one end as a portion of a sphere with a radius R_1 and the other end as a portion

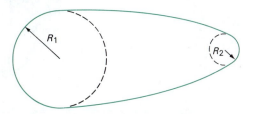

FIGURE 25.14

The electric field at the sharp end of the charged conductor is much greater than the electric field at the blunt end.

of a sphere with a radius R_2. The blunt end radius R_1 is much larger than the sharp tip radius R_2. Because the surface is an equipotential we can use Equation 25.12 to relate the fields at the two ends

$$E_1 = \frac{V}{R_1} \qquad E_2 = \frac{V}{R_2}$$

Evidently, the small radius at the sharp end produces a much larger electric field than the larger radius at the blunt end:

$$E_2 \gg E_1 \qquad R_2 \ll R_1$$

Corona discharges are prone to occur at sharp points, where the electric field reaches the 3×10^6 V/m value needed to ionize the surrounding air molecules.

The sharp tip of a lightning rod will set up a strong electric field at relatively low voltage. By ionizing the molecules of the air, the lightning rod transfers charge to the atmosphere and thereby lowers the potential difference between the building it is protecting and a charged cloud overhead. The function of the lightning rod is to reduce a dangerously high potential difference; it is not meant to attract lightning. Airplanes often have similar devices to eliminate excess charge.

If the electric field at the surface of a conductor is high enough, it can pull electrons out from the surface of the conductor. This phenomenon is called **field emission**. The electric field required for field emission to occur depends on the nature of the substance and the temperature. For a substance like tungsten at room temperature, the required field is a few million volts per centimeter, which is roughly a hundred times the electric field needed to ionize the molecules of oxygen and nitrogen in the air. Such high fields can be achieved in a vacuum at pointed surfaces. Field emission has been used in microscopes to study the nature of surfaces. By using magnification of over 1 million, we can see the outlines of heavy atoms on a pointed emitting surface (Figure 25.15).

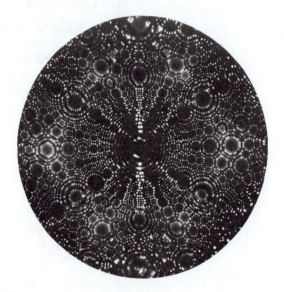

FIGURE 25.15
Field emission picture of atoms.

25.4
THE DERIVATIVE RELATION BETWEEN **E** AND V

The potential difference between two points is expressed as an integral of the electric field by the relation

$$V_B - V_A = -\int_A^B \mathbf{E} \cdot d\mathbf{s} \quad .$$

The fact that the change in V is obtained by integrating **E** suggests that **E** can be determined by differentiating V. Let's pursue this idea. The differential form for the potential difference is

$$dV = -\mathbf{E} \cdot d\mathbf{s} \qquad (25.13)$$

In a uniform electric field we can choose a coordinate system in which the x-axis is directed parallel to the electric field lines (Figure 25.16). The equipotentials will be perpendicular to this x-axis. We can write

$$\mathbf{E} = \mathbf{i}E_x$$

$$d\mathbf{s} = \mathbf{i}\, dx$$

whereupon Equation 25.13 becomes

$$dV = -E_x\, dx$$

This result shows that we can determine E_x by differentiating V:

$$E_x = -\frac{dV}{dx} \qquad (25.14)$$

In general, V depends on all three coordinates, and the y- and z-components of **E** are not zero. In this case the derivative in Equation 25.14 becomes a partial derivative:

$$E_x = -\frac{\partial V}{\partial x} \qquad (25.15)$$

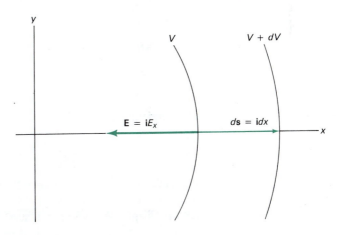

FIGURE 25.16
The x-axis can be chosen to lie along the direction of **E**.

We can also infer that E_y and E_z are likewise expressed as partial derivatives of V,

$$E_y = -\frac{\partial V}{\partial y} \tag{25.16}$$

$$E_z = -\frac{\partial V}{\partial z} \tag{25.17}$$

EXAMPLE 6

Electric Field of a Line of Charge

In Example 5 we determined the potential on the y-axis for a charge q distributed uniformly along the x-axis between $x = -a$ and $x = +a$ (Figure 25.10):

$$V(y) = \frac{1}{4\pi\epsilon_0}\left(\frac{q}{a}\right)\ln\left(\frac{a}{y} + \sqrt{1 + \left(\frac{a}{y}\right)^2}\right)$$

The electric field is in the y-direction and is given by

$$E_y = -\frac{\partial V}{\partial y}$$

Carrying out the differentiation gives

$$E_y = \frac{1}{4\pi\epsilon_0}\left(\frac{q}{y\sqrt{y^2 + a^2}}\right)$$

Observe that for $y \gg a$, the electric field approaches the point-charge field

$$E_y \approx \frac{1}{4\pi\epsilon_0}\frac{q}{y^2} \qquad y \gg a$$

The electric field of a point charge provides another illustration of the derivative relation between **E** and V. The Coulomb potential for a point charge q is

$$V = \frac{1}{4\pi\epsilon_0}\frac{q}{r}$$

The electric field is radial and, by analogy with Equation 25.15, is given by

$$E = -\frac{\partial V}{\partial r} \tag{25.18}$$

Differentiating we obtain the Coulomb electric field of a point charge

$$E = \frac{1}{4\pi\epsilon_0}\frac{q}{r^2}$$

We have used the derivative relation to evaluate the electric field. Alternatively, if we know the electric field, then we can integrate and find the potential. In Chapter 24 we used Gauss' law to determine the radial electric field set up by a charge Q spread uniformly throughout the volume of a sphere of radius R. We found

$$E = \frac{1}{4\pi\epsilon_0}\frac{Q}{r^2} \qquad r \leq R$$

$$E = \frac{1}{4\pi\epsilon_0}\frac{Qr}{R^3} \qquad r \geq R$$

From the derivative relation $E = -\partial V/\partial r$ we can write

$$V = -\int E\,dr$$

Carrying out the integrations gives

$$V = \frac{1}{4\pi\epsilon_0}\frac{Q}{r} + A \qquad\qquad r \geq R$$

$$V = \frac{1}{4\pi\epsilon_0}\left(\frac{Q}{R^3}\right)\left(B - \frac{r^2}{2}\right) \qquad r \leq R$$

where A and B are integration constants. In order for V to approach zero as r approaches infinity we must choose $A = 0$. The constant B is determined by requiring that both equations give the same value of V at $r = R$. This condition yields

$$B = \frac{3R^2}{2}$$

Thus we obtain for the potential

$$V = \frac{1}{4\pi\epsilon_0}\frac{Q}{r} \qquad\qquad r \geq R \qquad\qquad (25.19)$$

$$V = \frac{1}{4\pi\epsilon_0}\left(\frac{Q}{R}\right)\left(\frac{3}{2} - \frac{r^2}{2R^2}\right) \qquad r \leq R \qquad (25.20)$$

The electric field at the center of the sphere is zero. The potential is not zero at the center—it is one and one-half times as great as the potential at the surface of the sphere.

EXAMPLE 7

Electric Potential of a Lead Nucleus

The uniformly charged sphere gives a reasonable model of the proton charge distribution in an atomic nucleus. A nucleus of the lead isotope, lead 208, has a radius of 6.34×10^{-15} m and contains 82 protons, each with a charge of 1.60×10^{-19} C. Let's calculate the potential at the surface of this nucleus, assuming that the charge is distributed uniformly throughout a sphere of radius 6.34×10^{-15} m.

The total charge in the nucleus is $Q = 82(1.60 \times 10^{-19})$ C $= 1.31 \times 10^{-18}$ C. Hence from Equation 25.19 the potential at the surface $(r = R)$ is

$$V_{\text{surface}} = \frac{1}{4\pi\epsilon_0}\frac{Q}{R}$$

$$= (8.99 \times 10^9 \text{ N·m}^2/\text{C}^2)\cdot\frac{1.31 \times 10^{-18} \text{ C}}{6.34 \times 10^{-15} \text{ m}}$$

$$= 1.86 \times 10^6 \text{ V}$$

At the center of the nucleus, Equation 25.20 shows that the potential is higher than it is at the surface by a factor of 3/2. Hence

$$V_{\text{center}} = \frac{3}{2}(1.86 \times 10^6 \text{ V})$$

$$= 2.79 \times 10^6 \text{ V}$$

These large potentials reflect the small radii characteristic of atomic nuclei.

25.5
ELECTRIC POTENTIAL ENERGY

The field concepts of electric field and electric potential are useful when we want to describe how a set of electric charges modifies the surrounding space. However, when we want to study the **interaction** of electric charges, the concepts of force and potential energy are more appropriate. For example, an atom is composed of electric charges, a positively charged nucleus and one or more negatively charged electrons. These charges interact via electrical forces. The electrical potential energy of these charges is a more convenient variable than the electric potential. We can introduce the electrical potential energy most simply by showing how it is related to the electric potential.

Consider a charge q in an electric field **E**. The charge experiences a force **F** $= q$**E**. The electric potential energy is related to the work performed by this force. The difference in potential energy between two points A and B is defined by

$$U_B - U_A = -\int_A^B \mathbf{F} \cdot d\mathbf{s} = -q \int_A^B \mathbf{E} \cdot d\mathbf{s}$$

(See Chapter 8.) Comparing this with Equation 25.4 for the potential difference we see that

$$U_B - U_A = q(V_B - V_A)$$

We will associate the zero of electric potential energy ($U = 0$) with the zero of electric potential ($V = 0$). It follows that U and V are related by

$$U = qV \tag{25.21}$$

We learned how to determine V in Section 25.2. We can use $U = qV$ to determine the electric potential energy of a collection of charges.

Potential Energy of Point Charges

Consider a collection of N point charges $q_1, q_2, q_3, \ldots$. We assemble the collection by bringing each charge from infinity to its position in the collection. Let the symbol r_{ij} denote the distance between the charges q_i and q_j in the assembled collection (Figure 25.17). The first charge we bring in (q_1) sets up a Coulomb field. The potential at a point a distance r from q_1 is

$$V_1 = \frac{1}{4\pi\epsilon_0} \frac{q_1}{r}$$

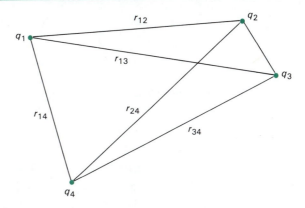

FIGURE 25.17

A collection of 4 point charges. The Coulomb potential energy of this collection is given by Equation 25.22 with $N = 4$.

Next, we bring in q_2 and position it a distance r_{12} from q_1. The potential energy of this 2-charge collection follows from Equation 25.21

$$U_{21} = q_2 V_1 = \frac{1}{4\pi\epsilon_0} \frac{q_1 q_2}{r_{12}}$$

When q_3 is added to the collection it experiences a potential set up by q_1 and q_2. The potential obeys a principle of superposition. The potential experienced by q_3 is the algebraic sum of the potentials set up by q_1 and q_2. The potential energy change brought about by adding q_3 is

$$U_{31} + U_{32} = q_3 V_1 + q_3 V_2$$
$$= \left(\frac{1}{4\pi\epsilon_0}\right)\left(\frac{q_3 q_1}{r_{31}} + \frac{q_3 q_2}{r_{32}}\right)$$

Each time we add another charge to the collection it interacts with all of the charges already present. The total potential energy of the N-charge collection can be expressed in the form of an array

$$\begin{aligned} U = U_{21} & \\ + U_{31} + U_{32} & \\ + U_{41} + U_{42} + U_{43} & \\ + \cdots & \\ + \cdots & \\ + U_{N1} + U_{N2} + \cdots + U_{NN-1} & \end{aligned} \qquad (25.22)$$

where

$$U_{ij} = \frac{1}{4\pi\epsilon_0} \frac{q_i q_j}{r_{ij}} \qquad (25.23)$$

denotes the mutual potential energy of the charges q_i and q_j. To illustrate these ideas we evaluate the potential energy of a helium atom.

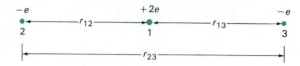

FIGURE 25.18

A model of the helium atom in which the electrons are diametrically opposite one another.

EXAMPLE 8

Helium Atom Potential Energy

The helium atom consists of two electrons and a nucleus. We want to evaluate the potential energy of the helium atom shown in Figure 25.18. The three charges shown in Figure 25.18 are

$$q_1 = 2e = 3.2 \times 10^{-19} \text{ C} \quad \text{(nucleus)}$$

$$q_2 = q_3 = -e = -1.60 \times 10^{-19} \text{ C} \quad \text{(electrons)}$$

The Coulomb potential energy is a sum of three terms. Two describe the interaction of the nucleus with the electrons. The third describes the electron-electron interaction. From Equation 25.22 we have

$$U = \frac{1}{4\pi\epsilon_0} \left(\frac{-2e^2}{r_{12}} - \frac{2e^2}{r_{13}} + \frac{e^2}{r_{23}} \right)$$

We take the electron-to-nucleus distance to be 2.64×10^{-11} m. With

$$r_{12} = r_{13} = 2.64 \times 10^{-11} \text{ m}$$

$$r_{23} = 5.28 \times 10^{-11} \text{ m}$$

we get for the potential energy

$$U = -3.05 \times 10^{-17} \text{ J}$$

This energy can be expressed in a more convenient unit called the **electron volt** (abbreviated eV):

$$1 \text{ eV} \equiv 1.60219 \times 10^{-19} \text{ J}$$

The name "electron volt" stems from the fact that one electron volt equals the work performed when an electron moves through a potential difference of one volt. In electron volt units the helium atom potential energy is

$$U = -3.05 \times 10^{-17} \text{ J} \cdot \left(\frac{1 \text{ eV}}{1.60 \times 10^{-19} \text{ J}} \right)$$

$$= -191 \text{ eV}$$

The fact that U is negative is a consequence of the opposite signs of the electron and nuclear charges.

Energy Conservation

The work-energy principle states that the work performed on a system equals its change in kinetic energy. If an electric force performs work W_{AB} on a charge q as it moves from A to B this principle states

$$K_B - K_A = W_{AB}$$

where K_A and K_B denote the kinetic energies at A and B. Our definition of potential difference (Equation 25.2) relates W_{AB} to the potential difference between points A and B:

$$W_{AB} = q(V_A - V_B)$$

The work-energy principle can be rewritten as

$$K_B + qV_B = K_A + qV_A \qquad (25.24)$$

showing that the quantity $K + qV$ is *conserved*. The quantity $U = qV$ is the electric potential energy of the charge q at a point where the potential is V. We can rewrite Equation 25.24 as

$$K_B + U_B = K_A + U_A \qquad (25.25)$$

showing that the sum of the kinetic energy and the electric potential energy is conserved.

EXAMPLE 9

Speed of a 1-eV Electron

An electron starts from rest and is accelerated through a potential difference of 1 volt. The work performed is 1 eV and so the electron acquires a kinetic energy of 1 eV. We want to find its speed. We use Equation 25.24 with $K_A = 0$, $V_B - V_A = 1$ volt, to get

$$K_B = q(V_A - V_B)$$

$$\tfrac{1}{2}mv_B{}^2 = (-e)(-1\text{ V})$$

This gives

$$v_B = \left[\frac{2(1.60 \times 10^{-19}\text{ C})(1\text{ volt})}{9.11 \times 10^{-31}\text{ kg}} \right]^{1/2}$$

$$= 5.93 \times 10^5\text{ m/s}$$

For comparison, this is about 20 times the speed of the earth around the sun.

Electric Dipole in a Uniform Electric Field

As a final example of electric potential energy, let's evaluate the potential energy of an electric dipole located in a uniform electric field. Recall that an electric dipole consists of two charges, $+q$ and $-q$, separated by a distance $2a$ (Figure 25.19). The dipole is positioned in a uniform electric field that is directed along the positive y-axis,

$$\mathbf{E} = \mathbf{j}E$$

The electric potential $V(y)$ for this field is

$$V(y) = -yE \qquad (25.26)$$

You can check this by using the derivative relation between field and potential, $E_y = -\partial V/\partial y$. The potential energy of the dipole is the sum of the potential energies of the two charges. With the two charges located at y_+ and y_- the potential energy is

$$U = qV(y_+) - qV(y_-) = -qE(y_+ - y_-)$$

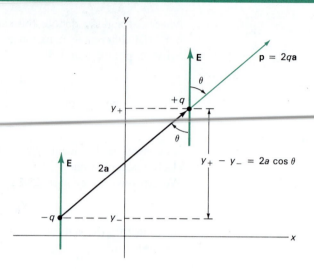

FIGURE 25.19

An electric dipole in an external electric field experiences a torque given by $\mathbf{p} \times \mathbf{E}$. The potential energy of the dipole is $-\mathbf{p} \cdot \mathbf{E} = -2qaE \cos \theta$.

From Figure 25.19 we see that

$$y_+ - y_- = 2a \cos \theta$$

Thus, the dipole potential energy is

$$U = -2qaE \cos \theta \qquad (25.27)$$

Notice that

$$p = 2qa$$

is the electric dipole moment. We define a vector electric dipole moment by

$$\mathbf{p} = 2q\mathbf{a} \qquad (25.28)$$

The vector $\mathbf{p}$ is directed from $-q$ to $+q$. The potential energy of the dipole can be expressed in terms of the scalar product of $\mathbf{p}$ and $\mathbf{E}$,

$$U = -\mathbf{p} \cdot \mathbf{E} = -pE \cos \theta \qquad (25.29)$$

where θ is the angle between $\mathbf{p}$ and $\mathbf{E}$.

EXAMPLE 10

Potential Energy of a Water Molecule in the Atmospheric Electric Field

In Example 1 we considered the atmospheric electric field near the surface of the earth. The magnitude of this field is approximately 100 N/C. The electric dipole moment of the water molecule is 6.17×10^{-30} C·m. Water molecules in the earth's atmosphere have potential energies given by Equation 25.29. The angle θ can range from $0°$ to $180°$. The corresponding values of U run from $-pE$ to $+pE$. Let's evaluate the range of U values.

$$U_{max} - U_{min} = 2pE = 2(6.17 \times 10^{-30} \text{ C·m})(100 \text{ N/C})$$
$$= 1.23 \times 10^{-27} \text{ J}$$

In electron volt units

$$U_{max} - U_{min} = 1.23 \times 10^{-27} \text{ J} \left(\frac{1 \text{ eV}}{1.60 \times 10^{-19} \text{ J}} \right)$$

$$= 7.7 \times 10^{-9} \text{ eV}$$

By comparison, the Coulomb potential energy of the molecule is several electron volts, roughly 10^9 times larger than the dipole energy. In short, the potential energy of the molecule's dipole moment is a very tiny fraction of the total molecular energy.

In a uniform electric field there is no net force on a dipole because the two charges experience equal but opposite forces, $-q\mathbf{E}$ and $+q\mathbf{E}$. In general, however, a net *torque* acts on the dipole. This torque tends to rotate the dipole until its dipole moment is aligned with the electric field. In Figure 25.19 we choose a torque reference point at the position of $-q$. The net torque is then

$$\boldsymbol{\tau} = \mathbf{r} \times \mathbf{F} = 2\mathbf{a} \times q\mathbf{E}$$

In terms of the vector electric dipole moment $\mathbf{p} = 2q\mathbf{a}$,

$$\boldsymbol{\tau} = \mathbf{p} \times \mathbf{E} \qquad (25.30)$$

In terms of the angle θ between $\mathbf{p}$ and $\mathbf{E}$,

$$\tau = pE \sin \theta \qquad (25.31)$$

If the dipole is free to rotate, this torque acts to align $\mathbf{p}$ and $\mathbf{E}$. With $\mathbf{p}$ parallel to $\mathbf{E}$ the torque is zero and the potential energy $-\mathbf{p} \cdot \mathbf{E}$ takes on its minimum value, $-pE$.

Torque: The Potential Energy Relation

There is a general relationship between torque and potential energy that we can infer by considering the electric dipole. We have just seen that an electric dipole in an external field experiences a torque given by

$$\tau = pE \sin \theta \qquad (25.31)$$

The potential energy of the dipole is

$$U = -pE \cos \theta \qquad (25.29)$$

We observe that*

$$\tau = -\frac{\partial U}{\partial \theta} \qquad (25.32)$$

This is not a coincidence; it is a general relation between torque and potential energy. In fact, this derivative relation between torque and potential energy is the angular analog of the derivative relation between force and potential energy that you first studied in Chapter 8.

$$F_x = -\frac{\partial U}{\partial x}$$

---

* Equation 25.32 gives $\tau = -pE \sin \theta$. The minus sign indicates that a positive torque corresponds to a negative value of $\sin \theta$.

WORKED PROBLEM

Three identical charges lie at the vertices of an equilateral triangle having sides 2 m long (Figure 25.20). Locate positions of electrostatic equilibrium within the triangle.

Solution

Electrostatic equilibrium positions are points where the electric field is zero. Symmetry suggests that the equilibrium positions lie on lines that pass through the charges and that bisect the opposite side. As Figure 25.20 shows, these three lines intersect at the point $(2 \text{ m})/\sqrt{3} = 1.155$ m from each vertex. This intersection is one position where the electric field is zero. But, there are three other positions where the electric field is zero—one along each of the three lines. We can figure out their locations by applying the condition

$$E_x = -\frac{\partial V}{\partial x} = 0$$

The potential at a point P on the x-axis is given by

$$V = \frac{q}{4\pi\epsilon_0}\left\{\frac{2}{r} + \frac{1}{\sqrt{3} - x}\right\}$$

where r and x are expressed in meters and are related by

$$r^2 = 1 + x^2$$

Setting

$$\frac{\partial V}{\partial x} = \frac{q}{4\pi\epsilon_0}\left\{\left(\frac{-2}{r^2}\right)\frac{\partial r}{\partial x} + \frac{1}{(\sqrt{3} - x)^2}\right\} = 0$$

and using $\partial r/\partial x = x/r$ gives the equation that locates equilibrium positions on the x-axis

$$\frac{2x}{r^3} = \frac{1}{(\sqrt{3} - x)^2}$$

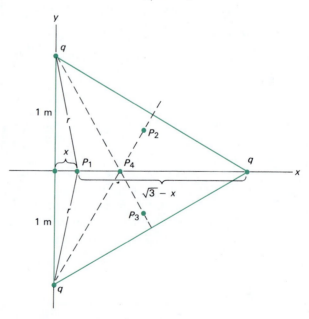

FIGURE 25.20

This can be rewritten as

$$f(x) \equiv 2x(\sqrt{3} - x)^2 - (1 + x^2)^{3/2} = 0$$

This equation is satisfied by two values of x corresponding to positions inside the triangle. To five-significant-figure accuracy, the solutions are

$$x = 0.57735 \text{ m} \quad \text{and} \quad x = 0.24859 \text{ m}$$

The solution $x = 0.57735$ m corresponds to the intersection of the three bisectors. The four equilibrium positions are identified in Figure 25.20 by P_1, P_2, P_3, and P_4.

One technique for finding roots is Newton's method. You can probably find a complete account of Newton's method in your calculus book. This method shows that if x is an approximate root of the equation

$$f(x) = 0$$

then a closer approximation is

$$x_1 = x - \frac{f(x)}{\dfrac{df}{dx}}$$

For the triangle problem

$$\frac{df}{dx} = 2(\sqrt{3} - x)^2 - 4x(\sqrt{3} - x) - 3x(1 + x^2)^{1/2}$$

If you have access to a microcomputer, try running the following BASIC program. It will locate the roots (one at a time), unless you accidentally make an initial guess for x that is near a zero, of df/dx.

```
10      REM ** ROOT3
20      INPUT "ENTER X (0 < X < 1.73205) "; X
30      X1=X−(2*X*(1.73205−X)^2−(1+X*X)*SQR(1+X*X))/
        (2*(1.73205−X)^2−4*X*(1.73205−X)−3*X*SQR(1+X*X))
40      IF ABS((X1−X)/X)<0.0001 GOTO 70
50      PRINT "NEW GUESS X = ";X1:PRINT
60      X=X1:GOTO 30
70      PRINT "ROOT = ";X1:END
```

EXERCISES

25.1 Potential Difference

A. A charge of 0.2 μC moves from the coordinate origin to the point (4, 4), with coordinates measured in meters (Figure 1). Calculate the work done by the uniform electric field $\mathbf{E} = 10^4\,\mathbf{i}$ N/C for each of the following three paths: (a) $(0, 0) \rightarrow (4, 0) \rightarrow (4, 4)$, (b) $(0, 0) \rightarrow (4, 4)$ along the diagonal $x = y$, (c) $(0, 0) \rightarrow (0, 4) \rightarrow (4, 4)$. (d) What is the potential difference $V(4, 4) - V(0, 0)$?

B. A potential difference of 10,000 V is established between two parallel metal plates, creating a uniform electric field between them. (a) Calculate the strength of the electric field if the plates are 10 cm apart. (b) When the electric field strength rises to 30,000 V/cm, the oxygen and nitrogen molecules of the air between

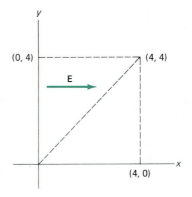

FIGURE 1

the plates are ionized and electrical breakdown occurs. What plate separation will give rise to this critical field?

C. Two parallel metal plates (Figure 2) are charged. The lower plate has a charge density $\sigma = +1.2\ \mu C/m^2$, and the upper plate has a charge density $\sigma = -1.2\ \mu C/m^2$. (a) Determine the electric field between the plates. (b) If the potential difference between the plates is 440 V, what is the plate separation?

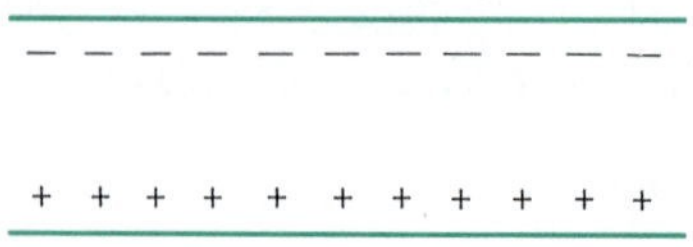

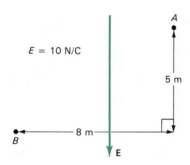

FIGURE 2

D. A uniform electric field of magnitude 10 N/C is directed vertically downward as shown in Figure 3. Determine the potential difference $V_B - V_A$ between the points A and B.

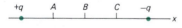

FIGURE 3

25.2   Electric Potential

E. Point charges $+q$ and $-q$ are located on the x-axis as drawn in Figure 4. Three points (A, B, C) are indicated on the line joining the two charges. If $V_B = 0$, and V_A is 12 V higher than V_B and 24 V higher than V_C, determine V_A and V_C.

FIGURE 4

F. Calculate the potential at the center of a uniform ring of charge with radius R and charge Q.

G. A molecule has an electric dipole moment $p = 10^{-30}$ C·m. Calculate the potential at a distance $r = 5.0$ nm from the molecule at angles (a) $\theta = 0$, (b) $\theta = 45°$, (c) $\theta = 90°$. (θ is the angle measured relative to the dipole axis.)

H. At a distance r from a particular isolated point charge Q the electric field is 1250 N/C. The potential is

5000 V. (a) Determine the distance r. (b) What is the magnitude of the point charge Q?

I. A point charge of 0.23 μC is at the origin (Figure 5). Calculate the potential difference between the points (a) $x = 4$ m, $y = 0$, $z = 0$ and $x = 2$ m, $y = 0$, $z = 0$, (b) $y = 4$ m, $x = 0$, $z = 0$ and $z = 2$ m, $x = 0$, $y = 0$.

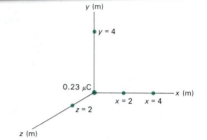

FIGURE 5

J. A $-1\ \mu C$ charge is located on the x-axis at $x = 1$ m and a $+2\ \mu C$ charge is located on the x-axis at $x = -2$ m. (a) Make a sketch of the electric potential for positions on the y-axis, (b) show that the electric potential is a maximum near $y = \pm 2$ m.

25.3   Equipotential Surfaces

K. The equipotential surfaces of an isolated point charge are concentric spheres. For an isolated point charge of 3.0×10^{-8} C, calculate the radius of the 100-V equipotential surface.

L. A long straight wire carries a charge per unit length of 10^{-8} C/m (Figure 6). A coaxial cylindrical equipotential surface, radius 1.0 m, has a potential of 100 V. (a) What is the radius of the 1000-V equipotential surface? (b) What is the potential of the coaxial equipotential surface of radius 2.0 m?

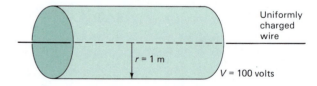

FIGURE 6

M. A cube, 1 m on a side, is centered on a $+50\ \mu C$ charge. How much work is done by the electric field set up by this charge when a $+2.0\ \mu C$ charge moves between any two corners of the cube?

25.4   The Derivative Relation Between E and V

N. The potential in a region between $x = 0$ and $x = 6$ m is given by:

$$V = a + bx \text{ volts}$$

where $a = 10$ V and $b = -7$ V/m. (a) Determine the potential at $x = 0$, 3 m, and 6 m. (b) Determine the electric field at $x = 0$, 3 m, and 6 m. Indicate the direction of E.

P. A charge Q is spread uniformly around a ring of radius a lying in the y–z plane (Figure 7). The potential on the x-axis is given by

$$V = \frac{1}{4\pi\epsilon_0} \frac{Q}{(x^2 + a^2)^{1/2}}$$

Determine the electric field on the x-axis.

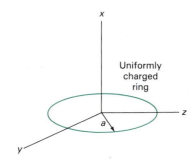

Uniformly charged ring

FIGURE 7

25.5 Electric Potential Energy

Q. A proton is fixed in place. A second proton is brought to within a distance r of the first proton and then released. When the potential energy of the protons is converted entirely into kinetic energy (at infinite separation), what speed will the second proton have for (a) $r = 1.0$ m, (b) $r = 3.0$ fm?

R. In a tandem Van de Graaff accelerator a proton is accelerated through a potential difference of 14×10^6 V. Assuming that the proton starts from rest, calculate its (a) final kinetic energy in joules, (b) final kinetic energy in MeV, (c) final speed.

S. Given two 2-μC charges, as shown in Figure 8, and a positive test charge $q = 1.28 \times 10^{-18}$ C at the origin, (a) what is the net force exerted on q by the two 2-μC charges? (b) What field E do the two 2-μC charges produce at the origin? (c) What is the potential V produced by the two 2-μC charges at the origin?

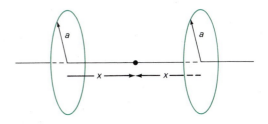

FIGURE 8

T. An electron is released from the negative plate of two parallel plates that are 2.0 cm apart. The electron is accelerated by a uniform electric field between the plates and strikes the positive plate with a speed of 3.0×10^6 m/s. (a) What is the potential difference between the plates? (b) What is the magnitude of the electric field?

PROBLEMS

1. Point charges ($+1$ μC and $+2$ μC) are located at $(-1, 0)$ and $(1, 0)$, respectively, with (x, y) coordinates specified in meters. (a) Calculate the potential at the points $(0, 2)$, $(0, 3)$, $(0, 1)$, $(-1, 2)$, and $(1, 2)$. (b) Using these potentials, estimate the electric field at the point $(0, 2)$.

2. A uniformly charged cylinder (radius $= a$, charge per unit volume $= \rho$ with $\rho > 0$) has its surface at zero potential. A negative point charge is launched outward at the surface with enough speed to move from $r = a$ to $r = 2a$ before coming to rest. The charge then retraces its path and passes through the axis of the cylinder to emerge on the opposite side. Show that the ratio of the launch speed to the speed at $r = 0$ is 0.762.

3. A negative point charge $(-q)$ is inside a uniformly charged cylinder (radius $= a$, volume charge density $= \rho$ with $\rho > 0$). (a) Sketch the potential energy of the point charge versus r, where r is the radial distance from the cylinder axis, for $r < a$. (b) Show that the point charge will execute simple harmonic motion in the radial direction with an angular frequency

$$\omega = \sqrt{q\rho/2\epsilon_0 m}$$

4. A uniformly charged cylinder (radius $= a$, volume charge density $= \rho$ with $\rho > 0$) has its surface at zero potential. A positive point charge at $r = 2a$ moves toward the axis of the cylinder, coming to rest at $r = \frac{1}{2}a$. Determine the initial speed of the charge.

5. Two charged rings having the same radius and the same linear charge density are coaxially mounted (Figure 9). Show that the potential at the center of either ring equals the potential on the axis halfway between them if their separation is 1.806 times their radius.

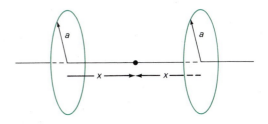

FIGURE 9

6. A spherical conductor of radius R carries a uniformly distributed charge $-Q$. A proton moving radially out

from the surface of the conductor is retarded by the electric field of the conductor, but reaches an arbitrarily large distance before its speed is reduced to zero $[v(r \rightarrow \infty) = 0]$. What is the proton's initial velocity, $v(r = R)$, in terms of the charge $-Q$ and the radius R? (*Note*: This is the electric analog of the escape velocity of a particle in a gravitational field.)

7. (a) Derive an expression for the electric field of a uniformly charged disk (total charge Q, radius R) along the rotational symmetry axis of the disk. (b) Show that the field has the expected behavior at $x = 0$, and for $x \gg R$.

8. From Gauss' law, the electric field set up by a uniform straight-line charge is

$$\mathbf{E} = \left(\frac{\beta}{2\pi\epsilon_0 r}\right)\hat{\mathbf{r}}$$

where $\hat{\mathbf{r}}$ is a unit vector pointing radially away from the line, and β is the charge per meter along the line. Derive an expression for the potential difference between $r = r_1$ and $r = r_2$.

9. Four equal positive charges, $q = 1.60 \times 10^{-19}$ C, are located on the vertices of a square of side length 1 Å. How much work, in eV, had to be done to assemble these four charges, bringing them in from infinity?

10. Figure 10a depicts an electric quadrupole located on the x-axis. (a) Show that at a position on the x-axis such that $x \gg a$ the electric potential is approximately

$$V = \frac{2k_e q a^2}{x^3}$$

(b) Apply the principle of superposition to the charge distribution in Figure 10b to show that at a position on the x-axis such that $x \gg a$, the electric potential is approximately

$$V = k_e q \left(\frac{2a^2}{x^3} - \frac{1}{x}\right)$$

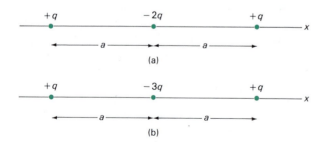

(a)

(b)

FIGURE 10

11. A point charge $+q$ is located at $x = -R$ and a point charge $-2q$ is located at the origin. Prove that the equipotential surface that has zero potential is a sphere centered at $(-4R/3, 0, 0)$ whose radius $r = 2R/3$.

12. The axis of rotational symmetry of a ring of charge dQ and radius r is taken to be the x-axis. On this axis the potential is

$$dV = \frac{1}{4\pi\epsilon_0}\frac{dQ}{(x^2 + r^2)^{1/2}}$$

(a) Determine the potential set up at x by a uniformly charged disk of radius R and charge Q. (b) Show that your result reduces to the potential set up by a point charge in the limit where $x \gg R$.

13. An electric quadrupole creates a potential $V = K/x^3$ along the x-axis. (K is a constant, x is large compared to the quadrupole dimensions.) Show that the electric field along the x-axis is $E_x = 3K/x^4$.

14. A charge $+q$ is at the origin. A charge $-2q$ is at $x = 2.0$ m on the x-axis. (a) For what finite value(s) of x is the electric field zero? (b) For what finite value(s) of x is the electric potential zero?

15. A charge $+4e$ is situated inside a square of side L. Charges $-e$ are located at the four corners of the square. Show that the minimum electric potential energy of the configuration is the one that has the charge $+4e$ at the center of the square.

16. A cylindrical wire (radius 1 mm) is surrounded by a concentric cylindrical conducting shell. If the charge per unit length on the wire is 10^{-8} C/m, determine (a) the electric field at the surface of the wire, (b) the electric field gradient, dE/dr, at the surface of the wire.

17. An electric dipole with a dipole moment p and a moment of inertia I undergoes small-amplitude oscillations in a uniform electric field E (Figure 11). Show that the frequency of the oscillations is

$$f = \frac{1}{2\pi}\sqrt{\frac{pE}{I}}$$

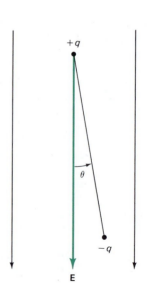

FIGURE 11

18. The x-axis is the symmetry axis of a uniformly charged ring of radius R and charge Q (Figure 12). A point charge Q of mass M is located at the center of the ring. When it is displaced slightly, the point charge accelerates along the x-axis to infinity. Show that the ultimate speed of the point charge is

$$v = \left(\frac{2k_e Q^2}{MR}\right)^{1/2}$$

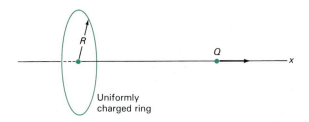

FIGURE 12

19. The electric potential is given by

$$V = ax^2 + bx + c$$

$$a = 12 \text{ V/m}^2 \quad b = -10 \text{ V/m} \quad c = 62 \text{ V}$$

(a) Determine the magnitude and direction of the electric field at $x = +2$ m. (b) Determine the position where the electric field is zero.

20. Figure 13 shows a pair of particles of mass m suspended from a common point by cords of length L. The particles carry equal but opposite charges. A uniform external electric field is directed horizontally to the right. The total potential energy of the system can be expressed as

$$U = 2mgL(1 - \cos\theta) - \frac{1}{4\pi\epsilon_0}\frac{q^2}{2L\sin\theta}$$
$$- 2qLE\sin\theta$$

(a) Explain the origin of each of the three terms in U. (b) Take $mgL = qLE = q^2/\pi\epsilon_0 L = 1$ J and show that there are two equilibrium configurations given by $\theta = 0.307$ rad and $\theta = 0.705$ rad. (c) Make a rough sketch of U versus θ and use it to decide which value of θ corresponds to a stable equilibrium.

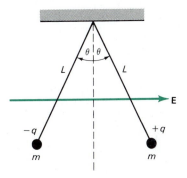

FIGURE 13

CHAPTER

26

CAPACITORS

AND CAPACITANCE

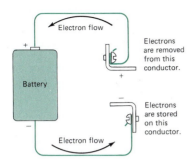

FIGURE 26.1

When a battery is connected to two conductors separated by an insulator, electric forces remove electrons from one conductor and store them on the other.

26.1
INTRODUCTION TO CAPACITORS AND CAPACITANCE

No single electronic component plays a more important role in our electronic age than a charge-storing device called a capacitor. Capacitors are used to store information in the memory elements of computers. They are used to establish electric fields, to minimize voltage variations in electronic power supplies, and to provide energy for certain types of nuclear particle accelerators. Capacitors are used in electronic circuits that detect and generate electromagnetic waves, and as components of electronic circuits used to measure time.

If two conductors separated by an insulator are connected to an energy source such as a battery, electric forces cause electrons to flow from the conductor connected to the positive terminal to the conductor connected to the negative terminal (Figure 26.1). This configuration of two conductors is called a **capacitor**. The symbol ⊣⊢ is used to represent a capacitor in circuit diagrams.

Electron flow stops when the potential difference between the conductors equals the potential difference across the battery. Conservation of charge demands that the amount of negative charge removed from one conductor (leaving this conductor with a net positive charge) equal the amount of negative charge accumulated on the other conductor. If the battery is removed, the charges remain stored on the two conductors.

The ability of a capacitor to store charge is measured by a quantity we call **capacitance**. We define the capacitance of a capacitor as

$$C = \frac{Q}{V} \tag{26.1}$$

where V is the potential difference between the two conductors and Q is the magnitude of the charge on either conductor.*

Capacitance has units of coulombs per volt, or farads (abbreviated F). Measurements of Q and V allow us to determine the capacitance of capacitors having any geometric shape. To compute the capacitance we determine the relation between Q and V, and then identify the ratio Q/V as the capacitance.

Spherical Capacitor

To illustrate the calculation of capacitance we consider two oppositely charged, concentric, hollow spherical conductors (Figure 26.2). The outer radius of the inner sphere is a and the inner radius of the outer sphere is b. The Coulomb electric field between the two spheres gives rise to a potential difference

$$V = \frac{Q}{4\pi\epsilon_0}\left(\frac{1}{a} - \frac{1}{b}\right)$$

Solving for the ratio Q/V gives the capacitance

$$C = \frac{Q}{V} = 4\pi\epsilon_0\left(\frac{ab}{b - a}\right) \tag{26.2}$$

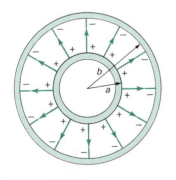

FIGURE 26.2

Two oppositely-charged concentric spheres separated by a vacuum. A charge $+Q$ is distributed uniformly on the surface of the inner sphere of radius a. A charge $-Q$ is distributed uniformly on the surface of the outer sphere of radius b.

---

* In Chapter 25 we used the symbol V to denote the *potential*. Rather than complicate the notation, we use the same symbol here to denote the *potential difference* between the two conductors.

Note that C depends on geometric factors through the lengths a and b and on the insulator (a vacuum in this case).

The effect of the outer sphere on the capacitance diminishes as its radius increases. As $b \to \infty$ the capacitance is determined by a, so we have, for an isolated spherical capacitor, $C = 4\pi\epsilon_0 a$. A spherical satellite in space is a good approximation to an isolated spherical capacitor.

EXAMPLE 1

Capacitance of the Echo Satellite

The *Echo I* satellite was a metal-coated plastic sphere, 30 m in diameter, that orbited the earth at a mean altitude of 1600 km. It had a capacitance of

$$C = 4\pi\epsilon_0 \cdot a = (8.99 \times 10^9 \text{ N·m}^2/\text{C}^2)^{-1} \cdot 15 \text{ m}$$
$$= 1.7 \times 10^{-9} \text{ F}$$

A capacitance of 1 F would require a radius of 8.85×10^9 m. This is about 1400 times the radius of the earth! Capacitances are usually many orders of magnitude smaller than a farad, and it is customary to use units of microfarads (μF) and picofarads (pF).

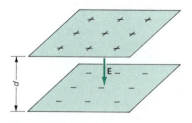

FIGURE 26.3

Two nonconcentric spheres separated by an insulator constitute a capacitor. Gauss' law cannot be used to determine the potential difference because the electric field lines do not extend radially out from the inner sphere.

A variation of this concentric sphere geometry involves shifting the spheres from the concentric position (Figure 26.3). The capacitance is still given by $C = Q/V$. We can always measure Q and V and take their ratio to determine the capacitance. Computing the capacitance, however, is difficult because the spherical symmetry is lost and the electric field lines are no longer radial. Because the electric field lines are no longer radial, we cannot use Gauss' law to obtain the electric field, which is needed for the calculation of the potential difference. We will consider only cases where symmetry or Gauss' law can be employed.

Parallel-Plate Capacitor

A parallel-plate capacitor consists of two plane-parallel conductors of equal area separated by an insulator (Figure 26.4). Let's assume that the insulator is a vacuum. If the capacitor plates were of infinite extent the electric field between the plates would be $E = \sigma/\epsilon_0$, where σ is the charge per unit area. In actual parallel-plate capacitors the length and width of the plates are large compared to the plate separation d. When this is the case, the electric field can be approximated as $E \approx \sigma/\epsilon_0$. Using this expression, we obtain

$$V = Ed = \left(\frac{\sigma}{\epsilon_0}\right)d \qquad (26.3)$$

for the potential difference between the plates of the capacitor. The magnitude of the total charge on either plate is $Q = \sigma A$, so the capacitance is

FIGURE 26.4

Two parallel conducting plates separated by an insulator (a vacuum here) constitute a capacitor. When the plate dimensions are large compared to the distance between the plates, the electric field between them is nearly uniform.

$$C = \frac{Q}{V} = \frac{\sigma A}{\dfrac{\sigma d}{\epsilon_0}} = \frac{A\epsilon_0}{d} \qquad (26.4)$$

Just as with spherical capacitors, C depends only on ϵ_0 and geometric factors. Note that C increases as d decreases and C increases as A increases.

Although Equation 26.4 pertains only to parallel-plate capacitors, its implications are much broader. The capacitance of any capacitor increases as the distance between the conductors decreases. Decreasing d is an effective way of increasing capacitance.

EXAMPLE 2

Computer Chip Capacitance

Each memory cell in a computer memory chip contains a capacitor. The charge stored in the capacitor represents the memory content of the cell. In order to pack more memory cells in a chip, solid-state physicists have developed **trench capacitors**. Instead of having the capacitor plates parallel to the surface of the chip they are mounted *vertically*, along the walls of narrow trenches. A typical trench capacitor has a capacitance of 50 femtofarads and has a plate area of 20×10^{-12} m^2. Let's estimate the plate separation, using the capacitance relation for a parallel-plate capacitor. From $C = \epsilon_0 A/d$ we get

$$d = \frac{\epsilon_0 A}{C} = (8.85 \times 10^{-12}\ \mathrm{C^2/N \cdot m^2}) \frac{20 \times 10^{-12}\ \mathrm{m^2}}{50 \times 10^{-15}\ \mathrm{F}}$$

$$= 3.5 \times 10^{-9}\ \mathrm{m}$$

This is about 35 times the diameter of an atom! The actual spacing in trench capacitors is somewhat greater because of the insulating material between the plates. We investigate this effect in Section 26.4.

Many electronic devices require a variable capacitance. For example, the selection of a radio station on a transistor radio often is done by varying a capacitance. A variable capacitor, symbolized by ⊬, exploits the dependence of capacitance on either the separation between the plates or the area of the plates. One type of variable capacitor maintains constant plate area but allows the distance between plates to be varied with a screw adjustment. A second type of variable capacitor (Figure 26.5) consists of a set of plates free to rotate between a second set of fixed plates. The effective plate area of the capacitor is varied by rotating the plates.

EXAMPLE 3

Comparison of a Capacitor with a Battery

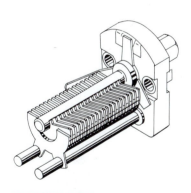

A 12-volt automobile storage battery is rated at 100 ampere hours (A·h), so it can transfer a total of 3.6×10^5 C of charge. Because capacitors are also charge-storing devices, you may wonder why they are not used in automobiles. The answer lies in their size. Let's determine the size of a parallel-plate capacitor that could store 3.6×10^5 C when its potential difference is 12 V. The capacitance would be $C = Q/V = 3 \times 10^4$ F. For a plate separation of 0.1 mm the area is

$$A = \frac{Cd}{\epsilon_0} = \frac{3 \times 10^4\ \mathrm{F} \cdot 10^{-4}\ \mathrm{m}}{8.85 \times 10^{-12}\ \mathrm{C^2/N \cdot m^2}}$$

$$= 3.39 \times 10^{11}\ \mathrm{m^2}$$

This area corresponds to a square of about 5.82×10^5 m (about 360 miles) to a side! Its size prevents this parallel-plate capacitor from being used in place of a storage battery. However, we will see later that capacitors perform many important functions that batteries cannot.

FIGURE 26.5

The capacitance of this capacitor is varied by adjusting the effective area of the plates.

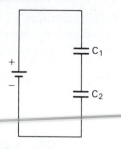

FIGURE 26.6

A *series* combination of two capacitors, connected to a battery. The symbol ⊣⊢ denotes the battery.

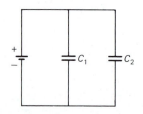

FIGURE 26.7

A *parallel* combination of two capacitors, connected to a battery.

26.2
CAPACITORS IN SERIES AND IN PARALLEL

Capacitors are connected and combined in electronic systems. If we understand how combinations of capacitors behave, we can understand how they affect the performance of these systems. Sometimes it is desirable to replace combinations of capacitors with a single equivalent capacitor. Sometimes we want to do just the opposite. Let's see how we determine *equivalent* capacitances.

Two combinations of capacitors are of special interest. Two or more capacitors connected to a battery (or other source of potential difference) in the way shown in Figure 26.6 are said to be connected in *series*. Two or more connected as shown in Figure 26.7 are connected in *parallel*.

Whether charging a single capacitor or a combination of capacitors, the principle of charge conservation is always satisfied. Charges are redistributed, but no charge is ever added to or subtracted from the total system. For example, consider the charging of two capacitors in series (Figure 26.8). Before connecting the battery by closing the switch, all conductors of the capacitors are electrically neutral. When the switch S is closed, the battery is connected and negative charge flows from the upper conductor of C_1 to the lower conductor of C_2. The insulators between the plates of the capacitors prevent any direct transfer of charge from one plate to the other. A separation of charge, however, occurs on the inner conductors while maintaining overall electrical neutrality. Capacitor C_2 charges *as if* electrons were removed from its upper conductor and stored on its lower conductor. Similarly, capacitor C_1 charges *as if* the electrons removed from its top conductor were stored on its bottom conductor. Because the connected inner conductors of the capacitors must be electrically neutral overall, the magnitude of the charge on both inner conductors must be equal. Thus each plate of the two capacitors contains the same magnitude of charge. Relating potential difference to charge and capacitance, we have

$$V_1 = \frac{Q}{C_1} \quad \text{and} \quad V_2 = \frac{Q}{C_2}$$

where V_1 is the potential difference across capacitor 1, which has capacitance C_1, and V_2 is the potential difference across capacitor 2, which has capacitance

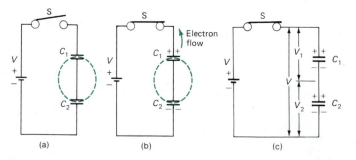

(a)   (b)   (c)

FIGURE 26.8

The charging of two capacitors in series by a battery. (a) The two capacitor plates outlined in dashed lines are connected electrically with a wire but there is no direct connection between these plates and the rest of the circuit. The *net* charge is always zero within the boundary shown. (b) Closing the switch (S) completes the electrical circuit, and electrons are extracted from the top plate of C_1 and transferred to the bottom plate of C_2. (c) The potential differences, V_1 and V_2, sum to V.

C_2. The charge Q is the same for both capacitors because they are connected in series. Potential differences V_1 and V_2 sum to V, the potential difference across the battery (Figure 26.8),

$$V = V_1 + V_2 = Q\left(\frac{1}{C_1} + \frac{1}{C_2}\right)$$

Suppose that we want to replace C_1 and C_2 with a single equivalent capacitor. "Equivalent" means that the capacitor stores the same charge Q when the potential difference across it is V. The equivalent capacitance C is related to Q and V by

$$V = \frac{Q}{C}$$

Comparing these two expressions for V shows

$$\frac{1}{C} = \frac{1}{C_1} + \frac{1}{C_2}$$

We describe this relation by saying that capacitors in series "add reciprocally." Note that the equivalent capacitance is less than either C_1 or C_2. For N series-connected capacitors we have the relation

$$\frac{1}{C} = \frac{1}{C_1} + \frac{1}{C_2} + \frac{1}{C_3} + \cdots + \frac{1}{C_N} \tag{26.5}$$

The reciprocal of the equivalent capacitance equals the sum of the reciprocals of the series capacitances.

EXAMPLE 4

Capacitors in Series

Two 10-μF capacitors are connected in series with a 10-V battery. Using symmetry arguments, let's determine the charge on each capacitor. Because both capacitors have the same capacitance, symmetry suggests that equal potential differences exist across each. Because the total potential difference across the combination is 10 V, the potential difference across each capacitor must be 5 V. Hence the charge on each capacitor is

$$Q = CV = 10 \times 10^{-6} \text{ F} \cdot 5 \text{ V}$$
$$= 50 \ \mu\text{C}$$

Alternatively, we can show that the equivalent capacitance of the combination is 5 μF:

$$\frac{1}{C} = \frac{1}{C_1} + \frac{1}{C_2} = \frac{1}{10 \ \mu\text{F}} + \frac{1}{10 \ \mu\text{F}} = \frac{2}{10 \ \mu\text{F}}$$

$$C = 5 \ \mu\text{F}$$

The charge on the equivalent capacitance is

$$Q = CV = 5 \times 10^{-6} \text{ F} \cdot 10 \text{ V}$$
$$= 50 \ \mu\text{C}$$

so the charge on each capacitor is also 50 μC.

When capacitors are connected in parallel, the potential difference across each capacitor is the same (Figure 26.7). The equivalent capacitance for N capacitors in parallel equals the sum of the capacitances in the combination.

$$C = C_1 + C_2 + C_3 \cdots + C_N \qquad (26.6)$$

The derivation of this relation for equivalent capacitance is straightforward and is presented as a problem.

EXAMPLE 5

Capacitors in Parallel

A combination of 10 identical capacitors connected to a 6-V potential difference causes a total of 120 μC of charge to be transferred. Let's determine the capacitance of each capacitor if the capacitors are all in parallel. The equivalent capacitance is

$$C = \frac{Q}{V} = \frac{120 \times 10^{-6}\ C}{6\ V} = 20\ \mu F$$

For a parallel combination, the equivalent capacitance is the sum of the individual capacitances. If C_1 is the capacitance of a single capacitor, then

$$10C_1 = 20\ \mu F$$

Thus, each capacitor has a capacitance of 2 μF.

When a combination of capacitors is connected to a potential difference we sometimes need to know the charge on each capacitor plate and the potential difference across each capacitor. The following example illustrates how knowledge of series and parallel combinations of capacitors allows us to determine these quantities.

EXAMPLE 6

Series and Parallel Combinations of Capacitors

Three capacitors are connected to a 10-V potential difference as shown in Figure 26.9a. We want to calculate the charge on each plate and the potential difference across each capacitor when fully charged. Toward this end, we first determine the equivalent capacitance of the combination.

Capacitors C_2 and C_3 are in parallel and can be replaced by a single capacitor having capacitance $C_4 = C_2 + C_3 = 10\ \mu F$, as shown in Figure 26.9b. As we saw in Example 4, two 10-μF capacitors in series are equivalent to a single 5-μF capacitor. Thus, the equivalent capacitance of the combination is $C_5 = 5\ \mu F$, as shown in Figure 26.9c.

The charge on C_5 is $Q_5 = C_5 V_5 = 5\ \mu F \cdot 10\ V = 50\ \mu C$. Because the charge is the same on each capacitor in a series connection, the charges on C_1 and C_4 (which are equivalent to C_5) are also 50 μC. The potential differences across C_1 and C_4 are

$$V_1 = \frac{Q_1}{C_1} = \frac{50\ \mu C}{10\ \mu F} = 5\ V$$

$$V_4 = \frac{Q_4}{C_4} = \frac{50\ \mu F}{10\ \mu F} = 5\ V$$

Note that $V_1 + V_4 = 10\ V$, as required.

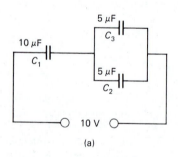

(a)

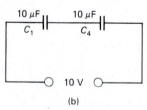

(b)

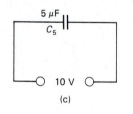

(c)

FIGURE 26.9

Because the potential difference is the same across each capacitor in a parallel connection, the potential difference across C_2 and C_3 (which together are equivalent to C_4) is 5 V. Hence the charges on C_2 and C_3 are

$$Q_2 = C_2V_2 = 5\ \mu F \cdot 5\ V = 25\ \mu C$$

$$Q_3 = C_3V_3 = 5\ \mu F \cdot 5\ V = 25\ \mu C$$

Note that $Q_2 + Q_3 = 50\ \mu C$, as required.

Understanding how capacitors combine when connected in parallel and in series is important if for no other reason than demonstrating the role of conservation of charge. However, there are important practical implications. For example, single capacitors cannot be made with arbitrarily large capacitance. An application may call for a capacitance for which a single capacitor is not available. The experimenter may achieve the desired capacitance by connecting capacitors in parallel. Electronic elements such as diodes and transistors have intrinsic capacitances. These elements may appear in parallel or in series with other capacitors, in which case the experimenter needs to know the effective capacitance of the combinations. The rules for combining capacitors in parallel and in series, then, are essential.

26.3
ENERGY STORED IN AN ELECTRIC FIELD

Work must be performed to charge a capacitor. The charges on the capacitor plates set up an electric field between the plates. We want to derive an expression for the work performed in charging a capacitor and relate it to the electric field set up between the capacitor plates. Figure 26.10 symbolizes a circuit consisting of a battery, a switch S, and a capacitor C. The battery maintains a constant potential difference V across its terminals and provides an energy source for charging the capacitor when the switch S is closed. Before the switch in Figure 26.10 is closed, there is no potential difference across the capacitor because there is no charge on the capacitor plates. When the switch is closed, charge begins accumulating on the plates. For each increment of positive charge dq transferred from the lower plate to the upper plate the potential energy of the capacitor increases by

$$dU = V_c\,dq = \frac{q}{C}\,dq$$

where q is the magnitude of the charge on each capacitor plate when the charge dq is transferred, and $V_c = q/C$ is the potential difference between the capacitor plates.

Charge transfer continues until the potential difference across the capacitor equals the potential difference across the battery. When fully charged, the charge on each plate is $Q = CV$. The total potential energy stored in the fully charged capacitor is

$$U = \int_0^Q \left(\frac{q}{C}\right) dq = \frac{Q^2}{2C} \tag{26.7}$$

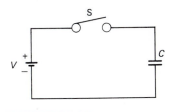

FIGURE 26.10

A scheme for charging a capacitor with a battery. When the switch is closed, electric forces cause a redistribution of charge on the capacitor plates.

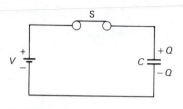

FIGURE 26.11
The work done by the battery in redistributing the charge Q is QV.

Because $Q = CV$, the potential energy can also be written as

$$U = \tfrac{1}{2}CV^2 \qquad (26.8)$$

or

$$U = \tfrac{1}{2}QV \qquad (26.9)$$

We use the most convenient of these three equations for solving the problem at hand.

Energy Aspects of Charging a Capacitor

When a charge Q is transferred from one conductor of a capacitor and stored on the opposite conductor, work equal to $\tfrac{1}{2}QV$ is transformed into potential energy. This work is performed by the energy source (a battery) connected to the capacitor. During the charging, is any energy expended other than this $\tfrac{1}{2}QV$?

Let's calculate the total work done by the battery. A total charge Q is moved through a *constant* potential difference V provided by the battery (Figure 26.11). Hence the total work done by the battery is

$$W = QV$$

But this is twice the energy stored on the capacitor. Where did the other $\tfrac{1}{2}QV$ of energy go? Work is also done in forcing the charges through the wires connecting the capacitor to the battery. This work produces heat, and accounts for the remaining $\tfrac{1}{2}QV$ of the work done by the battery.

In Example 2 we compared the charge-storing abilities of a capacitor and a battery. Now we compare the energy-storing capability of these two devices.

EXAMPLE 7

Comparison of a Capacitor with a Battery as an Energy-Storing Device

Twelve-volt, lead-acid batteries convert chemical energy into electric energy in the electrical systems of nearly all contemporary automobiles. Since capacitors also store electric energy, let us determine the feasibility of using capacitors rather than lead-acid batteries as electric energy sources in automobiles. The largest capacitance available in a 12-V commercial capacitor is about 0.1 F. Such a capacitor will store 7.2 J of energy. But a good 12-V lead-acid battery can deliver 4.3×10^6 J (see Example 3). This means that it would take 6×10^5 capacitors to provide the same amount of energy as one 12-V battery. Even if each capacitor cost only 10¢, the cost of 6×10^5 capacitors would be $60,000. Each of these capacitors would occupy about 20 cm^3 of space. The total volume of all 6×10^5 capacitors corresponds to a cube about 2 m on a side, which is roughly the size of a contemporary car. Thus, we conclude that capacitors cannot compete with lead-acid storage batteries as energy sources in contemporary automobiles.

Electric Field Energy Density

A charge q in an electric field $\mathbf{E}$ experiences an electric force $\mathbf{F} = q\mathbf{E}$. If the charge is free to move, then work is done on it by the electric force. If we think of the electric field as the environment that provides the force, then we may also identify the electric field as the energy source for the work done by the field. Let's pursue this interpretation of the electric field by considering the energy stored by a parallel-plate capacitor.

For any capacitor the total energy stored is

$$U = \tfrac{1}{2}CV^2$$

For a parallel-plate capacitor, $C = A\epsilon_0/d$ and $V = Ed$. Hence

$$U = \frac{1}{2}\left(\frac{A\epsilon_0}{d}\right)(Ed)^2 = \frac{1}{2}(Ad)\epsilon_0 E^2 \qquad (26.10)$$

Note that Ad is the volume between the plates of the capacitor and that it coincides with the volume where the electric field is nonzero. Dividing both sides of Equation 26.10 by this volume, we arrive at a quantity independent of the geometry of the capacitor:

$$u = \frac{U}{Ad} = \frac{1}{2}\epsilon_0 E^2$$

The quantity $u = \tfrac{1}{2}\epsilon_0 E^2$ has units of J/m^3, the units of energy per unit volume. We interpret $\tfrac{1}{2}\epsilon_0 E^2$ as the **energy density** associated with the electric field between the capacitor plates. This interpretation suggests that *there is energy stored wherever an electric field exists* and that the energy density is proportional to the square of the electric field. Although we have actually demonstrated this only for a parallel-plate capacitor, it is a general result and it is true of all electric fields,

$$u = \tfrac{1}{2}\epsilon_0 E^2 \qquad (26.11)$$

Because the strength of an electric field can vary with position, the energy density can also vary with position. This happens, for example, between the conductors of a spherical capacitor. To obtain the total energy in situations like these, we must integrate the energy density throughout the volume where E is nonzero, as we do in the following example.

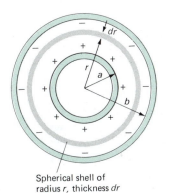

Spherical shell of
radius *r*, thickness *dr*

FIGURE 26.12

Between the conductors of a spherical capacitor there is a volume element in the form of a spherical shell of radius *r* and thickness *dr*. The volume of the shell is $4\pi r^2\, dr$ and the electric energy in the shell is $\tfrac{1}{2}\epsilon_0 E^2 \cdot 4\pi r^2\, dr$. The total electric energy stored in the electric field is obtained by integrating over the volume between the spherical conductors.

EXAMPLE 8

Energy of a Spherical Capacitor

We know that the energy stored in any capacitor is $U = \tfrac{1}{2}(Q^2/C)$. We also know that the capacitance of a spherical capacitor is

$$C = 4\pi\epsilon_0\left(\frac{ab}{b-a}\right)$$

and that the electric field between the conductors is

$$E = \frac{1}{4\pi\epsilon_0}\frac{Q}{r^2}$$

Let us take the energy density of a spherical capacitor to be $u = \tfrac{1}{2}\epsilon_0 E^2$, and show by direct integration that this leads to the known relation for the energy of a spherical capacitor.

We note that u depends only on the radial distance from the center of the capacitor. Therefore, if we pick a spherical shell (Figure 26.12) of thickness dr, volume $4\pi r^2\, dr$, the energy within this shell is

$$dU = u4\pi r^2\, dr = \frac{1}{2}\epsilon_0\left(\frac{1}{4\pi\epsilon_0}\frac{Q}{r^2}\right)^2 4\pi r^2\, dr$$

giving

$$U = \int dU = \frac{1}{2} \frac{Q^2}{4\pi\epsilon_0} \int_a^b \frac{dr}{r^2}$$

Integrating, we obtain

$$U = \frac{1}{2} \frac{Q^2}{4\pi\epsilon_0} \left(\frac{b-a}{ab} \right)$$

Because the capacitance is

$$C = 4\pi\epsilon_0 \left(\frac{ab}{b-a} \right)$$

it follows that

$$U = \frac{Q^2}{2C}$$

confirming the energy density formula for the spherical capacitor.

26.4
EFFECT OF AN INSULATOR ON CAPACITANCE

Earlier we noted that capacitance depends on the geometry, area, conductor separation, and the insulator between the conductors. We now discuss in more detail the role of the insulator. Figure 26.13 shows an apparatus for investigating capacitance changes when an insulator is inserted between the plates of a parallel-plate capacitor. The effect of the insulator is represented by the **dielectric constant**, κ (Greek kappa), which is defined by the following experiment.

A battery is connected to a parallel-plate capacitor in a vacuum. A constant potential difference V_0 exists across the capacitor when it is fully charged

FIGURE 26.13

A type of parallel-plate capacitor used to investigate the effects of an insulator on capacitance.

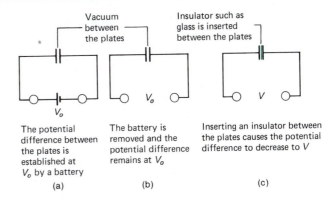

The potential difference between the plates is established at V_o by a battery

(a)

The battery is removed and the potential difference remains at V_o

(b)

Inserting an insulator between the plates causes the potential difference to decrease to V

(c)

FIGURE 26.14

Experimental procedure used to measure the effects of an insulator on the capacitance. (a) Vacuum between plates; battery connected. (b) Vacuum between plates; battery removed. (c) Insulator inserted between plates.

(Figure 26.14a). The energy source is then removed and the capacitor is isolated so that charge cannot escape from the plates (Figure 26.14b). If an insulator is inserted between the plates of the isolated capacitor, we observe that the potential difference across the plates decreases (Figure 26.14c). We call the new value of the potential difference V, and define the dielectric constant κ as

$$\kappa = \frac{V_0}{V} \qquad (26.12)$$

Although this ratio is affected to some extent by external influences such as temperature, systematic experiments reveal that it does *not* depend on plate size, plate separation, or initial potential difference. The ratio depends primarily on the type of insulator, and is a property of the insulator. The dielectric constant is unity for a vacuum and is a number greater than 1 for any material medium under static conditions. Representative dielectric constants are listed in Table 26.1.

TABLE 26.1

Representative Dielectric Constants of Some Common Materials*

Material	κ
Water	78.3
Air	1.000590
Lucite	2.84
Plexiglas	3.12
Polystyrene	2.55
Neoprene	6.60
Polyethylene	2.26
Pyrex	4–6[†]
Teflon	2.1
Titanates (Ba, Sr, Ca, Mg, and Pb)	15–12,000

* For a more complete listing see *Handbook of Chemistry and Physics,* CRC Press, 2000 N. W. 24th Street, Boca Raton, FL 33431.

[†] Depending on type.

Because potential difference and charge are related by $V = Q/C$, you may suspect that the reduction in V is due to a loss of charge. However, the plates are not connected electrically—remember, the battery has been removed—so there is no way for charge to leave the plates. Indeed, when the insulator is removed from the isolated capacitor the potential difference returns to the initial value, indicating that the presence of the insulator does *not* affect the charge stored on either plate. We conclude that the reduction in potential difference is due to an *increase in capacitance*. Combining the definition of capacitance and Equation 26.12, we find

$$C = \frac{Q_0}{V} = \frac{\kappa Q_0}{V_0} = \kappa C_0 \qquad (26.13)$$

The capacitance has increased by a factor of κ.

Now suppose we start with a charge Q_0 on the capacitor and then insert an insulator between the capacitor plates while the battery is still connected. If the battery maintains a constant potential difference V_0 across the capacitor, then the charge stored on the capacitor is

$$Q = CV_0$$

where C is the capacitance when the insulator is present. Because $C = \kappa C_0$ and $Q_0 = C_0 V_0$,

$$Q = \kappa C_0 V_0 = \kappa Q_0$$

Measurements confirm that the charge stored with the insulator present has increased by a factor of κ.

Equation 26.13 is a general equation. It shows that an insulator alters the capacitance of any capacitor. We can rewrite Equation 26.13 specifically for a parallel-plate capacitor as follows:

$$C = \kappa C_0 = \frac{\kappa \epsilon_0 A}{d} \qquad (26.14)$$

Because κ is dimensionless, $\kappa \epsilon_0$ has the same dimensions as ϵ_0. The quantity $\kappa \epsilon_0$, denoted by ϵ, is called the **permittivity**.

$$\epsilon = \kappa \epsilon_0 \qquad (26.15)$$

Permittivity is a property of the insulator. Because $\epsilon = \epsilon_0$ for a vacuum, ϵ_0 is appropriately called the **permittivity of free space**. In terms of permittivity, the capacitance of a parallel-plate capacitor becomes

$$C = \frac{\epsilon A}{d} \qquad (26.16)$$

Similarly, the energy density in an electric field in a material medium becomes

$$u = \tfrac{1}{2}\epsilon E^2 \qquad (26.17)$$

Dielectric Strength

Capacitors for electronic applications vary in capacitance from about 10^{-5} μF to about 10^5 μF. Manufacturers achieve this wide range of values by taking advantage of the dependence of capacitance on geometry and dielectric constant. A capacitor in a circuit is charged by an energy source that provides a potential difference across its terminals. This external energy source

TABLE 26.2
Dielectric Strengths of
Some Common Insulators*

Insulator	Dielectric Strength (kV/mm)
Air	3
Lucite	20
Plexiglas	20
Polystyrene	20
Neoprene	12
Polyethylene	18
Pyrex	14
Teflon	19
Titanates (Ba, Sr, Ca, Mg, and Pb)	2–12

* The dielectric strengths of plastics will vary with thickness. For this reason, there is some variation in the values quoted in the literature. Those listed here are representative. For a more complete listing see *Handbook of Chemistry and Physics*, CRC Press, 2000 N. W. 24th Street, Boca Raton, FL 33431.

establishes an electric field between the capacitor plates. For any insulator there is a maximum electric field that can be maintained without ionizing atoms in the insulator. This maximum electric field, called the **dielectric strength**, depends on the physical structure of the insulator. Representative values of the dielectric strength of some common insulators are shown in Table 26.2. Once the plate separation of a capacitor has been determined, there is a maximum potential difference that can be applied across its terminals to avoid breakdown of the insulator. For a parallel-plate capacitor the breakdown potential difference (V_b) and the dielectric strength (E_b) are related by

$$E_b = \frac{V_b}{d}$$

where d is the plate separation.

EXAMPLE 9

Breakdown Potential Difference of a Capacitor

A certain parallel-plate capacitor has a plate separation of 0.01 mm and uses Teflon® as an insulator. What is the maximum potential difference that can be applied to the terminals of the capacitor? From Table 26.2 the dielectric strength of Teflon is 19 kV/mm. Thus the maximum applied potential difference is

$$V_b = E_b \cdot d = (19 \text{ kV/mm}) \cdot 0.01 \text{ mm} = 0.19 \text{ kV}$$
$$= 190 \text{ V}$$

If a potential difference larger than 190 V is applied, this capacitor is in danger of being destroyed.

Electrolytic Capacitors

Electronic circuits often employ a type of capacitor termed **electrolytic**. Electrolytic capacitors are identified by + and − labels near the connections. This polarity must be honored when the capacitor is employed in a circuit.

Electrolytic capacitors have relatively large capacitance that is achieved with a very thin separation between the capacitor plates. One of the two conductors in an electrolytic capacitor is a metal foil, usually made from tantalum or aluminum. On the surface of this conductor is a very thin non-conducting oxide of the metal. This metal oxide serves as the insulator. The other conductor is a conducting paste or a liquid that makes intimate contact with the metal oxide. Because the metal oxide layer is very thin and has a large dielectric constant, capacitances as large as 500,000 μF can be attained.

26.5
POLARIZATION

The insertion of an insulator between the conductors of an isolated charged capacitor diminishes the potential difference between the conductors. We interpret this effect in terms of the atoms and molecules within the insulator. In doing so, we will show why the value of the dielectric constant is always greater than unity.

A sufficiently strong electric field applied to an insulator may free electrons from atoms and create an electric current, but ordinarily the electric fields used are only capable of distorting the electric structure of atoms and molecules.

The molecules in an insulator fall into two classes. In some molecules, such as H_2O and N_2O, internal electric fields cause a separation of the centers of positive and negative charges. This distortion of the electrical structure gives rise to a *permanent* electric dipole moment. Such molecules are termed **polar**.

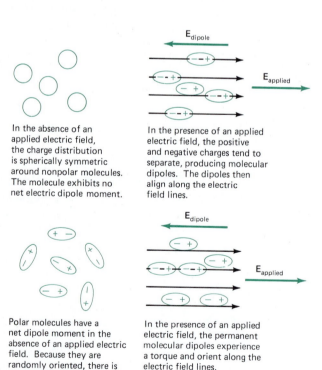

In the absence of an applied electric field, the charge distribution is spherically symmetric around nonpolar molecules. The molecule exhibits no net electric dipole moment.

In the presence of an applied electric field, the positive and negative charges tend to separate, producing molecular dipoles. The dipoles then align along the electric field lines.

Polar molecules have a net dipole moment in the absence of an applied electric field. Because they are randomly oriented, there is no net electric dipole field in a macroscopic size sample.

In the presence of an applied electric field, the permanent molecular dipoles experience a torque and orient along the electric field lines.

FIGURE 26.15
Alignment of nonpolar and polar molecules by an electric field.

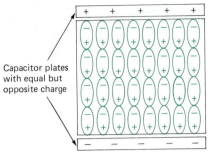

FIGURE 26.16

Alignment of molecular electric dipoles in an insulator between the plates of a capacitor.

Other molecules, like H_2, N_2, and O_2, do not possess a permanent electric dipole moment. But, when they are placed in an external electric field the ensuing charge separation gives rise to an **induced** electric dipole moment. Such molecules are called **non-polar**.

Insulators containing either polar or non-polar molecules are called **dielectrics** because the internal electric fields set up by their electric dipole moments *oppose* and weaken an applied electric field. Figures 26.15, 26.16, and 26.17 suggest how this action occurs. The applied electric field E_a tends to align the molecular dipole moments p parallel to E_a. This aligned orientation gives zero torque ($\tau = p \times E_a$) and minimum potential energy ($U = -p \cdot E_a$).

The electric dipoles establish an *internal electric field*, E_i. The electric field lines of the dipoles originate at the center of positive charge and terminate at the center of negative charge. As a consequence the dipole field E_i is directed opposite to the applied field.

The net field inside the dielectric—the vector sum of the applied and dipole fields—is thereby reduced. The weakened electric field in turn reduces the potential difference between the capacitor plates, as we can see from the relation between the electric field E and the potential difference V:

$$V = -\int E \cdot ds$$

Since the potential difference of the charged isolated capacitor (V_0) is always greater than the potential difference (V) produced when an insulator is inserted between the plates of the same isolated capacitor, the dielectric constant ($\kappa = V_0/V$) should be greater than unity. Experiments confirm this expectation.

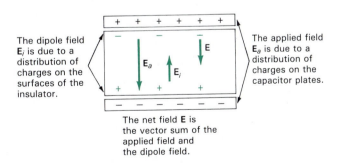

FIGURE 26.17

The net electric field E is the vector sum of the applied field E_a and the internal dipole field E_i. The net electric field is weaker than the applied field because the dipole field opposes the applied field.

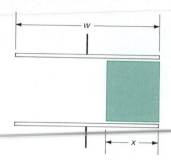

FIGURE 26.18

WORKED PROBLEM

A slab of material having a dielectric constant κ is placed between the plates of a parallel-plate capacitor (Figure 26.18). (a) Explain why this configuration is equivalent to two capacitors in parallel and show that its capacitance is

$$C = \left(\frac{A\epsilon_0}{d}\right)\left[\left(\frac{\kappa x}{W}\right) + \left(1 - \frac{x}{W}\right)\right]$$

(b) Show that this equation reduces to the expected results for $x = 0$, $x = W$, and $\kappa = 1$.

Solution

(a) Circuit elements connected in parallel have equal potential differences across their terminals. The two parallel metal plates of the capacitor are equipotential surfaces, giving equal potential differences across the slab-filled and empty segments.

Calling L the length of the plates perpendicular to the horizontal dimension, the plate areas of the two capacitors are $L(W - x)$ and Lx. Using the parallel-plate capacitor formula (Equation 26.16) the capacitances of the two segments are

$$C_1 = \frac{L(W - x)\epsilon_0}{d} \qquad C_2 = \frac{Lx\kappa\epsilon_0}{d}$$

The equivalent capacitance of two capacitors in parallel is the sum of their capacitances, so

$$C = C_1 + C_2 = \frac{L\epsilon_0}{d}[\kappa x + (W - x)]$$

$$= \frac{LW\epsilon_0}{d}\left[\left(\frac{\kappa x}{W}\right) + \left(1 - \frac{x}{W}\right)\right]$$

Recognizing that $A = LW$ is the total surface area of one plate gives the desired result.

(b) If $x = 0$, the entire space between the plates is a vacuum. Such a capacitor has a capacitance $C = A\epsilon_0/d$. Setting $x = 0$ in the final equation reproduces this result.

If $x = W$, the entire space between the plates is filled with material having a dielectric constant κ. Such a capacitor has a capacitance $C = \kappa A\epsilon_0/d$. Setting $x = W$ in the final equation reproduces this result.

If $\kappa = 1$, the entire space between the plates is a vacuum and we again expect $C = A\epsilon_0/d$. Setting $\kappa = 1$ in the final equation reproduces this result.

EXERCISES

26.1   Introduction to Capacitors and Capacitance

A.   Two separated metallic coffee cans (Figure 1) are con-

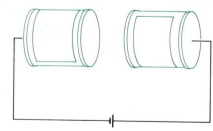

FIGURE 1

nected to the terminals of a 10-V battery. A sensitive electrometer indicates that each has accumulated a charge of 10^{-12} C. What is the capacitance of the configuration?

B.   A capacitor with a charge of 10^{-4} C has a potential difference of 100 V. What charge is required to produce a potential difference of (a) 1 V, (b) 25 V?

C.   Calculate the capacitance of an isolated sphere having a radius equal to that of the earth.

D.   The earth and the ionosphere may be considered to be a spherical capacitor with a separation of 100 km. Calculate the capacitance of this earth-ionosphere system.

E. A potential difference of 5 volts is established across a 40-femtofarad capacitor in a 1-megabit computer memory chip. Determine the number of electrons on the negative plate.

F. A 1-megabit computer memory chip contains many 60-femtofarad capacitors. Each capacitor has a plate area of 21 square micrometers (21×10^{-12} m^2). Determine the plate separation of such a capacitor (assume a parallel-plate configuration). The characteristic atomic diameter is 10^{-10} m = 1 Å. Express the plate separation in Å.

26.2 Capacitors in Series and in Parallel

G. Determine the equivalent capacitance for each of the capacitor configurations in Figure 2.

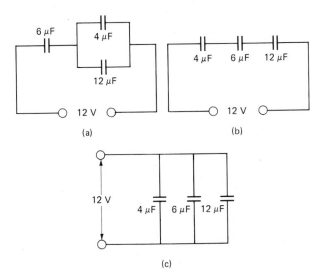

(a) (b)

(c)

FIGURE 2

H. Points A and B in Figure 3 are joined by an array of five capacitors. The numbers give the capacitances in microfarads. Find the equivalent capacitance between A and B.

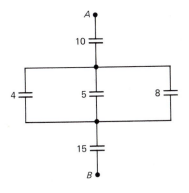

FIGURE 3

I. How many 0.25-pF capacitors must be connected in parallel in order to store 1.2 μC of charge when connected to a battery providing a potential difference of 10 volts?

J. A group of identical capacitors is connected first in series and then in parallel. The combined capacitance in parallel is 100 times larger than for the series connection. How many capacitors are in the group?

K. (a) Using a symmetry argument, determine the potential difference across each capacitor in Figure 4. (b) Determine the total amount of charge redistributed by the battery. (c) Calculate the equivalent capacitance of the group.

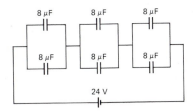

FIGURE 4

L. During the charging of the capacitors in Figure 5 the battery redistributes 3.1 μC of charge. Determine the equivalent capacitance of the configuration.

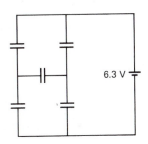

FIGURE 5

26.3 Energy Stored in an Electric Field

M. Determine the ratio of the energies (U_2/U_1) stored in the two capacitors shown in Figure 6 after the switch is closed and the capacitors have become fully charged.

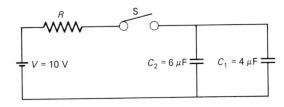

FIGURE 6

N. The energy stored in a particular capacitor is increased fourfold. What is the accompanying change in the (a) charge, (b) potential difference across the capacitor?

P. The energy stored in a 12-μF capacitor is 130 μJ. Determine the capacitor (a) charge, (b) potential difference.

Q. A 12-volt battery redistributes 8 μC between the plates of a capacitor connected across its terminals. (a) How much work is done by the battery? (b) How much electric energy is stored in the capacitor? (c) Why is the work done greater than the energy stored in the capacitor?

R. When considering the energy supply for an automobile, the energy per unit mass of the supply is an important point. Using the data below, compare the energy per unit mass (J/kg) for gasoline, lead-acid batteries, and capacitors.

> Gasoline: 126,000 Btu/gal; density = 670 kg/m^3
>
> Lead-acid battery: 12 volts; 100 ampere-hours; mass = 16 kg
>
> Capacitor: potential difference at full charge = 12 V; capacitance = 0.1 F; mass = 0.1 kg

S. Calculate the (a) electric field, (b) energy density at a distance of 2.8 fm from the center of a proton.

T. Water in a tank is allowed to drain into an adjacent tank (Figure 7a). (a) What determines when the flow stops? (b) What effect does the length and diameter of the connecting pipe have on the final levels of the water? (c) How is this situation analogous to the draining of charge from a capacitor to an initially uncharged capacitor connected to it (Figure 7b)?

26.4   Effect of an Insulator on Capacitance

U. A 1000-pF parallel-plate capacitor is to be made with

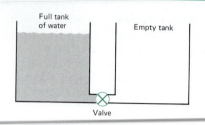

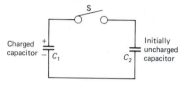

FIGURE 7

a Teflon dielectric 0.005 mm thick. Determine (a) the required plate area, (b) the breakdown potential.

V. A certain electronic circuit calls for a capacitor having a capacitance of 1.2 pF and a breakdown potential of 1000 V. If you have a supply of 6-pF capacitors having a breakdown potential of 200 V, how could you meet this circuit requirement?

W. A parallel-plate capacitor having air between its plates is charged to 31.5 V. The capacitor is then isolated from the charging source and the volume between the plates is filled with Plexiglas. Determine the new potential difference across the capacitor.

X. The manufacturer of a type of coaxial cable used for stereo systems quotes the capacitance per unit length as 18 pF/ft. Verify this value, knowing that the radii of the inner and outer conductors are 0.0254 inch and 0.242 inch, and that they are separated by polyethylene having a dielectric constant equal to 2.26.

PROBLEMS

1. (a) Consider both the inside and outside surfaces of a charged parallel-plate capacitor (Figure 8a). Indicate for each surface whether it is charged or uncharged and the sign of the charge. (b) A capacitor is constructed using three identical plates arranged to be parallel (Figure 8b). This capacitor is charged by connecting the central plate to the positive terminal of a battery and the two outer plates to the negative terminal. Indicate for each of the six surfaces whether it is charged or not, and the sign of the charge. (c) If each plate in Figure 8b has an area A and the plate separations are d, explain why the capacitance of the three-plate capacitor is twice that of a two-plate capacitor having a plate area A and plate separation d. (d) A capacitor is constructed using N identical plates arranged like those in Figure 8b, with alternate plates electrically connected together. Explain why the capacitance is $(N - 1)$ times the capacitance of the two-plate capacitor.

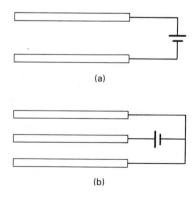

(a)

(b)

FIGURE 8

2. (a) A slab of insulating material is placed inside a parallel-plate capacitor, and the capacitor is charged. The battery is then disconnected and the slab is pulled from between the plates. Using an energy argument, explain why work is required to do this. (b) Derive an expression for the work required to remove the slab. Express your result in terms of the dielectric constant, plate area, plate separation, and charge.

3. A spring balance holds a metal plate of area A above an identical fixed plate (Figure 9). When there is a potential difference across the plates an additional force F is shown by the spring balance. Derive the following equations that show how F is related to the potential difference V, the charge Q on the plates, and the plate separation d:

$$V = d\sqrt{\frac{2F}{A\epsilon_0}} \qquad Q = \sqrt{2A\epsilon_0 F}$$

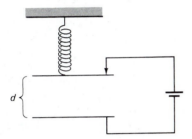

FIGURE 9

4. Calculate the equivalent capacitance of the combination shown in Figure 10.

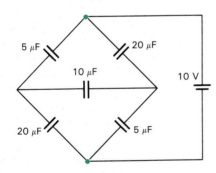

FIGURE 10

5. A 4.0-μF capacitor and a 6.0-μF capacitor in series are charged by a 200-V source. The charged capacitors are disconnected (without discharging them) and reconnected in parallel. (a) What is the charge on each capacitor while they are in series? (b) What is the final charge on each after they are connected in parallel?

(c) What is the potential difference across the capacitors when they are connected in parallel?

6. A 5-μF capacitor is charged by connecting it to a 10-V battery. The capacitor is removed and connected in parallel with an unknown, uncharged capacitor. At equilibrium, there is a potential difference of 1 V across each capacitor. Determine the unknown capacitance.

7. For a cylindrical capacitor the electric field between the cylinders is given by $E = Q/2\pi\epsilon_0 Lr$, where L is the length of the cylinder and r is the radial distance from the axis of the cylinders. Show that the total energy is $\frac{1}{2}Q^2/C$, thus supporting the concept that energy is stored in the electric field.

8. The potential difference across an isolated 5-μF capacitor is 10 V. The plates of this capacitor are then connected to the plates of an uncharged 5-μF capacitor. Determine the total electric energy of the combination and comment on any energy difference between the initial and final configurations.

9. The capacitance of two concentric spheres is

$$C = 4\pi\epsilon_0 \left(\frac{r_1 r_2}{d}\right)$$

where $d = r_2 - r_1$ is the separation of the spheres. If both radii become very large, their surfaces become locally flat. Show that if r_1 and r_2 are large compared to d, then the equation for C approaches that for a parallel-plate capacitor.

10. The capacitance of two spheres having radii a and b whose surfaces are a distance c apart (Figure 11) is given by

$$C = \frac{4\pi\epsilon_0}{\dfrac{1}{a} + \dfrac{1}{b} - \dfrac{2}{c}}$$

provided c is large compared to a and b. (a) Is the expression for C dimensionally correct? (b) Show that in the limit $c \to \infty$ the capacitance reduces to that of two spherical capacitors in series.

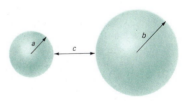

FIGURE 11

11. A parallel-plate capacitor with air as its dielectric is charged and then separated from the charging source. The plate separation is then doubled. Compare the new electric potential energy to the old and account for the change.

12. (a) Show that the electric energy of a charged parallel-plate capacitor can be written as

$$U = \frac{1}{2}\left(\frac{Q^2}{\epsilon_0 A}\right)x$$

where x is the plate separation. (b) Show that the pressure on a plate can be expressed by

$$P = -\tfrac{1}{2}\sigma E = -\tfrac{1}{2}\epsilon_0 E^2$$

13. Wires in electrical circuits give rise to stray capacitance. Treat a pair of parallel wires 10 cm long with diameters of 0.5 mm, with their centers separated by 2 mm, as a parallel-plate capacitor, and *estimate* their capacitance. Express your result in pF.

14. A parallel-plate capacitor is arranged so that the left plate is fixed in position while the right plate is free to move horizontally without friction (Figure 12). A string connects the right plate to a mass m hanging from a frictionless pulley. Show that if the charge Q is fixed, the net force acting on the right plate is given by

$$F = mg - \frac{Q^2}{2\epsilon_0 A}$$

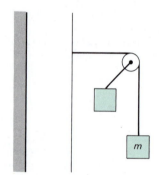

FIGURE 12

15. A parallel-plate capacitor is arranged so that the electric field is directed upward. The potential difference across the capacitor is adjusted until a charged particle moves horizontally through the capacitor undeflected under the combined influence of the electrical and gravitational forces. Calculate the ratio of the charge to mass of the particle if the potential difference is 5000 V and the plate separation is 1 cm.

16. Show that the capacitance of a coaxial cable is $2\pi\epsilon_0 L/\ln(b/a)$, where L is its length, b is the outer conductor radius, and a is the inner conductor radius.

17. Each capacitor in the combination shown in Figure 13 has a breakdown voltage of 15 V. What is the breakdown voltage of the combination?

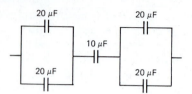

FIGURE 13

18. The capacitance of two parallel wires of length L and radius a whose axes are separated by a distance $d \gg a$ is

$$C = \frac{\pi\epsilon_0 L}{\ln\left(\dfrac{d}{a}\right)}$$

If the wires are uniformly charged, with a charge of β C/m on one and $-\beta$ C/m on the other, show that the potential difference between the wires is

$$V = \frac{\beta \ln\left(\dfrac{d}{a}\right)}{\pi\epsilon_0}$$

19. Some variable capacitors have nonparallel plates as shown in Figure 14. The capacitance is varied by changing the tilt angle of the upper plane. To calculate the capacitance we can divide the plates into strips of width dx and length L. This divides the capacitor into a large number of parallel-plate capacitors in parallel, each having a capacitance $dC = \epsilon_0 L dx/y$. (a) Show that the capacitance is

$$C = \frac{\epsilon_0 A}{s} \ln\left(1 + \frac{s}{d}\right)$$

where $s = w \tan\theta$ and A is the area of the bottom plate. (b) Show that in the limit $s \to 0$ this expression reduces to the parallel-plate result, $C = A\epsilon_0/d$.

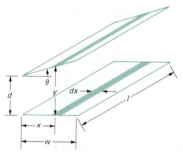

FIGURE 14

20. A slab of material having a dielectric constant κ is placed between the plates of a parallel-plate capacitor

(Figure 15). (a) Explain why this configuration is equivalent to two capacitors in series and then show that the capacitance is

$$C = \frac{\dfrac{\kappa A \epsilon_0}{d}}{\dfrac{x}{W} + \kappa \left(1 - \dfrac{x}{W}\right)}$$

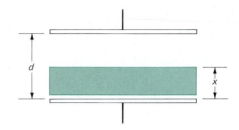

FIGURE 15

(b) Show that this equation reduces to the expected results for $x = 0$, for $x = W$, and for $\kappa = 1$.

CHAPTER

27

ELECTRIC

CURRENT

27.1
INTRODUCTION TO ELECTRIC CURRENT

Conduction of Electrons in Solids

Water molecules flowing down a river constitute a water current. Analogously, electric charges flowing in a wire or a transistor constitute an electric current. Electric currents that we deal with on an everyday basis occur in solid conductors, like light bulb filaments and transistors, although they can also exist in a near-vacuum. For example, protons circulating in the evacuated channels of the accelerator at the Fermilab in Batavia, Illinois, and electrons racing down the SLAC linear accelerator at Stanford University, constitute electric currents.

A current in a solid is caused by electrons because the nuclei are virtually immobile. Differences in the way the electrons are bound by electrical forces distinguish insulators from conductors. In an insulator, an insignificant fraction of the electrons are free to move in response to applied electrical forces. They are bound securely in the atoms. In a conductor, virtually every atom contributes an electron that is free to move in response to external electrical forces. Thermal energies are sufficient to free these electrons from their host atoms. These free electrons form a pool of **conduction electrons** that is then available for electric currents.

In the absence of an external electric field, electron motions are random. There is no net flow of electrons in any direction and thus no electric current. If an electric field $\mathbf{E}$ is established in the conductor, a force $\mathbf{F} = -e\mathbf{E}$ acts on each conduction electron, producing a net electron flow opposite to $\mathbf{E}$. This is equivalent to a net positive charge flow in the direction of $\mathbf{E}$.

A neutral copper atom has 29 electrons. In a copper wire, each atom contributes an average of about one electron to the conduction electron pool, producing about 10^{29} conduction electrons per cubic meter. In an insulator, the conduction electron number density may easily be 10 orders of magnitude smaller. Between these extremes is a class of materials called semiconductors. Semiconductors are neither good conductors nor good insulators. However, their special electrical characteristics form the basis of our computer-based technologies employing silicon chips in integrated circuits.

Electric current is measured by the net rate of flow of positive charge through a cross section of the conductor (Figure 27.1). A net charge dq flowing through a cross section in time dt produces a current (I)

$$I = \frac{dq}{dt} \tag{27.1}$$

Units of current are coulombs per second, called amperes* and abbreviated A. If the current is constant, Equation 27.1 reduces to

$$I = \frac{q}{t} \tag{27.2}$$

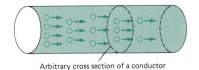

Arbitrary cross section of a conductor

FIGURE 27.1

Electric charges move along a conductor. The electric current is the rate of flow of charge (coulombs/second) through any cross section of the conductor.

* Although this is a meaningful interpretation of current and its units, the ampere is defined operationally in terms of the force exerted on a current-carrying wire in a magnetic field (Section 30.5).

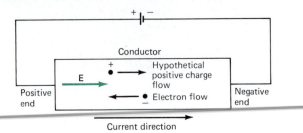

FIGURE 27.2

A battery establishes a potential difference between the ends of a conductor. With the left end of the conductor positive, an electric field is directed to the right. Electrons move to the left. The current is directed to the right, the direction in which positive charges would move.

where q is the net charge flowing through a cross section of the conductor in time t. A constant current of 1 A means that during each second, 1 C of charge flows through a cross section of the conductor. In this chapter we consider only constant currents.

Both positive and negative charges can produce electric currents. The current in an automobile storage battery involves the motion of both positive and negative charges in a liquid electrolyte. An energetic alpha particle produces both positive ions and free negative electrons in the gas of the detecting head of a Geiger counter, and both the positive charges and the negative charges give rise to electric currents. Charges of opposite sign move in opposite directions in a given electric field. An ordinary ammeter cannot determine whether the current detected is due to negative charges moving in one direction or positive charges moving in the opposite direction. With very few exceptions (such as the Hall effect, discussed in Section 29.3), the effects produced by negative charges moving in one direction are identical to those produced by positive charges moving in the opposite direction.

We define the electric current direction to be the direction of positive charge motion. According to this convention, if the charges are actually negative, they move opposite to the direction of the current (Figure 27.2).

Conservation of Charge

The flow of charge in a conductor is similar in some respects to the flow of water through pipes. In the cooling system of an automobile engine, the flow of water in kilograms per second is the same through any portion of the system (Figure 27.3). The plumbing may be circuitous, it may contain internal

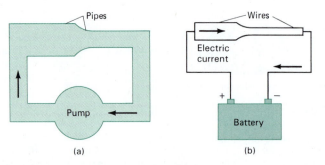

(a)   (b)

FIGURE 27.3

obstructions, and the cross section of the pipes may vary, but as long as there are no mechanisms for losing or gaining water, the rate of mass flow is constant (Figure 27.3a). Mass is conserved. Similarly, a wire may be bent in a peculiar shape, and may have a varying cross-sectional area, but the current is the same in any cross section of the conductor because there are no mechanisms for removing or adding charge (Figure 27.3b). Charge is conserved.

Drift Velocity

The work done by an electric field to establish a current in a conductor increases the kinetic energy of the conduction electrons. The electrons do not achieve unlimited speed because they collide with the ions or atoms that make up the conductor. On the average, kinetic energy is lost in such collisions. The conduction electrons acquire an average velocity in a direction opposite to the electric field. This resultant average electron velocity is called the **drift velocity** and is denoted by $\mathbf{v}_d$.

If the drift velocity is zero, the current is zero. We can relate the drift velocity to the current by considering a conductor of uniform cross-sectional area A (Figure 27.4). The electrons that move past any given point in a time t fill a cylindrical volume Av_dt. Let n denote the number of conduction electrons per cubic meter. The number (N) of conduction electrons in the volume Av_dt is

$$N = nAv_dt$$

The charge passing through the area A is

$$q = Ne = enAv_dt$$

where e is the magnitude of the electron charge. It follows that the current is

$$I = \frac{q}{t} = enAv_d \qquad (27.3)$$

Note that the current is directly proportional to the drift velocity.

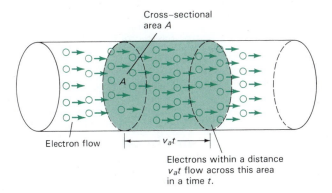

Cross–sectional
area A

Electron flow

v_dt

Electrons within a distance
v_dt flow across this area
in a time t.

FIGURE 27.4

By determining the amount of charge flowing through a cross section in some time interval we can calculate the current.

EXAMPLE 1

Drift Velocity of Electrons in a Metal

Let's estimate the drift velocity of electrons making up a 1-A current in a copper wire 1 mm in diameter.

Solving Equation 27.3 for the drift velocity v_d gives

$$v_d = \frac{I}{enA}$$

To evaluate v_d we need to estimate n, the density of conduction electrons.

Let's assume that each copper atom contributes one electron to the current. Then the number of conduction electrons per cubic meter equals the number of copper atoms per cubic meter. One mole of copper contains Avogadro's number of copper atoms and fills a volume V given by

$$V = \frac{\text{mass of one mole}}{\text{mass density}}$$

The gram molecular weight of copper is 63.5 grams and its mass density is 8.89×10^3 kg/m^3. The volume of 1 mole is

$$V = \frac{0.0635 \text{ kg}}{8.89 \times 10^3 \text{ kg/m}^3}$$
$$= 7.14 \times 10^{-6} \text{ m}^3$$

The number density of copper atoms and conduction electrons is therefore

$$n = \frac{N_A}{V} = \frac{6.02 \times 10^{23}}{7.14 \times 10^{-6} \text{ m}^3}$$
$$= 8.43 \times 10^{28} \text{ electrons/m}^3$$

The cross-sectional area of the conductor is

$$A = \left[\frac{\pi (10^{-3})^2}{4} \right] \text{m}^2 = 7.85 \times 10^{-7} \text{ m}^2$$

For the drift velocity we get

$$v_d = \frac{I}{neA}$$

$$= \frac{1 \text{ A}}{(8.43 \times 10^{28}/\text{m}^3)(1.60 \times 10^{-19} \text{ C})(7.85 \times 10^{-7} \text{ m}^2)}$$
$$= 9.4 \times 10^{-5} \text{ m/s}$$

At this speed, it takes about 3 hours for an electron to drift a distance of 1 meter!

Current Density

Equation 27.3 shows that the current I is directly proportional to the cross-sectional area A. The ratio I/A is an intrinsic measure of the rate of charge flow. We call this ratio the **current density** and denote it by J,

$$J = \frac{I}{A} \tag{27.4}$$

For example, the current density in the wire of Example 1 is

$$J = \frac{I}{A} = \frac{1 \text{ A}}{7.85 \times 10^{-7} \text{ m}^2} = 1.27 \times 10^6 \text{ A/m}^2$$

It follows from Equation 27.3 that the current density can also be expressed as

$$J = nev_d \qquad (27.5)$$

27.2
ELECTRICAL RESISTANCE AND OHM'S LAW

In the previous section we compared the flow of charge in a conductor to the flow of water in a pipe. Let's continue with this analogy.

An open pipe permits water to flow through it but offers resistance to the flow due to viscous friction. An electrical conductor also permits charge to flow through it, but again there is some opposition to the flow. This opposition is measured by a property called the **electrical resistance**, or simply the resistance.

For electrons to flow through the wires of a circuit, a potential difference must be maintained between the ends of the wires by a battery or some other source of electric energy. The electrical resistance (R) of a conductor is defined as the potential difference between the ends of the conductor (V) divided by the current (I) in the conductor,

$$R = \frac{V}{I} \qquad (27.6)$$

Electrical resistance has units of volts/amperes, called ohms, and symbolized by Ω. If a potential difference of 12.1 V across a conductor causes a current of 0.25 A in the conductor, its resistance is

$$R = \frac{12.1 \text{ V}}{0.25 \text{ A}} = 48.4 \ \Omega$$

For many conductors, particularly metals, electrical resistance is essentially constant, regardless of the applied potential difference. Such a conductor is said to obey **Ohm's law**. If we plot the current of a conductor that obeys Ohm's law against the potential difference, we get a straight line (Figure 27.5). The slope of the line is the reciprocal of the resistance. Resistors are conductors obeying Ohm's law. We incorporate resistors in circuits deliberately to impede or limit the current or to produce heat. In electrical circuit diagrams, resistors are represented by ⌇⌇⌇⌇.

In some conductors, the resistance is not constant (Figure 27.6). These conductors do not obey Ohm's law and are called **nonohmic**. Resistance as defined by Equation 27.6 has little practical usefulness for nonohmic conductors. In Section 27.4 we will discuss the concept of dynamic resistance, which does have utility for nonohmic conductors.

27.3
ELECTRICAL CONDUCTIVITY AND ELECTRICAL RESISTIVITY

The electrical resistance of a conductor of a uniform cross section depends on its length and cross-sectional area. As the length (L) increases, the resistance increases. As the cross-sectional area (A) increases, the resistance decreases. Additionally, the resistance of a conductor depends on the type of material

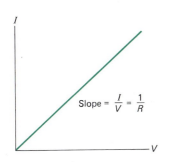

FIGURE 27.5

For a conductor obeying Ohm's law, a plot of current versus potential difference yields a straight line. The slope of the straight line, I/V, equals the reciprocal of the resistance.

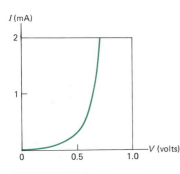

FIGURE 27.6

A plot of current versus potential difference for a solid state diode. The diode does not obey Ohm's law.

from which the conductor is made. Incorporating these ideas, we write the resistance of a conductor as

$$R = \rho\left(\frac{L}{A}\right) \qquad (27.7)$$

where ρ is called the **electrical resistivity**. Resistivity has units of ohm-meters ($\Omega \cdot m$) and is an intrinsic property of the material. It is independent of cross-sectional area and length. The reciprocal of electrical resistivity has units of reciprocal ohm-meters [$(\Omega \cdot m)^{-1}$], and is called **electrical conductivity**; it is denoted by σ. Evidently,

$$\sigma = \frac{1}{\rho} \qquad (27.8)$$

Using the relation $R = \rho(L/A)$, we can rewrite $V = IR$ in a form involving electrical resistivity, current density, and electric field.

For a conductor of uniform cross-sectional area A we can write

$$V = IR = I \cdot \rho\,\frac{L}{A}$$

Rearranging gives

$$\frac{V}{L} = \rho\left(\frac{I}{A}\right)$$

We observe that I/A is the current density (J) and V/L is the electric field (E) established by the potential difference V. Hence,

$$E = \rho J \qquad (27.9)$$

Replacing ρ by $1/\sigma$ we can write this as

$$J = \sigma E \qquad (27.10)$$

The quantities J, ρ, and E are characteristic of a point in the conductor; they are microscopic quantities. It is appropriate to think of $E = \rho J$ as a microscopic form of the relation $V = IR$.

Table 27.1 lists the conductivities of some common materials. Note that the largest and smallest conductivities listed differ by about 20 orders of magnitude. Conductors are customarily categorized as materials with conductivities greater than 10^5 $(\Omega \cdot m)^{-1}$ (Figure 27.7). Materials with conductivities between 10^{-5} and 10^5 $(\Omega \cdot m)^{-1}$ are called semiconductors. The elements germanium and silicon and the compound gallium arsenide are important semiconductors used in the fabrication of diodes, transistors, and integrated circuits. Materials whose conductivities are less than 10^{-5} $(\Omega \cdot m)^{-1}$ are classified as insulators.

EXAMPLE 2

Electrical Resistance of a Solenoid

A solenoid is constructed from 500 m of copper wire having a diameter of 1.5 mm. We can determine its electrical resistance by using Equation 27.7. Thre cross-sectional area of the wire is

$$A = \frac{\pi(1.5 \times 10^{-3}\ \text{m})^2}{4}$$

$$= 1.77 \times 10^{-6}\ \text{m}^2$$

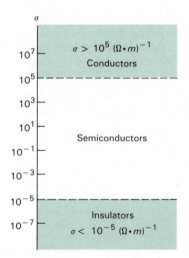

FIGURE 27.7

Classification of materials according to electrical conductivity. The boundaries between the categories are not sharply defined.

TABLE 27.1

Electrical Resistivity, Electrical Conductivity, and the Temperature Coefficient of Resistivity of Some Materials*

Material	Resistivity $(\Omega \cdot m)$	Conductivity $(\Omega \cdot m)^{-1}$	Temperature Coefficient $(/°C)$
Aluminum	2.824×10^{-8}	3.541×10^{7}	0.0039
Brass[†]	7×10^{-8}	1.4×10^{7}	0.002
Constantan[‡]	49×10^{-8}	2.0×10^{6}	0.000002
Copper	1.724×10^{-8}	5.800×10^{7}	0.00393
Gold	2.44×10^{-8}	4.10×10^{7}	0.0034
Iron	10.0×10^{-8}	1.00×10^{7}	0.0050
Nichrome[§]	100×10^{-8}	1×10^{6}	0.0004
Silver	1.59×10^{-8}	6.29×10^{7}	0.0038
Carbon	3500×10^{-8}	2.86×10^{4}	-0.0005
Polyethylene	$10^{8}-10^{9}$	$10^{-9}-10^{-8}$	
Polystyrene	$10^{7}-10^{11}$	$10^{-11}-10^{-7}$	
Neoprene	10^{9}	10^{-9}	
Teflon	10^{14}	10^{-14}	
Glass	$10^{10}-10^{14}$	$10^{-14}-10^{-10}$	
Porcelain	$10^{10}-10^{12}$	$10^{-12}-10^{-10}$	

* Measurements were made at 20°C. For a more complete listing see *Handbook of Chemistry and Physics*, CRC Press, 2000 N. W. 24th Street, Boca Raton, FL 33431.

† Average values; depends on type.

‡ An alloy of equal parts of nickel and copper having a very low temperature coefficient of resistivity.

§ An alloy of nickel and chromium often used to make heating elements.

The electrical resistivity of copper is given as 1.72×10^{-8} $\Omega \cdot$m in Table 27.1. The resistance of the solenoid is

$$R = \rho\left(\frac{L}{A}\right) = \frac{(1.72 \times 10^{-8}\ \Omega \cdot m)(500\ m)}{1.77 \times 10^{-6}\ m^2}$$

$$= 4.88\ \Omega$$

If the solenoid carries a current of 1.25 A, then the potential difference between the ends of the wire is

$$V = IR = 1.25\ A \cdot 4.88\ \Omega$$

$$= 6.10\ V$$

Electrical resistivity depends on temperature. The resistivity of most metals increases as the temperature increases. For temperature intervals of as much as a few hundred degrees, resistivity is related to temperature by a linear expression of the form

$$\rho = \rho_0[1 + \alpha(T - T_0)] \tag{27.11}$$

where ρ_0 is the resistivity at a temperature T_0, generally taken to be 20°C. The quantity denoted by the symbol α is called the **temperature coefficient of resistivity**. Given the resistivity and temperature coefficient of resistivity at temperature T_0, we can compute the resistivity for some different temperature T. The temperature coefficient of resistivity for several materials measured at 20°C is shown in Table 27.1.

The change of resistivity with temperature makes possible electrical thermometers. Platinum is commonly used in these thermometers. Commercial temperature-dependent resistors called thermistors are made of semiconducting materials. Thermistors rely on the fact that a small decrease in temperature can increase the resistivity of a semiconductor by an order of magnitude or more.

EXAMPLE 3

Change of Resistance with Temperature

The resistance of a conductor is given by $R = \rho L/A$. The resistivity, ρ, the length, L, and the cross-sectional area, A, all change if the temperature of the conductor changes. Which is more important, the temperature variation in resistivity, or thermal expansion that causes changes in L and A?

From Table 27.1 we see that the temperature coefficient of resistivity is *roughly* $10^{-3}/C°$. This means that a temperature change of 10 C° causes a change of roughly 1% in ρ and a corresponding change in R. The coefficient of linear expansion for solids is typically $10^{-5}/C°$ (see Table 19.1). A temperature change of 10 C° causes changes in linear dimensions of approximately 0.01%, about 100 times less than the change in resistivity. Thus, when determining the temperature variation in resistance we need to consider only the change in resistivity. The minor changes in L and A can be ignored.

Superconductivity

In 1911, the Dutch physicist J. Kamerlingh Onnes discovered that the resistivity of mercury decreased abruptly to zero at a temperature of 4.153 K (modern value). The resistivity remained zero as long as the temperature remained below 4.153 K (Figure 27.8). The total absence of resistivity is called **superconductivity** and materials showing this behavior are termed **super-**

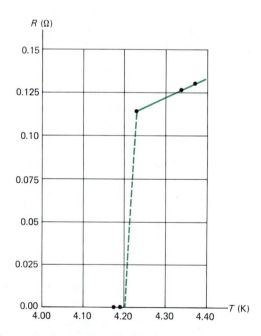

FIGURE 27.8

The resistance of mercury versus temperature as reported by Kamerlingh Onnes. The abrupt decrease at about 4.20 K marks the onset of superconductivity.

conductors. Once a current is initiated in a ring of superconducting material, the current will persist undiminished as long as the material is maintained in the superconducting state. Since Onnes' discovery of mercury's superconductivity, numerous other metals and alloys have been found to be superconductors (see Section 41.7).

Before 1986, superconductivity could be achieved only at temperatures maintained by liquid helium. This relatively expensive refrigerant relegated superconductivity to a laboratory phenomenon. Recent discoveries have led to the development of ceramic materials that become superconducting at temperatures that can be maintained by liquid nitrogen (77 K), a refrigerant that costs less than beer. The development of room-temperature superconductors would revolutionize the transmission of electrical energy by eliminating the wasteful generation of heat. Many experimental devices requiring large magnetic fields utilize superconducting materials in the magnets, as we will see in that chapter.

27.4
DYNAMIC RESISTANCE

The relation $R = V/I$ is useful because so many conductors obey Ohm's law. One determination of the resistance of a conductor establishes the relation between V and I, and the current can then be calculated for any given applied potential difference. Conductors that do not obey Ohm's law are *nonlinear*; the potential difference across them is not directly proportional to the current in them. Such is the case with a solid-state diode (Figure 27.9). The resistance as defined by $R = V/I$ has little practical use for nonlinear conductors. A more useful parameter for nonlinear conductors is the **dynamic resistance**, defined to be dV/dI:

$$\text{dynamic resistance} \equiv \frac{dV}{dI} \qquad (27.12)$$

The dynamic resistance reveals how the potential difference changes for small changes in current. Of particular interest are current ranges over which dV/dI

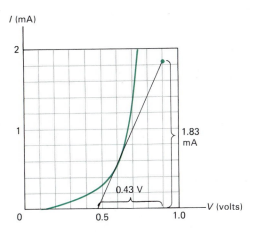

FIGURE 27.9

FIGURE 27.10

A plot of current versus potential difference for a tunnel diode. The dynamic resistance dV/dI is negative between points A and B.

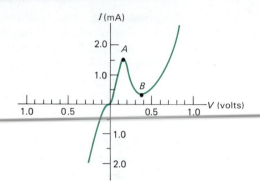

is negative (Figure 27.10). With a negative dynamic resistance, a decrease in potential difference produces an increase in current.

EXAMPLE 4

Determination of Dynamic Resistance

Figure 27.9 shows the relation between current and potential difference for a solid-state diode. Let us estimate the dynamic resistance when the potential difference across the diode is 0.6 V.

By definition, the dynamic resistance is dV/dI, the slope of the V versus I curve. In Figure 27.9 we have drawn a tangent to the curve at $V = 0.6$ volt. Measuring the slope, we find for the dynamic resistance.

$$\frac{dV}{dI} = \frac{0.43 \text{ V}}{1.8 \times 10^{-3} \text{ A}}$$

$$= 230 \ \Omega$$

To illustrate the negative resistance characteristic, we consider the neon glow lamp shown in Figure 27.11. (These reddish-glowing lamps are used as bedroom night-lights.) We connect the lamp in series with a resistor and a variable potential difference, and then record the potential difference across the glow lamp and the current in the glow lamp. We start with zero potential difference across the lamp. An increase in potential difference produces zero current until the critical value labeled V_f in Figure 27.12a is achieved. In the

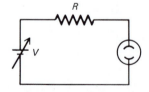

FIGURE 27.11

A neon glow lamp connected in series with a resistor and a potential difference that can be varied. The glow lamp consists of two electrodes in a glass envelope containing neon gas at low pressure. A potential difference of sufficient magnitude across the electrodes will produce an electric discharge in the gas. The discharge gives rise to a reddish glow often seen in bedroom night lights.

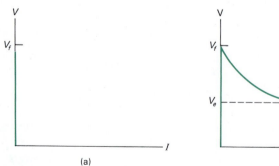
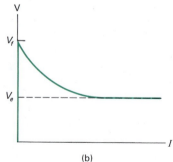

FIGURE 27.12

(a) As the potential difference across the glow lamp (Figure 27.11) increases, there is no current until a value V_f is achieved. (b) Once the potential difference V_f is reached, an electrical discharge occurs and the lamp becomes conductive. The potential difference across the lamp decreases rapidly to a value V_e.

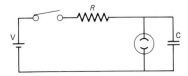

FIGURE 27.13

A modification of the circuit in Figure 27.11. Because the capacitor and neon glow lamp are connected in parallel, the potential difference across them is the same.

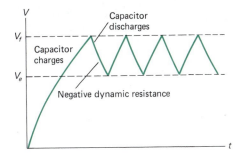

FIGURE 27.14

Once the potential difference V_f is reached, the glow lamp becomes conductive, short circuiting the capacitor. Charge leaves the capacitor plates and the potential difference decreases rapidly to V_e, whereupon the glow lamp becomes nonconductive. A cyclic charging and discharging action ensues.

potential difference range from 0 to V_f, $dV/dI \to \infty$; the dynamic resistance is infinite. Once V_f is achieved, the lamp suddenly becomes conductive and glows. The potential difference across the lamp decreases as the current in it increases. In the potential difference range $V < V_f$, $dV/dI < 0$; the dynamic resistance is small, negative, and rapidly approaches a near-zero value. The potential difference across the lamp decreases and approaches the value labeled V_e in Figure 27.12b. The circuit current is given by

$$I = \frac{V - V_e}{R}$$

where V is the potential difference provided by the battery.

The glow lamp can be made to blink on and off by connecting a capacitor in parallel with the lamp (Figure 27.13). Initially there is no charge on the capacitor. Closing the switch completes the circuit and charge accumulates on the capacitor plates. No charge flows through the lamp until the critical potential V_f is reached. Then it becomes conductive and its dynamic resistance becomes very low. The low resistance of the bulb allows the capacitor to discharge rapidly. When the potential difference across the capacitor falls below V_e, the bulb becomes non-conductive; its dynamic resistance becomes very large and the capacitor recharges, beginning a new cycle. The potential difference across the bulb and capacitor varies with time as shown in Figure 27.14.

27.5
ELECTRIC POWER

In a radio, electric energy is changed into sound energy; in a toaster, electric energy is transformed into heat; in the starter of an engine, electric energy is converted into mechanical energy; and in a headlamp, electric energy becomes light. The rate at which an electrical device transforms energy is practical information. With it, we can determine how much energy is used in a given time, and then, knowing the price of energy, we can compute the cost of operating the device.

Regardless of the nature of the electrical device or the manner of producing the potential difference, the electric energy converted by a charge dq moving through a potential difference V is

$$dU = V \, da \tag{27.13}$$

Equation 27.13 follows directly from the definition of potential difference (Section 25.1). The rate at which energy is converted is

$$\frac{dU}{dt} = V\frac{dq}{dt}$$

Replacing dq/dt by the current I gives

$$\frac{dU}{dt} = VI$$

The rate of converting energy is power, and is denoted as P. Thus, the **electric power** developed is

$$P = VI \tag{27.14}$$

Electric power can be expressed in the equivalent units of either volt·amperes, watts, or joules per second.

Equation 27.14 is a general relation for any type of device through which the current is I and across which the potential difference is V. If some special relation between V and I exists for a particular device, then this relation can be incorporated in Equation 27.14. For example, if the connected device is a resistor, we can replace V with IR and obtain

$$P = I^2R \tag{27.15}$$

or we can replace I with V/R and obtain

$$P = \frac{V^2}{R} \tag{27.16}$$

Electrons migrating through a resistor lose energy through collisions, thereby increasing the internal energy of the resistor. As a result, the temperature of the resistor rises. This effect is called the Joule effect and the thermal energy produced is called **joule heat**. Electric stoves, toasters, and hair dryers, for example, produce joule heat.

EXAMPLE 5

Electric Water Heater

Heaters used to warm small volumes of water consist of a coil of wire that is immersed in the water to be heated. One such heater carries a current of 5 A when connected to a 120-V source. Let us determine how long it would take for this heater to bring 0.5 liters (5×10^{-4} m^3) of water (enough for two cups of coffee) to a boil if the initial temperature of the water was 27°C, assuming that all the joule heat developed in the coil is absorbed by the water.

The energy required to warm m kg of water by ΔT degrees is

$$Q = mC\,\Delta T$$

where C is the specific heat of water. Inserting numerical values, we have

$$Q = 10^3 \text{ kg/m}^3 \cdot 5 \times 10^{-4} \text{ m}^3 \cdot 4.19 \text{ kJ/kg·C}° \cdot (100 - 27) \text{ C}°$$
$$= 153{,}000 \text{ J}$$

The power developed by the heater coil is

$$P = VI = 120 \text{ V (5 A)} = 600 \text{ W}$$
$$= 600 \text{ J/s}$$

The thermal energy produced by the heater in time t is Pt. Setting Pt equal to Q gives

$$t = \frac{Q}{P} = \frac{153{,}000 \text{ J}}{600 \text{ J/s}} = 255 \text{ s}$$

$$= 4.3 \text{ min}$$

What would be the cost of heating the water if the company that supplies the electric power charges 10¢ per kilowatt-hour? The energy, expressed in kilowatt-hours, is

$$Q = 0.6 \text{ kW} \cdot 4.3 \text{ min} \cdot \frac{1 \text{ hour}}{60 \text{ min}}$$

$$= 0.043 \text{ kWh}$$

At 10¢ per kilowatt-hour, the total cost is

$$\text{cost} = 0.043 \text{ kWh} \ (10\text{¢/kWh})$$

$$= 0.43\text{¢}$$

WORKED PROBLEM

A 2-ohm resistor and a solid-state diode are connected so that they carry the same current. The current-potential difference (I-V) relation for the diode is

$$I = I_0(e^{V/V_0} - 1) \qquad V_0 = 0.056 \text{ volt} \qquad I_0 = 200 \text{ nA}$$

(a) For what current is the dynamic resistance equal to 2 ohms?
(b) What are the potential differences across the 2-ohm resistor and the diode when they carry the current determined in part (a)?

Solution

(a) Start by deriving an expression relating the current I to the dynamic resistance, dV/dI. To introduce dV/dI, differentiate the I-V relation with respect to I:

$$1 = I_0 \left(\frac{1}{V_0} \right) e^{V/V_0} \left(\frac{dV}{dI} \right)$$

Solve this for $I_0 e^{V/V_0}$ and substitute back into the I-V relation. This gives,

$$I_0 e^{V/V_0} = \frac{V_0}{dV/dI}$$

and

$$I = V_0 \cdot \left(\frac{dV}{dI} \right)^{-1} - I_0$$

This equation can be used to calculate the current for any value of the dynamic resistance. With $dV/dI = 2 \ \Omega$,

$$I = \left(\frac{0.056 \text{ V}}{2 \ \Omega} \right) - (2 \times 10^{-7} \text{ A}) = 0.028 \text{ A}$$

$$= 28 \text{ mA}$$

(b) The potential difference across the resistor is

$$V_R = RI = 2 \ \Omega \ (0.028 \text{ A})$$

$$= 0.056 \text{ V}$$

The potential difference across the diode follows from the *I-V* relation,

$$V = V_0 \ln\left(1 + \frac{I}{I_0}\right) = (0.056 \text{ V}) \ln\left(1 + \frac{28 \times 10^{-3} \text{ A}}{2 \times 10^{-7} \text{ A}}\right)$$

$$= 0.66 \text{ volt}$$

EXERCISES

27.1   Introduction to Electric Current

A.   How many electrons per second flow through any cross section of a wire carrying a current of 1 ampere?

B.   In a high-energy particle accelerator, protons are moving at nearly the speed of light. If the electric current delivered to the target by the protons is 1 mA, how many protons hit the target in one second?

C.   Alpha particles from a Van de Graaff accelerator deposit 153 μC of charge in a target in 5 minutes. Determine (a) the number of alpha particles that strike the target, (b) the current produced by the beam.

D.   (a) What is the speed of electrons accelerated through a potential difference of 19 kV in a cathode-ray oscilloscope? (b) If the diameter of the electron beam is 1 mm and the current is 130 μA, what is the number density of electrons in the beam?

E.   A fuse in an electrical circuit is designed to open the circuit like a switch when the current exceeds some preset value. If a fuse is made of a material that melts when the current density reaches 400 A/cm^2, what diameter wire is needed to limit a current to 0.25 A?

27.2   Electrical Resistance and Ohm's Law

F.   An orbital electron in a hydrogen atom has a kinetic energy of 13.6 eV. If the radius of the orbit is 5.3×10^{-11} m, what is the electric current produced by the orbiting electron?

G.   A geologist measures a potential difference of 3 mV between two probes planted 6 m apart in the earth. The current between the probes is 10^{-12} A. Determine the resistance between the probes.

H.   An automobile starter delivers 200 A when driven by a 12-volt battery. Assuming that all electrical resistance in the circuit is in the starter, what is the starter resistance?

I.   Imagine an aluminum wire with a diameter of 4 cm circling the earth at the equator. (a) Determine the resistance of the wire. (b) Determine the weight of the wire in metric tons.

27.3   Electrical Conductivity and Electrical Resistivity

J.   A current of 1 A exists in a copper wire 1 m long and 0.1 mm in diameter. Determine (a) the electric field within the wire, (b) the potential difference between its ends.

K.   A Nichrome wire 2 m long and 0.2 mm in diameter is connected to a 100-V battery. Determine the current in the wire.

L.   Two copper wires having the same length but different cross-sectional areas are connected to the same battery. How does the current density in the two wires compare?

M.   The conductors that carry current from the generators of large electric power plants are often in the form of a hollow metallic pipe several inches in diameter. Calculate the resistance of an aluminum pipe 2 cm thick and 3 m long with an outside diameter of 20 cm.

N.   The copper wire leads for a 1/8 — W commercial resistor are 0.635 mm in diameter and 8 cm long. Determine the resistance of the leads. Does your calculation show that the lead resistance can normally be neglected in most applications?

27.4   Dynamic Resistance

P.   The current-potential difference relationship for a solid-state diode is shown in Figure 1. Estimate the resistance and dynamic resistance for currents of 0.5 and 1.5 mA.

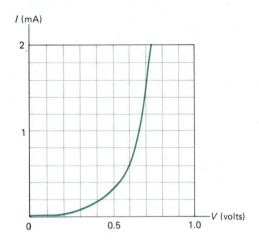

FIGURE 1

Q.   A device has a potential difference-current characteristic described by the equation

$$V = V_1 + \frac{P}{I}$$

where $V_1 = 9 \times 10^4$ V and $P = 10^5$ W. When the current $I = 10$ A, calculate (a) the ratio V/I, (b) the dynamic resistance. Express your answers in ohms.

R. Figure 27.10 shows the characteristic curve for a tunnel diode. Estimate the average dynamic resistance for the diode over the region between A and B.

27.5   Electric Power

S. An electron-beam machine delivers 200 A at 250 kV for 1.10 μs. (a) What is the power in watts during this time interval? (b) How much energy in joules is delivered in this time interval?

T. A 10-W light bulb is connected to a 12-V battery having negligible internal resistance. Determine (a) the current in the bulb, (b) the resistance of the bulb.

U. The electrical wiring in household kitchens normally incorporates #12 copper wires (diameter = 0.0808 in). The National Board of Fire Underwriters determined the maximum current-carrying capacity of a well-insulated #12 copper wire to be 25 A. For a #12 copper wire carrying 25 A determine (a) the current density in the wire, (b) the electric field in the wire, (c) the potential difference between the ends of 30 m of the wire, (d) the rate of joule heating in the wire.

V. A circuit breaker inserted in the wiring of an electrical circuit opens the circuit when the current achieves some critical value. An electrical circuit in an automobile powered by a 12-V battery utilizes a circuit breaker designed to open when the current reaches

3 A. What is the maximum electric power that can be delivered by the battery?

W. Suppose that you want to install a heating coil that will convert electric energy to heat at a rate of 300 W for a current of 1.5 A. (a) Determine the resistance of the coil. (b) The resistivity of the coil wire is 10^{-6} $\Omega \cdot$m. The diameter of the wire is 0.3 mm. Determine its length.

X. The bulb in a two-cell flashlight is rated at 2 W. If the flashlight is left on continuously, the batteries go dead in about 1 hr. (a) Determine the number of kilowatt-hours of energy produced during the lifetime of the batteries. (b) If the batteries cost 35¢ each, what is the cost per kilowatt-hour of energy? Compare this with a cost of 10¢ per kilowatt-hour from an electric utility company. What do you really buy when you use batteries to produce light?

Y. An electric heater with a resistance of 20 Ω requires 100 V across its terminals. A built-in switching circuit repetitively turns the heater on for 1 s and off for 4 s. (a) How much energy is produced by the heater in 1 hr? (b) What is the average power delivered by the heater over a period of one cycle?

Z. An electric water heater has a 60-gal capacity and a 5000-W heating element. Assuming that all the electric energy is transferred to the water, determine (a) the cost of heating 60 gallons of water from 70°F to 135°F if the electric energy costs 10¢ per kilowatt-hour, (b) the time required to do the heating.

PROBLEMS

1. Suppose that the southbound traffic on a straight north-south interstate highway moves with an average speed v. If the number of vehicles per mile of highway is n and the average mass of a vehicle is M, derive an expression for the rate at which mass moves past any point in the southbound lane. What is the relationship between this exercise and that leading to Equation 27.3?

2. A fluid of density ρ (kg/m^3) moves through a pipe of uniform cross section A (m^2) at a constant speed v (m/s). The rate of flow of energy through a unit area is $\frac{1}{2}\rho v^3$. (a) Using dimensional analysis, show that this expression has the characteristics of intensity. (b) Derive this relation by following the treatment in Section 27.1 for the flow of charge in a conductor.

3. The current in a wire varies with time according to the relation $I = 2.5e^{-at}$ mA where $a = 0.833$ s^{-1}. Determine the total charge that has flowed through the wire by the time the current has diminished to zero.

4. An electric discharge along a cylindrically shaped tube produces a cylindrically symmetric current. The cur-

rent density at a distance r from the axis of the tube can be represented by

$$J = -ar + b \qquad r \leq 0.01 \text{ m} \qquad a = 15 \text{ A/m}^3$$
$$J = 0 \qquad r > 0.01 \text{ m} \qquad b = 0.15 \text{ A/m}^2$$

Determine the total current in the discharge.

5. A wire can be stretched within limits without permanently deforming the shape. This strain changes the shape and resistance of the wire. Determine the change in resistance of a copper wire undergoing a strain of 0.001 if its unstrained resistance is 10.0 ohms. Take Poisson's ratio for copper to be 0.32.

6. An aluminum rod has a resistance of 1.234 ohms at 20°C. Calculate the resistance at 120°C by accounting for the change in both the resistivity and dimensions of the rod.

7. A student produces a net charge of 3 μC by running a comb through her hair. She then runs at a speed of 5 m/s toward a friend. If the charge on her head is spread over a distance of 12 cm (parallel to her motion), what current is created by its motion?

8. The proton beam of a 6-MeV Van de Graaff generator constitutes an electric current of 1.0 μA delivered to the target. (a) How many protons hit the target per second? (b) With each proton having 6-MeV of kinetic energy, how much energy (in million electron volts and in joules) is delivered to the target in 1 s? (c) What is the power input to the target?

9. A pan containing 1.2 liters of water rests on a 1000-W heating element of an electric stove. If the water is initially at a temperature of 25°C and is brought to a boil in 10.2 min, what is the energy efficiency of this arrangement for boiling water? (Here, energy efficiency is the ratio of energy required to heat the water to electrical energy supplied.)

10. A current density J exists in a conductor having a resistivity ρ and uniform cross-sectional area. (a) Show that the rate of joule heating per unit volume of the conductor is ρJ^2. (b) Calculate the rate of joule heating per unit volume in an aluminum wire 2 mm in diameter carrying a current of 0.83 A.

11. An aluminum wire 2 m long and 2 mm in diameter is surrounded by a cylindrical copper covering 1 mm thick. A 1-V potential difference is applied between the ends of the wire. Determine the current and thermal power in each component of the wire.

12. A common two-terminal electronic element called a *Zener diode* has a potential difference-current relationship closely approximated by Figure 2. Make a sketch of the resistance of the Zener diode versus the potential difference.

13. The resistance of a special type of wire decreases as the current through it increases. The equation describing the resistance-current relation is

$$R = \frac{100\ \Omega}{6 + bI^3}$$

where I is the current in amperes and $b = 2\ \mathrm{A}^{-3}$. (a) Determine the current which maximizes the power developed in the wire. (b) Determine the maximum power.

14. A resistor at 20°C has resistance R_0. (a) Show that if this temperature is changed by some small amount ΔT, its resistance changes by $\Delta R = \alpha R_0\,\Delta T$. (b) Resistors connected in series have a total resistance equal to the sum of the separate resistances. If two resistors in series have positive and negative temperature coefficients of resistivity, then the change in resistance of one resistor tends to be compensated by the opposite resistance change of the other. Show that the compensation for two series resistors R_1 and R_2 is exact if $R_1/R_2 = -\alpha_2/\alpha_1$. (c) Determine the values of a carbon resistor and an iron resistor if it is desired to maintain $R_{\text{carbon}} + R_{\text{iron}} = 100\ \Omega$.

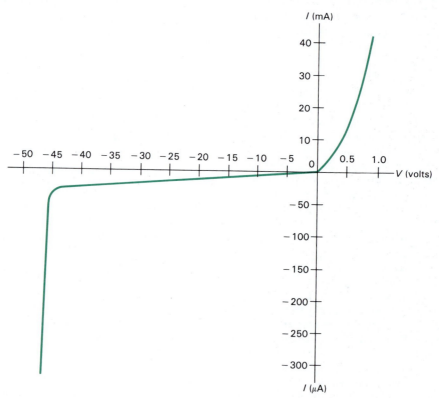

FIGURE 2

15. A more general definition of the temperature coefficient of resistivity is

$$\alpha = \frac{1}{\rho}\frac{d\rho}{dT}$$

where ρ is the resistivity at temperature T. (a) Assuming that α is constant, show that

$$\rho = \rho_0 e^{\alpha(T-T_0)}$$

where ρ_0 is the resistivity at temperature T_0. (b) Using the series expansion ($e^x \approx 1 + x; x \ll 1$), show that the resistivity is given approximately by $\rho = \rho_0[1 + \alpha(T - T_0)]$ for $\alpha(T - T_0) \ll 1$.

16. A student in an undergraduate laboratory is given the task of determining the resistivity of an unidentified length of wire. She makes measurements of the dimensions of the wire and current in the wire for a variety of applied potential differences between the ends of the wire. From the summary of the measurements given in Table 1 deduce the resistivity and identify the composition of the wire.

TABLE 1
Diameter = 0.512 mm
Length = 10.1 m

V (volts)	I (mA)
2.00	80
4.00	173
6.00	252
8.00	328
10.0	421
12.0	496

17. A fuse is constructed from aluminum. Assume the fuse is cylindrical and that the temperature rises at the rate of 100 C°/s when $I = 30$ A. Calculate the radius of the fuse.

18. A beam of 10-MeV alpha particles is collected in a copper cup (Faraday cup). The beam current is 5 μA. (a) How long does it take to collect 10^{15} alpha particles? (b) What temperature rise occurs during this time if the cup mass is 1 kg? Assume that no heat is lost through conduction, convection, and radiation.

19. The heating element of a coffee maker operates at 120 V and carries a current of 2 A. Assuming that all of the heat generated is absorbed by the water, how long does it take to heat 0.5 kg of water from room temperature (23°C) to the boiling point?

20. The wire windings in a magnet require 50 A from a 250-V supply. The wires are hollow to permit cooling water to flow through them. If the water inlet and outlet temperatures differ by 5 C°, what is the water flow rate expressed in liters/min?

21. A conductor of uniform cross section is constructed from a material whose resistivity varies along its length. If the resistivity can be represented by $\rho = \rho_0 + \rho_0 bx$, where x denotes the position in meters along the wire ($x = 0$ denotes one end) and b is a constant, derive an expression for the resistance of a wire of length L and cross-sectional area A.

22. A copper bar 10 cm long with a rectangular cross section of dimensions 1 cm × 2 cm is to be used to conduct 100 A of current between two opposite faces. (a) Through which two opposite faces should the charge flow in order to minimize the potential difference between the faces? (b) What is the current density in the bar? (c) What is the potential difference between the faces?

23. The resistivity of copper is 1.724×10^{-8} $\Omega \cdot$m; that of aluminum is 2.824×10^{-8} $\Omega \cdot$m. What diameter aluminum wire will have the same resistance per unit length as a copper wire 1 mm in diameter?

24. A wire is stretched to twice its original length by drawing it through a tiny hole. If the wire assumes a uniform cross-sectional area in the stretched configuration and its volume is unchanged by the stretching, find the relationship between the resistance of the stretched and unstretched wires.

25. A solenoid 10 cm long is formed by winding 25 layers of coil from a continuous length of copper wire 1.25 mm in diameter on a core 2 cm in diameter. Determine the resistance of the wire.

26. The potential difference across the terminals of a device is related to the current it carries by

$$V = aI - bI^2 \qquad a = 6 \text{ V/A} \qquad b = 2 \text{ V/A}^2$$

(a) Determine the current for which the power developed by the device is a maximum. (b) Determine the maximum power.

CHAPTER 28

DIRECT CURRENT CIRCUITS

28.1
SOURCES OF ELECTRIC ENERGY

A flashlight battery is a source of electric energy that maintains a potential difference between its terminals and that is capable of moving charges through a circuit. A battery is one example of a source of electric energy called a **source of emf** (pronounced like the series of letters "e," "m," "f"). The letters emf stand for **e**lectro**m**otive **f**orce, which is somewhat of a misnomer since the strength of a source of emf is not measured by a force.

A battery connected to a light bulb provides the electric energy to move charges through the filament of the bulb. Electric energy is converted to thermal energy and radiant energy when the charges move through the filament from the higher potential to the lower potential. The process is analogous to a ski lift (a source of energy) that raises skiers to an elevated position, thereby increasing their gravitational potential energy. The gravitational potential energy of the skiers is converted into kinetic energy and mechanical work when the skiers slide down the slope.

Some important types of sources of emf are (a) batteries used to convert chemical energy to electric energy; (b) biological sources that convert chemical energy to electric energy in nerve and muscle cells; (c) temperature-measuring thermocouples consisting of a junction of two dissimilar metals in which thermal energy is converted to electric energy; (d) photovoltaic cells that convert radiant energy into electric energy (used in light-metering systems of cameras and in satellite electric power supplies); (e) electric generators that convert mechanical energy to electric energy (used in electric power plants).

It is possible for a source of emf to generate currents that vary in both magnitude and direction. For example, the electric generator in a commercial power plant produces currents that change direction 120 times per second. However, a source of emf such as an automobile battery generates currents that do *not* change direction. These currents are called **direct currents**, abbreviated dc.

The strength ($\mathscr{E}$) of a source of emf is defined as the potential difference between its terminals when there is no current through it (open circuit potential difference). This potential difference is also called the emf of the source. Thus, the term "emf" is used in two related ways: to describe the energy source and to measure its terminal potential difference. Physically, the emf measures work per unit charge and is expressed in units of joules per coulomb, or **volts**. For example, a commercial type AA flashlight battery is a source of emf that maintains an emf of 1.5 volts between its terminals.

EXAMPLE 1

Work Done by a Source of emf

A flashlight battery maintains a constant emf of 1.50 volts between its terminals and supplies a current of 300 mA. Let's calculate the work done by this source of emf in one minute.

The charge moved from the negative terminal (lower potential) to the positive terminal (higher potential) in one minute is

$$q = It = 0.300 \text{ A} \cdot 60 \text{ s}$$
$$= 18.0 \text{ C}$$

FIGURE 28.1

Representation of a source of emf including its internal resistance. The circles at A and B denote the terminals of the source.

The work done by the battery in moving this charge is

$$W = \mathscr{E}q = (1.50 \text{ J/C})(18.0 \text{ C})$$
$$= 27.0 \text{ J}$$

Chemical energy is converted to electric energy to account for this work.

Not all of the energy converted to electric energy by a source of emf is available to devices connected to the source. Some of the electric energy is converted *internally* into thermal energy (joule heat). For example, you may have noticed that a battery feels warm to the touch when it is providing energy for a light bulb. This thermal energy is accounted for by attributing an **internal resistance** to a source of emf. Figure 28.1 shows the symbol used to denote the emf ($\mathscr{E}$) and the internal resistance (r) of a source of emf. The small circles at A and B denote the terminals of the source.

If the source of emf is connected to a resistor, the equivalent circuit is as shown in Figure 28.2. There is no mechanism for removing or storing charge anywhere in the circuit. Charge is conserved. Hence, the current is the same at all points in the circuit, and the current leaving the positive terminal A equals the current entering the negative terminal B.

The power delivered by a source of emf is

$$P = \mathscr{E}I \tag{28.1}$$

Electric energy is transformed into thermal energy through the external resistance R and the internal resistance, r. The rates of thermal energy generation are I^2R and I^2r. Because energy is conserved, we can equate the rate of thermal energy generation to the power delivered by the source of emf,

$$I^2R + I^2r = \mathscr{E}I \tag{28.2}$$

Solving for the current we have

$$I = \frac{\mathscr{E}}{r + R} \tag{28.3}$$

Equation 28.3 illustrates the important role of the internal resistance, r. If $r \ll R$, as it is for an automobile battery in good condition, then $I \approx \mathscr{E}/R$, and the internal resistance has little effect on the current. If the internal resistance increases to the point where $r \gg R$, then $I \approx \mathscr{E}/r$, and the current in the external circuit is determined largely by the internal resistance. When a battery goes dead, it is not necessarily because its emf decreases. Rather, it is probably because the internal resistance increases and prevents adequate current in the

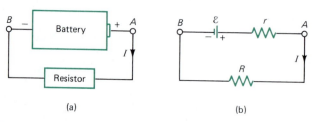

(a)   (b)

FIGURE 28.2

(a) A battery connected to a resistor such as a light bulb. (b) The circuit diagram for the battery and resistor.

external circuit. In subsequent discussions we assume that the emf ($\mathscr{E}$) and internal resistance (r) are constant.

With a current I in the circuit shown in Figure 28.2, the potential difference across the terminals of the source of emf is

$$V = \mathscr{E} - Ir$$

Only if there is no current delivered will the terminal potential difference equal the emf. If current is delivered, then $V < \mathscr{E}$. The difference between $\mathscr{E}$ and V increases as the current delivered increases.

EXAMPLE 2

Strength and Internal Resistance of a Battery

When a 5-Ω resistor is connected to a flashlight battery, a current of 0.25 A is produced. Replacing the 5-Ω resistor by a 9-Ω resistor produces a current of 0.15 A. We can determine the internal resistance and the emf of the battery from these measurements.

Using Equation 28.3, we can write $\mathscr{E} = Ir + IR$. This gives

$$\mathscr{E} = (0.25 \text{ A})r + 0.25(5 \ \Omega) \qquad \text{5-}\Omega \text{ resistor connected}$$

$$\mathscr{E} = (0.15 \text{ A})r + 0.15(9 \ \Omega) \qquad \text{9-}\Omega \text{ resistor connected}$$

Solving these two equations, we find

$$r = 1.0 \ \Omega$$

$$\mathscr{E} = 1.5 \text{ V}$$

28.2
KIRCHHOFF'S VOLTAGE RULE

Suppose that you are presented with several sources of known emf and known internal resistance and several additional resistors of known resistance. You then proceed to draw a diagram depicting how these components will be connected to form a circuit. Is it possible to calculate the expected currents in each component and the potential difference across each component when the circuit is actually constructed? In fact, this *can* be done with the help of two rules called Kirchhoff's voltage rule and Kirchhoff's current rule. Let's first discuss the voltage rule.

Consider the circuit shown in Figure 28.3. There is a definite electric potential

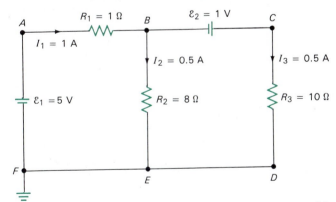

FIGURE 28.3
A circuit to illustrate Kirchhoff's voltage rule. Point *F* is grounded and so is at zero potential.

at each point in the circuit. For example, if the potential of point F is zero, the potential at point A is $+5$ volts. We can measure the potential at different points in the circuit and then evaluate the changes in potential from point to point. Regardless of how the potential changes as we move around the circuit, when we complete the loop by returning to point A, the potential there is still $+5$ volts. In other words, there can be no change in potential around a complete loop. This means that the *sum* of the *changes* in potential must be zero for a complete loop. The formal statement of this requirement is called **Kirchhoff's voltage rule**:

> The sum of potential changes is zero around any closed loop in an electrical circuit.

EXAMPLE 3

Verification of Kirchhoff's Voltage Rule

We can use the circuit of Figure 28.3 to verify Kirchhoff's voltage rule. Let's start at point F where the potential is zero and evaluate the potential changes as we move around the loop *FABEF*. Moving from F to A, or, in other words, from the negative to the positive terminal of the source with an emf $\mathscr{E}_1 = 5$ volts, the potential increases from zero to $+5$ volts. So, the potential change is $+5$ volts.

Moving with the current from A to B the potential decreases 1 volt. This result follows from Ohm's law,

$$I_1 R_1 = 1 \text{ A} \cdot 1 \text{ } \Omega = 1 \text{ V}$$

Because the voltage dropped 1 volt going from A to B the potential at B is $+4$ volts. Moving with the current from B to E the potential drops 4 volts. This result also follows from Ohm's law,

$$I_2 R_2 = 0.5 \text{ A} \cdot 8 \text{ } \Omega = 4 \text{ V}$$

Because the potential decreased by 4 volts going from B to E the potential at E is zero. There is no change of potential as we complete the loop by moving from E to F because E and F are at the same potential. Summing the potential changes, we verify Kirchhoff's voltage rule:

$$+5 \text{ V} - \quad 1 \text{ V} \quad - \quad 4 \text{ V} \quad + \quad 0 \text{ V} = \quad 0 \text{ V}$$
$$\text{F to A} \quad \text{A to B} \quad \text{B to E} \quad \text{E to F} \quad \text{Loop } FABEF$$

Kirchhoff's voltage rule applies to all loops. As a second illustration, consider the loop *BCDEB*. The potential at B is $+4$ volts. Moving from B to C gives a potential rise of 1 volt, so the potential at C is $+5$ volts. Moving with the current from C to D the potential falls by 5 volts:

$$I_3 R_3 = 0.5 \text{ A} \cdot 10 \text{ } \Omega = 5 \text{ V}$$

The potential at D is zero. There is no change in potential from D to E because these points are at the same potential. Moving *against* the current from E to B we experience a potential rise of 4 volts,

$$I_2 R_2 = 0.5 \text{ A} \cdot 8 \text{ } \Omega = 4 \text{ V}$$

Summing the potential changes, we again verify Kirchhoff's voltage rule:

$$+1 \text{ V} - \quad 5 \text{ V} \quad + \quad 0 \text{ V} \quad + \quad 4 \text{ V} = \quad 0 \text{ V}$$
$$\text{B to C} \quad \text{C to D} \quad \text{D to E} \quad \text{E to B} \quad \text{Loop } BCDEB$$

In general, the potential difference across a resistor (R) in which the current is I equals $\pm IR$. The potential change is $+IR$ if the movement is against the current—in a sense, uphill from lower to higher potential. The potential change is $-IR$ if the movement is with the current—downhill from higher to lower potential. A source with an emf $\mathscr{E}$ contributes a potential change of $\pm \mathscr{E}$. The positive or negative character ($\pm$) of these potential changes depends on the direction of the loop. If the loop chosen involves a movement from the negative to the positive terminal of a 10-volt source, the potential change is $+10$ volts. If the loop involves a movement from the positive to the negative terminal, the potential change is -10 volts.

EXAMPLE 4

The Wheatstone Bridge

The **Wheatstone bridge** is a dc circuit used to measure resistance (Figure 28.4). The circuit consists of a known resistance R_3, an unknown resistance R_x, and a uniform wire of total resistance $R_1 + R_2$. A movable contact C connects a detector to the junction between R_x and R_3 to a point along the wire. A source of emf supplies current to the bridge and maintains a fixed potential difference between the ends of the wire. For arbitrary positions of the contact, the points C and E, there will be a potential difference across the detector. However, there will be one position of the contact point C for which the points C and E will be at the same potential. In this case, we say the bridge is *balanced*.

When the bridge is balanced, there is no current through the detector. The current I_1 is the same in R_1 and R_2. Likewise, the current I_2 is the same in R_x and R_3. We apply Kirchhoff's voltage rule to the balanced bridge, around the loops $EACE$ and $BECB$. After rearranging we obtain

$$R_x I_2 = R_2 I_1 \qquad \text{loop } EACE$$
$$R_3 I_2 = R_1 I_1 \qquad \text{loop } BECB$$

The currents can be eliminated by forming ratios. Solving for R_x we find

$$R_x = R_3 \left(\frac{R_2}{R_1} \right)$$

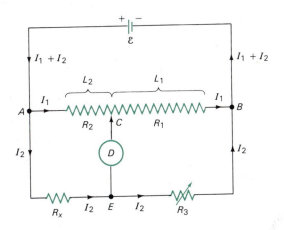

FIGURE 28.4

A balanced Wheatstone bridge. There is no current through the detector D when the bridge is balanced.

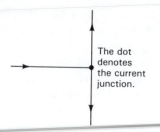

The dot denotes the current junction.

Because the wire is uniform, the ratio R_2/R_1 is the ratio of the lengths L_2/L_1 in Figure 28.4. Thus, the unknown resistance can be expressed in terms of the known resistance R_3 and the measured lengths L_1 and L_2.

$$R_x = R_3 \left(\frac{L_2}{L_1} \right)$$

FIGURE 28.5

A physical connection of two or more conductors forms a current junction.

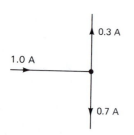

FIGURE 28.6

The total current entering the junction equals the total current leaving the junction. A current of 1 A enters the junction and currents of 0.3 A and 0.7 A leave the junction.

28.3
KIRCHHOFF'S CURRENT RULE

The intersection of electrical conductors such as metal wires is called a **junction** (Figure 28.5). Charges arrive and leave the junction through the conductors. Experiment shows that charge is conserved; the charge flowing into a junction per second must equal the charge flowing out of the junction per second. For example, in Figure 28.6, a current of 1.0 A enters the junction. A current of 0.30 A leaves the junction in the upward branch. In order for charge to be conserved, the current leaving the junction in the downward branch must be 0.70 A. This observation is the basis for **Kirchhoff's current rule**:

> The total current entering a junction equals the total current leaving a junction.

EXAMPLE 5

Charge Conservation and Kirchhoff's Current Rule

A battery having an internal resistance of 10 Ω and an emf of 18 V is connected to two 100-Ω resistors as shown in Figure 28.7. Measurements indicate that the currents at points A and B in the 100-Ω resistors are each 0.15 A. What are the currents at the points labeled C, D, E, and F?

Current is constant in any continuous segment of a circuit. Hence the current at C is the same as the current at A, and the current at D is the same as the current at B:

$$I_A = I_C = 0.15 \text{ A}$$
$$I_B = I_D = 0.15 \text{ A}$$

The current I_E enters junction 1. The currents I_A and I_B leave junction 1. Kirchhoff's current rule requires that at junction 1,

$$I_E = I_A + I_B = 0.15 \text{ A} + 0.15 \text{ A}$$
$$= 0.30 \text{ A}$$

The currents I_C and I_D enter junction 2, and the current I_F leaves it. The current rule gives

$$I_F = I_C + I_D = 0.15 \text{ A} + 0.15 \text{ A}$$
$$= 0.30 \text{ A}$$

Ammeters at points E and F will register the same values for the current: $I_E = I_F = 0.30$ A. This is expected from the principle of charge conservation.

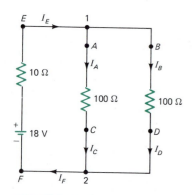

FIGURE 28.7

A parallel combination of a source of emf and two 100-ohm resistors.

28.4
APPLYING KIRCHHOFF'S RULES TO DC CIRCUITS

For all but the simplest circuits, applying Kirchhoff's rules can lead to frustration unless you follow a systematic procedure. The Guide below will serve you well when you apply Kirchhoff's rules.

A Guide for Applying Kirchhoff's Rules

- Draw a diagram of the circuit and label values of known quantities.
- Assign letters to identify key positions, especially the junctions.
- Start at some point and work your way around the circuit, assigning symbols to identify unknowns. As you work your way around the circuit, apply Kirchhoff's current rule at each junction to minimize the number of symbols representing unknown currents.
- Apply Kirchhoff's voltage rule to as many independent loops as you have unknowns.
- Solve the resulting equations to determine the unknowns.

We can illustrate the Guide by considering the circuit in Figure 28.8a. Let's assume that the resistances and the strengths of the sources of emf are known, and that we want to determine the unknown currents. In Figure 28.8b we have used letters to identify key points. Starting at the junction at B we have assigned labels I_1 and I_2 to the currents in the branches containing the resistors r_1 and

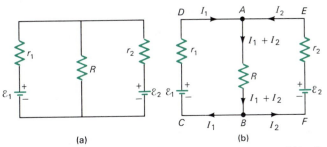

(a)
We start with the circuit diagram.

(b)
Then we label the junctions and identify the currents to be determined. Convince yourself, using Example 5 if you like, that the currents at junction B are the same as those at junction A.

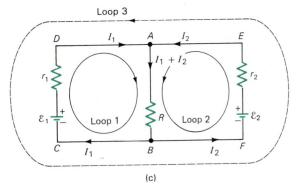

(c)
The next step is to identify the loops in the circuit.

FIGURE 28.8

Steps illustrating the application of Kirchhoff's rules.

r_2. How do we know what direction to assign to the currents? We need not know the actual directions of the currents. If we assign the wrong direction, the solution gives a *negative* value for the current, but the magnitude of the current will be correct. The total current leaving junction B is $I_1 + I_2$, so we label the current reaching B as $I_1 + I_2$. As you can verify, the total current reaching junction A is $I_1 + I_2$, so we apply Kirchhoff's current rule and assign the value $I_1 + I_2$ to the total current leaving A.

The currents I_1 and I_2 are the unknowns we seek to determine. Since we have assigned symbolic values to the resistances and emf's, our solutions for I_1 and I_2 will take the form of algebraic relations rather than numerical values.

To solve for I_1 and I_2 we must apply Kirchhoff's voltage rule to two loops. Figure 28.8c shows *three* possible loops. Let's write down Kirchhoff's voltage rule for all three loops. Then we can see that there are only *two independent* loop equations.

Loop 1 starts at A and proceeds to B, then to point C, then to point D, and returns to A. The other two loops are $ABFEA$ and $ADCBFEA$. The voltage rule equations for these loops are

$$-(I_1 + I_2)R + \mathscr{E}_1 - I_1 r_1 = 0 \qquad \text{for loop 1 } ABCDA$$

$$-(I_1 + I_2)R + \mathscr{E}_2 - I_2 r_2 = 0 \qquad \text{for loop 2 } ABFEA$$

$$I_1 r_1 - \mathscr{E}_1 + \mathscr{E}_2 - I_2 r_2 = 0 \qquad \text{for loop 3 } ADCBFEA$$

Note that the loop 3 equation can be reproduced by subtracting the loop 1 equation from the loop 2 equation. This shows that only two of three voltage equations are independent. We are left with only the required two independent equations involving the unknown currents I_1 and I_2.

Note also that the equations are linear; doubling the sizes of $\mathscr{E}_1$ and $\mathscr{E}_2$ causes all currents to double. Because the equations are linear, their solution is straightforward. You can verify the following equations:

$$I_1 = \frac{\mathscr{E}_1(r_2 + R) - \mathscr{E}_2 R}{r_1 r_2 + r_1 R + r_2 R} \qquad (28.4)$$

$$I_2 = \frac{\mathscr{E}_2(r_1 + R) - \mathscr{E}_1 R}{r_1 r_2 + r_1 R + r_2 R} \qquad (28.5)$$

The current in the resistor R is

$$I_1 + I_2 = \frac{\mathscr{E}_1 r_2 + \mathscr{E}_2 r_1}{r_1 r_2 + r_1 R + r_2 R} \qquad (28.6)$$

Let's see if the answers are reasonable. First, dimensions and units must be consistent. The left-hand side of each equation has units of amperes. The right-hand side of each equation has units of volts per ohm, which are also equal to amperes. Hence the dimensions are consistent.

Second, it may be possible to reduce the circuit to some situation where the answer is obvious or known from a previous calculation. For example, if loop 2 is open-circuited by severing the connection between A and E, then no charge flows through r_2 and the circuit reduces to that in Figure 28.2. The answer from a previous calculation is $I_1 = \mathscr{E}_1/(r_1 + R)$. Severing the wire is equivalent to letting r_2 approach infinity. Thus Equation 28.4 should yield $I_1 = \mathscr{E}_1/(r_1 + R)$ for the limiting case in which $r_2 \rightarrow \infty$. You can verify this by letting r_2 approach infinity in Equation 28.4.

Instead of loop 2, suppose that the wire connecting R to junction A is severed. Then the current in R must vanish. Severing the wire is equivalent to letting R approach infinity. Using Equation 28.6, convince yourself that $I_1 + I_2 \to 0$ if $R \to \infty$. With R removed, we expect

$$I_1 = -I_2 = \frac{\mathscr{E}_1 - \mathscr{E}_2}{r_1 + r_2}$$

Letting R approach infinity in Equations 28.4 and 28.5 verifies these forms for I_1 and I_2.

Kirchhoff's voltage and current rules always yield exactly the number of independent equations needed to solve for the circuit currents. However, it is a tedious job to sort out and solve the resulting simultaneous equations if there are more than three unknown currents. There are several systematic approaches to solving this problem that are beyond the scope of our treatment here. A particularly attractive method takes advantage of the ease of solving simultaneous linear algebraic equations with modern digital computers. Specially designed computer programs for circuit analysis are readily available.* You might want to become familiar with these methods if you have access to a computer.

EXAMPLE 6

More About Kirchhoff's Rules

We want to illustrate the application of Kirchhoff's rules to a circuit that includes a fully-charged capacitor. The circuit diagram is shown in Figure 28.9 where the values of the given quantities are indicated. The three unknowns, I, I_1, and Q, are labeled in Figure 28.9. I is the total current delivered by the battery, I_1 is the current in the 6-Ω resistor, and Q is the charge on the capacitor. To solve for these three unknowns we must apply Kirchhoff's voltage rule to three loops.

There can be no steady current through the branch containing the fully-charged capacitor. Accordingly, the current $I - I_1$ arriving at the junction E must continue through the 4-Ω resistor.

Kirchhoff's voltage rule for the loop $ABDJA$ gives

$$+5 \text{ V} - 2\,\Omega \cdot I - 6\,\Omega \cdot I_1 = 0$$

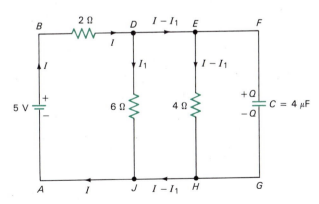

FIGURE 28.9

* See, for example, *Fortran for Engineering Physics—Electricity, Magnetism, and Light*, by Alan B. Grossberg. New York: McGraw-Hill, 1973.

For the loop *JDEHJ*, Kirchhoff's voltage rule gives

$$6\,\Omega \cdot I_1 - 4\,\Omega\,(I - I_1) = 0$$

Solving these two equations we find

$$I_1 = 0.455\ \text{A} \qquad I = 1.14\ \text{A}$$

To determine the charge on the capacitor we apply Kirchhoff's voltage rule to the loop *HGFEH*. The potential change from *E* to *H* is $-4\,\Omega\,(I - I_1)$. The positively charged capacitor plate is at a higher potential than the negatively charged plate. You simply choose one plate to be positive. If your choice is wrong, the solution will yield a negative value for the charge. The potential change from *G* to *F* across the capacitor is $+Q/C$. Kirchhoff's voltage rule for the loop *HGFEH* gives

$$-4\,\Omega\,(I - I_1) + \frac{Q}{C} = 0$$

Solving for *Q* we get

$$Q = C \cdot 4\,\Omega\,(I - I_1) = 4\ \mu\text{F} \cdot 4\,\Omega\,(0.682\ \text{A})$$
$$= 10.9\ \mu\text{C}$$

Kirchhoff's rules are applicable to all types of circuits, not just steady-state dc circuits. We will make frequent use of Kirchhoff's voltage rule in this and subsequent chapters.

Equivalent Resistance

In Chapter 26 we deduced rules for the equivalent capacitance of series and parallel combinations of capacitors. We can use Kirchhoff's rules to derive rules for the **equivalent resistance** of series and parallel combinations of resistors.

Resistors in Series

A series combination of resistors is shown in Figure 28.10. In a series combination there are no junctions at which the current can divide and so the current is the same in each resistor. We can use Kirchhoff's voltage rule to determine the equivalent resistance of a series combination. This is the resistance of the single resistor that could replace the series combination and leave the current unchanged.

Applying Kirchhoff's voltage rule to the circuit in Figure 28.10 and solving for the current, we find

$$I = \frac{\mathscr{E}}{r + R_1 + R_2 + R_3}$$

If a single resistor *R* replaced the series combination, the current would be

$$I = \frac{\mathscr{E}}{r + R}$$

The current would be unchanged provided the resistance of the single resistor equals the sum of the resistances in the series combination

$$R = R_1 + R_2 + R_3$$

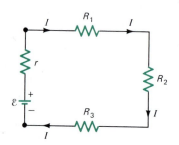

FIGURE 28.10

The three resistors, R_1, R_2, and R_3, are connected in series. The current is the same in each resistor.

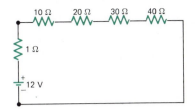

FIGURE 28.11

A series connection of four resistors and a battery. The internal resistance of the battery is 1 ohm.

In general, any series combination of resistors can be replaced by a single resistor having resistance equal to the sum of the resistances without changing the current in the circuit. Thus for N resistors in series the equivalent resistance (R) is

$$R = R_1 + R_2 + R_3 + \cdots + R_N \tag{28.7}$$

Note that determining the equivalent resistance of resistors in series is like computing the equivalent capacitance of capacitors in parallel.

EXAMPLE 7

Resistors in Series

A 10-Ω, a 20-Ω, a 30-Ω, and a 40-Ω resistor are connected in series to a battery having an emf of 12 V and an internal resistance of 1 Ω (Figure 28.11). What is the current in the circuit? The equivalent resistance of the resistors is

$$R = 10 \ \Omega + 20 \ \Omega + 30 \ \Omega + 40 \ \Omega$$
$$= 100 \ \Omega$$

The current is

$$I = \frac{\mathscr{E}}{r + R} = \frac{12 \text{ V}}{101 \ \Omega}$$
$$= 0.119 \text{ A}$$

Resistors in Parallel

Figure 28.12a shows a battery connected to a *parallel* combination of three resistors. In a parallel combination the potential difference across each resistor is the same. It is often helpful to know the *equivalent* resistance of a parallel combination of resistors. This is the resistance of the single resistor that could replace the parallel combination and leave the total current unchanged (Figure 28.12b).

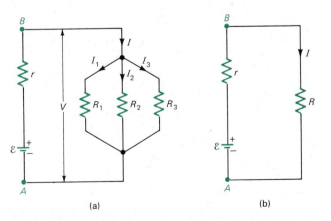

(a)   (b)

FIGURE 28.12

(a) Three resistors, R_1, R_2, and R_3, are connected in parallel to the terminals of a battery. (b) The single resistor R results in the same current as the parallel combination of the three resistors in (a).

Let V denote the common potential difference across the resistors. The total current I is the sum of the currents in each resistor,

$$I = I_1 + I_2 + I_3 = \frac{V}{R_1} + \frac{V}{R_2} + \frac{V}{R_3}$$

$$= \left(\frac{1}{R_1} + \frac{1}{R_2} + \frac{1}{R_3} \right) V$$

The equivalent resistance is defined by

$$I = \frac{V}{R}$$

Comparing these two expressions for I shows that the equivalent resistance R is related to R_1, R_2, and R_3 by

$$\frac{1}{R} = \frac{1}{R_1} + \frac{1}{R_2} + \frac{1}{R_3}$$

The generalization of this to any parallel combination of N resistors is

$$\frac{1}{R} = \frac{1}{R_1} + \frac{1}{R_2} + \cdots + \frac{1}{R_N} \tag{28.8}$$

We describe this relation by saying that resistors in parallel *add reciprocally*.

Note that resistors in parallel add like capacitors in series. In practice it is recommended that you use Equation 28.8 to evaluate $1/R$, and then take the reciprocal to determine R. For instance, the equivalent resistance of twenty 5-ohm resistors in parallel is found from

$$\frac{1}{R} = \underbrace{\frac{1}{5} + \frac{1}{5} + \frac{1}{5} + \cdots + \frac{1}{5}}_{20 \text{ terms}} \Omega^{-1} = 20 \left(\frac{1}{5} \right) \Omega^{-1}$$

which gives

$$= 4 \, \Omega^{-1}$$

$$R = \tfrac{1}{4} \Omega$$

From Equation 28.8 we can see that $1/R$ is *greater* than the reciprocal of the smallest resistance in the parallel combination. It follows that the equivalent resistance is *smaller* than the smallest resistance in the combination.

EXAMPLE 8

Resistors in Parallel

What is the equivalent resistance of a parallel combination of 5-Ω, 10-Ω, and 20-Ω resistors? Using Equation 28.8 we write

$$\frac{1}{R} = \left(\frac{1}{5} + \frac{1}{10} + \frac{1}{20} \right) \Omega^{-1} = \frac{7}{20} \Omega^{-1}$$

The equivalent resistance is

$$R = \frac{20}{7} \Omega = 2.86 \, \Omega$$

Note that the equivalent resistance is smaller than the smallest resistance in the combination.

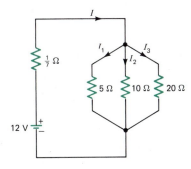

FIGURE 28.13

A parallel combination of three resistors (5 Ω, 10 Ω, and 20 Ω) connected to a 12-volt battery with an internal resistance of 1/7 ohm. The current delivered by the battery is 4 A.

If the parallel combination is connected to a 12-volt battery with an internal resistance of $\frac{1}{7}$ ohm, what current does the battery deliver (Figure 28.13)? Figure 28.14 shows the equivalent circuit. The current is

$$I = \frac{\mathcal{E}}{r + R} = \frac{12 \text{ V}}{\frac{1}{7}\Omega + \frac{20}{7}\Omega}$$

$$= 4 \text{ A}$$

A good battery provides a terminal voltage that is relatively insensitive to the resistance connected between its terminals. A parallel combination of resistors has a significant effect on the current delivered by such a battery, but has little effect on the battery's terminal voltage.

EXAMPLE 9

Turning on More Lights

Consider a house trailer supplied with a 12-volt battery. The lights each have the same resistance and are connected in parallel (Figure 28.15). How is the battery current affected when more lights are turned on?

If n lights are connected their equivalent resistance is R/n, where R is the resistance of one light. The current delivered by the battery is

$$I = \frac{\mathcal{E}}{r + \dfrac{R}{n}} = \frac{n\mathcal{E}}{nr + R}$$

If nr is negligible compared to R, then $I \approx n(\mathcal{E}/R)$, showing that the battery current is directly proportional to the number of lights turned on.

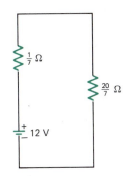

FIGURE 28.14

This circuit is equivalent to the one shown in Figure 28.13 in the sense that the battery delivers the same total current (4 A).

28.5
THE RC CIRCUIT

In Chapter 26 we examined the charging of a capacitor connected to a battery. Our primary interest was in the total charge stored by the capacitor. Now we want to learn which factors control the time required for the charge to accumulate and the manner in which the voltage changes across the plates as charge accumulates. Understanding the time behavior of the charging process is vital to understanding how capacitors function in electrical circuits.

Consider the circuit in Figure 28.16. With the switch in position 1 the capacitor is in series with the battery and the resistor, and charge flows to the capacitor. When the switch is in position 2 the capacitor discharges through the resistor. Let's first examine the charging.

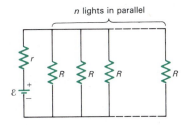

FIGURE 28.15

A parallel combination of *n* lights is connected to a battery. The resistance of each light is *R*.

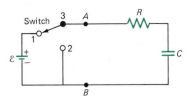

FIGURE 28.16

The capacitor can be charged by connecting terminals 1 and 3. It can be discharged by connecting terminals 2 and 3. We assume that terminals 2 and 3 have been connected long enough so that the capacitor is discharged. Then terminals 1 and 3 are connected and the capacitor begins charging.

Charging the Capacitor

We assume that the capacitor is completely discharged. This is assured if the switch is in position 2 for a sufficient length of time. Then the switch is moved to position 1 and the capacitor begins charging.

We can use Kirchhoff's voltage rule to determine how the charge on the capacitor and the current in the circuit vary with time. Moving in a clockwise fashion around the circuit in Figure 28.16, Kirchhoff's voltage rule gives

$$\mathcal{E} - IR - \frac{q}{C} = 0 \tag{28.9}$$

where q/C is the voltage across the capacitor. The current I is related to the charge by $I = dq/dt$. If Equation 28.9 is multiplied by $C\, dt$ and $I\, dt$ is replaced by dq, the resulting equation can be rearranged to give

$$\frac{dq}{q - C\mathcal{E}} = -\frac{1}{RC}\, dt$$

This equation must be integrated to express the charge q in terms of the time t. At $t = 0$ when the switch is closed, the charge on the capacitor is zero, so

$$\int_0^q \frac{dq}{q - C\mathcal{E}} = -\frac{1}{RC} \int_0^t dt$$

Carrying out the integrations leads to the result

$$q = C\mathcal{E}(1 - e^{-t/RC}) \tag{28.10}$$

Note that $q = 0$ at $t = 0$ in accordance with the requirement that there be no charge stored when the switch is closed.

The voltage V across the capacitor is q/C, and follows from Equation 28.10 as

$$V = \mathcal{E}(1 - e^{-t/RC}) \tag{28.11}$$

As t becomes large compared to the product RC, the exponential in Equation 28.11 goes to zero and the voltage approaches $\mathcal{E}$. We anticipate this because the maximum voltage across the capacitor equals $\mathcal{E}$, the emf of the battery.

RC Time Constant

The product RC has units of time and is called the **RC time constant**. If R is expressed in ohms and C is expressed in farads, then RC has units of seconds. The RC time constant measures the time to charge the capacitor. Setting $t = RC$ in Equation 28.10 we see that $q = C\mathcal{E}(1 - e^{-1}) = 0.632 C\mathcal{E}$. Thus about 63% of the maximum charge that can accumulate has done so in a time equal to the RC time constant. Figure 28.17 shows graphically how the charge accumulates as a function of time.

The current follows from Equation 28.10

$$I = \frac{dq}{dt} = \left(\frac{\mathcal{E}}{R}\right) e^{-t/RC} \tag{28.12}$$

Equation 28.12 shows that the current equals $\mathcal{E}/R$ at $t = 0$ when the switch is moved to position 1. This initial value of the current is in accord with Kirchhoff's voltage rule, Equation 28.9. At the moment the switch is closed, the capacitor

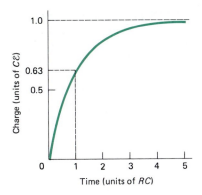

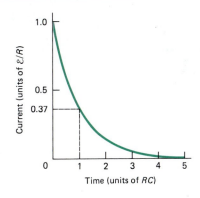

FIGURE 28.17

A plot of the charge stored by a capacitor connected in series with a resistor and a battery. Time is expressed in units of the time constant, *RC*. Charge is expressed in units of $C\mathscr{E}$, the maximum charge that can be stored. 63% of the maximum charge is stored in a time equal to *RC*.

FIGURE 28.18

A plot of the current in the *RC* circuit of Figure 28.16. Current is expressed in units of $\mathscr{E}/R$, the maximum current. The current drops exponentially after the switch is closed, reaching 37% of its initial value in a time equal to *RC*.

is uncharged and so the voltage across it is zero. With $q/C = 0$, Equation 28.9 gives $I = \mathscr{E}/R$.

The current in the circuit declines following the switch movement to position 1 because charge accumulating on the capacitor plates impedes the further flow of charge. As time progresses, the current decreases exponentially and approaches zero as the capacitor becomes fully charged. After a time $t = RC$ has elapsed, the current has decreased to $I = (\mathscr{E}/R)e^{-1} = 0.37\mathscr{E}/R$, or about 37% of its initial values. Figure 28.18 shows graphically how the current declines as a function of time.

EXAMPLE 10

Making Use of the Time Required to Charge a Capacitor

You may have played computer games where the position of some design on the screen is controlled by a knob called a joystick. In many systems this knob controls a variable resistor that is connected in series with a source of emf and a capacitor. The circuitry causes the capacitor to repeatedly charge and discharge.

The *RC* circuit is followed by an electronic circuit that relates the position coordinates of the design on the screen to the time required for the voltage of the capacitor to achieve 70% of its maximum voltage (Figure 28.19). As the player manipulates the joystick, the resistance changes. This changes the *RC* time constant, which in turn changes the time for the capacitor voltage to reach the 70% level. Finally, this causes the position of the figure to change.

Typically, the capacitor has a capacitance of 0.1 microfarads. Let's take $R = 100,000$ ohms and calculate the time required for the voltage across the capacitor to achieve 70% of the emf of the source.

The ratio of the capacitor voltage to the emf of the source follows from Equation 28.11:

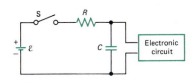

FIGURE 28.19

The *RC* circuit of Figure 28.16 is modified to include an electronic circuit connected in parallel with the capacitor. The electronic circuit is designed to respond when the potential difference across it reaches 70% of $\mathscr{E}$.

$$\frac{V}{\mathscr{E}} = 1 - e^{-t/RC}$$

We set $V/\mathscr{E} = 0.7$, and solve for the time t,

$$t = -RC \ln 0.3 = -(10^5 \; \Omega \cdot 10^{-7} \; F)(-1.2) = 1.20 \times 10^{-2} \; s$$

$$= 12 \; ms$$

Modern electronics associated with computer games can easily measure a time this brief. By comparison, the human eye perceives motion as continuous when the time between position changes is less than 40 ms. Furthermore, human reaction times are typically between 100 ms and 200 ms. Thus, the perceptions and motions of the player are slow compared to the time for the electronics to sense changes. For these reasons the motion appears to respond instantaneously to the player's actions.

Discharging the Capacitor

Suppose that the capacitor in Figure 28.16 is fully charged, so the voltage across its plates is $\mathscr{E}$. If the switch is moved to position 2, the battery is removed from the circuit. The capacitor acts like a short-lived battery and drives charge through the resistor. During discharge the current is directed opposite to its direction during the charging phase, and Kirchhoff's voltage rule gives

$$RI - \frac{q}{C} = 0$$

Because the capacitor charge is *decreasing*, the current-charge relation is $I = -dq/dt$, and Kirchhoff's voltage rule may be rearranged and integrated to give

$$\int_{C\mathscr{E}}^{q} \frac{dq}{q} = -\frac{1}{RC} \int_{0}^{t} dt$$

The lower limits correspond to the condition that the charge on the capacitor is $C\mathscr{E}$ when the switch is moved to position 2.

Carrying out the integrations we find

$$q = C\mathscr{E} e^{-t/RC} \tag{28.13}$$

$$V = \frac{q}{C} = \mathscr{E} e^{-t/RC} \tag{28.14}$$

$$I = -\frac{dq}{dt} = \left(\frac{\mathscr{E}}{R}\right) e^{-t/RC} \tag{28.15}$$

We expect the capacitor voltage to decline from $\mathscr{E}$ to zero as the capacitor discharges. We see that the decline is described by an exponential function and the RC time constant measures the discharge time. In a time RC the voltage falls to $1/e = 0.37$ of its initial value. We also expect the current to become zero as the capacitor discharges. Equation 28.15 shows that the decline of the current is also exponential.

Energy Stored by the Capacitor

During the charging of a capacitor, energy is converted by the source of emf into two forms; energy is converted into thermal energy (joule heat) in the resistor and energy is stored in the electric field between the plates of the capacitor. As we will now show, one-half of the energy converted by the source of emf is transformed into thermal energy and the other half is stored in the capacitor.

The maximum charge stored on the capacitor in the RC circuit is $q_{max} = C\mathscr{E}$, independent of the value of the series resistance. The total energy converted by the source of emf is

$$\text{energy converted by source of emf} = q_{max}\mathscr{E} = C\mathscr{E}^2$$

The energy stored by the fully-charged capacitor is

$$U = \tfrac{1}{2}C\mathscr{E}^2 \tag{28.16}$$

which is one-half of the energy converted by the source. The resistor affects the rate at which the capacitor charges but has no effect on the ultimate value of the charge or the energy stored.

To determine the thermal energy produced, we use Equation 28.12 for the current to express the thermal power developed in the resistor as

$$P = I^2 R = \left(\frac{\mathscr{E}^2}{R}\right) e^{-2t/RC}$$

The total thermal energy produced during the complete charging is the time integral of P,

$$\text{thermal energy} = \int_0^\infty P \, dt = \frac{\mathscr{E}^2}{R} \int_0^\infty e^{-2t/RC} \, dt$$
$$= \tfrac{1}{2}C\mathscr{E}^2 \tag{28.17}$$

Hence, during charging, the thermal energy produced is exactly equal to the energy stored in the capacitor. Of the total energy converted by the source, one-half is converted to thermal energy, and the other half is stored by the capacitor.

28.6
CURRENT AND VOLTAGE MEASUREMENTS

We have devised ways of calculating currents and voltages in electrical circuits but have said little about measuring these quantities. Not only do measurements give us confidence in the validity of Kirchhoff's rules, they sometimes provide the only way to determine currents and voltages. An instrument for measuring current is called an **ammeter**. To measure a current, charge must flow through the ammeter, so it has two terminals that allow the current to enter and to leave. The circuit must be broken at the point where the current is measured, and the terminals of the ammeter must be connected in series to recomplete the circuit (Figure 28.20).

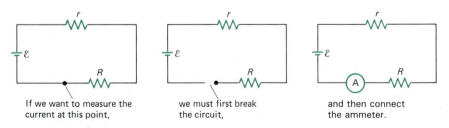

If we want to measure the current at this point,

we must first break the circuit,

and then connect the ammeter.

FIGURE 28.20
Steps involved in connecting an ammeter in a circuit.

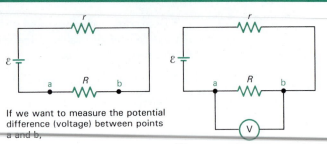

If we want to measure the potential difference (voltage) between points a and b,

We connect the two terminals of the voltmeter to these points.

FIGURE 28.21

Steps involved in connecting a voltmeter in a circuit.

An instrument for measuring the potential difference (voltage) between two points is called a **voltmeter**. A voltmeter has two terminals that can be connected in parallel to any two points in a circuit. The circuit does not have to be broken in order to connect a voltmeter (Figure 28.21). For this reason, it is more convenient to measure potential difference with a voltmeter than it is to measure current with an ammeter.

All ammeters and voltmeters rely on some effect that is produced by a current or a voltage. For example, one of the most versatile modern voltmeters, the oscilloscope, relates the amount of deflection of an electron beam to a voltage established between two parallel plates that the electron beam passed between. A digital ammeter or voltmeter that visually displays the numerical value of a current or voltage measurement is based on the principle that a transistor or electronic circuit will produce a response related to the current or voltage it is measuring. A versatile and widely used type of meter, which requires the experimenter to read the position of a pointer on a calibrated scale, relies on the principle that a current-carrying coil of wire between the poles of a permanent magnet will experience a torque that rotates the coil.

When an ammeter or voltmeter is used to make a measurement, the meter affects the circuit to some extent. This is because meters require energy for their operation, which they extract from the circuit. The energy requirement is associated with the electrical resistance (R_m) appearing between the two terminals of the meter (Figure 28.22). Resistance R_m is the equivalent of the resistance of the active element, and of any resistance that is connected in parallel or in series with the active element. For example, a conventional moving-coil ammeter has an intrinsic resistance associated with the wire constituting the moving coil (the active element) and a parallel resistance (commonly called a shunt) that allows current to bypass the moving coil (Figure 28.22).

To illustrate the effect of the measuring instrument, we consider the circuit in Figure 28.23 that shows an ammeter with resistance R_m connected to measure a current through the resistance designated R. The ammeter is connected in series with R because the charge that flows through R must also flow through the ammeter. In the absence of the ammeter the current I is

$$I = \frac{\mathscr{E}}{r + R}$$

With the meter present the current is

$$I_m = \frac{\mathscr{E}}{r + R + R_m}$$

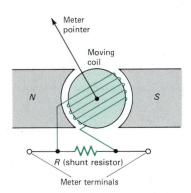

FIGURE 28.22

The components of a moving-coil ammeter. The active element is the moving coil. The deflection of the pointer is proportional to the current in the moving coil. The resistance of the meter is due to the resistances of the moving coil and the shunt.

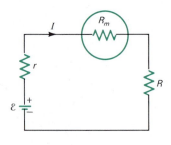

FIGURE 28.23

The ammeter introduces a resistance R_m in series with the resistors r and R.

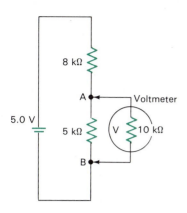

FIGURE 28.24

The 10,000-ohm voltmeter is connected to points *A* and *B* to measure the voltage across the 5,000-ohm resistor.

Clearly, $I \neq I_m$. However, if $R_m \ll R$, then $I \approx I_m$. When this is the case, R_m can be neglected. Generally, an ammeter will not significantly affect a current measurement, but we should always be aware of the possibility that it might, and be prepared to make corrections for it.

When a voltmeter is used to measure the potential difference across a resistor (Figure 28.21), the meter will not significantly affect the measurement if $R_m \gg R$.

EXAMPLE 11

Be Aware of the Resistance of a Voltmeter

A voltmeter having an internal resistance of 10,000 ohms is to be used to measure the voltage across the 5,000-ohm resistor in the circuit shown in Figure 28.24. Let's calculate the voltage before and after the voltmeter is connected to see the effect of the voltmeter.

The current before the voltmeter is connected is

$$I = \frac{5.0 \text{ V}}{8000 \ \Omega + 5000 \ \Omega} = 0.385 \times 10^{-3} \text{ A}$$

The voltage across the 5,000-ohm resistor is

$$V = IR = 0.385 \times 10^{-3} \text{ A} \cdot 5000 \ \Omega = 1.92 \text{ V}$$

When the volmeter is connected in parallel with the 5,000-ohm resistor, the resistance of the parallel combination is given by

$$\frac{1}{R} = \frac{1}{5000 \ \Omega} + \frac{1}{10000 \ \Omega} = \frac{3}{10000 \ \Omega}$$

or

$$R = 3330 \ \Omega$$

The current into this effective resistance is then

$$I = \frac{5 \text{ V}}{8000 \ \Omega + 3330 \ \Omega} = 0.441 \times 10^{-3} \text{ A}$$

and the voltage across the parallel combination is

$$V = IR = 0.441 \times 10^{-3} \text{ A} \cdot 3330 \ \Omega = 1.47 \text{ V}$$

It is this voltage that the voltmeter will record. Thus the difference between the voltages before and after the meter is connected is

$$1.92 \text{ V} - 1.47 \text{ V} = 0.45 \text{ V}$$

We conclude that the voltmeter records a voltage significantly different from the actual voltage.

WORKED PROBLEM

(a) Determine the current in each resistor in the circuit shown in Figure 28.25. (b) Determine the total resistance between the points *C* and *F*.

Solution

(a) The five unknowns are the currents in each of the resistors. To avoid introducing five symbols, apply Kirchhoff's current rule at the junctions at *D* and *G*. This step results in the assignment of currents shown in Figure 28.25.

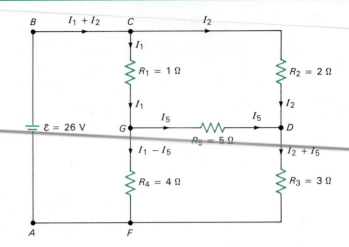

FIGURE 28.25

Use Kirchhoff's voltage rule for three independent loops to obtain equations that can be solved for I_1, I_2, and I_5.

1) For the loop *CDGC*,

$$-I_2 R_2 + I_5 R_5 + I_1 R_1 = 0$$

2) For the loop *GDFG*,

$$-I_5 R_5 - (I_2 + I_5)R_3 + (I_1 - I_5)R_4 = 0$$

3) For the loop *ABCGFA*,

$$\mathscr{E} - I_1 R_1 - (I_1 - I_5)R_4 = 0$$

We want to solve these equations for the three currents. Let's solve first for I_1 by successively eliminating I_2 and I_5. The current I_2 does not appear in the third equation, so it can be eliminated from the first two. From the first equation,

$$I_2 = \frac{I_5 R_5 + I_1 R_1}{R_2} = \frac{5}{2} I_5 + \frac{1}{2} I_1$$

Using this in the second equation gives

$$I_5 = \frac{(R_4 - \frac{1}{2}R_3)I_1}{R_4 + R_5 + \dfrac{7R_3}{2}}$$

$$= \frac{5}{39} I_1$$

Substituting this expression for I_5 into the third equation gives

$$I_1 = \frac{\mathscr{E}}{R_1 + R_4 - \dfrac{5R_4}{39}} = \frac{26 \text{ V}}{1\,\Omega + 4\,\Omega - \dfrac{20}{39}\,\Omega}$$

$$= 5.79 \text{ A}$$

The current in R_5 is

$$I_5 = \frac{5}{39} I_1 = \frac{5}{39}(5.79 \text{ A})$$

$$= 0.74 \text{ A}$$

For the current in R_2 we get

$$I_2 = \frac{5}{2}I_5 + \frac{1}{2}I_1 = \frac{5}{2}(0.74\text{ A}) + \frac{1}{2}(5.79\text{ A})$$
$$= 4.76\text{ A}$$

The current in R_3 is

$$I_2 + I_5 = 5.50\text{ A}$$

The current in R_4 is

$$I_1 - I_5 = 5.05\text{ A}$$

(b) The total current delivered by the battery is $I_1 + I_2 = 10.55$ A. The resistance of the network of resistors equals

$$\frac{\mathscr{E}}{I_1 + I_2} = \frac{26\text{ V}}{10.55\text{ A}}$$
$$= 2.46\ \Omega$$

EXERCISES

28.1   Sources of Electric Energy

A. A battery having an emf of 7 V produces a current of 0.5 A when connected to an external resistor. If 150 J of energy are produced in 1 min in the external resistor, determine the internal and external resistances.

B. An old 12-V automobile battery with an internal resistance of 1 Ω delivers 1 A of current to a resistor for 10 min. Determine (a) the chemical energy converted by the battery, (b) the energy produced in the resistor, (c) the resistance of the resistor.

C. A battery (emf = 12 V) of unknown internal resistance delivers 22 W to an external resistance when the current is I. If the battery current doubles, the external power delivered becomes 40 W. Determine (a) the original current I, (b) the internal resistance of the battery.

28.2, 28.3, and 28.4   Kirchhoff's Rules

D. (a) Show that the equivalent resistance of any series combination of resistors is greater than the resistance of any single resistor in the combination. (b) Show that the equivalent resistance of any number of resistors in parallel is always less than the smallest resistance in the combination.

E. Show how four 6-Ω resistors can be connected to produce equivalent resistances of 1.5, 6, 8, 15, and 24 Ω.

F. Two light bulbs, each having a resistance of 10 Ω, are connected in series with a 12-V storage battery having negligible internal resistance. When a third identical light bulb is connected in parallel with one of the original bulbs, which bulb will brighten and which will dim?

G. When two resistors are connected in series with a battery having an emf of 10 V and negligible internal resistance, a current of 1 mA is produced. When the resistors are connected in parallel to the same battery, a total current of 4 mA is produced. What are the resistances of the resistors?

H. Five 1000-Ω resistors are connected in series. Each of five identical resistors of unknown resistance is connected in parallel with the series combination of five 1000-Ω resistors. The equivalent resistance of the combination is 2500 Ω. What is the unknown resistance?

I. Two batteries and two resistors are connected in the single loop shown in Figure 1. Given that the potential at point D equals zero, determine the potentials at points (a) A, (b) B, (c) C.

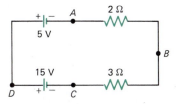

FIGURE 1

J. N identical resistors in parallel have an equivalent resistance of R/N where R is the resistance of a single resistor. (a) If an additional identical resistor is connected in parallel, show that the percentage change in the equivalent resistance is

$$\frac{100}{N+1}$$

(b) Evaluate the percentage change for $N = 1$, $N = 10$, and $N = 100$.

K. A dead battery is boosted by connecting it to the live battery of another car (Figure 2). Determine the current in the starter and in the dead battery.

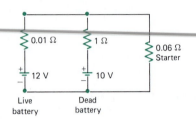

FIGURE 2

L. Determine the current I_2 in the circuit shown in Figure 3.

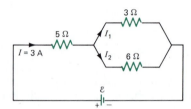

FIGURE 3

M. Determine the current in the 10-Ω resistor in Figure 4.

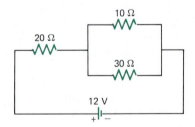

FIGURE 4

N. (a) Using Kirchhoff's voltage rule, show that the current in resistor R in the circuit shown in Figure 5 is

$$I = \frac{\mathcal{E}_1 - \mathcal{E}_2}{r_1 + r_2 + R}$$

(b) You can think of the current in R as the superposition of currents provided independently by $\mathcal{E}_1$ and $\mathcal{E}_2$. Calculate the current provided by $\mathcal{E}_1$ when $\mathcal{E}_2$ is short-circuited. Do the same for $\mathcal{E}_2$ assuming that $\mathcal{E}_1$ is short-circuited. Superpose these two currents and show that the result is identical to that obtained in part (a).

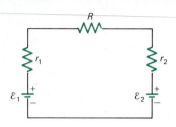

FIGURE 5

P. A circuit for an unbalanced Wheatstone bridge is shown in Figure 6. If $I_1 = 10$ mA, $I_2 = 6$ mA, and $I_3 = 2$ mA, use Kirchhoff's current rule to obtain (a) I_4, (b) I_5, and (c) I_6.

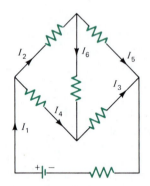

FIGURE 6

Q. A sealed box contains two wires leading to an unknown combination of resistors inside the box. When a potential difference of 1.51 V is connected across the leads, a current of 0.206 A is produced. (a) What is the effective resistance of the combination of resistors in the box? (b) Opening the box reveals the combination of resistors shown in Figure 7. Determine the equivalent resistance and verify the calculation in part (a).

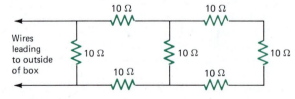

FIGURE 7

28.5   The RC Circuit

R. A 1-μF capacitor in a circuit has an initial charge of 10 μC. Suppose that we want another circuit to respond when the potential difference across the capacitor falls to 1 V as it discharges through a 100-kΩ

resistor. How much time is required for this circuit to respond?

S. Using Ohm's law and the definition of capacitance, show that RC has dimensions of time.

T. A circuit consists of a battery, a resistor, a capacitor, and a switch connected in series. The charge on the capacitor reaches about 63% of its maximum value in a time $t = RC$ after the switch is closed. At this time, what percentage of the maximum electric energy is stored by the capacitor?

U. A battery with negligible internal resistance charges a strobe light through an RC circuit in 2.3 s. (a) To what percentage of the battery's emf does the capacitor charge if the time constant of the circuit is 1 s? (b) After aging, the same battery requires 23 s to charge the same strobe light. Compare the internal resistance of the battery to R.

V. What time is required for a capacitor to lose 90% of its charge during the discharge of an RC circuit? Express your result as a multiple of the time constant of the circuit.

W. A fully-charged capacitor has 12 J of energy stored. How much energy remains stored when its charge has decreased to half its original value during a discharge?

X. Figure 8 shows an oscilloscope trace of the discharge of a 1-pF capacitor in an RC circuit. The vertical scale denotes the voltage across the capacitor. The horizontal scale measures time, with each scale division corresponding to 1 μs. Estimate the resistance of the resistor.

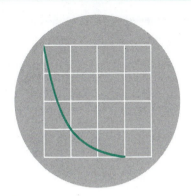

FIGURE 8

Y. Show by substitution that

$$R\frac{dq}{dt} + \frac{q}{C} = 0$$

is solved by $q = q_0 e^{-t/RC}$

28.6 Current and Voltage Measurements

Z. A battery having an emf of 12 V and internal resistance of 0.20 Ω is connected in series with two 1800-Ω resistors. A voltmeter is used to measure the potential difference across one of the two resistors. What potential difference will the voltmeter measure if its internal resistance equals 18,000 Ω?

PROBLEMS

1. In Section 28.4 we showed that the current in a series circuit consisting of a source of emf, a resistor, and a capacitor is $I = (\mathscr{E}/R)e^{-t/RC}$. If the power developed by the source is $\mathscr{E}I$, show that the energy converted by the source during the complete charging of the capacitor is $C\mathscr{E}^2$.

2. Determine I_2, I_4, I_5, and I_7 as shown in Figure 9.

3. For the circuit shown in Figure 10, (a) calculate the three currents I_1, I_2, and I_3. (b) Calculate the potential difference across each resistor. (c) Calculate the power dissipated in each resistor. (d) Show that the total thermal power developed in the resistors equals the electric power delivered by the two batteries.

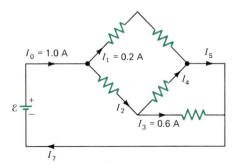

FIGURE 9

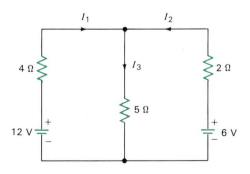

FIGURE 10

4. A potential difference V is applied across a network of four resistors (Figure 11). Table 1 gives three sets of values for the four resistors in terms of two resistances labeled R and ΔR. Calculate the potential difference, $V_A - V_B$, for each combination if $V = 5$ volts, $R = 1$ kΩ, and $\Delta R = 100$ Ω.

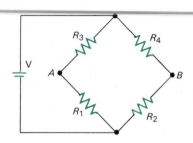

FIGURE 11

TABLE 1

R_1	R_2	R_3	R_4
R	R	R	R
$R + \Delta R$	$R - \Delta R$	R	R
$R + \Delta R$	$R - \Delta R$	$R - \Delta R$	$R + \Delta R$

5. Calculate the value of $\mathcal{E}$ in the circuit in Figure 12 that makes the current zero in the 4-ohm resistor.

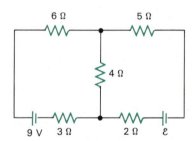

FIGURE 12

6. After the switch is closed in the circuit in Figure 13 the potential difference across the capacitor rises from

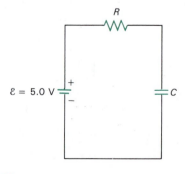

FIGURE 13

zero to 4.0 volts in 1.0 s. When a 0.001-μF capacitor is placed in parallel with the capacitor it takes 1.2 s for the potential difference across the capacitor to rise from zero to 4 volts. Determine the value of C.

7. (a) Determine the potential difference across the 5.0-kΩ resistor in Figure 14. (b) If the potential difference is measured with a voltmeter having an internal resistance of 10 kΩ, what potential difference would be recorded by the meter? (c) If a voltmeter reads 5 volts when measuring the potential difference across the 5-kΩ resistor, what must be the internal resistance of the meter?

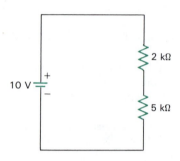

FIGURE 14

8. When two unknown resistors are connected in series with a battery, 225 W are dissipated with a total current of 5 A. For the same total current 50 W are dissipated when the resistors are connected in parallel. Determine the values of the two resistors.

9. An electrical circuit is somewhat like a jigsaw puzzle in which each independent loop outlines a piece of the puzzle. If you cut the puzzle into sections, each of which contains a single loop, then the number of sections equals the number of independent loop equations. For example, the circuit in Figure 15 has two sections. (a) Label the unknown currents in the circuit. (b) How many current equations are required in addition to the two voltage equations? (c) Apply Kirchhoff's current and voltage rules and solve for the circuit currents.

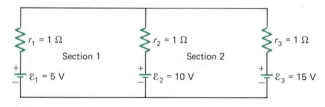

FIGURE 15

10. A 20-W bulb designed to operate with a 12-V potential difference across its terminals is connected to a 12-V battery having an internal resistance of 0.6 Ω.

(a) Determine the potential difference across the bulb. (b) Determine the power converted in the bulb. (c) Determine the change in the potential difference across the bulb and the change in the power converted by the bulb when a second 20-W bulb is connected in parallel with the battery.

11. A current I produces 25 mW of power in a resistor R_1. When R_1 is connected in series with a resistor R_2, the same current produces a total of 75 mW of power in the two resistors. Determine the total power when the same current I is delivered to a parallel combination of R_1 and R_2.

12. A battery with a constant emf $\mathcal{E}$ has an internal resistance r. It is placed in series with a variable resistor, as suggested in Figure 16. (a) Show that the power developed by the resistance R can be expressed by

$$P = \frac{R\mathcal{E}^2}{(R + r)^2}$$

(b) Make a sketch of P versus R for the range $R = 0$ to 10 Ω. Take $\mathcal{E} = 6$ V, $r = 3$ Ω. (c) Using the expression given for P in part (a), prove *analytically* that the maximum value of P occurs for $R = r$. Does your sketch in (b) show that P is a maximum at $R = 3$ Ω?

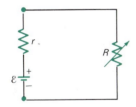

FIGURE 16

13. For a resistor, capacitor, and battery in series, we showed that

$$IR + \frac{q}{C} = \mathcal{E}$$

at any instant. (a) Differentiate this equation with respect to time to obtain a differential equation involving the current, I. (b) Solve the differential equation for the current as a function of time.

14. Four resistors are arranged in a rectangle (Figure 17). When a battery is connected between points a and b

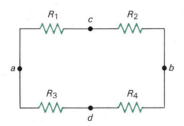

FIGURE 17

it is found that there is no potential difference between the points c and d. Prove that if the battery is then connected between points c and d, there will be no potential difference between points a and b.

15. Before the switch is closed in the circuit in Figure 18 there is no charge stored by the capacitor. Determine the current in R_1, R_2, and C (a) at the instant the switch is closed (that is, $t = 0$), and (b) after the switch is closed for a long period of time (that is, $t \to \infty$).

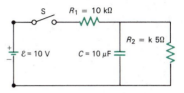

FIGURE 18

16. For the circuit shown in Figure 19, show that the resistance $R_{ab} = (27/17)$ ohms.

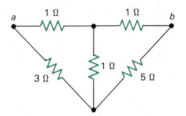

FIGURE 19

17. The circuit diagram shown in Figure 20 has *one* incorrect labeling of either a current or a potential. Identify which label is incorrect, and determine its correct value.

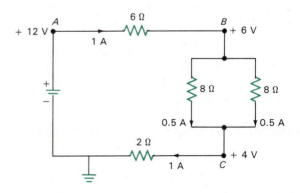

FIGURE 20

18. Figure 21 shows a portion of a microelectronic circuit used to create a flashing light. The capacitor C charges through the series combination of R_1 and R_2. When the voltage across the capacitor reaches 2.5 V it abruptly ceases to charge and begins to discharge through R_2 only. The capacitor discharges until the voltage across it drops to 1.25 V. Estimate the charge and discharge times.

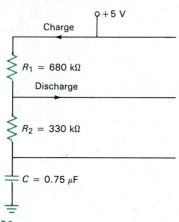

FIGURE 21

19. Determine the equivalent resistances of the configurations of the equal-valued resistors shown in Figure 22.

20. Each resistor in the circuit in Figure 23 has the same value (R). Show that the equivalent resistance of the combination is also R.

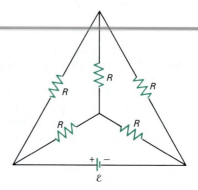

FIGURE 23

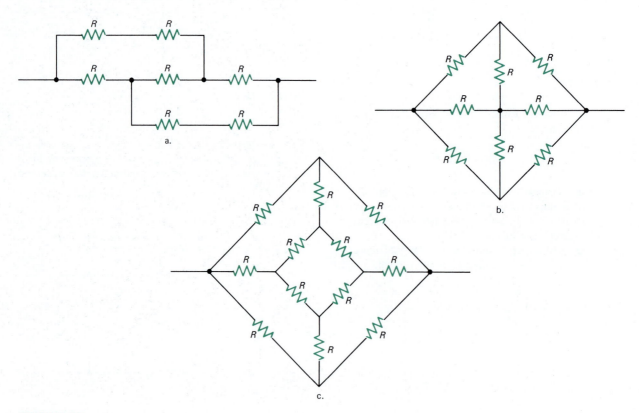

FIGURE 22

21. (a) Using symmetry arguments, show that the current through any resistor in the configuration of Figure 24 is either $I/3$ or $I/6$. (b) Show that the equivalent resistance between points A and B is $(5/6)r$.

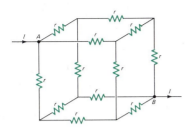

FIGURE 24

22. (a) For the circuit in Figure 25 show that

$$\frac{1}{I} = \frac{R}{\mathscr{E}} + \frac{r}{\mathscr{E}}$$

(b) The data below were recorded in the circuit shown in Figure 25 with an ammeter of negligible internal resistance. Using these data, make a plot of $1/I$ versus R and determine the emf and internal resistance of the battery.

$R\ (\Omega)$	$I\ (\mathrm{A})$
1	4.8
4	2.0
8	1.1
12	0.8
15	0.6

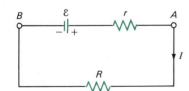

FIGURE 25

23. A symmetric infinite array of equal-valued resistors is shown in Figure 26. (a) Why is the equivalent resistance of this array equal to or greater than $2r$ but less than $3r$? (b) If the current in r_1 is I, why must the current in r_3 also be equal in magnitude to I? (c) If the current in r_4 is kI (k is a positive constant), why must the current in r_2 be $(1 - k)I$ and why must the current in r_5 be $k(1 - k)I$? (d) By applying Kirchhoff's voltage rule to the loop containing r_2, r_4, r_5, and r_6, show that $k = 2 - \sqrt{3}$. (e) Given that $k = 2 - \sqrt{3}$, apply Kirchhoff's voltage rule to the loop containing V, r_1, r_2, and r_3 and show that the equivalent resistance is $(1 + \sqrt{3})r$.

24. A resistor that does not obey Ohm's law is connected in series with a 10-ohm resistor and a 12-volt battery having negligible internal resistance (Figure 27). (a) If the potential difference V_x across the nonlinear resistor is 4.0 volts, what is the current in the circuit? (b) If it is found experimentally that the potential difference V_x is related to the current by $V_x = kI^2$, determine k.

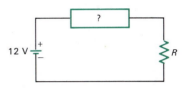

FIGURE 27

25. (a) Show that the ratio of the potential difference across the capacitor, V_C, and the current in the circuit in Figure 28.20 can be written as $V_C/I = R(e^{t/RC} - 1)$. (b) If we interpret V_C/I as the resistance of the capacitor, what is the resistance for $t = 0$ and $t \to \infty$? Do these results make sense physically in light of the way that the capacitor is charged by the battery?

26. An infinite number of resistors are connected in parallel. The resistance of the Nth resistor is $N!\ \Omega$. Thus, the first six resistances ($N = 0$ to $N = 5$) are $1\ \Omega$, $1\ \Omega$, $2\ \Omega$, $6\ \Omega$, $24\ \Omega$, and $120\ \Omega$. Determine the equivalent resistance of the combination to 4-figure accuracy. Try to solve the problem analytically. Otherwise, proceed with a brute-force solution.

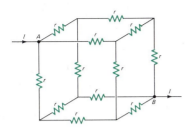

FIGURE 26

27. (a) Determine the charge on the capacitor in Figure 28 when $R = 10 \ \Omega$. (b) For what value of R will the charge on the capacitor be zero?

28. Determine the charge on the capacitor in Figure 29.

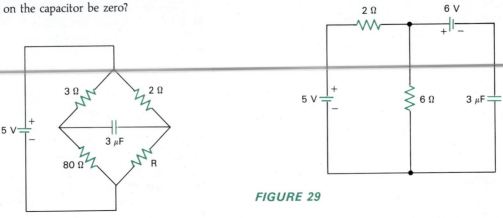

FIGURE 28

FIGURE 29

MAGNETIC FIELD

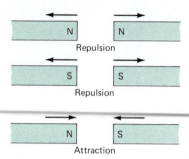

FIGURE 29.1

Two like poles repel each other, but unlike poles attract each other.

29.1
MAGNETISM

If you've ever played with a magnetic toy, used a compass, or attached notes to your refrigerator with a magnet, you've probably wondered how magnets work. In fact, magnets hold a certain fascination for many of us, perhaps because it is possible to *feel* magnetic force. For example, if you hold a magnet in your hand while standing near an iron or steel object, you feel your hand being pulled toward the object. If you hold a magnet in each hand, you sense forces exerted by one magnet on the other, even when the magnets are not in contact. If you place an insulating material such as glass between the two magnets, the forces persist. In fact, the forces exist even if the magnets are in a vacuum.

Experiments with magnets show that the sources of the magnetic force are concentrated in two regions called **poles** and that forces between magnetic poles can be attractive or repulsive. *Like poles repel* each other and *unlike poles attract* each other (Figure 29.1). Exhaustive experiments with magnets lead us to conclude that magnetic poles cannot be isolated. If we try to isolate a magnetic pole by cutting a magnet into two pieces, we do not obtain isolated poles. Instead, we get two smaller magnets (Figure 29.2).

If we sprinkle iron filings in the vicinity of a bar magnet, the filings become magnetized and produce a pattern similar to that produced by tiny bits of thread scattered in the vicinity of an electric dipole (Figure 29.3). The needle of a magnetic compass placed in the vicinity of the magnet will align itself with the iron filings (Figure 29.4). The poles of the bar magnet produce this alignment by exerting forces on the poles of the magnetized iron filings. We envision these forces as being transmitted to the iron filings and compass through a **magnetic field** created by the magnet.

Because isolated poles do not exist, we cannot explore magnetic fields in the way we studied electric fields, using isolated test charges. Instead we use the magnetic forces exerted on moving charges.

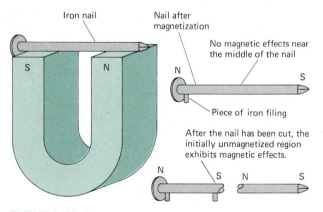

FIGURE 29.2

An iron nail becomes magnetized when laid across the pole faces of a magnet. A piece of iron filing is attracted to the ends of the magnetized nail but there is no attraction at the midpoint. Cutting the nail in half produces two magnets. The ends that were near the middle of the unsevered nail are now magnetic poles.

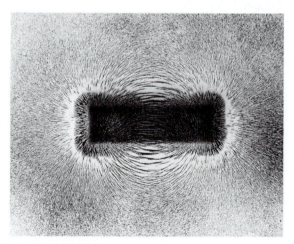

FIGURE 29.3

The magnetic field line pattern of a bar magnet, one form of magnetic dipole, as revealed by the alignment of tiny bits of iron.

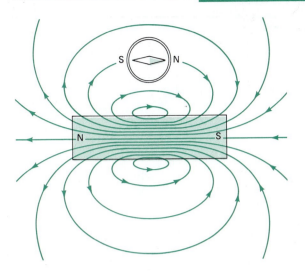

FIGURE 29.4
A compass needle aligns tangentially to a magnetic field line.

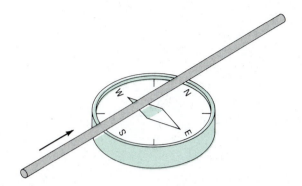

FIGURE 29.5
A compass needle beneath a current-carrying wire experiences a torque and orients perpendicular to the wire.

Although many similarities exist between electric phenomena and magnetic phenomena, no physical connection was inferred until July 12, 1820. During a lecture demonstration on this date, the Danish scientist Hans Christian Oersted accidentally discovered that *an electric current sets up a magnetic field.* During a lecture demonstration, Oersted noticed that a compass beneath a current-carrying wire orients itself perpendicularly to the wire (Figure 29.5). Oersted reversed the current in the wire, and observed that the compass needle reversed its orientation.

29.2
DEFINITION OF MAGNETIC FIELD
Magnetic Force on a Moving Charge

A properly oriented magnet experiences a force when placed in the magnetic field produced by an electric current. We might expect that a reaction force is experienced by the current. Indeed, experiment confirms this expectation. This fact enables us to set up an operational definition of magnetic field strength.

Moving charges can be produced in a variety of ways. Cathode-ray tubes similar to those used in oscilloscopes can be used to generate a beam of electrons. This technique is attractive because it permits the experimenter to control the speed of the charges by accelerating them through a known potential difference. Additionally, the electron deflections can be monitored by observing the spot of light produced by the charges as they strike a phosphorescent screen.

Let's use a beam of electrons produced by a cathode-ray tube to explore a uniform magnetic field. Figure 29.6a depicts a beam of electrons striking the center of the screen of a cathode-ray tube that is outside the magnetic field. When the tube is inserted into the magnetic field, as shown in Figure 29.6b, the electron beam is deflected. Electrons experience a force *perpendicular* to their direction of motion. Interchanging the magnetic poles reverses the deflection. The force increases as the speed of the electrons increases. The force also

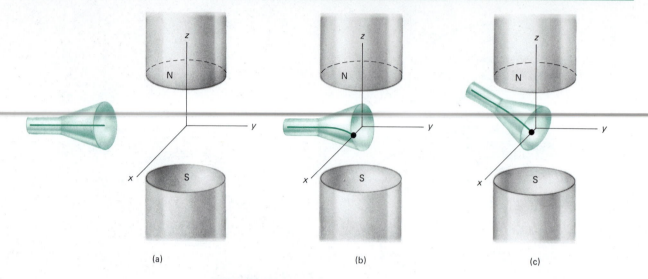

(a)   (b)   (c)

FIGURE 29.6

(a) The stream of electrons in the cathode ray tube experiences no magnetic force when the tube is outside the magnetic field produced by the magnet. (b) When the cathode ray tube is moved horizontally into the magnetic field, the electron stream is deflected by a magnetic force. (c) The magnetic force becomes weaker as the electron stream moves more nearly parallel to the z-axis.

depends on the direction of the beam of electrons. If the cathode-ray tube is tilted as shown in Figure 29.6c, the beam is still deflected but the deflection is diminished. The deflection vanishes if electrons move directly toward either pole.

Experiments also reveal that positive and negative charges traveling in the same direction experience magnetic forces in opposite directions.

In summary, these experiments disclose four facts about the magnetic force on a moving charge in a magnetic field:

1. There is a unique direction in space along which the moving charges experience no magnetic force. This direction, which we call the *zero-force direction*, lies along a line perpendicular to the pole faces (the z-axis in Figure 29.6).
2. The magnitude of the magnetic force is directly proportional to the product of the charge, the speed, and the sine of the angle between the velocity and the zero-force direction ($F \propto qv \sin \theta$).
3. The direction of the magnetic force is perpendicular to both the velocity and the zero-force direction.
4. Negative and positive charges moving in the same direction are deflected in opposite directions.

We can summarize these four facts by defining a magnetic field vector **B** to be along the zero-force direction, and by writing the force as

$$\mathbf{F} = q\mathbf{v} \times \mathbf{B} \tag{29.1}$$

In terms of the angle θ between **v** and **B**,

$$F = qvB \sin \theta \tag{29.2}$$

TABLE 29.1
Magnetic Fields

Neutron star, pulsar	10^8 T
Strongest man-made fields from brief bursts of electric current	10^3 T
Strong laboratory superconducting magnet	10 T
Strong bar magnet	10^{-1} T
Earth's magnetic field	10^{-5} T
Interplanetary magnetic fields	10^{-9} T
Magnetic field associated with electric currents in the human body	10^{-12} T

This definition of the magnetic field **B** is similar to the definition of the electric field **E** in that the field is defined in terms of the *force* it produces; this is an *operational* definition. To find the field, we must measure the force. But the magnetic field **B** is contained in a vector product, and its definition is therefore indirect. From Equation 29.1, the magnetic field **B** has units of

$$[B] = \frac{[\text{force}]}{[\text{charge} \cdot \text{velocity}]} = \frac{\text{N}}{\text{C} \cdot (\text{m/s})}$$

Because the ratio coulombs/second is equal to the unit of current, the ampere, the units of magnetic field can also be expressed as

$$\frac{\text{newtons}}{\text{ampere} \cdot \text{meter}} = \text{N/A} \cdot \text{m}$$

This unit is called a *tesla* and its symbol is T. The range of magnetic field strength encountered in nature is shown in Table 29.1. Many magnetic field strengths of interest are only a fraction of a tesla, and you may find magnetic field strengths expressed in the unit of *gauss* (G). For conversion purposes, $1\ \text{T} = 10^4$ G.

EXAMPLE 1

Measuring **B** *with Moving Protons*

Protons moving perpendicular to a magnetic field at a speed of 2.0×10^6 m/s experience an acceleration of 2.30×10^{14} m/s^2. Let's use Equation 29.2 to determine *B*.

$$B = \frac{F}{qv \sin \theta}$$

The force is given by $F = ma$, and θ is $90°$, so $\sin \theta = 1$.

$$B = \frac{ma}{qv}$$

$$= \frac{1.67 \times 10^{-27}\ \text{kg} \cdot 2.30 \times 10^{14}\ \text{m/s}^2}{1.60 \times 10^{-19}\ \text{C} \cdot 2.0 \times 10^6\ \text{m/s}}$$

$$= 1.2\ \text{T}$$

Measuring magnetic fields in this fashion is not practical because of the difficulty of measuring the acceleration. The Hall effect, which we will study in the next section, provides a practical method for measuring magnetic fields.

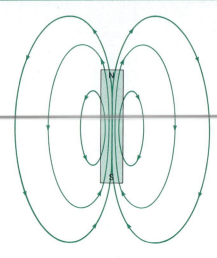

FIGURE 29.7

The magnetic field line pattern for a magnetic dipole. The magnetic field lines form closed loops.

Magnetic Field Lines

Electric field lines provide a visual picture of an electric field. The pattern of iron filings that forms around a bar magnet suggests that this field concept can be extended to magnetism. The direction of a magnetic field line at any point is the direction of **B** at that point. The spacing of field lines indicates the magnitude of **B**—that is, the closer together the lines are, the stronger the field is. Figure 29.7 shows the pattern of magnetic field lines for a bar magnet. Unlike electric field lines, *magnetic field lines do not begin or end*; magnetic field lines form continuous loops.

Magnetic Flux

We introduce a magnetic flux Φ_B in much the same way that we introduced electric flux. For a constant magnetic field **B** and a plane surface A, the magnetic flux is defined as the scalar product of the magnetic field and the area (Figure 29.8).

$$\Phi_B = \mathbf{B} \cdot \mathbf{A} \qquad (29.3)$$

The unit of magnetic flux is the tesla·meter^2, which is called a weber (Wb). In analogy with electric flux, *magnetic flux measures the number of magnetic field lines passing through (or threading) the area A*. For a non-uniform magnetic field and a curved surface we use a differential form of Equation 29.3:

$$d\Phi_B = \mathbf{B} \cdot d\mathbf{A} \qquad (29.4)$$

The *net* number of magnetic field lines, or net magnetic flux penetrating a surface, is obtained by integrating Equation 29.4 over the surface.

$$\Phi_B = \int \mathbf{B} \cdot d\mathbf{A} \qquad (29.5)$$

We found the net electric flux threading a closed surface to be given by Gauss' law:

$$\Phi_E = \int \mathbf{E} \cdot d\mathbf{A} = \frac{q}{\epsilon_0}$$

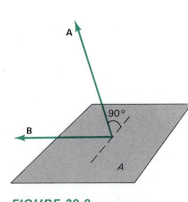

FIGURE 29.8

The magnetic flux equals the scalar product of **B** and the vector **A** representing the area A.

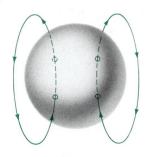

FIGURE 29.9
The magnetic flux through any closed surface is zero, because the magnetic field lines form continuous loops.

where q is the *net* charge inside the surface. There is no magnetic counterpart of electric charge. Unlike electric field lines, which originate on a positive charge and terminate on a negative charge, magnetic field lines have no origin and no termination. They form closed loops. Equal numbers of field lines enter and leave the surface (Figure 29.9). Thus, the net magnetic flux threading a *closed* surface is zero,

$$\int \mathbf{B} \cdot d\mathbf{A} = 0 \qquad (29.6)$$

Equation 29.6 is the magnetic analog of Gauss' law and is one of Maxwell's four equations describing electric and magnetic fields.

29.3
MOTION OF A CHARGED PARTICLE IN A MAGNETIC FIELD

The magnetic force on a charged particle is perpendicular to its velocity. This has some unusual and useful consequences. Let's investigate the motion of a charged particle in a uniform magnetic field and then look at some applications.

First, we show that the magnetic force does no work on the particle. The rate at which work is performed by a force $\mathbf{F}$ is given by the power P,

$$P = \mathbf{F} \cdot \mathbf{v}$$

The magnetic force $\mathbf{F} = q(\mathbf{v} \times \mathbf{B})$ is perpendicular to $\mathbf{v}$. It follows that the scalar product $\mathbf{F} \cdot \mathbf{v}$ is zero because the angle between $\mathbf{F}$ and $\mathbf{v}$ is 90° and $\cos 90° = 0$:

$$P = \mathbf{F} \cdot \mathbf{v} = Fv \cos 90° = 0 \qquad (29.7)$$

Zero power means that no work is performed. The work-energy principle states that the work done equals the change in kinetic energy. Because no work is performed, the kinetic energy of the particle remains constant and this in turn means that its speed is constant.

Even though the speed is constant, the velocity changes. The magnetic force causes an acceleration, which changes the direction of the velocity. If the velocity is perpendicular to the magnetic field, the particle will experience a centripetal acceleration toward a fixed point and undergo uniform circular motion.

Figure 29.10 shows a positive charge q moving in a uniform magnetic field. The field $\mathbf{B}$ is parallel to the y-axis. At the moment shown, the velocity $\mathbf{v}$ is directed along the positive x-axis. The resulting force and acceleration are along the z-axis. The particle moves in a circular orbit lying in the x–z plane, perpendicular to $\mathbf{B}$.

FIGURE 29.10
A particle with a positive charge q moves perpendicular to the magnetic field $\mathbf{B}$. The magnetic force $q(\mathbf{v} \times \mathbf{B})$ is toward the center of the circular path of the particle.

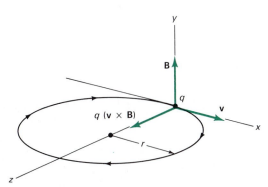

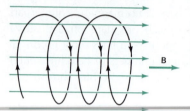

$v_\perp$ v

θ

$v_\parallel$

FIGURE 29.11

A charged particle in a uniform magnetic field travels along a helical path. The velocity component parallel to **B** is unaffected by the magnetic field. The velocity component perpendicular to **B** gives rise to a magnetic force perpendicular to the field.

Any velocity component parallel to **B** is unaffected by the magnetic field because **v** × **B** is zero when **v** and **B** are parallel. Thus motion parallel to **B** proceeds at a constant velocity, while movement perpendicular to **B** is uniform circular motion. The superposition of these two components results in a *helical* trajectory (Figure 29.11).

Orbit Radius, Linear Momentum, and Kinetic Energy

The radius of the circular orbit can be related to the magnitude of the linear momentum of the particle. Because the force is directed toward the center of the orbit we can set the acceleration equal to v^2/r in Newton's second law, $F = ma$. Setting the force equal to qvB we get

$$qvB = \frac{mv^2}{r}$$

The magnitude of the linear momentum, $p = mv$, follows as

$$p = mv = qBr \tag{29.8}$$

This result shows that a measurement of the orbit radius in a known magnetic field enables one to determine the particle linear momentum. This in turn makes it possible to relate the kinetic energy to the orbit radius. Thus,

$$K = \frac{p^2}{2m} = \frac{(qBr)^2}{2m} \tag{29.9}$$

EXAMPLE 2

Charge-to-Mass Ratio of the Electron

In an introductory laboratory experiment, electrons are accelerated through a potential difference of 400 volts. They move perpendicular to a magnetic field of 0.0095 T. The circular arc along which the electrons move is found to have a radius of 0.70 cm. We can use these data to determine the charge-to-mass ratio of the electron.

The work performed by the electric field associated with the 400-volt potential difference equals the kinetic energy acquired by the electrons. In equation form,

$$eV = \frac{p^2}{2m}$$

where e is the magnitude of the electron charge and V is the potential difference through which the electrons are accelerated. Using Equation 29.9 to replace $p^2/2m$ by $(eBr)^2/2m$ gives the charge-to-mass ratio,

$$\frac{e}{m} = \frac{2V}{B^2 r^2}$$

Inserting the data gives

$$\frac{e}{m} = \frac{2 \cdot 400 \text{ V}}{(0.0095 \text{ T})^2 (0.007 \text{ m})^2}$$
$$= 1.8 \times 10^{+11} \text{ C/kg}$$

High-precision experiments give a value of 1.758819×10^{11} C/kg.

Cyclotron Frequency

The magnetic force qvB results in uniform circular motion. The *period* T of the motion is related to the speed v and orbit radius r by

$$2\pi r = vT$$

Using this relation in Equation 29.8 gives for the period,

$$T = \frac{2\pi m}{qB}$$

The *frequency* of the motion f is the reciprocal of T,

$$f = \frac{qB}{2\pi m} \tag{29.10}$$

This frequency is called the **cyclotron frequency**. Note that the cyclotron frequency is independent of the radius of the particle orbit and the particle speed. This feature makes it possible to accelerate charged particles in a circular accelerator called a **cyclotron**. In a cyclotron (Figure 29.12) an electric field that oscillates at the cyclotron frequency does work on charged particles as they move in spiral-shaped trajectories. Cyclotrons are now used in materials research and in hospitals to produce radioactive elements employed in medical diagnostics and treatments.

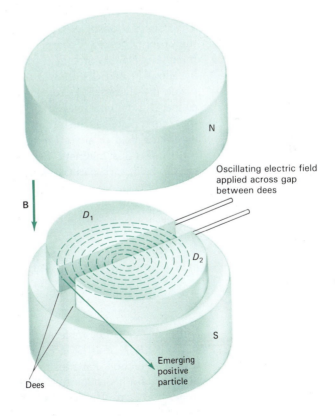

FIGURE 29.12

The cyclotron principle. An oscillating electric field accelerates charged particles across the gaps between the "dees" (labeled D_1 and D_2). Inside the dees the charges experience only the magnetic force $q(\mathbf{v} \times \mathbf{B})$. As they speed up they move in orbits of larger radius, but their orbital frequency equals the cyclotron frequency. Energy is efficiently transferred from the electric field to the particle motion provided the frequency of the oscillating electric field equals the cyclotron frequency.

EXAMPLE 3

Cyclotron Frequency of Ionospheric Electrons

The **ionosphere** is a layer of the earth's upper atmosphere in which ultraviolet radiation from the sun **ionizes** a small fraction of the atoms and molecules. Electrons are torn from atoms and molecules to form a partially ionized gas called a **plasma**. Electrons move in the earth's magnetic field at the cyclotron frequency. Let's calculate the cyclotron frequency for ionospheric electrons.

We take $B = 5 \times 10^{-5}$ T for the earth's magnetic field. The cyclotron frequency follows from Equation 29.10,

$$f = \frac{qB}{2\pi m} = \frac{1.60 \times 10^{-19} \text{ C} \cdot 5 \times 10^{-5} \text{ T}}{2\pi \cdot 9.11 \times 10^{-31} \text{ kg}}$$

$$= 1.40 \text{ MHz}$$

Radio transmissions at or near the cyclotron frequency can be strongly affected by a **cyclotron resonance** in which the electrons absorb energy from the radio wave in the same way that charges acquire kinetic energy in a cyclotron.

Charged-Particle Velocity Selector

An electric field can coexist with a magnetic field in a region of space. If a moving charge is in motion in such a region, it experiences an electric force $\mathbf{F}_E = q\mathbf{E}$, as well as a magnetic force $\mathbf{F}_B = q\mathbf{v} \times \mathbf{B}$. The total force is the superposition of the electric and magnetic forces, so

$$\mathbf{F} = \mathbf{F}_E + \mathbf{F}_B$$
$$= q(\mathbf{E} + \mathbf{v} \times \mathbf{B}) \tag{29.11}$$

The combination of electric and magnetic forces in Equation 29.11 is called the **Lorentz force**. A charged-particle **velocity selector** exploits the Lorentz force. A pencil-like stream of particles having a distribution of speeds is formed by allowing particles from a source to pass through holes in two separated plates. If a Cartesian coordinate system is oriented with the x-axis along a line connecting the two small holes, then the velocity of a particle passing through the second plate has the form $\mathbf{v} = \mathbf{i}v$ (Figure 29.13). If the particles pass through an electric field $\mathbf{j}E$ along the y-axis, then each particle experiences an electric force $\mathbf{j}(qE)$. If a magnetic field $\mathbf{k}B$ along the z-axis is superimposed on the electric field, each particle experiences a magnetic force

$$q\mathbf{v} \times \mathbf{B} = q(\mathbf{i}v) \times \mathbf{k}B = -\mathbf{j}(qvB)$$

The net force is

$$\mathbf{F} = \mathbf{j}q(E - vB)$$

Particles moving at a speed given by

$$v = \frac{E}{B} \tag{29.12}$$

experience zero net force and are undeflected. Particles with this speed pass through the second hole. Particles moving at slower speeds are deflected in the $+y$-direction. Particles moving at higher speeds are deflected in the $-y$-direction. Both groups of deflected particles miss the hole and are removed from the beam. The velocity that passes can be selected by adjusting the ratio E/B.

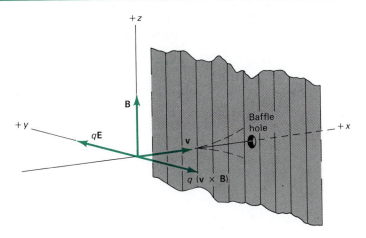

FIGURE 29.13

With both **E** and **B** at right angles to the velocity of the charged particle it is possible to balance the electric and magnetic forces for a particular speed. Particles moving at a speed $v = E/B$ experience zero net force and are undeflected. They travel through a small hole in the baffle. Other speeds result in a net force and cause the particle to be deflected. The deflected particles strike the baffle and are removed from the beam.

The Hall Effect

Figure 29.14a depicts a current in a thin, flat strip of metal connected to a battery. Electrons flow with drift speed v from the negative connection to the positive connection on the conducting strip. If a magnetic field is oriented perpendicular to the flat face of the strip, electrons feel a transverse force $-evB$ and are deflected from their previous course (Figure 29.14b). Because the electrons cannot escape from the conductor, negative charges accumulate on one side of the strip, leaving a net positive charge on the opposite side. This separation of charges produces a transverse electric field E_H, with H standing for Hall, for whom this effect is named. As a result of this electric field the electrons experience an electric force that opposes the magnetic force. Charges accumulate and E_H increases until the electric force on an electron cancels the

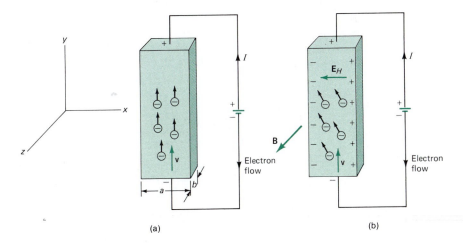

(a)   (b)

FIGURE 29.14

(a) A ribbonlike conductor is oriented with its long axis along the y-axis. The drift velocity of the electrons is in the positive y-direction. (b) A magnetic field is directed in the positive z-direction. Electrons experience a magnetic force in the negative x-direction.

magnetic force just like the action in the velocity selector. Equating the electric force on an electron to the magnetic force on an electron gives

$$-eE_H = -evB \tag{29.13}$$

The potential difference across the strip is called the **Hall voltage**, V_H. In terms of V_H the electric field can be written as

$$E_H = \frac{V_H}{a} \tag{29.14}$$

where a is the width of the strip. The drift speed v of the charges is related to the current density J and the number density of charges n by

$$J = nev$$

In terms of the cross-sectional area of the strip ab and the current I the current density (amperes per square meter) can also be written as

$$J = \frac{I}{ab} \tag{29.15}$$

Setting $J = nev$ in Equation 29.15 gives

$$v = \frac{I}{abne} \tag{29.16}$$

Combining Equations 29.13, 29.14, and 29.16 gives a relation between the Hall voltage and the magnetic field:

$$V_H = \left(\frac{I}{bne}\right) \cdot B \tag{29.17}$$

Here, $e = 1.60 \times 10^{-19}$ C. Measurements of I, B, b, and V_H allow us to calculate the number density n. Typically, V_H is in the microvolt range for metals and in the millivolt range for semiconductors. Its measurement requires a sensitive voltmeter. The Hall voltage is inversely proportional to the thickness of the sample; to enhance V_H, the sample is made quite thin.

Hall probes are widely used to measure magnetic fields. For a given strip of material, the Hall voltage is directly proportional to the magnetic field. By placing the strip in a known magnetic field, and by measuring V_H, we can determine the constant of proportionality between V_H and B. Unknown magnetic fields can then be determined by measuring the Hall voltage when the strip is placed in a magnetic field of interest.

EXAMPLE 4

Hall Voltage for a Copper Strip

The measured value of $1/ne$ for copper is 5.4×10^{-11} m^3/C. Let's calculate the Hall voltage in a copper strip 2 mm wide and 0.05 mm thick, if the strip carries a current of 100 mA and is placed in a magnetic field of 1 T. From Equation 29.17,

$$V_H = \frac{IB}{bne}$$

$$= \frac{(10^{-1}\ \text{A})(1\ \text{T})(5.4 \times 10^{-11}\ \text{m}^3/\text{C})}{5 \times 10^{-5}\ \text{m}}$$

$$= 1.1 \times 10^{-7}\ \text{V}$$

The Hall voltage is 0.11 μV. Measurement of this potential difference requires a sensitive measuring instrument.

Force on a Current-Carrying Conductor

By converting electric energy into mechanical energy, electric motors are able to perform countless tasks. Regardless of their size or complexity, most electric motors operate on the principle that a current-carrying conductor experiences a force when placed in a magnetic field.

At the atomic level this force originates with the $q\mathbf{v} \times \mathbf{B}$ force that acts on the electrons whose motions constitute the electric current. Because the electrons are confined to the conductor, the conductor as a whole experiences a net force.

To show this quantitatively, suppose that we establish a current I in a straight wire of uniform cross section, and place the wire in a uniform magnetic field $\mathbf{B}$ (Figure 29.15). In a time t the charge flowing through any cross section of the wire is $q = It$. This charge experiences a force

$$\mathbf{F} = q\mathbf{v} \times \mathbf{B}$$
$$= It\mathbf{v} \times \mathbf{B} \qquad (29.18)$$

where $\mathbf{v}$ represents the drift velocity of the charges. If t represents the time required for the charges to travel the length (L) of the wire, then it follows that

$$L = vt$$

By defining a vector $\mathbf{L} = \mathbf{v}t$ having the same direction as $\mathbf{v}$ (and the current) we can write

$$\mathbf{F} = I\mathbf{L} \times \mathbf{B} \qquad (29.19)$$

Because the charges are confined to the wire, this force is exerted on the wire as a whole. Note that if the magnetic field is parallel (or antiparallel) to the length of the wire, there is no force on the wire. If the magnetic field is perpendicular to the length of the wire, the force is a maximum, and has magnitude

$$F = ILB \qquad (29.20)$$

Next we show how this force acts to operate an electric motor.

Principle of a Direct-Current (dc) Electric Motor

Most electric motors operate on the principle that a current-carrying conductor experiences a force when placed in a magnetic field. Let's examine this principle further.

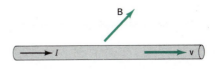

FIGURE 29.15
A straight wire of uniform cross section carrying a current *I* in a uniform magnetic field.

Figure 29.16 shows the origin of the torque that causes the shaft of the motor to rotate. A coil of wire and the shaft to which it is attached rotate as a result of a magnetic torque exerted on the current in the coil. We can determine the torque by examining a single loop of a current-carrying wire in the magnetic field of a permanent magnet. Sides 1 and 2, which are the same length (b) and carry the same current (I), experience equal and opposite forces ($\mathbf{F}_1 = -\mathbf{F}_2$) with the same lines of action. Hence they can neither translate nor rotate the loop. Sides 3 and 4 also experience equal but opposite forces, and therefore produce no translational motion. But because their lines of action are not colinear they exert a net torque on the loop. The magnitude of these equal forces follows from Equation 29.20,

$$F_3 = BIa = F_4$$

where a is the length of the side. Choosing a torque reference point on the line OO' passing through the midpoints of sides 1 and 2, we have for the net torque on the loop

$$\tau = F_3\left(\frac{b}{2}\right)\sin\theta + F_4\left(\frac{b}{2}\right)\sin\theta$$
$$= BIab\sin\theta \qquad (29.21)$$

This torque rotates the loop in a clockwise direction (Figure 29.17). In the upright position the torque is zero because the lines of action of $\mathbf{F}_3$ and $\mathbf{F}_4$ coincide ($\theta = 0$ in Equation 29.21).

The inertia of the loop carries it past the equilibrium position, at which time the current direction in the loop is reversed by the action of the commutator (Figure 29.18a). This current reversal at the two equilibrium positions of the coil provides a torque that is always clockwise (Figure 29.18b). Consequently, the coil rotates in a continuous fashion and produces a simple motor. This is called a direct-current (dc) motor because the current directions into and out of the motor never change. A practical motor contains many loops of wire in order to increase the torque on the coil.

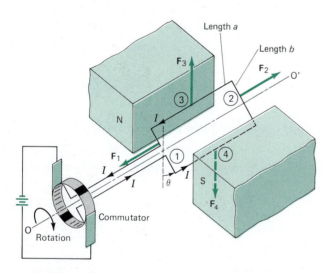

FIGURE 29.16
Viewed from the commutator end, the current in side 3 is directed toward the observer. In side 4, the current is directed away from the observer.

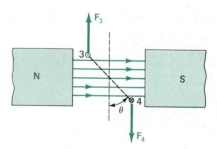

FIGURE 29.17
The magnetic force $\mathbf{F} = I\mathbf{L} \times \mathbf{B}$ is directed upward on side 3 and downward on side 4. The torques set up by the magnetic forces produce clockwise rotation as viewed from the commutator.

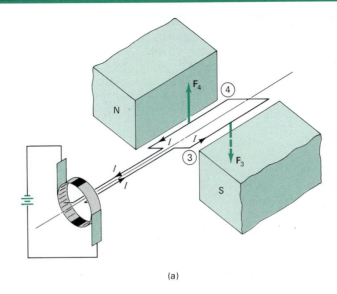

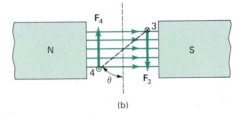

FIGURE 29.18

(a) When the coil rotates past the upright position, the commutator changes the direction of the current. (b) The current reversal maintains the direction of the torque. The coil continues to rotate clockwise.

29.4
MAGNETIC DIPOLE IN A MAGNETIC FIELD

Magnetic Dipole Moment

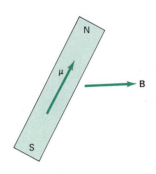

FIGURE 29.19

The magnetic dipole experiences a clockwise torque given by $\tau = \mu \times \mathbf{B}$. Measurements of the torque τ and the magnetic field $\mathbf{B}$ allow the magnetic moment μ to be determined.

An electric dipole experiences a torque in a uniform electric field. The torque on an electric dipole is given by

$$\tau = \mathbf{p} \times \mathbf{E}$$

where $\mathbf{p}$ is the electric dipole moment and $\mathbf{E}$ is the electric field. Magnetic structures ranging from bar magnets to electrons and atomic nuclei can experience a torque when placed in a magnetic field. We refer to such structures as **magnetic dipoles**. The torque τ on a magnetic dipole in a magnetic field $\mathbf{B}$ is related by a property of the dipole called the **magnetic dipole moment** (μ) that is *defined* by the relation

$$\tau = \mu \times \mathbf{B} \qquad (29.22)$$

In Figure 29.19, the bar magnet placed in a magnetic field experiences a torque described by Equation 29.22. Just as magnetic field is defined operationally by Equation 29.1, the magnetic dipole moment is operationally defined by Equation 29.22. We measure τ and **B**, and deduce μ. Magnetic moment has units of

$$\frac{\text{torque}}{\text{magnetic field}} = \frac{\text{newton} \cdot \text{meters}}{\text{newton/ampere} \cdot \text{meter}} = \text{J/T} \quad \text{or} \quad \text{A} \cdot \text{m}^2$$

The magnetic dipole moment of a proton is 1.41×10^{-26} J/T. A toy bar magnet might have a magnetic dipole moment of approximately 100 J/T.

Magnetic Dipole Moment of a Current-Carrying Loop

A current in a circular loop produces a magnetic field pattern like that produced by a magnetic dipole (Figure 29.20). When a current-carrying loop is placed in a magnetic field, the loop experiences a torque like that experienced by a magnetic dipole.

We can make this analogy between current loops and magnetic dipoles quantitative. Consider the expression for the magnitude of the torque on a magnetic dipole and the expression for the magnitude of the torque experienced by a current-carrying loop (Equation 29.21). The magnitude of the torque on a magnetic dipole in a uniform magnetic field follows from Equation 29.22,

$$\tau = \mu B \sin \theta \tag{29.23}$$

where θ is the angle between μ and **B**. The magnitude of the torque on a rectangular current loop in a uniform magnetic field is

$$\tau = IAB \sin \theta \tag{29.24}$$

where A is the area of the rectangular loop. Comparing Equations 29.23 and 29.24, we see that the current loop has a magnetic moment equal to the product of its area and the current,

$$\mu = IA \tag{29.25}$$

Here, as when we calculated magnetic flux in Section 29.2, we can treat area as a vector having magnitude A and direction perpendicular to the plane containing the area. If we take the direction of the current as the positive direction

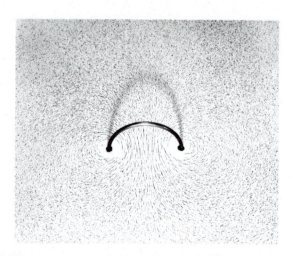

FIGURE 29.20

The arrangement of iron filings on a sheet of paper through which a circular loop of electric current passes shows a magnetic field pattern characteristic of a magnetic dipole.

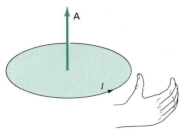

FIGURE 29.21

The magnetic moment of a current loop is given by the relation $\mu = I\mathbf{A}$, where I is the current and $\mathbf{A}$ is the vector representing the area of the current loop. The vector $\mathbf{A}$ has a magnitude equal to the area of the loop. To get the direction of $\mathbf{A}$, curl the right hand fingers in the direction of the current. The extended right thumb denotes the direction of $\mathbf{A}$ and thus also the direction of μ.

around the perimeter of the loop, the direction of the area vector can be determined by a right-hand rule (Figure 29.21). Thus we define the magnetic dipole moment of a loop to be

$$\mu = I\mathbf{A} \tag{29.26}$$

The equation $\tau = \mu \times \mathbf{B}$ for the magnetic torque is appropriate for either a magnetic dipole moment associated with a magnet such as a compass or a current-carrying loop. Although we assumed a rectangular current-carrying loop, the result of the derivation applies to a current loop of arbitrary shape.

Potential Energy of a Magnetic Dipole

The torque on a magnetic dipole performs work when the dipole rotates. Work performed on the dipole is stored as potential energy. We can write down the expression for the potential energy of a magnetic dipole by analogy with a relation derived in Section 25.5 for the electric dipole. There we showed that an electric dipole moment $\mathbf{p}$ in an external electric field $\mathbf{E}$ has a potential energy

$$U = -\mathbf{p} \cdot \mathbf{E}$$

The analogous relation for a magnetic dipole μ in an external magnetic field $\mathbf{B}$ is

$$U = -\mu \cdot \mathbf{B} \tag{29.27}$$

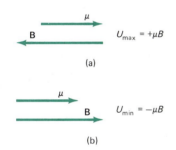

FIGURE 29.22

Orientations of magnetic moment and magnetic field for (a) maximum, (b) minimum potential energy.

In terms of the angle θ between μ and $\mathbf{B}$ we can express U as

$$U = -\mu B \cos \theta \tag{29.28}$$

Notice in Figure 29.22 that the potential energy U of the magnetic dipole is at a maximum when μ and $\mathbf{B}$ are antiparallel,

$$U_{max} = +\mu B$$

and at a minimum when μ and $\mathbf{B}$ are parallel,

$$U_{min} = -\mu B$$

The torque on the dipole ($\tau = \mu \times \mathbf{B}$) is zero whether μ and $\mathbf{B}$ are parallel or antiparallel. Therefore both orientations are equilibrium configurations. The parallel configuration is one of stable equilibrium, and the antiparallel configuration is one of unstable equilibrium. For example, the magnetic moment of a compass needle is in stable equilibrium when it is parallel to the earth's magnetic field. If the needle is rotated through $180°$, it remains in unstable equilibrium until some disturbance alters its alignment. The needle is then urged back toward the stable configuration.

A hydrogen atom has a magnetic dipole moment that can change from an unstable to a stable orientation, as described in the following example.

EXAMPLE 5

Magnetic Dipoles in Astrophysics

The hydrogen atom consists of an electron and a proton, both of which have an intrinsic magnetic moment. The magnetic moment of the electron sets up a magnetic field that affects the magnetic moment of the proton. There are two equilibrium configurations of the moments. In the stable configuration, the magnetic moment and

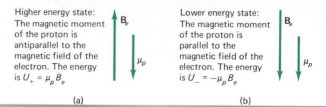

Higher energy state: The magnetic moment of the proton is antiparallel to the magnetic field of the electron. The energy is $U_+ = \mu_p B_e$

Lower energy state: The magnetic moment of the proton is parallel to the magnetic field of the electron. The energy is $U_- = -\mu_p B_e$

(a)   (b)

FIGURE 29.23

A schematic representation of the magnetic energy states in the hydrogen atom. Electromagnetic radiation is emitted when the atom changes from the higher energy state (a) to the lower energy state (b).

magnetic field are parallel (Figure 29.23a). In the unstable configuration, the magnetic moment and magnetic field are antiparallel (Figure 29.23b). The energies associated with these two configurations are

$$U_- = -\mu_p B_e \qquad \text{(stable configuration)}$$

$$U_+ = +\mu_p B_e \qquad \text{(unstable configuration)}$$

where μ_p is the magnetic moment of the proton and B_e is the magnetic field produced by the electron. The magnetic moment of the proton is

$$\mu_p = 1.41 \times 10^{-26} \text{ J/T}$$

The magnetic field set up by the electron is

$$B_e = 33.4 \text{ T}$$

There are vast clouds of hydrogen in our galaxy. Most of the individual hydrogen atoms are in the lower energy state. Occasionally, a collision will supply enough energy to boost an atom into the higher energy configuration. The energy required to do this is the energy *difference* between the stable and unstable states.

$$\text{energy required} = U_+ - U_- = 2\mu_p B_e$$
$$= 2(1.41 \times 10^{-26} \text{ J/T})(33.4 \text{ T})$$
$$= 9.42 \times 10^{-25} \text{ J}$$

This converts to 5.89×10^{-6} eV.

The atom can return to the stable configuration by *emitting* energy in the form of an electromagnetic wave with a wavelength of 21 cm. This radiation can be detected by a radio telescope. Radio astronomers have been able to map hydrogen-rich portions of our galaxy by measuring the intensity of this radiation.

Magnetic Force in a Non-Uniform Field

If the magnetic moment of a dipole is aligned with the magnetic field, the potential energy is given by $U = -\mu B$. In a uniform field the net force on the aligned dipole is zero. If the magnetic field varies in the x-direction there is a force given by $F_x = -\partial U/\partial x$, or

$$F_x = +\mu \frac{\partial B}{\partial x} \qquad (29.29)$$

The derivative $\partial B/\partial x$ is called the magnetic field gradient and measures how sharply B changes with position. The strength of the force is determined by the gradient of B rather than by B itself.

The magnetic force described by Equation 29.29 is the force that acts when a large magnet in a junkyard picks up an auto. This same magnetic force can be used to trap neutrons in storage rings.

EXAMPLE 6

A Neutron Storage Ring

The neutron has a magnetic moment of $\mu = 9.66 \times 10^{-27}$ J/T. Using specially shaped magnets, scientists have been able to use the magnetic force described by Equation 29.29 to hold slow-moving neutrons in circular orbits. One such neutron storage ring has a radius of 60 cm, and the neutrons move at speeds of 20 m/s.

Let's determine the magnetic field gradient, $\partial B/\partial x$. The neutrons experience a centripetal acceleration given by

$$v^2/r = \frac{(20 \text{ m/s})^2}{0.6 \text{ m}} = 667 \text{ m/s}^2$$

The neutron mass is $m = 1.67 \times 10^{-27}$ kg and so the radial force on the neutron is

$$F = ma = 1.67 \times 10^{-27} \text{ kg} \cdot 667 \text{ m/s}^2$$
$$= 1.12 \times 10^{-24} \text{ N}$$

The required magnetic field gradient follows from Equation 29.29,

$$\frac{\partial B}{\partial x} = \frac{F}{\mu} = \frac{1.12 \times 10^{-24} \text{ N}}{9.66 \times 10^{-27} \text{ J/T}}$$
$$= 116 \text{ T/m}$$

WORKED PROBLEM

A spring is connected to the end of a magnetic dipole that is free to pivot about its midpoint, as shown in Figure 29.24. The torque exerted on the dipole by the spring can be represented by $\tau = k\theta$, where k is a constant and the angle θ is expressed in radians. Assuming that only the spring and the magnetic field exert torques on the dipole, determine the equilibrium value of θ for the dipole. Take $\mu = 10$ J/T, $B = 0.025$ T, and $k = 1$ N·m.

Solution

The magnetic torque exerted on the dipole is $\mu B \sin \varphi = \mu B \cos \theta$. When the dipole is in equilibrium this torque is balanced by the spring torque $k\theta$:

$$k\theta = \mu B \cos \theta$$

The equilibrium value of θ satisfies

$$\theta = \left(\frac{\mu B}{k}\right) \cos \theta = \left(\frac{10 \text{ J/T} \cdot 0.025 \text{ T}}{1 \text{ N·m}}\right) \cos \theta$$
$$= \tfrac{1}{4} \cos \theta$$

This is a transcendental equation. Since $\cos \theta \leq 1$, the value of θ is less than $\frac{1}{4}$ radian. Take $\theta \approx \frac{1}{4}$ on the right side to get an approximate value for θ:

$$\theta \approx \tfrac{1}{4} \cos \left(\tfrac{1}{4}\right) = \tfrac{1}{4}(0.969) = 0.242$$

To get a better approximation, *iterate* this value in $\cos \theta$. This gives a result that is correct to three significant figures,

$$\theta \approx \tfrac{1}{4} \cos (0.242) = \tfrac{1}{4}(0.971) = 0.243$$

This converts to an angle of 13.9°.

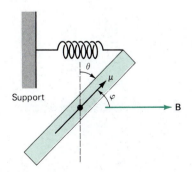

FIGURE 29.24

EXERCISES

29.2   Definition of Magnetic Field

A. An electron having velocity $10^6\mathbf{i}$ m/s experiences a maximum force of $1.6 \times 10^{-14}\mathbf{k}$ N when it enters a uniform magnetic field. What is the magnitude and direction of the magnetic field?

B. Electrons from outer space enter the earth's magnetic field with velocities on the order of 10^7 m/s. (a) Assuming the earth's magnetic field to be approximately 10^{-5} T, estimate the magnetic force on a cosmic electron. (b) Compare this force with the weight of an electron at the earth's surface.

C. A stream of particles, some positive, some negative, and some neutral, are directed along the east line in Figure 1. The magnetic field is directed into the plane containing the N–S, E–W lines (as represented by the crosses). Draw the possible trajectories for the three types of particles.

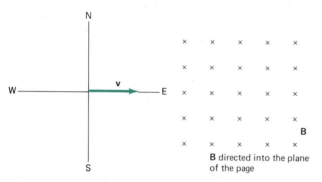

FIGURE 1

D. A positive charge ($q = 1.6 \times 10^{-19}$ C) moves through a magnetic field ($B = 2$ T). If the velocity is in the $\pm z$-direction, then the magnetic force equals zero. If the velocity is in the $+y$-direction, then the force is $\mathbf{F} = 6.4 \times 10^{-12}\mathbf{i}$ N. (a) What is the direction of $\mathbf{B}$? (b) If the charge moves parallel to the direction $(\mathbf{i} + \mathbf{j})$ with the same speed as in part (a), what is the magnitude and direction of the force?

E. A proton having velocity $\mathbf{v} = 10\mathbf{j}$ m/s enters a uniform magnetic field represented by $\mathbf{B} = 0.5\mathbf{i} - 0.8\mathbf{k}$ T. Determine (a) the force, (b) the acceleration experienced by the proton when it enters the field.

F. A charged particle moves with velocity $\mathbf{v}$ in a region where both an electric field $\mathbf{E}$ and a magnetic field $\mathbf{B}$ exist. What orientation of $\mathbf{v}$, $\mathbf{E}$, and $\mathbf{B}$ could make the net force on the charge equal zero?

G. Use the equation $\mathbf{F} = q\mathbf{v} \times \mathbf{B}$ to show that the units of magnetic field can be expressed in N/A·m.

29.3   Motion of a Charged Particle in a Magnetic Field

H. An electron passes through a velocity selector having $E = 10^4$ V/m, $B = 2 \times 10^{-2}$ T. Determine the electron speed.

I. **Cyclotron resonance** can occur when the cyclotron frequency of a charged particle equals the frequency of an incident electromagnetic wave. Determine the magnetic field strength required for plasma electrons to experience cyclotron resonance when they are subject to microwaves having a frequency of 6.3 GHz.

J. The electrons in a plasma composed of electrons and singly-charged positive sodium ions have a cyclotron frequency of 10.1 GHz. Determine the cyclotron frequency for the sodium ions.

K. A thin ribbon of copper is shown in Figure 2. The direction of the current and the applied magnetic field are shown in the figure. Make a diagram showing clearly and unambiguously: (a) the direction of motion and deflection of the conduction electrons; (b) the resulting polarity $(+, -)$ of the opposite edges of the ribbon; and (c) the Hall electric field, $\mathbf{E}_H$.

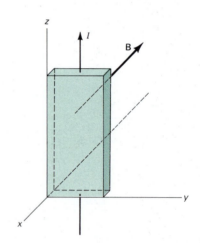

FIGURE 2

L. Protons from a cyclotron are steered between the pole faces of a magnet, as shown in Figure 3. We want to deflect the protons to the right of their path from the cyclotron—that is, out of the paper, toward you.

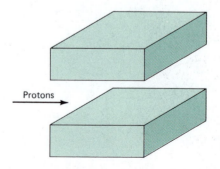

FIGURE 3

Determine the required direction of the magnetic field in the bending magnet and label the N and S poles of the magnet.

M. In a certain magnetic field, a proton and an alpha particle are observed to revolve in circles having the same radii. Compare their (a) linear momentum, (b) velocity, (c) kinetic energy.

N. A charged particle executes circular motion between the pole faces of a permanent magnet, as shown in Figure 4. Is the charge positive or negative?

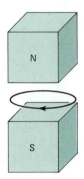

FIGURE 4

P. A horizontal length of copper wire (density = 8960 kg/m^3 and cross-sectional area = 8.37×10^{-6} m^2) experiences an upward magnetic force due to an imposed horizontal magnetic field ($B = 2$ T). What electric current is needed to support the weight of the wire exactly?

Q. A Hall probe used to measure magnetic fields has a thickness of 0.15 mm and is made from a material having a charge-carrier density of 10^{24}/m^3. For a current of 10 mA, a Hall voltage of 42 μV results when the probe is placed in a magnetic field of unknown strength. What is the value of the magnetic field?

29.4 Magnetic Dipole in a Magnetic Field

R. The needle of a Boy Scout compass experiences a maximum torque of 5.2×10^{-6} N·m when placed in a region where the earth's magnetic field has a strength of 2×10^{-5} T. Determine the magnetic dipole moment of the compass.

S. A certain non-uniform magnetic field is represented by $\mathbf{B} = 0.01x^2\mathbf{i}$ T, where x is the distance in meters from the origin of a Cartesian coordinate system. Determine the force on a magnetic dipole having a dipole moment of 2.8 J/T at a position 2 m from the origin.

T. A superconducting solenoid produces a field described by

$$\mathbf{B} = +16.4\mathbf{j} \text{ T}$$

In this field there is a magnetic dipole described by $\boldsymbol{\mu} = 3.18 \times 10^{-2}\mathbf{n}$ A·m^2, where $\mathbf{n} = 0.611\mathbf{j} - 0.792\mathbf{k}$. (a) Calculate the torque on this dipole. (b) Calculate

the potential energy of this dipole, taking the potential energy to be zero when the dipole and field are mutually perpendicular.

U. A coil consisting of 300 turns of wire has a mean radius of 10 cm and carries a current of 1.5 A. Initially its dipole moment is aligned parallel to a uniform magnetic field of 0.75 T. Calculate the minimum amount of work that must be done by the field in order to rotate the coil into a position where its magnetic moment is antiparallel to the magnetic field.

V. Determine the maximum and minimum magnetic potential energy of a proton in a magnetic field of 1 T. Express your result in electron volts.

W. The proton has a magnetic moment of $\mu = 1.41 \times 10^{-26}$ J/T. When a proton in a hydrogen atom flips from the unstable equilibrium position to the stable equilibrium position in the magnetic field of the electron, 9.47×10^{-25} J of energy are released as 21-cm electromagnetic radiation. Calculate the magnetic field experienced by the proton.

X. Typically, magnetic dipole moments of atoms and molecules are on the order of one **Bohr magneton** ($\mu_B = 9.2741 \times 10^{-24}$ J/T). These atomic and/or molecular magnetic moments may experience magnetic fields on the order of 1 T. Show that the corresponding magnetic potential energy is roughly 10^{-4} eV.

Y. The maximum torque on a rectangular current loop rotating in a magnetic field $B = 10^{-2}$ T is 4.6×10^{-4} N·m. The loop area is 2×10^{-3} m^2. Determine the current in the loop.

Z. A pattern of magnetic lines corresponding to a non-uniform magnetic field is shown in Figure 5. Several small bar magnets are shown in the figure and numbered for reference. (a) Ignoring forces between bar magnets, which magnets will experience a net force that urges them along a field line from left to right? (b) Which magnets experience the greatest torque?

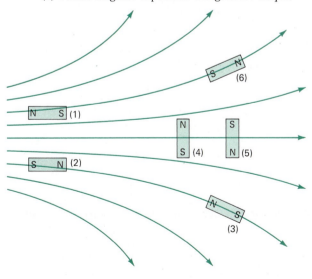

FIGURE 5

PROBLEMS

1. Magnetic flux is trapped in many situations where ionized material collapses. If the sun were to collapse from its present radius (7×10^8 m) to a white dwarf, it would trap much of its magnetic flux. The magnetic field of the sun is approximately 10^{-2} T. If the sun collapsed to a white-dwarf configuration with a radius equal to the earth's radius, estimate the resulting magnetic field strength.

2. A neutron has a magnetic moment of 9.66×10^{-27} J/T and a mass of 1.67×10^{-27} kg. Determine the strength of the magnetic field gradient $\partial B/\partial x$ required to cancel the gravitational force acting on a neutron.

3. Consider a copper rod 1 cm in diameter. A cross-sectional disk that is one atom in thickness contains about 6.2×10^{14} copper atoms. Each copper atom has 27 electrons, which means that there are about 1.7×10^{16} electrons in the disk. These electrons are tiny magnets, each having a magnetic moment $\mu = 9.3 \times 10^{-24}$ J/T. Suppose these elementary magnets are all parallel to the axis of the rod. (a) What is the disk's magnetic moment? (b) What circumferential current would give rise to the same magnetic moment for the disk?

4. A cosmic-ray proton traveling at half the speed of light is heading directly toward the center of the earth in the plane of the earth's equator. Will it hit the earth? As an estimate, assume that the earth's magnetic field is 5×10^{-5} T and extends out one earth diameter, or 1.3×10^7 m. Calculate the radius of curvature of the proton in this magnetic field.

5. In a Hall-effect experiment a potential difference will often develop across the potential difference contacts, even if no magnetic field is present. This is a potential difference resulting from current in a resistance produced by misalignment of the contacts. Although the polarity of the Hall voltage depends on the direction of the impressed magnetic field, the IR potential difference does not. Show that if two potential difference measurements are taken with oppositely directed magnetic fields, then the true Hall voltage is

$$V_H = \tfrac{1}{2}(V_1 - V_2)$$

6. Electrons in a color TV set are often accelerated through a potential difference of 10 kV. Calculate the maximum force that an electron experiences when it enters a magnetic field of 3000 G after being accelerated through a potential difference of 10 kV.

7. A free neutron at rest disintegrates into a proton, an electron, and an antineutrino (which has no charge). (a) Describe the subsequent motions shown in Figure 6 if the proton initially is directed along the $+x$-axis. and the electron initially is directed along the $-x$-axis. (b) Does the magnet experience a *net* reaction force?

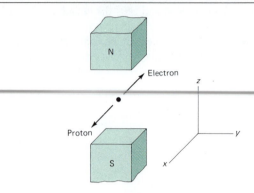

FIGURE 6

8. A proton having velocity $\mathbf{v} = 2 \times 10^8 \mathbf{i}$ m/s enters a uniform magnetic field having a strength of 1 T. The proton leaves the field with a velocity $\mathbf{v} = -2 \times 10^8 \mathbf{j}$ m/s. Determine (a) the direction of the magnetic field, (b) the radius of curvature of the path of the proton while in the magnetic field, (c) the distance traveled by the proton while in the magnetic field, (d) the time spent by the proton in the magnetic field.

9. Sodium melts at 210°F. Liquid sodium is an excellent thermal conductor and is used in some nuclear reactors to remove thermal energy from the reactor core. The liquid sodium can be moved through pipes by pumps that exploit the force on a moving charge in a magnetic field. The principle is as follows: imagine the liquid metal to be in a pipe having a rectangular cross section of width w and height h. A uniform magnetic field perpendicular to the pipe affects a section of length L (Figure 7). An electric current directed perpendicular to the pipe and to the magnetic field produces a current density J. (a) Explain why this arrangement produces a force on the liquid that is directed along the length of the pipe. (b) Show that the section of liquid in the magnetic field experiences a pressure increase equal to JLB. (c) Calculate the current density required to produce a pressure increase

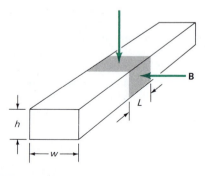

FIGURE 7

of 2 atm for a section of length 1 m and a field strength of 1 T.

10. (a) In the Bohr model of the hydrogen atom an electron executes circular motion about a proton fixed at the center of the orbit. The force on the electron is provided by the Coulomb force between the electron and proton. The radius of the smallest orbit is 0.529 Å. Determine the speed of the electron in this smallest orbit. (b) Another electron executes uniform circular motion in a plane perpendicular to a constant magnetic field. The speed of the electron and the radius of its orbit are the same as for the electron in part (a). Calculate the strength of the magnetic field.

11. A thin ribbon of a silver alloy 2.00 cm wide and 0.0150 mm thick carries a current of 6.98 A perpendicular to a magnetic field (Figure 2). The Hall voltage is found to be 1.24×10^{-3} V when the magnetic field strength is 25.0 T. (a) Starting with Figure 2, show the direction of motion of the charge carriers (electrons), the polarity of the opposite sides of the strip, and the direction of the Hall electric field, $\mathbf{E}_H$. (b) Calculate the drift speed. (c) Calculate n, the number of charge carriers per cubic meter.

12. A pendulum of length L has a bob made of a non-conducting material. The bob carries a charge q and swings in a plane perpendicular to a uniform magnetic field of strength B (Figure 8). (a) Prove that the frequency of oscillation is not changed by the charge on the pendulum. (b) In terms of B, g, L, and q, determine the difference in tension at the bottom of the swing for the two directions of motion of the bob.

horizontal circle in a magnetic field $\mathbf{B}$ directed downward. Show that when the angle θ is small ($\theta \ll 1$ radian), the two possible directions of motion in a horizontal plane have angular frequencies given by $\omega = \sqrt{g/L} \pm (qB/2m)$.

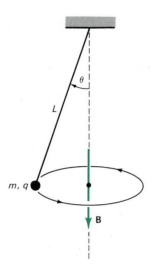

FIGURE 9

14. Table 1 shows measurements of a Hall voltage and corresponding magnetic field for a probe used to measure magnetic fields. (a) Make a plot of these data and deduce a relationship between the Hall voltage and magnetic field. (b) If the measurements were taken with a current of 0.2 A and the sample is made from a material having a charge-carrier density of $10^{26}/m^3$, what is the thickness of the sample?

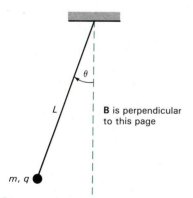

FIGURE 8

13. A conical pendulum of length L has a bob of mass m and charge q (Figure 9). The bob moves in a

TABLE 1

V_H (μV)	B (T)
0	0
11	0.1
19	0.2
28	0.3
42	0.4
50	0.5
61	0.6
68	0.7
79	0.8
90	0.9
102	1.0

CHAPTER 30

MAGNETIC FIELD OF ELECTRIC CURRENT

30.1
THE LAW OF BIOT AND SAVART

Soon after Oersted discovered that an electric current sets up a magnetic field, Jean Baptiste Biot and Felix Savart made quantitative studies of the relationship between the field and the current. Their results were subsequently refined by others, giving what is now known as the law of Biot and Savart. This law establishes a *differential* relationship between current and magnetic field.

Consider a conductor carrying a current I, and the magnetic field set up by this current. From a differential viewpoint, the magnetic field at any point is the vector sum (integral) of differential contributions, $d\mathbf{B}$, set up by the current in differential vector lengths, $d\mathbf{s}$ (Figure 30.1). The length $d\mathbf{s}$ is taken parallel to the current and the product $I\,d\mathbf{s}$ is called a **current element**. The law of Biot and Savart specifies the relation between the field $d\mathbf{B}$, and the current element $I\,d\mathbf{s}$:

$$d\mathbf{B} = \frac{\mu_0}{4\pi}\frac{I\,d\mathbf{s} \times \hat{\mathbf{r}}}{r^2} \tag{30.1}$$

The geometric relation between $d\mathbf{B}$, $d\mathbf{s}$, and $\hat{\mathbf{r}}$ is illustrated in Figure 30.1. In this figure, $\hat{\mathbf{r}}$ is a unit vector directed from the current element $I\,d\mathbf{s}$ (called the source point) toward the point P at which $\mathbf{B}$ is to be evaluated (called the field point). The differential magnetic field vector $d\mathbf{B}$ is perpendicular to $I\,d\mathbf{s}$ and to $\hat{\mathbf{r}}$. The proportionality factor in Equation 30.1 has the value

$$\frac{\mu_0}{4\pi} = 10^{-7} \text{ T·m/A}$$

The quantity μ_0 is called the **magnetic permeability** of free space. It is **not** a magnetic moment.

$$\mu_0 = 1.25664 \times 10^{-6} \text{ T·m/A}$$

The magnetic field produced by all current elements in a complete circuit is obtained by adding the $d\mathbf{B}$ vectors set up by each $I\,d\mathbf{s}$ current element.

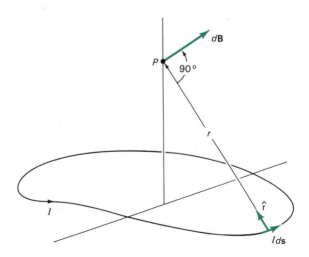

FIGURE 30.1

The current element $I\,d\mathbf{s}$ generates a contribution $d\mathbf{B}$ at the field point P.

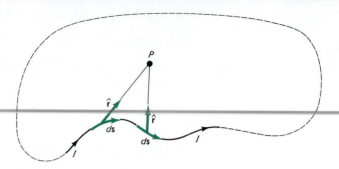

FIGURE 30.2
When the field point *P* is in the same plane as the current loop all current elements produce *d***B**'s in the same direction. For the counterclockwise current shown, *I d***s** × **r̂** and *d***B** are directed out of the page.

Mathematically, this addition amounts to the integration of Equation 30.1 over the entire circuit,

$$\mathbf{B} = \frac{\mu_0 I}{4\pi} \oint \frac{d\mathbf{s} \times \hat{\mathbf{r}}}{r^2} \tag{30.2}$$

The circle on the integral sign reminds us that the path of integration is a closed loop (Figure 30.2). The integral in Equation 30.2 is another type of **line integral**. To become familiar with such integrals we consider two important examples; the magnetic fields produced by a straight-line current and by a circular current loop.

Magnetic Field of a Straight-Line Current

Consider a straight-line segment of wire carrying a current *I*. We want to use the law of Biot and Savart to calculate the magnetic field a distance *R* from the center of the segment. Figure 30.3 shows the geometry. The segment of interest runs from $s = -a$ to $s = +a$. Each current element sets up a vector *d***B** directed out of the page in Figure 30.3 (the direction of *d***s** × **r̂**). Hence the magnitude of the vector **B** is the sum (integral) of the magnitudes of the *d***B**'s given by the law of Biot and Savart. The magnitude of *d***s** × **r̂** is *ds* sin θ, where θ is the angle between *d***s** and **r̂**. Thus,

$$B = \int dB = \frac{\mu_0 I}{4\pi} \int_{-a}^{+a} \frac{ds \sin \theta}{r^2}$$

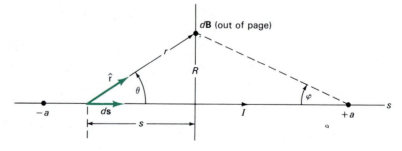

FIGURE 30.3
Geometry for evaluating the magnetic field of a straight-line current.

To evaluate the integral we express $\sin \theta$ and r^2 in terms of s. From Figure 30.3,

$$\sin \theta = \frac{R}{\sqrt{s^2 + R^2}} \qquad r^2 = s^2 + R^2$$

$$B = \frac{\mu_0 I R}{4\pi} \int_{-a}^{+a} \frac{ds}{(s^2 + R^2)^{3/2}}$$

The integral gives

$$\int_{-a}^{+a} \frac{ds}{(s^2 + R^2)^{3/2}} = \frac{s}{[R^2(s^2 + R^2)]^{1/2}} \bigg|_{-a}^{+a} = \frac{2a}{R^2 \sqrt{a^2 + R^2}}$$

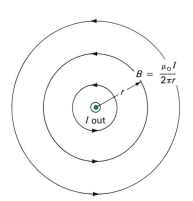

$B = \frac{\mu_0 I}{2\pi r}$

so that

$$B = \frac{\mu_0 I}{2\pi R} \cdot \frac{a}{\sqrt{a^2 + R^2}} \tag{30.3}$$

Observe that (Figure 30.3)

$$\cos \varphi = \frac{a}{\sqrt{a^2 + R^2}}$$

FIGURE 30.4

The straight-line current is directed out of the page. The concentric circles defined by r = constant are lines of constant B.

In terms of φ,

$$B = \frac{\mu_0 I}{2\pi R} \cdot \cos \varphi \tag{30.4}$$

A limiting case of Equations 30.3 and 30.4 is noteworthy. When the length of the segment ($2a$) is much larger than the distance R, then the angle φ is small and its cosine is nearly unity,

$$\cos \varphi = \frac{a}{\sqrt{a^2 + R^2}} \simeq 1 \qquad a \gg R$$

In this limit the field a distance $R = r$ from the wire is

$$B = \frac{\mu_0 I}{2\pi r} \tag{30.5}$$

We refer to the field given by Equation 30.5 as that of a long straight wire, it being clear that "long" means $a \gg r$. As Equation 30.5 shows, B has the same magnitude at all points on a circle concentric with the wire and in a plane perpendicular to the wire (Figure 30.4). The magnetic field lines form circles about the current.

A convenient way to remember the direction of the magnetic field of a current-carrying wire is to use a right-hand rule. If you let the extended thumb of your right hand point in the direction of the current, then the fingers of your right hand, when curled around the wire, give the direction in which the magnetic field lines encircle the wire (Figure 30.5).

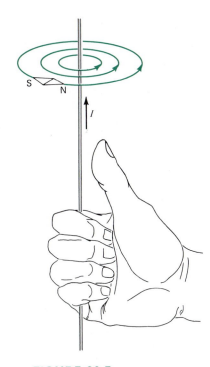

FIGURE 30.5

A right-hand rule relates the directions of the current and the magnetic field.

EXAMPLE 1

Magnetic Field at the Center of a Square Current Loop

Let's use Equation 30.3 to calculate the magnetic field at the center of a square loop that is 1 m on a side, and carries 1 A of current (Figure 30.6). First, we consider

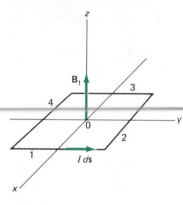

FIGURE 30.6

Each side makes an equal contribution to the magnetic field at the center of the square current loop.

the directions of the **B** vectors. Vector $\mathbf{B}_1$, produced by the current in line segment 1, is perpendicular to the plane containing line 1 and the origin. In other words, $\mathbf{B}_1$ is perpendicular to the x–y plane. Using the right-hand rule, we find that $\mathbf{B}_1$ is in the positive z-direction:

$$\mathbf{B}_1 = \mathbf{k}B_1$$

The same argument applies to the remaining three **B** vectors. All are in the positive z-direction. The four **B** vectors may be added algebraically,

$$\mathbf{B} = \mathbf{B}_1 + \mathbf{B}_2 + \mathbf{B}_3 + \mathbf{B}_4 = \mathbf{k}(B_1 + B_2 + B_3 + B_4)$$

Because all four sides are the same length and are the same distance from the origin, the four **B** vectors all have the same magnitude. Therefore

$$\mathbf{B} = \mathbf{k}4B_1$$

To find the magnitude of B_1 we substitute into Equation 30.3. With $a = 0.5$ m, $R = 0.5$ m, and $I = 1.0$ A, we get

$$B_1 = \frac{(4\pi \times 10^{-7}\ \text{T·m/A})(1\ \text{A}) \cdot (0.5\ \text{m})}{2\pi(0.50\ \text{m})[(0.5\ \text{m})^2 + (0.5\ \text{m})^2]^{1/2}}$$
$$= 2\sqrt{2} \times 10^{-7}\ \text{T}$$

The magnetic field at the center of this square loop is $4B_1$ and is directed in the positive z-direction:

$$\mathbf{B} = \mathbf{k}1.13 \times 10^{-6}\ \text{T}$$

Magnetic Field of a Circular Current Loop

Next, let's use the law of Biot and Savart to calculate the magnetic field set up by a circular current loop (Figure 30.7). We evaluate **B** at a point a distance z above the center of the loop. The vector $d\mathbf{B}$ in Figure 30.7 can be resolved into components parallel to and perpendicular to the z-axis. When we integrate around the loop, the components perpendicular to the z-axis cancel. The net field is the sum of the components parallel to the z-axis,

$$B = \oint dB_z = \oint dB \cdot \cos \alpha$$

The unit vector $\hat{\mathbf{r}}$ is perpendicular to the current element $I\,d\mathbf{s}$ so

$$|I\,d\mathbf{s} \times \hat{\mathbf{r}}| = I\,ds$$

Figure 30.7 shows that

$$\cos \alpha = \frac{a}{r}$$

where a is the radius of the loop. Thus,

$$B = \left(\frac{\mu_0 I}{4\pi}\right) \oint \frac{a\,ds}{r^3}$$

In the integration around the loop the distance r is the same for all current elements. Hence

$$B = \left(\frac{\mu_0 I a}{4\pi r^3}\right) \oint ds$$

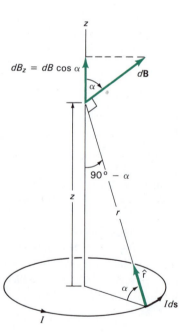

FIGURE 30.7

The circular current loop sets up a magnetic field given by the law of Biot and Savart. At points above the center of the loop the field is directed upward along the $+z$-axis.

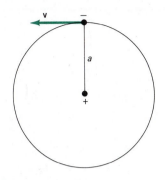

FIGURE 30.8

Planetary model of a hydrogen atom: an electron $(-)$ in orbit about a proton $(+)$.

The integral $\oint ds$ equals $2\pi a$, the circumference of the loop. Thus

$$B = \frac{\mu_0 I a^2}{2r^3}$$

In terms of the distance z, measured from the center of the loop,

$$B = \frac{\mu_0 I a^2}{2(z^2 + a^2)^{3/2}} \qquad (30.6)$$

We note a special case of Equation 30.6. At the center of the loop, $z = 0$ and B is a maximum and given by

$$B_{center} = \frac{\mu_0 I}{2a} \qquad (30.7)$$

EXAMPLE 2

Magnetic Field in a Hydrogen Atom

The planetary model of a hydrogen atom consists of an electron in a circular orbit about a proton (Figure 30.8). The motion of the electron constitutes an electric current. Let's calculate the magnetic field produced by the orbiting electron at the location of the proton.

The radius of the electron orbit is $a = 5.29 \times 10^{-11}$ m. The electron's orbital speed v is 2.19×10^6 m/s. This means that the time it takes to complete one orbit (the period) is

$$T = \frac{2\pi a}{v} = \frac{2\pi(5.29 \times 10^{-11} \text{ m})}{2.19 \times 10^6 \text{ m/s}}$$

$$= 1.52 \times 10^{-16} \text{ s}$$

The electric current I is equal to the electron charge e divided by this time,

$$I = \frac{e}{T} = \frac{1.60 \times 10^{-19} \text{ C}}{1.52 \times 10^{-16} \text{ s}}$$

$$= 1.06 \times 10^{-3} \text{ A}$$

The magnetic field at the center of the electron orbit follows from Equation 30.7 as:

$$B = \frac{\mu_0 I}{2a}$$

$$= \frac{(4\pi \times 10^{-7} \text{ T·m/A})(1.06 \times 10^{-3} \text{ A})}{2 \times 5.29 \times 10^{-11} \text{ m}}$$

$$= 12.6 \text{ T}$$

This magnetic field is quite strong. In fact, it is 30,000 times stronger than the earth's magnetic field at the equator.

30.2
AMPÈRE'S LAW

The work of Oersted and of Biot and Savart was continued by André Marie Ampère. Ampère formulated an integral relation between the magnetic field and the electric current. Ampère found that any line integral of the magnetic

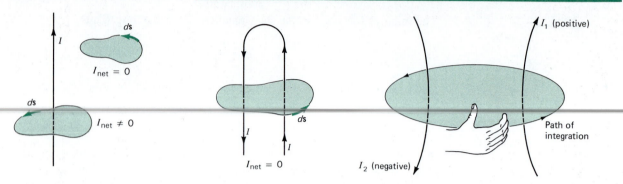

FIGURE 30.9

The line integral of **B** is determined by the net current encircled by the path of integration.

FIGURE 30.10

A right-hand rule is used to define the positive current direction.

field around a closed path was proportional to the net current encircled by the path. In SI units Ampère's law is

$$\oint \mathbf{B} \cdot d\mathbf{s} = \mu_0 I_{net} \tag{30.8}$$

A right-hand rule is used to associate a sign with the current (Figure 30.9). Curl the fingers of your right hand in the direction of the path of integration. Your extended thumb defines the positive current direction. In Figure 30.10, for example, the path of integration is counterclockwise. The current I_1 is positive and I_2 is negative. Ampère's law involves only the line integral of **B**. It is possible to use Ampère's law to evaluate **B** itself in situations where the field exhibits a high degree of symmetry. We illustrate this technique by using Ampère's law to determine the magnetic field set up by a long straight current-carrying wire.

Magnetic Field of a Long Straight Current-Carrying Wire

In the preceding section we used the law of Biot and Savart to determine the magnetic field set up by a long straight wire carrying a current I. We found the results given by Equation 30.5 and displayed in Figure 30.4. The field lines are circles, concentric with the current. This circular symmetry enables us to use Ampère's law to derive Equation 30.5. We exploit the symmetry of **B** by choosing a circular path to evaluate $\oint \mathbf{B} \cdot d\mathbf{s}$. Figure 30.11 shows that along a circle of radius r, concentric with the current, the vector **B** is parallel to the vector $d\mathbf{s}$ and so

$$\oint \mathbf{B} \cdot d\mathbf{s} = \oint B \, ds$$

Along the circular path, B is constant. The line integral that remains gives the circumference of the path

$$\oint B \, ds = B \oint ds = 2\pi r B \tag{30.9}$$

Using this result in Ampère's law (Equation 30.8) we obtain

$$2\pi r B = \mu_0 I$$

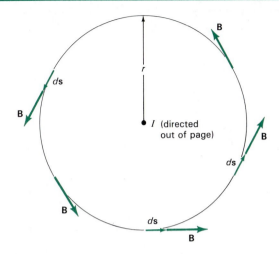

FIGURE 30.11

The magnetic field lines are circles concentric with the current. Along the circular path of integration, $\mathbf{B} \cdot d\mathbf{s} = B \, ds$.

This gives the same result for B obtained via the law of Biot and Savart,

$$B = \frac{\mu_0 I}{2\pi r} \tag{30.5}$$

We can also use Ampère's law to determine the behavior of B *inside* the wire as $r \to 0$. Equation 30.5 suggests that B becomes infinite as $r \to 0$. The following example shows that B tends toward zero as $r \to 0$ and provides a further illustration of Ampère's law.

EXAMPLE 3

Magnetic Field Inside a Current-Carrying Wire

A long round wire with radius a carries a total current I. To calculate B in the interior of the wire we apply Ampère's law around a circle of radius r. The line integral in Ampère's law is

$$\oint \mathbf{B} \cdot d\mathbf{s} = 2\pi r B$$

exactly as given by Equation 30.9. The path inside* the round wire (Figure 30.12) has the same circular symmetry that we have for the path outside the round wire. We assume that the current I is spread uniformly over the cross-sectional area (πa^2) of the wire. The net current encircled by the path of integration is

$$I_{\text{net}} = I\left(\frac{\pi r^2}{\pi a^2}\right)$$

Using these results in Ampère's law gives

$$2\pi r B = \frac{\mu_0 I r^2}{a^2}$$

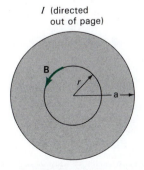

I (directed out of page)

FIGURE 30.12

A path of integration for calculating the magnetic field inside a cylindrical distribution of current. The current I is directed out of the page.

---

* In the interior of a material medium we should take the magnetic properties of the medium into account and replace μ_0 by μ, analogous to $\epsilon_0 \to \epsilon$ for dielectrics. Actually the effect is generally very small except in ferromagnetic media. For copper the effect is only a few parts per million.

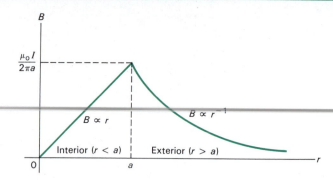

FIGURE 30.13

The magnetic field inside and outside a uniform cylindrical distribution of current.

and we find

$$B = \frac{\mu_0 I r}{2\pi a^2} \qquad r \leq a \tag{30.10}$$

Inside the round wire, the field is directly proportional to the distance from the center. This means that B is zero at the center of the wire—quite a contrast to the infinite field implied by Equation 30.5, which describes B *outside* the wire.

The strength of the magnetic field for the wire described in Example 3 is plotted in Figure 30.13 as a function of r. Equation 30.10 describes the linear portion $0 \leq r \leq a$ and Equation 30.5 yields a hyperbolic curve for $r \geq a$.

Magnetic Field of a Solenoid

A **solenoid** is a long current-carrying wire tightly wound into a helix, or coil (Figure 30.14). Usually the length of the coil is larger than the radius. Solenoids are useful because the magnetic field inside a solenoid can be very strong and nearly uniform over a large volume.

We want to determine the magnetic field of a solenoid. First, we develop an approximate picture to enable us to determine the symmetry that allows

FIGURE 30.14

A solenoid.

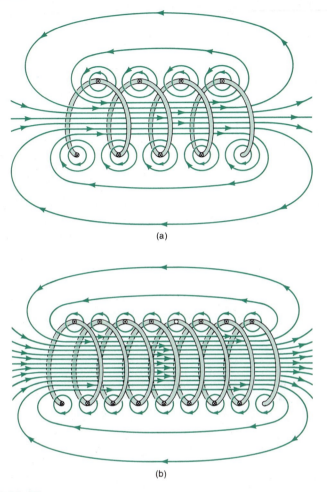

FIGURE 30.15

Magnetic field lines for (a) a loosely wound solenoid; (b) a tightly wound solenoid.

Ampère's law to be used. Then we calculate the magnetic field quantitatively, using Ampère's law.

Think of a solenoid as approximately a series of circular current loops. We calculated the field of a single circular current loop in Section 30.1. If the wire is loosely wound (Figure 30.15a), some magnetic field lines will encircle the individual turns of the coil. Other field lines will combine to form the axial magnetic field of the solenoid. Here again we use the principle of superposition, adding the vector fields of the individual circular current loops. If the wire is wound more tightly, the number of field lines encircling each individual turn is reduced and the desired axial field is strengthened (Figure 30.15b).

An ideal solenoid is infinitely long and so tightly wound that the current is essentially a cylindrical sheet. When applying Ampère's law, the problem is to find a path of integration over which the integral can be evaluated and that yields a useful result. We select the rectangular path shown in Figure 30.16.

The line segments EF and GH are parallel to the z-axis, and FG and HE are perpendicular to the z-axis. The integral around the closed path can be expressed as a sum of integrals along the four paths:

$$\oint \mathbf{B} \cdot d\mathbf{s} = \int_E^F B_z \, dz + \int_F^G B_r \, dr + \int_G^H B_z \, dz + \int_H^E B_r \, dr$$

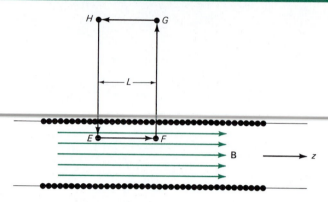

FIGURE 30.16

A rectangular path of integration is used to evaluate the axial magnetic field inside a solenoid.

We've taken the solenoid to be very long, essentially infinite. This means that since the ends are infinitely far away, nothing, including the radial field B_r, can depend on z. The fact that B_r is independent of z is an expression of another symmetry property: **translational symmetry**. If the solenoid is infinitely long, there is nothing physical to single out any one value of z from any other value of z. Then

$$B_r(z) = B_r(z + L)$$

since the two radial integrals ($\int_F^G B_r \, dr$ and $\int_H^E B_r \, dr$) are in opposite directions, they exactly cancel.

Next we must consider the integral

$$\int_G^H B_z \, dz$$

involving the value of B_z outside the solenoid. Let us assume that if there is a nonzero field B_z outside the solenoid, this field falls off with distance. We can take the line segment GH an arbitrarily large distance from the solenoid, which makes the field B_z arbitrarily small, and the integral over GH vanishes.

Inside the solenoid the magnetic field is **axial** (parallel to the *axis* of the solenoid). Because this is the z-axis, with B_z independent of z, the path integral is

$$\oint \mathbf{B} \cdot d\mathbf{s} = \int_E^F B_z \, dz = B_z \cdot L$$

For a current I, and a winding of N turns in the distance $EF = L$, the current enclosed by the path of integration is NI. Thus, Ampère's law for the solenoid gives

$$B_z \cdot L = \mu_0 NI$$

If we let $N = nL$, where n is the number of turns per meter, the field inside the solenoid may be written

$$B_z = \mu_0 nI \tag{30.11}$$

Note carefully that we have not specified where the line segment EF is located within the solenoid. This result holds inside the solenoid for any distance out from the z-axis, $0 \leq r < a$. Therefore the axial field B_z is uniform, constant across the cross section of the solenoid, and independent of r as well as of z. This is why solenoids are so useful in scientific work: they can produce strong, uniform magnetic fields. Also, because of the strength of the axial field, solenoids have important everyday applications. For example, solenoids are used in the electrical switches for starters in cars.

We can use a right-hand rule to determine the direction of the axial field. If the fingers of your right hand curl in the direction of the current in a solenoid, your extended thumb points in the direction of the field.

EXAMPLE 4

Magnetic Field of a Superconducting Solenoid

A solenoid 10 cm long has an inside diameter of 3 cm. The wire is a superconducting niobium alloy. The fine wire is wound in many layers, for a total of 4×10^4 turns per meter. What is the magnetic field produced when the current is 60 A?

From Equation 30.11

$$B_z = \mu_0 nI = (1.26 \times 10^{-6} \text{ T·m/A})(4 \times 10^4/\text{m})(60 \text{ A})$$
$$= 3.0 \text{ T}$$

Without superconducting wire you would need a large (and expensive) magnet to produce such a strong magnetic field over this volume. Many turns of fine wire, even if copper, would have a moderately high resistance. The large potential difference necessary to drive the 60-A current would require more electrical insulation (to avoid short circuits), which would result in a much bulkier piece of equipment. The heat produced (I^2R loss) would make it difficult, if not impossible, to cool the coil. For large solenoids the power demand would be exorbitant.

The Field of a Finite Solenoid

The analysis of the axial field B_z of an ideal solenoid (infinitely long) also applies to a *finite* solenoid if we stay away from the ends. Near the ends, some magnetic field lines go out between the turns and the axial field becomes weaker, as illustrated in Figure 30.17. Note that the finite solenoid does have an *external* axial field. This field exists because the magnetic field lines form

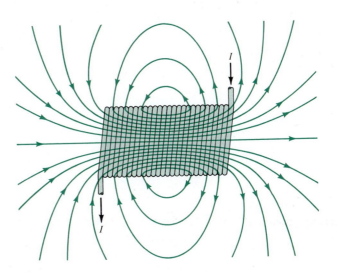

FIGURE 30.17
The magnetic field lines of a tightly wound solenoid.

closed loops. They exit from one end of the finite solenoid, curve back, and reenter the opposite end, just as they do in a bar magnet.

We can use a symmetry argument to show that the magnetic field on the axis at the ends of a solenoid is one-half the magnetic field at the center of the solenoid. By symmetry, the magnetic field on the axis has the same magnitude and direction at both ends. Imagine now that we construct a longer solenoid by connecting two identical solenoids, end to end. The magnetic field at the center of the longer solenoid is the superposition of the two equal magnetic fields at the adjacent ends of the shorter solenoids. It follows that the magnetic field at the end of a solenoid is one-half the magnetic field at the center. For a long solenoid the magnetic field at the center equals $\mu_0 nI$, and so the magnetic field at the ends equals $\frac{1}{2}\mu_0 nI$.

30.3
THE MAGNETIC FORCE BETWEEN PARALLEL CURRENTS

Imagine two long parallel conductors a distance d apart. One carries a current I_1 and the other carries a current I_2 (Figure 30.18). The first current creates a magnetic field

$$B_1 = \frac{\mu_0 I_1}{2\pi d}$$

at the position of the second conductor. As indicated in Figure 30.18, the field $\mathbf{B}_1$ is oriented counterclockwise (by the right-hand rule).

The force exerted on a length L of the current I_2 is

$$\mathbf{F}_2 = I_2 \mathbf{L} \times \mathbf{B}_1$$

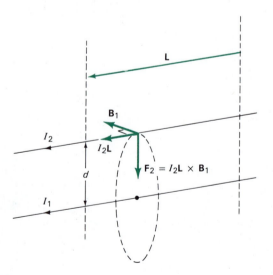

FIGURE 30.18

The current I_1 sets up a field $\mathbf{B}_1$ that exerts a force on the current element $I_2\mathbf{L}$. When I_1 and I_2 are parallel as shown here, the force attracts I_2 toward I_1. An equal but oppositely directed force attracts I_1 toward I_2.

where the vector **L** is in the direction of I_2. With **L** perpendicular to $\mathbf{B}_1$ the magnitude of the force is

$$F_2 = I_2 L B_1 = \frac{\mu_0 I_1 I_2 L}{2\pi d} \tag{30.12}$$

The current I_2 creates a magnetic field at the position of the first conductor. We can readily verify that the force exerted on I_1 by I_2 obeys Newton's third law—that is, the currents exert equal but opposite magnetic forces on each other. In Figure 30.18 the currents I_1 and I_2 are *parallel* and the wires are *attracted* to each other. When the currents are antiparallel the conductors experience mutually repulsive forces.

EXAMPLE 5

Welding by Lightning

Two parallel copper rods are 1 cm apart. Lightning sends a 10,000-ampere pulse of current along each conductor. Let's calculate the force per unit length on one conductor.

From Equation 30.12

$$\frac{\text{force}}{\text{meter}} = \frac{\mu_0 I_1 I_2}{2\pi d}$$

$$= \frac{1.26 \times 10^{-6} \text{ T·m/A} \times 10^4 \text{ A} \times 10^4 \text{ A}}{2\pi \times 10^{-2} \text{ m}}$$

$$= 2000 \text{ T·A}$$

$$= 2000 \text{ N/m}$$

This is about 140 pounds per foot. This force, together with the joule heat generated, might weld the two conductors together.

The mutual magnetic attraction of parallel currents occurs in plasma physics and is called the **pinch effect**. A plasma consists of positive and negative ions in motion. When a potential difference is established across a plasma, a current is set up. This current is distributed over the cross section of the plasma like a bundle of parallel-current filaments. These parallel-current filaments attract one another, causing a constriction—or pinching—of the plasma.

Definition of the Ampere

The force that one current-carrying conductor exerts on a second, parallel, current-carrying conductor is used to define the unit of current, the ampere. The same current is sent through two parallel conductors. The force per unit length on conductor 2 follows from Equation 30.12:

$$\frac{F_2}{L} = \frac{\mu_0 I^2}{2\pi d} \tag{30.13}$$

This force is measured with a very sensitive, accurate balance. The conductor length L and the separation distance d are also measured accurately. The constant μ_0 equals $4\pi \times 10^{-7}$ T·m/A. We can then solve Equation 30.13 for the current I in order to find the value of the current in amperes. The ampere is defined as follows:

> One ampere is the electric current in each of two parallel conductors separated by 1 m that gives rise to a force per unit length of 2×10^{-7} N/m on each conductor.

With the ampere defined we can now define the coulomb operationally as the quantity of electric charge that passes a given point in a conductor in 1 s when the current is 1 A. Thus,

$$1\,\mathrm{C} = 1\,\mathrm{A \cdot s}$$

The reason for this seemingly roundabout definition of the coulomb is strictly pragmatic: scientists can control and measure electric current far more precisely and accurately than they can control and measure electric charge.

WORKED PROBLEM

A ribbon-like conductor of width $2a$ and infinite length carries a current I. The conductor lies in the x–y plane with its center line on the x-axis, as shown in Figure 30.19. The current is distributed uniformly across the width of the ribbon. Determine the magnetic field at a point $y = R$ on the y-axis ($R > a$).

Solution

Treat the ribbon as a collection of narrow current strips. In Figure 30.19, the strip of width dy carries a current dI. Because the current is spread uniformly across the width $2a$, the fraction of the current carried by the strip is given by

$$\frac{dI}{I} = \frac{dy}{2a}$$

so

$$dI = \frac{I}{2a}\,dy$$

The current dI sets up a magnetic field dB given by Equation 30.5, the equation for the magnetic field of a long, straight current. The field point is a distance $R - y$ from the strip so

$$dB = \frac{\mu_0 dI}{2\pi r} = \frac{\mu_0 I\,dy}{4\pi a(R - y)}$$

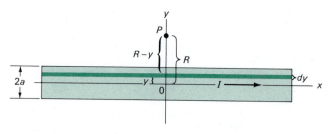

FIGURE 30.19

The total magnetic field is obtained by integrating the contributions from all strips. This is achieved by integrating from $y = -a$ to $y = +a$,

$$B = \frac{\mu_0 I}{4\pi a} \int_{-a}^{+a} \frac{dy}{R - y} = \frac{\mu_0 I}{4\pi a} \left(-\ln (R - y) \Big|_{-a}^{+a} \right)$$

$$= \frac{\mu_0 I}{4\pi a} \ln \left[\frac{R + a}{R - a} \right]$$

EXERCISES

30.1 The Law of Biot and Savart

A. An element of a conductor 1 mm long carries a current of 100 A. This current element is at the origin and is directed along the positive z-axis ($I\,d\mathbf{s} = \mathbf{k}I\,ds$). Find the contribution to the magnetic field $d\mathbf{B}$ of this current element at each of the following points (distances in meters): (a) (1, 0, 0), (b) (0, 0, 1), (c) (1, 1, 1).

B. (a) A square loop 1 m on a side conducts a current of 1 A (Figure 1). Calculate the magnetic field at the center of the square. Compare your result with the field produced at the center of a circular loop by the same current when the loop radius is (b) 0.5 m.

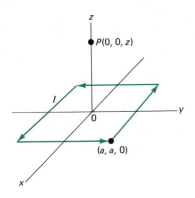

FIGURE 1

C. A circular loop has a radius of 3 cm and carries a clockwise current of 3 A. A concentric loop has a radius of 8 cm. What current must it carry to produce zero magnetic field at the center of the loops?

D. A circular coil of wire whose radius is 10 cm has 100 turns. The current in the coil is 0.3 A. (a) Calculate the magnetic field at the center of the coil. (b) How far above or below the center of the coil are the points where the magnetic field has dropped to half its value at the center of the coil?

E. Three concentric circular current loops have radii R, $2R$, and $3R$. The inner and outer loops carry clockwise currents of 5 A. Determine the magnitude and direction of the current in the middle loop if the net magnetic field at the center of the loops is zero.

F. Figure 2 shows a segment of a circuit consisting of two straight conductors and a semicircular conductor of radius a. The circuit carries a current I. (a) What magnetic field is produced at point P by the currents in the straight line segments? (b) Calculate the total magnetic field at P if $I = 10$ A and $a = 6$ cm.

FIGURE 2

G. A pair of long straight wires are parallel, and separated by a distance of 2 m. The wires carry parallel currents of 1 A. Superimpose the magnetic fields set up separately by the currents and determine the magnitude of the net magnetic field at a point in the plane containing the wires (a) midway between the wires, (b) at a position 2 m from the point midway between the wires.

30.2 Ampère's Law

H. A section of a circuit running to and from a battery consists of two parallel wires (Figure 3). The current in each wire is 2.0 A. Evaluate the loop integral $\oint \mathbf{B} \cdot d\mathbf{s}$ for (a) a path encircling the top wire only, as shown, (b) a path encircling only the bottom wire, and (c) a path encircling both wires.

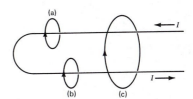

FIGURE 3

I. An evaluation of the loop integral in Ampère's law around a current-carrying conductor yields a value of 0.314×10^{-6} T·m. Determine the current.

J. A long wire lying on the line $x = a$ in the x–z plane carries a current I in the positive z-direction. A second long wire, parallel to the first on the line $x = -a$, carries a current I in the negative z-direction (Figure 4). (a) Sketch the magnetic field lines you expect in the x–y plane. (b) Calculate the net magnetic field at the point P on the y-axis. Express your result as a function of y.

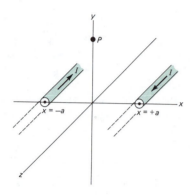

FIGURE 4

K. Lead becomes superconducting at 4.3 K. At this temperature the superconductivity is destroyed by a magnetic field of at least 0.052 T. Consider a wire of superconducting lead with a radius of 1.00 mm at 4.3 K. What current can it carry without destroying its own superconductivity with its own magnetic field?

L. At room temperature the superconducting solenoid of Example 4 has a resistance of 10,000 Ω. A 12-V potential difference is applied to the solenoid terminals. Calculate the resulting magnetic field inside the solenoid.

M. A 12-ft bubble chamber is inside a solenoid of niobium-tantalum coils. The interior field is equivalent to that of an ideal solenoid with 796 turns per meter, each carrying a current of 1800 A. Calculate the magnetic field inside the solenoid.

N. Two long solenoids are coaxially mounted. The solenoids are wound in opposite directions with the same number of turns per meter n and the same current I. If the radii are R_1 and $R_2 > R_1$, determine the magnetic field at a distance r from the axis for (a) $r < R_1$, (b) $R_1 < r < R_2$, (c) $r > R_2$.

P. How would the magnetic field inside a long solenoid change if (a) the current were doubled, (b) the number of turns per meter were doubled, (c) the length of the solenoid were doubled, cutting the number of turns per meter in half, (d) the length of the solenoid were doubled, holding the number of turns per meter constant, (e) the diameter of the solenoid were doubled?

30.3 The Magnetic Force Between Parallel Currents

Q. Three parallel straight wires lie in a plane and are equally spaced (Figure 5). The center wire is a distance of 10 cm from the outside wires. The current in each wire is 10 A. Give the direction and magnitude of the force per unit length on the center wire if its current is in the $+x$-direction, and (a) the currents in the outer wires are both in the $+x$-direction, (b) the current in wire A is in the $+x$-direction and the current in wire C is in the $-x$-direction.

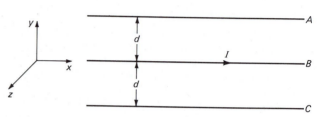

FIGURE 5

R. A rectangular conducting loop, 0.10 m × 0.80 m, carrying 2.0 A is placed 0.10 m away from a long, straight conductor carrying 6.0 A (Figure 6). (a) Calculate the net magnetic force on the loop. (b) Calculate the net torque on the loop about an axis through its center and parallel to the longer sides.

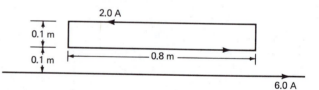

FIGURE 6

S. The movable conductor of a current balance in a freshman laboratory has an effective length of 0.80 m. The fixed parallel conductor is 0.010 m away. Both conductors carry the same current. If the balance is sensitive to 1 milligram, what is the minimum current that the balance can detect?

T. To get a feeling for the magnetic forces on the coils of a large solenoid, calculate the magnetic force on a 4-m length of a conductor that is carrying a current of 1800 A perpendicular to a field of 1.8 T.

U. A plasma inside a hollow cylinder carries a current and thereby produces a magnetic field. Assuming there is a uniform current density across the cylinder, what is the direction of the magnetic force on the outer edge of the plasma?

PROBLEMS

1. Consider a segment of current extending along the x-axis from 0 to s (Figure 7). Calculate the magnetic field that this current segment contributes to the magnetic field at point P on the y-axis.

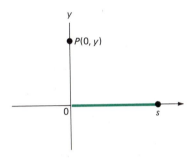

FIGURE 7

2. A flat current loop is in the shape of a rectangle. Prove that for a given fixed perimeter, the magnetic field at the center of the loop is smallest for a square-shaped loop.

3. A rectangle lies beside a long, straight wire, as shown in Figure 8. The rectangle and the wire are in the same plane. The wire carries a current of 4.90 A. Calculate the magnetic flux Φ_B threading the rectangle.

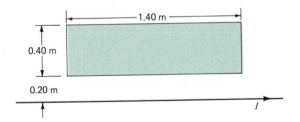

FIGURE 8

4. A circular coil (radius 0.50 m) consists of 180 turns. Each turn carries a current of 4.21 A. (a) Calculate the magnitude of the magnetic field at the center of the coil. At the center of the coil is a smaller circular coil of 20 turns with a radius of 1.00 cm. Each turn carries 0.680 A. (b) Calculate the magnitude of the magnetic moment of this smaller coil. (c) Assuming that the magnetic field of the larger coil is uniform over the smaller coil, calculate the magnitude of the maximum torque on the smaller coil. (d) Calculate the difference in potential energy between the unstable and stable orientations of the smaller coil.

5. Calculate the magnetic field for a point $P(0, 0, z)$ on the axis of a square loop of current (Figure 1).

6. The magnetic field produced by a straight wire at a perpendicular distance r from the center of the wire can be expressed as

$$B = \left(\frac{\mu_0 I}{2\pi r}\right) \sin \beta$$

where 2β is the angle subtended by the wire from the field point P. (a) Show that the field produced at the center of a current loop in the form of an N-gon (a regular polygon with N sides) can be expressed as

$$B_N = \left(\frac{\mu_0 I}{2r}\right) \frac{\sin(\pi/N)}{\pi/N}$$

(b) Show that for $N = 4$, the formula for B_N gives the result for a square loop derived in Example 1. (c) Obtain the field at the center of a circular loop of radius r by considering a limiting form of B_N.

7. An infinitely long, straight wire is bent, as shown in Figure 9, to form a circle 1 m in radius and a smaller circle of radius r. The direction of the current along the wire is shown in the diagram. The magnetic field at the common center is zero. Determine the value of r.

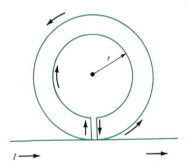

FIGURE 9

8. Helmholtz coils consisting of circular conductors, each of radius b, are placed parallel to each other with axes coinciding, a distance b apart (Figure 10). Each coil carries a current I in the same sense of circulation. (a) Sketch the magnetic field lines you expect for a cross-sectional plane containing the common axis (the x–y plane). (b) Calculate the magnetic field B_x on the common axis at the midpoint between the coils ($x = 0$). (c) Demonstrate that this field is relatively uniform by showing that

$$\left.\frac{dB_x}{dx}\right|_{x=0} = 0 \quad \text{and} \quad \left.\frac{d^2 B_x}{dx^2}\right|_{x=0} = 0$$

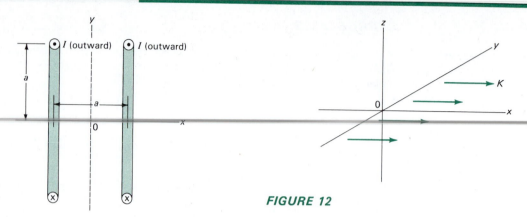

FIGURE 10

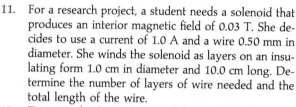

FIGURE 12

9. A coaxial cable (Figure 11) consists of an inner, solid conductor of radius a and an outer, concentric conductor of inner radius b and outer radius c. The inner conductor carries a current I directed into the page and the outer conductor carries an equal current directed out of the page. The current density is constant in each conductor. Determine the magnetic field as a function of the radial distance r for (a) $r < a$, (b) $a < r < b$, (c) $b < r < c$, (d) $r > c$.

11. For a research project, a student needs a solenoid that produces an interior magnetic field of 0.03 T. She decides to use a current of 1.0 A and a wire 0.50 mm in diameter. She winds the solenoid as layers on an insulating form 1.0 cm in diameter and 10.0 cm long. Determine the number of layers of wire needed and the total length of the wire.

12. Two circular loops are parallel, coaxial, and almost in contact, 1 mm apart (Figure 13). Each loop is 10 cm in radius. The top loop carries a current of 140 A clockwise. The bottom loop carries 140 A counterclockwise. (a) Calculate the magnetic force that the bottom loop exerts on the top loop. (b) The upper loop has a mass of 0.021 kg. Calculate its acceleration, assuming that the only forces acting on it are the force in part (a) and its weight.

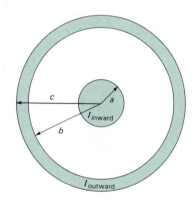

FIGURE 11

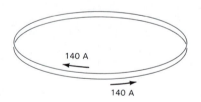

FIGURE 13

10. Electric charge flows along a wide sheet of metal (Figure 12). The sheet lies in the x–y plane and the charge flows in the $+x$-direction. The current is distributed uniformly over the sheet. Show that the magnetic field set up above or below the sheet by the current is given by

$$\mathbf{B} = \begin{cases} -\mathbf{j}\tfrac{1}{2}\mu_0 K & z > 0 \\ +\mathbf{j}\tfrac{1}{2}\mu_0 K & z < 0 \end{cases}$$

where K is the *surface* current density in amperes per meter. (Or, in other words, $K\,dy$ is the current carried by a strip of width dy.) Assume that $|z| \ll$ width and length of the current sheet.

13. The current density inside a cylindrical wire of radius R is given by $J = (3I/2\pi R^3)r$, where r is the radial distance from the axis of the wire and where I is the total current. (a) Verify that I is the total current by evaluating the integral

$$\int J\, dA$$

over the cross section of the wire. (b) Determine the magnetic field B as a function of r inside the wire.

14. A charge q is uniformly distributed around a ring of radius R. The ring rotates with a constant angular velocity ω about an axis perpendicular to the plane of

the ring and through its center. Show that the magnetic field at the center of the ring is

$$B = \frac{\mu_0 q \omega}{4\pi R}$$

15. A uniform distribution of charge in the shape of a thin disk of radius R rotates about an axis perpendicular to the plane of the disk and through its center with constant angular velocity ω. Show that the magnetic field at the center of the disk is

$$B = \frac{\mu_0 \sigma R \omega}{2}$$

where σ is the charge per unit area on the disk.

16. A long straight wire with a diameter of 3 mm carries a current of 20 A. The current density is constant. At what distance from the center of the wire does the magnetic field have its maximum value? Determine this maximum value.

17. A sphere of radius R has a constant volume charge density ρ. Determine the magnetic field at the center of the sphere when it rotates as a rigid body with angular velocity ω about an axis through its center.

18. A sphere of radius R has a constant volume charge density ρ. Determine the magnetic dipole moment of the sphere when it rotates as a rigid body with angular velocity ω about an axis through its center.

19. (a) A current in a long, cylindrical wire of radius a produces a constant current density J. Show that the magnitude of the magnetic field a distance r $(r > a)$ from the axis of the wire is

$$B = \frac{\mu_0 J a^2}{2r}$$

(b) Suppose that the wire is hollow with an inner radius b. Assuming that the current density is unchanged, show that the magnetic field a distance r $(r > a)$ from the axis of the wire is

$$B = \frac{\mu_0 J (a^2 - b^2)}{2r}$$

20. A current loop has the form of an equilateral triangle. The current in the loop is 30 A. The length of each side is 30 cm. Determine the magnetic field at the center of the triangle.

21. A square loop of side $2a$ carries a current I_1. The loop lies in the x–y plane with the origin at its center. Its sides are parallel to the x- and y-axes. A long, straight wire lying in the x–y plane and carrying a current I_2 is parallel to the x-axis and a distance d from it $(d > a)$. (a) Calculate the force exerted on the loop by I_2. (b) Prove that if $a \ll d$, your result reduces to the result you would find using $F = \mu \, \partial B/\partial x$.

22. Two long straight wires lie on the x- and y-axes, each carrying a current of 1000 A (Figure 14). (a) Show that at points (x, y) on a circle of radius $R = 1$ m the magnetic field set up by the currents is given by

$$B = \left(\frac{\mu_0 I}{2\pi}\right)\left(\frac{1}{x} \pm \frac{1}{\sqrt{R^2 - x^2}}\right)$$

(b) Determine the positions for which the magnitude of B is a minimum relative to adjacent points on the circle, and indicate the corresponding values of B.

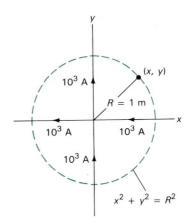

FIGURE 14

CHAPTER 31

ELECTROMAGNETIC INDUCTION

31.1
MOTIONAL EMF

Following Oersted's 1820 discovery that a steady electric current produced a steady magnetic field, experimentalists sought to demonstrate the reverse: that a steady magnetic field could create a steady electric current. All attempts failed.

In 1831 Michael Faraday accidentally discovered **electromagnetic induction**. When Faraday closed a switch to send a current through one circuit, a galvanometer needle in a nearby circuit moved and then returned to its original position. The motion of the galvanometer needle indicated that an electric current had been induced in the second circuit. It was a mark of Faraday's genius that he recognized the significance of his observation and followed it up with extensive experiments.

Let's consider an example of Faraday's law of electromagnetic induction that we can develop and analyze in detail. Figure 31.1 shows a conductor of length L sliding along parallel metal rails. The conductor moves with a constant velocity $\mathbf{v}$ in a direction perpendicular to its length and perpendicular to a constant magnetic field $\mathbf{B}$. A potential difference develops between the rails, indicating that an emf is being generated. We want to determine the origin of this emf.

The frame of reference must be specified carefully. The measuring instruments and the magnetic field $\mathbf{B}$ are stationary in the laboratory. The conductor is moving relative to the laboratory. Conduction electrons in the moving conductor move relative to the magnetic field $\mathbf{B}$ and therefore experience a force

$$\mathbf{F} = q\mathbf{v} \times \mathbf{B} \tag{31.1}$$

With $\mathbf{v}$ perpendicular to $\mathbf{B}$, the magnitude of $\mathbf{F}$ is

$$F = qvB \tag{31.2}$$

In response to this force, electrons move toward the lower conductor, segment PT in Figure 31.1, causing a net negative charge to accumulate on the lower rail. The equal positive charge that accumulates on the upper rail QS opposes continued flow of electrons. The electron motion continues but at a decreasing rate until equilibrium is achieved. At equilibrium, the magnetic force is balanced by the electrostatic force set up by the charge separation.

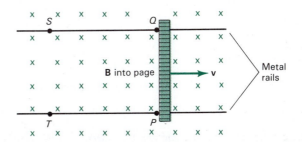

FIGURE 31.1

The conducting rod is sliding on the metal rails. Electrons accumulate on the lower rail, leaving the upper rail with an equal positive charge. A potential difference between the rails indicates an emf is being generated.

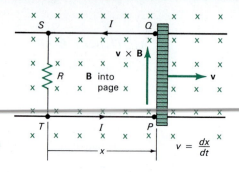

FIGURE 31.2

The moving conductor makes a sliding electrical contact at *P* and *Q*. The resistor *R* connects the rails, giving a complete circuit, *PQSTP*.

When a charge moves from *Q* to *P* through the distance *L*, the work performed by the force *F* is

$$W = FL = qBvL$$

The emf ($\mathscr{E}$) developed between *Q* and *P* is the work per unit charge W/q:

$$\mathscr{E} = \frac{W}{q} = BvL \qquad (31.3)$$

Note that $\mathscr{E} = 0$ when $v = 0$. This emf is a result of the motion of the conductor. Additionally, the emf depends on the length of the conductor ($\mathscr{E} \propto L$) and on the strength of the magnetic field ($\mathscr{E} \propto B$). These features are generally true of motional emf's, regardless of the nature or the shape of the conductors.

The moving conductor gives rise to an emf, but because we do not have a complete circuit, no current is present in the conductor after the initial redistribution of charge is complete. We complete the circuit by connecting a resistance *R* between the ends of the rails (Figure 31.2). The electrons driven across the moving conductor by the magnetic force are now able to move through the circuit.

The current direction is opposite to the electron flow, as indicated in Figure 31.2. In terms of potential differences, the induced emf creates a potential difference along the moving rod, $V_Q > V_P$. The upper rail acts like the positive terminal of a battery as it drives charge through the resistor. In a battery, chemical reactions maintain the current. In the system shown in Figure 31.2, an external force must act to maintain the motion of the conductor. Such a system is a device for converting mechanical energy into electrical energy.

EXAMPLE 1

A Moving Conductor in a Magnetic Field

The conductor in Figure 31.2 is 0.1 m long and moves with a speed of 10 m/s across a magnetic field of 1.5 T. The resistance is 100 Ω. Let's calculate the induced emf and the current.

From Equation 31.3,

$$\mathscr{E} = BvL = (1.5 \text{ T})(10 \text{ m/s})(0.10 \text{ m})$$
$$= 1.5 \text{ V}$$

The current is

$$I = \frac{\mathscr{E}}{R} = \frac{1.5 \text{ V}}{100 \ \Omega} = 15 \times 10^{-3} \text{ A}$$

$$= 15 \text{ mA}$$

31.2
FARADAY'S LAW OF INDUCTION

As the conductor moves, there is a change in the magnetic flux threading the area bounded by the circuit. From Figure 31.2, the magnitude of the magnetic flux is

$$\Phi_B = BA = BLx$$

The time rate of change of flux is

$$\frac{d\Phi_B}{dt} = BL \frac{dx}{dt} = BLv \tag{31.4}$$

Comparing this with Equation 31.3, we see that the induced emf equals the rate of change of magnetic flux,

$$\mathscr{E} = \frac{d\Phi_B}{dt}$$

With one small but important refinement—a minus sign—this result will become Faraday's law of electromagnetic induction.

To establish the general form of Faraday's law we express the emf in terms of the line integral of the induced electric field, **E**. We can introduce the electric field by changing from our laboratory frame of reference to a frame that moves with the conductor (Figure 31.3). An observer moving with the conductor measures the same current and emf that we measure from our laboratory perspective. However, for the observer moving with the conductor, there is no magnetic force; the velocity of the rod is zero in his frame of reference and so $q\mathbf{v} \times \mathbf{B}$ is zero. Instead, the moving observer measures an electric force $q\mathbf{E}$. The electric force for the moving observer equals the magnetic force sensed by the laboratory observer

$$q\mathbf{E} = q\mathbf{v} \times \mathbf{B}$$

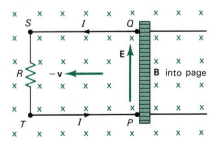

FIGURE 31.3

This is Figure 31.2 from the viewpoint of an observer sitting on the sliding rod. The rod is stationary. The rest of the circuit and the magnetic field are moving to the left at velocity −**v**.

The magnitude of this electric field is

$$E = Bv \qquad (31.5)$$

The induced emf is

$$\mathscr{E} = BvL = EL$$

The product EL equals the line integral of $\mathbf{E}$ from P to Q,

$$\mathscr{E} = EL = \int_P^Q \mathbf{E} \cdot d\mathbf{s}$$

There is no induced emf or field in any other portion of the circuit $PQSTP$, so we can express $\mathscr{E}$ as the line integral around the complete circuit

$$\mathscr{E} = \oint \mathbf{E} \cdot d\mathbf{s} \qquad (31.6)$$

Initially we showed that the induced emf equals the time rate of change of magnetic flux. Equation 31.6 shows that the induced emf equals the line integral of the electric field. But before we can equate these two expressions for $\mathscr{E}$ we must take care to establish a particular sign convention. This sign convention relates the direction of the path of integration in the emf line integral, Equation 31.6, and the direction of the area vector $d\mathbf{A}$ in the flux integral,

$$\Phi_B = \int \mathbf{B} \cdot d\mathbf{A} \qquad (31.7)$$

The path of the line integral in Equation 31.6 is closed. The area enclosed by this path is the area over which the flux integral is defined. For example, in Figure 31.4 the line integral is around the rectangular circuit $PQSTP$ and the flux integral is evaluated over the area enclosed by this circuit. We use a right-hand rule to relate the direction of the path to the direction of $d\mathbf{A}$. Figure 31.4 illustrates the rule: direct the fingers of the right hand in the direction of $d\mathbf{s}$. The extended right thumb defines the direction of $d\mathbf{A}$. When we apply this rule to the circuit of Figure 31.4 around the counterclockwise path indicated, we find that the line integral of $\mathbf{E}$ is positive because $\mathbf{E}$ and $d\mathbf{s}$ are parallel,

$$\mathscr{E} = \oint \mathbf{E} \cdot d\mathbf{s} = EL > 0$$

However, the flux integral is negative since the magnetic field $\mathbf{B}$ and the area vector $d\mathbf{A}$ are in opposite directions,

$$\Phi_B = \int \mathbf{B} \cdot d\mathbf{A} = -BLx$$

Our earlier result, $\mathscr{E} = d\Phi_B/dt$, correctly relates only the magnitudes of the emf and the time rate of change of magnetic flux. The general relationship is

$$\mathscr{E} = -\frac{d\Phi_B}{dt} \qquad (31.8)$$

or, in integral form,

$$\oint \mathbf{E} \cdot d\mathbf{s} = -\frac{d}{dt} \int \mathbf{B} \cdot d\mathbf{A} \qquad (31.9)$$

Equations 31.8 and 31.9 express **Faraday's law** of electromagnetic induction in different forms. Equation 31.8 emphasizes the emf and flux viewpoints;

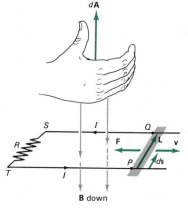

FIGURE 31.4
The right-hand rule is used to relate the directions of the area vector $d\mathbf{A}$ and the line integral displacement vector $d\mathbf{s}$. With the fingers of the right hand directed along $d\mathbf{s}$, the thumb defines the direction of $d\mathbf{A}$.

Equation 31.9 emphasizes the **field** viewpoint. The importance of the negative sign in Equations 31.8 and 31.9 will become apparent in the next section, in which we examine Lenz's law.

Faraday's discovery of induced emf's shifted attention from static phenomena to time-varying phenomena. Electromagnetic induction gave birth to **electrodynamics**.

The magnetic field was constant in the circuit we analyzed. The changing flux resulted because the moving conductor changed the area of the circuit. But Faraday's law as expressed by Equations 31.8 and 31.9 applies in general. *Any* change in flux induces an emf. For example, a time varying magnetic field gives rise to a changing magnetic flux through a stationary circuit having a fixed area. The flux $\Phi_B = \int \mathbf{B} \cdot d\mathbf{A}$ changes because $\mathbf{B}$ changes. In fact, this is the situation that Faraday initially observed. An increasing current through one coil created an increasing magnetic field in an adjacent coil. This increasing field produced a changing flux and thereby induced an emf. Let's look at another situation where a changing magnetic field induces an emf in a stationary circuit.

EXAMPLE 2

Electromagnetic Induction by a Time-Dependent Field

Consider a long solenoid with a cross-sectional area of 8 cm^2 (Figures 31.5 and 31.6). A time-dependent current in the wire creates a time-dependent magnetic field:

$$B(t) = B_0 \cos(2\pi ft)$$

Here B_0 is constant and equal to 1.2 T. The quantity f is the frequency of the oscillatory magnetic field. A conducting ring of radius r surrounds the solenoid. The time variation of the magnetic field produces a changing magnetic flux through the ring. With $f = 60$ Hz and the ring resistance $R = 1.0\ \Omega$, we want to calculate the emf and the current induced in the ring.

The magnetic flux is

$$\Phi_B = A \cdot B_0 \cos(2\pi ft)$$

The induced emf is

$$\mathscr{E} = -\frac{d\Phi_B}{dt} = (2\pi f)A \cdot B_0 \sin(2\pi ft)$$

$$= (2\pi \cdot 60\ \text{s}^{-1})(8 \times 10^{-4}\ \text{m}^2)(1.2\ \text{T})\sin(2\pi ft)$$

$$= 0.362 \sin(2\pi ft)\ \text{volt}$$

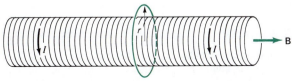

FIGURE 31.5

A long solenoid and a concentric ring outside the solenoid.

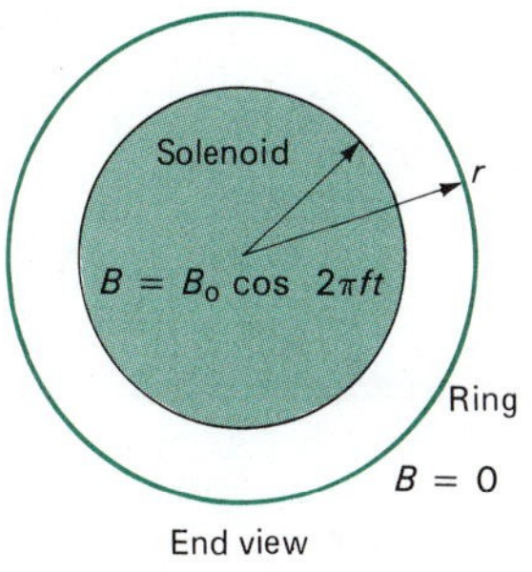

FIGURE 31.6

Solenoid and concentric ring: a cross-sectional view.

The current in the ring is $I = \mathscr{E}/R$. Therefore,

$$I = \frac{0.362 \sin{(2\pi ft)} \text{ V}}{1.0 \text{ }\Omega}$$

$$= 0.362 \sin{(2\pi ft)} \text{ A}$$

The induced current has a maximum value of 0.362 A and oscillates at a frequency of $f = 60$ Hz.

31.3
LENZ'S LAW

Let's look again at Faraday's law, Equation 31.8. We went to some length to account for the negative sign. What are its implications?

To answer this question, let's review the moving rod of Figure 31.4. As the rod moves across the magnetic field, an emf is induced. This emf sets up a current in the moving rod and the rod experiences a force given by

$$\mathbf{F} = I\mathbf{L} \times \mathbf{B} \tag{31.10}$$

The vector $\mathbf{L}$ has a magnitude equal to the length of the rod and has the direction of the current I. Using the right-hand rule you can see that the force $\mathbf{F}$ is opposite to the velocity $\mathbf{v}$ and so opposes the motion of the rod. The induced current interacts with the magnetic field, causing the rod to experience a force. This force opposes the motion that induced the current. This opposition is a consequence of the negative sign in Faraday's law and is formalized as **Lenz's law**:

> An induced emf gives rise to effects that oppose the changes that induced the emf.

In short, Lenz's law says: *Effect opposes Cause.*

The key idea is *opposition*. Lenz's law is a declaration that we are not going to get something for nothing. Suppose that Faraday's law did not contain the negative sign. Then if we started the rod moving in Figure 31.4, the induced current would experience an accelerating force. This force would cause the rod to move faster, which would cause the induced current to increase, which would increase the force, and so on. In short, we would have a system that generated kinetic energy—a perpetual motion machine. That is, we would have a device that violated the law of conservation of energy. However, the world doesn't work this way. Faraday's law *does* have a minus sign. Lenz's law guarantees that we can get electric energy out of this type of inductive system *only* by performing work.

EXAMPLE 3
Induced Current

A metal ring is released directly above a circular current loop. The ring falls toward the current loop, with its plane parallel to the plane of the current loop (Figure 31.7). Viewed from above the falling ring, the current in the loop circulates clockwise. We want to explain why an emf is induced in the falling ring and the direction of the current induced in the ring.

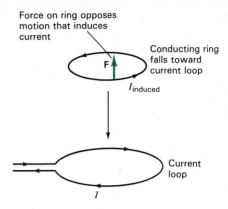

FIGURE 31.7

The conducting ring falls toward the current loop. A changing magnetic flux through the ring gives rise to an induced emf and an induced current. The induced current causes the ring to experience an upward force, in accordance with Lenz's law.

The current in the loop sets up a magnetic field that produces a magnetic flux through the falling ring. The magnetic flux changes as the ring falls toward the current loop because the magnetic field is stronger near the current loop. This changing magnetic flux induces an emf that induces a current in the ring.

The force of gravity is responsible for the motion that gives rise to the induced emf and current. In accord with Lenz's law, the induced current must move in a direction that gives rise to a force opposing gravity. We have seen that parallel currents attract and that antiparallel currents repel. To oppose gravity, the induced current must be counterclockwise, antiparallel to the current in the loop. The magnetic force exerted on the induced current in the ring is then upward, in opposition to gravity.

Eddy Currents

Eddy currents provide a striking illustration of Lenz's law. Whenever a conductor or portion of a conductor moves into or out of a magnetic field, an emf is generated. If this induced emf is not balanced by other emf's, charges will flow, creating a current. In low-resistance circuits these currents may be quite large. Such internal circulating currents are called **eddy currents** because of their resemblance to the eddies in the flow of fluids. In accord with Lenz's law, the eddy currents will circulate in such a way that the magnetic force exerted on them opposes the motion that induced them. The net effect is a drag, or slowing down.

The eddy current drag can be dramatically demonstrated by letting a sheet of copper swing into a magnetic field. Eddy currents are generated as shown in Figure 31.8. These eddy currents oppose the external magnetic field. We expect this if we use the right-hand rule to find the direction of the magnetic field produced by the induced current loops. But where does the retarding force, that is, the drag, come in? In Figure 31.8 the magnetic force on the induced current ($I\mathbf{L} \times \mathbf{B}$) acts to the left as the copper sheet swings *into* the field. This retarding force slows the sheet. As the sheet continues its swing *out of* the field the eddy currents reverse their direction of circulation and move clockwise. The magnetic force is still to the left—a retarding force. The eddy currents always circulate so that the magnetic force retards the inducing motion.

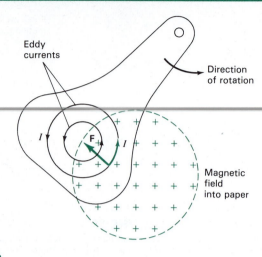

FIGURE 31.8

A copper pendulum in a magnetic field. As the pendulum swings into the field the eddy currents circulate couterclockwise. The portion of the eddy currents within the magnetic field gives rise to a retarding force. As the pendulum swings out of the magnetic field the eddy currents reverse direction and circulate clockwise. This also produces a retarding force.

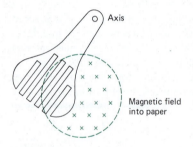

FIGURE 31.9

A slotted copper pendulum inhibits eddy currents and thereby reduces the retarding force.

The eddy currents in the sheet of copper can be minimized by interrupting their path. A series of slots, or rectangular holes, in the copper sheet does the job very effectively (Figure 31.9). The slots interrupt possible current paths, and thereby increase the resistance. An increase in the resistance means that the induced current will be smaller. This in turn results in a weaker retarding force.

Eddy currents in the iron cores of transformers are minimized by building the cores of thin sheets of iron called *laminations*. The laminated iron core retains the desired magnetic properties but inhibits eddy currents by preventing charge flow across the laminations.

31.4
APPLICATIONS OF FARADAY'S LAW

There are two ways of stating Faraday's law of electromagnetic induction. In terms of electromagnetic fields, Faraday's law says:

A changing magnetic field sets up an electric field.

We can also state Faraday's law as:

A changing magnetic flux induces an emf.

This statement of Faraday's law is formulated by Equation 31.8,

$$\mathcal{E} = -\frac{d\Phi_B}{dt} \tag{31.8}$$

In the examples that follow, we use this formulation of Faraday's law.

An Alternating Current Generator

If a loop of wire is rotated in a magnetic field (Figure 31.10) the flux threading the loop changes with time. It follows from Faraday's law that an

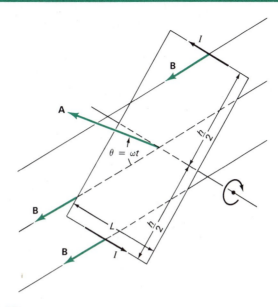

FIGURE 31.10

A rectangular coil is rotated in a uniform magnetic field. The magnetic flux through the coil changes, inducing an emf and a current in the coil.

emf is induced in the loop and that a current will result. Let's work out the details by considering a rectangular loop of length L and height h that is rotated about an axis perpendicular to a uniform field **B** (Figure 31.11). The

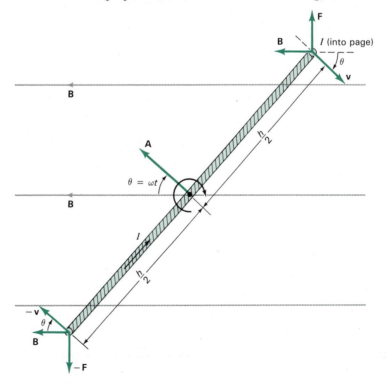

FIGURE 31.11

Side view of the rotating coil of Figure 31.10. At time $t = 0$ the area vector **A** and the magnetic field **B** are parallel. The two forces **F** and $-$**F** act to oppose the rotation of the coil. Mechanical work must be performed to maintain the rotation at a constant angular velocity.

loop rotates at a constant angular velocity ω. The angle θ between the area vector **A** and the field **B** is related to ω and the time t by

$$\theta = \omega t$$

The magnetic flux through the loop is

$$\Phi_B = \mathbf{B} \cdot \mathbf{A} = BA \cos \omega t \tag{31.11}$$

As the loop rotates, the angle $\theta = \omega t$ changes and thus the flux changes. The induced emf is

$$\mathcal{E} = -\frac{d\Phi_B}{dt} = \omega BA \sin \omega t \tag{31.12}$$

If the loop has a total resistance R the induced current is

$$I = \frac{\mathcal{E}}{R} = \left(\frac{\omega BA}{R}\right) \sin \omega t \tag{31.13}$$

This equation shows that the induced current is an alternating current that varies sinusoidally at the frequency at which the coil rotates.

We emphasized in connection with Lenz's law that external work must be performed in order to generate a motional emf. Let's compare the rate at which electrical energy is generated in the loop to the rate at which work is performed to keep it rotating at a constant angular velocity. The electric power developed by the induced emf is

$$P_{\text{elec}} = \mathcal{E}I \tag{31.14}$$

In Figure 31.11 the induced current travels into the page along the top edge of the loop and out of the page along the bottom edge. The edges experience equal and opposite magnetic forces; these are retarding forces, in accordance with Lenz's law. The velocities of the top and bottom edges are also equal and opposite, so the rate at which these two forces develop power is $2\mathbf{F} \cdot \mathbf{v}$. In order to maintain the rotation some external force must perform work at a rate

$$P_{\text{mech}} = -2\mathbf{F} \cdot \mathbf{v} = -2Fv \cos\left(\theta + \frac{\pi}{2}\right)$$

The individual factors in P_{mech} are

$$F = ILB$$

$$v = \frac{\omega h}{2}$$

$$\cos\left(\theta + \frac{\pi}{2}\right) = -\sin \theta = -\sin \omega t$$

The mechanical power delivered to the coil is

$$P_{\text{mech}} = IBLh\omega \sin \omega t \tag{31.15}$$

But Lh equals the area of the loop, and it follows from Equation 31.12 that

$$BLh\omega \sin \omega t = BA\omega \sin \omega t = \mathcal{E}$$

is the induced emf. Thus,

$$P_{\text{mech}} = \mathcal{E}I = P_{\text{elec}} \qquad (31.16)$$

This is the result to be expected on the basis of energy conservation. The rate at which electrical energy is generated equals the rate at which mechanical work is performed.

What we have just described are basic principles of the electric generator. The rotating loop may be opened, and the circuit completed to the outside world by using sliding contacts. When the circuit connection is made, the loop becomes a generator. Today almost all the electricity we use comes from generators operating on the principles of this rotating loop and on Faraday's law.

EXAMPLE 4

A Small Generator

Imagine the single loop of Figure 31.10 replaced by a coil of 240 turns (so the induced emf is multiplied by a factor of 240). The coil is square in shape (0.08 m on a side) and is rotated at a frequency $f = 5$ Hz. Let's calculate the maximum induced emf for a field $B = 0.25$ T.

From Equation 31.12 the induced emf for one loop is

$$\mathcal{E} = \omega BL^2 \sin \omega t$$

We multiply by 240 to get the emf for the entire coil. The angular frequency is

$$\omega = 2\pi f = 10\pi \text{ rad/s}$$

Setting $\sin \omega t = 1$, we obtain for the maximum emf

$$\mathcal{E}_{\text{max}} = 240(\omega BL^2) = 240(10\pi \text{ s}^{-1})(0.25 \text{ T})(0.08 \text{ m})^2 = 12 \text{ T·m}^2/\text{s}$$
$$= 12 \text{ V}$$

Note that the maximum induced emf is directly proportional to the frequency of rotation, the area of a loop, the number of turns, and the field strength. If any of these factors is changed the induced emf is changed in direct proportion.

Magnetic Mirrors

In Chapter 29 we studied the motion of a charged particle in a uniform magnetic field. The particle trajectories are **helical**—a superposition of uniform circular motion in a plane perpendicular to **B** and motion at a constant speed parallel to **B**. Now we want to analyze the motion when the field is not uniform. By carefully sculpturing the magnetic field it is possible to trap charged particles between a pair of magnetic mirrors.

We consider a field in which the field lines become concentrated at both ends of a uniform field region (Figure 31.12). Recall that the closer the spacing of the field lines, the stronger the field. We take the positive z-axis along the direction of **B** in the uniform field region. A graph of the magnitude of B versus z might look like Figure 31.13. The weaker field region is characterized by B_0. The stronger field region is identified by B_m.

Suppose we follow a particle that starts at $z = 0$ and drifts (spirals) into the region of stronger B. This particle experiences a retarding force that slows its motion parallel to the field. The retarding force may halt motion along the z-axis and then accelerate the particle back toward the origin. In doing so the

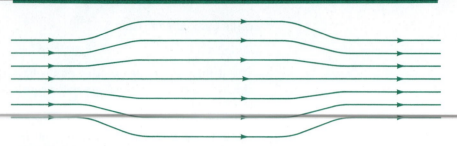

FIGURE 31.12
The magnetic field lines are closer together in the regions of stronger field.

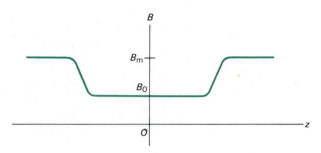

FIGURE 31.13
The magnetic field strength increases on both sides of the uniform field region.

regions of increasing field act like magnetic mirrors that reflect the charged particles.

To describe this mirror action in detail we will derive an equation for v_z, the z-component of the particle velocity. Reflections occur at points where $v_z = 0$. To derive an expression for v_z we use Faraday's law and a result established in Chapter 29. There we showed that the kinetic energy of a charged particle in a static magnetic field is constant. This result is a consequence of the fact that the magnetic force $\mathbf{F} = q\mathbf{v} \times \mathbf{B}$ does no work on the particle. We resolve the particle velocity into components v_z and $v_\perp$, parallel and perpendicular to the z-axis (Figure 31.14). The constant kinetic energy condition can be expressed as $\frac{1}{2}mv^2 = $ constant or, equivalently, as

$$v_z{}^2 + v_\perp{}^2 = \text{constant} \tag{31.17}$$

In Figure 31.12, at a point where the charged particle orbit radius is r, the orbit area is πr^2. The magnetic flux threading the particle orbit at this point is $\pi r^2 B_z$, where B_z denotes the z-component of the magnetic field. As the particle drifts into a region where B_z increases, the orbit radius decreases. This reaction to the increasing field is required by Lenz's law. The particle adjusts its orbit radius so that the flux remains constant. In this way it does not experience any induced emf and its energy remains constant. A constant flux means

$$\Phi_B = \pi r^2 B_z \approx \text{constant} \tag{31.18}$$

We want to recast this relation in terms of v and B_z. In Chapter 29 we showed how the particle orbit radius is related to the component of the linear momentum perpendicular to $\mathbf{B}$:

$$mv_\perp = qB_z r \tag{31.19}$$

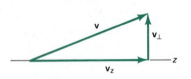

FIGURE 31.14
The particle velocity can be resolved into components parallel and perpendicular to the z-axis.

Using this relation to eliminate r from Equation 31.18 allows us to express the constant magnetic flux condition as

$$\frac{v_\perp^2}{B_z} = \text{constant} \qquad (31.20)$$

Let B_m denote the value of B_z at the mirror point where $v_z = 0$. When v_z is zero the motion is purely transverse to the field. Thus,

$$v_\perp = v \qquad \text{when} \qquad B_z = B_m$$

and we can express the constant in Equation 31.20 as v^2/B_m. This in turn lets us express $v_\perp$ in terms of B,

$$v_\perp^2 = v^2 \left(\frac{B_z}{B_m}\right) \qquad (31.21)$$

Using Equation 31.21 in Equation 31.17 to eliminate $v_\perp^2$ gives

$$v_z^2 = v^2 \left(1 - \frac{B_z}{B_m}\right) \qquad (31.22)$$

This is the desired result and shows that $v_z = 0$ when $B_z = B_m$. Note that the particles are not at rest at the mirror points. As B_z increases, v_z decreases and $v_\perp$ increases. At the mirror point, $v_z = 0$ and the particle motion is entirely transverse to the z-axis.

There are two prominent examples of magnetic mirrors: one is man-made, the other occurs naturally. In proposed thermonuclear fusion reactors it will be necessary to heat a plasma (an ionized gas) to temperatures in excess of 10^6 K. Magnetic mirrors are among the schemes under study to keep the hot plasma from coming into thermal contact with the reactor walls.

Another example of magnetic mirror action involves electrons moving along the earth's magnetic field lines. Figure 31.15 suggests how the earth's magnetic field acts to trap electrons between magnetic mirrors at the two magnetic poles. Many of these electrons originate in disturbances at the surface of the sun. The

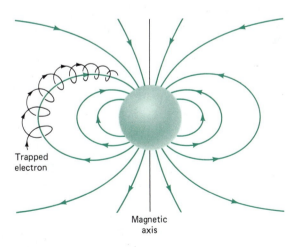

Trapped
electron

Magnetic
axis

FIGURE 31.15

Electrons entering the earth's magnetic field experience a magnetic force that traps them in spiraling paths around the magnetic field lines. The concentrated field lines near the poles function as magnetic mirrors that reflect the electrons.

magnetic mirror action does not trap all electrons. For example, electrons traveling parallel to the earth's magnetic field do not experience any magnetic force and leak through the mirror. These electrons strike atoms and molecules in the upper atmosphere. The collisions transform the kinetic energy of the electrons into internal *excitation energy* of the atmospheric atoms and molecules. The atoms and molecules subsequently rid themselves of this energy by emitting light. Occasionally, solar flares produce unusually large numbers of very energetic electrons. This causes a corresponding increase in the numbers and energies of electrons that leak through the magnetic mirrors. This in turn causes a great increase in the light emitted by the atmosphere in the polar regions. We observe this as the "northern lights" (*aurora borealis*) and the corresponding southern auroral displays (*aurora australis*).

WORKED PROBLEM

A flip coil with a cross-sectional area A and N turns has a resistance R. A constant magnetic field is originally perpendicular to the plane of the coil (Figure 31.16). Show that when the coil is rotated through $180°$, a total charge

$$q = \frac{2BNA}{R}$$

passes through the coil.

Solution

As the coil rotates, the magnetic flux through each turn changes. This induces an emf which in turn generates a current. The current is given by

$$I = \frac{\mathscr{E}}{R}$$

where $\mathscr{E}$ is the induced emf and R is the resistance of the coil. Calling t the time required to rotate the coil, the total charge that passes through the coil is

$$q = \int_0^t I \, dt$$

To determine the total change in magnetic flux during the same time interval, we integrate Faraday's law, $\mathscr{E} = -d\Phi_B/dt$,

$$\Delta\Phi_B = \int d\Phi_B = -\int_0^t \mathscr{E} \, dt$$

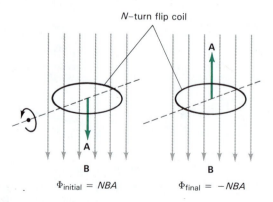

N–turn flip coil

$\Phi_{initial} = NBA$ $\Phi_{final} = -NBA$

FIGURE 31.16

Replacing $\mathscr{E}$ by IR we find,

$$\Delta\Phi_B = -R \int_0^t I\, dt = -Rq$$

Figure 31.16 shows that flux changes from $+NBA$ to $-NBA$ when the coil rotates through $180°$. Thus, an alternate expression for $\Delta\Phi_B$ is

$$\Delta\Phi_B = -NBA - (+NBA) = -2NBA$$

Equating the two expressions for $\Delta\Phi_B$ yields the desired result,

$$q = \frac{2NBA}{R}$$

EXERCISES

31.3 Motional emf

A. The conducting rod of Figure 1 is 82 cm long. It moves through a field of 0.58 T at 4.31 m/s. Calculate (a) the emf developed and (b) the potential difference between P and Q.

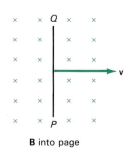

B into page

FIGURE 1

B. In Figure 2 the 10-cm rod PQ moves to the left with a speed of 8 m/s through a magnetic field of 1.3 T directed into the page. The rod PQ and the horizontal segments QS and TP are conductors with negligible resistance. The segment ST is made of Nichrome wire and has a resistance of 108 Ω. Calculate (a) the potential difference developed across the moving rod, (b) the magnitude and direction of the current in the loop $PQSTP$.

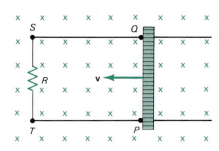

FIGURE 2

C. A car with a radio antenna 1 m long travels at 80 km/h in a locality where the earth's magnetic field is 5×10^{-5} T. What is the maximum possible induced emf in the antenna as a result of moving through the earth's magnetic field?

D. A circular loop of wire has a radius of 50 cm. The loop is in a magnetic field given by $B = 1.12(1 - at)$ T, where $a = 0.031$ s^{-1}. The positive normal to the loop and the magnetic field are parallel. Calculate (a) the magnetic flux Φ_B threading the loop at $t = 0$, (b) the time rate of change of magnetic flux, $d\Phi_B/dt$.

31.2 Faraday's Law of Induction

E. A bar magnet moves away from a metal ring as shown in Figure 3. What is the direction of the current induced in the ring?

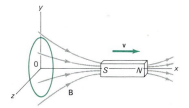

FIGURE 3

F. A 1000-turn coil is 10 cm in diameter. What emf is developed in the coil if the magnetic field through the coil is reduced from 6 T to 0 in (a) 1 s, (b) 0.01 s, (c) 0.0001 s?

G. The flux due to the axial magnetic field of a solenoid varies according to

$$\Phi_B = \Phi_0 \sin(2\pi ft)$$

where $\Phi_0 = 1.30 \times 10^{-4}$ T·m^2 and $f = 60$ Hz. Wrapped tightly around the outside of the solenoid is a small coil of 12 turns (Figure 4). Calculate the emf induced in the small coil at $t = (1/60)$ s.

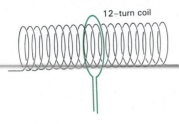

12-turn coil

FIGURE 4

H. In analogy to Example 1 a square loop of copper wire 15 cm on a side is pulled into, through, and out of a uniform but sharply bounded magnetic field, as indicated in Figure 5. The magnitude of the field is 1.4 T. The loop moves at 6.0 m/s and enters the field at time $t = 0$. Calculate the induced emf for (a) $t < 0$, (b) $0 < t < 0.025$ s, (c) 0.025 s $< t <$ 0.050 s, (d) 0.050 s $< t < 0.075$ s, (e) 0.075 s $< t$. (f) Plot the induced emf versus t. If the boundary of the magnetic field is not sharply defined, how will your plot be affected?

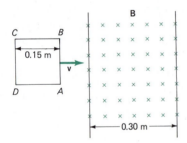

FIGURE 5

31.3   Lenz's Law

I. Two identical conducting rings, coaxially mounted, are free to move along the axis AB in Figure 6. Ring 1 carries a current I whose direction is defined to be positive. Give the direction ($+$ or $-$) of the current induced in ring 2 if (a) ring 1 moves toward A and ring 2 is at rest, (b) ring 2 moves toward A and ring 1 is at rest, (c) ring 1 moves toward B and ring 2 is at rest, (d) ring 2 moves toward B and ring 1 is at rest, (e) both rings move toward A with the same speed, (f) ring 1 moves toward B and ring 2 moves toward A with the same speed.

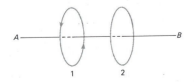

FIGURE 6

J. A bar magnet (Figure 7) moves vertically down, north pole first, approaching a circular conducting loop in the x–y plane. What is the direction of the induced current in the loop?

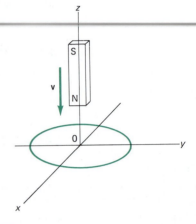

FIGURE 7

31.4   Applications of Faraday's Law

K. A metal rod is free to slide along a pair of conducting rails electrically connected by a 200-Ω resistor. If 50 W of mechanical power are expended in pushing the rod through a magnetic field, and all of this power is dissipated in the resistor, what is the induced current in the rails?

L. A rectangular loop generator has dimensions of 12 cm × 37 cm. It rotates in a uniform field of 0.682 T with an angular velocity of 120π radians/second. (a) Calculate the maximum magnetic flux threading the loop. (b) Calculate the maximum value of $d\Phi_B/dt$.

M. Let the rectangular loop generator of Section 31.4 be a square 10 cm on a side. It is rotated at a frequency of 60 Hz in a uniform field of 0.80 T. Calculate (a) the flux through the loop as a function of time, (b) the emf induced in the loop, (c) the current induced in the loop for a loop resistance of 1.0 Ω, (d) the power dissipated in the loop, (e) the torque that must be exerted to rotate the loop.

N. A 20-turn circular coil with a radius of 9 cm rotates with a constant angular velocity of 20π radians/second in a uniform magnetic field of 0.20 T. The axis of rotation lies in the plane of the coil (Figure 8) and is

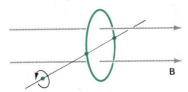

FIGURE 8

perpendicular to the magnetic field. The coil ends are joined to form a complete circuit with a resistance of 1.1 Ω. Calculate (a) the induced emf in the coil, (b) the current in the coil, (c) the rate of dissipation of electric energy.

P. A sheet of copper is partway out of a magnetic field (Figure 9). When you try to pull the sheet to the right, out of the field, there is a resisting force. Explain.

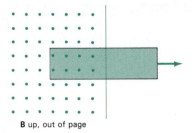

B up, out of page

FIGURE 9

PROBLEMS

1. A long wire carrying a current I and a rectangular wire loop of length L and width W lie on the surface of a nonconducting table. Figure 10 shows a top view of the wire and loop. The loop is pulled away from the wire along the tabletop at a velocity $\mathbf{v}$. (a) Why is there an induced emf around the loop? (b) Calling r the distance from the wire to the nearest edge of the loop, derive a relation for the induced emf.

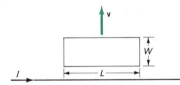

FIGURE 10

2. A circular loop of wire 10 cm in radius is in a uniform magnetic field, the plane of the circular loop perpendicular to the direction of the field. The magnetic field varies with time: $B = B_0 \cos \omega t$. (a) Calculate the magnetic flux through the loop. (b) Calculate the emf induced in the loop. (c) If the resistance of the loop is R, what current exists in the loop? (d) At what rate is electric energy being dissipated in the ring?

3. Figure 11 shows the magnetic flux Φ_B threading a con-

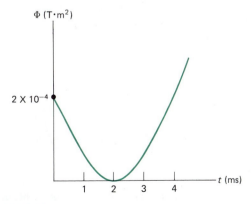

FIGURE 11

ducting loop as a function of time t in milliseconds. Assuming that the flux is proportional to $(t - 2\,\mathrm{ms})^2$, deduce a relation for the induced emf in the loop and evaluate the induced emf for $t = 0$, $t = 2$, and $t = 4$ ms.

4. A circular loop of wire 5 cm in radius is in a spatially uniform magnetic field, with the plane of the circular loop perpendicular to the direction of the field (Figure 12). The magnetic field varies with time:

$$B(t) = a + bt \qquad a = 0.20\ \mathrm{T} \qquad b = 0.32\ \mathrm{T/s}$$

(a) Calculate the magnetic flux through the loop at $t = 0$. (b) Calculate the emf induced in the loop. (c) If the resistance of the loop is 1.2 Ω, what is the induced current? (d) At what rate is electric energy being dissipated in the loop?

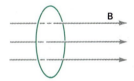

FIGURE 12

5. A changing magnetic field is confined to a cylindrical region of radius R. The field $\mathbf{B}$ varies with time but is spatially uniform. The direction of the field everywhere is parallel to the axis of the cylindrical region. Assume that at a radial distance r from the cylindrical axis the electric field is tangent to a circle of radius r. Show that the electric field is given by

$$E = -\frac{1}{2}\frac{R^2}{r}\frac{dB}{dt} \qquad r \geq R$$

$$E - -\frac{r}{2}\frac{dB}{dt} \qquad r \leq R$$

6. A wire is bent into the shape of an equilateral triangle with side length L. Each apex of the triangle

lies on a coordinate axis (Figure 13). A magnetic field described by $B = B_0 \sin \omega t$ is directed along the positive x-axis. Derive a relation for the emf induced in the wire.

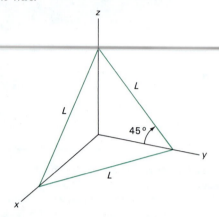

FIGURE 13

7. A wire 30 cm long is held parallel to and at a distance of 80 cm above a long wire carrying current of 200 A that rests on the floor of a room (Figure 14). The wire is released and falls, remaining parallel with the current-carrying wire as it falls. Assume that the falling wire accelerates at a constant rate of 9.80 m/s² and derive an equation for the emf induced. Express your result as a function of the time t after the wire was dropped. What is the induced emf 0.30 s after the wire is released?

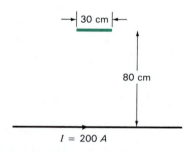

FIGURE 14

8. (a) A loop of wire in the shape of a rectangle of width w and length L and a long straight wire carrying a current I lie on a tabletop as shown in Figure 15.

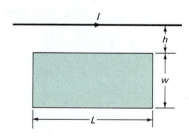

FIGURE 15

Determine the magnetic flux in the loop. (b) Suppose that the current is changing with time according to $I = a + bt$, where a and b are constant. Determine the induced emf in the loop if $b = 10$ A/s, $h = 1$ cm, $w = 10$ cm, and $L = 100$ cm.

9. A wire 1.0 m long rotates in a plane at 6.2 rev/s about an axis perpendicular to its length and through one end. A uniform magnetic field of 0.72 T directed perpendicular to the plane of rotation surrounds the wire. Determine the emf induced between the ends of the wire.

10. A device accelerates electrons in a circular path of fixed radius by changing the magnetic field through the electron orbit. Show that for the orbital radius R to be constant it is necessary that

$$B_R = 0.5 B_{av}$$

where B_R is the magnetic field at the orbital distance R and $B_{av} = \dfrac{1}{R} \int_0^R B_R \, dr$ is the spatial average of the magnetic field over the plane of the orbit.

11. A wire of mass m, length D, and resistance R slides without friction on parallel rails as shown in Figure 16. A battery that maintains a constant emf $\mathscr{E}$ is connected between the rails, and a constant magnetic field B is directed perpendicular to the plane of the page. If the wire starts from rest, show that at time t it moves with a speed

$$v = \frac{\mathscr{E}}{BD}\left(1 - e^{-B^2 D^2 t / mR}\right)$$

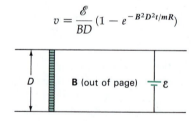

FIGURE 16

12. A bar magnet is aligned with the axis of a current loop as shown in Figure 17. The magnet moves with constant velocity from left to right. Sketch (a) the flux, (b) the current induced in the loop as a function of the position of the center of the magnet relative to the center of the loop.

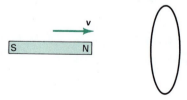

FIGURE 17

13. A steel rod 1.0 m long is dropped from a height of 2.2 m. It falls with its length horizontal at a location

where the component of the earth's magnetic field perpendicular to the plane swept out by the rod is 6.2×10^{-5} T. Determine the induced emf between its ends just before impact.

14. A small magnet is dropped through a circular coil with leads attached to the input terminals of an oscilloscope. The changing flux in the coil induces an emf that may be displayed by the oscilloscope. The coil has 200 turns and an area of 120 cm². The magnetic field rises from essentially zero to a maximum of 20 gauss in 0.1 s. Estimate the maximum induced emf in the coil.

15. A 2400-turn coil has a radius of 8 cm. A magnetic field perpendicular to the coil varies with time according to

$$B = B_0 e^{-at} \qquad B_0 = 0.16 \text{ T} \qquad a = 4 \text{ s}^{-1}$$

Calculate the induced emf at $t = 0$ and $t = 1.25$ s.

16. A rectangular motor coil has 400 turns, each with an area of 80 cm². The coil resistance is 0.24 ohm, and it rotates at 60 Hz. What magnetic field strength is required if the motor develops an average power of 6.3 hp?

17. The magnetic flux threading a metal ring varies with time t according to

$$\Phi_B = 3(at^3 - bt^2) \text{ T·m}^2 \qquad a = 2 \text{ s}^{-3} \qquad b = 6 \text{ s}^{-2}$$

The resistance of the ring is 3 Ω. Determine the maximum *current* induced in the ring during the interval from $t = 0$ to $t = 2$ s.

CHAPTER 32

INDUCTORS AND INDUCTANCE

32.1
INDUCTORS AND INDUCTANCE

A change of the magnetic flux threading a loop of conducting material induces an emf that in turn gives rise to a current in the loop. The current in the loop and the magnetic flux threading the loop are related by a property called the inductance of the loop.

There is a related concept called **mutual inductance** that involves magnetic flux from currents in nearby circuits. We do not treat mutual inductance formally in this chapter, but we consider its implications when we study transformers in Chapter 34.

While inductance is a property of all current-carrying structures, we are usually interested in particular circuit elements designed to exploit inductive effects. Such circuit elements, called **inductors**, are generally coils of wire of varied shapes and sizes. The symbol for an inductor is ⏛.

Inductance

Consider a conducting loop in which there is a current I. The current produces a magnetic field giving rise to a magnetic flux proportional to the field. According to the law of Biot and Savart, the magnetic field is proportional to the current. Hence the magnetic flux threading the loop is also proportional to the current in the loop

$$\Phi_B \propto I$$

The proportionality constant between magnetic flux and current is called the **inductance** and is denoted by the symbol L:

$$\Phi_B = LI \tag{32.1}$$

Any change in the current causes a change in the flux and induces an emf in accord with Faraday's law, $\mathcal{E} = -d\Phi_B/dt$ (Figure 32.1). Using Equation 32.1, the induced emf may be written

$$\mathcal{E} = -L\frac{dI}{dt} \tag{32.2}$$

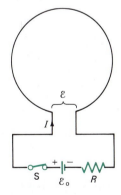

FIGURE 32.1

Closing the switch sets up a current in the loop. The current establishes a magnetic field and a magnetic flux through the loop. Any change in the current causes a change in the flux and induces an emf ($\mathcal{E}$) in the circuit.

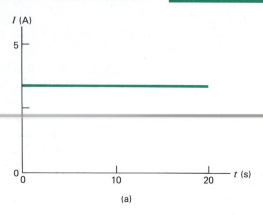

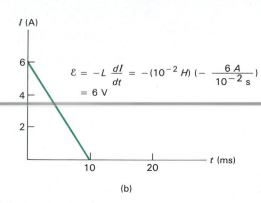

(a)   (b)

FIGURE 32.2

(a) The induced emf is zero if the current is constant. (b) A rapidly changing current in an inductor can produce a significant induced emf even though the current is small. In this example, the inductance is taken to be 0.01 H.

The induced emf is proportional to the time rate of change of current in the loop. This is the form of Faraday's law that we apply to electrical circuits.

Note that a constant current, no matter how large, induces zero emf, whereas even a small current, changing rapidly, induces a large emf (Figure 32.2). From Equation 32.2 it follows that the units of inductance are

$$\text{units of } L = \text{volt/(ampere/second)} = \text{ohm second}$$

An ohm second is called a **henry**, abbreviated H. For most applications, the henry is a rather large unit, and we use the more convenient units of millihenry (mH) and microhenry (μH).

EXAMPLE 1

Experimental Determination of Inductance

The current in a coil of wire decreases with time t according to the relation,

$$\frac{dI}{dt} = -0.10 \text{ A/s}$$

Measurements show that an induced emf of 0.13 mV is produced across the ends of the coil. What is the inductance of the coil?

Setting the emf $\mathscr{E}$ equal to 0.13 mV, we get for the inductance

$$L = \frac{\mathscr{E}}{-\dfrac{dI}{dt}} = \frac{0.13 \text{ mV}}{0.10 \text{ A/s}} = 1.3 \text{ mH}$$

The equation $\mathscr{E} = -L(dI/dt)$ is the basis for the measurement of inductance regardless of the geometry of the loop. Calculation of inductance is possible in cases where symmetry simplifies matters. The usual procedure is to calculate the total flux threading a loop for a given current and then evaluate the inductance as the ratio of flux to current ($L = \Phi_B/I$). We illustrate this technique for a solenoid and for a coaxial cable.

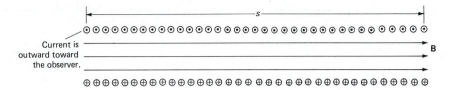

Current is
outward toward
the observer.

B

FIGURE 32.3
Wire wrapped tightly into a cylindrical form constitutes a solenoid. Shown here is a cross-sectional view of such a solenoid. The magnetic field is nearly uniform inside the solenoid.

Inductance of a Solenoid

Consider a long solenoid of length s containing N closely spaced identical turns of wire (Figure 32.3). We take the magnetic field B to be constant within the solenoid, with the field lines parallel to the solenoid axis. In Section 30.2 the magnetic field inside a long solenoid was shown to be

$$B = \frac{\mu_0 N I}{s}$$

The flux threading each loop of the solenoid is BA, where A is the cross-sectional area of each loop. The total flux threading the solenoid is NBA,

$$\Phi_B = NBA = \frac{\mu_0 N^2 I A}{s} \tag{32.3}$$

Dividing by I gives the inductance of the solenoid

$$L = \frac{\mu_0 N^2 A}{s} \tag{32.4}$$

Note that L depends on the geometry of the solenoid through A and s.

Inductance of a Coaxial Cable

A coaxial cable is a wire surrounded by a concentric cylindrical conductor (Figure 32.4). Currents from a personal computer to a monitor generally travel through a coaxial cable.

The inductance of a coaxial cable is an important characteristic because it influences the propagation of signals along the cable. Let's calculate the inductance of the coaxial cable shown in Figure 32.4. The radius of the inner conductor is a and the inner radius of the outer conductor is b. A source of emf is connected to one end of the cable and a resistor is connected to the other end, forming a complete circuit (Figure 32.4).

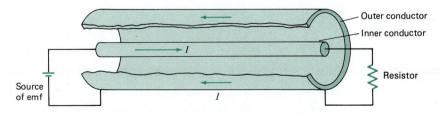

Outer conductor
Inner conductor
Resistor
Source of emf
I
I

FIGURE 32.4
A coaxial cable with a section of the outer conductor removed.

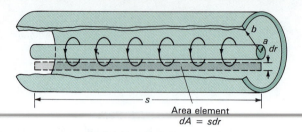

Area element
$dA = s\,dr$

FIGURE 32.5

The magnetic field lines inside a coaxial cable form circles concentric with the inner conductor. Magnetic flux threads the region between the two conductors.

To determine the inductance, we must first evaluate the magnetic flux produced by the current in the central conductor. The magnetic field set up by this current is the field of a long straight wire. The field a distance r from the center of the wire is given by

$$B = \frac{\mu_0 I}{2\pi r}$$

and the field lines are circles centered on the wire (Figure 32.5).

Because the magnetic field varies with the radial distance, we must perform an integration to determine the total magnetic flux. Consider an area element (dA) whose length equals the length (s) of the cable and whose width (dr) is measured radially outward from the conductor (Figure 32.5). The magnetic field lines are perpendicular to this area and so the flux threading dA is

$$d\Phi_B = B\,dA = \frac{\mu_0 I}{2\pi r}\,s\,dr$$

Integrating over the area between the conductors gives the total flux

$$\Phi_B = \frac{\mu_0 Is}{2\pi} \int_a^b \frac{dr}{r} = \frac{\mu_0 Is}{2\pi} \ln\left(\frac{b}{a}\right)$$

The inductance Φ_B/I is

$$L = \frac{\mu_0 s}{2\pi} \ln\left(\frac{b}{a}\right) \tag{32.5}$$

The inductance depends on the cable geometry through the length s and the radii a and b.

32.2
CIRCUIT ASPECTS OF INDUCTORS

To illustrate how an inductor in a circuit influences the current, we consider an LR series circuit—a series connection of a battery, a switch, an inductor, and a resistor (Figure 32.6). Unless stated otherwise, the resistance of the inductor and the internal resistance of the battery are neglected. When the switch is closed, charge begins flowing and the current increases ($dI/dt > 0$). The induced emf across the inductor opposes the current buildup.

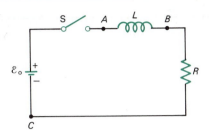

FIGURE 32.6

A series *LR* circuit. When the switch is closed, charge flows in the circuit. An emf is induced across the inductor whenever the current is changing.

The polarity of the induced emf is such that point *A* is positive relative to point *B* in Figure 32.6. That is, the inductor behaves like a battery of variable strength connected so as to oppose the action of the emf of the real battery. The induced emf changes as dI/dt changes, and drops toward zero as the current approaches a constant value. The ultimate value of the current is independent of the value of the inductance: the inductance only affects the time required for the current to reach its final value.

We can gain further insight into the role of inductance by applying Kirchhoff's voltage rule. An induced emf is subject to this rule, just like any other emf. Starting at point *C* in Figure 32.6, and proceeding clockwise, we get

$$\mathcal{E}_0 - L\frac{dI}{dt} - IR = 0$$

or

$$\mathcal{E}_0 = L\frac{dI}{dt} + IR \qquad (32.6)$$

At the moment the switch is closed in the circuit of Figure 32.6, the current is zero and Kirchhoff's voltage rule reads

$$\mathcal{E}_0 = L\frac{dI}{dt}$$

We can compare this equation with Newton's second law, $F = m\, dv/dt$. We see that the inductance L is the electrical analog of the mass m. Inductance measures electrical inertia. Just as a large mass prevents a rapid change in velocity, so too a large inductance inhibits a rapid change in current.

An emf is also induced when a current decreases. For example, if the switch controlling the current in the coil of an electromagnet is turned off, an emf is induced in the coil. This emf acts to maintain the existing current. When a switch is opened suddenly, the rate of change of current, dI/dt, can be very large. A combination of a large inductance and a large value of dI/dt can lead to dangerously high induced voltages. So, be careful how you turn off electromagnets!

EXAMPLE 2

How **Not** *to Turn Off an Electromagnet*

An electromagnet coil has an inductance of 10 henries and a resistance of 6 ohms. A 12-volt battery supplies a steady current during normal operation. The circuit is shown in Figure 32.6. With the switch closed, Equation 32.6 describes the behavior of the current. With $\mathscr{E}_0 = 12$ V and $R = 6 \; \Omega$, the steady current is

$$I = \frac{\mathscr{E}_0}{R} = \frac{12 \text{ V}}{6 \; \Omega} = 2 \text{ A}$$

If the switch is opened suddenly in a time of 2×10^{-3} s we can estimate the induced emf by taking for dI/dt its average value

$$\frac{dI}{dt} \approx \frac{\Delta I}{\Delta t} = \frac{-2 \text{ A}}{2 \times 10^{-3} \text{ s}} = -10^3 \text{ A/s}$$

The induced emf is $-L(dI/dt)$:

$$-L\frac{dI}{dt} = (-10 \text{ H})(-10^3 \text{ A/s}) = 10^4 \text{ H·A/s}$$

$$= 10,000 \text{ V}$$

This greatly exceeds the emf of the battery. Kirchhoff's voltage rule is still satisfied because there is an equal but opposite emf across the open switch. Such a potential difference is often sufficient to produce a dangerous discharge across the terminals of the switch. To avoid this, the current in the magnet coil should be reduced slowly to zero before the switch is opened.

Solution of the LR *Circuit Equation*

To determine how the current changes with time in an *LR* circuit, we rewrite Equation 32.6 in the form

$$\frac{dI}{I - \mathscr{E}_0/R} = -\frac{R \; dt}{L}$$

Next, we integrate from $t = 0$, when the current is zero, to an arbitrary time t. This gives

$$\ln \left(\left[I - \frac{\mathscr{E}_0}{R} \right] \Big/ \left[\frac{-\mathscr{E}_0}{R} \right] \right) = \frac{-Rt}{L}$$

which can be written as

$$I = \frac{\mathscr{E}_0}{R} \left(1 - e^{-Rt/L} \right) \tag{32.7}$$

In Figure 32.7 we show a graph of the current versus time for Equation 32.7. We note three points:

1. Equation 32.7 is dimensionally correct.
2. The current is zero at $t = 0$, as required by the initial conditions.
3. The current approaches the steady value $\mathscr{E}_0/R$ after the switch has been closed for a long time ($t \to \infty$).

Because the exponent Rt/L in Equation 32.7 is dimensionless, the ratio L/R has the dimension of time. The quantity L/R is called the **time constant** for the *LR* circuit. To interpret the time constant, consider the current in the cir-

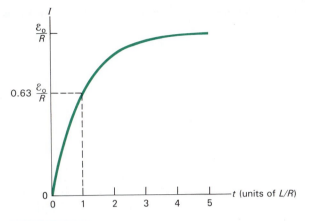

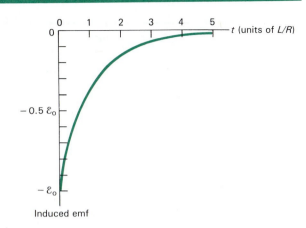

FIGURE 32.7

Time variation of the current in an *LR* circuit. Time is expressed in units of *L/R*. About 63% of the ultimate current $\mathscr{E}_0/R$ is achieved in a time equal to *L/R*.

FIGURE 32.8

Time variation of the induced emf in an *LR* circuit. Time is expressed in units of *L/R*. The induced emf tends toward zero as the current tends toward a constant value.

cuit after a time equal to *L/R* has elapsed. Setting $t = L/R$ in Equation 32.7 gives for the current

$$I = \frac{\mathscr{E}_0}{R}(1 - e^{-1}) = 0.632\frac{\mathscr{E}_0}{R}$$

This shows that 63.2% of the ultimate current $\mathscr{E}_0/R$ is achieved in a time equal to the time constant. The time constant measures the characteristic time for the current to change in an *LR* series circuit.

The way in which the time constant depends on *L* confirms that inductance measures electrical inertia. The larger *L*, the larger the time constant *L/R*, and the more slowly the current changes.

The potential difference across the inductor, $-L\,dI/dt$, follows from Equation 32.7,

$$-L\frac{d}{dt}\frac{\mathscr{E}_0}{R}(1 - e^{-Rt/L}) = -\mathscr{E}_0 e^{-Rt/L}$$

At $t = 0$ the potential difference across the inductor is $-\mathscr{E}_0$. This is in agreement with Kirchhoff's voltage rule. At $t = 0$, the current is zero and so there is no potential difference across the resistor. The emf of the battery ($\mathscr{E}_0$) must be balanced by the induced emf across the inductor. As *t* increases, the potential difference across the inductor rises from $-\mathscr{E}_0$ toward zero. After a time equal to one time constant ($t = L/R$) the potential difference across the inductor has reached $-\mathscr{E}_0 e^{-1} = -0.368\mathscr{E}_0$. The time evolution of $-L\,dI/dt$ is shown in Figure 32.8. The inertial effect of the inductor makes itself felt through the time constant. Thus, the larger *L*, the more slowly the potential difference approaches zero.

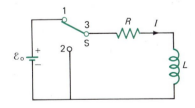

FIGURE 32.9

With the switch in position 1, charge flows through the *LR* combination and the current approaches a steady value of $\mathscr{E}_0/R$. When the switch is moved to position 2, the current decays exponentially according to $I = (\mathscr{E}_0/R)e^{-Rt/L}$.

Decline of Current in an Inductor

Let us replace the switch in Figure 32.6 with the one shown in Figure 32.9. This new switch is designed to connect terminals 2 and 3 as it breaks the connection between terminals 1 and 3. In position 1, the resistor and inductor are connected to the battery, as in Figure 32.6. In position 2, a closed circuit is formed, consisting of only the resistor and the inductor. With the switch in position 2 the battery is removed and Kirchhoff's voltage rule for the circuit becomes

$$L\frac{dI}{dt} + IR = 0$$

The solution is

$$I = I_0 e^{-Rt/L} \tag{32.8}$$

where I_0 is the current in the circuit at the instant $t = 0$ when the switch connects terminals 2 and 3. Assuming that terminals 1 and 3 were connected long enough for the current to reach its ultimate value, then I_0 equals $\mathscr{E}_0/R$, and Equation 32.8 becomes

$$I = \frac{\mathscr{E}_0}{R} e^{-Rt/L} \tag{32.9}$$

The current falls exponentially toward zero. In a time equal to the time constant L/R, the current drops to $1/e = 0.368$ of its initial value.

EXAMPLE 3

Initial and Ultimate Currents

A circuit consisting of a battery, a switch, an inductor, and two resistors is shown in Figure 32.10. Let's determine the current for the limiting cases $t \approx 0$ (a moment after the switch is closed) and $t \to \infty$ (after the switch has been closed for a long time).

Before the switch is closed the current is zero. Immediately after closing the switch there will be currents in R_1 and R_2. Electrical inertia keeps the current in the inductor near zero. Thus, at $t \approx 0$ the current through the inductor is essentially zero, although the potential difference across it ($L\, dI/dt$) is not zero. With no current in the inductor, the current is the same in R_1 and R_2. It follows from Kirchhoff's voltage rule that the current just after closing the switch is

$$I\,(t \approx 0) = \frac{\mathscr{E}_0}{R_1 + R_2}$$

When the switch has been closed for a long time, the current reaches a steady value. Assuming zero resistance for the inductor, there will be no current in R_2, so the circuit consists of the battery in series with R_1 and the inductor. The ultimate current is therefore

$$I\,(t \to \infty) = \frac{\mathscr{E}_0}{R_1}$$

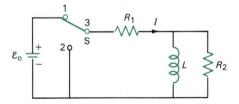

FIGURE 32.10

The *LR* circuit of Figure 32.9 is modified to include a resistor (R_2) in parallel with the inductor.

32.3
ENERGY STORED IN A MAGNETIC FIELD

An inductor can store magnetic energy. By examining the *LR* circuit shown in Figure 32.11 we can deduce an expression for the energy stored in the magnetic field of an inductor. We multiply Equation 32.6 for the *LR* circuit by the current *I* to obtain

$$\mathscr{E}_0 I = I^2 R + LI\left(\frac{dI}{dt}\right) \tag{32.10}$$

Each term has the dimensions of power. The term $\mathscr{E}_0 I$ is the rate at which energy is delivered by the battery. The quantity $I^2 R$ is the rate at which thermal energy is produced in the resistor. Evidently, the term $LI(dI/dt)$ is the power associated with the inductance. We note that this term can be written as the time derivative of $\frac{1}{2}LI^2$,

$$LI\frac{dI}{dt} = \frac{d}{dt}\left(\tfrac{1}{2}LI^2\right)$$

We introduce

$$U_M = \tfrac{1}{2}LI^2 \tag{32.11}$$

and interpret it as the magnetic energy stored in the inductor.

Why is $\frac{1}{2}LI^2$ *magnetic* energy? We call it magnetic energy because the current sets up a magnetic field. When the inductor has the form of a tightly-wound coil, or solenoid, the magnetic field is largely confined to the interior of the coil. If we remember that *L* and *I* are electrical analogs of mass and velocity, then we see that there is a strong analogy between the magnetic energy $\frac{1}{2}LI^2$ and kinetic energy $\frac{1}{2}mv^2$.

Interpreting $\frac{1}{2}LI^2$ as magnetic energy lets us extend the principle of energy conservation. In fact, Equation 32.10 becomes a statement of energy conservation:

Rate at which work is done by the emf		Rate at which thermal energy is developed in the resistor		Rate at which magnetic energy is stored or released
$\mathscr{E}_0 I$	$=$	$I^2 R$	$+$	$\dfrac{d}{dt}\left(\tfrac{1}{2}LI^2\right)$

$$\tag{32.12}$$

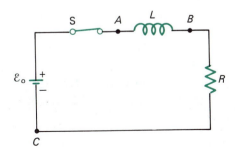

FIGURE 32.11

Magnetic Energy Density

Because energy is stored in an inductor it should be possible to express the magnetic energy in terms of the magnetic field. We can do this most simply for a solenoid. The magnetic field in the interior of an N-turn solenoid of length s is related to the current by

$$I = \frac{Bs}{\mu_0 N}$$

The inductance of the solenoid is

$$L = \mu_0 N^2 \, A/s$$

The magnetic energy U_M can be expressed as

$$U_M = \tfrac{1}{2}LI^2 = \tfrac{1}{2}(\mu_0 N^2 A/s)\left(\frac{Bs}{\mu_0 N}\right)^2 = \left(\frac{B^2}{2\mu_0}\right)As$$

Because the magnetic field is essentially confined to the core of the solenoid, the magnetic energy is confined to the same volume. The volume of the core is As, and so the magnetic energy density (magnetic energy per unit volume) is

$$u_M = \frac{U_M}{As} = \frac{B^2}{2\mu_0} \tag{32.13}$$

This relation, although derived for a solenoid, holds for all magnetic fields. To establish a magnetic field requires energy and we may consider the energy to be stored in the magnetic field. The magnetic field energy density $u_M = B^2/2\mu_0$ is similar in form to the electric field energy density, $u_E = \tfrac{1}{2}\epsilon_0 E^2$.

EXAMPLE 4

Magnetic Energy Density in a Superconducting Solenoid

The magnetic field in a particular superconducting solenoid is 3.0 T. The magnetic energy density of the field is

$$u_M = \frac{B^2}{2\mu_0} = \frac{(3.0 \text{ T})^2}{2 \cdot 1.26 \times 10^{-6} \text{ T·m/A}} = 3.58 \times 10^6 \text{ T · A/m}$$
$$= 3.58 \times 10^6 \text{ J/m}^3$$

For comparison, the energy density in gasoline is 3.5×10^{10} J/m³.

Strong forces are required to contain large magnetic energy densities, even though magnetic energy densities are well below the energy density of gasoline. In fact, magnets have been known to self-destruct as a result of magnetic forces.

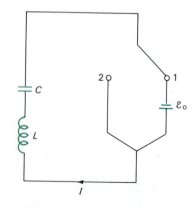

FIGURE 32.12

An *LC* circuit. With the switch in position 1, the capacitor charges. When the switch is moved to position 2 the capacitor discharges, initiating oscillations of the charge and current.

32.4

OSCILLATIONS IN LC AND LCR CIRCUITS

Let's consider the LC circuit shown in Figure 32.12. With the switch in position 1 the battery charges the capacitor. Kirchhoff's voltage rule states

$$\mathcal{E}_0 - L\frac{dI}{dt} - \frac{q}{C} = 0 \tag{32.14}$$

With the switch in position 2, the battery is removed from the circuit. Kirchhoff's voltage rule has the same form as Equation 32.14, but with $\mathscr{E}_0 = 0$,

$$L\frac{dI}{dt} + \frac{q}{C} = 0$$

Using the relation between current and charge ($I = dq/dt$) gives an equation governing the charge q:

$$L\frac{d^2q}{dt^2} + \frac{q}{C} = 0 \qquad\qquad (32.15)$$

This equation has the same form as Newton's second law for the frictionless spring-mass oscillator,

$$m\frac{d^2x}{dt^2} + kx = 0$$

The *LC* circuit is the electrical analog of the spring-mass system. Figure 32.13 displays the analogs between the *LC* circuit and the spring-mass system.

The spring-mass system undergoes simple harmonic motion at the frequency $f = (1/2\pi)\sqrt{k/m}$. The charge on the capacitor in the *LC* circuit oscillates at a frequency given by the analog of $(1/2\pi)\sqrt{k/m}$,

$$f = \left(\frac{1}{2\pi}\right)\sqrt{\frac{1}{LC}} \qquad\qquad (32.16)$$

In the absence of friction the mechanical energy (E) of the spring-mass system is conserved:

$$E = \tfrac{1}{2}mv^2 + \tfrac{1}{2}kx^2 = \text{constant}$$

The *LC* system ignores resistance, the electrical analog of friction. Consequently, the total energy (U) of the *LC* circuit is conserved. This energy is the sum of the energy stored in the magnetic field of the inductor and the energy stored in the electric field of the capacitor,

$$U = \tfrac{1}{2}LI^2 + \frac{q^2}{2C} = \text{constant} \qquad\qquad (32.17)$$

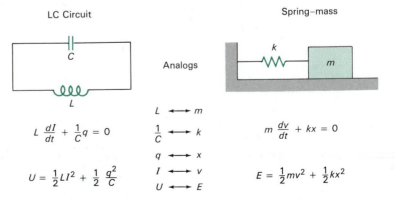

FIGURE 32.13

The *LC* circuit and the frictionless spring-mass oscillator are analogs of each other.

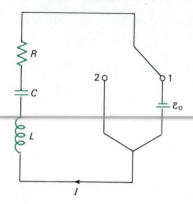

FIGURE 32.14

The series *LCR* circuit.

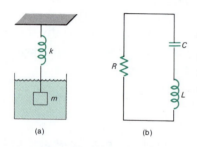

FIGURE 32.15

The resistance in the *LCR* circuit is analogous to the viscous (frictional) damping force acting on the moving mass.

The oscillations of the spring-mass system can be viewed as a cycle of energy transformations, from the kinetic energy of the moving mass to the elastic potential energy of the spring. A similar viewpoint reveals the *LC* circuit oscillations as a cycle of energy transformations between energy stored in the magnetic field of the inductor and energy stored in the electric field of the capacitor.

LCR Circuit

Friction in mechanical systems and resistance in electrical circuits are not always negligible. Both inevitably result in the production of thermal energy at the expense of other forms of energy. To examine the effects of resistance we analyze the *LCR* series circuit (Figure 32.14).

With the switch in position 1 the battery delivers charge, storing energy in the capacitor and inductor. Kirchhoff's voltage rule states

$$\mathscr{E}_0 - L\frac{dI}{dt} - RI - \frac{q}{C} = 0 \tag{32.18}$$

With the switch in position 2, the battery is removed from the circuit. Energy stored in the capacitor and inductor is transformed into thermal energy by the resistor. Kirchhoff's voltage rule has the same form as Equation 32.18, but with $\mathscr{E}_0 = 0$. Setting $I = dq/dt$ gives the equation governing the *LCR* circuit,

$$L\frac{d^2q}{dt^2} + R\frac{dq}{dt} + \frac{q}{C} = 0 \tag{32.19}$$

This equation has the same form as the equation describing the damped motion of a spring-mass system (Figure 32.15),

$$m\frac{d^2x}{dt^2} + \sigma\frac{dx}{dt} + kx = 0$$

To the analogs displayed in Figure 32.13 we add

$$R \leftrightarrow \sigma$$

The resistance R is the electrical analog of the frictional damping coefficient σ.

Having studied the spring-mass oscillator in detail in Chapter 13, we draw on that presentation to deduce the behavior of the *LCR* circuit. Assuming that the charge on the capacitor is q_0 at time $t = 0$, we have as a solution of Equation 32.19 for the condition $(1/LC) > (R/2L)^2$

$$q = q_0 e^{-Rt/2L}\cos(2\pi f't) \tag{32.20}$$

This solution shows that oscillations occur at a frequency,

$$f' = \frac{1}{2\pi}\sqrt{\left(\frac{1}{LC}\right) - \left(\frac{R}{2L}\right)^2} \tag{32.21}$$

This frequency is lower than the frequency for a circuit in which there is no damping ($R = 0$). The exponential term in Equation 32.20 steadily reduces the amplitude of the oscillations (Figure 32.16).

Increasing the resistance makes the damping more severe and reduces the frequency of the oscillations. At a critical value of R such that $(R/2L)^2 = 1/LC$,

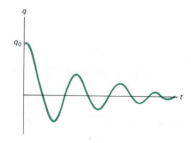

FIGURE 32.16

Resistance leads to damped oscillations of the charge on the capacitor of the *LCR* circuit.

the circuit no longer oscillates, and the capacitor simply discharges, converting its energy into thermal energy.

If $(R/2L)^2 > 1/LC$, the circuit is said to be **overdamped**. If $(R/2L)^2 = 1/LC$, the circuit is said to be **critically damped**. The behavior is completely analogous to the behavior of the damped spring-mass oscillator.

WORKED PROBLEM

An inductor and a resistor are connected so as to form a loop. The current in the loop falls exponentially from an initial value I_0 with a time constant L/R. If L/R equals 3.3 ms, how much time elapses before the energy converted to heat equals the magnetic energy stored at that moment?

Solution

The current decreases with time according to

$$I = I_0 e^{-Rt/L}$$

Total energy is conserved, with the decrease in magnetic energy appearing as heat. The magnetic energy stored at any time is $\frac{1}{2}LI^2$. When the magnetic energy has decreased to half its initial value the other half has been converted to heat. The time for this to occur is given by

$$\tfrac{1}{2}LI^2 = \tfrac{1}{2}LI_0{}^2 e^{-2Rt/L} = \tfrac{1}{2}(\tfrac{1}{2}LI_0{}^2)$$

This gives

$$-\frac{2Rt}{L} = \ln\left(\frac{1}{2}\right) = -0.693$$

Solving for the time t we find

$$t = \frac{L}{2R}\,0.693 = \frac{3.3\ \text{ms}}{2}\,0.693 = 1.1\ \text{ms}$$

EXERCISES

32.1 Inductors and Inductance

A. When the current in an inductor is 1.2 mA, the magnetic flux threading the inductor is 2.76×10^{-6} Wb. Determine its inductance.

B. A changing current in a 0.1-H inductor produces a constant potential difference of $+2$ V across the inductor. How does this current vary with time?

C. The current in a 10-mH inductor is observed to vary exponentially according to $I = 3.8e^{-0.21t}$ mA, where t is expressed in seconds. Calculate the maximum value of the induced emf.

D. The current in a 3-mH inductor is $2.2 \cos(60t)$ A, where t is expressed in seconds. Determine the maximum emf induced across the inductor.

E. Because of a misprint, a textbook equation for the inductance of a long solenoid reads $L = \mu_0 n^2 A$, where n is the number of turns per unit length and A is the cross-sectional area of the solenoid. Without actually calculating the inductance of a solenoid or looking up the expression for the inductance, show why this expression for inductance is incorrect.

F. Why should you expect the magnetic field to be zero outside a coaxial cable?

G. Present an argument in support of the following statement:

The inductance of a system increases as the linear dimensions of the system increase.

H. (a) What combination of inductance and resistance has the dimension of time? (b) What combination of inductance and capacitance has the dimension of time?

I. Potential difference and current measurements for an inductor in a circuit produce the graphs shown in Figure 1. Use the graphs to estimate the inductance.

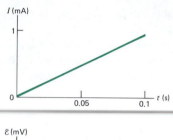

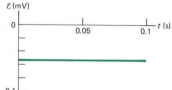

FIGURE 1

J. A type of commercial coaxial cable uses the prefix RG to designate a particular construction. For example, type RG-58/U has an inner wire diameter of 0.812 mm and an outer conductor diameter of 3.24 mm. Determine the inductance per unit length of type RG-58/U coaxial cable.

32.2 Circuit Aspects of Inductors

K. A 6-V battery, a switch that is initially open, and an inductor with $L = 100$ mH and $R = 1.5 \, \Omega$ are connected in series. Determine the current in the circuit 0.12 s after the switch is closed.

L. Consider the circuit shown in Figure 2. With $\mathscr{E}_0 = 6$ V, $R = 200 \, \Omega$, and $L = 3$ mH, and the switch closed, determine the current when the rate of change of the current is 800 A/s.

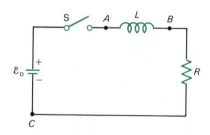

FIGURE 2

M. A circuit consisting of a 6-V battery, a 100-mH inductor, and two 1000-Ω resistors is shown in Figure 3. Determine the currents I_1 and I_2 for the limiting

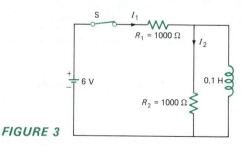

FIGURE 3

cases (a) $t \approx 0$ (switch is just closed) and (b) $t \to \infty$ (switch is closed for a long time).

N. Figure 4 shows a pair of separated electrodes attached to an inductor in a circuit. When the switch is opened, the induced emf across the electrodes can exceed $\mathscr{E}_0$. What makes this possible? (This principle is used to provide a short-lived large potential difference to turn on a fluorescent lamp.)

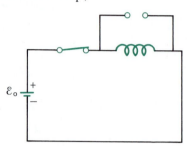

FIGURE 4

P. Measurements of the current I_1 in a circuit similar to that in Figure 3 reveal that it increases from a minimum of 0.25 A when the switch is first closed to a maximum of 1 A when the switch has been closed for a long time. Assuming that the emf of the battery is 6 V, what are the values of R_1 and R_2?

Q. The current in the LR circuit (Figure 5) drops to one-half its initial value in 0.24 s. How long will it take for the current to drop to one-tenth of its initial value?

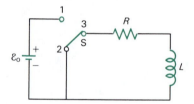

FIGURE 5

R. After the switch in the LR circuit of Figure 2 is closed, (a) how much time is required for the current to rise to 90% of its ultimate value? (b) How much time is required for the current to rise to 99% of its ultimate value? Express your answers as multiples of the L/R time constant.

32.3 Energy Stored in a Magnetic Field

S. The magnetic field inside a superconducting solenoid is 4.5 T. The solenoid has an inner diameter of 6.2 cm and a length of 26 cm. (a) Determine the magnetic energy density in the field. (b) Determine the magnetic energy stored in the magnetic field within the solenoid.

T. Consider the LR circuit shown in Figure 2. Compare the magnetic energy stored by the inductor with the thermal energy produced in the resistor after the switch has been closed for a time equal to L/R.

U. A parallel plate capacitor has a plate separation of
    0.01 mm and a potential difference of 2.7 V. (a) Deter-
    mine the electric energy density. (b) Compare this
    electric energy density with the magnetic energy den-
    sity in a solenoid having 125 turns per centimeter and
    carrying a current of 0.13 A.

32.4 Oscillations in *LC* and *LCR* Circuits

V. A 0.1-μF capacitor initially with a charge of 1 μC is
    connected to an inductor having a resistance of 2 Ω and
    an inductance of 0.1 μH (Figure 6). Make a graph of
    the charge on the capacitor as a function of time.

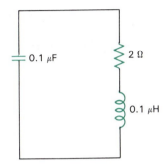

FIGURE 6

W. Consider the circuit shown in Figure 7 with $L = 2$ mH
    and $C = 4$ μF. When the switch is closed, determine
    the frequency of the oscillating current.

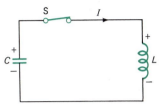

FIGURE 7

X. In an *LC* circuit the maximum energy stored in the
    capacitor equals 12 J. (a) What is the maximum energy
    stored in the inductor? (b) What is the total energy
    stored in the circuit?

Y. Design an *LCR* series circuit that will produce oscilla-
    tions having a frequency of 10 kHz with $L/R = 10$ μs.

PROBLEMS

1. A solenoid with tightly-wound turns of wire has a
    toroidal (doughnut) shape. (a) Following the analysis
    in Section 32.1, show that the inductance of a toroidal
    coil is $L = \mu_0 N^2 A/2\pi a$, where N is the number of turns,
    A is the cross-sectional area, and a is the mean radius.
    (b) Determine the inductance of a toroidal coil having
    1000 turns with a turn radius of 5 mm and a mean ra-
    dius of 10 cm.

2. The lead-in wires from a TV antenna are often con-
    structed in the form of two parallel wires (Figure 8).
    (a) Why does this configuration of conductors have
    an inductance? (b) What constitutes the flux loop for
    this configuration? (c) Neglecting any magnetic flux
    inside the wires, show that the inductance of a length
    x of this type of lead-in is

    $$L = \frac{\mu_0 x}{\pi} \ln\left(\frac{w-a}{a}\right)$$

    where a is the radius of the wires and w is the
    center-to-center separation of the wires.

3. (a) When the switch in the *LR* circuit (Figure 2) is
    closed a current is set up and thermal energy is gen-
    erated in the resistor. Derive an expression for the
    thermal energy produced in a time t. (b) Set $\mathscr{E}_0 = 5$ V
    and $R = 25$ Ω and let the time constant L/R equal 1 sec-
    ond. Plot the thermal energy developed versus time
    from $t = 0$ to $t = 10$ s. (c) Why does your graph ap-
    proach a straight line?

4. Calculate the ratio of the magnetic energy stored in-
    side the inner conductor to that stored between con-
    ductors in a *RG-58/U* coaxial cable (see Exercise J).

5. Consider a one-loop circuit where R, C, and L are
    provided by a length of *RG-58/U* coaxial cable (see
    Exercise J). Assume the resistance is due to the central
    wire and (a) prove that the natural frequency varies
    inversely with the length of the cable. (b) Show that
    a one-meter length has a natural frequency of 2.10 ×
    10^8 Hz. (c) Demonstrate that the circuit is under-
    damped. Take the resistivity to be 2.82 × 10^{-8} $\Omega \cdot$m.

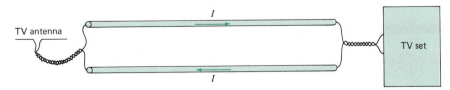

FIGURE 8

6. The switch in the circuit in Figure 9 is closed at $t = 0$. Before the switch is closed, the capacitor is uncharged and all currents are zero. Write down the Kirchhoff voltage rule equations for the loops containing $\mathcal{E}_0$, L, and C and $\mathcal{E}_0$, L, and R (a) the instant after closing the switch, (b) long after the switch is closed.

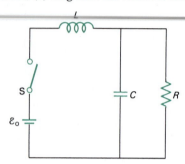

FIGURE 9

7. An electromagnet coil has an inductance of 3.2 H and a resistance of 40 Ω. It is connected to a 70-V emf as shown in Figure 10. (a) What is the current in the circuit? (b) Estimate the emf developed across the switch if it is suddenly opened, causing the current to fall to zero in 1.2 ms. (c) Why might it be dangerous to be the person who opens the switch in this fashion?

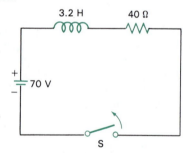

FIGURE 10

8. In an LR series circuit the magnetic energy stored drops to half its initial value in 3 ms. What additional time elapses before the energy stored drops (a) to one-fourth its initial value? (b) to one-eighth its initial value?

9. The capacitor shown in Figure 11 carries a charge of 8 μC when the switch is closed. (a) Determine the

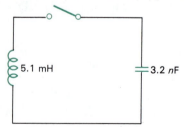

FIGURE 11

maximum current in the circuit. (b) Determine the current when equal amounts of energy are stored in the inductor and the capacitor.

10. A series LR circuit has $L = 0.1$ H and $R = 6\,\Omega$. At a certain moment the current is 5 A. (a) What is the potential difference across the inductor? (b) Calculate the time rate of change of current.

11. The inductor in the circuit in Figure 12 has negligible resistance. When the switch is opened after having been closed for a long time, the current in the inductor drops to 0.25 A in 0.15 s. What is the inductance of the inductor?

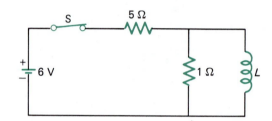

FIGURE 12

12. Using the analogs between mechanical and electrical quantities, deduce an expression that describes critical damping for the spring-mass system in Figure 32.15a.

13. The magnetic field outside a sphere of radius R is given by $B = B_0(R/r)^2$, where B_0 is a constant. Determine the total energy stored in the magnetic field outside the sphere and evaluate your result for $B_0 = 5 \times 10^{-5}$ T and $R = 6 \times 10^6$ m, values appropriate for the earth's magnetic field.

14. The magnetic field inside a long wire of radius a carrying a current I is given by $B = \mu_0 Ir/2\pi a^2$, where r is the radial distance from the axis of the wire. (a) Derive a relation for the total magnetic energy per unit length of the wire. (b) Evaluate the magnetic energy per meter inside a wire 2 mm in diameter that carries a current of 1.2 A.

15. (a) If the switch in the circuit shown in Figure 13 has been open for a long time, what is the current in the inductor? (b) Qualitatively, how do you expect the current in the inductor to change when the switch is closed? (c) Using Lenz's law as a guide, determine which of the points A or B is at the higher potential

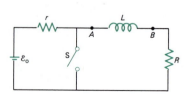

FIGURE 13

after the switch is closed. (d) Argue why the current in the inductor after the switch is closed is

$$I = \frac{\mathcal{E}_0}{R + r} e^{-Rt/L}$$

16. The switch in the circuit in Figure 14 is closed at $t = 0$. Before the switch is closed, the capacitor is uncharged and all currents are zero. Determine the currents in L, C, and R and the potential differences across L, C, and R (a) the instant after closing the switch, (b) long after the switch is closed.

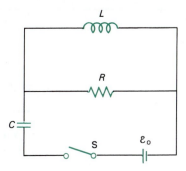

FIGURE 14

CHAPTER

33

MAGNETIC PROPERTIES OF MATTER

33.1
MAGNETIC CLASSIFICATION OF MATERIALS

Many materials experience a force when placed in a **nonuniform magnetic field**. For example, a paper clip is attracted to the pole of a magnet. In contrast, copper in the vicinity of the same magnet would be repelled. The magnitude and direction of the magnetic force experienced by different materials allow us to classify matter as diamagnetic, paramagnetic, or ferromagnetic (Figure 33.1). If the material is *repelled* by a magnet it is called **diamagnetic**. If the material experiences a relatively *weak attractive* force, it is called **paramagnetic**. Materials that experience much *stronger attractive* magnetic forces are called **ferromagnetic**. Table 33.1 describes these three categories of magnetic materials and lists examples of each.

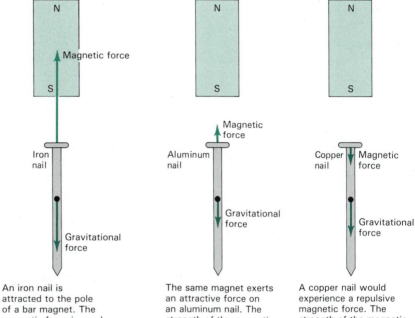

An iron nail is attracted to the pole of a bar magnet. The magnetic force is much greater than the weight of the nail.

The same magnet exerts an attractive force on an aluminum nail. The strength of the magnetic force is much less than the weight of the nail.

A copper nail would experience a repulsive magnetic force. The strength of the magnetic force is much less than the weight of the nail.

FIGURE 33.1

Comparison of the forces on nails made of iron, aluminum, and copper, in the field of a bar magnet.

TABLE 33.1

Magnetic Classification of Materials According to the Force Experienced in a Nonuniform Field

Magnetic Category of Matter	Interaction	Direction of Force	Examples
diamagnetic	weak	toward a region of weaker magnetic field	copper, bismuth, carbon, gold, lead, silver, zinc, hydrogen, helium
paramagnetic	weak	toward a region of stronger magnetic field	aluminum, chromium, palladium, platinum, potassium, sodium, tungsten (wolfram), manganese, magnesium
ferromagnetic	strong	toward a region of stronger magnetic field	iron, nickel, cobalt, gadolinium, dysprosium

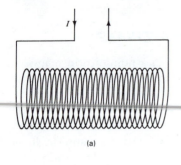

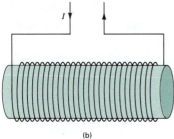

FIGURE 33.2

(a) An empty solenoid. A current in the solenoid sets up a nearly uniform magnetic field in the interior. (b) When a material fills the solenoid the magnetic field inside the material differs from the value of the field when the solenoid is empty.

An alternative scheme for classifying magnetic materials employs the magnetic field concept. When a material is placed in a magnetic field, the atomic components of the matter react and set up an internal magnetic field. The material becomes magnetically **polarized**. The situation is similar to the electric polarization that arises when a material is subjected to an electric field.

We view the magnetic field inside a material as a superposition of the applied field and the internal field, induced by the applied field. The field viewpoint of magnetic properties is based on the relationship between the induced magnetic field and the applied magnetic field.

Magnetic Susceptibility

A solenoid is useful for visualizing how to relate the induced and applied magnetic fields. A uniform magnetic field can be established inside a solenoid. Let B_0 denote the uniform magnetic field inside the *empty* solenoid shown in Figure 33.2a. When a sample is placed inside the solenoid, as shown in Figure 33.2b, the magnetic field is altered from B_0 to

$$B = B_0 + \chi_m B_0 \tag{33.1}$$

The quantity χ_m is called the **magnetic susceptibility**. It is a dimensionless quantity that measures the strength of the magnetic field induced within the material. The magnetic susceptibility need not be constant and may depend on the temperature and the applied magnetic field B_0.

In Equation 33.1, the field induced in the material is $\chi_m B_0$. For diamagnetic materials, the magnetic susceptibility is negative, which means that the induced magnetic field opposes the applied field. For paramagnetic and ferromagnetic materials, χ_m is positive, which shows that the induced magnetic field strengthens the applied field.

Magnetic Susceptibility Measurement

Equation 33.1 suggests that measurements of B and B_0 can be used to determine the magnetic susceptibility. A direct measurement of the magnetic field inside a material is difficult because ordinary probes cannot be inserted without modifying the field they are intended to measure. Therefore, we generally measure some quantity that is related to the magnetic field in a known way.

One measurable quantity is *magnetic flux*. To measure the magnetic flux in a sample we place a *sense coil* around a solenoid filled with the material. Figure 33.3 shows schematically how the sense coil makes it possible to determine the magnetic susceptibility. When the switch S is closed, a current is established in the solenoid, which in turn sets up a magnetic field. When the solenoid is empty this magnetic field is B_0. When the solenoid is filled with a sample the magnetic field is $B = B_0 + \chi_m B_0$.

The act of closing the switch and establishing the magnetic field gives rise to a change of magnetic flux threading the sense coil. Before the switch is closed there is no field and the flux is zero. After the switch is closed, the current builds up to a steady value. The magnetic field reaches a value B_0 when the solenoid is empty, or a value B when the solenoid is filled with a sample of the material of interest. The magnetic flux changes from zero to B_0NA or BNA, where A is the cross-sectional area threaded by the magnetic field lines and N is the number of turns of the sense coil. These flux changes induce emf's that drive electric charges through the sense coil. The magnetic

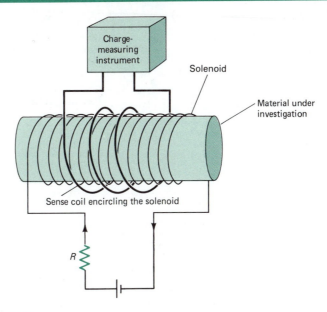

FIGURE 33.3

Schematic for arrangement used to measure the magnetic susceptibility. When the switch is closed, the ensuing magnetic flux change induces a current in the sense coil. Measurements of the charge passing through the sense coil when the solenoid is empty and when it is filled with a sample lead to a determination of the magnetic susceptibility.

susceptibility is determined by comparing the charge passing through the sense coil when the solenoid is empty and when it is filled with a sample of material.

The emf $\mathscr{E}$ induced in the sense coil is related to the change in magnetic flux by Faraday's law,

$$\mathscr{E} = -\frac{d\Phi_B}{dt}$$

The total flux change can be expressed as

$$\Delta\Phi_B = -\int \mathscr{E}\, dt \tag{33.2}$$

The integral is over the time interval during which the flux increases from zero to its final value. The emf is related to the resistance R and the current I in the sense coil circuit by

$$\mathscr{E} = IR$$

It follows that the flux change is given by

$$\Delta\Phi_B = -R\int I\, dt = -Rq \tag{33.3}$$

where q is the charge passing through the sense coil.

When the solenoid is empty, the flux change is $B_0 NA$. When the solenoid is filled with a sample, the flux change is

$$\Delta\Phi_B = BNA = (1 + \chi_m)B_0 NA$$

If we let q_0 and q denote the charges corresponding to B_0 and B we have

$$\frac{(1 + \chi_m)B_0 NA}{B_0 NA} = \frac{-Rq}{-Rq_0} = \frac{q}{q_0}$$

TABLE 33.2
Magnetic Susceptibilities for Some Paramagnetic and Diamagnetic Elements*

Element	Magnetic Susceptibility
Paramagnetic	
Aluminum	$+20.7 \times 10^{-6}$
Magnesium	$+11.8 \times 10^{-6}$
Potassium	$+5.82 \times 10^{-6}$
Diamagnetic	
Bismuth	-280.1×10^{-6}
Carbon	-14.1×10^{-6}
Copper	-10.8×10^{-6}
Silver	-23.8×10^{-6}

* The measurements reported here are for a temperature of 300 K, and were taken from a list found in *Handbook of Chemistry and Physics*, 61st ed. CRC Press, 2000 N. W. 24th St., Boca Raton, FL 33431.

and thus

$$\chi_m = \frac{q - q_0}{q_0} \tag{33.4}$$

A variety of ways exist for measuring the charges q and q_0; one is to allow the current to charge a capacitor, and then to measure the potential difference across the capacitor. The relation $q = CV$ can then be used to express χ_m in terms of the potential differences corresponding to q and q_0. Table 33.2 lists the magnetic susceptibilities of a number of substances.

The magnetic classification of materials has its basis in the atomic structure of matter. Only modern quantum physics is capable of giving a detailed explanation of the magnetic properties of matter. In the following sections we present qualitative accounts of the atomic origins of diamagnetism, paramagnetism, and ferromagnetism.

33.2
DIAMAGNETISM

The negative magnetic susceptibility of diamagnetic materials is a consequence of Lenz's law (Section 31.3). In diamagnetic materials, electric charges (positive ions and electrons) react to the induced emf that accompanies a change of magnetic flux. Their reaction always *opposes* the flux change. For example, consider a material in which the field changes from zero to some value **B**. Before the field changes, the net electric current is zero because of the random nature of the motions of the electric charges. When the field changes, each charge experiences a magnetic force

$$\mathbf{F} = q(\mathbf{v} \times \mathbf{B}) \tag{33.5}$$

This magnetic force gives rise to electric currents that are no longer random; these currents set up a magnetic field. The direction of this magnetic field always opposes the magnetic field that induced changes in the random motions of the charges.

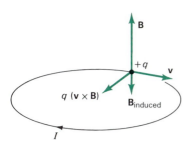

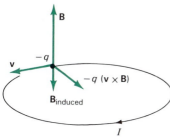

FIGURE 33.4

The diamagnetic character of charges free to move in a magnetic field. The circulating positive and negative charges constitute current loops that set up magnetic fields opposing the applied field, in accordance with Lenz's law.

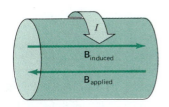

FIGURE 33.5

The induced current *I* on the surface of the superconductor produces a magnetic field $\mathbf{B}_{induced}$ that exactly cancels the applied magnetic field.

Figure 33.4 shows how both negative and positive charges moving in circular paths under the influence of the magnetic force $q(\mathbf{v} \times \mathbf{B})$ set up currents that produce magnetic fields opposing **B**. The positive and negative charges circulate in opposite directions, thereby producing currents in the same sense. The right-hand rule relating current direction and magnetic field direction shows that the induced magnetic field opposes the applied magnetic field.

Diamagnetic currents are induced in all types of materials. However, in paramagnetic and ferromagnetic materials, the weak diamagnetic effects are obscured by the stronger magnetic effects associated with permanent magnetic properties of the atomic components.

Diamagnetism in Superconductors

A superconductor is a material that loses all electrical resistance below a certain temperature. Superconductors have particularly interesting magnetic properties. For example, the magnetic field is zero in the interior of a superconductor. Figure 33.5 shows how this zero-field condition might be achieved. When a superconductor experiences an applied magnetic field, a surface current is induced. This surface current produces a magnetic field that exactly cancels the applied field.

Because $B = 0$ inside a superconductor, it follows from Equation 33.1 that the magnetic susceptibility of a superconductor is

$$\chi_m = -1$$

From the standpoint of Ohm's law, superconductors are perfect conductors; they have zero resistivity. From the standpoint of Lenz's law, superconductors are perfectly diamagnetic; they prevent any change in magnetic flux.

Figure 33.6 shows how superconductors can take advantage of Lenz's law to provide a magnetic levitation force. A small permanent magnet is shown suspended in liquid nitrogen above a disk of a ceramic (high temperature)

FIGURE 33.6

A permanent magnet levitates in a liquid nitrogen bath above a superconducting film. Currents induced in the superconductor take on whatever value is necessary to balance the weight of the magnet.

superconducting material. The weight of the magnet is balanced by a repulsive magnetic force arising from currents induced in the superconductor. Because there is no resistance, the current in the superconductor takes on whatever value is necessary to maintain a constant magnetic flux configuration.

EXAMPLE 1

Surface Current in a Superconductor

A superconducting cylinder 10 cm long is thrust into a magnetic field of 1 T. Let's estimate the current induced in its surface.

The surface current produces a field that is roughly equivalent to a one-turn solenoid. The field inside a solenoid of length s with N turns and carrying a current I is $\mu_0 I \, N/s$. The field induced by the surface current is therefore

$$B_{\text{ind}} = \frac{\mu_0 I}{s}$$

This induced field cancels the applied field of 1 T. Hence the surface current is

$$I = \frac{B \cdot s}{\mu_0} = \frac{(1 \text{ T})(0.10 \text{ m})}{4\pi \times 10^{-7} \text{ T·m/A}}$$
$$= 80,000 \text{ A}$$

In the superconductor, this enormous current can continue indefinitely because the electrical resistance is zero.

33.3
PARAMAGNETISM

At the atomic level we view a paramagnetic material as composed of a uniform distribution of atomic magnetic dipoles, sufficiently separated so that the magnetic field of any given dipole does not influence its neighbors. In the absence of an applied field, the net magnetic moment of a paramagnetic material is zero because the dipoles are randomly oriented as a result of thermal motions. If an external magnetic field is applied, the atomic dipoles tend to align themselves with the field, thereby producing a net magnetic moment for the material.

The alignment of atomic magnetic dipole moments enhances the applied field. As Figure 33.7 shows, the applied field $\mathbf{B_0}$ and the induced field $\chi_m \mathbf{B_0}$ due to the aligned dipoles are in the same direction, which means that χ_m is positive.

There are two ways to understand the tendency toward alignment. First, a magnetic dipole moment $\boldsymbol{\mu}$ in a magnetic field $\mathbf{B}$ experiences a torque $\boldsymbol{\tau}$ given by

$$\boldsymbol{\tau} = \boldsymbol{\mu} \times \mathbf{B} \tag{33.6}$$

This torque tends to align $\boldsymbol{\mu}$ and $\mathbf{B}$. When $\boldsymbol{\mu}$ and $\mathbf{B}$ are parallel (aligned) the torque is zero, giving rotational equilibrium. As a second way to understand the tendency toward alignment, we observe that the magnetic potential energy of a magnetic dipole is

$$U = -\boldsymbol{\mu} \cdot \mathbf{B} \tag{33.7}$$

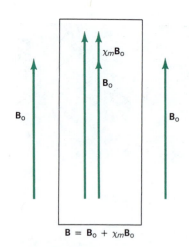

FIGURE 33.7

The induced magnetic moment of the paramagnetic material sets up an induced magnetic field $\chi_m \mathbf{B_0}$ that is parallel to the applied magnetic field $\mathbf{B_0}$. The net field inside the solid is the superposition of the applied field and the induced field.

The stable equilibrium configuration is the one for which U is a *minimum*. This minimum is achieved when **μ** and **B** are parallel.

The tendency toward magnetic alignment is opposed by thermal motions of the atoms. A useful measure of thermal energy per particle is kT, where k is the Boltzmann constant (Section 22.3). The ratio B/T reflects the relative strengths of magnetic alignment and thermal disordering. At room temperature, in a magnetic field of 1 T, the random thermal energy (kT) is approximately 100 times the magnetic alignment energy ($-$**μ** $\cdot$ **B**). Consequently, the thermal disruption is nearly complete. Magnetic alignment can be enhanced by increasing B and by lowering T.

Experiments can measure the total magnetic moment of a sample. A knowledge of the composition of a sample makes it possible to determine the number of atoms and thus the number of individual magnetic dipoles. Together these data yield a value of $\langle \mu \rangle$, the average magnetic dipole moment per particle (atom, molecule, ion). Figure 33.8 shows the measured values of $\langle \mu \rangle$ plotted versus the ordering parameter B/T. Figure 33.8 also shows the results of a theoretical calculation based on a quantum theory of paramagnetism.

The behavior shown in Figure 33.8 is typical: as B/T increases, $\langle \mu \rangle$ increases and then levels off. We can understand this behavior if we consider what happens to $\langle \mu \rangle$ if only B is increased while T is held constant. As the magnetic field is increased, dipole alignment becomes more and more complete, and thus $\langle \mu \rangle$ increases. Once the alignment is almost complete, increasing B has little effect and $\langle \mu \rangle$ levels off. This leveling off of $\langle \mu \rangle$ is called **saturation**.

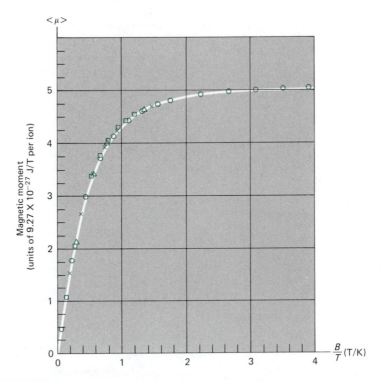

FIGURE 33.8

A plot of the average magnetic moment per particle $\langle \mu \rangle$ versus the alignment parameter B/T. The circles and diamonds denote experimental measurements. The solid line is the result predicted by a quantum theory of paramagnetism. (Data are taken from W. E. Henry, *Physical Review*, **88**, 592 [1952].)

For small values of B/T the average dipole moment per particle is directly proportional to B/T. The straight-line portion of a graph of $\langle \mu \rangle$ versus B/T can be represented by

$$\langle \mu \rangle = C\left(\frac{B}{T}\right) \tag{33.8}$$

where C is a constant whose value depends on the particular paramagnetic material. Equation 33.8 is called **Curie's law**, and C is called Curie's constant (after Pierre Curie). Curie deduced the form of Equation 33.8 on the basis of studies of a few compounds. By coincidence these compounds had very similar values of C, and Curie mistakenly believed that he had discovered a universal law, with a single value of C describing all paramagnetic materials.

33.4
FERROMAGNETISM

Ferromagnetism is exhibited by five elements—iron, nickel, cobalt, dysprosium, and gadolinium—and by some alloys, which usually contain one or more of these same five elements. Ferromagnetism relies strongly on the *mutual* interactions between the magnetic moments of electrons, or, the cooperative interactions that favor a high degree of parallel alignment of the magnetic dipole moments. There is no quantitative classical explanation for either the intrinsic electron magnetic moment or the alignment mechanism; they are quantum mechanical effects.

There are regions in every ferromagnetic sample that have nearly perfect alignment of magnetic dipole moments even when there is no applied magnetic field. These experimentally observable regions (Figure 33.9) are called **magnetic domains**. Depending on the structure and type of ferromagnetic material, the volumes of magnetic domains vary from about 10^{-18} to 10^{-12} m^3. Because each cubic centimeter of any solid contains roughly Avogadro's number ($\sim 10^{24}$) of molecules, domain volumes involve between 10^{12} and 10^{18} molecules. But even though the magnetic dipole alignment in a given domain is nearly complete, a ferromagnetic sample will not display a net magnetic dipole moment if the domains themselves are randomly oriented. A net magnetic dipole moment in a sample develops only when the domains are aligned by the application of a magnetic field.

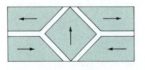

Magnetic domains in a crystalline whisker of iron. The crystal is 0.1 mm thick and there is no applied magnetic field.

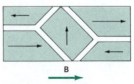

Application of a magnetic field causes some domains to grow at the expense of neighboring domains.

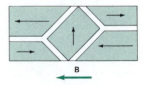

Domains having their intrinsic magnetic moments parallel to the applied magnetic field grow when the direction of the applied magnetic field is reversed. Note how the central domain has rotated.

FIGURE 33.9
Magnetic domains are altered when an external magnetic field is applied.

TABLE 33.3

*Curie Temperatures of the Ferromagnetic Elements**

Element	T_c (K)
Iron	1043
Cobalt	1404
Nickel	631
Gadolinium	289
Dysprosium	85

* All values except that for dysprosium were taken from *Handbook of Chemistry and Physics*, 61st ed. CRC Press, 2000 N. W. 24th St., Boca Raton, FL 33431.

There are two types of domain interaction that give rise to a net magnetic dipole moment:

1. Those domains with magnetic dipole moments parallel to the magnetizing field grow at the expense of neighboring domains (Figure 33.9). This effect is responsible for producing a net magnetic dipole moment in a weak applied magnetic field.
2. The magnetic dipole moments of the domains rotate toward alignment with the applied magnetic field. This is the mechanism of magnetic dipole alignment when the applied magnetic field is strong.

For a given ferromagnetic material, ferromagnetism vanishes at a sharply defined temperature called the **Curie temperature**. Above the Curie temperature, the material is only paramagnetic. For example, the Curie temperature of iron is 1043 K (Table 33.3). Because the interior of the earth is at a temperature of about 2000 K, we know that there can be no ferromagnetic contributions to the earth's magnetic field from its molten iron interior.

We saw that the magnetism of a paramagnetic or diamagnetic material vanishes once the material is removed from a magnetic field. However, this may not be the case if a substance is ferromagnetic. To illustrate, let us consider measurements of the magnetic field (B) in an iron sample as a function of the current in a solenoid arrangement like that discussed in Section 33.2. We start with the iron sample unmagnetized. When we increase the current (I) in the solenoid, the magnetic field in the iron (B) increases. We could display the data as a graph of B versus I. Because the magnetic field inside an empty solenoid is directly proportional to the current ($B_0 = \mu_0 n I$), we choose to display the measurements as a plot of B versus B_0. Figure 33.10 shows such a set of measurements plotted on a graph. Let's examine each part of this graph.

In Figure 33.10a we see that as the current increases, the magnetic field (B) also increases. The nonlinear relationship between B and B_0 means that the magnetic susceptibility is not constant—it depends on the value of B_0.

In Figure 33.10b we see that if B_0 is decreased from the value labeled 2, the magnetic field measurements are consistently higher than when the current was increased from the zero value. A **remanent magnetic field**, designated B_r, persists even when the current in the solenoid vanishes. The remanent magnetic field results from alignment of magnetic domains. This is one way to produce a permanent magnet.

Figure 33.10c shows that if we now reverse the direction of the current in the solenoid, the magnetic field (B) within the sample is reduced steadily from the remanent field value B_r. At a critical value of B_0, called the **coercive force** (B_c), the magnetic field is zero. The larger the coercive force B_c, the more difficult it is to demagnetize a ferromagnetic sample. Ferromagnetic materials having a large coercive force are said to be magnetically "hard"; those having a small coercive force are said to be magnetically "soft." Ordinary iron is magnetically soft and has a coercive force of about 10^{-4} T. A hard magnetic material used in the speaker of a high-fidelity system may have a coercive force 20–50 times that of ordinary iron.

In Figure 33.10d we find that if we keep the same orientation for the magnetic moment μ by maintaining the same current direction in the solenoid and by increasing the magnitude of the current, the magnetic field increases. However, the direction is now reversed from the starting direction. The

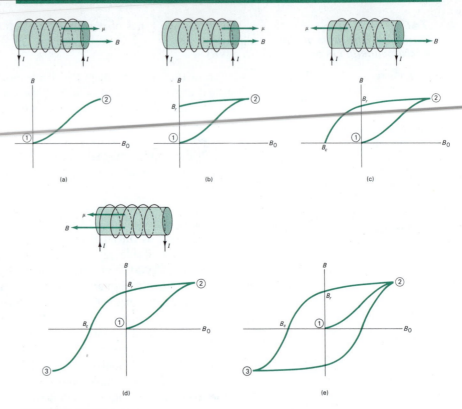

FIGURE 33.10

(a) As B_0 is increased, the field B inside the sample increases. The field **B** and the magnetic moment μ are in the same direction. The field saturates at point 2. (b) When B_0 is decreased the field inside the sample decreases but does not retrace its path from 1 to 2. (c) Reversing the current reverses **B**$_0$ and μ but does not immediately reverse **B**. At a critical value of B_0 denoted as B_c, the magnetic field inside the material reaches zero. (d) Increasing the current further causes **B** to again become parallel to **B**$_0$. The field saturates at point 3. (e) Decreasing the current to zero and then reversing its direction again we can return the magnetic field to its original saturation value at point 2.

magnetic field tends to saturate, achieving the value labeled 3. This indicates that the alignment of the magnetic domains is approaching completion.

Finally, Figure 33.10e shows that decreasing the magnetic intensity with an ultimate change in the initial direction brings the magnetic field back to its initial saturation value, labeled 2. Subsequent changes in the magnetic intensity retrace this closed loop. Measurements for the steel commonly used in transformers are shown in Table 33.4. Note that the magnetic susceptibility is not constant for a ferromagnetic material. It is generally 10^8 times larger than for a paramagnetic material and greatly depends on the value of B_0.

The failure of the B versus B_0 curve to retrace the initial magnetization curve, labeled 1 to 2 in Figure 33.10, is termed **hysteresis**. A closed curve representing measurements of B and B_0 is called a hysteresis loop. As a ferromagnetic sample is cycled around a hysteresis loop, frictionlike, irreversible changes occur in its domain structure.

Work is done by the magnetizing field in order to alter the domains, and the temperature of the sample increases. The greater the area of the hysteresis loop, the greater the amount of work required and the larger the corresponding temperature rise of the sample.

TABLE 33.4

Magnetic Measurements for Ordinary Transformer Steel*

B(T)	B_0(T)	χ_m
0.2	5.99×10^{-5}	3300
0.4	8.70×10^{-5}	4600
0.6	11.0×10^{-5}	5500
0.8	14.8×10^{-5}	5400
1.0	22.7×10^{-5}	4400
1.2	38.5×10^{-5}	3100
1.4	109×10^{-5}	1300
1.6	430×10^{-5}	370
1.8	1500×10^{-5}	120

* The data are taken from *Handbook of Chemistry and Physics*, 61st ed. CRC Press, 2000 N. W. 24th Street, Boca Raton, FL 33431.

WORKED PROBLEM

A coil of wire having 100 turns, a resistance of 5.5 ohms, and a cross-sectional area of 50 cm^2, encircles a long solenoid. A current is induced in the coil when the magnetic field in the solenoid changes from zero to its final steady value. The current induced in the coil during the time the magnetic field changes is described by

$$I = -(at - bt^3) \qquad a = 225 \text{ A/s} \qquad b = 10^8 \text{ A/s}^3$$

The current lasts for 1.5 milliseconds. Determine the final magnetic field in the solenoid.

Solution

In Section 33.1 it was shown that the change in magnetic flux is given by

$$\Delta\Phi_B = -Rq$$

where q is the total charge passing through the sense coil. Additionally, it was shown that the change in magnetic flux is related to the final value of the magnetic field by

$$\Delta\Phi_B = BNA$$

Equating these expressions and solving for B we find,

$$B = \frac{Rq}{NA}$$

In this problem we know explicitly how the current in the sense coil changes with time, so we need only integrate the current to determine the total charge,

$$q = \int I\,dt = -\int_0^{1.5 \text{ ms}} (at - bt^3)\,dt = -[\tfrac{1}{2}at^2 - \tfrac{1}{4}bt^4]\Big|_0^{1.5 \text{ ms}}$$

Substituting for a and b yields

$$q = -0.127 \times 10^{-3} \text{ C}$$

For the magnetic field we find,

$$B = \frac{-Rq}{NA} = \frac{-(5.5 \text{ } \Omega)(-0.127 \times 10^{-3} \text{ C})}{100(50 \times 10^{-4} \text{ m}^2)}$$

$$= 0.0014 \text{ T}$$

EXERCISES

33.1   Magnetic Classification of Materials

A. (a) A current of 1 A exists in a solenoid having 4000 turns per meter. Calculate the magnetic field within the solenoid when it is empty. (b) Determine the change in the magnetic field when the interior is filled with copper.

B. A current I in a long, tightly-wound solenoid produces a magnetic field B_0. When the solenoid is filled with aluminum, the same current produces a field B. Find the relative change, $(B - B_0)/B_0$.

C. The core of a solenoid having 250 turns per meter is filled with a material of unknown composition. When the current in the solenoid is 2 A, measurements reveal that the magnetic field within the core is 0.13 T. Determine the magnetic susceptibility of the material and classify it magnetically.

D. The sense coil surrounding a long solenoid has 120 turns of wire, a resistance of 0.13 Ω, and a mean diameter of 1.21 cm. When the flux in the solenoid is reduced to zero, a net charge of 1.3 mC flows through the sense coil. What was the magnetic field in the interior of the solenoid?

E. Figure 1 displays the current induced in the sense coil surrounding a solenoid similar to that shown in Figure 33.3. Estimate the total charge moved through the sense coil.

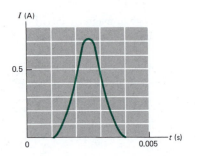

FIGURE 1

33.2   Diamagnetism

F. Consider a circular superconducting disk (diameter = 1 cm) placed in an external magnetic field ($B_0 = 2 \times 10^{-3}$ T) parallel to the axis of the disk. What surface current around the disk perimeter would be needed to make $B = 0$ at the center of the disk?

33.3   Paramagnetism

G. Given N magnetic dipoles, each having a dipole moment μ, (a) organize these N dipoles into an arrangement having a net magnetic dipole moment of 500 μ. (b) Is there any restriction on the size of N? (c) Is your arrangement the only possibility?

H. The magnetic dipole moment of an atom in a paramagnetic gas is 10^{-23} J/T. Determine the temperature at which the average thermal energy of an atom is equal to the magnitude of its magnetic potential energy when the dipole moment is aligned with a magnetic field of 0.6 T.

33.4   Ferromagnetism

I. Measurements of the total magnetic field (B) and of the magnetic field (B_0) due to solenoid current are presented in Table 1. Determine the magnetic susceptibility χ_m for the values of B_0 recorded in the table.

TABLE 1

B(T)	B_0(T)
0.094	3.14×10^{-4}
0.38	6.28×10^{-4}
0.70	9.42×10^{-4}
0.90	12.60×10^{-4}
0.99	15.70×10^{-4}
1.05	18.90×10^{-4}

J. Show that the product of B and B_0/μ_0 has units of joules per cubic meter.

K. Determine the remanent magnetic field B_r and coercive force B_c for the material whose B versus B_0/μ_0 properties are shown in Figure 2.

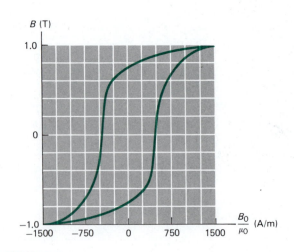

FIGURE 2

L. The area bounded by a hysteresis loop represents the work required to take the material through a hysteresis cycle. Figure 3 shows an approximation to the real hysteresis loop shown in Figure 2. Determine the work required in joules per cubic meter to take t'is material through one hysteresis cycle.

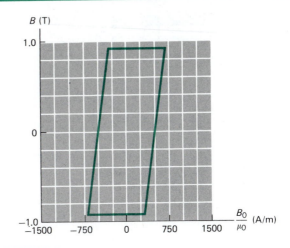

FIGURE 3

PROBLEMS

1. A rod of a superconducting material 2.5 cm long is placed in a uniform magnetic field of 0.54 T with its cylindrical axis along the magnetic field lines. (a) Sketch the directions of the applied field and the induced surface current, and (b) estimate the magnitude of the surface current.

2. For an electromagnet to pick up a solid piece of material, the upward magnetic force must exceed the downward gravitational force. Show that the variation of magnetic field must be at least $dB/dy = mg/\mu$, where m is the mass and μ is the magnetic dipole moment.

3. A coil of wire having 50 turns, a resistance of 10.5 ohms, and a cross-sectional area of 75 cm^2 encircles a long solenoid. A current is induced in the coil when the magnetic field in the solenoid changes. The current in the sense coil is described by

$$I = -(a - bt)t \qquad a = 2000 \text{ A/s} \qquad b = 10^6 \text{ A/s}^2$$

The current lasts for 2 milliseconds. Determine the change in magnetic field in the solenoid.

4. The table below records the magnetization of a ferromagnetic material. (a) Construct a magnetization curve from the data. (b) Determine the ratio B/B_0 for each pair of values of B and B_0 and construct a graph of B/B_0 versus B_0. (B/B_0 is called the relative permeability, and it is a measure of the induced magnetic field.)

5. A linear array of identical magnetic dipoles tends to have one of two configurations, (a) all dipoles parallel or (b) every other dipole parallel, as shown in Figure 4. Consider only interactions between nearest neighbors and show that both configurations result in zero net force and zero net torque on any interior dipole.

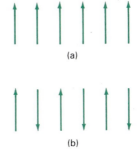

FIGURE 4

TABLE 2

B(T)	B_0(T)
0.2	4.8×10^{-5}
0.4	7.0×10^{-5}
0.6	8.8×10^{-5}
0.8	1.2×10^{-4}
1.0	1.8×10^{-4}
1.2	3.1×10^{-4}
1.4	8.7×10^{-4}
1.6	3.4×10^{-3}
1.8	1.2×10^{-1}

6. A bar magnet (mass = 39.4 grams, magnetic moment = 7.65 J/T, length = 10 cm) is connected to the ceiling by a string. A uniform external magnetic field is applied horizontally, as shown in Figure 5. The magnet is in equilibrium, making an angle θ with the horizontal. If $\theta = 5°$, determine the strength of the applied magnetic field.

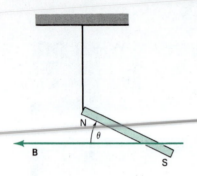

FIGURE 5

7. A superconducting aluminum ring of radius $R = 10$ cm, and weight $W = 0.02$ N lies in a horizontal plane. A magnetic field is turned on at $t = 0$. The field is given by

$$B = \frac{B_0 R^3}{z^3} \qquad B_0 = 0.1 \text{ T}$$

where the $+z$-axis is up. If the field is just strong enough to support the ring, determine the current induced in the ring when the field is turned on. Take $z = R$ to be the center of the ring.

8. A particle (charge Q, mass M) orbits in a circular path about a fixed point charge. A magnetic field **B** is directed parallel to the orbital plane. The circular motion of the charge constitutes a current loop. As such it has a magnetic moment. Show that this orbital magnetic moment precesses about **B** at an angular velocity given by,

$$\omega = \frac{QB}{2m}$$

CHAPTER 34

ALTERNATING CURRENTS

34.1
AC GENERATORS

An **alternating current (ac) generator** is a source of electrical energy that maintains a time-varying potential difference (voltage) of the form

$$\mathcal{E} = \mathcal{E}_0 \cos \omega t \qquad (34.1)$$

across its terminals. The amplitude $\mathcal{E}_0$ is constant and the angular frequency ω equals $2\pi f$, where f is the generator frequency expressed in hertz. A magnet rotating inside a stationary coil is one type of ac generator. Radios, TV sets, and computers incorporate ac generators having no moving components. Commercial electric power plants employ ac generators to produce electric energy for homes and industries. In America, the frequency f is 60 Hz and $\mathcal{E}_0$ is 170 volts. In Europe, the frequency f is 50 Hz and $\mathcal{E}_0$ is 340 volts.

Kirchhoff's voltage and current rules govern electrical circuits. Because these rules are expressed as linear equations, the current in a circuit *driven by* an ac generator is sinusoidal* and varies at the same frequency as the generator. An ac generator is represented by the symbol $(\sim)$. In this chapter we focus on the response of simple circuits that are driven by an ac generator.

34.2
RMS VALUES

If we connect a resistor to an ac generator as shown in Figure 34.1, an alternating current is established in the circuit. Kirchhoff's voltage rule for the circuit gives

$$\mathcal{E}_0 \cos \omega t - IR = 0$$

The current can be expressed as

$$I = I_m \cos \omega t \qquad (34.2)$$

where

$$I_m = \frac{\mathcal{E}_0}{R} \qquad (34.3)$$

is the maximum current in the circuit.

Two cycles of $\mathcal{E}$ and I are shown in Figure 34.2 for a circuit with $f = 60$ Hz, $\mathcal{E}_0 = 170$ V, and $R = 10\ \Omega$. Both $\mathcal{E}$ and I are proportional to $\cos(2\pi f t)$, so their maxima occur at the same time. We say $\mathcal{E}$ and I are **in phase**.

The current represented by Equation 34.3 is called the steady-state current. In addition, there are transient currents that arise when the generator is first connected. We ignore transient currents throughout this chapter; they decrease exponentially with time and have no bearing on the ac response of the circuit.

In many situations we are interested in the energy transferred to a circuit element such as a resistor. The power developed in a resistor R carrying a current I is $P = RI^2$. If the power is constant, the energy delivered to the resistor in a time t is Pt. If the current varies with time, then the power also

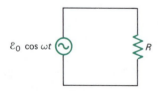

FIGURE 34.1

An ac generator connected to a resistor.

---

* Quantities that are directly proportional to $\sin \omega t$ or $\cos \omega t$ are referred to as **sinusoidal**.

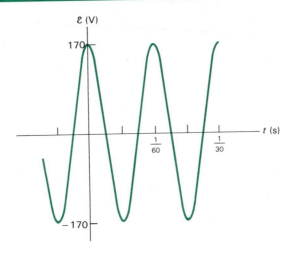

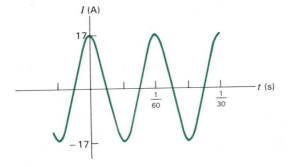

FIGURE 34.2
The generator voltage and the current are in phase for the circuit shown in Figure 34.1.

varies and the energy delivered is the *time integral* of the instantaneous power:

$$\text{Energy delivered in time } t = \int_0^t P \, dt \qquad (34.4)$$

We need to develop the relationship between the time integral of the power and the **average power**.

Time Averages

The time average of a quantity $F(t)$ during the time interval from $t = 0$ to $t = T$ is defined by

$$\langle F \rangle = \frac{1}{T} \int_0^T F(t) \, dt \qquad (34.5)$$

For example, if $F(t) = \cos^2 \omega t$ and the interval is one cycle of the cosine, we have

$$\langle \cos^2 \omega t \rangle = \frac{1}{T} \int_0^T \cos^2 \omega t \, dt$$

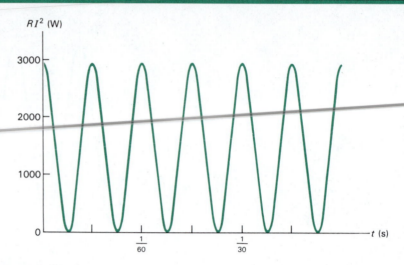

FIGURE 34.3

Time variation of the thermal power developed in a resistor.

where T is the period of the cosine function. The integration gives

$$\langle \cos^2 \omega t \rangle = \frac{1}{T} \left(\frac{1}{2} t + \frac{1}{2} \cos(2\omega t) \right) \Big|_0^T = \frac{1}{2}$$

Thus, the average value of $\cos^2 \omega t$ over a cycle is $\frac{1}{2}$.

If $F(t)$ is the instantaneous power, the energy delivered in a time t can be expressed as $\langle P \rangle \cdot t$. Thus, if the power varies with time we use the average power to determine the energy delivered.

The instantaneous power developed in a resistor R carrying a current $I_m \cos(2\pi ft)$ is given by

$$P = RI^2 = RI_m^2 \cos^2(2\pi ft)$$

Figure 34.3 shows a graph of RI^2 for $I_m = 17$ A, $f = 60$ Hz, and $R = 10\ \Omega$. The time variation of RI^2 is contained in the $\cos^2(2\pi ft)$ factor. The average power $\langle RI^2 \rangle$ depends on the average value of $\cos^2(2\pi ft)$, which equals $\frac{1}{2}$. Thus,

$$\langle P \rangle = \tfrac{1}{2}RI_m^2 \tag{34.6}$$

RMS Value

Comparing Equation 34.6 with the expression $P = RI^2$ for the power developed by a constant current, we see that the sinusoidal current $I_m \cos(2\pi ft)$ produces the same average power as a constant current $I_m/\sqrt{2} = 0.707\ I_m$. The quantity $I_m/\sqrt{2}$ is called the **root-mean-square** (rms) value of the sinusoidal current $I_m \cos(2\pi ft)$. It is the square root of the mean value of the square of the current

$$I_{\text{rms}} = \sqrt{\frac{1}{T} \int_0^T I^2\, dt} \tag{34.7}$$

We frequently deal with sinusoidal currents and emf's. Keep in mind that their rms values are 0.707 times their maximum values,

$$I_{rms} = \frac{I_m}{\sqrt{2}} = 0.707\ I_m \qquad (34.8)$$

and

$$\mathscr{E}_{rms} = \frac{\mathscr{E}_0}{\sqrt{2}} = 0.707\ \mathscr{E}_0 \qquad (34.9)$$

In American homes where $\mathscr{E}_0 = 170$ V, the rms value of the voltage is $0.707(170\text{ V}) = 120$ V.

The average power developed in the resistor $\frac{1}{2}RI_m{}^2$ can be expressed in terms of the rms current by

$$\langle P \rangle = RI_{rms}{}^2 \qquad (34.10)$$

By using the rms current we maintain a correspondence between the alternating current and constant current relations for power.

EXAMPLE 1

Resistance and rms Current for a Light Bulb

A 60-watt light bulb operates at 120 volts (rms). Let's treat it as a resistor in series with an ac generator and determine the rms current in it and its resistance.

The rms voltage across a resistor in series with an ac generator is related to the rms current in it by

$$V_{rms} = RI_{rms}$$

The average power developed is given by Equation 34.10,

$$\langle P \rangle = RI_{rms}{}^2$$

The resistance is

$$R = \frac{V_{rms}{}^2}{\langle P \rangle} = \frac{(120\text{ V})^2}{60\text{ W}} = 240\ \Omega$$

The rms current is

$$I_{rms} = \frac{V_{rms}}{R} = \frac{120\text{ V}}{240\ \Omega} = 0.500\text{ A}$$

If you measure the resistance of a 60-watt bulb with an ohmmeter, you will get a much *smaller* value than we have just calculated. Knowing that the bulb gets very warm, can you explain why?

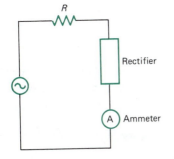

FIGURE 34.4

The rectifier is placed in series with the ac ammeter.

The inertia of a moving coil galvanometer-type ammeter prevents using it to measure the instantaneous current. It can be modified to record rms currents by connecting it in series with a **rectifier** (Figure 34.4). The rectifier permits charge to pass in only one direction, so the meter records the average value of the current for one direction of the current. As we now show, the rms value and the average value of the rectified current are related by a constant factor, making it possible to compute the rms value from the average value.

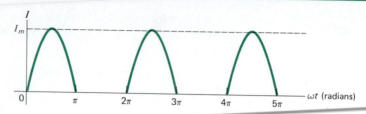

FIGURE 34.5

Rectified ac current. The average current is *not* zero.

EXAMPLE 2

How to Make a dc Ammeter Record an rms Current

The ac generator in Figure 34.4 delivers a sinusoidal current $I = I_m \sin \omega t$. The rectifier eliminates the negative half of the sinusoidal current. The current reaching the meter corresponds to the graph shown in Figure 34.5. The meter records the average value of this rectified current.

The length of one cycle of the sinusoidal current is $T = 2\pi/\omega$. Because the current is zero over half of the cycle the average current is

$$\langle I \rangle = \frac{1}{T} \int_0^T I\, dt = I_m \frac{\omega}{2\pi} \left(\int_0^{\pi/\omega} \sin \omega t\, dt + \int_{\pi/\omega}^{2\pi/\omega} 0\, dt \right)$$

$$= \frac{I_m}{\pi}$$

The rms value of $I_m \sin \omega t$ is $I_m/\sqrt{2}$. Hence

$$I_{\text{rms}} = \langle I \rangle \frac{\pi}{\sqrt{2}}$$

Multiplying the measured average current by $\pi/\sqrt{2}$ gives the rms current. Most ac meters incorporate this factor in their calibration so the rms value can be read directly.

In earlier chapters we learned how various combinations of resistors, capacitors, and inductors functioned in circuits energized by a constant emf such as a battery. In the next three sections we analyze the response of circuits which have an ac generator connected to different combinations of a resistor, a capacitor, and an inductor.

34.3
THE RC CIRCUIT

Consider a resistor and a capacitor connected to an ac generator, as shown in Figure 34.6. We will analyze the RC circuit in considerable detail to illustrate the physical principles involved and to develop techniques that we will use later on.

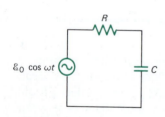

FIGURE 34.6

An ac generator drives the *RC* series combination.

Kirchhoff's voltage rule for the RC circuit has the form

$$\mathcal{E}_0 \cos \omega t - RI - \frac{q}{C} = 0 \qquad (34.11)$$

Because the ac generator maintains a sinusoidal emf, the steady-state current in the circuit also varies sinusoidally at the generator frequency. Accordingly we write

$$I = I_m \cos(\omega t - \varphi) \tag{34.12}$$

where I_m is the maximum current, and the **phase angle** φ allows for the possibility that the current and the emf of the ac generator may not be in phase.

To solve Equation 34.11, we must express I_m and φ in terms of the given quantities $\mathscr{E}_0$, R, C, and ω.

The charge on the capacitor is related to the current by $dq/dt = I$. You can verify that

$$q = \frac{I_m}{\omega} \sin(\omega t - \varphi) \tag{34.13}$$

satisfies the derivative relation between charge and current. Inserting I and q as expressed by Equations 34.12 and 34.13 into Equation 34.11 gives

$$\mathscr{E}_0 \cos \omega t - RI_m \cos(\omega t - \varphi) - I_m \left(\frac{1}{\omega C}\right) \sin(\omega t - \varphi) = 0 \tag{34.14}$$

The combination $1/\omega C$ is called the **capacitive reactance**, and has the units of ohms. We denote it by the symbol X_C

$$X_C \equiv \frac{1}{\omega C} \tag{34.15}$$

Using trigonometric identities for $\cos(\omega t - \varphi)$ and $\sin(\omega t - \varphi)$, Equation 34.14 can be rewritten in the form

$$\sin \omega t \{-RI_m \sin \varphi - X_C I_m \cos \varphi\}$$
$$+ \cos \omega t [\mathscr{E}_0 - RI_m \cos \varphi + X_C I_m \sin \varphi] = 0 \tag{34.16}$$

For this equation to be satisfied at all times, the coefficients of $\sin \omega t$ and $\cos \omega t$ must equal zero. This gives two equations relating the unknowns I_m and φ to the known quantities $\mathscr{E}_0$, R, C, and ω.*

Setting the coefficient of $\sin \omega t$ equal to zero gives

$$-RI_m \sin \varphi = X_C I_m \cos \varphi \tag{34.17}$$

This gives a solution for the phase angle φ,

$$\tan \varphi = -\frac{X_C}{R} \tag{34.18}$$

Figure 34.7 shows how the phase angle can be related to X_C, R, and the combination

$$Z \equiv \sqrt{R^2 + X_C^2} \tag{34.19}$$

called the **impedance**.

---

* An equivalent way of obtaining the two equations is to evaluate Equation 34.16 at the times for which $\omega t = 90°$ and $\omega t = 0°$.

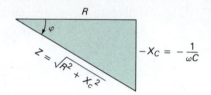

FIGURE 34.7
Phase angle triangle for the *RC* circuit.

Equating the coefficient of cos ωt in Equation 34.16 with zero gives

$$I_m(R \cos \varphi - X_C \sin \varphi) = \mathscr{E}_0 \tag{34.20}$$

From the triangle of Figure 34.7 we see that

$$\cos \varphi = \frac{R}{Z} \qquad \sin \varphi = -\frac{X_C}{Z}$$

Inserting these expressions into Equation 34.20 gives for the maximum current

$$I_m = \frac{\mathscr{E}_0}{Z} = \frac{\mathscr{E}_0}{\sqrt{R^2 + X_C^2}} \tag{34.21}$$

The rms current is related to the rms voltage by

$$I_{rms} = \frac{\mathscr{E}_{rms}}{Z} \tag{34.22}$$

This is a generalization of the dc relation $I = \mathscr{E}/R$ and shows that *the impedance Z is a generalization of the resistance.*

EXAMPLE 3

rms Current in an RC Circuit

A 100-μF capacitor is connected in series with an 18-Ω resistor and a 60-Hz ac generator. The rms voltage across the generator is 120 V. We want to determine the rms current in the circuit.

The capacitive reactance is

$$X_C = \frac{1}{\omega C} = \frac{1}{(2\pi \cdot 60 \text{ s}^{-1} \cdot 100 \times 10^{-6} \text{ F})} = 26.5 \ \Omega$$

The impedance is

$$Z = \sqrt{R^2 + X_C^2} = \sqrt{(18 \ \Omega)^2 + (26.5 \ \Omega)^2} = 32.0 \ \Omega$$

The rms current is

$$I_{rms} = \frac{\mathscr{E}_{rms}}{Z} = \frac{120 \text{ V}}{32 \ \Omega} = 3.75 \text{ A}$$

The impedance is larger than R so the rms current is less than it would be if the capacitor were absent.

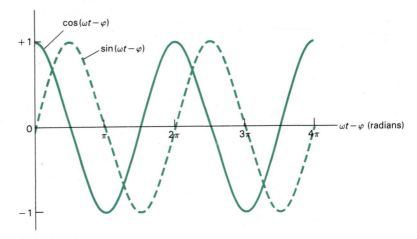

FIGURE 34.8
This illustrates that the cosine function *leads* the sine function by 90°.

Equations 34.18 and 34.21 express φ and I_m in terms of $\mathscr{E}_0$, R, C, and ω. In terms of the impedance Z the steady-state current is

$$I = \frac{\mathscr{E}_0}{Z} \cos(\omega t - \varphi) \qquad (34.23)$$

The voltage across the capacitor is

$$\frac{q}{C} = \frac{1}{\omega C}\frac{\mathscr{E}_0}{Z} \sin(\omega t - \varphi) = X_C I_m \sin(\omega t - \varphi) \qquad (34.24)$$

Note that the current in the circuit varies as $\cos(\omega t - \varphi)$, but the voltage across the capacitor varies as $\sin(\omega t - \varphi)$. The cosine and sine functions differ in phase by 90°, as illustrated in Figure 34.8. The current reaches its maximum value *before* the capacitor voltage reaches its maximum. We describe this situation by the statement:

The current *leads* the voltage across the capacitor by 90°.

Filter Action

The maximum voltage across the capacitor is

$$\frac{\mathscr{E}_0}{\omega C Z} = \frac{\mathscr{E}_0}{\sqrt{1 + (\omega RC)^2}} \qquad (34.25)$$

Figure 34.9 displays $\mathscr{E}_0/\sqrt{1 + (\omega RC)^2}$ versus ωRC. For fixed values of R and C, the maximum voltage across the capacitor *decreases* as the angular frequency ω increases. This behavior is exploited in filter circuits.

In a circuit such as a high-fidelity amplifier, the ac voltage across one component often acts as an ac generator that drives another stage of the circuit.

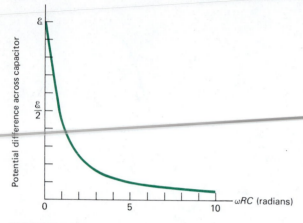

FIGURE 34.9

The maximum voltage across the capacitor in the *RC* circuit is plotted versus the dimensionless parameter ωRC.

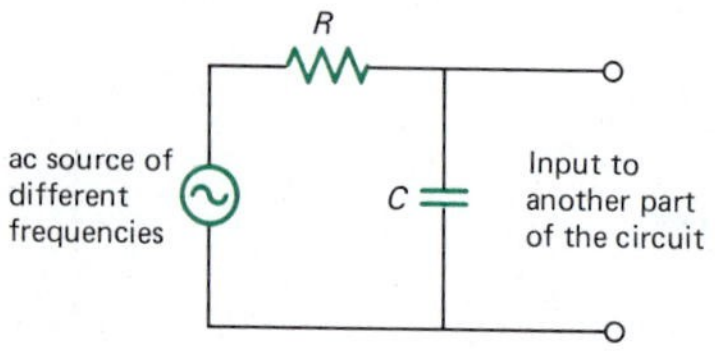

FIGURE 34.10

The *RC* circuit can be used as a high-frequency filter.

In Figure 34.10 the voltage across the capacitor in the *RC* circuit acts as an ac generator that provides the input to another part of the circuit. The maximum voltage across the capacitor decreases as the frequency increases. If an ac source operating over a wide range of frequencies provides the emf to the *RC* series combination, then the capacitor voltage will be relatively small at high frequencies. We say that the high-frequency components have been *filtered* from the output.

Reactive Power

The instantaneous power developed by a circuit element carrying a current *I* and across which the voltage is *V* is given by

$$P = VI$$

The power developed by the capacitor in the *RC* circuit is

$$P_C = X_C I_m^2 \sin(\omega t - \varphi)\cos(\omega t - \varphi) \tag{34.26}$$

No energy is dissipated by the capacitor. This follows from the fact that the time average of P_C is zero,

$$\langle P_C \rangle = X_C I_m^2 \langle \sin(\omega t - \varphi)\cos(\omega t - \varphi)\rangle = 0$$

The average power is zero for reactive circuit elements (capacitors and inductors) because they store energy during one quarter of a cycle and return energy to the circuit during the following quarter of a cycle. This ebb and flow of reactive power for a capacitor is shown in Figure 34.11. The capacitor is charging during the quarter of the cycle for which $\omega t - \varphi$ ranges from zero to $\pi/2$, and the reactive power is positive. As the capacitor charges, it stores energy in the electric field between its plates. The charge on the capacitor and the energy stored in it reach a maximum at $\omega t - \varphi = \pi/2$, at which time the current is momentarily zero. The reactive power is negative during the next quarter cycle, as $\omega t - \varphi$ increases from $\pi/2$ to π. During this quarter cycle, the charge and energy stored in the capacitor decrease. At $\omega t - \varphi = \pi$, the charge is momentarily zero and the energy stored in the capacitor reaches zero.

Even though the average of P_C is zero, its maximum value is significant. The maximum value of the reactive power is a measure of the energy available

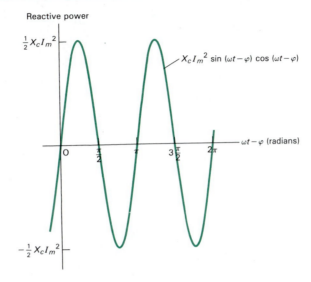

FIGURE 34.11

Reactive power in the *RC* circuit. The average power over a cycle is zero.

in the circuit. The energy available under steady-state conditions can greatly exceed the energy transformed in the resistor during one cycle.

The form of P_C given by Equation 34.26 can be expressed as

$$P_C = \tfrac{1}{2} X_C I_m^2 \sin(2\omega t - 2\varphi)$$

showing that its maximum value is

$$(P_C)_{max} = \tfrac{1}{2} X_C I_m^2 = X_C I_{rms}^2 \tag{34.27}$$

EXAMPLE 4

Reactive Power and Thermal Power in an RC Circuit

An *RC* circuit consists of a 47-μF capacitor and a resistance of 4 Ω in series with a 115-volt rms, 60-Hz ac generator. We want to compare the maximum reactive power $X_C I_{rms}^2$ to the average power $R I_{rms}^2$ developed in the resistor. Their ratio is

$$\frac{X_C I_{rms}^2}{R I_{rms}^2} = \frac{X_C}{R} = \frac{1}{\omega C R}$$

Using

$$\omega = 2\pi \cdot 60 \text{ s}^{-1} = 377 \text{ s}^{-1}$$

$$C = 47 \ \mu\text{F} = 4.7 \times 10^{-5} \text{ F}$$

$$R = 4 \ \Omega$$

gives

$$\frac{1}{\omega C R} = \frac{1}{(377 \text{ s}^{-1} \cdot 4.7 \times 10^{-5} \text{ F} \cdot 4 \ \Omega)} = 14$$

This 14:1 ratio shows that the energy made available by the generator each cycle is substantially greater than the electrical energy transformed into thermal energy in the resistor each cycle.

Power Factor

The average power developed in the resistor of the RC circuit is

$$\langle P \rangle = RI_{rms}{}^2 = R\left(\frac{\mathscr{E}_{rms}}{Z}\right)I_{rms}$$

Using $R/Z = \cos\varphi$ we can write

$$\langle P \rangle = \mathscr{E}_{rms}I_{rms}\cos\varphi \qquad (34.28)$$

The factor $\cos\varphi = R/Z$ is called the **power factor**. It is a measure of the current *in phase* with the ac generator. Equation 34.28 is a generalization of the dc relation, $P = \mathscr{E}I$.

EXAMPLE 5

Power Factor for an RC Circuit

Let's calculate the power factor and thermal power for the RC circuit of Example 4. The capacitive reactance is

$$X_C = \frac{1}{\omega C} = \frac{1}{(377\ \text{s}^{-1} \cdot 4.7 \times 10^{-5}\ \text{F})} = 56.4\ \Omega$$

The impedance is

$$Z = \sqrt{R^2 + X_C{}^2} = \sqrt{(4\ \Omega)^2 + (56.4\ \Omega)^2} = 56.5\ \Omega$$

The power factor is

$$\cos\varphi = \frac{R}{Z} = \frac{4\ \Omega}{56.5\ \Omega} = 0.0708$$

The power developed in the resistor is

$$\langle P \rangle = \mathscr{E}_{rms}I_{rms}\cos\varphi = \mathscr{E}_{rms}\left(\frac{\mathscr{E}_{rms}}{Z}\right)\cos\varphi$$

$$= \frac{(115\ \text{V})^2(0.0708)}{(56.5\ \Omega)}$$

$$= 16.6\ \text{W}$$

34.4
THE RL CIRCUIT

Replacing the capacitor in the RC circuit with an inductor produces an RL circuit (Figure 34.12). The current and potential differences of an RL circuit can be found using the same technique employed for the RC circuit. Kirchhoff's voltage rule has the form

$$\mathscr{E}_0 \cos\omega t - RI - L\frac{dI}{dt} = 0$$

Proceeding as we did when solving for the current in the RC circuit, we assume a solution of the form $I = I_m \cos(\omega t - \varphi)$. Kirchhoff's voltage rule becomes

$$\mathscr{E}_0 \cos\omega t - RI_m \cos(\omega t - \varphi) + \omega L I_m \sin(\omega t - \varphi) = 0 \qquad (34.29)$$

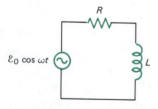

FIGURE 34.12

An ac generator drives the RL series combination.

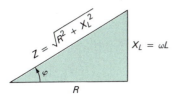

FIGURE 34.13

Phase angle triangle for the *RL* circuit.

Comparing this equation with Equation 34.15 shows they are identical in form. In fact, if we replace $1/\omega C = X_C$ in Equation 34.15 by $-\omega L$ we get Equation 34.29. This shows that the solution of Equation 34.29 has the same mathematical form as the solution of the *RC* equation, but with $-\omega L$ replacing $1/\omega C$.

We introduce the **inductive reactance** X_L defined by

$$X_L \equiv \omega L \tag{34.30}$$

X_L is measured in ohms. The current in the *RL* circuit can be expressed as

$$I = \left(\frac{\mathscr{E}_0}{Z}\right) \cos(\omega t - \varphi) \tag{34.31}$$

where the phase angle φ is given by (Figure 34.13)

$$\tan \varphi = \frac{X_L}{R} \tag{34.32}$$

The impedance Z is

$$Z = \sqrt{R^2 + X_L{}^2} \tag{34.33}$$

Equations 34.32 and 34.33 are formally equivalent to Equations 34.18 and 34.19 for the *RC* circuit.

The current-voltage relations in the *RL* and *RC* circuits are similar, but there are notable differences. In the *RC* circuit the current *leads* the voltage across the capacitor. In the *RL* circuit, the current reaches its maximum *after* the voltage across the inductor reaches its maximum. The current and inductor voltage differ in phase by 90°. We describe this situation by the statement:

The current *lags* the voltage across the inductor by 90°.

The frequency response of *RL* and *RC* circuits also differ. A high-frequency filter can be achieved using the output voltage across the capacitor in an *RC* circuit. The voltage across the inductor in the *RL* circuit decreases as ω decreases, allowing it to function as a **low-frequency filter**.

The inductor stores energy over parts of a cycle and feeds energy back into the circuit over other parts of a cycle. The average power developed by the inductor over a cycle is zero. The maximum power developed by the inductor is an important parameter because it measures the power available to the circuit. The maximum power developed by the inductor follows from Equation 34.27 by replacing X_C by the inductive reactance X_L,

$$(P_L)_{max} = \tfrac{1}{2} X_L I_m{}^2 = X_L I_{rms}{}^2 \tag{34.34}$$

The average power developed in the resistor of the *RL* circuit has the same form as Equation 34.28, derived for the *RC* circuit,

$$\langle P \rangle = \mathscr{E}_{rms} I_{rms} \cos \varphi \tag{34.35}$$

EXAMPLE 6

Determining R, L, and φ with an AC Meter

A meter records an rms current of 0.5 A and an rms generator voltage of 104 V in an *RL* circuit. A wattmeter shows that the average thermal power developed in the resistor is 10 W. We want to determine *R, L,* and the phase angle φ.

From Equation 34.35,

$$\cos \varphi = \frac{P}{\mathscr{E}_{rms} I_{rms}} = \frac{10\ W}{(104\ V \cdot 0.5\ A)} = 0.192$$

The phase angle is

$$\varphi = \cos^{-1}(0.192) = 78.9°$$

The impedance is

$$Z = \frac{\mathscr{E}_{rms}}{I_{rms}} = \frac{104\ V}{0.5\ A} = 208\ \Omega$$

The resistance follows from

$$R = Z \cos \varphi = 208\ \Omega \cdot (0.192) = 39.9\ \Omega$$

We can determine the inductance by first calculating the inductive reactance

$$X_L = Z \sin \varphi = 208\ \Omega \cdot \sin(78.9°) = 208\ \Omega \cdot (0.978) = 203\ \Omega$$

The inductance follows from

$$L = \frac{X_L}{\omega} = \frac{203\ \Omega}{377\ s^{-1}} = 0.538\ H$$

34.5
THE RLC CIRCUIT

If we connect a resistor, an inductor, and a capacitor in series with an ac generator we get the RLC circuit shown in Figure 34.14. For the series RLC circuit, Kirchhoff's voltage rule gives

$$\mathscr{E}_0 \cos \omega t - RI - L\frac{dI}{dt} - \frac{q}{C} = 0$$

The same techniques we applied to the RC and RL circuits lead to the following solution for the current:

$$I = I_m \cos(\omega t - \varphi) = \frac{\mathscr{E}_0}{Z} \cos(\omega t - \varphi) \tag{34.36}$$

where the impedance is

$$Z = \sqrt{R^2 + (X_L - X_C)^2} \tag{34.37}$$

The reactances are $X_L = \omega L$ and $X_C = 1/\omega C$. The phase angle φ is given by

$$\tan \varphi = \frac{(X_L - X_C)}{R} = \frac{\left(\omega L - \dfrac{1}{\omega C}\right)}{R} \tag{34.38}$$

Consistency checks can be made by considering the special cases $X_L = 0$, which gives the RC circuit results, and $X_C = 0$, which gives the RL circuit results. In general, relations derived for the RL and RC circuits are special cases of the RLC circuit results.

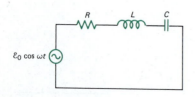

FIGURE 34.14

An ac generator drives the series RLC circuit.

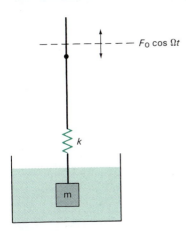

$- - +- - +- - - F_0 \cos \Omega t$

k

m

FIGURE 34.15

The damped spring-mass system is driven by a force varying sinusoidally with time. This system is the mechanical analog of the series *RLC* circuit of Figure 34.14.

Resonance

The series *RLC* circuit is of special importance because it can exhibit *resonance*. We encountered resonance in Chapter 13 in connection with a spring-mass system driven by a sinusoidal force (Figure 34.15). The *RLC* circuit is the electrical analog of this driven oscillator. The analogy can be displayed by setting $I = dq/dt$ in Kirchhoff's voltage rule for the *RLC* circuit. This gives

$$L \frac{d^2q}{dt^2} + R \frac{dq}{dt} + \frac{q}{C} = \mathcal{E}_0 \cos \omega t$$

Newton's second law for the spring-mass system driven by a sinusoidal force $F_0 \cos \Omega t$ can be written as (Section 13.4)

$$m \frac{d^2x}{dt^2} + \sigma \frac{dx}{dt} + kx = F_0 \cos \Omega t$$

Comparison of these two equations shows that the analogs are

$$L \leftrightarrow m \qquad R \leftrightarrow \sigma \qquad C \leftrightarrow \frac{1}{k}$$

The spring-mass system exhibits resonance when the angular frequency of the driving force Ω equals $\sqrt{(k/m)}$, the resonant angular frequency. In the series *RLC* circuit, the analogous resonance occurs when the angular frequency ω of the ac generator equals $\sqrt{1/LC}$, the resonant angular frequency of the *LC* combination.

Mechanical resonance can have dramatic effects. The collapse of the Tacoma Narrows suspension bridge in 1940 was caused by a wind-driven resonance. We can use the series *RLC* circuit to illustrate the effects of electrical resonance.

Resonance in the series *RLC* circuit occurs when the impedance is a minimum. Equation 34.37 shows that Z is a minimum for $X_L = X_C$, which requires $\omega L = 1/\omega C$. Thus, resonance occurs when the angular frequency is

$$\omega_r = \sqrt{\frac{1}{LC}}$$

The resonant frequency f_r expressed in Hz is given by

$$f_r = \frac{1}{2\pi} \sqrt{\frac{1}{LC}} \tag{34.39}$$

Resonance has a marked effect on the power developed in the resistor. At resonance $Z = R$, and $\cos \varphi = 1$, so the phase angle is zero. This means that the current is in phase with the ac generator. The average power developed at resonance is

$$\langle P \rangle = \mathcal{E}_{\text{rms}} I_{\text{rms}}$$

Figure 34.16 is a graph of the average power developed in the resistor of a series *RLC* circuit. The mountain-like appearance of the graph inspires its name, the **resonance peak**. The increased power developed in the resistor at resonance occurs because the rms current is a maximum at resonance. This follows from the relation $I_{\text{rms}} = \mathcal{E}_{\text{rms}}/Z$, and the fact that Z is a minimum at resonance. A large current provides large energy storage in the inductor and capacitor.

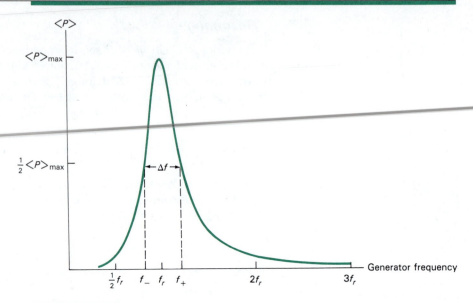

FIGURE 34.16

Average power delivered to the series *RLC* circuit is plotted versus the frequency of the ac generator. The power is a maximum at the resonant frequency f_r. At the frequencies labeled f_- and f_+, the power is one half of its maximum value. The frequency difference $f_+ - f_-$ is called the half-width of the resonance peak.

EXAMPLE 7

Resonance in a Radio Tuning Circuit

A radio is tuned to a particular station by adjusting a capacitor to achieve resonance at the broadcast frequency of the station. Let's determine the rms current produced at resonance when the antenna acts like an ac generator and provides an rms voltage of 0.05 V. We take $R = 150\ \Omega$, $L = 15$ mH, and $C = 3.45$ pF. The resonant frequency is

$$f_r = \frac{1}{2\pi}\sqrt{\frac{1}{LC}} = \frac{1}{2\pi}\sqrt{\frac{1}{(15 \times 10^{-3}\ \text{H} \cdot 3.45 \times 10^{-12}\ \text{F})}}$$
$$= 7.0 \times 10^5\ \text{Hz} = 700\ \text{kHz} \tag{6.28}$$

This is in the AM broadcast band. At resonance the impedance equals the resistance: $Z = R = 150\ \Omega$. Hence, the rms current is

$$I_{\text{rms}} = \frac{\mathcal{E}_{\text{rms}}}{Z} = \frac{0.05\ \text{V}}{150\ \Omega} = 333\ \mu\text{A}$$

For comparison we calculate the current developed when the same circuit is driven at 720 kHz by the signal from another station. We find

$$X_L = \omega L = 2\pi(7.20 \times 10^5\ \text{s}^{-1})(15 \times 10^{-3}\ \text{H}) = 6.79 \times 10^4\ \Omega$$

and

$$X_C = \frac{1}{\omega C} = \frac{1}{(2\pi \cdot 7.20 \times 10^5\ \text{s}^{-1} \cdot 3.45 \times 10^{-12}\ \text{F})} = 6.41 \times 10^4\ \Omega$$

The impedance at this frequency is

$$Z = \sqrt{(150\ \Omega)^2 + (67900\ \Omega - 64100\ \Omega)^2} = 3800\ \Omega$$

The rms current for this nonresonant frequency is

$$I_{rms} = \frac{0.05\ V}{3800\ \Omega} = 13.1\ \mu A$$

Changing the frequency by about 3% has led to a decrease in the rms current by a factor of about 25. This is a sharply-tuned resonant circuit, capable of discriminating against signals that differ even slightly from the resonant frequency.

Another dramatic effect of resonance is that large voltages can be established across the inductor and capacitor. In a mechanical system, resonance enables a weak driving force to produce large-amplitude oscillations. In the series *RLC* circuit, resonance enables a modest ac generator voltage to produce much greater voltages across the inductor and capacitor. For example, consider the voltage across the inductor,

$$-L\frac{dI}{dt} = \omega L I_m \sin(\omega t - \varphi)$$

The maximum rms value of this voltage is

$$\omega L I_{rms} = \omega L \frac{\mathscr{E}_{rms}}{Z}$$

At resonance, $Z = R$, and the rms voltage across the inductor is $\omega L/R$ times the rms generator voltage. With ω equal to its resonance value, $\sqrt{1/LC}$,

$$\frac{\omega L}{R} = \sqrt{\frac{L}{CR^2}}$$

For the *RLC* circuit described in Example 7, $\sqrt{L/CR^2}$ equals 440. Thus, the rms voltage across the inductor is 440 times the rms generator voltage.

This result may seem perplexing in view of Kirchhoff's voltage rule. How can any of the voltages across individual circuit elements exceed the voltage across the generator when Kirchhoff's voltage rule requires that their *sum* equal the generator voltage? The answer is that the voltages across the inductor and capacitor are $180°$ out of phase—they have opposite signs. At resonance their magnitudes are equal and their sum is zero,

$$L\frac{dI}{dt} + \frac{q}{C} = 0$$

Separately, each of these voltages can have much larger magnitudes than that of the ac generator that drives the circuit.

There is a mechanical analog of this situation: at resonance in the driven spring-mass system

$$m\frac{d^2x}{dt^2} + kx = 0$$

Separately, the spring term (kx) and the inertial term ($m\ d^2x/dt^2$) can have much larger magnitudes than the driving force.

The Q

The Q of a circuit is a dimensionless parameter that compares the maximum power delivered to reactive elements (inductor, capacitor) to the power developed in the resistive elements. We define the Q by

$$Q \equiv \frac{\text{maximum reactive power}}{\text{average thermal power}} \qquad (34.40)$$

For the reactive power we found the results

$$(P_L)_{\text{max}} = X_L I_{\text{rms}}^2$$

$$(P_C)_{\text{max}} = X_C I_{\text{rms}}^2$$

The average thermal power is

$$\langle P \rangle = R I_{\text{rms}}^2$$

Since they are equal at resonance, we can use either $(P_L)_{\text{max}}$ or $(P_C)_{\text{max}}$ for the maximum reactive power for a series RLC circuit near resonance. Choosing $X_L I_{\text{rms}}^2 = \omega L I_{\text{rms}}^2$ for the maximum reactive power gives

$$Q = \frac{\omega L}{R} \qquad (34.41)$$

A large Q means that the circuit stores energy efficiently. Only a small fraction ($\approx 1/Q$) of the energy stored leaks out each cycle as thermal energy. This leakage is restored continuously by the ac generator that thereby maintains the steady-state condition of the circuit.

Resonance Half-Width

The **half-width** of the average-power resonance peak (Figure 34.16) is *defined* as the difference between the two frequencies for which the average power is half of its resonance value.* We want to deduce a relation between the half-width and the Q.

The average power can be expressed as

$$\langle P \rangle = \frac{R \mathscr{E}_{\text{rms}}^2}{Z^2}$$

where

$$Z^2 = R^2 + \left(\omega L - \frac{1}{\omega C} \right)^2$$

At resonance, $Z^2 = R^2$, and the average power has its maximum value. At frequencies for which $Z^2 = 2R^2$, the average power is half its maximum value. The condition $Z^2 = 2R^2$ is satisfied if

$$\left(\omega L - \frac{1}{\omega C} \right) = \pm R$$

---

* The quantity we call the half-width answers to many different names, depending on the discipline. These names include the **bandwidth** and the **full width at half-maximum power**.

This condition gives a quadratic equation for ω whose positive roots are the two angular frequencies for which the average power is half its resonance maximum. The roots are

$$\omega_+ = \frac{1}{2}\frac{R}{L} + \sqrt{\left(\frac{1}{2}\frac{R}{L}\right)^2 + \frac{1}{LC}}$$

and

$$\omega_- = -\frac{1}{2}\frac{R}{L} + \sqrt{\left(\frac{1}{2}\frac{R}{L}\right)^2 + \frac{1}{LC}}$$

The half-width of the resonance peak is

$$f_+ - f_- = \frac{(\omega_+ - \omega_-)}{2\pi} = \frac{R}{2\pi L} \qquad (34.42)$$

The *ratio* of the half-width to the resonant frequency is a dimensionless measure of the width of the resonance peak,

$$\frac{(f_+ - f_-)}{f_r} = \frac{R}{2\pi f_r L} = \frac{R}{\omega_r L}$$

Because Q equals $\omega_r L/R$, we have

$$\frac{(f_+ - f_-)}{f_r} = \frac{1}{Q} \qquad (34.43)$$

This relation shows that a large Q corresponds to a sharply-defined resonance.

EXAMPLE 8

Q and Half-Width of a Tuning Circuit

Let's illustrate Equation 34.43 by calculating the Q and the half-width for the tuning circuit of Example 7. The Q at resonance is

$$Q = \frac{\omega_r L}{R} = \frac{2\pi(7.0 \times 10^5 \text{ s}^{-1})(15 \times 10^{-3} \text{ H})}{150 \ \Omega} = 440$$

This large Q implies a sharply-defined resonance. The half-width follows from Equation 34.43,

$$f_+ - f_- = \frac{f_r}{Q} = \frac{700 \times 10^3 \text{ Hz}}{440} = 1.59 \text{ kHz}$$

In Example 7 we found a significant decrease in the rms current when the frequency changed from the resonant frequency of 700 kHz to the nonresonant frequency of 720 kHz. This is consistent with the fact that the frequency change of 20 kHz is many times the half-width of 1.59 kHz.

34.6
IMPEDANCE MATCHING

There are many situations where a source delivers power to a load. An efficient transfer of power may require a means of matching the source and load. For example, a 10-speed bicycle provides different gear ratios to match the power output of the rider and the load presented by different road conditions. An automobile transmission performs a similar matching function. The efficient transfer of electrical power from an ac generator to a load requires an **impedance matching**. Let's see exactly what is meant by this terminology.

Figure 34.17a shows an idealized representation of an ac generator. It is characterized by an rms voltage $\mathscr{E}_G$ and an **internal impedance** Z_G. The impedance of the generator is characterized by a resistance R_G and a reactance X_G. Consider the parameters $\mathscr{E}_G$, R_G, and X_G to be fixed; they cannot be changed to improve power transfer. Figure 34.17b shows an idealized generator-load circuit; the load is characterized by a resistance R and a reactance X. It is useful to think of R and X as variables that can be adjusted to achieve an efficient power transfer to the load. The load and generator impedances are in series. The combined resistance is $R + R_G$; the combined reactance is $X + X_G$. The rms current delivered to the load by the generator is

$$I_{rms} = \frac{\mathscr{E}_G}{Z}$$

where

$$Z^2 = (R + R_G)^2 + (X + X_G)^2 \tag{34.44}$$

The average power delivered to the load is

$$\langle P \rangle = RI_{rms}{}^2 = \frac{R\mathscr{E}_G{}^2}{(R + R_G)^2 + (X + X_G)^2} \tag{34.45}$$

One condition that will help to maximize P is evident from the sum of squares $(R + R_G)^2 + (X + X_G)^2$ in the denominator. Resistances are always positive, but the series reactance can be positive, negative, or zero. This feature of reactance arises because the voltages across a capacitor and an inductor in series are $180°$ out of phase, which means they have opposite signs. As a consequence their total reactance is the *difference* of their separate reactances, $X_L - X_C$. This difference can be positive, negative, or zero. The condition

$$X = -X_G \tag{34.46}$$

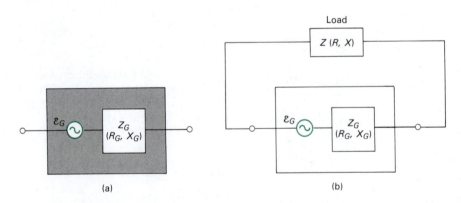

(a)   (b)

FIGURE 34.17

(a) The ac generator has an rms voltage $\mathscr{E}_G$ across its terminals and an internal impedance described by the resistance R_G and reactance X_G. (b) The load connected to the ac generator has an impedance that is characterized by a resistance R and a reactance X.

eliminates the term $(X + X_G)^2$ from the denominator of $\langle P \rangle$ and gives

$$\langle P \rangle = \frac{R\mathscr{E}_G{}^2}{(R + R_G)^2} \tag{34.47}$$

Matching reactances $(X = -X_G)$ is similar to achieving the resonance condition $(\omega L = 1/\omega C)$ in a series RLC circuit; in both instances the total reactance is zero, and the current is in phase with the generator voltage.

The value of R that maximizes $\langle P \rangle$ (as given by Equation 34.47) is determined by setting $d\langle P \rangle/dR$ equal to zero. This leads to an equation showing that $\langle P \rangle$ is a maximum for

$$R = R_G \tag{34.48}$$

Equations 34.46 and 34.48 describe an **impedance match** between the generator and load: *equal resistances* and *equal but opposite reactances*.

What can be done in a situation where the generator and load impedances do not match? The following example illustrates how power transfer can be improved by inserting an impedance-matching element.

EXAMPLE 9

Matching Impedances

An ac generator maintains an rms voltage of 120 volts at 60 Hz. The generator resistance is 100 Ω. The generator reactance is *inductive* with a value of 60 Ω. The load is an inductance with a resistance of 40 Ω and an inductive reactance of 100 Ω. We want to change the load so as to achieve an impedance match.

We can insert in series with the load a circuit element that gives an impedance match. The following table shows that the impedance-matching element should have a resistance of 60 Ω and a reactance of -160 Ω.

	Generator		Load		Matching Element
R	100 Ω	$=$	40 Ω	$+$	(60 Ω)
X	60 Ω	$=$	100 Ω	$+$	(-160 Ω)

One way of achieving this match would be to use a 60-Ω resistor in series with a capacitor for which

$$-\frac{1}{\omega C} = -160\ \Omega$$

For a 60-Hz generator ($\omega = 377$ s^{-1}) the required capacitance is

$$C = \frac{1}{(377\ \text{s}^{-1} \cdot 160\ \Omega)} = 16.6\ \mu\text{F}$$

With the impedance-matching element in the circuit, the power delivered to the load is 36 W. Without the impedance-matching element, the power delivered is 27.2 W.

34.7
TRANSFORMERS

A power company would like to deliver to its customers all of the electrical energy that is produced at the plant. Unfortunately there are energy losses between the generators and the consumers. Foremost is the energy converted

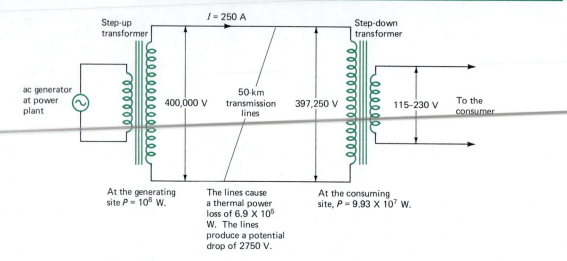

FIGURE 34.18
Transformers are employed at both ends of electric power transmission lines.

to heat in the transmission lines because of their electrical resistance. To minimize thermal energy losses, the transmission line current and resistance are kept as low as possible. The resistance of the copper or aluminum transmission lines can be lowered by increasing their cross-sectional area. But increasing the cross-sectional area increases the weight and cost of the lines, and eventually sets a limit.

The I^2R power losses in a transmission line can be lowered by reducing the current. In order to minimize the current in the transmission line and still transmit the desired amount of electrical energy, power companies use **transformers**. A transformer transfers electric energy from one circuit to another through magnetic flux that links the circuits.

An electric power system consists of a transformer to step up the emf at the generating site, a transformer to step down the emf at the consumer end, and a transmission line connecting the two facilities (Figure 34.18). At a power plant, voltages are made large, typically 400 kV, and currents are made small to reduce I^2R losses in power transmission. Then at the consumer end of the transmission lines, the voltages are lowered to 230 V for safe, practical use. It is the ease with which ac voltages can be changed by transformers that makes ac power particularly useful.

EXAMPLE 10

The Cost of Transmitting Electric Power

The I^2R power losses in the transmission lines bringing electricity from an electric power plant to a community are very costly. Let's see why.

At the generating site, the rms voltage across the secondary of the step-up transformer is typically 400,000 volts. A typical level of average-power transmission is 100 MW. Assuming no losses, the rms current in the secondary coil is

$$I_{rms} = \frac{\langle P \rangle}{\mathscr{E}_{rms}} = \frac{10^8 \text{ W}}{4 \times 10^5 \text{ V}} = 250 \text{ A}$$

Suppose the total length of two wires connecting a generating plant to the community is 50 km and that the wires are made of solid copper 1 cm in diameter. Using the resistivity for copper given in Table 27.1 we obtain a resistance of 11 ohms for the lines. For a resistance of 11 ohms, the power produced in the transmission line is

$$RI_{rms}^2 = 11\ \Omega(250\ \text{A})^2 = 0.69\ \text{MW}$$

At 10 cents per kWh this is an hourly cost of $69. Multiply this by 24 and then again by 365 and you will appreciate the expense of transmitting electric power.

The symbol for a transformer is $\underline{\overline{}}$. A diagram of the conventional arrangement for a transformer is shown in Figure 34.19. In a transformer, a changing flux in one coil, called the **primary**, induces an emf in a second coil, called the **secondary**. The changing flux in the primary winding is produced by an ac generator connected to its leads. The transformer core enhances the flux produced by the current in the primary and guides this flux through the secondary. Ideally, all of the flux produced by the primary threads the secondary. Such a condition is realized approximately if the core is made of a magnetic material such as soft iron that can steer magnetic flux and minimize hysteresis and eddy current losses.

The primary is an RL circuit whose resistance is that of the primary coil. In practice the IR voltage across the primary resistance is negligible compared to the emf induced by the flux change in the primary coil. For a primary coil having N_P turns and a magnetic flux Φ_B threading each turn, the induced emf is $N_P\ d\Phi_B/dt$. The ac generator that drives the primary circuit is represented by $\mathscr{E}_P \cos \omega t$. Ignoring the IR voltage across the resistor, Kirchhoff's voltage rule for the primary gives

$$\mathscr{E}_P \cos \omega t = N_P \frac{d\Phi_B}{dt} \tag{34.49}$$

We assume initially that no load is connected to the secondary. Thus, the secondary is an open circuit in which the current is zero (Figure 34.20a). However, in accord with Faraday's law, a changing magnetic flux in the secondary will induce an emf. This emf is called the open-circuit voltage.

We assume the same flux threads each turn of the primary and secondary coils. For a secondary coil having N_S turns, the secondary emf is $N_S\ d\Phi_B/dt$.

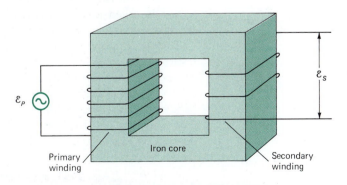

FIGURE 34.19

The current in the primary coil establishes a magnetic flux in the transformer core. The transformer core guides this flux through the secondary coil. Flux changes in the primary induce currents in the secondary, and result in an energy transfer from the primary to the secondary.

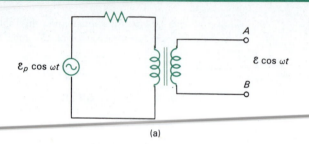

(a)

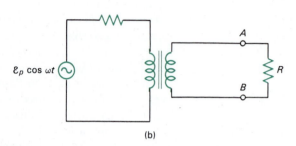

(b)

FIGURE 34.20

(a) The secondary circuit is open. There is no current in the secondary coil, but there is an induced voltage between the terminals A and B. (b) With a load connected between the terminals there is a current in the secondary. This current is in phase with the voltage of the ac generator in the primary. This allows an efficient transfer of power from the primary to the secondary.

Using Equation 34.49 to eliminate $d\Phi_B/dt$ shows that the secondary emf can be expressed as $\mathscr{E}_S \cos \omega t$, where

$$\mathscr{E}_S = \left(\frac{N_S}{N_P}\right)\mathscr{E}_P \qquad (34.50)$$

Note that the secondary emf is in phase with the ac generator that drives the primary circuit.

The ratio N_S/N_P is called the **turns ratio** of the transformer. If the turns ratio is greater than unity, then the induced emf in the secondary is greater than the voltage of the ac generator in the primary circuit. Such a transformer is called a **step-up transformer**. If the turns ratio is less than unity, the secondary emf is less than the voltage of the primary generator. Such a transformer is called a **step-down transformer**. Thus, the induced emf in the secondary coil may be raised or lowered by choosing the proper turns ratio.

To transfer power from the primary to the secondary we connect a load resistor R to complete the secondary circuit (Figure 34.20b). The potential drop across the load resistor equals the induced emf in the secondary coil. The current in the secondary and the voltage across the ac generator are in phase because both are proportional to $d\Phi_B/dt$, the time rate of change of magnetic flux. This is a key to transformer action, permitting an efficient transfer of power from the ac generator that drives the primary to the load connected to the secondary.

In an ideal transformer, the power delivered to the load equals the power developed by the primary generator. In practice, thermal energy losses prevent a complete transfer of power. Assuming ideal behavior, the equality of primary and secondary powers can be expressed by

$$\mathscr{E}_P I_P = \mathscr{E}_S I_S \qquad (34.51)$$

where I_P and I_S denote the rms currents in the primary and secondary. In practice, the efficiency of power transfer may approach 99%, in which case Equation 34.51 is an excellent approximation.

If a step-up transformer is used to increase the potential difference across the load in the secondary circuit, then Equation 34.51 shows that the current in the secondary is smaller than the primary current. This is the scheme used to reduce the current in electric power transmission lines, which in turn lowers I^2R losses during transmission.

Effective Resistance of a Transformer-Load Combination

We noted that the current in the load resistance is in phase with the ac generator driving the primary. This is the same condition that existed in the circuit we studied in Section 34.2, consisting of an ac generator in series with a resistor. The transformer along with its load resistor appears as a resistor to the ac generator. We can use Equations 34.50 and 34.51 to derive an expression for the effective resistance of the transformer-plus-load-resistor combination.

The average power developed in a load resistor R connected across the secondary of a transformer is

$$\langle P \rangle = R I_S^2$$

where I_S is the rms current in the secondary. Using Equations 34.51 and 34.50 we can express the load power in terms of the rms current in the primary I_P,

$$\langle P \rangle = R I_S^2 = R \left(\frac{\mathscr{E}_P I_P}{\mathscr{E}_S} \right)^2 = R \left(\frac{N_P}{N_S} \right)^2 I_P^2 \qquad (34.52)$$

If an ac generator delivers an rms current I_P to a resistor, it develops a power equal to I_P^2 times the resistance of the resistor. Equation 34.52 shows that the combination of the transformer and load resistor present an *effective* resistance equal to $(N_P/N_S)^2$ times the actual load resistance,

$$R_{\text{eff}} = R \left(\frac{N_P}{N_S} \right)^2 \qquad (34.53)$$

Impedance Matching With a Transformer

We showed in Section 34.6 that the most efficient transfer of power from an ac generator to a load occurs when there is an impedance match. For an optimum impedance match, the combined reactance of the generator and load is zero. This causes the load current to be in phase with the generator voltage, and so the load resistance equals the internal resistance of the generator.

Using the fact that the effective resistance of a transformer and load resistor combination equals $(N_P/N_S)^2$ times the actual load resistance, we can match impedances between an ac generator and a load resistor by using a transformer with the proper turns ratio, N_P/N_S. The transformer makes it possible to increase the secondary current and thereby increase the power delivered to the load resistor. An impedance match is achieved when the internal resistance of the generator R_G equals the effective resistance of the transformer-load resistor combination,

$$R_G = R \left(\frac{N_P}{N_S} \right)^2 \qquad (34.54)$$

With this impedance match, a maximum of half the power developed by the ac generator can be delivered to the load.

WORKED PROBLEM

A 60-watt light bulb is designed to operate at 120 volts (rms). It is to be used with a 240-volt (rms), 60-Hz ac generator. By connecting an inductor in series with the bulb, the rms voltage across the bulb can be reduced to 120 volts. Determine the inductance required. Neglect the resistance of the inductor.

Solution

The bulb alone acts as a resistor. As shown in Example 1, the resistance of the bulb is given by

$$R = \frac{V_{rms}^2}{\langle P \rangle} = \frac{(120 \text{ V})^2}{60 \text{ W}} = 240 \text{ } \Omega$$

The bulb and inductor constitute a series RL combination with an impedance

$$Z = \sqrt{R^2 + (\omega L)^2}$$

When connected to an ac generator delivering an rms voltage $\mathscr{E}_{rms}$, the rms current is

$$I_{rms} = \frac{\mathscr{E}_{rms}}{Z} = \frac{\mathscr{E}_{rms}}{\sqrt{R^2 + (\omega L)^2}}$$

The rms voltage across the bulb is $I_{rms}R$. We want the rms voltage to be 120 volts. Hence,

$$120 \text{ V} = \frac{\mathscr{E}_{rms}R}{\sqrt{R^2 + (\omega L)^2}}$$

With $\mathscr{E}_{rms} = 240$ V, $R = 240$ Ω, and $\omega = 377$ s^{-1} ($f = 60$ Hz), we get

$$L = \frac{\sqrt{3}R}{\omega} = \frac{1.73(240 \text{ } \Omega)}{377 \text{ s}^{-1}}$$

$$= 1.10 \text{ H}$$

EXERCISES

34.1   AC Generators

A.   A 10-turn coil having an average cross-sectional area of 100 cm^2 is rotated 20 rev/s in a uniform magnetic field of 0.5 T. Calculate the maximum voltage induced across the coil.

B.   Consider the RLC parallel circuit shown in Figure 1. Show that the three currents can be expressed as

$$I_R = \left(\frac{\mathscr{E}_0}{R}\right) \cos \omega t$$

$$I_C = -(\mathscr{E}_0 \omega C) \sin \omega t$$

$$I_L = \left(\frac{\mathscr{E}_0}{\omega L}\right) \sin \omega t$$

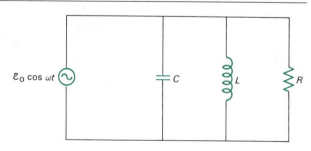

FIGURE 1

34.2   RMS Values

C.   Determine the average and rms current in a 6-kW clothes dryer connected to a 230-V (rms), 60-Hz source.

D. An electric light bulb dissipates 100 W when con-
    nected to an ac generator producing a voltage $V =$
    170 sin ωt volts. (a) What is the light bulb's resistance?
    (b) What is the maximum current in the light bulb?

34.3   The *RC* Circuit

E. The current in a 2.5-μF capacitor connected to an ac
    generator is represented by

$$I = -4.71 \sin{(377t)} \; \mu A$$

Determine the maximum voltage across the capacitor.

F. Determine the maximum and rms currents in a 15-pF
    capacitor connected to a 10-V (rms), 50-MHz
    generator.
G. Figure 2 shows a plot of the current in a 1.2-nF ca-
    pacitor connected to an ac generator. Determine the
    frequency and peak voltage of the generator.

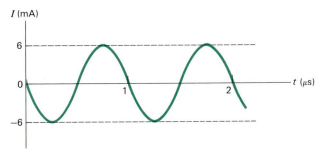

FIGURE 2

H. A 200-Ω resistor, in series with a 0.55-μF capacitor,
    is in series with an ac generator with a peak voltage
    of 170 volts. (a) For what frequency is the capacitive
    reactance equal to the resistance? (b) What is the peak
    current at the frequency of part (a)?
I. A 2.5-V (rms), 100-Hz generator is connected in series
    with a 100-nF capacitor and a 2500-Ω resistor. (a)
    Determine the impedance of the circuit, (b) the rms
    current in the circuit.
J. Show that for sufficiently large frequencies the cur-
    rent in the *RC* circuit can be represented by
    $I = (\mathscr{E}_0/Z) \cos \omega t$.
K. In an *RC* circuit having a 10.5-Ω resistor and a 1.5-μF
    capacitor, at what frequency will the impedance equal
    50 Ω?
L. Determine the rms voltage across a 15-Ω resistor in
    series with a 15-μF capacitor and a 10-V (rms), 1-kHz
    generator.
M. A capacitor has a capacitive reactance of 300 Ω at
    60 Hz. It is connected in series with a 100-Ω resistor
    and a 120-V (rms), 60-Hz generator. Determine the
    average power developed in the resistor.

34.4   The *RL* Circuit

N. Figure 3 shows a plot of the current in an ideal

inductor ($R = 0$) connected to an ac generator that
produces a peak voltage of 1 V. Determine (a) the
frequency of the generator, (b) the inductance of the
inductor.

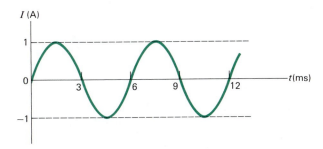

FIGURE 3

P. Calculate the maximum and rms currents in a 22-μH
    inductor connected across a 5-V (rms), 75-MHz
    generator.
Q. An *RL* series combination consisting of a 1.5-Ω resistor
    and a 2.5-mH inductor is connected to a 12.5-V (rms),
    400-Hz generator. Determine (a) the impedance of the
    circuit, (b) the rms current, (c) the rms voltage across
    the resistor, and (d) the rms voltage across the inductor.
R. As a way of determining the inductance of a coil used
    in a research project, a student first connects the coil
    to a 12-V battery and measures a current of 0.63 A.
    The student then connects the coil to a 24-V (rms),
    60-Hz generator and measures an rms current of
    0.57 A. What value does the student calculate for the
    inductance?

34.5   The *RLC* Circuit

S. Compute the rms current in a series *RLC* circuit having
    $R = 10 \; \Omega$, $C = 10 \; \mu F$, and $L = 1.5$ mH when con-
    nected to a 25-V (rms), 100-Hz generator.
T. A series *RL* combination and a series *RC* combination
    have equal impedances of 47 Ω at a frequency of 2 kHz.
    The resistances are equal and $C = 3.3 \; \mu F$. Determine
    the resistance and the inductance.
U. A series *RLC* circuit with $R = 1500 \; \Omega$ and $C = 15$ nF
    is connected to an ac generator whose frequency can
    be varied. When the frequency is adjusted to 50.5 kHz,
    the rms current in the circuit reaches a maximum at
    0.14 A. Determine (a) the inductance, (b) the rms value
    of the generator voltage.
V. A series *RLC* circuit has $L = 1.2$ mH and $C = 1.6$ nF.
    It is desired to change the resonant frequency by
    changing the capacitance. What percentage change in
    the capacitance is required to increase the resonant
    frequency by 5%?
W. An FM radio frequency of 92.1 MHz produces a res-
    onance in a series *RLC* circuit having $R = 2.5 \; \Omega$,

$L = 1.5\ \mu H$, and $C = 1.99$ pF. If the rms voltage of 2.5 mV is provided to the circuit by the antenna, determine the rms current at resonance.

34.6   Impedance Matching

X. A 12-V (rms), 60-Hz generator has an internal resistance of 600 Ω and an internal capacitive reactance of $-6\ \Omega$. The load resistance is 400 Ω. (a) How can you alter the load to maximize the power developed in the load resistor without changing the load resistance? (b) What is the maximum power developed for your solution?

Y. Cite two or more everyday examples of impedance matching, not involving electrical phenomena.

34.7   Transformers

Z. In Europe, households and businesses are provided with 240-V, 50-Hz current, rather than 120-V, 60-Hz current as in the United States. What type of transformer would allow you to use an American appliance in Europe? (The frequency is not crucial for this application.)

PROBLEMS

1. A type of rectifier called a bridge rectifier can transform a sinusoidal voltage to that shown in Figure 4. (a) Show that the average rectified voltage is

$$\langle V \rangle = \frac{2\mathscr{E}_0}{\pi} = 0.637\mathscr{E}_0$$

(b) Show that the rms value of the rectified voltage is

$$V_{\text{rms}} = \frac{\mathscr{E}_0}{\sqrt{2}} = 0.707\mathscr{E}_0$$

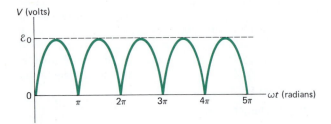

FIGURE 4

2. A series RLC circuit has $R = 10\ \Omega$, $L = 2$ mH, and $C = 4\ \mu F$. (a) Determine the impedance at a frequency of 60 Hz. (b) Determine the resonant frequency in Hz. (c) Determine the impedance at resonance. (d) Determine the impedance at a frequency of one-half the resonant frequency.

3. A series RLC circuit having $R = 20\ \Omega$, $L = 0.10$ mH, and $C = 10$ nF is connected to a 100-V (rms) ac generator whose frequency can be varied. Determine the thermal power developed in the resistor at (a) the resonant frequency and at (b) nine-tenths of the resonant frequency.

4. Calculate the half-width (in Hz) of the resonance peak of a series RLC circuit that has $R = 100\ \Omega$, $L = 20\ \mu H$, and $C = 0.1$ pF.

5. A series RLC circuit in which $R = 1\ \Omega$, $L = 1$ mH, and $C = 1$ nF, is connected to an ac generator

delivering 1 V (rms). Make a careful plot of the power delivered to the circuit as a function of the frequency and verify that the half-width of the resonance peak is $R/2\pi L$.

6. A series RLC circuit is to have a resonant frequency of 1 MHz and a resonance half-width of 10 kHz. Determine R, L, and C if the maximum average power developed is 10 W for a generator delivering 1 V (rms).

7. Show that the angular frequency for which the voltage across the inductor in a series RLC circuit is a maximum is given by

$$\omega = \sqrt{\frac{1}{LC}\left(1 - \frac{R^2 C}{2L}\right)^{-1/2}}$$

8. A plot of average-power-versus-angular-frequency for an RLC series circuit is shown in Figure 5. The parameters used were $R = 10\ \Omega$, $L = 0.10$ H, and $\mathscr{E}_{\text{rms}} = 100$ V. (a) Using the graph as a source of information, determine the capacitance. (b) Determine the half-width of the resonance peak and show that your result is consistent with $f_+ - f_- = R/2\pi L$.

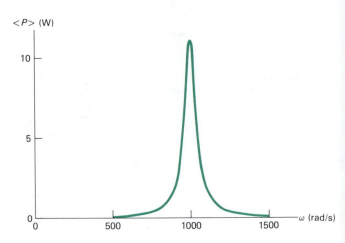

FIGURE 5

9. The generator current in the parallel RLC circuit (Exercise B) can be written

$$I = \mathcal{E}_0 \sqrt{\left(\frac{1}{R}\right)^2 + \left(\omega C - \frac{1}{\omega L}\right)^2} \cos \omega t \equiv I_0 \cos \omega t$$

Show that I_0 is a minimum when $\omega^2 = 1/LC$.

10. (a) Calculate the rms value of a sawtooth voltage that rises linearly from 0 to 1 volt in a time of 1 μs and falls linearly back to zero in 1 μs. (b) Is your result larger or smaller than the rms value of a sinusoidal voltage that reaches a peak value of 1 volt and has a period of 4 μs?

11. Show that the potential difference between the points 1 and 2 in the circuit of Figure 6 is zero provided $L/R_2 = R_1C$.

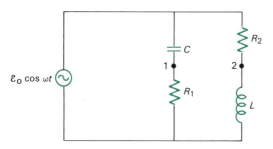

FIGURE 6

12. A series RLC circuit is driven by a 12-volt, 200-Hz ac generator. The circuit parameters are as follows: $R = 5 \ \Omega$, $L = 0.250$ H, and $C = 3 \ \mu$F. (a) Determine the maximum reactive power delivered to the inductor. (b) Determine the average power developed in the resistor.

13. A 4-nF capacitor is in series with a 5-mH coil having a resistance of 15 ohms. The RLC combination is driven by an ac generator. (a) Determine the resonant frequency. (b) When the generator drives the circuit at its resonant frequency, the maximum voltage across the capacitor is 20 volts. Determine the rms voltage of the ac generator.

14. An amplifier with an internal resistance of 600 Ω acts as an ac generator, driving a load having a resistance of 8 Ω. The average power developed by the load resistor is 1.12 W. A transformer is to be used to match the impedances of the generator and load. (a) What transformer turns ratio should be used to give the best impedance match? Assume that the combined reactance of the generator and load is zero and (b) determine the average power developed in the resistor for an optimum impedance match.

CHAPTER
35

MAXWELL'S EQUATIONS AND ELECTROMAGNETIC WAVES

35.1
MAXWELL'S EQUATIONS

In 1864, James Clerk Maxwell brought together the four basic laws governing electric and magnetic fields. He modified one of them and thereby achieved a unified theory of electric and magnetic phenomena. The resulting set of four basic laws are now known as **Maxwell's equations**.

Maxwell found that these four equations have wavelike solutions and showed that the speed of electromagnetic waves equals the speed of light. This led him to conclude that light is an electromagnetic wave. Thus, Maxwell's research not only united electricity and magnetism, it also laid a theoretical foundation for many optical phenomena.

Maxwell's theory seemed esoteric at the time of its publication in 1864. But times change. Today we can look back more than 100 years at a growing parade of scientific and technological developments that are outgrowths of Maxwell's theory. Radio, television, and fiber optics communications are all based on Maxwell's theory.

In this section we review the four basic equations of electricity and magnetism and show how Maxwell modified Ampère's law by adding what is called the **displacement current.**

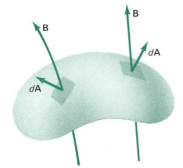

FIGURE 35.1

Magnetic Flux

The magnetic flux Φ_B is defined as the surface integral of the magnetic field (Section 29.2),

$$\Phi_B = \int \mathbf{B} \cdot d\mathbf{A} \tag{35.1}$$

Φ_B can be interpreted as the net number of magnetic field lines threading the surface (Figure 35.1).

Two of the basic laws of electromagnetism involve magnetic flux. The first of these states that the net magnetic flux through any *closed* surface is zero:

$$\Phi_B = \int_{\substack{\text{closed} \\ \text{surface}}} \mathbf{B} \cdot d\mathbf{A} = 0 \tag{35.2}$$

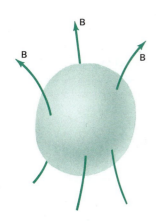

FIGURE 35.2

Geometrically, $\Phi_B = 0$ means that equal numbers of magnetic field lines enter and leave the volume enclosed by the surface (Figure 35.2). Physically, $\Phi_B = 0$ is a consequence of the fact that there are no isolated magnetic poles on which magnetic field lines originate or terminate.

Faraday's Law of Induction

The second basic law involving magnetic flux is Faraday's law of induction (Section 31.2),

$$\mathscr{E} = \oint \mathbf{E} \cdot d\mathbf{s} = -\frac{d}{dt} \int \mathbf{B} \cdot d\mathbf{A} = -\frac{d\Phi_B}{dt} \tag{35.3}$$

The line integral of the electric field extends around the periphery of the surface through which the flux of the magnetic field is evaluated (Figure 35.3). Faraday's law describes how a changing magnetic flux induces an electric field. Faraday's law is of special significance because it describes a coupling of **E** and **B** and recognizes that this coupling requires a time variation of the flux. Only when Φ_B changes is there an induced electric field.

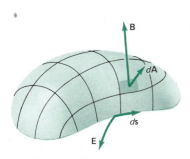

FIGURE 35.3

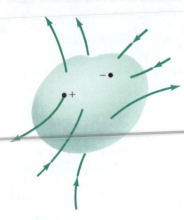

Electric Flux, Gauss' Law

The third basic law describing the fields is Gauss' law (Section 24.4), which relates the electric flux (Φ_E) through a closed surface to the net charge enclosed by that surface:

$$\Phi_E = \int_{\substack{\text{closed} \\ \text{surface}}} \mathbf{E} \cdot d\mathbf{A} = \frac{q}{\epsilon_0} \tag{35.4}$$

Gauss' law is a quantitative expression of the idea that electric field lines originate on positive charges and terminate on negative charges. If a surface encloses a net-positive charge, more electric field lines leave the surface than enter it ($\Phi_E > 0$). If the surface encloses no net charge, equal numbers of electric field lines enter and leave and $\Phi_E = 0$ (Figure 35.4).

FIGURE 35.4

Ampère's Law

The fourth basic law describing electromagnetic phenomena is Ampère's law (Section 29.3), which recognizes that a current sets up a magnetic field. In equation form, Ampère's law is

$$\oint \mathbf{B} \cdot d\mathbf{s} = \mu_0 I \tag{35.5}$$

The line integral of **B** is around a closed path, and I is the net current encircled by that path (Figure 35.5).

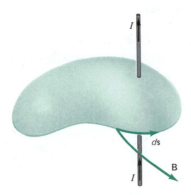

As formulated in Equation 35.5, Ampère's law is adequate only for phenomena involving *steady* currents. We can show that Ampère's law is incomplete when the current is not steady by considering the region near a parallel-plate capacitor while the capacitor is charging (Figure 35.6). Charge flows onto one plate and charge flows off the other plate. A current I enters one plate and a current I leaves the other plate. There is no current between the plates. That is, no charge flows across the space separating the plates.

FIGURE 35.5

However, experiment shows that a magnetic field exists in the region between the plates as well as on either side of the plates. To a good approximation, the magnetic field lines are circles, concentric with the current, as shown in Figure 35.7. Experiment shows that the value of the line integral $\oint \mathbf{B} \cdot d\mathbf{s}$ is the same around the loops labeled 1, 2, and 3 in the figure. The loops

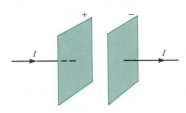

FIGURE 35.6

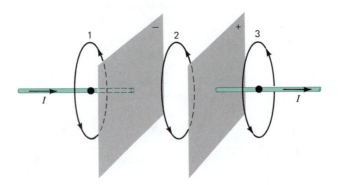

FIGURE 35.7
Experiment shows that $\oint \mathbf{B} \cdot d\mathbf{s}$ is the same for the loops labeled 1, 2, and 3. Loops 1 and 3 encircle a current I, and Equation 35.5 is satisfied. Loop 2 does not encircle any conventional current. The fact that $\oint \mathbf{B} \cdot d\mathbf{s}$ is not zero for loop 2 shows that Equation 35.5 is incomplete.

labeled 1 and 3 encircle a current I and Equation 35.5 is satisfied. According to Equation 35.5, $\oint \mathbf{B} \cdot d\mathbf{s}$ should vanish for loop 2, because no conventional current is encircled. Evidently, Ampère's law, as expressed by Equation 35.5, is incomplete.

Maxwell's Displacement Current

Maxwell remedied the shortcoming of Equation 35.5 by adding what is called the **displacement current**. The displacement current is related to the **changing electric field** between the plates of the capacitor. To establish this relation for ourselves, we apply Gauss' law to the closed surface shown in Figure 35.8. The surface encloses a net charge q, the charge on one plate of the capacitor. This charge is the source of the electric field between the plates. The charge q is related to the electric flux through the surface by Gauss' law (Equation 35.4):

$$\frac{q}{\epsilon_0} = \int \mathbf{E} \cdot d\mathbf{A} = \Phi_E$$

The charge q changes with time because the capacitor is charging. As q changes the electric field also changes. The rate at which q changes is

$$\frac{dq}{dt} = \epsilon_0 \frac{d}{dt} \int \mathbf{E} \cdot d\mathbf{A} \qquad (35.6)$$

and equals the current in the wire leading to the capacitor plate. Maxwell realized that the quantity

$$\epsilon_0 \frac{d}{dt} \int \mathbf{E} \cdot d\mathbf{A} \equiv I_d \qquad (35.7)$$

that he called a displacement current was equivalent to the conventional current in the wires on either side of the capacitor plates. The displacement current is equivalent to a conventional current because it produces the same magnetic field. Maxwell modified Ampère's law, Equation 35.5, by adding a

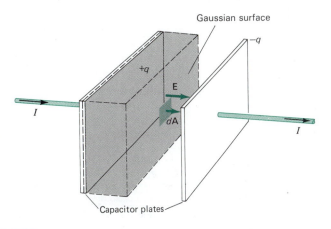

FIGURE 35.8

The shaded Gaussian surface encloses a charge q, which gives rise to the electric field $\mathbf{E}$ between the plates. Both q and $\mathbf{E}$ change with time as the capacitor charges and discharges.

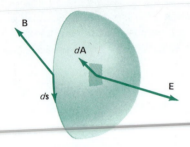

FIGURE 35.9

displacement current to the conventional current. That is, in Equation 35.5 he replaced I by $I + I_d$. The modified version of Ampère's law is

$$\oint \mathbf{B} \cdot d\mathbf{s} = \mu_0 I + \mu_0 \epsilon_0 \frac{d}{dt} \int \mathbf{E} \cdot d\mathbf{A} \qquad (35.8)$$

The line integral $\oint \mathbf{B} \cdot d\mathbf{s}$ is evaluated around the periphery of the surface over which the electric flux integral $\int \mathbf{E} \cdot d\mathbf{A}$ is evaluated (Figure 35.9). In the region between the capacitor plates there is no conventional current, only a displacement current, and Equation 35.5 reduces to

$$\oint \mathbf{B} \cdot d\mathbf{s} = \mu_0 \epsilon_0 \frac{d}{dt} \int \mathbf{E} \cdot d\mathbf{A} \qquad (35.9)$$

This special form of Ampère's law describes the fact that a changing electric flux induces a magnetic field. Recall that Faraday's law,

$$\oint \mathbf{E} \cdot d\mathbf{s} = -\frac{d}{dt} \int \mathbf{B} \cdot d\mathbf{A} \qquad (35.3)$$

describes how a changing magnetic flux induces an electric field. Collectively, Equations 35.3 and 35.9 emphasize the coupled or interlocking nature of time-varying electric and magnetic fields. It is because of this coupling that we speak of the **electromagnetic field**.

Maxwell's genius enabled him to discover the underlying unity of electric and magnetic phenomena. To honor his contribution, the set of four basic electromagnetic equations are now called **Maxwell's equations**:

$$\int_{\substack{\text{closed} \\ \text{surface}}} \mathbf{B} \cdot d\mathbf{A} = 0 \qquad (35.2)$$

$$\int_{\substack{\text{closed} \\ \text{surface}}} \mathbf{E} \cdot d\mathbf{A} = \frac{q}{\epsilon_0} \qquad \text{(Gauss)} \qquad (35.4)$$

$$\oint \mathbf{E} \cdot d\mathbf{s} = -\frac{d}{dt} \int \mathbf{B} \cdot d\mathbf{A} \qquad \text{(Faraday)} \qquad (35.3)$$

$$\oint \mathbf{B} \cdot d\mathbf{s} = \mu_0 I + \mu_0 \epsilon_0 \frac{d}{dt} \int \mathbf{E} \cdot d\mathbf{A} \qquad \text{(Ampére)} \qquad (35.8)$$

We have written Maxwell's equations in integral form; they can also be expressed in differential form. We will find it useful to use the differential forms of Faraday's law and Ampère's law in our treatment of electromagnetic waves.

35.2
ELECTROMAGNETIC WAVES

In this section we draw together ideas developed earlier to create a picture of electromagnetic waves. After we develop a qualitative picture of electromagnetic waves we will show that Maxwell's equations have wavelike

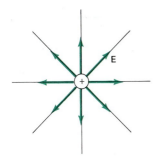

FIGURE 35.10

Electric field lines of a positive charge at rest.

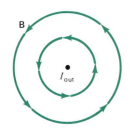

FIGURE 35.11

Magnetic field lines surrounding a steady current *I* directed out of the page.

solutions and show that electromagnetic waves travel at the speed of light. The implication is that light is an electromagnetic wave.

To begin, we recall the electric field set up by an isolated positive charge at rest. The electric field lines are directed radially away from a positive charge (Figure 35.10). So long as the charge is at rest it does not set up a magnetic field. If the charge moves, however, it constitutes an electric current and produces a magnetic field as well as an electric field. You may recall that the direction of the magnetic field is related to the direction of the current by the right-hand rule. In a plane perpendicular to a steady current the magnetic field lines are circles, concentric with the current (Figure 35.11).

To establish a rudimentary picture of an electromagnetic wave, we envision a single electric charge oscillating up and down. As a matter of convenience we consider a positive charge, although in practice the charge could be negative. Figure 35.12 suggests what happens to *one* of the electric field lines originating on the charge. As the charge oscillates, the electric field line is carried up and down like the end of a rope attached to a moving hand. Ripples in the electric field move outward as a wavelike disturbance.

A test charge located some distance from the oscillating charge feels the effects of this wave and experiences a time-varying electric force. Note that the electric field of the wave lies in the plane of motion of the charge. Figure 35.12 suggests that an oscillating charge generates an electromagnetic wave. However, the actual electric field set up by an oscillating charge is more complicated than that shown in Figure 35.12 and cannot be represented faithfully by a few lines.

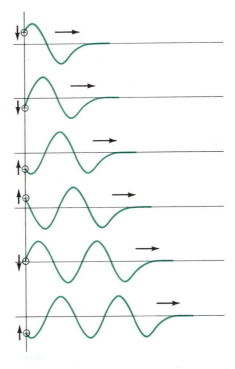

FIGURE 35.12

As the charge oscillates up and down, changes in the electric field travel outward as a wave. Only one field line is shown.

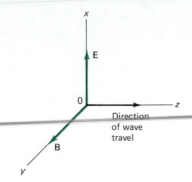

FIGURE 35.13

In an electromagnetic wave, the fields **E** and **B** are perpendicular to each other and perpendicular to the direction of wave travel. The direction of wave travel is the same as the direction of the vector product **E** × **B**.

The magnetic field produced by the oscillating charge is also time varying. As the charge oscillates up and down it constitutes a current that changes direction repeatedly. The magnetic field set up by this oscillating current travels through space as a wave. The magnetic field wave accompanies the electric field wave. In fact, the two wave fields are inseparably linked. Accordingly, we often speak of an **electromagnetic wave** rather than of two separate waves.

The magnetic field wave lies in a plane perpendicular to the oscillating current, just as the magnetic field lies in a plane perpendicular to a steady current. On the other hand, the electric field wave lies in a plane parallel to the plane of the current. Thus, the two planes containing the electric field and magnetic field of an electromagnetic wave are mutually perpendicular. Furthermore, both the electric field and the magnetic field of the electromagnetic wave are perpendicular to the direction of wave travel. This means that electromagnetic waves are **transverse**. Figure 35.13 illustrates this transverse character of electromagnetic waves.

The Speed of Electromagnetic Waves

We now want to show that Maxwell's equations have wavelike solutions. Specifically, we look for solutions that represent plane waves traveling through a vacuum in which **E** lies along the x-axis and **B** lies along the y-axis (Figure 35.13). Such waves propagate along the z-axis. This means that **E** and **B** will vary spatially only along the z-axis. We can write

$$\mathbf{E} = \mathbf{i}E_x(z, t) \tag{35.10}$$

$$\mathbf{B} = \mathbf{j}B_y(z, t) \tag{35.11}$$

The subscripts on E_x and B_y are useful reminders of the directions of **E** and **B**.

In place of the *integral* formulation of Maxwell's equations we use the equivalent *differential* form. The differential forms of Maxwell's equations can be obtained most elegantly by using vector techniques that are beyond the scope of this text. The differential form of Faraday's law (Equation 35.3) for the geometry employed here is

$$\frac{\partial E_x}{\partial z} = -\frac{\partial B_y}{\partial t} \tag{35.12}$$

This is one of two equations linking the space and time variations of **E** and **B**. The second such equation is Ampère's law, Equation 35.8. For free space* the differential form of Equation 35.8 is

$$-\frac{\partial B_y}{\partial z} = \mu_0 \epsilon_0 \frac{\partial E_x}{\partial t} \tag{35.13}$$

We can obtain the wave equation for E_x by forming the space derivative of Equation 35.12 and the time derivative of Equation 35.13. We find

$$\frac{\partial^2 E_x}{\partial z^2} = -\frac{\partial}{\partial z}\left(\frac{\partial B_y}{\partial t}\right)$$

$$\mu_0 \epsilon_0 \frac{\partial^2 E_x}{\partial t^2} = -\frac{\partial}{\partial t}\left(\frac{\partial B_y}{\partial z}\right)$$

The right sides of these two equations are equal because the z- and t-derivatives commute,

$$\frac{\partial}{\partial z}\left(\frac{\partial B_y}{\partial t}\right) = \frac{\partial}{\partial t}\left(\frac{\partial B_y}{\partial z}\right)$$

It follows that the left sides of the two equations are also equal. This gives us the wave equation satisfied by E_x

$$\frac{\partial^2 E_x}{\partial z^2} = \mu_0 \epsilon_0 \frac{\partial^2 E_x}{\partial t^2} \tag{35.14}$$

By forming the time derivative of Equation 35.12 and the space derivative of Equation 35.13 you can derive the wave equation satisfied by B_y. It has the same form as Equation 35.14:

$$\frac{\partial^2 B_y}{\partial z^2} = \mu_0 \epsilon_0 \frac{\partial^2 B_y}{\partial t^2} \tag{35.15}$$

The standard form of the wave equation is (Section 17.1)

$$\frac{\partial^2 \psi}{\partial z^2} = \frac{1}{c^2}\frac{\partial^2 \psi}{\partial t^2} \tag{35.16}$$

The wave equation describes a wavelike disturbance $\psi(z, t)$ traveling in the z-direction at a speed c.

Clearly, both Equation 35.14 and Equation 35.15 are of the standard wave equation form. Collectively they describe a traveling electromagnetic wave. Comparing Equations 35.14 and 35.15 with Equation 35.16 shows that electromagnetic waves travel at a speed given by

$$c = \frac{1}{\sqrt{\mu_0 \epsilon_0}} \tag{35.17}$$

where μ_0 is the magnetic permeability of empty space and ϵ_0 is the electric permittivity of free space. Inserting the values

$$\epsilon_0 = 8.85419 \times 10^{-12} \text{ C}^2/\text{N} \cdot \text{m}^2$$

$$\mu_0 = 1.25664 \times 10^{-6} \text{ N/A}^2$$

---

* *Free space* means a vacuum, a region *free* of matter.

into Equation 35.17 gives

$$c = 2.99793 \times 10^8 \text{ m/s}$$

Equation 35.17 was first derived by Maxwell. The value of $1/\sqrt{\mu_0\epsilon_0}$ agrees with the measured value of the speed of light. This is remarkable because Equation 35.17 expresses c in terms of μ_0 and ϵ_0, which are magnetic and electric constants previously unrelated to light. This was one of the triumphs of Maxwell's theory—the implication that light is an electromagnetic wave. The electric and magnetic fields of the wave are inseparably linked. Their coupling is described quantitatively by Maxwell's equations.

The implications of Maxwell's theory stretch far beyond the suggestion that light is an electromagnetic wave. Maxwell's theory does not restrict the wavelength to the visible spectrum. In fact, Equation 35.17 predicts that the speed of the waves is independent of wavelength. In principle, then, there could be electromagnetic waves of any wavelength. In 1887, Heinrich Hertz generated and detected electromagnetic waves with wavelengths of approximately 5 m. He was able to verify Maxwell's predicted speed by measuring both frequency (f) and wavelength (λ). Their product gave the speed $f\lambda = c$. Hertz's results were in substantial agreement with the speed of visible light. Thus, they confirmed Maxwell's theory.

Maxwell died in 1879 at age 47, and Hertz died in 1894 at age 36. Fate deprived both men of the satisfaction of watching their labors bear fruit. Maxwell's theory still finds new applications, most recently in lasers and fiber optics. Hertz's 5-m waves were soon stretched to 50 m by Marconi and transmitted across the Atlantic Ocean. Now they link global communications and even extend into the reaches of outer space.

Sinusoidal Plane Waves

Let's verify that the wave equations for E_x and B_y have sinusoidal solutions. Sinusoidal waves have the form

$$E_x = E_0 \sin(kz - \omega t) \tag{35.18}$$

and

$$B_y = B_0 \sin(kz - \omega t) \tag{35.19}$$

where ω/k is the wave speed. Equations 35.18 and 35.19 describe a transverse plane electromagnetic wave that advances in the positive z-direction at a speed given by ω/k. Figure 35.14 suggests how E_x and B_y vary along the z-axis. Keep in mind that E_x and B_y fill space. That is, they are not restricted to points on the z-axis. If we substitute Equation 35.18 into Equation 35.14 we get

$$-k^2 E_0 \sin(kz - \omega t) = -\mu_0\epsilon_0\omega^2 E_0 \sin(kz - \omega t)$$

which is satisfied provided the wave speed $\omega'k$ is given by

$$\frac{\omega}{k} = \frac{1}{\sqrt{\mu_0\epsilon_0}} \tag{35.20}$$

As we have seen, $1/\sqrt{\mu_0\epsilon_0}$ equals the speed of light. You should verify that

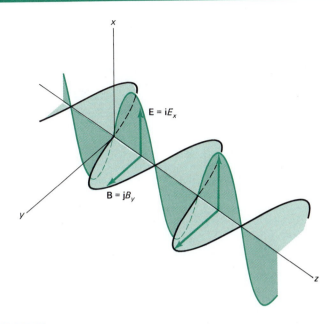

FIGURE 35.14

E_x and B_y vary from point to point along the z-axis. But keep in mind that the wave fields **E** and **B** fill space.

the sinusoidal form of B_y given by Equation 35.19 also satisfies Equation 35.15, provided Equation 35.20 is satisfied.

The amplitudes E_0 and B_0 can be related using Faraday's law. Substituting from Equations 35.18 and 35.19 into Equation 35.12 gives

$$+ kE_0 \cos(kz - \omega t) = \omega B_0 \cos(kz - \omega t)$$

which shows that

$$B_0 = \left(\frac{k}{\omega}\right) E_0$$

Because $\omega/k = c$ is the speed of light we have the amplitude relation

$$B_0 = \frac{E_0}{c} \qquad (35.21)$$

EXAMPLE 1

Magnetic Field Amplitude for a Plane Wave

A plane electromagnetic wave has an electric field strength of $E_0 = 800$ V/m. This value of E_0 is approximately equal to the amplitude of the electric field of a plane wave having the same intensity as the sunlight reaching the earth. We wish to determine the strength of the magnetic field. From Equation 35.21 we find

$$B_0 = \frac{E_0}{c} = \frac{800 \text{ V/m}}{3 \times 10^8 \text{ m/s}}$$

$$= 2.67 \times 10^{-6} \text{ T}$$

For comparison, the magnetic field strength of the earth is approximately 10^{-4} T.

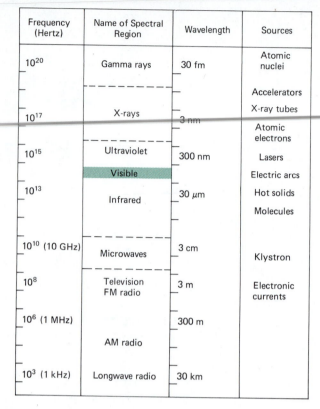

Frequency (Hertz)	Name of Spectral Region	Wavelength	Sources
10^{20}	Gamma rays	30 fm	Atomic nuclei
			Accelerators
10^{17}	X-rays	3 nm	X-ray tubes
			Atomic electrons
10^{15}	Ultraviolet	300 nm	Lasers
	Visible		Electric arcs
10^{13}	Infrared	30 μm	Hot solids
			Molecules
10^{10} (10 GHz)	Microwaves	3 cm	Klystron
10^{8}	Television FM radio	3 m	Electronic currents
10^{6} (1 MHz)		300 m	
	AM radio		
10^{3} (1 kHz)	Longwave radio	30 km	

FIGURE 35.15

The electromagnetic spectrum and typical sources. The narrow range of visible wavelengths runs from red light at about 670 nm to violet light at about 430 nm.

The Electromagnetic Spectrum and Its Sources

In 1665, at age 23, Isaac Newton used a glass prism to disperse a beam of sunlight into a rainbow of colors. He concluded that white light is a mixture of colors. The range of color, or **spectrum**, as Newton called it, ran from deep violet to blue, green, orange, and finally dark red. The full electromagnetic spectrum extends beyond both ends of the visible spectrum and includes radio and television waves, infrared radiation, ultraviolet radiation, X rays, and gamma rays. Figure 35.15 shows the electromagnetic spectrum. The borders separating the various regions of the spectrum are not sharply defined. The range of frequencies and wavelengths explored to date is enormous, but the spectrum has not yet been completely spanned.

Sources of Electromagnetic Waves

If you charge a glass rod by rubbing it, and then wave it up and down periodically, you will generate electromagnetic waves. The frequency of the waves equals the frequency at which you wave the rod. Radiation is produced because the charges are *accelerated*. Charges at rest or in uniform motion do not radiate, even though they may produce electric and magnetic fields. The current in a radio antenna produces electromagnetic waves because the moving charges experience accelerations. When the electron beam in a color TV set hits the screen, some X rays are emitted. The electrons are decelerated via collisions and convert their kinetic energy into X rays. An electron moving along a curved path in a magnetic field is accelerated, and it too emits electromagnetic radiation. The energy carried away by the waves is supplied by the kinetic energy of the electron.

Nature provides us with many examples of the conversion of kinetic energy into electromagnetic energy via the acceleration of electric charges. None is more spectacular than the Crab nebula, the remnant of a supernova (an exploding star). Electrons in the Crab nebula convert kinetic energy into electromagnetic energy as they move through the magnetic field associated with the nebula. The rate at which the Crab nebula radiates electromagnetic energy is enormous—nearly 100,000 times the rate at which our sun radiates.

There are many technological aspects of generating, transmitting, and receiving electromagnetic waves. Different techniques are used to generate X rays, visible light, and AM radio waves. For example, small crystals are used to control the frequency of the current in a radio-transmitter antenna. The frequency of the current equals the frequency of the electromagnetic wave that is radiated. The crystal is stimulated to vibrate at one of its natural (or resonant) frequencies, and the mechanical vibrations are then used to control the antenna current. The size of such a crystal can be estimated by recalling that the fundamental frequency of a vibrating structure is given by

$$f_{fund} \approx \frac{\text{speed of wave}}{\text{largest dimension of vibrating structure}} = \frac{v}{\mathscr{L}} \qquad (35.22)$$

(See Section 17.4.) Crystal oscillators can be used to generate wavelengths as small as several centimeters. Still shorter electromagnetic waves can be generated by klystrons, devices that rely on the coherent accelerated motions of electrons.

EXAMPLE 2

Size of an Oscillator Crystal

We want to estimate the characteristic dimension of a crystal that vibrates at 1 MHz (the middle of the AM radio band). We take the speed of sound in the crystal to be 5×10^3 m/s, a typical value for many solids. With $f_{fund} = 10^6$ Hz and $v = 5 \times 10^3$ m/s, Equation 35.22 gives

$$\mathscr{L} \approx \frac{v}{f_{fund}} = \frac{5 \times 10^3 \text{ m/s}}{10^6 \text{ s}^{-1}} = 5 \times 10^{-3} \text{ m} = 5 \text{ mm}$$

Thus, a thin wafer about 5 mm thick would oscillate at the desired frequency.

For wavelengths less than about 1 mm, the "classical" sources—accelerated charges—give way to various "quantum" sources: atomic nuclei, atoms, and molecules. Molecular rotations are one source of infrared radiation, for example. Electronic transitions in atoms and incandescent solids are sources of visible light. Changes in nuclear structure often result in the emission of gamma rays. Figure 35.15 indicates the more prominent sources of electromagnetic radiation across the spectrum.

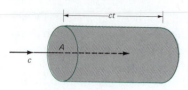

FIGURE 35.16

The energy U crossing the area A in time t is contained in a volume Act.

35.3
ENERGY TRANSFER VIA ELECTROMAGNETIC WAVES

Anyone who stays in the sun too long knows that electromagnetic waves can transfer energy. The rate at which sunlight and other electromagnetic waves transport energy is described by the **wave intensity**. The intensity (I) of a wave is defined as the rate at which the wave transfers energy divided by the area over which the energy is spread.

If an electromagnetic wave carries energy U through an area A perpendicular to the direction of energy flow in a time t, the wave intensity is

$$I = \frac{U}{tA} \tag{35.23}$$

(See Figure 35.16.) The units of intensity are watts per square meter (W/m^2).

Solar energy reaches each square kilometer of the earth's surface at a rate of roughly 1000 MW. Contemporary power plants deliver electric energy at approximately this same rate. This favorable comparison explains why solar energy is attractive. But although solar energy is plentiful and arrives free of charge, there are considerable costs in collecting it and converting it to thermal and electric energy.

EXAMPLE 3

Solar Energy and the Intensity of Sunlight

The sun radiates electromagnetic energy at the rate of 3.90×10^{26} W. Its radius is $R_\odot = 6.96 \times 10^8$ m, and so its surface area is

$$A_\odot = 4\pi R_\odot{}^2 = 6.09 \times 10^{18} \text{ m}^2$$

The intensity of sunlight at the solar surface is therefore

$$I = \frac{3.90 \times 10^{26} \text{ W}}{6.09 \times 10^{18} \text{ m}^2} = 5.60 \times 10^7 \text{ W/m}^2$$

As light waves travel outward from the sun their energy is spread over larger and larger areas, causing the intensity to decrease. At a distance of 150 million km from the sun (the distance from the sun to the earth) the energy has spread over an area

$$A = 4\pi(1.50 \times 10^{11} \text{ m})^2 = 2.83 \times 10^{23} \text{ m}^2$$

and the intensity has decreased to

$$I = \frac{3.90 \times 10^{26} \text{ W}}{2.83 \times 10^{23} \text{ m}^2} = 1.38 \times 10^3 \text{ W/m}^2$$

This intensity is referred to as the **solar constant**. The intensity at the surface of the earth is lower because about 40% of the incident solar energy is reflected back into space.

Lasers are usually rated in terms of the power they radiate. When used in this sense, *power* refers to the total electromagnetic energy per second carried away by the laser beam. The power ratings of the helium-neon lasers used in classroom demonstrations typically range from 0.1 mW to 1.0 mW. This power seems inconsequential when compared with that of the sun, 3.90×10^{26} W. However, the intensity of a laser beam may exceed the intensity of sunlight.

EXAMPLE 4

Laser Intensity

The power of a particular helium-neon laser is rated at 1 milliwatt. It is possible to focus a laser beam on an area of approximately one square wavelength. We want to estimate the corresponding intensity of the laser beam.

The wavelength of the visible He-Ne laser light is 632.8 nm. The cross-sectional area of the focused laser beam is approximately

$$A = (632.8 \times 10^{-9} \text{ m})^2 = 4.00 \times 10^{-13} \text{ m}^2$$

The beam intensity is

$$I = \frac{\text{power}}{\text{area}} = \frac{10^{-3} \text{ W}}{4.00 \times 10^{-13} \text{ m}^2}$$
$$= 2.5 \times 10^9 \text{ W/m}^2$$

This is over 20 times the intensity of sunlight at the surface of the sun, and nearly a million times the intensity of sunlight that reaches the earth.

The Intensity-Field Relationship

To see how intensity is related to the electric field and magnetic field, we multiply the numerator and denominator of Equation 35.23 ($I = U/tA$) by c, the speed of light. This gives

$$I = \left(\frac{U}{ctA} \right) c = \left(\frac{U}{V} \right) c \qquad (35.24)$$

The quantity

$$ctA = V$$

is the volume containing the energy U that has crossed the area A in time t (Figure 35.16). The ratio U/V, the electromagnetic energy per unit volume, is called the **energy density**,

$$\frac{U}{V} = \text{energy density} \equiv u \qquad (35.25)$$

The intensity can be expressed as the product of the energy density and the speed of light:

$$I = uc \qquad (35.26)$$

In an electromagnetic wave, both the electric field and the magnetic field contribute to the energy density (Sections 26.3 and 32.3). The total energy density is the sum of the electric-field energy density u_E and the magnetic-field energy density u_B:

$$u = u_E + u_B = \tfrac{1}{2}\epsilon_0 E^2 + \frac{B^2}{2\mu_0} \qquad (35.27)$$

Because of the wave nature of E and B, the values of E^2 and B^2 vary from point to point and from moment to moment. The effective values of E^2 and B^2 are their time averages. For sinusoidal plane waves (Equations 35.18 and 35.19) the electric field and magnetic field make equal contributions and the time-average energy density u_0 can be expressed as

$$u_0 = \tfrac{1}{2}\epsilon_0 E_0{}^2 \tag{35.28}$$

where E_0 is the amplitude of the sinusoidal electric field. It follows from $I = uc$ that the time average intensity I_0 is

$$I_0 = \tfrac{1}{2}c\,\epsilon_0 E_0{}^2 \tag{35.29}$$

Equation 35.29 is the desired relation between average intensity and electric field amplitude.

EXAMPLE 5

"Exit" Intensity

An "EXIT" light is covered with a filter that transmits red light. The electric field of the emerging beam is represented by a sinusoidal plane wave

$$E_x = 36 \sin{(kz - \omega t)}\ \text{V/m}$$

Let's calculate the time average intensity of the beam. The amplitude of the electric field is

$$E_0 = 36\ \text{V/m}$$

Using Equation 35.29 we have

$$
\begin{aligned}
I_0 &= \tfrac{1}{2}c\,\epsilon_0 E_0{}^2 \\
&= \tfrac{1}{2}(3.00 \times 10^8\ \text{m/s}) \cdot (8.85 \times 10^{-12}\ \text{C}^2/\text{N}\cdot\text{m}^2) \cdot (36\ \text{V/m})^2 \\
&= 1.72\ \text{C}^2\cdot\text{V}^2/\text{s}\cdot\text{N}\cdot\text{m}^3 = 1.72\ \text{J/s}\cdot\text{m}^2 \\
&= 1.72\ \text{W/m}^2
\end{aligned}
$$

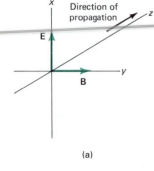

(a)

(b)

FIGURE 35.17

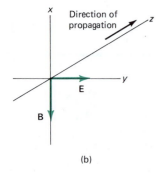

FIGURE 35.18

The superposition of many light waves with randomly polarized electric fields gives unpolarized light.

35.4
POLARIZATION

All types of transverse waves can exhibit a preferred direction or alignment, which we call **polarization**. The plane of polarization of an electromagnetic wave is defined as the plane containing the electric field and the direction of propagation. In Figure 35.17a the plane of polarization is the x–z plane. In Figure 35.17b the plane of polarization is the y–z plane.

The light streaming from an ordinary light source, such as an incandescent bulb, is a superposition of the waves emitted by an enormous number of atoms and molecules. The electric fields of these many waves are randomly oriented, and as a result, the light is unpolarized (Figure 35.18). In general, microscopic sources composed of many independently radiating atoms, such as an incandescent light, produce unpolarized waves. Macroscopic sources such as radio and TV generally produce waves with a definite polarization.

A beam of unpolarized light may be polarized, or partially polarized, in several different ways. Reflection, scattering, refraction, and absorption are all

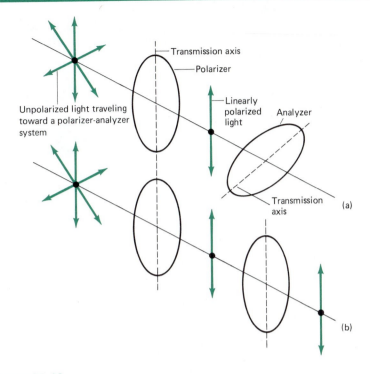

FIGURE 35.19

(a) With transmission axes crossed, no light passes through the analyzer. (b) With transmission axes parallel, the analyzer transmits the incident linearly polarized light.

capable of affecting the polarization of light. Polaroid ™ *H*-sheet* converts unpolarized light by **selective absorption**. The *H*-sheet is formed from polyvinyl alcohol chemically treated with iodine. The long-chain molecules of the plastic are aligned by stretching the sheet during its fabrication. The aligned molecules absorb almost completely the component of the electric field that is parallel to the axis of molecular alignment. The component of the electric field that is perpendicular to the axis of molecular alignment is virtually unaffected by absorption. The transmission axis of the sheet is therefore perpendicular to the direction in which the sheet was stretched.

A Polaroid sheet acts almost like an ideal polarizer, a filter that transmits only the component of the electric field that is parallel to the characteristic transmission axis. Figures 35.19a and b illustrate the effect of a Polaroid sheet on the electric field of a beam of unpolarized light, and the way in which the electric field can be analyzed with a second sheet of Polaroid. The first sheet (the **polarizer**) converts the unpolarized beam into a plane-polarized beam. If the second sheet (the **analyzer**) has its polarizing axis at right angles to the polarizing axis of the first sheet, essentially no light is transmitted. But when the polarizer and analyzer axes are aligned, there is a nearly complete transmission by the analyzer (Figure 35.19b). Reflection reduces slightly the intensity of the transmitted light.

---

* Edwin H. Land invented a synthetic polarizing material, which he called Polaroid *J*-sheet, in 1928. Ten years later he invented an improved polarizing material, Polaroid *H*-sheet. We generally refer to these simply as Polaroids or sheets of Polaroid. Land is perhaps better known for his invention of the Polaroid camera.

FIGURE 35.20

Two pictures showing how Polaroid ™ sunglasses reduce glare. (a) A scene filmed without a Polaroid filter. (b) The same scene filmed through a Polaroid filter, showing a great reduction in glare.

Longitudinal waves, such as sound waves in air, do not single out a plane of polarization. Only transverse waves can be polarized. Hence, the demonstration of the polarization of light waves is a convincing proof of their transverse character.

When unpolarized light strikes a smooth surface, the reflected light is partially polarized. Specifically, the amplitude of the electric field component that is parallel to the reflecting surface is larger than the amplitude of the component that is perpendicular to the surface. Polaroid sunglasses take advantage of this fact to reduce reflected light, or glare (Figures 35.20a and b). Most surfaces that cause glare are horizontal (for example, automobile hoods and water surfaces). The glare is therefore composed predominantly of light polarized in a horizontal plane. The transmission axes of Polaroid sunglasses are vertical, which enables them to absorb over half of the reflected light. You can demonstrate this yourself by viewing reflected glare through one lens of Polaroid glasses. Rotating the lens 90° will cause the intensity of this transmitted light to increase noticeably.

The vector character of the electric field is important to any quantitative discussion of polarized light. The electric field **E** of the wave can be resolved into two mutually perpendicular components (Figure 35.21). The fact that we can resolve **E** into components means that any transverse electric field can be built up by superposing two mutually perpendicular electric fields. Thus, when we write the equation for the field **E** of Figure 35.21,

$$\mathbf{E} = \mathbf{i}E_x + \mathbf{j}E_y$$

we recognize that the field **E** is superposition of a field $\mathbf{i}E_x$ polarized in the x–z plane, and a field $\mathbf{j}E_y$ polarized in the y–z plane.

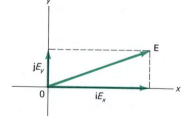

FIGURE 35.21

The electric field of an electromagnetic wave can be built up by superposing the electric fields of two waves polarized along mutually perpendicular axes.

Intensity

Suppose a beam of plane-polarized light strikes a Polaroid sheet such that the angle between the transmission axis and the plane of polarization is θ. Let's determine the relationship between the angle θ and the intensities of the incident and transmitted beams.

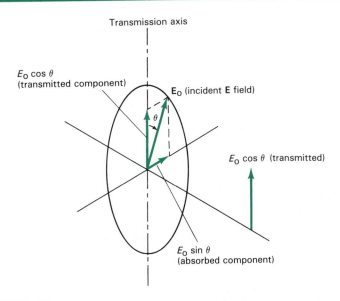

FIGURE 35.22

The incident electric field may be resolved into components parallel and perpendicular to the transmission axis. The parallel component $E_0 \cos \theta$ is transmitted. The perpendicular component $E_0 \sin \theta$ is absorbed by the Polaroid sheet.

As Figure 35.22 suggests, we can resolve the incident electric field into two components. If E_0 is the amplitude of the incident electric field, $E_0 \cos \theta$ is the amplitude of the transmitted field. The intensity of the incident beam (I_0) is given by Equation 35.29, and is proportional to $E_0{}^2$,

$$I_0 = \tfrac{1}{2} c \epsilon_0 E_0{}^2 \qquad (35.29)$$

It follows that the intensity of the transmitted wave (I_t) is given by

$$I_t = I_0 \cos^2 \theta \qquad (35.30)$$

When the axes are aligned, $\theta = 0^0$, and we have complete transmission ($I_t = I_0$). When the axes are perpendicular, $\theta = 90°$, and we have complete extinction ($I_t = 0$).

EXAMPLE 6

Polaroids Crossed at 45°

A beam of light passes through a Polaroid sheet (a polarizer). The emergent beam of plane-polarized light strikes a second Polaroid (analyzer). The transmission axis of the analyzer is oriented at 45° with respect to the plane of polarization of the incident beam (Figure 35.23a). Using Equation 35.30, we can show that the transmitted intensity is one-half that of the incident intensity. Thus, with $\theta = 45°$, Equation 35.30 gives

$$I_t = I_0 \cos^2 45° = \tfrac{1}{2} I_0$$

Half the incident intensity is absorbed and half is transmitted. It should be noted that the transmitted beam remains plane polarized; the plane of polarization now coincides with the transmission axis of the analyzer.

If the emergent light is now sent through a second analyzer, with its transmission axis oriented at 45° with respect to the first analyzer (and at 90° with respect to the transmission axis of the original polarizer), a transmitted beam will emerge

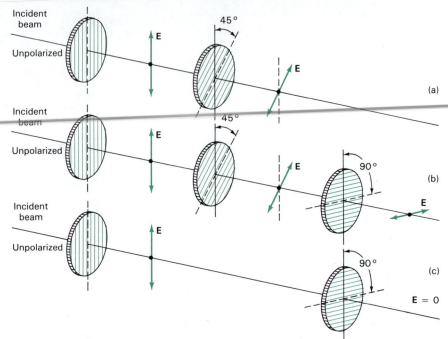

FIGURE 35.23

(a) Two polaroids with transmission axes inclined at 45°. The transmitted beam is plane polarized. (b) When a third polaroid is put in place with its transmission axis perpendicular to the transmission axis of the first polaroid, there is still a transmitted beam. (c) When the middle polaroid is removed, we are left with two polaroids with their transmission axes perpendicular and there is no transmitted light.

(Figure 35.23b). Surprisingly, if the first analyzer is now removed, the transmitted intensity drops to zero because we are left with two Polaroids with their transmission axes at 90° (Figure 35.23c).

Equation 35.30 can also be used to show that if a beam of unpolarized light strikes a Polaroid sheet, the intensity of the transmitted light is one-half that of the incident light. To prove this statement we first recall that the incident wave is the superposition of a multitude of randomly oriented electric fields. Equation 35.30 applies to each of the component electric fields, but the angle θ ranges from 0° to 360°. Because the orientation is random, all values of θ will occur equally often. As a result, Equation 35.30 will give the transmitted intensity if we use the average value of $\cos^2 \theta$, which is 1/2. Replacing $\cos^2 \theta$ by 1/2 in Equation 35.30 shows that one-half of the incident intensity is transmitted.

EXAMPLE 7

A Polarization Puzzle

A beam of light is composed of a mixture of unpolarized and plane-polarized waves. The beam is analyzed with a Polaroid sheet. As the transmission axis is rotated, the transmitted intensity varies from a maximum I_{max} to a minimum of $(1/3) I_{max}$. We want to show that the intensities of the polarized and unpolarized beams are equal.

Let I_0 and I_p denote the intensities of the unpolarized (I_0) and polarized (I_p) beams that reach the analyzer. The transmitted intensity is

$$I_t = \tfrac{1}{2}I_0 + I_p \cos^2 \theta \tag{35.31}$$

where θ is the angle between the plane of polarization and the transmission axis. The transmitted intensity is a maximum at $\theta = 0°$ and 180°, where $\cos^2 \theta = 1$. Thus

$$I_{max} = \tfrac{1}{2}I_0 + I_p \tag{35.32}$$

The intensity is a minimum at $\theta = 90°$ and $270°$, where $\cos^2 \theta = 0$. Thus

$$I_{min} = \tfrac{1}{2}I_0$$

Setting $I_{min} = (1/3)I_{max}$ gives

$$I_0 = (2/3)I_{max}$$

Inserting this result into Equation 35.32 shows that

$$I_p = (2/3)I_{max}$$

which establishes the desired result, $I_p = I_0$. The two intensities are equal.

WORKED PROBLEM

The electric field vector of a circularly polarized plane electromagnetic wave has the form

$$\mathbf{E} = \mathbf{i}E_0 \cos \omega t + \mathbf{j}E_0 \sin \omega t$$

in the $z = 0$ plane. Show that the magnitude of $\mathbf{E}$ is E_0, and that the vector $\mathbf{E}$ rotates in the x–y plane with an angular velocity ω.

Solution

The x- and y-components of $\mathbf{E}$ are

$$E_x = E_0 \cos \omega t$$

$$E_y = E_0 \sin \omega t$$

Hence the magnitude of $\mathbf{E}$ is

$$E = \sqrt{E_x^2 + E_y^2} = \sqrt{E_0^2 \cos^2 \omega t + E_0^2 \sin^2 \omega t}$$
$$= E_0 \sqrt{\cos^2 \omega t + \sin^2 \omega t}$$
$$= E_0$$

Figure 35.24 shows a vector of magnitude E_0 that rotates in the x–y plane with an angular velocity ω. If we let $\theta = \omega t$ denote the angle between the rotating vector and the positive x-axis, then the x-component of the rotating vector is

$$E_0 \cos \theta = E_0 \cos \omega t$$

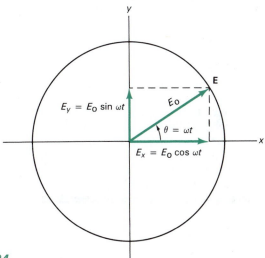

FIGURE 35.24

and the y-component is

$$E_0 \sin \theta = E_0 \sin \omega t$$

These are precisely the components of the vector **E**, which confirms that it rotates in the x–y plane with an angular velocity ω.

EXERCISES

35.1 Maxwell's Equations

A. Radio astronomers learn much about the structure of our galaxy by observing radiation emitted by hydrogen atoms at a frequency of 1420 MHz. What is the wavelength of this radiation?

B. (a) Red light has a wavelength of 650 nm. What is its frequency? (b) Blue light has a wavelength of 450 nm. What is its frequency?

C. The electric field of a sinusoidal plane wave traveling in a vacuum is represented by

$$E_x = 250 \sin(kz - \omega t) \text{ V/m}$$

$$k = 1.21 \times 10^7 \text{ rad/m}$$

$$\omega = 3.63 \times 10^{15} \text{ rad/s}$$

(a) Determine the frequency of the wave. (b) Determine the wavelength. Express the wavelength in nanometers. If it corresponds to visible light, indicate its color.

D. Police radar operates in the microwave range at a wavelength of about 3 cm. What is the frequency of a microwave whose wavelength is 3 cm?

E. From Figure 35.13 it can be seen that the direction of propagation of an electromagnetic wave is in the direction of the vector $\mathbf{E} \times \mathbf{B}$. The vectors **E** and **B** reverse their directions every half cycle of the wave. Does $\mathbf{E} \times \mathbf{B}$ change direction when both **E** and **B** reverse direction? If **E** is directed along the positive x-axis, along what (perpendicular) axis must **B** lie for a wave traveling along the positive z-axis?

F. Write down expressions for the electric and magnetic fields of a sinusoidal plane electromagnetic wave with a frequency of 3 GHz traveling in the positive x-direction. The amplitude of the electric field is 300 V/m.

G. At a track meet the starter is located at one end of the 100-meter track and the timers are at the other end. When the gun is fired the timers see the gun smoke and then hear the sound of the gun firing, what is the time interval between sight and sound of the gun?

H. Show that the ordinary current (I) and Maxwell's displacement current $\epsilon_0 \, d/dt \int \mathbf{E} \cdot d\mathbf{A}$ have the same dimensions.

I. Prove that $\epsilon_0 E^2$ and B^2/μ_0 have dimensions of energy per unit volume.

35.2 Electromagnetic Waves

J. In a certain plane electromagnetic wave the maximum electric field is 2.1 microvolts/meter. (a) Calculate the maximum electric force (eE) on an electron. (b) Assume the electron of part (a) is moving with velocity **v** perpendicular to **B**. Taking c as an upper limit to the electron's speed, calculate the magnitude of the maximum magnetic force [$q(\mathbf{v} \times \mathbf{B})$] on the electron.

35.3 Energy Transfer via Electromagnetic Waves

K. A 10-mW laser has a beam diameter of 1.6 mm. (a) What is the intensity of the light, assuming it is uniform across the circular beam? (b) What is the average energy density of the laser beam?

L. Suppose that the intensity of sunlight reaching the outermost layers of our atmosphere (Example 3) were due to a monochromatic plane wave. Calculate the amplitude of the electric field of the wave.

M. Assume that 60% of the arriving solar energy reaches the earth's surface. *Estimate* the amount of solar energy you absorb in a 60-minute sunbath, assuming you absorb 50% of the incident energy.

N. In 1965, Penzias and Wilson discovered the cosmic microwave radiation left over from the Big Bang expansion of the universe. The energy density of this radiation is 4×10^{-14} J/m³. Determine the corresponding electric field amplitude.

P. Assume that 60% of the arriving solar energy reaches the earth's surface. Calculate the rate at which energy reaches a surface area of one square kilometer (10^6 m²). Would all of the incident energy be absorbed? Could all of the absorbed energy be converted into electric energy?

Q. The magnetic field of a sinusoidal plane electromagnetic wave is represented by

$$B_y = 1.42 \times 10^{-4} \sin(kz - \omega t) \text{ T}$$

$$k = 1.43 \times 10^7 \text{ rad/m}$$

$$\omega = 4.29 \times 10^{15} \text{ rad/s}$$

Determine the time average intensity of the electromagnetic wave.

R. A high-power laser has a power output of 1.0 MW. The laser beam is focused onto an area of 10^{-6} m². Calculate (a) the intensity, (b) the average energy density, (c) the maximum electric field, (d) the maximum magnetic field.

S. Pluto is approximately 39 times as far from the sun as is the earth. Use the results of Example 3 to estimate the intensity of sunlight on Pluto.

35.4 Polarization

T. A plane-polarized electromagnetic wave with an electric field amplitude of $E_0 = 100$ V/m strikes an ideal polarizer. The incident **E** vector makes an angle of $62°$ with the transmission axis of the polarizer. Determine the amplitude of the transmitted electric field.

U. In an undergraduate laboratory experiment, a beam of plane-polarized light strikes a Polaroid sheet such that the angle between the transmission axis and the plane of polarization is θ. A student measures the ratio of intensities of the transmitted and incident beams and records the data shown in Table 1. By plotting the ratio of intensities versus $\cos^2 \theta$, show that the experimental results are consistent with the relation $I_t = I_0 \cos^2 \theta$.

TABLE 1

I_t/I_0	θ (degrees)
0.97	10
0.87	20
0.76	30
0.59	40
0.40	50
0.26	60
0.12	70
0.03	80

PROBLEMS

1. Differential forms of Faraday's law and Ampère's law for empty space and an appropriate geometry are as follows:

$$\frac{\partial E_x}{\partial z} = -\frac{\partial B_y}{\partial t} \qquad \text{(Faraday)}$$

$$\mu_0 \epsilon_0 \frac{\partial E_x}{\partial t} = -\frac{\partial B_y}{\partial z} \qquad \text{(Ampère)}$$

Show that the plane wave forms of E_x and B_y given by Equations 35.18 and 35.19 satisfy these differential equations provided that

$$\left(\frac{\omega}{k}\right)^2 = \frac{1}{\mu_0 \epsilon_0}$$

2. A horizontal beam of light with an intensity of 100 W/m^2 is 100% polarized in the vertical direction. The beam passes through three polarizers in succession. The first polarizer's transmission axis is $30°$ from the vertical, the second is $60°$ from the vertical, and the third is $90°$ from the vertical. What is the intensity and what is the polarization of the beam emerging from the third polarizer?

3. A beam of light is known to be a mixture of unpolarized light (intensity I_0) and plane-polarized light (intensity I_p). When the beam is analyzed with a Polaroid sheet, the transmitted intensity varies from a maximum I_{max} to a minimum of $0.60\, I_{max}$. The total intensity, $I_0 + I_p$, is 12 W/m^2. Determine I_p and I_0.

4. A beam of unpolarized light of intensity 200 W/m^2 strikes a Polaroid sheet (a polarizer). (a) What is the intensity of the transmitted beam? (b) The transmitted beam passes through a second Polaroid sheet. The transmission axes of the two sheets are oriented at $45°$. What is the intensity of the beam transmitted by the second sheet? (c) The beam continues, passing through a third sheet, whose transmission axis is inclined at an angle of $45°$ with respect to the second sheet and $90°$ with respect to the first sheet. Determine the intensity transmitted by the third sheet. (d) The third sheet is rotated so that its transmission axis is at right angles ($90°$) with respect to the second sheet. Determine the transmitted intensity.

5. An electron initially at rest at the origin ($x = y = z = 0$) experiences the electric field of a light wave,

$$E_x = E_0 \sin(kz - \omega t)$$

where $\omega = 2\pi \times 10^{15}$ rad/s and $E_0 = 10^3$ V/m. Calculate (a) the maximum speed, (b) the maximum displacement of the electron caused by the electric field.

6. A SYNCOM satellite transmits signals to earth with a power of 15 kW. If the beam reaching the earth is spread over a circle with a radius of 400 km, determine the amplitude of the electric field. Assume sinusoidal fields.

7. The momentum p carried in an electromagnetic wave is related to its energy U by

$$p = \frac{U}{c}$$

Calculate the momentum-per-second (and thus the force) due to sunlight striking the earth and compare this with the gravitational force between the sun and the earth.

8. Assuming that the antenna of a 10-kW radio station

radiates electromagnetic energy uniformly in all directions, compute the maximum value of the magnetic field at a distance of 5 km from the antenna, and compare this value with the magnetic field of the earth at the surface of the earth.

9. The electric field vector of an *elliptically polarized* plane electromagnetic wave has the form

$$E = ia \cos \omega t + jb \sin \omega t \qquad (a, b \text{ constant})$$

in the $z = 0$ plane. Show that the tip of the vector E traces an ellipse as it rotates in the xy–plane with an angular velocity ω.

10. (a) For a parallel-plate capacitor having a small plate separation compared with the length or width of a plate, show that the displacement current is given by

$$I_d = C \frac{dV}{dt}$$

where dV/dt is the time rate of change of potential difference across the plates. (b) Calculate the time rate of change of potential difference required to produce a displacement current of 1 ampere in a 1-microfarad capacitor.

11. Many homes throughout the country are using solar radiation to heat water for domestic use. The following data is reasonable for many areas of the country. Use the data and answer the questions posed to estimate the size and cost of a domestic solar hot water system.

number of occupants = 4

incoming water temperature = 15°C

hot water temperature = 57°C

usage rate = 20 gallons per person per day

average solar energy available = 14×10^6 J/m²·day

collector efficiency for converting solar energy to thermal energy = 65%

thermal energy per m² per day available for heating water = _____

thermal energy per day required to heat water = _____

collecting area required = _____

installed system cost at \$400/m² of collecting area = _____

12. (a) A straight wire of length L and radius a is connected to a battery providing a potential difference of V volts. Show that the thermal power per unit surface area is consistent with the relation

$$\frac{P}{A} = \left(\frac{1}{\mu_0} \right) EB$$

where E and B are the magnitudes of the electric field and magnetic field at the surface of the wire. (b) Calculate the power per unit surface area for a copper wire 1.0 m long, 0.1 mm in diameter when the wire is connected to a 1.5-volt battery.

CHAPTER 36

GEOMETRIC OPTICS

36.1
INTRODUCTION TO GEOMETRIC OPTICS

Optics is at one of the frontiers of science—swept there by laser beams and by the integration of solid-state physics and optical physics. New technologies have sprung up, and both theoretical and applied developments continue at an exhilarating pace. We see daily reminders of the revolution in optics, such as liquid crystal and light-emitting diode displays in watches and calculators, and laser scanners at supermarket checkout counters. Many optical innovations also affect our lives even though we are not aware of them, remote sensing devices such as weather satellites, for example. The versatile laser is powerful enough to weld auto bodies, yet delicate enough to perform eye surgery.

36.2
REFLECTION AND REFRACTION

The observation that an object can cast a sharp shadow leads to the notion of a **light ray**, a narrow stream of light that travels without diverging or converging. We can think of a broad beam of light as a bundle of parallel rays. A shadow is formed when an object reflects or absorbs some of the incident rays.

Geometric optics is the study of light in terms of light rays. Geometric optics is adequate provided that the wavelength of light is much smaller than the characteristic dimensions of the optical system. For example, geometric optics gives an adequate description of lenses as long as the wavelength of the light is much smaller than the diameter of the lens.

The laws of reflection and refraction describe the behavior of light rays. Both laws can be derived from Maxwell's wave theory of light and both are amply confirmed by experiment.

Reflection

Consider a light ray that strikes a smooth surface. In general, both a reflected ray and a transmitted (refracted) ray are formed, as shown in Figure 36.1. The angle between the incident ray and a line perpendicular to the surface

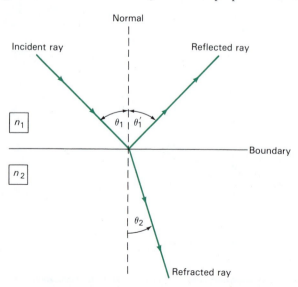

FIGURE 36.1

The light ray strikes the boundary and is divided into a reflected ray and a refracted ray.

(the *normal*) is called the **angle of incidence**. The angle between the normal and the reflected ray is called the **angle of reflection**. The **law of reflection** states:

The angles of incidence and reflection are equal.   (36.1)

This means that $\theta_1 = \theta_1'$ in Figure 36.1.

Index of Refraction

When light travels from one medium to another its speed changes. The speed of light in water is approximately three-quarters its speed in air. The **index of refraction** (n) is used to compare the speed of light in a medium to its speed in vacuum, and is defined by

$$n = \text{speed in vacuum/speed in medium} = \frac{c}{v} \qquad (36.2)$$

Table 36.1 lists values of n for several substances. Notice that $n > 1$ in every case. The speed of light in a material medium is less than its speed in a vacuum. The reduction in speed when light enters a medium is a consequence of the interaction between the electromagnetic field and the atomic electrons in the medium.

TABLE 36.1
Index of Refraction (reference wavelength = 589.3 nm)*

Substance	n	Substance	n
Water (ice)	1.31	Amber	1.56
Water (liquid)	1.33	Emerald	1.58
Acetone	1.36	Crown glass	1.59
Fluorite	1.43	Flint glass	1.65
Glycerol	1.47	Calcite	1.66
Benzene	1.50	Ruby	1.77
Sodium chloride	1.54	Diamond	2.42

* The index of refraction can be measured more precisely than indicated here. Over the visible spectrum it is generally possible to achieve five-figure precision. Values of n vary with the temperature of the substance as well as with the wavelength of light.

Law of Refraction

When light strikes a surface, the transmitted rays may travel in a direction different from that of the incident ray. The change in direction, or bending, is called **refraction**, and this gives rise to many fascinating optical effects (Figure 36.2).

Snell's law of refraction relates the angle of incidence and the angle of refraction to the indices of refraction. Referring to Figure 36.1, Snell's law states

$$n_1 \sin \theta_1 = n_2 \sin \theta_2 \qquad (36.3)$$

Note that θ_1 and θ_2 are measured relative to a line perpendicular to the boundary. In Figure 36.1, the angle of refraction θ_2 is smaller than the angle of incidence θ_1. Snell's law shows that this occurs for $n_2 > n_1$. Thus, Figure

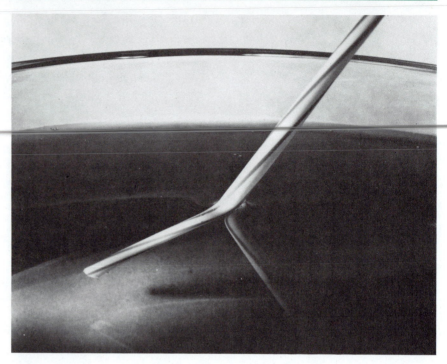

FIGURE 36.2

Refraction at the air–water boundary causes the partially submerged straw to appear bent.

36.1 depicts a situation where the refracted (transmitted) light travels at a slower speed than the incident light.

EXAMPLE 1

Refraction at an Air-Water Boundary

Light from a fish reaches the surface of a pond at an angle $\theta_2 = 36.3°$ (Figure 36.3). What is the angle of refraction θ_1 in the air above?

With $n_{air} = 1.00$ and $n_{water} = 1.33$, Snell's law gives

$$1.00 \sin \theta_1 = 1.33 \sin 36.3°$$

from which we find

$$\sin \theta_1 = 0.787$$

and

$$\theta_1 = 51.9°$$

As Figure 36.3 shows, light reaching the observer's eye seems to come from position B, above the actual position of the fish.

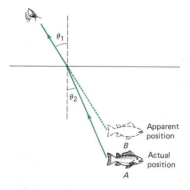

FIGURE 36.3

The apparent position of the fish is above its actual position because of refraction at the air–water boundary.

Qualitatively, the larger the change in the index of refraction across the boundary, the greater the bending of the light as it crosses the boundary. The index of refraction varies with wavelength, causing a **dispersion** of white light (Figure 36.4). The different wavelengths in white light are refracted through different angles, causing a fanning out, or a dispersal, of the beam. Dispersion in a lens produces a defect called **chromatic aberration**. Different wavelengths are focused at different points. Nature provides a spectacular example of the dispersion of light—the rainbow.

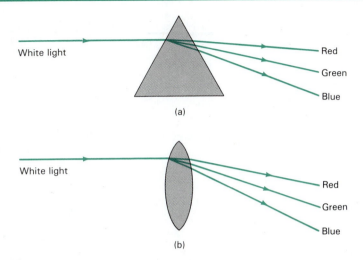

FIGURE 36.4

White light is dispersed when it passes through (a) a prism or (b) the edge of a lens. Dispersion occurs because the index of refraction varies with wavelength.

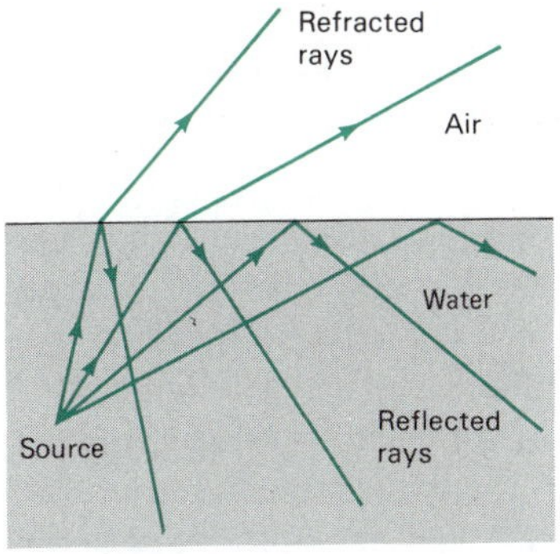

Total Internal Reflection

In Figure 36.5 we see a series of light rays that travel through water and strike a water-air boundary. For some light rays the light reaching the boundary divides into a reflected ray and a refracted ray, which passes into the air. But other rays exhibit **total internal reflection**. All of the light of these rays is reflected; no light escapes from the water to the air.

The condition that determines whether light undergoes total internal reflection follows from Snell's law. If $n_1 > n_2$, then $\theta_2 > \theta_1$. As Figure 36.6 shows, a limit for refraction is reached when the value of $\sin \theta_2$ reaches $+1$, corresponding to $\theta_2 = 90°$. The angle of incidence that corresponds to $\theta_2 = 90°$ is called the **critical angle** and is denoted by θ_c. With $\theta_2 = 90°$ and $\theta_1 = \theta_c$, Snell's law takes the form,

$$\sin \theta_c = \frac{n_2}{n_1} \tag{36.4}$$

Light rays incident at angles equal to or exceeding the critical angle undergo total internal reflection. For a water-air boundary the critical angle is

$$\theta_c = \arcsin\left(\frac{1.00}{1.33}\right) = 48.6°$$

FIGURE 36.5

Rays from a submerged source reach the water–air boundary. For some rays there is no refracted light. Such rays exhibit total internal reflection.

FIGURE 36.6

Rays originating beneath the surface that strike the water–air boundary at angles greater than the critical angle of 48.6° undergo total internal reflection.

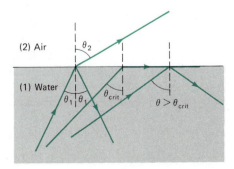

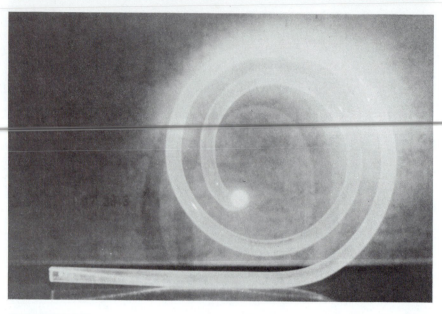

FIGURE 36.7

A laser beam enters the left end of a light pipe. Light travels around the curved sections via total internal reflections. Some rays strike the surface at angles less than the critical angle and are refracted out of the pipe. The circular bright spot is the emerging light.

Total internal reflection results in many beautiful optical effects, including the sparkle of a diamond. Most light rays entering the top of a diamond undergo total internal reflection one or more times and then come back out. There are also important practical applications of total internal reflection. Figure 36.7 shows a clear plastic light pipe through which a laser beam travels by internal reflections. Some rays strike at angles less than the critical angle and are refracted out of the pipe. Compared to commercially available optical fibers, the pipe of Figure 36.7 leaks badly. Commercial fibers have diameters ranging from a few micrometers to several hundred micrometers. They are very flexible and can be tied into knots without noticeable diminution in the ability to transmit light. These fibers are therefore able to carry beams of light around corners, and are also small and lightweight.

Fiber optics have made possible many innovative surgical and diagnostic techniques in medicine. Fiber optics communication systems are now in operation, most notably in the telephone industry.

Rainbows

The rainbow is nature's most beautiful illustration of dispersion. We can understand the formation of a rainbow by applying the laws of reflection and refraction to a spherical raindrop. Figure 36.8 shows the two refractions and one reflection that divert a ray of light from the sun into an observer's eye. The angle of deviation D is the total angle through which a ray is turned by refraction and reflection. In Figure 36.8, the triangle ABC and CBE are isosceles. The deviation of the ray at A is $\theta_1 - \theta_2$. At B the reflected ray is deviated through $180° - 2\theta_2$. The ray refracted at E is deviated through the angle $\theta_1 - \theta_2$. The total deviation is the sum of these three angles:

$$D = 2(\theta_1 - \theta_2) + (180° - 2\theta_2)$$

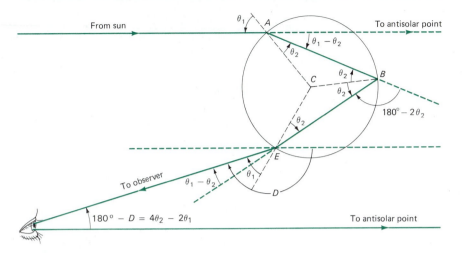

FIGURE 36.8

Two refractions and one internal reflection deviate the ray through an angle $D = 2(\theta_1 - \theta_2) + (180° - 2\theta_2)$. Light reaches the observer's eye at $4\theta_2 - 2\theta_1$, measured with respect to the antisolar point.

In Figure 36.9 the angle of the red bow is shown as 42.2°. The violet bow angle is shown as 40.6°. These are the angles for which the light is most intense. The intensity of the light is greatest for the angle of incidence (θ_1) that bunches the refracted rays. This turns out to be the angle of incidence for which the deviation D is a minimum. Setting $dD/d\theta_1 = 0$, and using Snell's law, leads to equations for θ_1 and θ_2 that produce the minimum angle of deviation,

$$\tan \theta_1 = \sqrt{\frac{4 - n^2}{n^2 - 1}}$$

$$\tan \theta_2 = \tfrac{1}{2} \tan \theta_1$$

(36.5)

The dispersive action of the raindrop enters through the index of refraction. For violet light, $n = 1.3435$, and for red light $n = 1.3318$. These slightly different values of n result in slightly different values of the minimum angle of

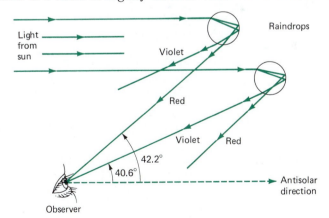

FIGURE 36.9

Different colors of light refracted by the raindrops reach the observer's eye from different directions, producing a rainbow. Red light forms the upper edge of the bow, 42.2° above the antisolar direction. Violet light forms the lower edge of the bow, 40.6° above the antisolar direction.

deviation. Different colors emerge at different angles. An observer viewing the emerging light (Figure 36.9) receives different colors from slightly different directions and thus sees a rainbow.

EXAMPLE 2

Locating the Rainbow

Figure 36.8 shows that light reaches the observer's eye at the angle $180° - D$, measured with respect to the direction of the antisolar point. This angle equals $4\theta_2 - 2\theta_1$.

Setting $n = 1.3318$ in Equation 36.5 gives the angles for red light:

$$\theta_1 = 59.48°$$

$$\theta_2 = 40.30°$$

The angle at which the red bow appears is

$$4\theta_2 - 2\theta_1 = 42.2°$$

With $n = 1.3435$, Equation 36.5 gives the angles for violet light:

$$\theta_1 = 58.80°$$

$$\theta_2 = 39.54°$$

The angle at which the violet bow appears is

$$4\theta_2 - 2\theta_1 = 40.6°$$

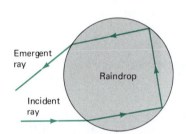

FIGURE 36.10

A secondary bow is formed by rays that undergo two internal reflections.

The rainbow we ordinarily view is one of a series of bows. A secondary bow is formed by rays that undergo two internal reflections, as suggested by Figure 36.10. The secondary bow appears higher in the sky than the primary bow (between angles ranging from 50.7° to 53.6°). A tertiary bow is produced by rays that undergo three internal reflections. Tertiary bows are seldom seen because the observer must face the sun to view them. The normal atmospheric scattering of sunlight ordinarily outshines the triply reflected light of the bow.

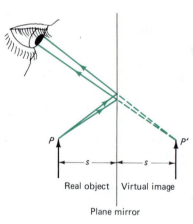

FIGURE 36.11

The upright arrow a distance s in front of the mirror is the object whose image appears behind the mirror. The two rays leaving the point P at the tip of the object are reflected and enter the observer's eye. To the eye the rays appear to come from the point P'. The point P' is the image of the point P. For a plane mirror the image appears to be as far behind the mirror as the object is in front of the mirror.

36.3
MIRRORS

Mirrors and lenses gather and redirect light rays, thereby forming an **image** of a particular object. Image formation by mirrors involves only the law of reflection; the angles of incidence and reflection are equal.

Plane Mirror

The way in which a *plane* mirror forms an image is depicted in Figure 36.11. The full image can be inferred by locating the images of a few key points on the object. The object in Figure 36.11 is an upright arrow. Consider two rays leaving the point P at the tip of the arrow. Diverging rays from P strike the mirror and are reflected to the eye of an observer. The rays appear to diverge from P'. The image of P is the point from which the reflected rays appear to come. Thus, the point P' in Figure 36.11 is the image of the point P. The image is opposite the object and appears to be as far behind the mirror as the object

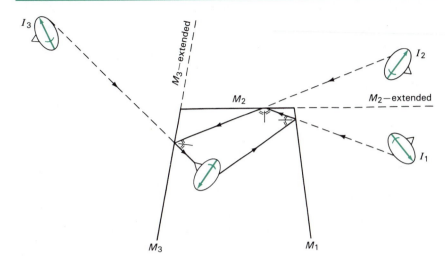

FIGURE 36.12

A triple mirror in a clothing store lets you see your back.

is in front. The image P' is called a **virtual image** because no light energy actually passes through that point.

The image formed by a plane mirror is upright and reversed. The reversal of right and left by a mirror is the reason ambulances often have "ƎƆИA⅃UꓭMA" painted across their front. When seen in a rearview mirror, the word "AMBULANCE" is immediately evident.

Clothing stores often provide triple mirrors that allow you to view your side and back while looking straight ahead. Figure 36.12 shows the path of a light ray as it travels from the back of the head to the eye. Virtual images (I_1, I_2, I_3) are formed at each of the three reflections.

Spherical Mirrors

Let's consider the image-forming action of spherical mirrors, which are formed from portions of a spherical surface and usually have circular edges. Consequently, spherical mirrors have an axis of symmetry, and we can conveniently analyze the image formed by such a mirror by tracing rays in any plane containing the symmetry axis. In many situations, most light rays travel nearly parallel to the spherical mirror axis. A ray is said to be **paraxial** if the angle φ between the ray and the symmetry axis is small enough to allow us to use the small-angle approximations,

$$\sin \varphi \approx \varphi \approx \tan \varphi$$

As we now show, most paraxial rays originating from a given point on an object are focused at a common image point. We restrict ourselves to paraxial rays in our study of mirrors and lenses.

Point Object

Consider first the image of a point object. In Figure 36.13 we illustrate how the image of a point object is located for a concave spherical mirror. A ray from P is reflected at Q, the incident and reflected rays making equal angles

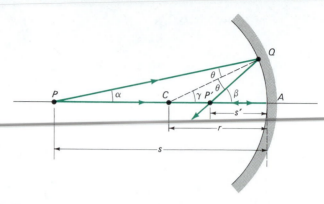

FIGURE 36.13

Paraxial rays from a point object P are focused by a concave mirror to give a real image at P'.

(θ) with the normal CQ (C denotes the center of curvature for the spherical mirror surface). A second ray travels from P along the mirror axis to A and is reflected back along itself. The two reflected rays intersect at P', forming a **real image** of P. An image is called real if light energy passes through the image position. In contrast, no light energy passes through the position of a **virtual image**.

The basic equation governing spherical mirrors relates three lengths, each measured from the mirror surface along the axis. These are the object distance s, the image distance s', and the radius of curvature of the mirror, r. In Figure 36.13 we see how two paraxial rays, PQP' and PAP', intersect at P' to form an image of P. We can use the geometric proposition that states that the sum of the opposite interior angles of a triangle equals the exterior angle to derive the spherical mirror equation. Figure 36.14 shows the geometry of Figure 36.13. The three angles γ, α, and θ are related by

$$\gamma = \alpha + \theta$$

The three angles β, α, and 2θ are related by

$$\beta = \alpha + 2\theta$$

Eliminating the angle θ from this pair of equations gives

$$\alpha + \beta = 2\gamma \tag{36.6}$$

The angle γ is measured from C and is related to the radius of curvature r and the arc length $\overline{QA}$ by

$$\gamma = \frac{\overline{QA}}{r}$$

When the angles α and β are small (as they must be for paraxial rays) the length $\overline{QB}$ is very nearly equal to the length $\overline{QA}$. Thus

$$\tan \alpha \approx \frac{\overline{QA}}{s}$$

and

$$\tan \beta \approx \frac{\overline{QA}}{s'}$$

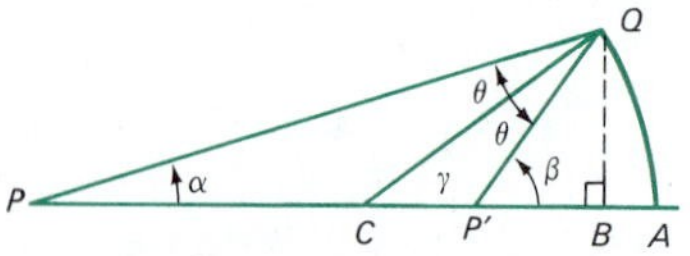

FIGURE 36.14

A point object at P produces a point image at P'.

Further, by using the small-angle approximation for the tangent

$$\tan \alpha \approx \alpha \qquad \tan \beta \approx \beta$$

we can write

$$\alpha \approx \frac{\overline{QA}}{s} \qquad \beta \approx \frac{\overline{QA}}{s'}$$

Inserting these expressions for α, β, and γ into Equation 36.6 and then canceling the common factor $\overline{QA}$ gives the **spherical mirror equation**:

$$\frac{1}{s} + \frac{1}{s'} = \frac{2}{r} \tag{36.7}$$

The fact that the angles α, β, γ, and θ do not appear in the mirror equation shows that all (paraxial) rays from a given point on the object form an image at the same point.

Incident rays traveling parallel to the mirror axis converge at a common point called the **focal point**. The **focal length** f is defined as the image distance when the object is at infinity ($s \to \infty$ corresponds to incident rays traveling parallel to the mirror axis). For a concave mirror,

$$\frac{1}{\infty} + \frac{1}{f} = \frac{2}{r}$$

which shows that the focal length of a concave spherical mirror is half the radius of curvature,

$$f = \tfrac{1}{2}r \tag{36.8}$$

We can rewrite the mirror equation in terms of f as

$$\frac{1}{s} + \frac{1}{s'} = \frac{1}{f} \tag{36.9}$$

Extended Object

So far we have considered only a point object. An extended object can be treated as a set of point objects. Each point on an object that faces a mirror gives rise to an image, and collectively the image points form an extended image. Equation 36.9 applies for each point on the object. If the object distance is essentially the same for all object points, then Equation 36.9 shows that the image distance is the same for all image points. Stated somewhat differently, if the object points lie in a plane perpendicular to the symmetry axis, then the image points also lie in a plane perpendicular to the symmetry axis. The object plane is mapped onto the image plane.

FIGURE 36.15
The candle is at P. The image of the candle at P' is real and inverted.

EXAMPLE 3

Locating the Image of a Candle

A concave mirror with a radius of curvature of 1 m is illuminated by a candle located on the symmetry axis 3 m from the mirror (Figure 36.15). We want to locate the image of the candle.

From Equation 36.8 the focal length is

$$f = \tfrac{1}{2}r = 0.50 \text{ m}$$

Equation 36.9 gives

$$\frac{1}{s'} = \frac{1}{f} - \frac{1}{s} = \frac{1}{0.50 \text{ m}} - \frac{1}{3 \text{ m}} = \frac{5}{3} \text{ m}^{-1}$$

showing that

$$s' = \frac{3}{5} \text{ m} = 60 \text{ cm}$$

This is the mirror-to-image distance. A real image is formed 60 cm in front of the mirror. The image is inverted.

Optical Reversibility

The symmetric way in which s and s' enter the mirror equation is noteworthy:

$$\frac{1}{s} + \frac{1}{s'} = \frac{1}{f}$$

This symmetry illustrates the **principle of optical reversibility**, which states that if any ray is reversed, it will retrace its path through an optical system. Mathematically, we can interchange s and s' and still satisfy the mirror equation. Physically, this interchange means that if the candle in Figure 36.15 is placed at P', its image will be formed at P.

Image Construction

The image of an extended object can be located by following two of the three so-called "easy" rays shown in Figure 36.16. You should learn the routes followed by these rays.

① *A ray traveling parallel to the mirror axis*: This ray appears to come from infinity and is therefore reflected through the focal point.

② *A ray passing through the focal point*: This ray is reflected and travels away parallel to the mirror axis.

③ *A ray passing through the center of curvature*: This ray strikes the spherical mirror along a normal and is reflected back along the same line.

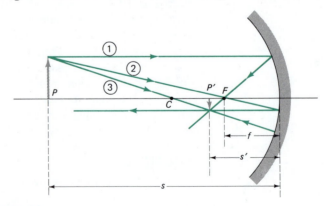

FIGURE 36.16

The three "easy" rays ①, ②, and ③ can be used to locate the image. With the object distance (s) greater than the focal length (f), the image distance (s') is positive, which corresponds to a real image at P'.

The three easy rays in Figure 36.16 intersect at a common point—the image point P' of the tip of the arrow. All three rays are shown, even though the intersection of any two is sufficient to locate the image.

Notice that in Figure 36.16 $s > f$. In other words, the object is *outside* the focal point of the mirror. It follows from the mirror equation, Equation 36.9,

$$\frac{1}{s} + \frac{1}{s'} = \frac{1}{f}$$

that s' is positive when $s > f$. A positive value of s' means that the image is real. As we can see in Figure 36.16, the image is also *inverted*. Figure 36.17 shows a different situation. Now the object is between the focal point and the mirror surface. Mathematically this means that $s < f$, and the mirror equation is satisfied by a negative value of the image distance s'. The negative image distance corresponds to a virtual image.

Construction of the easy rays shows that they diverge after reflection. The image is virtual and is located at a point P' behind the mirror. As indicated in Figure 36.17, the easy rays show that the virtual image is *upright* and *magnified*. Easy ray ① travels toward the mirror parallel to the mirror axis. It is reflected and passes through the focal point. It appears to come from a point behind the mirror surface. Easy ray ② travels toward the mirror along a line passing through the focal point, F. It is reflected, and then travels away parallel to the mirror axis, appearing to come from a point behind the mirror. Easy ray ③ strikes the mirror at right angles and is reflected back along its incident direction, passing through the center of curvature, C. Like rays ① and ②, ray ③ appears to come from a point behind the mirror. The intersection of the three easy rays defines the image of the tip of the arrow. All three easy rays are shown in Figure 36.17, although the intersection of any two is enough to locate an image point. Concave mirrors with long focal lengths magnify the image and are sold as vanity mirrors—the type that might be used when shaving whiskers or applying make up.

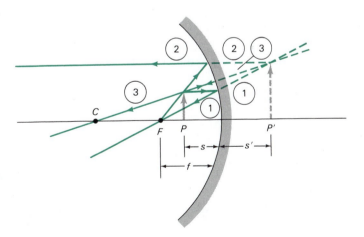

FIGURE 36.17

When the object distance (s) is less than the focal length (f), a concave mirror forms an upright virtual image. The virtual image corresponds to a negative value of the image distance (s') in the mirror equation.

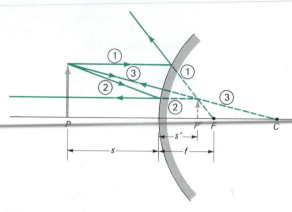

FIGURE 36.18

The easy rays show that a convex mirror always produces an upright virtual image.

Convex mirrors are also properly described by the mirror equation, Equation 36.9, provided that both s' and r (and $f = \frac{1}{2}r$) are assigned negative values. Figure 36.18 shows the easy rays for a convex mirror. Easy ray ① travels parallel to the mirror axis and is reflected along a line whose extension passes through the focal point, F. Easy ray ② travels toward the focal point. After reflection it travels away parallel to the mirror axis. Easy ray ③ travels toward the center of curvature, C, and is reflected back along itself. All three easy rays are shown in Figure 36.18. The intersection of any two rays is sufficient to locate the image, which is *virtual* ($s' < 0$) and *upright*. Convex mirrors are often used as rearview mirrors, particularly by truck drivers and other drivers towing wide loads, because they allow the driver to view a wide area.

With a spherical mirror not all paraxial rays are reflected through the same point, and therefore the image of a point is not a point. In other words, the image is not sharp. This lack of image sharpness caused by the spherical shape of the mirror surface is called spherical aberration. In reflecting telescopes, spherical aberration is avoided by using a mirror with a parabolic surface (Figure 36.19). Parallel rays traveling parallel to the symmetry axis are reflected and converge at a point, the *focus* of the parabola. The reflecting surfaces for automobile headlamps are parabolic. With the lamp at the focus of the parabola, the rays travel outward as a nearly parallel beam.

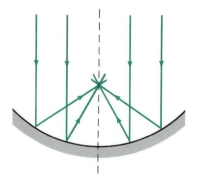

FIGURE 36.19

Parallel rays converge at the focus of a parabolic mirror. If a light source is placed at the focus the reflected rays travel outward as parallel rays.

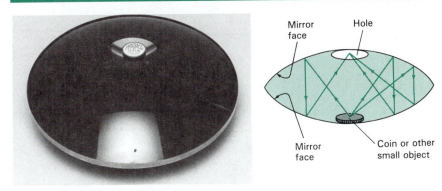

FIGURE 36.20
(a) The Illusion® demonstration. (b) Rays from the object form an image in the hole.

EXAMPLE 4

An Illusion

An intriguing optical demonstration called Illusion® consists of two parabolic mirrors (Figure 36.20). The mirrors are **confocal**; the focal point of each mirror lies on the surface of the other. The upper mirror has a hole at its center. The hole is the location of the focal point of the lower mirror. If a coin is placed at the center of the lower mirror, rays from the coin strike the upper mirror and travel along parallel paths. They strike the lower mirror, are reflected, and converge to produce a real image in the hole. The lower mirror is also imaged in the hole, so the image of the coin seems to be resting on a solid mirror surface.

36.4
LENSES

A lens forms an image by refraction. A real or virtual image is formed, depending on the type of lens and on the object-to-lens distance. We use the words *convex* and *concave* to describe the surface of the lens as viewed from a position outside the lens. Figure 36.21 shows three different combinations of lens surfaces.

The Thin-Lens Equation

The equation describing image formation by a lens is derived by applying Snell's law at the two refracting surfaces. When the thickness of the lens is ignorably small compared to the other three relevant lengths—image distance, object distance, and focal length—the equation takes on a particularly simple form known as the **thin-lens equation**:

$$\frac{1}{s} + \frac{1}{s'} = \frac{1}{f} \tag{36.10}$$

In Figure 36.22, s denotes the object distance (point object at P), s' denotes the image distance, and f is the focal length. Mathematically, the thin-lens equation has the same form as the mirror equation, Equation 36.9. In Figure 36.22 a real image is formed at P'. In general, the character of the image (real, virtual,

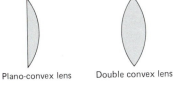

Plano-convex lens   Double convex lens

Plano-concave lens

FIGURE 36.21
Three types of lenses, classified according to the shapes of their surfaces.

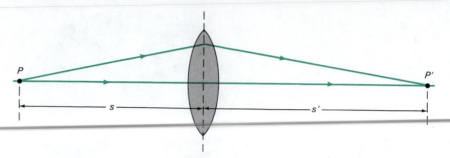

FIGURE 36.22
Object-image distances for a convex lens.

upright, inverted, magnified, minified) depends on the focal length and the object distance. In the thin-lens equation a positive value of s' indicates a real image. When s' is negative the image is virtual.

EXAMPLE 5

Real and Virtual Images

Figure 36.23a shows an object placed 15 cm from a convex lens whose focal length is 6 cm. The image distance follows from the thin-lens equation

$$\frac{1}{s'} = \frac{1}{f} - \frac{1}{s} = \frac{1}{6 \text{ cm}} - \frac{1}{15 \text{ cm}} = \frac{9}{90 \text{ cm}}$$

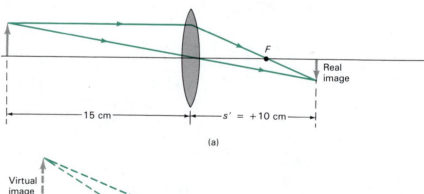

(a)

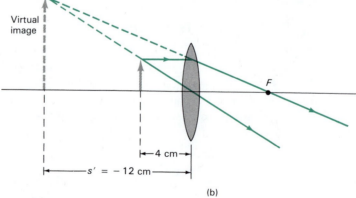

(b)

FIGURE 36.23
(a) A real image results when $s > f$. (b) A virtual image results when $s < f$.

which gives

$$s' = 10 \text{ cm}$$

The positive value of s' indicates that a real image is formed 10 cm beyond the lens. In Figure 36.23b the object has been moved closer to the lens, to a position where $s = 4$ cm. The image distance is

$$\frac{1}{s'} = \frac{1}{f} - \frac{1}{s} = \frac{1}{6 \text{ cm}} - \frac{1}{4 \text{ cm}} = -\frac{2}{24 \text{ cm}}$$

which gives

$$s' = -12 \text{ cm}$$

The negative value of s' indicates that a virtual image is formed 12 cm in front of the lens.

For a lens formed from two convex spherical surfaces (see Figure 36.22) the focal length f is determined by the **lensmaker's formula**:

$$\frac{1}{f} = \left(\frac{n_2}{n_1} - 1\right)\left(\frac{1}{r_A} + \frac{1}{r_B}\right) \tag{36.11}$$

In Equation 36.11, r_A and r_B are the radii of the surfaces, n_2 is the index of refraction of the lens, and n_1 is the index of refraction of the surrounding medium. In most situations the surrounding medium is air, and n_1 can be set equal to 1.

EXAMPLE 6

How to Cut the Focal Length in Half

A plano-convex lens has a radius of curvature of 6 cm. Its index of refraction is $n = 1.55$. The focal length in air follows from Equation 36.11,

$$\frac{1}{f} = (1.55 - 1.00)\left(\frac{1}{\infty} + \frac{1}{6} \text{ cm}\right)$$

giving $f = 10.9$ cm. If two identical plano-convex lenses are cemented together, back to back, the resulting double convex lens has a focal length that is one-half that of the plano-convex lenses. Thus, with $r_A = r_B = 6$ cm,

$$\frac{1}{f} = (1.55 - 1.00)\left(\frac{1}{6} \text{ cm} + \frac{1}{6} \text{ cm}\right) = \frac{1.10}{6 \text{ cm}}$$

This gives

$$f = 5.45 \text{ cm}$$

which is just half the value for a single plano-convex lens. We can see from Equation 36.11 that this is a general result. With $r_A = r_B$ (double convex), $1/f$ is twice as great as with $r_A = \infty$ (plano-convex), and thus f is cut in half. Physically, this means that equal ray bending occurs at the two identical refracting surfaces.

A lens has two focal points, located at equal distances on either side of the lens. The focal length f is the distance from the lens to a focal point. The focal point is defined as the image distance for an object at infinity ($s' \rightarrow f$

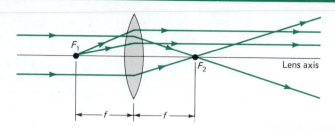

FIGURE 36.24

A double convex lens. Rays from infinity are focused at a focal point. Rays originating at a focal point are refracted and emerge traveling parallel to the lens axis.

for $s \to \infty$), or as the object distance for an image at infinity ($s' \to f$ for $s' \to \infty$). An image or object at infinity corresponds to rays traveling parallel to the lens axis. Figure 36.24 shows one set of parallel rays being focused at one focal point (F_2) and another set of divergent rays leaving the other focal point (F_1) and traveling parallel to the lens axis after being refracted.

Concave Lens

For a thin lens formed by two concave shperical surfaces the thin-lens equation, Equation 36.10, still applies, but the focal length is *negative*. The focal length for a double concave lens is given by the lensmaker's formula, Equation 36.11, with r_A and r_B taken to be *negative*.

A lens with a negative focal length causes light rays to **diverge**. A lens with a positive focal length causes light rays to **converge**. By using different combinations of lenses and mirrors it is possible to tailor the image-forming characteristics of an optical system.

Easy Rays

The types of images formed by thin lenses can be determined by using a ray diagram. Just as we did with mirrors, we use a set of easy rays. Figures 36.25a and b show the three easy rays for concave and convex lenses:

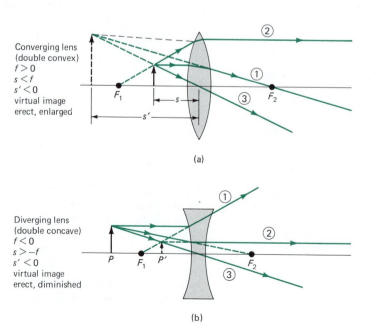

FIGURE 36.25

The easy rays can be used to locate image points. (a) Converging lens. (b) Diverging lens.

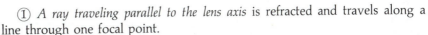

① *A ray traveling parallel to the lens axis* is refracted and travels along a line through one focal point.

② *A ray traveling along a line through the other focal point* is refracted and travels parallel to the lens axis.

③ *A ray passing through the center of the lens* undergoes no net refraction.

The Fresnel Lens

In certain applications, a large-diameter, short-focal-length lens is required. For example, large lens arrays are needed in certain types of solar energy collectors. Simple plano-convex lenses are not well suited for such arrays because they are heavy and costly. The Fresnel lens is a cheap, lightweight substitute. The surface of a Fresnel lens is a series of concentric circular ridges, as shown in the cross-sectional view in Figure 36.26. The focusing action of a plano-convex lens occurs at the curved surface. Figure 36.26 illustrates a process in which segments of the curved surface of a plano-convex lens are removed and placed on a thin flat surface. Incident light normally meets the same curved refracting surface as it would encounter on the much bulkier plano-convex lens. Fresnel lenses are also used in lighthouses and in overhead projectors. They collect diverging rays and combine them into a well-defined beam. Fresnel lenses with a negative focal length are often used in the rear windows of recreational vehicles or motor homes. They offer the driver a wide-angle view, much of which would otherwise be blocked by the vehicle.

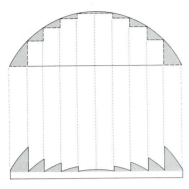

FIGURE 36.26

Cross-sectional view showing the principle of the Fresnel lens.

36.5
OPTICAL SYSTEMS

To illustrate and reinforce the ideas developed in the preceding sections, we consider three important optical systems: the human eye, a simple magnifier, and the telescope.

The Human Eye

Light enters the eye through the **cornea**, a transparent bulge on the **sclera**, the otherwise opaque outer surface of the eye (Figure 36.27). Most of the converging action of the eye occurs at the air–cornea boundary, where the index of refraction changes from 1.00 to 1.38. The light rays travel inward through a liquid called the **aqueous humor**, passing through the **pupil**, a variable-diameter opening in the opaque **iris**, the colored part of the eye. The rays undergo further refraction in the **crystalline lens**. The curvature and thus the focal length of the crystalline lens can be varied by eye muscles. The capacity to vary the curvature is called **accommodation**, and allows the eye to focus on objects at different distances. The light rays then travel onward through the **vitreous humor**, and are at last focused on a thin layer of photosensitive cells that make up the **retina**.

The retinal cells are of two kinds: the **rods**, numbering over 100 million, are very sensitive to low-intensity light, but do not distinguish colors or provide sharp images; the **cones**, numbering between 5 and 10 million, sharpen images and are sensitive to color, but are ineffective at low intensities. Thus, when you view a magazine in a semi-dark room, the cones are not activated. You may be able to see that the pages carry words and pictures, but you will not be able to distinguish the words. Nor will you be able to tell whether

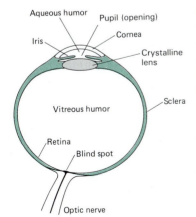

FIGURE 36.27

The human eye.

X   L   R

FIGURE 36.28

How to locate your blind spot. Close your left eye and look at the "X." With the page about 15 inches from your eye, slowly move the page toward your eye. The "R" will disappear at some point. Moving the page closer to your eye causes the "R" to reappear and the "L" to disappear.

the pictures are in color or in black and white. Increasing the intensity of the light brings the cones into play.

The retinal signals are relayed to the brain through the **optic nerve**. There are no rods or cones in the region where the optic nerve is attached to the retina. Accordingly, light focused on this spot is not perceived. This region is called the **blind spot**. Figure 36.28 shows you how to recognize your blind spot.

In a normal human eye, light from a distant object is focused on the retina with the lens muscle relaxed. As an object is brought closer, the lens contracts, becoming more convex. This shortens the focal length (f) of the lens. The necessity of shortening f follows from the thin-lens equation, $1/s + 1/s' = 1/f$. In order to keep the image focused on the retina, the image distance s' must remain constant. It follows that a decrease in the object distance s must be accompanied by a decrease in the focal length.

The ability of the eye to vary its focal length is limited. Objects closer than a certain distance appear blurred. The **near point** is the closest point on which the eye can focus. The distance from the eye to the near point ranges from less than 10 cm for youngsters to over 100 cm for older people.

Two common eye disorders are nearsightedness (**myopia**) and farsightedness (**hyperopia**). In the relaxed myopic eye, parallel rays are focused before they reach the retina. This defect can be corrected by a spectacle or contact lens with a negative focal length (Figure 36.29). Such a corrective lens diverges the rays slightly, thereby compensating for the excessive refraction by the eye. In the farsighted eye, rays are converged toward a focal point behind the retina. This condition is alleviated by a lens with a positive focal length. Such a lens converges rays, compensating for the lack of refraction by the eye.

Two thin lenses in contact, with focal lengths f_1 and f_2, behave like a single thin lens with a focal length f given by

$$\frac{1}{f} = \frac{1}{f_1} + \frac{1}{f_2}$$

(36.12)

Contact lenses are thin glass or plastic lenses that adhere to the cornea. We can get a good idea of how contact lenses alter rays by assuming that the eye is described by the thin-lens equation,

$$\frac{1}{s} + \frac{1}{s'} = \frac{1}{f_E}$$

where f_E is the focal length of the unaided eye. This allows us to treat the eye-plus-contact-lens as a pair of thin lenses in contact. If we denote the focal length of the contact lens by f_L, then the equivalent focal length (f) of the eye-plus-lens combination is given by

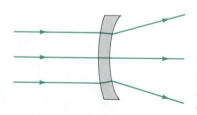

FIGURE 36.29

Section of a spectacle lens for a myopic (nearsighted) eye.

$$\frac{1}{f} = \frac{1}{f_E} + \frac{1}{f_L}.$$

(36.13)

EXAMPLE 7

Contact Lens Focal Length for a Myopic Eye

To apply Equation 36.13 we consider a myopic eye in which incident parallel rays are focused at a distance $f_E = 1.69$ cm. The correct focal length, which will focus rays on the retina, is $f = 1.71$ cm.

The contact lens must have a focal length f_L, given by Equation 36.13. Thus,

$$\frac{1}{f_L} = \frac{1}{f} - \frac{1}{f_E} = \frac{1}{1.71 \text{ cm}} - \frac{1}{1.69 \text{ cm}} = -0.00692 \text{ cm}^{-1}$$

This gives

$$f_L = -144 \text{ cm}$$

In this case a weak diverging contact lens will remedy the myopic condition.

Our study of optics has been restricted to lenses that are symmetric. If a symmetric lens is rotated about its symmetry axis, the image will not change. If the lens is asymmetric, however, rotation will distort the image. Very often the cornea of the human eye becomes asymmetric, causing what is called **astigmatism**. If you wear glasses, there is a simple way to determine whether or not you have astigmatism. Remove your glasses and rotate the lenses about a horizontal axis while looking through them. If objects appear to change shape and become distorted as you rotate the lenses, you are astigmatic; your cornea and the correcting lens are not symmetric about the rotation axis.

The Magnifying Glass

The size of the retinal image is determined by the angle subtended by the object. As an object is moved closer to the eye its apparent size increases (Figure 36.30). Psychologists have developed a test that shows that the eye detects only the angular size of objects. Using one eye, a subject views two automobiles. One is a full-sized vehicle, placed rather far from the subject, and the other is a miniature model. The model is placed much closer to the subject, with the result that both the model and the full-sized car subtend the same angle. With proper lighting, most people perceive both cars to be full-sized and far away.

When an object is brought closer than the near point, its apparent size continues to increase, but the image blurs. A simple magnifier consists of a single positive lens $(f > 0)$ that permits an object to be brought closer than the near point and still remain in focus. In addition to magnification, it is desirable

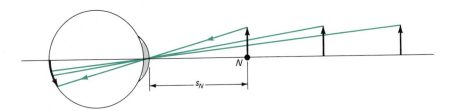

FIGURE 36.30

The size of the retinal image increases as the object is moved closer to the eye. The largest clear image is formed with the object at the near point, N.

that a magnifier give an upright image. These conditions require that the object distance be less than the focal length.

The **angular magnification** M (also called the **magnifying power**) is the ratio of the angles subtended by the magnified image and by the unmagnified image when the object is at the near point. Figure 36.31a shows that the angular size of the unmagnified image is

$$\alpha_0 = \frac{y}{S_N}$$

where y is the height of the object and S_N is the distance from the eye to the near point. The greatest usable magnification occurs when the image is at the near point; bringing the image closer makes it larger, but blurred. Figure 36.31b shows the image-object relation when the virtual image is at the near point, a distance S_N from the magnifier. At this position the angular size of the image is

$$\alpha = \frac{y}{s}$$

where s is the object-to-magnifier distance. The angular magnification is

$$M = \frac{\alpha}{\alpha_0} \tag{36.14}$$

It follows that

$$M = \frac{S_N}{s}$$

Using the thin-lens equation with $s' = -S_N$ for the virtual image distance gives

$$\frac{1}{s} = \frac{1}{f} - \frac{1}{s'} = \frac{1}{f} + \frac{1}{S_N}$$

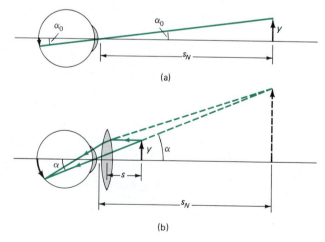

(a)

(b)

FIGURE 36.31

The size of the retinal image determines how big an object appears. (a) With the object at the near point, the angular size of the unmagnified image is α_0. (b) Maximum magnification is achieved when the magnified image is at the near point.

The magnification follows as

$$M = 1 + \frac{S_N}{f} \tag{36.15}$$

Equation 36.15 shows that a large magnification requires a lens with a short focal length. Customarily, S_N is taken to be 25 cm in the design of magnifiers and other commercial optical instruments. For example, to determine the focal length of an "8 power" or "8 X" magnifier, we use Equation 36.15 with $S_N = 25$ cm and $M = 8$. This gives

$$f = \frac{S_N}{M - 1} = \frac{25 \text{ cm}}{8 - 1} = 3.57 \text{ cm}$$

The Telescope

Credit for inventing the telescope is given to a Dutchman, Hans Lippershey. When Lippershey applied for a patent for his invention in 1608, the Dutch government refused to grant it. Instead, the rights to the telescope were purchased by the government and Lippershey was hired to perform research aimed at improving the instrument. The Dutch government recognized the great military significance of the telescope and wanted the device to remain its military secret. But then, as now, scientific secrets were difficult to keep. When the knowledge that Lippershey had fashioned a telescope reached Italy, Galileo was able to figure out independently how a combination of two lenses could be arranged to magnify distant objects. Galileo then constructed his own telescope and used it to discover four of Jupiter's moons, sunspots, and Saturn's rings.

The simplest refracting telescope (Figure 36.32) consists of two lenses that form a magnified image of a distant object. The objective lens gathers light and focuses rays in a plane. The image formed by the objective lens acts as an object for the eyepiece lens that produces the final magnified image.

The object distance (s_0) is nearly infinite by comparison to the objective lens focal length (f_0). It follows from the thin-lens formula that the image distance for the objective lens (s_0') is just slightly greater than f_0. The eyepiece is positioned so that the image formed by the objective coincides with the focal point of the eyepiece. This makes the object distance for the eyepiece equal to the eyepiece focal length $(s_E = f_E)$ and gives an inverted virtual image at infinity $(s_E' = -\infty)$.

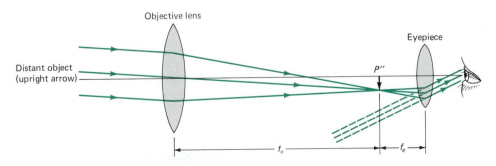

FIGURE 36.32

Ray paths for a refracting telescope. Parallel rays are focused in the plane passing through P'', which is also a focal plane of the eyepiece. The image at P'' acts as an object for the eyepiece.

The magnification of the telescope is given by

$$M = \frac{f_0}{f_E} \tag{36.16}$$

A large magnification can be achieved by using an objective with a large focal length. In order to gather light effectively from faint sources at great distances the objective lens must also have a large diameter. Happily, these two requirements are compatible: It is possible to make a large-diameter lens that has a slight curvature and thus a large focal length.

The basic optical features of the magnifying glass, the human eye, and the telescope can be explained by using geometric optics. In general, the ray picture of geometric optics is satisfactory for phenomena in which relevant dimensions of the optical system—such as the diameter of a lens—are large by comparison to the wavelength of light. The ray picture is inadequate for describing interference phenomena, such as the diffraction of light. Such phenomena must be described in terms of waves. In Chapters 37 and 38 we will develop physical optics, which is based on the wave picture of light.

WORKED PROBLEM

A thin convex lens ($f = 10$ cm) is placed 25 cm in front of a concave mirror ($f = 5$ cm). An object is positioned 15 cm in front of the lens. (a) Draw a ray diagram to locate the image and determine if the image is real or virtual, upright or inverted. (b) Use the mirror and lens equations to locate the position of the image.

Solution

(a) Ray tracing in this problem requires an accurate drawing because two optical systems are involved. Let's look first at the action of the lens alone. In Figure 36.33a, the easy ray parallel to the symmetry axis and the easy ray through the focal point intersect about 30 cm from the lens. In the absence of the mirror, a real and inverted image would form at this distance from the lens.

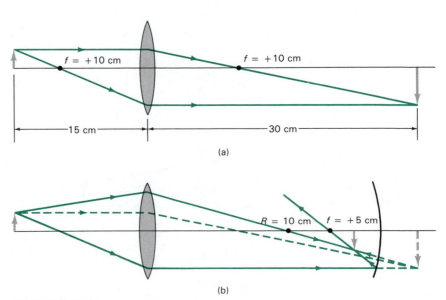

(a)

(b)

FIGURE 36.33

When the mirror is inserted 25 cm from the lens, it reflects rays that are converging (Figure 36.33b). One easy ray for the mirror is the ray emerging from the lens parallel to the symmetry axis. This ray reflects from the mirror and passes through the focal point of the mirror. To locate the image formed by the mirror we need a second reflected ray. We pick the ray emerging from the lens that passes through the center of curvature of the mirror. This ray strikes the mirror and reflects back along itself. The intersection of the two reflected rays locates the image formed by the mirror. As drawn in Figure 36.33b, the image is approximately 3 cm to the left of the mirror. The image is real and inverted.

(b) We can use the lens formula to locate the position of the image formed by the lens alone. With $s = 15$ cm and $f = 10$ cm, we find for the image distance s',

$$\frac{1}{s'} = \frac{1}{f} - \frac{1}{s} = \frac{1}{10 \text{ cm}} - \frac{1}{15 \text{ cm}} = \frac{1}{30 \text{ cm}}$$

and thus,

$$s' = 30 \text{ cm}$$

Because this image falls to the right of the mirror it becomes a virtual object for the mirror. This means that the object distance for the mirror equation is negative. Applying the mirror equation we have $s = -5$ cm and $f = 5$ cm. The image distance follows from

$$\frac{1}{s'} = \frac{1}{f} - \frac{1}{s} = \frac{1}{5 \text{ cm}} - \frac{1}{-5 \text{ cm}} = \frac{2}{5 \text{ cm}}$$

Thus,

$$s' = 2.5 \text{ cm}$$

The image formed by the lens and mirror system is located 2.5 cm to the left of the mirror.

EXERCISES

36.2 Reflection and Refraction

A. A ray of light in air making an angle of 40° with the vertical strikes a horizontal water surface. Part of the light is reflected. (a) What is the angle of the reflected ray relative to the vertical? (b) Calculate the angle of the refracted ray relative to the vertical.

B. (a) A coin rests on the bottom of a pool of water 80 cm deep (Figure 1). When viewed from positions nearly above the coin, the angles of the incident and refracted rays are small. Use the small-angle approximations for $\sin \theta_1$ and $\sin \theta_2$ in Snell's law to show that the apparent depth of the water is 60 cm. Take the index of refraction of water to be 4/3. (b) For larger angles, does the apparent depth increase or decrease? (c) Can the apparent depth exceed the actual depth?

C. Light rays travel upward through the glass bottom of a fish bowl. The index of refraction of the glass is 1.52. Determine the critical angle for the glass-water boundary. Is it larger or smaller than the critical angle for a water-air boundary?

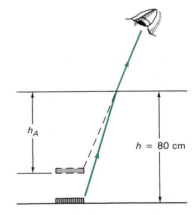

FIGURE 1

D. Suppose that you are under water near the edge of a swimming pool. Does someone standing above water at poolside look shorter or taller to you than when viewed from the same position when the pool is empty?

36.3   Mirrors

E.   Sketch the palm of your left hand. Then sketch the mirror image of the palm of your right hand. Compare the two sketches.

F.   No light energy passes through the position of a virtual image. Could you photograph a virtual image?

G.   Many trucks are equipped with two rearview mirrors; one plane, and the other convex. What special function does each mirror serve that the other cannot?

H.   The radius of curvature of a convex mirror is 10 cm. (a) What is its focal length? (b) The image of an object lies halfway between the mirror surface and the focal point. Use the mirror equation to determine the position of the object. Make a ray diagram.

I.   A point object lies at one end of the diameter of a hemispherical mirror (Figure 2). Paraxial rays are focused at a point F located $\frac{1}{2}R$ from the mirror surface. Do rays reflected through large angles cross the mirror axis at points closer to or farther from the mirror than F? Draw one ray that confirms your answer.

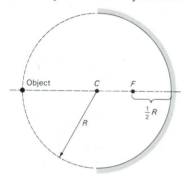

FIGURE 2

J.   A concave mirror has a radius of curvature of 1.2 m. If you stand 5 m in front of the mirror, where is your image formed? Can you look into the mirror and see your image? Explain.

K.   A concave mirror has a focal length of 2.0 m. If you stand 1.0 m in front of it, where is your image formed? Is the image real or virtual? Is it upright or inverted?

L.   An object is placed 7.5 cm in front of a concave mirror with a 10 cm radius of curvature. (a) Using the three easy rays, construct the image. (b) Is the image real or virtual; erect or inverted? (c) Calculate the image distance.

M.   An object is 20 m from a convex mirror with a focal length of -0.20 m. Determine the position of the image. Is it upright or inverted; real or virtual?

36.4   Lenses

N.   Use a ray diagram to show how a raindrop on the windshield of a car gives an inverted image.

P.   Suppose that you were handed a convex lens and a ruler and told to determine the focal length of the lens. How would you proceed?

Q.   A thin lens has a focal length of 22 cm. An object is placed 8 cm to the left of the lens. Describe the image.

R.   The surfaces of a symmetric double convex lens are ground to a curvature of 20.3 cm. The index of refraction of the glass is 1.64. What is the focal length of the lens?

S.   The image and object for a thin convex lens are on opposite sides of the lens, each 28 cm away from the lens. What is the focal length of the lens?

T.   A symmetric double convex lens is designed to have a focal length of 6.0 cm under water. The index of refraction of the glass is 1.54. Calculate the radius of curvature of the convex surfaces.

U.   An object is placed 12 cm to the left of a thin lens with a focal length of 7.5 cm. (a) Use easy rays to construct the image. (b) Use the thin-lens equation to determine the image distance.

V.   A point source of light is placed 8 cm from a convex glass surface ($r = 2$ cm, $n = 1.60$). Determine the image location. Is it a real image or a virtual image? Repeat for an object distance of 1 cm. Sketch rays for both situations.

W.   (a) Using glass with an index of refraction 1.52, what radius of curvature is needed to produce a plano-convex lens with a focal length of 30 cm? (b) What radius of curvature (the same for both faces) would be required to produce a double convex lens with a focal length of 30 cm?

36.5   Optical Systems

X.   The objective lens of a small portable telescope has a focal length of 1.20 m. What focal length eyepiece should be used to achieve 200 power magnification?

Y.   Determine the magnification of a simple magnifier with a focal length of 2.5 cm.

PROBLEMS

1.   If the distance from the cornea to the retina is 2.12 cm, by how much must the focal length of the eye *change* to follow an object from the near point (25 cm) to infinity? Treat the eye as a thin lens in which all of the refraction occurs at the cornea.

2.   (a) Paraxial rays emerge from a point source and are refracted from a thin lens ($f = +5$ cm) that is 10 cm away. Where is the image located? Is it real or virtual? (b) The rays travel onward through a second lens ($f = -5$ cm) located 20 cm beyond the first. Where is the image located? Is it real or virtual?

3.   A thin convex lens ($f = 10$ cm) is placed 25 cm in

front of a concave mirror ($f = 5$ cm; Figure 3). An object is positioned 15 cm in front of the lens. (a) Draw a ray diagram to locate the image. Is the image real or virtual; upright or inverted? (b) Use the mirror and lens equations to locate the position of the image.

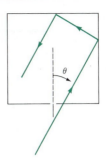

FIGURE 5

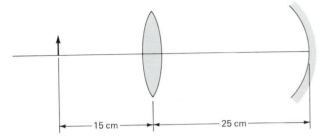

FIGURE 3

4. (a) Use a ray diagram to show that a plane mirror of length $\frac{1}{2}L$ is long enough for a person of height L to view her entire reflected image. (b) Is there any restriction on the position of the top edge of the mirror? Would the minimum length be increased or decreased if the mirror were (c) slightly concave, (d) slightly convex?

5. The high index of refraction of diamond enables a gem to sparkle via internally reflected rays. Such rays would be refracted out of materials with lower values of n, thereby reducing the retroreflected rays. Compare the critical angle of incidence for diamond and flint glass surrounded by air. For the rays indicated in Figure 4, show that the second reflection is totally internal for diamond, but that refraction occurs for glass.

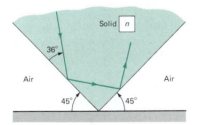

FIGURE 4

6. Figure 5 shows a top view of a square enclosure. The inner surfaces are plane mirrors. A ray of light enters a small hole in the center of one mirror. (a) At what angle θ must the ray enter in order to exit through the hole after being reflected once by each of the other three mirrors? (b) Are there other values of θ for which the ray can exit after multiple reflections? If so make a sketch of one of the ray's paths.

7. Five identical plane mirrors form 5 sides of a hexagon enclosure (Figure 6). Describe what happens to a ray that enters the enclosure parallel to and just above the face of the mirror to the right of the open side.

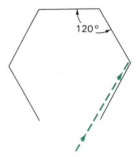

FIGURE 6

8. Students in an undergraduate laboratory allow a narrow beam of laser light to strike a water surface. They arrange to measure the angle of refraction for selected angles of incidence and record the data shown in the accompanying table. Use the data to verify Snell's law by plotting the sine of the angle of incidence versus the sine of the angle of refraction. Use the resulting plot to deduce the index of refraction of water.

Angle of Incidence (degrees)	Angle of Refraction (degrees)
10.0	7.5
20.0	15.1
30.0	22.3
40.0	28.7
50.0	35.2
60.0	40.3
70.0	45.3
80.0	47.7

9. A ray of light passes through a prism as shown in Figure 7. (a) Calling D the angle of deviation, show that

$$\theta + \beta = A + D$$

where θ is the angle of incidence and β is the angle of refraction for the emerging ray. (b) If the ray inside the prism is parallel to the base of the prism show that

$$\sin[\tfrac{1}{2}(A + D)] = n \sin(\tfrac{1}{2}A)$$

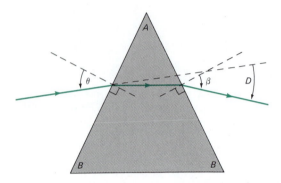

FIGURE 7

10. An object 1 cm high is placed 20 cm in front of a concave lens having a focal length of -10 cm. A convex mirror having a focal length of -10 cm is placed 20 cm beyond the lens. (a) Make a scale drawing of the system and locate the image by tracing rays. (b) Confirm the location and nature of the image using the lens and mirror equations.

11. An object and a screen are separated a fixed distance L. Between the screen and object is a converging lens having a focal length f. (a) Use the symmetry of the thin-lens equation to argue why there are two positions of the lens for which a real image is formed. (b) Determine the location of the two positions in terms of L and f.

12. Consider a concave mirror. Plot s versus s' and confirm graphically that there are no solutions to the mirror equation for which both s and s' are negative.

13. Consider two mirrors intersecting at 90° (Figure 8).

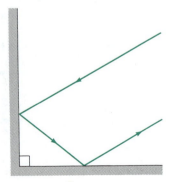

FIGURE 8

Prove that for any angle of incidence the emerging ray is parallel to the incident ray. (Three-dimensional versions, called corner cubes, were placed on the moon and are used to reflect a laser beam used to make precise earth-moon distance measurements.)

14. (a) A ray of light is incident on a pane of glass 9.0 mm thick (Figure 9). The light makes an angle of 35° with the vertical. The glass has an index of refraction of 1.60. The top and bottom surfaces of the glass are parallel and both give rise to reflected rays. Calculate the separation δ of the two reflected rays. (b) Show that when the angle of incidence θ is small, the separation δ is given by

$$\delta = t\theta\left(1 - \frac{n_{air}}{n_{glass}}\right)$$

where t is the thickness of the glass. (c) Use the data from part (a) to calculate δ using the formula derived in part (b).

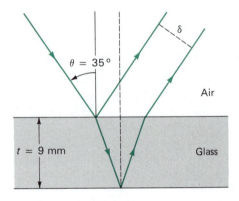

FIGURE 9

15. The prism shown in Figure 10 has an index of refraction of 1.55. Light is incident at an angle of 20°. Determine the angle θ at which the light emerges.

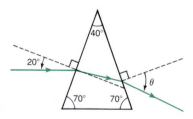

FIGURE 10

16. (a) It is desired to have light enter one face of a prism, undergo total internal reflection, and emerge

from an adjacent face as shown in Figure 11. What is the minimum index of refraction of the glass? (b) What is the minimum index of refraction when the prism is submerged in water?

17. A laser beam strikes one end of a slab of material, as shown in Figure 12. The index of refraction of the slab is 1.48. Determine the number of internal reflections of the beam before it emerges from the opposite end of the slab.

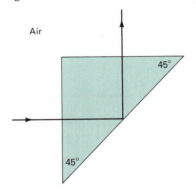

Air

45°

45°

FIGURE 11

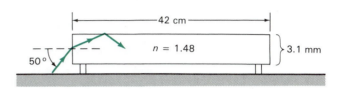

42 cm

$n = 1.48$

3.1 mm

50°

FIGURE 12

CHAPTER 37

PHYSICAL OPTICS: INTERFERENCE

37.1
INTERFERENCE

Light is treated as a wave in physical optics. This enables us to apply the principle of superposition for waves that we first introduced in Section 16.4:

A meeting of two or more waves produces a waveform that is the sum of the waveforms produced by each wave acting separately.

The superposition of waves can result in **interference**, as shown for water waves in Figure 37.1. Circular water waves are generated at two points. At certain positions the two waves meet and cancel each other and the surface of the water at these positions is undisturbed. At other positions the two waves reinforce one another and produce a maximum disturbance of the surface.

In general, interference is said to be **constructive** at points where the intensity of the superposed waves is greater than the sum of the intensities of the two separate waves. At points where the intensity is less than the sum of the intensities of the two separate waves, the interference is called **destructive**.

Ordinarily, light does not display interference. For example, if two flashlights illuminate the same surface, their combined intensity is simply the sum of the two separate intensities. Likewise, the many floodlamps used to illuminate a stadium produce a total intensity that is the sum of the individual lamp intensities.

Under the proper conditions, however, light does display interference. The interference of light was first demonstrated in 1801 by Thomas Young, a

FIGURE 37.1

The superposition of water waves results in interference.

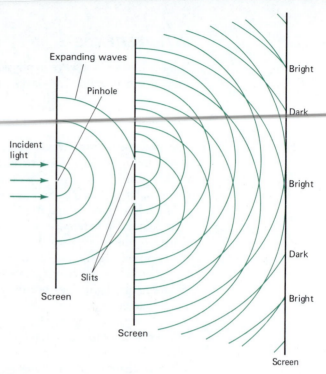

FIGURE 37.2

Thomas Young's experiment. Monochromatic light from a sodium flame reaches a pinhole in a card. Wavefronts from the pinhole expand and reach a pair of slits in a second card. Waves passing through the slits travel on to the screen where they intefere.

brilliant English physician. Young's experiment, the basic features of which are shown in Figure 37.2, is important because it established the wave picture of light. In Young's experiment, monochromatic light from a sodium flame reaches a small hole in a screen. Waves from the hole expand and reach two closely spaced slits in a second screen.* Waves from these two slits spread out and superpose, interfering constructively at some places and destructively at others. When displayed on a third screen, the regions of constructive and destructive interference appear as an alternating series of bright and dark bands called **interference fringes**. Figure 37.3 shows the interference fringes produced by two closely-spaced slits.

We can explain the results of Young's experiment if we describe each wave in terms of its electric field, $E(x, t)$. We let E_1 and E_2 denote the electric fields of the waves from the two slits. For monochromatic light we can represent E_1 and E_2 in terms of sinusoidal waveforms like the one shown in Figure 37.4a. A mathematical description of the sinusoidal waveform of Figure 37.4a is given by

$$E_1(x_1, t) = E_0 \sin\left[2\pi\left(\frac{x_1}{\lambda} - ft\right)\right] \tag{37.1}$$

---

* In Young's original experiment there were two pinholes in the second screen rather than slits. The use of narrow slits increases the intensity of the interference pattern.

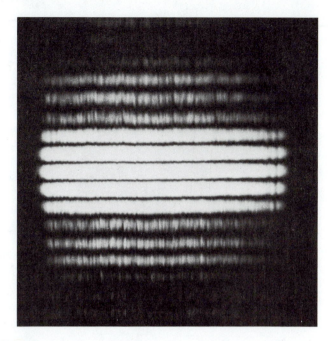

FIGURE 37.3

Interference fringes produced by light passing through two narrow slits.

where λ is the wavelength (the crest-to-crest distance in Figure 37.4a), and f is the frequency. The quantity in brackets is the **phase** of the wave and is represented by φ_1:

$$\varphi_1 = 2\pi\left(\frac{x_1}{\lambda} - ft\right) \qquad (37.2)$$

The field E_1 can be expressed in terms of the phase as

$$E_1 = E_0 \sin\varphi_1 \qquad (37.3)$$

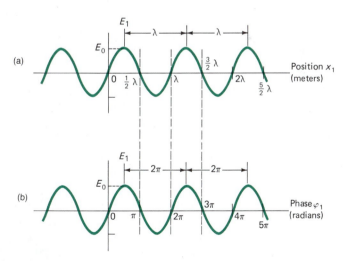

FIGURE 37.4

(a) Graph of sinusoidal waveform versus position. (b) The same sinusoidal waveform graphed versus the phase angle φ.

Likewise,

$$E_2(x_2, t) = E_0 \sin \varphi_2 \qquad (37.4)$$

with

$$\varphi_2 = 2\pi \left(\frac{x_2}{\lambda} - ft \right) \qquad (37.5)$$

The distances x_1 and x_2 are measured from origins located at slits 1 and 2 (Figure 37.5). The fields E_1 and E_2 have equal amplitudes (E_0) because we assume that the two slits are identical.

The principle of superposition states that the resultant electric field at points where the two waves superpose is the sum $E_1 + E_2$. At the point P where the two waves superpose, the resultant electric field is

$$E = E_1 + E_2 = E_0 \sin \varphi_1 + E_0 \sin \varphi_2 \qquad (37.6)$$

The phases of the two wavelets differ at P because the slit-to-screen distances (x_1 and x_2) differ. The phase difference $\varphi_1 - \varphi_2$ depends on the difference in the path lengths, $x_1 - x_2$,

$$\varphi_1 - \varphi_2 = \left(\frac{2\pi}{\lambda} \right)(x_1 - x_2) \qquad (37.7)$$

Referring to Figure 37.5, let's establish the conditions under which the point P is an intensity maximum (a point of constructive interference). If the **path difference** $x_1 - x_2$ is an integral number of wavelengths, the waves superposed at P are **in phase**. Crests of waves from slit 1 meet crests of waves from slit 2, giving a maximum wave amplitude and intensity at P. Thus, P will be a position of **maximum intensity** provided that

$$x_1 - x_2 = m\lambda \qquad m = 0, 1, 2, 3, \ldots \qquad (37.8)$$

We can transform Equation 37.8 into an equation that locates the angular position of P (the angle θ in Figure 37.5) or the linear position of P (the distance y in Figure 37.5). Let a denote the slit separation and let x_0 denote the distance

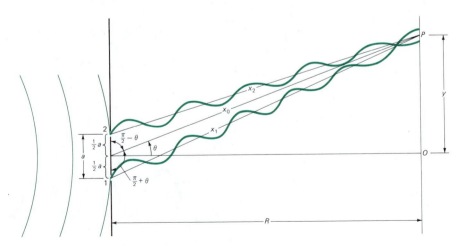

FIGURE 37.5

Top view of the geometry for the double slit. If the path difference $x_1 - x_2$ is an integral number of wavelengths, the waves interfere constructively at P. Destructive interference results if the path difference is an odd number of half-wavelengths.

from a point midway between the two slits to the point P. In practice x_1 and x_2 are nearly equal to x_0, and x_0 is much larger than a, the slit separation. To a very good approximation, the path difference is given by

$$x_1 - x_2 = a \sin \theta$$

Inserting this result into Equation 37.8, we get a relation that locates the angular positions of the **intensity maxima**,

$$a \sin \theta = m\lambda \qquad m = 0, 1, 2, \ldots \qquad (37.9)$$

The integer m is called the **order** of the intensity maxima. Each value of m corresponds to a definite value of θ for which the interference is constructive. The positions of constructive interference are intensity maxima. The central maximum ($m = 0$) occurs at $\theta = 0$. Note that $\theta = 0$ corresponds to points directly opposite points midway between the slits (Figure 37.5). Young's wave theory correctly predicts that these are points of maximum intensity; they are brighter than regions to either side. In contrast, the ray picture of light predicts that $\theta = 0$ should correspond to a shadow region. Young's experiment thus supports the wave theory of light.

A first-order maximum corresponds to $m = 1$ in Equation 37.9 and occurs at an angle given by

$$a \sin \theta = \lambda \qquad m = 1 \qquad (37.10)$$

This equation shows that the first-order maximum occurs at different angles for different wavelengths, unlike the central maximum, which occurs at $\theta = 0$ for all wavelengths. For example, if white light illuminates the slits, then the central maximum is also white. But because the first-order maximum occurs at different positions for different wavelengths, the first-order maximum is a colored fringe pattern. The higher-order maxima also occur at different positions for different wavelengths.

The linear position of an intensity maximum y is related to the angular position θ by

$$\sin \theta = \frac{y}{\sqrt{R^2 + y^2}}$$

Thus, the linear positions of **intensity maxima** are given by

$$\frac{ay}{\sqrt{R^2 + y^2}} = m\lambda \qquad m = 0, 1, 2, \ldots \qquad (37.11)$$

Equation 37.11 shows that the wavelength of light can be determined from measurements of a, y, and R.

EXAMPLE 1

Wavelength Measurement Using a Double Slit

Red light illuminates a double-slit system with $a = 0.20$ mm. The first-order maximum occurs at $y = 3.3$ mm on a screen 1.0 m from the slits. We can determine the wavelength by using Equation 37.11 with $m = 1$,

$$\lambda \approx a \cdot \frac{y}{R} = \frac{(2 \times 10^{-4} \text{ m})(3.3 \times 10^{-3} \text{ m})}{1 \text{ m}} = 6.6 \times 10^{-7} \text{ m}$$

$$= 660 \text{ nm}$$

The remarkable aspect of the double-slit system is that it enables us to measure the wavelength of light—a length far too small to be gauged by any conventional measuring rod. To appreciate just how small 660 nm is, consider this: a string of 1500 wavelengths, each 660 nm long, would not span one millimeter. In Chapter 38 we will see how a multiple-slit diffraction grating enables us to make precise measurements of wavelength.

At points on the screen where waves from slits 1 and 2 arrive one-half wavelength out of step there is an **intensity minimum**. When we say that waves are a half-wavelength out of step we mean that wave crests from slit 1 meet wave troughs from slit 2 and result in destructive interference. In general, a path difference that is any odd number of half-wavelengths results in destructive interference. The angular positions of the intensity minima can be found by using the same procedure used to locate the intensity maxima, Equation 37.9. The minima are located at angles given by

$$a \sin \theta = (m + \tfrac{1}{2})\lambda \qquad m = 0, 1, 2, \ldots \qquad (37.12)$$

The linear positions of the minima are given by

$$\frac{ay}{\sqrt{R^2 + y^2}} = \left(m + \frac{1}{2}\right)\lambda \qquad m = 0, 1, 2, \ldots \qquad (37.13)$$

EXAMPLE 2

Casting Shadows with a Double-Slit System

Light from a helium-cadmium laser ($\lambda = 442$ nm) passes through a double-slit system with $a = 0.40$ mm. Let's determine how far away a screen must be placed in order that a dark fringe appear directly opposite both slits (Figure 37.6).

Taking $m = 0$ and $y = 0.2$ mm in Equation 37.13 gives

$$R \approx \frac{2ay}{\lambda}$$

$$= \frac{2(0.4 \times 10^{-3}\ \text{m})(0.2 \times 10^{-3}\ \text{m})}{442 \times 10^{-9}\ \text{m}} = 0.36\ \text{m}$$

$$= 36\ \text{cm}$$

Geometric optics incorrectly predicts bright regions opposite the slits and darkness in between. But as this example shows, interference can produce just the opposite.

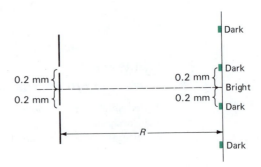

FIGURE 37.6
With a slit-to-screen distance of $R = 36$ cm, there will be dark fringes directly opposite the slits.

37.2
COHERENCE

When we use a light source such as the sun or an ordinary light bulb to demonstrate the double-slit experiment shown in Figure 37.2, we find that the intensity of the interference pattern is quite low. We can increase the intensity greatly by removing the first card in Figure 37.2. This allows light from a broad source (the sun or a light bulb) to fall directly on the two slits. However, such an arrangement destroys the interference pattern—the intensity on the screen no longer shows the maxima and minima characteristic of interference. Evidently, the waves traveling away from the slits have a property that waves coming directly from a broad source do not. This critical property is called **coherence**. The superposition of waves from coherent sources results in interference.

Sources, and the waves they emit, are said to be *coherent* if they

1. have *equal frequencies,*
2. maintain a *phase difference* that is *constant in time.*

If either property is lacking, the sources are incoherent, and the waves do not exhibit interference. Some examples of coherent waves and their sources are:

1. sound waves from two loudspeakers driven by the same audio oscillator,
2. electromagnetic waves from two microwave horns driven by the same oscillator,
3. sound waves from audio headsets, such as those used on commercial airliners,
4. light waves generated by Young's double-slit system,
5. light waves from a laser.

By contrast, the sound waves from two speakers are incoherent if the loudspeakers are driven by audio oscillators tuned to different frequencies. The incoherent light streaming from a neon sign is the result of the superposition of waves that do not maintain a constant phase difference.

To study coherence in detail we must examine the intensity pattern produced by the superposition of two waves. In Section 35.3 we showed that the average intensity of a monochromatic plane wave is given by

$$I = c\epsilon_0 \langle E^2 \rangle \tag{37.14}$$

where $\langle \ \rangle$ denotes a time average over an integral number of cycles of the wave. We can use this expression for the intensity I to illustrate the key aspects of interference.

Consider the superposition of light waves from two sources. The combined electric field is the sum

$$E = E_1 + E_2$$

where E_1 and E_2 are the electric fields produced by the two sources. The time-averaged value* of E^2 is

$$\langle E^2 \rangle = \langle E_1{}^2 \rangle + \langle E_2{}^2 \rangle + 2\langle E_1 E_2 \rangle$$

---

* The time-averaging operation is an integration over time and is therefore linear—the average of a sum equals the sum of the averages.

The intensity is

$$I = c\epsilon_0\langle E_1{}^2\rangle + c\epsilon_0\langle E_2{}^2\rangle + 2c\epsilon_0\langle E_1 E_2\rangle$$

If we compare this equation to Equation 37.14, we see that the individual intensities of the sources are

$$c\epsilon_0\langle E_1{}^2\rangle = I_1$$

and

$$c\epsilon_0\langle E_2{}^2\rangle = I_2$$

Therefore we can write,

$$I = I_1 + I_2 + 2c\epsilon_0\langle E_1 E_2\rangle \tag{37.15}$$

The product term $2c\epsilon_0\langle E_1 E_2\rangle$ is called the **interference term**. It represents the interference between the two sources. If

$$\langle E_1 E_2\rangle = 0 \tag{37.16}$$

the sources are **incoherent** and there is no interference. With $\langle E_1 E_2\rangle = 0$, the intensity is

$$I = I_1 + I_2 \tag{37.17}$$

The intensities simply add when the sources are incoherent. For example, when the headlights of an automobile illuminate the same area, their combined intensity is simply the sum of the two separate intensities. The headlights are incoherent sources and there is no interference.

When light sources are coherent, the intensities do not simply add. Coherence alters the intensity. Let's consider two cases where the interference term is not zero. In both cases two coherent light waves of equal amplitude are superposed. The interference is constructive in one instance, and destructive in the other.

1. Let $E_1 = E_2$. This means that the two electric fields have the same amplitude, frequency, and phase. The intensities of the individual waves are equal,

$$I_1 = I_2 = c\epsilon_0\langle E_1{}^2\rangle$$

and the interference term is

$$2c\epsilon_0\langle E_1 E_2\rangle = 2I_1$$

The total intensity is

$$I = I_1 + I_2 + 2c\epsilon_0\langle E_1 E_2\rangle = 4I_1$$

which is twice the intensity that the two sources would produce if they were incoherent. In this instance the coherent waves interfere **constructively**.

2. Let $E_1 = -E_2$. This describes the superposition of two waves with the same amplitude and frequency, but with a constant phase difference of $180°$. Their superposition results in complete destructive interference. The resultant wave amplitude and intensity are zero:

$$E = E_1 + E_2 = 0 \qquad I = 0$$

The superposition of incoherent light waves can never result in zero intensity. The situation in which $I = 0$ requires coherent waves and marks the limit of complete destructive interference.

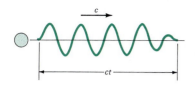

FIGURE 37.7

An atom that radiates for a time t generates a wave train of length ct, which is the coherence length.

Coherence Length

A wave front is defined as a surface of constant phase. Thus, the notion of a wave front implies one type of coherence—a lateral phase coherence across the wave front. A beam of light also displays **longitudinal** coherence—a phase coherence *along* the direction of wave travel. The distance along the direction of propagation over which such phase coherence exists is called the **coherence length**.

If an atom radiates for a time t, the emitted wave train has a length ct, where c is the speed of light (Figure 37.7). The coherence length L_c is given by

$$L_c = ct \qquad (37.18)$$

For a typical incoherent light source the coherence length is roughly 1 mm. For example, a sodium atom typically radiates for 4×10^{-12} s. This time interval is called a **coherence time**, because the waves emitted during this interval are coherent. Using Equation 37.18 we find for the coherence length

$$L_c = ct = (3 \times 10^8 \text{ m/s})(4 \times 10^{-12} \text{ s}) = 1.2 \text{ mm}$$

The light emitted by individual atoms forms short wave trains because the radiation process lasts only a short time. The coherence length is the average length of these wave trains. In a typical interference experiment the wave trains are divided, either physically (by the use of slits) or optically (by reflection). The two sets of wave trains are then sent along different paths that ultimately reunite them. If the paths differ by more than the coherence length, there is no interference because the reunited waves are not parts of the same group of wave trains. They are incoherent.

Light reflected from an oil film can exhibit interference (Figure 37.8). Waves are reflected from both the front and the back of the film, and are superposed after reflection. The superposed waves can interfere provided that the thickness of the film is small compared to the coherence length. If the film thickness exceeds the coherence length, then the reflected waves are not parts of the

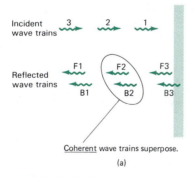
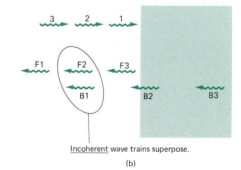

FIGURE 37.8

(a) Wave trains (1,2,3) of finite length strike an oil film. Coherence extends over the length of each wave train. There is no coherence between different wave trains. Each reflected wave is a superposition of wave trains reflected from the front and rear of the film. If the thickness of the film is small compared to the length of the wave train, the superposed waves are coherent and interfere. (b) If the thickness of the film is large compared to the length of the wave train, the superposed waves are incoherent and do not interfere.

same group of wave trains; there is no phase coherence between the reflected waves and thus no interference.

37.3
THIN-FILM INTERFERENCE

Light reflected from oil films, soap bubbles, and other thin films of material often exhibits beautiful interference effects. In practice such interference is most pronounced when the films are very thin—no more than a few times the wavelength of light. Accordingly, we call it **thin-film interference**.

The concept of **optical path length** is useful in our study of thin-film interference. If light travels a distance x in a medium of index of refraction n, its optical path length is defined as nx.

$$\text{Optical path length} = (\text{index of refraction})(\text{physical path length}) \qquad (37.19)$$

To show why optical path length is a useful concept, we consider the phase difference between two coherent waves that are superposed after traveling different distances (x_1 and x_2) through a medium with an index of refraction n. Their phases differ by

$$\delta = \left(\frac{2\pi}{\lambda_n}\right)(x_1 - x_2) \qquad (37.20)$$

where λ_n is the wavelength of light in the medium. It is convenient to refer all wavelengths to their vacuum value. We can replace λ_n by λ, the corresponding vacuum wavelength, if we also replace ($x_1 - x_2$) by $n(x_1 - x_2)$, the difference in optical path lengths. To prove this, we first note that when a wave crosses a boundary from one medium to another there is a change in wave speed and wavelength, *but not in wave frequency*. The wave frequency equals the number of wave crests per second reaching the boundary in one medium, and it also equals the number per second traveling away from the boundary in the second medium. No wave crests can get lost at the boundary. Thus, the wave frequency remains unchanged when a wave crosses a boundary.

Next we compare the basic kinematic relation for a vacuum and for the medium. In a vacuum the frequency f and wavelength λ are related to the speed of light c by

$$f\lambda = c$$

In a material medium the speed of light v differs from c, and the wavelength λ_n differs from its vacuum value, but the frequency remains unchanged.

$$f\lambda_n = v$$

The ratio

$$n \equiv \frac{c}{v} = \frac{\lambda}{\lambda_n}$$

introduces the index of refraction n and shows that

$$\lambda_n = \frac{\lambda}{n} \qquad (37.21)$$

Inserting this result into Equation 37.20 shows that we can replace λ_n by λ provided that we also replace the path length difference by the optical path length difference,

$$\delta = \left(\frac{2\pi}{\lambda}\right) \cdot \overbrace{n(x_1 - x_2)}^{\text{optical path length difference}} \qquad (37.22)$$

vacuum wavelength

Note that when the optical path lengths differ by one vacuum wavelength, the phases differ by 2π.

EXAMPLE 3

Optical Path Length in a Soap Film

A soap bubble ($n = 1.333$) is 300 nm thick. Green light ($\lambda = 530$ nm) strikes the surface and is partially reflected and partially transmitted. The transmitted wave travels through the film, where it is again partially reflected and partially transmitted. The wave that crosses and returns travels a path of length 600 nm. We determine its optical path length as follows:

$$\text{Optical path length} = n\lambda = (1.333)(600 \text{ nm})$$
$$= 800 \text{ nm}$$

The change in phase for the round trip is $2\pi/\lambda$ times the optical path length,

$$\delta = \left(\frac{2\pi}{530 \text{ nm}}\right) \cdot 800 \text{ nm} = 9.48 \text{ rad}$$

Phase Change Accompanying Reflection

Equation 37.20 gives the phase difference arising from differences in optical path lengths. Additional phase changes may occur as a result of **reflection**. Specifically, we assert that reflection has the following effects:

1. A reflected wave undergoes a phase change of π rad if it travels *faster* than the transmitted wave.
2. A reflected wave undergoes no phase change if it travels slower than the transmitted wave.

The criterion for reflected-wave change can also be expressed in terms of indices of refraction. *If the incident wave strikes a surface of higher index of refraction, the reflected wave undergoes a phase change of π radians.* This follows from the original statement of the criterion and the definition of n,

$$n = \frac{c}{v}$$

Thus, a higher n corresponds to a slower wave speed.

Likewise, if the incident wave strikes a surface of lower index of refraction, there is no phase change in the reflected wave. For example, consider a light

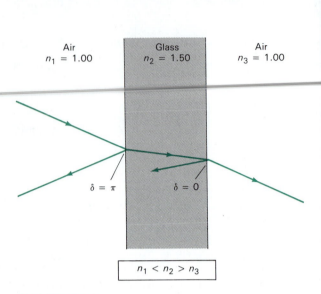

Air
$n_1 = 1.00$

Glass
$n_2 = 1.50$

Air
$n_3 = 1.00$

$\delta = \pi$

$\delta = 0$

$n_1 < n_2 > n_3$

FIGURE 37.9

At the left (air-to-glass) interface the reflected wave undergoes a phase change of π radians relative to the incident wave. At the right (glass-to-air) interface the reflected wave undergoes no phase change.

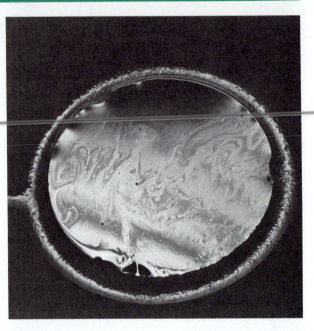

FIGURE 37.10

A soap film suspended by a wire loop viewed by reflected light. The upper portion of the film appears black because of destructive interference between waves reflected from the front and rear surfaces of the film.

wave that strikes a window pane (Figure 37.9). The wave reflected from the outer surface experiences a phase change of π radians relative to the incident wave because the index of refraction for glass is higher than that of air. If we follow the transmitted wave through the glass, there will be a second reflection at the inner surface, where the light emerges. This reflection produces no phase change because the wave encounters an air surface having a lower index of refraction.

A rigorous proof of the phase-change effects of reflection involves Maxwell's equations. However, there are several ways to simply *demonstrate* the phase change that accompanies reflection. For example, we can consider a thin soap film suspended vertically in a wire loop (Figure 37.10). Gravity establishes a variable thickness in the film; the thinnest portion is at the top of the loop. In fact, the thickness of the film at the top of the loop is small compared to the wavelength of visible light. The round trip across the film and back introduces an insignificant phase change because the film is so thin. Light reflected from such a thin film exhibits destructive interference, as shown by Figure 37.10. The origin of the destructive interference is shown in Figure 37.11. Reflected light entering the eye is a superposition of waves reflected from the two air-film boundaries. Light reflected at the front surface undergoes a phase change of π radians. Light reflected from the rear surface experiences no phase change. Thus, a phase difference of π radians is established between the two reflected waves, and they interfere destructively.

Thin-film interference results from the combined effects of optical path length differences and reflection phase shifts. Consider the situation shown in Figure 37.12. White light is incident normally on a soap bubble. The light

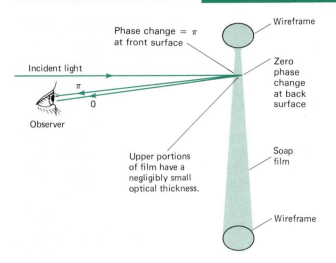

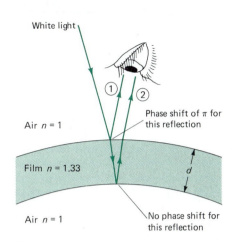

FIGURE 37.11

The reflection at the front surface introduces a phase shift of π radians (faster to slower). The reflection at the rear surface does not introduce any phase change (slower to faster). The phase difference of π radians between the two reflected waves results in destructive interference.

FIGURE 37.12

White light is incident normally on a soap film. The light entering the eye is a superposition of waves reflected from the two air-to-film boundaries.

that enters the eye is a superposition of waves reflected from the two air-film boundaries. If white light is incident, the reflected light is not white. Certain reflected wavelengths interfere destructively, and these wavelengths are removed from the reflected light. Likewise, certain reflected wavelengths interfere constructively, and the reflected light is richer in colors corresponding to these wavelengths.

For a film of thickness d the total phase difference between the two waves is

$$\delta = \left(\frac{2\pi}{\lambda}\right)n(2d) \qquad + \qquad \pi$$

$$\underbrace{\hphantom{\delta = \left(\frac{2\pi}{\lambda}\right)n(2d)}}_{\substack{\text{due to optical} \\ \text{path difference}}} \qquad \underbrace{\hphantom{\pi}}_{\substack{\text{due to air-film reflection} \\ \text{from top surface}}} \tag{37.23}$$

The reflected waves interfere constructively when δ is an integral multiple of 2π,

$$\delta = (m + 1)2\pi \qquad m = 0, 1, 2, 3, \ldots \tag{37.24}$$

By combining Equations 37.23 and 37.24, we get the conditions for constructive interference of the reflected light,

$$2nd = (m + \tfrac{1}{2})\lambda \qquad m = 0, 1, 2, 3, \ldots \tag{37.25}$$

Wavelengths satisfying Equation 37.25 are reflected strongly. These same wavelengths are missing from the transmitted light.

The reflected waves interfere destructively when the phase difference δ is an *odd* multiple of π. Setting

$$\delta = (2m + 1)\pi \qquad m = 0, 1, 2, \ldots$$

in Equation 37.23, the conditions for destructive interference of the reflected light take the form,

$$2nd = m\lambda \qquad m = 0, 1, 2, \ldots \qquad (37.26)$$

Wavelengths satisfying Equation 37.26 are missing from the reflected light. This means they are completely transmitted.

When the film thickness d is much less than the wavelength λ, Equation 37.26 is satisfied to a good approximation by $m = 0$. This situation prevails with the soap film shown in Figure 37.10. The thickness of the top of the film is less than any visible wavelength. The top of the film appears black because all visible wavelengths are transmitted and there is *no reflected light*.

EXAMPLE 4

Reflected and Transmitted Colors

A soap film has a thickness of 300 nm and an index of refraction of 1.33. What colors are strongly reflected and transmitted by this film? Wavelengths for which interference is constructive are most prominent in the reflected light. Using Equation 37.25 we can show that these wavelengths are

$$\lambda = \frac{2nd}{(m + \frac{1}{2})} = \frac{2(1.33)(300 \text{ nm})}{(m + \frac{1}{2})}$$

Taking $m = 0$, 1, and 2 we get $\lambda = 1600$ nm, 533 nm, and 320 nm. Larger values of m give still shorter wavelengths. A wavelength of 1600 nm lies in the infrared (IR) region of the spectrum, outside the visible range (410–700 nm). The 320-nm wavelength is in the ultraviolet (UV) region, outside the visible spectrum. The only wavelength within the visible spectrum that interferes constructively is 533 nm, which corresponds to the color green.

Equation 37.26 allows us to determine wavelengths that interfere destructively. These wavelengths are missing in the reflected light and hence are strongly transmitted. The $m = 0$ solution is ruled out because we are concerned with wavelengths in or near the visible spectrum. From Equation 37.26 we find that for $m = 1$, 2, and 3, the wavelengths are $\lambda = 800$ nm, 400 nm, and 267 nm. None of these wavelengths lies in the visible range, although the first two are close to the red and violet ends of the visible spectrum.

The film will have a greenish color when viewed by reflected light because green is strongly reflected and red and blue are transmitted. When viewed in transmitted light the film will appear purple—a mixture of red and blue with very little green.

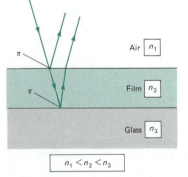

FIGURE 37.13

Phase changes of π radians occur at both reflecting surfaces.

Antireflection Coatings

A thin film of material can be used to reduce the intensity of reflected light. Camera lenses are often coated with thin films to reduce reflection. Such antireflection coatings result in destructive interference of the reflected waves. A single-layer coating can achieve zero reflectance for only a very narrow range of wavelengths. We will assume that the index of refraction of the coating is intermediate between those of the air and lens. The two most commonly used coatings have indexes of refraction of 1.38 (magnesium fluoride) and 1.35 (cryolite). In these circumstances, reflections at both surfaces produce phase changes of π rad (Figure 37.13). Therefore, the overall phase difference between the two reflected waves depends on the difference of their optical path lengths.

For a coating of thickness d and index of refraction n, the optical paths differ by $2nd$ and the phase difference between the two reflected waves is

$$\delta = 2\pi\left(\frac{2nd}{\lambda}\right) \tag{37.27}$$

The reflected waves interfere destructively when $\delta = \pi$; that is, when the two waves are one-half of a wavelength out of step. Setting $\delta = \pi$ in Equation 37.27 gives,

$$d = \frac{\lambda}{4n} \tag{37.28}$$

which makes the film one-quarter of a wavelength thick.* Optical components sometimes incorporate a "quarter-wave plate" as an antireflection device.

EXAMPLE 5

Thickness of an Antireflection Coating

Let us determine the thickness of an antireflection coating of magnesium fluoride, for which $n = 1.38$. We want the film to give maximum transmission at 550 nm, a wavelength near the middle of the visible spectrum. For the thickness, Equation 37.28 gives

$$d = \frac{550 \text{ nm}}{(4 \cdot 1.38)}$$
$$= 99.6 \text{ nm}$$

Thin-film interference can also be used to produce highly reflective coatings. A film thickness of one-half wavelength produces a path difference of one wavelength and results in constructive interference for the reflected light. Such coatings are particularly useful in fabricating laser mirrors, where high reflectivity for only a single wavelength is desired.

37.4
OPTICAL INTERFEROMETERS

An optical interferometer is an instrument that allows us to measure small changes in optical path length by observing an interference pattern. Basically, an interferometer operates by splitting a beam of light, sending the two beams along different paths, and then recombining the beams to produce an interference pattern. There are many varieties of optical interferometers. We describe two, the Rayleigh interferometer and the Michelson interferometer.

Rayleigh Interferometer

A diagram of the Rayleigh interferometer is shown in Figure 37.14. A slit source of light (S) is placed at the focal point of a lens (L_1). Light emerges from the lens in the form of plane waves that strike the double-slit system

---

* Recall that λ is the wavelength in a vacuum; λ/n is the wavelength inside the film.

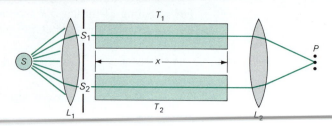

FIGURE 37.14
The Rayleigh interferometer.

(S_1, S_2). The waves emerging from S_1 and S_2 are coherent and travel along parallel paths through two identical sealed tubes (T_1, T_2). A second lens (L_2) superposes the waves that interfere and produce a fringe pattern (P).

The Rayleigh interferometer can be used to measure the index of refraction of a gas. Suppose that one tube is filled with a gas having an index of refraction n, and the other tube is evacuated. The optical path length through the gas is nx, where x is the length of the tube. The optical path length through the evacuated tube is simply x. The difference between the two optical path lengths is $(n - 1)x$.

If the gas-filled tube is slowly evacuated, the difference in optical path lengths drops to zero and the interference fringe pattern changes. Each time $(n - 1)x$ changes by one wavelength, the fringe pattern returns to its original form. If N denotes the number of times the fringe pattern returns to its original form (N is called the number of **fringe shifts**), then $N\lambda$ equals the change in optical path length,

$$N\lambda = (n - 1)x \qquad (37.29)$$

In the following example we use Equation 37.29 to determine the index of refraction of air.

EXAMPLE 6

Index of Refraction of Air

The tube length of a Rayleigh interferometer is 10.0 cm. Initially, one tube is filled with air and the other is evacuated. Using light with a wavelength of 656.3 nm we observe 44.5 fringe shifts as the air is removed. Using Equation 37.29,

$$n - 1 = \frac{N\lambda}{x} = \frac{44.5(656.3 \times 10^{-9} \text{ m})}{0.100 \text{ m}}$$

$$= 0.000292$$

The index of refraction of air for a wavelength of 656.3 nm is

$$n = 1.000292$$

Michelson Interferometer

The Michelson interferometer (Figure 37.15) was developed by the American physicist Albert A. Michelson. Figure 37.16 shows how the Michelson interferometer produces an interference pattern. Light from a source strikes a half-silvered mirror (HSM), a glass plate with a thin metallic coating on its back. The

FIGURE 37.15

A Michelson interferometer.

thickness of the coating is such that half of the incident intensity is reflected and half is transmitted. The half-silvered mirror thereby serves as a **beam splitter**, producing two coherent light waves of equal amplitude. The reflected wave travels to a movable mirror (MM) and then retraces its path back to the half-silvered mirror. The transmitted wave passes through a compensator plate (C), strikes a fixed mirror (FM), and then retraces its path to the half-silvered mirror.

The compensator is an unsilvered twin of the half-silvered mirror. It is inserted so that both waves make three passes through equal thicknesses of glass. With the compensator in position, any difference in optical path lengths

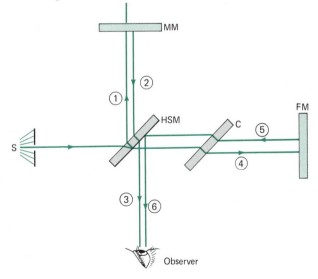

FIGURE 37.16

Light paths for the Michelson interferometer. The half-silvered mirror (HSM) divides an incident wave into two coherent waves of equal intensity.

comes about as a result of differences in physical path lengths between the half-silvered mirror and the other two mirrors. The two reflected waves return to the half-silvered mirror where both are again split into two waves of equal amplitude. The transmitted portion of one wave and the reflected portion of the other wave are superposed and produce an interference pattern.

The interference pattern appears as a series of concentric interference fringes. The target-like pattern of bright and dark fringes shows that positions of constructive and destructive interference are located symmetrically about the axis of the interferometer.

If the movable mirror is displaced a distance D, the optical path length of the waves traveling to and from it is changed by $2D$. Each time the optical path length changes by one wavelength the fringe pattern returns to its original form. If the fringe pattern reproduces itself N times when the mirror is moved a distance D, we have

$$2D = N\lambda \tag{37.30}$$

In order to obtain interference it is necessary that the difference in optical path lengths not exceed the coherence length. For laser light this is seldom a problem, since the coherence length is typically 30 cm or more.

EXAMPLE 7

Helium-Neon Laser Wavelength

In an undergraduate laboratory experiment, light from a helium-neon laser is beamed into a Michelson interferometer. The movable mirror is displaced 0.0567 mm, causing the fringe pattern to reproduce itself 180 times. The wavelength follows from Equation 37.30,

$$\lambda = \frac{2D}{N} = \frac{2(0.0567 \times 10^{-3} \text{ m})}{180} = 6.30 \times 10^{-7} \text{ m}$$

$$= 630 \text{ nm}$$

Because D and N are measured to only three-significant-figure accuracy, the precision of the resulting wavelength is limited. A more precise value of the helium-neon laser wavelength is 632.8 nm.

You should not conclude from Example 7 that Michelson's interferometer is an instrument of low precision. Michelson obtained very high precision when he used his interferometer to compare the standard meter (then defined as the length between two marks on a rod) with the wavelength of the red light emitted by cadmium atoms. His value for the wavelength of the light had seven significant figures, which shows that his interferometer had a precision of one part in 10 million. Using interferometers, scientists now routinely measure distances accurate to a fraction of one wavelength of light.

Experiments performed with this interferometer by Michelson and E. W. Morley helped establish an experimental basis for Einstein's special relativity theory (Chapter 39). In 1887 Michelson and Morley attempted to measure the speed of the earth relative to the hypothetical medium (the ether) that propagated light waves. The expected fringe shifts in the interference pattern were never observed. Einstein's theory explains the null result of the Michelson-Morley experiment by recognizing that there is no such medium. There is no preferred frame of reference for light waves.

WORKED PROBLEM

A soap bubble appears green when exposed to normally incident white light. Determine the minimum thickness of the soap film that can produce the observed effect. Take the wavelength of green light to be 500 nm. The index of refraction of the film is 1.41.

Solution

When light strikes the front surface, some is reflected and some penetrates into the film. The reflected light is a superposition of waves reflected at the front surface and waves reflected at the rear surface that are then transmitted when they return to the front surface.

The fact that the soap film appears green indicates that the green component of the white light experiences constructive interference. We need to formulate the conditions for constructive interference.

The optical path difference for the interfering waves is $2nd$ where d is the thickness of the soap film. The factor 2 accounts for two passes through the film. The phase difference corresponding to the optical path difference is

$$\left(\frac{2\pi}{\lambda}\right)(2nd)$$

Because there is air on both sides of the film, only waves reflected from outside the front surface undergo a phase change of π radians. The total phase difference between the superposed waves is

$$\left(\frac{2\pi}{\lambda}\right)(2nd) + \pi$$

For constructive interference this phase difference must equal an integral multiple of 2π. Therefore, the condition for constructive interference in the reflected light is

$$\left(\frac{2\pi}{\lambda}\right)(2nd) + \pi = 2\pi(m + 1) \qquad m = 0, 1, 2, 3, \ldots$$

Solving for the film thickness we get

$$d = \frac{(2m + 1)\lambda}{4n}$$

The minimum thickness corresponds to $m = 0$. Substituting $m = 0$, $n = 1.41$, and $\lambda = 500$ nm, we find

$$d = 88.7 \text{ nm}$$

EXERCISES

37.1 Interference

A. In a double-slit experiment the wavelength is 600 nm. Waves reaching a particular point on the screen from one slit travel 621 nm farther than waves reaching the same point from the other slit. Determine the phase difference between the two waves that superpose at that point on the screen.

B. Monochromatic light illuminates a double-slit system with $a = 0.30$ mm. The second-order maximum occurs at $y = 4.0$ mm on a screen 1 m from the slits. (a) Determine the wavelength. (b) Determine the position (y) of the third-order maximum. (c) Determine the angular position (θ) of the $m = 1$ minimum.

C. Monochromatic light passes through two slits 0.26 mm apart. The $m = 1$ maximum is observed at an angle of 0.14°. Calculate the wavelength of the light.

D. Plane waves with a wavelength of 500 nm are incident on two narrow slits separated by 0.25 mm. The diffraction pattern is viewed on a screen 5 m from the slits. Determine the separation of the central intensity maximum and the first-order intensity minimum.

E. The helium-cadmium laser of Example 2 is replaced by a helium-neon laser (wavelength = 632.8 nm). What slit-to-screen distance will place the $m = 1$ dark fringe directly opposite the slits?

37.2   Coherence

F. Give two reasons why the following waves are not coherent:

$$E_1 = A \sin (2kx - 5\pi ft)$$

$$E_2 = A \sin (kx - 2\pi ft + \sin [0.01ft])$$

G. Two coherent waves are described by

$$E_1 = E_0 \sin \left(\frac{2\pi x_1}{\lambda} - 2\pi ft + \frac{\pi}{6} \right)$$

$$E_2 = E_0 \sin \left(\frac{2\pi x_2}{\lambda} - 2\pi ft + \frac{\pi}{8} \right)$$

Determine the relationship between x_1 and x_2 that produces constructive interference when the two waves are superposed.

H. Light waves are reflected from the interior and exterior faces of a department store window. Why don't the two sets of reflected waves show interference effects?

I. The coherence time for a low-power laser used in a classroom demonstration is approximately 10^{-7} s. Estimate the coherence length of the light.

37.3   Thin-Film Interference

J. If the antireflection coating of Example 5 is 90 nm thick, what wavelength will receive maximum transmission?

K. Light with a wavelength of 620 nm travels through a film of water 720 nm thick. Determine (a) the optical thickness of the film, (b) the phase change of the light across the film.

L. A thin film of cryolite ($n = 1.35$) is applied to a camera lens ($n = 1.50$). The coating is designed to reflect wavelengths at the blue end of the spectrum and to transmit wavelengths in the near infrared. What minimum thickness will give high reflectivity at 450 nm and high transmission at 900 nm?

37.4   Optical Interferometers

M. Using the following data for a Rayleigh interferometer, determine the index of refraction of the gas (hydrogen):

    tube length = 11.2 cm
    wavelength = 589.2 nm
    number of fringe shifts = 25.1

N. Light from a helium-cadmium laser is beamed into a Michelson interferometer. The movable mirror is displaced 0.382 mm, causing the interferometer pattern to reproduce itself 1700 times. Determine the wavelength of the laser light. What color is it?

P. A microwave version of the Michelson interferometer uses waves whose wavelength is 4 cm. The detector receiving the two interfering microwave beams is sensitive to phase differences corresponding to path differences of one-tenth of a wavelength. What is the smallest displacement that can be detected with the interferometer?

Q. How far must the movable mirror on a Michelson interferometer be moved to produce 30 fringe shifts with light having a wavelength of 606 nm?

PROBLEMS

1. A thin film of soap and water is held in a wire frame so that its surface is vertical (Figure 1). Gravity causes

FIGURE 1

the thickness of the film to be smallest at the top. With white light incident on the film, the very top of the film appears black in reflected light. Lower portions of the film appear as colored bands. (a) Explain why the upper portion of the film appears black. (b) The three bands marked 1, 2, 3 in Figure 1 are red, blue, and green—not necessarily in that order. Identify the color of each band and indicate the reasoning for your ordering. (c) Calculate the thickness of the green band. The index of refraction of the film is 1.35. Take the wavelength of green light to be 500 nm.

2. A thin film of oil ($n = 1.38$) floats on water. (a) Show that the equation

$$\lambda = \frac{2nd}{(m + \frac{1}{2})} \qquad m = 0, 1, 2, \ldots$$

where d is the film thickness, gives the wavelengths that exhibit constructive interference for reflected light. (b) What color is the film at points where $d = 300$ nm? (The film is viewed from above in white light.)

3. A double-slit system having a slit separation of 0.1 mm is illuminated with two sources of light having different wavelengths. Each source produces its own diffraction pattern. If the first-order intensity maxima for the two patterns are separated by 0.001 radian, determine the difference in the wavelengths of the two sources.

4. An oil film having an index of refraction of 1.48 coats the surface of a pan of water. The film is illuminated from above. How thick must a single film be in order to prevent reflections of wavelengths of 550 nm and 750 nm?

5. A plano-convex lens resting on a flat glass plate is illuminated from above with light of wavelength λ (Figure 2). When viewed from above one sees a series of concentric rings (Newton's rings) formed from interference of light reflected from the spherical surface of the lens and from the glass plate. (a) Show that the radius (r) of the rings can be determined from

$$R - \sqrt{R^2 - r^2} = \tfrac{1}{2}(m - \tfrac{1}{2})\lambda \qquad m \geq 1$$

where R is the radius of curvature of the lens and m is a positive integer. (b) If $r \ll R$ show that a good approximation is

$$r \approx \sqrt{R\lambda(m - \tfrac{1}{2})}$$

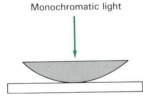

Monochromatic light

FIGURE 2

6. A wedge-shaped film of air is formed by two glass plates separated at one side by an optical fiber 1 μm in diameter (Figure 3). The film is illuminated and

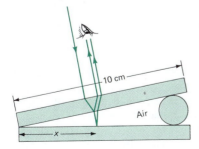

10 cm

Air

x

FIGURE 3

viewed from above. (a) Determine the thicknesses of the air film that give constructive interference for yellow light (wavelength = 590 nm). (b) Determine

the distances from the end (x) at which constructive interference occurs for yellow light. (c) The air is replaced by water. Does the spacing of the yellow bands increase or decrease? How many yellow bands are there?

7. A film of air is formed between two glass plates, as shown in Figure 3. The diameter of the fiber is 1 μm. How many bright fringes will be observed over a horizontal distance of 5 cm when the film is illuminated with red light of wavelength 630 nm?

8. Identical coherent sources S_1 and S_2 are symmetrically located on the x-axis at $x = \pm d$ (Figure 4). (a) Explain why no intensity variations occur at points on the y–z plane. (b) Consider a plane parallel to the y–z plane. Explain why all points on a circle centered on the x-axis should exhibit the same intensity. (c) Assume that $d \ll x$, and show that intensity minima occur for angles θ given by

$$4d \cos \theta = (2n + 1)\lambda \qquad n = \text{integer}$$

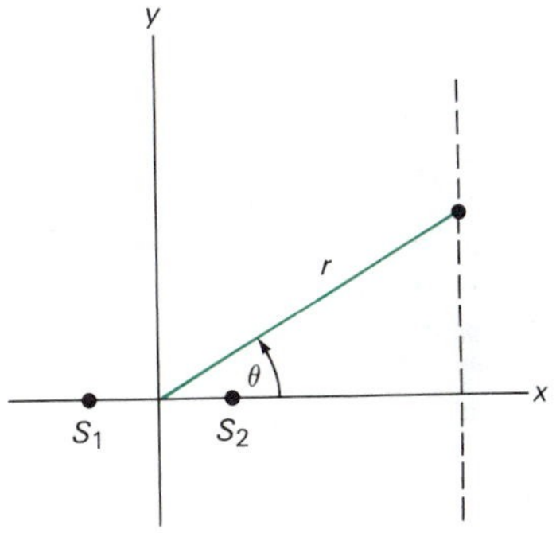

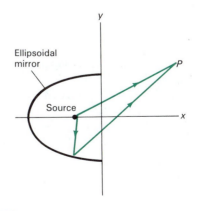

FIGURE 4

9. A half-ellipsoid is coated, making the inside surface perfectly reflecting (Figure 5). A source placed at one focus emits waves that can reach an external point P directly, as well as after being reflected, as shown. Show that the locus of points for which the phase difference between the two waves is constant is a hyperbola.

Ellipsoidal mirror

Source

FIGURE 5

10. A sheet of plate glass has its faces not quite parallel (Figure 6). To determine the angle θ, light of wavelength 500 nm is reflected off each face, producing interference fringes separated by 0.20 mm. If the index of refraction of the glass is 1.40, determine the angle θ.

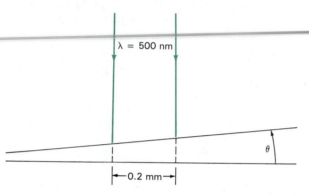

$\lambda = 500$ nm

$\leftarrow$ 0.2 mm $\rightarrow$

θ

FIGURE 6

11. (a) Write the equations describing constructive interference for light reflected from the thin films shown in Figure 7. (b) Calculate the visible wavelengths for which reflected waves interfere constructively for the two films. Indicate the colors corresponding to each.

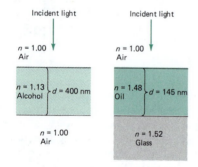

Incident light

n = 1.00
Air

n = 1.13
Alcohol $d = 400$ nm

n = 1.00
Air

Incident light

n = 1.00
Air

n = 1.48
Oil $d = 145$ nm

n = 1.52
Glass

FIGURE 7

12. The sealed tubes of a Rayleigh interferometer are each 5.00 cm in length. When light with a wavelength of 430.8 nm is used, 34.5 fringe shifts are counted as the air is removed from one tube. Determine the index of refraction of air.

13. Plane mirrors M_1 and M_2 make an angle θ with each other (Figure 8). Light from a source S is reflected from each mirror and arrives at the point P from the virtual sources S_1 and S_2. The coordinates of S, P,

and S_2 are given in the figure. Assuming that θ is very small, show that the condition for the first minimum to be at P is,

$$\theta = \left(\frac{\sqrt{5}}{4}\right)\left(\frac{\lambda}{d}\right)$$

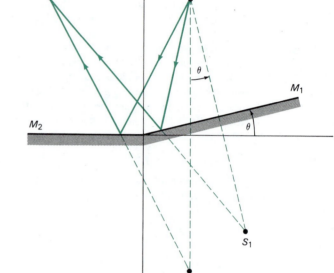

y

$P(-2d, 3d)$

$S(d, 3d)$

θ

M_1

M_2

θ

x

S_1

$S_2(d, -3d)$

FIGURE 8

14. The time average $\langle f \rangle$ of a function $f(t)$ over the time interval from 0 to T is defined by

$$\langle f \rangle = \left(\frac{1}{T}\right)\int_0^T f\, dt$$

Two plane wave electric fields are given by

$$E_1 = E_0 \sin(2kx - 2\pi ft)$$

$$E_2 = E_0 \sin(kx - 6\pi ft)$$

Verify that at $x = 0$, in the limit as $T \to \infty$,

$$\langle (E_1 + E_2)^2 \rangle = \langle E_1{}^2 \rangle + \langle E_2{}^2 \rangle$$

This shows that the two waves are incoherent and do not interfere.

15. Both sides of a thin film ($n = 1.30$) are surrounded by air. Determine the minimum film thickness that gives maximum transmission for a wavelength of 625 nm and maximum reflection for a wavelength of 500 nm.

CHAPTER 38

PHYSICAL OPTICS: DIFFRACTION

38.1
DIFFRACTION

The superposition of coherent waves results in interference. In this chapter we study an aspect of interference called diffraction that can result when a wave strikes an obstacle.

When a wave encounters an obstacle, portions of the wave may be absorbed. The remaining portions travel away in all directions, and we say that the wave has been *scattered* by the obstacle. Water waves on a pond are scattered when they strike a partially submerged rock, and light waves are scattered by interstellar dust grains.

If the scattered portions of the wave meet they interfere. The interference of the scattered waves is called **diffraction**. The interference pattern is called a **diffraction pattern**.

We discuss diffraction in terms of light, a type of electromagnetic wave. However, all types of waves exhibit diffraction, and so the ideas and techniques developed here apply to all types of waves.

Figure 38.1 shows the diffraction pattern produced when light encounters a circular obstacle. Waves scattered by the edges superpose on the screen, exhibiting constructive interference at some points and destructive interference at other points. The shadow cast is a circular interference pattern of alternating bright and dark rings. At the center of the shadow is a bright circular spot, indicating constructive interference of the scattered waves reaching that region. The explanation of this bright spot is quite simple: the scattered waves start from the edge of the obstacle with equal phases. By symmetry, waves reaching the *center* of the shadow travel equal distances. Consequently, they arrive in phase and interfere constructively.

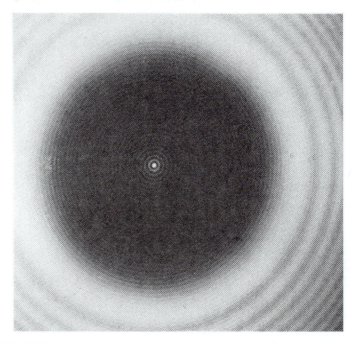

FIGURE 38.1

Diffraction pattern of a circular disk. Observe the bright spot at the center of the shadow region.

FIGURE 38.2

The space between the fingers forms a single slit and can be used to view a diffraction pattern.

It is easy to create your own diffraction pattern. Hold two fingers together so as to form a narrow slit a few inches from your eyes, and view a source of light (Figure 38.2). You will notice one or more dark lines parallel to the edges of the slit. These dark lines mark regions of destructive interference.

The general approach to determining the diffraction pattern is indicated in Figure 38.3. The wave incident on an aperture A is divided into small sections. Each section acts as a source of scattered waves. The scattered waves reach the viewing screen S where they interfere.

The most general treatment of diffraction, called **Fresnel diffraction**, takes into account the curvature of the wave fronts. Under certain conditions the curvatures of the wave fronts are ignorably small and the waves can be treated as plane waves. Such special conditions produce what is called **Fraunhofer diffraction**. Figure 38.4 shows that the curvature of the wave fronts can be made ignorably small by placing the source and screen at substantial distances from the aperture. We will investigate Fraunhofer diffraction because it reveals the full range of diffraction phenomena with fewer mathematical complexities than Fresnel diffraction.

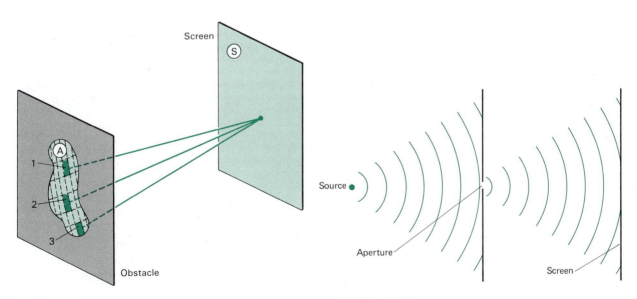

FIGURE 38.3

General approach to diffraction. The wave front entering the aperture A is divided into rectangular sections. Each section is treated as a source of scattered waves. The scattered waves interfere when they reach the screen S.

FIGURE 38.4

Fraunhofer diffraction conditions can be achieved by placing the source and screen at large distances from the aperture.

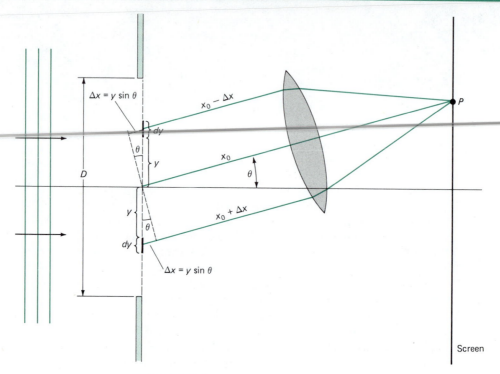

FIGURE 38.5

Geometry for single-slit diffraction, viewed from above the slit. The length y measures distance from the center of the slit. The lens is used to achieve Fraunhofer conditions.

38.2
SINGLE-SLIT DIFFRACTION PATTERN

Consider the diffraction of light by a single narrow slit. In Figure 38.5 the geometry is shown from above the slit. Each point on the wave front within the slit acts like a source of scattered waves. Waves reaching the screen from different sections of the slit differ in phase because they travel different distances. These phase differences result in interference and produce a characteristic diffraction pattern. Our goal is to derive an expression for the light intensity on the screen as a function of the angular position on the screen (the angle θ of Figure 38.5).

We take the incident and scattered electromagnetic waves to be sinusoidal with a wavelength λ and frequency f. The phase of a scattered wave at the screen has the form

$$\varphi = (kx - \omega t) \tag{38.1}$$

where x is the distance from a point in the slit to the screen. The wave number k is related to the wavelength by $k = 2\pi/\lambda$. The angular frequency ω is related to the frequency by $\omega = 2\pi f$.

The width of the slit is denoted by D. Figure 38.6 shows a portion of the slit as seen looking in a direction parallel to the direction of the incident light wave. We can think of the slit as being divided into narrow strips parallel to the length of the slit. Each strip is the source of a scattered wave. We take advantage of symmetry by first adding the electric fields reaching P from two

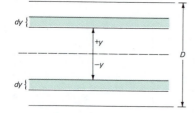

FIGURE 38.6

The slit width is D. The slit is conceptually divided into narrow strips of width dy parallel to the length of the slit.

strips at positions $\pm y$, equidistant from the center of the slit. These strips are shaded in Figure 38.6. The intensity of the light reaching the screen can be expressed in terms of the electric field. We assume that the electric fields reaching P from the strips are proportional to the product of three factors:

1. E_0, the amplitude of the wave entering the slit,
2. dy/D, the fraction of the slit filled by a strip of width dy,
3. $\sin[k(x_0 \pm \Delta x) - \omega t]$, the sinusoidal phase factor for waves starting from $\mp y$. As Figure 38.5 indicates, $x_0 \pm \Delta x$ are the distances from the strips at $\mp y$ to the point P on the screen.

The sum of the two electric fields reaching P from the strips at $+y$ and $-y$ is denoted by dE. We write

$$dE = bE_0 \left(\frac{dy}{D}\right) \sin[k(x_0 + \Delta x) - \omega t]$$
$$+ bE_0 \left(\frac{dy}{D}\right) \sin[k(x_0 - \Delta x) - \omega t] \tag{38.2}$$

where b is a dimensionless proportionality factor. Calling $u = kx_0 - \omega t$ and $v = k\,\Delta x$, we can use the trigonometric identity $\sin(u + v) + \sin(u - v) = 2 \sin u \cos v$ to convert Equation 38.2 to the form

$$dE = 2bE_0 \left(\frac{dy}{D}\right) \sin(kx_0 - \omega t) \cos(k\,\Delta x) \tag{38.3}$$

As Figure 38.5 shows, Δx is related to the positions of strips by

$$\Delta x = y \sin\theta \tag{38.4}$$

The angle θ is the angular position of the point P on the screen. Inserting $\Delta x = y \sin\theta$ into Equation 38.3 gives the electric field at P produced by waves originating from the strips at $\pm y$:

$$dE = \left(\frac{2bE_0}{D}\right) \sin(kx_0 - \omega t) \cos[(k \sin\theta)y]\,dy$$

The contributions from all the strips are summed by integrating from $y = 0$ to $y = D/2$. The resultant electric field at P is

$$E = bE_0 \sin(kx_0 - \omega t) \cdot \frac{\sin\Phi}{\Phi} \tag{38.5}$$

where

$$\Phi = \frac{1}{2} kD \sin\theta = \left(\frac{\pi D}{\lambda}\right) \sin\theta \tag{38.6}$$

The intensity at P is the time average of $c\epsilon_0 E^2$; this is denoted by I_θ and can be expressed as

$$I_\theta = \frac{1}{2} c\epsilon_0 b^2 E_0{}^2 \left[\frac{\sin^2\Phi}{\Phi^2}\right] \tag{38.7}$$

The intensity I_θ gives a quantitative description of the interference. The factor $\frac{1}{2}c\epsilon_0 b^2 E_0{}^2$ is the intensity at $\theta = 0$, the point opposite the center of the slit. In terms of I_0 the diffraction pattern is given by

$$I_\theta = I_0 \left[\frac{\sin^2 \Phi}{\Phi^2} \right] \tag{38.8}$$

This expression is the result we have been seeking—the intensity on the screen as a function of angular position. We refer to I_θ as the single-slit diffraction pattern. The dependence of I_θ on the angle θ enters through the parameter Φ, defined by Equation 38.6. Equation 38.8 can be rewritten to display the explicit dependence on θ,

$$\frac{I_\theta}{I_0} = \frac{\sin^2 \left[\dfrac{\pi D \sin \theta}{\lambda} \right]}{\left[\dfrac{\pi D \sin \theta}{\lambda} \right]^2} \tag{38.9}$$

The quantity I_θ/I_0 is the relative intensity. We can display the form of the diffraction pattern by plotting the relative intensity versus the angle θ. Figure 38.7 is a plot of I_θ/I_0 versus θ for the case where the slit width is ten times the wavelength, $D/\lambda = 10$. The quantity Φ is also indicated in Figure 38.7. Two significant features of the diffraction pattern can be understood as interference effects:

1. The intensity is sharply peaked in the forward ($\theta = 0$) direction. The optical path length from all segments of the slit to the screen is the same for $\theta = 0$. Thus, scattered waves from all segments arrive in phase, giving complete constructive interference and an intensity maximum.
2. The intensity drops to zero at certain angles. As θ increases away from $\theta = 0$, the optical path length from the slit to the screen becomes different for each segment of the slit. The waves from each segment arrive

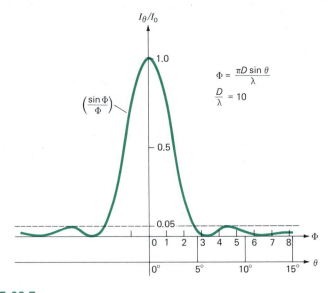

FIGURE 38.7

Relative intensity I_θ/I_0 for the single-slit diffraction pattern. The slit width D is ten times the wavelength λ.

at the screen with different phases. At certain angles their relative phases result in complete destructive interference and zero intensity.

Let's confirm these features by analyzing the relative intensity,

$$\frac{I_\theta}{I_0} = \frac{\sin^2 \Phi}{\Phi^2}$$

We determine the angular positions of zero intensity by setting $\sin \Phi = 0$. The zeros of $\sin \Phi$ occur at

$$\Phi = m\pi \qquad m = \pm 1, \pm 2, \pm 3, \ldots$$

indicating that the intensity falls to zero at angles θ_m given by

$$\sin \theta_m = m\left(\frac{\lambda}{D}\right) \qquad m = \pm 1, \pm 2, \ldots \qquad (38.10)$$

The plus-or-minus signs show that the zeros of intensity are located symmetrically about the central peak at $\theta = 0$. In particular, the intensity falls to zero on either side of the central maximum, at an angle given by setting $m = \pm 1$ in Equation 38.10,

$$\sin \theta_1 = \pm\frac{\lambda}{D} \qquad \text{(first-order minimum)} \qquad (38.11)$$

In Figure 38.7, $\lambda/D = 1/10$, and the values of θ_1 that satisfy $\sin \theta_1 = \pm 1/10$ are $\theta_1 = \pm 5.74°$.

The Central Peak

The equation $\sin \theta_1 = \lambda/D$ defines an angular radius or spread of the central peak of intensity. The fact that the intensity of the central peak is so much greater than the secondary peaks indicates that the bulk of the wave intensity is scattered into a narrow angular region about the forward direction. If $\lambda \ll D$, we use the small-angle approximation $\sin \theta_1 \approx \theta_1$ to show that the intensity falls to zero on either side of the central peak at an angle

$$\theta_1 = \frac{\lambda}{D} \qquad (38.12)$$

There are two immediate implications of the relation $\theta_1 = \lambda/D$: (1) For a fixed slit width, longer wavelengths are diffracted over a wider range of angle than are shorter wavelengths. For example, red light has a longer wavelength than other colors in the visible spectrum; as a result, when white light is diffracted, the edges of the central peak have a reddish color. (2) For a given wavelength, the angular width of the central peak is increased when the slit width is decreased. In other words, if the slit is narrowed, the energy diffracted into the central peak spreads over a wider range of angle.

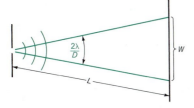

FIGURE 38.8

Single-slit diffraction. W denotes the distance between adjacent positions of zero intensity, and $2\lambda/D$ measures the angular spread of the central maximum.

EXAMPLE 1

An Estimate of the Wavelength of Laser Light

A beam of red light from a helium-neon laser is diffracted by a slit 0.50 mm wide. An interference pattern is formed on a wall 2.12 m beyond the slit (Figure 38.8). The distance between the positions of zero intensity on either side of the forward peak is

5.1 mm. Let's estimate the wavelength of the laser light. The angular separation of the zero-intensity points is $2\theta_1 = 2\lambda/D$. At a distance L from the slit their linear separation is approximately

$$W = L\left(\frac{2\lambda}{D}\right)$$

The wavelength is

$$\lambda = \frac{D \cdot W}{2L} = \frac{(0.50 \times 10^{-3}\ \text{m})(5.1 \times 10^{-3}\ \text{m})}{4.24\ \text{m}} = 6 \times 10^{-7}\ \text{m}$$

$$= 600\ \text{nm}$$

A more precise value is 632.8 nm. The single-slit diffraction pattern provides an estimate of the wavelength. A more precise measurement, obtained by using a diffraction grating, is described in Section 38.4.

38.3
DIFFRACTION AND ANGULAR RESOLUTION

The relatively large intensity of the central peak of the single-slit pattern is typical of many diffraction patterns. Furthermore, the relation $\theta_1 = \lambda/D$ for the single slit illustrates a very general feature of diffraction:

> When waves of wavelength λ encounter an obstruction or aperture whose characteristic dimension D is large by comparison with λ, most of the diffracted intensity is channeled into an angular range 2θ given by

$$2\theta \approx \frac{2\lambda}{D} \tag{38.13}$$

This relation applies to all kinds of diffraction phenomena, not just those involving light waves.

An important consequence of diffraction is that the image of a point source is *not* a point. Instead, it is a diffraction pattern. For example, when light from a star passes through the lens of a refracting telescope it is diffracted. The lens is a circular aperture. Figure 38.9 shows the targetlike diffraction pattern for a circular aperture. A circular lens produces this type of pattern. The focusing action of the lens cannot compensate for the interference that produces the diffraction pattern.

The intensity is greatest over the central region of the pattern, called the Airy disk. The intensity of the Airy disk falls to zero at an angle θ given by $\sin \theta = 1.22\lambda/D$, where λ is the wavelength and D is the diameter of the lens. The image formed at the focal point of the lens consists of the relatively bright Airy disk surrounded by much fainter concentric rings (Figure 38.9).

Angular Resolution

We can see the limitations imposed on image-forming systems by diffraction if we consider the images formed by two point sources. Figures 38.10 a, b, and c show three situations in which the light of wavelength λ from two point sources of equal intensities passes through a lens of diameter D. The waves from each source form diffraction patterns with angular spreads of approximately λ/D. The angle between the two sources (φ) as viewed from the

FIGURE 38.9

The diffraction pattern of a circular aperture. The bright circular area in the center is the Airy disk.

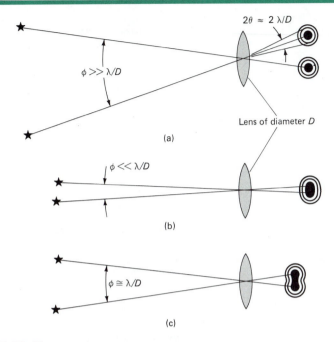

FIGURE 38.10

The diffraction limit on angular resolution. (a) The angular separation of the objects (φ) is much larger than λ/D; the objects are well resolved. (b) The angular separation of the objects is much smaller than λ/D; the objects are not resolved because their diffraction patterns overlap. (c) The angular separation of the objects is approximately λ/D; the objects are just resolved.

lens is called their **angular separation**. In Figure 38.10a the angular separation of the two sources is much greater than λ/D and the two diffraction patterns do not overlap. The two images are clearly separated, and we say that the two objects are **resolved**. In Figure 38.10b the angular separation of the sources is smaller than λ/D and the two diffraction patterns overlap significantly. It is not possible to tell whether there are two sources or only one. In this case, we say that the objects are **not resolved**. Figure 38.10c shows the borderline situation where the angular separation is approximately equal to λ/D. The diffraction patterns overlap slightly. It is now barely possible to infer that there are two sources, and we say that the objects are **just resolved**.

The **angular resolution** of an image-forming system is the smallest angular separation it can distinguish (resolve). **Rayleigh's criterion** is widely used to establish the angular resolution of a circular lens. Rayleigh observed that when the angular separation (φ) of the two sources is such that

$$\sin \varphi = 1.22 \left(\frac{\lambda}{D} \right) \qquad (38.14)$$

the central intensity peak of one source falls on the first intensity zero of the other source. The result is a slight dimple in the total intensity pattern, from which we can infer the presence of two sources. Figures 38.11 a, b, and c show the intensity patterns for two sources separated by angles for which $\sin \varphi = 0.61(\lambda/D)$, $1.22(\lambda/D)$, and $2.44(\lambda/D)$.

When $\lambda/D \ll 1$, Rayleigh's criterion gives

$$\text{angular resolution} = \varphi = 1.22 \left(\frac{\lambda}{D} \right) \qquad (38.15)$$

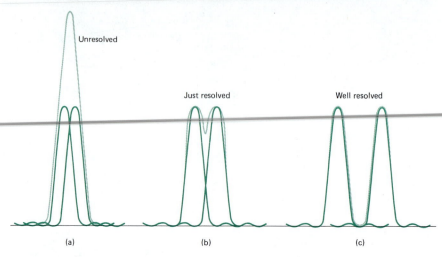

FIGURE 38.11

Rayleigh's criterion. Overlapping diffraction patterns for two objects whose angular separations are given by (a) $\sin \varphi = 0.61 \ (\lambda/D)$, (b) $\sin \varphi = 1.22 \ (\lambda/D)$, (c) $\sin \varphi = 2.44 \ (\lambda/D)$. Darker lines indicate the intensity patterns for each of the two objects. The fainter line is the sum of the two intensities, which is the observed intensity. The slight dimple in the total intensity for (b) marks the limit of resolution according to Rayleigh's criterion.

The smaller the angular resolution, the better the image resolution of an instrument. The world's largest radio telescope, at Arecibo, Puerto Rico, has a diameter of 305 m. At a wavelength of $\lambda = 21$ cm the angular resolution is 8.4×10^{-4} radians. By comparison, the angular resolution of an amateur astronomer's 6-inch (lens diameter) telescope viewing green light ($\lambda = 500$ nm) is 4×10^{-6} radians. Contrary to what we might expect, the amateur's telescope is superior with respect to angular resolution. However, the Arecibo telescope affords several advantages. It is able to detect objects far beyond the range of the 6-inch telescope and can scan a wide range of radio wavelengths far beyond the visible range.

The human eye is another optical system that is subject to diffraction effects. Let's determine the diffraction limit for the angular resolution of the human eye.

EXAMPLE 2

Angular Resolution of the Human Eye

The diameter of the pupil of a human eye varies from 2 mm in bright light to 5 mm in the dark. If yellow light with a wavelength of 600 nm reaches the eye, its wavelength in the vitreous humor is 450 nm. The angular resolution for the light-adapted eye is

$$1.22\left(\frac{\lambda}{D}\right) = 1.22\left(\frac{4.5 \times 10^{-7} \text{ m}}{2 \times 10^{-3} \text{ m}}\right)$$
$$= 3 \times 10^{-4} \text{ rad}$$

The millimeter markings on a meter stick subtend an angle of 3×10^{-4} rad at a distance of approximately 3 m. Therefore, if you view a meter stick more than 3 m away, you won't be able to resolve the individual millimeter marks. Images of the individual millimeter marks will overlap because of diffraction and you will probably see only a gray band. Try it.

Angular resolution is also limited by the spacing of the light-sensitive cells on the retina. Most people are unable to resolve the millimeter marks at distances greater than 3 m.

Other factors also affect the ability of an image-forming system to resolve objects. Turbulence causes fluctuations in the index of refraction of the earth's atmosphere. Starlight traveling through the atmosphere is refracted many times in an irregular, fluctuating fashion. We see this refraction as the twinkle of starlight. The direction of a star appears to us to change very slightly from one moment to the next. For the astronomer, twinkling also limits the angular resolution of a telescope. Twinkling smears images, and if two objects are close enough, their smeared images may overlap and thereby limit resolution.

It should be emphasized that the diffraction limitations on angular resolution are not restricted to lenses. Any device that collects waves and senses the direction of their origin is subject to the limits imposed by diffraction. Thus, the angular resolution of reflecting telescopes and of radio telescopes are subject to diffraction limitations.

38.4
DIFFRACTION GRATINGS

The angular resolution of a single slit, a mirror, or a radio telescope dish is limited by the diameter D. The angular resolution can be improved by using a series of slits, mirrors, or dishes. Any arrangement of elements equivalent to an array of parallel slits is called a **diffraction grating**. A diffraction grating produces a series of sharply defined intensity peaks, and can be used to make precision measurements of wavelength.

High-quality *transmission* gratings are made by scribing a series of closely spaced parallel lines on a transparent plate. A grating may have over 10,000 lines per centimeter. Needless to say, formidable technological problems must be overcome to produce high-quality gratings. A modified version of the Michelson interferometer can be used to control the spacing of the grating lines. The lines scatter the light diffusely and are effectively opaque. The spaces between lines act as the slits.

The Diffraction Pattern

Consider a grating composed of N identical slits. Each slit has a width D and a center-to-center distance of a. Let's determine the diffraction pattern for the grating—an expression that describes the light intensity. In Figure 38.12 we see that waves that interfere to produce the diffraction pattern travel different distances. The path difference for waves from adjacent slits is $a \sin \theta$, just as it was for the double-slit interference studied in Section 37.2. The corresponding phase difference between waves from adjacent slits is

$$\delta = \left(\frac{2\pi a}{\lambda}\right) \sin \theta$$

As Figure 38.12 indicates, the electric field at a point on the screen is the superposition of the electric fields originating from the N slits. The electric

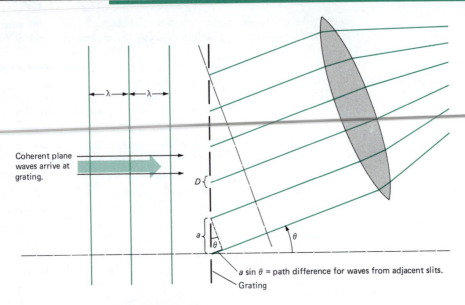

a sin θ = path difference for waves from adjacent slits.

Grating

FIGURE 38.12

Geometry for a diffraction grating. The optical path difference for waves from adjacent slits is $a \cdot \sin \theta$.

field from the first (top) slit ($E_{1\theta}$) is given by the expression derived in Section 38.2 for a single slit (Equation 38.6),

$$E_{1\theta} = bE_0 \sin(kx_0 - \omega t) \cdot \frac{\sin \Phi}{\Phi}$$

Waves from the second slit travel a distance $a \sin \theta$ farther than those from the first slit. The electric field for the second slit ($E_{2\theta}$) therefore has the same form as $E_{1\theta}$, but with x_0 replaced by $x_0 + a \sin \theta$,

$$E_{2\theta} = bE_0 \sin(kx_0 + ka \sin \theta - \omega t) \cdot \frac{\sin \Phi}{\Phi}$$

The quantity

$$\Phi = \left(\frac{\pi D}{\lambda}\right) \sin \theta$$

was introduced for the single slit in Section 38.2. The factor $\sin \Phi / \Phi$ is the same for all slits because D and the angle θ have the same values for all slits. The electric fields from successive slits differ from $E_{1\theta}$ only in having $x_0 + 2a \sin \theta$, $x_0 + 3a \sin \theta$, and so on, in place of x_0. Thus, for the Nth slit,

$$E_{N\theta} = bE_0 \sin[kx_0 + (N-1)ka \sin \theta - \omega t] \cdot \frac{\sin \Phi}{\Phi} \tag{38.16}$$

The total electric field at the screen is the sum of the fields arriving from the N slits and is given by

$$E_0 = E_{1\theta} + E_{2\theta} + \cdots + E_{N\theta}$$

$$= \left\{\frac{bE_0 \sin \Phi}{\Phi}\right\} \{\sin \varphi_0 + \sin(\varphi_0 + \delta) + \cdots \tag{38.17}$$

$$+ \sin[\varphi_0 + (N-1)\delta]\}$$

where $\varphi_0 = kx_0 - \omega t$, and

$$\delta = ka \sin \theta = \left(\frac{2\pi a}{\lambda}\right) \sin \theta \qquad (38.18)$$

To use E_θ to compute the intensity, we convert the sum of sines in Equation 38.17 into a product. The mathematical relation needed is

$$\sin \varphi_0 + \sin (\varphi_0 + \delta) + \cdots + \sin [\varphi_0 + (N - 1)\delta]$$

$$= \left(\frac{\sin \frac{1}{2}N\delta}{\sin \frac{1}{2}\delta}\right) \sin [\varphi_0 + \tfrac{1}{2}(N - 1)\delta] \qquad (38.19)$$

Using Equation 38.19 in Equation 38.17 gives

$$E_\theta = bE_0 \left(\frac{\sin \Phi}{\Phi}\right) \left(\frac{\sin \frac{1}{2}N\delta}{\sin \frac{1}{2}\delta}\right) \cdot \sin [\varphi_0 + \tfrac{1}{2}(N - 1)\delta] \qquad (38.20)$$

The intensity is the time average of $c\epsilon_0 E_\theta^2$ and can be expressed as

$$I_\theta = I_0 \left[\frac{\sin^2 \Phi}{\Phi^2}\right] \left(\frac{\sin \frac{1}{2}N\delta}{\sin \frac{1}{2}\delta}\right)^2 \qquad (38.21)$$

This equation describes the **N-slit diffraction pattern**. The angular position of a point on a viewing screen is specified by the angle θ. Both Φ and δ, which control the value of I_θ, are functions of θ. Thus,

$$\Phi = \left(\frac{\pi D}{\lambda}\right) \sin \theta \qquad (38.7)$$

$$\delta = \left(\frac{2\pi a}{\lambda}\right) \sin \theta \qquad (38.18)$$

Note that I_θ, given by Equation 38.21, is the product of a single-slit intensity factor $I_0[(\sin^2 \Phi)/\Phi^2]$, and an **interference factor**, $(\sin \frac{1}{2}N\delta/\sin \frac{1}{2}\delta)^2$. The interference factor describes interference *between* adjacent slits and depends on the slit separation (a). The factor $[(\sin^2 \Phi)/\Phi^2]$ describes interference arising *within* the slits and depends on the width (D) of the slits.

Gratings with Large N

The dependence of I_θ on the number of slits, N, is contained in the interference factor $(\sin \frac{1}{2}N\delta/\sin \frac{1}{2}\delta)^2$. We can see the effect of increasing the number of slits by comparing plots of the interference factor for different values of N. Figure 38.13 is a plot of $(\sin \frac{1}{2}N\delta/\sin \frac{1}{2}\delta)^2$ for $N = 2$, 4, and 6. The large peaks are called **principal maxima**. The effect of increasing N is to increase the intensity of the principal maxima at the expense of the regions in between.

The principal maxima correspond to angles (θ) for which waves from adjacent slits are out of step by an integral number of wavelengths and therefore interfere constructively. The path difference for waves from adjacent slits is $a \sin \theta$. The condition for a principal maximum is therefore

$$a \sin \theta = m\lambda \qquad m = 0, 1, 2, 3, \ldots \qquad (38.22)$$

Setting $a \sin \theta = m\lambda$ in $\delta = (2\pi a/\lambda) \sin \theta$ shows that the principal maxima correspond to values of δ that are integral multiples of 2π:

$$\delta = 2\pi m \qquad m = 0, 1, 2, 3, \ldots \qquad (38.23)$$

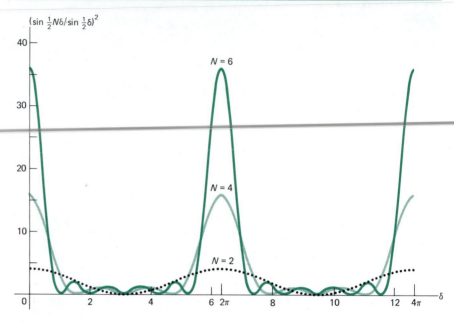

FIGURE 38.13
The interference factor $(\sin\frac{1}{2}N\delta/\sin\frac{1}{2}\delta)^2$ for $N = 2, 4,$ and 6. The principal maxima occur at angles where δ is an integral multiple of 2π. The interference factor has the value N^2 at a principal maximum.

The integer m is called the order of the maximum.

The interference factor $(\sin\frac{1}{2}N\delta/\sin\frac{1}{2}\delta)^2$ has the value N^2 at a principal maximum. This is a consequence of complete constructive interference of the waves from all N slits. Figure 38.13 confirms numerically that the interference factor has the value N^2 at principal maxima.

EXAMPLE 3

Determining the Wavelength of Laser Light

In Example 1 we estimated the wavelength of light from a He-Ne laser by using a single slit. A multiple-slit grating provides a much more precise method for measuring the wavelength.

Light from the laser strikes a grating with 5220 lines per centimeter. The zeroth- and first-order principal maxima ($m = 0, 1$) are separated by a distance of $S = 0.639$ m on a wall that is a distance $L = 1.83$ m from the grating. From Figure 38.14 it follows that

$$\sin\theta = \frac{S}{\sqrt{L^2 + S^2}} = \frac{0.639 \text{ m}}{\sqrt{(1.83 \text{ m})^2 + (0.639 \text{ m})^2}}$$

$$= 0.330$$

The slit spacing is the reciprocal of the number of lines per centimeter

$$a = \frac{1}{(5220 \text{ cm}^{-1})} = 1.920 \times 10^{-6} \text{ m}$$

$$= 1920 \text{ nm}$$

From Equation 38.22, with $m = 1$,

$$\lambda = a\sin\theta = (1920 \text{ nm})(0.330) = 633 \text{ nm}$$

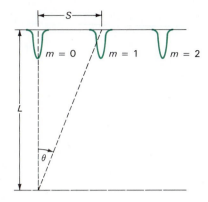

FIGURE 38.14

The angle θ is the angular separation of the $m = 0$ and $m = 1$ principal maxima.

Intensity Zeros

The positions of zero intensity correspond to angles at which complete destructive interference occurs. The interference factor (and the intensity) is zero provided that

$$\sin \tfrac{1}{2}N\delta = 0 \qquad \text{and} \qquad \sin \tfrac{1}{2}\delta \neq 0 \qquad (38.24)$$

If both $\sin \tfrac{1}{2}N\delta$ and $\sin \tfrac{1}{2}\delta$ are zero, we get a principal maximum. At a principal maximum, Equation 38.23 shows that $\delta = 2\pi m$, where m is an integer. The value of $\tfrac{1}{2}N\delta$ at a principal maximum is $\pi m N$. The intensity zero adjacent to a principal maximum occurs when $\tfrac{1}{2}N\delta$ increases by π, from $\pi m N$ to $\pi(mN + 1)$. The change in δ between a principal maximum and an adjacent zero is

$$\Delta\delta = \frac{2\pi}{N} \qquad (38.25)$$

Recall that

$$\delta = \left(\frac{2\pi a}{\lambda}\right) \sin \theta \qquad (38.18)$$

The range of angle ($\Delta\theta$) corresponding to $\Delta\delta = 2\pi/N$ can be found by forming the differential of Equation 38.18 and replacing $d\delta$ and $d\theta$ by $\Delta\delta$ and $\Delta\theta$. Thus,

$$\Delta\delta = \left(\frac{2\pi a}{\lambda}\right) \cos \theta \cdot \Delta\theta$$

Setting $\Delta\delta = 2\pi/N$ gives

$$\Delta\theta = \frac{\lambda}{Na \cos \theta} \qquad (38.26)$$

This is a key result—it specifies $\Delta\theta$ as a function of N. The quantity $\Delta\theta$ is the difference between the values of θ at a principal maximum and at an adjacent zero. As such, $\Delta\theta$ measures the angular spread of the intensity peaks. Equation 38.26 shows that the angular spread of the intensity peaks decreases as N is increased (see also Figure 38.13). Therefore, as the number of slits is increased, light of a given wavelength is concentrated into a smaller range of angle. This makes it possible for a diffraction grating to separate light with different wavelengths. The separation becomes more clear-cut as the value of N is increased. Gratings with thousands of lines per centimeter enable us to measure visible wavelengths with a precision of 0.01 nm—a length that is about one-tenth the characteristic dimension of an atom!

Resolving Power

The precision with which wavelengths can be measured using a grating spectroscope can be related to $\Delta\theta = \lambda/Na \cos\theta$. Suppose that light consisting of a mixture of two wavelengths, λ and $\lambda + \Delta\lambda$, strikes the grating. If λ and $\lambda + \Delta\lambda$ are nearly equal, the intensity maxima for the two wavelengths overlap, and it is not possible to determine that different wavelengths are present. In this instance we say that the wavelengths are *not resolved*.

In our discussion of Rayleigh's criterion in the preceding section, resolution was a question of detecting two sources that emit the same wavelength, but are located at slightly different angular positions. Here, on the other hand, the sources coincide with respect to direction, but have slightly different wavelengths. When the difference $\Delta\lambda$ is large enough, the principal maxima do not overlap, and it is clear that two different wavelengths are present. In this instance we say the wavelengths are *clearly resolved*. The arbitrary criterion used to decide the issue of wavelength resolution is this:

The difference $\Delta\lambda$ must be large enough to cause the principal maximum for one wavelength to fall on the first position of zero intensity of the other wavelength.

This criterion results in a slight dimple in the overall intensity, which shows that two wavelengths are present.

The equation governing resolvability is

$$\frac{\Delta\lambda}{\lambda} = \frac{1}{mN} \tag{38.27}$$

According to Equation 38.27 the larger N is, the smaller $\Delta\lambda/\lambda$ and the greater the precision will be. This is one incentive for having a large number of lines on the grating. A high-quality grating with $N \approx 100{,}000$ can resolve wavelength differences as small as $\lambda/100{,}000$. For visible light (430−700 nm), this amounts to a precision of better than 0.01 nm.

Great precision in measurements of wavelengths is desirable for several reasons. For example, the set of wavelengths (spectrum) of light emitted by each element is unique. No two elements have the same optical spectrum. The spectrum acts like a set of optical fingerprints that can be used to identify the emitter, provided that precision measurements of wavelength can be made.

A measure of the precision with which a grating can measure wavelengths is the quantity

$$R \equiv \frac{\lambda}{\Delta\lambda} \tag{38.28}$$

called the *resolving power* of the grating. A large value of R means that the grating can resolve small differences of wavelength. For the mth order of a grating with N slits, Equation 38.27 shows that

$$R = mN \tag{38.29}$$

Notice that the resolving power is directly proportional to the number of slits. Clearly, a large grating is desirable.

Equation 38.29 shows that the resolving power increases with the order m. Two wavelengths that overlap in the first-order spectrum may be resolved in the second- or third-order spectra. In practice, spectra from different orders begin to overlap for values of m larger than 2 or 3. Thus, the resolving power $R = mN$ is not improved by using a very high-order spectrum.

EXAMPLE 4

Resolving the Sodium Doublet

The yellow-orange light from a sodium vapor lamp consists primarily of a mixture of two wavelengths, $\lambda = 589.0$ nm and $\lambda + \Delta\lambda = 589.6$ nm. These two wavelengths are called the *sodium doublet*. Let's determine the minimum number of slits required to resolve these wavelengths in the second-order spectrum. We use Equation 38.27,

$$\frac{\Delta\lambda}{\lambda} = \frac{1}{mN}$$

with $\Delta\lambda = 0.6$ nm, $\lambda = 589.0$ nm, and $m = 2$, the minimum number of slits is

$$N = \left(\frac{589 \text{ nm}}{0.6 \text{ nm}}\right)\left(\frac{1}{2}\right) \approx 491$$

The very modest resolving power required for such a grating is

$$R = mN = 982$$

38.5
Holography

A photographic negative and the positive prints made from it are two-dimensional records of light intensity. The negative contains no information about the phase of the light waves that produced it. Because the phase information is lost, an ordinary photograph gives us a two-dimensional image of a three-dimensional object. If, on the other hand, we record both phase and intensity, we can obtain a three-dimensional image. The process of **holography** allows us to make such images, called **holograms**.

The superposition of coherent light waves produces an interference pattern. Constructive interference occurs at points where the waves have phases differing by integral multiples of 2π radians. If the interfering light waves fall on a photographic film, we can record the interference pattern. Such a photographic record contains information about the relative phases of the two waves. It also contains information about the amplitudes of the individual waves.

A hologram is a photographic record of the interference pattern formed by the superposition of two coherent light waves. It can be used to reconstruct the wave fronts that produced it, and this reconstruction gives us a three-dimensional image.

Figure 38.15 shows how the simplest type of hologram is produced. Coherent light from a laser is scattered by a point object, P. The scattered waves then interfere with other waves from the laser, while the film at S records the interference pattern. The pattern is a series of concentric rings that mark alternating regions of constructive and destructive interference. This ring pattern is called

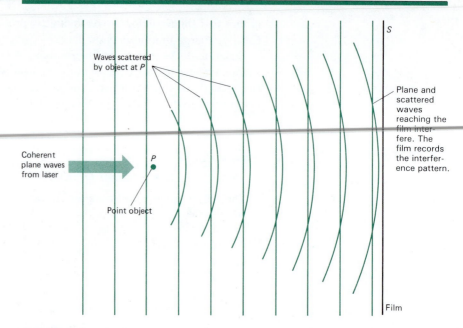

FIGURE 38.15

A hologram of a point object (P) is formed when waves scattered by the object interfere with unscattered waves. The interference pattern is recorded in the film. For a point object the hologram consists of a set of circular interference fringes called a Gabor zone plate.

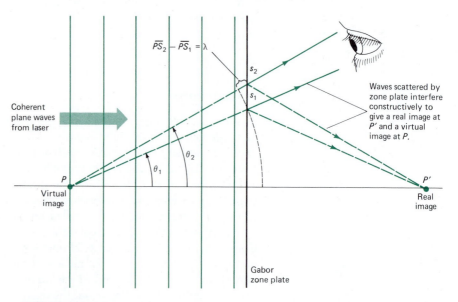

FIGURE 38.16

When the Gabor zone plate is illuminated with coherent light, the scattered waves interfere constructively in certain directions. These waves reconstruct the waves that formed the zone plate and thereby produce an image of the object.

a **Gabor zone plate**, in honor of Dennis Gabor, who conceived the principles of holography in 1947. In Figure 38.16 we see that when the Gabor zone plate is illuminated with the same coherent laser light used in its production, it acts like a diffraction grating with circular slits. Secondary waves from the slits interfere constructively in certain directions. In Figure 38.16, the points

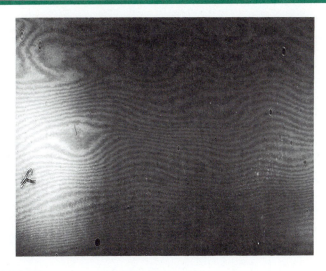

FIGURE 38.17
A piece of film used to record the holographic interference pattern produced by an extended object. The scale of the holographic interference pattern is too small to be seen. The visible interference fringes are caused by minor imperfections in lenses used to expand the laser beam.

S_1 and S_2 mark adjacent positions at which waves from P interfere constructively with the waves arriving directly from the laser. The locations of S_1 and S_2 are such that $\overline{PS_2}$ and $\overline{PS_1}$ differ by one wavelength, λ,

$$\overline{PS_2} - \overline{PS_1} = \lambda$$

Rings of destructive interference occur at positions where the path difference equals an odd number of half-wavelengths. The geometry of Figures 38.15 and 38.16 shows that waves diffracted by adjacent slits through the angles θ_1 and θ_2 interfere constructively and form a real image at P'. A second image—this one virtual—is formed behind the hologram at P, the original site of the point object. An observer sees light waves diverging from P. The relative phase of these waves is the same as for the scattered waves that produced the hologram. In effect, then, the diffracted waves are **reconstructions** of the waves that produced the hologram.

An extended object produces a complicated-looking hologram (Figure 38.17) that bears no resemblance to the object. Nevertheless, the diffracted laser light reconstructs the wave fronts that produced the hologram and creates two images—one real and one virtual. Figures 38.18 and 38.19 show how a hologram can be produced and used to display the real and virtual images of an extended object. Advances in holography have made it possible to view a hologram with ordinary white light and not just coherent laser light.

In Chapter 37 we introduced the concept of **coherence length**, that is, the length of the wave trains emitted by a source. There is a definite phase relationship between different portions of a single wave train; the waves that make up the train are coherent. Therefore, the superposition of different portions of the same train results in interference. On the other hand, there is no correlation of phases between different wave trains, and the superposition of portions of different wave trains consequently does not produce interference. In the production of a hologram, wave trains are divided and later superposed. One portion becomes a reference wave and the other becomes a wave scattered by the object. In order for the two waves to interfere, they must be

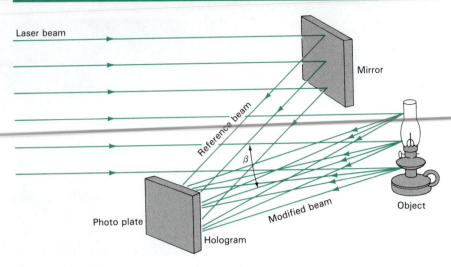

FIGURE 38.18

Hologram production. Waves reflected from the mirror and from the object interfere at the photo plate. This developed plate is the hologram. (From *Fundamentals of Optics*, 4th ed., by F. A. Jenkins and H. E. White; McGraw-Hill, New York, 1976. Used with permission of McGraw-Hill Book Company.)

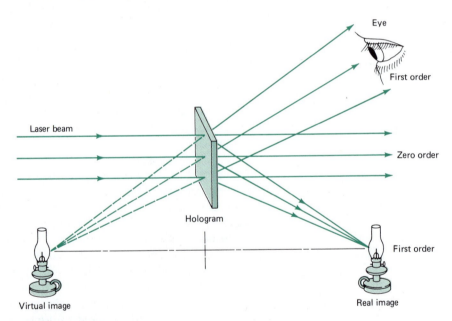

FIGURE 38.19

Reconstruction of wave fronts by a hologram. The hologram diffracts light, producing a real image and a virtual image. (From *Fundamentals of Optics*, 4th ed., by F. A. Jenkins and H. E. White; McGraw-Hill, New York, 1976. Used with permission of McGraw-Hill Book Company.)

portions of the same wave train. The two waves will interfere provided that their optical path lengths differ by less than the coherence length. If the optical path difference exceeds the coherence length, the superposed waves do not interfere—they are incoherent.

Before the development of the laser in 1960 the maximum coherence length that could be achieved with a visible light source was approximately 1 mm, and consequently holograms could be produced only for very small objects.

The coherence length for lasers, however, can be many meters and allows us to produce holograms of life-sized subjects.

WORKED PROBLEM

The moon is approximately 400,000 km away. Can two lunar craters 50 km apart be resolved by a telescope with a mirror 15 cm in diameter? Can craters 1 km apart be resolved? Take the wavelength to be 700 nm and justify your answers with approximate calculations.

Solution

We regard the craters as point objects. The observer using the telescope will see a diffraction pattern formed by light from the craters. According to the Rayleigh criterion, the craters will be resolved if their angular separation, measured from the position of the telescope, is greater than $1.22\lambda/D$. In this problem, D is the diameter of the mirror and λ is the wavelength of the light. If we call S the separation of the craters and L the distance from the telescope to the moon, then the angular separation of the craters is S/L. The criterion for the craters to be resolved is expressed by

$$\frac{S}{L} > \frac{1.22\lambda}{D}$$

or

$$S > \frac{1.22L\lambda}{D}$$

Substituting $L = 400,000$ km, $\lambda = 700$ nm $= 7 \times 10^{-7}$ m, and $D = 0.15$ m, we obtain

$$S > 1.22(400,000 \text{ km})\left(\frac{7 \times 10^{-7} \text{ m}}{0.15 \text{ m}}\right) = 2 \text{ km}$$

We conclude that craters separated by 50 km would be resolved, but craters separated by 1 km would not be resolved.

EXERCISES

38.2   Single-Slit Diffraction Pattern

A. Monochromatic light is diffracted by a single slit. The slit width is twenty times the wavelength of the light. What is the smallest angle of diffraction for which the intensity is zero?

B. A beam of green light from a helium-cadmium laser is diffracted by a slit 0.55 mm wide. The diffraction pattern forms on a wall 2.06 m beyond the slit. The distance between the positions of zero intensity ($m = \pm 1$) is 4.1 mm. Estimate the wavelength of the laser light. The helium-cadmium laser can produce light at wavelengths of 441.2 nm or 537.8 nm.

C. The relative intensity I_θ/I_0 for a single slit exhibits peaks between the zeros located at $\Phi = m\pi$. Assuming that the peaks occur midway between zeros ($\Phi = (m + \frac{1}{2})\pi$), evaluate I_θ/I_0 at the first three secondary peaks ($m = 1, 2, 3$).

D. Hold two fingers close together so as to form a narrow slit. View a source of light through this slit and explain what you see. Try varying the slit width. (Two pencils or two pieces of chalk can also be used to form a slit.)

E. Light with a wavelength of 589 nm strikes a single slit 0.50 mm wide, and the diffraction pattern is displayed on a screen 1.8 m away. Determine the distance between the two first-order dark fringes on either side of the central maximum.

38.3   Diffraction and Angular Resolution

F. Radar waves ($\lambda = 2.6$ cm) emerge through a circular opening (Figure 1). The diameter of the opening is

FIGURE 1

22 cm. Determine the approximate angular diameter of the emerging radar beam.

G. A student constructs a reflecting telescope with a 15-cm diameter mirror. Using this telescope she photographs a star (point source) in blue light ($\lambda = 450$ nm). Determine the angular radius of the Airy disk

H. A helium-neon laser emits light with a wavelength of 632.8 nm. The circular aperture through which the beam emerges has a diameter of 0.50 cm. Estimate the diameter of the beam at a distance of 10 km from the laser.

I. When Mars is nearest the earth, the distance separating the two planets is 88.6×10^6 km. Mars is viewed through a telescope whose mirror has a diameter of 30 cm. (a) If the wavelength of the light is 590 nm, what is the angular resolution of the telescope? (b) What is the smallest distance that can be resolved on Mars (the distance between two points that are just resolved)?

J. Grote Reber was a pioneer in radio astronomy. He constructed a radio telescope with a 10-m diameter receiving dish. What was the telescope's angular resolution for radio waves with a wavelength of 2 m?

K. Estimate the angular resolution of an ordinary camera lens having a diameter of 5.5 cm. Take the wavelength to be 600 nm. Could the lens resolve the millimeter markings on a meter rod placed 40 m away?

L. Two disks, each 0.5 cm in diameter, lay side-by-side. You view them at various distances. At what distance will they blend into one image according to the Rayleigh criterion? Assume a pupil diameter of 3 mm and a wavelength of 450 nm.

M. A whisper is composed of higher-frequency sound waves than ordinary conversational speech. Why is it more difficult to hear someone facing away from you when he whispers than it is to hear him when he talks in a normal tone, even though the intensity levels are the same in both cases?

N. The angular resolution of a radio telescope is to be $0.1°$ when it operates at a wavelength of 3 mm. Approximately what minimum diameter is required for the telescope's receiving dish?

38.4   Diffraction Gratings

P. Light from an argon laser strikes a diffraction grating with 5310 lines per centimeter. The central and first-order principal maxima are separated by a distance of 0.488 m on a wall that is 1.72 m from the grating. Determine the wavelength of the laser light.

Q. A diffraction grating with 38,380 lines has a width of 8.211 cm. Determine the resolving power for the second-order spectrum when the grating is used to study light with a wavelength of 583 nm.

R. Imagine that you are designing a large diffraction grating that must be able to resolve $\Delta\lambda = 0.01$ nm for $\lambda = 500$ nm in the second-order spectrum. How many lines must your grating have?

PROBLEMS

1. An earth satellite designed to inventory crops is in orbit at an altitude of 705 km. If the satellite's camera has an angular resolution of 42.5×10^{-6} radians, what is the minimum separation of two objects which are resolved?

2. A source emits light with wavelengths of 531.62 nm and 531.81 nm. (a) What is the minimum number of lines required for a grating that resolves the two wavelengths in the first-order spectrum? (b) Determine the slit spacing for a grating 1.32 cm wide that has the required minimum number of lines.

3. A spy is accused of reading, with his naked eyes, a sensitive document 10 m away. Assuming that the printed matter has the same structure as newspaper print, explain why the accusation is invalid.

4. (a) Show that values of Φ corresponding to maxima in a single-slit diffraction pattern are determined by the relation

$$\tan \Phi = \Phi$$

(b) To an accuracy of three figures, use a calculator or computer to find the three smallest positive values of Φ that solve $\tan \Phi = \Phi$. You may find it helpful to sketch the functions Φ and $\tan \Phi$ versus Φ in order to locate approximate solutions.

5. Light from a helium-neon laser (wavelength = 632.8 nm) is directed toward a diffraction grating oriented perpendicular to the laser beam. On a screen 2.0 m from the grating the first-order principal maximum is 35 cm from the central maximum. (a) Determine the number of lines per centimeter on the grating. (b) What is the order of the largest principal maximum that can be observed with this arrangement.

6. Light from a helium-neon laser (wavelength = 632.8 nm) illuminates a single slit and a diffraction pattern is formed on a screen 1.0 m from the slit. The following data record measurements of the relative intensity as a function of distance from the line formed by the undiffracted laser beam.

Relative Intensity	Distance (mm)
0.95	0.8
0.80	1.6
0.60	2.4
0.39	3.2
0.21	4.0
0.079	4.8
0.014	5.6
0.003	6.5
0.015	7.3
0.036	8.1
0.047	8.9
0.043	9.7
0.029	10.5
0.013	11.3
0.002	12.1
0.0003	12.9
0.005	13.7
0.012	14.5
0.016	15.3
0.015	16.1
0.010	16.9
0.0044	17.7
0.0006	18.5
0.0003	19.3
0.003	20.2

Make a plot of the relative intensity versus distance. Choose an appropriate value for the slit width D and plot the theoretical expression for the relative intensity

$$\frac{I_\theta}{I_0} = \frac{\sin^2 \Phi}{\Phi^2}$$

on the same graph used for the experimental data. What value of D gives the best fit of theory and experiment?

7. A telescope in a spy satellite in orbit at an altitude of 100 miles must resolve objects separated by a distance

of 1 foot. If ultraviolet light (wavelength = 200 nm) is used, what diameter must the telescope have?

8. In the acoustic analog of diffraction-limited resolution described by Figure 38.10, the lens might be replaced by a doorway of width D. Consider sound waves with a frequency of 1 kHz and a speed of 330 m/s. What is the approximate minimum angular separation of the two speakers which can be resolved when the waves are diffracted by a doorway 1 m wide?

9. William L. Bragg scattered X rays off crystals and observed diffraction effects similar to those produced by a grating. Consider a parallel beam of X rays of wavelength λ incident at the angle θ to a set of crystal planes separated by a distance d (Figure 2). Show that the diffracted beams interfere constructively, provided that

$$2d \sin \theta = m\lambda \qquad (m = \text{integer})$$

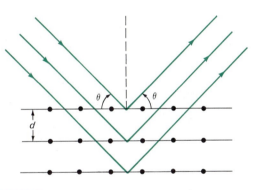

FIGURE 2

10. Consider a double slit, taking into account both diffraction and interference effects. What condition must be satisfied if the broad central maximum is to have exactly 21 fringes?

CHAPTER 39

SPECIAL RELATIVITY

39.1
EINSTEIN'S POSTULATES OF SPECIAL RELATIVITY

Albert Einstein's theories of special relativity (1905) and general relativity (1915) represent one of mankind's greatest intellectual achievements. Relativity is concerned with the laws of physics as formulated by observers in relative motion. In this chapter, we develop the theory of special relativity. Special relativity is *special* because it is restricted to **inertial frames of reference** (see Section 6.1). We present special relativity as a generalization of Galilean relativity. This generalization requires us to reexamine the basic concepts of length and time. You must reeducate your common sense to bring it into agreement with experimental facts.

Nineteenth-century scientists recognized the wave properties of light and they searched for the medium through which light waves propagated. At every turn the proposed medium seemed to elude the investigators. The French mathematician Henri Poincaré spoke of a "conspiracy of nature" to conceal the medium. It remained for the genius of Albert Einstein to realize that this apparent conspiracy was actually evidence for a new law of nature. In 1905, Einstein established the special theory of relativity on the basis of two postulates. The first postulate is called the **principle of relativity**:

The laws of nature have the same form in all inertial frames of reference.

The second postulate concerns the speed of light as measured by observers in relative motion:

The speed of light is constant, the same in all inertial frames, independent of any relative motion of the source and the observer.

In his first postulate Einstein asserted the existence of a single principle of relativity that is valid for mechanics, for electromagnetism, and for all of nature.

In his second postulate Einstein dispensed with the medium; light does not need a medium to transmit it. Einstein's second postulate seems to contradict common sense because it is incompatible with Galilean velocity addition (Section 5.3). To illustrate this incompatibility, imagine that you are stationary as a friend walks toward you. You can increase her velocity relative to you by walking *toward* her. You can decrease her velocity relative to you by walking *away* from her. This is just the common sense of Galilean relativity. But, if your friend is replaced by a light wave traveling toward you, then Einstein's second postulate means that the speed of light relative to you is the same when you are moving toward or away from the light source as it is when you are stationary. This seems to defy common sense. But common sense is wrong when it disagrees with experiment. To quote Einstein, "Common sense is that thin veneer of prejudice we acquire before the age of 17" (Figure 39.1).

Einstein's postulates are judged on a very pragmatic basis. We ask, are their consequences in agreement with experiment? The answer is, "Yes." The postulates lead to conclusions that are verified by experiment, so we accept them as valid. Also, we should remember that our common sense has evolved in a world where velocities of most objects are on the order of meters per second, whereas light travels over a million times faster. The extension of everyday concepts to light is a giant extrapolation. In this case, extrapolated common sense fails. The consequences of Einstein's postulates are verified by the experimental evidence.

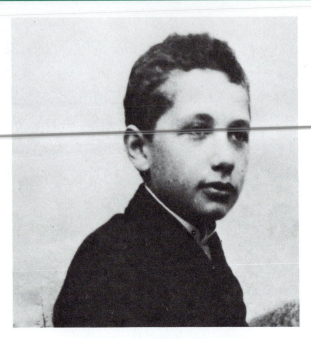

FIGURE 39.1
Einstein as a boy.

This pragmatic judgment of Einstein's postulates on the basis of experiment is not the whole story. Einstein's postulates have great scope and generality. The "laws of nature" referred to in the first postulate include all the laws of our natural world. This universality appealed to many scientists who expected perfection in the natural laws, much as the Greeks expected perfection in natural phenomena. Quite literally, the great esthetic appeal of Einstein's theory gained it a sympathetic hearing. Its ultimate acceptance as valid, however, is based on its agreement with experimental evidence.

39.2
THE RELATIVITY OF SIMULTANEITY

We classify physical quantities as **relative** or **absolute**, depending on their values as determined by observers in relative motion. Velocity is among the relative quantities we have studied. Observers in relative motion will measure different values for the velocity of an object. For example, suppose you stand at a street corner and measure the velocity of an approaching car to be 50 mph directed westward. The velocity of the car relative to its driver is zero. You and the driver are observers in relative motion and you measure different velocities. As an example of an absolute quantity, suppose you and the driver both count the number of people in the car. You count four people. The driver will also count four people. The number of people is an absolute quantity—its value is the same for observers in relative motion.

The concept of time was thought to be absolute until 1905, when Einstein showed it to be a relative concept. Einstein pointed out that observers in relative motion will disagree about the **simultaneity** of events. Because time measurements involve the concept of simultaneity, time is also a relative quantity.

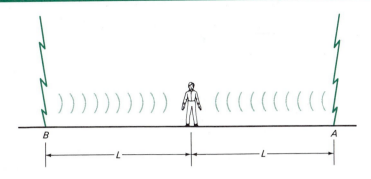

FIGURE 39.2

A definition of simultaneity. The observer is stationed midway between the points where the lightning flashes originated. He says the flashes were simultaneous events if they reach him at the same moment.

Simultaneity

We will start by defining simultaneity. We will then demonstrate its relative character. Two events are defined to be simultaneous if an observer stationed midway between them sees them occur at the same moment. Figure 39.2 illustrates this definition: an observer is stationed midway between points where two lightning bolts strike the ground. If he sees the two flashes at the same moment, he reasons that the light traveled equal distances at equal speeds and hence required equal times to travel to his position. He then concludes that the two flashes occurred at the same instant. If the flashes arrive at different times he can apply the same reasoning to conclude that the light flashes occurred at different moments—they were not simultaneous events.

Consider the same pair of events, that is, the arrival of lightning bolts, from frames of reference in relative motion. Without invoking anything beyond critical thinking we will be forced to the conclusion that simultaneity is a relative concept. Observers in relative motion will not agree as to the simultaneity of events. Figure 39.3 shows two observers in relative motion. One observer, a railroad repairman, stands on the ground beside a railroad track. The second observer, the railroad president, rides aboard a passing train. The president is seated at the midpoint of the train. The train is struck by lightning at its front and in the rear. The lightning jumps to the track, marking it also.

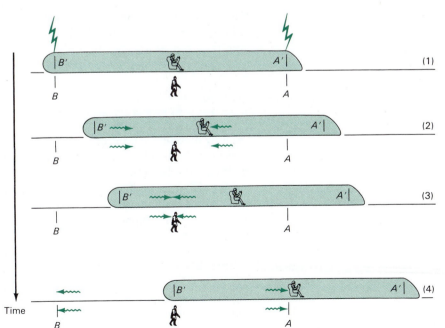

FIGURE 39.3

(1) Lightning strikes the train at A' and B' and the track at A and B. (2) The light from A' reaches the president. (3) The light from A and B reach the repairman. (4) The light from B' reaches the president.

Time

The flash from each lightning bolt is seen by the president inside the train and by the repairman beside the track. We suppose that the repairman sees light from the bolts at the same moment and that he carefully measures the distance from his position to the marks on the track where the bolts originated and finds that he was midway between them. He correctly concludes that the bolts struck simultaneously.

The president also sees two flashes, but even though she is situated midway between the two points where they originated she does not see the light from them arrive simultaneously. As Figure 39.3 shows, the light coming from the front of the train passes her before it reaches the repairman. The light coming from the rear of the train passes the repairman before it catches up with the president. Because the light pulses reach the repairman simultaneously, they can not reach the president at the same moment. In fact, the president sees the light from the front of the train before she sees the light from the rear of the train.

This may seem strange, but there is no paradox. Both observers give correct descriptions of the sequence of lightning bolts. We must conclude that **simultaneity is a relative concept**. Observers in relative motion will disagree as to the simultaneity of events. The relative nature of simultaneity is at the heart of special relativity. Measuring time requires judgments about simultaneity. Hence time becomes a relative quantity. Length is also a relative concept. To measure the length of an object, an observer must locate its endpoints simultaneously. For example, you would not associate the length of an airplane with the distance between its nose when it is in New York and its tail when it reaches California a few hours later. If one observer makes a length measurement by *simultaneously* locating the endpoints of an object, an observer in relative motion can measure a different length because of the relative character of simultaneity.

As a student you can test your understanding of the relativity of simultaneity by changing the premise. We supposed that the repairman saw the lightning flashes simultaneously. You can change the supposition and let the lightning bolt flashes arrive simultaneously for the president aboard the train. You should then be able to argue that the lightning bolt flashes are not simultaneous for the repairman. Which flash does the repairman see first, the flash from the front or the flash from the rear?

39.3
THE LORENTZ TRANSFORMATION EQUATIONS

We now present the transformation equations implied by Einstein's postulates. These equations relate the position and time measurements made by inertial observers in relative motion.

Consider the two inertial frames of reference, S and S', shown in Figure 39.4. Their relative motion is parallel to their x- and x'- axes at a speed v. At the moment the origins coincide, observers located at $x = 0$ and $x' = 0$ set their identical clocks to read zero. Mathematically, this means that the origins coincide when

$$t = t' = 0$$

Note that we must allow for separate time scales for the two frames because time is a relative quantity in special relativity.

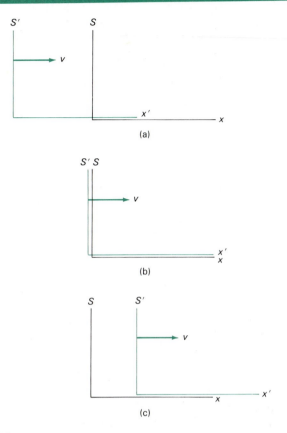

FIGURE 39.4

The inertial frames S and S' are in relative motion parallel to their x- and x'-axes. Their relative speed is v. The frame of reference S might be a coordinate system on the ground and S' a coordinate system in a moving car. The sequence of figures shows the two coordinate origins (a) approach, (b) coincide, and then (c) recede.

For Galilean relativity the relationships between position and time measurements in the two frames are given by

$$x' = x - vt$$

$$x = x' + vt'$$

$$t' = t$$

These Galilean transformation equations are not consistent with Einstein's postulates because they *assume* time is absolute. We look for modifications of the form

$$x' = \alpha(x - vt) \tag{39.1}$$

$$x = \alpha(x' + vt') \tag{39.2}$$

In addition, since we have seen that time is not an absolute concept, we expect to find that

$$t' \neq t$$

The quantity α is a scale factor which is unknown at this point. This scale factor cannot depend on position or time because that would single out special positions or times. We will find that α depends only on v, the relative speed of S and S'. The only other restriction on α is that it equal unity when $v = 0$. This is because for $v = 0$ there is no relative motion, and Equations 39.1 and 39.2 must then reduce to $x = x'$.

To determine the scale factor, we consider the position-time relation for a pulse of light emitted from the origins of S and S' when they coincide at $t = t' = 0$ (Figure 39.5). According to Einstein's second postulate, observers in S and S' see the pulse move at speed c. Their descriptions of the position of the pulse are given by

$$x = ct \tag{39.3}$$

$$x' = ct' \tag{39.4}$$

We use Equations 39.3 and 39.4 to eliminate t and t' from Equations 39.1 and 39.2. Setting $t' = x'/c$ in Equation 39.2 gives

$$x = \alpha x'\left(1 + \frac{v}{c}\right) \tag{39.5}$$

Setting $t = x/c$ in Equation 39.1 gives

$$x' = \alpha x\left(1 - \frac{v}{c}\right) \tag{39.6}$$

Next, we use Equations 39.5 and 39.6 to form the product xx' (multiply equals by equals). This gives

$$xx' = \alpha^2 xx'\left(1 - \frac{v}{c}\right)\left(1 + \frac{v}{c}\right)$$

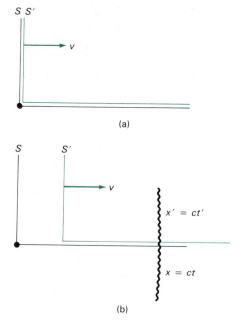

(a)

(b)

FIGURE 39.5

(a) The origins of S and S' coincide at the moment $t = t' = 0$. A light pulse is emitted from the origin at that moment. (b) Observers in S and S' both see the pulse travel at speed c.

Because x and x' are not zero we can cancel the factors xx' from both sides, leaving

$$1 = \alpha^2\left(1 - \frac{v}{c}\right)\left(1 + \frac{v}{c}\right) = \alpha^2\left(1 - \frac{v^2}{c^2}\right) \tag{39.7}$$

It follows that the scale factor α is given by

$$\alpha = \frac{1}{\sqrt{1 - \dfrac{v^2}{c^2}}} \tag{39.8}$$

We choose the positive root to make $\alpha = +1$ when $v = 0$. With α given by Equation 39.8, the transformation equations for x and x' become

$$x' = \frac{x - vt}{\sqrt{1 - \dfrac{v^2}{c^2}}} \tag{39.9}$$

$$x = \frac{x' + vt'}{\sqrt{1 - \dfrac{v^2}{c^2}}} \tag{39.10}$$

The corresponding time transformations follow by using $x = ct$ and $x' = ct'$ in Equations 39.9 and 39.10. The results are

$$t' = \frac{t - \dfrac{vx}{c^2}}{\sqrt{1 - \dfrac{v^2}{c^2}}} \tag{39.11}$$

and

$$t = \frac{t' + \dfrac{vx'}{c^2}}{\sqrt{1 - \dfrac{v^2}{c^2}}} \tag{39.12}$$

Equations 39.9 and 39.11 are the **Lorentz transformation equations**. They amount to a recipe that lets us transform measurements made by an observer in S into the corresponding measurements that would be obtained by an observer in S'. Equations 39.10 and 39.12 are the *inverse* Lorentz transformation equations—they let us convert from S' values to S values.

EXAMPLE 1

Position-Time Data Aboard a Space Tug

A space tug (S') moves past the earth (S) at a speed of $v = 0.6c$ (Figure 39.6a). Clocks on earth and on the tug are synchronized at $t = t' = 0$ as they pass ($x = x' = 0$). The tug subsequently passes a star located 6 light-years* from the earth, and at rest

---

* Recall that one light-year equals the distance light travels in one year; 1 ly $= c \cdot 1$ year.

relative to the earth. Clocks at the star, at rest relative to the earth, are synchronized with the earth clocks. The tug's passage of the star is an event that can be described in terms of the S data, x and t, and in terms of the S' data, x' and t'. We want to compare the position-time data describing this event.

According to the earth observer, the tug requires 10 years to travel 6 light-years at a speed of 0.6c. Thus, the coordinates of the event for the earth observer are

$$x = 6 \text{ ly}, \qquad t = 10 \text{ yr}$$

Let's use the Lorentz transformation equations to determine the corresponding quantities (x', t') aboard the tug. We can anticipate the value of x', without using fancy equations. The tug is located at the origin of the S' system, and so $x' = 0$. Equation 39.9 verifies this,

$$x' = \frac{x - vt}{\sqrt{1 - \dfrac{v^2}{c^2}}} = \frac{6 \text{ ly} - (0.6c)(10 \text{ yr})}{\sqrt{1 - \dfrac{v^2}{c^2}}} = 0$$

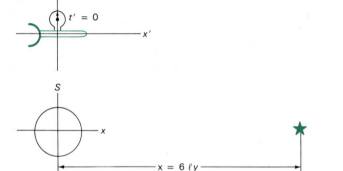

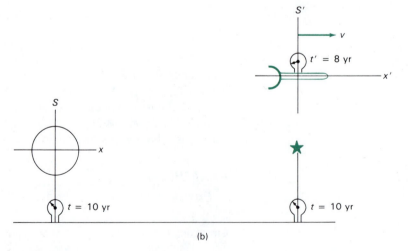

FIGURE 39.6

(a) The tug passes the earth at $t = t' = 0$ at a speed of 0.6c. It takes 10 years, earth time, for it to reach the star 6 light-years from the earth. (b) When the tug passes the star its clock reads $t' = 8$ years. The clock at the star, which remains synchronized with the earth clock, reads $t = 10$ years.

The clocks on board the tug read the time t' when it reaches the star. From Equation 39.11 we find

$$t' = \frac{t - \dfrac{vx}{c^2}}{\sqrt{1 - \dfrac{v^2}{c^2}}}$$

$$= \frac{10 \text{ yr} - \dfrac{(0.6c)(6 \text{ ly})}{c^2}}{\sqrt{1 - (0.6)^2}}$$

$$= 8 \text{ yr}$$

Whereas 10 years elapse on the earth clocks, we find that only 8 years elapse on clocks aboard the tug. Although the clocks in S and S' were synchronized when the tug passed the earth, they do not remain synchronized. Clocks aboard the tug read $t' = 8$ years when the tug reaches the star. Clocks at the position of the star and at rest relative to the earth remain synchronized with earth clocks and read $t = 10$ years when the tug passes (Figure 39.6b).

The fact that the earth clocks and tug clocks do not remain synchronized means that clocks in relative motion run at different rates. This is a consequence of the relativity of simultaneity. Time is a relative quantity.

Symmetry Relations

The Lorentz transformation equations exhibit a symmetry related to the fact that only relative motion is detectable. If an observer at rest in S sees S' move to the right at speed v, then an observer at rest in S' sees S move to the left at speed v (velocity $-v$). Interchanging x and x' and t and t' and replacing v by $-v$ is equivalent to interchanging S and S'. The transformation equations are left unchanged in form by these interchanges. This invariance of form is required by Einstein's first postulate. The laws of nature—the Lorentz transformation equations in this instance—must have the same form in all inertial frames.

Low-Speed Limit

For speeds that are much less than the speed of light we expect the Lorentz transformation equations to reduce to the Galilean transformation equations. Why? Because the Galilean transformation equations have been amply verified in low-speed experiments.

For speeds that are small compared to the speed of light ($v \ll c$), the factor $\sqrt{1 - v^2/c^2}$ approaches unity and Equations 39.9 and 39.10 reduce to their Galilean form ($x' = x - vt$).

Light travels a distance ct in a time t. For distances x that are small compared to ct, we have the inequalities

$$\frac{vx}{c^2} \ll \frac{v(ct)}{c^2} = \frac{vt}{c} < t$$

Ignoring vx/c^2 compared to t and replacing $\sqrt{1 - v^2/c^2}$ by 1 reduces Equation 39.11 to its Galilean form, $t' = t$.

We see that Galilean relativity, amply verified by experiment, is the low-speed limit of Einstein's special relativity. Einstein's relativity becomes essential for phenomena involving speeds comparable to the speed of light.

39.4
THE EINSTEIN VELOCITY ADDITION FORMULA

All motions considered in this chapter take place parallel to the x- and x'-axes. If this is accepted, we can speak of the "addition of velocities." We now derive the Einstein velocity addition formula and show that it is consistent with the postulates of special relativity.

We let the observers in S and S' record the motion of some object moving parallel to their x- and x'-axes (Figure 39.7). For S the object undergoes a displacement Δx in a time interval Δt. For S' the same motion is recorded as a displacement $\Delta x'$ over a time interval $\Delta t'$. The displacements and time intervals are related by the Lorentz transformation equations. From Equation 39.10 we find the relation

$$\Delta x = \frac{\Delta x' + v\,\Delta t'}{\sqrt{1 - \dfrac{v^2}{c^2}}} \tag{39.13}$$

From Equation 39.12 we obtain a second relationship,

$$\Delta t = \frac{\Delta t' + \dfrac{v\,\Delta x'}{c^2}}{\sqrt{1 - \dfrac{v^2}{c^2}}} \tag{39.14}$$

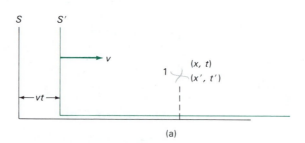

(a)

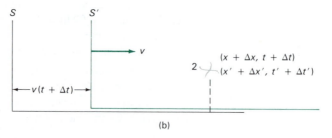

(b)

FIGURE 39.7

Observers in S and S' move with a velocity v relative to each other. They measure the displacement of the object relative to their reference frames. During a short time interval (Δt in S, $\Delta t'$ in S') the object moves from 1 to 2 and the observers record its displacement as Δx and $\Delta x'$. Using these data they can determine the velocity of the object.

The velocity of the object as observed in S' is

$$u' = \frac{\Delta x'}{\Delta t'} \qquad (39.15)$$

while in S the velocity of the object is

$$u = \frac{\Delta x}{\Delta t} \qquad (39.16)$$

Dividing Equation 39.13 by Equation 39.14 and performing a bit of algebraic manipulation gives

$$\frac{\Delta x}{\Delta t} = \frac{\Delta x' + v\,\Delta t'}{\Delta t' + \dfrac{v\,\Delta x'}{c^2}}$$

$$= \frac{\dfrac{\Delta x'}{\Delta t'} + v}{1 + \dfrac{v\left(\dfrac{\Delta x'}{\Delta t'}\right)}{c^2}}$$

Using the definitions of u and u' converts this into the desired Einstein **velocity addition formula**

$$u = \frac{u' + v}{1 + \dfrac{u'v}{c^2}} \qquad (39.17)$$

For speeds small compared to the speed of light the denominator on the right side of Equation 39.17 approaches unity and we recover the Galilean velocity addition relation, $u = u' + v$.

First, let's illustrate the Einstein velocity addition formula by showing that it is consistent with the second postulate of relativity. This is the postulate that seems to defy common sense by stating that the speed of light is the same for all inertial observers, regardless of their motion relative to the light source.

Suppose an observer in S' is at rest relative to a light source and records the velocity of an oncoming light pulse as $u' = c$ (Figure 39.8). An observer in S is in motion with respect to S' at speed v and so is also in motion relative to the light source. According to Galilean relativity, the observer in S would see the light approaching at speed $c + v$. Using the Einstein velocity addition

FIGURE 39.8

A light source is at rest in S', which moves at speed v relative to S. Light travels toward the observer in S' at speed $u' = c$. Galilean relativity incorrectly predicts that the observer in S would find a light speed of $c + v$. Einstein's velocity addition formula correctly shows that the speed of light for the observer in S is also c.

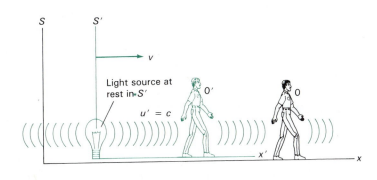

formula, Equation 39.17, we get a different result. With $u' = c$, Equation 39.17 gives the velocity of the light pulse relative to S,

$$u = \frac{c + v}{1 + \dfrac{cv}{c^2}} = c$$

Thus, $u = c$ and observers in both S and S' see the same speed of light regardless of any motion relative to its source.

Notice that it is the denominator of the Einstein velocity addition formula that distinguishes it from the Galilean velocity addition formula. In solving problems involving relative velocity it is a good practice to start by first working the problem by applying the Galilean velocity addition relation. This will help you identify u, u', and v. Then you can obtain the correct relativistic result by dividing the Galilean result by the Einstein denominator.

EXAMPLE 2

Klingon-Enterprise Chase

A Klingon battle cruiser moves away from the earth at a speed of 0.8c. The Enterprise pursues at a speed of 0.9c. Both speeds are measured relative to the earth. In Figure 39.9, the earth is the reference frame S and the Klingon ship is the frame S'. The Enterprise is the object whose motion is followed from S and S'.

We want to find the speed at which the Enterprise overtakes the Klingons. In velocity terms this is the velocity of the Enterprise relative to the Klingons. We identify the velocities as

$$u = \text{velocity of Enterprise relative to the earth} = 0.9c$$

$$v = \text{velocity of Klingons relative to the earth} = 0.8c$$

$$u' = \text{velocity of Enterprise relative to Klingons}$$

We get the common sense result of Galilean relativity by taking

$$u' = u - v = 0.9c - 0.8c = 0.1c$$

The Einstein velocity addition formula **ALWAYS** modifies the Galilean result by dividing it by

$$1 + \frac{\text{product of the two Galilean numerator velocities}}{c^2}$$

$$= 1 + \frac{(u)(-v)}{c^2}$$

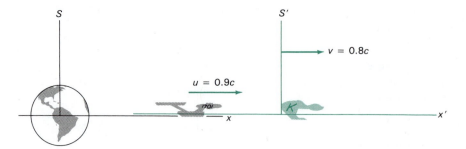

FIGURE 39.9

The earth is frame S. The Klingon ship is frame S'. The Enterprise is the object whose motion is followed from S and S'.

Rearranging Equation 39.17 gives for u',

$$u' = \frac{u - v}{1 - \dfrac{uv}{c^2}} = \frac{0.9c - 0.8c}{1 - \dfrac{(0.9c)(0.8c)}{c^2}}$$

$$= \frac{0.1c}{0.28}$$

$$= 0.357c$$

The procedure used in this example is recommended to help you avoid algebraic and conceptual errors. The Einstein velocity addition formula does not change your *qualitative* concepts involving relative motion. It changes only the *quantitative* evaluation of relative velocities.

39.5
TIME DILATION AND THE TWIN EFFECT

We have seen that time is relative. Now we want to use the Lorentz transformation equations to find the relationship between time intervals measured by observers in relative motion. The result we will find is that a clock in motion records a *shorter* time interval than an identical clock at rest. This effect is called **time dilation**.

Figure 39.10 shows the arrangement used to compare times. We place one clock at the origin ($x' = 0$) of the moving frame, S'. This clock remains fixed at $x' = 0$. We refer to it as the moving clock because the frame S' moves at speed v relative to the frame S. In S we place a number of identical clocks, at rest and synchronized with each other. Observers in S compare times recorded by their clocks with the times recorded on the moving clock as it passes. The Lorentz transformation equation we need is Equation 39.12,

$$t = \frac{t' + \dfrac{vx'}{c^2}}{\sqrt{1 - \dfrac{v^2}{c^2}}} \tag{39.12}$$

Let t_1 and t_2 denote the times recorded by clocks in S when the moving clock passes. The corresponding times on the moving clock are t_1' and t_2', and these

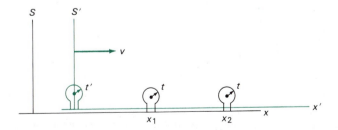

FIGURE 39.10
The moving clock remains fixed at the origin of the moving frame S'. The clocks at rest in S at x_1 and x_2 are synchronized. Observers in S compare the readings of their clocks with those of the moving clock as it passes.

times are related by Equation 39.12. Because the moving clock is located at the fixed position $x' = 0$ we have

$$t_1 = \frac{t'_1}{\sqrt{1 - \dfrac{v^2}{c^2}}} \tag{39.18}$$

$$t_2 = \frac{t'_2}{\sqrt{1 - \dfrac{v^2}{c^2}}} \tag{39.19}$$

The quantity $t_2 - t_1$ is the time interval recorded by the clock at rest in S,

$$\Delta t_R = t_2 - t_1 \Rightarrow \text{time interval recorded by clock at rest}$$

The quantity $t'_2 - t'_1$ is the time interval recorded by the moving clock in S',

$$\Delta t_M = t'_2 - t'_1 \Rightarrow \text{time interval recorded by the moving clock}$$

Subtracting Equation 39.18 from Equation 39.19 gives

$$\Delta t_R = \frac{\Delta t_M}{\sqrt{1 - \dfrac{v^2}{c^2}}}$$

Solving for Δt_M gives

$$\Delta t_M = \Delta t_R \sqrt{1 - \frac{v^2}{c^2}} \tag{39.20}$$

This is the result noted earlier. Because the factor $\sqrt{1 - v^2/c^2}$ is less than unity, the time interval measured by the moving clock (Δt_M) is *shorter* than the time interval measured by the clock at rest (Δt_R). This effect is called **time dilation**. To make it easier to remember we say, "A moving clock runs slow." Equation 39.20 is a general result of special relativity. It has nothing to do with the structure of any particular clock. Time dilation is a consequence of the relativity of simultaneity. Time *is* relative—it runs at different rates for observers in relative motion.

In Example 1 we saw a specific example of time dilation. The earth clocks recorded a time interval of 10 years. Clocks aboard the tug moved at a speed of $v = 0.6c$ relative to the earth. The elapsed time on the moving clock aboard the tug is given by Equation 39.20,

$$\Delta t_M = \Delta t_R \sqrt{1 - \frac{v^2}{c^2}} = 10 \text{ yr } \sqrt{1 - (0.6)^2} = 8 \text{ yr}$$

as found in Example 1.

Our everyday experiences leave us unaware of time dilation because the factor $\sqrt{1 - v^2/c^2}$ is nearly unity for most motions we encounter. Even though time dilation is not readily apparent to our senses, it has been confirmed by experiment. In 1971 four cesium atomic clocks were flown around the world on commercial jet flights. The times recorded by these clocks were compared with reference atomic clocks at the U.S. Naval Observatory.* The clocks were

---

* J. C. Hafele and R. E. Keating, *Around-the-World Atomic Clocks: Predicted Relativistic Time Gains*, and *Around-the-World Atomic Clocks: Observed Relativistic Time Gains*, Science **177**, 166; 168 (1972).

flown around the world along an eastward course, in the same direction that the earth's rotation carried the reference clocks. Then they were flown around the world along a westward course. The interpretation of the measurements are complicated by two circumstances:

1. The earthbound reference clocks were not at rest relative to a nonrotating (inertial) frame anchored at the center of the earth; they were moving in an easterly direction at a speed comparable to the airspeed of the flying clocks. When the clocks flew eastward they ran slow relative to the earthbound reference clocks. When the clocks flew westward, they ran fast relative to the reference clocks. To put it differently, during the westward flight the reference clocks moved faster than the flying clocks, *relative to the inertial frame*, and so ran more slowly than the flying clocks.
2. In addition to the time dilation resulting from the motion of the clocks, there is also a *gravitational* effect; the weaker gravitational acceleration experienced by the flying clocks tended to make them run faster than the reference clocks for both the eastbound and westbound flights. The gravitational effect on clockrate is independent of speed, and was verified with great precision in a 1960 experiment conducted by Pound and Rebka. Interestingly, the two time dilation effects nearly canceled each other during the eastward flight.

The results of the Keating-Hafele experiment confirm the predicted time dilation. Table 39.1 summarizes the predicted and observed time dilations. The tabulated values are $\Delta t_{Fly} - \Delta t_{Ref}$, the difference between the times elapsed on the flying clocks and the reference clocks, expressed in nanoseconds. A negative value means that the flying clocks ran slower than the reference clocks. There is a vast body of *indirect* experimental evidence that supports time dilation. The Keating-Hafele experiment is special because it used actual clocks.

The Twin Effect

Time dilation is an unusual and unexpected effect. So much so that thought experiments have been invented to challenge the idea. We consider one of these, sometimes called the twin paradox. Because there is no paradox we refer to it as the twin effect.

The Home twins are a brother named Levi and a sister named Happy. They are separated at birth. Happy Home remains on earth. Levi Home travels off in a spaceship. Levi's ship quickly accelerates to $v = 0.8c$, and travels to Alpha Centauri, 4 light-years away. Levi's ship turns around at Alpha Centauri and returns to earth at the same constant speed, $v = 0.8c$.

TABLE 39.1

Effect	$\Delta t_{Fly} - \Delta t_{Ref}$ (ns)	
	Eastward	Westward
gravitational	144 ± 14	179 ± 18
special relativity	-184 ± 18	96 ± 10
net (predicted)	-40 ± 23	275 ± 21
observed	-59 ± 10	273 ± 7

Happy sees Levi moving at $v = 0.8c$ over a round-trip distance of 8 light-years. According to Happy's earthbound clocks the round trip takes 10 years,

$$\Delta t_R = \frac{\text{distance}}{\text{speed}} = \frac{8 \text{ ly}}{0.8c} = 10 \text{ yr}$$

Happy calculates the time that elapses on her brother's moving clocks. They move at speed $v = 0.8c$ and so record a round-trip time of

$$\Delta t_M = \Delta t_R \sqrt{1 - \frac{v^2}{c^2}} = 10 \text{ yr} \sqrt{1 - (0.8)^2} = 6 \text{ yr}$$

This time dilation affects all clocks aboard Levi's ship—including Levi's biological clock. When Levi returns he is 6 years old, 4 years younger than his twin sister.

This unusual result may seem paradoxical if we argue from Levi's point of view. Levi argues that only relative motion is observable. He claims that from his frame of reference the earth receded at $v = 0.8c$ until it was 4 light-years away, whereupon it turned around and returned. He goes through the same chain of reasoning used by his sister. Levi calculates that the round trip takes 10 years as recorded on his clocks and that his sister's clock runs slow by the factor $\sqrt{1 - v^2/c^2} = 0.6$. Levi argues that he ages 10 years over the round trip and that his sister is only 6 years old when he returns.

If Levi's argument were true there would be a genuine paradox. But Levi's argument is false. We don't have symmetry between the two trips. Happy Home, who remains on earth, is always in the same inertial frame. Levi Home is in one inertial frame on the way to Alpha Centauri, but changes to a different inertial frame during turnaround. In the turnaround Levi undergoes an acceleration. During the acceleration he is not in an inertial frame and he cannot apply the time-dilation relation.

Happy Home's time-dilation calculation is valid. Levi Home's claim for reciprocity is not valid. We do not have symmetry and we do not have a paradox. We have the twin effect—the traveling twin returns younger than his stay-at-home sister.

If you cannot bring yourself to believe in the twin effect, remember the around-the-world atomic clocks. One set of clocks made a round trip and returned 273 nanoseconds younger than their stay-at-home twins. Common sense is wrong when it disagrees with experiment.

39.6
THE LORENTZ-FITZGERALD LENGTH CONTRACTION

The German scientist and philosopher Leibniz once defined space as a "relationship between simultaneous events." Considering that Leibniz lived 200 years before the advent of special relativity, his remark was prophetic. To determine the length of an object, you first locate its endpoints *simultaneously* and then measure the distance between those points. As we have seen, simultaneity is a relative concept, and so length becomes relative.

We want to use the Lorentz transformation equations to find the relationship between lengths measured by observers in relative motion. Figure 39.11 shows the arrangement used to compare lengths as measured by observers in

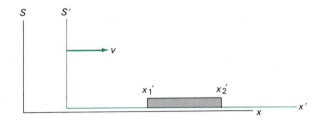

FIGURE 39.11

A rod at rest in S' moves at speed v relative to S.

relative motion. The frame S' moves at speed v relative to S. A rod is at rest in S', its endpoints being located at the points x'_1 and x'_2. As measured in S' the length of the rod is

$$L_R = x'_2 - x'_1$$

The subscript R is a reminder that L_R denotes the length of the rod as measured by an observer *at rest* relative to the rod. An observer in S also measures the length of the rod as it passes. To do so he locates the positions (x_1, x_2) of the endpoints of the rod. He locates x_1 and x_2 *simultaneously*; in other words, at the same time as recorded by his clocks. The length of the *moving* rod, measured by the observer in S, is

$$L_M = x_2 - x_1$$

Using Equation 39.9 we can relate x_1 and x_2 to x'_1 and x'_2 and the time t when S makes his measurements. We have

$$x'_2 = \frac{x_2 - vt}{\sqrt{1 - \dfrac{v^2}{c^2}}}$$

$$x'_1 = \frac{x_1 - vt}{\sqrt{1 - \dfrac{v^2}{c^2}}}$$

Subtracting we get

$$x'_2 - x'_1 = \frac{x_2 - x_1}{\sqrt{1 - \dfrac{v^2}{c^2}}}$$

Notice that the terms involving t cancel because the measurements of x_1 and x_2 occur at the same time t, simultaneously in S. Multiplying through by $\sqrt{1 - v^2/c^2}$ gives

$$x_2 - x_1 = (x'_2 - x'_1)\sqrt{1 - \frac{v^2}{c^2}} \tag{39.21}$$

But

$$x_2 - x_1 = L_M \Rightarrow \text{length of moving rod} \tag{39.22}$$

$$x'_2 - x'_1 = L_R \Rightarrow \text{length of rod at rest} \tag{39.23}$$

In terms of L_M and L_R, Equation 39.21 becomes

$$L_M = L_R \sqrt{1 - \frac{v^2}{c^2}} \qquad (39.24)$$

Suppose an observer at rest relative to the rod measures its length to be 1 m. Then,

$$L_R = 1 \text{ m}$$

Because $\sqrt{1 - v^2/c^2} < 1$, Equation 39.24 shows that the measured length of the rod in motion is less than 1 m. In general, Equation 39.24 shows that a rod in motion parallel to its length is shortened by the factor $\sqrt{1 - v^2/c^2}$. The result described by Equation 39.24 is called the **Lorentz-FitzGerald length contraction**. As we have seen, the factor $\sqrt{1 - v^2/c^2}$ is very close to unity for phenomena we encounter at the sensory level. Consequently, our senses do not make us aware of the Lorentz-FitzGerald length contraction.

EXAMPLE 3

A Moving Klingon Ship Contracts

A Klingon D7-class battle cruiser measures 216 earth meters in length when it is at rest. What length would a Federation observer measure it to be as it flashes past at $v = 2.0 \times 10^8$ m/s?

Using Equation 39.24 with $L_R = 216$ m, the length measured by the Federation observer (reference frame S) is

$$L_M = 216 \text{ m} \sqrt{1 - \left[\frac{2.0 \times 10^8 \text{ m/s}}{3.0 \times 10^8 \text{ m/s}}\right]^2} = 161 \text{ m}$$

Example 3 illustrates the direct calculation of the Lorentz contraction for a high speed. If the speed v is small compared to c, we can use the binomial theorem to determine the difference between the lengths L_R and L_M. Using Equation 39.24 we have,

$$L_R - L_M = L_R \left[1 - \sqrt{1 - \frac{v^2}{c^2}}\right] \qquad (39.25)$$

By the binomial theorem

$$\left(1 - \frac{v^2}{c^2}\right)^{1/2} \approx 1 - \frac{1}{2}\left(\frac{v^2}{c^2}\right)$$

The difference in lengths is

$$L_R - L_M = \frac{1}{2} L_R \left(\frac{v^2}{c^2}\right) \qquad (39.26)$$

This is a useful approximation so long as $v \ll$ c.

Now let's apply Equation 39.26 to a meter stick. Suppose the meter stick flies past you at 40 m/s (89.5 mi/h). What is the contraction, $L_R - L_M$? From Equation 39.26 we have

$$L_R - L_M = \frac{\frac{1}{2}(1.0 \text{ m})(40 \text{ m/s})^2}{(3 \times 10^8 \text{ m/s})^2}$$

$$= 8.89 \times 10^{-15} \text{ m}$$

For this speed the contraction is tiny; it is on the order of the diameter of an atomic nucleus.

The Lorentz-FitzGerald length contraction contradicts common sense. But remember, time is relative and length measurements require simultaneous observations of the endpoints. This is why observers in relative motion measure different lengths. Their notions of simultaneity are relative and so also are their measurements of time and length.

WORKED PROBLEM

Two jets of material fly away from the center of a radio galaxy in opposite directions (Figure 39.12). Both jets move at a speed of 0.75c relative to the galaxy. Determine the speed of one jet relative to the other.

Solution

In Figure 39.12, we let S denote a reference frame attached to the left jet and we let S' denote a reference frame attached to the galaxy. The right jet becomes the object whose motion is observed from S and S'. The velocities are

$$v = 0.75c = \text{velocity of galaxy } (S') \text{ relative to left jet } (S)$$

$$u' = 0.75c = \text{velocity of right jet relative to galaxy}$$

We want to determine

$$u = \text{velocity of right jet relative to left jet}$$

If Galilean relativity were valid we would have $u = u' + v = 1.5c$. Instead, we use Einstein's velocity addition formula, Equation 39.17, to find

$$u = \frac{u' + v}{1 + \dfrac{u'v}{c^2}}$$

$$= \frac{0.75c + 0.75c}{1 + 0.75 \cdot 0.75}$$

$$= 0.96c$$

This confirms our expectation that u must be less than the speed of light.

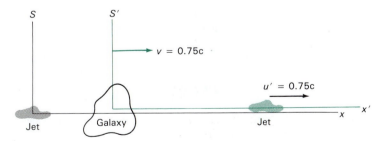

FIGURE 39.12

Two jets move away from the radio galaxy. The galaxy is frame S'. The jet moving to the left is frame S. The jet moving to the right is the object whose motion is viewed from S and S'. The Einstein velocity addition formula shows that the jets move apart at a speed of 0.96c. Galilean relativity gives an incorrect relative speed of 1.5c.

EXERCISES

39.2   The Relativity of Simultaneity

A. Reconsider the example of the train struck by lightning bolts. Assume that the light flashes arrive simultaneously for the president aboard the train. Use a sketch to show that the flashes do not arrive simultaneously for the repairman.

39.3   The Lorentz Transformation Equations

B. (a) Show that the Lorentz transformation equation, Equation 39.9, reduces to the Galilean transformation, $x = x' + vt$, in the limit $v \ll c$. (b) What do you have to neglect to get the Galilean equation?

C. Show that if the speed of light were infinite, then the Lorentz transformation equations (Equations 39.9–39.12) would reduce to the corresponding Galilean transformation equations, and we would have no special relativity.

39.4   The Einstein Velocity Addition Formula

D. Calculate the Einstein sum of the velocities (a) $u' = 0.4c$ and $v = 0.4c$; (b) $u' = 0.9c$ and $v = 0.9c$.

E. Can the Einstein velocity addition relation give a velocity greater than c? Present an example to illustrate your answer.

F. Two spaceships approach the earth from opposite directions. Each travels at a speed of 0.5c relative to the earth. Determine their speed relative to each other.

39.5   Time Dilation and the Twin Effect

G. An astronaut circling the earth has a speed of 18,000 mph. For 1 s of earth time, consider the astronaut to be an inertial system. (a) As seen by a ground observer, by how much time will the astronaut's watch run slow in this 1 s of time? (b) How many seconds slow would this be in one year?

H. An atomic clock moves at 1000 km/h for one hour as measured by an identical earthbound clock. How many nanoseconds slow will the moving clock be at the end of the one-hour interval?

I. Muons are components of cosmic rays that rain down on the earth from outer space. Muons are unstable particles and decay with an average lifetime of 2.2 μs when they are at rest. Imagine muons approaching the

earth at speeds of 0.99c. (a) For a clock on the earth, what is the average lifetime of the muons and how far does a muon travel in this time? (b) An observer on earth sees the muon travel a vertical distance of 4800 m through the atmosphere. Relative to a frame of reference on a muon, what is the thickness of this slice of the atmosphere? (c) As recorded by a clock on a muon, how long does it take for this thickness of the atmosphere to pass the muon?

J. Suppose that a rocket leaves the earth traveling at 0.6c toward a star moving toward the earth at 0.8c. At launch time the star is 14 light-years away, as measured by earthbound observers. Neglecting effects of accelerations, how long does it take for the rocket to reach the star (a) according to earth clocks, (b) according to a clock moving with the star?

39.6   The Lorentz-FitzGerald Length Contraction

K. For $v \ll c$, show that the Lorentz-FitzGerald contraction may be approximated by $L_M = L_R(1 - \frac{1}{2}v^2/c^2)$.

L. In the Stanford Linear Accelerator (SLAC), electrons are accelerated to an energy of 20 GeV. Their speed is so close to the speed of light that $1 - v/c = 3.3 \times 10^{-10}$. The evacuated tube through which they move is 3.2 km long. How long will the tube appear to an observer "sitting" on a 20-GeV electron?

M. The length of an airplane is measured to be 30 m in a reference system in which the plane is at rest. If observers on earth could make a precise measurement of the plane's length as it flies past at 1000 mph, by what fraction of a nanometer would they find it had contracted?

N. Our galaxy has a diameter of 10^5 light-years. (a) As measured by a clock at rest in the galaxy, how long does it take a particle moving at 0.95c to cross its diameter? (b) In the frame of reference of the particle, what is the diameter of the galaxy and how long does it take the particle to cross it?

P. The picture tube in a color TV set is 40 cm long. An electron travels from one end to the other at a speed of 0.30c. (a) How much time elapses in the electron frame of reference? (b) What is the length of the picture tube in the electron frame of reference?

PROBLEMS

1. A Klingon scout ship and a Federation scout ship approach the earth from opposite directions. Both ships travel at 0.5c relative to the earth. Initially, they are each 1.5 light-years from earth, as measured from earth. How long does it take them to meet as measured by clocks (a) on earth, (b) on board the scout ships?

2. In the example of the train struck by lightning, the

president sees first the flash from the front. The flash from the rear reaches the president a short time later. For a train 180 m long traveling at 40 m/s, calculate this short time.

3. A rod with a rest-frame length L moves at speed v parallel to its length. A flash of light leaves one end, travels to a mirror at the other end, and returns.

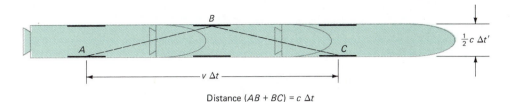

$$\text{Distance } (AB + BC) = c\,\Delta t$$

FIGURE 1

The round-trip time is $2L/c$ for an observer moving with the rod. (a) Calculate the round-trip time for an observer who sees the rod move at speed v. (b) If the length of the rod is Lorentz-contracted to $L(1 - v^2/c^2)^{1/2}$, show that the time of flight equals $2L/c(1 - v^2/c^2)^{1/2}$.

4. Observer A travels past a laboratory observer O at half the speed of light. Observer B catches and passes observer A, moving at half the speed of light relative to A. All three observe a sequence of events and record the elapsed time. Observers A and O calculate that 6 seconds elapse on B's clock. Determine the elapsed times for A and O.

5. On a rocket ship a pulse of light bouncing up and down between two mirrors constitutes a clock (Figure 1). The light path seen by observers in the rocket is perpendicular to the motion of the rocket. An observer in space sees the rocket moving at speed v. He sees the light pulse follow a diagonal path as shown in Figure 1. With the help of the figure, rederive the time dilation formula that relates $\Delta t'$ and Δt.

6. The Klingons in Example 2 were being overtaken by the Enterprise. The Klingons departed 1 year before the Enterprise (earth-clock time). (a) How much time had elapsed on the Klingon clocks when the Enterprise departed? (b) As viewed from earth the Klingons are 0.8 light-year away when the Enterprise leaves. What was the distance between the earth and the Klingons when the Enterprise departed, as measured by the Klingons? (c) The Klingons saw the Enterprise overtaking them at a speed of 0.357c. How long does it take for the Enterprise to overtake the Klingons, as measured on Klingon clocks?

7. In a particular inertial frame a rod is at rest. Its length is measured to be 10 meters by an observer at rest relative to the rod. Locating the endpoints constitutes a pair of simultaneous events that occur at different positions. As observed from a second inertial frame the positions appear to be separated by 100 meters.

Determine the time interval between the two events as recorded by an observer in the second frame of reference.

8. A platform moves horizontally at a speed of 0.5c relative to the earth. The platform is the bottom member of a series of N such platforms, all of which move in the same direction (Figure 2). Each platform moves at a speed of 0.5c relative to the platform beneath it. Show that the speed of the topmost platform relative to the earth does not exceed c, but approaches c in the limit as N approaches infinity.

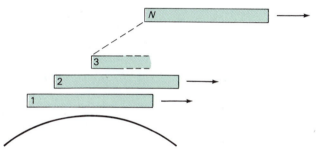

FIGURE 2

9. This problem is an extension of the preceding problem. If you have access to a personal computer, write a short program that evaluates v_N/c in terms of v_{N-1}/c, where v_N is the speed of the topmost platform in Problem 8. Then run your program for values of N ranging from 1 to 10.

10. Two students walk through a classroom doorway 10 s apart. Their passages comprise two events that take place at the same position at different times. An observer in motion relative to the classroom sees the two events occur one minute apart. Because of the relative motion, this observer also sees the two events take place at different positions. What is the spatial separation of the events for this observer?

CHAPTER

40

RELATIVISTIC

DYNAMICS

40.1
RELATIVISTIC MECHANICS

Newtonian mechanics, as developed in Chapters 6–13, is consistent with *Galilean* relativity, but it is not consistent with *special* relativity. In this chapter we develop a relativistic mechanics consistent with special relativity, and which has been amply confirmed by experiment. This relativistic mechanics incorporates Newtonian mechanics in the limiting case where speeds are small compared to the speed of light.

We begin by introducing expressions for linear momentum and kinetic energy. Two considerations guide the development.

1. Relativistic linear momentum and relativistic energy are defined in such a way that linear momentum and energy are conserved for an isolated system. We recognize the conservation laws as fundamental and we want to maintain them as part of the framework of special relativity.
2. For speeds that are small compared to the speed of light ($v \ll c$), the relativistic linear momentum expression must reduce to the Newtonian expression $\mathbf{p} = m\mathbf{v}$ and the relativistic kinetic energy expression must likewise reduce to $K = \frac{1}{2}mv^2$. The reason for this restriction is simply that Newtonian mechanics works admirably for speeds that are small compared to the speed of light.

40.2
RELATIVISTIC LINEAR MOMENTUM AND FORCE

The form of relativistic linear momentum was originally developed by Einstein on a purely theoretical basis as part of a comprehensive relativity theory for mechanics (Figure 40.1). Modern experiments measuring the deflection of charged particles moving at speeds close to the speed of light in magnetic fields yield values that are in excellent agreement with Einstein's expression

FIGURE 40.1
An Israeli banknote features the likeness of Albert Einstein and a conception of the nuclear atom.

for relativistic momentum. On the basis of theory and experiment we define the relativistic form of linear momentum as

$$\mathbf{p} = \frac{m\mathbf{v}}{\left(1 - \dfrac{v^2}{c^2}\right)^{1/2}} \tag{40.1}$$

Figure 40.2 compares this relativistic form of linear momentum and the Newtonian form $p = mv$, plotted against v/c. Note how the factor $(1 - v^2/c^2)^{1/2}$ in the denominator raises the value of the relativistic momentum above the Newtonian momentum and drives the relativistic momentum toward infinity as v approaches c.

For $v/c < 0.5$, the two curves of Figure 40.2 are nearly identical, indicating that the relativistic momentum reduces to the Newtonian momentum. The formal mathematical reduction of Equation 40.1 to the Newtonian form is accomplished by letting v/c be so small that v^2/c^2 is negligible compared to 1. Then the denominator goes to 1, so $\mathbf{p} = m\mathbf{v}$, showing that for $v/c \ll 1$ the special relativistic linear momentum reduces to the Newtonian form.

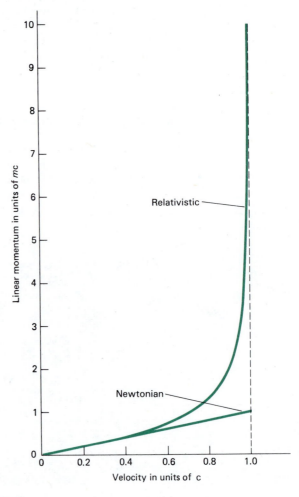

FIGURE 40.2

A comparison of momentum in relativistic mechanics and Newtonian mechanics. Note how the relativistic momentum approaches infinity as v approaches c.

One further point should be made regarding Equation 40.1. In Newtonian mechanics the mass of a particle is taken to be a constant, independent of the state of motion. We use this interpretation here. The mass m is the mass measured when the particle is at rest; m does not change with velocity.

In Newtonian mechanics, the net force acting on a particle is related to the linear momentum of the particle by Newton's second law,

$$\frac{d\mathbf{p}}{dt} = \mathbf{F} \tag{40.2}$$

This same *form* of Newton's second law applies in relativistic mechanics, provided we use the relativistic form of linear momentum defined by Equation 40.1.

Motion with Constant Force

Consider a particle of mass m that is accelerated from rest by a force of constant magnitude and direction. According to Newtonian mechanics its velocity increases without limit. We have $v = at = (F/m)t$. As time t approaches infinity, so does the velocity v.

Using the relativistic version of Newton's second law we can show that the particle speed approaches c in the limit as $t \rightarrow \infty$. With $F =$ constant, we can integrate Equation 40.2,

$$\frac{dp}{dt} = F = \text{constant}$$

to obtain

$$p = Ft$$

Substituting

$$p = \frac{mv}{\left(1 - \dfrac{v^2}{c^2}\right)^{1/2}}$$

and solving for v/c gives

$$\frac{v}{c} = \frac{\dfrac{Ft}{mc}}{\left[1 + \left(\dfrac{Ft}{mc}\right)^2\right]^{1/2}} \tag{40.3}$$

Figure 40.3 compares v/c as given by Equation 40.3 with the result predicted by Newtonian mechanics, $v/c = Ft/mc$. For small values of v/c the two curves in Figure 40.3 merge, showing that the relativistic results reduce to the form predicted by Newtonian mechanics at speeds small compared to the speed of light.

We can also check Equation 40.3 algebraically by looking at limiting values. For small values of time we have $Ft/mc \ll 1$, and v/c approaches the limit

$$\frac{v}{c} = \frac{Ft}{mc}$$

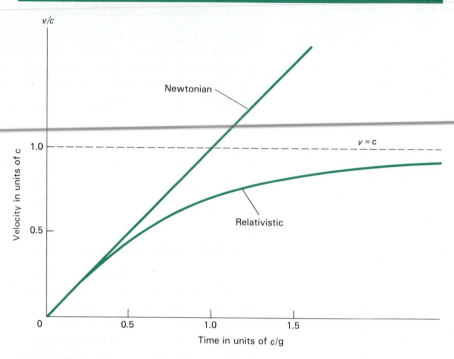

FIGURE 40.3

A comparison of speed-versus-time for a constant force, *mg*. The straight line describes the motion according to Newtonian mechanics. The curved line describes the motion according to relativistic mechanics. Relativistic mechanics shows that the speed of light is an upper limit to the speed of the particle.

In the nonrelativistic limit the ratio F/m equals a, the acceleration, and we have

$$v = at$$

in agreement with Newtonian mechanics. In the limit of very long times $Ft/mc \gg 1$, and Equation 40.3 reduces to

$$\frac{v}{c} = 1$$

Unlike the nonrelativistic situation, a constant force does not imply a constant acceleration. No matter how long the force is applied, the speed of the particle will never exceed c.

EXAMPLE 1

Acceleration Toward the Stars

Imagine a rocket of mass m that experiences a constant force mg ($g = 9.80$ m/s^2) for a time of 2 years (6.31×10^7 s). What speed does the rocket achieve, starting from rest?

We calculate the speed first according to Newtonian kinematics and then relativistically using Equation 40.3.

According to Newtonian mechanics, if a particle accelerates from rest at a constant rate $a = g$ for a time t it acquires a speed,

$$v = at$$
$$= (9.80 \text{ m/s}^2)(6.31 \times 10^7 \text{ s})$$
$$= 6.18 \times 10^8 \text{ m/s} = 2.06c$$

We see that Newtonian kinematics gives a speed that is more than twice the speed of light. The result $v > c$ is incorrect and shows the inadequacy of Newtonian mechanics.

However, using Equation 40.3, which is derived using the relativistic dynamics, we get

$$\frac{Ft}{mc} = \frac{gt}{c} = \frac{(9.80 \text{ m/s}^2)(6.31 \times 10^7 \text{s})}{3.00 \times 10^8 \text{ m/s}} = 2.06$$

and

$$\frac{v}{c} = \frac{2.06}{[1.0 + (2.06)^2]^{1/2}} = 0.900$$

Thus, special relativity correctly predicts a speed less than the speed of light.

Equation 40.3 can be integrated to obtain the distance traveled. With $v = dx/dt$, Equation 40.3 gives

$$\frac{1}{c} \int_0^x dx = \int_0^t \frac{\left(\dfrac{Ft}{mc}\right) dt}{\left[1 + \left(\dfrac{Ft}{mc}\right)^2\right]^{1/2}}$$

Integrating, we obtain

$$x = \left(\frac{mc^2}{F}\right)\left\{\left[1 + \left(\frac{Ft}{mc}\right)^2\right]^{1/2} - 1\right\} \tag{40.4}$$

Again, we consider the limiting forms. For small values of time, $Ft/mc \ll 1$, the binomial expansion leads to

$$x \approx \frac{1}{2}\frac{Ft^2}{m} = \frac{1}{2}at^2$$

in full agreement with the Newtonian result. For $Ft/mc \gg 1$,

$$x \approx ct$$

For $Ft/mc \gg 1$, the mass m spends most of the time traveling at almost $v = c$, so the distance x is approximately ct.

EXAMPLE 2

Are We There Yet?

How far does the rocket in Example 1 travel in 2 yr? From Equation 40.4 with $F/m = 9.80$ m/s^2,

$$x = \frac{(3.00 \times 10^8 \text{ m/s})^2\{[1 + (2.06)^2]^{1/2} - 1\}}{9.80 \text{ m/s}^2}$$

$$= 1.18 \times 10^{16} \text{ m}$$

The meter is an inconvenient unit for interstellar distances. Let's convert to light-years. One light-year (ly) is the distance light travels in 1 year,

$$1 \text{ ly} = (3.00 \times 10^8 \text{ m/s})(3.16 \times 10^7 \text{ s})$$

$$= 9.48 \times 10^{15} \text{ m}$$

Then

$$x = \frac{1.18 \times 10^{16}\text{m}}{9.48 \times 10^{15} \text{ m/ly}} = 1.24 \text{ ly}$$

The nearest star beyond our sun is about 4 light-years away, and so our rocket is less than one-third of the way there.

40.3
RELATIVISTIC ENERGY

The kinetic energy of the particle (K) is defined as the work done in accelerating the particle from rest. The work is defined by the integral of the force. Thus,

$$K = \int_0^x F \, dx \qquad (40.5)$$

In Newtonian mechanics this definition leads to $K = \frac{1}{2}mv^2$. As we now show, relativistic mechanics leads to a different result.

Using $F = dp/dt$ and $dx = v \, dt$ gives

$$K = \int_0^t \left(\frac{dp}{dt}\right) v \, dt = \int_0^p v \, dp$$

We express v in terms of p in order to carry out the integration. Solving the relativistic relation, $p = mv/(1 - v^2/c^2)^{1/2}$ for v gives

$$v = \frac{\dfrac{p}{m}}{\left[1 + \left(\dfrac{p}{mc}\right)^2\right]^{1/2}}$$

The kinetic energy is

$$K = \int_0^p \frac{\left(\dfrac{p}{m}\right) dp}{\left[1 + \left(\dfrac{p}{mc}\right)^2\right]^{1/2}}$$

which gives the result

$$K = mc^2 \left\{ \left[1 + \left(\frac{p}{mc}\right)^2\right]^{1/2} - 1 \right\} \qquad (40.6)$$

Using Equation 40.1 to express p in terms of v, we can rewrite Equation 40.6 as

$$K = \frac{mc^2}{\left(1 - \dfrac{v^2}{c^2}\right)^{1/2}} - mc^2 \qquad (40.7)$$

This is the relativistic generalization of $K = \frac{1}{2}mv^2$. Let's check to see if it reduces to $K = \frac{1}{2}mv^2$ for speeds small compared to the speed of light.

For $v/c \ll 1$, the $(1 - v^2/c^2)^{-1/2}$ term in Equation 40.7 can be expanded by the binomial theorem, to give for the kinetic energy

$$K = mc^2 \left[1 + \frac{1}{2}\left(\frac{v^2}{c^2}\right) + \frac{3}{8}\left(\frac{v^2}{c^2}\right)^2 + \dots - 1 \right]$$

or

$$K = mc^2 \left[\frac{1}{2}\frac{v^2}{c^2} + \frac{3}{8}\left(\frac{v^4}{c^4}\right) + \dots \right]$$

If v^2/c^2 is small enough that the v^4/c^4 term can be neglected, then

$$K = \tfrac{1}{2}mv^2$$

in agreement with the Newtonian form. Figure 40.4 shows graphically how the Newtonian and special relativistic kinetic energy curves merge for small v.

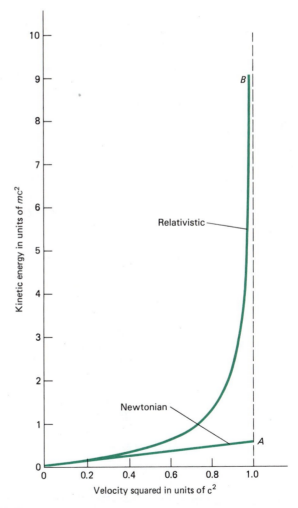

FIGURE 40.4

A comparison of kinetic energy in relativistic mechanics and of kinetic energy in Newtonian mechanics.

Mass Energy and Total Energy

Let's return to the relativistic expression for kinetic energy:

$$K = \frac{mc^2}{\left(1 - \dfrac{v^2}{c^2}\right)^{1/2}} - mc^2 \tag{40.7}$$

The second term, mc^2, is independent of the velocity; it is called the **mass energy** (or **rest energy**) and is denoted by E_0:

$$E_0 = mc^2 \tag{40.8}$$

Following Einstein, we interpret $mc^2/(1 - v^2/c^2)^{1/2}$ as the **total energy** and denote it by E:

$$E = \frac{mc^2}{\left(1 - \dfrac{v^2}{c^2}\right)^{1/2}} \tag{40.9}$$

In terms of E, we can rewrite Equation 40.7 as

$$E = E_0 + K \tag{40.10}$$

This relation is interpreted as a statement that the total energy E is the sum of the mass energy E_0 and the kinetic energy K.

In Section 40.4 we will see that it is this total relativistic energy E that is conserved in particle interactions.

EXAMPLE 3

Kinetic Energy and Mass Energy

At what speed will the kinetic energy of a particle be equal to its mass energy? To answer this question we set

$$K = E_0$$

or

$$\frac{mc^2}{\left(1 - \dfrac{v^2}{c^2}\right)^{1/2}} - mc^2 = mc^2$$

Cancelling the factor mc^2 leads to

$$\frac{1}{\left(1 - \dfrac{v^2}{c^2}\right)^{1/2}} = 2$$

which gives

$$\frac{v}{c} = \frac{\sqrt{3}}{2} = 0.866$$

Thus, the particle must travel at nearly 87% of the speed of light to make its kinetic energy equal to its mass energy. Substituting $c = 3 \times 10^8$ m/s, we get $v = 2.60 \times 10^8$ m/s for the speed at which the particle's kinetic energy is equal to its mass energy.

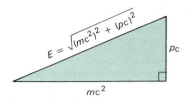

FIGURE 40.5

The total energy is the hypotenuse of a right triangle. The other two sides are pc and the mass energy mc^2.

An Energy-Momentum Relation

Relativistic energy and relativistic linear momentum are related by the useful identity

$$E^2 = (pc)^2 + (mc^2)^2 \qquad (40.11)$$

This is derived directly from Equation 40.1 and can be verified by substituting for p from Equation 40.1 and for E from Equation 40.9 into Equation 40.11. The identity is in the form of a Pythagorean equation and can be interpreted as shown in Figure 40.5.

The energies E and E_0 are customarily expressed in electron volt units (eV) or in metric multiples thereof, such as MeV (10^6 electron volts), GeV (10^9 electron volts), or TeV (10^{12} electron volts). The conversion relation is

$$1\ \text{MeV} = 1.60219 \times 10^{-13}\ \text{J}$$

For example, the proton mass is $m_p = 1.67265 \times 10^{-27}$ kg. Its mass energy is

$$m_p c^2 = 1.67265 \times 10^{-27}\ \text{kg}\ (2.99792 \times 10^8\ \text{m/s})^2$$

$$= 1.50330 \times 10^{-10}\ \text{J} \left(\frac{1\ \text{MeV}}{1.60219 \times 10^{-13}\ \text{J}} \right)$$

$$= 938.3\ \text{MeV}$$

When a particle is in motion it possesses kinetic energy in addition to its mass energy. Certain types of particles possess only kinetic energy. These are particles that have zero mass.

Zero-Mass Particles

Mass is a fundamental property of a particle. A particle of mass m at rest possesses a mass energy, mc^2. A particle in motion possesses kinetic energy in addition to its mass energy.

There are three particularly important particles that have zero mass. These are the photon, the neutrino, and the as yet unobserved graviton.

The photon is associated with light. We can view a beam of light as a stream of photons, each with a definite energy, and each traveling at the speed of light. The photon has zero mass—it does not exist as an object at rest.

The neutrino is produced in certain types of radioactive decay processes. Like the photon, the neutrino is a zero-mass particle, and it does not exist as an object at rest.

The graviton is a hypothetical zero-mass particle associated with the universal force of gravitation. Photons and neutrinos have been detected, but gravitons have not.

The photon, neutrino, and graviton share two important characteristics. Each has zero mass and each travels at the speed of light. We can prove that a zero-mass particle carrying energy must travel at the speed of light.

We first rewrite Equation 40.9 for the total energy as

$$E \cdot \left(1 - \frac{v^2}{c^2} \right)^{1/2} = mc^2$$

If $m = 0$, we are left with

$$E \cdot \left(1 - \frac{v^2}{c^2} \right)^{1/2} = 0$$

Before

π^{o}

At rest

After

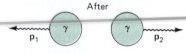

p_1 γ γ p_2

FIGURE 40.6

The decay of a neutral pion into two gamma photons.

But because E is not zero, $(1 - v^2/c^2)^{1/2}$ must equal zero, which means that $v = c$. Zero-mass particles carrying energy must travel at the limiting speed, the speed of light. Conversely, a particle moving at the speed of light (but with only a finite amount of energy) must have zero mass.

For a zero-mass particle the energy-momentum relation, Equation 40.11, reduces to

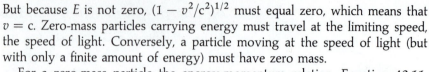

$$E = pc \tag{40.12}$$

Because a zero-mass particle has no rest energy, the energy E is purely kinetic.

EXAMPLE 4

Decay of a Neutral Pion

The *pion* is an unstable subatomic particle. Pions have a mass intermediate between that of an electron and that of a proton. Pions may carry a positive or a negative electric charge, or be electrically neutral.

A neutral pion at rest transforms (decays) into two photons, as suggested by Figure 40.6. Conservation of linear momentum requires that the photon linear momenta be equal in magnitude and opposite in direction. It follows from Equation 40.12 that the two photons have the same energy. Hence the mass energy of the pion is equally divided between the two photons.

The neutral pion mass m_π is 264 times the electron mass m_e. In terms of electron mass energy units, the energy of each photon is

$$E = 132m_ec^2 = 132(0.511 \text{ MeV})$$
$$= 67.5 \text{ MeV}$$

From Equation 40.12 the magnitude of the momentum of each photon is

$$p = \frac{E}{c} = 132m_ec$$

Can you imagine zero-mass particles carrying energy and momentum? The sale of suntan lotions testifies to the transport of energy from the sun to the skin by photons. The recoil of an object absorbing a burst of laser light offers a dramatic demonstration of the linear momentum of photons.

40.4
RELATIVISTIC ENERGY AND MOMENTUM CONSERVATION

One spectacular aspect of the interaction of very high-energy subatomic particles is the creation and destruction of particles. In these interactions kinetic energy is transformed into mass energy and vice versa. An example of this phenomenon is the creation of a proton antiproton pair in a collision between two high-energy protons. We will calculate the minimum (threshold) energy necessary for this proton-antiproton-pair production.*

Modern high-energy accelerators use **storage rings** to achieve an efficient transformation of kinetic energy into particle mass energy. The accelerated particles are stored in oppositely directed beams before they are allowed to

---

* The creation of a single proton would violate the conservation of electric charge. The production of a pair of oppositely charged particles does not produce any *net* electric charge.

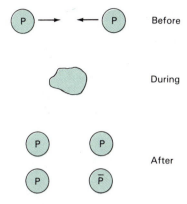

FIGURE 40.7

The two colliding protons have equal but opposite linear momenta. They momentarily form a complex particle that fragments into three protons and one antiproton.

collide. The two colliding beams carry particles with equal but opposite linear momenta. When the two particles collide they momentarily form a complex particle whose total linear momentum is zero.

We can apply the principles of total energy conservation and linear momentum conservation to calculate the minimum kinetic energy needed to produce a proton-antiproton pair in a head-on collision between two protons. The before- and after-collision situations are shown in Figure 40.7. The two colliding protons form a complex particle that then decays into four particles—three protons and one antiproton.

If the four particles are all at rest after the decay, then the total energy will be a minimum, given by

$$E_a = 3m_pc^2 + m_{\bar{p}}c^2$$

The proton and antiproton have equal masses, so

$$E_a = 4m_pc^2$$

Because their linear momenta are equal in magnitude, the two colliding protons have equal energies. Using Equation 40.9, their combined energy before the collision can be expressed as

$$E_b = 2(m_pc^2 + K)$$

where K is the kinetic energy of each proton. The total energy is conserved in the collision. Setting $E_a = E_b$ gives

$$2(m_pc^2 + K) = 4m_pc^2$$

and thus

$$K = m_pc^2 = 938.3 \text{ MeV}$$

This is the minimum, or **threshold**, kinetic energy needed by the protons. The colliding protons transform *all* of their *kinetic* energy into particle mass energy. The total kinetic energy of the protons before the collision was $2m_pc^2$ and this was transformed into the mass energy of the proton-antiproton pair.

Before the advent of storage rings, energy-to-mass transformations employed accelerators in which a single beam of particles collided with particles at rest in a laboratory target. This is an inefficient arrangement because the total linear momentum of the colliding particles is not zero. The conservation of linear momentum requires that the products of the collision have the same total linear momentum as the colliding pair. This means that the products must be moving and hence they must possess kinetic energy. This kinetic energy is supplied by the accelerator beam particle. It is wasted energy in the sense that it is not available to be transformed into mass energy.

In colliding beams experiments, kinetic energy is transformed into mass energy. The following example illustrates the inverse transformation in which mass energy is transformed into kinetic energy. We again apply the principles of total energy conservation and linear momentum conservation.

Before

At rest

After

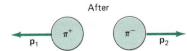

FIGURE 40.8

The decay of a neutral kaon into two pions.

EXAMPLE 5

Radioactive Decay of a Neutral Kaon

The neutral *kaon* is an unstable particle that can be produced by high-energy accelerators. A neutral kaon at rest can decay into a pair of oppositely charged pions, as shown in Figure 40.8. The kaon mass is 3.566 times the mass of a charged pion.

We seek the kinetic energy of each pion, and we want to know whether it is necessary to use relativistic mechanics rather than Newtonian mechanics.

The pions must be oppositely directed, as shown in Figure 40.8. This follows from linear momentum conservation. For the kaon, $\mathbf{p} = 0$, and hence $\mathbf{p}_1 + \mathbf{p}_2 = 0$ for the pion pair. The linear momenta of the pions must be equal in magnitude and opposite in direction.

From Equation 40.11 the total energy of a particle is given by

$$E = [(pc)^2 + (mc^2)^2]^{1/2}$$

Because the pions have equal masses and linear momenta of equal magnitude, they have equal energies. The total energy of the two pions can be written as

$$E_\pi = 2(K_\pi + m_\pi c^2)$$

where K_π is the kinetic energy of each pion. The energy of the kaon is

$$E_K = m_K c^2 = 3.566 m_\pi c^2$$

Conservation of energy requires that E_K equal E_π, or

$$3.566 m_\pi c^2 = 2(K_\pi + m_\pi c^2)$$

which gives

$$K_\pi = 0.783 m_\pi c^2$$

Is relativistic dynamics required? If we try to use $K_\pi = \frac{1}{2}m_\pi v^2$, the nonrelativistic expression for kinetic energy, we get

$$\tfrac{1}{2}m_\pi v^2 = 0.783 m_\pi c^2$$

This gives the incorrect result $v = 1.25c$. It is clearly wrong to use $K = \frac{1}{2}mv^2$ in this instance. Relativistic dynamics is required. The relatively large transformation of mass energy into kinetic energy gives the pions kinetic energies comparable to their mass energies and speeds close to the speed of light.

40.5
MASS ENERGY TRANSFORMATIONS

Example 5 illustrates the transformation of mass energy into kinetic energy. The total mass of the two product pions is less than the mass of the original kaon. Thus,

$$m_K c^2 = 3.566 m_\pi c^2 > 2 m_\pi c^2$$

shows that $m_K > 2 m_\pi$. It is the difference in mass energy that is transformed into the kinetic energy of the pions.

If a reaction *releases* energy, the mass of the products is less than the mass of the reactants. The energy release ΔE equals the difference in mass times c^2,

$$\Delta E = (M_{reactants} - M_{products})c^2 = \Delta Mc^2 \tag{40.13}$$

We describe such processes as **mass energy transformations**. It is another example of the transformation of one form of energy into another form of energy.

A small amount of mass corresponds to a large amount of energy. With $M = 1$ kg,

$$\Delta E = 1 \text{ kg } (3 \times 10^8 \text{ m/s})^2 = 9 \times 10^{16} \text{ J}$$

TABLE 40.1

The Fractional Mass Transformation in Various Reactions

Reaction	$\Delta M/M$
Chemical	$\sim 10^{-9}$
Nuclear Fission	$\sim 10^{-3}$
Nuclear Fusion	$\sim 6 \times 10^{-3}$
Electron-positron annihilation	1
π° decay	1

If 1 kg of mass were transformed over a period of one year the average power developed would be

$$P = \frac{E}{t} = \frac{9 \times 10^{16} \text{ J}}{3.16 \times 10^7 \text{ s}} = 2.86 \times 10^3 \text{ MW}$$

For comparison, the electric generating capacity of the United States is about 400×10^3 MW.

Rewriting Equation 40.13 as $\Delta M = \Delta E/c^2$, shows that the change in mass is small when the energy release is small. The energy release will be directly proportional to the number of reactions and thus also directly proportional to the total mass M of the reactants. The *fractional change in mass*, $\Delta M/M$, is a measure of how efficiently the reaction converts mass energy into other forms of energy.

In chemical reactions the fractional change in the mass of the particles is not directly measurable, being approximately one part in a billion (Table 40.1).

EXAMPLE 6

Mass Energy Transformation in a Chemical Reaction

The combination of 2 moles of hydrogen with 1 mole of oxygen is described by the reaction equation

$$2 H_2 + O_2 \rightarrow 2 H_2O$$

This reaction releases 573 kJ of energy. That is, it transforms 573 kJ of mass energy into other forms of energy,

$$\Delta Mc^2 = 573 \text{ kJ}$$

The mass of 2 moles of molecular hydrogen is 4 grams. The mass of 1 mole of molecular oxygen is 32 grams. The mass energy of the reactants is

$$Mc^2 = (0.036 \text{ kg})(3 \times 10^8 \text{ m/s})^2 = 3.24 \times 10^{15} \text{ J}$$

The fractional change in mass for the reaction is

$$\frac{\Delta M}{M} = \frac{\Delta Mc^2}{Mc^2} = \frac{5.73 \times 10^5 \text{ J}}{3.24 \times 10^{15} \text{ J}} = 1.77 \times 10^{-10}$$

In nuclear reactions, the fractional mass change is much larger than it is in chemical reactions. In nuclear fission a heavy element such as uranium splits into fragments. The fractional mass change is about 0.001, roughly one million times greater than it is for a chemical reaction. This factor of one million explains the great appeal of nuclear power. By burning a few hundred *pounds* of

FIGURE 40.9

An Orion vehicle would be propelled by using a hemispherical thruster to absorb the linear momentum of the blast debris.

uranium nuclei in a fission reactor we can release the same energy as by burning hundreds of thousands of *tons* of coal.

Nuclear fission is the source of the electrical energy produced by nuclear reactors. In the stars, nuclear *fusion* reactions operate to release energy by combining nuclei of light elements, usually forms of hydrogen, to build up heavy elements. Here, too, the fractional mass change $\Delta M/M$ is small (about 0.006), but enormously greater than it is for chemical reactions. Whereas a practical nuclear fission reactor is a working reality, a nuclear fusion reactor is only a hope.

At present the only application of nuclear fusion reactions on earth is in the form of hydrogen bombs. Project Orion was aimed at showing that even hydrogen bombs can have a benevolent aspect.

EXAMPLE 7

The Orion Spaceship

Project Orion was a research effort aimed at proving the feasibility of a spaceship that would be propelled by repeated hydrogen bomb explosions (Figure 40.9).

The ultimate speed that can be achieved by any rocket propulsion system can be estimated by assuming that all of the energy transformed in the propulsion engine is transformed into the kinetic energy of the vehicle. Thus,

$$\text{kinetic energy acquired} = \text{energy transformed}$$

or

$$\tfrac{1}{2}Mv^2 = \Delta Mc^2$$

This relates the maximum achievable speed to the fraction of the mass transformed by the propulsion reactions,

$$v = \sqrt{2\left(\frac{\Delta M}{M}\right)}\,c$$

For chemical reactions, Table 40.1 shows that $\Delta M/M \approx 10^{-9}$, so

$$v \approx \sqrt{2 \times 10^{-9}}(3 \times 10^8 \text{ m/s}) \approx 13 \text{ km/s} \qquad \text{(chemical reaction)}$$

For hydrogen fusion reactions, $\Delta M/M \approx 0.006$, so

$$v \approx \sqrt{2(0.006)}\ c \approx 0.11c \qquad \text{(hydrogen fusion reaction)}$$

In principle, an Orion vehicle could reach speeds on the order of one-tenth the speed of light—over 2000 times the speed typical of chemical propulsion systems.

In extreme cases, the fractional mass change $\Delta M/M$ may be unity; 100% of the mass is converted into radiant energy. For example, an electron and a positron can annihilate each other, giving rise to radiant energy in the form of gamma photons. The decay of the pion π° into two gamma photons was considered in detail in Example 4, and also illustrates the complete conversion of mass energy into radiant energy.

WORKED PROBLEM

A high-speed proton collides with a stationary proton. The result of their interaction is two protons plus a pair of oppositely charged pions. Determine the minimum kinetic energy of the incident proton necessary for this reaction to occur. The mass energy of a charged pion is 139.6 MeV.

Solution

The fact that there are four particles after the interaction appears to make this a complicated, many-body problem. To avoid dealing with four separate particles, we view the system of colliding particles as a *single* complex particle, which subsequently breaks up, forming two protons and two pions.

Let

$$E_p = m_p c^2 \qquad E_\pi = m_\pi c^2$$

denote the mass energies of a proton and a pion. The minimum mass energy of the complex particle that will allow it to break up into two protons and two pions is

$$E_C = 2E_p + 2E_\pi$$

Energy is conserved in the interaction. Before the collision, the total energy of the system is

$$E_p + \sqrt{E_p^2 + p^2 c^2}$$

where p is the linear momentum of the bombarding proton. The total energy of the complex particle is

$$\sqrt{E_C^2 + p^2 c^2}$$

The total linear momentum of the complex particle equals that of the colliding protons because linear momentum is conserved. Energy conservation is expressed by

$$E_p + \sqrt{E_p^2 + p^2 c^2} = \sqrt{E_C^2 + p^2 c^2}$$

Squaring both sides and cancelling the $p^2 c^2$ terms gives

$$2E_p^2 + 2E_p\sqrt{E_p^2 + p^2 c^2} = E_C^2 = (2E_p + 2E_\pi)^2$$

The kinetic energy of the bombarding proton is

$$K = \sqrt{E_p^2 + p^2 c^2} - E_p$$

Replacing $\sqrt{E_p^2 + p^2 c^2}$ by $K + E_p$ and solving for K yields

$$K = 4E_\pi + \frac{2E_\pi^2}{E_p} = 4(139.6 \text{ MeV}) + \frac{2(139.6 \text{ MeV})^2}{938.3 \text{ MeV}}$$

$$= 600 \text{ MeV}$$

If a proton with a kinetic energy of 600 MeV induces such a reaction, the net mass energy produced is $2E_\pi = 279$ MeV. The remaining 321 MeV is carried as kinetic energy by the reaction products.

EXERCISES

40.2 Relativistic Linear Momentum and Force

A. A high-energy proton has a speed of 0.9999995c. Calculate the ratio of its relativistic momentum to its Newtonian momentum.

B. If the relativistic momentum of a particle of mass m is given by $p = 10mc$, what is the ratio v/c?

C. (a) What constant F/m ratio (force/mass) would get a space probe to Alpha Centauri (4.3 light-years away) in 10 years (earth time)? (b) How fast would the space probe be going as it flashed past Alpha Centauri?

D. An electron experiences a constant force of 1.60×10^{-13} N in a linear accelerator. (a) If it starts from rest, how fast is it moving 10 ns later? Express your result as a fraction of c. (b) How far does it move in the first 10 ns?

E. Calculate the speed for which the relativistic expression for momentum differs from the nonrelativistic expression by 1 percent.

F. In Newtonian mechanics, doubling the velocity of a particle will cause the momentum to be doubled. Discuss the effect on the relativistic momentum of doubling the velocity of a particle. (Take $v/c < 0.5$).

40.3 Relativistic Energy

G. Calculate the kinetic energy in joules of an electron and a proton moving with $v = 0.866c$.

H. Determine the speed of a proton whose kinetic energy is 400 GeV. Express your result as a fraction of c.

I. A particle moves at a speed that is one-third the speed of light. Determine the ratio of its kinetic energy to its mass energy (K/mc^2) by using (a) Newtonian mechanics, (b) relativistic mechanics.

J. A particle of mass m has a kinetic energy that is eight times its mass energy. Calculate the ratio of its speed to the speed of light using (a) Newtonian mechanics, (b) relativistic mechanics.

K. An electron has a linear momentum mc, where m is the electron mass. Determine the total energy of the electron in MeV.

L. An electron moves at a speed such that its total energy equals the mass energy of a proton. Determine the speed of the electron in units of c. Keep 9 significant figures.

M. In Newtonian mechanics, doubling the velocity of a particle causes its kinetic energy to increase four fold. What would be the effect on the relativistic kinetic energy of doubling the velocity? (Take $v/c < 0.5$.)

N. A proton in the Fermilab accelerator has a speed of 0.999998c. Calculate the ratio of its total relativistic energy to its mass energy.

P. Prove the statement, "A particle of nonzero energy moving at the speed of light must have zero mass."

Q. A fast-moving pion has a total energy of 400 MeV and a value of pc equal to 375 MeV. Determine its mass. Express your result in terms of electron mass units. The mass energy of an electron is 0.511 MeV.

R. A proton has a value of pc equal to 1500 MeV. (a) Calculate the total energy of the proton in MeV. (b) Determine v/c for the proton.

S. Assume that the neutrino is not a zero-mass particle, but that it has a mass energy equal to 10 eV. Calculate $1 - v/c$ for the neutrino when its total energy is 1 MeV.

40.4 Relativistic Energy and Momentum Conservation

T. An electron and a positron at rest annihilate each other, creating two photons in the process. Prove that the photons must emerge in opposite directions and have equal energies.

U. How much mass would have to be converted to other forms of energy each second to produce 10^9 watts?

PROBLEMS

1. The star Alpha Centauri is 4.3 light-years away. (a) Determine how long it would take a rocket to reach Alpha Centauri if it started from rest and experienced a constant force equal to its weight at the surface of the earth. (b) Repeat for a constant force equal to ten times its weight at the surface of the earth.

2. The oxidation of glucose is a chemical reaction that converts mass into energy. It is described by

$$C_6H_{12}O_6 + 6\,O_2 \rightarrow 6\,CO_2 + 6\,H_2O$$

On a molar basis, 1 mole of glucose and 6 moles of oxygen combine and release (convert) 2.82 MJ of energy. Determine $\Delta M/M$ for this chemical reaction.

3. A hydrogen atom emits a photon with an energy of 10.2 eV. Determine $\Delta M/M$ for the reaction.

4. Show that the velocity, momentum, and total energy of a particle are related by

$$v = \frac{pc^2}{E}$$

Show that this equation reduces to the Newtonian form in the nonrelativistic limit where the kinetic energy is negligibly small compared to the mass energy.

5. The Λ^0 is an unstable neutral particle that decays into a proton and a negative pion (Figure 1). Determine the kinetic energies of the proton and the pion if the Λ^0 is at rest when it decays. Notice that the less

massive particle gets the larger share of the kinetic energy. The mass of the Λ^0 is $7.993m_\pi$, where m_π is the pion mass, and the mass of the proton is $6.722m_\pi$.

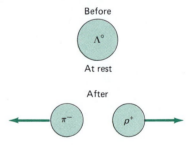

FIGURE 1

6. A particle of mass m and speed v undergoes a perfectly inelastic collision with a particle of mass M at rest. The total energy of the system is conserved. Express the speed of the resulting particle in terms of m, M, and v.

7. A proton collides with another proton at rest. The composite particle formed has a mass three times the proton mass. Determine the kinetic energy of the incident proton. Express your result as a multiple of the mass energy of a proton.

8. A particle of mass M at rest decays into two particles. One particle has zero mass and the other has a mass m. Use energy and momentum conservation to

show that the kinetic energy of the emerging particle of mass m is given by

$$K = \frac{(M - m)^2 c^2}{2M}$$

9. A particle whose mass energy equals its kinetic energy is accelerated until its kinetic energy has doubled. Show that the particle speed is increased by approximately 9%.

10. A ^{57}Fe nucleus at rest emits a 14-keV photon. Use the conservation of energy and momentum to deduce the kinetic energy of the recoiling nucleus in eV. Use $Mc^2 = 8.6 \times 10^{-9}$ J for the final state of the ^{57}Fe nucleus.

11. Two protons are on a head-on collision course. Each has a kinetic energy of 10 times its mass energy. Prove that the kinetic energy of one proton relative to the other is 240 times its mass energy.

12. A pion at rest decays into a muon and a neutrino. The pion mass is 273 times the electron mass. The muon mass is 207 times the electron mass. Determine the kinetic energies of the muon and the neutrino. Express your results in units of the electron mass energy and in MeV. The electron mass energy is 0.511 MeV.

13. The velocity of a particle of mass m that starts from rest at time $t = 0$ and experiences a constant force F is given by

$$v = \frac{\dfrac{Ft}{m}}{\left[1 + \left(\dfrac{Ft}{mc}\right)^2\right]^{1/2}}$$

(a) Show that the acceleration of the particle is given by

$$a = \frac{\dfrac{F}{m}}{\left[1 + \left(\dfrac{Ft}{mc}\right)^2\right]^{3/2}}$$

(b) Show that the acceleration approaches F/m as t approaches zero, and interpret this result. (c) Show that the acceleration approaches zero as t approaches infinity, and interpret this result.

CHAPTER 41

QUANTUM PHYSICS

41.1
THE ORIGINS OF QUANTUM PHYSICS

Toward the end of the nineteenth century it seemed that Newton's mechanics and Maxwell's electromagnetic theory provided a firm basis for the complete understanding of all branches of physics. Newton's laws adequately described a breadth of motions, from falling apples to orbiting planets. The theories of heat and sound had been successfully extended into the microscopic domain by using a kinetic theory of matter based on Newtonian mechanics. Maxwell's research not only unified electrical and magnetic phenomena, but also described the propagation of light and other electromagnetic waves. To be sure, there were many unanswered questions, but most scientists were confident that the theories of Newton and Maxwell would lead to satisfactory answers.

They were mistaken. It was Max Planck (Figure 41.1) who ushered in a revolution in scientific and philosophical thought in 1900. Albert Einstein and Niels Bohr were the champions of this revolution. Their research revealed that quantities such as energy and angular momentum are **quantized** at the atomic level; that is, they can take on only certain discrete values, like the times displayed by a digital clock.

All objects radiate energy in the form of electromagnetic waves, at a rate that depends strongly on the object's temperature. This energy is called **thermal radiation** (Section 20.3). Near the end of the nineteenth century, experiments had revealed the key features of thermal radiation. These experiments led to the concept of a **blackbody**, an ideal radiating body that absorbs all incident electromagnetic radiation.

Quantum physics originated in 1900 when Max Planck developed his theory of blackbody radiation. Planck envisioned the surface of a blackbody as containing innumerable microscopic oscillators capable of absorbing and emitting

FIGURE 41.1

Max Planck (right). Quantum theory originated in 1900 with Planck's theory of blackbody radiation. Niels Bohr is on the left.

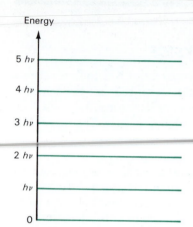

FIGURE 41.2

In Planck's theory of blackbody radiation the quantized energies of an oscillator are integral multiples of the energy quantum $h\nu$.

electromagnetic radiation. Each oscillator had a characteristic frequency at which it emitted and absorbed electromagnetic energy.

Planck found he could explain the observed features of blackbody radiation by assigning *discrete* energies to the oscillators. For the first time, energy was *quantized*. In Planck's theory the energy E of an oscillator of frequency ν must be an integral multiple of a quantum of energy, $h\nu$ (Figure 41.2),

$$E = 0, h\nu, 2h\nu, 3h\nu, \ldots, nh\nu, \tag{41.1}$$

where n is an integer. The constant h is called Planck's constant. Planck determined h by fitting his theory to experimental results. The modern value for h is

$$h = 6.6262 \times 10^{-34} \text{ J·s}$$

When energies are measured in electron volt (eV) units it is more convenient to express h in units of electron volt seconds (eV·s),

$$h = 4.1357 \times 10^{-15} \text{ eV·s} \tag{41.2}$$

Despite the success of his theory, Planck initially believed that quantization was only a mathematical trick that led to the correct blackbody spectrum.

Einstein and Photons

Albert Einstein readily accepted the idea of energy quantization. He went beyond Planck and argued that light quanta were a necessary consequence of Planck's theory. For Einstein, the fact that an oscillator absorbs and emits electromagnetic energy in packets of size $h\nu$ meant that electromagnetic radiation could be interpreted as a collection of energy parcels, or **photons** as they are now called*. Thus, Einstein extended Planck's quantization of energy to the radiation field. The energy emitted by a light source of frequency ν is carried by photons of energy

$$E = h\nu \tag{41.3}$$

EXAMPLE 1

The Energy of a Blue Photon

A source emits blue light having a wavelength of 460 nm. To determine the photon energy we use the kinematic relation $\nu\lambda = c$ to obtain

$$E = h\nu = \frac{hc}{\lambda} = \frac{(4.14 \times 10^{-15} \text{ eV·s})(3 \times 10^8 \text{ m/s})}{460 \times 10^{-9} \text{ m}}$$

$$= 2.70 \text{ eV}$$

The Photoelectric Effect

When ultraviolet light strikes a metal surface, electrons are ejected from the metal. This is the **photoelectric effect**, discovered in 1887 by Hertz in experiments that also verified Maxwell's theory of electromagnetic waves.

In 1905, Einstein used the photon picture of light to explain the photoelectric effect. Einstein assumed that when a photon is absorbed by a metal surface

---

* The name *photon* was suggested by G. N. Lewis in 1926.

its energy is transformed into the work needed to free an electron from the metal and into the kinetic energy of the ejected electron,

$$h\nu = \text{work to free electron} + \text{kinetic energy of electron}$$

The amount of energy required to free an electron depends primarily on the type of metal surface. The needed energy also depends on the depth from which the electron originates. If W is the *minimum* energy required to free an electron, the *maximum* kinetic energy of the ejected electrons K_{max} is related to W by

$$h\nu = W + K_{max} \tag{41.4}$$

The quantity W is called the **work function** of the surface. Typically, W is a few electron volts. To measure K_{max} we can subject the electrons to a repelling electrostatic force (Figure 41.3). The electrons are decelerated as they move toward the negatively charged plate. If the potential difference is large enough, then not even the most energetic electrons will cross the gap, and the electron current will drop to zero. The **stopping potential** V_s is the potential that reduces the current to zero. An electron ejected with the maximum kinetic energy K_{max} and brought to rest momentarily just short of the negatively charged plate converts its kinetic energy into electric potential energy (eV_s). Thus,

$$K_{max} = eV_s \tag{41.5}$$

Using this result in Equation 41.4 gives

$$V_s = \left(\frac{h}{e}\right)\nu - \frac{W}{e} \tag{41.6}$$

This equation reveals the key prediction of Einstein's theory: a graph of the stopping potential V_s versus the frequency ν should be a straight line whose slope (h/e) is the same for all materials.

Experiments carried out in 1916 by Millikan verified Einstein's theory. Millikan demonstrated that the value of h determined from experiments on the photoelectric effect agreed with the value of h obtained by Planck from

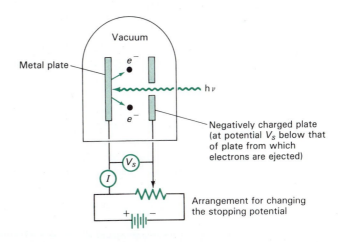

FIGURE 41.3

The photoelectric effect. Photons eject electrons from the metal plate. The electrons can be prevented from reaching the plate at the right by applying a sufficiently large stopping potential.

blackbody experiments. Thus, the photoelectric effect provided an independent confirmation of the quantum hypothesis.

The lowest frequency that ejects electrons from a metal is called the **threshold frequency**. The threshold frequency v_0 is related to the work function by

$$hv_0 = W \qquad (41.7)$$

This equation can be regarded as a limiting case of Equation 41.4, in which K_{max} approaches zero. Thus, a photon with an energy exactly equal to W has barely enough energy to free an electron.

EXAMPLE 2

Threshold Frequency for Silver

The work function for silver is 4.73 eV. Let's calculate the threshold frequency. We have

$$v_0 = \frac{W}{h} = \frac{4.73 \text{ eV}}{4.14 \times 10^{-15} \text{ eV·s}} = 1.14 \times 10^{15} \text{ Hz}$$

The corresponding wavelength is

$$\lambda = \frac{c}{v_0} = \frac{3 \times 10^8 \text{ m/s}}{1.14 \times 10^{15} \text{ s}^{-1}} = 2.63 \times 10^{-7} \text{ m}$$
$$= 263 \text{ nm}$$

The wavelength of 263 nm lies in the ultraviolet (UV) portion of the electromagnetic spectrum. Photons of visible light have longer wavelengths and lower frequencies and therefore less energy than UV photons. Photons of visible light cannot eject electrons from silver.

In addition to energy, a photon possesses other particle-like properties. We now examine two of these: the photon mass and the photon linear momentum.

Photon Mass

Any particle that travels at the speed of light and has energy must be a zero-mass particle (Section 40.3). The photon possesses energy and travels at the speed of light and so is a zero-mass particle:

$$m_{\text{photon}} = 0 \qquad (41.8)$$

Zero-mass particles have no existence as objects at rest. For example, when a photon is emitted by an atom, energy stored in the atom is transformed into radiant energy. The photon has no existence prior to its emission. Likewise, when an atom absorbs a photon, the photon is *destroyed*—it does not survive bottled up inside the atom.

Photon Linear Momentum

The linear momentum of a photon is parallel to the direction of motion of the photon. The magnitude of the linear momentum follows from the general relation between energy, linear momentum, and mass derived in Section 40.3:

$$E^2 = (pc)^2 + (mc^2)^2 \qquad (41.9)$$

(Before)

MV_i

(After)

MV_f

FIGURE 41.4

Head-on collision in which a photon is absorbed by an atom. The linear momentum of the atom decreases, causing it to slow down.

Taking the photon mass to be zero shows that the photon linear momentum is directly proportional to the photon energy

$$p = \frac{E}{c} \qquad (41.10)$$

Another useful expression for the photon linear momentum is obtained by using $E = h\nu$ and the kinematic relation $\nu\lambda = c$. The expression $p = E/c$ then becomes

$$p = \frac{h}{\lambda} \qquad (41.11)$$

EXAMPLE 3

Cooling Atoms with Photons

Figure 41.4 shows the head-on collision of an atom of mass M and a photon. The photon is absorbed, transferring its energy and linear momentum to the atom. The momentum transfer causes the speed of the atom to decrease from V_i to V_f.

We apply the conservation of linear momentum principle to determine the change in the speed of the atom. Using $p = h/\lambda$ for the photon momentum we have*

$$MV_f = MV_i - \frac{h}{\lambda}$$

The change in the speed of the atom is

$$V_i - V_f = \frac{h}{M\lambda}$$

One experiment uses sodium atoms and laser light with a wavelength of 596 nm. The change in speed is

$$\frac{h}{M\lambda} = \frac{6.62 \times 10^{-34} \text{ J·s}}{3.84 \times 10^{-26} \text{ kg} \cdot 596 \times 10^{-9} \text{ m}}$$

$$= 0.029 \text{ m/s}$$

Although this change is small compared to the typical initial atomic speed of 100 m/s, there are many absorptions per second. Using photon absorption, physicists have been able to slow groups of atoms to average speeds of less than 1 m/s.

A decrease in average speed amounts to a decrease in temperature because

$$\text{average speed} \propto (\text{kelvin temperature})^{1/2}$$

Temperatures below 10^{-3} K have been achieved using multiple laser beams to trap atoms.

Compton Scattering

Deep inside the core of the sun, thermonuclear fusion reactions transform mass energy into kinetic energy (Section 42.5). Much of this kinetic energy is carried by gamma photons. As these photons struggle outward they collide repeatedly with electrons that comprise part of the solar interior. The collisions

---

* The mass of the atom increases slightly because of the energy of the absorbed photon. This effect is negligible here.

cause the photons to move along erratic, zigzag paths. Progress toward the surface is slow; even though the photons travel at the speed of light in between collisions, it takes an average of 10,000 years for them to reach the surface.

The scattering of photons by electrons is called **Compton scattering**, after A. H. Compton, who developed its theory in 1922. Compton used his theory to explain the results of scattering of X rays and gamma rays by atomic electrons. Compton found that he could explain his observations by treating the X rays and gamma rays as photons. Thus, Compton scattering demonstrates the particle-like character of light.

Compton applied the principles of energy conservation and linear momentum conservation to the scattering. Figure 41.5a shows the geometry. A photon with energy E_0 and linear momentum $\mathbf{P}_0$ strikes an electron at rest. The photon scatters through an angle θ and the electron recoils. After the collision the scattered photon has energy E and linear momentum $\mathbf{P}$. The recoiling electron has energy e and linear momentum $\mathbf{p}$.

The conservation of linear momentum is expressed by

Total linear momentum before scatter = Total linear momentum after scatter

Linear momentum conservation is expressed in symbolic form as

$$\mathbf{P}_0 = \mathbf{P} + \mathbf{p} \qquad (41.12)$$

Figure 41.5b shows the triangular relation between $\mathbf{P}_0$, $\mathbf{P}$, and $\mathbf{p}$. Energy conservation is described by

Total energy before scatter = Total energy after scatter

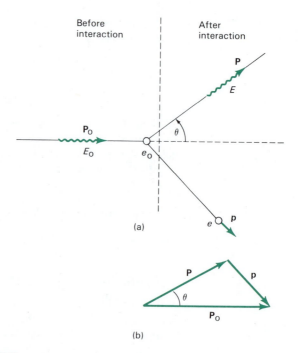

(a)

(b)

FIGURE 41.5

Compton scattering. (a) A photon of energy E_0 interacts with an electron at rest. The photon scatters through an angle θ and the electron recoils. (b) Triangle describing the conservation of linear momentum.

Symbolically,

$$E_0 + e_0 = E + e \tag{41.13}$$

where $e_0 = mc^2$ is the mass energy of the electron.

We eliminate the electron momentum and energy from this pair of equations by noting that $\mathbf{p} \cdot \mathbf{p} = p^2$ and using Equation 41.9 to form $e_0{}^2 + (pc)^2$, which equals the square of the total electron energy

$$e^2 = e_0{}^2 + (pc)^2$$

This step gives

$$e^2 = e_0{}^2 + E_0{}^2 + E^2 - 2E_0 E \cos \theta$$

A second expression for e^2 follows from energy conservation, Equation 41.13:

$$e^2 = (e_0 + E_0 - E)^2$$

Eliminating e^2 leads to

$$-2E_0 E \cos \theta = -2E_0 E + 2mc^2(E_0 - E) \tag{41.14}$$

Dividing through by $2E_0 E$ and introducing the wavelengths λ and λ_0 through

$$E = h\nu = \frac{hc}{\lambda} \qquad E_0 = h\nu_0 = \frac{hc}{\lambda_0}$$

results in Compton's equation for the change in wavelength

$$\lambda - \lambda_0 = \frac{h}{mc}(1 - \cos \theta) \tag{41.15}$$

EXAMPLE 4

Compton Scattering

Compton verified Equation 41.15 by observing the scattering of X rays with $\lambda_0 = 0.708$ Å. He observed that the X rays scattered through an angle of $90°$ had a wavelength of 0.730 Å. Thus, the observed value of $\lambda - \lambda_0$ equaled 0.022 Å. For $\theta = 90°$, $\cos \theta = 0$, and the Compton formula predicts

$$\lambda - \lambda_0 = \frac{h}{mc} = \frac{6.62 \times 10^{-34} \text{ J·s}}{9.11 \times 10^{-31} \text{ kg} \cdot 3 \times 10^8 \text{ m/s}} = 2.4 \times 10^{-12} \text{ m} = 0.024 \text{ Å}$$

The length h/mc is called the "Compton wavelength." In a subsequent series of experiments Compton verified his theory using gamma rays with wavelengths of 0.022 Å. When scattered by atomic electrons in iron, aluminum, and paraffin, the observed change in wavelength agreed closely with that calculated from Equation 41.15.

In the sun, the gamma photons have much greater energies than the electrons with which they collide. On average, the electrons gain energy and the photons lose energy. The energy transfer from the photons to the electrons maintains the high temperature of the solar interior. The relation

$$E = h\nu = \frac{hc}{\lambda}$$

shows that as the photon energy decreases, the wavelength increases. By the time the photons reach the outer layers of the sun many of them have wavelengths in the visible range. The photons undergo a final series of

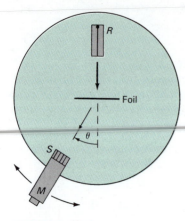

FIGURE 41.6

Top view of the alpha particle scattering apparatus of Geiger and Marsden. The radioactive source (R) gives a collimated beam of alpha particles. The microscope (M) is focused on the screen (S). The foil and source are fixed. The microscope can rotate about the foil.

interactions with atoms and ions in the outer layers before leaving the solar surface.

41.2
RUTHERFORD AND THE NUCLEAR ATOM

The research of Planck and Einstein introduced quantization into physics. The next thread in the fabric of quantum physics was Rutherford's discovery of the atomic nucleus.

In 1909, Rutherford's assistants, Geiger and Marsden, began analyzing the deflection of alpha particles passing through thin metal foils. The alpha particles were emitted by radioactive substances and had kinetic energies of several MeV. An alpha particle was known to have a positive charge and a mass several thousand times that of an electron. It was subsequently identified as a doubly-ionized helium atom.

Figure 41.6 shows the experimental arrangement used by Geiger and Marsden. A movable eyepiece was focused on a small screen that gave off brief flashes of light when struck by alpha particles. Most alpha particles suffered very small deflections. But surprisingly, a tiny fraction of the alpha particles was scattered backward, through angles as great as 150°, the largest angle observable with their apparatus. Such large deflections were completely unexpected.

In 1911, Rutherford described an atomic model that accounts quantitatively for the experimental observations. Rutherford argued that the alpha particle could bounce back only if it struck an object more massive than itself. This could be the case if the positive charge and most of the mass of the atom were concentrated in a small portion of the atom, rather than spread diffusely throughout the atom. Rutherford called this concentrated region the **nucleus**. The fact that backward scattering occurred infrequently suggested that the nucleus was a small target—much smaller than the length that characterizes the diameter of the atom.

Rutherford analyzed the scattering of alpha particles by a nucleus, assuming that the only force between the two was the Coulomb force of repulsion between their positive charges. The experiments of Geiger and Marsden verified the pattern of scattering predicted by Rutherford (Figure 41.7).

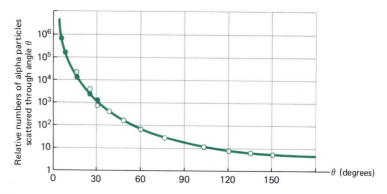

FIGURE 41.7

Rutherford's theory of the nuclear atom (solid line) predicted the relative number of alpha particles that will scatter through different ranges of angle. Rutherford's theory explained the measurements of Geiger and Marsden, shown as solid and open circles.

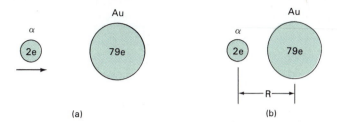

FIGURE 41.8

(a) An alpha particle approaches a stationary nucleus. (b) The distance between the nucleus and the alpha particle is a minimum at the moment when the alpha particle comes to rest.

Knowing that the force between a nucleus and an alpha particle is electrical, Rutherford was able to estimate the radius of the nucleus. His calculation was based on energy conservation. Figure 41.8 shows a head-on collision between an alpha particle and a gold nucleus. When the alpha particle is outside the gold atom, the repulsive force of the nucleus is shielded by the atomic electrons. Initially, the energy of the system (alpha plus nucleus) equals the kinetic energy of the alpha particle. As the alpha particle approaches the nucleus it experiences a repulsive Coulomb force and slows down, and its kinetic energy is transformed into Coulomb potential energy.

In a head-on collision the speed of the alpha particle is reduced to zero, at which point the energy of the system is stored as Coulomb potential energy.* If R is the center-to-center distance between the alpha particle (charge $2e$) and a nucleus containing Z protons (charge Ze), the potential energy is

$$U = \frac{1}{4\pi\epsilon_0} \frac{2e \cdot Ze}{R}$$

The conservation of energy is expressed by

$$K = U$$

where K is the kinetic energy of the alpha particle before it encounters the nucleus. Solving for R gives

$$R = \frac{2Ze^2}{4\pi\epsilon_0 K} \tag{41.16}$$

EXAMPLE 5

Estimate of the Radius of a Gold Nucleus

The value of R given by Equation 41.16 is an upper limit to the radius of the nucleus. Let's estimate the radius of a gold nucleus, using data from Geiger and Marsden's experiment.

The kinetic energy of the alpha particles used by Geiger and Marsden was 7.7 MeV. This converts to

$$K = 7.7 \text{ MeV} \cdot 1.60 \times 10^{-13} \text{ J/MeV} = 1.23 \times 10^{-12} \text{ J}$$

The atomic number of gold is 79. With $K = 1.23 \times 10^{-12}$ J and $Z = 79$, Equation 41.16 gives

$$R = \frac{158(1.60 \times 10^{-19} \text{ C})^2(8.99 \times 10^9 \text{ N·m}^2/\text{C}^2)}{1.23 \times 10^{-12} \text{ J}}$$

$$= 2.96 \times 10^{-14} \text{ m}$$

$$= 29.6 \text{ fm}$$

* This assumption ignores the kinetic energy of the recoiling nucleus.

This overestimates the nuclear radius because the alpha particle and the nucleus are not touching when the alpha particle comes to rest.

The atomic electrons spread themselves over a volume roughly 1 Å in diameter. Our estimate shows that the diameter of the atomic nucleus is approximately 10,000 times smaller than the atomic diameter. The nucleus is concentrated into a tiny fraction of the atomic volume, which explains why large-angle deflections of alpha particles are infrequent. Only rarely will an alpha particle suffer the head-on collision needed to produce a large deflection. Most alpha particles will miss the nucleus and undergo small deflections when they are scattered by the atomic electrons.

The experiments of Geiger and Marsden, and Rutherford's analysis, established the nuclear model of the atom. The nucleus carries the full positive charge of the atom and most of the mass. The atom's electrons contribute an equal negative charge and an insignificant fraction of the atomic mass.

The positions of the atomic electrons were not revealed by alpha particle scattering, but it was clear that the electrons were not distributed in a static configuration. The strong attractive force of the positively charged nucleus would lead to the collapse of any static configuration. However, if the electrons moved about the nucleus in orbits they could achieve a dynamic stability, just as the motion of the planets about the sun prevents the gravitational collapse of the solar system. It was Niels Bohr who combined a dynamic model of the nuclear atom with the photon concept and quantization to develop a successful model of the hydrogen atom.

41.3
BOHR AND THE HYDROGEN ATOM

The idea of an atom in which the electrons whirl about a tiny nucleus intrigued Niels Bohr, a young Danish physicist. Early in 1913 Bohr was alerted to an empirical relation discovered in 1885 by J. J. Balmer, a Swiss schoolteacher. Balmer's equation described the measured wavelengths of light emitted by hydrogen atoms

$$\lambda_n = 364.56 \, \frac{n^2}{n^2 - 2^2} \, \text{nm} \qquad n = 3, \, 4, \, 5, \, 6, \, \dots \qquad (41.17)$$

Balmer's formula described the experimental data exceptionally well, but it had no theoretical basis; Balmer deduced it by trial and error.

Bohr constructed a theoretical model of the hydrogen atom that tied together the light quanta of Einstein and the nuclear atom of Rutherford, and which led to a prediction of the hydrogen wavelengths in agreement with Balmer's empirical equation. Bohr envisioned the hydrogen atom as an electron moving in a circular orbit about a much more massive nucleus (Figure 41.9). The electron is held in orbit by the attractive Coulomb force exerted by the nucleus.

Bohr's model raised a troublesome point. According to Maxwell's theory of electromagnetism, an accelerated charge radiates. The electron in orbit experiences a centripetal acceleration and therefore should radiate. The continuous loss of energy via radiation would cause the electron to spiral inward toward the nucleus, resulting in a collapse of the atom. Furthermore, the collapse time

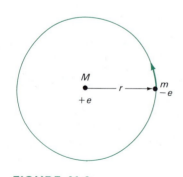

FIGURE 41.9

Bohr model of the hydrogen atom. The electron moves in a circular orbit about a stationary proton.

was calculated and found to be very brief, about 10^{-10} s. Bohr concluded that Maxwell's electromagnetic theory was not adequate to describe fully the electrodynamics of atomic electrons.

With a bold intellectual leap, Bohr went beyond the limited domain of classical physics and postulated the following:

1. **Stationary State Postulate.** There exist *stationary states* of the atom in which the orbiting electron does not radiate.
2. **Radiation Postulate.** When the electron changes from one stationary state to another stationary state of lower energy, a photon is emitted. The energy carried away by the photon equals the decrease in the energy of the atom.

If the atom makes a transition from a state of energy E_a to a state of lower energy E_b, the energy decrease is $E_a - E_b$. The radiation postulate can be expressed as

$$h\nu = E_a - E_b \tag{41.18}$$

3. **Quantization Postulate.** The angular momentum of the atom is quantized in units of $h/2\pi$.

If L denotes the angular momentum of the electron, the quantization postulate can be expressed as

$$L = n\frac{h}{2\pi} \qquad n = 1, 2, 3, \ldots \tag{41.19}$$

Note that the quantization postulate *imposes* quantization on the atom. Quantization did not follow from Bohr's theory as a natural consequence of the classical physics of Newton and Maxwell; instead, it was inserted into the theory.

The quantization of angular momentum leads to the quantization of two properties of the hydrogen atom: the radii of electron orbits and the energies of the stationary states.

These quantization relations follow from an analysis of the total energy of the hydrogen atom, which is the sum of the kinetic energy and the Coulomb potential energy,

$$E = \tfrac{1}{2}mv^2 - \frac{1}{4\pi\epsilon_0}\frac{e^2}{r}$$

The kinetic energy can be expressed in terms of the angular momentum,

$$\tfrac{1}{2}mv^2 = \frac{(mvr)^2}{2mr^2} = \frac{\left(\dfrac{nh}{2\pi}\right)^2}{2mr^2}$$

Thus,

$$E = \frac{\left(\dfrac{nh}{2\pi}\right)^2}{2mr^2} - \frac{1}{4\pi\epsilon_0}\frac{e^2}{r} \tag{41.20}$$

The stationary states for a given value of n are fixed by requiring that E be a minimum. Setting $dE/dr = 0$ leads to the equation describing the quantized values of r:

$$r \equiv r_n = n^2 a_0 \qquad n = 1, 2, 3, \ldots \qquad (41.21)$$

where a_0 is a length called the **Bohr radius**,

$$a_0 = \frac{4\pi\epsilon_0}{me^2}\left(\frac{h}{2\pi}\right)^2 = 0.529 \text{ Å} \qquad (41.22)$$

The smallest orbit radius is $r_1 = a_0$. The corresponding diameter of the hydrogen atom is $2a_0$, which is approximately 1 Å, the length that characterizes all atoms. Bohr's theory thereby explains that the length scale of atoms is a result of quantization.

EXAMPLE 6

A Quantum Explanation of Elastic Moduli

Table 14.1 shows that for many solids the three elastic moduli (Y, B, μ) have values in the range 10^{10} N/m² to 10^{12} N/m². This clustering of values is not accidental. Two facts can explain the observed values to within an order of magnitude:

1. The binding forces in solids are electrical.
2. Quantum considerations set the characteristic atomic dimension to be a few times the Bohr radius.

From their definitions (Chapter 14), Y, B, and μ have the dimensions of force per unit area,

$$\text{Modulus} = \frac{\dfrac{\text{Force}}{\text{Area}}}{\text{Strain}}$$

The strain is dimensionless, and it is typically small. We *overestimate the strain* by taking it to be unity,

$$\text{Strain} = 1$$

For an estimate of the force we use the electrical force between a proton and an electron separated by a distance r,

$$\text{Force} \approx \frac{1}{4\pi\epsilon_0} \frac{e^2}{r^2}$$

Although this is a good estimate of the force that binds electrons and protons in atoms, it is an overestimate of the force *between atoms* in a solid. Because we overestimate both the force and the strain, we can hope that the ratio of force/strain is not too far off the mark.

The area over which the binding force acts is approximately r^2, where r is a characteristic atomic length,

$$\text{Area} \approx r^2$$

For the elastic modulus we get

$$\text{Modulus} \approx \frac{1}{4\pi\epsilon_0} \frac{e^2}{r^4}$$

We take $r \approx 1.5$ Å, the atomic radius deduced in Chapter 22. This is about three times the Bohr radius. The corresponding modulus is

$$\text{Modulus} \approx \frac{(8.99 \times 10^9 \ \text{N·m}^2/\text{C}^2)(1.60 \times 10^{-19} \ \text{C})^2}{(1.5 \times 10^{-10} \ \text{m})^4}$$

$$\approx 5 \times 10^{11} \ \text{N/m}^2$$

This is in reasonable agreement with the values of Y, B, and μ listed in Table 14.1.

A small change in the value of r strongly affects the modulus. This strong dependence on r is one reason the observed values of elastic moduli range over two orders of magnitude.

The fact that our dimensional analysis leads to a reasonable estimate supports our initial assumptions: the elastic forces are electrical and the atomic dimension imposed by quantum considerations is a few angstroms.

Energy Quantization

The quantized energies of the stationary states, which follow by setting $r = n^2 a_0$ in Equation 41.20, can be expressed as

$$E_n = -\frac{E_H}{n^2} \qquad n = 1, 2, 3, \ldots \qquad (41.23)$$

where E_H is a characteristic energy of the hydrogen atom,

$$E_H = \frac{1}{4\pi\epsilon_0} \frac{e^2}{2a_0} = 13.6 \ \text{eV} \qquad (41.24)$$

The quantized energies for the hydrogen atom are displayed in Figure 41.10. The quantum number n labels the stationary energy states of the atom and the corresponding energy levels. In Figure 41.11 each energy level is represented by a horizontal line. The state of lowest energy ($n = 1$) is called the **ground state**, and the other states are referred to as **excited states**.

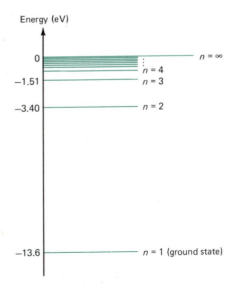

FIGURE 41.10

Quantized energies of the hydrogen atom. Each energy is represented by a horizontal line.

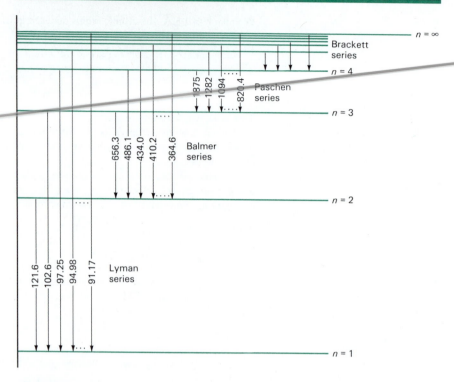

FIGURE 41.11

Spectral series for hydrogen. Wavelengths are indicated in nanometers.

Derivation of Balmer's Formula

When the hydrogen atom makes a transition from an excited state to a state of lower energy, a photon is emitted. Energy is conserved in the overall process. The photon energy ($h\nu$) equals the decrease in the energy of the atom—as prescribed by Bohr's radiation postulate, $h\nu = E_a - E_b$. When the transition terminates on the $n = 2$ energy level, the energy of the emitted photon is

$$h\nu_n = -\frac{E_H}{n^2} + \frac{E_H}{2^2} \qquad n = 3, 4, 5, \ldots$$

Replacing ν_n by c/λ_n and solving for the wavelength gives

$$\lambda_n = \frac{4hc}{E_H} \frac{n^2}{n^2 - 2^2} \tag{41.25}$$

Evaluating $4hc/E_H$, we find that

$$\lambda_n = 364.51 \frac{n^2}{n^2 - 2^2} \text{ nm} \qquad n = 3, 4, 5, 6, \ldots \tag{41.26}$$

Comparing this equation with Balmer's formula, Equation 41.17, we see that the two are in substantial agreement. Thus, Bohr's theory provided a theoretical basis for Balmer's empirical formula.

The explanation of Balmer's formula was a great triumph; *it revealed that the atomic world is quantized*. Bohr's original theory of the hydrogen atom has since been replaced, but the quantization of physical properties such as energy and angular momentum survives as an inherent feature of modern quantum theory.

Bohr's theory inspired many experimental investigations. The series of transitions that produced the Balmer wavelengths terminated on the $n = 2$ level (Figure 41.11). It seemed likely that other spectral series should be possible, corresponding to transitions that end on the $n = 1, 3, 4, \ldots$ levels. Experiment soon confirmed these expectations. In 1913 Theodore Lyman observed the series of wavelengths that now bear his name. The Lyman series arises from transitions that terminate on the $n = 1$ level (ground state). The wavelengths in the Lyman series lie in the ultraviolet region of the spectrum.

EXAMPLE 7

Wavelength of the Lyman α Radiation

The wavelengths of the Lyman series are labeled $L_\alpha, L_\beta, L_\gamma, \ldots$, with L_α being the longest. The L_α radiation is emitted in the $n = 2$ to $n = 1$ transition (Figure 41.11). The wavelength follows from the radiation condition, Equation 41.18:

$$h\nu = E_a - E_b$$

Setting $h\nu = hc/\lambda$, $E_a = -E_H/2^2$, and $E_b = -E_H/1^2$, we find

$$\lambda = \frac{4hc}{3E_H}$$

We noted earlier that $4hc/E_H = 365$ nm, and so

$$\lambda = 122 \text{ nm} \qquad (L_\alpha \text{ wavelength})$$

The shortest wavelength in the Lyman series, which corresponds to the $n = \infty$ to $n = 1$ transition, is 91.2 nm. The entire Lyman series lies in the ultraviolet portion of the spectrum. The atmosphere of our sun, composed primarily of hydrogen, emits Lyman radiation. The earth is surrounded by a tenuous cloud of hydrogen and also has a halo of Lyman radiation, largely L_α.

FIGURE 41.12

Louis Victor de Broglie. In his 1924 Ph.D. thesis, de Broglie recognized the wave-particle duality as a universal principle.

41.4
DE BROGLIE AND THE WAVE-PARTICLE DUALITY

In our study of optics we treated light as a wave. In this chapter we have seen how light can also behave like a particle. The wave and particle interpretations of light are not contradictory, but rather, complementary. In fact, taken together, they enable us to explain the full range of optical phenomena. The wave picture provides a simple explanation of diffraction and other optical interference phenomena. The particle picture provides a simple explanation of the photoelectric effect and the scattering of X rays by the electrons in atoms. We now recognize that light is neither a wave nor a particle, but displays aspects of both. The dual nature of light is one example of the overall wave-particle duality exhibited by nature.

In 1924, a young Frenchman Louis Victor de Broglie (Figure 41.12) argued that wave-particle duality is a universal principle not restricted to light. Not only do waves exhibit particle-like behavior, he argued, but objects that we traditionally regard as particles must also exhibit wavelike properties. For example, we are accustomed to thinking of an electron as a particle but, as we shall see, there are interference experiments involving electrons that call for a wave interpretation.

Two quantities used to describe a wave are frequency (ν) and wavelength (λ). Two quantities used to describe a particle are energy (E) and linear momentum (p). The relationships between the wave and particle attributes of light are

$$E = h\nu \tag{41.3}$$

$$p = \frac{h}{\lambda} \tag{41.11}$$

De Broglie postulated Equations 41.3 and 41.11 as **universal relationships** between wave (ν, λ) and particle (E, p) attributes.

De Broglie tested his hypothesis by applying it to the electron in a hydrogen atom and found that it leads to the same quantization of angular momentum imposed by Bohr, and, in turn, to the same quantized energy levels and spectrum.

The Davisson-Germer Experiment

De Broglie's theory had two points in its favor: it predicted the same quantum structure of the hydrogen atom as did the Bohr theory, and it offered a different viewpoint, which was potentially broader and more general than Bohr's approach. But there were also shortcomings. There was nothing to indicate the *nature* of the de Broglie waves. Specifically, there was no wave equation describing how the waves travel. Furthermore, there was no direct experimental evidence to support de Broglie's hypothesis of the wave nature of matter.

Both of these shortcomings were soon remedied. First, in 1925 Erwin Schrödinger set forth a wave equation that described the behavior of de Broglie's waves. Then in 1927 Clinton Davisson and Lester Germer performed an experiment that confirmed the wave character of electrons. Their experiment showed that electrons could exhibit interference—a clear indication of wave behavior. A beam of electrons accelerated through a potential difference of 54 V was directed toward a nickel crystal. Electrons were scattered in all directions, but their intensity showed a strong maximum at a certain angle. The regularly spaced atoms in the crystal acted like a diffraction grating. The direction of maximum intensity of the scattered electrons corresponded to the direction of constructive interference of the electron waves. Davisson and Germer were able to relate the atomic spacing to the electron wavelength, and found that $\lambda = 1.65 \times 10^{-10}$ m. The de Broglie wavelength can also be calculated by using Equation 41.11 and our knowledge that the measured kinetic energy of the electrons is 54 eV. Thus,

$$\lambda = \frac{h}{p} = \frac{h}{\sqrt{2mK}} \tag{41.27}$$

where K is the kinetic energy and m is the electron mass. With $K = 54$ eV, the calculated de Broglie wavelength is

$$\lambda = \frac{6.62 \times 10^{-34} \text{ J} \cdot \text{s}}{\sqrt{2(9.11 \times 10^{-31} \text{ kg})(54 \text{ eV})(1.60 \times 10^{-19} \text{ J/eV})}}$$
$$= 1.67 \times 10^{-10} \text{ m}$$

The difference between the measured and calculated values of λ can be traced to the fact that the crystal contributes to the overall potential difference experienced by the electrons. Thus, the potential difference differs slightly from the applied 54 V.

41.5
SCHRÖDINGER'S EQUATION AND PROBABILITY WAVES

In 1925, Erwin Schrödinger presented an equation that describes de Broglie's waves. The Schrödinger equation is similar in some respects to the equation describing waves on a string,

$$\frac{\partial^2 \psi}{\partial t^2} = v^2 \frac{\partial^2 \psi}{\partial x^2}$$

For waves on a string, the wave function $\psi(x, t)$ describes the displacement of the string from equilibrium. The variable x labels positions along the direction of wave travel and t denotes time. The **Schrödinger wave equation** for stationary states of the hydrogen atom (states of constant energy, E) has the form

$$\left(E + \frac{e^2}{4\pi\epsilon_0 r}\right)\psi = -\frac{h^2}{8\pi^2 m}\left(\frac{\partial^2 \psi}{\partial x^2} + \frac{\partial^2 \psi}{\partial y^2} + \frac{\partial^2 \psi}{\partial z^2}\right) \qquad (41.28)$$

The variables x, y, z mark the position of the electron relative to the nucleus, and $r = \sqrt{x^2 + y^2 + z^2}$ is the distance between the electron and the nucleus. The physical significance of the Schrödinger wave function ψ was not immediately recognized. Yet despite the lack of a physical interpretation of ψ, Schrödinger found that only certain *discrete* (quantized) values of E yield proper solutions of the wave equation. A proper solution is one for which the wave function satisfies prescribed boundary conditions.* Remarkably, the quantized energies obtained from the Schrödinger equation for hydrogen were the same as those predicted by the Bohr theory.

Schrödinger's wave equation caused great excitement, especially since it was not restricted to the hydrogen atom. Techniques were quickly developed for describing the absorption and emission of radiation when an atom undergoes a transition from one energy state to another. The frequency of the emitted radiation is given by Bohr's radiation condition, $h\nu = E_a - E_b$. The great beauty of Schrödinger's wave equation is that quantization emerges in a natural way— as a consequence of **boundary conditions**. It is not necessary to impose quantization, as Bohr did with the quantization postulate.

The Schrödinger wave equation was subsequently solved for numerous atomic and molecular systems, and the optical spectra of many atoms and molecules were successfully interpreted. Schrödinger's wave equation gave birth to a *new* quantum theory—a theory that is known today as **quantum mechanics**.

---

* The role of boundary conditions on the classical wave equation is discussed in Section 17.4. For a wave that represents an electron localized in the vicinity of a nucleus, one boundary condition requires that $\psi \to 0$ as $r \to \infty$.

Despite its immediate successes and recognition as the key equation governing quantum structure, the Schrödinger equation at first presented one very troublesome point. Neither the equation nor its mathematical solutions reveals the physical meaning of the wave function, ψ. The question physicists asked themselves was "What does ψ measure?" The answer was given in 1926 by Max Born. Schrödinger's wave is a **probability wave**. Specifically, the square of the magnitude of $\psi(x, y, z)$ for the electron in a hydrogen atom is proportional to the probability that the electron will be located in the vicinity of the point with coordinates (x, y, z)*,

$$|\psi(x, y, z)|^2\, dV = \frac{\text{probability that the particle is in the}}{\text{volume } dV \text{ centered at } (x, y, z)} \qquad (41.29)$$

The essence of Born's probability interpretation of ψ is this: The *deterministic* statements of physical laws that characterize classical physics, such as Newton's laws of motion, are replaced at the atomic level by *probability* statements. For example, the classical description of an electron at rest is quite deterministic. An electron at rest, free of any external forces, will remain at rest. But the quantum description of the electron is different. It says that an electron that is free of any external forces, and within a specified region of space at one moment, need not remain in that region. There is a probability that it will leave the region, and quantum theory prescribes how to calculate this probability.

"Has physics lost touch with reality?" you may wonder. Actually, the concept of physical reality is as elusive as Schrödinger's wave function. Our conception of physical reality has changed as our perception of nature has widened and deepened.[†] Quantum theory adequately describes reality in the sense that its predictions agree with experiment.

Quantum Mechanical Tunneling

One of the most striking of all quantum phenomena is **quantum tunneling**, the penetration of a potential energy barrier. To illustrate the idea of tunneling, consider an electron that moves in response to the potential energy function shown in Figure 41.13.[††] The total energy of the electron (E) is the sum of its potential energy (U) and its kinetic energy (K). The total energy E is less than the maximum potential energy (U_B). The peak in the potential energy presents a barrier in the sense that the kinetic energy of an electron decreases as it approaches the point $x = x_A$. According to classical physics, an electron traveling to the right must come to rest momentarily at the turning point at $x = x_A$. At the turning point $E = U$, and the kinetic energy is zero. The electron will then move away to the left. In short, the electron is reflected by the potential energy barrier. In the words of classical physics, the region between the turning points (x_A and x_B) is inaccessible because it corresponds to a negative kinetic energy, $K = (E - U) < 0$. Classically, this is impossible. The kinetic energy, $K = \frac{1}{2}mv^2$, cannot be negative.

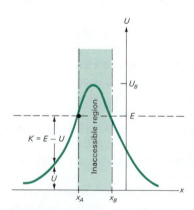

FIGURE 41.13

The total energy (E) of the particle is less than the potential energy maximum (U_B). The region between x_A and x_B is inaccessible according to classical physics, because it corresponds to negative kinetic energy.

---

* This interpretation has profound philosophical implications. See *Atomic Physics*, 6th ed., by Max Born, Hafner, New York, 1956 (pp. 94–97); and *Physics for Poets*, 2nd ed., by Robert H. March, McGraw-Hill, New York, 1978 (Chapter 17).

† See *The Nature of Physical Reality*, by Henry Margenau; McGraw-Hill, New York, 1950.

†† The potential energy function of Figure 41.13 is not typical of the potential energy of an electron in an atom. It has some of the features of the potential energy of electrons in certain types of solid-state circuit elements.

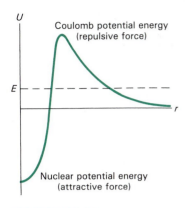

FIGURE 41.14

The potential energy barrier facing colliding protons.

Quantum mechanics, on the other hand, describes the electron in terms of its wave function, ψ, which is not zero in the barrier region between x_A and x_B. There is a definite probability—related to ψ—that an electron can travel from x_A to x_B and continue on. We say that the electron tunnels through the region that is inaccessible by classical mechanics.*

Quantum tunneling has many practical applications. For example, many solid-state electronic circuits incorporate so-called tunnel diodes, whose operation relies on the quantum tunneling of electrons. One of the earliest applications of quantum mechanics to the nucleus invoked quantum tunneling to explain the type of radioactivity known as alpha particle decay.

The very existence of life on earth depends on quantum tunneling. The energy that nourishes life on our planet is released via nuclear fusion reactions inside our sun. One important nuclear fusion reaction unites two protons. Figure 41.14 shows that the protons face a Coulomb potential energy barrier as they approach one another. The barrier is a consequence of the repulsive Coulomb force between the positive charges of the two protons. To get close enough to experience the attractive nuclear force, a proton must tunnel through the Coulomb barrier. There is only a small probability that a proton will tunnel through the barrier, and therefore only a tiny fraction of proton–proton collisions result in a fusion reaction. However, this tiny fraction is enough to supply the energy output of the sun, which sustains life on earth. If quantum tunneling were not a physical reality, our sun would never have been able to ignite fusion reactions, nor kindle life.

41.6
QUANTUM OPTICS AND LASERS

The quantum theory of light is called **quantum optics**. As we will now see, the quantum theory of matter and quantum optics are tied together.

Absorption and Emission of Electromagnetic Radiation

One of the basic equations linking quantum optics to the quantum theory of matter is the **Bohr radiation condition**. The radiation condition relates the photon energy to the energy change of an atom or molecule or nucleus:

$$h\nu = E_a - E_b \tag{41.18}$$

In 1917, Albert Einstein showed that three basic processes are described by Equation 41.18. These are *absorption, spontaneous emission,* and *stimulated emission.* Figure 41.15 depicts these three processes. In Figure 41.15a an atom in a state of energy E_b absorbs a photon and is thereby raised to a state of energy E_a. In Figure 41.15b an isolated atom in an excited state of energy E_a spontaneously emits a photon. This leaves the atom in a state of lower energy, E_b. The process of stimulated emission is shown in Figure 41.15c. A photon of energy $h\nu = E_a - E_b$ interacts with an atom in a state of energy E_a and triggers the birth of another photon of energy $h\nu$. The two photons travel off together, leaving the atom in a level of lower energy. From the wave viewpoint, the excited atom emits a wave that is *coherent* with the stimulating

---

* There are classical analogs of quantum tunneling. For example, light waves can exhibit tunneling.

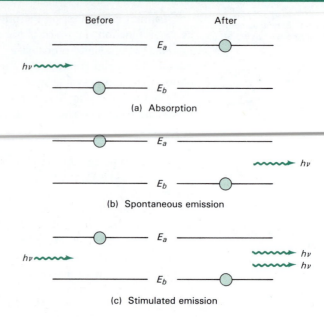

Before   After

E_a

$h\nu$

E_b

(a) Absorption

E_a

$h\nu$

E_b

(b) Spontaneous emission

E_a

$h\nu$ $h\nu$
                                         $h\nu$

E_b

(c) Stimulated emission

FIGURE 41.15

Three basic mechanisms by which atoms interact with photons. (a) Absorption: A photon of energy $E_a - E_b$ is absorbed, raising the energy of the atom from E_b to E_a. (b) Spontaneous emission: An atom spontaneously emits a photon of energy $E_a - E_b$. The energy of the atom decreases from E_a to E_b. (c) Stimulated emission: A photon of energy $E_a - E_b$ triggers the emission of its twin. The energy of the atom decreases from E_a to E_b.

wave. The emitted wave has the same frequency, direction of propagation, and polarization as the stimulating wave. Their superposition results in an amplification of the stimulating wave.

Lasers

A laser* is a device that produces a highly directional and intense beam of coherent light. Ordinarily, when a beam of light travels through matter, the effects of absorption and spontaneous emission overpower the effects of stimulated emission, and there is a net reduction in the intensity of the beam. In a laser, however, conditions are arranged so that stimulated emissions are dominant over absorptions and spontaneous emissions. As a result, the intensity of the beam grows as the beam moves through the matter. Let's first examine the conditions that reduce the intensity of a light beam traveling through matter, and then describe the conditions that increase the intensity in a laser.

Consider a narrow beam of photons entering a slab of material that contains atoms capable of absorbing the photons (Figure 41.16). When a photon is absorbed, the intensity of the beam is reduced. Each atom that absorbs a photon is left in an excited state. Its energy is increased by $h\nu$, the photon energy. Ordinarily, the atom remains in an excited state only very briefly— typically about 10^{-8} s. It can return to its former state by emitting a photon of the same frequency as the photon it absorbed. If every photon emitted this way were emitted in the direction of propagation of the beam, no overall change in the beam intensity would result. Absorptions followed by

* The word *laser* is an acronym for **l**ight **a**mplification by **s**timulated **e**mission of **r**adiation.

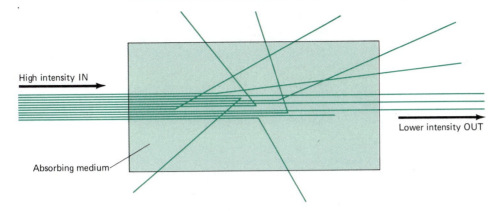

FIGURE 41.16

A light beam enters a medium capable of absorbing photons. Ordinarily, photon absorption is followed quickly by spontaneous emission. The emitted photons emerge traveling in all directions. Only a tiny fraction end up traveling with the beam. The intensity decreases as the beam moves through the medium.

spontaneous emissions along the beam would not change the intensity. But the direction of a spontaneously-emitted photon is rarely along the direction of the beam. Instead, photons are emitted in all directions, and only a tiny fraction end up traveling with the beam. Thus, absorptions followed by spontaneous emissions result in a *net loss of photons* from the beam and thereby *reduce* the beam intensity.

Next, consider a photon in the beam that encounters an atom in an appropriate excited state. If the incident photon should cause the excited atom to emit a photon—*stimulated* emission—then the two photons travel off hand in hand in the direction of the beam (Figure 41.15c). Stimulated emissions thereby *increase* the beam intensity.

As you can see, we have two competing effects: Absorptions reduce the beam intensity; stimulated emissions increase the beam intensity. Normally, the net result of these two processes is a reduced intensity, because there are many more absorptions than there are stimulated emissions. The reason that absorptions normally outweigh stimulated emissions is that most atoms are ordinarily in their lowest energy state, where they can absorb but not emit. In order to have stimulated emissions there must be atoms in excited states, and ordinarily only a tiny fraction of atoms are in excited states.

Population Inversion

There are several methods that make it possible to invert the usual population of energy states. A **population inversion** creates a situation where there are relatively many atoms in excited states, capable of emitting light of a particular wavelength, and relatively few in states capable of absorbing light of the same wavelength. In particular, it is possible to add energy selectively to atoms, thereby pumping them into a particular excited state. A beam of photons passing through an inverted population is amplified because there are more stimulated emissions than absorptions. Each photon of the incident beam can trigger the emission of others, and each newborn photon can travel on to stimulate further emissions. If a sufficient number of atoms are in excited states, an avalanche of photons can result. Figure 41.17 illustrates how the intensity of a photon beam can grow as it moves through a medium with an inverted population.

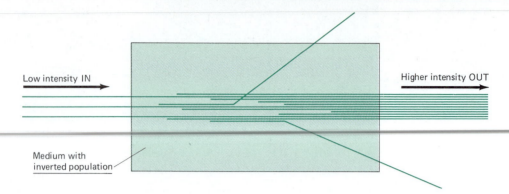

FIGURE 41.17

A beam of photons enters a medium in which there is an inverted population of atoms. When stimulated emission outweighs the effects of absorptions, the intensity increases as the beam travels through the medium.

The possibility of amplifying light by stimulated emission became evident after Einstein's theoretical study (in 1917) of the basic processes of absorption and emission. The possibility became a reality in 1954, when Gordon, Zeiger, and Townes built a device that amplified microwave radiation via stimulated emission. This type of microwave amplifier is called a **maser** (**m**icrowave **a**mplification by **s**timulated **e**mission of **r**adiation). Then, in 1960 Theodore Maiman first demonstrated the amplification of visible light via stimulated emission. Maiman's laser produced red light with a wavelength of 694.3 nm.

Laser Operation

The two essential components of a laser are an **active medium** and an **optical resonator**. In the helium-neon (He-Ne) laser the active medium is a gaseous mixture of helium and neon. An electric discharge causes some of the atoms to ionize. Thus, the active medium also contains positive ions and electrons. The optical resonator usually consists of a pair of mirrors made highly reflective at the laser wavelength via thin-film interference. One of the mirrors transmits about 1% of the incident laser intensity. This transmitted light constitutes the laser output beam.

The He-Ne laser is referred to as a four-level system. Figure 41.18 shows how four different energy levels are used to achieve a population inversion in a four-level system. Atoms* are pumped (Ⓐ) from the ground state (1) to a band of excited states (2). Transitions (Ⓑ) from the band of excited states populate the upper laser level (3). The transitions from the band of excited states must be selective. Transitions to the upper laser level must be more probable than transitions to the lower laser level. In the absence of pumping, both laser levels are essentially empty. Therefore, any selective population of the upper laser level results in a population inversion. Stimulated emissions (Ⓒ) from level 3 to level 4 result in laser action.

Figure 41.19 shows the energy levels for the He-Ne system. The helium plays a vital role in the pumping scheme. An electric discharge provides energetic electrons that bombard the helium atoms. Inelastic collisions between electrons and helium atoms in the ground state transfer energy to the helium atoms, boosting them into excited states. The energy of the excited state

---

* Lasers may employ molecules or ions as well as atoms.

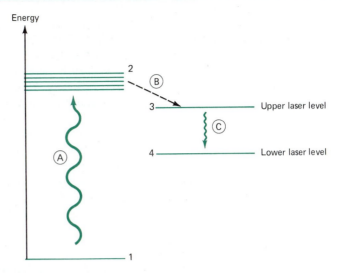

FIGURE 41.18

A four-level scheme for achieving an inverted population. Ⓐ pump transition; Ⓑ selective transitions to upper laser level; Ⓒ laser transition.

shown for the helium atom is a few hundredths of an electron volt greater than the energy needed to excite the upper laser level of the neon atom. Because of this, an inelastic collision between an excited helium atom and a ground-state neon atom can transfer energy to the neon atom. The neon atom is raised to the upper laser level and the helium atom returns to its ground state. The transfer of energy through inelastic collisions is more probable than energy loss through spontaneous emission, making the pumping

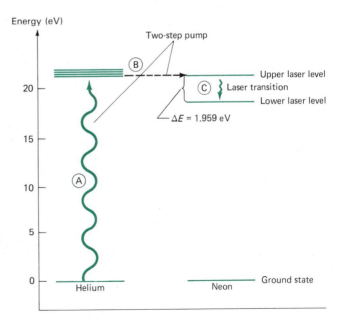

FIGURE 41.19

Steps involved in producing an inverted population and laser action in the helium-neon laser. Ⓐ An inelastic collision between an electron and a helium atom pumps the helium atom to an excited state. Ⓑ An inelastic collision between a helium atom in an excited state and a neon atom transfers energy to the neon atom, thereby pumping it to an excited state. Ⓒ The laser transition, in which a neon atom is stimulated to emit a photon.

action efficient. The energy transfer between helium and neon atoms is favored by the very small difference in their excitation energies. The small excess of energy appears as translational kinetic energy between the colliding atoms. This collisional energy transfer process is selective. The excitation energy of the helium atom is nearly 2 eV higher than the excitation energy of the lower level of the laser transition. This relatively large difference makes it very unlikely that an excited helium atom will transfer its energy to a neon atom and thereby populate the terminal state of the laser transition.

Of course, collisions between the swift electrons and neon atoms in the ground state will populate both laser levels. However, such collisions are not selective. They do not tend to produce a population inversion. The key to the pumping scheme is the tiny energy difference between the two excited states of helium and neon, which favors the upper laser level.

EXAMPLE 8

Helium-Neon Laser Wavelength

Using the energies shown in Figure 41.19 for the upper and lower laser levels, we can calculate the frequency and wavelength of the radiation. The frequency follows from the Bohr radiation condition, Equation 41.18:

$$h\nu = E_a - E_b$$

For the He-Ne laser, $E_a - E_b = 1.9593$ eV. Thus,

$$\nu = \frac{E_a - E_b}{h} = \frac{1.9593 \text{ eV}}{4.1357 \times 10^{-15} \text{ eV} \cdot \text{s}} = 4.7375 \times 10^{14} \text{ Hz}$$

The laser wavelength λ follows from $\nu\lambda = c$,

$$\lambda = \frac{c}{\nu} = \frac{2.9979 \times 10^8 \text{ m/s}}{4.7375 \times 10^{14} \text{ s}^{-1}} = 6.328 \times 10^{-7} \text{ m} = 632.8 \text{ nm}$$

This wavelength gives the laser light a fiery red color.

The laser output is highly directional. This stems from the fact that the amplitude of the light wave builds up via repeated trips through the active medium. Only waves that start out nearly parallel to the axis of the laser cavity are trapped and have the opportunity to grow in amplitude. Waves starting in other directions are refracted or reflected out of the active medium. The light that is transmitted is also diffracted as it passes through the exit mirror, and this causes a slight divergence of the laser beam. The high directionality of laser light was demonstrated spectacularly in an experiment in which laser light was beamed to the moon. The light was reflected back to earth by an array of reflectors left by the *Apollo 11* astronauts. Optical components expanded the laser beam to give it a diameter of 3 m when it left earth. After traveling nearly 400,000 km to the moon, its diameter was approximately 2 km. In traveling to the moon the beam diameter grew by only 1 m every 200 km.

The lunar reflectors have made it possible for scientists to measure the earth-to-moon distance accurately. These measurements, which have a precision of a few centimeters, confirm the prediction that the moon is slowly spiraling away from the earth because of the effects of tidal friction (Section 11.3). Measurements of continental drifts on earth have also been achieved by using laser signals between the moon and selected points on earth.

Laser Applications

The laser touches our lives in more ways than we realize. The denim in your blue jeans may have been cut by a laser beam. Your compact disc player uses a laser beam to help produce high fidelity sound. Powerful lasers are used to weld auto frames and anneal large metal pipes. Delicate lasers are used to remove tattoos and weld detached retinas. Low-power lasers are now used routinely to scan merchandise at the supermarket.

Laser research pushes frontiers in many directions—toward higher powers and shorter wavelengths. Sophisticated information storage and retrieval systems using lasers are under development. Multiple-laser systems operated in tandem are being tested as part of a broad effort aimed at the development of thermonuclear power reactors. New methods of pumping lasers are also being developed. The nuclear pumped laser converts the kinetic energy of fission fragments into the energy of excitation of laser atoms. A nuclear pumped laser would make it possible to beam power from one space vehicle to another. The laser seems destined to become an important aspect of future engineering systems, both on earth and in space.

41.7
SUPERCONDUCTIVITY AND SQUIDS

Quantum engineering has given us a class of ultrasensitive instruments called **squids**. The word *squid* is an acronym for **s**uperconducting **q**uantum **i**nterference **d**evice. Basically, a squid measures changes in magnetic flux, but it can be converted to an ammeter or voltmeter by using auxiliary circuitry.

The physics underlying squids is almost as old as quantum theory itself. We begin by sketching the early physics that culminated in the invention of the squid.

Superconductivity

In 1908 Heike Kamerlingh Onnes succeeded in liquefying helium by lowering its temperature to 4.2 K. The liquid helium bath furnished Onnes with a new frontier. He could immerse materials in the frigid liquid and study their properties at temperatures never before achieved. Onnes conducted systematic studies of the properties of different electrical conductors. In 1911 he discovered that mercury loses all electrical resistance below a temperature of 4.153 K. This phenomenon is called **superconductivity**. The temperature below which a material is superconducting is called the **transition temperature**. Since the initial discovery by Onnes, several thousand compounds and many elements have been found to exhibit superconductivity. Recent discoveries (1986), have led to the development of ceramic materials that become superconducting at transition temperatures above 90 K.

Meissner Effect

In 1933 Meissner and Ochsenfeld performed an experiment that revealed that a superconductor is not only a perfect conductor, but a perfect diamagnet as well (Section 33.2). They observed that when a tin cylinder has (a) been cooled below its superconducting transition temperature and then (b) been subjected to a magnetic field, the lines of magnetic flux do not penetrate the

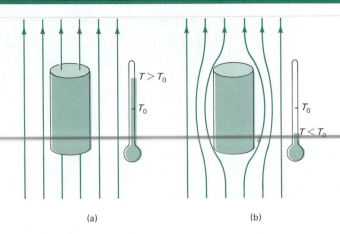

(a)   (b)

FIGURE 41.20

The Meissner effect. (a) Lines of magnetic flux penetrate the material when it is above its transition temperature. (b) Flux is expelled when the temperature falls below the transition temperature.

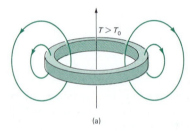

(a)

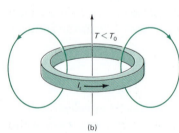

(b)

FIGURE 41.21

Flux trapping. (a) Flux produced by an external magnetic field penetrates a ring when the ring is above its transition temperature. (b) If the external field is removed when the ring is below its transition temperature, the flux threading the ring is trapped. Supercurrents induced in the ring maintain the trapped flux.

cylinder. This is the behavior expected for a perfect diamagnet. As the magnetic field is turned on, currents are induced that tend to prevent any change in the magnetic flux through the sample (Lenz's law). In the absence of any resistance, this current assumes whatever value is necessary to prevent flux penetration.

Meissner and Ochsenfeld then reversed the order of (a) and (b). The magnetic field was first applied to the sample in its normal (nonsuperconducting) state (Figure 41.20). Because the magnetic properties of the normal state are very weak, the magnetic flux lines fully penetrated the sample. The material was then cooled below its transition temperature. If zero resistivity were the only attribute of superconductivity, the flux threading the specimen would have remained unchanged as the transition temperature was passed. What Meissner and Ochsenfeld observed, however, was something quite different: they found that the magnetic flux was completely expelled from the sample as it became superconducting (Figure 41.20b). This perfect diamagnetism means that the magnetic field inside a superconductor is zero.*

If the solid cylinder is replaced by a ring-shaped superconductor, a phenomenon called *flux trapping* can be observed (Figure 41.21). When the ring is above its transition temperature, an external magnetic field causes flux to penetrate the ring. If the ring is cooled below its transition temperature and the external field is then removed, the flux threading the ring remains unchanged—the magnetic flux is trapped.

Cooper Pairs

A quantum theory of superconductivity was developed in 1957 by Bardeen, Cooper, and Schrieffer (Figure 41.22). The BCS theory shows that in a superconductor the motions of a small fraction of the electrons are strongly correlated.† A pair of correlated electrons is treated as a single entity and is called a **Cooper pair**.†† When a material becomes superconducting a small fraction

---

* The magnetic field does penetrate the surface layers to a depth of about 10^{-7} m.

† Quantum mechanically, it is the electron wave functions that are correlated.

†† In 1956, L. N. Cooper showed how such pairs might form in a superconductor.

FIGURE 41.22
John Bardeen, Leon N. Cooper, and J. Robert Schrieffer received the 1972 Nobel Prize in physics for their theory of superconductivity.

of its electrons condense, forming a group of Cooper pairs. In a superconductor it is the ordered drift motions of Cooper pairs that results in the supercurrent. The correlated motion of Cooper pairs proceeds without resistance, thereby shortcircuiting any normal electric current.

Magnetic Flux Quanta

In 1957 A. A. Abrikosov predicted the existence of magnetic flux quanta. The quantum of magnetic flux is called the **fluxon**. The fluxon (Φ_0) is the ratio of Planck's constant (h) and the magnitude of the electric charge of a Cooper pair ($2e$):

$$\Phi_0 = \frac{h}{2e} = 2.0678538 \times 10^{-15} \text{ T·m}^2 \qquad (41.30)$$

The existence of such flux quanta was verified experimentally in 1961 by Fairbank and Deaver and independently by Doll and Nabauer. More recently, the arrangement of flux tubes has been studied directly by using a novel photographic technique. The quantization of magnetic flux is a special property of superconductors. The magnetic flux threading an ordinary solenoid or transformer coil is not quantized.

Josephson Junctions

In Section 41.5 we saw that the wave properties of electrons enable them to penetrate potential energy barriers. Several types of electron tunneling can occur in solids. For example, electrons can tunnel through a very thin layer of insulating material separating two metals. A significant tunnel current can exist only when the de Broglie wavelength of the electron is comparable to or greater than the barrier thickness. In practice, this requires very thin barriers, generally with thicknesses less than 50 Å.

In 1962, an English graduate student, Brian Josephson (Figure 41.23), developed a theory describing the tunneling of supercurrents through a junction separating two superconductors. These tunneling junctions between two superconductors are called **Josephson junctions**. At a Josephson junction, Cooper pairs tunnel through the oxide layer separating the two superconductors.

FIGURE 41.23
Brian Josephson developed a theory describing the tunneling of supercurrents. Josephson shared the 1973 Nobel Prize for physics.

A novel feature of the tunneling supercurrent is that a steady supercurrent can exist without any potential difference across the junction. For a given junction area there is a maximum tunnel supercurrent (I_c) that can exist without setting up a potential difference across the junction. If the current through the junction exceeds I_c a potential difference appears across the junction. The tunneling is then due to individual electrons. In other words, if $I > I_c$, the Cooper pairs are broken.

The maximum supercurrent that can tunnel is strongly influenced by an applied magnetic field. More specifically, the supercurrent exhibits interference effects related to the quantization of magnetic flux.

Supercurrent Interference Effects

Josephson showed that a constant magnetic field, acting at right angles to the supercurrent through the junction, results in interference that limits the critical current. Specifically, the critical current has the form

$$I_c = I_0 \frac{\left| \sin\left(\frac{\pi\Phi}{\Phi_0} \right) \right|}{\frac{\pi\Phi}{\Phi_0}} \tag{41.31}$$

where Φ is the magnetic flux through the junction and Φ_0 is the fluxon, the quantum of magnetic flux. Figure 41.24 is a graph of I_c versus Φ/Φ_0. The graph shows that I_c is zero whenever the flux threading the junction is an integral number of fluxons. The mathematical form of I_c is identical to that of the magnitude of the electric field for single-slit diffraction. There are certain analogies between the optical and supercurrent interference effects. In both cases, interference results from the superposition of coherent waves having different phases. In optical interference, the waves are electromagnetic and the phase difference results because of differences in optical path lengths. In supercurrent interference, the waves are the de Broglie waves of Cooper pairs and the phase difference is a consequence of the applied magnetic field. In a typical junction, the area penetrated by the applied magnetic field is roughly

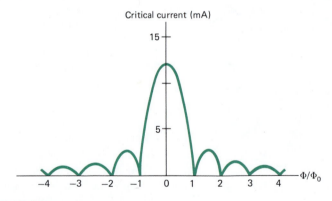

FIGURE 41.24

Critical current versus magnetic flux (in units of the flux quantum, Φ_0) for a Josephson junction. The quantum interference of the de Broglie waves of the Cooper pairs produces a pattern similar to the optical diffraction pattern for a single slit.

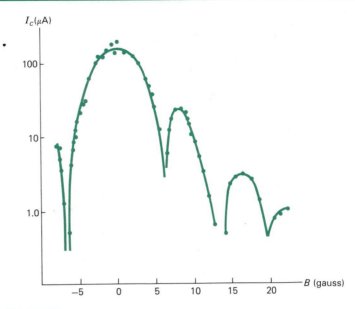

FIGURE 41.25

The single-slit interference pattern for the critical current in a Josephson junction operated at $T = 1.3$ K. [After J. M. Rowell, Phys. Rev. Lett. 11, 200 (1963)].

10^{-11} m^2. A change in the field of $\Delta B = 2 \times 10^{-4}$ T gives a flux change of

$$\Delta \Phi = \Delta B \cdot A = 2 \times 10^{-15} \text{ T·m}^2 \approx \Phi_0$$

The $|\sin(\pi\Phi/\Phi_0)|$ factor in Equation 41.31 shows that the critical current goes from one zero to the next each time the flux changes by Φ_0. If the magnetic field through the junction is varied by several gauss (1 gauss $= 10^{-4}$ tesla), the critical current should exhibit variations having the form described by Equation 41.31. In 1963 John Rowell studied the critical current through a lead/lead–oxide/lead junction operated at a temperature of 1.3 K. The critical current varied with the applied magnetic field strength in the manner displayed in Figure 41.25. Rowell's experiment was a striking confirmation of the quantum interference predicted by Josephson.

In 1964, a research group headed by James Mercereau performed the supercurrent analog of the double-slit interference experiment (Young's experiment). A supercurrent is divided (Figure 41.26) and sent along two different paths. The supercurrents tunnel through separate junctions and are reunited. Josephson's theory predicts that the phase difference between the reunited currents is directly proportional to the magnetic flux through the loop formed by the two supercurrent paths. This loop area, the area A in Figure 41.26, can be made much larger than the junction area. The change in magnetic field necessary to cause a flux change of one fluxon is therefore much smaller. Thus, a double-junction device has a greater sensitivity than one using a single junction. Areas on the order of 10^{-6} m^2 were typical of some of the early devices. An area of 10^{-6} m^2 requires a field change of approximately 2×10^{-9} T to cause the flux to change by one fluxon.* Sensitivities approaching 10^{-15} T are possible with more sophisticated devices.

---

* $\Delta B = \Phi_0/A \approx \dfrac{2 \times 10^{-15} \text{ T·m}^2}{10^{-6} \text{ m}^2} = 2 \times 10^{-9}$ T

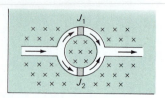

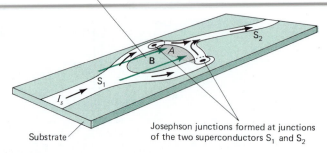

Applied magnetic
field is parallel
to substrate surface

Josephson junctions formed at junctions
of the two superconductors S_1 and S_2

Substrate

FIGURE 41.26

The double-slit interference of supercurrents. A supercurrent (I_s) is divided and sent along two different paths. The coherent supercurrents tunnel through Josephson junctions and exhibit interference when they reunite. An external magnetic field B is applied parallel to the surface of the glass substrate. The phase difference between the superposed supercurrents is proportional to the magnetic flux through the area A. A schematic representation of the double-junction device is shown at the upper right.

Observing the interference of supercurrents allows us to count flux quanta in a direct way and has led to the development of ultrasensitive measuring instruments called squids.

Squids

The heart of a squid is a *superconducting ring*, a single loop of superconducting material. The magnetic flux threading a superconducting ring is quantized. The total flux Φ must be an integral number of fluxons.

$$\Phi = n\Phi_0 \qquad n = \text{integer} \qquad (41.32)$$

In order for a superconducting ring to be a useful part of the squid, it must be able to exchange energy with other parts of the instrument. This is accomplished by introducing a **weak link** in the ring. A weak link is a region that has a much lower critical current than the rest of the superconducting ring. The weak link is also a region into which an applied magnetic field can penetrate. Figure 41.27 shows one way of producing a weak link.

Because of the constrictive nature of the weak link, the current density is greater in the link than in any other portion of the ring. If the current in the link exceeds the critical value, the link becomes normal—it is no longer superconducting. This break in the superconducting path around the ring allows fluxons to move through the link. The weak link thus acts as a gate through which fluxons enter and leave the ring. The weak link can be tailored to let a single fluxon pass through before reverting to the superconducting state. When a fluxon passes through the link, the link is resistive, and the potential difference across it can be measured with a voltmeter. This makes it possible to count individual fluxons with a squid.

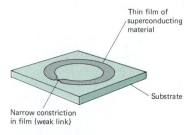

Thin film of
superconducting
material

Substrate

Narrow constriction
in film (weak link)

FIGURE 41.27

A thin circular film of superconducting material is deposited on a substrate. A narrow constriction in the film—the weak link—is created by scraping away portions of the film.

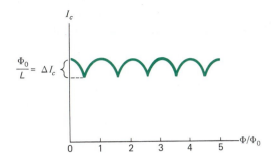

FIGURE 41.28

The critical current in the weak link varies periodically with the total flux through the squid area (area enclosed by the superconducting ring). The current executes one cycle each time the flux changes by one fluxon (Φ_0). The maximum variation in the critical current is given by $\Delta I_c = \Phi_0/L$. The potential difference across the link shows the same periodic variations as does I_c.

In practice, the weak link of the squid is operated in the *resistive mode*, in which the link is in the normal state while the rest of the ring is supercon- ducting. Flux can therefore pass *continuously* through the link. The counting of fluxons then depends on the periodic relationship between the flux and the critical current. The critical current repeats as the applied flux passes through the values $\Phi = 0$, Φ_0, $2\Phi_0$, $3\Phi_0$, ... (Figure 41.28). The potential difference across the link shows the same periodic behavior of the critical squid current and potential difference, and converts these into counts of fluxons. The po- tential difference across the link can be measured directly and used to drive an electronic counting circuit. Such a circuit is used to count the number of peri- odic variations in the squid current, each of which corresponds to a flux change of 1 fluxon (Φ_0). Counting rates of 2000 fluxons/s have been achieved with such digital squids. The squid current can be sensed by a nearby coil (Figure 41.29). Periodic variations in the squid current set up a changing magnetic field that produces similar variations in the flux threading the coil. The in- duced current and emf in the coil have the same period as the squid current. The emf induced in the coil acts to drive electronic counting circuitry.

As we have seen, the squid measures flux changes. The relationship between flux and magnetic field strength,

$$\Phi = BA \tag{41.33}$$

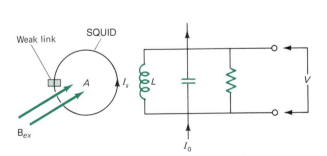

FIGURE 41.29

As the external field of (B_{ex}) is turned on, the flux (Φ) through the squid area (A) changes, causing periodic changes in the squid current, I_s. Variations in the squid current induce an emf in the adjacent coil. The induced emf (V) has the same flux-dependent period as the squid current.

makes it possible to convert flux measurements into B-field measurements. The sensitivity of squids is determined by the largest area (A) that can be used and the minimum flux change that can be detected. If $\Delta\Phi$ is the smallest detectable change in Φ, then the smallest change in the magnetic field (B) follows from Equation 41.33 as

$$\Delta B = \frac{\Delta\Phi}{A} \tag{41.34}$$

The smallest flux change that can be detected has steadily decreased with advances in the design of squids. Changes as small as $\Delta\Phi \approx 0.01\Phi_0$ have been detected. The largest area A that can be used is limited by the inductance (L) of the squid. The larger A, the larger the inductance L.* The largest inductance that can be used is approximately 10^{-8} H. This sets a limit for A of about 1 cm². The corresponding magnetic field sensitivity follows from Equation 41.34. With

$$\Delta\Phi = 0.01\Phi_0 = 2 \times 10^{-17}\ \text{T·m}^2$$
$$A = 10^{-4}\ \text{m}^2$$

we find

$$\Delta B = \frac{2 \times 10^{-17}\ \text{T·m}^2}{10^{-4}\ \text{m}^2} = 2 \times 10^{-13}\ \text{T}$$

The earth's B-field is on the order of 10^{-4} T. Fluctuations in the earth's field of about 10^{-9} T are caused by irregularities in the ionosphere. In order to achieve a sensitivity of $B \approx 10^{-13}$ T, it is essential that the squid be shielded from fluctuations in the earth's field.†

The Squid as an Ammeter

We can convert a squid into a sensitive ammeter by coupling it to a superconducting coil placed near the squid (Figure 41.30). A current I in the coil produces a field that tends to change the flux through the squid. The expression $LI = \Phi$ lets us relate the smallest detectable change in current (ΔI) and the minimum flux change that the squid can sense ($\Delta\Phi$):

$$\Delta I = \frac{\Delta\Phi}{L} \tag{41.35}$$

With

$$\Delta\Phi \approx 0.01\Phi_0 = 2 \times 10^{-17}\ \text{T·m}^2$$
$$L = 10^{-8}\ \text{H}$$

we find

$$\Delta I \approx 2 \times \frac{10^{-17}}{10^{-8}} = 2 \times 10^{-9}\ \text{A}$$

---

* Remember, inductance measures "electrical inertia." The larger the squid, the larger its electrical inertia.

† Very elaborate shielding from stray electric fields is also required. The first generation of squids was very fragile; even a slight electrostatic spark between an experimenter's fingertip and the apparatus could burn out the weak link.

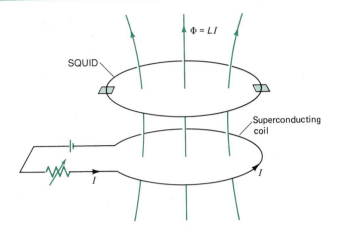

FIGURE 41.30

The current I in the superconducting coil produces a flux $\Phi = LI$ that threads the coil and the squid. Changes in the current cause changes in the flux. The smallest detectable change in current is related to the smallest detectable change in flux.

Squids have revolutionized the technology of electrical and magnetic measurements. Geologists are adapting quantum magnetometers to measurements of rock magnetism and continental drift. Applications of squids to the fields of biology, medicine, and psychology are being pursued actively. The heart and the brain generate tiny electric currents, which in turn set up weak magnetic fields. Magnetic fields of about 10^{-14} T are generated by the human heart, and the human brain generates magnetic fields of about 10^{-15} T. Squids are able to detect even these feeble fields. Squids may soon supplement the equipment now used to make electrocardiograms (EKG's) and electroencephalograms (EEG's).

The squid owes its existence to the quantization of magnetic flux in a superconducting ring. It is a product of quantum engineering, the branch of engineering that also produced the transistor and the laser. It is safe to predict that the squid will find many applications that promote our well-being.

Ceramic Superconductors

Prior to 1986, most superconducting materials were elements or alloys, and the highest transition temperature was less than 24 K. Virtually all superconductivity research required cooling materials with liquid helium, a relatively expensive refrigerant.

In 1986, studies begun by K. A. Mueller and J. G. Bednorz led to the development of ceramic materials that become superconducting at transition temperatures above 90 K. The discovery of these high temperature superconductors has spurred worldwide efforts by thousands of scientists, intent on pushing transition temperatures to the room temperature range. Nitrogen liquefies at 77 K, making it possible to use this relatively inexpensive refrigerant (it costs less than beer) to maintain materials in the superconducting state. The development of room temperature superconductors would revolutionize the transmission of electrical energy by eliminating the wasteful generation of heat.

The chemical makeup of the ceramic superconductors can be described by the chemical formula $RBa_2Cu_3O_{7-x}$. The letter "R" denotes yttrium or one of the rare earth elements, such as lanthanum, and x is a number less than one whose value depends on the way in which the material is prepared. The crystal structure and the positions of the oxygen atoms strongly affect the transition temperature. The superconducting paths are believed to be along linear chains of copper and oxygen atoms.

Ceramic superconductors can carry much larger current densities than conventional superconductors. This may make it possible to use ceramic superconductors to fabricate powerful electromagnets.

WORKED PROBLEM

A photon with an energy of 50 keV is scattered by an electron at rest. Determine the kinetic energy of the recoiling electron if the photon scatters through an angle of 135°.

Solution

We use the notation of Section 41.1 and Figure 41.5. The electron is at rest before the collision, so its kinetic energy after the collision (K) equals the change in photon energy ($E_0 - E$),

$$K = E_0 - E$$

We can use the Compton formula, Equation 41.15, and $E = h\nu = hc/\lambda$, to express the energy of the scattered photon in terms of the angle of scatter (θ) and the photon energy before the collision. Thus, from Equation 41.15 we have,

$$\lambda - \lambda_0 = \frac{h}{mc}(1 - \cos\theta)$$

The energy of the scattered photon is,

$$E = \frac{hc}{\lambda} = \frac{hc}{\lambda_0 + \dfrac{h}{mc}(1 - \cos\theta)}$$

$$= \frac{\dfrac{hc}{\lambda_0}}{1 + \dfrac{hc}{\lambda_0 mc^2}(1 - \cos\theta)}$$

$$= \frac{E_0}{1 + \dfrac{E_0}{mc^2}(1 - \cos\theta)}$$

With $\theta = 135°$, $E_0 = 50$ keV, and $mc^2 = 511$ keV, we find

$$E = \frac{50 \text{ keV}}{1 + \dfrac{50 \text{ keV}}{511 \text{ keV}}(1 + 0.707)}$$

$$= 42.8 \text{ keV}$$

The kinetic energy of the recoiling electron is

$$K = E_0 - E = 50 \text{ keV} - 42.8 \text{ keV}$$

$$= 7.2 \text{ keV}$$

EXERCISES

41.1   The Origins of Quantum Physics

A. The work function of platinum is 6.30 eV. Determine the photoelectric threshold frequency and the corresponding wavelength.

B. The photoelectric threshold frequency for gold is 1.16×10^{15} Hz. Determine the work function for gold. Express your results in electron volts.

C. What is the wavelength associated with a photon with an energy of 1 eV?

D. Our atmospheric ozone layer protects us from ultraviolet (UV) photons emitted by our sun. Why might UV photons be harmful to human beings, whereas visible light photons are not?

E. Compton used gamma radiation with a wavelength of 0.022 Å to study the Compton scattering by electrons in iron, aluminum, and paraffin. He observed the scattered gamma photons at angles of $\theta = 45°$, 90°, and 135°. The observed wavelengths are shown in the table. Complete the table and comment on the agreement between theory and experiment.

θ	Observed Wavelength	Theoretical Wavelength
45°	0.030 Å	
90°	0.043 Å	
135°	0.068 Å	

41.2   Rutherford and the Nuclear Atom

F. A bowling ball weighs 16 pounds. A copper pellet weighs 0.012 oz. Compare the mass ratio of a bowling ball to a copper pellet with the mass ratio of an alpha particle to an electron. Is the mass ratio for alpha particle scatter by an electron comparable to the scatter of a bowling ball by a copper pellet?

G. An alpha particle with a kinetic energy of 8.2 MeV collides head-on with a uranium nucleus (which carries a charge 92 times that of the proton). (a) Assume that the uranium nucleus does not recoil, and determine the distance of closest approach (R) between the alpha and the nucleus. (b) Would the calculated value of R increase or decrease if you took recoil into account?

H. Use dimensional considerations to find a combination of h, c, and the neutron mass m, that has units of length. Compare this length to the characteristic nuclear length, 10^{-15} m.

41.3   Bohr and the Hydrogen Atom

I. Two hydrogen atoms collide head-on and end up with zero kinetic energy. Each then emits a photon with a wavelength of 121.6 nm ($n = 2$ to $n = 1$ transition). At what speed were the atoms moving before the collision?

J. Calculate the energy of a Lyman alpha photon. Express your result in electron volts.

K. A hydrogen atom undergoes a transition from the $n = 105$ level to the $n = 104$ level. Show that the wavelength of the emitted radiation is a few centimeters. (This places it in the radio-frequency region of the electromagnetic spectrum.)

L. The three energy levels for a hypothetical atom are shown in Figure 1. Identify all possible transitions that result in the emission of light. Which transition produces the longest wavelength? If the atom is in the $n = 2$ level, how many different wavelengths can it absorb?

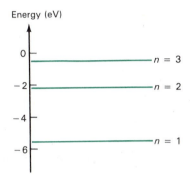

FIGURE 1

M. Liquid oxygen has a bluish color. This means that the liquid preferentially absorbs light with wavelengths toward the red end of the visible spectrum. Although the oxygen molecule (O_2) does not strongly absorb radiation in the visible spectrum it does absorb strongly at a wavelength of 1269 nm, which is in the infrared region of the spectrum. Research has shown that it is possible for *two* colliding O_2 molecules to absorb a *single* photon, sharing its energy equally. The transition that both molecules undergo is the same transition that results when they absorb radiation with a wavelength of 1269 nm. What is the wavelength of the single photon that causes this double transition? What is the corresponding color?

41.4   De Broglie and the Wave-Particle Duality

N. The neutron has a mass of 1.67×10^{-27} kg. Neutrons emitted in nuclear reactions can be slowed down via collisions with matter. They are referred to as thermal neutrons once they come into thermal equilibrium with their surroundings. The average kinetic energy ($3kT/2$) of a thermal neutron is approximately 0.04 eV. Calculate the de Broglie wavelength of a neutron with a kinetic energy of 0.04 eV. How does it compare with the characteristic atomic spacing in a crystal? Would you expect thermal neutrons to exhibit diffraction effects when scattered by a crystal?

P. Take the electron de Broglie wavelength of 1.65×10^{-10} m measured by Davisson and Germer to be correct. Calculate the kinetic energy of the electrons (in eV). If the applied potential difference is 54.0 V, what potential difference is contributed by the crystal?

Q. If an electron and a photon have the same kinetic energy, which has the shorter de Broglie wavelength?

R. If an electron and a photon have the same de Broglie wavelength, which has the greater kinetic energy?

41.5 Schrödinger's Equation and Probability Waves

S. Figure 2 shows the Schrödinger wave function, ψ, for a particle confined between the points $x = 0$ and $x = a$. About what point is the particle most probably located?

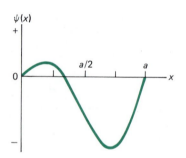

FIGURE 2

T. A particle starts inside the potential energy well shown in Figure 3. Its total energy is E. (a) Describe its motion on the basis of classical physics. (b) From a quantum viewpoint, which side of the well is it most likely to tunnel through? (Make a guess.) (c) On which side of the well would it move with the greater speed?

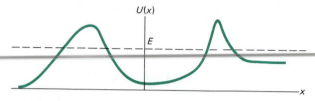

FIGURE 3

41.6 Quantum Optics and Lasers

U. The helium-neon system is capable of lasing at several different infrared wavelengths. The most prominent infrared wavelength is 3.3913 μm. Determine the energy difference (in electron volts) between the upper and lower levels for this wavelength.

V. The carbon dioxide (CO_2) laser is one of the most powerful lasers developed. The energy difference between the two laser levels is 0.117 eV. Determine the frequency and wavelength of the radiation. In what portion of the electromagnetic spectrum is this radiation?

41.7 Superconductivity and Squids

W. The quantum of magnetic flux, the fluxon, is defined by

$$\Phi_0 = \frac{h}{2e}$$

Show that h/e has the units of magnetic flux ($T \cdot m^2$).

PROBLEMS

1. The following data were obtained in a low-precision photoelectric experiment.

Wavelength	Stopping Potential
$\lambda_1 = 552$ nm	$V_1 = 1.8$ V
$\lambda_2 = 410$ nm	$V_2 = 2.8$ V

(a) Show that Planck's constant is given in terms of these data by

$$h = \frac{e}{c}(V_2 - V_1)\frac{\lambda_1 \lambda_2}{\lambda_1 - \lambda_2}$$

(b) Evaluate h and compare it with the more precise value 6.62×10^{-34} J·s.

2. A gold atom has a radius of 2.6×10^{-10} m. The gold nucleus has a radius of 8.6×10^{-15} m. (a) What fraction of the total cross-sectional area of the atom is blocked by the nuclear cross-sectional area? The gold film used by Geiger and Marsden was approximately 2000 atoms thick. (b) Estimate the fraction of the incident alpha particles that suffered a head-on collision with a gold nucleus.

3. Estimate the area of a ring that would fit one of your fingers, and calculate the magnetic flux through the ring due to the earth's magnetic field (take $B_{earth} = 5.8 \times 10^{-5}$ T). If this flux were quantized, how many fluxons would the ring enclose?

4. A squid operating as an ammeter can detect a flux change of $0.03\Phi_0$. The inductance of the superconducting coil is 0.06 μH. What is the minimum change in current that the ammeter can detect?

5. Water is pumped from a reservoir G to a tank A at a rate of 600 liters/min (Figure 4). The water leaks out of A through holes. One hole returns water to G at

a rate of $6N_A$ liters/min, where N_A is the number of liters in A. The other hole leaks water into tank B at a rate of $54N_A$ liters/min. Water leaks from B to C at a rate of $2N_B$ liters/min and from C to G at a rate of $30N_C$ liters/min. Tanks A, B, and C are all empty when pumping begins. (a) Determine the steady-state content of tank A (the number of liters in A when it reaches a constant—steady—value). (b) Determine the steady-state contents of B and C. (c) Draw a diagram showing the steady-state rate of transfer into and out of A, B, C, and G.

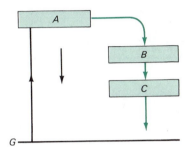

FIGURE 4

6. (a) Show that Balmer's empirical formula (Equation 41.17) can be written in the form

$$\frac{1}{\lambda} = a\left(\frac{1}{2^2} - \frac{1}{n^2}\right)$$

where a is a constant. (b) Wavelengths corresponding to $n = 3, 4, 5$, and 6 are presented in Figure 41.11. Using these data, make a plot of $1/\lambda$ versus $1/n^2$ and deduce the value of a. Compare your value of a with that derived from Balmer's empirical formula.

7. Compare the de Broglie wavelengths of a 30-eV photon and a 30-eV electron, and comment on which would provide a more appropriate probe of crystal structure.

8. (a) Using data presented in Figure 41.10, derive a formula for the Lyman series similar to the formula presented for the Balmer series (Equation 41.26). (b) Compare predictions from your formula with the data presented for the Lyman series.

9. Consider a singly ionized helium ion in its ground state. Using the Bohr theory calculate the minimum energy of a photon that can be absorbed by the ion.

10. Consider an electron with linear momentum of magnitude p moving in a circular orbit of radius r about a proton. Prove that the Bohr quantization rule for angular momentum is satisfied if the de Broglie wavelength equals an integral multiple of the circumference of the electron orbit.

11. (a) Show that the classical frequency of motion of an electron in the nth Bohr orbit of a hydrogen atom is given by $v_c = \left(\dfrac{me^4}{4h^3\epsilon_0^2}\right)(1/n^3)$. (b) Using the Bohr theory show that, for $n \gg 1$, the classical frequency v_c equals the frequency of the radiation emitted when the atom makes a transition from the nth state to the $(n - 1)^{th}$ state.

12. Positronium is a hydrogen-like system consisting of an electron and a positron. Apply Bohr's theory to determine the ground state energy and compare your result with Equation 41.24. (You may wish to start with the fact that the particles rotate about their center of mass.)

13. The negative hydrogen ion (H^-) consists of a proton and two electrons. It is an important constituent in the atmosphere of the sun. Develop a Bohr theory for the H^- ion. Assume that the electrons move in the same orbit, diametrically opposite one another. Further, assume that the angular momentum of each electron is quantized according to $mvr = n(h/2\pi)$; $n = 1, 2, 3, \ldots$. (a) Derive an equation for the quantized energies. (b) Calculate the binding energy of the ground state. This is defined to be the difference between the ground state energy of the H^- ion and the ground state energy of a neutral hydrogen atom. Compare your answer with the experimental value of 0.75 eV. (c) Show that the ground state is the only bound state of the H^- ion.

14. A gamma photon with a wavelength of 0.0110 Å suffers a head-on collision with an electron at rest. The photon scatters through 180°. Determine (a) the energy of the incident photon in MeV, (b) the wavelength of the scattered photon, (c) the kinetic energy of the recoiling electron in MeV.

15. How many head-on Compton scatters are required to double the wavelength of a photon with an initial wavelength of 2.60×10^{-11} m? (The photon scatters through 180° in a head-on collision.)

16. A photon with an energy of 100 keV is scattered by an electron at rest. The photon has an energy of 87 keV after the collision. Determine the angle through which the photon scatters.

17. A gamma photon with an energy of 0.511 MeV strikes an electron at rest and scatters through 90°. Determine its wavelength (a) before, and (b) after being scattered.

18. Photons carry linear momentum and so they exert a force when they are absorbed or reflected by a surface. A solar sail presents an absorbing area of 6×10^5 m² to the sun. The sail material has a mass per unit area of 0.10 kg/m². The intensity of the sunlight striking the sail is 1.38 kW/m². (a) Estimate the acceleration of the sail, assuming it *absorbs* all photons striking it. (b) Starting from rest, how many years would be required for it to reach a speed of one-tenth the speed of light, assuming the constant acceleration of part (a)?

CHAPTER 42

NUCLEAR STRUCTURE AND NUCLEAR TECHNOLOGY

42.1
THE NEUTRON-PROTON MODEL OF THE NUCLEUS

James Chadwick discovered the neutron in 1932 while studying the bombardment of beryllium with alpha particles. The neutron is electrically neutral and has a mass slightly greater than that of the proton. In atomic mass units* the neutron and proton masses are

$$m_n = 1.008665 \text{ u}$$

$$m_p = 1.007277 \text{ u}$$

Also in 1932, John Cockcroft and Ernest Walton used their newly constructed proton accelerator to bombard the element lithium and confirmed the Einstein mass-energy relation, $\Delta E = \Delta mc^2$. They accounted for the kinetic energy of the alpha particles produced in the proton-lithium interactions by assuming that some of the mass of the reactants was converted to energy according to $\Delta E = \Delta mc^2$. Guided by these experimental results, Werner Heisenberg introduced the **neutron-proton model of the nucleus**.

According to this nuclear model, neutrons and protons are bound together by a **strong nuclear force**, the strongest of the four fundamental forces of nature. This force is called a **short-range** force because it essentially vanishes if the neutrons and protons are separated by distances greater than about 3×10^{-15} m. This feature is in sharp contrast to the **long-range** gravitational and electrical forces, whose influence extends to infinity.

Protons in the nucleus account for the nuclear charge. We call the number of protons in the nucleus the **atomic number** and give it the symbol Z. The number of neutrons is denoted by N. The total number of protons and neutrons, $A = Z + N$, accounts for the nuclear mass, and is called the **mass number**. A neutral atom has Z extranuclear electrons and Z protons in the nucleus. It is customary to attach the mass number A as a superscript to the chemical symbol that identifies the atom. For emphasis, the atomic number is sometimes designated as a subscript. Thus, $^{235}_{92}U$ identifies a uranium atom with a nucleus containing

$A = 235$ neutrons and protons
$Z = 92$ protons, corresponding to the element uranium
$N = A - Z = 143$ neutrons

Atoms having the same atomic number but a different neutron number are called **isotopes**. Isotopes form a family of atoms sharing the same chemical symbol. For example, the three most abundant isotopes of uranium are ^{234}U, ^{235}U, and ^{238}U. These isotopes contain 142, 143, and 146 neutrons.

Unless stated otherwise, the notation $^A X$ denotes the neutral atom of the element having the chemical symbol X and mass number A. On occasion we are interested in the properties of bare nuclei, that is, atoms with the electrons removed. In cases where the distinction between atoms and bare nuclei is important we indicate explicitly that we are dealing with bare nuclei.

When designating masses, we use a lowercase letter to denote the mass of a bare nucleus and an uppercase letter to indicate the mass of a neutral atom.

---

* 1 atomic mass unit = 1 u = 1.660531×10^{-27} kg; it is one-twelfth of the mass of a neutral ^{12}C atom.

For example, m_N symbolizes the mass of a bare nitrogen nucleus; M_N denotes the mass of a neutral nitrogen atom. The relative difference in the mass of a bare nucleus and a neutral atom is small. For nitrogen, $(M_N - m_N)/M_N = 0.0003$.

Energy Conservation

The equivalence of mass and energy and the conservation of energy play central roles in understanding how neutrons and protons are bound together in a nucleus. To illustrate, deuterium (^{2}H) is a hydrogen isotope with one proton and one neutron. When a neutron and a proton meet and interact, they can produce a deuterium nucleus, called a deuteron, and a photon having 2.23 MeV of energy. The process can be represented as

$$n + p \rightarrow d + \gamma \tag{42.1}$$

The energy of the gamma photon (γ) arises from the conversion of some of the mass of the neutron and proton. Conservation of energy requires the deuteron mass to be less than the combined masses of the proton and neutron.

A deuteron can be separated into a proton and neutron by adding energy to the deuteron to make up the mass difference. One way of adding the required energy is to bombard deuterium nuclei with gamma photons. This is the inverse of the reaction described by Equation 42.1,

$$\gamma + d \rightarrow p + n \tag{42.2}$$

Both the formation of a deuteron according to Equation 42.1 and the photodisintegration of a deuteron according to Equation 42.2 are routine events in many physics laboratories.

EXAMPLE 1

Conservation of Energy in a Nuclear Reaction

A photon is emitted in the capture of a neutron by a proton to form a deuteron. Using conservation of energy at the nuclear level we can calculate the photon energy. The reaction is

$$n + p \rightarrow d + \gamma$$

The kinetic energies of the neutron, proton, and deuteron are ignorably small. Hence only the mass energy of these particles need be considered. The energy of the gamma photon is provided by the difference between the mass energies of the reactants (proton + neutron) and the product deuteron,

$$\Delta E = \Delta m \, c^2 = (m_n + m_p - m_d)c^2$$

The mass of the deuteron is

$$m_d = 2.013553 \text{ u}$$

The mass energy equivalent of 1 atomic mass unit ($1 \text{ u} \cdot c^2$) is 931.48 MeV, so the photon energy is

$$E = (1.008665 + 1.007277 - 2.013553) \text{ u} \cdot (931.48 \text{ MeV/u})$$
$$= 2.23 \text{ MeV}$$

Measurement of the photon energy confirms the conservation of energy principle. Once the conservation principle is established, it can be used to deduce nuclear masses from measurements of photon energies. For example, if the neutron mass were unknown, then a knowledge of the masses of the proton and deuteron and a measurement of the photon energy would permit a determination of the mass of the neutron.

Nuclear Binding Energy

The attractive forces that bind neutrons and protons in a nucleus perform work when a nucleus is formed. These forces can be overcome by doing work against the attractive forces. We speak of the **binding energy** that must be added to a nucleus to separate it into its component neutrons and protons.

The binding energy (*BE*) of a nucleus of mass m containing Z protons and N neutrons is defined as

$$BE = (mc^2)_{\text{pieces}} - (mc^2)_{\text{nucleus}} = Zm_pc^2 + Nm_nc^2 - mc^2 \qquad (42.3)$$

In practice it is easier to measure the mass of the neutral atom than it is to measure the mass of the bare nucleus. Equation 42.3 is therefore converted to atomic masses by adding and subtracting the mass energy of Z electrons. This gives

$$BE = Zm_pc^2 + Zm_ec^2 + Nm_nc^2 - (mc^2 + Zm_ec^2)$$

which reduces to

$$BE = ZM_Hc^2 + Nm_nc^2 - Mc^2 \qquad (42.4)$$

where $M_H \approx m_p + m_e$ is the mass of the hydrogen atom and $M \approx m + Zm_e$ is the mass of the atom. This final expression is approximate because it neglects the binding energies of the atomic electrons (the energies required to remove the electrons from the atom). Typically, atomic binding energies are small compared to the nuclear energies, and Equation 42.4 is sufficiently accurate. For example, the atomic binding energy of deuterium is 13.6 eV, whereas the nuclear binding energy for the deuteron is 2.23 MeV.

The binding energy increases systematically as the number of **nucleons** (neutrons and protons) in a nucleus increases. But, if we divide the binding energy by the number of nucleons to get the average binding energy per nucleon, the result is approximately the same for most nuclei. Figure 42.1 shows that the binding energy per nucleon is approximately 8 MeV/nucleon.

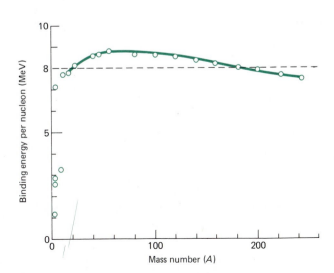

FIGURE 42.1

The binding energy per nucleon for representative nuclei, plotted versus mass number. The binding energy per nucleon has a broad maximum of 8.7 MeV/nucleon for various isotopes of iron, nickel, and cobalt.

The average binding energy per nucleon is approximately 10^5 times the average atomic binding energy per electron. This huge ratio reflects the enormous difference in strength of the nuclear force and the electrical force.

The binding energy per nucleon reaches a maximum of approximately 8.7 MeV per nucleon (Figure 42.1). For large and small mass numbers the binding energy per nucleon is significantly less than the nominal value of 8 MeV per nucleon. This is an important consideration in nuclear technology. A conventional nuclear reactor derives energy from **nuclear fission**, by splitting nuclei having mass numbers around 235. The binding energy per nucleon of the fragments is larger than that of the original nucleus. The energy difference is given to the fragments in the form of kinetic energy. In other words, there is a conversion of mass energy (binding energy) into kinetic energy. From Figure 42.1 we see that the difference in binding energy per nucleon for $A = 235$ and $A = 120$ is about 1 MeV per nucleon. Hence the splitting of a single nucleus with $A = 235$ liberates about 235 MeV of energy. This energy appears as the kinetic energy of the fission fragments.

By comparison, a chemical reaction generally liberates only a few electron volts of energy. This million-to-one ratio explains the appeal of nuclear energy. "Burning" a few pounds of atomic nuclei in a fission reactor releases the same energy as the burning of thousands of tons of coal.

At the other end of the scale, it is energetically favorable for nuclei with low mass numbers to combine, or *fuse*, forming a more massive nucleus. The product nucleus has a greater binding energy per nucleon than the reacting nuclei. The difference in binding energies is liberated as kinetic energy of the products. This process is called **nuclear fusion**. The energy released in the core of the sun and other stars is the result of nuclear fusion reactions. On the earth, fusion has yet to be developed as a practical energy source, but as we will see in Section 42.5, there is intense interest in controlled fusion reactions.

EXAMPLE 2

Binding Energy of ^{235}U

The uranium used in contemporary nuclear reactors is ^{235}U. To see how binding energies are computed, let's determine the binding energy of ^{235}U.

There are 92 protons and 143 neutrons in a ^{235}U nucleus. Hence the binding energy as determined from Equation 42.4 is

$$BE = 92M_H c^2 + 143 m_n c^2 - M_U c^2$$

Using Table 42.1 we find

$$BE = [92(1.007825) + 143(1.008665) - 235.043945]\, u \cdot (931.48\ \text{MeV/u})$$

$$= 1780\ \text{MeV}$$

This is the minimum energy that must be given to a ^{235}U nucleus to separate it into 92 protons and 143 neutrons. The binding energy per nucleon is

$$\frac{BE}{A} = \frac{1780\ \text{MeV}}{235\ \text{nucleons}}$$

$$= 7.59\ \text{MeV/nucleon}$$

The fact that the binding energy per nucleon is less than the binding energy per nucleon of the fission fragments accounts for the energy release in nuclear fission.

TABLE 42.1
Selected Atomic Masses (Neutral Atoms) in Atomic Mass Units*

Isotope	Z	Mass	Isotope	Z	Mass	Isotope	Z	Mass
n	0	1.008665	^{16}O	8	15.994915	^{210}Po	84	209.982883
^{1}H	1	1.007825	^{17}O	8	16.999133	^{234}Th	90	234.043635
^{2}H	1	2.014103	^{27}Al	13	26.981540	^{233}U	92	233.039654
^{3}H	1	3.016050	^{40}Ar	18	39.962383	^{234}U	92	234.040976
^{3}He	2	3.016030	^{40}K	19	39.963999	^{235}U	92	235.043945
^{4}He	2	4.002603	^{40}Ca	20	39.962591	^{238}U	92	238.050817
^{6}Li	3	6.015124	^{60}Co	27	59.933809	^{239}U	92	239.054326
^{7}Li	3	7.016005	^{60}Ni	28	59.930778	^{239}Np	93	239.052954
^{7}Be	4	7.016930	^{75}As	33	74.921598	^{238}Pu	94	238.049583
^{8}Be	4	8.005305	^{90}Sr	38	89.907751	^{239}Pu	94	239.052177
^{9}Be	4	9.012183	^{91}Kr	36	90.923241	^{240}Pu	94	240.053828
^{11}B	5	11.009306	^{95}Sr	38	94.918903	^{252}Cf	98	252.081654
^{12}B	5	12.014354	^{137}Cs	55	136.907072			
^{11}C	6	11.011434	^{137}Ba	56	136.905812			
^{12}C	6	12.000000	^{140}Ba	56	139.910636			
^{13}C	6	13.003355	^{151}Nd	60	150.923886			
^{14}C	6	14.003242	^{197}Au	79	196.966547			
^{14}N	7	14.003074	^{210}Bi	83	209.984130			

* A complete list may be found in *Nuclear Data Tables, 9*, 267 (1971), by A. H. Wapstra and N. B. Gove.

Nuclear Size

As a first approximation we view a nucleus as a sphere of constant density. In this model the volume of the nucleus is proportional to the number of nucleons,

$$V \propto A$$

Introducing the radius R, it follows that

$$R^3 \propto A$$

We write

$$R = aA^{1/3}$$

where a is a proportionality constant with the dimension of length. Experiments reveal that the length a has the value

$$a = 1.2 \times 10^{-15} \text{ m} = 1.2 \text{ fm}$$

The femtometer, also called the **fermi**, is the characteristic length associated with nuclei. With this notation the nuclear radius can be expressed as

$$R = 1.2A^{1/3} \text{ fm} \tag{42.5}$$

Because the nuclear radius is proportional to the cube root of the mass number A, the difference between the radii of light and heavy nuclei is fairly small. The mass number of $^{239}_{94}Pu$ is about 34 times the mass number of $^{7}_{3}Li$, but their radii differ only by a factor of about 3.

EXAMPLE 3

Neutron Star Radius

Under certain circumstances the outer layers of a dying star can explode to produce a supernova. A spectacular supernova appeared in 1987. The accompanying implosion of the core of the star may result in a neutron star. A neutron star is essentially a giant nucleus, composed almost entirely of neutrons. Let's use Equation 42.5 to estimate the radius of a neutron star whose mass is 7.96×10^{30} kg, four times the mass of the sun.

The volume of the star is $(4\pi/3)r_*^{\ 3}$, where r_* is its radius. The number of neutrons in the star is

$$N = \frac{\text{mass of star}}{\text{mass of neutron}} = \frac{7.96 \times 10^{30} \text{ kg}}{1.67 \times 10^{-27} \text{ kg}}$$

$$= 4.77 \times 10^{57}$$

The volume occupied by one neutron is $(4\pi/3)R^3$, where R is given by Equation 42.5. For the neutron, $A = 1$ and $R = 1.2$ fm. The volume of N neutrons is the volume of the star. Thus,

$$\left(\frac{4\pi}{3}\right)r_*^{\ 3} = \left(\frac{4\pi}{3}\right)R^3 \cdot N$$

The neutron star radius is

$$r_* = R \cdot N^{1/3} = 1.2 \times 10^{-15} \text{ m} \cdot (4.77 \times 10^{57})^{1/3}$$

$$\approx 2 \times 10^4 \text{ m}$$

The radius of the neutron star is approximately 20 km, or about 12 miles. Several neutron stars have been detected through their associated **pulsar** activity.

42.2
NUCLEAR STABILITY

Atoms and nuclei occur in stable and unstable configurations. The light produced by neon signs and lasers is a result of photon emission by energetically unstable atoms. Unstable nuclei can change to more stable configurations by emitting photons or particles that carry away energy. Both stable and unstable nuclei play important roles in nuclear physics (Figure 42.2). We begin by examining the relationship between the numbers of neutrons and protons in stable nuclei.

We find empirically that light stable nuclei tend to have nearly equal numbers of neutrons and protons. Helium has two stable isotopes: ^{4}He and ^{3}He. Oxygen has three stable isotopes: ^{16}O, ^{17}O, and ^{18}O. These nuclei are characterized by this feature of $Z \approx N$. But as the mass number increases, stable nuclei have an increasing excess of neutrons (Figure 42.3). In the heaviest stable nucleus $^{209}_{83}$Bi the neutron excess is $N - Z = 43$.

A proton or neutron in a nucleus is attracted to another proton or neutron by the strong nuclear force. Experiment shows that the strong nuclear force between two neutrons is essentially the same as the nuclear force between two protons or between a neutron and a proton. The neutron excess in nuclei reduces the destabilizing effect of repulsive electrical forces between protons.

FIGURE 42.2 — Chart of Nuclides

Z \ N	0	1	2	3	4	5	6	7	8	9	10
6				^{9}C 0.1265 s, e$^+$, EC, E 16.50	^{10}C 19.2 s, e$^+$ 1.865, 0.72, 1.02, E 3.651	^{11}C 20.38 m, e$^+$ .961, no γ, E 1.982	^{12}C stable	^{13}C stable	^{14}C 5730 y, e$^-$ .1565, no γ, E .1565	^{15}C 2.449 s, e$^-$ 9.82,4.51, γ 5.30, E 9.772	^{16}C 0.75 s, e$^-$, E 8.011
5				^{8}B 0.769 s, e$^+$, E 17.98	^{9}B 0.85 x 10^{-18} s	^{10}B stable	^{11}B stable	^{12}B 0.0204 s, e$^-$ 13.4,8.9, γ 4.43, E 13.37	^{13}B 0.0174, e$^-$ 13.4, γ 3.68, E 13.44	^{14}B 0.016 s, e$^-$, E 20.64	
4			^{6}Be	^{7}Be 53.3 d, EC, γ .478, E .862	^{8}Be 7 x 10^{-17} s, 2α, E 0.0919	^{9}Be stable	^{10}Be 1.6 x 10^6 y, e$^-$ .556, no γ, E .556	^{11}Be 13.8s, e$^-$ 11.5,9.3, E 11.51	^{12}Be 0.0114 s, e$^-$, E 11.66		
3			^{5}Li	^{6}Li stable	^{7}Li stable	^{8}Li 0.84 s, e$^-$ 13,12.5, E 16.01	^{9}Li 0.178 s, e$^-$ 13.5,11.0, E 13.61		^{11}Li 0.0085 s, e$^-$, E 20.76		
2		^{3}He stable	^{4}He stable	^{5}He	^{6}He 0.808 s, e$^-$ 3.51, no γ, E 3.51		^{8}He 0.122 s, e$^-$, E 10.66				
1	^{1}H stable	^{2}H stable	^{3}H 12.33 y, e$^-$ 0.0186, no γ, E .0186								
0		n 10.6 m, e$^-$ .782, E .782									

Z (atomic number) ↑

N (neutron number) → 1 2 3 4 5 6 7 8 9 10

FIGURE 42.2

A portion of a chart of nuclides. This chart summarizes the properties of nuclei and is very useful for following pictorially the disintegration of unstable nuclei. The entire chart includes over 1000 nuclei having atomic number (Z) ranging from 0 to 105.

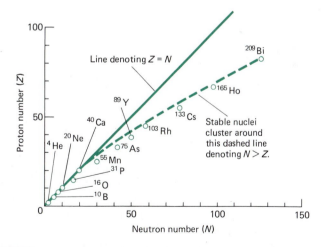

FIGURE 42.3

Plot of the proton and neutron numbers of selected stable nuclei. Notice that the neutron number exceeds the proton number for all but the lightest nuclei.

One measure of the destabilizing effect of electrical forces is the electric potential energy of the protons in a nucleus. The electric potential energy of a pair of protons separated by a distance r is

$$U = \frac{1}{4\pi\epsilon_0} \frac{e^2}{r} \qquad (42.6)$$

Each proton interacts with every other proton. The total electric potential energy is a sum of terms like U, one for each pair of protons. For a nucleus with Z protons there are $\frac{1}{2}Z(Z-1)$ pairs of protons and the total electric potential energy has the form

$$U_t = \frac{Z(Z-1)e^2}{8\pi\epsilon_0 R} \qquad (42.7)$$

where R is approximately equal to the nuclear radius as given by Equation 42.5. Note that U_t is nearly proportional to Z^2 for large values of Z.

EXAMPLE 4

Electric Potential Energy of Nuclei

To illustrate the rapid increase of the electric potential energy with the atomic number, let's calculate U_t for helium and uranium, elements at opposite ends of the periodic table.

For helium, $Z = 2$, $A = 4$, $R = 1.9$ fm, and

$$U_t = \frac{(8.99 \times 10^9 \text{ N·m}^2/\text{C}^2)(1.60 \times 10^{-19} \text{ C})^2}{1.9 \times 10^{-15} \text{ m}}$$

$$= 1.2 \times 10^{-13} \text{ J}$$

Converting to MeV gives

$$U_t = 1.2 \times 10^{-13} \text{ J} \cdot \left(\frac{1 \text{ MeV}}{1.60 \times 10^{-13} \text{ J}}\right) = 0.75 \text{ MeV}$$

This energy is not large on a nuclear scale. The $Z(Z-1)$ dependence of U_t causes it to become a major factor in nuclear stability for large Z. For uranium, $Z = 92$, $A = 238$, $R \approx 7.4$ fm, and we get

$$U_t = \frac{(8.99 \times 10^9 \text{ N·m}^2/\text{C}^2)\frac{1}{2}(92)(91)(1.60 \times 10^{-19} \text{ C})^2}{7.4 \times 10^{-15} \text{ m}}$$

$$= 1.30 \times 10^{-10} \text{ J} \left(\frac{1 \text{ MeV}}{1.60 \times 10^{-13} \text{ J}}\right)$$

$$= 810 \text{ MeV}$$

The nuclear potential energy associated with the attractive nuclear force is negative. However, because of the short-range nature of the strong nuclear force, each neutron or proton interacts only with its nearest neighbors. As a consequence, the total nuclear potential energy is directly proportional to the mass number, the total number of neutrons and protons. This contrasts with the Z^2 dependence of the electric potential energy. The competition between the stabilizing nuclear potential energy and the destabilizing electric potential energy establishes an upper limit on the proton number for stable nuclei. There is no stable nucleus in nature having Z greater than 83, corresponding to bismuth.

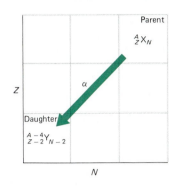

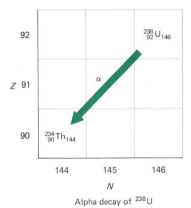

Alpha decay of ^{238}U

FIGURE 42.4

On a chart of nuclides, alpha emission is represented by a diagonal arrow pointing from the parent nucleus having Z protons and N neutrons to a daughter nucleus having $Z - 2$ protons and $N - 2$ neutrons.

Alpha, beta, and gamma radiations from unstable nuclei play important roles in nuclear physics and nuclear technology. These three radiations provide clues to the detailed structure of nuclei. Their properties have been exploited in a wide variety of technological applications that we briefly describe in this chapter.

Alpha-Particle Emission

Nuclear instability may lead to a spontaneous transformation of the nucleus with the emission of one or more particles. When this occurs, the nucleus is said to **decay**, or disintegrate. Alpha-particle emission is a spontaneous nuclear decay of this type. An alpha particle is the nucleus of a helium atom of mass number 4, $^4_2$He. According to the neutron-proton model of the nucleus, alpha-particle emission involves the escape of a bound configuration of two protons and two neutrons from the nucleus. For example, $^{238}_{92}$U undergoes alpha decay as follows:

$$^{238}_{92}\text{U} \rightarrow {}^4_2\text{He} + {}^{234}_{90}\text{Th} \tag{42.8}$$

Note that charge is conserved; initially there are 92 protons in the uranium nucleus, and finally there are 2 protons in the alpha particle and 90 protons in the thorium nucleus (Figure 42.4).

Mass energy plays a central role in alpha decay. When a uranium nucleus decays, kinetic energy is imparted to the alpha particle and to the residual thorium nucleus. The mass energy of the uranium nucleus exceeds the combined mass energies of the alpha particle and the thorium nucleus. The difference in mass energies is converted into kinetic energy, which is shared by the alpha particle and the thorium nucleus. The thorium nucleus may be created in an *excited* energy state. The thorium nucleus normally loses this excess energy by emitting energetic photons (Figure 42.5).

For the alpha decay of uranium described by Equation 42.8, the requirement that the mass of the parent nucleus (U) exceed the combined masses of the alpha particle and the daughter nucleus (Th) can be formulated in terms of atomic masses by

$$M_U > M_{He} + M_{Th} \tag{42.9}$$

From Table 42.1 we find,

$$M_U = 238.050817 \text{ u} \qquad M_{He} = 4.002603 \text{ u} \qquad M_{Th} = 234.043635 \text{ u}$$

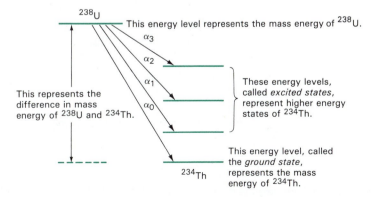

FIGURE 42.5

Energy level diagram for the alpha decay of ^{238}U. Alpha particle decay features discrete energies and a discrete spectrum.

and thus,

$$M_U - (M_{He} + M_{Th}) = 0.004579 \text{ u}$$

This mass difference corresponds to an energy difference of

$$0.004579 \text{ u } (931.48 \text{ MeV/u}) = 4.27 \text{ MeV}$$

This energy is available to be shared by the alpha particle and the thorium nucleus.

When there are just two decay products, as in alpha decay, the conservation of energy and linear momentum uniquely determine their kinetic energies. Let K_α and K_N denote the kinetic energies of the alpha particle and the other product nucleus. Energy conservation is expressed by

$$K_\alpha + K_N = \Delta M c^2 \qquad (42.10)$$

where ΔM is the mass difference between the parent nucleus and the decay products.

Assuming that the parent nucleus is at rest when the decay occurs, the conservation of linear momentum is expressed by

$$\mathbf{P}_\alpha + \mathbf{P}_N = 0$$

The kinetic energies of the decay products are related to their linear momenta by

$$K_\alpha = \frac{P_\alpha{}^2}{2M_\alpha} \qquad K_N = \frac{P_N{}^2}{2M_N}$$

where M_α and M_N are the mass of the alpha particle and its decay partner. Using linear momentum conservation we find

$$K_N = \frac{P_\alpha{}^2}{2M_N} = \left(\frac{M_\alpha}{M_N}\right) K_\alpha$$

Using this result in Equation 42.10 shows that the alpha particle energy is uniquely determined; that is, every alpha particle emerges with the same kinetic energy,

$$K_\alpha = \frac{\Delta M c^2}{1 + \dfrac{M_\alpha}{M_N}} \qquad (42.11)$$

EXAMPLE 5

Two-Particle Decay

We can use the alpha particle decay of ^{238}U to illustrate Equation 42.11. The energy release in the decay

$$^{238}\text{U} \rightarrow {}^4\text{He} + {}^{234}\text{Th}$$

is $\Delta M c^2 = 4.27$ MeV. The alpha particle's share of this energy follows from Equation 42.11

$$K_\alpha = \frac{\Delta M c^2}{1 + \dfrac{M_\alpha}{M_N}} = \frac{4.27 \text{ MeV}}{1 + \dfrac{4}{234}} = 4.19 \text{ MeV}$$

The remaining 0.08 MeV is carried off as kinetic energy by the thorium nucleus.

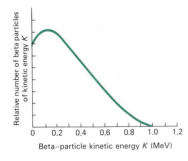

FIGURE 42.6

The distribution of beta particle
energies for beta decay. This is a
continuous spectrum, from zero
energy up to a maximum energy.
The detailed shape of the curve
depends on the specific beta decay
reaction. The upper limit on energy
(1 MeV above) represents the
total energy released in the beta
decay.

Mass considerations severely limit the number of alpha-particle emitters.
There are a few naturally-occurring nuclei having mass numbers (A) between
140 and 190 that are alpha-particle emitters. The majority of alpha-particle
emitters have $A > 200$.

Beta-Particle Emission

Nature provides two types of beta particles. They are identical in mass but
differ in the sign of electric charge. The negative beta particle is identical to
an electron (e^-). The positive counterpart is called a **positron** (e^+).

The character of beta-particle emission is quite different from that of alpha-
particle emission. Alpha particles are emitted with discrete energies, but beta
particles are emitted with a distribution of energies. For example, ^{210}Bi emits
negative beta particles having a continuous distribution of energy varying from
zero to 1.16 MeV (Figure 42.6). This distribution of energies is a consequence
of the simultaneous emission of a companion particle, an **antineutrino**.

Neutrinos and antineutrinos are electrically neutral, zero-mass particles carrying energy and momentum, but interacting very weakly with matter. The mass
energy available in beta decay is shared by the beta particle, the neutrino (or
antineutrino), and the nucleus remaining after the decay. For example, in the
beta decay of $^{14}_{6}$C, the products are a negative beta particle, an antineutrino,
and a nitrogen nucleus:

$$^{14}_{6}\text{C} \rightarrow ^{14}_{7}\text{N} + e^- + \bar{\nu} \qquad \text{(bare nuclei)}$$

Although there is a fixed total energy available to the decay products, the energy and linear momentum received by each product particle are not uniquely
fixed by energy and momentum conservation. In some decays the beta particle
gets virtually zero kinetic energy; in others it receives the maximum available
energy. Figure 42.6 shows a typical distribution of beta particle energies in
beta decay.

Beta decay is mediated by the **weak nuclear force**, the fourth and final
fundamental force we shall encounter. At the time of emission of an electron,
the weak interaction operates to transform a neutron into a proton, an electron,
and an antineutrino (Figure 42.7). The proton remains in the nucleus, but the
electron and antineutrino are ejected. Symbolically, this process is

$$n \rightarrow p + e^- + \bar{\nu} \qquad (42.12)$$

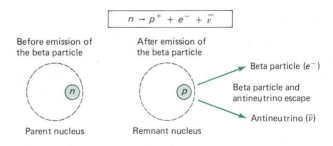

FIGURE 42.7

Schematic representation of the beta decay process. A neutron transforms into a proton that
remains in the nucleus, and a beta particle and an antineutrino that escape.

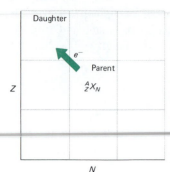

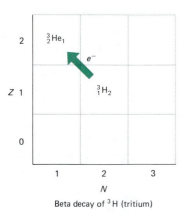

Beta decay of ^{3}H (tritium)

FIGURE 42.8
On a chart of nuclides, negative beta particle emission is represented by a diagonal arrow pointing from a parent nucleus having Z protons and N neutrons to a daughter nucleus having $Z + 1$ protons and $N - 1$ neutrons. The mass number A does not change.

This does not mean that a neutron consists of a proton, an electron, and an antineutrino. The product particles are created and the neutron is annihilated through the weak interaction at the moment of decay.

In beta-particle emission, a nucleus transforms itself into a different element. If an electron is created in the decay, the nuclear charge Z increases by one (Figure 42.8). Tritium is an unstable isotope of hydrogen (^{3}H) and undergoes electron beta decay:

$$^3_1H \rightarrow \, ^3_2He + e^- + \bar{\nu} \qquad \text{(bare nuclei)} \qquad (42.13)$$

The product nucleus ^{3}He is a stable isotope of helium that is rare in nature. Because ^{3}He is in great demand for experiments at cryogenic temperatures it is produced, one atom at a time, by the beta decay reaction described by Equation 42.13. So, the transmutation of one element into another, once the alchemist's dream, is now done routinely.

Net electric charge is conserved in beta decay. In the decay of tritium, Equation 42.13, there is $+1$ unit of electric charge in the tritium nucleus initially. In the decay products there are $+2$ units in the helium nucleus and -1 unit in the electron, for a total of $+1$ unit.

The mechanisms for alpha-particle emission and beta-particle emission are very different. However, the energy available for both types of reactions comes from the difference in mass energies of the parent nucleus and the decay products.

EXAMPLE 6

Kinetic Energy Available in Beta Decay

One of the most useful techniques for determining the age of very old carbon-based materials such as wood is based on the radioactive decay of ^{14}C. Using energy conservation, we want to determine the kinetic energy available to decay products in the reaction

$$^{14}_6C \rightarrow \, ^{14}_7N + e^- + \bar{\nu} \qquad \text{(bare nuclei)}$$

The available energy (E) equals the mass energy difference between the parent carbon nucleus and the decay products,

$$E = (m_C - m_N - m_e)c^2$$

We add and subtract $6m_ec^2$ to deal with atomic masses:

$$E = (m_C + 6m_e - m_N - 7m_e)c^2 = (M_C - M_N)c^2$$

From Table 42.1 we find

$$E = (14.003242 - 14.003074) \, u \cdot (931.48 \, \text{MeV/u})$$
$$= 0.156 \, \text{MeV}$$

This decay energy represents the maximum kinetic energy of the beta particle. In general, the beta particle, the antineutrino, and the nitrogen nucleus share the total energy. By observing many beta decays, we determine the distribution of beta-particle energies. The graph in Figure 42.6 shows a typical distribution of energies.

Carl Anderson discovered the positron in 1932 while investigating cosmic rays. The mechanics of positron emission are analogous to those of electron emission. A proton in a nucleus may be transformed into a neutron, a positron, and a neutrino. The neutron remains in the nucleus, but the positron and

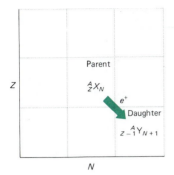

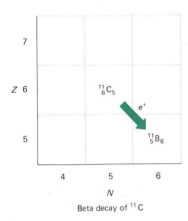

Beta decay of ^{11}C

FIGURE 42.9

On a chart of nuclides, positron emission is represented by a diagonal arrow pointing from a parent nucleus having Z protons and N neutrons to a daughter nucleus having $Z - 1$ protons and $N + 1$ neutrons. The mass number A does not change.

neutrino are ejected (Figure 42.9). The $^{11}_{6}$C nucleus undergoes positron decay according to

$$^{11}_{6}\text{C} \rightarrow ^{11}_{5}\text{B} + e^{+} + v \qquad \text{(bare nuclei)} \qquad (42.14)$$

As with electron beta decay, charge is conserved and the product nucleus is chemically different from the parent nucleus.

If a proton within a nucleus can transform itself into a neutron, a positron, and a neutrino, we might wonder why a free proton doesn't simply disintegrate according to $p \rightarrow n + e^{+} + v$. The reason is that such a transformation is energetically impossible. The mass of the proton is less than the mass of the neutron. In order to decay within a nucleus, a proton must borrow energy from its nuclear companions.

On the other hand, a free neutron can undergo beta decay into a proton, an electron, and an antineutrino. The mass of the neutron is greater than the combined mass of the proton and the electron. The beta decay of a free neutron releases a total of 0.42 MeV of kinetic energy.

Gamma Emission

Quantization of energy in atoms is revealed by the discrete wavelengths of light they emit. The energies in nuclei are also quantized. The primary difference between the quantized atomic and nuclear energies is one of *scale*. Atomic energies are on the order of electron volts. Nuclear energies are approximately a million times greater—on the order of millions of electron volts. The larger nuclear energies reflect the greater strength of the strong nuclear force.

Figure 42.10 shows the four lowest energy levels of the scandium isotope $^{45}_{21}$Sc. Like an atom, a nucleus can make a transition to a higher energy state by absorbing energy. Once in a higher energy state it can change to a lower energy state by giving up energy. As in the atomic case, this energy release is often in the form of a photon. The photon energy (hv) is related to the change in the energy of the nucleus by Bohr's radiation relation:

$$hv = E_{\text{higher}} - E_{\text{lower}} \qquad (42.15)$$

To illustrate, consider in Figure 42.10 the transition for the scandium nucleus from the state of energy 974.5 keV to the state of energy 543.0 keV. The energy of the emitted photon is 431.5 keV. This places the photon in the gamma

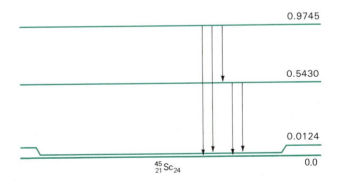

FIGURE 42.10

The energy states of a nucleus are discrete. Shown here are some of the measured energy states for ^{45}Sc, with energy expressed in MeV. Transition from a higher energy state to a lower energy state involves the emission of a gamma photon.

region of the electromagnetic spectrum, with a wavelength over 100,000 times shorter than visible light.

The Bequerel and Curie

The SI unit for disintegration rate is the **bequerel**, symbolized by Bq. One Bq equals one disintegration per second. The curie, symbolized by Ci, is an older but widely used unit of disintegration rate. One Ci equals 3.70×10^{10} disintegrations per second.

Gamma radiation sources like those used in cancer treatment, or the spent fuel elements from nuclear reactors, are generally on the order of hundreds or thousands of curies. Sources encountered in nuclear physics research are usually in the millicurie range. Demonstration sources used in classrooms are typically microcuries. The radioactive content of samples of material from our environment is typically on the order of picocuries ($1 \text{ pCi} = 10^{-12} \text{ Ci}$). Sensitive instruments have made possible the measurement of extremely small amounts of radioactivity.

Radioactive Decay Law

For a given nucleus, any disintegration process consistent with the conservation principles is possible. But the conservation principles say nothing about the rate of decay of a group of unstable nuclei. Given a collection of tritium atoms, for example, we do not find that all of their nuclei simultaneously undergo beta decay. Some decay in fractions of a second. Others survive for years before disintegrating.

When we detect beta particles with a Geiger counter we hear a series of random clicks, indicating that the time interval between beta decays is unpredictable. The decays, whether by alpha, beta, or gamma emission, have only a *statistical* regularity. We can assign an *average* lifetime against decay, but cannot predict when any particular nucleus will decay.

When describing nuclear disintegrations we generally measure the number of disintegrations Δn that occur in a time interval Δt. The ratio $\Delta n/\Delta t$ is the average disintegration rate over the time interval Δt. A succession of measurements give us a set of data representing the decay rate as a function of time. Generally, the data are displayed as a plot of the logarithm of the decay rate versus time, as shown in Figure 42.11.

The decay rate for a radioactive substance can be derived by assuming that nuclei decay independently of each other, and independently of how long a given nucleus has existed. Let n denote the number of nuclei present at time t. The instantaneous rate of decay is dn/dt. Because n decreases with time, dn/dt is negative and we can write

$$\frac{dn}{dt} = -\lambda n \qquad (42.16)$$

The proportionality factor λ represents the probability per unit time that any particular nucleus will decay. The value of the constant λ is different for different radioactive nuclei. Equation 42.16 can be integrated to yield

$$n = n_0 e^{-\lambda t} \qquad (42.17)$$

where n_0 is the number of nuclei present at some initial time ($t = 0$). The **activity** a is defined as the rate of decay, $|dn/dt|$. It follows from Equation

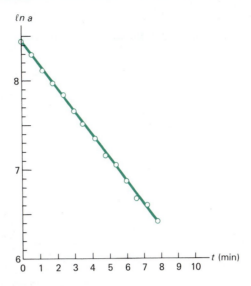

FIGURE 42.11
A plot of the natural logarithm of the activity versus time for a radioactive species, ^{137}Ba.

42.17 that the activity decreases exponentially with time, mirroring the decrease in the number of nuclei.

$$a \equiv \left| \frac{dn}{dt} \right| = a_0 e^{-\lambda t} \tag{42.18}$$

Activities are generally displayed on a logarithmic scale. The logarithmic form of Equation 42.18 is

$$\ln a = \ln a_0 - \lambda t \tag{42.19}$$

This equation predicts that the logarithm of the activity decreases linearly with time. The data graphed in Figure 42.11 confirm the predicted linear decrease of $\ln a$.

Half-Life

The half-life ($\tau_{1/2}$) is the time required for the activity of a sample to decrease by a factor of 2. If a_0 is the initial activity, then the activity one half-life later is $\frac{1}{2}a_0$. Substituting into Equation 42.19 we have

$$\ln \left(\tfrac{1}{2}a_0 \right) = \ln a_0 - \lambda \tau_{1/2}$$

from which we find

$$\tau_{1/2} = \frac{0.693}{\lambda} \tag{42.20}$$

The half-life can be determined by an inspection of the activity data. Or if λ is determined by fitting Equation 42.19 to the data, the half-life can be calculated from Equation 42.20, as in the case of the data in Figure 42.11.

There are few physical parameters in nature spanning a greater range of values than the half-life for nuclear disintegration. The half-life of only 2×10^{-21} s for ^{5}He is not much greater than the time required for light to travel a distance equal to the diameter of a nucleus. An extremely important

TABLE 42.2
Some Natural Alpha-Particle Emitters and their Half-Lives*

Uranium Isotopes		A = 140–190		Technically Significant Emitters	
Mass Number	Half-life	Isotope	Half-life	Isotope	Half-life
227	1.1 min	^{146}Sm	1.03×10^8 yr	^{238}Pu	87.74 yr
228	9.1 min	^{142}Ce	$>5 \times 10^{16}$ yr	^{239}Pu	2.44×10^4 yr
229	58 min	^{144}Nd	2.1×10^{15} yr	^{239}Np	2.35 days
230	20.8 days	^{150}Gd	1.8×10^6 yr	^{232}Th	1.41×10^{10} yr
232	72 yr	^{152}Gd	1.1×10^{14} yr	^{226}Ra	1.60×10^3 yr
233	1.592×10^5 yr	^{156}Yb	24 s	^{210}Po	138.38 days
234	2.45×10^5 yr	^{174}Hf	2.0×10^{15} yr		
235	7.038×10^8 yr	^{185}Au	4.3 min		
236	2.342×10^7 yr				
238	4.468×10^9 yr				

* Column 1 indicates the enormous range of half-lives encountered in the uranium isotopes. Column 2 lists some of the alpha-particle emitters in the mass region from $A = 140$ to $A = 190$. Column 3 shows some alpha-particle emitters of particular interest in nuclear technology. The data are from *Table of Isotopes*, 7th ed., C. Michael Lederer and Virginia S. Shirley, eds., Wiley, New York, 1978.

nucleus in nuclear technology, ^{239}Pu, has a half-life of 24,390 years. A sampling of alpha decay half-lives is presented in Table 42.2.

EXAMPLE 7
Student Determination of the Half-Life of ^{137}Ba

The radioactive barium isotope (^{137}Ba) has a relatively short half-life and can be easily extracted from a solution containing radioactive cesium (^{137}Cs). This barium isotope is commonly used in an undergraduate laboratory exercise for demonstrating the radioactive decay law. The data presented in Figure 42.11 were taken by undergraduate students using modest experimental equipment. We want to determine the half-life for the decay of ^{137}Ba using their data.

The logarithmic plot shown in Figure 42.11 is fitted by

$$\ln a = 8.44 - 0.262t$$

where t is the time measured in minutes. Comparing this with Equation 42.19 shows that the decay constant λ is 0.262 per minute. The half-life is

$$\tau_{1/2} = \frac{0.693}{\lambda} = \frac{0.693}{0.262/\text{min}} = 2.64 \text{ min}$$

The reported half-life of ^{137}Ba is 2.55 min. The discrepancy reflects experimental uncertainties.

42.3
RADIOACTIVE DATING

Archaeologists are interested in determining the age of artifacts, and geologists are interested in determining the age of geologic samples. Both use dating techniques based on the half-life of radioactive atoms. We consider first the carbon–14 method used by archaeologists.

The Carbon–14 Method

The earth's atmosphere is 78% nitrogen, most of it in the form of ^{14}N. Neutrons produced in the atmosphere by cosmic rays interact with ^{14}N nuclei, producing ^{14}C and a proton according to the reaction

$$^{14}N + n \rightarrow {}^{14}C + p \tag{42.21}$$

^{14}C is unstable and disintegrates in the beta decay reaction

$$^{14}C \rightarrow {}^{14}N + e^- + \bar{v} \tag{42.22}$$

The half-life for this decay is 5730 years. These production and decay processes have been operating for much longer than 5730 years. As a consequence the rate of production of ^{14}C now equals its rate of disintegration, and the amount of ^{14}C in our environment has reached a steady value*. The steady-state concentration of ^{14}C in the atmosphere amounts to approximately one ^{14}C atom for every 10^{12} carbon atoms.

   ^{14}C enters into chemical reactions in the same way as the stable carbon isotopes, ^{12}C and ^{13}C. In particular, ^{14}C forms gaseous carbon dioxide just as readily as do the stable isotopes. The ratio of ^{14}C to total carbon in a plant or animal consuming CO_2 remains constant during its lifetime. But the intake of CO_2 ceases when the plant or animal dies. In the solid parts of the organism, bones for example, the amount of ^{14}C subsequently decreases with a half-life of 5730 years, whereas the stable carbon isotope content remains constant. By measuring the ratio of ^{14}C atoms to total carbon atoms we can determine the time that has elapsed since the organism's death.

   Let N denote the number of ^{14}C atoms per gram of total carbon present in the sample. We define the specific disintegration rate I to be the rate of decay per gram of total carbon†. I is related to N by

$$I = -\lambda N \tag{42.23}$$

where λ is the disintegration constant. Using the radioactive decay law for N, Equation 42.17, we can write for I

$$\frac{I}{I_0} = e^{-\lambda t} \tag{42.24}$$

If we assume that the initial rate of decay per gram (I_0) for the sample is the same as it is for a present-day living sample, then measurements of I and I_0 and a knowledge of the decay constant allows us to determine the age of the sample,

$$t = -\frac{1}{\lambda} \ln\left(\frac{I}{I_0}\right)$$

In terms of the half-life,

$$t = -1.44\tau_{1/2} \ln\left(\frac{I}{I_0}\right) \tag{42.25}$$

The measured value of I_0 is 13.6 disintegrations per minute per gram.

---

* The rate of production of ^{14}C fluctuates about its average value because the cosmic ray bombardment varies in an irregular way.

† In Section 42.2 we used the symbol a (for activity) to denote the rate of decay of a radioactive sample. Here, the symbol I denotes the rate of decay *per gram*. Thus, the activity a and the specific disintegration rate I are closely related, but they have different units.

EXAMPLE 8

The ^{14}C Dating Game

The disintegration rate of ^{14}C in a wooden archaeological artifact is measured to be 10.5 disintegrations per hour per gram. From this measurement, let's estimate the age of the artifact.

For I_0 we have

$$I_0 = [13.6/\text{min} \cdot \text{gram}](60 \text{ min/hour}) = 816/\text{hour} \cdot \text{gram}$$

The ratio I/I_0 is

$$\frac{I}{I_0} = \frac{10.5}{816} = 0.0129$$

Using 5730 yr for the half-life of ^{14}C in Equation 42.25 we find

$$t = -1.44(5730 \text{ yr}) \ln(0.0129) = 36{,}000 \text{ yr}$$

So, the tree died 36,000 years ago.

Although simple in principle, the ^{14}C dating scheme is tedious in practice. Generally, the carbon content is oxidized to produce gaseous carbon dioxide. The radioactivity of the carbon dioxide gas is measured with an instrument, such as a Geiger counter, that detects the beta particles emitted in the decay of ^{14}C. It is necessary to shield the counting apparatus from extraneous background sources of radiation in our environment. Even in a comparatively young sample the counting rates are low, requiring on the order of 10 hours of counting time. These complications limit this form of ^{14}C dating to ages of less than about 50,000 years.

An alternative ^{14}C dating scheme counts atoms rather than beta particles emitted during the decay of nuclei. The method uses a cyclotron, an accelerator that can be used to separate positive ions of atoms. (A positive ion is produced by stripping one or more electrons from an atom.) The cyclotron imparts equal accelerations to ions having equal charge-to-mass ratios. When the cyclotron is adjusted to accelerate ions of ^{14}C, it will not accelerate ions of ^{12}C and ^{13}C, which have charge-to-mass ratios different from that of ^{14}C. When the atoms of a sample are ionized, the cyclotron separates them according to their mass and they can be counted as they emerge. By sequentially adjusting the cyclotron to accelerate ^{14}C ions, ^{13}C ions, and ^{12}C ions, the researcher can determine the ^{14}C-to-total-carbon ratio.

Because the method counts atoms rather than disintegrations, it takes advantage of the still rather large reservoir of ^{14}C atoms in a sample. Milligram samples are generally required, and the range of age determinations can be extended to approximately 100,000 years. The precision of the method is limited by the presence of other ions, notably ^{14}N, having very nearly the same charge-to-mass ratio as ^{14}C. The cyclotron technique also lends itself to other long-lived radioactive nuclei, such as the beryllium isotope, ^{10}Be, which has a half-life of 1.6 million years.

Geologic time scales are much larger than archaeological ages. A radioactive dating scheme for geologic samples therefore requires unstable nuclei with correspondingly long half-lives. Several naturally occurring radioactive nuclei are used. We will discuss a method using a radioactive rubidium isotope, ^{87}Rb.

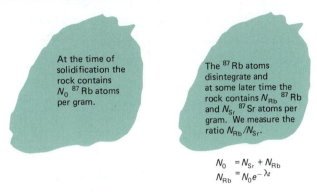

FIGURE 42.12

Age determination via the Rb-Sr technique.

At the time of solidification the rock contains N_0 ^{87}Rb atoms per gram.

The ^{87}Rb atoms disintegrate and at some later time the rock contains N_{Rb} ^{87}Rb and N_{Sr} ^{87}Sr atoms per gram. We measure the ratio N_{Rb}/N_{Sr}.

$$N_0 = N_{Sr} + N_{Rb}$$
$$N_{Rb} = N_0 e^{-\lambda t}$$

The Rubidium-Strontium Method

The rubidium-strontium method is based on the beta decay of ^{87}Rb,

$$^{87}\text{Rb} \rightarrow {}^{87}\text{Sr} + e^- + \bar{\nu}$$

The half-life for the decay is 4.8×10^{10} years, about ten times the age of the earth. At the time a rock solidified, suppose it contained N_0 rubidium atoms per gram. A time t later, the surviving number per gram is

$$N_{\mathbf{Rb}} = N_0 e^{-\lambda t} \tag{42.26}$$

where the decay constant is

$$\lambda = 1.44 \times 10^{-11} \text{ yr}^{-1}$$

Because each ^{87}Rb decay forms a stable ^{87}Sr nucleus, the number of ^{87}Sr atoms per gram accumulating in the time t is

$$N_{\mathbf{Sr}} = N_0 - N_{\mathbf{Rb}} = N_{\mathbf{Rb}}(e^{\lambda t} - 1) \tag{42.27}$$

where we have used Equation 42.26 to write N_0 in terms of the number of rubidium atoms. Solving for the time t we get

$$t = 6.92 \times 10^{10} \cdot \ln\left(1 + \frac{N_{\mathbf{Sr}}}{N_{\mathbf{Rb}}}\right) \text{ yr} \tag{42.28}$$

By measuring the ratio $N_{\mathbf{Sr}}/N_{\mathbf{Rb}}$ we can calculate the time since solidification of the rock (Figure 42.12). Generally the ratio $N_{\mathbf{Sr}}/N_{\mathbf{Rb}}$ is determined with a mass spectrometer. This radioactive dating method is one of several used for dating geologic materials.

EXAMPLE 9

Dating a Meteorite

One technique using the ^{87}Rb dating method measures the ratio of the concentrations of ^{87}Rb and ^{86}Sr and the ratio of the concentrations of ^{87}Sr and ^{86}Sr. Since ^{86}Sr is common to both of these measurements, the ratio $N_{\mathbf{Sr}}/N_{\mathbf{Rb}}$ needed to determine the age of the sample follows as the ratio of the measured ratios:

$$\frac{N(^{87}\text{Sr})}{N(^{87}\text{Rb})} = \frac{N(^{87}\text{Sr})/N(^{86}\text{Sr})}{N(^{87}\text{Rb})/N(^{86}\text{Sr})}$$

In this way a certain meteorite sample is found to have

$$\frac{N(^{87}\text{Sr})}{N(^{87}\text{Rb})} = \frac{N_{\mathbf{Sr}}}{N_{\mathbf{Rb}}} = 0.0664$$

The age of the meteorite follows from Equation 42.28

$$t = 6.92 \times 10^{10} \ln \left(1 + \frac{N_{Sr}}{N_{Rb}} \right) \text{yr}$$

$$= 6.92 \times 10^{10} \ln (1.0664) \text{ yr}$$

$$= 4.45 \times 10^9 \text{ yr}$$

42.4
NEUTRON ACTIVATION ANALYSIS

When an atom emits a photon, the energy of the photon equals the difference between the energies of the initial and final atomic states. By measuring photon energies it is possible to infer the chemical nature of the emitting atoms. Spectroscopic analysis exploits this principle.

An analogous situation exists with nuclei. Like an atom, the quantized energy states of a nucleus are unique. When a nucleus emits a photon the energy of the photon equals the difference between the energies of the initial and final nuclear states. By measuring the photon energies it is possible to identify the emitting nuclei. The primary difference between the atomic and nuclear analyses is one of energy scale; photons emitted by atoms have energies in the electron volt range. Photons emitted by nuclei have energies a million times greater, in the MeV range. This places the nuclear photons in the *gamma* region of the electromagnetic spectrum.

Neutron activation analysis takes advantage of the unique relation between photon energy and the emitting nucleus. *Neutron activation* means that nuclei have been made radioactive by bombardment with neutrons. *Analysis* refers to the measurement of the energies of the gamma photons released in the radioactive decay processes.

There are three steps involved in neutron activation analysis:

1. A nucleus captures a neutron, forming an unstable isotope.
2. The unstable isotope undergoes beta decay, producing a product nucleus in an excited state.
3. The excited state decays by gamma photon emission.

For example, the capture of a neutron by ^{75}As produces an unstable arsenic isotope, ^{76}As:

$$^{75}\text{As} + n \rightarrow {}^{76}\text{As}$$

The unstable arsenic isotope undergoes beta decay, forming a selenium isotope according to the reaction

$$^{76}\text{As} \rightarrow {}^{76}\text{Se*} + e^- + \bar{\nu}$$

The selenium nucleus is formed in an excited state (denoted by an asterisk), which then decays by emitting a gamma photon:

$$^{76}\text{Se*} \rightarrow {}^{76}\text{Se} + \gamma$$

Figure 42.13 shows this sequence of events on an energy diagram. The experiment involves the following steps: (1) We first identify the emitting nucleus by measuring the gamma photon energies (Figure 42.14). (2) Knowing that the nucleus was formed by beta decay, we then identify the isotope that led to

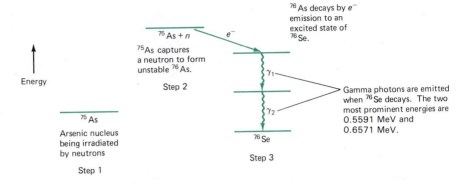

FIGURE 42.13

The activation of ^{75}As by neutron capture (1) leads to the formation of an unstable ^{76}As nucleus (2), which then decays (3) by gamma photon emission. The analysis procedure examines these three steps in reverse.

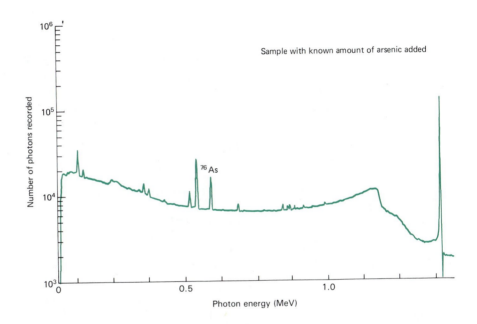

FIGURE 42.14

Recordings of gamma photons emitted from neutron-activated sediment samples from a lake. The samples are being analyzed for their arsenic content. Each peak identifies a gamma photon source in the sample. One photon peak is identified as being initiated by ^{76}As. The relative sizes of the ^{76}As peaks in the standard and unknown samples enable us to deduce the arsenic content in the unknown.

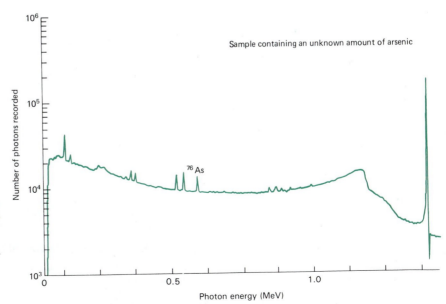

the beta decay. (3) Knowing that this isotope was formed by neutron capture, we finally work backward to identify the nucleus that initiated the sequence.

Neutron activation analysis makes it possible to determine the *concentration* of the initial isotope. For example, suppose we want to determine the concentration of a particular element in the sediments of a river. To do this, we prepare two identical samples of the sediment. Each sample contains C atoms per gram of the element of interest. To one of these samples we add an additional *known* concentration of C_0 atoms per gram. Both samples are irradiated simultaneously with a rain of neutrons from a nuclear reactor. We then measure the gamma photon activities for the two samples. The ratio of the activities of the gamma photons is equal to the ratio of the concentrations of the isotopes,

$$\frac{I_0}{I} = \frac{C_0 + C}{C} = 1 + \frac{C_0}{C} \tag{42.29}$$

I_0 denotes the activity of the sample containing $C_0 + C$ atoms per gram, and I denotes the activity of the unaltered sample containing C atoms per gram. The only unknown in Equation 42.29 is C, the concentration of the element of interest.

EXAMPLE 10

Accountability

A metal-working company claims that the arsenic content of its wastewater is less than 75 μg/liter. Let's see how the neutron activation analysis technique can check the company's accountability.

Two 5-liter samples of the effluent are prepared. Into one is dissolved 0.00136 gram of arsenic pentoxide (As_2O_5), having a molecular weight of 229.1. The atomic weight of arsenic is 74.96. The mass of arsenic added is thus

$$\left[\frac{2(74.96)}{229.1}\right] \cdot 0.00136 \text{ gram} = 8.90 \times 10^{-4} \text{ gram} = 890 \text{ } \mu g$$

The added arsenic concentration is

$$C_0 = 890 \text{ } \mu g/5 \text{ liter} = 178 \text{ } \mu g/\text{liter}$$

After irradiating 150 milliliters of both samples with neutrons, the selenium gamma activities of the two samples are

$$\text{standard } I_0 = 531 \text{ disintegrations/min}$$

$$\text{specimen } I = 93 \text{ disintegrations/min}$$

Using Equation 42.29 to solve for C we find

$$C = \frac{I}{I_0 - I} C_0 = \frac{93}{531 - 93} 178 \text{ } \mu g/\text{liter}$$

$$= 38 \text{ } \mu g/\text{liter}$$

In this case the measured concentration verifies the company's statement.

Note that by exposing two samples to the same neutron source and by taking the ratio of the induced activities, we need *not* know (1) how many neutrons actually impinged on the samples, nor (2) the probability that the neutrons are captured by the nuclei of interest. These are very important features of neutron activation analysis.

It is the great sensitivity of neutron activation analysis that makes the technique so useful. In many instances it is possible, using only a tiny sample, to determine concentrations equivalent to 1 atom of interest in the presence of 1 billion other atoms. Environmentalists, criminologists, and analytical chemists, all of whom must often analyze samples for trace amounts (very small quantities) of impurities, use neutron activation analysis to great advantage.

42.5
NUCLEAR ENERGY

Nuclear Fission

The separation of a nucleus into two nearly equally-massive fragments is termed **nuclear fission**. In a nuclear reactor, energy is liberated by inducing nuclear fission of ^{235}U with neutrons. A typical neutron-induced fission reaction that liberates 174 MeV of energy is

$$^{1}_{0}n + ^{235}_{92}U \rightarrow ^{143}_{56}Ba + ^{90}_{36}Kr + 3\,^{1}_{0}n \qquad (42.30)$$

Products other than ^{143}Ba and ^{90}Kr may result from ^{235}U fission, but the energy released is always about 200 MeV. For comparison, the chemical oxidation of a carbon atom in coal according to the reaction $C + O_2 \rightarrow CO_2$ liberates 4.1 eV, about 50 million times less energy than that obtained from the uranium fission reaction. This large factor is the major attraction of nuclear energy.

The fission energy is distributed among several particles, but the kinetic energy of the fission fragments (^{143}Ba and ^{90}Kr in Equation 42.30) accounts for more than 80% of it (Table 42.3). The fission fragments are relatively massive and charged. In a nuclear power reactor the fission fragments readily transfer their kinetic energy to atoms in the fuel elements, thereby heating the fuel elements (Figure 42.15). The fuel elements in turn heat the water circulating around them, producing steam. In this fashion nuclear reactors convert nuclear energy into thermal energy. The thermal energy carried by the steam is transformed into rotational kinetic energy of a steam turbine driving an electric generator that produces electricity.

Although simple in principle, few technological accomplishments rival the nuclear reactor. It could not have been developed without an understanding of nuclear physics principles.

TABLE 42.3
Energy Distribution in a Nuclear Fission Reaction

	MeV
Kinetic energy of fission fragments	165
Instantaneous gamma-ray energy	7
Kinetic energy of fission neutrons	5
Beta particles from fission products	7
Gamma rays from fission products	6
Neutrinos from fission products (not available for power production)	10
Total energy per fission	200

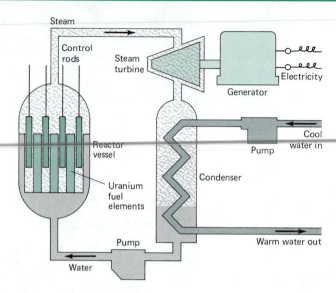

FIGURE 42.15

Schematic illustration of energy conversion and energy transfer in a nuclear fission reactor. Energy released from fission reactions causes the temperature of the uranium fuel elements to rise. Water circulating around the hot fuel elements is vaporized and channeled to a steam turbine in which thermal energy is transformed into rotational kinetic energy. After passing through the turbine, the steam is condensed and returned to the reactor, completing the cycle.

On the average, the fission of ^{235}U produces 2.5 neutrons. Therefore, on the average, ^{235}U fission reactions not only replenish the neutrons used to initiate fissions, but also provide neutrons for succeeding reactions. In this way a **self-sustaining chain reaction** can be achieved, making the nuclear reactor possible.

Achieving a self-sustaining chain reaction involves a number of considerations. When a neutron impinges on a fissionable nucleus, there is no guarantee that it will induce fission. Other processes compete with fission. For instance, the neutron may simply scatter elastically, like a golf ball bouncing off a bowling ball. It may scatter and transfer energy to the target nucleus, leaving it in one of its higher energy states. Or it may induce a nuclear reaction that transmutes the target nucleus into a different nuclear species. Even if the target nucleus captures the neutron, there is no guarantee that the intermediate nucleus will fission; it may simply lose energy and achieve stability by emitting one or more gamma photons. Each of these possible results occurs with a particular probability, determined by the details of the structure of the target nucleus and the kinetic energy of the neutron.

Figure 42.16 shows the measured neutron-induced fission probabilities for ^{235}U and ^{238}U. In both cases the probabilities depend sensitively on neutron energy. For ^{238}U the probabilities are negligibly small for neutron energies less than about 1 MeV; we say that ^{238}U fissions only under *fast* neutron bombardment. In contrast, ^{235}U is most likely to fission under bombardment by low-energy, or *slow*, neutrons. The effective minimum energy of neutrons in a reactor is set by the temperature. For a temperature near 300 K the minimum energy is approximately 1/40 eV. Neutrons having energies in this range are called **thermal** neutrons.

Note that the scale of relative probabilities in Figure 42.16 is **logarithmic**. The probability for fission of ^{235}U under slow neutron bombardment is

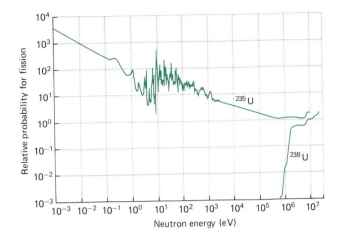

FIGURE 42.16

The relative probability for the neutron-induced fission of ^{235}U and ^{238}U. Note that the scale of relative probabilities is logarithmic.

about 1000 times the probability for the fission of ^{238}U under fast neutron bombardment.

Note the large increases in the fission probability for certain energies between 1 eV and 1000 eV. These sharp increases are referred to as **resonances**. At a resonance energy, the nucleus readily absorbs the incident neutron, enhancing the likelihood of fission.

Neutrons emerging from a fission reaction are very fast: their kinetic energies are in the MeV range. The probability that they will induce fissions is relatively low. Therefore, most nuclear reactors provide a means of slowing down, or **moderating**, the fission neutrons to promote a self-sustaining chain reaction.

The rate of fission reactions in a reactor is controlled by inserting rods between the fuel elements. These rods contain boron or cadmium, both of which have a very high affinity for capturing neutrons. The positions of the rods determine the number of neutrons available for inducing fission reactions. In turn, the number of neutron-induced fission reactions determines the energy output of the reactor.

Nuclear Fusion

Nuclear fusion reactions between light nuclei release energy because the reaction products have a smaller total mass than the reacting nuclei. The difference in mass energy is released as the kinetic energy of the products.

In terms of the binding energy per nucleon curve (Figure 42.1), it is energetically favorable for heavy nuclei to fission and for light nuclei to fuse. In both fission and fusion, the product nuclei are more tightly bound and hence more stable.

In the sun and other stars, a key reaction involves the fusion of two protons into a deuteron according to the reaction

$$p + p \rightarrow d + e^+ + \nu \tag{42.31}$$

The kinetic energy released is shared by the deuteron, the positron, and the neutrino.

EXAMPLE 11

Energy Release in Proton-Proton Fusion

Our existence depends on energy from the sun. The proton-proton fusion reaction is the key to a variety of nuclear reactions involved in energy production in the sun. Let's determine the energy release in the proton-proton fusion reaction described by Equation 42.31.

By adding two electrons to both sides of the reaction we can use the tabulated neutral atom masses. In terms of neutral atoms, Equation 42.31 becomes

$$^1H + {}^1H \rightarrow {}^2H + e^- + e^+ + v$$

The energy released is the mass energy difference between the reactants and the products:

$$\Delta E = \Delta M c^2 = (2M_H - [M_D + 2m_e]) c^2$$

where M_D is the mass of the deuterium atom and $2m_e$ is the combined mass of an electron and a positron. From Table 42.1 we find

$$\Delta E = \Delta M c^2 = (2.015650 - 2.015210) \text{ u} \cdot (931.48 \text{ MeV/u})$$
$$= 0.42 \text{ MeV}$$

The fusion of the two protons yields 0.42 MeV of energy. In addition, the positron produced in the reaction quickly undergoes an electron-positron annihilation reaction. This reaction converts the 1.02 MeV mass energy of the electron-positron pair into the kinetic energy of gamma photons. Thus the total energy release in the proton-proton fusion reaction is 1.44 MeV.

Numerous other fusion reactions are possible, but three involving hydrogen isotopes are of special importance. Shown with their energy releases in parentheses they are

$$^2H + {}^2H \rightarrow {}^3He + n \qquad (3.3 \text{ MeV})$$
$$^2H + {}^2H \rightarrow {}^3H + p \qquad (4.0 \text{ MeV})$$
$$^2H + {}^3H \rightarrow {}^4He + n \qquad (17.6 \text{ MeV})$$

It appears likely that the first generation of controlled nuclear fusion reactors will be based on these reactions. The first two reactions are called DD reactions because they involve a pair of deuterons. The third reaction between a deuteron and a *triton* (the nucleus of a tritium atom) is called the DT reaction.

Producing these reactions on an individual basis is not difficult. Deuterons from an accelerator can be made to bombard deuterium atoms in a suitable target, and fusion reactions will result. But such a scheme is not practical as a nuclear reactor because the investment in energy to accelerate the deuterons far exceeds the energy derived from the fusion reactions. If nuclear fusion reactions are to provide a useful energy source, a technology must be devised that derives more energy than is used to initiate the reactions.

For a fusion reaction to occur, the reacting nuclei must come close enough to feel the attractive nuclear force. Because of the electrical repulsion between the positively charged nuclei, the nuclei can approach one another closely enough to fuse only if they have sufficient kinetic energy. The potential energy barrier that the nuclei face is just the electric potential energy of the two interacting nuclei evaluated at a separation where the attractive nuclear force comes into play. This distance is roughly 3 fm. The electric potential energy of two hydrogen nuclei separated by 3 fm is

$$U = \frac{1}{4\pi\epsilon_0} \frac{e^2}{r}$$

$$= \frac{(8.99 \times 10^9 \ \text{N·m}^2/\text{C}^2)(1.60 \times 10^{-19} \ \text{C})^2}{3 \times 10^{-15} \ \text{m}}$$

$$= 7.7 \times 10^{-14} \ \text{J}$$

Converting to keV gives

$$U = 7.7 \times 10^{-14} \ \text{J} \left(\frac{1 \ \text{keV}}{1.60 \times 10^{-16} \ \text{J}} \right) = 480 \ \text{keV}$$

In order for the reacting nuclei to achieve this energy *thermally*, a very high temperature is required. In stars, the high temperatures are achieved through enormous gravitational compression. We speak of **thermonuclear fusion** because thermal motions supply the energy required to ignite the reactions. We can estimate the temperature by recognizing that a significant reaction rate can be achieved when only a small fraction of all nuclei have the kinetic energy required to surmount the electric potential energy barrier. The characteristic thermal energy per particle is kT, where T is the temperature and k is the Boltzmann constant (Section 22.3). We take 10kT as the energy characteristic of a small but significant fraction of the reacting particles. Setting 10kT equal to 481 keV gives a temperature of 6×10^8 K.

The actual **ignition temperature** for fusion is lowered because of **quantum tunneling** (Section 41.5). For DD reactions the ignition temperature is about 4×10^8 K. For the DT reaction ignition occurs at a temperature of 4×10^7 K. Because the ignition temperature is lower for the DT reaction, and the energy release per reaction is larger, the DT reaction is receiving the most attention in nuclear fusion energy research.

At the ignition temperature, the thermal energy of a particle greatly exceeds the atomic binding energy of electrons. Hence the reacting matter is completely ionized. It forms an electrically neutral gaseous mixture of negative electrons and positively charged nuclei called a **plasma**. If the plasma comes into contact with the reactor structure, it cools and is unable to participate in fusion reactions. To achieve a controlled thermonuclear fusion reaction requires a containment scheme to prevent the hot plasma from contacting the walls of the reactor. Some schemes attempt to do this magnetically. Others rely on *inertial* confinement; the matter is compressed, heated, and the reactions ignited before the matter has enough time to disperse.

To ignite the fusion reactions, energy must be added to heat the plasma to the ignition temperature. The energy invested is directly proportional to the number density (n) of the reacting nuclei.

$$\text{energy invested} = Bn$$

where B is a proportionality constant.

The rate at which energy is released by fusion reactions is proportional to the rate at which nuclei collide. The collision rate is proportional to the product of the densities of the reacting nuclei. In a DD-type reaction in which the reacting nuclei are identical, the reaction rate is proportional to the square of the number density of the hydrogen nuclei (n^2). The rate of energy release is likewise proportional to n^2,

$$\text{rate of energy release} = Cn^2$$

where C is a proportionality constant. The energy released in a time τ is $Cn^2\tau$. A net energy production requires that the energy released be greater than the energy invested:

$$Cn^2\tau > Bn$$

or

$$n\tau > \frac{B}{C}$$

The factors B and C include the details of the specific fusion reactions. The quantity τ is called the **containment time**, and $n\tau$ is called the **confinement parameter**. For the DT reaction a theoretical analysis shows that B/C has a value of 10^{14} s/cm^3. Thus the confinement parameter for the DT reaction is given by

$$n\tau = 10^{14} \text{ s/cm}^3 \tag{42.32}$$

This condition is termed the **Lawson criterion**, after its originator, British physicist John D. Lawson.

In summary, a practical energy-producing nuclear fusion system must satisfy two basic criteria: (1) an ignition temperature in the 10^7 K to 10^8 K range that ensures sufficient kinetic energy for the reacting nuclei to overcome the repulsive electric potential energy barrier; (2) achievement of the Lawson criterion that insures a net energy output from the fusion reactions.

Although the Lawson criterion is simple, achieving it has proved to be elusive, despite 40 years of worldwide research.

WORKED PROBLEM

James Chadwick observed that radiation produced by the interaction of alpha particles and beryllium caused protons to be ejected from paraffin with an energy of about 6 MeV. Conceivably this ejection could have been the result of a collision of a photon with a proton. In such a collision the photon will transfer the maximum amount of energy to the proton when the collision is head-on and the photon scatters backward. (a) Assuming a head-on collision of a photon with a proton initially at rest, treat the proton nonrelativistically and show that

$$E = \sqrt{\tfrac{1}{2}Km_pc^2} + \tfrac{1}{2}K$$

where E is the photon energy, K is the kinetic energy of the proton, and m_p is the proton mass. (b) Determine the photon energy needed to eject a proton with a kinetic energy of 6 MeV.

Solution

(a) Figure 42.17 shows the before-collision and after-collision pictures. Energy conservation is expressed by,

$$E = E' + K$$

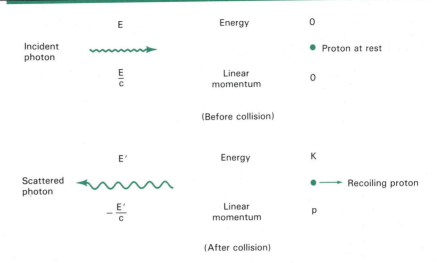

FIGURE 42.17

where E' is the photon energy after the collision. The linear momentum of a photon of energy E is E/c, where c is the speed of light. Linear momentum conservation is expressed by,

$$\frac{E}{c} = -\frac{E'}{c} + p$$

Using the linear momentum relation to eliminate E', the energy conservation relation becomes

$$E = \tfrac{1}{2}pc + \tfrac{1}{2}K$$

The kinetic energy of the proton is related to its linear momentum by

$$\frac{p^2}{2m_p} = K$$

Using this relation to eliminate p gives the desired result,

$$E = \sqrt{\tfrac{1}{2}Km_pc^2} + \tfrac{1}{2}K$$

(b) Setting $K = 6$ MeV and $m_pc^2 = 938$ MeV gives the energy the photon needs to eject a 6-MeV proton,

$$E = \sqrt{\tfrac{1}{2}(6 \text{ MeV})(938 \text{ MeV})} + \tfrac{1}{2}(6 \text{ MeV})$$
$$= 53 \text{ MeV} + 3 \text{ MeV}$$
$$= 56 \text{ MeV}$$

EXERCISES

Atomic mass data for Exercises and Problems may be found in Table 42.1.

42.1   The Neutron-Proton Model of the Nucleus

A. Determine and compare the binding energy per nucleon of ^{40}Ar, ^{90}Sr, ^{137}Ba, and ^{238}U.

B. Determine the nuclear radii of ^{60}Co, ^{131}I, and ^{235}U.

C. Assume that a nucleus is a sphere having a radius given by Equation 42.5. Determine the number of nucleons per cubic fermi in a nucleus.

D. A beryllium nucleus has four protons. Label these protons p_1, p_2, p_3, and p_4. Determine the number (n) of possible pairs of protons and show that your result is consistent with the formula $n = \tfrac{1}{2}Z(Z - 1)$.

42.2   Nuclear Stability

E. Beryllium 8 (^{8}Be) is a very special case among light nuclei because it disintegrates by alpha-particle emission. (a) Write the nuclear equation for this disintegration. (b) The atomic masses of ^{8}Be and an alpha particle are 8.005305 u and 4.002603 u, respectively. Show that this disintegration is energetically possible.

F. Determine the maximum beta-particle energy in the beta decay of the neutron.

G. Plutonium 239 decays by alpha-particle emission according to

$$^{239}_{94}Pu \rightarrow {}^{235}_{92}U + {}^{4}_{2}He$$

Determine the energy available for this decay.

H. The activity of a certain radioactive sample declines by a factor of 10 in 4 min. What is the half-life of this radioactive species?

I. Determine the disintegration rate (activity) of 1 gram of ^{60}Co. The half-life of ^{60}Co is 5.24 yr.

J. Figure 42.10 shows some of the transitions in which ^{45}Sc emits gamma photons. Calculate the wavelength of the gamma photon emitted in the transition from the excited state at 543.0 keV to the ground state.

K. What prevents a ^{12}C nucleus from decaying spontaneously into three alpha particles according to the following reaction?

$$^{12}C \rightarrow {}^{4}He + {}^{4}He + {}^{4}He$$

L. When a nucleus at rest decays spontaneously into two fragments, why does the fragment of smaller mass acquire the greater amount of kinetic energy?

M. A radioactive nucleus above the line of stability in Figure 42.3 (the dashed curve) may undergo beta decay, producing a more stable nucleus closer to the line of stability. Would the decay be by positron emission or by electron emission?

42.3 Radioactive Dating

N. The ^{14}C specific disintegration rate in an artifact is measured to be 23.1 disintegrations/hr·g. Determine the age of the artifact.

P. The specific disintegration rate in an artifact is measured to be 35.1 ± 0.5 disintegrations/hr·g. Estimate the age and the uncertainty in the age by using the ^{14}C dating method.

Q. Suppose that the natural ^{14}C disintegration rate per gram of total carbon 5000 yr ago was in fact less than what it is now. If we want to date an artifact that is known to be 5000 yr old, would the assumption that the natural disintegration rate per gram of total carbon was the same as now yield an age greater than or less than 5000 yr?

42.4 Neutron Activation Analysis

R. If the sodium activity in a neutron-irradiated material is 59,300 disintegrations per second, how many sodium atoms are in the sample? (The half-life of the sodium isotope is 15.0 hr.)

S. In a neutron activation analysis experiment involving aluminum, a measurement of the activity in a sample in which 1.25 mg of aluminum had been inserted was 386 disintegrations per second. Ten minutes later the activity in an identical sample having no aluminum purposely inserted was found to be 5.90 disintegrations per second. Determine the unknown amount of aluminum (^{28}Al has a half-life of 2.30 min.)

42.5 Nuclear Energy

T. How much energy (in MeV) is released in the following neutron-induced fission reaction?

$$n + {}^{233}_{92}U \rightarrow {}^{91}_{36}Kr + {}^{140}_{56}Ba + 3n$$

U. Determine the difference in excitation energy for the capture of a neutron by ^{233}U and ^{238}U. Which of these two reactions would most likely lead to nuclear fission?

V. A modern nuclear power plant produces about 1000 MW of electric power. The overall energy conversion efficiency of the plant is generally close to 33 percent. If the plant were operated at full capacity for a year and all the energy were derived from the fissioning of ^{235}U (each reaction producing 190 MeV of energy), how many kilograms of ^{235}U would be required?

W. If all the energy from the fissioning of 1 g of ^{235}U could be converted to electrical energy, and each reaction produces 190 MeV of energy, how long could this energy keep a 200-W bulb operating?

X. About 200 MeV of energy is liberated in the fission of a ^{235}U nucleus. (a) Show that the complete fissioning of 1 kg of ^{235}U liberates about 8.2×10^{13} J. (b) On the average, a city of 1 million people requires 2000 MW of electric power. If nuclear energy is converted to electrical energy at an efficiency of 30%, determine the mass of ^{235}U that would have to be fissioned to furnish the daily energy of a city of 1 million people.

Y. About 1 of every 6800 water molecules contains one deuterium atom. (a) If all the deuterium atoms in 1 liter of water are fused in pairs according to the DD reaction $^{2}H + {}^{2}H \rightarrow {}^{3}He + n + 3.3$ MeV, how many joules of energy would be liberated? (b) Burning gasoline produces about 3.4×10^7 J/liter. Compare the energy obtainable from the fusion of the deuterium in a liter of water with the energy liberated from the burning of a liter of gasoline.

Z. Assuming that a deuteron and a triton are at rest when they fuse according to $^{2}H + {}^{3}H \rightarrow {}^{4}He + n + 17.6$ MeV, determine the kinetic energy acquired by the neutron.

PROBLEMS

1. A ^{239}Pu nucleus at rest undergoes alpha decay, leaving a ^{235}U nucleus in its ground state. Determine the kinetic energy of the alpha particle.

2. In positron decay, a nuclear proton is transformed into

a neutron, a positron, and a neutrino. (a) Complete the following nuclear equation for the positron decay of the $^{11}_6C$ nucleus.

$$^{11}_6C \rightarrow$$

(b) The nuclear equation in part (a) involves only nuclear masses. Maintaining a balance of charge on both sides of the equation, change the nuclear equation into one involving masses of the neutral atoms by adding the same number of electrons to both sides of the equation. (c) Determine the maximum kinetic energy of the positron.

3. A disintegrating nucleus is often referred to as a parent, and the end nuclear product is called the daughter. For example, in the decay of tritium, the parent is ^3H and the daughter is ^3He. If the decay of the parent is described by $n = n_0 e^{-\lambda t}$, what is the corresponding equation for the growth of the daughter? Make a simple sketch of the decay and growth curves assuming the daughter is stable.

4. The disintegration of a radioactive sample is described by $n = n_0 e^{-\lambda t}$. The average lifetime of a nucleus is defined by

$$\bar{t} = \frac{\int_{n_0}^{0} t \, dn}{\int_{n_0}^{0} dn}$$

Show that $\bar{t} = 1/\lambda$.

5. If the daughter (remnant) nucleus in a radioactive decay process is stable, then the rate of decay of the parent nuclei $|dn_p/dt|$ is exactly equal to the rate of growth of the daughter nuclei dn_d/dt. But if the daughter is radioactive, the rate of change of the daughter is a combination of growth and decay rates. Show that the rate of change of the daughter is zero when $\lambda_d n_d = \lambda_p n_p$, where λ_d and λ_p are the decay constants of the daughter and parent, respectively.

6. (a) Calling R the ratio of daughter to parent atoms, show that the age in the rubidium-strontium method is given by

$$t = \frac{1}{\lambda} \ln (1 + R)$$

(b) Under what conditions would the approximate relation

$$t \approx \frac{R}{\lambda}$$

be valid? (c) The age determination in Example 9 utilized the expression in part (a). Show that the approximate equation in part (b) gives essentially the same result.

7. In a piece of rock from the moon the ^{87}Rb content is assessed to be 1.82×10^{10} atoms per gram of material. In a piece of the same rock, the ^{87}Sr content is found

to be 1.07×10^9 atoms per gram. (a) Determine the age of the rock. (b) Could the material in the rock actually be much older? What assumption is implicit in using the radioactive dating method?

8. In a contemporary carbon-based material the ^{14}C specific activity is 13.6 disintegrations/min·g of total carbon. (a) Show that there are 5.9×10^{10} ^{14}C atoms for each gram of total carbon. (b) How many ^{14}C atoms will have disintegrated for each gram of total carbon in a sample that is 5730 yr old?

9. Suppose that a neutron-induced fission reaction produces two neutrons, and that the two neutrons in each reaction proceed to induce fission reactions. If the products of the first reaction are called the first generation, derive a relation for the total number of neutrons produced in N generations.

10. Both ^{235}U and ^{238}U are radioactive. The half-life of ^{235}U is 7.04×10^8 yr and the half-life of ^{238}U is 4.5×10^9 yr. In natural uranium there is one ^{235}U atom for every 140 ^{238}U atoms. (a) What would this ratio have been 10^9 yr ago? (b) What will it be 10^9 yr from now?

11. (a) Neutrons are reduced in energy through elastic collisions with appropriate nuclei, such as protons. The energy lost depends on how the neutron strikes the scatterer. If a neutron collides with a more massive nucleus at rest and scatters at an angle of $90°$ measured relative to its incident direction, use nonrelativistic conservation of linear momentum and energy to show that the ratio of its kinetic energy after the collision (K) to its kinetic energy before the collision (K_0) is

$$\frac{K}{K_0} = \frac{M - m}{M + m}$$

where m is the mass of the neutron and M is the mass of the scatterer. (b) How many $90°$ scatterings must a neutron make with ^{12}C nuclei in order to reduce its energy from 2 MeV to 0.2 eV?

12. When a material of interest is irradiated by neutrons, radioactive atoms are produced continually and some decay according to their given half-life. (a) If radioactive atoms are produced at a constant rate R and their decay is governed by the conventional radioactive decay law, show that the number of radioactive atoms accumulated after an irradiation time t is

$$n = \frac{R}{\lambda} (1 - e^{-\lambda t})$$

(b) What is the maximum number of radioactive atoms that can be produced?

13. Figure 1 shows—in an idealized way—data recorded for the gamma-ray activities of two neutron-irradiated samples. One sample has been doped with 5.23×10^{16} lanthanum atoms. The data for the doped sample required 1 min of counting time. The data for the other

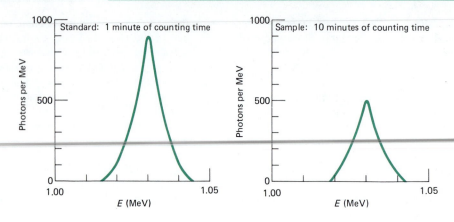

FIGURE 1

sample required 10 min of counting time. Determine the number of atoms in the unknown sample.

14. Students in an undergraduate laboratory recorded the data shown below for the activity of a radioactive source emitting gamma photons. Determine the half-life of the nuclei involved.

Time (minutes)	Gamma Photons Detected per Second
0	88.5
0.5	75.8
1.0	71.8
1.5	57.8
2.0	52.8
2.5	47.1
3.0	38.8
3.5	34.4
4.0	31.7
4.5	30.5
5.0	25.3
5.5	19.7
6.0	19.1
6.5	14.7
7.0	14.2
7.5	13.3
8.0	9.9
8.5	9.6
9.0	8.5
9.5	7.3
10.0	7.3

15. A nucleus at rest emits a gamma photon. The measured energy of the photon is 14.4 keV. Take the mass of the nucleus to be 57 u and determine (a) the kinetic energy of the recoiling nucleus, (b) the energy of the excited state in the nucleus from which the photon originated.

16. A large nuclear power reactor produces about 3000 MW of thermal power in its core. Three months after a reactor is shut down, the thermal power in the core is 10 MW, due to radioactive byproducts. Assuming that each emission delivers 1 MeV of energy to the thermal power, estimate the activity in bequerels three months after the reactor is shut down.

17. Consider a nucleus with $A = 176$, $R = 6.7$ fm, and $Z = 71$. (a) How much electric energy is transformed when the nucleus is split into its constituent neutrons and protons? (b) How much total energy input is required to effect the splitting?

18. Consider a model of the nucleus in which the positive charge (Ze) is uniformly distributed throughout a sphere of radius R. By integrating the energy density, $\frac{1}{2}\epsilon_0 E^2$, over all space, show that the electrostatic energy may be written

$$U = \frac{3Z^2 e^2}{20\pi\epsilon_0 R}$$

19. The number of radioactive daughter atoms accumulated in a time t is

$$n = \frac{R}{\lambda}(1 - e^{-\lambda t})$$

(a) Show that if $t \ll 1/\lambda$, then n increases linearly with t. (b) Show that R is the time rate of change of n at $t = 0$.

20. Free neutrons decay with a half-life of 10.6 minutes. Suppose you start with 2 free neutrons and add 2 new neutrons every 10.6 minutes. Show that your pile of neutrons can never contain more than 4 neutrons.

ANSWERS TO

ODD-NUMBERED PROBLEMS AND

EXERCISES A, C, E, G, I, K, M, N, S, U, AND V

CHAPTER 1

A. a) 1 microyear = 0.527 min;
    b) 1 microcentury = 0.878 hour
C. A sidereal day would be approximately 4 min longer than the solar day.
E. 6.38×10^{11} m, which is 1680 times the actual distance
G. 6.02×10^{23}
I. a) LT^{-2} b) MLT^{-2} c) $M^2 L^2 T^{-2}$
    d) MLT^{-1}
K. h/mc
M. 0.242
N. a) 0.0714 gal/s   b) 2.70×10^{-4} m^3/s   c) 1.03 h
S. 9.81 m/s^2

1. $(ax)^{\frac{1}{2}}$
3. 2.70×10^{12} days (1 day = time for 1 revolution of earth)
5. a) 30 days   b) 1.07×10^4 kg
7. Reynolds number = $\rho u D/\eta$
9. a) (proof)   b) 4 kg
11. 24 hours, 50 minutes

CHAPTER 2

A. $\mathbf{A} + \mathbf{B} = 7.65$ N,   113° above positive x-axis; $\mathbf{A} - \mathbf{B} = 18.5$ N, 22.5° below negative x-axis (analytic result)
C. Yes. The three vectors have equal magnitudes and form the sides of an equilateral triangle.
E. 50 km/h, 53.1° south of east
G. $E_x = 31$ V/m, $E_y = 52$ V/m, $E_z = -18$ V/m
    $E = 63.2$ V/m
I. $v_{\text{horiz}} = 12.3$ m/s $v_{\text{vert}} = 17.0$ m/s
K. 5.66
M. $-7\mathbf{i} + 16\mathbf{j} - 10\mathbf{k}$
N. $20.4\,\mathbf{k}$ kg·m^2/s

1. The driver ends up 2.34 km from the starting point, along a direction 11.6° west of north.
3. $\mathbf{r}_{AB} = (2, 2, -1)$ m
5. 7
7. (proof)

9. $(1.98 \times 10^{-15}\,\mathbf{i} - 2.08 \times 10^{-14}\,\mathbf{j})$ N
11. $\mathbf{B} = -0.2\mathbf{i} + 0.4\mathbf{j}$
13. a) $\mathbf{r}_{AB} = \mathbf{j}L$ $\mathbf{r}_{AC} = \dfrac{(\sqrt{3}\,\mathbf{i} + \mathbf{j})L}{2}$

    $\mathbf{r}_{AD} = \dfrac{(\mathbf{i} + \sqrt{3}\,\mathbf{j} + 2\sqrt{2}\,\mathbf{k})L}{2\sqrt{3}}$

    b) (proof) both areas $= \dfrac{\sqrt{3}L^2}{4}$
15. a) (proof)   b) $\mathbf{A} = \dfrac{2\pi}{\sqrt{3}L}(\mathbf{i} + \sqrt{3}\,\mathbf{j})$

    $\mathbf{B} = \dfrac{2\pi}{\sqrt{3}L}(-\mathbf{i} + \sqrt{3}\,\mathbf{j})$ $\mathbf{C} = \dfrac{2\pi}{W}\mathbf{k}$
17. (proof—Show dot product equals zero.)
19. $\dfrac{(\mathbf{i} + \mathbf{j} + \mathbf{k})}{\sqrt{3}}$ or $\dfrac{-(\mathbf{i} + \mathbf{j} + \mathbf{k})}{\sqrt{3}}$

CHAPTER 3

A. 4760 N
C. Equilibrium requires zero net force. The two forces acting on the bob are not oppositely directed and so cannot possibly add to zero.

E. 0.01 N
G. $T_1 = T_2 = 5200$ N $T_3 = 9000$ N
I. $F = 10.0$ N, up   torque = 6.00 N·m, counterclockwise

K. a) 20 N·m, counterclockwise   b) 10 N
M. zero torque for V and H 200 N·m clockwise for
    W 1200 N·m clockwise for T_2
    1400 N·m counterclockwise for T_1
N. a) 444 N   b) 784 N
S. $\bar{x} = 4$ in $\bar{y} = 3.46$ in
U. $\bar{x} = 0$ $\bar{y} = 3.09$ cm
V. $\bar{x} = 0.75$ cm $\bar{y} = 1$ cm $\bar{z} = 0.75$ cm

1. 100 N
3. $P = 200$ N $T = 283$ N
5. a) $T_1 = T_2 = 3.46$ N   b) $T_1 = 3.98$ N
    $T_2 = 2.92$ N

7. a) $P_1 = P_3 = 1.64$ N $P_2 = 3.28$ N
    b) contact force magnitude = 2.32 N
9. tension = compression = 577 N
11. $\tan \theta = a/b$
13. $L = R = 200$ N
15. a) (proof)   b) $60°$
17. tension = 327 N   frictional force = 258 N
19. 6.34 m
21. left wedge force = 3.14 N
    right wedge force = 4.99 N

CHAPTER 4

A. displacement = -2 m for both observers
C. estimates based on Figure 1:   a) 0.58 m/s
    b) 1.2 m/s
E. a) -0.81 m/s   b) -1.22 m/s   c) -1.54 m/s
G. approximately 0.8 m/s
I. 420 m (analytic result)
K. a) 29.5 m/s^2 b) 24.5 m/s^2
M. first gear, 13 mph/s
N. a) 70 mi/h   b) -10 mph/s
S. a) 8.05 m/s^2 b) 80.5 m/s

1. a) 2.94 m/s   b) 0.441 m

3. 180 frames/s
5. 15 yards
7. a) 9.80 m/s^2 b) 46.5 m/s   c) 110 m
9. a) Expect G to increase as v increases, so
    $G = A + Bv$ is reasonable.   b) 0.597 gal
11. a) 15.7 s   b) 257 m
13. a) $v = \sqrt{(10 \text{ m/s}^2)x}$ b) $x = (2.5 \text{ m/s}^2)t^2$
    c) $v = (5 \text{ m/s}^2)t$
15. a) 3.84 m   b) 0.997 s
17. 10 light-years
19. 20 s, 278 m

CHAPTER 5

A. (sketch)
C. speed = 8.94 m/s, $\theta = 63.4°$
E. 24.2 m/s
G. $2.99°, 87.01°$
I. 1.41 hours
K. 35.9 m/s^2
M. 377 m/s^2
N. 2.48 km
S. A and C, 16 m/s

1. a) 0.474 s   b) 2.32 m/s   c) 2.32 m/s

3. 19.2 m/s, $63.4°$ above horizontal
5. (proof)
7. a) 20.7 m/s   b) 1.24 s
9. a) 1.03 s   b) $v_{\text{horiz}} = 8.67$ m/s, $v_{\text{vert}} = 4.20$ m/s
    c) $25.8°$ above horizontal
11. a) 1.40×10^6 m/s^2 b) approximately 18 s
13. a) $y = c - d(x/b)^2$, parabola   b) $\mathbf{v} = b\mathbf{i} - 2dt\mathbf{j}$
    c) 1.73 s
15. a) 2 km/h   b) (proof)
17. (proof)

CHAPTER 6

A. Not force free. Average velocity changes.
C. 8.66 N, east
E. 4000 N
G. 5.15 m/s^2, $14°$ south of east
I. a) 2 m/s^2 b) 160 N   c) 3.2 m/s^2
K. 3 N

M. The vehicles exert forces of equal magnitude on each
    other. Occupants of the more massive trucks experi-
    ence smaller accelerations and routinely escape injury.
    Occupants of the less massive cars experience larger
    accelerations and frequently suffer severe injuries.
N. $ma = mg$ implies $a = g$ for all m

S. 39.9 km
U. a) 544 N   b) 2.7 m/s^2, upward
V. a) 100 N/m   b) 1 m/s^2

1. a) 2.5 m/s^2 b) 2500 N
3. a) $F_{sun}/F_{earth} = 6.01 \times 10^{-4}$
    b) $F_{moon}/F_{earth} = 3.38 \times 10^{-6}$ c) 0.483 N
5. $r_{IV} = 3.30 \times 10^6$ m
7. a) 2.35×10^{-7} m/s^2 b) 6.00 N   c) 8.1 N
9. (proof)
11. a) 9 m/s^2 b) 9000 N

13. a) 1240 km   b) 69.3 m
15. a) 14.3°   b) 4.42 s
17. a) 2.36 s   b) 5.32 s
19. 248 years
21. (proof)
23. a) 491 N   b) 50 kg   c) 2.00 m/s^2
25. a) $T = 787$ N, $T_{horiz} = 68.6$ N, $T_{vert} = 784$ N
    b) 0.857 m/s^2
27. a) 4 m   b) 3.72 m/s
29. 24.5 m/s^2
31. (proof)

CHAPTER 7

A. a) 0.0328 J   b) -0.0328 J
C. a) zero   b) 78.4 J   c) 78.4 J   d) -78.4 J
    e) -78.4 J
E. Work by gravity is zero when same height is achieved.
    Work by gravity is positive when height decreases.
G. a) 0.53 J   b) 0.20 J   c) 0.60 J   d) zero
I. a) positive   b) negative   c) zero
K. a) -100 J   b) zero   c) 100 J   d) zero   e) zero
M. a) 16.2 m/s, 0, -16.2 m/s   b) 65.6 J, 0, 65.6 J
N. 4.47 m/s
S. a) 3.92 kJ   b) 980 W
U. $r_2 = 0.235$ m

1. a) zero   b) -51.3 kJ   c) 51.3 kJ
3. 6.08×10^{-13} J
5. -1.37×10^{-21} J

7. a) 28 m/s   b) 2.85 kJ
9. a) 0.098 J   b) -0.098 J   c) zero
11. a) 4.43 m/s   b) 7.67 m/s   c) 7.67 m/s
13. a) 1.34×10^7 m/s   b) 1.23×10^7 m/s
    c) 1.03×10^7 m/s   d) 3.79×10^{-14} m
15. a) $(2.5$ m/s$^2)t$ b) 25 kW   c) $(3125$ J/s$^2)t^2$
    d) 200 kJ
17. a) (proof)   b) 8.49×10^5 kg/s   c) 7.34×10^7 m^3
    d) 1.53 km
19. a) 314   b) 3.18 lb
21. 314 J
23. $W_1 = 31.4$ J, $W_2 = 62.7$ J, $W_{gravity} = -94.1$ J
25. a) -147 J   b) 8.08 m/s
27. a) (proof)   b) 1.38 kW
29. a) 0.15 J   b) 2.25 J

CHAPTER 8

A. a) -2.2 J   b) 2.2 J
C. -825 J
E. a) 5 J   b) 0.279 J
G. a) 19.6 MJ   b) -19.6 MJ   c) zero   d) 10.3 MJ
I. 0.447 m/s
K. 8.32 m/s
M. 11.2 km/s
N. (proof)
S. a) -588 N   b) -588 N
U. $-3Cx^2$
V. a) 2–3, 4–6, 7–8   b) 2, 3, 4, 6, 7, 8
    c) 1–2, 3–4, 6–7, 8–9   d) 3, 6, 8   e) 2, 4, 7

1. 45.6 km/s
3. a) 9.07 m/s   b) $\theta_1 = 180°$, $\theta_2 = \theta_3 = 90°$
5. a) ± 3.34 m   b) 6.53 m/s
7. a) (proof)   b) 6.24×10^{10} J   c) 1.27×10^7 m
9. 468 MJ
11. (proof)
13. 8.22 J
15. a) $v_{esc} = \sqrt{\dfrac{2GM}{R_E + H}}$ b) $v_{orb} = \sqrt{\dfrac{GM}{R_E}}$
    c) $R_E = H$

CHAPTER 9

A. a) $-\omega mA$ b) zero
C. 60 N·s
E. a) 12 N·s   b) 2.40 m/s
G. 2.56×10^5 m/s
I. 8.64 m/s, 40.6° above positive x-axis
K. a) 33.3 km/h   b) 16.7 km/h

M. a) (proof)   b) converted into work done by forces
    that deform cars
N. 14.5°
S. 1.1 units above middle of bottom of torso
U. $x_{cm} = y_{cm} = 0.1$ m
V. $x_{cm} = 4.67 \times 10^6$ m, toward moon

1.   No. Kinetic energy is not conserved.
3.   (proof)
5. $(23.6\mathbf{i} - 2.6\mathbf{j})$ kg·m/s
7. 1.25×10^4 N
9.   a) 4.67 m/s, north   b) 6.67 m/s, north
11.   (proof)
13.   3
15.   a) both move at 1 m/s   b) $2M$ moves at 1 m/s, M moves at 5 m/s

17.   (proof)
19.   5.19 m/s (0.3 kg) and 7.78 m/s (0.2 kg)
21.   a) $\dfrac{u_0}{2}$ b) $\dfrac{7u_0}{12}$
23.   (proof)
25. $\frac{a}{4}$ above center of base
27. $(0, 4.70, 3.15) \times 10^{-11}$ m
29.   2.45 cm

CHAPTER 10

A.   a) $8.11°$ b) 0.142 rad
C.   0.628 rad/s
E.   a) 0.447 rad/s   b) -0.0318 rad/s^2 c) 14.1 s
G.   a) 25 rad/s   b) 39.8 rad/s^2 c) 0.628 s
I. 2.14×10^{29} J
K.   2.47 kJ
M. $I = (\frac{1}{12})ML^2 + Mx^2$
N.   cube is largest $\left(\dfrac{2MR^2}{3}\right)$, sphere is smallest $\left(\dfrac{2MR^2}{5}\right)$

1.   a) 36.8 s   b) 736 rad
3.   310 rad/s
5. $10MR^2$
7. $I_x = \frac{1}{2}MR^2\left(1 - \dfrac{8}{\pi^2}\right)$ $I_y = \frac{1}{2}MR^2$
    $I_z = \frac{1}{2}MR^2\left(1 - \dfrac{4}{\pi^2}\right)$
9. $\dfrac{2}{5}\dfrac{(R_2^5 - R_1^5)}{(R_2^3 - R_1^3)}M$
11.   a) (proof)   b) 20 hours

CHAPTER 11

A. 5.31×10^7 kg·m^2/s
C. $3.53\,\hbar$
E. 3×10^4 m/s
G. -22.0 **k** kg·m^2/s
I.   a) No. A net torque acts.   b) No. **L** reverses direction each time the bob comes to rest.   c) No.
K.   2400 kg·m^2/s, the same for origins at A, B, and C.
M.   (sketch)
N.   **L** $= 16$ **k** kg·m^2/s for all three origins
S.   a) 1.06×10^{-34} J·s
    b) 1.59×10^{-34} J·s or 0.53×10^{-34} J·s

1.   (proof)
3.   a) 26 **k** kg·m^2/s   b) 6 **k** kg·m^2/s
5.   18.8 ms. No, the earth could not withstand the stress.
7.   (proof)
9.   a) 2.56×10^7 kg·m/s   b) 5.98×10^{16} kg·m^2/s
    c) 1.99×10^{14} m^2/s
11. 3.57×10^{30} kg·m^2/s
13.   833 kg·m^2/s
15.   a) 4.12 rad/s   b) 6.56 revolutions
17.   a) 31.4 rad/s   b) 126 kg·m^2/s   c) 1.5 s

CHAPTER 12

A.   20.8 rad/s^2
C.   The frictional force opposes the relative motion of the spinning ball and the floor, causing the center of mass of the ball to move sideways. The frictional force sets up a torque that can reverse the direction of the spin (angular momentum). This can give rise to a series of bounces, each causing the ball to reverse its direction of spin and its direction of horizontal motion.
E.   a) 14.7 rad/s^2 b) 14.7 m/s^2
G.   0.209 rad/s
I.   a) 60 N·m   b) 94.2 J
    c) 12.1 rad/s

K.   1630 rad/s
M.   Uniform disk will win. Smaller I_c gives larger v_{cm}.
N.   2.19 kJ

1.   a) (proof)   b) (proof)
3.   (proof)
5.   a) (proof)   b) (proof)   c) (proof)
7.   a) $\frac{v}{2}$ b) $\frac{6v}{5L}$ c) $\frac{3}{10}$
9.   a) 2.5 s   b) 5 kW
11.   1.25
13.   a) 223 J   b) 5.70 m/s   c) 2.23 rad/s

15. a) (proof)   b) (proof) $a = \dfrac{(2m - M \sin \theta)g}{\dfrac{3M}{2} + 2m}$

17. a) 5.36×10^5 kg·m²   b) 3.33 hours   c) 8190 N·m

19. a) 0.153 m/s   b) 9.55 N·m
21. (proof)
23. $7.43R$

CHAPTER 13

A. period = 0.25 s, frequency = 4 Hz, angular frequency = 25.1 rad/s
C. a) 38.6 N/m   b) 1.32 kg
E. a) $x = \pm 7.3$ cm   b) 3.14 s
G. 0.384 J
I. 3.36 N/m
K. (proof)
M. a) 0.31 m   b) 1.57 s
N. a) 1.53 rad/s   b) 4.11 s   c) 2220 s

3. a) 2 s   b) one oscillation for one particle and two oscillations for the other
5. (verification)
7. 1.68 N
9. (proof)
11. a) 0.5 s   b) new mass = four times old mass
13. (proof)
15. a) 2.38 rad/s   b) 0.477 rad   c) -0.817 rad   d) 0.817 rad

1. $k = 2.45$ N/m, $x_0 = -4.08$ cm (directed opposite to acceleration)

CHAPTER 14

A. In units of 10^3 kg/m³, Mercury = 5.5, Venus = 5.22, Earth = 5.52, Mars = 3.96, Jupiter = 1.34, Saturn = 0.70, Uranus = 1.56, Neptune = 2.16. The densities of the terrestrial planets are all close to the density of Earth. The densities of the Jovian planets are comparable to the density of Jupiter and are distinctly different from the density of the terrestrial planets.
C. a) 43.8 psi   b) 2.98 atm   c) 302 kPa
E. 4.99×10^7 N/m²
G. 120 N/m²
I. 1.23×10^4 Pa, 1.78 psi, 0.121 atm
K. 2.65 mm
M. a) 2260 N   b) 1.70 J
N. tensile stress = 1.25×10^8 N/m²
   tensile strain = 7.81×10^{-4}
   $Y = 16 \times 10^{10}$ N/m²

S. a) 1.25×10^8 N/m²   b) Damage will occur.
U. a) 3.5×10^7 N/m²   b) 3.5×10^4 N
V. 1/2

1. (Answers given to one significant figure.)
   a) 5×10^{14} m²   b) 5×10^{19} N   c) 5×10^{18} kg
   d) 10^{44} molecules
3. a) (proof)   b) 1.2×10^{-3} c) (proof)
5. a) 262 N/m   b) 46.1 m
7. (proof)
9. a) (proof)   b) Estimates based on graph. Elastic energy density = 1.3×10^6 J/m³, for tensile strain = 0.015.   Elastic energy density = 1.1×10^6 J/m³ for compressive strain = -0.015.
11. 4.21 cm
13. (proof)

CHAPTER 15

A. 2.99×10^{-29} m³
C. 149 atm
E. approximately 200 km
G. No change. Buoyant force of air on bucket and contents is same before and after because volume of air displaced does not change. Scale reading is (true weight − buoyant force), so scale reading is unchanged.
I. 2.7 m/s
K. 2.37×10^{-3} m³/s
M. 1.42×10^5 N/m²
N. a) 16.6 m/s   b) 3.5 m
S. 3.77×10^{-4} m³/s = 5.98 gal/min
U. a) 90,000   b) turbulent
V. 3.5 mm

1. a) (proof)   b) 66.4 N
3. density of oil = 1.25×10^3 kg/m³
   density of sphere = 500 kg/m³
5. 7.9 km (Mt. Everest rises 8.89 km above sea level.)
7. (proof)
9. (proof)
11. a) $\dfrac{\rho g W H^2}{2}$ (ρ = density of water)   b) $\dfrac{\rho g W H^3}{6}$ c) $\dfrac{H}{3}$
13. a) spring tension = $W - \rho V g$
    b) $\rho_{\text{liquid}} = \rho_{\text{water}} \dfrac{[\text{tension(liquid)} - W]}{[\text{tension(water)} - W]}$
15. (proof)
17. $k = 4.02 \times 10^{-3}$ N·s³/m³
19. nickel

CHAPTER 16

A. a, b, and c
C. amplitude = 1.80 cm, wavelength = 1.02 m
E. 30 cm in positive x-direction
G. period = 0.6 s, frequency = 1.67 Hz, wave number = 1.05 rad/m, angular frequency = 10.5 rad/s
I. 5.8 km/s
K. (proof)
M. 65 nm
N. The x' origin is 0.70 m to the left of the x origin.
S. 1.59 beats/s

U. 444 Hz
V. a) 532 Hz   b) 447 Hz

1. (proof)
3. 0.673 m/s
5. a) (proof)   b) clockwise
7. a) (sketch)   b) linearly polarized along the direction $x = y$
9. a) (sketch)   b) 10 cm/s
   c) $\psi = 10 \exp\{-a(x - vt)^2\}$ cm, where $v = 10$ cm/s

CHAPTER 17

A. 77.5 m/s   b) 0.103 s
C. a) $T_{top} = 85.9$ N, $T_{bottom} = 80$ N
   b) $v_{top} = 29.3$ m/s, $v_{bottom} = 28.3$ m/s
E. 2.4×10^{11} N/m^2
G. Yes. 16.5 m and 1.65 cm
I. 5.58 m/s. Surface tension contribution is insignificant.
K. $B = 6.00 \times 10^{11}$ N/m^2, $\mu = 1.84 \times 10^{10}$ N/m^2
M. 0.052 mm
N. a) 3×10^{15} W/m^2 b) 10^7 J/m^3
S. a) 3.82×10^{23} W/m^2 b) 3.24 MJ
U. 4 m
V. a) (proof)   b) approximately 800 Hz, which is in audible range

1. a) (proof)   b) (proof)

3. a) 6.02 dB   b) 4   c) No, change is 3.52 dB.
5. $I_{orchestra} = 31.6 I_{violin}$
7. (proof)
9. 2750 Hz
11. $\psi = 0$ at $x = L$ and $\dfrac{\partial \psi}{\partial x} = 0$ at $x = 0$, where $x = L$ is the end attached to wall   b) (sketch)
   c) $f_{fund} = \dfrac{1}{4L}\sqrt{\dfrac{T}{\mu}}$
13. a) (proof)   b) $A = -10$ cm, $B = 1.2$ cm
15. a) (proof)   b) (sketch)   c) The two pulses are identical in shape and travel in opposite directions at equal speeds. At $t = \frac{a}{v}$ they cancel exactly.

CHAPTER 18

A. a) 375.3 K, 374.2 K, 373.7 K
   b) 373.1 K (linear extrapolation)
C. 390 J
E. a) 1.68×10^{15} kJ   b) approximately 52 years
G. a) silver   b)1.95 C$^\circ$
I. 21 C$^\circ$
K. 1.92 MJ
M. a) 12 MJ   b) -12 MJ
N. a) (sketch)   b) $W_{isobaric} > W_{isothermal}$
S. 22 MJ
U. 39.4 MJ

V. a) 1.42×10^4 kJ   b) zero   c) -4.25×10^3 kJ
   d) 9.93×10^3 kJ

1. a) (proof)   b) $R_0 = 0.0119\ \Omega$ B = 3570 K
   c) 318 K = 45°C
3. 1.41 MJ
5. a) 1.01 kJ   b) 1.01×10^3 N/m^2 c) 7.00 kJ
7. 0.817 kJ/kg·C$^\circ$
9. 252 quarts
11. 0.801 kJ/kg·C$^\circ$
13. 4.64 MJ

CHAPTER 19

A. 2.171 cm
C. 89.6 J
E. a) increase   b) 1.603 cm
G. a) 0.165 m   b) 1.5 μm
I. a) 0.0111 m^3 b) 0.0197 m^3
K. a) 0.0120 m^3 b) 0.0240 m^3 c) 0.0182 m^3
   d) 222 K

M. 80.5 K. This is below the boiling point of oxygen, implying that some oxygen in the air might liquefy.
N. a) helium   b) Yes, MC_v for H_2 is slightly smaller than MC_v for He, so H_2 would undergo larger temperature change.
S. Adiabat crosses fusion line.

U. 55.4 kJ

V. 554 J

1. Actual distance is 0.19 mm greater than tape indicates (4.75 times width of scale mark).

3. $T_{final} = 62.4$ K a) Cooling is greater for monatomic gas. b) As $\gamma \to 1$, $T_{final} \to T_{initial} = 300$ K

5. 17.9 W

7. a) 225 MJ b) Yes c) Yes d) Available energy is roughly 80 times the energy required to vaporize the comet, so a volume of earth several times that of the comet could be vaporized. In short, complete devastation over an area of a few hundred square miles.

9. a) Transition temperature is greater than typical house temperature ($70°F = 21°C$). b) 523 kg c) salt volume $= 0.358$ m^3, less than 1/5 volume of refrigerator

11. 3.36

13. 285 C°

15. sample melts

17. 1.37

19. (proof)

CHAPTER 20

A. 20.4 kW

C. 16 cm

E. 6.25% lower at 20°C

G. 90 W

I. It absorbs radiation emitted by the collector, raising the steady state temperature of the collector. It also inhibits convection currents, thereby reducing convective heat loss.

K. 10.4 MJ, very nearly equal to 2500 kilocalories

M. 518 kcal

N. 3.5 kW

1. a) (proof) b) 279 K

3. a) $287 b) $57

5. a) 18,300 K b) 1.14×10^{13} m to 1.72×10^{13} m

7. a) (proof) b) (proof) c) 337°C

9. (proof)

11. a) (proof) b) approximately 3 hours (9190 s)

13. $2.8 R_{earth}$

15. 0.110 W/m·C°

17. a) (sketch) b) (sketch) c) $\rho C L^2/k$
d) $T(x, t)$ approaches linear temperature variation characterizing the steady state, in a time of order $\rho C L^2/k$.

CHAPTER 21

A. Don't invest! Claimed performance violates law of energy conservation.

C. a) 60 kJ b) 0.40

E. (proof)

G. 18.9 kWh

I. a) 120 kJ b) 4

K. Larger temperature difference between high- and low-temperature reservoirs gives greater efficiency in North Atlantic.

M. a) (proof) b) There is some result other than the conversion of heat into work: the gas ends in an expanded, lower-pressure state.

N. 0.331

S. 0.059

U. 500 K to 510 K gives greater increase in efficiency.

V. 967 K

1. a) (sketch)

b) area = net heat absorbed per cycle = net work done per cycle

3. Both gases spread throughout larger volumes.

5. a) 0.5 b) 0.285

7. a) 4 kJ b) $\frac{2}{7} = 0.283$

9. (proof)

11. a) 3 kJ b) -1 kJ c) 2 kJ d) 1 kJ e) 1 kJ
f) $\frac{1}{3}$

13. a) (proof) b) (proof) c) 17.6

15. a) $20°C = 293$ K b) (proof)
c) $\Delta S = 4.88$ J/K > 0 d) Mixing is irreversible, as implied by $\Delta S > 0$ result of c).

17. 90 K, 1.49 J/K

19. a) 31.9°C b) -1.4 J/K c) Spontaneous change with $\Delta S < 0$ is not possible. Event claimed by guest violates second law.

CHAPTER 22

A. a) 1.99×10^{-26} kg b) 1.67×10^{-27} kg

C. approximately 0.04 mm

E. 0.05 N/m^2

G. The mean free path is halved.

I. approximately 4 mm

K. $v_{mean} = 2$ km/s $v_{rms} = 2.16$ km/s

M. approximately 20 ms (assumes a 10 m displacement traveling at 515 m/s)

N.   a) 1.96×10^7 m/s   b) 2.82×10^3 m/s

S. $\dfrac{v_{\text{oxygen}}}{v_{\text{smoke}}} = 14{,}000$

1.   a) 300 K   b) 100 K
3.   The Maxwellian distributions are the same for all 4 because all have the same value of m/T.
5.   a) 83   b) 610   c) 82

7.   approximately $10^{28}/\text{m}^3$. For water, $n = 3.3 \times 10^{28}/\text{m}^3$

9. $\left(\dfrac{M}{N_A \rho}\right)^{1/3} = 3.55$ Å

11. 0.224 (analytic result)
13. 0.4 ns
15. 0.215
17. a) 0.51 m/s   b) approximately 20 ms

ANSWERS TO

ODD-NUMBERED PROBLEMS AND

EXERCISES A, C, E, G, I, K, M, N, S, U AND V

CHAPTER 23

A. (proof)
C. 2.54×10^{11}
E. a) 8.22×10^{-8} N   b) 2.19×10^6 m/s, roughly 1% of the speed of light
G. a) midway between charges   b) Yes
   c) Equilibrium would be unstable.

1. $+5.82 \ \mu C$ and $-2.31 \ \mu C$ or $-5.82 \ \mu C$ and $+2.31 \ \mu C$

3. a) $0.120 \ \mu C$ b) $5°$ for both
5. (proof)
7. a) $F_x = 0, F_y = -2k_e \beta Q/R$ b) 36.3 cm
9. $-4 \ \mu C$
11. a) (sketch)   b) Net $\mathbf{F} = 0.0508 \ \mathbf{i}$ N
13. a) (proof)   b) (proof)
15. (proof)
17. (proof)
19. (proof)

CHAPTER 24

A. 5.58×10^{-11} N/C, down
C. 5×10^5 N/C, directed away from the screen
E. $\mathbf{E} = 5.08 \times 10^4 \mathbf{i}$ N/C
G. (sketch)
I. 28.2 N·m²/C
K. 5220 N·m²/C
M. 2.26×10^5 N·m²/C
N. a) zero   b) 135 N/C
S. 2 alpha particles and 1 electron
U. 2.56 cm

1. a) (proof)   b) (proof)
3. a) 2.93×10^{21} N/C   b) 1.18×10^{17} N/C
5. a) (proof)   b) (proof)
7. (proof)
9. a) 9.57 cm   b) 1.96 N   c) 1.39×10^6 N/C

11. a) $E(x = 2 \ m) = 2.2 \times 10^4$ N/C
       $E(x = 0) = 1.8 \times 10^4$ N/C
       $E(x = -2 \ m) = -3.4 \times 10^4$ N/C
    b) (proof)   c) 0.171 m
13. a) 1.15×10^{14} m/s²   b) 3.03×10^7 m/s
    c) 0.264 μs
15. (proof)
17. 613 N/C
19. a) A positive test charge on the y-axis ($0 < y < a$) will be attracted by $-Q$ and repelled by the two $+Q$s, so E_y cannot be zero.   b) A positive test charge on the x-axis will experience a net vertical force, exerted by $-Q$, so E_y cannot be zero.
    c) $y = 3.88a$ and $y = -0.330a$
21. 1.71 cm
23. a) (proof)   b) $x = \frac{R}{\sqrt{2}}$

CHAPTER 25

A. a) 8×10^{-3} J   b) 8×10^{-3} J   c) 8×10^{-3} J
   d) -4×10^4 V
C. a) 1.36×10^5 V/m   b) 3.25 mm
E. $V_A = 12$ V, $V_C = -12$ V
G. a) 3.6×10^{-4} V   b) 2.54×10^{-4} V   c) zero
I. a) 517 V   b) 517 V
K. 2.70 m
M. zero

N. a) $V(x = 0) = 10$ V $V(x = 3 \ m) = -11$ V
       $V(x = 6 \ m) = -32$ V
   b) $E = 7$ V/m, in positive x-direction at all three positions
S. a) net force is zero   b) $E = 0$ c) 4.50×10^4 V

1. $E_x = -600$ V/m $E_y = 5300$ V/m
3. (proof)

5. (proof)
7. $E_x = \pm (Q/2\pi\varepsilon_0 R^2)\{1 - x/(x^2 + R^2)^{1/2}\}$, with $+$ for $x > 0$ and $-$ for $x < 0$ b) $E_x \to \pm Q/4\pi\varepsilon_0 x^2$ for $x \gg R$, and $E_x(x = 0) = \pm\sigma/2\varepsilon_0$ (Gauss' law result)
9. 77.9 eV

11. (proof)
13. (proof)
15. (proof)
17. (proof)
19. a) -38 V/m b) 0.417 m

CHAPTER 26

A. 0.1 pF
C. 709 μF
E. 1.25×10^6
G. a) 4.36 μF b) 2 μF c) 22 μF
I. 480,000
K. a) 8 V b) 128 μC c) 5.33 μF
M. $U_2/U_1 = 3/2$
N. Both Q and V are doubled.
S. a) 1.83×10^{20} V/m b) 1.49×10^{29} J/m³
U. a) 2.69 cm² b) 95 V
V. Five 6-pF in series meet requirements.

1. a) Outer faces are virtually uncharged and inner faces carry a charge of the same sign as the battery terminal to which they are connected. b) Central plate has positive charges on both faces. Top and bottom have virtually all charge on inner faces. c) Configuration is equivalent to 2 identical capacitors in parallel. d) Configuration is equivalent to $N - 1$ identical capacitors in parallel.
3. (proof)
5. a) 480 μC b) $Q_4 = 384$ μC, $Q_6 = 576$ μC c) 96 V
7. (proof)
9. (proof)
11. Energy doubled. Force that increases spacing does work.
13. approximately 1 pF
15. 1.96×10^{-5} C/kg
17. 45 V
19. a) (proof) b) (proof)

CHAPTER 27

A. 6.25×10^{18}/s
C. a) 4.78×10^{14}
   b) 5.1×10^{-7} A
E. 0.282 mm
G. 3×10^9 Ω
I. a) 900 Ω
   b) 136,000 metric tons
K. 1.57 A
M. 7.49×10^{-6} Ω
N. 4.36×10^{-3} Ω, ignorably small
S. a) 50 MW b) 55 J
U. a) 7.56×10^6 A/m² b) 0.130 V/m c) 3.91 V
   d) 97.7 W
V. 36 W

1. Rate $= Mnv$
3. 2.08 mC
5. 0.0036 Ω
7. 0.125 mA
9. 0.616
11. In Al, $I = 55.6$ A, $P = 55.6$ W. In Cu, $I = 273$ A, $P = 273$ W.
13. a) 1.82 A b) 18.3 W
15. a) (proof) b) (proof)
17. 1.01 mm
19. 11.2 min
21. $R = \rho_0 L/A + \rho_0 bL^2/2A$
23. 1.28 mm
25. 4.52 Ω

CHAPTER 28

A. $R_{int} = 4$ Ω, $R_{ext} = 10$ Ω
C. a) 2 A b) 0.5 Ω
E. All four in parallel gives 1.5 Ω. Two in parallel, in series with the other two in parallel, gives 6 Ω. Three in parallel, in series with the fourth, gives 8 Ω. Two in parallel, in series with the other two in series, gives 15 Ω. All four in series gives 24 Ω.
G. Both are 5-kΩ resistors.
I. a) -5 V b) -9 V c) -15 V
K. starter: 171 A, battery: 0.28 A
M. 0.327 A

N. a) (proof) b) (proof)
S. (proof)
U. a) 90 percent b) Internal resistance is 9 times R.
V. 1.15 RC

1. (proof)
3. a) $I_1 = 1.42$ A, $I_2 = -0.158$ A, $I_3 = 1.26$ A
   b) 5.68 V (4 Ω), 6.32 V (5 Ω), 0.316 V (2 Ω)
   c) 8.07 W (4 Ω), 7.98 W (5 Ω), 0.05 W (2 Ω)
   d) Sum of RI^2 for all 3 resistors equals 16.1 W, which equals sum of VI for the two emfs.

5. 7 V
7. a) 7.1 V   b) 6.3 V   c) 3.3 kΩ
9. b) None or one, depending on labeling in part a). c) $I_1 = -5$ A (counterclockwise through r_1), $I_2 = 0$, $I_3 = 5$ A (counterclockwise through r_3).
11. 16.7 mW
13. a) (proof)   b) (proof)
15. a) $I_1 = 1$ mA, $I_2 = 0$, $I_C = 1$ mA
    b) $I_1 = I_2 = 0.67$ mA, $I_C = 0$
17. The potential at point C is 2 V.
19. $(7/5)R$, $(2/3)R$, $(3/4)R$

21. a) (proof)   b) (proof)
23. a) Array consists of series combination of $r + r$, in series with r in parallel with infinite array.   b) by symmetry (reversing direction of current won't change magnitudes.)   c) Kirchhoff's current rule   d) (proof)   e) (proof)
25. a) (proof)   b) "resistance" = 0 at $t = 0$ and $\to \infty$ as $t \to \infty$. This makes sense because uncharged $(t = 0)$ capacitor readily accepts charge. Fully-charged $(t \to \infty)$ capacitor cannot accept more charge.
27. a) 1.96 μC   b) 53.3 Ω

CHAPTER 29

A. 0.10 T, in negative y-direction
C. (sketch)
E. a) $-(8.0\,\mathbf{k} + 12.8\,\mathbf{i}) \times 10^{-19}$ N
   b) $-(4.79\,\mathbf{k} + 7.66\,\mathbf{i}) \times 10^8$ m/s^2
G. (proof)
I. 0.225 T
K. (diagram)
M. a) $p_{alpha} = 2\,p_{proton}$ b) $v_{alpha} = (1/2)v_{proton}$
   c) $K_{alpha} = K_{proton}$
N. negative
S. 0.112 $\mathbf{i}$ N
U. 21.2 J
V. $\pm 5.8 \times 10^{-5}$ eV

1. approximately 120 T
3. a) 1.6×10^{-7} A·m^2 b) 2.0 mA
5. (proof)
7. Initially, both experience forces in positive y-direction. They move along curved paths in opposite senses. The ratio of their accelerations is $a_p/a_e = (1836)^2$
   b) yes
9. $\mathbf{F} = I\,\mathbf{L} \times \mathbf{B}$ is along length of pipe.   b) (proof)
   c) 2.03×10^5 A/m^2
11. a) (proof)   b) 2.48 mm/s   c) 5.86×10^{28} /m^3
13. (proof)

CHAPTER 30

A. a) $10\,\mathbf{j}$ nT   b) zero   c) $1.92\,(\mathbf{j} - \mathbf{i})$ nT
C. 8 A
E. 13.3 A
G. a) zero   b) 2.67×10^{-7} T
I. 0.250 A
K. 260 A
M. 1.80 T
N. a) zero   b) $\mu_0 nI$ c) zero
S. 0.78 A
U. radially inward

1. $(\mu_0 I/4\pi)\{s/y(s^2 + y^2)^{1/2}\}$
3. 1.51 μT/m^2
5. $B = 2\mu_0 Ia^2/\pi(a^2 + z^2)(2a^2 + z^2)^{1/2}$, $a = 0.5$ m, $I = 1$ A

7. 0.759 m
9. a) $\mu_0 Ir/2\pi a^2$ b) $\mu_0 I/2\pi r$
   c) $(\mu_0 I/2\pi r)\{(c^2 - r^2)/(c^2 - b^2)\}$ d) zero
11. 12 layers, total length approximately 121 m
13. a) (proof)   b) $\mu_0 Ir^2/2\pi R^3$
15. (proof)
17. $\mu_0 \omega \rho R^2/2\sqrt{2}$
19. a) (proof)   b) (proof)
21. a) $2\mu_0 I_1 I_2 a^2/\pi(d^2 - a^2)$ b) (proof)   points, $x = -y = \frac{R}{\sqrt{2}} = 0.707$ m, and at $x = -y = \frac{-R}{\sqrt{2}} = -0.707$ m. At these positions, the fields set up by the two currents have equal magnitudes and opposite directions.

CHAPTER 31

A. a) 2.05 V   b) 2.05 V
C. 1.11 mV
E. counterclockwise as viewed from the magnet
G. -0.588 V
I. a) +   b) −   c) −   d) +   e) no current
   f) −

K. 0.5 A
M. a) $0.008 \cos(377t)$ T·m^2 b) $3.02 \sin(377t)$ V
   c) $3.02 \sin(377t)$ A   d) $9.10 \sin^2(377t)$ W
   e) $0.0241 \sin^2(377t)$ N·m
N. a) $6.40 \sin(62.8t)$ V   b) $5.81 \sin(62.8t)$ A
   c) $37.2 \sin^2(62.8\,t)$ W

1. a) The flux through the loop decreases because B decreases as the loop moves away.
   b) $\mu_0 ILWv/2\pi r(r + W)$
3. 0.2 V, zero, and -0.2 V at $t = 0$, 2 ms, and 4 ms
5. (proof)
7. $\mathcal{E} = 118t/(0.8 - 4.9t^2)\ \mu V$, with t in seconds
   $\mathcal{E} = 98.3\ \mu V$ at $t = 0.3$ s

9. 14.0 V
11. (proof)
13. 4.07×10^{-4} V
15. 30.9 V ($t = 0$), and 0.208 V ($t = 1.25$ s).
17. 6 A

CHAPTER 32

A. 2.3 mH
C. 7.98 mV
E. 1) wrong dimensions 2) independent of length, contrary to the idea that L measures electrical inertia and should therefore increase with length of solenoid
G. $L \sim \Phi/I \sim (B/I)A \sim (1/\text{length})(\text{length})^2 \sim \text{length}$
I. approximately 6 mH
K. 3.34 A
M. a) $I_1 = I_2 = 3$ mA b) $I_1 = 6$ mA, $I_2 = 0$
N. If the switch is opened quickly, the current drops to zero suddenly. The rapidly-changing current induces a large $L(dI/dt)$ emf. The potential difference across the electrodes equals $L(dI/dt)$ because they are in parallel with the inductor.
S. a) 8.06 MJ/m^3 b) 6.32 kJ
U. a) $u_E = 0.323$ J/m^3 b) $u_M/u_E = 5.14$
V. (sketch) Circuit is critically damped.

1. a) (proof) b) 1.57×10^{-4} H
3. a) joule heat $= (\mathcal{E}_0^2/R)\{t - (2L/R)(1 - e^{-Rt/L}) + (L/2R)(1 - e^{-2Rt/L})\}$ b) (graph) c) The current approaches a constant value in a time of order $L/R = 1$ s, and the joule heat produced in time t approaches $RI^2 t$.
5. a) (proof) b) (proof) c) (proof)
7. a) 1.75 A b) approximately 4700 volts c) The person might become part of the switch across which there is a 4700-volt potential difference.
9. a) 1.98 A b) 1.40 A
11. 96 mH
13. 2.70×10^{18} J
15. a) $I = \mathcal{E}_0/(R + r)$ b) Current will decrease. c) B is at higher potential. d) L and R are in series. Current falls exponentially from initial value, $\mathcal{E}_0/(R + r)$.

CHAPTER 33

A. a) 5.03×10^{-3} T b) decrease by 5.43×10^{-8} T
C. 206, ferromagnetic
E. 1.1 mC (approximate result deduced from graph)
G. a) $N/2 + 250$ in one direction, $N/2 - 250$ in opposite direction
   b) N must be greater than or equal to 500.
I. 300, 600, 740, 710, 630, 560

K. $B_r = 0.72$ T, $B_c = 9.4 \times 10^{-4}$ T (approximate results deduced from graph)

1. a) (sketch) b) 10,700 A
3. 0.0373 T
5. (proof)
7. 0.212 A

CHAPTER 34

A. 4.44 V
C. $\langle I \rangle = 0$, $I_{rms} = 26.1$ A
E. 5.00 mV
G. $f = 1$ MHz, $\mathcal{E}_0 = 0.796$ V
I. a) 16.1 kΩ b) 0.155 mA
K. 2.17 kHz
M. 14.4 W
N. a) 167 Hz b) 0.955 mH
S. 0.158 A
U. a) 0.661 mH b) 210 V

V. 9.3% decrease

1. a) (proof) b) (proof)
3. a) 500 W b) 237 W
5. (sketch)
7. (proof)
9. (proof)
11. (proof)
13. a) 35.6 kHz b) 0.190 V

CHAPTER 35

A. 21.1 cm
C. a) 5.78×10^{14} Hz b) 519 nm, green

E. No, the direction of $\mathbf{E} \times \mathbf{B}$ is unchanged.
   $\omega = 1.88 \times 10^{10}$ rad/s, $k = 62.8$ rad/m

G. approximately 0.29 s, the time for sound to travel 100 m
I. (proof)
K. a) 4.97 kW/m^2 b) 1.66 × 10^{-5} J/m^3
M. approximately 2.4 MJ (assumes absorbing area of 0.8 m^2)
N. 0.095 V/m (This is E-field amplitude, not rms value.)
S. 0.9 W/m^2
U. (graph)

1. (proof)
3. $I_p = 3$ W/m^2, $I_0 = 9$ W/m^2
5. a) 2.80 cm/s   b) 4.45 × 10^{-18} m
7. Force by sunlight = 5.9 × 10^8 N. Force of gravity = 3.5 × 10^{22} N
9. (proof)
11. 9.1 MJ/m^2·day, 34.3 MJ/day, 3.8 m^2, $1520

CHAPTER 36

A. a) 40°   b) 28.9°
C. 61.0° larger (water-air critical angle is 48.8°)
E. They are identical.
G. Convex gives wide-angle view. Plane gives true-distance view.
I. Closer to mirror than F
K. 2.0 m behind mirror; virtual, upright
M. 0.198 m behind mirror; upright, virtual
N. (sketch)
S. 14 cm
U. a) (sketch)   b) 20 cm
V. For $s = 8$ cm, a real image is formed 9.14 cm inside glass. For $s = 1$ cm, a virtual image is formed 2.29 cm in front of glass-air interface.

1. 0.17 cm

3. a) (diagram) real, inverted   b) 2.5 cm in front of mirror
5. (sketch) $\theta_c = 37.3°$ (flint glass), $\theta_c = 24.4°$ (diamond)
7. The ray reflects and travels parallel to the mirror opposite the open side. It reflects again and travels parallel to the mirror at the left side of the opening. This lets it escape from the enclosure, its total angle of deviation being 240°.
9. a) (proof)   b) (proof)
11. a) Interchanging s and s' leaves lens equation unchanged.   b) Use $s + s' = L$ and lens equation to eliminate s' and show $s = L/2 \pm \sqrt{(L/2)^2 - fL}$ (requires $L > 4f$)
13. (proof)
15. Light emerges 45.2° below the normal to the exit face.
17. 82

CHAPTER 37

A. 6.50 radians
C. 635 nm
E. 25.3 cm
G. $x_2 - x_1 = \lambda (m + 1/48)$
I. approximately 30 m
K. a) 958 nm   b) 9.70 radians
M. 1.000132
N. 449 nm, blue

1. a) Waves reflected from the front and rear surfaces of the film differ in phase by π and therefore interfere destructively.   b) 1 (blue), 2 (green), 3 (red)
   c) 92.6 nm
3. 100 nm
5. a) (proof)   b) (proof)
7. 2
9. (proof)
11. a) alcohol: $(m - 1/2)\lambda = 904$ nm, $m = 1, 2, \ldots$; oil: $m\lambda = 429$ nm, $m = 1, 2, \ldots$ b) alcohol: 603 nm (orange); oil: 429 nm (violet)
13. (proof)
15. 481 nm

CHAPTER 38

A. 2.87°
C. 0.045 (m = 1), 0.0162 (m = 2), 0.00827 (m = 3)
E. 4.24 mm
G. 3.66 × 10^{-6} radian
I. a) 2.4 × 10^6 b) approximately 210 km
K. 4.6 × 10^{-5} radian, markings not resolved
M. Higher-frequency waves are diffracted through smaller angles.

N. 2.1 m

1. 30 m
3. Newsprint characters are not resolved at 10-m distance.
5. a) 2730 lines/cm   b) 5
7. 12.9 cm
9. (proof)

CHAPTER 39

A. (proof)
C. (proof)
E. Yes, but only if one of the added velocities is greater than c.
G. a) 0.361 ns b) 0.0114 s
I. a) 15.6 μs, 4.63 km b) 677 m c) 2.28 μs
K. (proof) Use the binomial expansion.
M. 0.0333 nm
N. a) 105,000 years b) 3.12×10^4 light-years in diameter; 32,800 years

1. a) 3 years b) 2.60 years
3. a) $2L/c(1 - v^2/c^2)$ {assuming no Lorentz-FitzGerald contraction of rod length L} b) (proof)

5. (proof)
7. 0.332 μs
9.

N	v_N/c
1	0.5
2	0.8
3	0.92857
4	0.97561
5	0.99180
6	0.99726
7	0.99909
8	0.99970
9	0.99990
10	0.99997

CHAPTER 40

A. 1000
C. a) 1.00 m/s^2 b) 2.18×10^8 m/s
E. $0.140c = 4.21 \times 10^7$ m/s
G. electron, 0.511 MeV $= 8.18 \times 10^{-14}$ J proton, 938 MeV $= 1.50 \times 10^{-10}$ J
I. a) 0.0556 b) 0.0607
K. 0.723 MeV
M. The kinetic energy is increased by more than a factor of 4.
N. 500
S. $1 - v/c = 5 \times 10^{-11}$

U. 1.11×10^{-8} kg

1. a) 1.64×10^8 s $= 5.18$ years
    b) 1.39×10^8 s $= 4.40$ years
3. 1.09×10^{-8}
5. $K_{proton} = 5.37$ MeV, $K_{pion} = 32.5$ MeV
7. $5mc^2/2$
9. (proof)
11. (proof)
13. a) (proof) b) (proof) c) (proof)

CHAPTER 41

A. 1.52×10^{15} Hz, 197 nm
C. 1240 nm
E. Theoretical wavelengths are 0.029 Å, 0.046 Å, and 0.063 Å, in respectable agreement with experiment.
G. a) 32.3 fm b) R would increase because the total kinetic energy would not be zero in the closest-approach configuration.
I. 4.43×10^4 m/s
K. 5.22 cm
M. 635 nm, red
N. 1.44 Å, comparable to atomic spacing. Thermal neutron can exhibit diffraction when scattered by crystal.
S. The probability is greatest for x slightly less than $3a/4$.
U. 0.36560 eV
V. 2.83×10^{13} Hz, 10.6 μm; infrared

1. a) (proof) b) 8.5×10^{-34} J · s
3. 8.8×10^6
5. a) 10 liters b) 270 liters in B and 18 liters in C c) (sketch)
7. 4.14 μm (photon), 2.24 Å (electron); electron more appropriate
9. 40.8 eV
11. a) (proof) b) (proof)
13. a) $E_n = -(9/8)E_H/n^2$, $n = 1, 2, 3, \ldots$
    b) $E_{binding} = E_H/8 = 1.7$ eV c) (proof) The $n = 2$ state energy is greater than that of a neutral hydrogen atom in the ground state and a free electron.
15. 6 head-on collisions
17. 0.0243 Å b) 0.0486 Å

CHAPTER 42

A. 8.59 MeV, 8.70 MeV, 8.39 MeV, 7.57 MeV
C. 0.138 nucleon/cubic fermi
E. a) $^8Be \rightarrow {}^4He + {}^4He$
    b) $8.005305 - 2(4.002603) > 0$

G. 5.24 MeV
I. 1140 Ci
K. The mass of 3 alpha particles is greater than the mass of a ^{12}C nucleus.

M. position emission (Z decreases.)

N. 29,500 yr

S. 0.564 mg

U. 6.84 MeV (^{233}U), 4.80 MeV (^{238}U). ^{233}U is more likely to fission.

V. 1220 kg

1. 5.16 MeV

3. $n_{daughter} = n_0(1 - e^{-\lambda t})$

5. (proof)

7. a) 4.11×10^9 yr
   b) No escape of Sr or Rb. No decay of Sr.

9. number of neutrons produced $= 2^{N+1} - 2$

11. a) (proof) see Equation 9.25   b) approximately 97

13. 3×10^{15} (approximate result based on graphical data)

15. a) 1.95×10^{-3} eV
    b) 14.4 keV (to 3 significant figures)

17. a) 650 MeV   b) approximately 1400 MeV (assumes binding energy per nucleon = 8 MeV)

19. a) (proof)   b) (proof) Evaluate dn/dt at $t = 0$.

INDEX

The entire Index is provided for the two-volume set of *University Physics*, as well as for the combined volume. Entries on pages 1–467 can be found in Volume One; entries on pages 468–938 can be found in Volume Two.

Phase Change Data

Substance	Normal Melting Point (K)	Latent Heat of Fusion (kJ/kg)	Normal Boiling Point (K)	Latent Heat of Vaporization (kJ/kg)
Water	273.16	334	373.16	2,260
Silver	1234	111	2485	2,360
Aluminum	931.7	399	2600	10,500
Gold	1336.16	64.4	2933	62,300
Cesium	301.9	15.7	963	514
Copper	1356.2	205	2868	4,800
Germanium	1232	479	2980	3,920
Lithium	459	416	1640	19,600
Sodium	371	115	1187	4,260
Lead	600.6	23.1	2023	859
Silicon	1683	1650	2750	10,600
Zinc	692.7	102	1180	1,760
Helium-4	——	——	4.2	20.5
Hydrogen (H_2)	14.0	57.8	20.3	452
Neon	24.5	13.8	27.1	85.8
Nitrogen (N_2)	63.2	25.8	77.4	199
Argon	83.8	29.6	87.3	163
Oxygen (O_2)	54.4	13.9	90.2	213
Xenon	161	17.6	165	396

Specific Heat Capacity

Substance	C_p(kJ/kg·C°)
Solids	
Silver (Ag)	0.234
Aluminum (Al)	0.900
Gold (Au)	0.130
Bismuth (Bi)	0.123
Copper (Cu)	0.385
Iron (Fe)	0.448
Ice (H_2O, 0°C)	2.10
Brick	0.837
Concrete	0.879
Magnesium (Mg)	1.04
Sodium (Na)	1.23
Nickel (Ni)	0.431
Lead (Pb)	0.130
Zinc (Zn)	0.352
Wood (Pine)	2.81
Liquids	
Benzene (C_6H_6)	1.62
Bromine (Br)	0.448
Ethyl alcohol (C_2H_5OH)	2.43
Methyl alcohol (CH_3OH)	2.52
Water (H_2O)	4.19
Gases	
Steam (100°C)	2.05
Argon (Ar)	0.523
Carbon dioxide (CO_2)	0.95
Chlorine (Cl_2)	0.481
Helium (He)	5.24
Hydrogen (H_2)	14.2
Nitrogen (N_2)	1.04
Neon (Ne)	1.03
Oxygen (O_2)	0.917

Thermal Conductivity

Material	k (W/m·C°)
Gases	
Air	0.0237
Carbon dioxide	0.0145
Oxygen	0.0246
Hydrogen	0.167
Helium	0.142
Methane	0.0305
Liquid	
Water	0.569
Solids	
Styrofoam	0.040
Fiberglass	0.045
Corrugated cardboard	0.064
Rubber	0.12
Wood	0.19
Concrete	1.0
Glass	0.75–1.2
Ice	2.21
Steel	15–50
Iron	73
Brass	120
Aluminum	228
Copper	386

Astrophysical Data

Quantity	Value
Earth	
Mass	5.98×10^{24} kg
Radius (equatorial)	6.38×10^{6} m
Radius (polar)	6.37×10^{6} m
Sidereal day	8.6164×10^{4} s
Solar day	8.6400×10^{4} s
Sidereal year	3.1558150×10^{7} s
Average distance from sun	1.50×10^{11} m
Moon	
Mass	7.35×10^{22} kg
Radius	1.74×10^{6} m
Sidereal period	2.3606×10^{6} s
Average distance from earth	3.84×10^{8} m
Sun	
Mass	1.99×10^{30} kg
Radius	6.96×10^{8} m
Sidereal period (at equator)	2.1868×10^{6} s
Distance from center of galaxy	3×10^{20} m
Time for sun to complete one orbit of galaxy	7.5×10^{15} s (250 million years)
Milky Way Galaxy	
Mass	2.8×10^{41} kg (1.4×10^{11} solar masses)
Diameter (disk)	8×10^{20} m (25,000 parsecs)
Thickness (disk)	6×10^{19} m (2,000 parsecs)
Universe	
Mass	3×10^{51} kg
Radius	10^{26} m
Average mass density	2×10^{-28} kg/m^3
Age (upper limit)	18×10^{9} years
Hubble constant	1.8×10^{-18} s$^{-1} = 17 \dfrac{\text{km/s}}{\text{Mly.}}$

Index of Refraction

(reference wavelength = 589.3 nm)

Substance	n	Substance	n
Water (ice)	1.31	Amber	1.56
Water (liquid)	1.33	Emerald	1.58
Acetone	1.36	Crown glass	1.59
Fluorite	1.43	Flint glass	1.65
Glycerol	1.47	Calcite	1.66
Benzene	1.50	Ruby	1.77
Sodium chloride	1.54	Diamond	2.42

Dielectric Constants

Material	κ
Water	78.3
Air	1.000590
Lucite	2.84
Plexiglas	3.12
Polystyrene	2.55
Neoprene	6.60
Polyethylene	2.26
Pyrex	4–6
Teflon	2.1

Conversion Factors

Length

1 fermi (fm) $= 10^{-15}$ m

1 angstrom (Å) $= 10^{-10}$ m

1 nanometer (nm) $= 10^{-9}$ m

1 inch $= 0.0254$ m

1 foot $= 0.3048$ m

1 mile $= 1.609$ km

1 astronomical unit (AU) $= 1.496 \times 10^{11}$ m

1 light year (ly) $= 9.461 \times 10^{15}$ m

1 parsec (pc) $= 3.086 \times 10^{16}$ m

Area

1 in^2 $= 6.4516 \times 10^{-4}$ m^2

1 ft^2 $= 9.2903 \times 10^{-2}$ m^2

Volume

1 gallon $= 3.785 \times 10^{-3}$ m^3

1 ft^3 $= 2.832 \times 10^{-2}$ m^3

1 ft^3 $= 7.477$ gallons

1 liter $= 10^3$ cm^3 $= 10^{-3}$ m^3

Time

1 solar day $= 86{,}400$ s

1 sidereal day $= 86{,}164$ s

1 year $= 3.156 \times 10^7$ s

Speed

1 ft/s $= 0.3048$ m/s

1 mile/hour $= 0.4470$ m/s

Mass

1 $u = 1.66053 \times 10^{-27}$ kg

Force

1 pound $= 4.448$ N

1 dyne $= 10^{-5}$ N

Pressure

1 pascal (Pa) $= 1$ N/m^2

1 lb/in^2 $= 6.895 \times 10^3$ N/m^2

1 atmosphere $= 1.013 \times 10^5$ N/m^2

1 atmosphere $= 14.70$ lb/in^2

Energy

1 kWh $= 3.600 \times 10^6$ J

1 calorie $= 4.186$ J

1 Btu $= 1.055 \times 10^3$ J

1 ft·lb $= 1.356$ J

1 erg $= 10^{-7}$ J

1 electron volt $= 1.602 \times 10^{-19}$ J

1 megaton $= 4 \times 10^{15}$ J

Power

1 ft·lb/s $= 1.356$ W

1 hp $= 746$ W

Magnetic Field

1 gauss (G) $= 10^{-4}$ tesla (T)

Angle

2π radian $= 360$ degrees

1 radian $= 57.30$ degrees

1 degree $= 0.01745$ radian

Temperature

Absolute zero $= -273.15°$C

Mathematics

$\pi = 3.14159$

$\pi^2 = 9.86960$

$e = 2.71828$

Quadratic Equation/Roots

$ax^2 + bx + c = 0$

$$x = \frac{-b \pm \sqrt{b^2 - 4ac}}{2a}$$

Binomial Expansion

$(1 + x)^n = 1 + nx + n(n-1)x^2/2 + \cdots$

Series Expansions

$\sin x = x - x^3/6 + \cdots$

$\cos x = 1 - x^2/2 + \cdots$

$e^x = 1 + x + x^2/2 + \cdots$

Law of Cosines

$c^2 = a^2 + b^2 + 2ab \cos \theta$

SI Prefixes

Power of 10	Prefix	Symbol
10^{18}	exa	E
10^{15}	peta	P
10^{12}	tera	T
10^{9}	giga	G
10^{6}	mega	M
10^{3}	kilo	k
10^{-2}	centi	c
10^{-3}	milli	m
10^{-6}	micro	μ
10^{-9}	nano	n
10^{-12}	pico	p
10^{-15}	femto	f
10^{-18}	atto	a